7th International Conference on Advanced Materials and Structures (AMS 2018)

IOP Conference Series: Materials Science and Engineering
Volume 416

Timisoara, Romania
28 - 31 March 2018

Editors:

Viorel-Aurel Serban
Liviu Marsavina
Ion-Dragos Utu
Emanoil Linul

ISBN: 978-1-5108-7380-3
ISSN: 1757-8981

Printed from e-media with permission by:

Curran Associates, Inc.
57 Morehouse Lane
Red Hook, NY 12571

Some format issues inherent in the e-media version may also appear in this print version.

Printed by Curran Associates, Inc. (2018)

For permission requests, please contact the Institute of Physics at the address below.

Institute of Physics
Dirac House, Temple Back
Bristol BS1 6BE UK

Phone: 44 1 17 929 7481
Fax: 44 1 17 920 0979

techtracking@iop.org

Additional copies of this publication are available from:

Curran Associates, Inc.
57 Morehouse Lane
Red Hook, NY 12571 USA
Phone: 845-758-0400
Fax: 845-758-2633
Email: curran@proceedings.com
Web: www.proceedings.com

Table of contents

Volume 416

7th International Conference on Advanced Materials and Structures - AMS 2018

28–31 March 2018, Timisoara, Romania

Accepted papers received: 12 August 2018 Published online: 26 October 2018

Preface

Papers

IOP Conf. Series: Materials Science and Engineering **416** (2018) 011001 doi:10.1088/1757-899X/416/1/011001

Preface

The 7th International Conference on Advanced Materials and Structures - AMS 2018 is organized by the University Politehnica Timişoara with the support of the Politehnica Foundation in collaboration with Óbuda University from Budapest Hungary and Lublin University of Technology Poland.

The AMS 2018 conference celebrates the 95th anniversary of the Strength of Materials Laboratory from University Politehnica Timişoara.

The conference Advanced Materials and Structures (AMS'18) was held between 28 to 31 March 2018 at Timisoara, Romania and it is the seventh edition of the scientific event organized by the Department of Materials Engineering and Manufacturing and Department of Mechanics and Strength of Materials in collaboration with the Politehnica Foundation.

The main topics of the AMS 18 were:

- Advanced materials - amorphous, nano-structured materials, composites, cellular materials, biomaterials etc.
- Surface engineering
- Modern fabrication, joints and recycling technologies
- Materials damage under time-dependent-actions (fatigue, creep, impact, corrosion)
- Computational techniques for advanced engineering materials and structures

Proceeding of 7th International Conference on Advanced Materials and Structures - AMS 2018 is a collection of papers presented by researches in area of advanced materials, with all important aspects: technology of fabrication, property characterization, advanced applications, investigation and mathematical modelling.

The organizers are grateful to all the people that have made possible this conference. Special thanks to the POLITEHNICA Foundation.

The proceedings editors are

Viorel-Aurel Serban
Liviu Marsavina
Ion-Dragos Utu
Emanoil Linul

IOP Conf. Series: Materials Science and Engineering **416** (2018) 011002 doi:10.1088/1757-899X/416/1/011002

Peer review statement

All papers published in this volume of *IOP Conference Series: Materials Science and Engineering* have been peer reviewed through processes administered by the proceedings Editors. Reviews were conducted by expert referees to the professional and scientific standards expected of a proceedings journal published by IOP Publishing.

IOP Conf. Series: Materials Science and Engineering **416** (2018) 012001 doi:10.1088/1757-899X/416/1/012001

Microstructure and Phase Composition of NiCrBSi-TiB_2 Vacuum Furnace Fused Flame-Sprayed Coatings

N Kazamer[1,2,*], P C Vălean[1,2], D T Pascal[2,*], R Muntean[2], G M Mărginean[2] and V A Şerban[1]

[1]Materials and Manufacturing Engineering Department, Politehnica University Timişoara, Piaţa Victoriei 2, Timişoara, 300006, Romania

[2]Materials Science and Testing Department, Westphalian University of Applied Sciences, Neidenburgerstr. 43, Gelsenkirchen, 45897, Germany

*Corresponding author: dragos.pascal@w-hs.de, norbert.kazamer@w-hs.de.

Abstract. Several studies have revealed that NiCrBSi reinforced with different intermetallic compounds exhibits characteristics that allow the material to be successfully used as a thermally sprayed coating for applications that require wear and corrosion protection. One of such materials is TiB_2, as it has an elevated melting point, high elasticity modulus, increased hardness and good thermal stability. NiCrBSi was mechanically mixed with 5%, 10%, 15% and 20% vol. TiB_2, flame sprayed and subsequently fused in a vacuum furnace. A remelting process under controlled atmosphere offer important advantages like good gas extraction and precise adjustment of temperature. The best performing batch of samples were selected for further analyse of microhardness and porosity measurements. The experimental results show that the thermal cycle created with the aid of the data offered by differential thermal analysis lead to a reduced porosity without affecting the hardness. X-ray diffractometry was used to characterize the phase composition while scanning electron microscopy was employed to investigate the microstructure and phase distribution. Further theoretical aspects regarding the influence of the phase composition on the coatings properties, like hardness or sliding wear conclude the work.

1. Introduction

Patented in 1909 by the Swiss inventor Max Ulrich Schoop [1], thermal spraying developed as a widely used technology in applications like heat exchangers, extruders, pistons, pipes or laminators [2-4]. The research in the area of Ni-based alloys began almost five decades ago [5] and shows that this type of materials can be used for a large range of technologies like flame spraying [6], Atmospheric Plasma Spray (APS) [7] or High Velocity Oxy Fuel (APS) [8].

Among the most well-known Ni-based alloys is NiCrBSi, a material used with industrial applications in respect of wear and corrosion protection [9-11]. Regarding its chemical composition, Ni as the main element influences the alloy's mechanical properties, mainly its ductility and hardness. The role of Cr is to improve the corrosion resistance while B and Si lowers the melting temperature of the material [12]. The content of B helps improving the hardness. One of the possible reinforcements for the Ni-based alloys is TiB_2, which because of its characteristics, is a suitable candidate for thermal spraying depositions. This material exhibits a high hardness, 3460 HV, low density, 4.52 g cm^{-3}, and an excellent resistance to oxidation, which goes up to 1000°C [13].

The NiCrBSi thermally sprayed products show moderate adhesion to the substrate, relatively high porosity and unwanted trapped gasses in the coatings [14]. A fusing process is therefore necessary to enhance the characteristics of the depositions and mitigate the above mentioned defects. Besides the properties previously described, B and Si also reduce the melting interval of the alloy [12], which has a positive impact for the coating post-treatment. Vacuum furnaces have recently started to be applied for such post-treatments as they offer reproducibility, the possibility of maneuvering products with complex geometries and a relative low cost with a short recoupment period [15].

The aim of the research work was to investigate the microstructure and phase composition of NiCrBSi-TiB_2 flame sprayed and vacuum fused coatings. NiCrBSi powder was mechanically mixed with TiB_2 in different volumetric concentrations, flame-sprayed and post-fused. Powder analysis, heat-treatment, microhardness measurement, porosity calculation, microstructure and phase composition examinations were all considered in the elaboration of the present work.

2. Experimental procedure

The Ni-based alloy was provided by Karl Schumacher GmbH and the NiCrBSi-TiB_2 powder was deposited on an S355JR carbon steel with 0.22% C and 1.6% Mn at the same German company. A successful deposition on a widely used steel in industry will offer the coating a larger applicability.

2.1. Feedstock powder and methods for characterisation

As one can see, Figure 1(a) shows the SEM micrograph of the NiCrBSi mixed with TiB_2. The spheroidization of Ni-based powder produced through water atomization helps the flowability through the spraying equipment. The EDX spectrum in Figure 1(b) shows the main element Ni, the hard carbide forming element Cr and Si, which promotes the wettability of the coating during deposition. The spectrum in Figure 1(c) shows Ni and the hard element Ti. The chemical composition and the size ranges of the two powders can be found in Table 1. Four sets of mechanically mixed powder with 5% (NTB5), 10% (NTB10), 15% (NTB15) and 20% (NTB20) vol. TiB_2 were prepared.

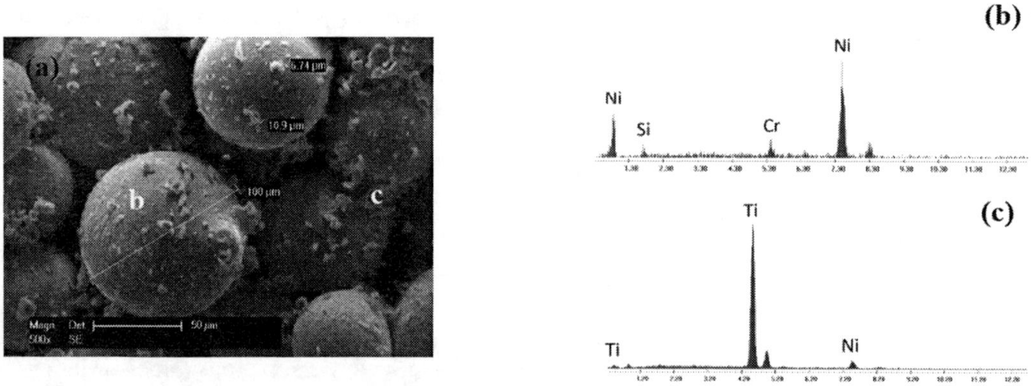

Figure 1. SE micrograph (a) of the NiCrBSi- TiB_2 mechanically mixed powder along with two EDX spectra demonstrating the chemical composition of the (b) NiCrBSi and (c) TiB_2 powder.

Table 1. Chemical composition and size range of the NiCrBSi and TiB_2 powder

	Ni [%]	Ti [%]	Cr [%]	B [%]	Si [%]	Fe [%]	C [%]	O_2 [%]	Size range [µm]
NiCrBSi	bal.		6	1	4	1.5	0.3		45-106
TiB_2		bal.		min. 30		max. 0.1	0.5	max. 1.1	6.5-10

2.2. Coating deposition and remelting

A careful thermal analysis of the powder was performed with a Netzsch STA 449F1 instrument. The DTA curve shows that all the examined powders exhibit a similar melting temperature at about 1030°C. This result has to be taken into account when selecting the fusion cycle as a temperature close to the melting point provides a densification of the coating without spilling it [16]. The surface was activated through grit blasting prior the deposition, this being a highly recommended step for a good adherence of the coating [17]. The fusion of the deposition was realised in a HITERM 80-200 vacuum furnace. Figure 3 represents the performed heating cycle which lasted a total of 288 min. with a maximum heating temperature of 1090°C with a holding time of 45 min. The ramps were set up to avoid changes during the heating and cooling steps that could have induced stress or cracks.

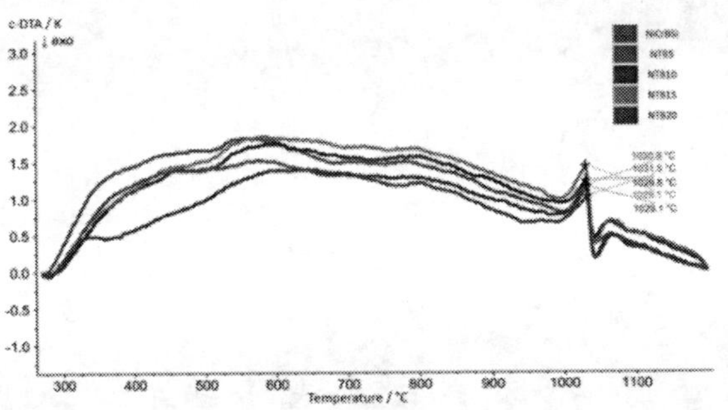

Figure 2. DTA curve highlighting the liquidus temperature of the NiCrBSi and NTB powders.

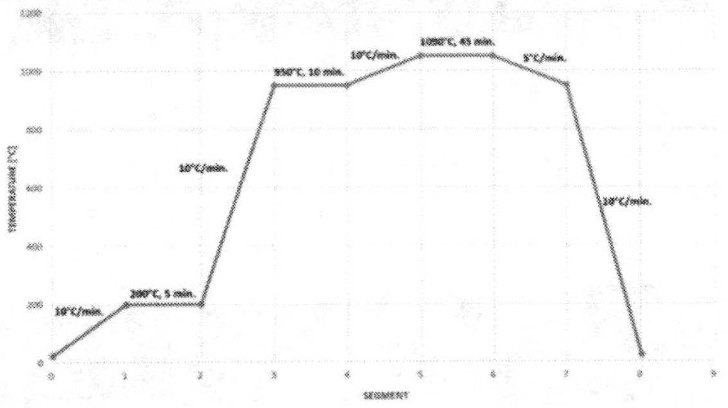

Figure 3. Vacuum furnace heating cycle of the as-sprayed samples.

2.3. Sample microhardness measurement and porosity calculation

The improvement of the post-treatment of the applied coatings was done considering the porosity and the microhardness of the samples. The porosity of the coatings was calculated on five different regions on micrographs using the image analysis software ImageJ. The microhardness tests were performed on the coating region at a load of 0.3 kgf using a Zwick/Roel ZHVμ-S machine.

2.4. Microstructure and phase composition

The quality of the coating was measured through several methods. The micrographs taken for the preliminary porosity calculation test were taken with a Leica DM-RME light microscope. The morphology, microstructure and chemical composition was studied using a Philips XL 30 SEM combined with EDX spectroscopy. The phase composition was determined using a Philips X'Pert X-ray diffractometer using a Cu K-α radiation source scattering the 2θ angle between 20° and 100°.

3. Results and discussions

3.1. Sample porosity and microhardness

The porosity level and the microhardness were chosen as the two eliminating criteria after the post-treatment of the samples. Both factors are commonly employed as excluding responses in thermal spraying [18,19] as they are directly influencing the mechanical properties of materials. Figure 4(a) represents an as-sprayed sample where a porosity level of approximately 25% in all coating batches prior the post-treatment process was measured. Figure 4(b) shows the post-fused NTB15 sample etched in a 3% HNO_3 solution. One can observe a drastic decrease of porosity, with an improved roughness, a good adhesion and no visible major defects. Porosity calculations of post-treated NTB5, NTB10, NTB15 and NTB20 can be consulted in Figure 4(c), Figure 4(d), Figure 4(e) and Figure 4(f). The NTB10 and NTB20 exhibited close mean values at 6.5±0.5% and 7.5±0.5%, respectively. The NTB5 and NTB15 performed better, showing average values of 1.5±0.5% and 0.8±0.5%, comparable

IOP Conf. Series: Materials Science and Engineering **416** (2018) 012001 doi:10.1088/1757-899X/416/1/012001

results to APS or HVOF processes. The decrease of porosity in the case of the fused samples may be attributed to the characteristics of the furnace to control and maintain the temperature, leading to the wetting of the surface offering a good gas extraction and void closure. The porosity of the coating is affected as well through the control of B and Si as they offer wettability [2] and act as fluxing agents.

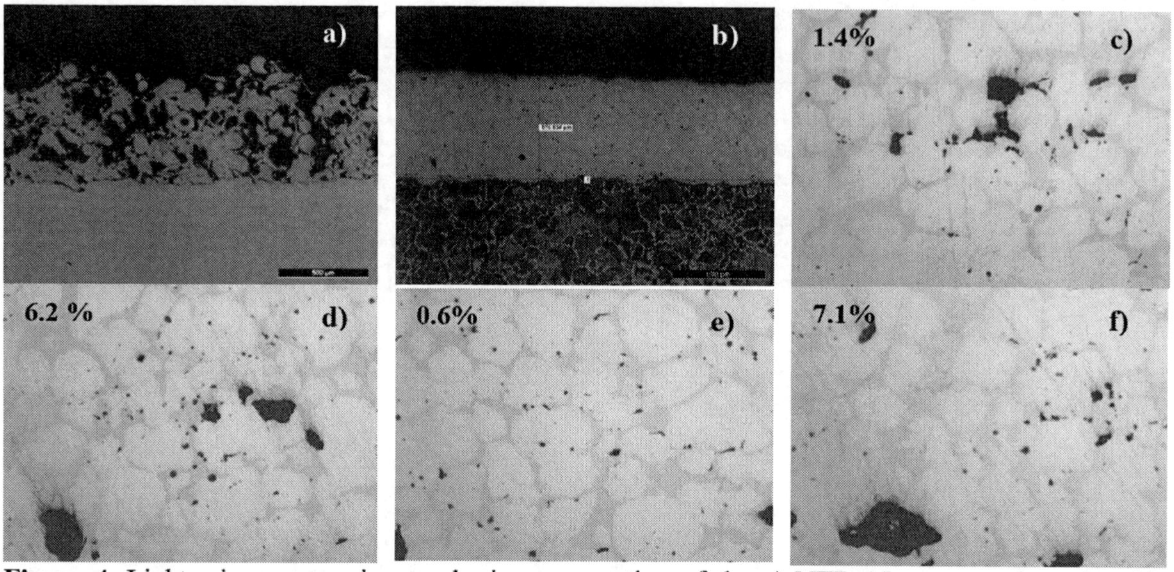

Figure 4. Light microscopy micrographs in cross-section of the a) NTB 15 as-sprayed coating, b) NTB15 post-treated coating and c) NTB5, d) NTB10, e) NTB15, f) NTB20 porosity measurements.

The microhardness of the deposited coatings was evaluated realizing seven imprints along the coating, the distribution being presented in Figure 5. The NTB10 and NTB20 coatings have both showed differences of over 30% between the extreme points of the measurement. The NTB5 coating exhibited a mean value of 225HV. The NTB15 samples showed smaller variations, the mean calculated value being 310HV. The variations of the measurements may be attributed to the different containing phases in the coating. The topography of a microindentation of the NTB15 sample presented in Figure 6 shows plastic deformation with no crack appearance, denoting a good phase bonding and low residual stresses.

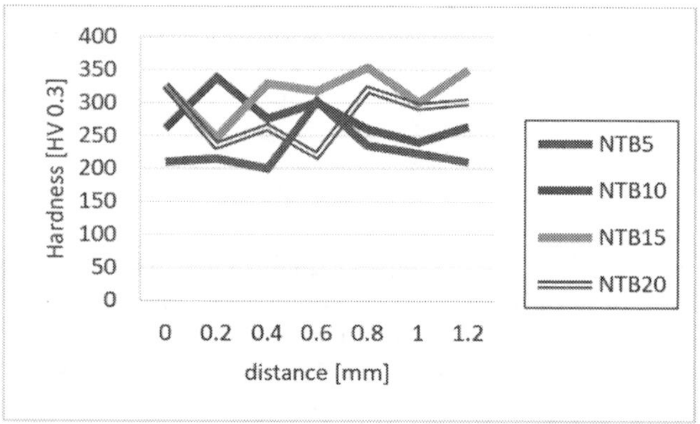

Figure 5. Graph highlighting the variation of the HV0.3 microindentation along the coating.

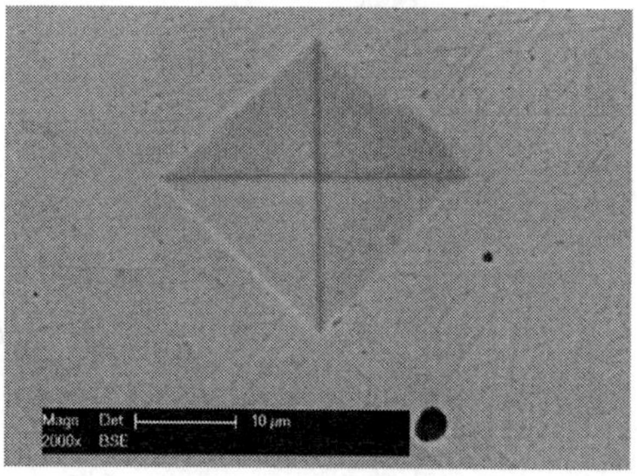

Figure 6. HV0.3 microindentation of the NTB15 sample.

IOP Conf. Series: Materials Science and Engineering **416** (2018) 012001 doi:10.1088/1757-899X/416/1/012001

Considering the porosity of under 1% and the rather constant microhardness with crack-free imprints, the NTB15 post-fused coating was chosen to be further analysed regarding its microstructure and phase composition.

3.2. Microstructure and phase composition

The NTB15 sample presented in Figure 7 exhibits a heterogeneous distribution of the dark phases TiB_2 phases. The reason for a non-homogeneous distribution lies in the mixing process of the powders and the high melting temperature of TiB_2 [13] which resists to plastic deformation and solubility in solid solutions even at high temperatures [21]. An EDX spot analysis at the interface demonstrates diffusion of Ni into the substrate creating a metallurgical bonding instead of the commonly reached mechanical hooking [22], aiding the adhesion.

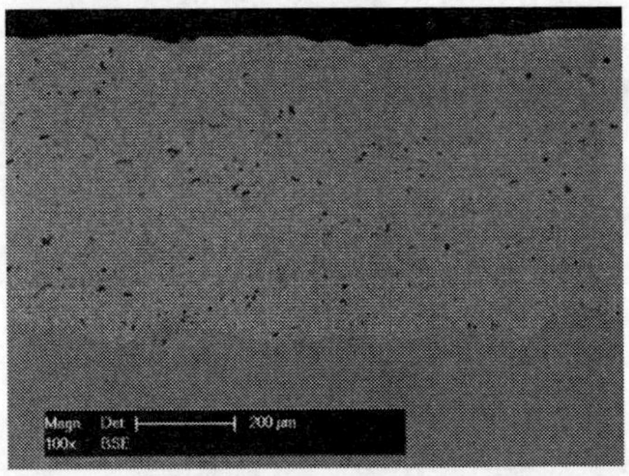

Figure 7. Cross-section SEM micrographs at lower magnification of the NTB15 sample.

Figure 8(a) shows four different regions with different chemical composition as detected by means of EDX spot analyses in Figure 8(b), Figure 8(c) and Figure 8(d). Corroborating the XRD spectra from Figure 9 with the EDX spot analyses, the matrix marked with b in Figure 8(a) mainly consists of the γ-Ni and BNi_2, similar eutectic phases being previously reported in Ni-based thermal sprayed coatings research [23-26]. Another component of the matrix is the σ-$Cr_3Ni_5Si_2$ phase, that can be found in the ternary Cr-Ni-Si diagram [27]. As previously reported by Schuster et al [28], the low Si content (4% in the present work) favours forming the σ phase.

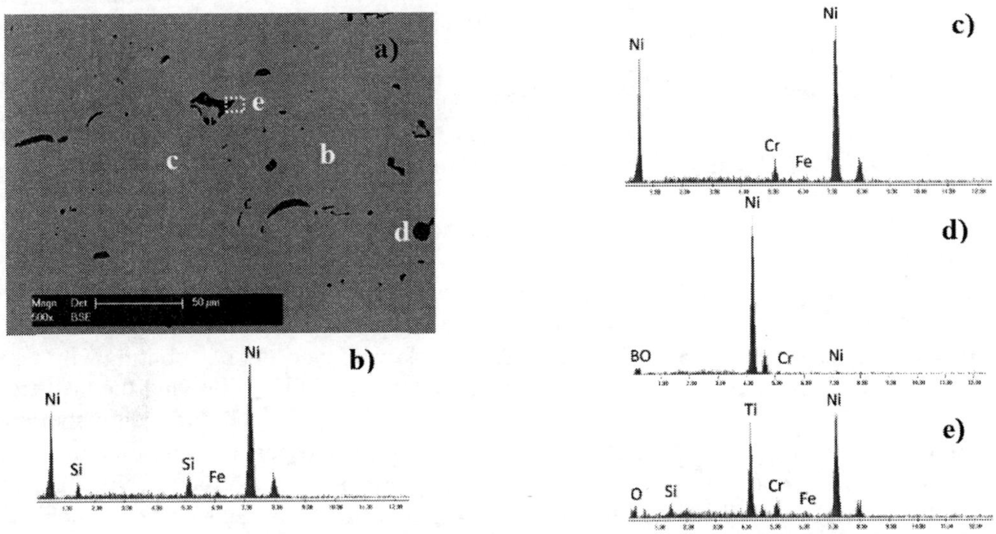

Figure 8. a) SEM micrograph of NTB15 sample along with the corresponding b), c), d), e) EDX spectra.

A confirmation that the heating cycle parameters were well chosen comes from the fact that it was not reached a temperature which transforms the Cr containing σ-$Cr_3Ni_5Si_2$ hard phase into a Ni solution along with minor constituents [29]. Observing both EDX analysis XRD results, the region c in Figure 8(a) corresponds to the rest γ-Ni/BNi_2 phase. The area d on the SEM micrograph is represented by the hard TiB_2 phase which is expected to aid the wear resistance of the coating and partly the Ti_3O. The EDX spectrum of the e area is showing a diffusion at the interface between the matrix and the TiB_2. The diffusion process forms on the one side Ti_3SiC_2, conferring high temperature stability and high fracture toughness [30], and Ti_2FeNi. The crack free microindentations may be attributed to the presence of the ductile Ti_3SiC_2 phase which is binding the matrix and the TiB_2 particles together leading to a coating with low residual stresses. The Ti_2FeNi phase with the more known empirical formula $FeNiTi_2$ may be consulted in the ternary Fe-Ni-Ti diagram [27]. The data provided by Duarte el al [31] shows that the solid solubility of Ti in (Fe,Ni) varies in dependence with the Fe:Ni ratio and lowers with decreasing temperature.

As the maximum operating pressure of the furnace vacuum pump is 10^{-6} mbar and the elimination of a large number of oxide and gases is successfully performed, Ti and Si have both according to the Ellingham diagram a high susceptibility of forming oxides until an oxygen partial pressure of about 10^{-23} mbar [32] and the formation of Ti_3O and SiO_2 is therefore inevitable. Nevertheless, in the current case the presence of Ti_3O during sliding wear tests it is proved to activate a tribofilm on the contact surfaces, resulting a decrease of the wear rate and the friction coefficient of the tribological pair [33].

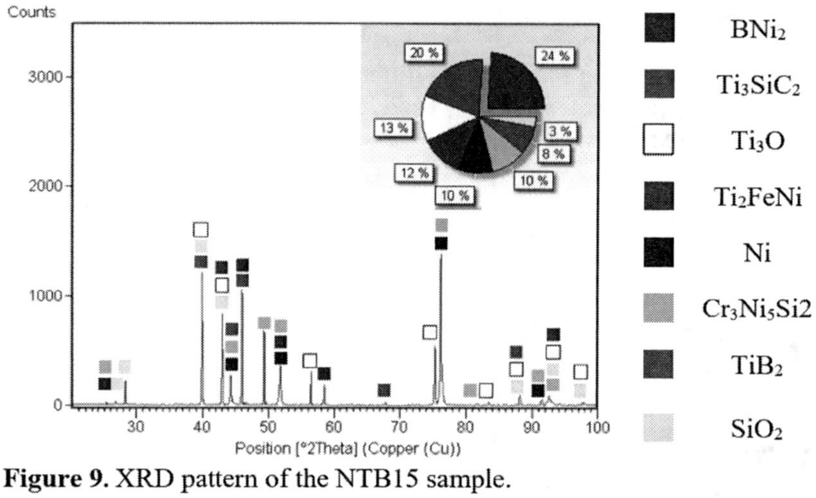

Figure 9. XRD pattern of the NTB15 sample.

4. Conclusions

Reinforcements and the improvement of self-fluxing alloys are a continuous topic in the coating industry. NiCrBSi-TiB_2 with 5%, 10%, 15% and 20% vol. concentration of TiB_2 have been successfully flame sprayed and vacuum furnace fused on a low alloyed steel. A coating with no visible major defects and a strong metallurgical bonded interface was obtained. Following the post-treatment process, considering the results of the microhardness tests and porosity measurements, the NiCrBSi-TiB_2 85-15 was further analyzed. The microstructure exhibits a heterogeneous distribution of the TiB_2 in the coating. A complex structure, with a matrix mainly composed of γ-Ni and BNi_2 was observed. The hard TiB_2 particles are partly diffused in the matrix creating Ti-based phases which offer the coating ductile properties, resisting to deformations without generating cracks. Recent research show that the presence of Ti_3O might represent a positive aspect as during sliding wear tests it activates a tribofilm decreasing the wear rate of the tribological pair. Further research could be performed in the direction of examining the deposited coating in respect for their corrosion and wear behaviour.

IOP Conf. Series: Materials Science and Engineering **416** (2018) 012001 doi:10.1088/1757-899X/416/1/012001

5. References

[1] Siegmann S and Abert C 2012 *Surface & Coating Technology* **220** 3

[2] Davis JR 2004 *Handbook of Thermal Spray Technology* vol 1, (Ohio, ASM International) p 10

[3] Lin MC, Chang LS, Lin HC, Yang CH and Lin KM 2006 *Surface & Coating Technology* **201** 3193

[4] Battez F, Viesca J and Gonzales R 2010 *Wear* **268** 325

[5] Kühn H and Morach E 1974 *Chemie Ingenieur Technik* **46** 281

[6] Chaliampalias D, Vourlias G, Pavlidou E, Skolianos S, Chrissafis K and Stergioudis G 2009 *Applied Surface Science* **255** 3605

[7] Wang Y, Stella J, Darut G, Poirier T, Liao H and Plance M-P 2017 *Journal of Alloys and Compounds* **699** 1095

[8] Kong D and Zhao B 2017 *Journal of Alloys and Compounds* **705** 700

[9] Kim HJ, Hwang SY, Lee CH and Juvanon P 2003 *Surface & Coating Technology* **172** 262

[10] Singh S and Kaur M 2016 *Surface Engineering* **32** 464

[11] Yao SH M 2014 *Materials Research Innovations* **18** S2-332

[12] Houdkova S, Vostrak M, Hruska M, Riha J, Smazalova E, Cesanek Z and Schubert J, 2013 *Proc. Int. Metal Conf. (Brno)* p 1407

[13] Wu Y, Qiu W, Yu H, Zhong X, Liu Z, Zeng D and Li S, 2011 *Applied Surface Science* **257** 10224

[14] Navas C, Colaco R, de Damborenea J and Vilar R, 2006 *Surface & Coating Technology* **200** *6854*

[15] Konakov SP and Lyapunov AI 2000 *Metal Science and Heat Treatment* **42** 84

[16] Bergant Z and Grum J 2011 *Image Analysis & Stereology* **30** 53

[17] Pawlowski L 2008 *The Science and Engineering of Thermal Spray Coatings* (Chichester: Wiley & Sons) p 53

[18] Lih WC, Yang Sh, Su CY, Huang SC, Hsu IC and Leu MS 2000 *Surface & Coating Technology* **133-134** 54

[19] Bertrand G, Bertrand P, Roy P, Rio C and Mevrel R 2007 *Surface & Coating Technology* **202** *1994*

[20] Schlichting KW, Padture NP, Jordan EH and Gell M, 2003 *Materials Science and Engineering A* **342** 120

[21] Zhu HBZ and Li ZX 2013 *Surface & Coating Technology* **235** 620

[22] Packham DE 2005 *Handbook of Adhesion,* Second Ed., (Chichester: Wiley & Sons) p 276

[23] Gonzales R, Garcia MA, Penuelas I, Cadenas M, del Rocia Fernandez Ma, Hernandez Battez A and Felgueroso 2007 *Wear* **263** 619

[24] Mrdak MR 2012 *Vojnotehnicki glasnik* **60(1)** 183

[25] Singh S and Kaur M 2016 *Surface Engineering* **32(7)** 464

[26] Karagoz M, Islak S, Buytoz S and Kurt B 2011 *Proc. 6th Int. Advanced Technol. Symposium* (Elazig, TUrkey) p13

[27] SGTE 2015 *Thermodynamic Properties of Inorganic Materials Compiled* ed A Watson and A Markus (Heidelberg: Springer) Vol. 2, Part 2 p 73 and p 194

[28] Schuster JC and Du Y 2000 *Metallurgical and Material Transactions A* **31(7)** 1795

[29] Sampath S, Neiser RA and Herman H 1993 *Journal of Materials Research* **8(1)** 78

[30] Barsoum MW, El-Raghi E, Rawn CJ, Porter WD, Wang H, Payzant EA and Hubbard CR 1999 *Journal of Physics and Chemistry of Solids* **60** 429

[31] Duarte LI, Klotz UE, Leinenbach C, Palm M, Stein F and Löffler JF 2010 *Intermetallics* **18** 374

[32] Atkins P and de Paula J 2006 Physical Chemistry : Thermodynamics and Kinetics ed Heidi Bamatter 8th Ed (New York: WH Freeman and Company) p 275

[33] Umanskyi O, Hussainova I, Storozhenko M, Terentyev O and Antononv M 2014 *Key Engineering Materials* **604** 16

IOP Conf. Series: Materials Science and Engineering **416** (2018) 012002 doi:10.1088/1757-899X/416/1/012002

Investigations on the Characteristics of Thermally Sprayed NiCrBSi Coatings Fused by Flame and Inductive Processing

P C Vălean[1,2], N Kazamer[1,2,*], R Muntean[1], D T Pascal[1], Y Kilic[1], G Mărginean[1] and V A Șerban[2]

[1]Westphalian University of Applied Sciences, Neidenburger Str. 43, 45897 Gelsenkirchen, Germany
[2]Politehnica University Timisoara, Piața Victoriei 2, Timișoara 300006, România

*Corresponding author: norbert.kazamer@yahoo.com

Abstract. Ni-based materials are some of the most used hardfacing alloys to improve wear and corrosion resistance for parts functioning under severe conditions. For the current study, a NiCrBSi self-fluxing alloy powder with particle size in range of 45-106 μm was employed as feedstock material and deposited onto a steel substrate material using the oxyacetylene thermal spraying process. The as-sprayed components have been subsequently subjected to thermal post-processing in order to increase the adhesion to the substrate and to reduce the inherent porosity of the coatings. Consequently, two types of heating sources were chosen. One of them was the neutral oxyacetylene flame and the second one was based on high frequency currents. The aim of this work was to study the influence of the heat source type of the post treatment on the final characteristics of the coatings. Evaluation of the morphology and microstructure of the developed coatings was performed after the remelting process by means of scanning electron microscopy. Hardness tests were carried out in cross-section complying with the ASTM E384 standard. In both cases, after fusing, the microstructure was refined, with a drastic decrease in porosity. Evaluation of the hardness distribution along the coating cross-section revealed similar values of microhardness.

1. Introduction

In order to compensate the rising personal and raw material costs, machines and technical installations, which are working in high-technology industrialised countries necessitate a constant rise in productivity and performance [1]. Recently, the requirements on components of machines that are utilised in offshore environments have become so stringent that the demands cannot be any more satisfied by the base material alone [2]. Thermally sprayed coatings are a practical and economical solution for the protection of parts against wear and corrosion. These processes are divided according to the type of heating source (laser, combustion gas, electric arc, plasma) and the feedstock material used for the deposition (powder, rod, wire) [3, 4, 5]. In this work a low velocity oxygen fuel (LVOF) process, known in the industry as the combustion process with oxygen-acetylene flame was used. This spraying method has been well-known in industry for many years being currently used for repairing damaged parts as well as for depositing new protective coatings, which can provide good resistance to wear and corrosion. In the process of repairing or depositing new coatings with oxyacetylene flame, the particles are accelerated and also partially melted inside of a turbulent flame and are deposited on a surface prepared in advance, having high roughness, thus forming a coating with a lamellar and

heterogeneous structure [6]. In the thermal spraying process, the most important mechanism of adhesion of the coating to the substrate is mechanical interlocking; the process creates a mechanical bonding between the particles and the asperities resulted after the sand blasting process. Following this deposition process, the adhesion of the coating is very low, due to this fact further treatment is necessary [7].
In the current work, to solve the initial problems regarding the increased porosity of the as-sprayed coatings and to minimise the induced micro-cracks, two types of post treatment processes are applied. The first one is the flame remelting process, where, typically NiCrBSi coatings are remelted by oxyacetylene flame during or after the deposition process. Due to the difficulty of maintaining a constant temperature during the fusion process and also interfering with the operator's experience and skills, this remelting process is not reproducible and also difficult to control [8]. The second post-treatment was performed using high frequency currents for heating and remelting the coating, increasing in this way, for the both applied processes, the adhesion of the coating to the substrate and significantly reducing the porosity.

2. Experimental procedure

2.1. Materials
The feedstock powder used in this study (480-FS6, LSN Diffusion, England) obtained through water atomisation, is a nickel-based alloy with a chemical composition shown in Table 1, and a particle size between 106-45μm.

Table 1. Chemical composition of Nickel powder and substrate.

	Composition, wt. %										
	Ni	Cr	B	Si	Fe	C	Mn	P	S	Mo	Al
Powder NiCrBSi	Base	13	2.5	3.5	3.4	0.5	-	-	-	-	-
Substrate 42CrMo4	-	1.13	-	0.26	Base	0.42	0.73	0.012	0.023	0.21	0.027

The high amount of chromium in the powder composition, increases the wear resistance of the coating, and the presence of Si and B favours the wettability and deoxidation during the remelting process [9].

2.2. Samples preparations
The geometry of substrate was cylindrical with a diameter of 160 mm and a length of 200 mm. Prior to deposition the samples were cleaned with alcohol, sandblasted with cast iron particles having a nominal size between 0.8 and 1.25 mm in order to reach a surface roughness higher than 75 μm. An increased roughness is strongly necessary because after the LVOF process the coatings adhere only mechanically to the substrate [1]. The coating procedure was performed by Karl-Schumacher GmbH, Germany using a flame spraying gun produced by Metatherm, Germany.
After the deposition, the coatings exhibit very high porosity and require a subsequent thermal treatment in order to close the pore and to achieve an increased adhesion to the substrate. Considering these essential aspects, two types of thermal treatments using different heat sources to reach the necessary remelting energy for the system were applied. First sample was remelted by flame using an oxyacetylene gas process with a neutral stoichiometry. The results for this type of process are directly connected and depend on the experience of the worker. The sample is slowly heated to the melting temperature, the heat dissipates inside the material and therefore it is no risk for crck inintiation after the remelting process. The main disadvantage of this method is that it is not suitable for large parts.

IOP Conf. Series: Materials Science and Engineering **416** (2018) 012002 doi:10.1088/1757-899X/416/1/012002

The second sample was remelted using an induction heater (EKOHEAT 200/30, Ambrell, The Netherlands) This induction heating unit requires 15 kHz to 40 kHz frequency range. According to a similar research made by Hemmati and the team [4], the remelting process consisted of a preheating step, comprising several passes at a lower power and a relatively faster velocity of the inductor in comparison with than the one used for the remelting procedure.
For further analysis, metallographic samples were prepared according to the standard guide for metallographic preparation of thermal sprayed coatings [10].

3. Results and discussions

3.1. Porosity of the NiCrBSi coatings
The micrographs were acquired with a Leica DM-RME light microscope and further processed. Figure 1 exhibits a considerably reduction of the porosity from 15% in the case of as sprayed coatings to a mean value of 3% after the flame remelting and to 1% after electromagnetic remelting treatment respectively. It is already known that heat treatment of NiCrBSi coatings will reduce their volume up to 20%. [1, 3]
The porosity degree was calculated with the aid of ImageJ processing Software from metallograpfic specimens in as-polished cross-section, presented in figure 1.

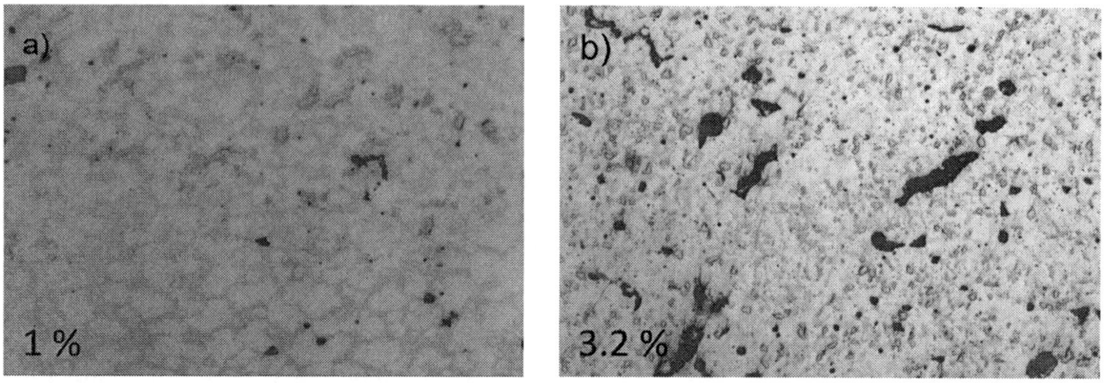

Figure 1. SEM micrograph of the a) inductive remelted and b) flame remelted NiCrBSi coatings

Previous works of the authors as well as the experimental results of this work showed that a complete densification cannot be achieved. In the case of flame remelting process, one may not assure an homogeneous temperature distribution in the coating material, fact which will induce some difference concerning the degree of porosity.
The flame remelted coating was more compact at the surface than at the interface (coating/base material), phenomenon which in the case of electromagnetic induction remelting was not observed. The induction fused coating revealed a very compact structure due to the influence of frequency on the penetration depth, which assured a constant temperature distribution maintained during the whole process.

3.2. Microhardness and adhesive properties of the NiCrBSi coatings
Microhardness investigations were carried out on a Zwick Microhardness tester according to ISO 6507 / ASTM E 384. The indentation were performed in different spots along the cross-section of the NiCrBSi coating, the load used for determining the Vickers microhardness was 3 N applied for 15 seconds on each indentation and the distance between measurements was minimum 0.15 mm. The investigations, revealed for the inductively remelted samples an average value of around 600 HV 0.3, and for flame-remelted ones, the hardness mean value was around 490 HV 0.3. This difference can be very goodassociated with the degreeof coating porosity.

Another very important property for the thermal sprayed coatings is their adhesion to the substrate. In order to determine if the coating has a good adhesion to the substrate, the samples were investigated concerning their adhesive properties

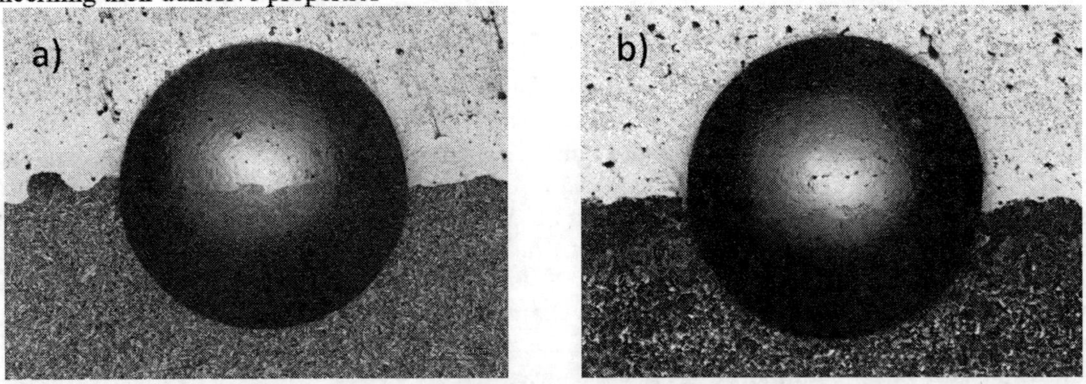

Figure 2. HBW2,5/187,5 indentation for a) inductive remelted and b) flame remelted NiCrBSi coatings

The test was carried out with a Hardness Testing Instrument equipped with a tungsten carbide Brinel type indentation ball with 2.5 mm nominal diameter and an applied force of 187.5 kgf. The indentations were made in different points along the coating/substrate interface (in cross-section). The results (figure 2) have shown that both remelting processes offer a good adhesion of the coating and do not present delamination at the interface

3.3. *Corrosion behavior*

The corrosion behavior of the NiCrBSi coatings was investigated by electrochemical testing in 3.5% NaCl aqueous solution. A saturated calomel electrode (SCE) was used as reference electrode, a platinum disk as auxiliary electrode and the NiCrBSi samples represented the working electrode. The samples were polarized in a potential interval from - 1000mV to + 1000 mV, with a scan rate of 0.16 mVs^{-1}

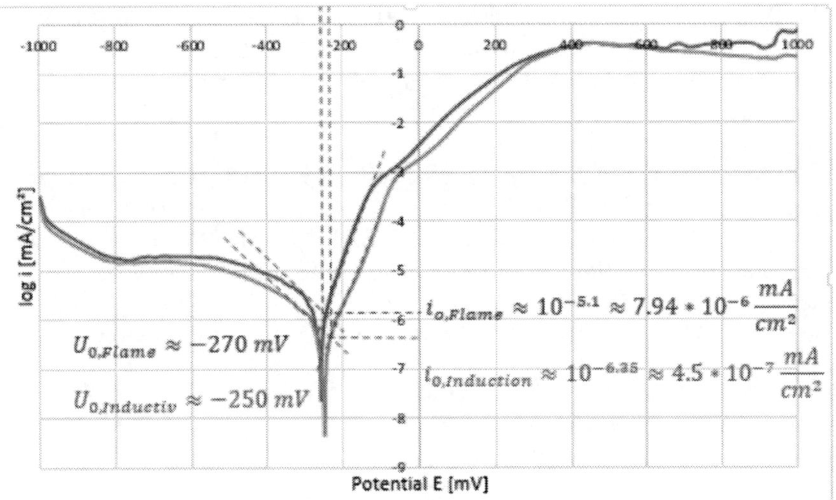

Figure 3. Polarisation curves of the flame and inductively fused NiCrBSi coatings tested in 3.5% NaCl aqueous solution.

The polarisation curves illustrated in figure 3 reveal no differences concerning the values for the corrosion potential of the tested coatings. As can be noticed, the corrosion current density i_{Corr} is 20 times lower in the case of the inductive remelted sample compared to the flame remelted one. This observation indicates a better corosion resistance of the inductive fused coatings exposed to a 3,5 % NaCl aqueous solution.

4. Conclusions

In the preliminary state, flame sprayed NiCrBSi coatings present high porosity, low cohesion between particles and poor adhesion to the substrate, as well as an irregular surface geometry. Therefore, the as sprayed coatings do not fulfil the technical requirements to be used in the industry and necessitate a remelting post-treatment. Two different types of remelting methods were applied and analyzed, namely manual oxyacetylene flame remelting and inductive remelting process. The obtained samples were further analysed regarding the microstructure and mechanical properties.

The inductive remelted coatings are more compact (lower porosity), the structures are much finer and better dispersed compared to the oxyacetylene flame remelted ones. After the remelting process, porosity decreased from 15% as in the case of the as-sprayed samples to 3.2% for flame remelted samples and as low as 1% for inductive remelted ones.

The microhardness investigations revealed a higher value (up to 100 Vickers units) for the inductive remelted coatings in comparison that remelted with oxyacetylene flame. This aspect was attributed to the more compact microstructure respectively lower porosity

The electrochemical corrosion tests showed that inductive remelted specimens exhibit better corrosion resistance than the flame remelted one.

In conclusion, inductively remelting process can be successfully applied for thermal sprayed coatings in order to improve the mechanical characteristics and the corrosion resistance, since the inductive remelted specimen exhibit a higher hardness, and an improved corrosion resistance, in comparison to the samples remelted with oxyacetylene flame, these aspects making them suitable for of shore applications.

5. References

[1] Methesius H-A and Krömmer W 2014 *Practice of Thermal Spraying vol. 10* (Unterscheissheim:DVS Media GmnH)
[2] Vieira LFS, Voorwald HJC and Cioffi MOH 2015 *Procedia Engineering* **114** pp 606-612
[3] Bergant Z, Trdan U Grum J 2014 *Corrosion Science* **88** pp 372-286
[4] Hemmati I, Ocelik V and Hosson JTM 2013 *Physics Procedia* **41** pp 302-311
[5] Şerban VA and Răduţă A 2012 Ştiinţa şi Ingineria Materialelor 3rd Ed. (Timişoara: Editura Politehnica)
[6] Davis JR 2004 *Handbook of Thermal Spray Technology* vol 1, (Ohio, ASM International)
[7] Bergant Z and Grum J 2009 *Journal of Thermal Spray Tehnology* **18(1059-9630)** pp 380-391
[8] Gonzales R, Garcia M, Penuelas I, Cadenas M and Felgueroso D 2007 *Wear* **263** pp 619-624
[9] ASM International 2016 *Alloy Phase Diagrams* vol. 3 ed. Okamoto H, Schlesinger ME and Mueller EM (Ohio: ASM International)
[10] ASTM Standard E1920-03 2014 (West Conshohocken: ASM International)

IOP Conf. Series: Materials Science and Engineering **416** (2018) 012003 doi:10.1088/1757-899X/416/1/012003

Electrochemical Corrosion Behavior of High Temperature Vacuum Brazed WC-Co-NiP Functional Composite Coatings

D T Pascal[1,*], N Kazamer[1,2,*], R Muntean[1], P C Valean[1,2], G M Marginean[1] and V A Serban[2]

[1]Department of Materials Science and Testing, University of Applied Sciences Gelsenkirchen, Neidenburger Str. 43, 45877 Gelsenkirchen, Germany, EU
[2]Department of Materials Science and Engineering, Politehnica University Timisoara, Piata Victoriei Nr.2, 300006 Timisoara, Timis, Romania, EU

*Corresponding author: norbert.kazamer@w-hs.de, dragos.pascal@w-hs.de

Abstract. The present study aims to investigate the electrochemical corrosion behavior of high temperature vacuum brazed WC-Co-NiP coatings deposited onto 16MnCr5 (1.7131) case hardening steel in 3.5% NaCl (pH 7) electrolyte solutions. Corrosion resistance is appraised using the potentiodynamic polarization method at room temperature in a three-electrode cell configuration. The coating characteristics like morphology, microstructure and chemical composition before and after the corrosion tests have been analyzed by means of scanning electron microscopy (SEM) combined with energy-dispersive X-ray (EDX) spectroscopy. This coating technology makes possible the deposition of high quality hardfacings, with minimal influence on the metallic substrate. The electrochemical measurements resulted in a lower corrosion current density, correlated with a significant improvement in the corrosion behavior of the WC-Co-NiP coated samples, compared to the uncoated ones.

1. Introduction

In nearly every area of industry, corrosion and wear are phenomena of paramount concern. It is estimated that the costs of corrosion in the US surpassed 1.1 trillion dollars in 2016 [1]. Recent surveys express that the worldwide (direct) cost of corrosion is between 1.3 and 1.4 trillion dollars, that is 3.1% to 3.5% of a developed country' annual GDP. These figures represent only the direct costs of corrosion basically materials, equipment, and services involved with repair, maintenance, and replacement. They do not comprise the environmental damage, waste of resources, loss of production, or personal injury caused by corrosion. Both NACE and NASA corrosion experts have determined that a net of 15% to 35% of the annual corrosion-related expense can be reclaimed by applying presently accessible corrosion control methods. According, even though corrosion is a natural process, thus it cannot be completely stopped, five primary techniques namely: material selection, coatings, inhibitors, cathodic or anodic protection and design can be applied in order to control it [1-3].

Although the brazing process is usually employed as a joining technique, in this study the possibility of using this method as a coating technology was tested, with the purpose of bringing fresh and innovative solutions in the field of functional coatings. Applied as a coating technique, high temperature vacuum brazing of flexible tapes concludes in increased cohesion, and adhesion of the coating to the substrate. These coatings are usually employed as protection against wear in a wide variety of applications, most of which requiring at the same time high corrosion resistance [4].

Metallic matrix functional composite coatings are generally utilized to isolate the base material from the corrosive environment to protect it. Therefore, they are more noble than the substrate and offer protection by way of anodic or barrier action. In order to possess adequate corrosion resistance, the structure has to be as dense as possible, without pores, cracks or any other defects. If the deposit is porous or it contains micro-cracks, the corrosive media can reach down to the base material. In this case, the coating might detach, and the substrate will come in direct contact with the working environment [5]. The reinforcing material selected for the current study was cobalt cemented tungsten carbide (WC-Co). Cemented tungsten carbides possess a heterogeneous structure, where corrosion occurs predominantly in the metallic binder, because it generally has a lower corrosion resistance compared to the ceramic phase (WC). For example, cobalt (often used as binder) has relatively low corrosion resistance in sulfuric, hydrochloric, nitric and other strong acids. The corrosion behavior of cemented carbides is highly dependent on the nature and proportions of metallic binder constituents. For instance, the presence of nickel significantly improves the corrosion resistance, possessing lower corrosion rates compared to cobalt. Addition of chromium will improve the resistance to oxidation, nevertheless it can also severely affect particular mechanical properties through the formation of chromium-rich brittle carbides [6, 7].

2. Experimental

Manufacturing of the currently employed polymer bond hardfacing tapes involves mixing of 75 wt.% WC-Co (PA2), comprising of 7.5 wt.% Co, 5.7 wt.% C, and W as balance, recycled cermet powder with 25 wt.% NiP (Ni6), consisting of ≃ 11 wt.% P, max. 0.06 wt.% C and Ni as balance, brazing filler metal powder and an additional 2 wt.% water-based fugitive organic binder until the latter one fibrillates, and doing so, anchors the powder particles, allowing further processing steps. Both types of powders are commercialized by the Swedish company Höganäs AB. The mixture is subsequently rolled to a predetermined thickness, forming a flexible cloth, afterwards cut to the required shape and size.

The high temperature vacuum brazing process was carried out in a HITERM 80-200 cold wall vertical vacuum furnace at the stable pressure of around $3.0 \cdot 10^{-4}$ mbar and a brazing temperature of 1000 °C for a soaking time of 30 min. During the high temperature vacuum brazing process, the NiP brazing filler alloy melts and infiltrates the WC-Co particles. Metallurgical reactions, governed by diffusion and alloying occur at the interfaces between the reinforcing particles (WC-Co hard phase) and metallic matrix (NiP alloy), as well as at the coating/substrate one. Consequently, the WC-Co-NiP/16MnCr5 systems possess a strong metallurgical bond [4]. The detailed procedure of manufacturing such hardfacings was reported in a different study [8]. The morphology and microstructure of the coating and coating-substrate interface is illustrated in the BSE cross-section micrograph of figure 1.

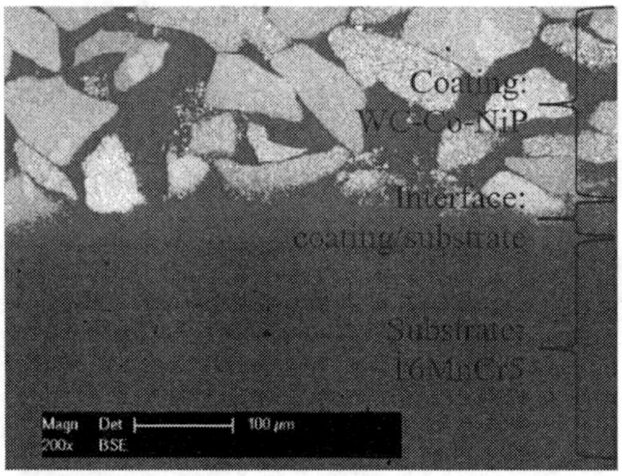

Figure 1. BSE micrographs of WC-Co-NiP/16MnCr5 system.

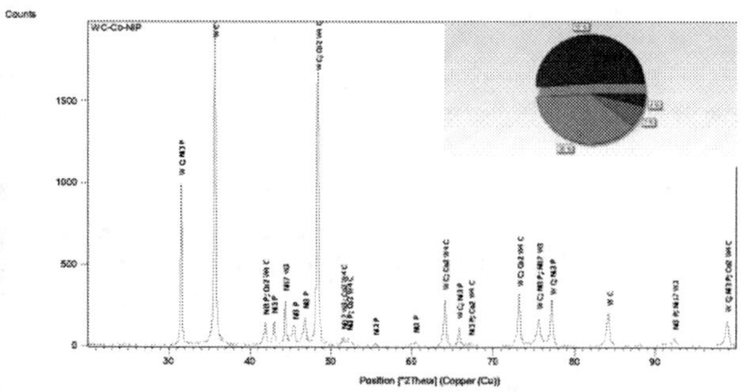

Figure 2. XRD pattern of WC-Co-NiP coating.

The high temperature vacuum brazed coatings display a dense structure, with uniform distribution of the cermet particles. One can clearly observe that the molten self-fluxing alloy infiltrated between the WC-Co particles, cementing them together, and at the same time bonding the entire system with the metallic substrate.

The X-ray diffractogram displayed in figure 2 reveals that the hardfacings consist of 51 wt.% WC, 38 wt.% Ni_3P, 7 wt.% $Ni_{17}W_3$ and 4 wt.% Co_2W_4C. Presence of other phases (in undetectable quantities) is plausible, because of the high number of elements involved.

The electrochemical corrosion behavior is evaluated through potentiodynamic polarization, employing a VoltaLab PGP201 potentiostat/galvanostat in a three-electrode cell configuration. For comparison purposes, working electrodes (WE) were prepared from the 16MnCr5 substrate material, WC-Co-NiP functional coatings developed throughout the present study, and two types of thermal sprayed coatings with similar chemical composition, namely WC-Co-Mo and WC-Co respectively. The chemical composition of the thermal sprayed coatings is only similar and not identical because at the present time there are no available methods of depositing thick coatings with the chemical composition of the vacuum brazed coatings, namely WC-Co-NiP. A platinum disk was used as counter electrode and a saturated calomel electrode (SCE) as reference. Prior to the investigations, the WEs have been ground with metallographic paper in order to remove potential oxides and to ensure as close as possible the same surface roughness (R_a 0.03). Subsequently, the specimens are rinsed with ethanol and dried under a stream of warm air. Polarization studies were performed at room temperature in the potential interval between -1000 mV and +1000 mV vs. SCE with a scan rate of 10 mV min^{-1}. The 1 cm^2 geometrical surface area of the sample allowed current readings to be directly expressed as current density (A cm^{-2}). The cathodic branch in the polarization curves corresponds predominantly to hydrogen evolution reaction, while the anodic branch features the most important information related to the corrosion resistance.

Figure 3 illustrates typical polarization plots for the discussed samples subjected to potentiodynamic electrochemical corrosion in 3.5% NaCl solution (pH 7). As expected, the corrosion current density (i_{corr}) of the WC-Co-NiP brazed coating is substantially lower than that of 16MnCr5.

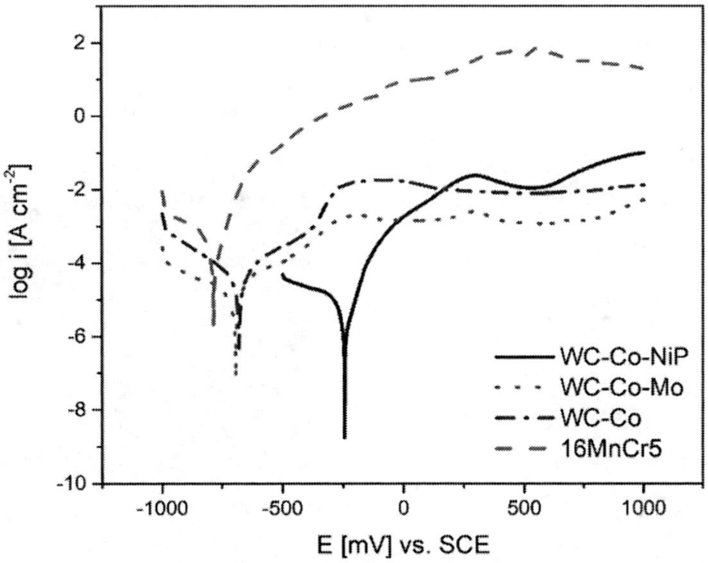

Figure 3. Polarization behavior of vacuum brazed WC-Co-NiP, thermal sprayed WC-Co-Mo, WC-Co and 16MnCr5 steel in 3.5% NaCl solution.

Correspondingly, the corrosion potential (E_{corr}) of the hardfacing is shifted to more positive potentials (-245 mV) in comparison to the uncoated sample (-780 mV), aspect associated with the

Ni-based chemical composition. The anodic branch (see figure 3) of the polarization curve shows an inclination to passivate, due to the high amount of Ni present in the brazing filler alloy. Adversely, the uncoated steel sample points no tendency towards passivation, due to the Cr content lower than 12 wt.%. Both corrosion current density (i_{corr}) and corrosion potential lean toward a lower reaction rate, and consequently, better corrosion resistance of the coated samples compared to the uncoated ones is to be expected [9].

The polarization curves of WC-Co-NiP coated samples show passive plateaus at values between 300 mV and 550 mV, indicating the formation of a passive oxide film. Further increasing the potential in the anodic direction, the passive film previously formed on the surface of the sample, brakes down and the nickel rich phase of the coating is subsequently dissolved. The values presented in table 1 indicate a corrosion current density around 45 times lower for the brazed coating in comparison to substrate material. Although both types of thermal sprayed coatings show a more stable behavior in the anodic branch, with a more pronounced tendency of passivation and a wider passive region, the brazed coating exhibits lower corrosion currents and the corrosion potential is shifted to more positive values, indicating a higher resistance of the material in this specific testing conditions. This fact can be attributed to the presence of the corrosion resistant Ni_3P phase revealed from the XRD patterns, which is formed during the vacuum brazing process [10].

Average corrosion potential and corrosion current density of the analyzed samples obtained from the electrochemical polarization investigations performed in 3.5% NaCl aqueous solution are summarized in table 1.

Table 1. Corrosion potential and corrosion current density values in 3.5 wt.% NaCl.

Sample	E_{corr} [mV] vs. SCE	i_{corr} [A cm^{-2}]
WC-Co-NiP	-245	$0.4 \cdot 10^{-5}$
WC-Co-Mo	-695	$0.6 \cdot 10^{-5}$
WC-Co	-685	$1.8 \cdot 10^{-5}$
16MnCr5	-780	$18 \cdot 10^{-5}$

The micrographs illustrated in figure 4 offer an overview (top-view) of the coatings' surface (testing area) a) before, and b) after the electrochemical investigation. It can be observed that the metallic binder is severely affected, while the reinforcing phase (WC) retains its structural integrity.

Figure 4. Optical micrographs of: a) coating surface before electrochemical investigation, and b) after electrochemical investigation.

Because transition metal carbides have high resistance to chlorine containing solutions, corrosion of WC-based composite coatings manifests predominantly through the dissolution of the metallic binder. Subsequently, the metallic matrix is depleted, leaving behind only a skeleton of WC particles with severely affected properties, as shown in the scanning electron micrograph of figure 5.

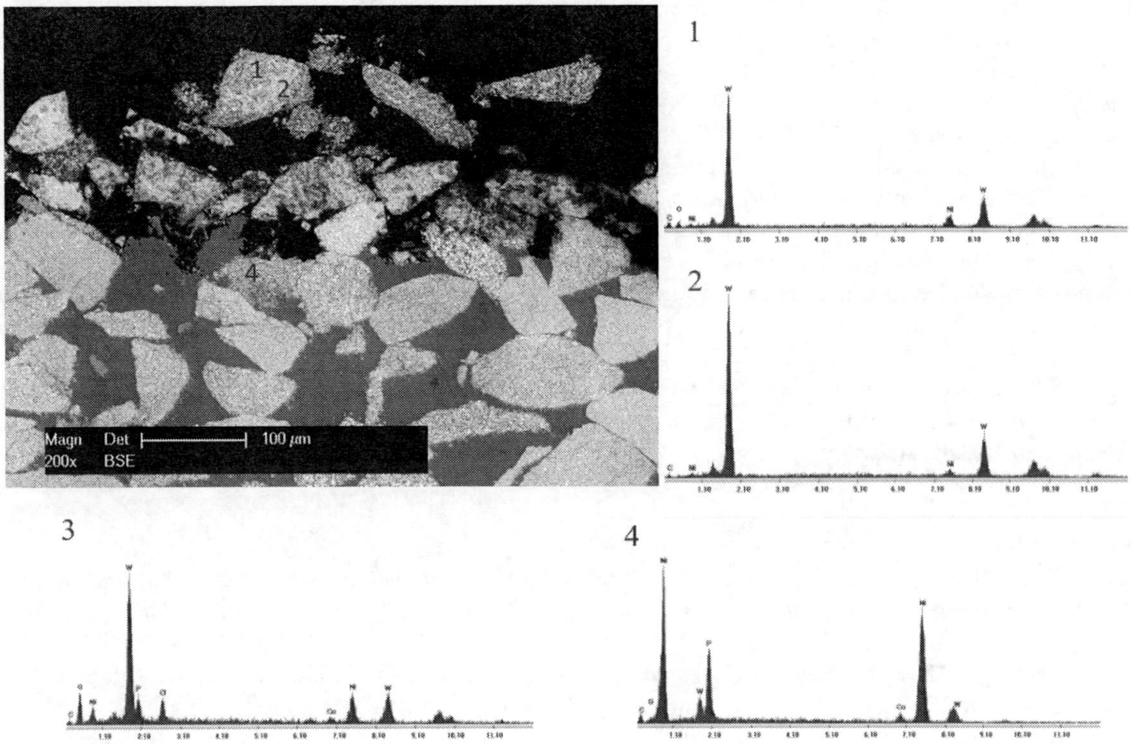

Figure 5. SEM cross-section micrograph and EDX spectra corresponding to the marked regions, after electrochemical corrosion.

The selective corrosion mechanism is experienced by the majority of (metal-matrix) composite coatings and it was observed in several studies conducted by other researchers [11-15]. The superior corrosion resistance of WC-Co-NiP coatings compared to the 16MnCr5 case hardening steel substrate is attributed to the formation of the nickel passive layer. This film forms willingly in contrast to the development of the passive films on iron. Dissimilarities in the nature of the oxide layer grown on Fe and Ni are accountable for this phenomenon. The layer thickness on Ni is between 0.9 mm and 1.2 mm, while the Fe oxide layer is situated in the interval 1 mm and 4 mm and has an elevated predisposition to chemical dissolution. There are two theories regarding the morphology and chemical composition of the nickel passive film. One theory stipulates that the layer is completely NiO with a small quantity of nonstoichiometric nickel, producing Ni^{3+} cation vacancies. In the case of the second one, it comprises of an inner film of NiO and an outer one of anhydrous $Ni(OH)_2$. Once generated, the oxide film of nickel, neither cathodic treatment nor chemical dissolution can easily remove it [16]. The passive film of nickel will not offer protection against corrosion in strongly oxidizing media, for instance in nitric acid. Alloying with chromium will generate a denser, more stable film, increasing the corrosion resistance to a variety of oxidizing environments. However, these alloys are susceptible to attack in atmospheres containing chlorides or other halides, particularly if oxidizing compounds are present. Corrosion will manifest mostly in the form of pitting and occasionally as stress corrosion cracking. Addition of molybdenum, titanium and/or tungsten should improve the behavior under all working conditions [17]. The EDX spectra of figure 5 confirm the formation of Ni oxides and possibly W oxides as well as the dissolution of the metallic binder. The corrosion behavior of the present high

IOP Conf. Series: Materials Science and Engineering **416** (2018) 012003 doi:10.1088/1757-899X/416/1/012003

temperature vacuum brazed WC-Co-NiP functional coatings was found to be in good compliance with those of WC-based claddings tested under similar conditions by other researchers [11, 18].

3. Conclusion

The corrosion behavior of high temperature vacuum brazed WC-Co-NiP functional composite coatings, potentiodynamic polarized in 3.5% NaCl was found to be similar to the one observed in the case of thermal sprayed WC-Co-Mo and WC-Co respectively. The corrosion current density (i_{corr}) of the WC-Co-NiP brazed coating is lower than that of thermal sprayed WC-Co-Mo and WC-Co. The corrosion potential (E_{corr}) of the brazed hardfacing is shifted to more positive values in comparison to that of thermal sprayed coatings, which is associated with a lower reaction rate. The electrochemical polarization curves illustrated a 45 times lower corrosion current density in the case of WC-Co-NiP coatings compared to 16MnCr5 case hardening steel samples in the 3.5% NaCl electrolyte solution. The superior corrosion resistance of WC-Co-NiP functional coatings is assigned to the chemical composition of the metallic matrix, namely the Ni-P alloy (Ni_3P phase), which is able to readily generate a passive layer, stable up to the potential of around ≈ 600 mV vs. SCE. The corrosive attack onto vacuum brazed WC-Co-NiP coatings is characterized by selective dissolution of the metallic matrix, compared to the generalized corrosion observed in the case of the unprotected substrate material. Accordingly, the present work proved vacuum brazed WC-Co-NiP coatings are highly effective in the anodic protection of the 16MnCr5 substrate material against the 3.5% NaCl corrosive media.

4. References

[1] Koch G, Varney J, Thompson N, Moghissi O, Gould M and Payer J 2016 *International Measures of Prevention, Application, and Economics of Corrosion Technologies Study* (Houston: NACE International)

[2] Herridge L 2006 *Combating Corrosion* (Florida: NASA's John F. Kennedy Space Center)

[3] Davis J R 2000 *Corrosion: Understanding the Basic* (Ohio: ASM International)

[4] Bobzin K 2013 *Oberflächentechnik für den Maschinenbau* (Weinheim: Wiley-VCH Verlag)

[5] Cramer S D and Covino B 2003 *Corrosion: Fundamentals, Testing, and Protection* (Ohio:ASM International)

[6] Federal Carbide Company 2018 *Corrosion resistance of Tungsten Carbide Grades* (Tyrone: Federal Carbide Company)

[7] Robert J K Wood 2007 *J. Phys. D: Appl. Phys.* **40** 5502

[8] Pascal D T, Serban V A and Marginean G 2016 *Solid State Phenom* **254** 164-9

[9] Ranganatha S and Venkatesha T V 2012 *Phys. Scr.* **85** 035601

[10] Zhang D Q, Gao L X and W Huan 2012 *Adv Mat Res* **347-353** 1605-8

[11] Zhang D Q, Liu T, Joo H G, Gao L X and Lee K Y 2012 *Int J Refract Metals Hard Mater* **35** 246-50

[12] Joo H G, Zhang D Q, Gao L X, Liu T and Lee K Y 2014 *Int J Surf Sci Eng* **8** 380-91

[13] Kazamer N, Pascal D T, Marginean G, Serban V A, Codrean C and Utu I D 2016 *Solid State Phenom* **254** 71-6

[14] Gao L X, Zhou T, Zhang D Q and Lee K Y, 2014 *Corros Eng Sci Techn* **49** 204-8

[15] Kazamer N, Pascal D T, Marginean G, Serban V A, Brandl W and Valean P C 2016 *Nanocon* (Brno: TANGER Ltd.)

[16] Schweitzer P A 2004 *Encyclopedia of Corrosion Technology, Second Edition* (CRC Press Taylor & Francis Group)

[17] Schweitzer P A 2007 *Fundamentals of Metallic Corrosion: Atmospheric and Media Corrosion of Metals* (CRC Press Taylor & Francis Group)

[18] M L Benea and L P Benea 2015 IOP Conf. Ser.: Mater. Sci. Eng. **85** 012004

Studies on Pulse Electrodeposition of Pt-Ni binary Alloy For Electrochemical Cell Applications

R Muntean, D T Pascal, U Rost, P Podleschny, M Schumacher and G Mărginean

Westfälische Hochschule Gelsenkirchen Bocholt Recklinghausen, Neidenburger Straße 43, D-45877 Gelsenkirchen, Germany

*Corresponding author: roxana.muntean@w-hs.de

Abstract. The possibility to prepare a Pt-Ni bimetallic catalyst for electrochemical cells applications using a two-step electrochemical process was tested. The binary alloy catalyst was deposited on a graphitic substrate covered with carbon nanofibers previously treated in oxygen plasma. This support system provides a high surface area for the alloy particles, improving considerably the catalytic activity. A simultaneous reduction reaction of Pt and Ni was performed by strictly controlling the metals ratios in the solution and by choosing suitable deposition parameters. A pulsed current technique was applied. Varying the Pt: Ni ratio in the electrolyte and adjusting the deposition main parameters lead to a homogeneous distribution of the catalyst. A comparison with pure platinum nano-scaled supported electrodes is also presented, regarding hydrogen/oxygen evolution reaction. Although a significant decrease of platinum amount was achieved by alloying, a higher activity of the catalyst was obtained. Morphology, structure, and chemical composition of the catalysts was investigated by means of scanning electron microscopy combined with energy dispersive X-ray spectroscopy. X-ray diffraction was used to identify the alloy phase composition. The amount of deposited alloy determined using thermogravimetric measurements correlated with the cyclic voltammetry performed in 0.5 M H_2SO_4 solution to estimate the electrochemical surface area offered important information regarding the catalytic activity.

1. Introduction

The most common catalyst material for both fuel cells and electrolysers is platinum. Improving the efficiency of platinum-based catalysts for the main electrochemical reactions, such as oxygen reduction/oxidation reaction, has always been a priority for research and development [1]. As this material is a very scarce metal, numerous studies have been conducted to reduce the platinum consumption, while maintaining the essential characteristics. Alloying has been demonstrated to be a suitable approach to obtain advanced catalytic materials [2]. Recently, platinum alloys have attracted attention due to their great electrocatalytic properties and applicability in various fields, ranging from catalytic applications in fuel cells, electrolysers, batteries, and sensors to medical and aerospace industry. Lately, Pt-based binary and ternary alloys have been developed using various routes, in order to achieve a better control of particle size and phase composition, aspects which are considered significant for an optimum catalyst [3]. Among the various deposition methods, electrodeposition exhibits several distinctive and unique characteristics [4]. Electrodeposition is an appropriate method for catalyst synthesis because it is both easy to apply and requires a low-cost technology. Moreover, using this technique, a good control of the catalyst properties can be achieved [5]. By reducing the catalyst particle size, an increase in the active surface area is attained, and moreover, a significant decrease in the Pt loading may be reached, correlated with faster reaction rates. Alloying Pt with non-noble metals like Ni or Co reduces the Pt content of the electrocatalysts and significantly increases the electrochemical activity [6]. Likewise, the addition of a second metal has been shown to improve the surface chemistry and the reactivity of Pt and confers the catalyst a bi-functional mechanism [2, 7].

Particularly, Pt-transition metal (Pt-M) bimetallic alloys, such as Pt–Co, Pt–Ni, Pt–Fe, Pt–Cr, and Pt–Ti were found to exhibit a lower overpotential for the oxygen reduction reaction, presenting a higher catalytic activity [8- 10]. Pt_3Ni alloy phase composition was found to enhance the ORR activity by two orders of magnitude (in comparison to Pt/C catalysts) by changing the electronic structure of Pt surface atoms [8]. Furthermore, Ni alloyed with Pt was found to reduce the electronic binding energy of Pt, which also contributes to a higher activity of the catalyst [2].

In addition, an appropriate support material for the catalyst particles, with a great surface area, high electrical conductivity, and stability, offers a good dispersion of Pt and Pt alloy nanoparticles, reaching a maximum utilization degree of the noble material. Carbon based materials with delocalized π-electrons, are considered to be an excellent support for the catalyst particles, due to the extremely large surface area, high electrical conductivity, good mechanical resistance and chemical and thermal stability [2, 11].

Considering the advantages of electrodeposition, such as high purity of deposits and relatively simple procedure of deposition, in the present study, platinum–nickel (Pt-Ni) bimetallic alloy nanoparticles with a uniform mean diameter of 10 nm, were anchored onto a carbon support material using the electrodeposition method. The resulting Pt-Ni/CNF catalysts have been characterized by Scanning Electron Microscopy (SEM), X-ray diffraction (XRD), energy dispersive X-ray spectroscopy (EDX). The electrocatalytic properties of the Pt-Ni/CNF electrodes have been investigated by cyclic voltammetry (CV).

2. Experimental

2.1 Materials and methods

Graphene related materials (GAqUA, carbon nanofibers-water suspension) with high specific surface area and good electrical conductivity purchased from Grupo Antolin Ingineria, Spain were used as support material for the catalyst alloy particles. CNF are firstly deposited on the surface of a commercially available gas diffusion layer (GDL, H2315 I2C8, Freudenberg FCCT SE & Co. KG) by air brushing (USI Prism BT 500 System), having a uniform 10 $\mu g\ cm^{-2}$ CNF loading. Other reagents used for the Pt-Ni alloy deposition are hexachloroplatinic acid (H_2PtCl_6) (Galvatron Platinbad, 10 g Pt L^{-1}, Wieland Edelmetalle GmbH, Germany), nickel (II) chloride hexahydrate (97%-Ni, Merck KGaA, Darmstadt, Germany), sulfuric acid (H_2SO_4) (98%, Carl Roth GmbH + Co. KG, Germany) and distilled water. Three electrodeposition baths were prepared, with different concentrations of nickel, as presented in table 1.

Table 1. Chemical composition of PtNi deposition baths

Bath	1:2	1:4	1:6
Pt content [g L^{-1}]	10	10	10
Ni content [g L^{-1}]	20	40	60

The main challenge of the catalyst nanoparticles electrodeposition on this specific support material is the hydrophobic character of the carbon. For a homogenous and fine dispersed deposition of the catalyst, a functionalization treatment of the substrate surface is advantageous. An oxygen plasma

IOP Conf. Series: Materials Science and Engineering **416** (2018) 012004 doi:10.1088/1757-899X/416/1/012004

treatment of the carbon surface before the electrodeposition significantly improves the catalyst distribution. The detailed procedure and parameters are described elsewhere [7,12]. Moreover, the effects of the electrolyte composition and electrodeposition parameters have been investigated in detail. For electrocatalytic applications, nucleation step should be predominant in order to generate numerous particles and consequently a large electrochemical active surface area. Various techniques, including CV, SEM, XRD were used in this study to investigate the catalytically relevant parameters of the developed electrodes. The morphology, size and distribution of catalyst particles on the CNF support were examined by SEM. XRD method was used to investigate the phase structure of the developed catalyst materials and further to calculate the crystallite size according to Debbie-Scherrer equation.

2.2 *Electrode preparation*

The synthesis process of Pt–Ni electrodes using electrodeposition consisted of two steps. The first step was the surface modification of the carbon support material, creating a hydrophilic layer with functional groups attached onto the surface of the CNF. The second step was the galvanostatic electrodeposition of Pt–Ni and Pt on the pretreated carbon substrates.

As the first step was developed and described in a previous work, the focus was directed on the second step, namely the Pt–Ni co-deposition. The electrodeposition was performed in a 250 ml plating bath containing 10 g L^{-1} M H_2PtCl_6 and 20-60 g L^{-1} M $NiCl_2 \cdot 6H_2O$, in a three-electrode system, consisting of working electrode (as-prepared carbon support), counter electrode (platinum disk), and reference electrode-saturated calomel electrode (SCE), using an electrochemical workstation (Potentiostat/Galvanostat Ivium Technologies Vertex). The active area of the working electrode was 1 cm^2. Deposition parameters such as current density, applied time (t_{on} and t_{off}), and electrolyte composition were varied in order to obtain an optimum electrochemical performance of Pt–Ni alloy catalysts. Moreover, a good dispersion of the catalyst on the carbon support is an important condition for attaining a superior electrocatalytic activity for fuel cells reactions.

3. Results and Discussions

3.1 *Influence of the deposition current density*

The influence of the applied current density on morphology and chemical composition of the Pt-Ni alloy has been investigated by applying different current densities in the range of 50 to 200 mA cm^{-2}. The finest structures were obtained for three of the applied values, namely 100 mA cm^{-2}, 125 mA cm^{-2}, 150 mA cm^{-2} according to the SEM micrographs presented in figure 1.

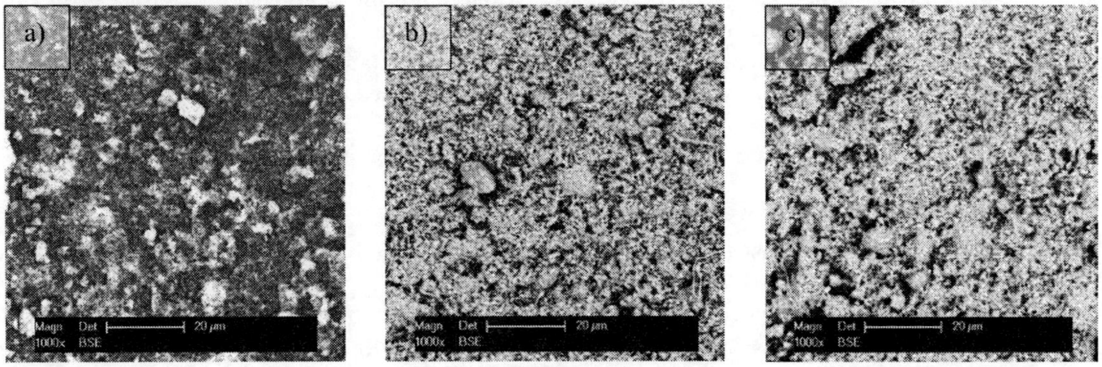

Figure 1. SEM micrographs of PtNi/CNF samples deposited with: a) 100 mA cm^{-2}; b) 125 mA cm^{-2}; c) 150 mA cm^{-2} from electrolyte 1:2, with t_{on} = 12 ms, d.c = 10%, 2000 cycles .

IOP Conf. Series: Materials Science and Engineering **416** (2018) 012004 doi:10.1088/1757-899X/416/1/012004

The other variables like bath composition, t_{on} = 12 ms, d.c = 10%, and number of cycles (2000) were maintained constant. As the current density affects the nucleation process of the catalyst particles, an increase in the current density over 125 mA cm^{-2} intensifies the hydrogen evolution reaction (HER), favouring the particles growth mechanism by achieving the diffusion limiting current density. According to the preliminary linear voltammetry performed with the working electrolyte, the electrodeposition of Pt-Ni alloy using a current density higher than 125 mA cm^{-2} takes place under mass transfer limitations for platinum and the excess current density is used for the nickel deposition, simultaneously with HER. As HER proceeds with a faster rate than nickel deposition, a larger fraction of the excess current is used towards the proton reduction reaction and consequently the CNF layer may be damaged. As it can be observed in SEM micrographs illustrated in figure 1, when a higher current density is applied, the particles start to grow, and dendrites start to form. An optimum nucleation and size particle is obtained for a deposition conducted with 125 mA cm^{-2}. Furthermore, the SEM investigation are correlated with the cyclic voltammetry measurements performed in 0.5 M H_2SO_4 solution, in order to highlight the electrochemical activity of the catalysts. The electrodes were scanned for 200 cycles with 100 mV s^{-1} scan rate and the results showed in figure 2 represent the last scan of each sample.

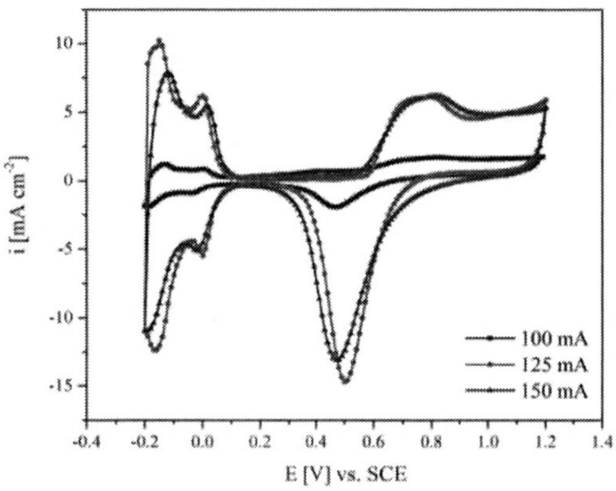

Figure 2. CV polarization curves for three PtNi/CNF electrodes deposited with different current densities (100 mA cm^{-2}, 125 mA cm^{-2} and 150 mA cm^{-2}) in 0.5 M H_2SO_4 solution with a scan rate of 100 mV s^{-2}.

CV plots performed on the three investigated samples exhibit, as expected, the same shape as for a typically pure platinum electrode. In the potential range of -0.2 – 0.1 V vs. SCE, the peaks are associated to the adsorption/desorption processes of hydrogen on Pt-Ni surface. In the anodic branch, in the potential range of 0.1 – 0.6 V vs. SCE, the curve corresponds to the charge of the double layer due to the oxygenated groups on the carbon support surface [9]. Starting from 0.6 V vs SCE, the oxidation process of Pt takes place. In the cathodic polarisation, the correspondent reduction peak from 0.5 V is linked to the reduction process of platinum oxides previously formed. In the presented voltammograms, the higher activity is observed for the Pt-Ni sample deposited with 125 mA cm^{-2} and this aspect may be attributed to the amount of catalyst and size of the particles in the alloy [9].

3.2 Influence of the t_{on} and t_{off}

By changing the main deposition parameters, it is possible to modify the alloy composition, current efficiency and nucleation rate [13]. Regarding the deposition time, generally, a short t_{on} pulse makes the electrodeposition to be dominated by nucleation. The t_{off} pulse has also a significant role in the deposition kinetics since many effects like replenishment of concentration gradients and recrystallisation occur in this step. The plots presented in figure 3 display the influence of t_{on} on the ECSA of the catalysts. A higher value is obtained for the sample deposited with 12 ms (16 m^2 g^{-1} Pt).

The effect of t_{off} pulse on the nucleation rate is less substantial. If the deposition current density is maintained for a longer period of time, the deposition is under high crystal growth conditions, the diffusion layer is depleted in the platinum ions, generating a concentration gradient between the bulk electrolyte and the electrolyte/electrode interface and the electrodeposition changes to diffusion limited

IOP Conf. Series: Materials Science and Engineering **416** (2018) 012004 doi:10.1088/1757-899X/416/1/012004

step [13]. Co-deposition of Pt with Ni can be achieved only under diffusion limiting conditions for Pt, resulting also in hydrogen evolution [4]. Consequently, the deposition of nickel, which is less noble than platinum takes place easier in these specific conditions. In the case of smaller t_{on} (less than 5 ms), the deposition consists predominantly of platinum particles. When a higher t_{on} is applied, nickel particles are co- deposited forming the alloy structure. A further increase of t_{on} (more than 15 ms) leads to a decrease in the nucleation rate and the system is controlled by the crystal growth and the alloy particles start to increase. Figure 4 presents the structure of the catalytic PtNi material obtained with 12 ms pulse time.

3.3 Influence of the number of cycles

Number of cycles has a strong influence on the catalyst loading of the electrodes. Since the total deposition time is higher, according to Faraday's Law, the amount of the alloy is correspondingly higher. TGA measurements confirm the increase in the amount of the metallic catalyst, proportional with the applied number of cycles. Regarding the morphology and the chemical composition, no structural changes can be observed when increasing the number of applied cycles. The size of the particles remains constant according to XRD measurements.

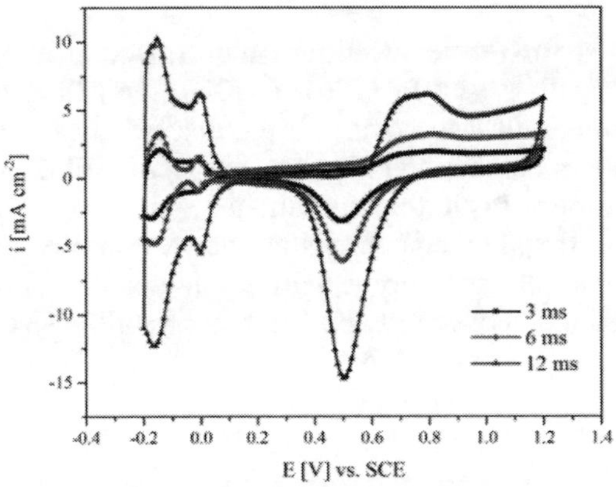

Figure 3. CV polarization curves for three PtNi/CNF electrodes deposited with different t_{on} (3 ms, 6 ms, 12 ms), 125 mA cm^{-2}, in 0.5 M H_2SO_4 solution with a scan rate of 100 mV s^{-2}.

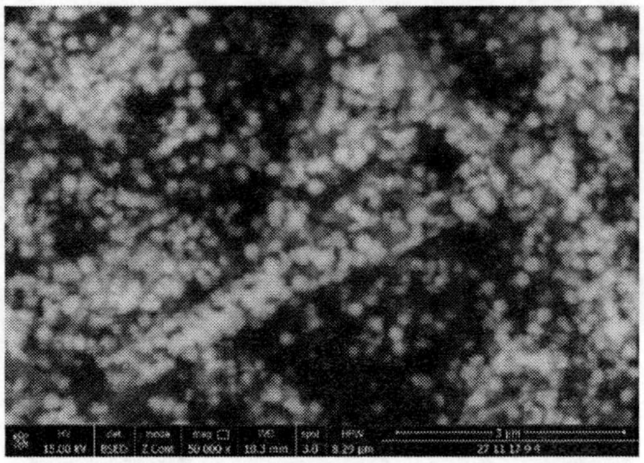

Figure 4. SEM micrographs of PtNi/CNF samples deposited with 12 ms; from electrolyte no. 1, with a current density of 125 mA cm-2, constant d.c of 10%, and 2000 cycles (mag. 50000x).

3.4 Influence of the bath concentration

The chemical composition of the working electrolyte has been shown to influence the alloy phase composition of the catalyst particles. Different ratios of Ni could be co-deposited with Pt by keeping the concentration of H_2PtCl_6 constant (10 g L^{-1}) and varying that of $NiCl_2$ in the plating bath (from 20 g L^{-1} to 60 g L^{-1}). As XRD measurements confirm, increasing the Ni ions concentration in the deposition bath determines an increase in the Ni content found in the catalyst alloy structure, correlated with a higher amount of NiPt3 alloy phase. Moreover, a higher concentration of nickel in the electrolyte determines an increase in the nucleation rate and additionally, creates changes in the catalyst structure. The deposition is acicular when higher Ni concentration is applied. Consequently, smaller particle sizes are obtained for Pt-Ni alloys deposited from the deposition bath with a higher nickel concentration (figure 5).

IOP Conf. Series: Materials Science and Engineering **416** (2018) 012004 doi:10.1088/1757-899X/416/1/012004

Figure 5. SEM micrographs of PtNi/CNF samples deposited from different Ni content deposition baths a) 1:2; b) 1:4; c) 1:6, with 125 mA cm^{-2}, t_{on} = 12 ms, constant d.c of 10%, 2000 cycles.

In figure 6, the CV measurements for three PtNi/CNF electrodes deposited from different deposition baths (1:2, 1:4, 1:6), with 125 mA cm^{-2}, in 0.5 M H_2SO_4 solution with a scan rate of 100 mV s^{-2} are presented.

Figure 7 shows the XRD patterns of the Pt/CNF and PtNi/CNF electrodes with different nickel contents. The intensity of Pt characteristic peaks such as Pt (111), Pt (200), Pt (220), and Pt (311) decreases when the Ni content increases in the deposition bath. Moreover, all the detected peaks for PtNi samples are shifted toward higher values of 2θ angle in comparison to standard spectra of Pt.

During the co-deposition process, nickel atoms randomly substitute platinum atom positions in the metallic lattice. As it can be remarked from the diffractograms, no characteristic lines for pure Ni can be observed [14]. As Ni atoms are considerably smaller that platinum atoms, lattice parameters become also smaller, fact which clearly emphasizes the metallic bond between Pt and Ni. The crystallite size is calculated from the Pt_3Ni (111) reflexion using Scherrer Equation.

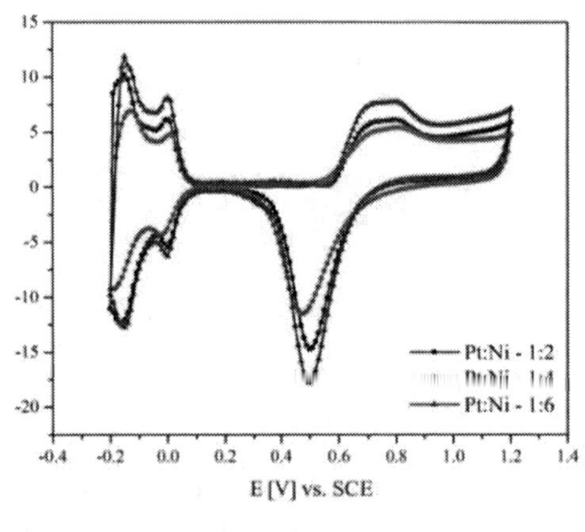

Figure 6. CV polarization curves for three PtNi/CNF electrodes deposited from different deposition baths (1:2, 1:4, 1:6), with 125 mA cm^{-2}, in 0.5 M H_2SO_4 solution with a scan rate of 100 mV s^{-2}.

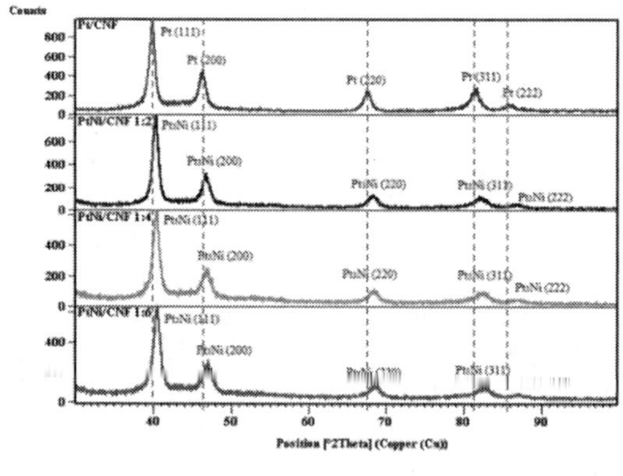

Figure 7. XRD patterns for Pt/CNF (reference electrode) and PtNi/CNF electrodes deposited from different bath compositions (1:2, 1:4, 1:6)

3.5. HER/OER on the PtNi/CNF developed electrodes

As it can be observed in figure 8, HER is significantly intensified and the onset potential decreases considerably for PtNi/CNF samples in comparison to a reference Pt/CNF electrode.

The HER depolarisation observed in the linear voltammetric plots is influenced by the presence of nickel in the structure of the catalyst. Beyond the HER onset potential, the cathodic current densities are considerably higher for PtNi/CNF electrodes at the same overpotential value. Around -0.5 V vs. SCE, the cathodic current density for PtNi/CNF (2000 cycles) reaches significantly higher current densities (400 mA cm^{-2}) in comparison to the reference Pt/CNF sample. The enhanced HER activity can be attributed mainly to the increased ECSA.

OER for platinum containing electrodes is known to be slow in acidic medium due to catalyst surface oxidation and Ni lower stability. Although the Pt/CNF shows a sluggish OER process (figure 9) with an onset potential around 1.5 V vs. SCE, the PtNi/CNF electrode prepared with 2000 cycles exhibits a lower onset potential of around 1.4 V vs. SCE, developing higher current densities in comparison to Pt/CNF sample at the same overpotential.

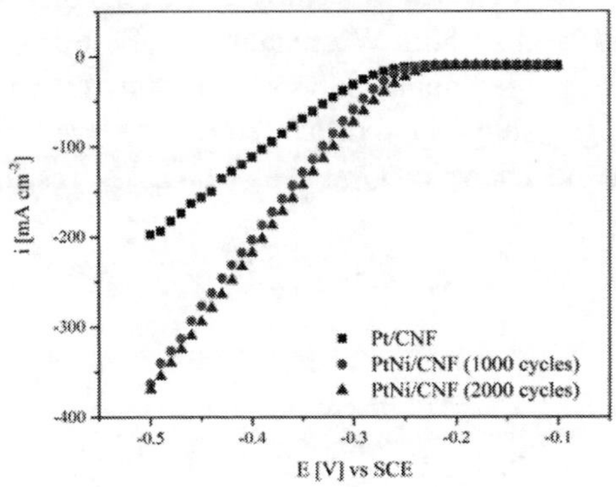

Figure 8. HER on selected Pt/CNF and PtNi/CNF electrodes in 0.5 M H_2SO_4 solution, acquired with a polarization rate of 5 mV s^{-1}

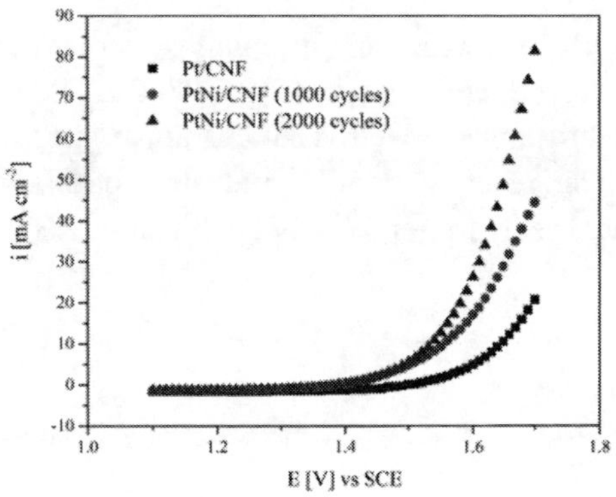

Figure 9. OER on selected Pt/CNF and PtNi/CNF electrodes in 0.5 M H_2SO_4 solution, acquired with a polarization rate of 5 mV s^{-1}

4. Conclusions

Pt-based alloys are considered a promising alternative to pure platinum catalyst for HOR/ORR in fuel cells or HER/OER in electrolysers, due to their low Pt loading and high activity. The present work suggests a new route of electrodepositing Pt-Ni alloy nanoparticles on CNF, using a two-step deposition process. The influence of the most important deposition parameters like current density, time and bath composition were studied in order to obtain PtNi catalysts with high electrochemical activity, which perform well in electrochemical systems. Improved results were obtained applying 125 mA cm^{-2} current density for 12 ms (2000 cycles), from the PtNi 1:4 electrolyte. The SEM micrographs are in good agreement with CV measurements as well as with TG and XRD investigations. XRD analysis confirms Pt_3Ni alloy phase formation on the CNF structure, since the Pt characteristic peaks as identified by the diffraction patterns shift from that of pure Pt.

5. References

[1] Rao R K C and Pushpavanam M 2006 *Trans. Inst. Met. Finish.* **84** 79-82
[2] Xiu R, Zhang F, Wang Z, Yang M, Xia J, Gui R and Xia Y 2015 *RSC* **5** 86578-83
[3] Li M, Lei Y, Sheng N and Ohtsuka T 2015 *J. Power Sources* **294** 420-29
[4] Zangari G 2015 *Coatings* **5** 195-218
[5] Petrii O A 2015 *Russ. Chem. Rev.* **84** 159-93

[6] Escano M C, Gyenge E, Nakanishi H and Kasai H 2011 *J. Nanosci. Nanotechnol.* **11** 2944-51
[7] Muntean R, Pascal T D, Mărginean G and Vaszilcsin N 2017 Int. J Electrochem. Sci. **12** 4597-609
[8] Wang C, Chen Z L, Tao A W and Zhang H 2016 *J. Electrochem. En. Conv. Stor.* **13**
[9] Do L C, Pham S T, Nguyen P N, Tran Q V and Pham H H 2015 *Adv. Nat. Sci.: Nanosci. Nanotechnol.* **6**
[10] Hwang J S, Yoo J S, Jang S, Lim T H, Hong S A and Kim S K 2011 *J. Phys. Chem*, **115** 2483-88
[11] Littlejohn S D 2014 *Springer International Publishing* 1-38
[12] Rost U, Muntean R, Mărginean G, Merino C, Diez R, Vaszilcsin N and Brodmann M 2016 *Int. J. Electrochem. Sci.* **11** 9110-22
[13] Egetenmeyer A, Baumgärtner M, Linckh T, Durneata D, Natter H, Hempelmann R, Radev I and Peinecke V 2017 Trans. IMF **95** 9-19
[14] Zhao Y, Yifeng E, Fan L, Qiu Y and Yang S 2007 Electrochimica Acta **52** 5873-78.

Acknowledgements

Part of this research has been carried out in the frame of an ongoing research project (Entwicklung von Low Cost Gasdiffusionselektroden auf Basis von CNT/CNF für den Einsatz in PEM-Brenstoffzellen (LOCOPEM)), which is financially supported by the state of North Rhine Westphalia and the European Union (funding number EFRE-0800111). Furthermore, the financial support of the Deutsche Bundesstiftung Umwelt (DBU; grant no. 20016/456) is highly appreciated. Moreover, we gratefully acknowledge the work of our industrial partners Zoz GmbH and Thermico GmbH as well as the research institute the hydrogen and fuel cell center ZBT GmbH.

Analysis of Cavitation Erosion Resistance of Grey Cast Iron EN-GJL-200 by the Surface Induction Hardening

E Riemschneider*, I Bordeasu, I Mitelea and I D Utu

„Politehnica“ University of Timisoara, Timisoara, Romania, EU

*Corresponding author: riemschneider.eduard@gmail.com

Abstract. By surface hardening of the cast iron having metallic matrix consisting of pearlite and fine lamellar graphite separations, it has been aimed an increase in hardness and wear resistance. Testing of cavitation erosion resistance was done in the laboratory in accordance with the standard ASTM G32 2010. The mass losses curves, depending on the duration of the cavitational attack by induction hardened and tempered at 220 °C samples, were analysed in comparison with those samples obtained after the stress relief annealing at 525 °C. The hardness measurements performed on the longitudinal section of the cavitation samples beside the microstructural investigations of the eroded surfaces allowed the explanation of the wear mechanism both by the action of the graphite stress concentrator and also by the sensitivity of the metallic mass to the notch effect.

1. Introduction

All machineries bodies (pressure valves, distribuitors, devices and debit regulators), used in the operating hydraulic plants for command, distribution and regulation, are casted from grey cast iron with lamellar or globular graphite [2, 3, 5]. When exploitation conditions impose great working speeds of the liquid environment, in the contraction area from these machineries, the pressure can decrease under a critic value (vaporization pressure) and, consequently, it is initiated the cavitation hydrodynamic process, with all its effects, the most dangerous being the erosion of the flow limits [5, 10, 11].

In Figure 1 it is schematically presented the area exposed to cavitation erosion obtained by the cavitational flow of the liquid from a pressure valve.

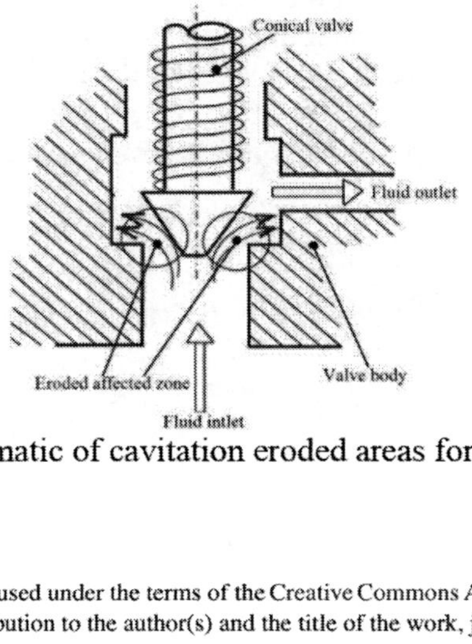

Figure 1. Schematic of cavitation eroded areas for conical valves.

The more the liquid viscosity is lower, close to the water one, the higher is the cavity operating danger [1, 3].

During the maintenance works, on the margins of the devices bodies, in the areas marked in the Figure 1, there have been noticed damages as pitting, specific to the erosion by cavitation. Destructions of these areas is dangerous and unwanted for hydraulic machineries, because it affects the operating of a plant. Moreover, in case of releasable pressure valves [2, 5, 11], as those from the hydraulic presses, pressure forming installations, their blocking cannot be provided and can represent a danger for the operator.

Consequently, specialists started to search for methods in order to increase the resistance of those areas exposed to cavitation erosion. In this direction there are presented the results of this paper, which were obtained from the researches performed on the surface of grey cast iron with pearlite metallic matrix and lamellar graphite separations, hardened by induction currents.

2. Material and experimental procedures

The investigated material (casted bar with diameter $\phi = 25$ mm), is a grey cast iron with metallic matrix composed from pearlite and lamellar graphite separations, subjected to stress relieving annealing, followed by induction hardening and tempering at a temperature of 220 °C for 90min., air cooled (Figure 2).

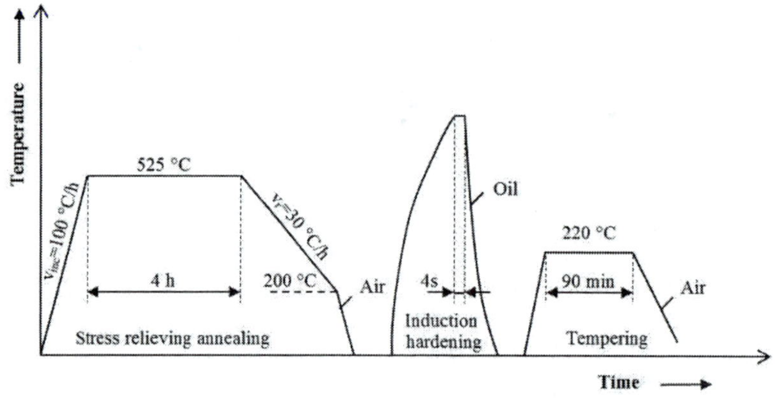

Figure 2. Cyclograms of the applied heat treatments.

Induction hardening process parameters are: specific capacity AP=0.9 kW/cm^2, heating period t = 4s; frequency, f=32 kHz; cooling was made in the oil. The structure of the heat treated samples was investigated by optical microscopy and scanning electron microscopy. The intensity of microstructural changes produced in the substrate coating system was evaluated by Vickers hardness measurements.

The chemical composition of the grey cast iron has been: 3.26 %C; 1.94 %Si, 0.90 %Mn, 0.12 %P; 0.11 %S. After the stress relieving annealing the tensile strength of the material was $R_m = 247$ N/mm^2 and the hardness of 192 HV.

Cavitation resistance investigations of the samples surface after the annealing and induction hardening treatment, have been performed by sets of three samples, using the standard vibrating machinery with piezoceramic crystals, presented in Figure 3, from the Cavitation Laboratory of the Polytechnics University from Timisoara.

IOP Conf. Series: Materials Science and Engineering **416** (2018) 012005 doi:10.1088/1757-899X/416/1/012005

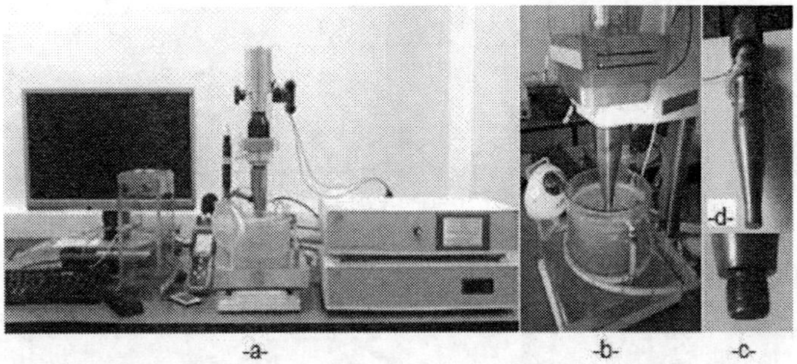

Figure 3. Standard vibrating equipment with piezoceramic crystals.

a-overall image; b- image during the cavitation attack period; c - sonotrode head where it is fixed the cavitation sample; d - mechanical vibrating system with the sample fixed in the sonotrode

Laboratory regulation, regarding the investigation procedure (total period and the intermediary periods of the cavitation attack, cleaning and washing the cavitation sample, Fig. 2d, evaluation of the mass loss, samples preserving) is in conformity with the standardASTM G32-2010 [3], [4], [5], [9], [12] .

During the entire period of the cavitation test, functional parameters of the equipment (vibrations amplitude and frequency, electric capacity of the ultrasounds electronic generator) and the distilled water temperature (22±1°C) [3], [4], [5] have been maintained under the values recommended by the ASTM G32-2010standard.

3. Cavitation test results. Discussions

In Figure 4 and 5 are indicated the experimental values and the approximation analytical curves of the mass loss and of the related erosion speeds, obtained in conformity to the cavitation test for the hardened grey cast iron and for the stress relieving annealed, at 525 °C used in order to highlight the resistance increase, obtained by induction surface hardening.

These diagrams are drawn up based on the mass loss, m and of the erosion rate registered at the end of each period of the cavitation attack, t = 5, 10 and 15 minutes, which have been summed while the cavitation attack period increased until the final value of 165 minutes. The calculation relations which have been used are [4, 5, 6, 7, 8]:

- for the cumulated mass, lost by erosion during the cavitation attack

$$M_i = \frac{1}{3}\sum_{i=1}^{12}\Delta m_i \tag{1}$$

- for the erosion rate, related to each exposing period to the vibrating cavitation

$$v_i = \frac{1}{3}\sum_{i=1}^{12}\frac{\Delta m_i}{\Delta t_i} \tag{2}$$

The mediation/approximation curves of the experimental values are drawn using the following analytical relations [5.9]:

- for the mass lost by erosion:

$$M(t) = A\cdot t\cdot(1-e^{-B\cdot t}) \tag{3}$$

- for the erosion rate:

$$v(t) = A\cdot(1-e^{-B\cdot t}) + A\cdot B\cdot t\cdot e^{-B\cdot t} \tag{4}$$

IOP Conf. Series: Materials Science and Engineering **416** (2018) 012005 doi:10.1088/1757-899X/416/1/012005

where: parameters A and B are statisticallyestablished according to the method used within the cavitation laboratory of UP Timişoara [5].

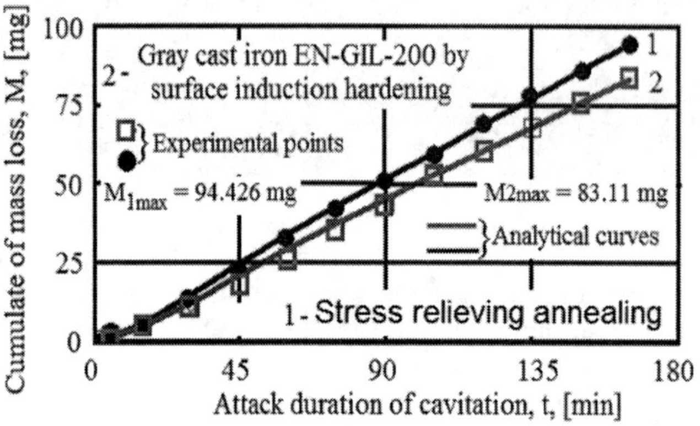

Figure 4. Mass loss variation with the vibrating cavitation attack period.

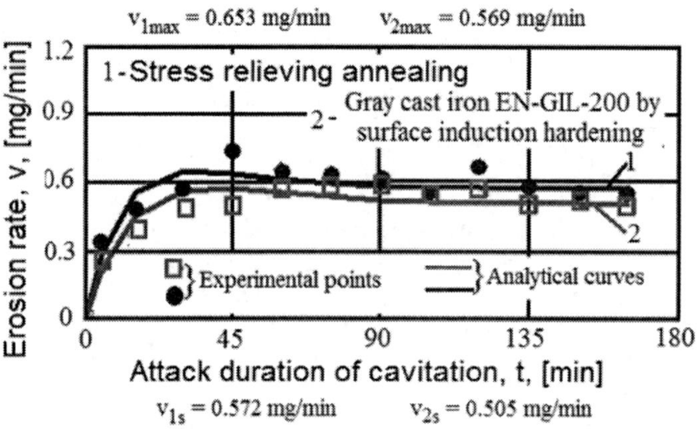

Figure 5. Erosion rate variation with the cavitation attack period.

The experimental points dispersions, of the mass losses related to the approximation curves, Figure 4, are very reduced, which indicates the fact that the structure of hardened surface layer is fine and homogeneous. The comparison with the annealed cast iron indicates that by performing the surface hardening, the mass loss, within 165 minutes of erosion by vibrating cavitation, is diminished by approximatively 13.6 %.

The experimental points distribution, of erosion speeds, Figure 5, indicates that, during the cavitation attack, at certain moments (90, 105, 150 and 165 minutes), the behaviours are identical. From the mechanical damage point of view of the cavitated layer, the explanation is offered by identical values of the expelled metal mass, composed by equal dimensions, respectively summed, of graphite particles.

But, as seen in the experimental values distribution, after 30 minutes of cavitation attack, when the process is effectively running in the metallic structure, generally, it is brought a resistance increase, the approximation curve of the hardened cast iron being under the one of the annealed one for tempering. This growth is due, mainly, to the strength increase obtained by the surface hardening. Consequently, according to the maximum value of the erosion rate, the increase is of 14.7 %, and after the value tends to stabilize at 13.2 %.

It is interesting that in both cases, the approximation curves reach to the maximum value after nearly equal times of the cavitation attack (30-35 minutes). Similar evolutions of approximation curves v(t) suggest that, from point of view of the cast iron behavior to the vibrating cavitation, by this surface hardening method, no substantial modification was obtained, but, by hardness, as indicated the correlation between the specific parameters, the surface resistance to the cavitational microjets impact increased.

4. Sclerometer examinations and microstructural analysis

The intensity of the microstructural modifications produced in the layer-substrate system has been evaluated by the Vickers hardness measurements. In Figure 6 it is shown the hardness variation curve recorded on the cross section of the induction surface hardened samples followed by tempering at low temperature. As it was expected, the hardness gradient curve corresponds to the distribution of the structural constituents present in the parts subjected to this composed heat treatment.

The layer hardened to martensite, having the highest carbon concentration interstitially dissoluted in the lattice of the solid solution α, respectively a high density of lattices defects, will have the highest hardness values.

The recorded hardness values were HV = 480 - 490 daN/mm^{2}in the marginal zone and 190 - 195 HV in the core of the sample. The working depth, defined by the depth where after the heat treatment of hardening and tempering at low temperature, it is obtained a hardness of over 480 HV, is 1.7 – 1.8 mm (Figure 6).

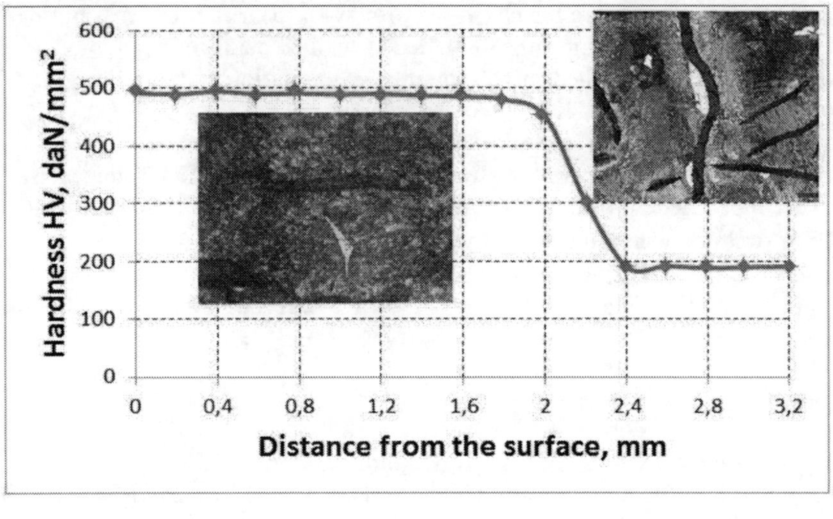

Figure 6. Hardness gradient curve on the cross section of the induction hardened sample.

By surface hardening of grey cast iron parts with pearlitic matrix it is desired to obtain a significant increase of the wear resistance. The existence of lamellar graphite separations in the metallic mass structure lowers the transformation stability of the subcooled austenite and initiates its decomposition in ferrite or pearlite. For this reason, the austenite contained by the cast iron is more inhomogeneous regarding the carbon content. These elements justify the lower hardness values obtained in the surface layer and a limited increase of the cavitation erosion resistance. Moreover, the presence of graphite reduces the metallic matrix which takes the mechanical requests during cavitation and concentrates the mechanical loads in the separations peak, due to their notch effect. The sharper the graphite separations peaks, the more powerful the notch effect and cavitation resistance.

Typical topographies of the cavitated zones of the different heat treated samples (Figure 7), highlights a more pronounced surface degradation in case of stress relieving annealed samples compared with the induction hardened one.

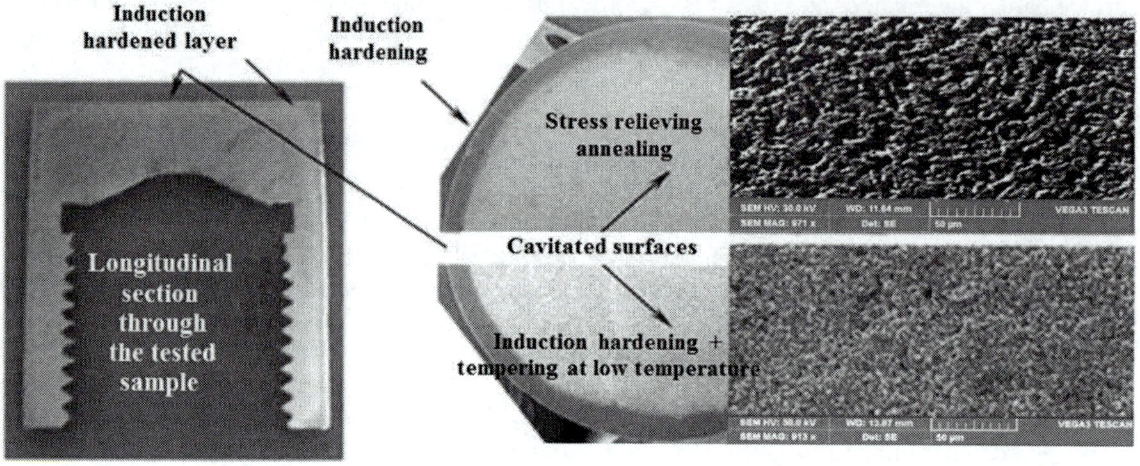

Figure 7. Macroscopic aspect of the hardened layer and of surface tested to cavitation.

5. Conclusions

Induction surface hardening of grey cast iron parts with pearlitic matrix and lamellar graphite, subjected first to the stress relieving annealing, leads to increasing of cavitation erosion resistance, expressed by the mass loss determined in 165 minutes of cavitation exposure, respectively by the final value of the erosion rate of about 13 %.

Limited increase of the cavitation resistance is justified by:

- the graphite effect, non-metallic structural constituent, which reduces the transformation stability of subcooled austenite, favouring its partial decomposition into ferrite or pearlite;
- the role of stress concentrator given by the graphite;
- matrix sensibility to the stresses concentration.

Typical topographies of the cavitated areas show a lower degradation of the induction surface hardened samples.

6. References

[1] Anton I 1985 *Cavitaţia* vol. II (Bucureşti: Editura Academiei RSR)
[2] Mitelea I, Bordeasu I, Katona SE and Craciunescu CM 2017 *International Journal of Materials Research* **108(12)** 1099
[3] Bordeasu I 2006 *Cavitation Erosion of Materials* (Timisoara Polytechnic Publishing House)
[4] Bordeasu I, Micu LM, Mitelea I, Utu ID, Pirvulescu LD and Sirbu NA 2017 *Revista de Materiale Plastice* **53(4)** 781-786
[5] Ghera C 2017 *Rolul tratamentelor duplex în creşterea rezistenţei la cavitaţie a oţelurilor pentru aparatura sistemelor hidraulice* (Teza de doctorat Timisoara)
[6] Krella A and Czyzniewski A 2007 *Wear* **263** 395
[7] Krella A 2009 *Surf. Coat. Technol.* **204(3)** 263
[8] Hattori S and Ishikura R 2010 *Wear* **268** 109
[9] Micu LM, Bordeasu I and Popoviciu MO 2017 *Revista de Chimie* **68(4)** 894
[10] Mitelea I, Micu LM, Bordeasu I and Craciunescu CM 2016 *Journal of Materials Engineering and Performance* **25(5)** 1939
[11] Raszga C 1998 *Fenomenul de cavitaţieîndistribuitoare cu sertar cilindric* (Timisoara:Phd work)
[12] ** *Standard method of vibratory cavitation erosion test* 2010 ASTM Standard G32-10.

IOP Conf. Series: Materials Science and Engineering **416** (2018) 012006 doi:10.1088/1757-899X/416/1/012006

The Particularities of the CMT Braze-Welding Process Between Ferritic Stainless Steel - Magnesium Alloy

N Simon*, I Mitelea, M Burca and I D Utu

Politehnica University of Timisoara, Timisoara, Romania, EU

*Corresponding author: nati_simon19@yahoo.com

Abstract. Jointing of dissimilar materials with the CMT braze-welding process, ferritic stainless steel - magnesium alloy, offers important technical and economic advantages to the automotive industry. In the actual research paper we analyse the opportunity of realizing heterogeneous joints by the CMT process, using a layer of copper as a layer to be deposited firstly on ferritic stainless steel. Macro - and micrographic investigations, corroborated with hardness testing on (Cu-stainless steel, Mg - Cu) deposition by TIG, aim to highlight the process specificities and problems occurring in the interface areas of the joints made

1. Introduction

In the automotive industry, welding of dissimilar materials is of great interest [1, 2]. They have attracted significant interest when compared to the joints of similar materials. Components which have corresponding ductility, high strength, good corrosion resistance, and desirable thermal conductivities have been achieved through dissimilar materials welding [3, 4]. Several welding techniques have been reported, including Welding-Brazing, Arc, Friction, Friction stir, ultrasonic, laser, and hybrid laser-arc welding [1 - 6].

There is an increased risk of encountering problems during dissimilar welding, like the carbon migration from materials with higher carbon content to the materials with less carbon content, thermal expansion coefficient variation of materials, and challenges in the execution of the post-welding heat treatments [1].

The ferritic stainless steels are generally considered to have poor weldability when compared to the austenitic stainless steels. Consequently, few systematic studies have been conducted on their weldability other than to determine the effect of welding on their mechanical properties and corrosion resistance. In the first phase of this program the effects of elements such as C, N, Cr, Si, Mn, Mo, Ni, P, S, Ti, Nb, and Ta on the hot cracking susceptibility of Type 430 and Type 444L were studied [2].

Magnesium and its alloys are increasingly used in automobile industry and other fields for their lower density, higher specific strength, higher specific stiffness, etc. [3].

Copper and its alloys are also widely used in many fields due to their outstanding properties, such as good ductility, corrosion resistance, electric conductivity and thermal conductivity.

Hybrid structure of Mg/Cu dissimilar materials not only satisfies the requirements of electric conductivity, thermal conductivity and corrosion resistance but also meets the demand of light weight and high strength. However, joining of magnesium and copper has a metallurgical challenge because of the differences in their chemical and physical properties, and mass of brittle Mg_2Cu and $MgCu_2$ intermetallic compounds, which seriously decrease the mechanical properties of the joints. Therefore, it is mandatory to control the formation and growth of Mg–Cu intermetallic compounds [4].

This paper highlights the process specificities and problems occurring in the interface areas of the joints made of a ferritic stainless steels with a copper alloy and of copper alloy with a magnesium alloy.

2. Experimental procedure

In order to obtain a good joint between magnesium and stainless steel, with desired properties, due to their metallurgical incompatibility it is necessary to use a material as an intermediate layer between the two, with the condition to have a corresponding degree of compatibility. After consulting the literature the best candidate is copper as an intermediate layer.

Therefore, the materials used in the study include a 3 mm thick ferritic stainless steel sheets (30x30 mm) and a 3 mm thick T2 pure copper sheets (30x30 mm). As a filler material was used AZ31 magnesium wire and a T2 pure copper wire. The chemical composition of the materials used, per the manufacturer's data sheet, is presented in Table 1.

Table 1. Chemical composition.

Material	**Elements, % wt.**											
	Fe	**Cr**	**Mn**	**Mg**	**Ni**	**S**	**Zn**	**Al**	**P**	**Cu**	**Si**	**O**
Ferritic steel W1.4713	Balance	7.1	0.68	-	-	0.011		0.72	0.026	-	0.74	
Mg-AZ31	0.005	-	1.0	Balance	0.005	-	1.3	3.5		0.01	0.05	-
T2 Copper (ECu-58)					0.021		0.025			Balance		0.034

Before welding, the oil on substrates (Cu–T2) surface were removed by a wire brush and acetone. The experimental setup is presented in Figure 1.

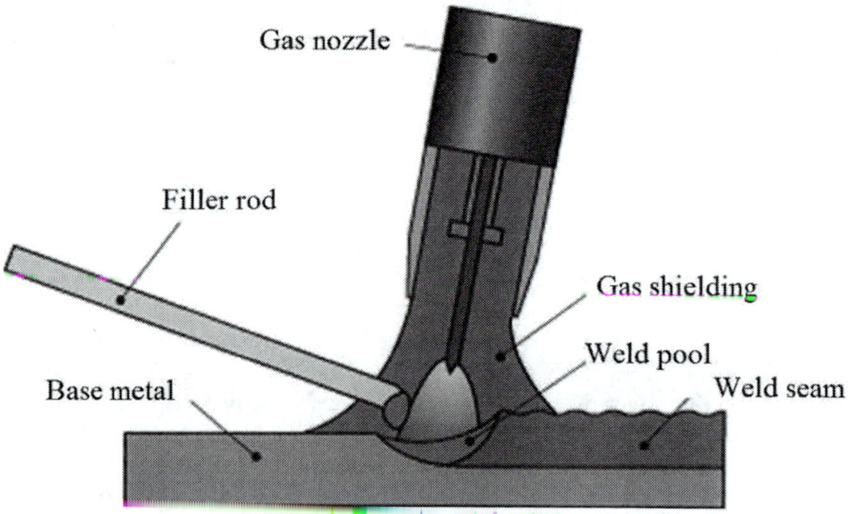

Figure 1. Experimental setup.

In order to study the peculiarities between the ferritic stainless steels, magnesium alloy and the intermediate layer of copper, the following depositions were carried out:
- Deposition of a copper layer on a stainless steel sheet;
- Deposition of a magnesium layer on a copper sheet.

Figure 2 a, b shows weld appearance of the Mg–AZ31 to Cu–T2 and the weld appearance of Cu-T2 to the ferritic stainless steel.The TIG welding was carried out using Fronius Magicwave 3000 welding machine. The main welding parameters, such as welding speed, wire feed speed, welding current and welding voltage, were presented in Table 2.

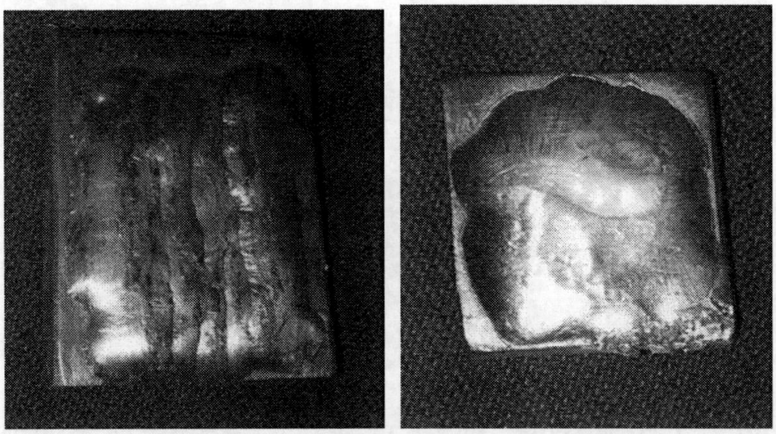

a) b)

Figure 2. Welding deposition
a) - Cu-T2 to the ferritic stainless steel; b)-Mg–AZ31 to Cu–T2.

In addition, 99% argon shielding gas with a flowrate of 15 L/min was used throughout the experiments .

Table 2. Welding parameters.

Material	Polarity	Welding speed	Wire feed speed	Welding current and welding voltage
Cu-T2 to the ferritic stainless steel	DC -	1 mm/s	3mm/s	180/20.5
Mg–AZ31 to Cu–T2	AC	1 mm/s	2mm/s	215 / 21.5

The weld geometry was studied at the weld zones to observe the changes in the seam profile, and weld joint microstructures through optical microscopy (Leica, DM IL M LED).
Metallographic investigations were conducted on a region of 30 mm × 10 mm × 2 mm which involved all of the weldment regions (base metal, heat-affected zone, and weld). The samples were etched for a period of 20 s. The compound weldment regions were polished with alumina and distilled water, using sheets of SiC of varied sand sizes ranging from 240 to 1200. The polishing process followed regular metallographic techniques to achieve a perfect mirror finishing of approximately 1 μm.
Figure 3 shows the weld seam dimensions (width of weld seam, depth of penetration), while the micro-hardness was measured at 1 mm from the top surface of the welded joint. Vickers micro-hardness technique was used to study the mechanical behavior by using a hardness tester (ZWICK 3212). Hardness measurements were studied across 3 zones the interface. Standard HV2 – hardness method was used, at consistent intervals of 1 mm.

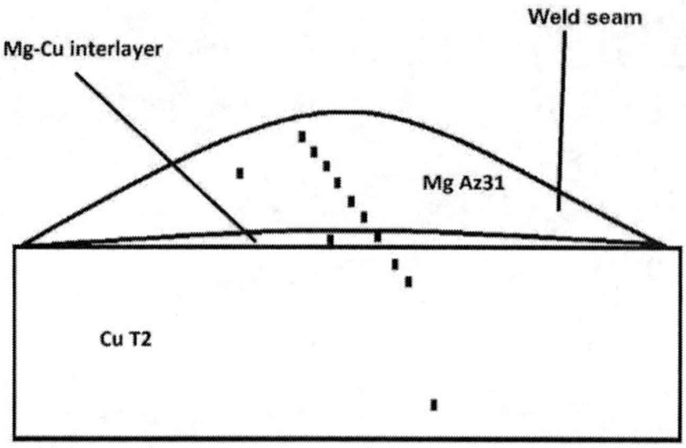

Figure 3. Weld seam profile and the micro-hardness measurements direction.

3. Results and discussions

In order to understand the mechanisms of WIG deposition of Mg–AZ3 to Cu–T2 and Cu-T2 to the ferritic stainless steel W14713, the microstructures of the joints were analysed. Figure 4 shows the macroscopic cross-section of Mg-Cu WIG joints

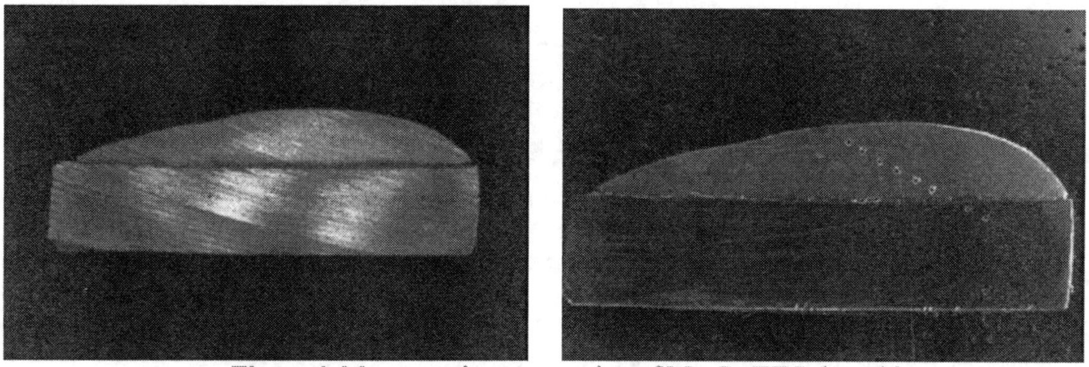

Figure 4. Macroscopic cross-section of Mg-Cu WIG deposition.

As shown in Figure 4, the joint was composed of three zones:

- Mg fusion zone, formed between Mg base metal and weld metal;
- weld metal zone;
- brazing interface zone formed by molten weld metal and local molten Cu base metal.

Figure 5 shows the macroscopic cross-section of Cu deposited on the ferritic stainless steel

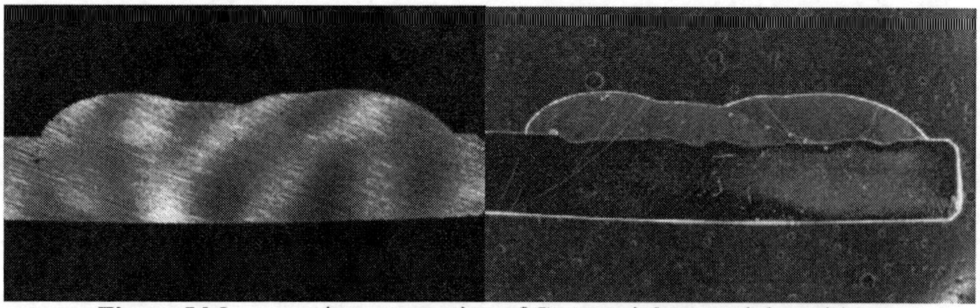

Figure 5. Macroscopic cross-section of Cu to stainless steel deposition.

Figure 6 a, b, c, shows the microstructures of various regions of Mg-Cu WIG joint

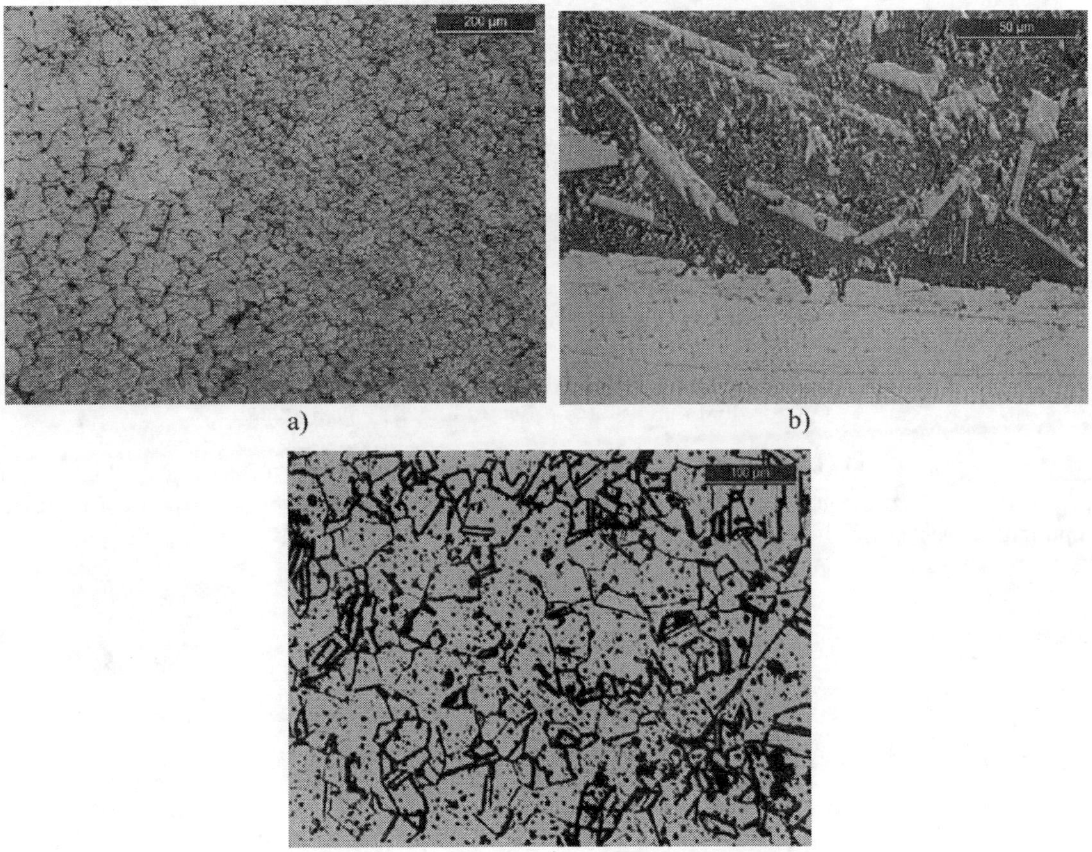

a) b)

c)

Figure 6. Microstructure of the joint
a) – deposited material; b) – fusion line; c) – base metal-Cu.

Figure 7 a, b, c, show the microstructures of various regions of Cu deposited on the ferritic stainless steel

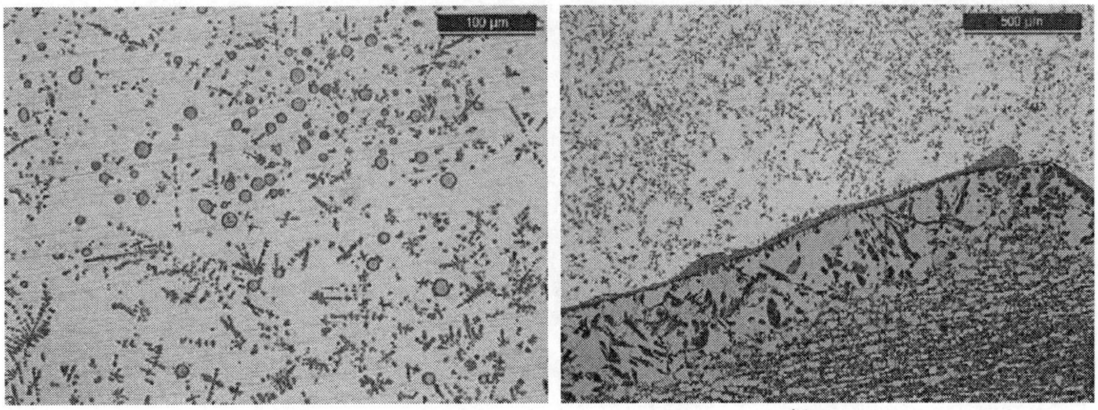

a) b)

IOP Conf. Series: Materials Science and Engineering **416** (2018) 012006 doi:10.1088/1757-899X/416/1/012006

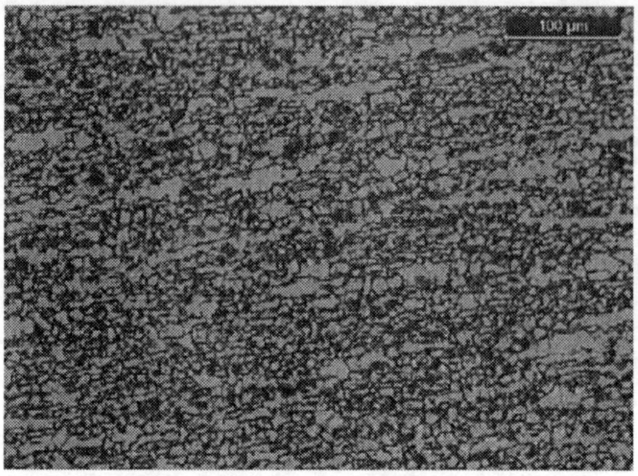

c)

Figure 7. Microstructure of the weld
a) - deposited material; b) – fusion line; c) – base metal.

Figure 8 presents the hardness distribution of the Mg deposition on Cu. From Fig. 7, high hardness appears in the brazing interface layer (Fusion line - FL), which is composed of series intermetallic compounds, consistent with Figure 6.

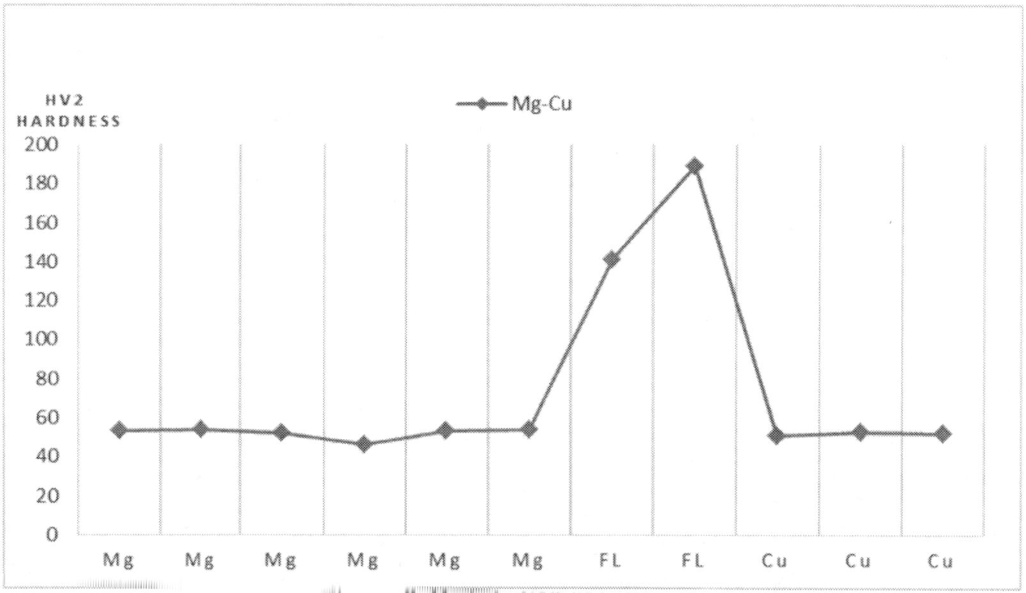

Figure 8. Distribution of hardness.

Figure 9 presents the hardness distribution of the Cu-T2 deposition on the ferritic stainless steel.

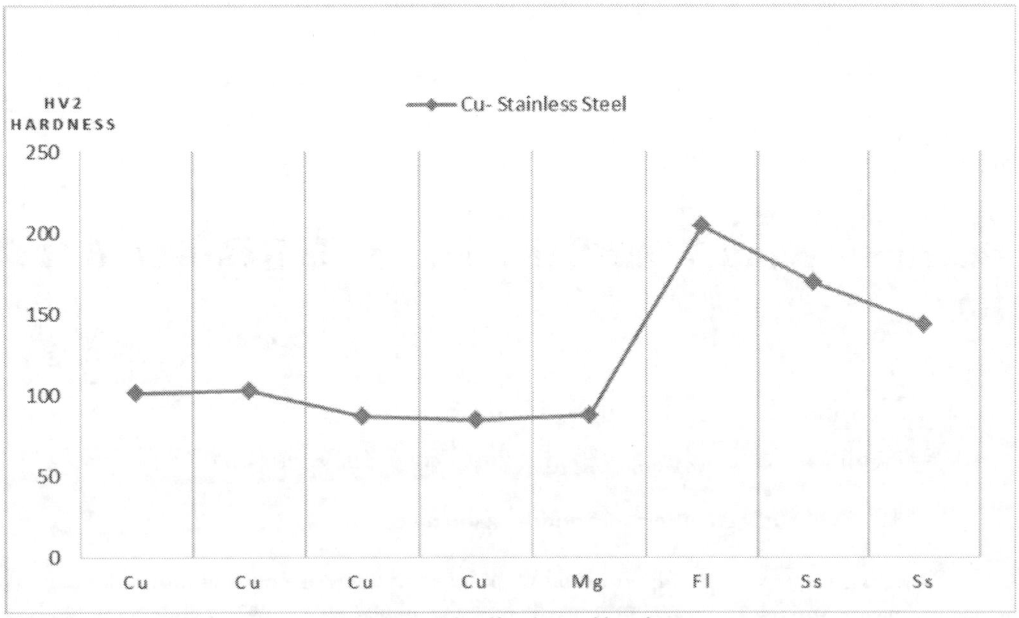

Figure 9. Distribution of hardness.

4. Conclusions

1. Joining of ferritic stainless steels with magnesium alloys remains a topical issue because it can not be achieved with the usual welding techniques.
2. The application of the CMT brazing process using an intermediate Cu layer can lead to solving the incompatibility problems between the two dissimilar materials.
3. Changing the chemical composition of the welding bath avoids the formation of brittle intermetallic phases between Fe and Mg, and the working environment of the process limits the rapid oxidation of Mg.

5. References

[1] Liu LM, Wang JF and Song G 2004 *Mater. Science Eng. A* **381** 129
[2] Ren DX and Liu LM 2014 *Mater. Des.* **59** 69
[3] Miao Y, Han D, Yao J and Li F 2010 *Mater. Des.* **31** 3121
[4] Elthalabawy W and Khan T 2011 *Mater. Sci. Technol.* **27** 22
[5] Kumar S and Shahi AS 2011 *Mater. Des.* **32** 3617
[6] Wahba M and Katayam S 2012 *Mater. Des.* **35** 701

IOP Conf. Series: Materials Science and Engineering **416** (2018) 012007 doi:10.1088/1757-899X/416/1/012007

Microstructure and Wear Behaviour of Ti Reinforced HVOF Coating

A Szabo*, I D Utu, I Hulka, I Bordeasu and I Mitelea

„Politehnica" University of Timisoara, Timisoara, Romania, EU

*Corresponding author: alexandru.szabo@upt.ro

Abstract. In order to improve the wear resistance and biocompatibility of titanium, which is well known as a common orthopaedic and dental implants material, aluminium oxide (Al2O3)-hydroxyapatite (HA) composite coatings were deposited onto titanium substrate using high velocity oxygen fuel (HVOF) thermal spraying process.
The microstructure of the Ti reinforced HVOF coating was investigated by scanning electron microscopy (SEM) and phase identification was performed using X-Ray diffraction techniques (XRD). The surface properties of the titanium substrate before and after Al_2O_3-HA coating deposition in terms of roughness and wear resistance, using pin on disk method, were determined.
The obtained results showed that Al_2O_3-HA coatings can extremely improve the wear resistance of the titanium substrate. A higher roughness of the coating evidently has a positive effect onto the adhesion of the implant osseointegration.

1. Introduction

Titanium and its alloys are the most used materials in the biomedical domain gradually replacing other metallic biomaterials like stainless steel and cobalt-chromium alloys in applications where high mechanical strength requirements [1]. They have been successfully used in the field of orthopaedic and dental implants due to their excellent mechanical properties, corrosion resistance, and biocompatibility [2]. Usually these characteristics of the implants in terms of surface texture, bioadhesion, non-toxicity, wear resistance, etc can be enhanced by depositing materials [1].

Coating deposition of aluminium oxide (Al_2O_3) and hydroxyapatite (HA, $Ca_{10}(PO_4)_6(OH)_2$) mixtures may improve the bioactivity and corrosion resistance of the originally bio-inert material [3,4]. Hydroxyapatite coated implants are being widely used to improve osteoconductivity of metallic implants, its structure being close to that of the human bone mineral component, while alumina is a ceramic material which possesses high values for hardness and good wear resistance properties in physiological environment [5].

Many studies are also focused in recent years on the manufacturing of composite coatings, using different deposition methods (thermal spraying, laser, sol-gel, etc), consisting of metallic biocompatible materials and bioactive ceramic materials based on biodegradable calcium phosphates [6].

The paper is focus on manufacturing of different Al_2O_3-HA coatings using the high velocity oxygen fuel (HVOF) spraying method onto the surface of a titanium substrate in order to improve its wear resistance and biocompatibility.

IOP Conf. Series: Materials Science and Engineering **416** (2018) 012007 doi:10.1088/1757-899X/416/1/012007

2. Materials and experimental procedure

As substrate material a commercially titanium bar was used which was cut in specimens of 60 mm diameter. Before depositions the surface of the substrate was sand blasted and cleaned with acetone (Fig.1a).The coatings were obtained by HVOF spraying method using a powder mixture of Al_2O_3-HA (Fig.1b). The powders were mixed by ball milling with agate balls with a Al_2O_3/HA (wt%) ratio of two chemical composition: 80 %Al_2O_3+20%HA and 50 %Al_2O_3+50%HA.

a)

b)

Figure 1. Macroscopic image of the titanium surface substrate: a) - sand blasted, b) - coated by HVOF sprayed Al_2O_3-HA powder.

For deposition, a 2700 DJM HVOF gun from Sulzer Metco was used. The spraying parameters were: 430 l/min oxygen, 60 l/min propane and 470 l/min air, the spraying distance being kept by 170 mm.

The morphology of the sprayed coating has been characterized by scanning electron microscopy (SEM: Philips XL-30 scanning electron microscope equipped with EDAX analyzer) and the phase composition has been investigated by X-ray diffraction analysis using a Cu Kα radiation (Philips X'Pert Diffractometer).

The roughness, R_a [μm], of the deposited materials was measured using a SJ-201 equipment from Mitutoyo.

In order to determine the sliding wear resistance of the coating and of the titanium substrate a pin-on disk tribometer (Tr-20 Micro from DUCOM-Material Characterization System) was used. The wear rate of the materials was calculate according to the variation of the wear track depth with applied load. The testing conditions were: normal load 20 N, the relative velocity between the ball (WC) and surface v=20 cm/s, and the testing distance 2000 m (the trajectory was a circle with a radius of 15 mm).

3. Results and discussion

3.1 Coatings morphology and phase composition

Figure 2 presents the surface morphology of the sprayed coating using the Al_2O_3/HA powder compositions. Analysing the SEM images from Fig. 1b and Fig.2 it can be seen that no specific thermal spraying defects (cracks, exfoliations, etc) were noticed. The EDAX analysis from Figure 3 confirms the differences between the chemical compositions of the coatings.

In Figure 4 and 5 are shown the XRD patterns of the deposited coatings. In both cases the identified phases were $Ca_5(PO_4)_3(OH)$, $Ca_3(PO)_4$, Al_2O_3 and $CaAl_2O_4$. Beside the commonly phases hydroxyapatite and tricalcium phosphate which are usually identified, the presence of alumina and monocalcium aluminate can be noticed. $CaAl_2O_4$ is the main constituent of calcium aluminate cements.

IOP Conf. Series: Materials Science and Engineering **416** (2018) 012007 doi:10.1088/1757-899X/416/1/012007

a)

b)

Figure 2. SEM micrographs of the deposited coatings:
a) 50 %Al_2O_3+50%HA, b) 80 %Al_2O_3+20%HA.

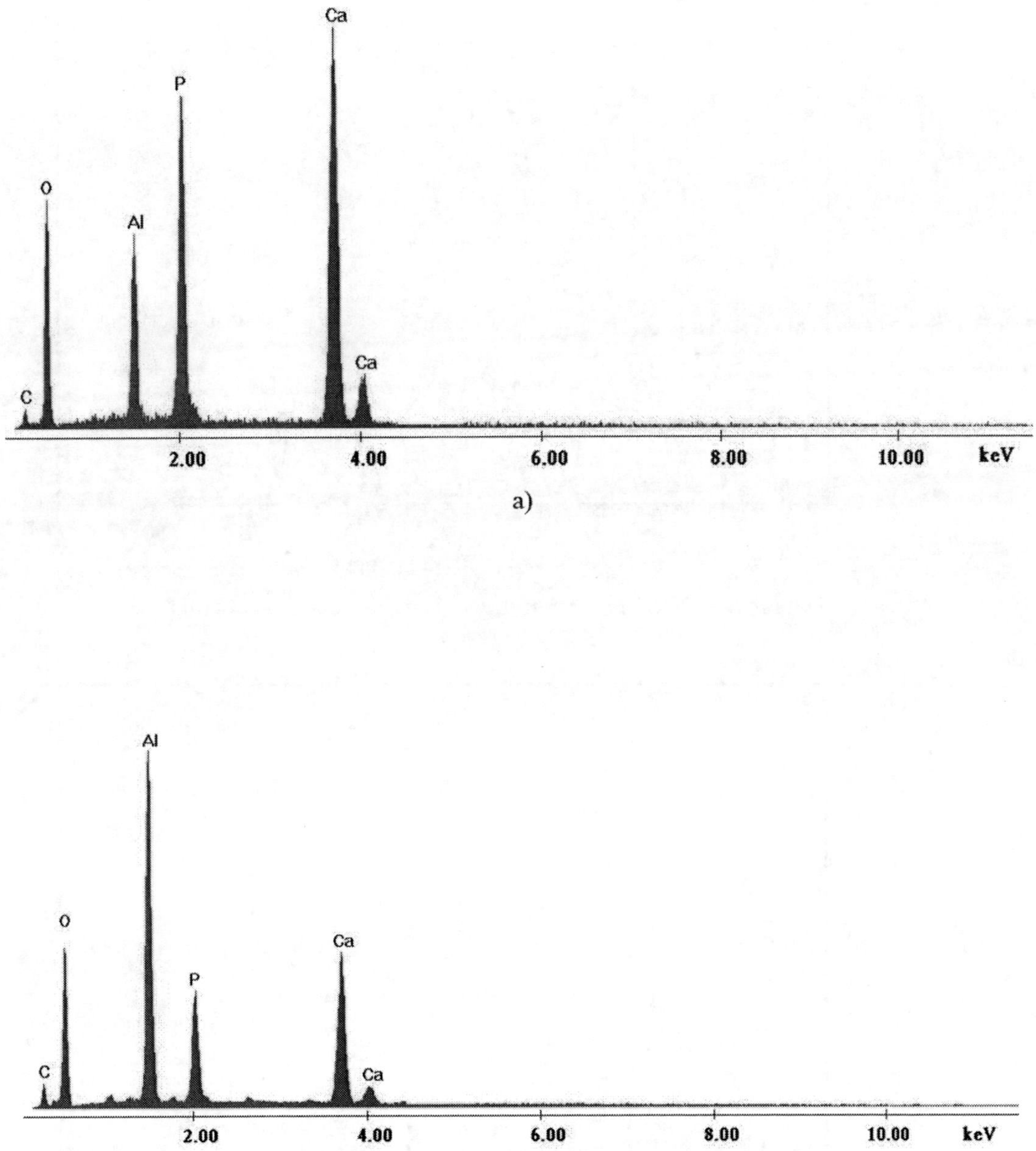

Figure 3. EDAX analyses of the the deposited coatings:
a) 50 %Al_2O_3+50%HA, b) 80 %Al_2O_3+20%HA.

IOP Conf. Series: Materials Science and Engineering **416** (2018) 012007 doi:10.1088/1757-899X/416/1/012007

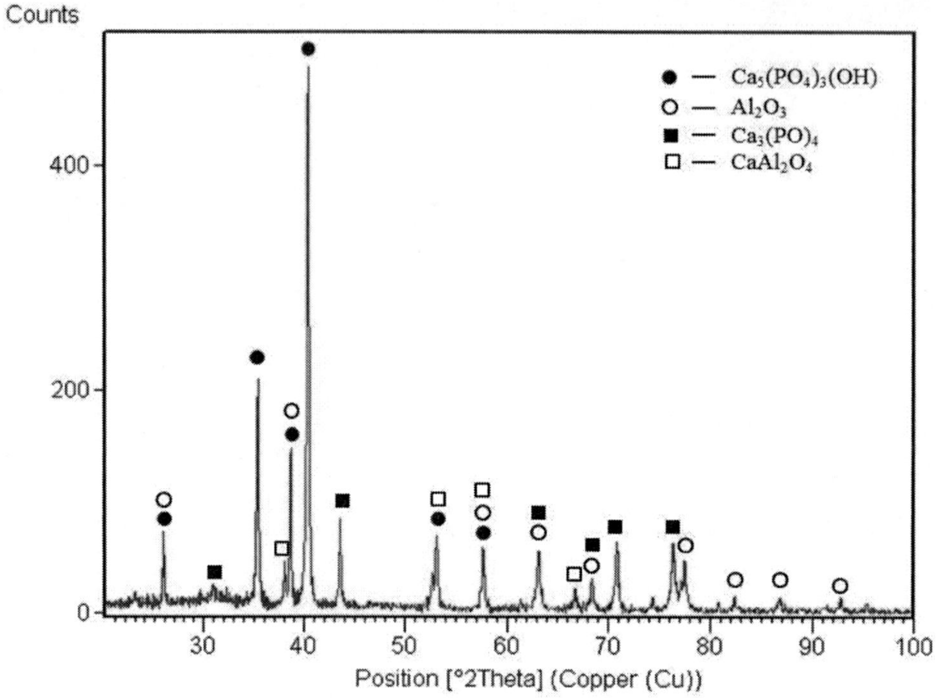

Figure 4. XRD pattern of the 50 %Al_2O_3+50%HA coating.

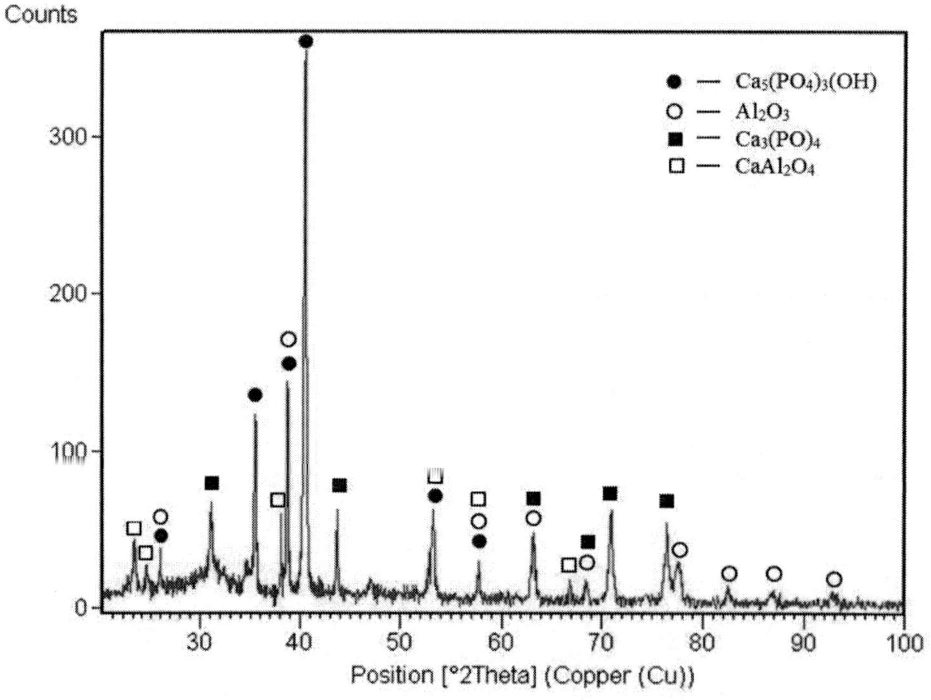

Figure 5. XRD pattern of the 80 %Al_2O_3+20%HA coating.

IOP Conf. Series: Materials Science and Engineering **416** (2018) 012007 doi:10.1088/1757-899X/416/1/012007

3.2 Roughness measurements of the coatings

The roughness of a surface is an important parameter that characterizes the quality of the coatings. In most biomedical applications, where hydroxyapatite coating are used a rough surface is required. This enhances the initial attachment of osteoblast cells and mineralization phenomena.

In the experiments the surface of the Al_2O_3-HA coatings was measured in five different areas, the value, Ra, being taken as an arithmetic mean between the five measured values of the roughness. The values obtained are shown in Table 1.

Table 1. Roughness of the Al_2O_3-HA sprayed coatings.

Coating	**Ra, µm**
50%Al2O3+50%HA	5.64
80%Al2O3+20%HA	5.48

In both cases the average roughness value was about Ra=5.5 µm.

3.3 Sliding wear behavior of the coatings

In order to evaluate the tribological behaviour of the materials the wear resistance of titanium substrate before and after coating deposition was tested. The evolution of the coefficient of friction for the tested materials is presented in Fig. 6 and the average values are shown in Table 2.

Table 2. Average values of the coefficient of friction.

Material	μ_{med}
Titanium	0.42
50%Al_2O_3+50%HA coating	0.52
80%Al2O3+20%HA coating	0.67

In case of the Al_2O_3-HA coatings higher values for the friction coefficient were recorded compared with the Ti substrate. The different tribological behaviour of the tested materials was influenced by the wear of the counterpart. For titanium there was quite no wear trace on the surface of the counterpart compared with the Al_2O_3-HA coatings where the wear of the worn WC ball was higher, so that the contact surface between the wear partners was greater and the friction coefficient increased.

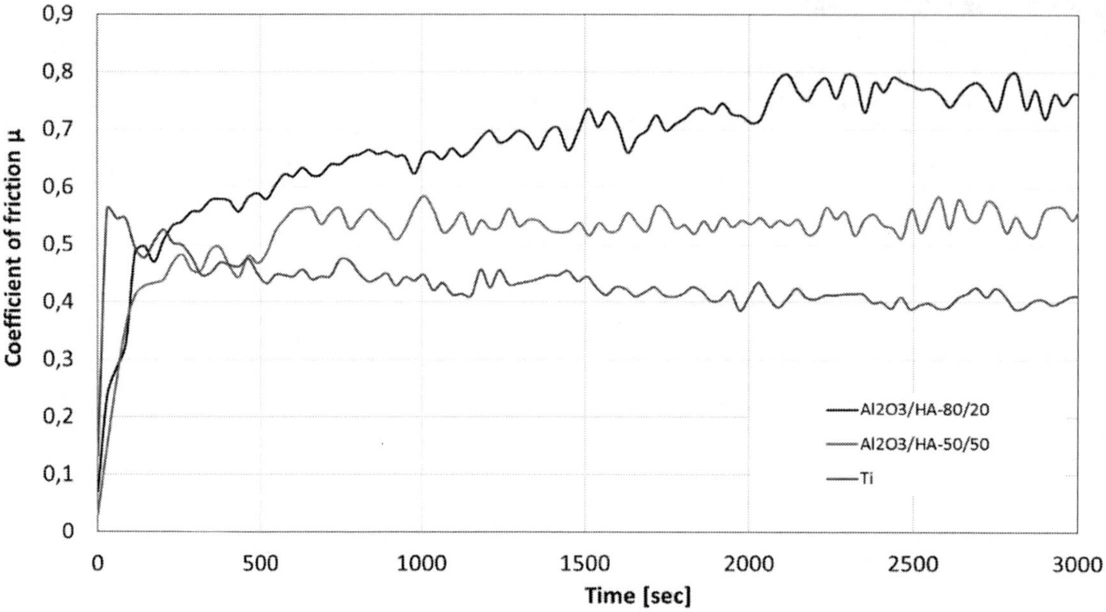

Figure 6. Evolution of the friction coefficient in time for the tested materials.

The wear resistance of the materials was determined according with the volume of the loss material (Fig.8). As it was expected the deposition of Al_2O_3-HA coatings had a positive effect onto the sliding wear behaviour, the wear rate of the coated titanium exhibited the lower degree. The $80\%Al_2O_3$+20%HA coating had the best wear resistance because of the higher amount of the alumina phase.

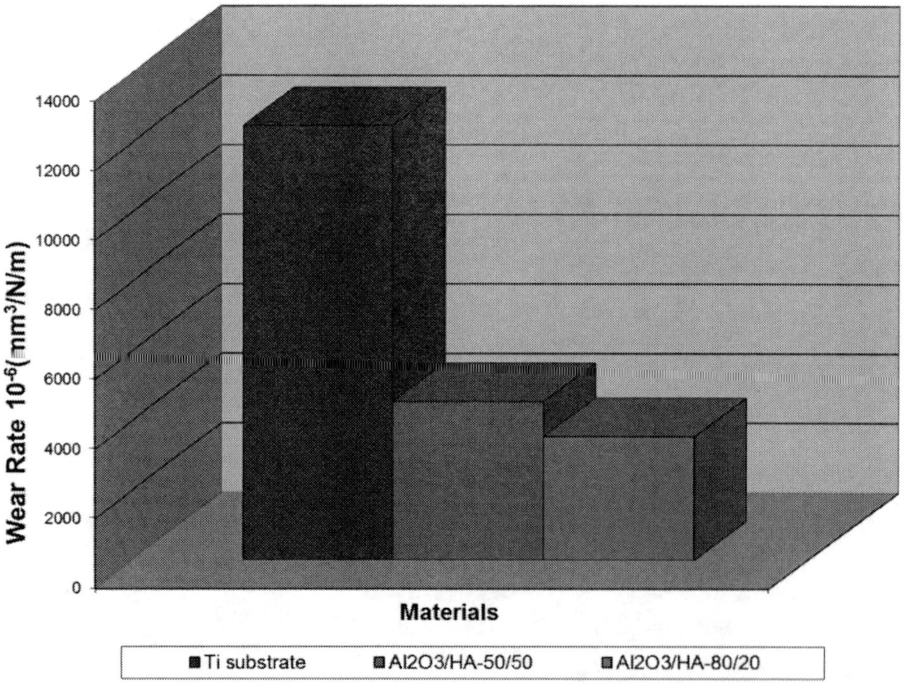

Figure 7. Wear rates of the tested materials.

IOP Conf. Series: Materials Science and Engineering **416** (2018) 012007 doi:10.1088/1757-899X/416/1/012007

4. Conclusions

High velocity oxygen fuel (HVOF) spraying method can be used for obtaining of Al2O3-HA coatings deposited onto the surface of a titanium substrate in order to improve its wear resistance and biocompatibility.

The XRD patterns showed that the hydroxyapatite structure has suffered no important modifications, which could decrease the restauration of the bone tissue. Moreover the measured surface roughness revealed high values (Ra≈5.5 μm) which provide a good osteointegration of the implant in the human body because of a greater contact surface.

The presence of alumina in the deposited coatings improved the wear resistance of coated titanium by approximatively 2-3 times.

5. References

[1] Tengwall P and Lundstrom I 1991 *Clinical Materials* **9** 115
[2] Chen Y et al 2017 *Journal of the Mechanical Behavior of Biomedical Materials* **75** 169
[3] Utu ID and Mitelea I 2015 *Materials Testing* **57(3)** 262
[4] He X et al 2017 *Ceramics International* **43(18)** 161
[5] Vahabzadeh S, Roy M, Bandyopadhyay A and Bose S 2015 *Acta Biomaterialia* **17** 47
[6] Rosu R, Serban VA, Bucur AI and Utu D 2012 *Applied Surface Science* **258** 3871

IOP Conf. Series: Materials Science and Engineering **416** (2018) 012008 doi:10.1088/1757-899X/416/1/012008

Mechanical Behaviour of Narrow Gap MAG Welding of API 5L X65M Steel Pipeline

D Simionescu*, I Mitelea, M Burcă and I D Uțu

[1]University Politehnica Timisoara, Piata Victoriei no. 2, Timisoara, Romania

*Corresponding author: dsimionescu@ccc.ae

Abstract. The paper analyzes the main mechanical characteristics of pulsed MAG welded joints obtained from API 5L X65M thermo mechanical treated steel. The results of the tensile strength measurements corroborated with those of toughness tests and metallographic investigations, highlighted the effect of structural changes in the heat affected zone (HAZ) and weld (W) area, on the mechanical behavior of the welded joints.

1. Introduction

Currently, the oil and gas transport are in a constant growth and requires the development of new underground and submarine pipeline networks. Since the costs of achieving these types of pipelines are quite high, the researchers conducted worldwide are oriented towards the following directions [1], [2], [3]:

- selection of low-alloy thermo mechanical treated steels with high mechanical strength characteristics which will allow a reduction in the thickness of the pipe wall and consequently a lower consumption of welding consumables
- implementation of high productivity welding techniques
- use of automated methods for monitoring the quality of welded joints

The matrix requirements for such properties of the steel pipe comprising:
- high mechanical strength characteristics
- good tenacity
- high resistance to fatigue and corrosion
- good metallurgical and technological welding behaviour
- a reasonable cost price

Such steels raise problems that mainly depend on the negative effect of some thermal cycles during the welding operation on the mechanical properties obtained by thermo mechanical treatment.
It is also taken into account that, in order to obtain mechanical characteristics at the base metal level, the welded seam must have an equivalent carbon, Cech. significantly higher than it [3], [4].
A particular aspect is the weakening of the welded joint as a whole due to the softening of some portions of the heat-affected zone, HAZ. The existence of such a portion in the ZIT may reduce the mechanical strength characteristics of the welded joint and this effect becomes more important when the relative width of the softened area is greater; this width is defined as the ratio of the softened portion width and the thickness of the welded element [1].
Taking into consideration the welding process features of thermo mechanically treated steels, this paper studies the performances of the mechanical strength of the joints achieved by applying a high productivity technique.

2. Experimental procedure

Pipelines with a diameter of 42 " (1066.8mm) and a thickness of 31.75mm, made of micro alloyed high-strength steel API5L X65M thermo mechanically treated steel were welded by MAG spray arc the root & hot pass and pulsed current for the fill layers of the welded joint using CRC Evans welding equipment.

The equipment used is composed by an internal welding machine, IWM, for the root pass and two external ones, P625, each one with two welding torches for the fill layers, where the pipe is fix and the welding equipment make an downward vertical orbital movement around the pipe [4].

Welding consumables selected were ER70S-G for the root pass and ER70S-6 for the fill layers.

The geometry of the selected joint is shown in Figure 1.

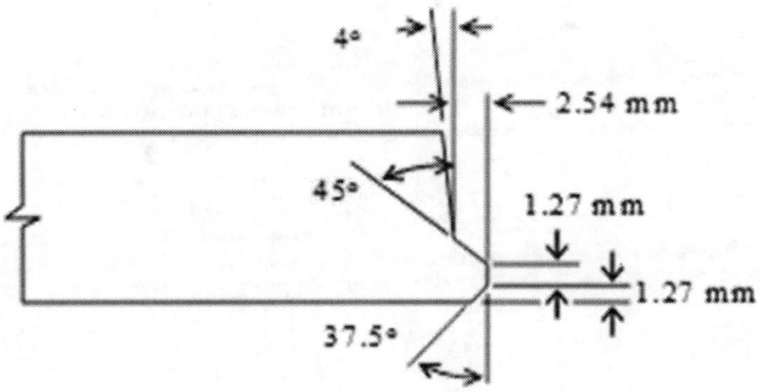

Figure 1. Joint configuration.

The optimal welding parameters of the thermal welding regime were determined experimentally and are presented in Table 1.

Table 1. Welding parameters.

Weld Layers	Current (A)	Voltage (V)	Travel Speed (cm/min)	Heat Input (KJ/cm)
Root	198	20	80	2.97
Hot	282	24.5	128	3.23
Fill	215	22	50	5.67
Cap	132	22.5	48	3.71

From the joints made, samples were taken for mechanical tests and metallographic examinations.

3. Evaluation of experimental results

3.1 Tensile strength test

The purpose of this test is to determine the strength of the welded joint and the location of the fracture.

The specimens used were oxy-gas flame cut followed by machining so that the sides are smooth and parallel.

IOP Conf. Series: Materials Science and Engineering **416** (2018) 012008 doi:10.1088/1757-899X/416/1/012008

The location of the tensile specimens is shown in Fig.2 and the shape and dimensions of the specimens (according to API 1104, 2013) are shown in the Fig.3.

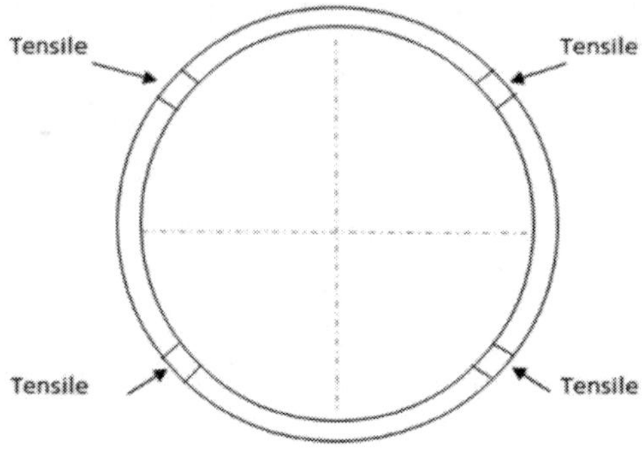

Figure 2. Location of tensile specimens.

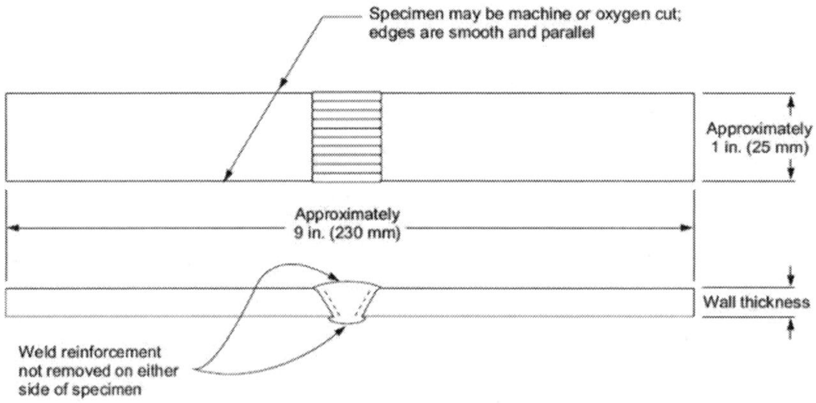

Figure 3. Tensile strength test specimen.

Tensile test was performed using a universal hydraulic machine, type Satec 5500R 200BTE (fig.4).

Figure 4. Universal machine type Satec 5500R 200BTE.

The results of the tensile tests are shown in Table 2 and the appearance of a test specimen is shown in fig. 5.

Table 2 Experimental results.

Specimen ID	So (mm²)	Fmax (N)	Rm (N/mm²)	Fracture Location
TT1	800.87	476226	595	Base metal
TT2	796.45	476126	598	Base metal
TT3	801.24	469729	586	Base metal
TT4	799.22	460440	576	Base metal

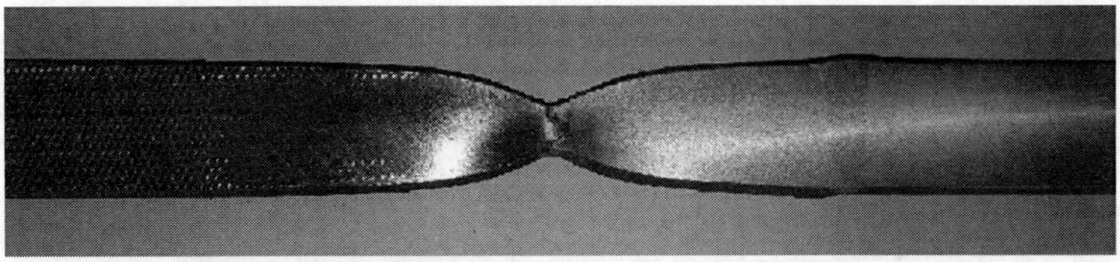

Figure 5. Fracture location of the tensile strength test specimen.

The analysis of the obtained data show that, each time, the fracture occurred in the base material (MB) and the values of the tensile strength for all 4 sets of tensile specimens are superior to the minimum required for the basic material (in accordance with API 5L, 2012), which is 535 N/mm^2. According with API 1104, 2013, the tensile strength of the weld, including the fusion zone of each specimen has to be greater than or equal to the specified minimum tensile strength of the pipe material.

3.2 Charpy Impact Testing

The charpy impact test seeks the appreciation of the welded joint toughness. The Charpy method uses prismatic specimens with a U-shaped or V-shaped notch in the middle of the specimen. V- notch specimens are preferred to those with U-shaped notches because the absorbed energy is mostly accredited for crack propagation; in U-shape notch specimens the absorbed energy mainly characterizes the crack initiating.

The experiments were conducted on samples with a V-shaped notch (ASTM E23, 2012), and the test temperatures were: + 20 ° C, 0 ° C, -30 ° C and -50 ° C.

For the tests performed at low temperature the used media was a refrigerant (dry ice) and a moderating liquid (ethanol). This combination (dry ice and ethanol) is used for the range: + 20 °C…-75° C.

The specimens were kept at the test temperature, T ± 1 ° C, for a period of 5 minutes. The Charpy impact machine was a Charpy pendulum with hammer, type PSW30 which had an available energy of 300J.

Impact test specimens location is shown in fig.6.

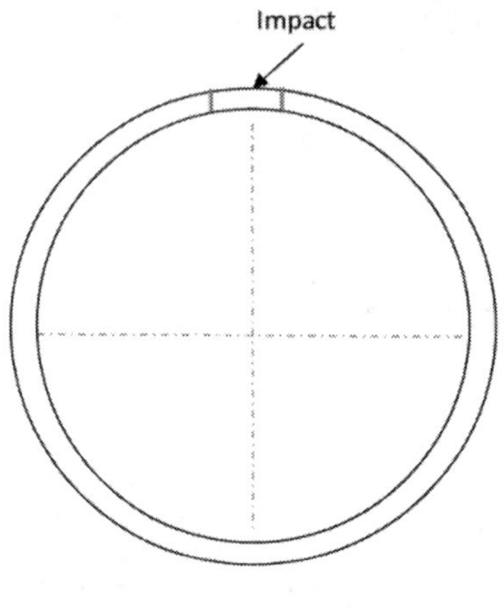

Figure 6. Location of impact specimens.

The notch axis was normal to material surface, as shown in fig.7.

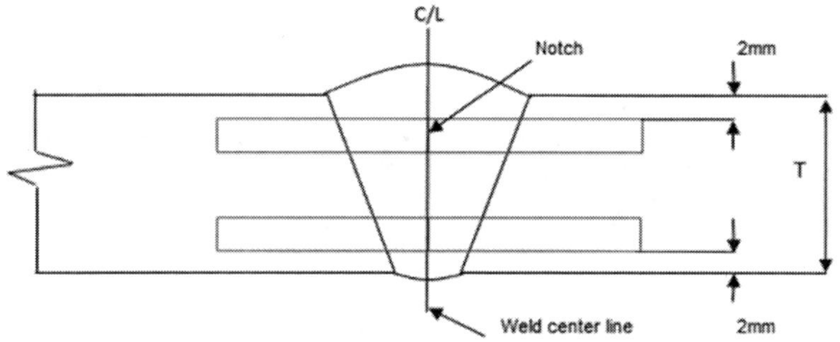

Figure 7. Notch location of impact specimens.

For each test temperature and each location of the notch, at least 3 specimens were used. The results obtained are shown in Table 3 and graphically represented in fig.8.

Table 3. The values of the breaking energy KV.

Notch Location	Specimen size, mm	Temp.°C	Absorbed Energy, Joules			
			A	B	C	Average
WCL Cap	10 x 10 x 55	+20°	178	176	190	181
WCL Root	10 x 10 x 55	+20°	220	220	208	216
WCL Cap	10 x 10 x 55	0°	150	138	150	146
WCL Root	10 x 10 x 55	0°	198	188	226	204
WCL Cap	10 x 10 x 55	-30°	102	110	116	109
WCL Root	10 x 10 x 55	-30°	130	150	138	139
WCL Cap	10 x 10 x 55	-50°	80	50	62	64
WCL Root	10 x 10 x 55	-50°	44	56	50	50

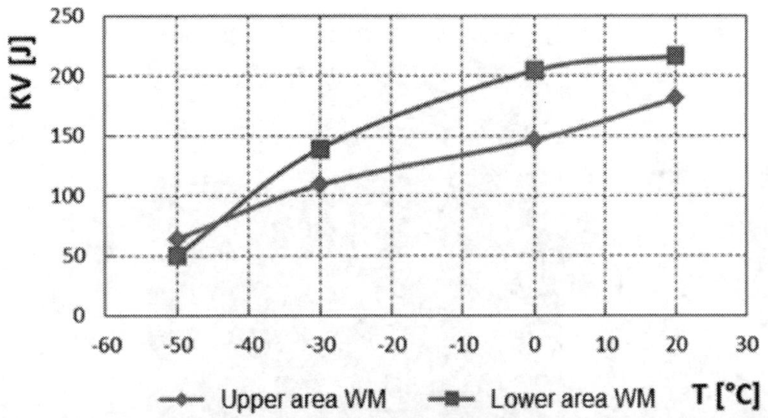

Figure 8. Absorbed energy variation KV with the test temperature T [°C].

Analysing the variation of the absorbed energy KV with the test temperature (Fig. 8), it can be observed that the welded joint has a high resistance to brittle fracture, its behaviour being ductile in a wide range of temperatures (-50 ° C - + 20 ° C).
The appearance of the fracture surface (fig. 9) is matte - fibrous, characteristic of the ductile fracture, which are preceded by large plastic deformations, respectively high values of the absorbed energy for crack propagation.
Towards the edges of the fracture surface where the shear stress is predominant, the fracture develops on the inclined surface to the core of the fracture surface.

Figure 9. Halves of broken Charpy V-Notch Impact specimen.

3.3 Metallographic examinations
Investigation of the weld macrostructure (Fig.10), reveals a full penetrated weld which interpenetrates correctly in the base metal without the appearance of metallic continuity defects.

The weld microstructure (fig.11a) consists of columnar grains oriented in the thermal gradient direction, and a heat-affected zone, HAZ, (Figure 11b) of ferrite, bainite and fine precipitations of secondary phases which impede the dimensional growth of crystalline grains in a purely mechanical manner. The base metal has a ferrite-bainitic microstructure (Fig. 11c).

IOP Conf. Series: Materials Science and Engineering **416** (2018) 012008 doi:10.1088/1757-899X/416/1/012008

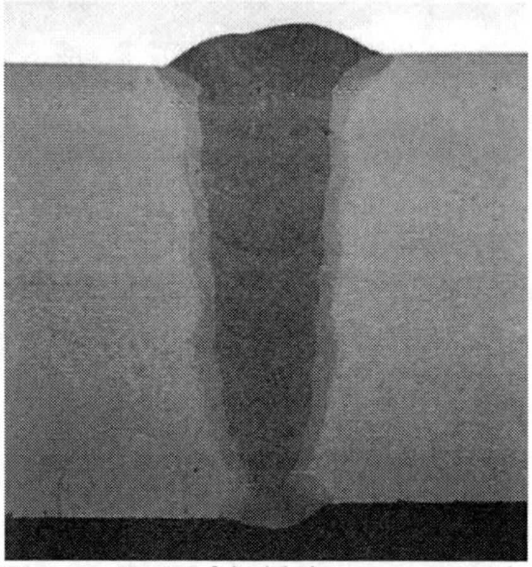

Figure 10. Welded joint macrograph.
Chemical Reagent: NITAL (10cm^3 HNO_3, 100cm^3 Ethyl Alcohol)

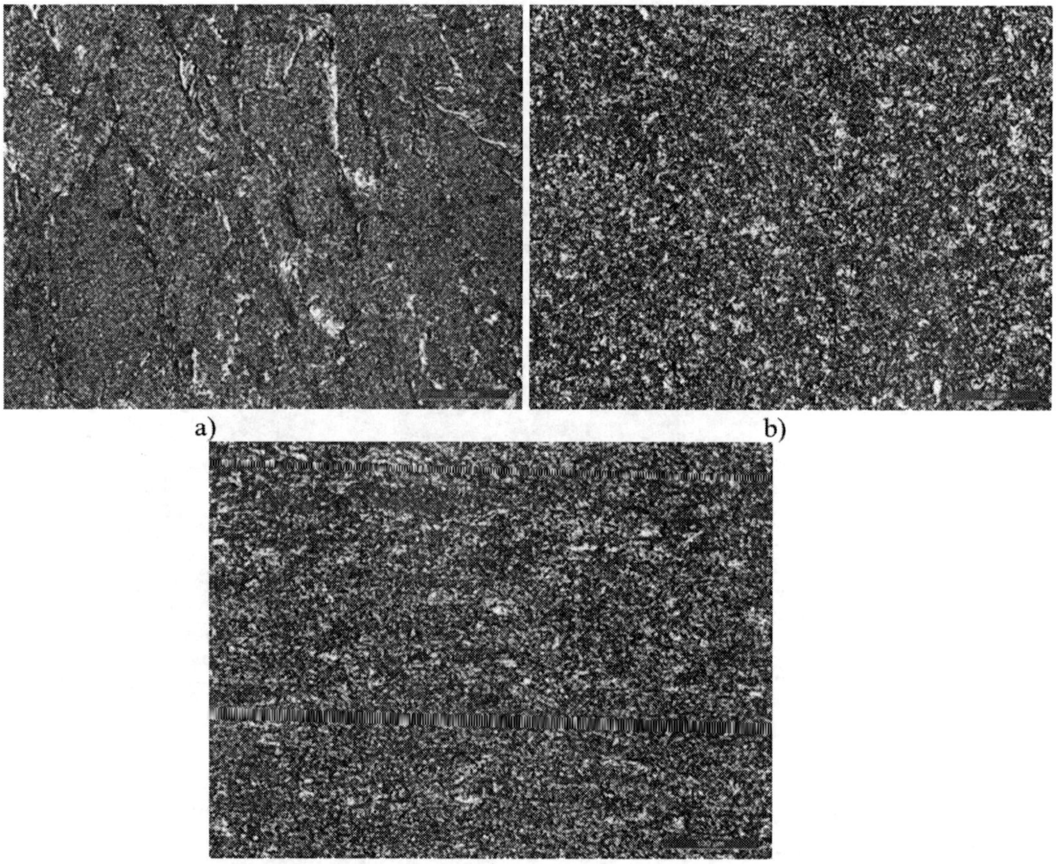

a) b)

c)

Figure 11. Microstructure of welded joint zones: a) weld metal; b) HAZ; c) base metal.
Chemical Reagent: NITAL (10cm^3 HNO3, 100cm^3 Ethyl Alcohol)

IOP Conf. Series: Materials Science and Engineering **416** (2018) 012008 doi:10.1088/1757-899X/416/1/012008

4. Conclusions

MAG welded joints of low-alloy thermo mechanical treated steels API 5L X65M, presents tensile strengths values Rm = 576…598 N/mm^2 higher than those required for base metal Rm ≥ 535 N/mm^2.

At the lowest charpy impact test temperature, (-50 ° C), the absorbed energy KV from the upper and lower metal deposited exceeds the minimum value of 38 J imposed by the DNV-OS-F 101 standard for this base material.

5. References

[1] Zhou M, Du L, Liu X 2011 *Journal of Iron and Steel Research, International* **18(3)** 59
[2] Coelho SR *et al* 2013 *Journal Materials Science & Engineering A* **578** 125
[3] Zhang T. *et al* 2017 *International Journal of Hydrogen Energy* 42(39) 25102
[4] Simionescu D, Mitelea I and Burcă M 2017 *Conference Proceeding Metal* **2017** (under printing)

IOP Conf. Series: Materials Science and Engineering **416** (2018) 012009 doi:10.1088/1757-899X/416/1/012009

Mechanical Behaviour of MAG Pulsed Welded Joints from 6082 T6 Aluminium Alloy

CP Lucian*, I Mitelea, M Burcă and I D Uţu

University Politehnica Timisoara, Piata Victoriei no. 2, Timisoara, Romania

*Corresponding author: ciprian.pavel.lucian@gmail.com

Abstract. One of the difficulties encountered by aluminum alloys fusion welding consist in reducing of the mechanical resistance characteristics of the weld seam and of the heat-affected zone (HAZ) compared to those of the base metal. The paper investigates the microstructural and mechanical properties changes of pulsed MAG welded joints made from the 6082-T6 alloy delivered in the form of 5 mm thick sheets. The dissolving phenomena of the hardenable phases in the sub-zones near the molten metal bath and the overaging of alloy in HAZ sub-zones at temperatures lower than those of precipitates redissolving, justifies the decrease of the mechanical resistance in the HAZ zone.

1. Introduction

Due to attractive usage properties obtained by heat treatments and reduced specific mass, Al-Mg-Si alloys have a growing expansion for structural applications and for the automotive industry [1],[2]. Although they have relatively low mechanical properties compared to steels, they are characterized by an excellent specific strength (R_m/ρ) .

Therefore, they are widely used in the aerospace and automotive industry. The most common aluminium alloys are: Al-Cu, Al-Si, Al-Mg, Al-Cu-Mg, Al-Cu-Mg-Si, Al-Mg-Si, and Al-Zn-Mg-Cu.In state of equilibrium, this alloys form a poorly alloyed solid solution and intermetallic phases $CuAl_2$ (phase θ), Mg_2Si, Al_2CuMg (phase S), Al_6CuMg_4 (phase T), Al_3Mg_2, $Al_2Mg_3Zn_3$ (phase T) etc. [3],[4].

A series of these alloys can be hardened by applying solution treatment followed by natural or artificial aging. The hardening ability is related to the variation in the solubility of the components (Mg, Si, etc.) in Al, depending on the temperature. Heating leads to the dissolution of excess phases and to the formation of an over saturated solid solution after solution treatment. The aging process is accompanied by the separation of the hardenable phases. The nature, shape, size, and distribution of the precipitated phases has a strong influence on growth of the ultimate tensile strength of Al-Mg-Si alloys. One of the difficulties encountered in the use of welded structures in these alloys is the overall reduction in the mechanical properties of the welded joint areas compared to the base material. This is due to the lower mechanical strength of the weld seam and the deterioration of the initial structure of the HAZ under the thermal welding cycles. Other papers [4], [5] highlight the microstructural stability of the 6082 - T6 alloy and the mechanical strength losses in welded seam and HAZ. Studies related to the effects of the heat of welding on precipitation phenomena and liquationcracking, which are responsible for the worsening of the mechanical properties of welded joints, are incompletely analysed [4],[6],[7].

IOP Conf. Series: Materials Science and Engineering **416** (2018) 012009 doi:10.1088/1757-899X/416/1/012009

2. Examined material and working procedure

Aluminium alloy, 6082 - T6 (AlSi1MgMn according to EN 573), delivered in the form of 5 mm thick sheets, was solution treated at 535 ± 5 ° C/25 min./water followed by artificial aging, 175 ± 10 °C for 8 h/air .

The nominal chemical composition of the alloy sheets used to make the full penetrated butt welded joint is: Si = 1.18 %, Fe = 0.39 %, Cu = 0.065 %, Mn = 0.70 %, Mg = 0.80 %, Cr = 0.10 %, Ni = 0.015 %, Zn = 0.044 %, Ti = 0.011 %, Ga = 0.01 %, V = 0.023 %, Al = Rest.

As filler material was selected the electrode wire AlSi 5 (Alloy 4043) according to ISO 18273 and EN 573 – 3 having a diameter of 1.2 mm , which has the following chemical composition prescriptions: Si = 4.5 – 6.0 %, Fe ≤ 0.8 %, Cu ≤ 0.30 %, Mn ≤ 0.05 %, Mg ≤ 0.05 %, Zn ≤ 0.10 %, Ti ≤ 0.20 %, Be ≤ 0.0003 %, Al = Rest.

The references [1], [3], [5] recommend using this wire to obtain a molten area without cracks.

As shielding gas was used Argon 4.8 (purity ≥ 99.998%), Linde, with a flow rate of Q = 14 - 15 l/min. The welding was made in a horizontal position, the position PA/SR EN ISO 6943, the welding direction being to the left with a 85° electrode wire inclination. The joint preparation and positioning of the components is shown in Figure 1.

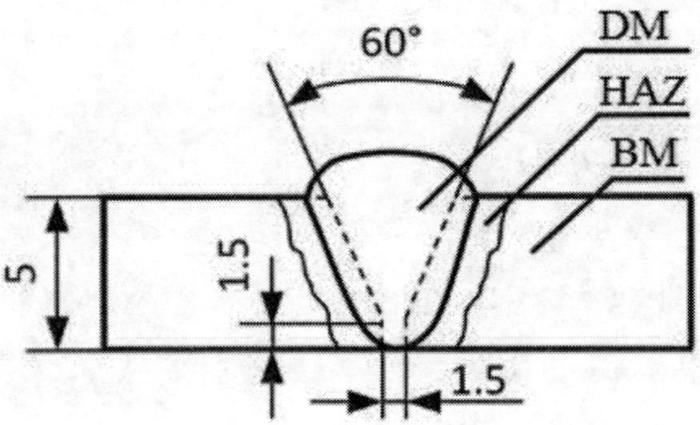

Figure 1. Geometry of the butt welded joint.

The welded joints were made in one pass with the following welding technological parameters:

- Wire feed speed: 6.2 m/min.
- Welding medium current: ≈ 130 A
- Arc voltage: 23 V (ΔUa = 0)
- Pulse current:220 A
- Pulse time: 2 ms
- Background current: 64 A
- Pulse frequency: 210 Hz
- Slope time: 9
- Self-regulating coefficients : k_a = 36 %; k_i = 0 %
- Welding speed : 25 cm/min

The used welding equipment (ESAB ARISTO LUD 450 W) allows a synergic control of process parameters. Their determination was made by properly selecting the wire feed speed,v_{as} = 6.2 m/min., Fig.2. In Fig. 3, 4 and 5 it is presented the recording of pulse current welding parameters under the conditions described above.

IOP Conf. Series: Materials Science and Engineering **416** (2018) 012009 doi:10.1088/1757-899X/416/1/012009

Figure 2. Setting of the wire speed feed.

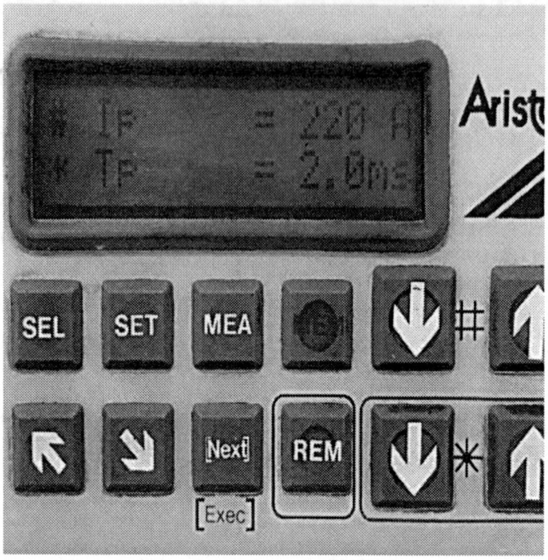

Figure 3. Record of the pulse parameters.

Figure 4. Record of pulse frequency.

Figure 5. Measured value of arc voltage.

The corresponding value of the arc voltage for these pulse current conditions is shown in Fig.5.The obtained welded joints were performed to tensile testing, Vickers microhardness and metallographic investigations.

3. Evaluation of experimental results

3.1Macrography of welded joints

For the examination of the overall structure and highlight the heterogeneities occurred in welded joints, samples with transverse faces (normal to the longitudinal welding axis), in accordance with standard techniques,were taken and prepared.
The macroscopic image of a cross section through a welded joint is shown in Fig.6.

IOP Conf. Series: Materials Science and Engineering **416** (2018) 012009 doi:10.1088/1757-899X/416/1/012009

Figure 6. Macrography of welded joint areas.

Although the welding operation provides the continuity of the material, the welding area does not have a homogeneous structure. As a result of high-speed heating, melting of the filler material and a small portion of the base material occurs, and subsequent cooling in the molten zone causes a series of structural transformations. Thereby, the liquid of the molten zone first reaches the solidification temperature in the areas in contact with the edges of the weld, where the solid grains of the two welding components are located. Consequently, the solid begins to grow from these grains in a columnar manner. The development of crystalline grains in the molten area from pre-existing grains is called epitaxial growth. The appearance of the outer surface of the weld is shown in fig.7.

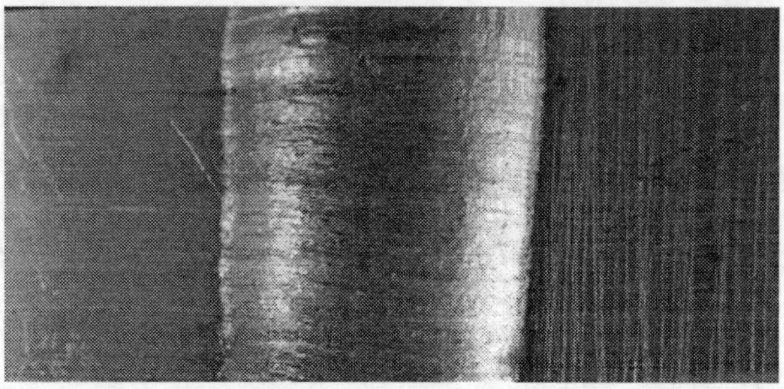

Figure 7. Macroscopic image of the outer surface of the weld seam.

It is noted that there are no surface defects such as cracks and cap undercuts and it's looks like there is a good moistening of deposits to the base metal. The correct positioning of the electrode wire in the joint at the execution of a single pass is evidenced by the symmetry of the joint and a corresponding excess of the weld metal height. The selected filler material led to the following chemical composition for the deposited metal: Si = 4.95 %, Fe = 0.15 %, Cu = 0.010 %, Mn = 0.0016 %, Mg = 0.0008 %, Zn = 0.002 %, Ti = 0.010 %, Al = Rest.

3.2 Micrography of welded joints

Figures 8…10 show some microstructural images of the characteristic areas of the welded joints made with parameters established by the experimental tests. The chemical reagent used etching has the following chemical composition: 1 ml of 40%HF + 1.5 ml of concentrated HCl + 2.5 ml of concentrated HNO_3 + 95% of H_2O_2.
Crystallization of the molten metal bath is initiated by the precipitation of relatively large grains of solid solution with aluminium base, α, having a dendritic character, after which, upon reaching the eutectic transformation line,starts the formation of mechanical mixtures of solid solutions with aluminium base and intermetallic phases intergranularly arranged. The further decrease of the temperature to that of the room causes separation from the solid solution α of secondary phases distributed inside and on the limits of the crystalline grains (Fig. 8).

The bonding zone of the metal deposited with the base metal is achieved through a very narrow range of fine grains of solid solution with intermetallic phase particles having a punctual appearance (Fig. 9).
The base metal microstructure consists of a matrix of α solid solution and secondary phase particles with a hardening role (fig.10).

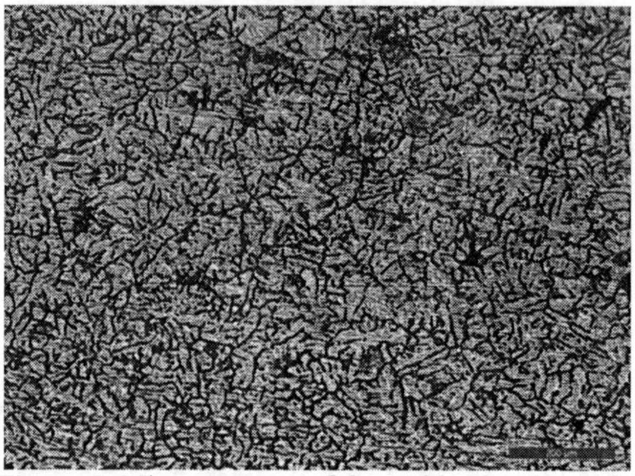
Figure 8. Microstructure of the welded bead(x 100).

Figure 9. Base metal microstructure(x100).

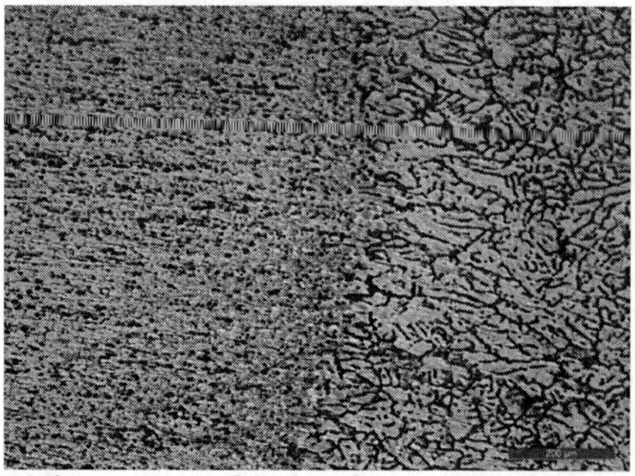
Figure 10. Microstructure of the interface welding - base metal(x100).

3.3 Static traction tests

Tensile testing was performed on samples taken from welded joints as well as samples taken from the base metal. The results are centralized in Table 1 and they highlight the following aspects:

- A decrease in ultimate tensile strength from approx. 328 N/mm^2, characteristic of the base metal, at ca. 202 N / mm^2, characteristic of the welded samples.
- A reduction in elongation at break from approx. 16% (unwelded samples), at ca. 9.8% (welded samples).
- At all tested samples the rupture occurred at the interface between weld and heat-affected area (HAZ) or in weld.

The data obtained demonstrates that the welding process eliminates the effects of the heat treatment applied to the base metal through structural changes in the weld and HAZ. The coarse microstructure resulted from primary and secondary crystallization along with the formation of light fusible eutectic mechanical mixtures and excess phase precipitation justifies the decrease of the mechanical characteristics of the welded joints.

Table 1. Tensile testing on welded joints samples.

No.	Source	Rm, N/mm^2	A5, %	Location of rupture
1	Base Material, BS	329	18,5	-
2	BS	327	14,5	-
3	BS	328	16,0	-
4	Welded Joint, WJ	200	9,6	HAZ
5	WJ	202	9,8	HAZ
6	WJ	201	9,8	HAZ
7	WJ	204	10,2	WELD
8	WJ	200	9,6	HAZ

3.4 Vickers microhardness tests

Since hardness is the mechanical characteristic most sensitive to the structural changes that occurred during the welding process, cross-faced samples subjected to such tests were taken from the joints. Based on the results obtained, the graph of Fig. 11 was drawn. The main observations from this graph are:

- The hardness of the base metal has values of 100-118 HV, corresponding to the applied heat treatment.
- The welded seam, without heat treatment after the welding operation, has values of 66-75 HV.
- At a distance of about 2 mm from the welded joint axis, the lowest hardness values (58-61 HV) occur due to the over-aging of the alloy in the strip of material that was heated to a temperature lower than redissolving the secondary phases in the solid solution.

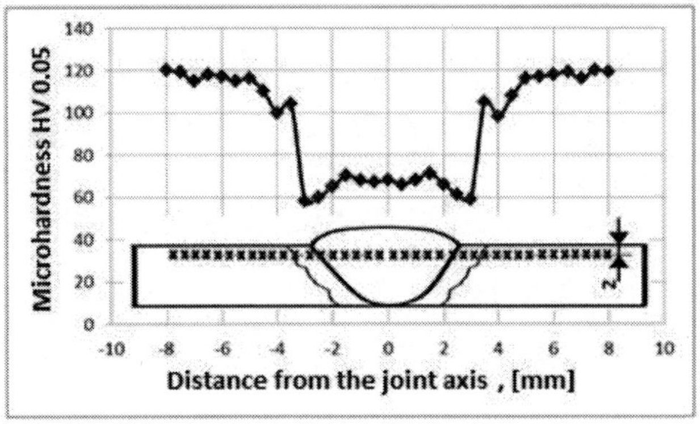

Figure 11. The hardness gradient curve on the cross section of the welded joint.

4. Conclusions

For used experimental conditions, the ultimate tensile strength of welded joints decreases with approx. 39% and elongation at break with approx. 40% compared to the base metal nominal values.
In the heat affected zone, at a distance of approx. 2 mm from the joint axis there is a pronounced softening of the material (HV hardness decreases with about 45%) as a result of the over-age phenomenon induced by the thermal welding cycle.

5. References

[1] Malin V1995 *Welding Journal* **74 (9)** 305
[2] Miyazaki M, Nishio K, Katoh M, Mukae S and Kerr W 1990 *WeldingJournal* **69 (9)** 362
[3] Kluken AO and Bjoerneklett B 1997 *Welding Journal* **76 (2)** 39
[4] Huang Cand Kou S 2004 *Welding Journal* **83** 111
[5] SinghG, Kumar S and Singh A 2013 *Int. J. Res. Mech. Eng. Technol.***3(2)** 143
[6] Utu ID, Mitelea I, Urlan SD, Craciunescu CM 2016 *Materials* **9** 606
[7] Mitelea I, Utu ID, Urlan SD, Craciunescu CM 2017 *Mat.-wiss. u. Werkstofftech.* 2017 **48** 1040

IOP Conf. Series: Materials Science and Engineering **416** (2018) 012010 doi:10.1088/1757-899X/416/1/012010

On Surface Topography and Microstructure in Cavitation Erosion Tests of Alloy 80 A

C Belin*, I Mitelea, I Bordeasu and C Craciunescu

„Politehnica" University of Timisoara, Timisoara, Romania, EU

*Corresponding author: cosmin.belin@student.upt.ro

Abstract. From practical and economic reasons for the design of components for Diesel engines, the selection of materials and /or heat treatments is essential to provide a high cavitation erosion resistance. This paper studies cavitation behavior of the Nimonic 80A alloy designed for execution of Diesel engine exhaust valves. The cavitation tests where done on an ASTM G32 -2010 compliant vibrating device with piezoceramic crystals.
The metalographic examinations performed with optical microscope and electronic scanning microscopy revealed the microstructural changes occurring in the region of the cavitation affected material and the path of the crack propagation

1. Introduction

The phenomenon of cavitation consists of the formation of bubbles/cavities filled with gases and vapors in a liquid as a result of the local pressure decrease below the vapor pressure, known as the critical value. When these bubbles come in contact with high local pressure, they will implode and generate microjets and/or shock waves [1], [2], [3]. If bubble implosion occurs near a solid surface, the microjet or shock waves create high pressure on the solid surface. When repeating such events, the surface area under such stress will suffer a local fatigue phenomenon, accompanied by elasto-plastic deformations leading to fracture, with solid material losses [4], [5]. This effect is known as cavitation erosion and is manifested in many engineering components such as hydraulic turbines and pumps, marine and river propellers, aerospace vessels, exhaust valves and combustion chambers of Diesel engines, etc. Frequently the phenomenon occurs in the parts exposed to fluids with high velocity, but also to those that vibrate in static liquids [2]. Studies in this field show that cavitation erosion can be countered by applying thermal treatments [6] to improve the mechanical properties of metallic materials or by various surface hardening techniques.

This paper highlights the cavitation erosion performance of the Nimonic 80 A alloy which has been subjected to solution quenching.

2. Experimental procedure

The considered alloy, Nimonic 80 A (EN NiCr20TiAl, UNS No. 7080, W. No. 24952 & 24631), is part of the deformable material category with high mechanical strength up to service temperatures of 815 °C [3], [4]. Vacuum refinements are commonly used in the manufacture of exhaust valves of Diesel engines and gas turbine components (vane, rotors, discs) [5]. The chemical composition of the investigated material is: 20.5% Cr, 1.98% Ti, 1.54% Al, 2.14% Fe, 1.61% Co, 0.63% Mn, 0.48% Si 0.12% Cu 0.08% % Zr, and % Ni balance. Cavitation samples were made from cylindrical rolled bars,

IOP Conf. Series: Materials Science and Engineering **416** (2018) 012010 doi:10.1088/1757-899X/416/1/012010

Ø20 x 300 mm, which were thermally treated by solution quenching, at 1080 °C for 8h and air cooled, having the shape and dimensions shown in Figure 1.

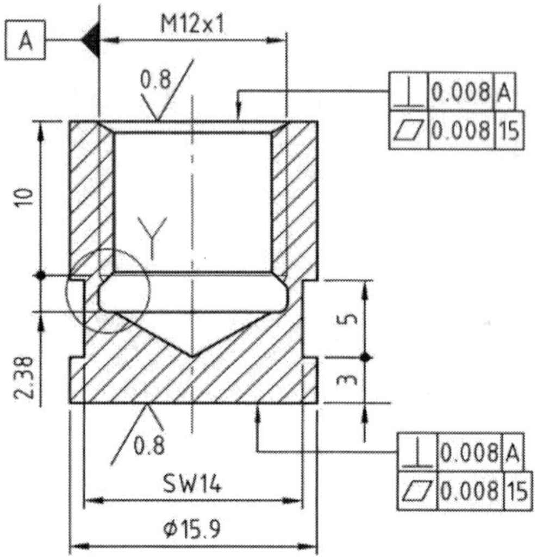

Figure 1. Cavitation sample.

Cavitation tests were performed on a vibrator device with piezoceramic crystals, which complies with the ASTM G32-2010 standard requirements (see figure 2) [6]. The functional parameters which determine the cavitation erosion intensity are: vibration amplitude of 50 μm peak-to-peak, 20000 ± 2% Hz frequency and 22 °C ±1°C cavitating liquid temperature (double distilled water). To determine the wear rate through cavitation erosion, the test was interrupted at regular intervals; the samples were washed in acetone, dried in a hot air stream and weighed with an analytical balance that can weigh with an accuracy of five decimals.

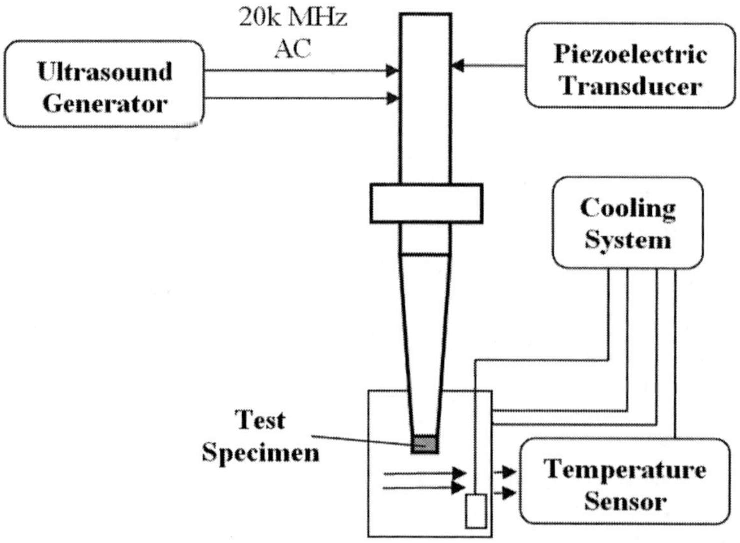

Figure 2. The ultrasonic equipment used in the cavitation erosion experiments.

The surface topography of cavitation tested samples was investigated by scanning electron microscopy.

IOP Conf. Series: Materials Science and Engineering **416** (2018) 012010 doi:10.1088/1757-899X/416/1/012010

3. Results and discussions

The microstructure obtained after the heat treatment is composed of γ solid solution grains of polyhedral form, some of which have annealing twins. Within the grains and on the separation boundaries there are secondary phase particles [3].

In Figure 3 are presented the characteristic cavitation erosion curves for the Nimonic 80 A alloy which was solution quenched at 1080°C, for 8h and cooled in air. They indicate the variation of the mass losses and their velocity, respectively the variation of the average penetration depth of the erosion and its velocity, with the duration of the cavitation attack generated by the vibratory apparatus.

The specific curves represent the cumulative mass loss (m(t)), the mean depth of erosion (MDE (t)), the specific velocities (v(t)) and the mean depth erosion rate (MDER(t)), and have the following forms [1] [2]:

$$m(t) = A1 \cdot t \cdot \left(1 - e^{-B1t}\right) \quad \text{or} \quad MDE(t) = A2 \cdot t \cdot \left(1 - e^{-B2 \cdot t}\right) \qquad [1]$$

$$v(t) = \frac{dm}{dt} = A1 \cdot \left[1 - e^{-B1 \cdot t}\right] + A1 \cdot t \cdot B1 \cdot e^{-B1 \cdot t}$$

respectively [2]

$$MDER(t) = \frac{dMDE}{dt} = A2 \cdot \left[1 - e^{-B2 \cdot t}\right] + A2 \cdot t \cdot B2 \cdot e^{-B2 \cdot t}$$

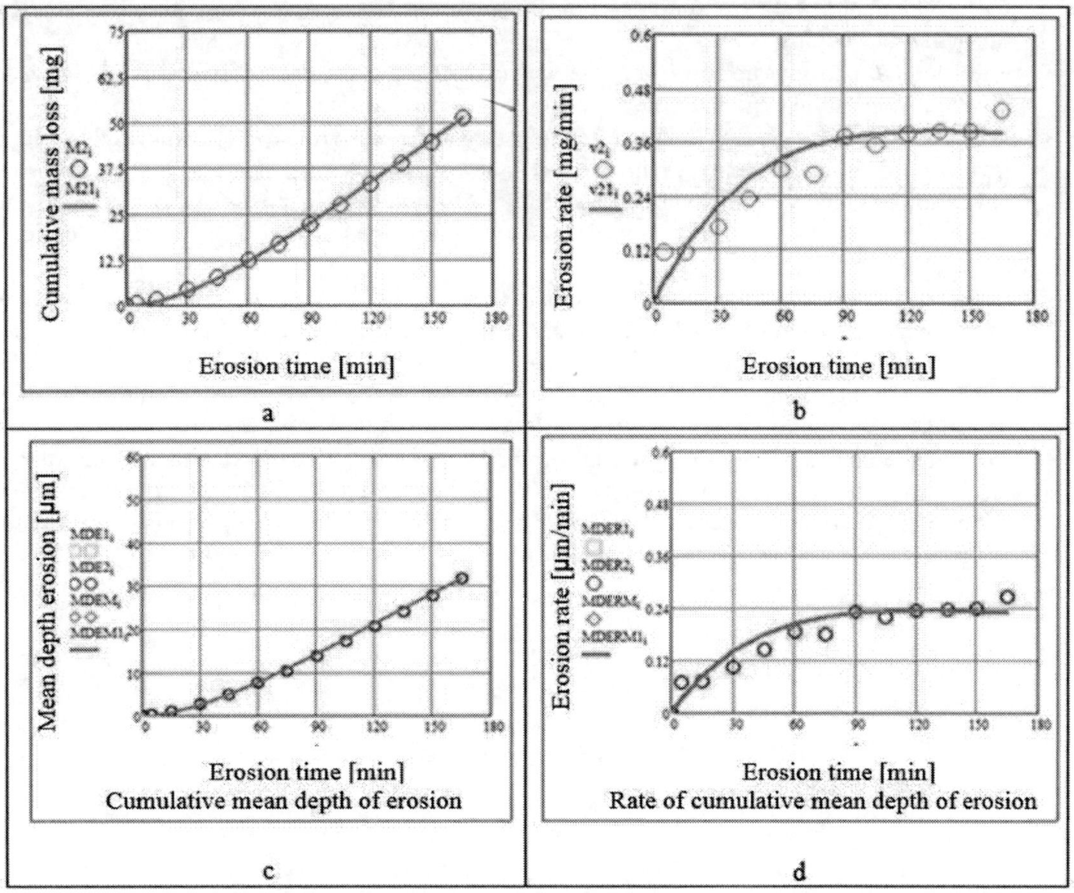

Figure 3. Characteristic cavitation erosion curves for the Nimonic alloy
a - Cumulative mass losses [mg]; b - Erosion rate [mg/min]; c - Mean depth of erosion, MDE [µm]; d - Mean depth erosion rate, MDER [µm/min].

The form of the approximation curves of the experimental points, as well as the dispersions of the points against these curves, (fig. 3), demonstrate the following:

- a small degree of dispersion of the experimental points compared to the approximation curves, specific to the materials with a good resistance to cavitation erosion due to high mechanical resistance characteristics;
- roughly linear variations of the curves m (t), Figure 3a, MDE (t), Figure 3 c respectively, after about 90 minutes of cavitation exposure, which results in the stabilization of the erosion rate to the maximum value ($\cong$ 0.39 mg /min, respectively $\cong$ 0.24 μm/min), phenomenon encountered only in materials with increased cavitation resistance.

The images shown in Figure 4. obtained with the Canon Power Shot A480 camera, the resolution of which, furthermore, highlights the extent of surface degradation and less in depth at various times of the cavitation attack, shows the evolution of surface degradation as a function of the resistance provided by the thermal treatment applied. These images (Fig. 4) represent the values and dispersions of the experiments points versus the erosion speed curves in Fig. 3b and 3d. Thus, it can be seen that:

- During the 5 minute-10 minutes of vibratory cavitation (the first two experimental points), the erosion speeds have very close values, reflected by craters and ring formation from the periphery of the cavitated surface. During these periods, the roughness peaks resulting from the mechanical grinding and abrasive dust remaining after the washing are eliminated before the actual test begins;
- after this attack time, up to 60 minutes (inclusive) the erosion rate increases, which leads to the extension of the degradation on the entire exposed surface;
- Within 60-75 minutes, the erosion rate has a slight decrease due to the compressing cavitation layer. By its repetitive impact with microjets and shock waves generated by the implosions of cavitation bubbles, stellar caverns are starting to form at the periphery of the surface due to the expulsion of grains, which bonds were cracked by the impact force between the microjets and the exposed surface;
- in the interval of 90-150 minutes, the erosive process is carried out at close speeds values due to the behavior of the layer, which, beyond a natural hardening by cold deformation, provides a constant attenuation of the impact force through the air pockets that penetrate the microcracks, blocking the propagation of cracks across the whole area / with large expulsions of material, but emphasizing the growth of caverns from the peripheral area;
- local fatigue takes place during the last cavitation period (150-165 minutes), when the speed has a jump (the last experimental point) as a result of the expulsion of a larger amount of material generated by the crack network formed earlier, which, uniting, leads to the fracture of boundaries, but also within the grains interior.

IOP Conf. Series: Materials Science and Engineering **416** (2018) 012010 doi:10.1088/1757-899X/416/1/012010

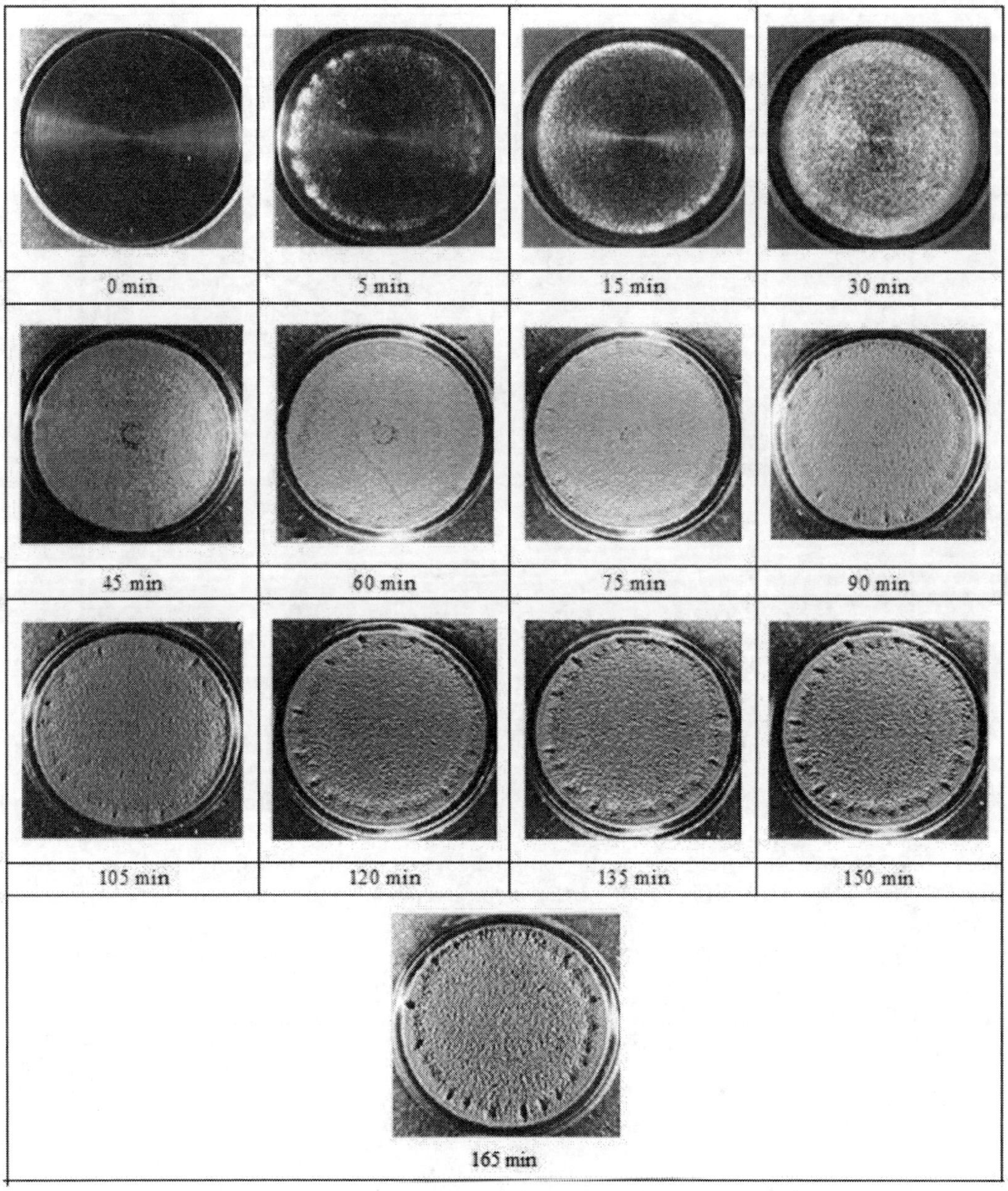

Figure 4. Aspect of the surface of samples tested at cavitation.

The statistical processing of the experimental results, with the determination of the tolerance interval in which the experimental point's dispersion fall (Figure 5.), demonstrates the accuracy of the measurements performed. The upper and lower limits of the dispersion band were constructed according to the procedure described in [1], using the standard estimation error σ and the exponential regression curve of the form (1), which approximates the experimental values of the three tested samples.

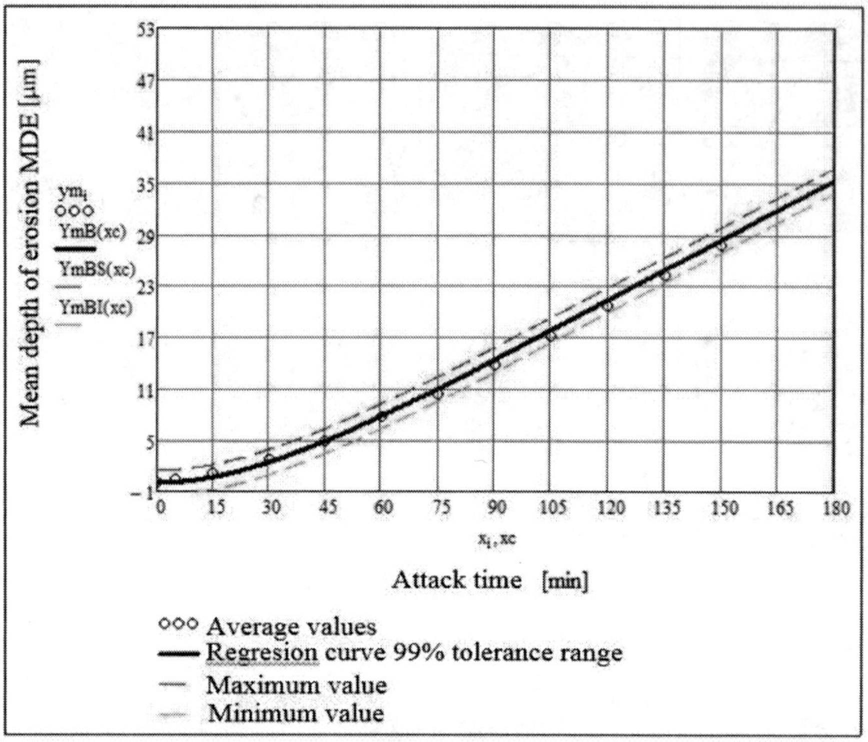

Figure 5. Sample dispersion band tested at cavitation.

The 99% tolerance range shows the low dispersion band and the uniform surface behavior of the cavitation attack. In fact, it is a reflection of the homogeneous distribution of mechanical properties (hardness, tensile strength, yield strength, toughness) that define the cavitation resistance of the material [2], [6]. The small differences in the error band are natural due to the complexity of the cavitation erosion process.

Figure 6 illustrates the topography of the samples tested at ultrasonic cavitation for 165 minutes. It is observed that the removal of material particles is uniform, being characteristic of the f.c.c. crystalline network (face-centered cubic), and that some differences in cavity depths occur.

IOP Conf. Series: Materials Science and Engineering **416** (2018) 012010 doi:10.1088/1757-899X/416/1/012010

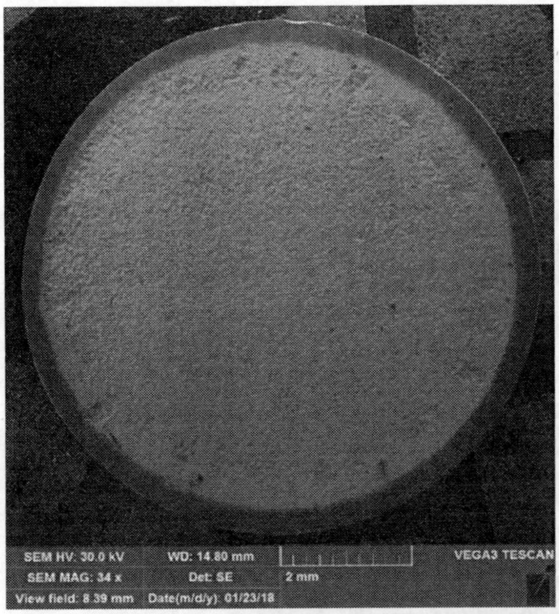

a)

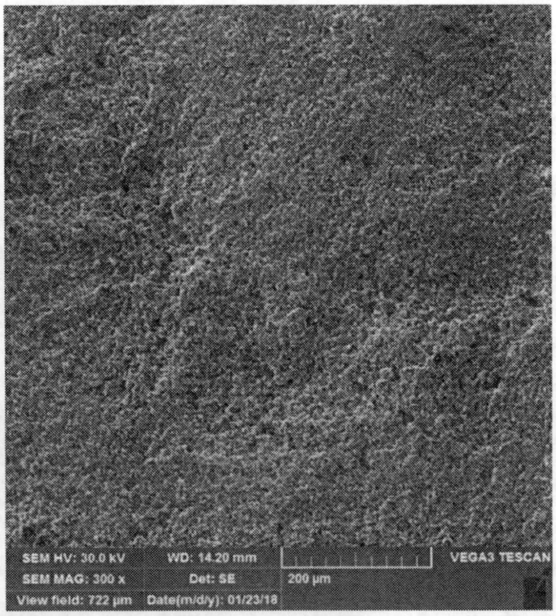

b)

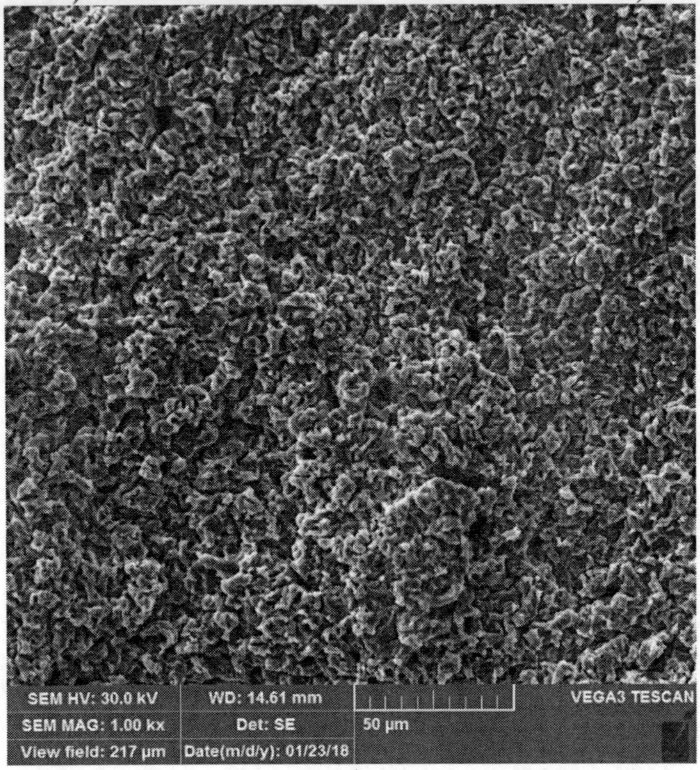

c)

Figure 6. Macro (a) and micrograph imagines (b, c) of the cavitation exposed surface for 165 min. after solution treatment.

The formed cavities were initiated on the crystalline grains boundaries, where there are phases of chemical combinations, and as the erosion progresses, the attack occurs in the γ solid solution matrix.

4. Conclusions

The cavitation behaviour of the investigated material is characteristic to the alloys with an austenitic microstructure with a f.c.c. crystalline network and secondary phases placed inside and on the grain boundaries.

For the experimental conditions used, after 165 min. of cavitation attack, the mean depth of erosion, MDE, is about 32 μm and the mean depth erosion rate, MDER, tends to stabilize at 0.24 μm / min.

The shape of the erosion speed curves, with maximum capping, and the low degree of dispersion of the experimental points from them, show that the studied Nimonic alloy has a good behavior in the vibrational cavitation erosion, specific to the materials with high mechanical characteristics.

The analysis of surface degradation over various cavitation times (Fig.4), correlated with the experimental values of the erosion speeds (Fig.3b and 3d), allowed the description of the damage mechanism, which occurs due to the impact between the samples surface and the shock waves, respectively by the microjets resulted from the implosion of cavitation bubbles.

The surface wear phenomena are triggered on the grain boundaries and advances into the γ solid solution matrix.

5. References

[1] Bordeaşu I, Patrascoiu C, Badarau R, Sucitu L, Popoviciu M and Balasoiu 2006 *FME Transactions Faculty of Mechanical Engineering, New Series* **34(1)** 39

[2] Bordeaşu I 2006 *Eroziunea cavitaţională a materialelor* (Editura Politehnica, Timişoara)

[3] Baumgaertner T, Plath A, Kempf B, Bothe K and Gerold V 1989 *Strength of Metals and Alloys (ICSMA 8)* **3** 1045

[4] Matthew DM, Singh V, Chen W and Wahi PR 1991 *Acta Metallurgica et Materialia* **39(7)** 1507

[5] Di Martino FS, Faulkner GR, Hogg CS, Vujic S and Tassa O 2014 *Materials Science & Engineering A* **619** 77

[6] Mitelea I, Bordeasu I, Katona SE and Craciunescu CM 2017 *International Journal of Materials Research* **108(12)** 1099

IOP Conf. Series: Materials Science and Engineering **416** (2018) 012011 doi:10.1088/1757-899X/416/1/012011

Roughness Parameters during Cavitation Exposure of Nodular Cast Iron with Ferrite-Pearlite Microstructure

T Bena*, I Mitelea, I Bordeaşu and C M Crăciunescu

[1] „Politehnica“ University of Timisoara, Timisoara, Romania, EU

*Corresponding author: traian2461@yahoo.com

Abstract. The cavitation erosion resistance of EN-GJS-400-15 nodular cast iron was tested in the laboratory according to ASTM G32- 2010 standard. The erosion average penetration depth values, MDE, were correlated with the roughness parameters and the microstructure of the samples subjected to volume heat treatments consisting from stress relieving and for softening, annealing, normalization and quenching followed by tempering. The results obtained showed that the changes in the surface roughness of the samples tested at cavitation can be used to predict the resistance of the material to this wear phenomenon.

1. Introduction

The evaluation of the cavitation erosion resistance of a material is done by standardized tests in order to simulate this process under laboratory conditions. Such tests provide the relevant results that can be used to compare the strength of a material in different structural states or more to test materials under similar conditions [1, 4, 5, 6]. The equipment used allows the assessment of cavitation erosion resistance of a material in terms of erosion rate, thus allowing a classification of materials based on this property [3, 4, 6]. One of the difficulties encountered in such examinations is the multiple unsuccessful or only partially successful attempts to correlate cavitation erosion resistance with one or a combination of metallic properties [7, 8].

This paper aims to find a correlation between the depth of erosion penetration, MDE at the end of the test period and the characteristic parameters of the surface roughness of the EN-GJS-400-15 nodular cast iron surface subjected to volume heat treatments.

2. Investigated material, experimental stand and working procedure

The investigated material is a ferrite-perlite matrix nodular cast iron, EN-GJS-400-15, having the following chemical composition: C = 3.57%, Si = 2.51%, Mn = 0.23%, P = 0.044%, S = 0.010% and Fe = balance.

The microstructure of the base metallic mass consists of approximately 60% F si 40% P (Figure 1).

IOP Conf. Series: Materials Science and Engineering **416** (2018) 012011 doi:10.1088/1757-899X/416/1/012011

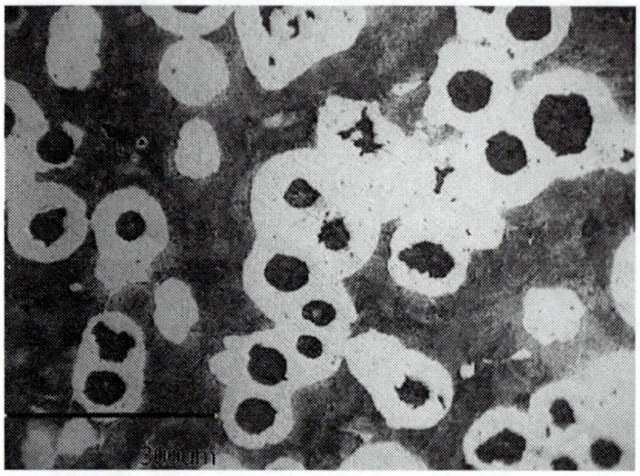

Figure 1. Image of the foundry nodular casting microstructure (x100).

Cast iron bars with the dimensions Φ 25 x 40 mm were subjected to the following volume heat treatments:

- Stress relief annealing (500 ± 10 °C/120 min./ oven);
- Softening annealing (710 ±10 °C/30 min./ oven);
- Normalizing (860 ±10 °C/60 min./air);
- Martensitic quenching followed by tempering (860 ±10 °C/30 min./oil + 450 ±10 °C/60 min./air).

Subsequently, samples were prepared for cavitation tests [3-6] and for the investigation of surface roughness. Cavitation attempts were conducted on an oscillator with piezoceramic crystals, performed in accordance with the requirements of ASTM G32-2010 [2].

The assessment of the cavitation erosion resistance was done by determining the mean erosion depth, MDE and erosion rate MDER curves with attack time [5]

At various attack times, mass losses were determined, and eroded surfaces were examined by optical and scanning electron microscopy.

The roughness measurements were made in 10 directions, on the surface of each sample using the Mitutoyo apparatus [3,9]

3. Evaluation of experimental results

After each cavitation attack period, images of the surface tested were made using a Canon Power Shot SX200 IS camera. The results obtained are shown in (Figure 2) and they highlight the way in which the surface degradation phenomena extend. Their analysis confirms that in the first 30 - 45 minutes of testing only the surface of the heat-treated samples is sensibly affected by softening annealing (Figure 2b). At samples subjected to other volume heat treatments, degradation of the surface is minimal up to 45 minutes (Figure 2a, 2c, 2d), after which there is a continuous increase in the density of material pinches and their depth with attack time. The phenomenon is explained by low particle expulsion the material (graphite and matrix particles) at the beginning of the wear process. At higher attack times, there are large differences in the degree of degradation of the sample surface depending on the applied treatment. The geometric quality of the surface is significantly improved by the application of heat treatments for normalization (Figure 2c) or quenching followed by tempering (Figure 2d).

IOP Conf. Series: Materials Science and Engineering **416** (2018) 012011 doi:10.1088/1757-899X/416/1/012011

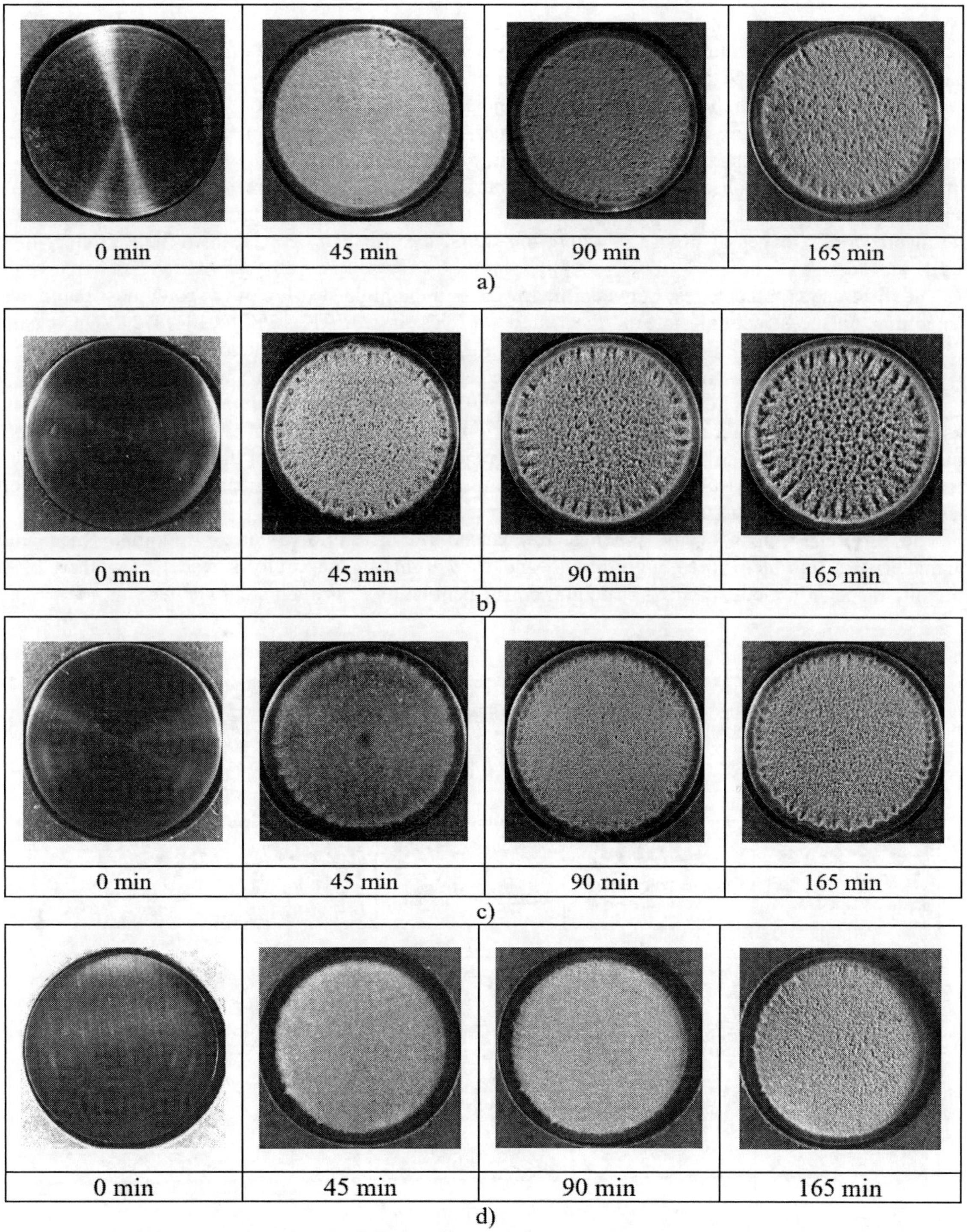

Figure 2. The macroscopic aspect of the surface of the tested samples at cavitation variable times depending on the thermal treatment condition: a) – stress relief annealing; b) - softening annealing; c) - normalizing; d) - quenching –tempering.

In (Figure 3a ... 3d), for each type of heat treatment, one significant image of the roughness profiles is displayed, with the corresponding values of the Ra, Rz and Rt parameters measured with the Mitutoyo

IOP Conf. Series: Materials Science and Engineering **416** (2018) 012011 doi:10.1088/1757-899X/416/1/012011

equipment. The profiles of these images and the values of the roughness parameters Ra, Rz and Rt were recorded in 10 arbitrarily chosen directions in the cavity surface, from the periphery to the central area. They show the complicated dynamics of surface degradation under the local fatigue demands created by the repeated impact of microjets produced by cavitation bubble implosion.

The substantial differences between the values of these parameters, recorded in different areas, show the intensity of the shock caused by the impact with the hydrodynamic microjet, but also the different structure response of the required material. Smaller values are specific to areas near the center of the cavity, and larger values are recorded to the periphery. The phenomenon is supported by the macroscopic images of Figure 2, where the stellar caverns can be seen from the periphery of the cavity surface.

The difference (percentages) between the lowest and the highest value of the roughness parameters, depending on the structural state of the tested samples, is from 13.13% to 138.27%. The smallest differences (percentages) are for softening annealing, and the highest for the quenching - tempering.

These values prove the complexity of the cavitation mechanism destruction, based on the hydrodynamics of the phenomenon (cavitation bubble formation and implosion) and on the iron response mechanics (deformation, cracks, expulsions after impact with microjets). At the same time, this difference also has an explanation in the formation of gas pockets in the eroded surface caverns (much larger in samples subjected to softening annealing), acting as attenuators of the pressure force developed at impact with microjets and shock waves.

The different shapes of the profiles are caused by different values of the main mechanical characteristics (hardness, breaking strength and flow limit) in the cavity surface, as well as by the graphite dispersion mode, the first structural constituent being destroyed and expelled.

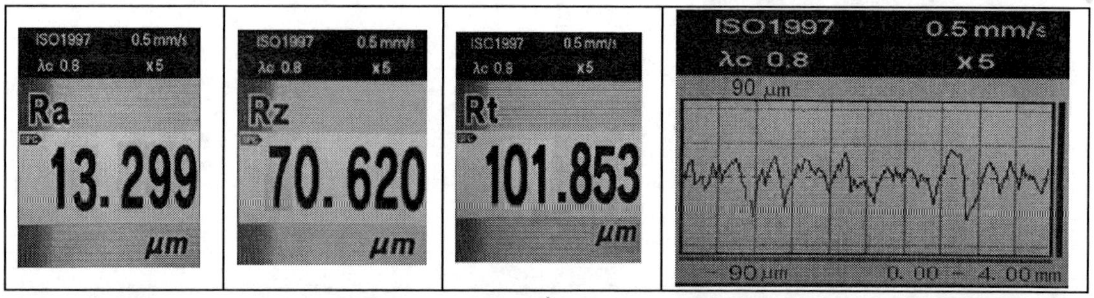

a)

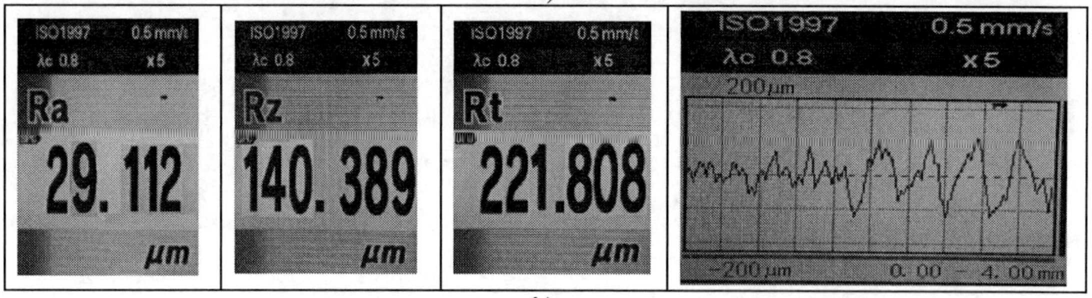

b)

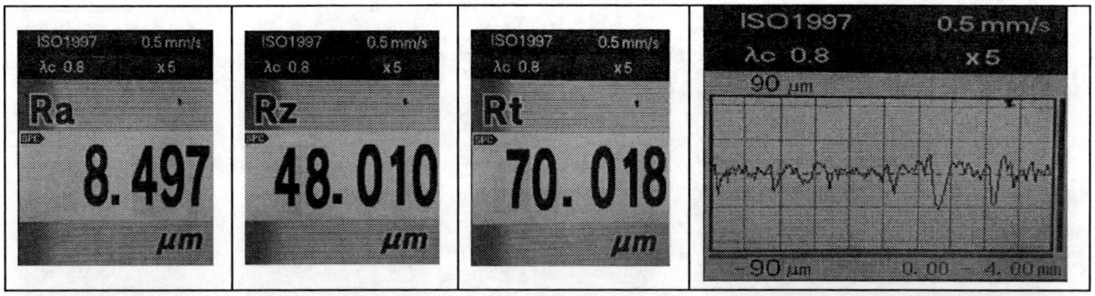

c)

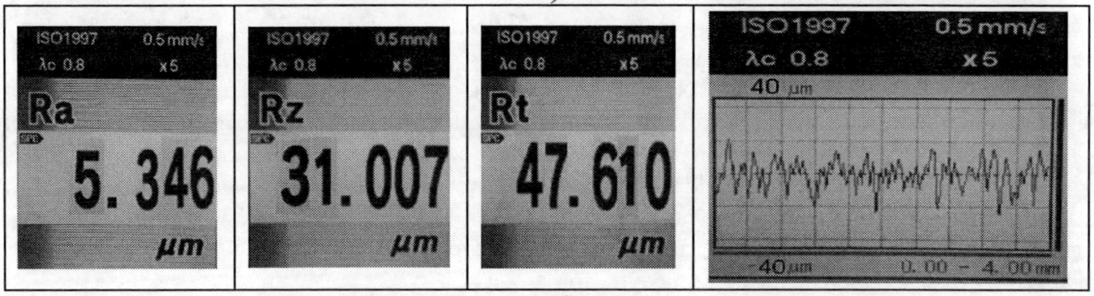

d)

Figure 3. Roughness and profile profiles specific to cavity surfaces 165 min. after volume heat treatments: a) – stress relief annealing; b) – softening annealing; c) - normalization; d) - quenching + tempering.

Table 1 compares the values of roughness parameters and MDE erosion depths obtained at the end of the cavitation test, using as reference the one obtained from stress relief annealing. The data in this table represents the arithmetic mean of roughness measured in three directions. They reflect the resistance to vibration cavity acquired by the exposed surface. It is noted that the best resistance to cavitation erosion after the MDE parameter is obtained by the heat treatment of quenching-tempering (increase of about 160%) and the worst by softening annealing (decrease is about 84%).

Since the ASTM G32-2010 standard recommends cavitation resistance assessment based on the average MDE erosion depth, and the laboratory's practice is to use this parameter, in **Table 1** are shown the MDE values for the four heat treated states. The comparison was performed with the mean values of Rz.

Table 1. Comparison of roughness parameters with MDE (165 minutes).

The treatment state of the sample	$MDE_{(165minute)}$ [µm]	Ra [µm]	Rz [µm]	Rt [µm]
Softening annealing	125.336	27.746	134.848	233.917
Stress relief	67.802	14.307	74.010	101.974
Normalizing	39.619	8.68	49.725	74.976
Queching tempering	26.054	5.976	34.092	45.411
Treatment effect (the reference state: stress relief annealing)				
Softening annealing	Increase with 84.86%	Increase with 93.93%	Increase with 82.76 %	Increase with 129.38%
Normalizing	Decrease with 71.13 %	Decrease with 64.82 %	Decrease with 48.83 %	Decrease with 36. 00%
Quenching tempering	Decrease with 160.23%	Decrease with 139.40 %	Decrease with 117.08%	Decrease with 124.55 %

Comparison between $MDE_{(165min)}$ and Rz				
The treatment state of the sample	**$MDE_{(165\ minute)}$ [µm]**	**Rz [µm]**	**Increase Rz**	**Rz difference between $Val._{min}$ and $Val._{max}$**
Softening annealing	125.336	134.848	7.58 %	28.03%
Stress relief annealing	67.802	74.010	9.15 %	40.08%
Normalizing	39.619	49.725	25.50 %	90.61%
Queching tempering	26.054	34.092	30.85 %	102.03%

It is found that the differences (from 7.58% to 30.85%) are slightly lower than the minimum and maximum values of Rz (see Figure 4) ranging from 28.03% to 102.3%.

This finding confirms that the Rz parameter is a correct indicator for assessing cavitation resistance. The histogram of (Figure 4) shows that the value of the MDE parameter at the end of the cavitation test is within the range of the Rz roughness parameter, confirming its importance as an important indicator in comparing surface resistance to cavitation erosion.

At the same time, this histogram shows that the volumetric heat treatment and queching-tempering thermal treatments can be successfully applied to the nodular graphite cast iron, which can substantially increase their resistance to cavitation erosion.

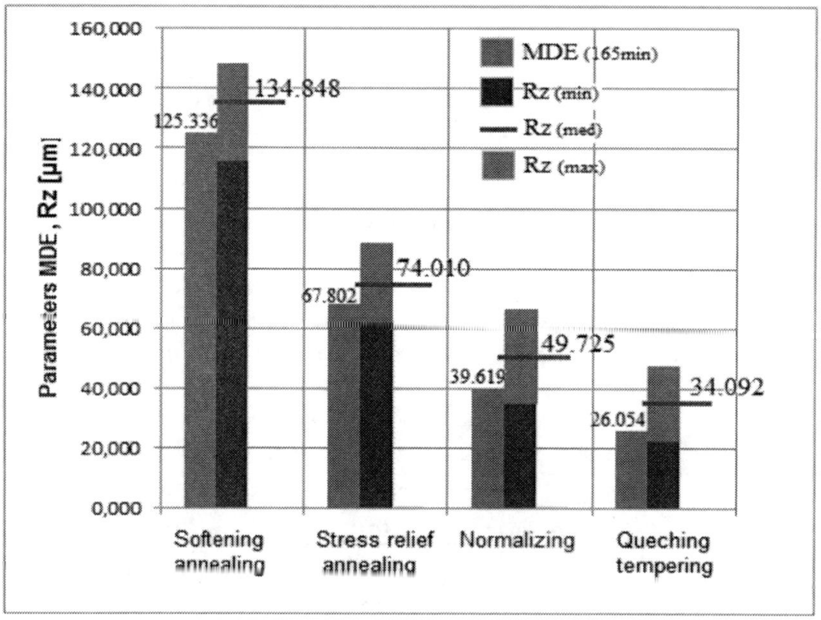

Figure 4. Comparison of parameter Rz with MDE (165 minutes).

4. Conclusions

1. The average surface roughness, Ra, of the samples tested to cavitation decreases from 14.307 µm (stres relief annealing) to 5.976 µm (quenching-tempering); Rz roughness decreases from 74.01 µm (stress relief annealing), to 34.092 µm (quenching-tempering); the Rt roughness decreases from 101.974 µm (stres relief annealing) to 45.411 µm (the quenching-tempering state);

2. The different shape of the profiles regardless of the type of thermal treatment recorded in various areas of the exposed area and the large differences between the values of the three parameters Ra, Rz and Rt express the complexity of degradation of surfaces exposed to cavitation.
3. The comparison of the MDE cavitation resistance (165 minutes) and the roughness Rz parameter shows that the last is a good indicator of cavitation resistance and which, in the future, can be considered for the evaluation of this property.

5. References

[1] Hattori S and Ishikura R 2010 *Wear* **268** 109
[2] *** Standard test method for cavitation erosion using vibratory apparatus **ASTM G32-2010**
[3] Bena T, Mitelea I, Bordeasu I and Crăciunescu C 2016 *Conference Proceeding Metal* **2016** 653-658
[4] Mitelea I, Bordeasu I, Pelle M and Craciunescu CM 2015 *Ultrasonics Sonochemistry* **23** 385
[5] Micu LM, Bordeasu I and Popoviciu MO 2017 *Revista de Chimie* **68(4)** 894
[6] Bordeasu I, Mitelea I, Salcianu L and Craciunescu CM 2016 *Journal of Tribology-Transaction of the Asme* **138(3)**
[7] Franc JP *et al* 1995 *La cavitation. Mécanismes physiques et aspects industriels* (Press Universitaires de Grenoble, Grenoble, France)
[8] Garcia R, Hammitt FG and Nystrom RE 1960 *Correlation of cavitation damage with other material and fluid properties, Erosion by Cavitation or Impingement* (ASTM, STP 408 Atlantic City, New Jersey, U.S.A.)
[9] Katona SE 2017 *Cavitational erosion of stainless steels with indirect martensitic transformation*, (PhD thesis, Timisoara)

IOP Conf. Series: Materials Science and Engineering **416** (2018) 012012 doi:10.1088/1757-899X/416/1/012012

Evaluation of Different pH Solutions on the Mechanical Behaviour of Two Hot Pressed Dental Ceramics

R D Vasiliu[1,*], I D Uţu[2] and L Porojan[1]

[1] V. Babes University of Medicine and Pharmacy Timisoara, Faculty of Dentistry, Department of Prostheses Technology, Specialization Dental Technology, 9 Revolutiei1989 Blvd, Timişoara,Romania
[2]Universitatea POLITEHNICA Timişoara Departamentul Ingineria Materialelor şi Fabricaţiei Bd. Mihai Viteazul nr.1, 300222 Timişoara, Romania

*Corresponding author: roxana.vasiliu@umft.ro

Abstract: From all the materials that are used in modern dentistry to restore the natural dentition, ceramics have the best features to mimic the tooth structure. Although they are aesthetically superior, their main disadvantages are brittleness and low fracture resistance. Some processes used in manufacturing the ceramic, as well as the environment, can affect it's structure and it's mechanical properties. The aim of this study is to illustrate how the surface roughness and hardness, of one feldspathic glass ceramic and a lithium disilicate glass ceramic, are influenced by the surface processing (glazing and polishing) and after immersion in solutions with different pH and temperatures (tea, carbonated acidic beverage and distilled water) for a given amount of time. The research goals are calculating the Ra parameter and the Vickers microhardness of the tested ceramic disc-shaped samples. This study results showed that the surface treatments (glazing and polishing) affected the Ra parameter and the Vickers microhardness. The glazed samples from both types of ceramic proved to have better hardness .From the selected immersion solutions, the tea affected the most the roughness and especially affected the polished surfaces of the tested ceramics.

1. Introduction

Dental ceramics have been used extensively as a restorative material for all kinds of dental restorations, because they provide good aesthetical and mechanical properties. In order to be successful a material has to resist against the intrinsic characteristic of the material and the environment to which it is exposed [1]. Oral cavity is a complex environment in which the ceramic is in contact with saliva first of all and with other different beverages and foods, that have different pH and different temperatures. The pH found in the oral cavity ranges from 1 to 10 [2]. The pH of the ingested beverages ranges from 3 to 8 and the ones that have low pH values are carbonated acidic beverages.

Another important aspect is the temperature found in the oral cavity that varies from 0 °C to 67 °C. Some in vitro studies showed that increased temperature is accelerating the reactions between the liquids and the ceramics surface [3,4,5]. These factors are known to influence the mechanical

IOP Conf. Series: Materials Science and Engineering **416** (2018) 012012 doi:10.1088/1757-899X/416/1/012012

properties, such as surface roughness and microhardness of the dental restorations [6]. The surface roughness has been shown to influence the long term prognosis of the restoration by decreasing it's flexural strength [7]. Another aspect that affects the surface roughness is the preparation of the surfaces. Polishing and glazing the ceramic surfaces to a high surface lustre are able to reduce surface roughness, fracture risk and staining [8,9]. Glazing of ceramic restorations is a routine laboratory procedure and it is demonstrated that it is a way of strengthening the glass [10]. Surface roughness can be calculated in terms of average roughness [Ra] and the maximum roughness [Rz], using a profilometer [11]. Vicker's microhardness is based on indentation toughness measurement and the results are obtained by measuring the cracks that emanate from the vertices of the Vickers indentation [12,13].

This study aimed to investigate the influence of the pH and temperature of different solutions on the mechanical properties of pressed dental ceramics after different surface treatments.

2. Material and methods

Two commercially available hot pressed dental ceramic materials were used in the present study: a feldspathic glass ceramic and a lithium disilicate glass ceramic (Table. 1).

24 ceramic specimens, 12 of each type of ceramic with a thickness of 1.5 mm and 1.6 mm diameter were prepared .The specimens were fabricated according to the manufacturer's guidelines. The wax-ups were invested in a phosphate-bonded investment material (Bellavest SH; BEGO GmbH & Co. KG, Bremen, Germany) (Table.2). Prefabricated ingots with medium translucency were pressed into 24 ceramic specimens using the lost-wax technique.

The pressing process was performed using a press furnace (Multimat 2 Touch+ Press Dentsply; Salzburg, Österreich) according to the manufacturer's guidelines for each material (Table. 3).

Table 1. Materials used in this study.

Ceramic system	Shade	Manufacturer	Type	Composition
Vita PM 9	A2 Medium Translucency	Vita Zahnfabrik H. Rauter GmbH&Co KG	Feldspathic glass ceramic	Feldspar with aluminium oxide
IPS e.max Press	A2 Medium Translucency	Ivoclar Vivadent Schaan, Lichtenstein.	Lithium disilicate glass ceramic	57-80% wt% SiO_2,11-19 wt% Li_2O and other oxides such as K_2O, MgO, ZnO, Al_2O_3, P_2O_3 .

Each ceramic disk has 2 surfaces, resulting 48 surfaces. 24 were glazed and 24 were polished. First of all the samples were cleaned ultrasonically. All the surfaces were prepared using abrasive paper (220, 320, 500, 800, 1000, 2000-grit silicon carbide papers) and diamond paste with 40 μm particles (Zirkopol; Feguramed, Germany) for 60 seconds. After this step half of the surfaces were glazed. The glaze was applied using manufacturer's instructions for each material (Table 3,4). Two layers of glaze were applied to simulate fabrication processes of the restoration.

The specimens of each material (n=12) were randomly divided into three groups (n=4) according to the immersion solution: Tea (Black Tea, Vallery)(pH=7), carbonated acidic beverages (Coca- Cola) (pH=2,5), Distilled Water (D)(pH=7) as the control group. The beverages were chosen according to their pH and temperature. The selected time was 24 hours that simulates one year and a half clinical use.

Table 2. Parameters used for the investment material.

Setting time after investing	20-30 minutes
Insertion temperature	850°C
Holding time	60 minutes
Final temperature	850°C

Table 3. Pressing programs according to each manufacturer.

Type of ceramic	Starting temperature °C	Vacuum level hPa	Heat rate °C/ min	Press temperature °C	Hold time minutes	Press time minutes
Vita PM 9	700	47	50	1000	20:00	10:00
IPS e.max Press	700	45	60	930	25:00	05:00

The tea was prepared using 1g of tea for 100 g of water. The specimens were immersed in a closed individual vial containing 20 ml of each solution. The vials were kept in the incubator for 24 hours at temperature of 37 °C for distilled water, 55° C for tea and at 5°C temperature for carbonated acidic beverage.

Table 4. Specific glaze for each ceramic.

Type of ceramic	Powder	Liquid
Vita PM 9	Vita Akzent plus glaze LT powder 5g	Vita Akzent Liquid
IPS e.max Press	IPS e.max Ceram Glaze Powder 5 gr Ivoclar	IPS Inline System Glaze Liquid 15ml Ivoclar

Table 5. Ceramic programs for glazing.

Type of ceramic	High temperature °C	Low temperature °C	Predry minutes	Preheat minutes	Vacuum level hPa	Heating minutes	Vacuum hold minutes	Firing time minutes
Vita PM 9	820	500	00:00	04:00	0	04:00	00:00	01:30
IPS e.max Press	820	403	00:00	06:00	50	07:00	07:00	02:00

Microhardness of the samples was measured using a DM 8 / DM 2 (Yang Yi Technology Co., Lt, Tainan City 70960, Taiwan) micro-hardness tester. A load from 1 kgf was applied for 10 seconds with a pyramidal indenter. After removing the indenter, the indentation dimensions were microscopically recorded at a 40x magnification. Five indentations were made on each specimen on the glazes and polished surfaces post immersion.

The roughness of both surfaces glazed and polished was measured using a profilometer (SJ 201, Mitotoyo, Japan, Kawaski). The values of interest were Ra and Rz, Ra meaning arithmetical mean roughness and Rz meaning the maximum roughness of the analysed surface. These roughness parameters were measured after immersing of the samples in specific solutions.

3. Results

3.1 Microhardness

For the glazed specimens that were immersed in distilled water, the Vickers microhardness of the ceramics ranged from 9.56 GPa to 11.80 GPa for lithium disilicate ceramic and for the feldspathic glass ceramic from 8.60 GPa to 9.02 GPa.

For the polished specimens immersed in distilled water, Vickers micro hardness ranged from 8.30 GPa to 9.10 GPa for the lithium disilicate ceramic and from 7.30 GPa to 8.30 GPa for the feldspathic ceramic.

After immersing the glazed samples in tea Vickers microhardness ranged from 9.45 GPa to 10.24 GPa for lithium disilicate and from 8.30 GPa to 8.90 GPa for the feldspathic ceramic (Figure 1). Immersing the samples in carbonated acidic beverages proved to change the microhardness parameter and ranged from 7.50 GPa to 9.2 GPa for the feldspathic ceramic for the glazed samples and from 6.67 GPa to 7.50 GPa for the polished samples. The samples made from lithium disilicate proved to have a microhardness from 9 GPa to 10.07 GPa the glazed samples and from 7.60 to 8.90 GPa the polished samples.

The immersion solutions and the temperature do not influence so much the Vickers microhardness. The important factor that makes a difference is the surface processing method. The glazed samples proved to have higher microhardness values. The means are the same ($p>0.05$) (ANOVA test) meaning that are not noticeable differences (Table 6).

Table 6. Microhardness (Pa) of the feldspathic and lithium disilicate ceramic.

	Feldspathic ceramic		Lithium disilicate	
	Glazed	Polished	Glazed	Polished
Distilled water	8.6—9.02GPa	7.3-8.3 GPa	9.56-11.80GPa	8.3-9.1GPa
Tea	8.3-8.9GPa	7.1-8.1GPa	9.45-10.24GPa	7.3-8.4GPa
Carbonated acidic beverage	7.5-9.2GPa	6.67-7.50GPa	9-10.3	7.6-8.9GPa

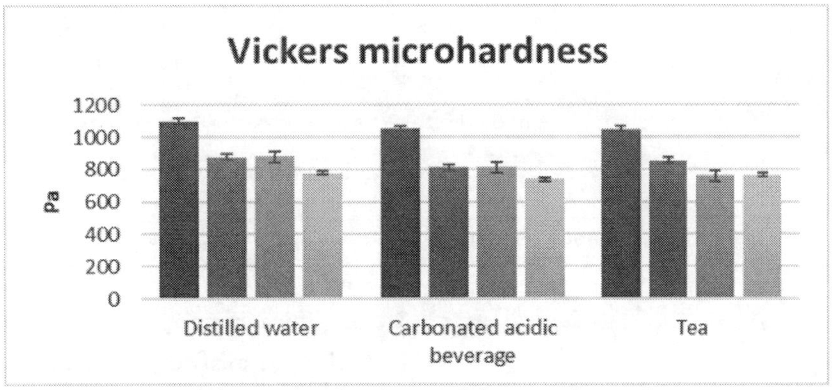

Figure 1. Vickers microhardness of the tested samples.

3.2 Roughness

Ra values of the glazed specimens immersed in **distilled water** were 0.14 μm for the glazed feldspathic glass ceramic and 0.20 μm for the polished samples. For the lithium disilicate glazed samples the values were 0.1 μm and for the polished samples 0.09 μm.

After immersion in **tea, Ra** values of the glazed samples from feldspathic glass ceramic were higher (0.19) than for the glazed samples from lithium disilicate ceramic (0.10 µm). For the polished feldspathic samples the Ra values 0.30 µm and for the polished lithium disilicate 0.08 µm.
After immersion in **carbonated acidic beverage** the **Ra** values were for the glazed samples 0.10 µm for feldspathic and for lithium disilicate 0.18 µm. For the polished samples 0.08 µm for feldspathic and for the lithium disilicate 0.111 µm. The differences between the samples are not significant.
The smoothest obtained surface was the glazed surface from the lithium disilicate glass ceramic **after distilled water immersion** (Ra=0.02) and roughest obtained surface was the polished feldspathic surface after **tea** immersion (Ra=0.38).
The maximum obtained roughness (Rz) was 2.31 µm for the polished feldspathic samples that were immersed in **tea** (Figure 2, 3)(Table7, 8).
The ANOVA test showed that there were no noticeable differences between the values($p>0.05$), means were the same.

Table 7. Ra parameter for the tested samples.

	Feldspathic ceramic		Lithium disilicate ceramic	
	Glazed	Polished	Glazed	Polished
Distilled water	0.14	0.2	0.1	0.09
Tea	0.19	0.30	0.10	0.08
Carbonated acidic beverages	0.10	0.08	0.18	0.111

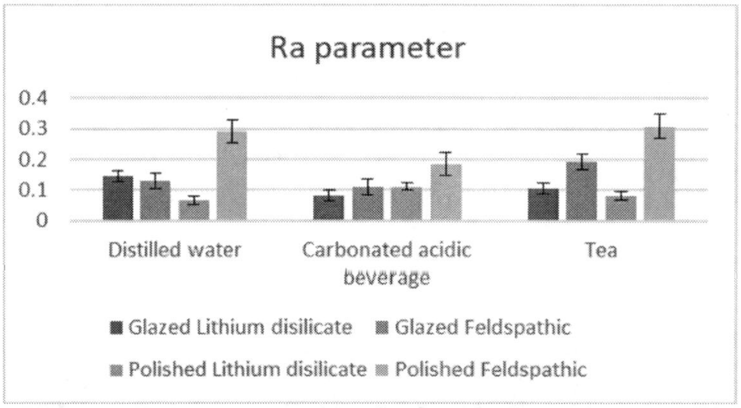

Figure 2. Ra parameter values for the tested samples.

Table 8. Rz parameter values after imersing the samples in distilled water.

	Feldspathic ceramic		Lithium disilicate ceramic	
	Glazed	Polished	Glazed	Polished
Distilled water	1.09	2.053	1.375	0.513
Tea	1.72	2.31	0.715	0.585
Carbonated acidic beverages	0.847	1.55	0.578	0.936

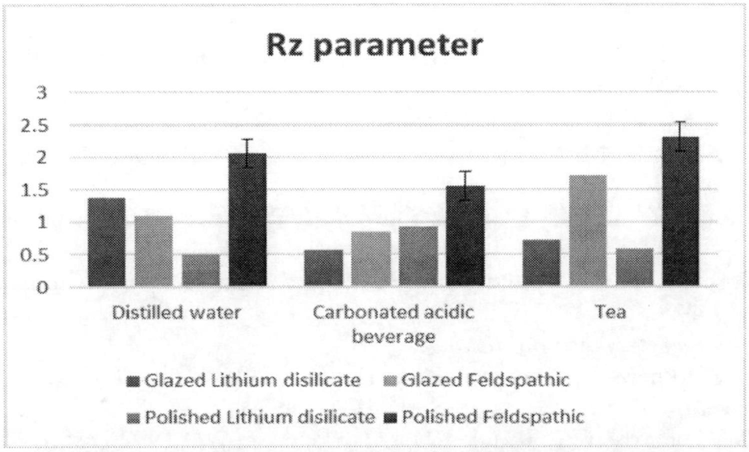

Figure 3. Rz parameter values for the tested samples.

4. Discussion

The results of this test revealed that the surface roughness of the samples immersed in distilled water are the lowest compared to the samples immersed in other liquids. However, the Vickers microhardness proved to be slightly influenced by the immersion solutions, both by the temperature and the pH. Several in vitro studies [14,15] show that the ceramics are susceptible to corrosion in range of oral fluid pH. The dissolution potentials of ceramics in acidic and alkaline liquids such as carbonated acidic beverages [15] and fruit juices differ considerably based on the patients saliva and diet [16]. Moreover, ceramic corrosion can affect the fracture strength of the materials and can increase their roughness [16]. It is known that patients experience discomfort, when the value of Ra is more than 0.50 μ [17]. All the registered roughness values after immersion in distilled water, tea and carbonated acidic beverages where lower than this value. Previous in vivo studies proved that clinical degradation and roughening of ceramic surfaces appear after 1-2 years after cementing them [18, 19, 20], this is why this study period was 1 year and a half. In this study the Ra parameter slightly increase after the period, but the Rz parameter doubled in tea for feldspathic ceramics. Indentation hardness is a predictor of the wear resistance of an material . The thickness of the samples used in this study was 1.5 mm and simulated the clinical practice. An increase in Vickers microhardness represents an increase in resistance to abrasion for the ceramic surfaces. On the basis of the results found here, it can be concluded that the high temperature of the tea produced a small decrease of the microhardness experienced better on the polished samples of both types of ceramics. The glaze layer protected both of the ceramics and the values of Vickers microhardness were higher than the polished samples. The glaze protected the ceramic samples against low and high temperatures.

5. Conclusions

Within the limitation of this study some conclusions can be drawn:

1. The pH and the temperature have no influence on the mechanical behaviour on the two types of ceramics at the chosen time.
2. The mechanical behaviour and the roughness is influenced by the surface protocols (glazing and polishing) and by the type of ceramic (feldspathic and lithium disilicate).
3. The values of both roughness and microhardness proved that the feldspathic ceramic should be glazed rather than polished.

IOP Conf. Series: Materials Science and Engineering **416** (2018) 012012 doi:10.1088/1757-899X/416/1/012012

6. References

[1] Lsquivel-Upshaq J, Dieng F, Clark AE, Neal D and Anusavice KJ 2013 *Journal of Dental Restorations* **92** 467..

[2] Angker L. Swain MV, Wong L and Sissons C 2011 *New Zeeland Dental Journal* **107** 12.

[3] Ccahuana VzS, Ozcan M, Mesquita AMM, Nishioka RS, Kimpara ET and Bottino MA 2010 *Journal of Applied Oral Science* **18** 155.

[4] Palmer DS. Barco T and Billy EJ 1992 *Journal of Prosthetic Dentistry* **67** 325.

[5] Al-Wahab ZN 2009 *Mustansiria Dentistry Journal* **6** 14.

[6] Albero A, Pascual A. Camps I and Grau-Benitez M 2015 *Journal of Clinical and Experimental Dentistry* **7** 49.

[7] Binns D 1983 *Quintessence Publishing*.

[8] Yilmaz C, Korkmaz T, Demirkoprulu H, Ergun G and Ozkan Y 2008 *Journal of Prosthodontics* 17 20.

[9] Fairhurst CW, Lockwood PE, Ringle RD and Thompson WO 1992 *Dental Materials* **8** 203.

[10] Alao AR, Stoll R, Song XF, Abbott JR, Zhang Y, Abduo J and Yin *L* 2017 *Journal of the Mechanical Behavior of Biomedical Materials* **74** 251.

[11] A. Sakar- Deliormanli and M. Guden 2006 *Journal of Biomedical Materials Research B* **76** 257.

[12] J.J. Kruzic, D.K. Kim, K.J. Koester and R.O. 2009 *Journal of the Mechanical Behavior of Biomedical Materials* **2** 384.

[13] Jain P, Nihil P, Sobkowski J and Augustin MZ 2007 *General Dentistry* **55** 150.

[14] Bartlett DW, Fares J, Shirodaria S, Chiu K, Ahmad N and Sheriff M 2011 *Journal of Dentistry* **39** 811.

[15] Milledin P, Gerdes S, Holmberg K and Karlsson S 1999 *European Journal of Oral Science* **107** 384.

[16] Jones CS, Billington RW and Pearson GJ 2004 *British Journal of Dentistry* **196** 42.

[17] Esquivel- Upshaw JF, Young H, JonesJ, Yang M and Anusavice KJ 2006 *International Journal of Prosthodontics* **19** 391.

[18] Esquivrl- Upshaw JF, Rose W.F., Barett AA, Oliveira ER, Yang MC and Clark AE 2009 *Dental Materials* **28** 615.

[19] Esquivel-Upshaw J1, Rose W, Oliveira E, Yang M, Clark AE, Anusavice K. 2013 *Journal of Prosthodontics* **22** 166.

Long-term Stable Electrodes Based on Platinum Electrocatalysts Supported on Titanium Sintered Felt for the Use in PEM Fuel Cells

U Rost*, P Podleschny, M Schumacher, R Muntean, D T Pascal, C Mutascu, J Koziolek, G Marginean and M Brodmann

Westfälische Hochschule Gelsenkirchen Bocholt Recklinghausen, Neidenburger Straße 43, D-45877 Gelsenkirchen, Germany

*Corresponding author: ulrich.rost@w-hs.de

Abstract. In this work a long-term stable electrode structure, suitable for polymer electrolyte membrane fuel cell (PEMFC) applications is presented. Investigated electrodes are prepared by using sintered titanium felt as porous and current conducting gas diffusion layer, due to its corrosion resistance in acidic environment. Platinum is deposited onto the surface of the titanium fibres by an electrochemical method. This carbon-free electrode structure has been studied regarding stability and performance ex-situ and in-situ by cyclic voltammetry, corrosion tests, i-U polarisation measurements, and accelerated stress tests. Furthermore, the electrode morphology of the catalytic layer on titanium sintered felt has been investigated by means of scanning electron microscopy. Results of this study claim high corrosion stability for the use of these electrodes in PEMFCs. A small-scale laboratory prototype has been realised, which is also presented in this work.

1. Introduction

In autonomously operating sensor systems, batteries are typically used as energy source. Niche applications such as for the determination of environmental conditions in underground deposits (e.g. uranium wastes), often demand low energy and power but require an autonomous operation for decades. Batteries are able to provide sufficient electrical energy at very high efficiency. However, due to degradation mechanisms, these suffer from self-discharge, which consequently inhibits the use of batteries for extended operation periods far beyond ten years without being recharged. As an alternative energy source for a long-term autonomous operation, polymer electrolyte membrane fuel cell (PEMFC) systems with integrated hydrogen and oxygen tanks are proposed as continuous electrical energy generator.

The main principle of a fuel cell is based on the anodic hydrogen oxidation reaction (HOR) and the cathodic oxygen reduction reaction (ORR) resulting in an overall electrochemical reaction expressed by the equation $H_2 + ½ O_2 \rightarrow H_2O$ + Energy [1].

Typically, relevant components in PEMFCs are based on carbon material or stainless steel with high electrical conductivity. Pole plates made from graphite compound or stainless steel provide necessary gas channels within the cell for the process gases hydrogen and oxygen. These gases are finely distributed to the anodic and cathodic catalysts, respectively, by porous gas diffusion layers

(GDL), which are typically graphitic. As electro catalysts in PEMFCs usually chemically or electrochemically reduced nano-scaled platinum particles supported on graphitic nano-structures are employed. In order to manufacture gas diffusion electrodes (GDE), the catalyst layer is applied onto the GDL. [1]

Although PEMFCs provide high power densities at reasonable efficiencies, carbon corrosion occurs at cell potentials above 0.65 V. This leads to continuous cell degradation mainly due to a loss of catalyst support material which, again, leads to an electrochemical surface area (ECSA) loss and a reduced cell performance.

As electrolyte in PEMFCs functionalized polymer membranes with high proton conductivity such as perfluorinated sulfonic acid (PFSA) based membranes are used. Further relevant degradation in fuel cells arises from poisoning of these membranes due to surface corrosion of metal bipolar plates exposed to the acidic PEMFC environment. Metal cations can enter the polymer by water flow and block the negatively charged sulfuric acid groups within the membrane, which leads to a continuous loss in proton conductivity and, therefore, in decreased cell performance over the operation time. Furthermore, chemical degradation of the membrane provoked by metal cations (radical attack) occurs. For a corrosion resistant protection layer, metal bipolar plates are often coated with noble metals such as gold. Accounting for the costs of noble metal coatings, a deposited gold layer for PEMFC pole plates typically has a thickness of far below 1 micron resulting in a corrosion stability for a few thousand operation hours. [2]

Titanium on the contrary is well known to be nearly completely corrosion stable under acidic environment. E.g. for PEM electrolyzers porous current distributors made from sintered titanium particles or sintered titanium fibers are used as anodic electrode material operating long-term stable at an electrode potential above 1.5 V [3]. In this work we describe an approach for the manufacturing of titanium-based long-term stable GDE for the use in PEMFCs. Two electrodes are hot pressed with proton conductive polymer membrane forming a titanium-based membrane electrode assembly (MEA).

Technical fuel cells have been developed for a subsequent use of prepared long-term stable electrodes. The cell design was engineered according to a sensor application that needs to fit into a bore hole with a diameter of 100 mm. Pole plates made from stainless steel have been plated with an above-average thick anti-corrosive gold coating of about 2 microns. Two pole plates are clamped together having the titanium-based MEA in between to form a single cell. A PEMFC stack has been assembled with 20 single cells that are electrically connected in series being operated with hydrogen and oxygen.

2. Methodology and Experimental Equipment

In order to overcome degradation mechanisms in PEMFCs that result from carbon or stainless-steel corrosion, titanium-based electrodes and pole plates with an anti-corrosive coating have been prepared for a long-term stable PEMFC application. Titanium is well known to be corrosion resistant in acidic environments even at high potentials as e.g. in polymer electrolyte membrane electrolyzers. In this study, titanium substrate made from sintered titanium fibers in micro scale was coated with platinum as electrocatalyst for the hydrogen oxidation reaction as well as for the oxygen reduction reaction.

For this work, prepared electrodes have been investigated regarding their morphology using scanning electron microscope (SEM) as well as their electrochemical activity in sulfuric acid by determining the ECSA. Furthermore, an accelerated stress test has been performed regarding the analysis of the prepared electrodes' corrosion stability.

The realized technical cells and the manufactured PEMFC stack have been operated in a laboratory test bench under realistic environmental conditions referred to their designated underground application. In-situ tests of the laboratory device are also reported in this work.

2.1. *Electrode preparation*

Regarding the preparation of long-term stable electrodes, sintered titanium felts have been used as substrate for platinum catalyst deposition. The utilized specimens are circular shaped with a diameter of 30 mm having a thickness of 1 mm and a high porosity above 70%.

In order to remove oxides from the surface for a subsequent platinum deposition, the utilized sintered titanium felts have been pre-treated in two steps. Initially, titanium felts have been cleaned in an electrolytic pickling solution from Wieland Edelmetalle, Germany, based on nitric acid and hydrofluoric acid with a pH-value of 0.5 at room temperature. Voltage has been applied to the felts to provoke anodic dissolution of the titanium/titaniumdioxide surface. A counter electrode based on titanium expanded mesh plated with platinum from Wieland Edelmetalle has been used for this preparation step. The anodic voltage versus the counter electrode was set to 2.5 V.

After rinsing with deionized water, the surface of the titanium work pieces has been plasma treated in argon, in order to remove still remaining contaminations on the titanium surface. A plasma reactor type Pink V 15-G from PINK Thermosysteme GmbH, Germany, was used and the parameters were set to an argon flow rate of 100 ml min^{-1}, at a gas pressure of 60 Pa and a treatment time of 30 min at a microwave power of 400 W.

Consecutively to the plasma physical cleaning process and further rinsing with deionized water, the titanium fibers have been immediately plated with platinum by electrochemical method under argon atmosphere to inhibit the formation of oxides on the titanium surface after the electrode preparation process. A commercially available plating bath type Galvatron Platinbad from Wieland Edelmetalle based on hexachloroplatinic acid has been used. The bath parameters were set to a pH-value of 8 and a temperature of 50 °C. The plating process was performed for 10 min at a constant cathodic voltage of -3.2 V versus the counter electrode (titanium/platinum from Wieland Edelmetalle). During the plating process, the titanium electrode was placed between two counter electrodes that are connected electrically. Subsequently, for an increase in the catalysts electrochemical active surface area by the deposition of micro and nano-scaled platinum particles, the plating mode has been switched to a pulsed plating at a cathodic voltage of -3.0 V versus the counter electrode without interrupting the process. The on-time was set to 10 ms and the off-time was set to 56.7 ms, resulting in a 15% duty cycle. The second part of the plating process was performed for further 10 min.

Regarding the later assembly of polymer membranes with the prepared specimens to form MEAs, electrodes have been impregnated with 0.5 ml proton conductive ionomer solution type Nafion® (5 wt.% in ethanol) subsequently to the platinum plating. The solution has been applied by an air brush casting, while the titanium-based electrode was attached to a heated sample holder. The temperature was set to 60 °C to speed up the evaporation of the ethanol from the electrodes surface.

It was already mentioned that the investigated electrodes consist of two parts, a titanium-based felt with catalyst coating and ionomer impregnation as well as a stainless-steel (type 1.4404/SS 316) pole plate having a labyrinth flow field formed by channels with 1 mm depth and 1 mm width. Regarding the corrosion protection, an ultra-thick gold coating has been applied to these workpieces after preliminary removal of oxides by a plasma physical pre-treatment in argon. The deposition has been performed by an electrochemical method. The commercially available plating solution type Haftgoldbad AC 3 SSF from Wieland Edelmetalle based on potassium dicyanoaurate was utilized at room temperature. A constant cathodic voltage of -3.0 V versus the platinum counter electrode was applied for 10 min. Deposition of the anti-corrosive coating has been performed according to the specification sheet of the electrolyte manufacturer.

2.2. *Ex-situ analyses*

Surface morphology and cross sections of selected specimens have been investigated using a scanning electron microscope (SEM) type Phillips ESEM XL 30 with an integrated detector for energy dispersive X-ray (EDX) analysis. The focus of this microscopy work has been on the observation of the corrosion protection layers of the titanium-based electrodes (platinum) as well as of the pole plates

IOP Conf. Series: Materials Science and Engineering **416** (2018) 012013 doi:10.1088/1757-899X/416/1/012013

(gold) and the nano/micro structure of the prepared catalyst layer. Furthermore, punctual and area material analyses have been performed by EDX measurements. For understanding the crystalline structure of the deposited platinum catalyst, further material characterization has been performed by means of X-ray diffraction measurements (XRD type Philips X´Pert).

For the electrochemical determination of the ECSA a galvanic cell has been prepared in order to entirely fit prepared electrodes into a sample holder. A platinum foil has been used as a counter electrode, having a circular shape with a diameter of 33 mm. The distance between sample holder and counter electrode was designed to be 30 mm. A saturated calomel electrode has been utilized as a reference. Cyclic voltammetry (CV) has been performed in 0.5 M H_2SO_4 using a potentiostat/galvanostat type Ivium Technologies Vertex (in three electrodes mode). CV starting potential and vertex potential were set to -0.2 V versus SCE and of 1.2 V versus SCE, respectively, and scan rate was set to 100 mV sec^{-1}. ECSA has been calculated after 200 cycles between starting and vertex potential. After determination of the initial ECSA of an investigated sample, an accelerated stress test (AST) that is known for testing automotive PEMFC electrodes has been performed. In analogy to an earlier work [4], samples have been cycled 1,000 times between 0.4 V versus SCE and 1.4 V versus SCE at a scan rate of 1,000 mV sec^{-1}. Subsequently to the AST, ECSA has been determined again following the above-mentioned parameters to analyse the change in catalytic activity due to high electrode potentials occurring especially at start-stop operation.

2.3. Cell and stack manufacture

Technical PEMFCs have been assembled using the prepared titanium-based felts with catalytic layer and stainless-steel pole plates with anti-corrosive coating. As a polymer membrane Nafion® type 115 has been used. This rather thick membrane (125 µm) inhibited relevant gas cross-over through the membrane, which was necessary for a long-term PEMFC operation. Regarding the prevention of gas leakage, two gaskets made of polytetrafluoroethylene (PTFE) were located between pole plate and membrane outside the electrochemical active electrode area, which is a typical design for PEMFC applications. Two pole plates, two long-term stable electrodes and a membrane with PTFE gaskets formed a single fuel cell.

Figure 1. Macrograph of the manufactured lab-scale PEMFC stack operated in its test bench environment.

A fuel cell stack has been manufactured using 20 single cells. To save material and therefore costs, pole plates were designed bipolarly having a flow field on each side of the plate, as it is common for PEMFCs. Each single cell was gas technically connected in parallel and electrically connected in series. The stack was designed for dead-end operation, meaning gas was fed without having a relevant gas flow through the cells. In order to compress the inner components, especially to decrease contact resistance between the different MEA layers, tie bolts were utilized to clamp the first with the last cell of the PEMFC stack. Furthermore, the stack provided connectors to apply hydrogen and oxygen feed gases as well as a waste water container.

2.4. Fuel cell testing

In order to provide the necessary infrastructure for the simulation of the underground environment, a test bench has been constructed, in which the manufactured PEMFC test stack has been operated under relevant testing conditions. The test bench was operated with pure hydrogen and oxygen (quality 5.0) from Linde, Germany. It was equipped with the necessary fuel processing components (pressure regulator, safety valves, mass flow meter etc.) as well as with an oven type Memmert UF55 to simulate the underground temperature. Regarding the determination of electrical specifications of the manufactured PEMFC stack, polarization measurements have been performed by the use of an electronic load type Höcherl & Hackl: DS400 SV20. The fuel cell stack was tested at 80 °C. Gas pressure was adjusted to 1.1 bar absolute. Polarization plots have been determined in constant current mode from open circuit voltage (OCV) to the maximum power point in forward direction.

3. Results and Discussions

Presented results indicate high corrosion stability of the prepared titanium-based GDEs as well as of the technical fuel cell that has been manufactured for this study. Microscope images show well distributed platinum nano-particles on the titanium fibers that have been used as a substrate. In cross section analyses dense corrosion protection layers of platinum (on titanium fibers) and gold (on stainless-steel pole plates) have been observed. Furthermore, electrochemical characterization result in high ECSA due to the roughness of the platinum surface of the prepared titanium-based electrodes. From in-situ testing performed by the aid of the assembled fuel cell under simulated underground conditions it can be assumed that the developed components exhibit the required electrochemical performance for the proposed application.

3.1. Microscopic investigations

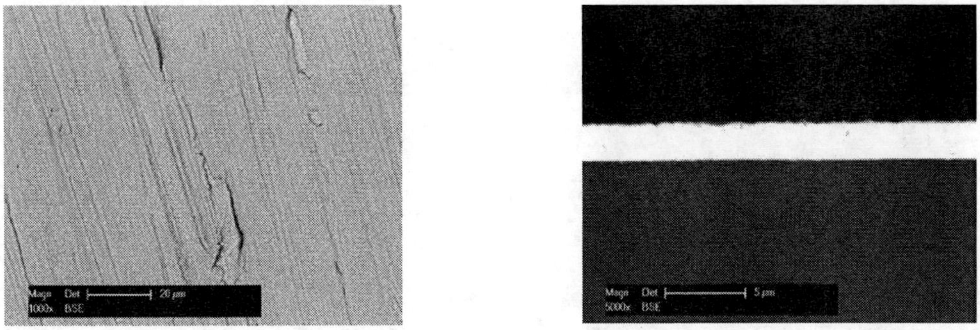

Figure 2. SEM micrographs of: left) top-view of the gold coating surface, and right) cross-section of the coating/substrate system.

IOP Conf. Series: Materials Science and Engineering **416** (2018) 012013 doi:10.1088/1757-899X/416/1/012013

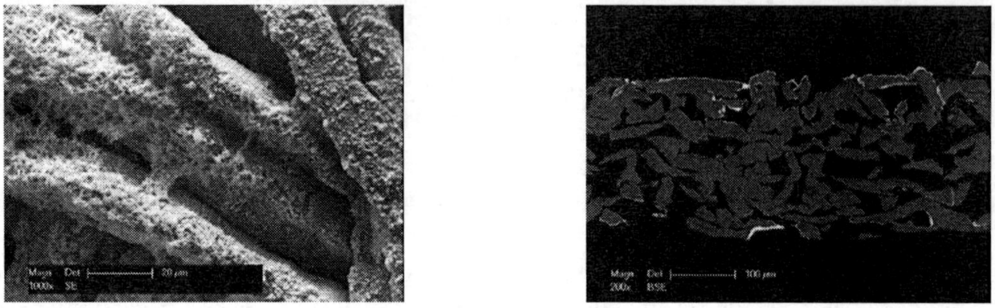

Figure 3. SEM micrographs of: left) top-view, and right) cross-section of the platinum coated titanium fibres.

Cross-sections of prepared titanium-based electrodes as well as on stainless-steel based pole plates have been investigated by SEM/EDX. Observing the SEM micrographs of the top view and cross-sections of the coated pole plates illustrated in Figure 2 it can be agreed that dense anti-corrosive coatings have been obtained with a thickness of about 2 μm. No oxide layer has been found between coating and stainless-steel substrate resulting in increased adhesion.

The microscopic images of Figure 3 show sintered titanium fibers with platinum coating and particles, respectively. It becomes obvious that only for these fibers situated at the outside of a utilized felt, an entire coating can be claimed. This effect is explained by the electrical field distribution within the galvanic cell (working electrode in between and parallel to two counter electrodes). However, the achieved coating is sufficient for the long-term stable adhesion of the platinum nano particles, which were decorated subsequently. Furthermore, micrographs have been analyzed of platinum particles that have been deposited on the plated titanium fibers by pulsed electro plating process. Nano and micro scaled platinum particles are visible, resulting in a high specific surface area due to a high roughness factor. In addition to that, EDX analysis (see Figure 4) showed only a small amount of oxides having been present on the titanium fibers decorated with platinum.

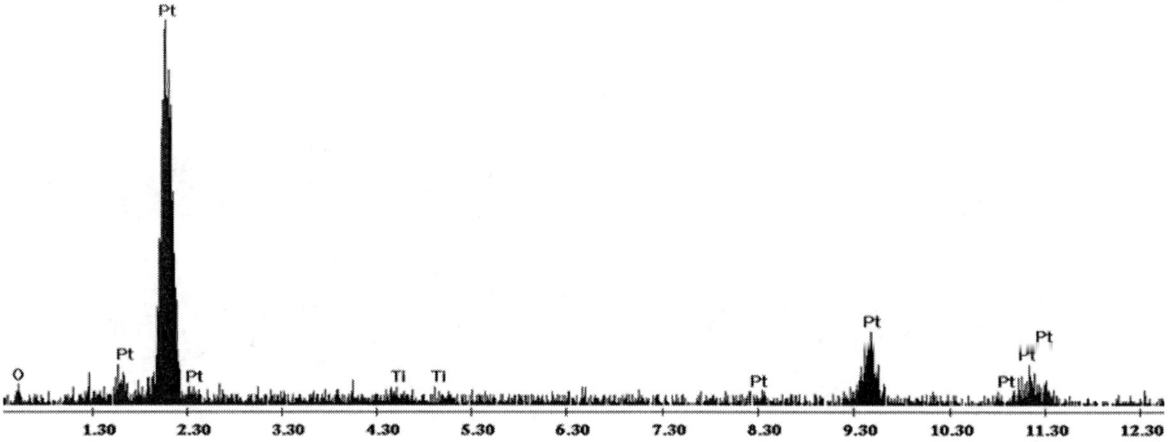

Figure 4. EDX spectrum of the platinum decorated titanium fibres.

3.2. X-ray analysis

The performed X-ray analysis confirmed the presence of Pt particles with a face-centered cubic crystal structure. Moreover, physical properties such as the average crystallite size were obtained from the X-ray diffraction results. The experiments were conducted in the 2θ range from 20° up to 100° using CuKα radiation. The average crystallite size was determined, employing the Debye-Scherrer equation, using the full width at half maximum of the Pt (111), Pt (200) and Pt (220) reflection peaks [5]. The X-ray diffractogram presented in Figure 5 shows a typical XRD pattern of the titanium GDL decorated with platinum catalyst particles. The resulting average size of the platinum crystallites was 10 nm underlining the high specific surface area of the prepared catalyst layer.

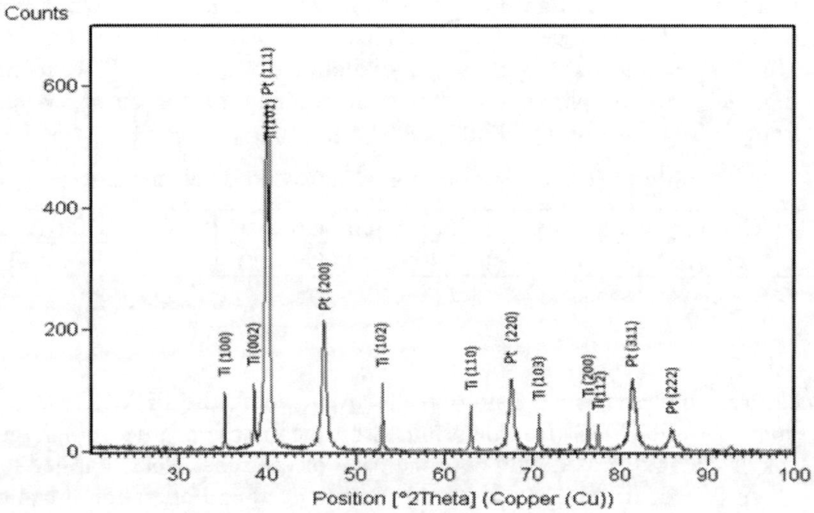

Figure 5. XRD pattern of the Ti fibres (GDL) coated and decorated with Pt nanoparticles.

3.3. CV and ECSA measurements

For a better understanding of the electrochemical processes that take place on the Pt/Ti catalyst, cyclic voltammetry measurements have been performed on selected samples in a manner similar to that of other studies [1, 4]. Figure 6 presents a typical cyclic voltammetry curve obtained from a Pt/Ti electrode with a geometrical surface area of 8.55 cm^2 immersed in a 0.5 M H_2SO_4 solution in the potential range of -0.2 V to +1.2 V vs. SCE with a constant scan rate of 100 mV s^{-1} after 200 cycles. Region A characterizes the hydrogen adsorption ($i < 0$) and desorption ($i > 0$) area, region B corresponds to the double layer capacity, and in region C, the platinum oxidation ($i > 0$) and reduction ($i < 0$) processes occur [6] [4].

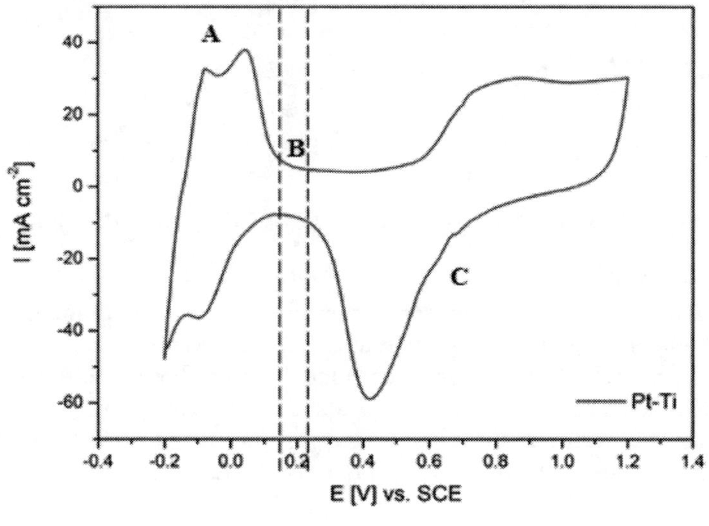

Figure 6. Cyclic voltammogram of the Pt decorated Ti GDL.

ECSA has been determined by calculation of charge related to the hydrogen desorption peaks applying a numerical integration algorithm. Taking into account the charge for a polycrystalline platinum surface (210 μC cm^{-2}) [6], ECSA was obtained by the division of the calculated charge with this value (results shown in Table 1). After the above described AST procedure ECSA has been determined again. Tests revealed a slight increase of the active surface area, which may result from a further cleaning of the catalyst surface from contaminations such as chlorides (arising from the used electrolyte) due to the potentiostatic cycling (Pt oxidation followed by Pt reduction). This result underlines the high stability of prepared samples in contrast to carbon-based electrodes, which typically suffer from ECSA loss after this kind of AST [4].

Table 1. Surface area of representative Pt-Ti electrodes

	Geometrical SA [cm^2]	**ECSA [cm^2 cm$_{geo}^{-2}$]**	**ECSA after AST [cm^2 cm$_{geo}^{-2}$]**
Pt-Ti electrode	8.55	≈ 38.60	≈ 42.22

3.4. In-situ tests

From in-situ measurements performed using an electronic load, the PEMFC stack performance at elevated temperatures of 80 °C has been determined by means of recorded polarization curves. The manufactured stack has been operated with hydrogen and oxygen that were supplied to the cells with a slight overpressure of 0.1 bar (dead-end operation). Due to dead-end operation it can be assumed, that the process gases are nearly 100% humidified. From i-U polarization an OCV of 19.5 V and a mean OCV of 0.975 V per cell have been measured (Figure 7). By increasing the stack current, voltage drops rapidly, which may be an effect of high ohmic resistance due to the poor electrical conductivity of titanium components in comparison to typically used graphitic material in combination with the poor ionic conductivity of the rather thick membrane. However, at a stack voltage of about 8.5 V the maximum electrical power has been determined to be in the range of 4 W (0.025 W cm^{-2}), which is sufficient for the aspired sensor application.

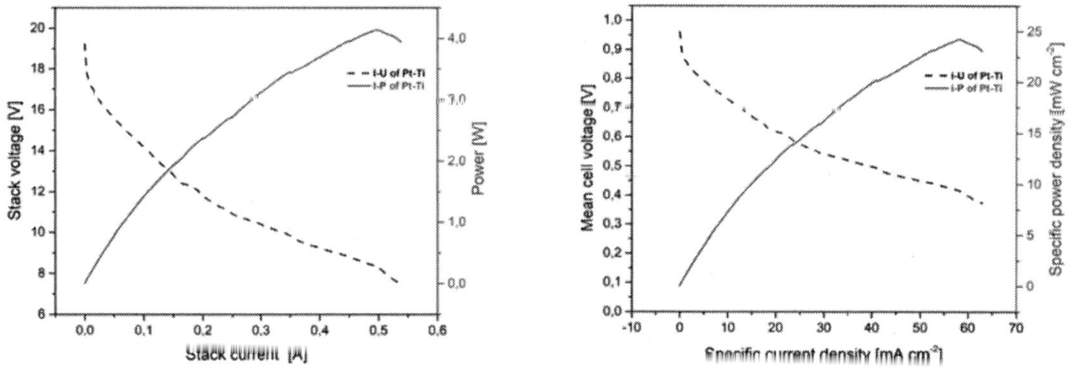

Figure 7. Polarization curves of the manufactured lab-scale PEMFC left) stack and right) single cell.

4. Conclusions and outlook

In this work a preparation route for long-term stable electrodes based on titanium sintered felt has been presented. Material characterization of prepared electrodes revealed high corrosion stability and sufficient electrochemical activity in acidic environment. From these electrodes, a lab-scale PEMFC stack has been manufactured that offered sufficient electrical power for underground sensor applications. However, further in-situ testing is necessary in order to proof long-term stability of the realized technical PEMFC stack as well as to integrate the device into the aspired sensor system.

5. References

[1] Rost U, Muntean R, Marginean G, Merino C, Diez R, Vaszilcsin N, Brodmann M, 2016, Int. J. Electrochem. Sci., **11** pp. 9110 – 9122

[2] Koça M, Mahabunphachai S, Dundar F, 2009, ECS Transactions, **25** (1) pp. 1773-1782

[3] Wirkert F J, Roth J, Rost U W, Brodmann M, 2017, Int. J. Smart Grid Clean Energy, **6** (3) pp. 171-176

[4] Podleschny P, Rost U, Muntean R, Marginean G, Heinzel A, Peinecke V, Radev I, Muhler, M, Brodmann M, 2018, Investigation of Carbon Nanofiber-Supported Electrocatalysts with Ultra-Low Platinum Loading for the Use in PEM Fuel Cells, Fuel Cells – *In press*.

[5] Zhong C-J, Luo J, Fang B, Wanjala B N, Njoki P N, Loukrakpam R, Yin J, 2010, Nanotechnology **21** 062001

[6] Trasatti S, Petrii O A, 1991, Pure Appl. Chem. **63** (5) pp. 711-734

Acknowledgments

Presented results have been obtained in the frame of a research and development project that was funded by the German Federal Ministry for Economic Affairs and Energy (funding number KF2697004ZG4). Moreover, we gratefully acknowledge the work of our industrial project partner IBeWa-Ingenieurpartnerschaft für Bergbau, Wasser und Deponietechnik, Wilsnack & Partner, Germany.

Surface Characterization of DLC Layers PVD Coated on AISI 52100 Steel Substrate

D I Feldiorean[1], D Cristea[2], L Jakab-Farkas[3] and M H Tierean[1,*]

[1] Materials Engineering and Welding Department, Transilvania University of Brasov, 29 Eroilor Blvd., 500036 Brasov, Romania
[2] Materials Science Department, Transilvania University of Brasov, 29 Eroilor Blvd., 500036 Brasov, Romania
[3] Faculty of Engineering, Sapientia University, 540485 Tg. Mures, Romania

*Corresponding author: mtierean@unitbv.ro

Abstract. The purpose of this study is to evaluate the influence of substrate and layer deposition parameters on the morphology, topography and adhesion of the DLC layer. The carbonic layer was deposited on the rotary hook body made of AISI 52100 steel. Using the PVD magnetron sputtering technique, the deposition was performed at three different temperatures. Information about the coating was collected using optical, SEM and AFM microscopy and XRD. The DLC hardness and adhesion to substrate was determined. Comparing substrate hardness values, DLC adhesion and topography, it turned out that the best wear resistance was exhibited by the layers deposited at 180 °C.

1. Introduction

The hook body, the bobbin case, the cover ring, the needle guard and the leaf spring are the main components of the rotary hooks. The hook's role is to form and to realize the stitch (Figure 1).

The large number of rotations that rotary hooks runs during the sewing process, bigger than 2500 rpm, leads to a strong wear stress between the parts being in movement and also between the thread and the hook. During the sewing, the hook comes in constant contact with the thread, thus forming a friction coupler (Figure 2). Some of the special sewing machines, for example those used for automotive airbags are designed to run without lubrication, or a minimal lubrication in the rotary hook area.

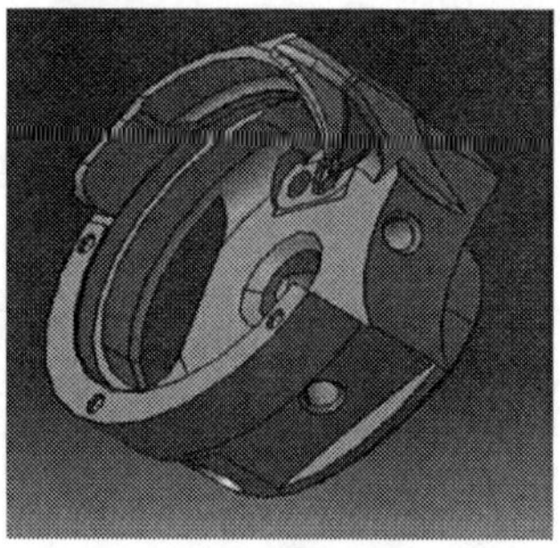

Figure 1. 3D model of the rotary hook.

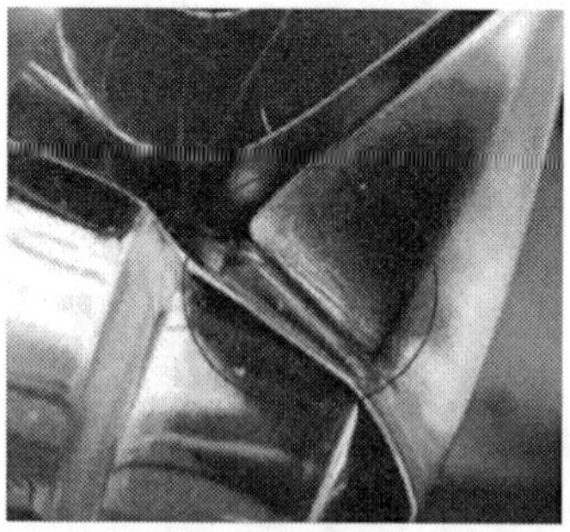

Figure 2. Wear traces left by thread on the hook.

IOP Conf. Series: Materials Science and Engineering **416** (2018) 012014 doi:10.1088/1757-899X/416/1/012014

Currently, one of the common steels used to made the hook body is AISI 52100, due to its high hardenability properties and wear resistance [1]. For this application AISI 52100 it is used in hardened condition, 61-63 HRC (C-Rockwell units).

This hardness provides a satisfactory wear resistance. To increase the wear resistance, a a-c:H DLC layer thin film was deposited using the magnetron sputtering PVD method on the surface of the hook. DLC coatings are well known for their tribological properties (high hardness low friction coefficient) and are used in a wide range of industrial applications [2, 3]. The amorphous carbon layer consists of a mixture of sp^2 structures with sp^3 structures, the proportion in which they are found determines the physical and chemical characteristics of the material [2]. The presence of hydrogen in the DLC layer gives a better behaviour reducing the friction coefficient [4], and increases the adhesion to the substrate [5]. Hardness characteristics of the substrate are influenced by the DLC coating temperature. A high deposition temperature leads to lower substrate hardness, phenomenon that can reverberate on the adhesion of the coating to the substrate [6, 7].

It has been analysed how the DLC deposition temperature influences the substrate properties, with the purpose of identifying the combination between de substrate hardness and DLC layer properties, which can ensure the optimal tribological behaviour of the hook body.

2. Materials and method

For the heat treatment of AISI 52100 steel, a batch of 30 samples (Ø 40 x 15 mm) divided into 3 series of 10 pieces was prepared. The chemical composition of the samples was verified using an OBLF QS 750 spectrometer (Table 1).

Table 1. Chemical composition of AISI 52100 samples.

Standard	Symbol	C	Si	Mn	P	S	Cr	Mo	Ni
		Average values %							
ASTM A295-98	AISI 52100	0.875	0.247	0.388	0.011	0.025	1.588	0.037	0.163

Heat treatment was performed in a CBUT 10 batch atmosphere furnace, electrically heated. Samples were hardened and tempered for a hardness of 62-63 HRC (Rockwell Units).

Hardening parameters:
- austenitizing temperature: 820 °C;
- austenitizing time: 60 min;
- Carbon potential: 0.9 – 1.0%;
- quenching media: oil at 90 °C.

Tempering parameters:
- temperature: 180 °C;
- holding time: 60 min.

Figure 3. a-c:H DLC coated sample.

For the DLC coating, the specimens surface has been grinded, sand blasted with 40-60 µm diam. glass pearls, polished, cleaned and ultrasonic degreased, thus attempting to reproduce how the hook body is made under industrial conditions. The a-c:H DLC coating was performed using the magnetron sputtering (PVD) deposition technique in the Eifeler Vacotec Alpha 400C system at three different temperatures: 180 °C, 200 °C and 250 °C (Figure 3). For each deposition temperature there were used 10 samples, named series. Working pressure 3 mPa, argon consumption 45 l/charge. A small negative voltage (-35 V) has been applied to the substrates to increase ion assistance during coating deposition.

To investigate the morphology, and adhesion to the substrate, the DLC thin film was SEM studied in cross-section using the JEOL JSM-5200 microscope. The surface porosity of the DLC deposition was assessed, by atomic-force microscopy (AFM) measurements. The sp^2/sp^3 ratio was investigated by X-ray Diffraction – XRD and GXRD (grazing incidence) method. Adhesion of DLC coating to substrate was determined using a Rockwell tester (HRC) with a conical indenter with 150 kg load, according to the German standard "VDI Richtlinien 3198" [8].

3. Results and discussion

The results of the applied heat treatment were evaluated using OTTO WOLPERT HT1 Rockwell hardness tester (HRC) with 150 Kg load. The microstructure obtained after the heat treatment was investigated by optical metallographic microscopy using the optical METTALUX II microscope. Surface preparation for metallographic characterization was done using the LECO PR 22 cold mounting press and the JEAN WIRTZ sample polishing machine. For etching was used NITAL in concentration of 3% for 3-5 seconds. A fine-grained tetragonal martensitic microstructure, with uniformly distributed carbides, was observed (Figure 4).

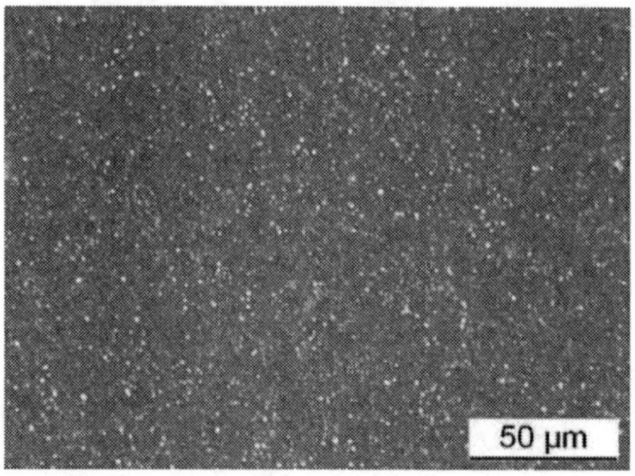

Figure. 4 Hardened AISI 52100 micrograph of the sample 5.

In order not to contaminate the results and to avoid overheating, the samples were not cut with the cut-off machine, they were cut to the size of 15x15x3 mm by electro-erosion machining.

Amorphous aspect of the carbonic layer was observed in SEM cross-section microscopy (Figure 5). DLC layers are clearly defined, adhesion with substrate is generally very satisfactory, a slight fracture in adhesion to the substrate was observed in the sample coated at 180°C. In the case of the DLC sample coated at 200 °C, SEM microscopy highlighted an area where the interface between the substrate and the carbon layer shows a substrate defect resulting from the sectioning and preparation of the sample (Figure 5b).

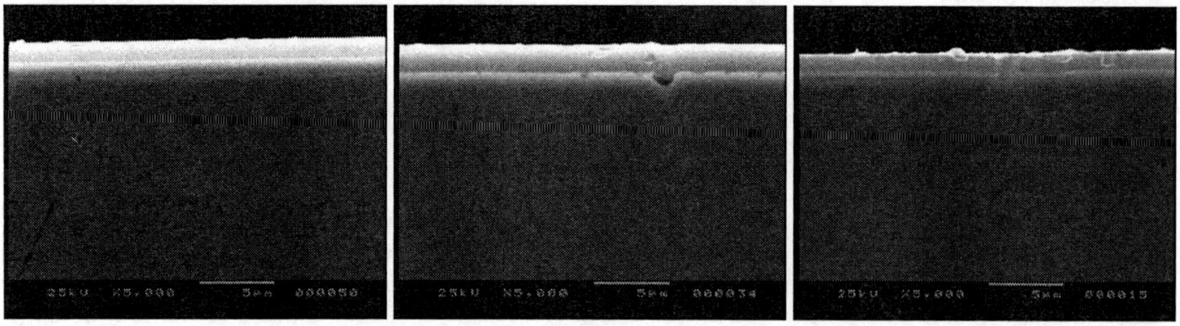

a) b) c)

Figure 5. SEM cross-section microscopy of the sample X, DLC layer:
a) coating at 180 °C; b) coating at 200 °C; c) coating at 250 °C.

The surface roughness and the morphology of the carbon layers was determined by atomic force microscopy-AFM shown in Figure 6. It can be seen that the film surfaces are covered with oriented needle-like peaks. The surface of a-c:H film, deposited at 180 °C is more compact, uniform and smooth than the others.

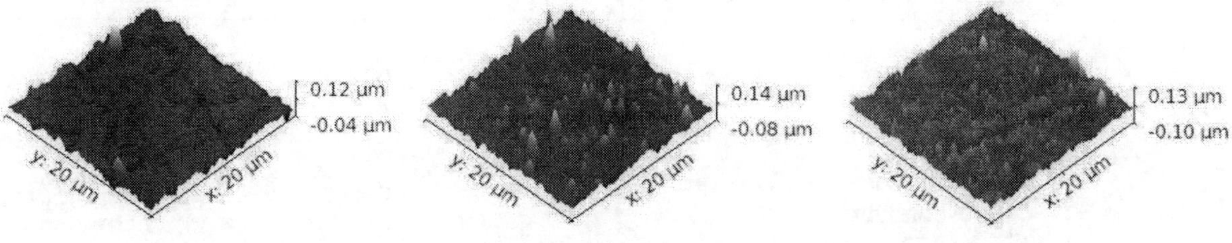

a) b) c)

Figure 6. AFM micrograph of the sample 4:
a) coating at 180 °C; b) coating at 200 °C; c) coating at 250 °C.

The results of the determinations can be seen in table 2. It is observed that the roughness values of the DLC layer deposited at 200 °C and 250 °C as well as the general morphological aspect are very close. Better roughness is shown by the DLC coating at 180 °C.

Table 2. Roughness parameters of the DLC layer.

Sample	Surface	RMS (nm) (root-mean-square roughness)	Ra (nm) (roughness average)	Surface Skewness (degree of bias of the roughness shape (asperity))	Surface kurtosis (sharpness of the roughness profile)
DLC coating at 180 °C	20x20	12.64	9.40	0.9656	10.5636
DLC coating at 200 °C	20x20	17.88	12.99	0.9147	6.7271
DLC coating at 250 °C	20x20	16.48	12.17	0.1947	7.0981

Prior to the thin layer coating, the average value of the Ra roughness parameter on the surface of the steel substrate, measured with the Mitutoyo Surftest SJ-210 roughness tester, was 0.246 µm. The topography of a-c:H DLC coating is highly dependent on the surface roughness of the substrate and deposition parameters [9]. A significant improvement in roughness was found after DLC coating, phenomenon explained by the fact that during the coating process, at the beginning the deposit rate is higher on the peaks and valleys of the rough surface, then as the deposited layer gains in thickness, the surface roughness Ra begins to decrease because the deposition is more pronounced on the slopes and tends to fill the valleys [10].

To determine the DLC coating adhesion to substrate, a Rockwell tester (HRC) with a conical indenter and 150 kg load and was used to imprint the samples surface. It was measured the samples hardness (table 3), and the imprint shape was compared with the reference images designating six adherence indices symbolized from HF 1 to HF 6 [8, 11]. The method has some restrictions:

- the substrate hardness must be at least 54 HRC;
- the thickness of the layer may not exceed 5 µm.

For a thin layer to be considered as adhering to the substrate, traces left on the samples should not exhibit detachments and cracks of the layer around the imprint made by the diamond cone. Samples with adhesion index HF 1, HF 2 HF 3 or HF 4 are considered acceptable (figure 7).

Table 3. Hardness values before and after DLC coating.

	Substrate hardness AISI 52100 (HRC)			
Samples	Hardened and tempered at 180 °C	DLC at 180 °C (series 1)	DLC at 200 °C (series 2)	DLC at 250 °C (series 3)
1.	30 pcs. corresponding to series 1, 2, and 3	62	61	60
2.		62	62	59
3.		62	61	60
4.		61	61	60
5.		61	61	60
6.		61	61	59
7.		62	61	60
8.		62	62	60
9.		61	61	60
10.		62	61	59
Av. value:	62.5	61.67	61.22	59.78
Std. dev.:	0.527	0.516	0.422	0.483

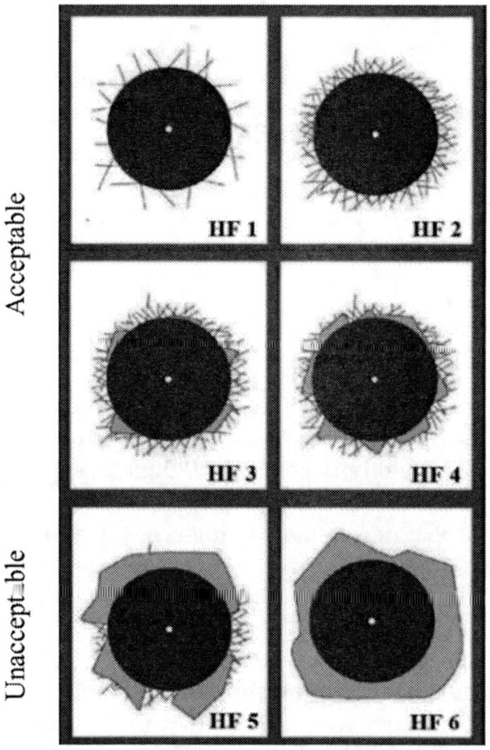

Figure 7. Adhesion classification according to the six imprint shapes - VDI standard [11].

Imprint determinations according VDI standard, were performed on DLC layer deposited at 180 °C, 200 °C and 250 °C (Figure 8). For this purpose, the Rockwell tester was set at a load of 150 kgf and equipped with a diamond cone penetrator having a peak angle of 120°.

IOP Conf. Series: Materials Science and Engineering **416** (2018) 012014 doi:10.1088/1757-899X/416/1/012014

As it can be seen in Figure 8, the obtained imprints after indentation can be categorized as HF1 or HF2 with acceptable adhesion indices. Due to the amorphous nature of DLC deposition, no cracks or combinations of cracks and delamination of the thin layer were observed after the indentation. Microscope images showed only small delamination at the edges of the imprints, proving a good adhesion of the DLC layer to the substrate. Better adhesion to the substrate present the layers deposited at 180 °C, corresponding to HF1 adhesion index.

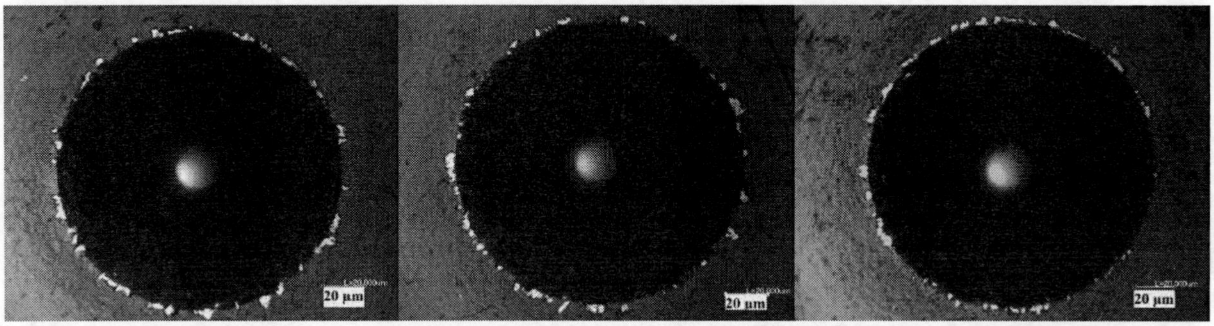

a) b) c)

Figure 8. Imprint shapes of the sample 3:
(a) coating at 180 °C; (b) coating at 200 °C; (c) coating at 250 °C.

The crystalline phase composition of the carbonic layers was determined by the X-ray diffraction (XRD). X-ray diffraction obtained on the DLC thin films using conventional θ/2θ scanning methods generally produced a weak signal from the film and an intense signal from the substrate, it can be noticed that X rays have penetrated the DLC layer and the substrate is seen; the peaks presented in Figure 9 come as a result. To avoid intense signal from the substrate and get stronger signal from the film, a GIXRD (grazing incidence) scan was performed, with a lower incidence angle for total reflection of the film without the substrate interference.

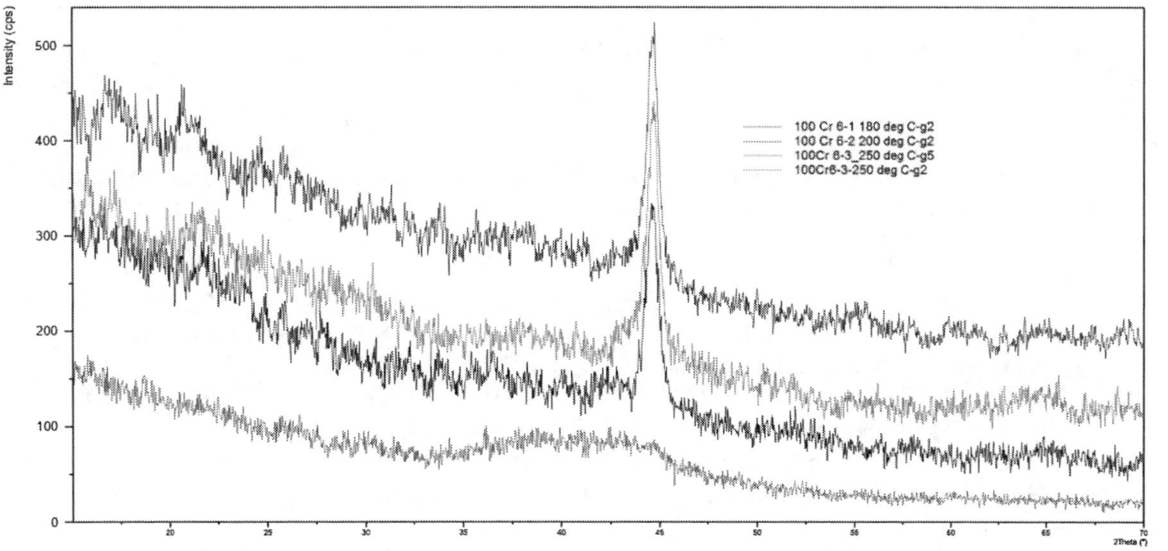

Figure 9. X-ray diffractogram of the 10 sample.

After the GIXRD, no obvious peaks can be observed (Figures 9 and 10), the peaks from the substrate have fallen (brown colour curve). Due to few crystallites in the DLC film, one discernible peak was noticed at 2Θ~ 45° that may correspond to a sp^3 diamond plane.

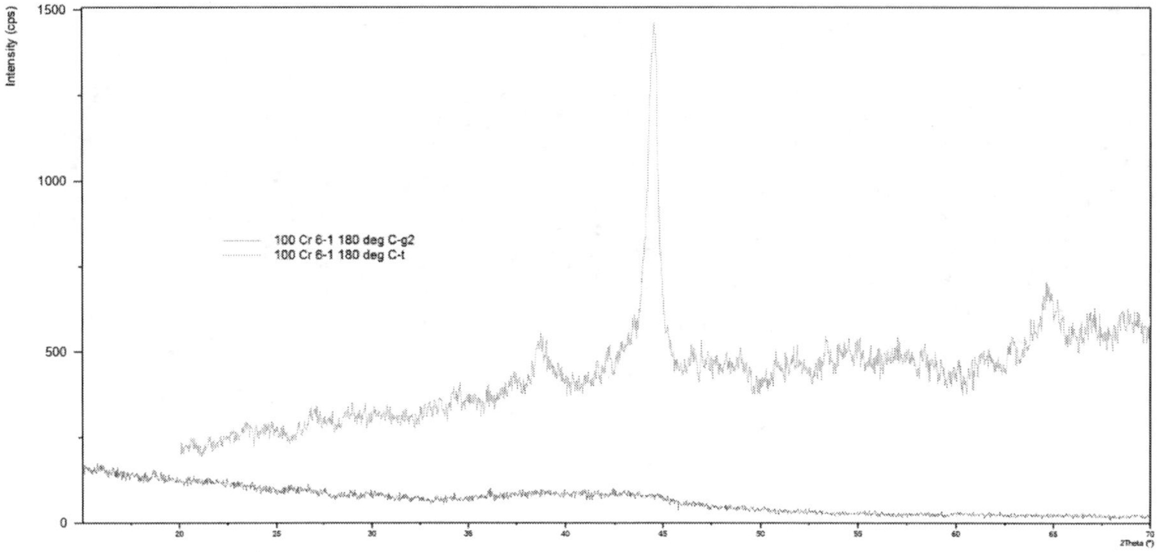

Figure 10. Comparison between GIXRD and symmetrical XRD patterns of the 10 sample - series 1.

4. Conclusions

The microstructure obtained after heat treatment consisting on fine-grained tetragonal martensite with uniformly distributed carbides highlighted by optical microscopy and hardness higher than 54 HRC provide wear and mechanical stresses resistance, being a good substrate for the subsequent DLC deposition.

A secondary tempering effect was observed after DLC, hardness values decrease depending on the coating temperature. SEM and XRD analyses revealed the amorphous character of DLC deposition. Surface roughness of the DLC layer is much lower compared to the roughness of the substrate prior to the coating.

Comparing substrate hardness values, DLC adhesion and topography, it turned out that the best wear resistance is exhibited by the layers deposited at 180 °C.

5. References

[1] Bhadeshia H K D H 2012 Steels for bearings *Progress in Materials Science* **57(2)** 268-435

[2] Erdemir A and Donnet C 2006 Tribology of diamond-like carbon films: recent progress and future prospects *Journal of Physics D: Applied Physics* **39(18)** R311-R327

[3] Cicek H 2018 Wear Behaviors of TiN/TiCN/DLC Composite Coatings in Different Environments *Ceramics International* **44(5)** 4853-4858

[4] Gao F, Erdemir A and Tysoe W T 2005 The Tribological Properties of Low-friction Hydrogenated Diamond-like Carbon Measured in Ultrahigh Vacuum *Tribology Letters* **20(3-4)** 221-227

[5] Paskvale S, Kahn M, Čekada M, Panjan P, Waldhauser W and Podgornik B 2011 Tribological properties of diamond-like carbon coatings prepared by anode layer source and magnetron sputtering *Surface and Coatings Technology* **205(2)** S99-S102

[6] Ebrahimi M, Mahboubi F, and Naimi-Jamal M R 2015 Wear behavior of DLC film on plasma nitrocarburized AISI 4140 steel by pulsed DC PACVD: Effect of nitrocarburizing

temperature *Diamond and Related Materials* **52** 32-37

[7] Waseem B, Alam S, Irfan M, Shahid M, Farooqa M, Soomro B D, Hashim S and Iqbal R 2015 Optimization and characterization of adhesion properties of DLC coatings on different substrates *Materials Today: Proceedings* **2** 5308 – 5312

[8] Vidakis N, Antoniadis A and Bilalis N 2003 The VDI 3198 indentation test evaluation of a reliable qualitative control for layered compounds *Journal of Materials Processing Technology* **143-144** 481-485

[9] Feldiorean D I and Tierean M H 2015 Overview on DLC layers used to increase the tribological properties *Bulletin of the Transilvania University of Braşov. Series I* **8(57)1** 43-50

[10] Singh R K, Xie Z H, Bendavid A, Martin P J, Munroe and R Hoffman M 2008 Effect of substrate roughness on the contact damage of DLC coatings *Diamond and Related Materials* **17(6)** 975-979

[11] Normen, V.D.I., VDI 3198. VDI Verlag Dusseldorf 1991 *Coating (CVD, PVD) of cold forging tools*

IOP Conf. Series: Materials Science and Engineering **416** (2018) 012015 doi:10.1088/1757-899X/416/1/012015

Research Regarding Capacitor Discharge Stud Welding with Tip Ignition on Galvanized Thin Sheets

A Magda[1,*], M Burca[1] and M Lego[2]

[1]Politehnica University Timisoara, Materials and Manufacturing Engineering Department, 1 Mihai Viteazu Blvd., 300222, Timisoara, Romania
[2]Technological Transportation High School, 1 Ardealul Str., 300556, Timisoara, Romania

*Corresponding author: aurelian.magda@upt.ro

Abstract. Arc stud welding is a widely used welding process used to fix joints like studs, screws, pins, or similar elements on a wide variety of products. The process is mainly characterized by simplicity, high welding productivity, very good welding joint quality, low cost price, easy welding equipment and easy to handle, flexibility. The process successfully replaces other assembly methods such as: threaded assembly, riveting, classic arc welding or pressure welding and soldering. The paper proposes to establish welding technology for M5 x 35 mm steel stud with tip ignition and electric charge stored in capacitors on thin galvanized steel (s <1 mm thick). The main problem of welding galvanized steel is vaporization of zinc on the surface of the sheet which produces pores, splashes and very toxic fumes. This vaporization produces also the corrosion hazard in the surface of the sheet. Welding technology will be applied to joint stud on the cover of an air filter.

1. Introduction

The principle of stud welding with tip ignition is relatively simple. The stud is initially fixed by the welder operator in the elastic muff of the special welding torch and positioned on the surface of the workpiece. By pressing the torch trigger, the welding current passes through the stud producing the heat of the tip stud. By easy lifting and for a short time the stud on the surface an electric arc is primed that produces the melting of the stud tip and the local melting of the piece. At the interruption of the electric arc, the stud is pushed by a relatively small force, by means of a spring placed in the head of the torch, in the welding pool, and by joining and solidifying the common bath results the welded joint [1]

Depending on the spring arc mode, the duration of the welding time, the type of metal bath protection, the type of the source, the arc stud welding process can be performed in several variants.

For this application is of interest arc stud welding with tip ignition and electric charge stored in capacitors [2]. The particularities of this welding process are:

- The diameter of the stud d_b= 3…10mm;
- Welding current: $I_s \leq 5000$ A;
- Welding time: $t_s \leq 10$ ms (1 .. 3 ms);
- Electric arc and welding pool protection: It does not require protection;
- Minimum thickness of the workpiece: 1/10 of the diameter of the stud (s_{min} = 0,6 mm);

- Stud material: carbon steel, high alloy steel (stainless steel), aluminum or brass on metal surfaces.

Due to a penetration depth of only 0.1 mm, the welding of the studs by this process leaves no trace, staining or deformation even on galvanized or painted surfaces on the opposite side of the sheet even at small thicknesses.

Figure 1 shows the principle of arc stud welding with tip ignition and electric charge stored in capacitors.

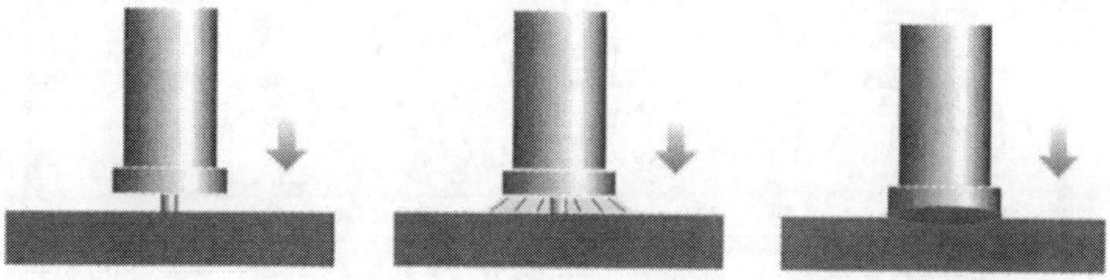

Figure 1. The principle of arc stud welding with tip ignition [1].

2. Welding parameters

In the case of arc stud welding, the quality of welded joints depends decisively on the correct setting of welding technology, more precisely the choice of technological welding parameters and less the skill of the welding operator. This is explained by the fact that the welding takes place in a very short time (usually less than 1.5 s), and the process takes place without the intervention or the participation of the welder operator.

The main technological parameters in arc stud welding are the welding current (I_s) and the welding time (t_s), which act on the power of the electric arc, respectively on the welding energy, with implications of welding pool volume, the penetration, the degassing or the danger of defects occurrence in the weld joint [6].

The determination of the main technological parameters of welding can be done by using empirical relations presented in table 1 [6].

Table 1. Relationships of calculation of the main welding parameters.

Welding current	$I_s = 80{*}d_b$	For stud diameter < 12 mm
I_s (A)	$I_s = 90{*}d_b$	For stud diameter 12–25 mm
Welding time t_s (s)	$t_s=0.04{*}d_b$	For all stud diameters

Arc welding of studs can be carried out both in direct current and in alternating current. Continuous current is preferred for increasing the arc stability. In alternating current, the arc blow is eliminated. When direct current welding is used, direct polarity is preferred which provides easier priming of the electric arc [3].

An important technological aspect in the welding of studs is the danger of a strong arc blow that can affect the quality of the weld joint.

Due to the very short welding time and process without the participation of the welder operator, it is not possible to control the arc blower by operator intervention as can be done with the classic manual welding. It is therefore necessary to take additional measures to reduce or avoid the magnetic arc blow [4].

The most important measure to avoid the magnetic arc blow is the use of symmetric workpiece connection, knowing that the arc blow always manifests in the opposite direction to the workpiece contact, figure 2 [6].

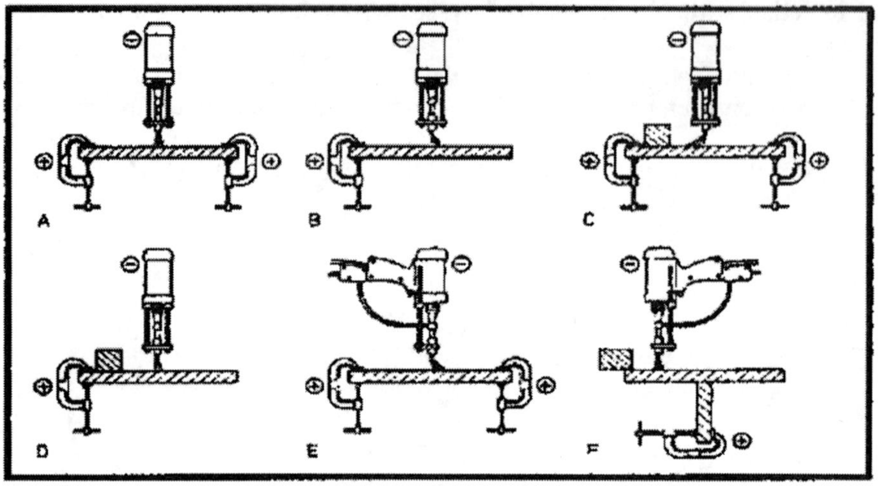

Figure 2. Aspects of magnetic arc blow manifestation in stud welding.

In this way new welding sources are provided with two workpiece connections. Positioning of the torch is also very important in this case, in order to cancel the magnetic arc blow, the torch should be oriented perpendicularly to the joining.

3. Technological aspects of galvanized sheets welding

The main problem is vaporization of zinc on the surface of the sheet which produces pores, splashes and very toxic smoke. Welding process can also produce the corrosion of the sheet in the surface area because the zinc is burned from the surface [2].

MAG welding can be done in CO_2 or gas mixtures Ar + CO_2, and with a welding speed of less than 20 cm / min by comparison of steel sheets welding. Oxy-fuel welding is applied at thicknesses below 3 mm and has the disadvantage that the zinc layer is destroyed on a very large area. Manual metal arc welding can be done at low welding speed recommended, high opening of joints, oscillation of the electrode and the use of slag electrodes that slowly solidify to allow the pore out of the welding pool. TIG welding does not apply because vaporization of zinc causes electric arc instability by coating the tungsten electrode with zinc oxide [6].

Regardless of the process used, it is recommended that at welding to have ventilation to remove smoke (zinc oxide).

For welding studs there are mainly 3 welding processes as follows:

- drawn arc stud welding with ceramic ferrule or shielding gas;
- short-cycle drawn arc stud welding;
- capacitor discharge drawn arc stud welding or capacitor discharge stud welding with tip ignition [6].

Starting from the particularities of these processes and the characteristics of the given application, it results that the possible welding process to be applied is capacitor discharge stud welding with tip ignition.

Characteristic of this process are the shape and dimensions of the fusible tip having a cylindrical shape with a diameter of 0.5-0.75 mm and a length of less than 1 mm. The welding process can be applied to the stud with a diameter between 2 and 10 mm.

4. Experimental research

Research has been done to establish the welding technology of capacitor discharge stud welding with tip ignition for M5x35mm studs on thin sheets (s <1 mm thick) covered by galvanizing. The suitable welding technology from a qualitative point of view will be applied to the welding of studs on the air filter cover.

The difficulty of stud welding in this case is determined by the following:

- low thickness of the sheet (s = 0.75 mm) leading to the danger of penetration of the workpiece during the welding process;
- coating the sheet by galvanizing with a thin layer of zinc leading to the danger of material expulsion and pores in welding area due to zinc evaporation;
- absence of information in the literature on welding of galvanized thin sheets by capacitor discharge stud welding.

For welding the stud, it is necessary before marking the welding spot on the surface of the workpiece. This operation can be done by drawing with the vernier caliper or by using a jig. This marking is presented in figure 3 and 4.

Figure 3. Marking with vernier caliper.

Figure 4. Marking with jig.

In the first case it is observed that the location of the stud is determined by a line, and in the second case it is marked by a point obtained by plastic deformation using a pointing device.

The first marking method is simple but has the disadvantage of scratching the piece and removing the zinc layer which can lead to local corrosion in time, respectively the wrong positioning of the stud over the drawing as a result of sliding the stud on the piece when the torch is fixed. The second method eliminates the disadvantages presented above because the tip of the stud is guided in the local plastic deformation that appears on the piece.

The disadvantage of this method is that plastic deformation can lead to changing the welding process conditions by reducing the length of the electric arc. This can lead to influencing the quality of the welded piece, therefore the application of this method should avoid a local plastic deformation with a depth greater than 0.1 - 0.15 mm.

4.1 Verification of welding technology

Manufacturers of capacitor discharge stud welding equipment give technological recommendations regarding the capacitor loading voltage depending on the diameter of the stud, respectively the capacity of the capacitor. Capacitor capacitance differs from one equipment manufacturer to another. Therefore, a welding technology recommended by a manufacturer cannot always be applied to welding equipment of another manufacturer because it would lead to different welding energies. Also recommendations are valid for welded steel stud on the not galvanized sheet.

Research was carried out with a HBS KES 2100 equipment. This source has a capacitor capacitance of 88 000 μF. Following the calculations resulted 100 V of a capacitor voltage.

Figure 5 shows the welded stud appearance following bending technological test [5].

Figure 5. The appearance of the stud after bending.

It is to be noticed that the welded stud resistance is adequate, since the stud has been bent at an angle greater than 60 degrees, without detachment of the stud and without cracks. With this technology were made welds on 0.75 mm thick galvanized sheet [5].

Four weldings were made using different capacitance loading voltages (figure 6), respectively two ways of marking the welding place by drawing or plastic deformation as shown in figures 3 and 4.

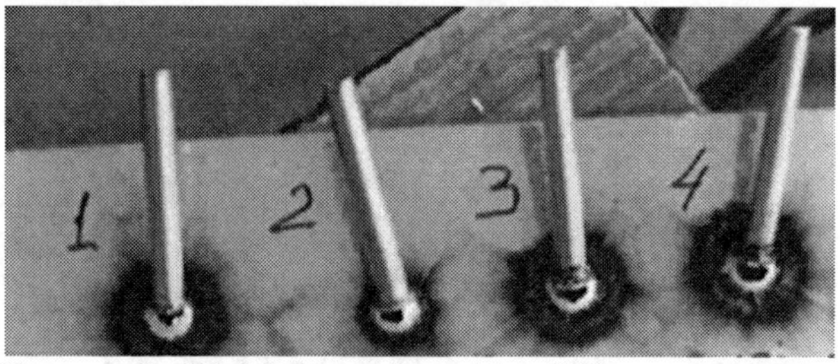

Figure 6. Welded studs with different capacitor loading voltages; 1- 100V; 2-100V; 3- 115V; 4-130V.

The test voltage for the first two studs was 100 V following the influence of the marking method on the stud welding quality. The studs were tested by bend according to ISO 14555 [5]. No major qualitative differences are observed regarding the influence of the marking method on the resistance of the welded point. However, in both situations the melting does not occur on the entire section of the stud.

This may lead to the conclusion that in the case of a galvanized sheet, a higher loading voltage is necessary. Therefore studs 3 and 4 were welded with 115V or 130V. We notice the beneficial effect of increasing the voltage by increasing the tear section of studs.

These tests lead to the conclusion that the stud welding on the finished product is recommended to use a capacitor loading voltage of 130 V.

Experimental investigations continued on samples of 0.75 mm galvanized sheet using Soyer BMS 8H source. The welding quality has been checked for different values of the capacitor loading voltage corresponding to the voltages of 90 V, 115 V, 130 V and 150 V. The appearance of the welds is shown in the figure 7.

In order to highlight the influence of the capacitor loading voltage on the welding resistance, initially 90 V and 115 V voltages were used, characterized by insufficient energy for an M5 stud. This is observed following the technological torque test when weld failure (figure 8). Increasing the voltage at 130 V, respectively 150 V determines the best quality of the welded point because the welding has broken off the sheet without failure.

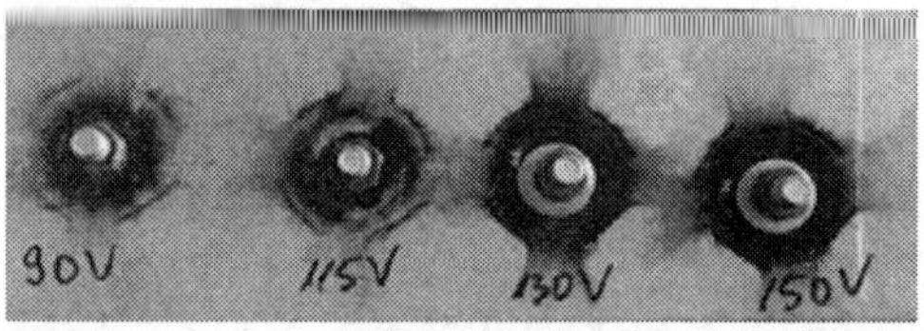

Figure 7. The appearance of the welds.

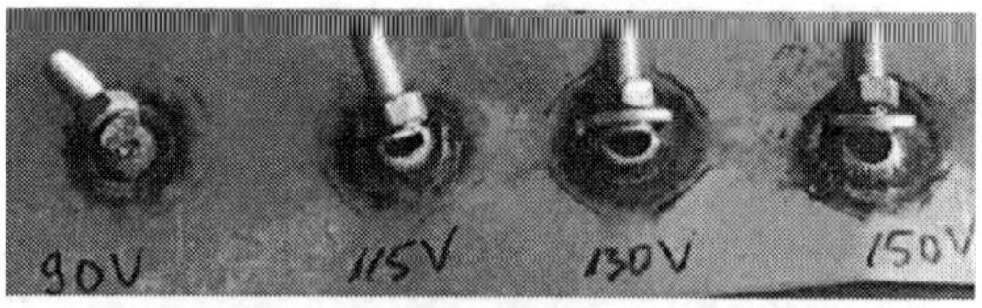

Figure 8. Torque test of the welds.

After the attempts made on the galvanized sheet were welded studs on the air filter cover trying different welding technologies.

The charging voltage used for welding on galvanized sheet of 115V, 130V turns out to be inappropriate when welding on the air filter cover, partial welding of the stud is observed. From the analysis it was found that the cover of the sheet was made by galvanizing and the cover of the air filter

IOP Conf. Series: Materials Science and Engineering **416** (2018) 012015 doi:10.1088/1757-899X/416/1/012015

was made by thermal zinc coating. The result is that in the case of zinc galvanizing, the thickness of the zinc layer is much smaller and more uniform than in the case of thermal coating.

5. Conclusion

Welding technologies have been made with welding installations from different companies were implemented, namely the KES 2100 equipment from HBS and the Soyer BMS-9 welding equipment.

Welding technologies proposed by experimental research are:

- for the KES 2100 installation, optimal welding technology was 135 V and capacitor capacitance 88 000 μF;
- for the Soyer BMS-9 installation, optimal welding technology was 160 V and capacitor capacitance 66 000 μF;
- the two welding technologies are equivalent in terms of energy stored in the capacitor, the differences being determined by the capacity of the capacitors in the two installations, namely 88 000 μF at the HBS installation and 66 000 μF at the Soyer installation. In both cases, the accumulated energy is 800 J.

The appearance of welded studs on the finished product is shown in figures 9 and 10.

Figure 9. The appearance of the welded studs on the air filter cover.

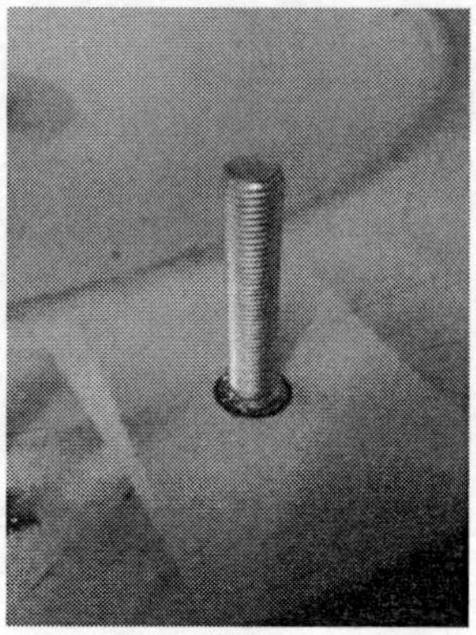

Figure 10. Stud appearance after avoiding coating with oxide.

From the analysis of the tests it can be seen that capacitor discharge stud welding with tip ignition for M5x35mm studs on the surface of the air filter with results corresponding to the quality, resistance and aesthetics being the only welding process possible to apply for such an application. In addition, it is noted that it has been solved to avoid covering the area adjacent to the stud with oxide (figure 10).

6. References

[1] Trillmich R 2001 *Welding Journal ASR* **2** 15

[2] Burca M 2001 *Research on the realization of heterogeneous joints in arc stud welding ASR conference* (Cluj-Napoca, Romania) p 186

[3] Hsu C Mumaw J 2011 *Welding Journal* **90** 45S

[4] Samardzic I Kladaric I and Klaric S 2009 *Metalurgija* **48** 181

[5] *** EN ISO - 14555/2006 *Welding — Arc stud welding of metallic materials*

[6] *** 2013 *Welding Handbook ninth edition* **2** AWS

IOP Conf. Series: Materials Science and Engineering **416** (2018) 012016 doi:10.1088/1757-899X/416/1/012016

Design of Porous Structures in Shape Memory Alloy Systems

R Sprincenatu*, A Novac, G Chilnicean, V Bolocan and C Craciunescu

[1]Department of Materials and Manufacturing Engineering, Faculty of Mechanical Engineering, Politehnica University Timisoara, Bd. Mihai Viteazul 1, 300006 Timisoara, Romania

*Corresponding author: sprincenatu.roxana@gmail.com

Abstract. The fabrication of porous shape memory alloys is analyzed based on the particularities of the alloy families and the spacer materials potentially to be used for pores generation. The analysis is made for two distinct set of shape memory alloy families Cu-based and Ti-based, taking into account solid state sintering technologies. Sodium chloride, magnesium and carbamide are comparatively analysed for the use in the fabrication of the porous shape memory materials. Particularities in using spark plasma sintering technique for the fabrication of porous shape memory alloys is discussed. Experiments performed on alloys belonging to the shape memory system are analyzed from macro and microstructural point of view in order to assess the influence on the resulting functionality. The results allowed the selection of the optimal processing parameters for the fabrication of porous alloys.

1. Introduction

Shape memory alloys are materials that are known to exhibit shape recovery following a plastic deformation by heating above a certain temperature that is specific to the chemical composition of the alloy.

Porous materials are materials with possible use in the fabrication of lightweight structures, filters, damping components, heat exchanges and even biomedical applications especially due to the presence of a large volume of porosity [1-5]. Porous shape memory alloys have been manufactured via various techniques: self-propagating high temperature synthesis (SHS), conventional sintering (CS), hot isostatic pressing (HIP), metal or powder injection molding (MIM), spark plasma sintering (SPS) and the space holder technique (SHT) [6]. Sodium chloride, magnesium and carbamide are the main materials used as spacers. Spark plasma sintering is a technique that allows the rapid sintering of materials in one single process that combines pressing and sintering, thus avoiding extensive oxidation, generation of uniform structures with limited grain growth during the fabrication process and high reproducibility [7].

Porous shape memory alloys are gaining importance in practical applications especially in two fields: medical implant devices and as energy absorption structural materials. While for biomedical applications the Ti-based alloys are the one with real potential, for energy absorbing applications other alloy system can also be considered. The main target for porous NiTi remains biomedical applications because of the following properties good biocompatibility, comparable to conventional porous stainless steel and titanium (Ti) implant materials; a combination of high strength (important to prevent deformation or fracture), relatively low stiffness (useful to minimize stress shielding effects) and high toughness (essential to avoid brittle failure); and shape-recovery behavior facilitating implant insertion and ensuring good mechanical stability within the host tissue [8].

Copper based shape memory alloys are among the materials with very high damping capacity, especially in martensite and in the martensite austenite transformation range. Among copper based shape memory alloys the mostly used ones are CuZnAl and CuAlNi alloys. CuAlNi alloys are more expensive, but they are more resistant to degradation of functional properties due to undesired aging effects [9].

The aim of this paper is to present our work on the fabrication and characterization of porous shape memory alloys using spark plasma sintering. The focus is on CuZnAl shape memory alloys that could be used for damping and filtering applications.

2. Experimental details

The samples of CuZnAl were made by sintering following mechanical alloying of the powders using a planetary ball mill. The milling was realized for 8 hour with a speed of 300 rpm. After this the powder was mixed with salt (NaCl) of tow granulation (fine and coarse) with a proportion of 2/3 powder -1/3 salt. The mixing was made in a cylindrical recipient with a rotation speed of 50-60 rpm for 5 hours. The sintering was realized with a spark plasma sintering machine (FCT Systeme GmbH), (Figure 1), with a 20 mm diameter tungsten carbide (WC) mould (Figure 2).

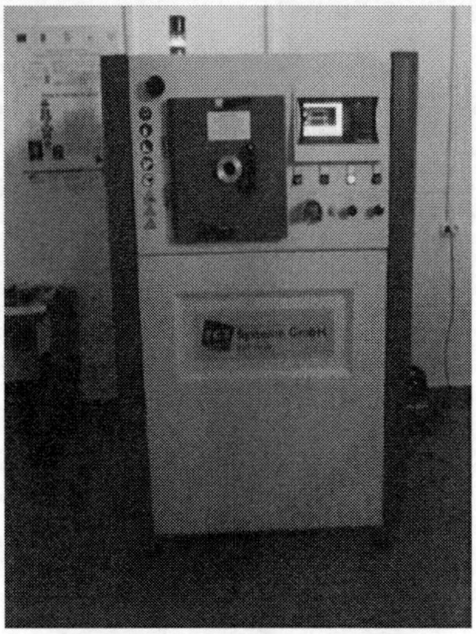

Figure 1. Spark plasma sintering machine.

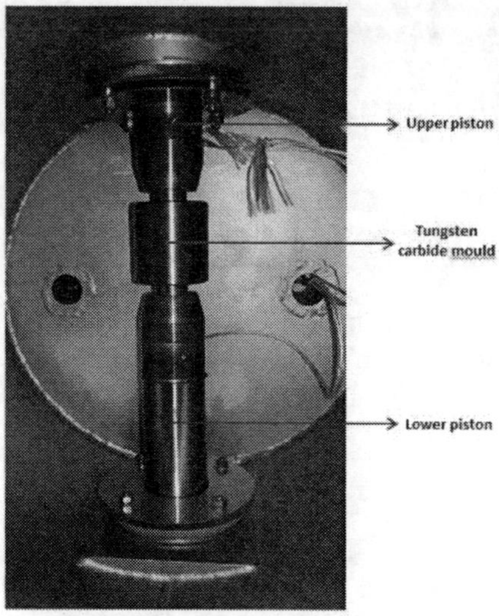

Figure 2. Tungsten carbide mould.

The sintering was made was made in two steps using 50°C/min heating speed: the first step was heating until 350°C, with a holding time of 3 minutes; the second step was heating until 450°C with a holding time of 10 minutes. The pressing force used was 20 kN. The cooling was controlled until 50°C, after that it was cooled in the chamber of the machine until room temperature. The samples were heated at 850°C and maintained for 15 minutes, followed by water cooling. For the sintered samples, the salt was dissolved following the sintering process.

A Ni-Ti was also made by conventional sintering, using mechanically alloyed Ni and Ti powders and fine salt, followed by conventional sintering in a vacuum sealed quartz ampoule.

Microstructural observations of the samples was made by scanning electron microscopy (SEM) in a TESCAN Vega 3 LM scanning electron microscope (SEM), equipped with a Bruker Quantax 200 Energy Dispersive X-ray Spectroscopy (EDX) system with Peltier-cooled XFlash 410M silicon drift detector.

3. Results and discussions

The structure of the of the NiTi sample produced by vacuum sintering following salt dissolution shows typical pores and a very limited quantity of salt in places where it was not fully removed. The composition was analyzed and was in the compositional range of shape memory alloys, with a composition slightly higher in Ti, making it Ti-rich (figure 3).

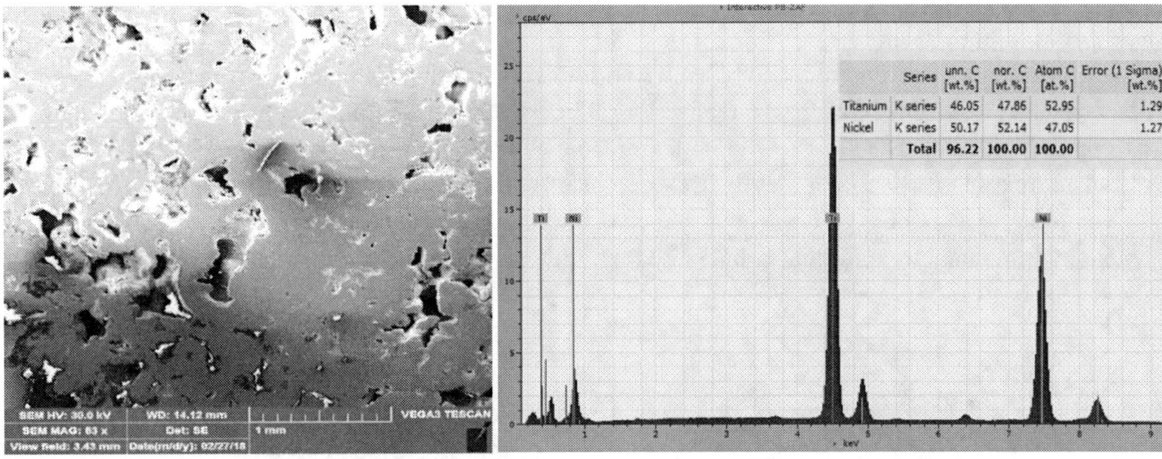

Figure 3. NiTi shape memory alloys produced by vacuum sintering.

The Cu-Zn-Al porous shape memory alloy produced by spark plasma sintering were of disk shape with a thickness in the range of 1-5 mm and dense (figure 4a), the SEM investigation also revealed a fine NaCl cubical particle in the structure, prior to the dissolution process (figure 4b).

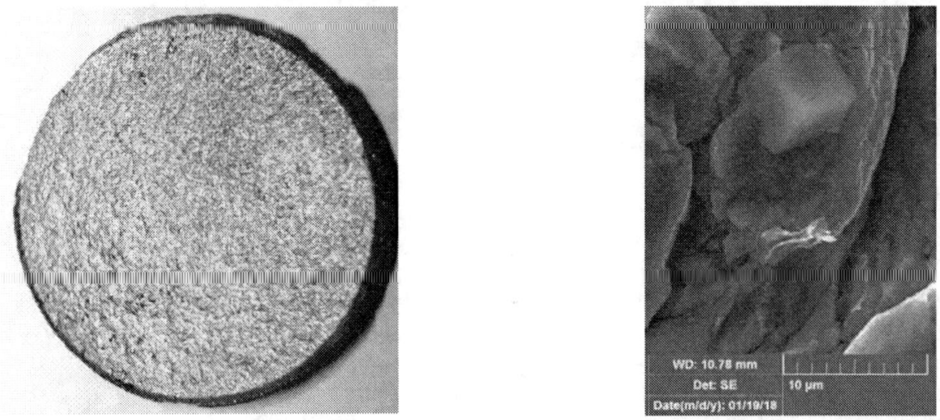

a) as sintered sample b) fine NaCl particle prior to dissolution

Figure 4. CuZnAl with NaCl spark plasma sintered sample.

The composition of the manufactured alloy as it resulted from the EDX investigation is 62 at% Cu, 20 at% Zn, 18 at% Al (figure 5).

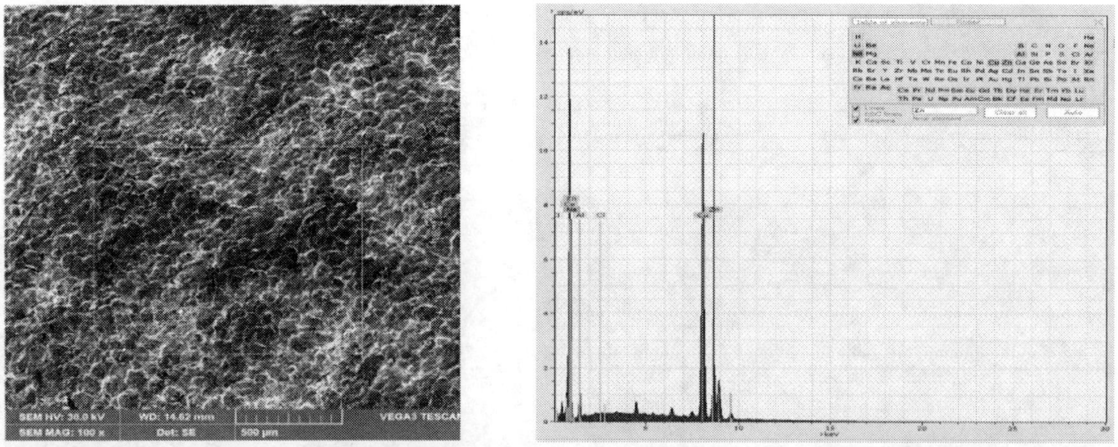

Figure 5. Surface analysis of the Cu-Zn-Al alloy sintered with fine NaCl particles.

Figure 6 a, b shows the microstructure of the pores obtained after the dissolution of the Cu-Zn-Al samples sintered with fine NaCl particles. It can be seen that the pores are small and most of them have almost the same shape, this is because the NaCl particles used had the same dimensions and shape. The distribution of pores in the surface of sample is almost equal, there aren't pores agglomerated in one place.

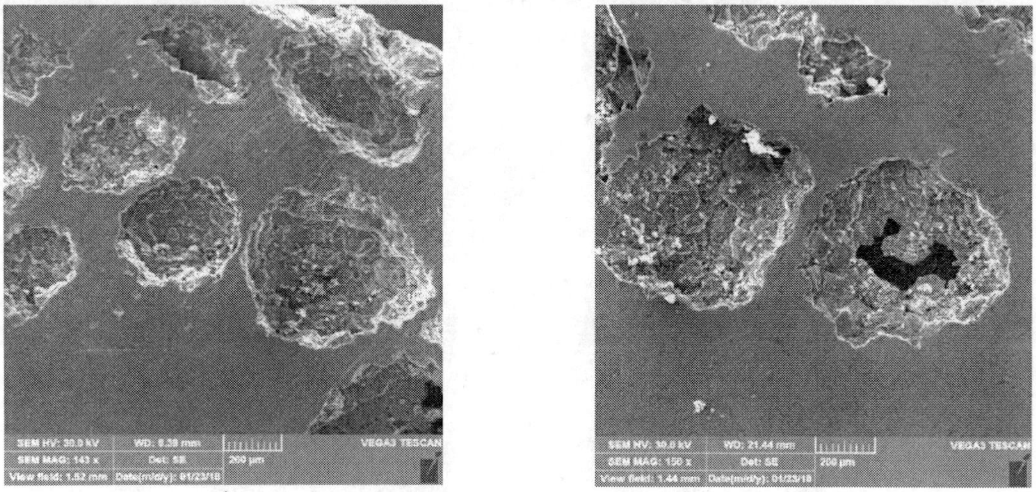

a) b)

Figure 6. Microstructure of small pores.

For the CuZnAl porous shape memory alloy sintered with large NaCl particles, the pores have irregular shapes as it can be seen in figure 7 a, as the shape of the large particles. Because of the bigger dimensions of the NaCl particles used in this sample the pores tend to agglomerate in one place in the surface of the sample resulting in an irregular distribution of pores in the surface. On the other hand, the bottom of the pore shows that the material is well sintered.

IOP Conf. Series: Materials Science and Engineering **416** (2018) 012016 doi:10.1088/1757-899X/416/1/012016

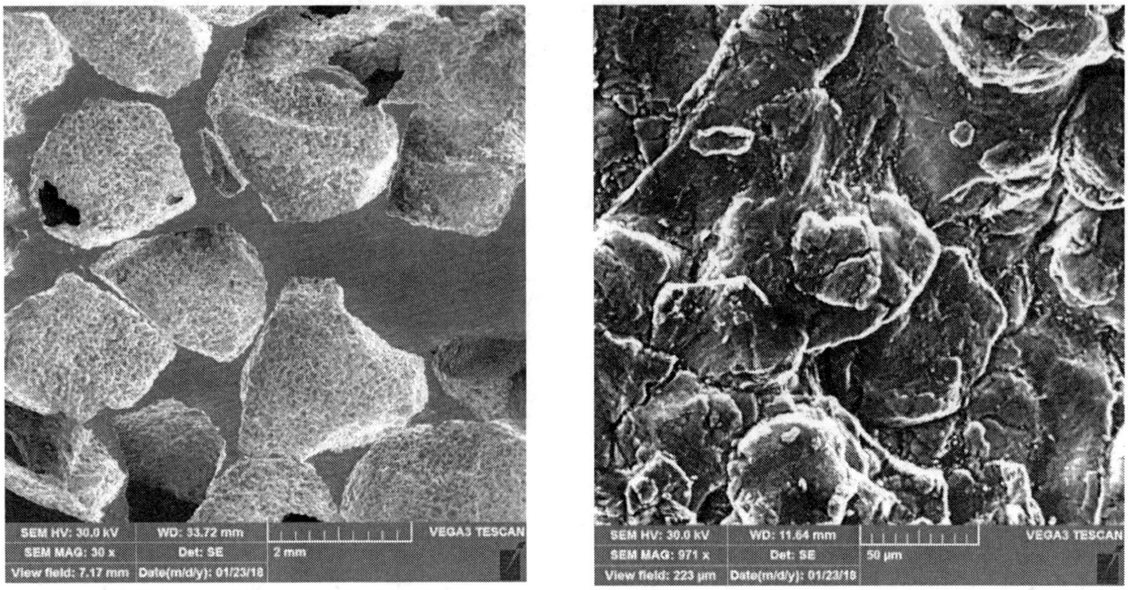

a) b)

Figure 7. Microstructure of large pores in the Cu-Zn-Al alloy: a) -Large pores, b)- The bottom of a large pore.

The analysis of the quenched sample show the formation of a typical martensitic structure, thus confirming that the manufactured alloy belong to the shape memory system not only by composition but also by structure.

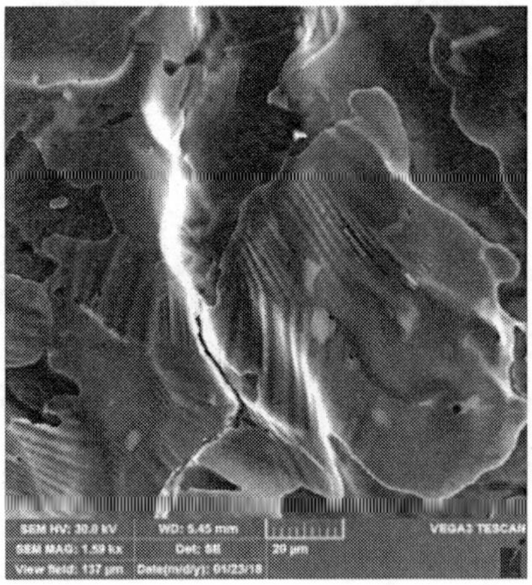

Figure 8. Martensite structure observed in the quenched Cu-Zn-Al alloy.

Further experiments are made to characterize the martensitic transformation in porous structures.

4. Conclusions

NaCl was successfully used to manufacture alloys belonging to the systems were shape memory effect and in their corresponding compositional range. It can be used in conventional as well as in spark plasma sintering, as long as the sintering temperature is lower than the melting temperature of the salt.

The resulting pores after the salt dissolution copy the features of the salt crystals. For the Cu-Zn-Al sintered alloy the large salt particles tend to agglomerate on the bottom of the sample during the sintering process in a larger amount compared to the fine particles.

The martensitic structure was observed in the Cu-Zn-Al sintered alloy following quenching.

5. References

[1] Aydoğmuş T and Bor Ş 2009 *Journal of Alloys and Compounds* **478** 705
[2] Bansiddhi A and Dunand D C 2007 *Intermetallics* **15** 1612
[3] Zhao Y Taya M Kang Y S and Kawasaki A 2005 *Acta Materialia* **53** 337
[4] Zhu S L Yang X J Fu D H Zhang LY Li C Y and Cui Z D 2005 *Materials Science and Engineering A* **408** 264
[5] Utsunomiya H Matsumoto R 2014 *Procedia Materials Science* **4** 245
[6] Aydoğmuş T and Bor A Ş 2011 *Turkish J. Eng. Env. Sci.* **35** 69
[7] Suárez M Fernández A Menéndez J L Torrecillas R Kessel H U Hennicke J Kirchner R and Kessel T 2013 *Sintering Applications* (book edited by Burcu Ertuğ ISBN 978-953-51-0974-7)
[8] Bansiddhi A Sargeant T D Stupp S I and Dunand D C 2008 *Acta Biomaterialia* **4** Issue 4 773
[9] Kneissl A C Unterweger E Bruncko M Lojen G Mehrabi K Scherngell H 2008 *Metalurgija - journal of metallurgy* 89

Acknowledgements
The authors acknowledge the support through the PIRSES-GA-2013-612585-MIDAS International Research Staff Exchange Grant of the PEOPLE MARIE CURIE ACTIONS FP7-PEOPLE-2011-IRSES. Equipment for experiments acquired through the Romania-Republic of Serbia IPA Cross-border Cooperation Programme, Project MIS ETC no 1328 "Pole of Collaboration in New Functional Alloys".

IOP Conf. Series: Materials Science and Engineering **416** (2018) 012017 doi:10.1088/1757-899X/416/1/012017

Ti-Ta High Temperature Shape Memory Alloys

G Chilnicean*, A Novac, R Sprincenatu, V Bolocan and C Craciunescu

Department of Materials and Manufacturing Engineering, Faculty of Mechanical Engineering, Politehnica University Timisoara, Bd. Mihai Viteazul 1, 300006 Timisoara, Romania

*Corresponding author: george.chilnicean@yahoo.com,

Abstract. The actuation based on shape memory alloys relies on the martensitic phase transformation that allows the reversible transition from the "soft" martensite phase to the "hard" austenite phase. Usual shape memory alloys usually have transformation temperatures around room temperature. While these are suitable for biomedical devices a higher temperature range is needed for applications that are designed to work in aerospace applications. High temperature shape memory alloys belonging to the Ti-Ta system manufactured by arc melting are investigated for their potential to be used in applications requiring the actuation based on martensitic transformation. The microstructure is analyzed based on X-ray diffraction and scanning electron microscopy whereas the martensitic transformation is investigated based on differential scanning calorimetry experiments.

1. Introduction

The shape memory effect was observed on numerous alloy systems, among which the so-called Ni-free alloys, based on Ti (like Ti-Ta, Ti-Mo, Ti-Nb) are ones that can provide a large range of transformation temperatures for a variation of the composition [1-4].

A number of applications requiring high temperature actuation cannot use the typical shape memory alloys because their actuation temperatures are limited to about 100 °C. Hot environments like the one required by home appliances, automotive or aerospace industry or power generation systems need significantly higher transformation temperatures.

Among the high temperatures shape memory alloys (HTSMA) the Ti-X (X=Pd, Au, Zn) and Ru-X (X=Nb, Ta) are the ones capable of leading to Ms transformation temperatures above 400 °C [5-8]. The β-type Ti-base alloys can show interesting shape memory and associated properties (like superelasticity and shape memory properties by taking advantage of the β↔α" reversible martensitic transformation. The parent β-phase in Ti alloys is obtained by stabilization via the addition of β stabilizing elements. Tantalum is a stabilizing element that also has the potential to increase the martensitic transformation temperature and to limit the formation of the ω phase during aging at intermediate temperatures, a phase that can supress the martensitic transformation [9].

The fabrication and microstructure of a Ti-Ta alloy was investigated in this paper in order to develop the shape memory alloys with higher transformation temperatures.

2. Experimental details

Ti-Ta alloys have been manufactured by arc melting technique using an Arcmelter equipment (Edmund Buhler GmbH), in Ar atmosphere, after the chamber was high-vacuumed to 3x10-5 mbar. Several re-melting were made to ensure the homogeneity of the resulting alloys. Heat treatments were performed for samples transversally cut from the as-cast buttons encapsulated in quartz tubes vacuumed at 3x10-5 mbar and sealed. The samples were heated at 1000 °C and maintained for 36 ks, followed by furnace cooling. Further on the vacuum sealed samples were heated at 900 °C and quenched in room temperature water by breaking the quartz tube in water.

Structural analysis was performed by X-ray diffraction in an X'Pert³ Powder (PANalytical) at room temperature. Monochromatic Cu Kα was used as radiation, with a 1.541840 Å wavelength, 40 mA power at 40 KV, 0,04 mm step with a 0,5 s/step, for a 20° – 90° 2θ range. Microstructural observations of the samples was made by scanning electron microscopy (SEM) in a TESCAN Vega 3 LM scanning electron microscope (SEM), equipped with a Bruker Quantax 200 Energy Dispersive X-ray Spectroscopy (EDX) system with Peltier-cooled XFlash 410M silicon drift detector.

3. Experimental results

The buttons made out of Ti-Ta alloys by arc melting showed a relatively good homogeneity following seven to eight re-melting and homogenization annealing. An analysis of the phase diagram corroborated with the EDS investigation of the composition of the alloy recorded on a sample cut from the button is shown in figure 1.

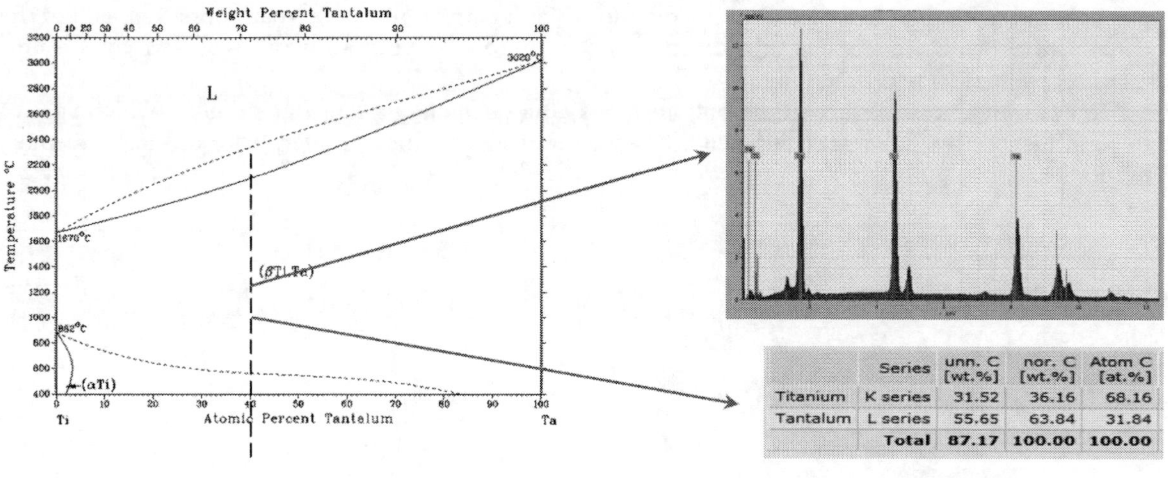

	Series	unn. C [wt.%]	nor. C [wt.%]	Atom C [at.%]
Titanium	K series	31.52	36.16	68.16
Tantalum	L series	55.65	63.84	31.84
	Total	87.17	100.00	100.00

Figure 1. Analysis of the alloy composition based on the phase diagram and the EDS results. (Ti-Ta phase diagram after Phase Diagrams of Binary Titanium Alloys, 1987).

The composition of the alloy falls in the range where the shape memory effect was reported for the Ti-Ta system. The microstructural observation of the free-solidification part of the as-cast button (figure 2) shows a dendritic structure (with dendrites and interdendritic spaces), typical for Ti-alloys [9]. According to the phase diagram (figure 1) an α+β structure is expected to be present at room temperature over a large compositional range in the as-cast alloy. Further EDS investigations showed that the solidification surface is Ti-rich, indicating that the Ti tends to segregate during the solidification process.

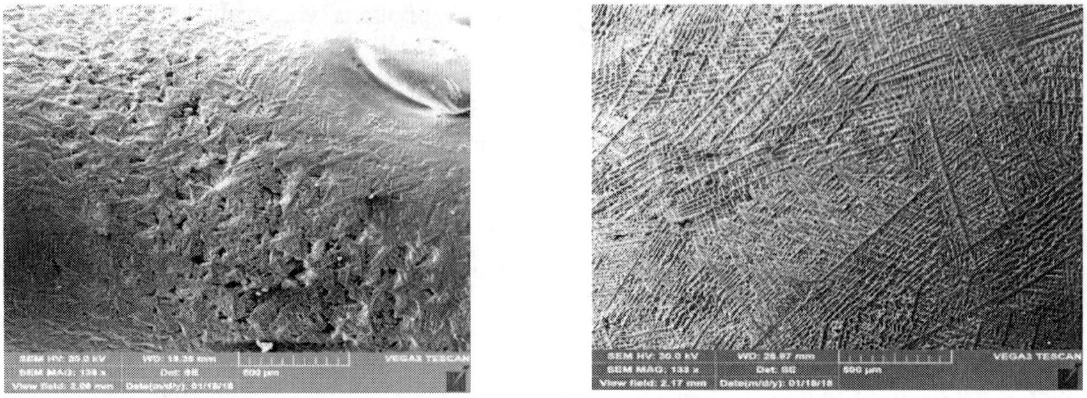

a) View of the as-cast solidification surface b) Area with denidritic structure

Figure 2. Microstructural aspects of the as-cast surface.

The martensitic transformation that appears in Ti-Ta shape memory alloy is influenced by the composition, with the Ms temperature tending to reach temperatures lower than room temperature for compositions with Ta content higher than 40%. Buttons quenched were analyzed by electron microscopy and X-ray diffraction.

The Ti-rich as-cast surface of the button shows after quenching a lamellar structure which appears to be influenced by the pre-existing dendritic structure and is of limited interest for the shape memory properties.

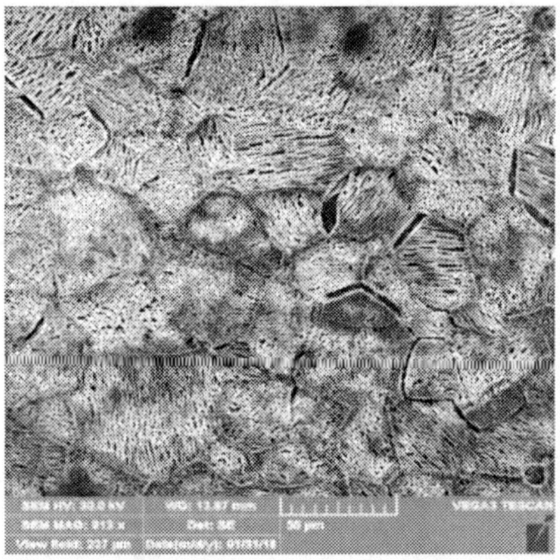

Figure 3. Detailed microstructural aspects of the as-cast surface.

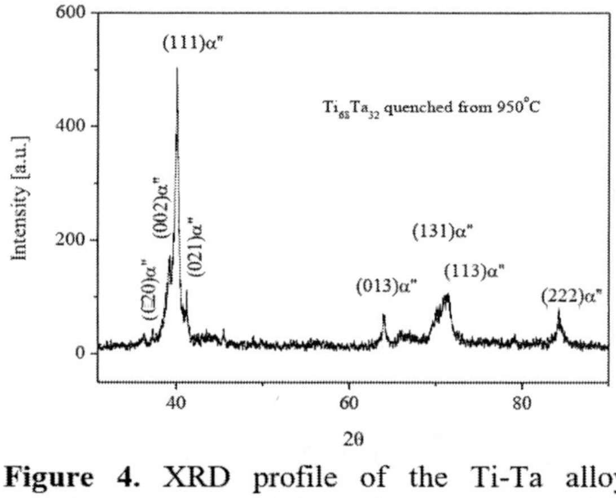

Figure 4. XRD profile of the Ti-Ta alloy solution treated at 900 °C and water quenched.

The quenching of the alloy leads to the formation of a martensitic structure that identified in the X ray spectra shown in figure 4. The structure is composed of α" martensite that was stabilized at room temperature by quenching and has an orthorombic crystal structure, which is consistent with the data reported in literature.

IOP Conf. Series: Materials Science and Engineering **416** (2018) 012017 doi:10.1088/1757-899X/416/1/012017

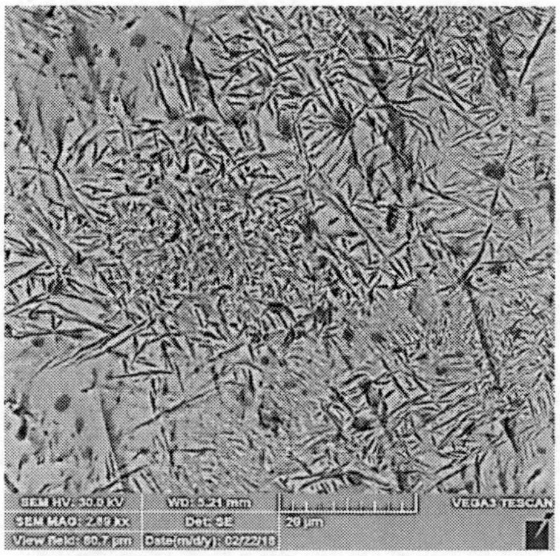

a) Martensite structure of the alloy

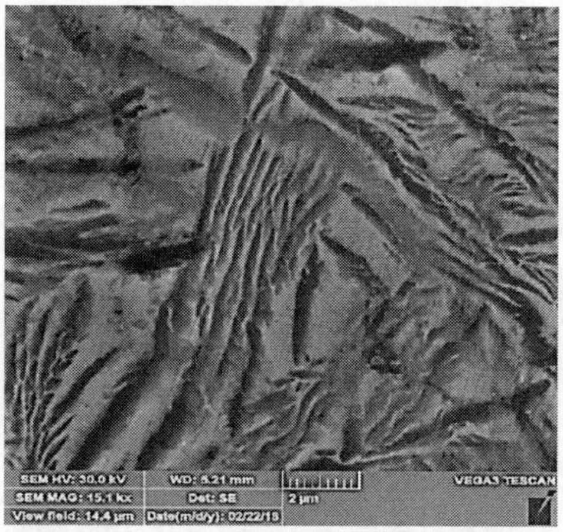

b) Details of the martensite structure

Figure 5. SEM image of the Ti-Ta alloy following quenching (center of the cross sectional sample).

The presence of the martensite in the microstructure is a premise for the existence of shape memory properties in the alloy.

4. Conclusions

The structure of a Ti-Ta alloy analyzed in the as-cast and quenched state reveals the formation of a martensitic structure following quenching from 900°C in water. The corresponding martensite is α" orthorombic as it resulted from the X-ray diffraction profile.

A segregation of Ti-rich phase was observed on the surface of the as cast sample, thus several re-melting are needed in order to obtain a homogenous composition throughout the thickness of the arc melted button.

5. References

[1] Kim H S, Kim W Y and Lim S H, 2006 *Scripta Mater.* **54** 887.
[2] Ping D H, Mitarai Y and Yin Y X, 2005 *Scripta Mater.* **52** 1287.
[3] Kim J I, Kim H Y, Hosoda H and Miyazaki S, 2005 *Materials Transactions.* **46** 852.
[4] Kim H Y, Hashimoto S, Kim J I, Inamura T, Hosoda H and Miyazaki S, 2006 *Mater. Sci. Eng. A* **417** 120
[5] Yamabe-Mitarai Y, Arockiakumar R, Wadood A, Suresh K S, Kitashima T, Hara T, Shimojo M, Tasaki W, Takahashi M, Takahashi S and Hosoda H, 2015 *Materials Today: Proceedings 2S* S517
[6] Shinohara Y, Tahara M, Inamura T and Miyazaki S, 2015 *Materials Transactions* **56(3)** 404.
[7] Xue D et al. 2016 *Sci. Rep.* **6**, 28244; doi: 10.1038/srep28244
[8] Abdul Wadood and YokoYamabe-Mitarai, 2014 *Materials Science and Engineering: A* **601** 106
[9] He G, Eckert J, Löser W and Schultz L, 2003 *Nature Materials* **2** 33

Acknowledgements

The authors acknowledge the support through the PIRSES-GA-2013-612585-MIDASInternational Research Staff Exchange Grant of the PEOPLEMARIE CURIE ACTIONSFP7-PEOPLE-2011-IRSES. Equipment for experiments acquired through the Romania-Republic of Serbia IPA Cross-border Cooperation Programme, Project MIS ETC no 1328 "Pole of Collaboration in New Functional Alloys".

Martensitic Transformation in Fe-Pd Ferromagnetic Shape Memory Alloy Wires

A Novac[*], R Sprincenatu, G Chilnicean, V Bolocan and C Craciunescu

Department of Materials and Manufacturing Engineering, Faculty of Mechanical Engineering, Politehnica University Timisoara, Bd. Mihai Viteazul 1, 300006 Timisoara, Romania

*Corresponding author: andreinovac2000@yahoo.com

Abstract. Shape memory alloys are materials with the functionality controlled by a temperature change that leads to a reversible transformation from austenite into martensite. When magnetic field can also be used to influence the functionality, either by magnetic induced transformation or by magnetic induced reorientation. Each of mechanism is analyzed in the paper and their prevalence in the ferromagnetic shape memory alloy systems is discussed. FePd alloys manufactured by arc melting are characterized in the as cast, wire drawn and heat treatment state in view of their integration into functional applications with thermal and magnetic control of the actuation.
Structural and microstructural analysis is used to assess the particularities of this alloys in view of their integration into sensor and actuator-based applications.

1. Introduction

Shape memory alloys are materials that are known to show shape recovery following a plastic deformation by heating the alloy above a certain temperature that is specific to the chemical composition of that alloy.

Ferromagnetic shape memory Heusler alloys, are receiving increasing attentions due to their multifunctional properties: magnetocaloric effect (MCE), magnetic-field-induced strain (MFIS), magnetoresistance, etc. Ferromagnetic shape memory alloys when exposed to a magnetic field they show shape memory effects. The deformation induced by being exposed to a magnetic field is called magnetoelastic deformation. Examples of FSMA are: Ni-Mn-Ga, Fe-Pd, Fe-Pt, Co-Ni-Ga etc [1-5].

Ferromagnetic shape memory alloys are well-known to exhibit short response time and large reversible strains owing to high stability of ferromagnetic martensite, modulated structure, high magnetocrystalline anisotropy (MCA) and high saturation magnetization. For high temperature applications, high TC is also an important requirement [5-6].

The mode of deformation in ferromagnetic shape memory alloys (FSMAs) is the movement of twin boundaries under the influence of an applied magnetic field. Strain in shape memory ferromagnets takes place either through nucleation and growth of favorably oriented twin variants or by the growth of already present variant which are oriented favorably to the applied magnetic field [7].

Long range ordered intermetallics obtained based on Fe or Co and noble metals like Pt or Pd exhibit high uniaxial magnetocrystalline anisotropy with potential large energy products, good mechanical and corrosion properties. They are attractive for both thin film device applications (high density magnetic recording media) and permanent magnets [8-9].

The current work uses meltsuction method following arc melting for the fabrication of ferromagnetic shape memory alloy rods. The focus is on Fe-Pd shape memory alloys that could be used for actuators, valves, dumpers and biomedical applications.

2. Experimental details

Melt-suction was the method used to produce the Fe-Pd alloy, with this method the arc melted alloy is cast by suction into a mold channel due to chambers pressure differences, the alloy is then fast cooled when it gets in contact with the walls of the mold that are water-cooled (this leads to fast cooling of the alloy).

Fe-Pd shape memory alloys were manufactured by melt-suction, the alloy was melted in an arc melting equipment (Edmund Buhler GmbH), followed by melt-suction casting in a copper mold, resulting in a 3 mm diameter rod. The arc melting was performed in Ar atmosphere, after the chamber was high-vacuumed to $3x10^{-5}$ mbar. Several re-melting were made to ensure the homogeneity of the resulting alloys. The rod obtained by melt-suction was vacuum sealed at $3x10^{-5}$ in a quartz tube and was homogenized at 1000 °C for 10 hours, followed by furnace cooling. The homogenized samples were further annealed for 15 minutes at 1000 °C and quenched in room temperature water by breaking the quartz tube in water.

The composition and the microstructure of the samples used in the experiments was investigated in a TESCAN Vega 3LM electron microscope, equipped with a Bruker Quantax 200 Energy Dispersive X-ray Spectroscopy (EDX) system with Peltier-cooled XFlash 410M silicon drift detector.

The martensitic phase transformation of the manufactured rod was studied by the 4 wires electric resistivity method in a Picotest multimeter, on cooling, using liquid nitrogen as cooling agent. Further investigations of the martensitic transformation were performed in an Olympus optical microscope, equipped with a Peltier heating and cooling system (Figure 1).

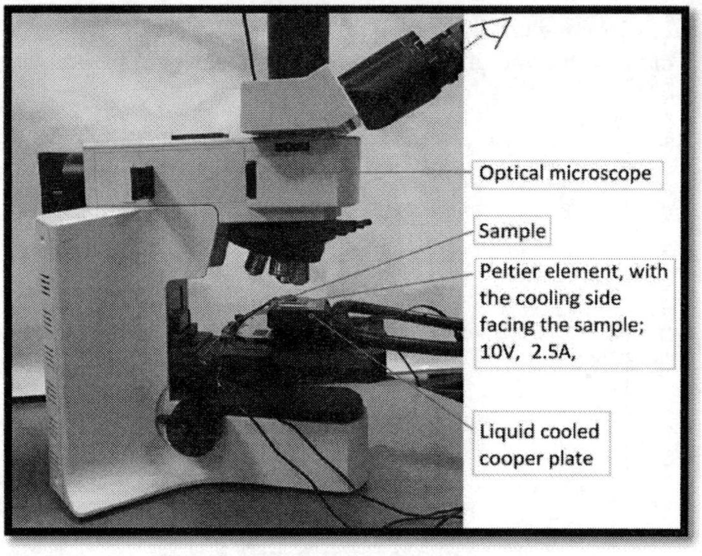

Figure 1. The martensitic transformation of the Fe-Pd alloy was observed using the optical microscope, the sample was placed on the cold side of a Peltier element, that was placed on a liquid cooled cooper plate.

3. Experimental results and dicussions

The EDX analysis of the alloy (in Figure 2a) indicated a composition of 68.26 at.% Fe and 31.74 at.% Pd, putting the alloy in the compositional range of the Fe-Pd where the shape memory properties have been reported and where the ordered FePd face centered tetragonal structure can be stabilized.

IOP Conf. Series: Materials Science and Engineering **416** (2018) 012018 doi:10.1088/1757-899X/416/1/012018

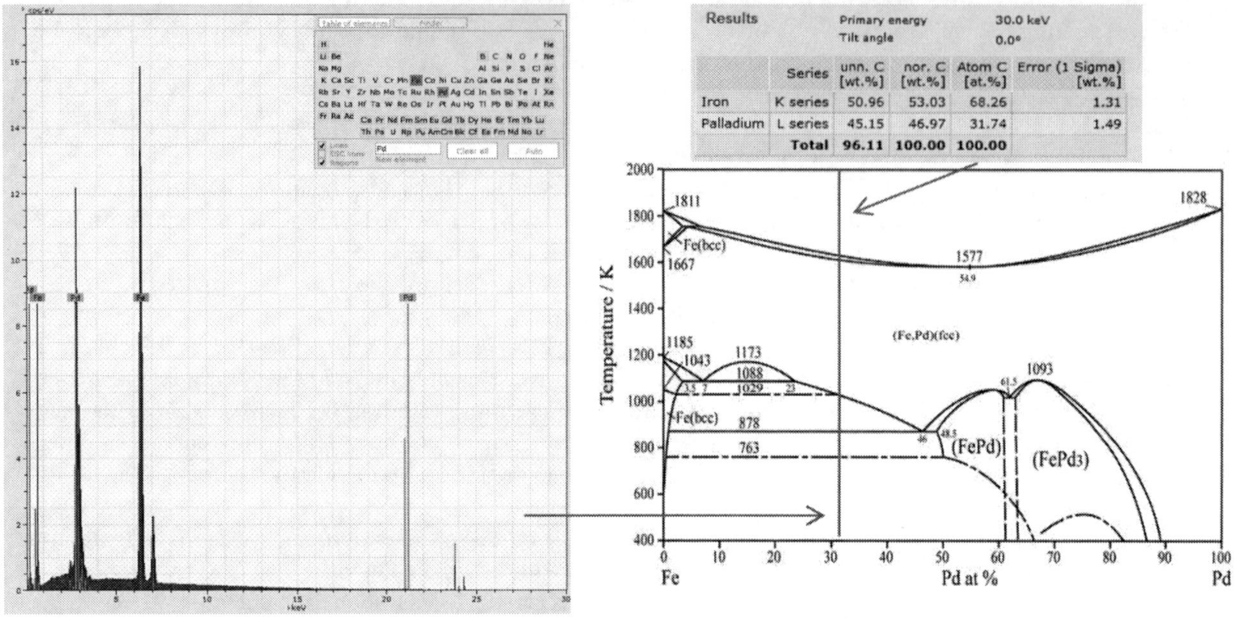

a) EDX compositional analysis of the Fe-Pd sample b) EDX results (top) and Fe-Pd phase diagram [10] (bottom)

Figure 2. Analysis of the compositional range of the Fe-Pd shape memory alloy.

Further investigations by X-ray diffraction at room temperature allowed us to identify the corresponding peaks that set the structure into a predominantly austenite and a small fraction of martensite (Figure 3).

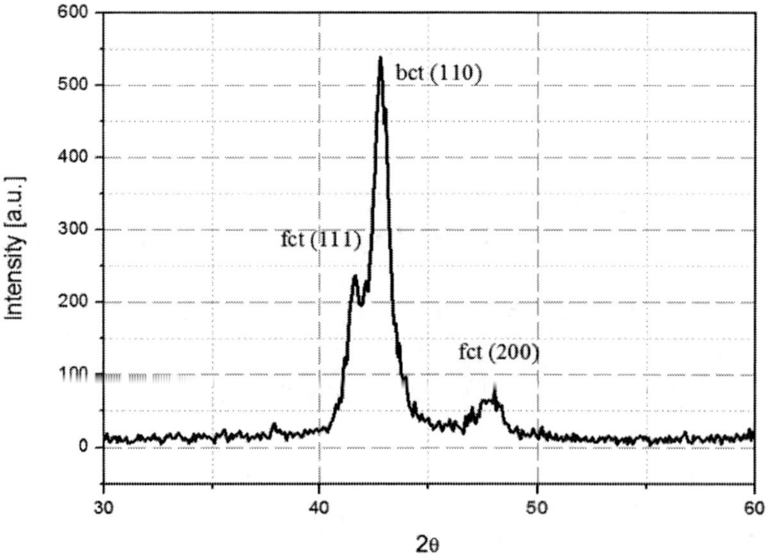

Figure 3. X ray diffraction data for the Fe-Pd alloy.

The X-ray diffraction observations were further confirmed by the temperature dependent optical microscopy, where the phase transformation was observed and recorded, with a typical martensitic structure observed at a temperature of -10°C vanishes on heating at +5 °C.

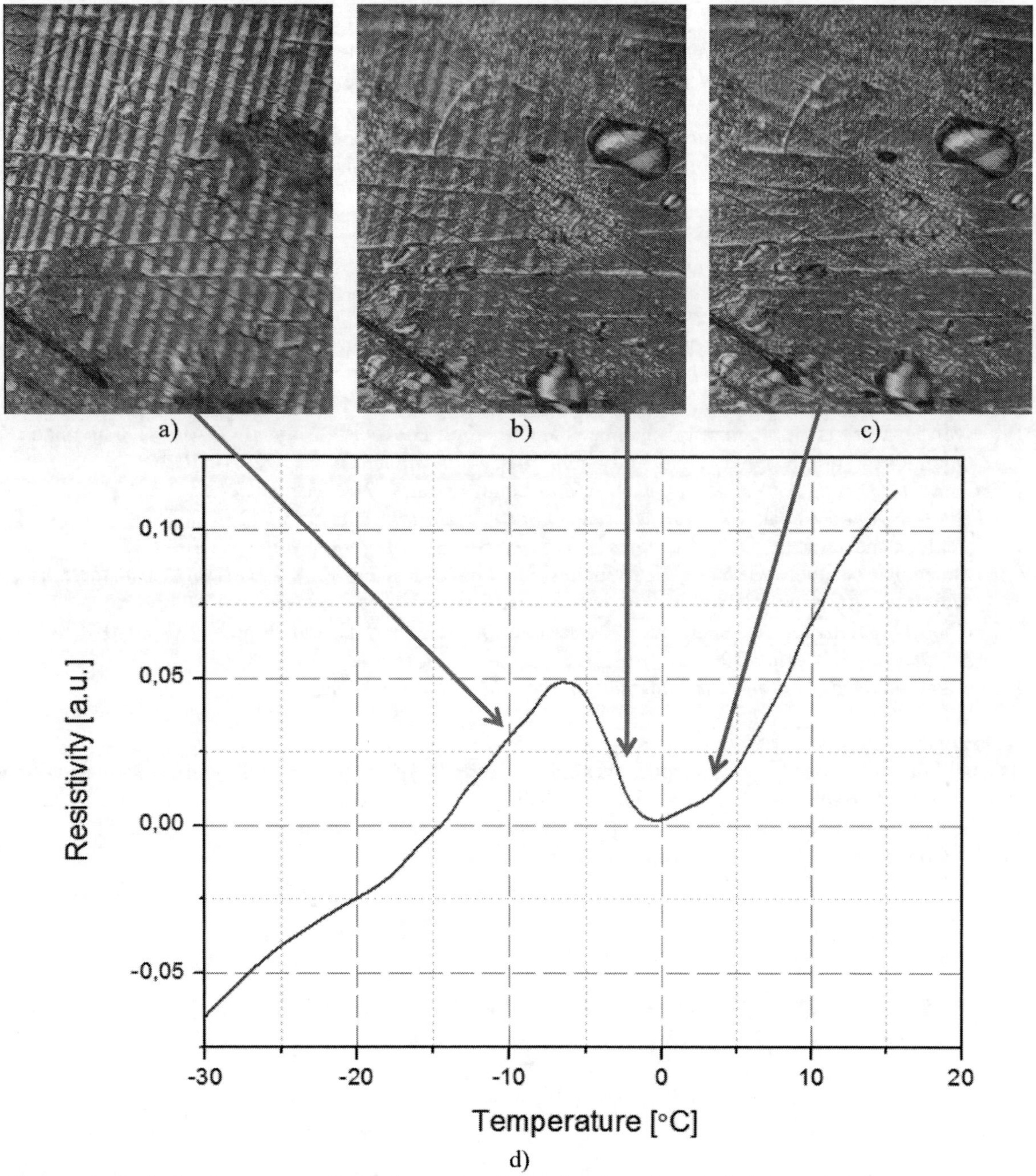

Figure 4. Optical microscopy of the phase transformation (a - c) and electric resistivity measurement as a function of temperature for the Fe-Pd sample (d).

A further confirmation of the presence of the martensitic transformation was obtained from the electric resistivity measurements as shown in figure 4d, with the martensite and austenite slopes defining the transformation in the same range where the microstructural observations located it.

4. Conclusions

Fe-Pd alloys manufactured by melt suction as 3 mm diameter rods in the compositional range 68.26 at% Fe and 31.74 at. % Pd were austentic at room temperature and showed a martensitic phase transition upon cooling below 10°C.

The martensitic transformation on heating lead to the disappearance of the martensitic plates formed on cooling by heating above 5°C. The observations were also confirmed by electric resistance measurements.

5. References

[1] Lin H P, Chen D and Kuo J C 2015 *Materials 2015* **8(6)** 3254
[2] Vlasova N I, Popov A G, Shchegoleva N N, Gaviko V S, Stashkova L A, Kandaurova G S and Gunderov D V 2013 *Acta Materialia* **60(7)** 2560
[3] Wada T, Tagawa T and Taya M 2003 *Scripta Materialia* **48** 207
[4] Sven H, Markus G, Stephan I, Buschbeck J, Christoph B, Kock I, Mayr S G, Alan S, Thienhaus S, Quandt E, Fähler S, Peter E and Ludwig A 2010 *Acta Materialia* **58** 5949 10.1016
[5] Cong D Y, Wang S, Wang Y D, Ren Y, Zuo L and Esling C 2008 *Mater Sci Eng A*, **473** 213
[6] Khalid F A and Abbas S Z 2011 *Materials characterization* **62** issue 12 1134
[7] Tickle R and James R D 1999 *Journal of Magnetism and Magnetic Materials* **195** 627
[8] Skuza J R, Clavero C, Yang K, Wincheski B and Lukaszew R A 2010 *IEEE Transactions on Magnetics* **46** issue 6 1886
[9] Chbihi A, Sauvage X, Genevois C , Blavette D, Gunderov D and Popov A G 2010 *Advanced Engineering Materials* **12**
[10] Okamoto H 1993 *Materials Information Soc. Materials Park* Ohio.

Acknowledgements

The authors acknowledge the support through the PIRSES-GA-2013-612585-MIDAS International Research Staff Exchange Grant of the PEOPLE MARIE CURIE ACTIONS FP7-PEOPLE-2011-IRSES. Equipment for experiments acquired through the Romania-Republic of Serbia IPA Cross-border Cooperation Programme, Project MIS ETC no 1328 "Pole of Collaboration in New Functional Alloys".

Shape Memory Alloy Films Manufactured by Magnetron Sputtering

V Bolocan*, R Sprincenatu, A Novac, G Chilnicean and C Craciunescu

[1]Department of Materials and Manufacturing Engineering, Faculty of Mechanical Engineering, Politehnica University Timisoara, Bd. Mihai Viteazul 1, 300006 Timisoara, Romania

*Corresponding author: vlad.bolocan@yahoo.ro

Abstract. By transferring the functional properties from bulk alloys to thinner structures in the range of micrometers it becomes possible to manufacture devices that can be used in micro-opto-electro-mechanical systems. Shape memory alloy possess a wide range of functional properties among which the shape memory effect, superelasticity and the high damping are influenced by the martensitic transformation. The current paper shows the results obtained on manufacturing shape memory alloy films by magnetron sputtering and the analysis of the martensitic transformation in the films attached to the substrate.

1. Introduction

The deposition of shape memory alloy films is a technique that is considered for both the fabrication of sensors and actuators [e.g. 1], as well as for the exploration of new and optimized composition based on combinatorial exploration [2]. The common substrate used for the deposition is silicon or glass. Glass is readily available and inexpensive. Over the time, several other substrates have been explored in order to assess the possible use in various applications, based on shape memory alloy films deposited on substrates, such as metallic substrates [3] or substrates / photoresist films that can be dissolved following the deposition or put of which the film may be removed, leading to freestanding films [4].

In most cases, the shape memory alloy films deposited at room temperature need to be annealed in order to show the shape memory effect. In these cases an annealing temperature in the range of 600°C or higher may be needed. For such annealing temperatures it is difficult to find adequate polymeric substrates. Another way is to deposit the films on heated substrates, a technique requiring lower heating temperatures for the substrate. It was shown that for NiTi films deposited on Si substrates, the crystallization starts at temperatures above 250-300°C [5] and for Ni-Mn-Ga freestanding films the sequence of the occurrence of ferromagnetic above and shape memory properties occurs during crystallization by heating.

Ishida et al. [6] was successful in depositing crystalline 8 μm films out of Ti, Ni and Cu elemental targets on heated polyimide substrates, using a carrousel-type magnetron-sputtering system, thus opening the path for the use of such substrates for manufacturing shape memory alloy thin film actuators. The peeled-off films out of glass substrates also showed the B2-B19 reversible transformation.

IOP Conf. Series: Materials Science and Engineering **416** (2018) 012019 doi:10.1088/1757-899X/416/1/012019

The aim of this work is to further explore the possibility to analyze the deposition of shape memory alloy films Ni-Ti-Cu films using alloyed target and to study the occurrence of the martensite in the as-deposited and further annealed films, respectively.

2. Experimental details

A 2" Ti-Ni-Cu target was used for depositing films using a custom made DC magnetron sputtering equipment on glass substrates, resulting in bimorph film/substrate structures. The deposition was made at room temperature (RT) with the following parameters: $2x10^{-6}$torr preliminary vacuum, 10 mtorrAr pressure and 100W power, for a total duration of 10.8 ks. One set of samples was further on annealed at 400°C in high vacuum by placing the glass substrate with the film on it on a K.J.Lesker ultra high vacuum heater module. The temperature was controlled with an OMRON temperature controller. The thickness of the film deposited was measured with a Sloan-Dektak-IID-Profilometer with 5Å vertical resolution.

The composition of the target and the film was analyzed in TESCAN Vega 3LM electron microscope, equipped with a Bruker Quantax 200 Energy Dispersive X-ray Spectroscopy (EDX) system with Peltier-cooled XFlash 410M silicon drift detector. Structural analysis was performed by X-ray diffraction in an X'Pert³ Powder (PANalytical) at room temperature. Monochromatic Cu Kα was used as radiation, with a 1.541840 Å wavelength, 40 mA power at 40 KV, 0,04 mm step with a 0,5 s/step, for a 20° – 55° 2θ range.

3. Results and discussions

The as deposited films had a measured thickness in the range of 1μm, as detailed in Figure. 1a, resulting in a deposition rate of about 1Å/s.

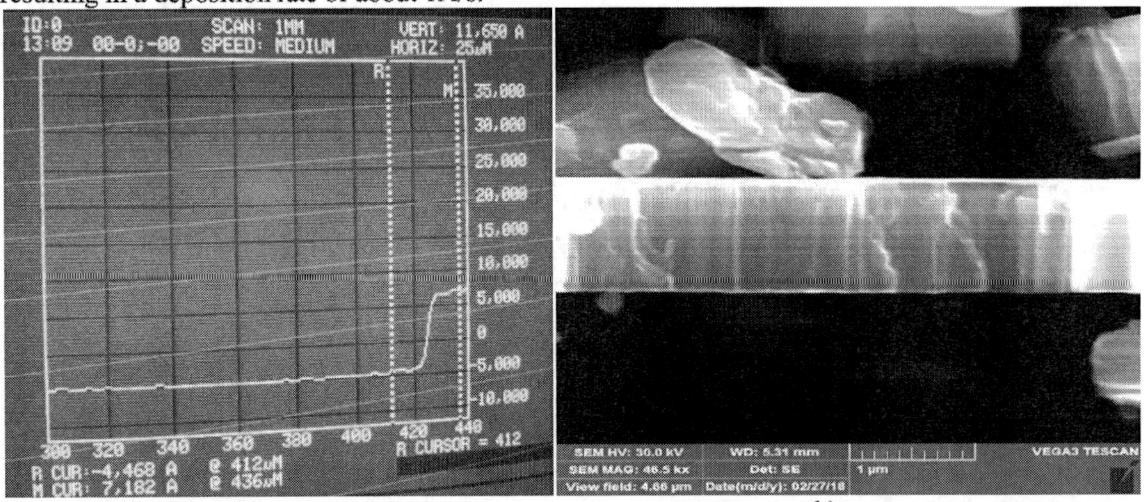

a) b)

Figure 1. Crosssectional details of the Ni-Ti-Cu films deposited at RT on glass substrate. a) Film thickness profile measured with the stylus profiler, b) Microstructure of the film deposited art RT showing a columnar growth.

The thickness was also confirmed by microstructural analysis of the cross section of the film detached from the substrate. A typical columnar growth is also observed as well as the uniformity of the film thickness.

The analysis of the composition of the film deposited at RT (Figure 2a) shows a composition that has about 12.5 at % Cu, thus positioning the alloy in the compositional range of the structural phase diagram (Figure 2b) where orthorhombic martensite (B19) is present.

IOP Conf. Series: Materials Science and Engineering **416** (2018) 012019 doi:10.1088/1757-899X/416/1/012019

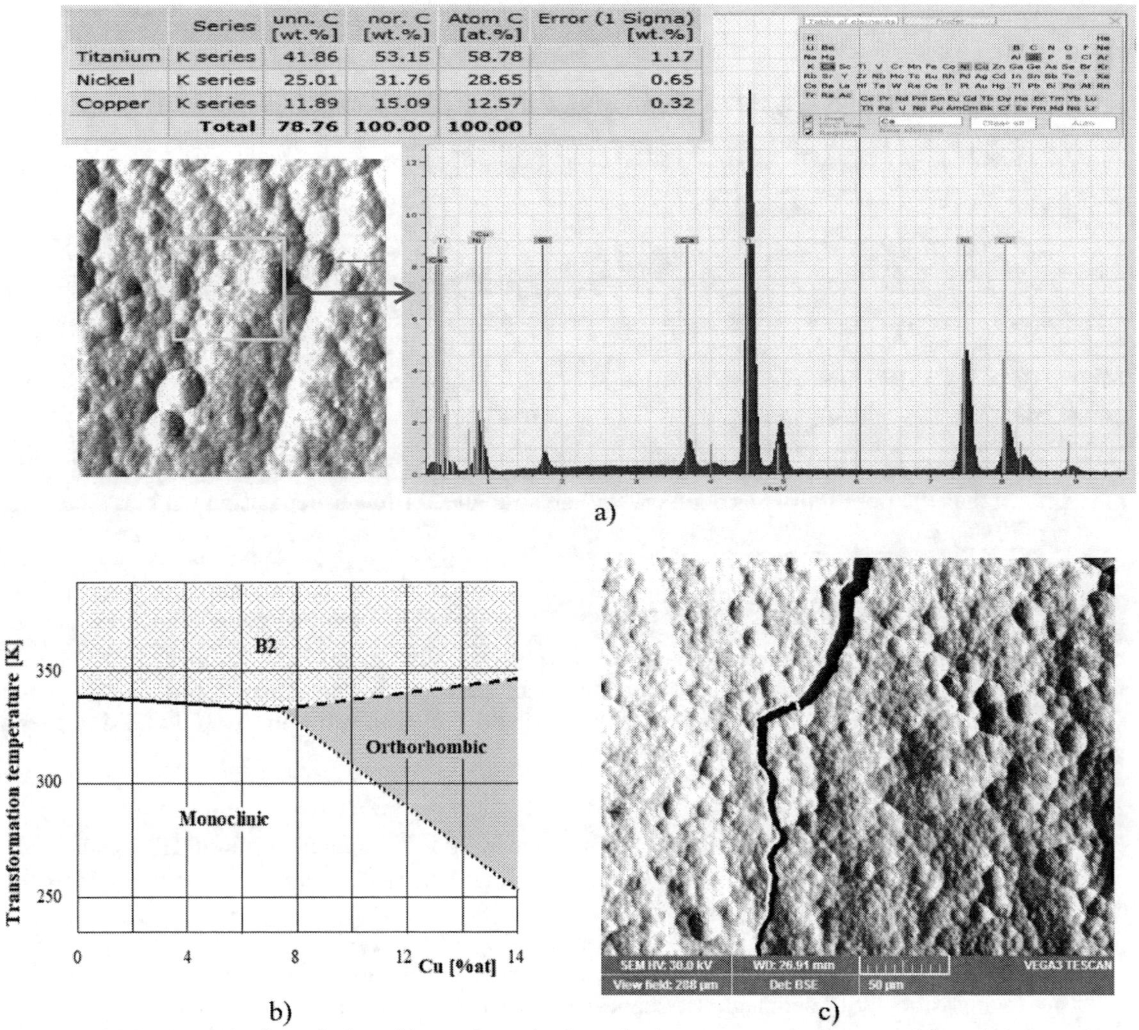

	Series	unn. C [wt.%]	nor. C [wt.%]	Atom C [at.%]	Error (1 Sigma) [wt.%]
Titanium	K series	41.86	53.15	58.78	1.17
Nickel	K series	25.01	31.76	28.65	0.65
Copper	K series	11.89	15.09	12.57	0.32
	Total	78.76	100.00	100.00	

Figure 2. EDX analysis of the films deposited and the corresponding position in the phase diagram(structural diagram redrawn after [7]), a) EDX results of the Ti-Ni-Cu film, b) Structural diagram for Ni-Ti-Cu alloys, c) Surface of the as-deposited film.

Depending on the Cu content, the Ti-Ni-Cu shape memory alloys show different paths for phase transitions from the B2 high temperature phase. According to Figure 2b, for contents below7.5 at% Cu the phase transformation is B2↔B19' (monoclinic martensite), while for higher contents the transition sequence is B2↔B19'↔B19 (orthorhombic martensite).

The formation of the orthorhombic martensite is further confirmed by the X-ray diffraction data (Figure 3), where B19 peaks are identified.

In fact, the presence of the B19 martensite at room temperature is justified by the fact that the B2↔B19' takes place in a temperature range above 300K for 12 at% Cu content.

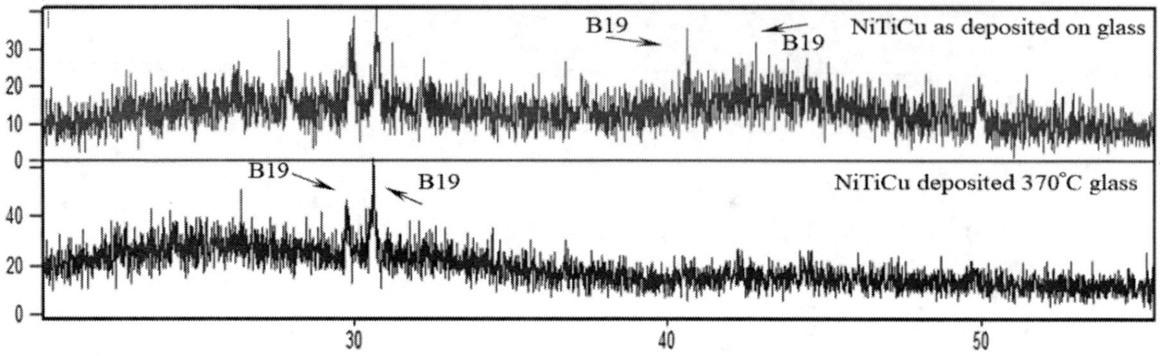

Figure 3. X-ray diffraction profiles for the Ti-Ni-Cu films deposited at RT (top) and subsequently annealed at 370°C.

Also, it is observed that even the film deposited at RT is at least partially crystalline and not amorphous, thus the possibility to obtain crystalline structure for lower deposition temperatures exists.

4. Conclusions

The EDX analysis of the film compositions and of the target shows changes that occurred during the plasma transfer from the target to the substrate.

Films deposited at room temperature showed to be partially crystalline – as investigated by room temperature X-ray diffraction - with B19 orthorhombic martensite and did not show the presence of B19', expected to appear at lower temperature for 12at% Cu. Annealing up to 370°C did not lead to significant changes in the crystallinity of the film.

5. References

[1] Fu Y, Huang W, Du H, Huang X, Tan J and Gao X 2001 Characterization of TiNi shape memory alloy thin films for MEMS applications *Surface & Coatings Technology* **145** 107-112.

[2] Takeuchi I, Famadou O O, Read J C, Aronova M A, Chang K S, Craciunescu C, Lofland S E, Wuttig M, Wellstood F C, Knauss L and Orozco A 2003 Identification of novel compositions of ferromagnetic shape-memory alloys using composition spreads *Nature Materials* **2** 180-184.

[3] Winzek B and Quandt E 1999 Shape-memory Ti-Ni-X-Films (X=Cu, Pd) under constraint *Zeitschrift fuer Metallkunde* **90** 796-802.

[4] Rumpf H, Craciunescu C M, Feydt J, Gilles A, Wuttig M, and Quandt E 2003 Fabrication and characterization of free-standing NiMnGa thin films *Materials and Devices for Smart Systems, Materials Research Society* **785**

[5] Crăciunescu C M, Li J and Wuttig M 2003 Thermoelastic Stress-Induced Thin Film Martensites *Scripta Materialia* **40/1** 65-70.

[6] Ishida A and Martynov V 2002 Sputter-deposited shape-memory alloy thin films: properties and applications *Materials Research Society Bulletin* **27** 111–114.

[7] Miyazaki S, Komatsubara K and Otsuka K 1998 Collected Abstracts of Annual Fall Meeting of Japan Institute of Metals Sapporo **408**.

Acknowledgements

The authors acknowledge the support through the PIRSES-GA-2013-612585-MIDAS International Research Staff Exchange Grant of the PEOPLE MARIE CURIE ACTION SFP7-PEOPLE-2011-IRSES. Equipment for experiments acquired through the Romania-Republic of Serbia IPA Cross-border Cooperation Programme, Project MIS ETC no 1328 "Pole of Collaboration in New Functional Alloys".

IOP Conf. Series: Materials Science and Engineering **416** (2018) 012020 doi:10.1088/1757-899X/416/1/012020

Mechanical Properties of Fe/Zr Dual Amorphous Phase Bulk Metallic Glasses

M Vodă[1], C Codrean[1,*], D Toşa[2] and V A Şerban[1]

[1] Politehnica University Timişoara, Piaţa Victoriei, No.2, 300006, Timişoara, Romania
[2] ELBA SA, 135 Paul Morand, Freidorf Industrial Park Timişoara, Romania

*Corresponding author: cosmin.codrean@upt.ro

Abstract. Dual amorphous phase bulk metallic glasses (DAPBMGs) are a new class of bulk metallic glasses (BMGs) containing several amorphous phases in order to bring together all the favorable properties of each phase. A Fe/Zr dual phase bulk metallic glasses of cylindrical form with the diameter of 10 mm and 5 mm in height, were successfully prepared by hot-pressing of Fe – based and Zr – based glassy alloy powder at different mixing times. The samples obtained were structural investigated by X-Ray diffraction and mechanically characterized by hardness and compression tests. It was found that increasing the mixing time of powders leads to a decrease in hardness and mechanical compressive strength.

1. Introduction

The bulk amorphous metallic materials with dual phases are characterized by the presence of two distinct metallic phases with amorphous structures, exhibiting a reunion of properties specific to each phase and thus attracting a great interest in the prospects of obtaining new materials [1]. The dual amorphous phase bulk metallic glasses (DAPBMGs), similar to a composite material, are expected to exhibit the properties of both amorphous alloys of which they are composed.

In the case of Fe / Zr dual phase bulk amorphous metallic materials, the Fe-based amorphous alloy has good magnetic properties, high hardness, and high compressive strength but has a low glass forming ability and high brittleness [2]. By introducing high-toughness Zr-based amorphous alloy, with excellent glass forming ability, it is desirable that this dual amorphous alloy retain its magnetic properties but also to improve toughness and ductility [3].

A viable method for obtaining dual bulk amorphous alloys is powder metallurgy [4]. It involves the following steps [3, 4]: (i) obtaining amorphous metal powders by atomization or milling; (ii) mixing of this two glassy alloy powders by mechanical milling for homogenization; (iii) compacting of the mixed glassy alloy powder in the supercooled liquid region using hot- pressing technique.

In this paper, has been studied the influence of mixing time, and thus the homogenization degree, on the mechanical properties of a Fe/Zr-based dual bulk amorphous alloy.

2. Experimental procedures

The glassy powders of the $Fe_{74}Mo_4P_{10}C_{7.5}B_{2.5}Si_2$ and $Zr_{52.5}Cu_{17.9}Ni_{14.6}Al_{10}Ti_5$ were mixed in a volume ratio of 50%, using RETSCH PM400 planetary ball mill at 100 rpm for 1h, 10h, and 20h with a ball-to-powder mass ratio of 10:1. Subsequently, the powders were compacted by hot pressing in the

WEBER-PRESSEN pressing device. The hot pressing was performed in the supercooled liquid region, at 420°C under an applied force of 60 kN, for 10 minutes.

The amorphous structure of the samples was identified by X-ray diffraction with an X'Pert³ Powder diffraction system, with the radiation of a Cu anode with a wavelength λ = 1.54 Å.

The mechanical properties were determined by Vickers method hardness test and compressive test. The hardness tests were performed using Volpert Micro - Vickers Hardness Tester with a 0.3 kgf load. The compressive tests were done at room temperature, at a loading speed of 1mm/min on a Zwick/Roell- machine.

3. Results and discussion

The hot pressed samples of 10 mm diameter and 5 mm in height are shown in Figure 1. The sample obtained after 1h mixing time was marked with S1, the sample obtained after 10h mixing time was marked with S2 and the sample obtained after 20h of mixing was marked with S3.

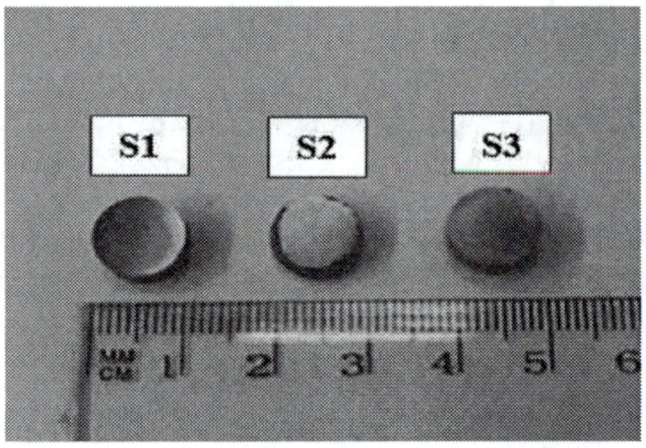

Figure 1. The hot pressed samples mixed with different times.

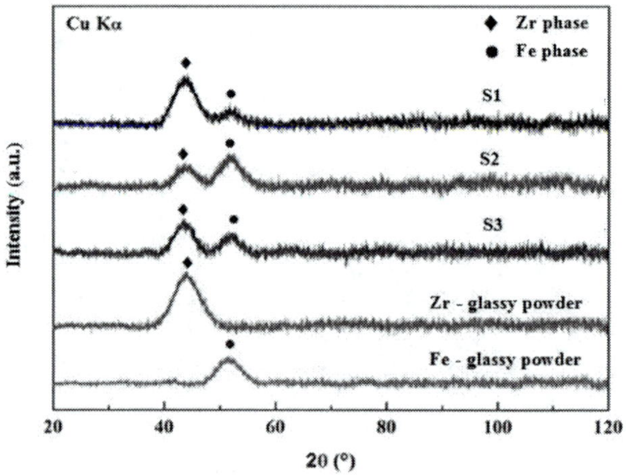

Figure 2. XRD patterns of the glassy powders and of the hot pressed samples.

The samples were subjected to structural analysis by X-ray diffraction both the Fe and Zr based glassy powders and the samples obtained by hot pressing.

The XRD patterns show that the amorphous structure is preserved also in compacted powders even after longer mixing times (Figure 2). It is also observed that in the case of hot pressed samples, the two amorphous phases are clearly distinguishable: one Fe-based amorphous phase and one Zr-based amorphous phase.

Therefore, X-ray diffraction analysis certifies the obtaining of a dual bulk amorphous alloy, which combines two distinct metallic amorphous phases, thus it can be considered a composite material.

The hardness tests were performed both on the Fe phase and on the Zr phase. The imprints left by the indenter on the surface of each examined sample are shown in Figure 3. Five tests were performed on each sample and the averages of the resulting values are listed in Table 1.

It can be seen that the Fe - phase has a higher hardness than the Zr - phase. It is also found that with increasing of powder mixing time, the hardness of the Fe - phase decreases and the hardness of the Zr - phase increases.

This can be explained by the fact that with increasing mixing time increase the homogeneity of the material and, after hot pressing of the mixed powder, diffusion processes occur that lead to the homogenization of the structure.

At the same time, the hardness of the material, calculated as the average hardness measured on the two phases, decreases with increasing mixing time as a result of the tendency to homogenize the structure.

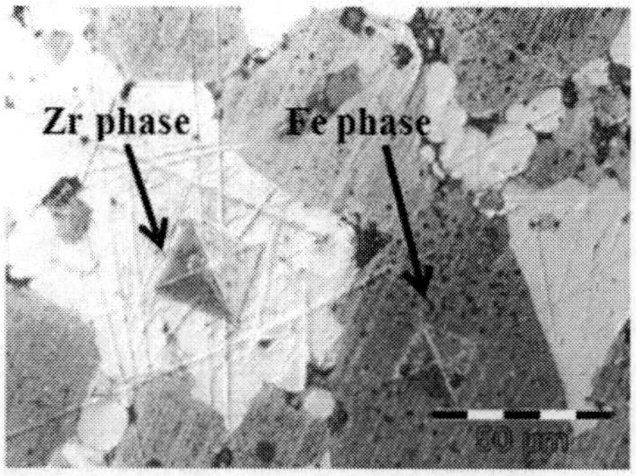

Figure 3. The imprints left by the indenter on the dual bulk amorphous alloy.

Table 1. The hardness values obtained by the Vickers method.

Sample	HV0.3 on Fe-phase	HV0.3 on Zr-phase	Average
S1	724.5	423.1	573.8
S2	697.1	424.9	561
S3	672.9	435.3	554.1

Figure 4 presents the compressive stress-strain curves of the hot pressed samples. The results of the compression tests are shown in Table 2.

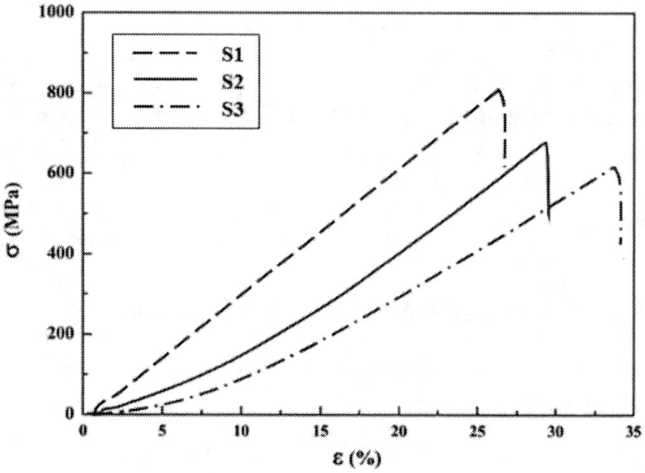

Figure 4. The stress-strain curve of the hot pressed samples.

Table 2. Compressive strength of the hot pressed samples.

Samples	**S1**	**S2**	**S3**
Compressive strength, σ_f [MPa]	824	683	612

It can be observed that unlike Fe - based amorphous bulk alloys, which have a high compressive strength (over 2000...2500 MPa) and a reduced deformation (below 2...3%) [5, 6], the amorphous alloys have a lower compressive strength (under 1000 MPa), but a higher elasticity, the deformation exceeding 30%. The presence of the Zr –based amorphous phase led to a significant increase in material elasticity, but also to a decrease in compressive strength.

It can be seen that with increasing mixing time, the fracture strength decreases from 824 MPa, corresponding to the sample with a mixing time of 1h to 612 MPa, corresponding to the sample with a mixing time of 20h. Thus, by increasing the mixing time and implicitly the degree of homogenization of the structure, the influence of the Zr- based amorphous phase on the mechanical properties of the dual amorphous alloy is more pronounced.

4. Conclusions

Fe/Zr dual phase bulk metallic glasses of cylindrical form with the diameter of 10 mm and 5 mm in height have been successfully obtained by hot-pressing method.

With increasing the mixing time of Fe and Zr amorphous powders, decreases the hardness and compressive strength of the dual phase bulk amorphous alloy, but increases the elasticity. Consequently, augmentation the degree of homogenization by increasing the mixing time leads to a magnification of the Zr – amorphous phase influence on the mechanical properties of the dual phase bulk amorphous alloy.

Therefore, the mixing time of the amorphous powders in order to obtain the dual phase amorphous alloys can be optimized so that, on the one hand to preserve the amorphous structure and on the other hand to obtain the desired mechanical properties.

5. References

[1] Li F, Zhang T, Guan S and Shen N 2005 A novel dual-amorphous-phased bulk metallic glass with soft magnetic properties *Materials Letters* **59** 1453–7

[2] Inoue A and Takeuchi A 2002 Recent Progress in Bulk Glassy Alloys *MATERIALS TRANSACTIONS* **43** 1892–906

[3] Toşa I, Serban V.A., Codrean C., Buzdugan D., Stoica M., Dual phase bulk metallic glasses fabricated by hot pressing using two different types of glassy alloy powder, *Journal of Optoelectronics and Advanced Materials* **17** 7-8

[4] Lee P-Y, Chung S-T, Lin C-K, Tsai W-T and Lin H-M 2007 Preparation and Characterization of Dual-Phase Bulk Metallic Glasses through Powder Metallurgy Route *MATERIALS TRANSACTIONS* **40** 1595–9

[5] Suryanarayana C and Inoue A 2013 Iron-based bulk metallic glasses *International Materials Reviews* **58** 131–66

[6] Stoica M 2017 *Fe-based bulk metallic glasses: understanding the influence of impurities on glass formation* (Wiesbaden: Springer)

Acknowledgements

Acknowledgements to Leibniz-Institute for Solid State and Materials Research (IFW) Dresden, Institute for Complex Materials which has provided the raw materials and required equipment for this research.

Simulation of Thermal Field in Bulk Amorphous Steels

C Codrean, B Radu, D Buzdugan* and C Opriş

Politehnica University Timişoara, Piaţa Victoriei, No.2, 300006, Timişoara, Romania

*Corresponding author: dragos.buzdugan@upt.ro

Abstract. Bulk amorphous steels (BASs) have shown great potential for some structural applications due to very attractive physical, thermal and mechanical properties. This paper presents some results for numerical simulation of the thermal field developed during rapid cooling of Fe-Cr-Co-Mo-Y-C-B alloy, to obtain bulk amorphous steels. Rapid cooling technology gives the possibility to obtain bulk metallic alloys with an amorphous structure under conventional industrial conditions, using, for example, commercial-grade raw materials, low vacuum, conventional casting techniques, etc. Once the model being validate, it can be used to optimize the rapid cooling process and casting mold geometry to be able to investigate the possibility to obtain larger bulk amorphous steel components.

1. Introduction

The increasing use of amorphous alloys is due to the combinations of excellent properties: hardness, mechanical strength and ductility superior to metallic crystalline materials, high magnetic permeability and low coercive field, electrical resistivity almost independent of temperature and very high corrosion resistance. Therefore, in the last decades, the research has been focused on producing amorphous alloys with thicknesses greater than 1 mm, also called bulk amorphous alloys.

Compared to most other bulk amorphous alloys, such as Zr or Pd based alloys, Fe-based bulk amorphous alloys offer several important advantages: much lower raw material costs, higher mechanical strength, better corrosion resistance, and greater thermal stability [1, 2, 3]. A major obstacle to the development of Fe-based bulk amorphous alloys is their low glass forming ability (GFA) [1, 4, 5]. Also, another disadvantage of the Fe-based bulk amorphous alloys is their relatively high fragility.

The first bulk Fe-based amorphous alloy with the chemical composition $Fe_{73}Al_5Ga_2P_{11}C_5B_4$ was developed in 1995 by a group of Inoue-led researchers [6, 7] using high purity components. Recent research has shown the possibility of obtaining so-called bulk amorphous steels (BAS), which are actually C containing bulk amorphous Fe-based alloys, by melting pure components in a vacuum electric arc furnace and casting in the form of cylindrical rods in a copper mold in a protective atmosphere of argon [8, 9, 10, 11]. These BASs are very attractive materials for making, lighter automobiles, tall buildings, corrosion-resistant coatings, ship hulls and surgical instruments, in the future [8].

The GFA increase of Fe-based alloys was achieved by adding to their chemical composition yttrium and / or rare earths [1, 9, 10, 11]. Therefore, the aim of this work is to evaluate the cooling rate by thermal field analysis of Fe-Cr-Co-Mo-Y-C-B alloy in form of disc quenched in a water cooled copper mold.

2. Simulation

Multicomponent $Fe_{48}Mo_{14}Cr_{11}Co_4B_6P_8Si_7Y_2$ BAS in disk form were elaborated by water cooled copper mold casting using mostly ferroalloys and pure elements. The elaborated samples were structurally analyzed by X-ray diffraction using an X'Pert³ Powder diffraction system with a Cu anode. Finite Element Analysis method was used to investigate the thermal field evolution during copper mold casting in order to estimate the cooling rate needed to obtain an amorphous structure. A 3D geometric model of the Cu mold and disk were used for the transient analysis. The Cu mold, having a diameter of 120 mm and 60 mm in length, is made up of two symmetrical parts which are screwed together, thus resulting the cavity with a diameter of 10 mm and thickness of 2 mm. The boundary conditions are the following:

- the melt temperature is 1150 °C and the initial mold temperature is 22 °C
- the melt already fills the mold cavity at the initial moment
- all the surfaces separating the mold and the disk from the outside environment have been considered as surfaces through which the heat is transferred by convection into the external environment
- the contact surface between the cast alloy (disk) and the mold was considered a surface, through which heat is transferred by conduction, as a result of contact between the cast alloy and the copper mold.
- the mold is cooled with water flowing through the channels and the heat is transferred by the forced convection.

3. Results and discussion

The BAS disk obtained by water cooled copper mold casting is presented in Figure 1. Figure 2 shows the diffraction pattern of the elaborated disk in which can be observed only a broad peak specific to the amorphous structure.

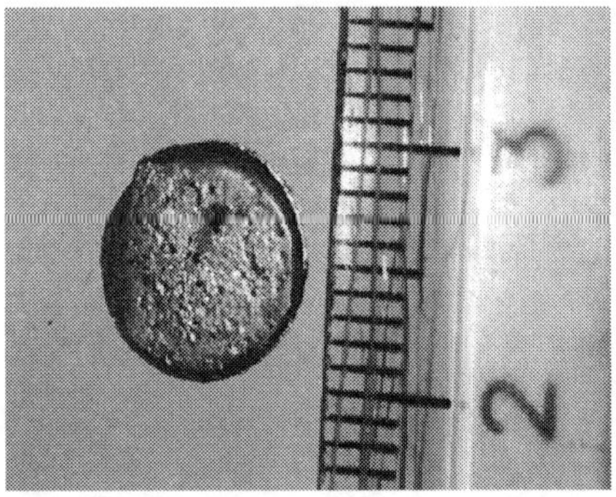

Figure 1 $Fe_{48}Mo_{14}Cr_{11}Co_4B_6P_8Si_7Y_2$ BAS disc.

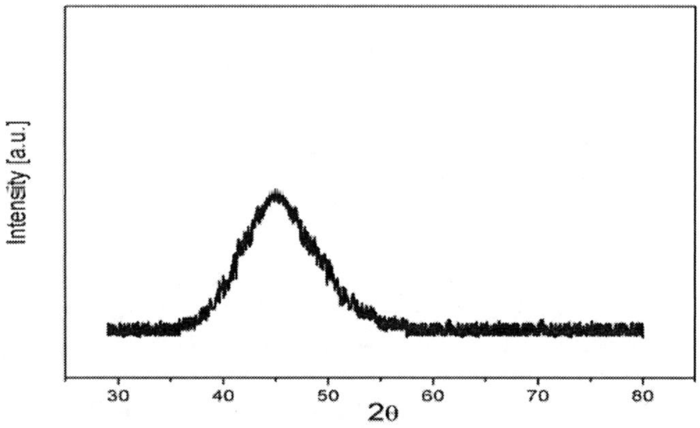

Figure 2 XRD pattern of the $Fe_{48}Mo_{14}Cr_{11}Co_4B_6P_8Si_7Y_2$ BAS disc.

The BAS disk obtained by water cooled copper mold casting is presented in Figure 1. Figure 2 shows the diffraction pattern of the elaborated disk in which can be observed only a broad peak specific to the amorphous structure. The accuracy of the finite element results were assured by using a fine tetrahedral mesh with higher number of nodes in areas where the temperature gradient was estimated to be the highest. The mesh contains 169171 elements and 247836 nodes and is presented in Figure 3.

Transient analyze performed on the entire assembly cast disk – mold provided data for temperature evolution during cooling time, allowing to estimate the cooling rate to obtain a bulk amorphous disk. Figure 4 shows the temperature evolution in cross section of the mold and disk until cooling at room temperature.

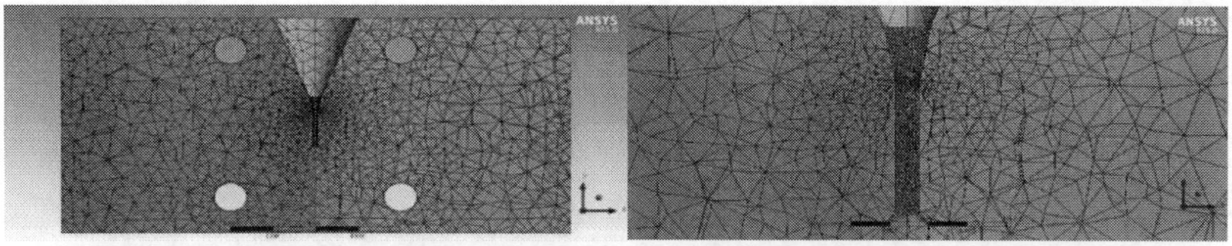

Figure 3. Mesh of the mold and detail in the disc area.

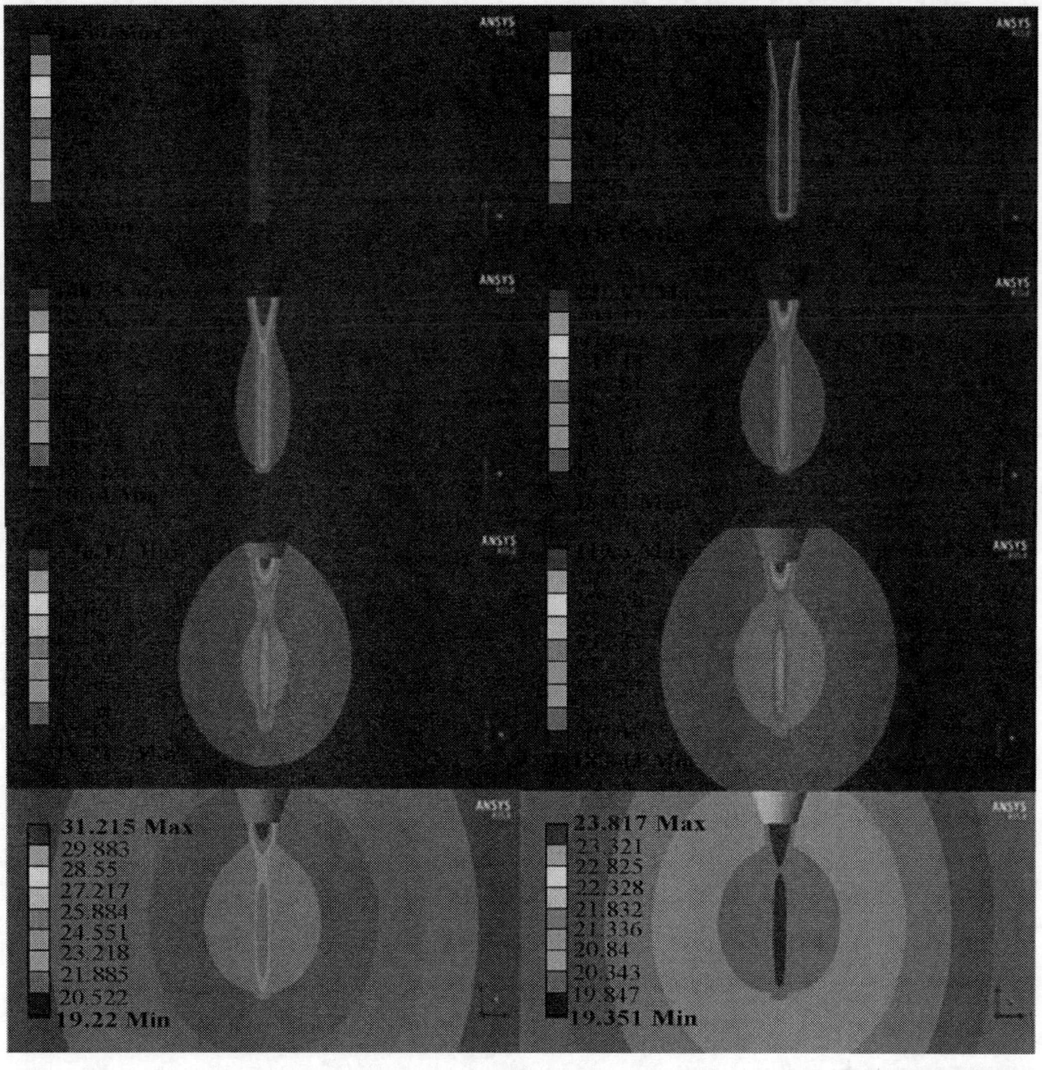

Figure 4. Thermal field distribution in cast alloy at 0s; 0.02s; 0.1s; 0.3s; 0.5s; 0.8s; 1s; 1.5s.

Figures 5 and 6 present the temperature evolution and cooling rates in disk cross section at different distances from the center of the disk. It can be noted that the cooling rate decreases from 3739 K/s in the surface to 1872 K/s in the core. The values of cooling rate are higher than the values reported in the literature, approximately 100 K/s [7] for Fe-based bulk metallic glasses.

IOP Conf. Series: Materials Science and Engineering **416** (2018) 012021 doi:10.1088/1757-899X/416/1/012021

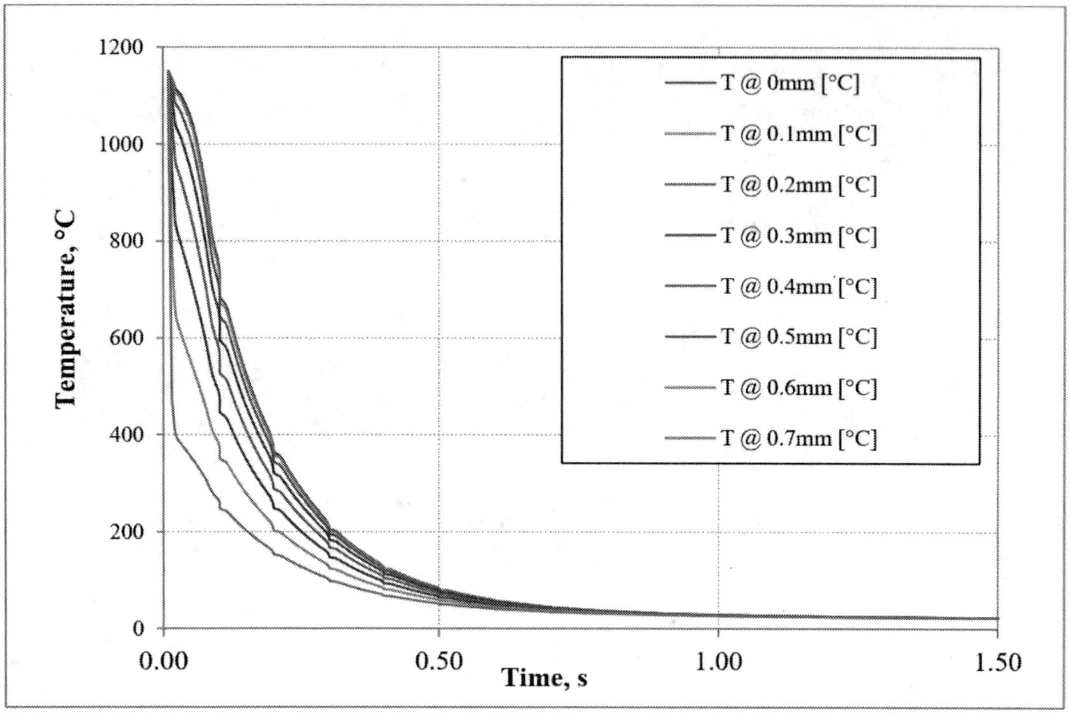

Figure 5 Temperature evaluation in disk cross section at 0, 0.1, 0.2, 0.3, 0.4, 0.5, 0.6 and 0.7 mm from the center of the disk.

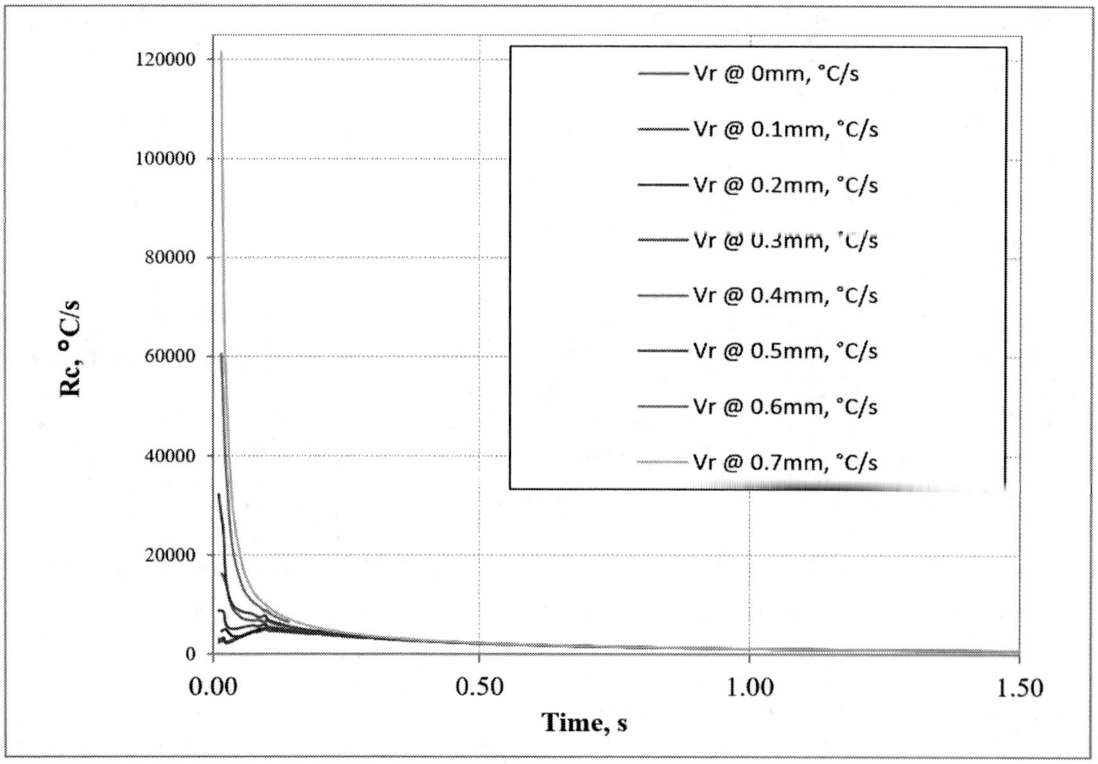

Figure 6 Cooling rate in disk cross section at 0, 0.1, 0.2, 0.3, 0.4, 0.5, 0.6 and 0.7 mm from the center of the disk.

4. Conclusions

Multicomponent $Fe_{48}Mo_{14}Cr_{11}Co_4B_6P_8Si_7Y_2$ BAS in disk form was elaborated by copper mold casting using mostly ferroalloys and pure elements. The finite element analysis model is verified since it was obtained a minimum cooling rate of 1872 K/s which has led to an amorphous structure. This model helps to estimate the critical cooling rate in processing the bulk metallic glasses by copper mold casting since it is difficult to determine it by experimental analysis.

5. References

[1] Babilas R., Nowosielski R. 2010 Iron - based bulk amorphous alloys, *Archives of Materials Science and Engineering* **44** 5-27.

[2] Rao B.R. 2009 Bulk Metallic Glasses: Materials of Future, *DRDO Science Spectrum*. 212-8.

[3] Wang W H, Dong C and Shek C H 2004 Bulk metallic glasses *Materials Science and Engineering: R: Reports* **44** 45–89

[4] Inoue A., Takeuchi A. 2002 Recent progress in bulk glassy alloys, *Materials Transactions* **43** 1892–06.

[5] Kui H W, Greer A L and Turnbull D 1984 Formation of bulk metallic glass by fluxing *Applied Physics Letters* **45** 615–6

[6] Basu J., Ranganathan S. 2003 Bulk metallic glasses: A new class of engineering materials, *Sadhana-* **28** 783-798.

[7] Suryanarayana C and Inoue A 2011 *Bulk metallic glasses* (Boca Raton, FL: CRC Press)

[8] Fang H, Hui X and Chen G 2008 Effects of Mn addition on the magnetic property and corrosion resistance of bulk amorphous steels *Journal of Alloys and Compounds* **464** 292–5

[9] Iqbal M, Akhter J I, Zhang H F and Hu Z Q 2008 Synthesis and characterization of bulk amorphous steels *Journal of Non-Crystalline Solids* **354** 3284–90

[10] Gostin P F, Gebert A and Schultz L 2010 Comparison of the corrosion of bulk amorphous steel with conventional steel *Corrosion Science* **52** 273–81

[11] Tsai P H, Xiao A C, Li J B, Jang J S C, Chu J P and Huang J C 2014 Prominent Fe-based bulk amorphous steel alloy with large supercooled liquid region and superior corrosion resistance *Journal of Alloys and Compounds* **586** 94–8

Acknowledgements

This work was supported by research grants PCD-TC-2017 No. 16179/21.11.2017.

Determination of Young's Modulus of Surface Coating Paint with Impulse Excitation Technique and Analysis of Laser Signals

E Nyaguly, I Crâştiu, D Vlădaia, G Drăgănescu, S Stanciu and L Bereteu*

Mechanics and Materials Strength Department, Politehnica University Timişoara, Bd. Mihai Viteazul, nr. 1, 300222, Timişoara, Romania

*Corresponding author: liviu.bereteu@upt.ro

Abstract. Polymer coatings of metallic surfaces offer them not only aesthetic value but at the same time provide anticorrosive protection or even improvement of some mechanical properties of the composite structure. The mechanical properties of paints are often negligible although the two aspects considered: aesthetics and durability are related to the mechanical properties of both the base material and the paint. The durability of the paint depends on how it adheres to the substrate, but at the same time a jump in the bending stress value occurs in the separation area of the two materials, which leads to the appearance of shear stress. The shear stress value depends on the ratio of the modulus of elasticity of the base material and the paint. In this paper it will follow, starting from the mechanical properties of the substrate, the metallic sheet in general, to determine the new properties of the assembly of substrate and the two painting layers, also the determination of mechanical properties of the layers. From the analysis of laser signals obtained by the impulse excitation of the samples, one can determine the elastic modulus. These results come to validate the results based on Finite Element Analysis (FEA) of the same samples. From the analysis of Bernoulli's equation for the two situations: the uncovered beam and the painted beam, the ratio of natural frequencies is determined as a function depending on the modulus of elasticity, density and geometric dimensions of the substrate, respectively the same elements of the coating paint. From this ratio, Young's modulus of the paint can be determined.

1. Introduction

By covering the work piece with different layers of paint, it changes its mechanical properties into a composite structure. The new properties of the composite structure will be different from the properties of any of the component materials. Painting is done for both metallic and non-metallic materials. As far as metallic coatings are concerned, it should be borne in mind that ferrous metals are subject to rusting in contact with corrosive environments, and their coating is a vital issue.

At the same time, a bending composite structure exhibits a jump of tension stress at the metal-paint interface, which results in a shear stress [1]. That is why it can be said that the cracking, the most common defect in painting metal surfaces, is due to the different loading of the two components. For this reason, the primer layer applied between paint and metal has an essential role in reducing this jump of tension stress. Therefore, the rigidity of the adhesive film has an important role in preventing cracks. To prevent shear cracking, it would be ideal that the flexural stiffness of the base material be

IOP Conf. Series: Materials Science and Engineering **416** (2018) 012022 doi:10.1088/1757-899X/416/1/012022

equal to the rigidity of the coating [2]. With the polymeric coating of metals, the elastic characteristics of the structure change [3].

To determine the mechanical properties of materials over time, several techniques have been developed, grouped into two large categories: static methods and dynamic methods [4], [5]. The first category includes traction and shear tests. The second category of tests includes acoustic resonance spectroscopy [6-8] and vibration resonance techniques [9-10].

The latest methods, techniques and patents for determining the elastic characteristics of materials are presented in the papers [11-12].

This paper can be included in the category of research on the determination of the mechanical characteristics of some materials by analyzing the vibroacoustic signals obtained from a mechanical impulse and recorded by an acquisition board, having a microphone as a sensor, and by analyzing the vibration motion signals recorded by a laser vibrometer, respectively [13-16].

This paper aims to determine the Young's modulus of paints used for coating steel sheets in refrigeration industry or automotive industry. For this purpose the resonance frequencies of the vibrations of some samples, clamped at one end and free at the other end, before and after painting, are determined. The frequencies of the signals given by a laser vibrometer after applying a mechanical impulse to the samples, by Fast Fourier Transform (FFT) are determined.

2. Vibrational modelling

In the analysis of the bending vibrations of a slender cantilever metal beam, considered one-dimensional and covered on the two surfaces with the paint, the equation Euler-Bernoulli will be considered (Figure 1). Neglecting the effects of shear and rotational inertia, vibration movements are described by the differential equation [17]

$$(EI)_{cb}\frac{\partial^4 Y(x,t)}{\partial x^4}+(\rho A)_{cb}\frac{\partial^2 Y(x,t)}{\partial t^2}=0, \tag{1}$$

where Y(x, t) is the displacement of neutral axis of the beam at x distance to the clamped end, $(EI)_{cb}$ is the flexural rigidity of the entire composite beam, and $(\rho A)_{cb}$ is the mass of the composite beam length unit.

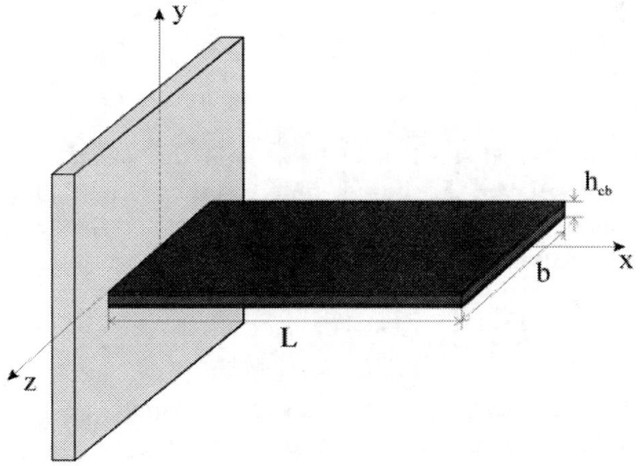

Figure 1. Schematic representation of a three layers beam.

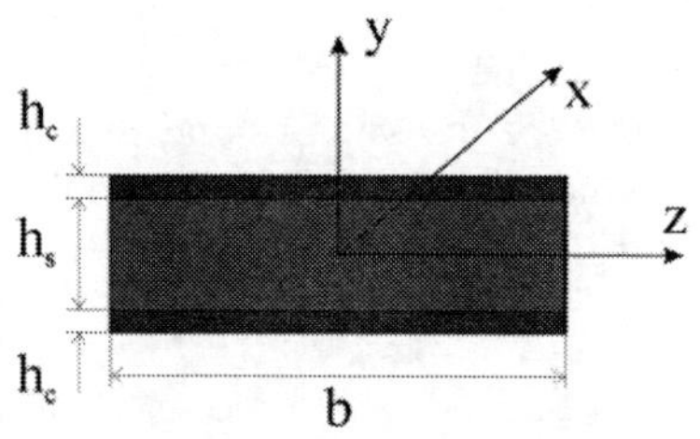

Figure 2. Cross section and its dimensions.

The three layers composite bar has a bending stiffness equal to the sum of the stiffness's of each component material. The paint layers are assumed to be homogeneous, of the same thickness and perfectly adhered to the surface of the metal substrate.

IOP Conf. Series: Materials Science and Engineering **416** (2018) 012022 doi:10.1088/1757-899X/416/1/012022

$$(EI)_{cb} = (EI)_s + 2(EI)_c, \tag{2}$$

where the indexes cb, s and c refer to the composite beam, substrate, and paint coats, respectively.

The mass per unit length of the three layers beam is given by the following formula:

$$(\rho A)_{cb} = (\rho A)_s + 2(\rho A)_c. \tag{3}$$

Analytical solving the differential equation (1) can determine the modal shapes and natural frequencies of vibration modes. The resonance frequencies are determined from the characteristic equation corresponding to the clamped–free boundary conditions

$$1 + chX \cos X = 0, \tag{4}$$

where

$$X = L\left((2\pi f_{cb})_i^2 \frac{(\rho A)_{cb}}{(EI)_{cb}} \right)^{1/4}, \tag{5}$$

L is the length of composite beam and $(f_{cb})_i$ (i=1,2,…..,n) is one of its natural frequencies.

If only the substrate metal beam is considered, the same characteristic equation is obtained, so it can be written:

$$X = L\left((2\pi f_s)_i^2 \frac{(\rho A)_s}{(EI)_s} \right)^{1/4}. \tag{6}$$

From equations (5) and (6), taking into account relations (2) and (3), as well by replacing I_s and I_c of cross section (Figure 2), with their correspondent expressions [18], the ratio of the frequencies of one vibration mode for the three layers beam and the substrate beam is obtained

$$\left(\frac{f_{cb}}{f_s}\right)_i^2 = \frac{2(EI)_c + (EI)_s}{(EI)_s} \frac{(\rho A)_s}{2(\rho A)_c + (\rho A)_s} = \frac{1 + \frac{E_c}{E_s}\left[6\left(\frac{h_c}{h_s}\right) + 12\left(\frac{h_c}{h_s}\right)^2 + 8\left(\frac{h_c}{h_s}\right)^3\right]}{1 + 2\left(\frac{\rho_c}{\rho_s}\right)\left(\frac{h_c}{h_s}\right)}. \tag{7}$$

Using the ratios between the natural frequencies of the coated beam and the substrate beam, between the thicknesses and densities of the coating layer and the substrate given by $(f_r)_i = (f_{cb}/f_s)_i$, $h_r = h_c/h_s$, and $\rho_r = \rho_c/\rho_s$, where the subscriptions c and s correspond to the coating paint and the substrate, then from equation (7) one can determine Young's modulus for coating, so

$$E_c = \frac{E_s}{6h_r + 12h_r^2 + 8h_r^3}\left[f_r^2(1 + 2\rho_r h_r) - 1\right]. \tag{8}$$

Assuming that the paint layers have very small thicknesses compared to the substrate, i.e. the ratio $h_r \ll 1$, then the cube and square of this ratio can be neglected compared to itself. With this observation equation (8) can be written in the form:

$$E_c = \frac{E_s}{6h_r}\left[f_r^2(1 + 2\rho_r h_r) - 1\right]. \tag{9}$$

Relationships similar to those given by equations 8 and 9 are obtained by Lopez in the paper [19] and Berill in the works [20-21].

If in addition to the above observation, it is considered that the covering material has a density much lower than the substrate material, then it is obtained a simplified form for the determination of the modulus of elasticity. It can be used to determine the Young module for thin and low density coating (TLDC)

$$E_c = \frac{E_s}{6h_r}\left(f_r^2 - 1\right) = \frac{E_s}{3h_r}\frac{\Delta f}{f_s}. \tag{10}$$

3. Numerical simulation

For the numerical simulation of the dynamic behavior of a three layers composite beam, clamped at one end and free at the other end, Finite Element Analysis (FEA) was performed using ANSYS 14.5 software. To construct a 3D model, hexahedral solid layered elements SOLID 186 with three degrees of freedom for each node were used. Element SOLID 186 is defined by 20 nodes that exhibit quadratic displacement behavior. These layered elements allow the definition of layer-by-layer properties, so it is suitable for the multi-layer material model. The geometric model consists of three parts: the substrate and the two paint coating layers. The substrate material is ordinary steel S 235 JR having the following properties E_s=200GPa, ρ_s=7850 kg/m^3, μ_s=0.3, with a rectangular cross section with a length L=240 mm, a width b=40 mm, and a thickness h_s=2mm. For the layers of paint two types of samples were used. The first type had the layer of paint deposited through a single electrostatic crossing and the thickness h_c=90μm, and for the second type of the sample, the deposition was made by two passes having the thickness of the paint layer h_c=180μm. The meshing of coated sample is shown in Figure 3, and the shape of the first vibration mode and corresponding frequency is shown in Figure 4.

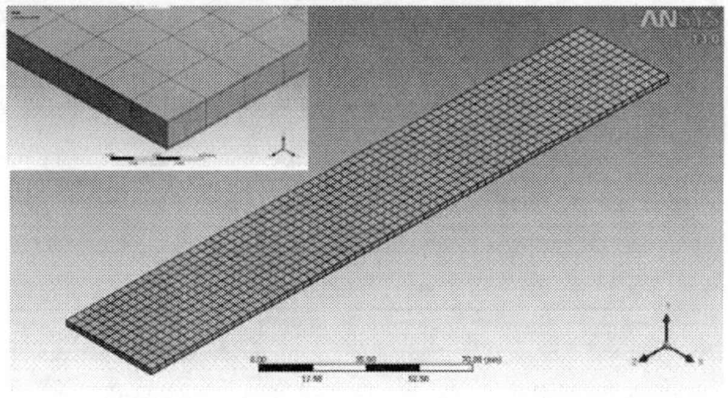

Figure 3. Finite element discretization of beam.

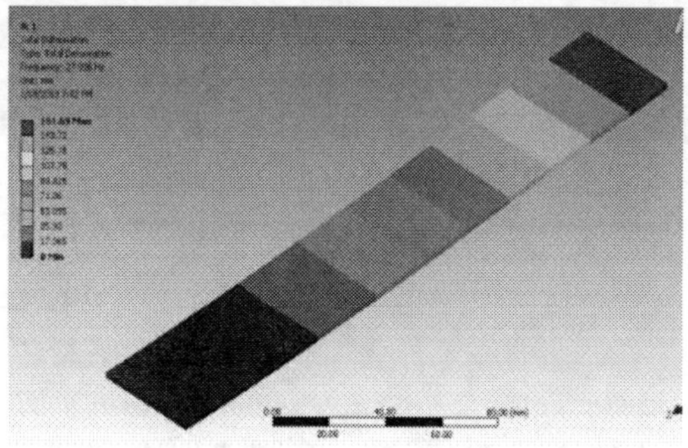

Figure 4. First vibration mode.

4. Measurement of fundamental frequency

The determination of the length L, width b and thickness of the substrate h_s was made by six measurements at six different points. The average of these measurements is for the length of the clamped sample, L=239.42 mm, for the width b=39.25 mm, and for the thickness h_s=1.94 mm.

To determine the thickness of the paint coatings, ten measurements were made in ten different points. The average of these ten measurements gave a thickness of h_{1c}=89.1μm, for one coat of paint and thickness h_{2c}=188.2μm, for coating with two layers of paint.

The purpose of the experimental technique developed in this paper is to determine the natural frequencies of the sample, before and after coating with paint layers. For the experimental determinations a stand (Figure 5) consisting of sample **1,** clamped in the vise **3,** and excited by a mechanical impulse given by a mini hammer **2** was used. The recording of a signal, proportional to the speed of a sample point, is based on the working principle of a Doppler Vibrometer Laser **4,** which emits a luminous signal to the sample vibratory motion and the receiver is a photo detector. The demodulated signal can be viewed on a PC display **5,** and through an acquisition board and computer **6** it is digitalized and used in the Fast Fourier Transform (FFT) program to determine the Fourier Spectrum.

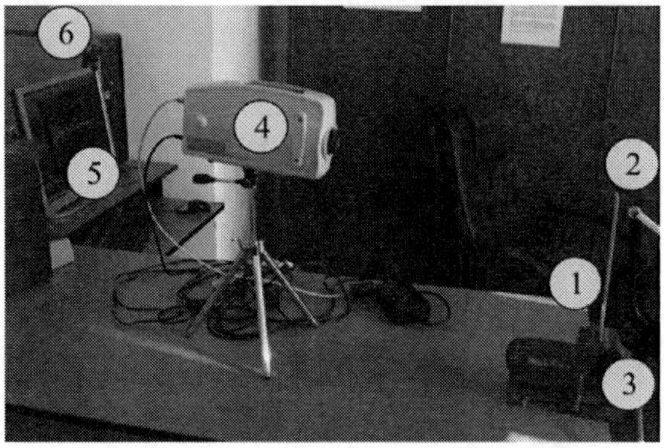

Figure 5. Vibratory set-up employed.

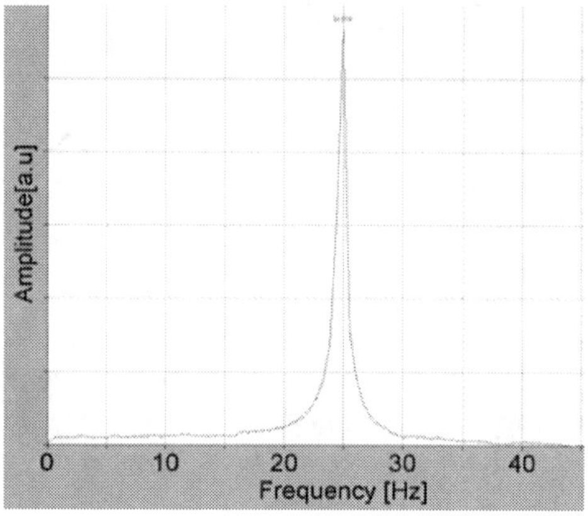

Figure 6. Fundamental frequency from FFT.

5. Results and conclusions

In order to determine the Young's modulus, the signals given by the laser vibrometer are analyzed, and from Fourier Spectrum (Figure 6), the fundamental frequency for each sample is obtained. The numerical determinations of the elasticity modulus of the coating paint for the three analyzed methods equations (8), (9) and (10) are used. The results are given in Table 1 and Table 2. For determining the steel elastic modulus, as a substrate sample, the standardized method presented in the ASTM E 1876-15 was used [22].

In Table 1 are presented the results obtained by experimental measurements on samples that had a single pass through the electrostatic field of paint deposition.

Table 1. Young's modulus of paint, after one pass of the sample through the electrostatic field

Sample frequency (Hz)	Composite sample frequency (Hz)	E_s (GPa) Young's modulus	E_c (GPa) TLDC method	E_c (GPa) Berry method	E_c (GPa) Lopez method
23.398	24.248	199.74	4.580	5.823	4.712

In Table 2 are presented the results of the elastic modulus of the paint determined experimentally from the sample covered by passing twice through the electrostatic field.

Table 2. Young's modulus of paint, after two passes of the sample through the electrostatic field

Sample frequency (Hz)	Composite sample frequency (Hz)	E_s (GPa) Young's modulus	E_c (GPa) TLDC method	E_c (GPa) Berry method	E_c (GPa) Lopez method
23.398	24.824	199.74	4.599	5.987	4.921

The values of the elastic modulus given in the paint catalogs are in the range of 4.7 GPa and 6.3 GPa. In conclusion, by comparing the results obtained experimentally with those given in the specialty catalogs, one can find that the proposed method for determining the Young's module is valid and gives good results for determining the modulus of elasticity of coating paints.

Numerical simulation results, using Modal Analysis (MA), come to confirm once again the validity of the method based on laser measurements.

6. References

[1] Lambourne R and Strivens T A 1999 *Paint and Surface Coatings. Theory and practice*, Second Edition (Cambridge, Woodhead Publishing Ltd)

[2] Kubisztal M, Haneczok G, Chrobak A, Rams B, Miara D and Rasek J 2009 Elastic properties of epoxy system coatings on aluminum substrate *Surface and Coatings Technology* **204** p 120

[3] Bellan C and Dhers J 2004 Evaluation of Young modulus of CVD coatings by different techiches. *Thin Solid Films* **469-470** pp 214-220

[4] Radovic M, Lara-Curzio E and Riester L 2004 Comparison of different experimental techniques for determination of elastic properties of solids. *Mater. Sci. Eng.* **A 368** pp 56–70

[5] Thomasová M, Sedlák P, Seiner H, Janovská M, Kabla M, Shilo D, and Landa M 2015 Young's moduli of sputter-deposited NiTi films determined by resonant ultrasound spectroscopy: Austenite, R-phase, and martensite. *Scripta Materialia.* **101** pp 24–27

[6] Tan Y, Shyam A, Choi W B, Lara-Curzio E and Sampath S 2010 Anisotropic elastic properties of thermal spray coatings determined via resonant ultrasound spectroscopy *Acta Materialia,* **58** pp 5305–15

[7] Sedmák P, Seiner H, Sedlák P, Landa R, Mušálek R and Matejícek J 2013 Application of resonant ultrasound spectroscopy to determine elastic constants of plasma-sprayed coatings with high internal friction, *Surface and Coatings Technology* **232(1)** pp 747–757

[8] Mazot P and Pautrot S 1998 Détermination du module d'young de dépôts par flexion dynamique: Application aux systèmes bicouche et tricouche. *Ann. Chim. Sci. Mat.* **23** pp 821–827.

[9] Gadaud P and Pautrot S 2004 Application of the dynamical flexural resonance technique to industrial materials characterisation. *Mater. Sci. Eng.* **A 370** pp 422–426

[10] Gadaud P, Milhet X and Pautrot S 2009 Bulk and coated materials shear modulus determination by means of torsional resonant method. *Mater. Sci. Eng.* **A 521–522** p 303

[11] Pagnotta L 2008 Recent progress in identification methods for the elastic characterization of materials *International Journal of Mechanics* **4(2)** pp 129-140

[12] Alfano M and Pagnotta L 2009 Recent developments and selected patents on vibration based methods for elastic characterization of izotropic materials *Recent Pat. on Mec. Eng.* **2** pp 130-143

[13] Suciu O, Bereteu L, Drăgănescu G and Ioanovici T 2013 Determination of the elastic modulus of hydroxiapatite doped with magnesium through nondestructive testing *Advenced Materials Research* **814** pp 115-122

[14] Bereteu L, Medgyesi T, Popescu M ,Draganescu G, Rosu R and Simoiu D 2013 Vibration tests at thin stainless steel sheets joined by CMT pulse with CuSi3 filler material for determination of the elasticity modulus *Advenced Materials Research* **814** pp 135-140

[15] Moisa R, Medgyesi T, Bereteu L, Draganescu G, Simoiu D and Sava M 2013 Vibration tests for determination of longitudinal elasticity modulus and shear modulus of some structures welded with tubular wire *Acoustics & Vibration of Mechanical Structures* **430** pp 101-107

[16] Neamtu A, Simoiu D, Nyaguly E, Crastiu I and Bereteu L 2018 Analysis of whole–body vibration on rheological models for tissues *Materials, Sciene and Engeeniring* **294** pp 1-8

[17] Meirovitch L, 2010 *Foundamentals of vibration*, Waveland Press, Inc.

[18] Nyaguly E, Crastiu I, Deac S, Gozman-Pop C, Draganescu G and Bereteu L 2018 Determination of elastic mechanical characteristics of surface coatings from analysis of signal obtained by impulse excitation *Materials, Sciene and Engeeniring* **294** pp 1-10

[19] López-Puerto A and Oliva A I 2014 A vibrational approach to determine the elastic modulus of individual thin films in multilayers *Thin Solid Films* **565** pp 228–236

[20] Berry B S and Pritchet W C 1975 Vibrating reed internal friction apparatus for films and foils *IBM J Res Dev* **19** pp 334-343

[21] Berry B S,(Gupta D, and Ho P S Eds.) 1988 *Diffusion Phenomena in Thin Films and Microelectronic Materials*, Noyes, Park Ridge, New York pp 73-141

[22] ASTM E 1876-15 2015 Standard test method for dynamic Young's modulus, shear modulus and Poisson's ratio by impulse excitation of vibration

IOP Conf. Series: Materials Science and Engineering **416** (2018) 012023 doi:10.1088/1757-899X/416/1/012023

Design and Dynamic Analysis of Fixture Behaviour Used for Vibration Testing of some Automotive Components

A B Bârgău, D Simoiu, C Gozman-Pop, D Vlădaia, L C Berce and L Bereteu*

Mechanics and Materials Strength Department, Politehnica University Timişoara, Bd. Mihai Viteazul, nr. 1, 300222, Timişoara, Romania

*Corresponding author: liviu.bereteu@upt.ro

Abstract. The purpose of this paper is to develop a project for fixtures used in the automotive industry for fixing of certain components such as side airbags, curtain airbags, driver airbag, airbags in the seats airbags knee protection, devices for securing the seat belts, steering wheels, in order to test to vibration. Due to the multitude of elements which are in continuous movement on the vehicle or the high forces that occur during vehicle running, mechanical vibrations phenomena are produced. These cause aging products mounted on the car, lead to fatigue of materials, and to lower initial mechanical properties and finally the complete destruction of subassemblies. To prevent such situations, car manufacturers always ask before approval to make the simulation of vibration and then to recover destructive tests and prepare a report of the simulation correctly. Therefore, after carrying out the project it will be 1odelled by applying the Finite Element Method (FEM), and then using Finite Element Analysis (FEA) to determine modal parameters. A fixture for the locking of the seat belt socket is the object of the analysis in this paper.

1. Introduction

Among the methods most used in machine monitoring, due to the development of sensor systems, signal acquisition systems and real-time processing, vibration analysis has become a very powerful tool. This analysis is used in dynamics qualification of newly designed structural components, defects prediction, structural diagnosis and structural aging issues [1].

Subjecting these devices to some aging tests involves knowing some mechanical characteristics that govern their dynamic behavior in order to select parameters that can lead to aging within a prescribed time. These dynamic features are given by the natural frequencies and the natural modes of the mechanical structure [2].

Progress in automotive technology has included devices that provide increased protection for vehicle occupants in the event of accidents or vehicle collisions. Primary safety features of the vehicle have always been considered as safety belts and airbags. The seat belt socket is considered one of the most important components of the seat belt [3].

Determining the regimes to which the seat belt socket is subjected by aging or destruction tests is the purpose of this paper. For this purpose, in the Catia V5 software a fixture for the seat belt socket dynamic testing was designed.

2. Finite element modelling

The finite element method is a numerical method for solving engineering problems or physics-mathematics. The analytical solution given for such problems requires the solving of differential equations with partial derivatives, with inputs of some initial conditions and border conditions. MEF includes, in a first stage, mesh generation techniques to divide a complex problem related to the whole body's behavior into small parts called finite elements. The simple equations governing the behavior of these finite elements are assembled into a large system of equations that govern the behavior of the whole. Therefore this numerical method leads to solving a set of algebraic equations. This part of the problem is known in engineering practice as FEA, meaning it is a computational tool for performing engineering analyzes by using coded software, based on specific algorithms. In the dynamic behavior of some mechanical structures, it is very important to know the natural frequencies and natural modes of dynamic deformation, and the algorithm of this application was first given by [4] and is known as Fast Fourier Transform (FFT) or Modal Analysis (MA).

Fixture design for seat belt socket dynamic testing, two elements, Solid 186 and Solid 187 were used to generate mesh (Figure 1). The first type of finite element is shown in Figure 2.

This type of upper order 3D element is defined by 20 nodes, each having three degrees of freedom corresponding to the three translational displacements along x, y and z nodal directions. The element supports plasticity, hyper elasticity, creep, stress stiffness, large deformation and large displacement capabilities [5-6].

Figure 1. Finite element mesh for a fixture.

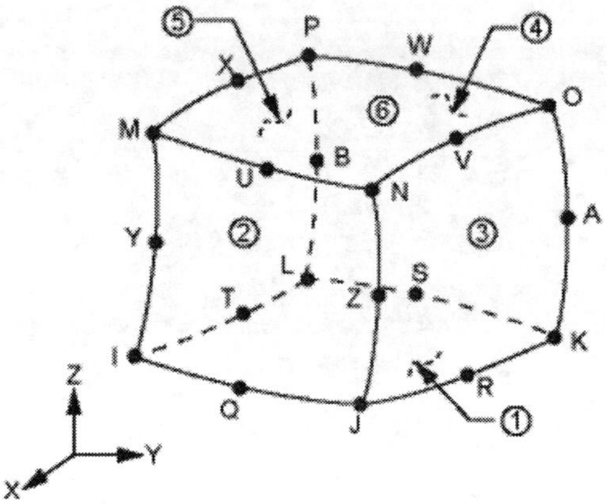

Figure 2. Spatial element SOLID186.

The second type of finite element, Solid 187, has a tetrahedral geometric shape with only 10 nodes but allows for quadratic displacement behavior as SOLID 186. This type of element is well for modeling irregular meshes, such as those made from different Catia V5 and CAD/CAM systems.

3. Modal Analysis

To determine the natural frequencies and the shape of the natural vibration modes of a fixture for a seat belt socket, the software provided by ANSYS 14.5 was used, and the finite element modeling was followed by the finite element analysis (FEA). In the construction of the mathematical model, based on finite elements, a three-dimensional model (3D), having a hexahedral geometric shape (Solid 186) and a model type having a tetrahedral geometric shape (Solid 187) were used. The first element has a number of 20 nodes, and the second one has 10 nodes, each node having three degrees of freedom. For simulation, 5 cases were analyzed. Depending on the dimensions of the geometric element, 29 767 nodes, 33 250, nodes, 38 527 nodes, 43 187 nodes and 47 383 nodes were obtained. For the modal analysis of fixture vibrations, the ANSYS module solves numerically a matrix differential equation of an undamped vibrant system given in the form:

$$[M]\{\ddot{u}\}+[K]\{u\}=\{0\}, \tag{1}$$

where [M] and [K] are mass and rigidity matrices, respectively {u} are nodal displacements. The number of equations in the matrix system was three times greater than the number of nodes. In the absence of damping, the vibrations of the nodes can be expressed by harmonic functions such as:

$$\{u\} = \{\phi\}_i \cos(p_i t + \varphi_i), \tag{2}$$

where $\{\phi\}_i$ are the eigenvectors and p_i are corresponding natural circular frequencies.

By imposing the solution (2) to verify the differential equation (1), this leads to a problem of eigenvalues and eigenvectors

$$\left(-p^2[M] + [K]\right)\{\phi\} = \{0\}, \tag{3}$$

from which one can determine the unknown eigenvalues p_i by imposing the condition that the equation (3) has nontrivial solutions, that means to solve the determinant:

$$\left|-p^2[M] + [K]\right| = 0. \tag{4}$$

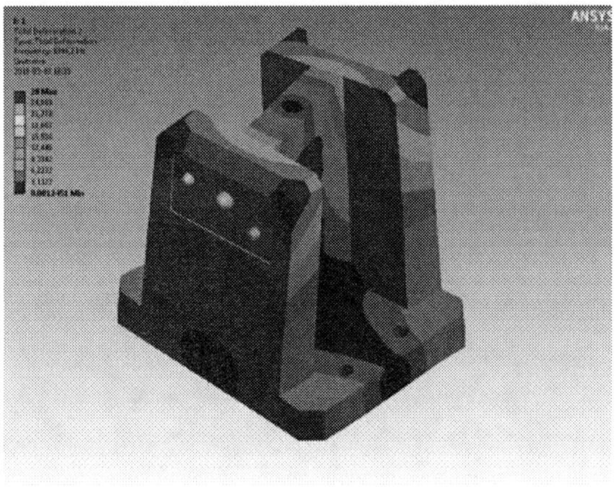

Figure 3. The first mode shape at natural frequency 1996.2 Hz.

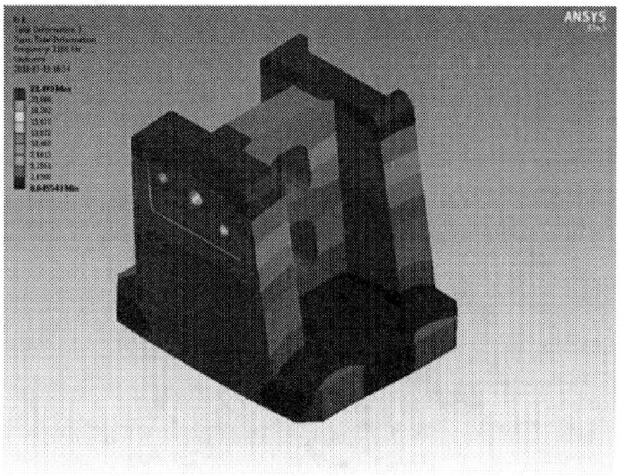

Figure 4. The second mode shape at natural frequency 2110 Hz.

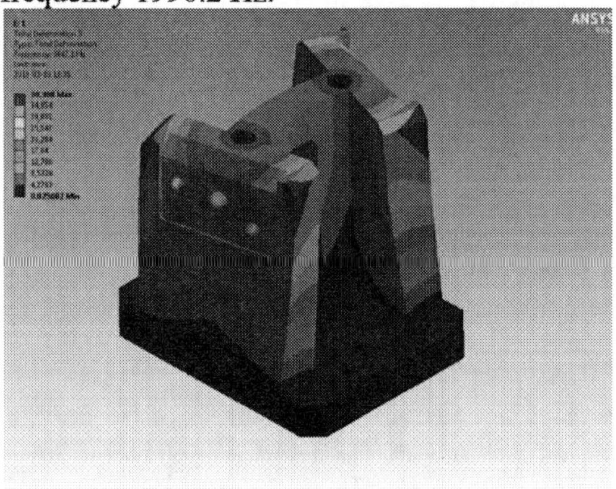

Figure 5. The fourth mode shape at natural frequency 3947.1 Hz.

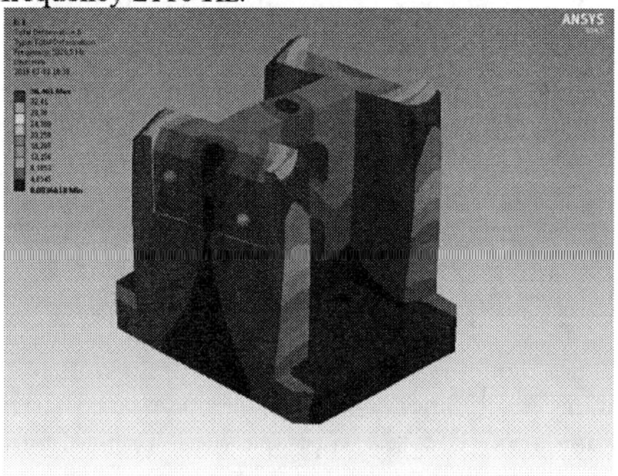

Figure 6. The fifth mode shape at natural frequency 5326.5Hz.

Figures 3 and 4 give the forms and frequencies of the first two modes of vibration, and Figures 5 and 6 give the forms and frequencies of modes four and five.

IOP Conf. Series: Materials Science and Engineering **416** (2018) 012023 doi:10.1088/1757-899X/416/1/012023

4. Results and conclusions

The modal analysis of fixtures, used to fix auto parts subjected to dynamic tests, is made for knowing the natural frequencies, that is, those frequencies in which resonance occurs. A first objective proposed in the paper was to determine how it influences the number of elements and implicitly the number of nodes on the convergence of the ANSIS algorithm in the determination of natural frequencies. It can be seen (Figure 7) that with the increase of the number of nodes the differences obtained in the frequency of a natural vibration mode is 8-12 Hz for the low modes (first, second and third mode), respectively 26-35 Hz for the other two higher modes. In conclusion for the practical purposes of determining the frequencies to which the head shaker extender is required to act, no finite elements with dimensions of less than 5 mm are necessary. This has an effect on reducing the time allocated to dynamic analysis.

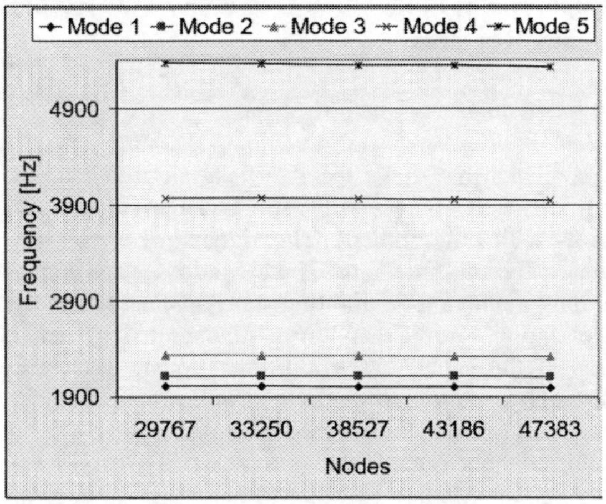

Figure 7. Variation of natural frequencies according to the number of nodes.

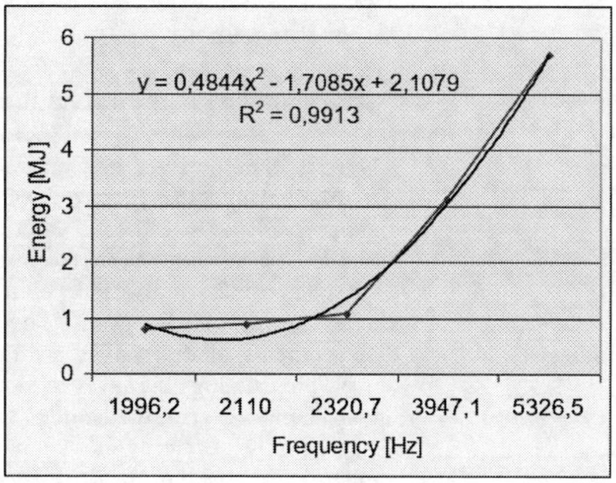

Figure 8. Variation of sample energy according to the frequencies of natural modes.

A second aspect to be taken into account in dynamic testing is that the energy transmitted to the samples is proportional to the square of the vibratory machine frequencies. The graphical representation (Figure 8) of energy variation according to the natural frequencies confirms this, and the interpolation curve is a parabola having a correlation coefficient R^2 equal to 0.9913, i.e. the mean square deviation, between the results obtained by the FEA and the theoretical ones are very good.

5. References

[1] Albarbar A.and Teay S.H. 2017 MEMS Accelerometers Testing and Practical Approach for Smart Sensing and Machinery Diagnostics. In: Zhang D., Wei B. (eds) Advanced Mechatronics and MEMS Devices II. Microsystems and Nanosystems. Springer, Cham, pp 19-40

[2] Al-Bahkali E A, Elkenani H, and Souli M 2014 Fatigue life estimate of landing Gear's leg using modal analysis *Int. Jnl. of Multiphysics* **8(2)** pp 231-244

[3] Alic C I; Miklos C C and Miklos I Zs 2011 Finite element analysis of a seat belt buckle device, *Annals of the Faculty of Engineering Hunedoara - International J* **9** **(4)** p135

[4] Cooley J W and Tukey J W 1965 An algorithm for the machine calculation of complex Fourier series, *Math. Comput.* **19** pp 297–301

[5] Kuk M G, Kwon S J and Tak T O 2008 Dynamic analysis of seatbelt systems with antiinertial release mechanisms, *International Journal of Automotive Technology,* **9(5)** pp 593-599
Moffatt E A, Thomas T M and Cooper E R 1995 Safety Belt Buckle Inertial Responses in Laboratory and Crash Tests." *SAE International Congress & Exposition*, Detroit, MI, pp 210-218

IOP Conf. Series: Materials Science and Engineering **416** (2018) 012024 doi:10.1088/1757-899X/416/1/012024

Dynamic Analysis of Satellites Used for Side-Airbag System

C Gozman-Pop, D Simoiu, I Crâştiu, E Nyaguly, E N Oanţă and L Bereteu*

Mechanics and Materials Strength Department, Politehnica University Timişoara, Bd. Mihai Viteazul, nr. 1, 300222, Timişoara, Romania

*Corresponding author: liviu.bereteu@upt.ro

Abstract. Satellites are mechanical components that make the interface between the outside environment and the Electronic Central Unit (ECU). The Electronic Central Unit receives signals from the satellites it interconnects with environment. The signals are given, after a collision, by capacitor-type accelerometers. The multiple caps of the condenser are connected to a seismic mass and elastic element, thus achieving a vibrating mechanical system with a single degree of freedom. The deceleration produced during the collision generates displacements of the movable armatures of the capacitor, proportional to the value of this decay. This analogy signal is modulated and transmitted to the ECU. For correct processing of the accelerator signal, the bearer's frequency must be not allowed. For these reasons, a dynamic analysis of the accelerometer support satellite is performed. Dynamic analysis involves a finite element mesh for the use of the Modal Analysis Method in order to determine the resonance frequencies. In this paper we will analyze the influence of the different geometric types of finite elements, which, combined with the effects given by the nature of the component materials and the disagreement of the satellite, will lead to the requirements imposed in the limitations of the low resonance frequency values.

1. Introduction

In a broad sense, a sensor is a device or module that detects and responds to physical input phenomena from the outside environment characterized by specific sizes. These input sizes can be given by movement (displacement, velocity, acceleration, temperature, light, pressure, humidity or others in the multitude of physical phenomena. The answer is sent in the form of electrical signals to electronic device, usually ECU or the processor of a computer.

With the need to increase passenger safety, more means of protection have developed. The airbags are an important roll of these. During collisions, large decelerations occur in a short period of time [1]. Therefore, the command of opening the airbags gives it a category of sensors called accelerometers. There are several physical phenomena that underlie the operation of accelerometers. The most known are piezoelectric accelerometers. In the last three decades, with the development of technological processes, the production of Micro-Electro-Mechanical Systems has taken place. Based on these technologies, condenser type accelerometers could be produced [2]. The silicon-based capability of a MEMS capacitive sensor enables mass production at low cost due to mature technology on surface micromachining [3]. In recent years, various techniques have been developed to study MEMS capacitive accelerometers to optimize their performance and robustness [4]. The purpose of the paper is to perform an FEA analysis of a type of satellite that incorporates a MEMS capacitor accelerometer and validates the results based on the dynamic analysis of its behavior.

2. Capacitive MEMS accelerometer

The simplest capacitive sensor consists of two parallel-plate between which is a dielectric material, which can even be the air. Capacitance C of the capacitor formed by the two armatures is expressed by the known relationship:

$$C_0 = \varepsilon_0 \varepsilon_r \frac{A}{d} = \varepsilon \frac{A}{d} \tag{1}$$

where ε_0 is the permittivity of the air, ε_r is the relative permittivity of the dielectric, A is the area of the plates, and d is the distance between them. Changing any of the three parameters leads to capacity modification and consequently each of the three variables can be used as sensitive capacitor elements.
A MEMS condenser type accelerometer is schematically represented in the figure 1. The central part is a proof mass and when acceleration is applied to the reference frame, the proof mass has an opposite movement. This means that between the fixed plates and the movable plate, two capacitors of variable capacities C_1 and C_2 are formed as in the figure 1. In fact, every accelerometer has lots of condenser sets. In the equilibrium position of the mobile plate, x = 0, therefore the two capacitors will have the same capacitance C_0. By neglecting the end effects, the two variable capacitors will have the capacities given by the following relations [6]:

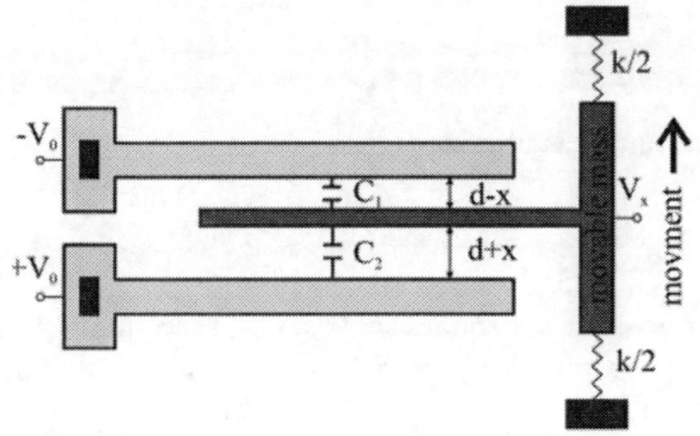

Figure 1. Simple structure of accelerometer.

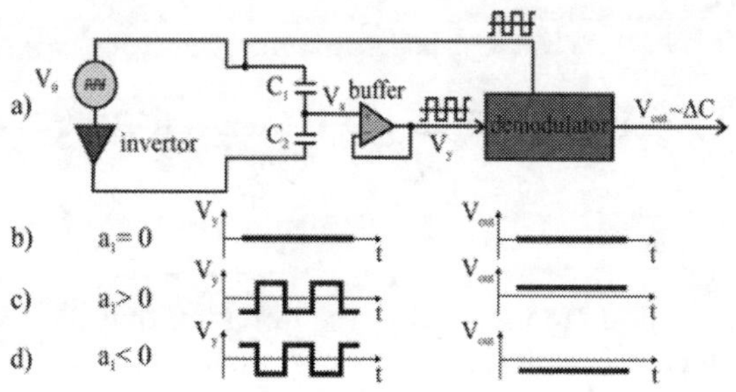

Figure 2. Electrical circuit.

$$C_1 = \varepsilon_0 \varepsilon_r \frac{A}{d-x} = \varepsilon \frac{A}{d-x} = C_0 + \Delta C; \tag{2}$$

$$C_2 = \varepsilon_0 \varepsilon_r \frac{A}{d+x} = \varepsilon \frac{A}{d+x} = C_0 - \Delta C. \tag{3}$$

The movable plate x displacement results due to acceleration movable mass and, there is a difference between the two capacities, which is

$$C_1 - C_2 = 2\Delta C = 2\frac{x}{d^2 - x^2}\varepsilon A. \tag{4}$$

If the difference capacities ΔC is measured, the displacement x can be determined from the following algebraic equation of the second degree

$$\Delta C x^2 + \varepsilon A x - \Delta C d^2 = 0. \tag{5}$$

In micromechanical systems the capacities are very low, then for small movements the term ΔCx^2 can be neglected. From the equation (4) can determine the displacement x [7-8]

$$x \approx \frac{\Delta C d^2}{\varepsilon A}. \tag{6}$$

This means that the displacement is approximately proportional to the capacitance difference ΔC. On the other hand, if one takes account of Hook's law for elastic element, and write Newton's second law for moving mass

$$ma = kx = k\frac{\Delta C d^2}{\varepsilon A}. \tag{7}$$

It is found that acceleration is a measure of the difference ΔC in capacities of the two variable capacitors.

The total electrical charge on the movable plate is zero, which gives to the next relationship between the potentials and capacities

$$(V_x + V_0)C_1 + (V_x - V_0)C_2 = 0. \tag{8}$$

From this relationship, the potential of the mobile armature V_x is determined

$$V_x = \frac{C_1 - C_2}{C_1 + C_2}V_0 = \frac{\Delta C}{C_0}V_0. \tag{9}$$

Replacing the difference ΔC from equation (7) to equation (9) is obtained

$$V_x = \frac{m}{k}\frac{a}{d}V_0. \tag{10}$$

It can be seen from the equation (10) that if an oscillator gives the square waves V_0, at the input of the electric circuit, represented in the figure 2, then the output signal V_x is proportional to the acceleration and has the same waveform as V_0. This output voltage gives not only a signal proportional to acceleration but also its sign. The Electronic Central Unit can evaluate this signal and make decision how to act based on this information.

3. Finite element analysis

In order to determine the natural frequencies of a G-type satellite, a satellite that incorporates an acceleration sensor, the software provided by ANSYS 13 was used, and the finite element modeling was followed by the Finite Element Analysis (FEA).

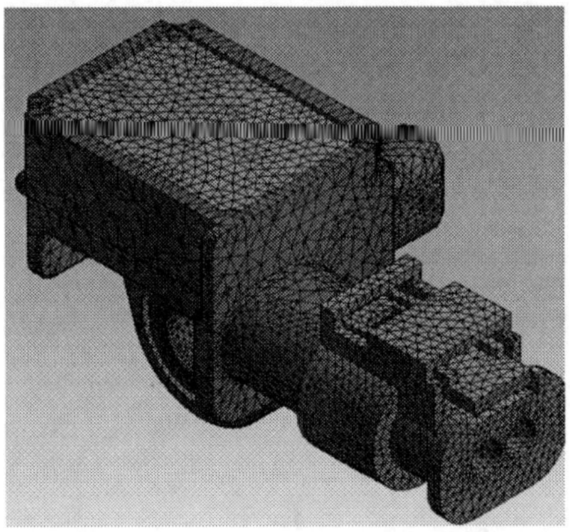

Figure 3. Finite element discretization of satellite.

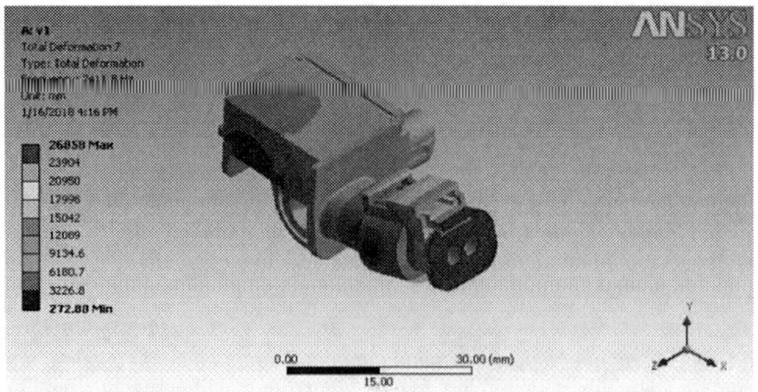

Figure 4. First vibration mode in OXY plane for free-free vibration.

IOP Conf. Series: Materials Science and Engineering **416** (2018) 012024 doi:10.1088/1757-899X/416/1/012024

In the construction of the mathematical model based on finite elements a three-dimensional (3D) model was used, with geometric elements having the hexahedral shape and each node having three degrees of freedom. For the modal analysis of free satellite vibrations, i.e. the determination of natural frequencies and natural vibration modes, the ANSYS module solves numerically a matrix differential equation of an undammed vibrant system given in the form:

$$[M]\{\ddot{u}\}+[K]\{u\}=\{0\}, \tag{11}$$

where [M] and [K] are mass and rigidity matrices, respectively {u} are nodal displacements. In the absence of damping, the vibrations of the nodes can be expressed by harmonic functions such as:

$$\{u\}=\{\phi\}_i\cos(p_i t+\varphi_i), \tag{12}$$

where $\{\phi\}_i$ are the eigenvectors and p_i are corresponding natural circular frequencies. The eigenvalues problem leads to solve the determinant:

$$\left|-p^2[M]+[K]\right|=0. \tag{13}$$

In these case, for the construction of the 3D model, it was chosen the geometry of a solid element named SOLID 186, which is defined by 20 nodes, each having three degrees of freedom representing displacements in three perpendicular directions. This type of finite element exhibits quadratic displacement behavior. A G-type satellite is meshed into finite elements as can be seen in the figure 3 and includes: a housing, a bushing, a Printed Circuit Board (PCB) and a polyurethane potting.

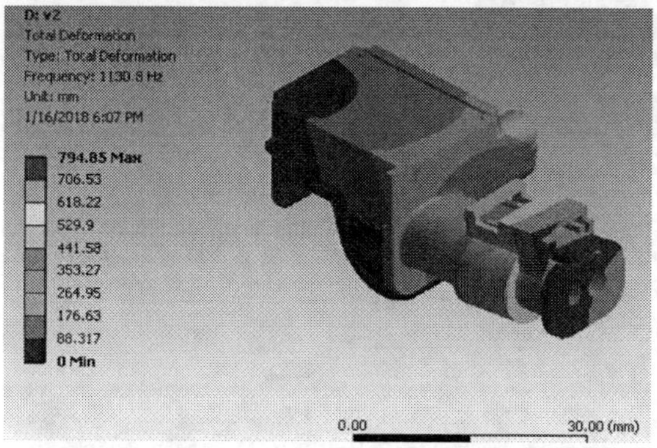

Figure 5. First mode in OXY plane for fixed bushing.

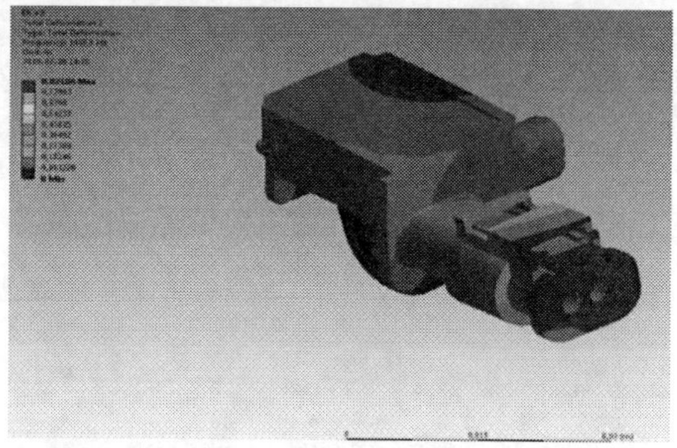

Figure 6. First mode in OXZ plane for fixed bushing.

The mechanical characteristics of the component materials are as follows: for housing with harness connector the density is 1550 kg/m^3, the longitudinal elastic modulus is 9.7 GPa, the Poisson coefficient is 0.3; -for bushing the density is 7,850 kg/m^3, the modulus of longitudinal elasticity is 200 GPa and Poisson ratio is 0.3; -for potting the density is 1100 kg/m^3, the modulus of longitudinal elasticity is 10 MPa, and the Poisson ratio is 0.3; for PCB, the density is 2 500 kg/m^3, the longitudinal elastic modulus is 18.4 GPa, the Poisson coefficient is 0.39.

Figure 5 shows the first mode of vibration in free-free border conditions. This mode has the frequency f_1=2411.8Hz and the dynamic deformations occur in the OXY plane. Figures 6, and 7 show two bending modes of vibrations. The first one is located in the OXY plane and has its natural frequency f_1=1130.8 Hz and the other one is located in the OXZ plane and has its natural frequency f_2=1419.8Hz.

4. Experimental technique

To validate the results obtained by the numerical modeling of the dynamic behavior of a satellite, the technique of the resonance diagram method was used. For this purpose an experimental setup was made consisting of the following components: **1** laser vibrometer, **2** shaker, **3** specimen, **4** accelerometer, **5** charge amplifier, **6** function generator, **7** power supply, **8** oscilloscope, **9** computer display and **10** data acquisition board and PC. The function generator has the roll of producing a sinusoidal function with the required frequencies and is placed in the experimental chain at the input of the power amplifier to give excitation to the shaker. The vibration generator receives the signal from the power amplifier via the function generator and vibrates the specimen at the given frequency range. To view the excitation signal of the specimen through the shaker, an oscilloscope is used. The signal from the piezoelectric accelerometer enters the charge amplifier and then is applied to the oscilloscope. Given that the mass of the satellite is smaller than the mass of a piezoelectric accelerometer, which would affect the measurement accuracy, a laser vibrometer is used to determine the specimen vibration amplitudes.

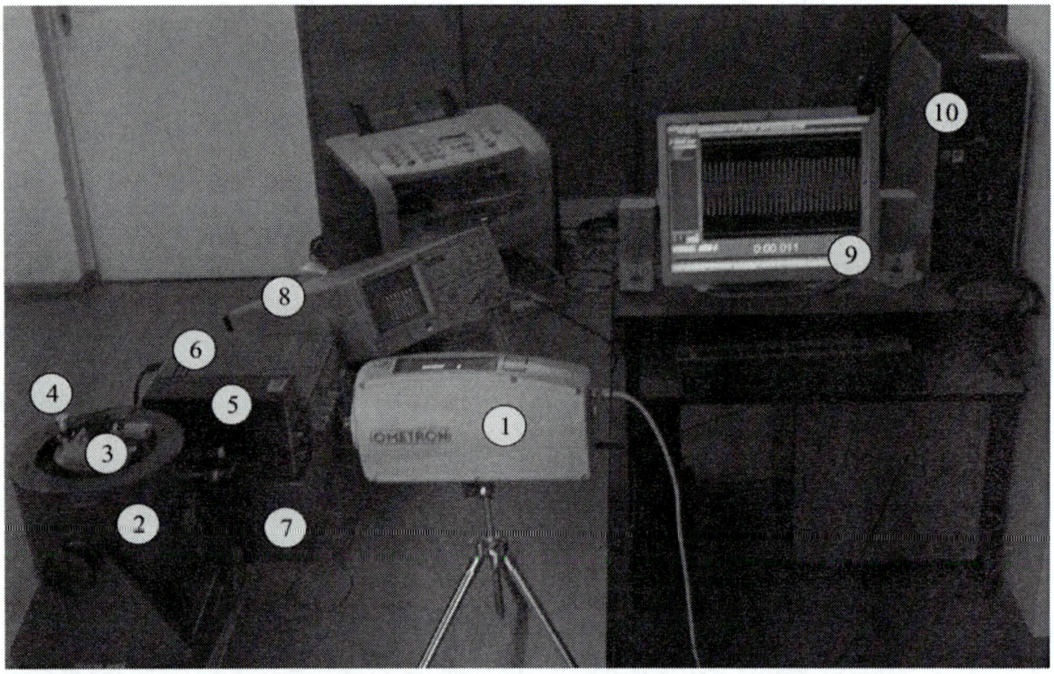

Figure 7. Experimental stand for vibration measurements.

5. Experimental results and conclusions

To obtain the experimental results, the satellite was attached to a shaker in the same way it is attached to the car. An anti-rotation device was also used, so the specimen was vibrated in the vertical direction in the 250-1800 Hz range. The measurement of the amplitudes of the forced vibrations was given by a Ometron type Vibrometer. By rotating the shaker to 90°, measurements were made with the vibrating satellite in the horizontal direction.

The graphical representation of the amplitude values given by the laser vibrometer for the excitation frequencies generated by the function generator is shown in Figure 8. It is the resonance diagram of the satellite. From the measurements, but also from the graph, it results that it has two peaks in the range 250-1800 Hz. The two peaks of the amplitudes determine the frequencies at which the resonances occur. In the OXZ plane the resonance takes place at the frequency f_1=1087.5 Hz, and in the OXZ plane it takes place at the frequency f_2 =1437.5 Hz.

Modal Analysis (MA) based on the Finite Element Method (FEM), for the satellite that was the similar boundary conditions to those in the dynamic testing, gives for the two frequencies modes f_1=1130.8Hz and f_2=1419.8 respectively. Experimental validation of the numerical evaluations obtained by FEA is found.

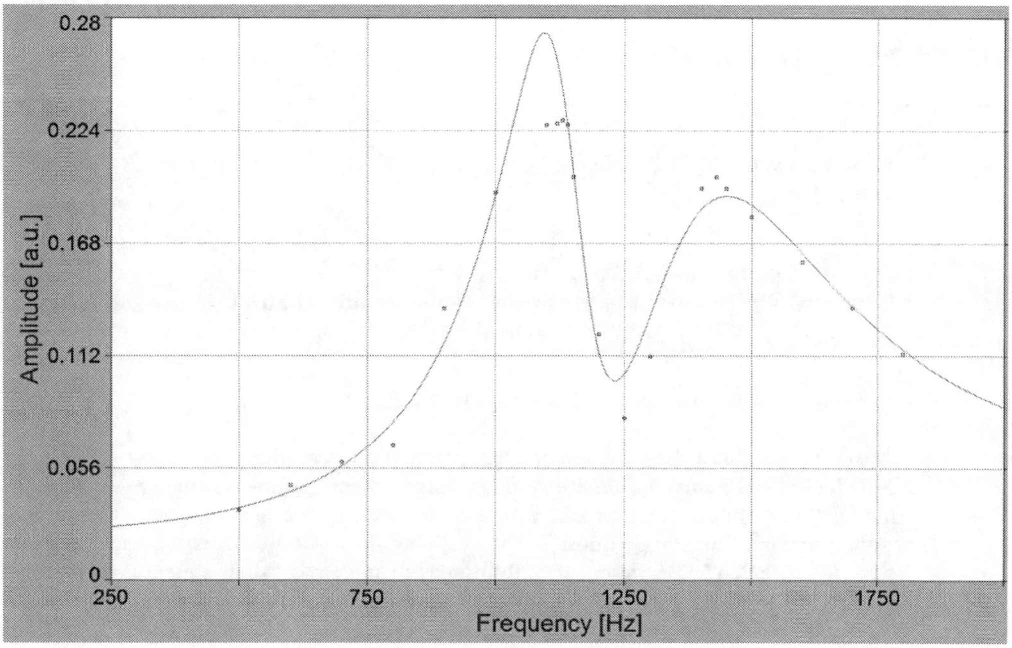

Figure 8. Resonant diagram.

6. References

[1] Deac S, Perescu A, Simoiu D, Nyaguly E, Crastiu I and Bereteu L 2018 Modeling and simulation of cars in frontal collision *Materials, Sciene and Engeeniring* **294** pp 1-12

[2] Albarbar A and Teay S.H. 2017 *Advanced Mechatronics and MEMS Devices II, Microsystems and Nanosystems* ,MEMS Accelerometers: Testing and Practical Approach for Smart Sensing and Machinery Diagnostics. Zhang D., Wei B. (Basel:Springer International Publishing Switzerland) pp 19-4

[3] Zhang D and Wei B 2017 *Advanced Mechatronics and MEMS Devices II, Microsystems and Nanosystems,* ed (Basel:Springer International Publishing Switzerland) ed

[4] Albarbar A, Sinha J and Starr 2009 Performance evaluation of MEMS accelerometers J Meas **42 (5)** pp 790-795

[5] Analog Devices 2015 MEMS accelerometers. Analog Devices. http://www.analog.com/en/products/mems/mems-accelerometers.html.

[6] Cioara T Gh 1999 *Tehnici Experimentale in Ingineria Mecanica-Traductoare si Senzori* (Timisoara, Editura Politehnica Romania)

[7] Lyshevski S E 2002 *MEMS and NEMS: systems, devices and structures*, CRC Press LLC, USA.

[8] Giurgiutiu V and Lyshevski S E 2009 *Micromechatronics, Modeling, Analalysis and Design with MATLAB^R*, CRC Press LLC, USA.

IOP Conf. Series: Materials Science and Engineering **416** (2018) 012025 doi:10.1088/1757-899X/416/1/012025

Some Laboratory Tests Regarding the X20Cr13 Martensitic Stainless Steel Behaviour

E Birtărescu[1], M D Nedeloni[1,*], P C Pedrali[2], C V Câmpian[1], L Nedeloni[1], T Ene[1] and S L Bogdan[1]

[1]Eftimie Murgu University of Reşiţa, Faculty of Engineering and Management, Piaţa Traian Vuia, No. 1-4, 320085, Reşiţa, Romania
[2]Regional University of Northwestern Rio Grande do Sul, Department of Exact Sciences and Engineering, Rio Grande do Sul, Brazil

*Corresponding author: m.nedeloni@uem.ro

Abstract. In this paper, beside the documentation regarding the many laboratory tests of X20Cr13 steel, the cavitation erosion, dry sliding wear, hardness respectively corrosion of this martensitic stainless steel was investigate. As cavitation erosion aspects, the authors present some preview results and continue to investigate the cavitation erosion resistance of X20Cr13 steel for an extended duration of 900 minutes of the cavitational attack, duration divided into 30 periods of 30 minutes. All this research through the stationary specimen method was done by using a cavitation stand and the obtained results were presented through cumulative mass loss and cavitation erosion rate versus time curves. As dry sliding wear aspects, the research was done with a CSM Tribometer through the pin-on-disk method and the obtained results were presented by means of the coefficient of friction and the wear rate. As hardness aspects, the measurement was made through the Vickers method using a Zwick/Roell microhardness tester. Finally, as corrosion aspects the research was done with a three-electrode cell using potentiodynamic polarization method and the obtained results were presented as corrosion curve in a logarithmic form.

1. Introduction

Hydraulic machines (hydraulic pumps and turbines) have a certain lifetime, which extends from operation to laying off of works [1,2]. In the case of hydraulic turbines, they can operate over 30 years (over 100.000 hours) [3]. Therefore, during the operation of the hydraulic turbines components is given maintenance and various repairs until the capital repair or refurbishment [4].

A hydraulic turbine requires different components, which are made of different materials, materials that encounter different types of wear and corrosion [5-7]. According to [8], in choosing the materials correctly, they are based on several criteria and requirements as follows: functional, technological, economic and social. For the manufacture of large components of hydraulic turbines, especially steel, cast iron and bronze are used [2]. For example, the rotor of a hydraulic turbine is cast from stainless steel [9]. Several grades of stainless steels are the following 12Cr-4Ni, 13Cr-1Ni, 13Cr-4Ni, 16Cr-5Ni, 18Cr-8Ni, 18Cr-10Ni, steels which must have very good properties for a good cavitation behavior and implicitly against cavitation erosion [10-13].

Like a type of wear, cavitation erosion occurs as a result of the cavitation phenomenon encountered in the liquid medium where the hydraulic turbine rotors work. Different research on the cavitation erosion of materials respectively on the cavitation phenomenon has been carried out by several authors

[14-21]. Also, researches on the hydraulic turbine components reconditioning were made due to its aspects on economics of raw materials, materials and energy [22,23]. For example, in the case of hydraulic turbine rotors, repairs of their damaged surfaces can be done in situ in a hydropower plant, where a modern method is the 3D scanning of the affected area and replaced the empty space with the one deposited by the Rapid Prototyping technique [24]. There are also other different techniques for the rehabilitation of areas affected by different wear types which most often are thermal treatments, coatings deposition, welding repairs and redesign of the components geometry [2].

Welding repair of rotor blades affected by certain imperfections, cracks and cavitation erosion includes several stages, such as: polishing of areas with imperfections of up to 20 mm; eliminate them by milling, drilling or grinding of areas with imperfections greater than 20 mm; application of heat treatment (where it was found that a stress-relief and annealing heat-treatment after welding repair and a heat treatment of the joining is necessary [25-29]; such thermal treatments can be increased by the hardness of the used steels) and finally the visual control with penetrating liquids.

In addition to the use of stainless steels in the casting of hydraulic turbine rotors, they also have a variety of other applications [30-33], such as steam turbines, water and steam valves, or machines working in corrosive environments. In all these areas, wear (cavitation erosion and sliding wear) occur as well the corrosion, corrosion which means the materials deterioration and degradation under the action of chemical agents from the environment, starting from the surface the material [34-37].

In this way, given that wear and corrosion are inevitable phenomena that cause industrial economic losses and in agreement with many research which have been published [38-42], this paper, through many laboratory tests, reports the experimental results regarding the cavitation erosion, dry sliding wear, hardness respectively corrosion of the X20Cr13 martensitic stainless steel.

Regarding the cavitation erosion tests, based on the reference [16] where a part of the authors tested two samples of this X20Cr13 steel for 180 minutes (with periods of 5, 10 and 15 minutes) and obtained for the four surfaces the following material losses: 9.32, 9.25, 10.25 and 8.68 mg; in this paper the authors continue the cavitation erosion tests for an extended time of 900 minutes of cavitation attack only for sample no. 2, surface 4 (surface with less loss) and investigates for the X20Cr13 steel, the dry sliding wear behavior, microhardness and corrosion.

2. Experimental procedure

The laboratory tests were based on the following ASTM and ISO standards: G32-92 and G32-10 standards for cavitation erosion; ASTM G99-95a standard for dry sliding wear; ASTM E384 for the microhardness tests respectively ISO 13129 for electrochemical measurement.

The experimental conditions for the cavitation erosion tests on the cavitation stand (vibratory cavitation apparatus) [43] were: frequency value must set of 20 ±0.5 kHz respectively amplitude of 50 µm; the water container (with distilled water) must be cylindrical with the depth of 100 ±10 mm; the water temperature shall be 25 ± 2°C; the working samples shall have a diameter of 15.9 ± 0.05 mm and the sample surface must be smooth. Cavitation erosion resistance testing on the experimental stand can be done by two methods: direct cavitation (for a short time) or indirect - stationary specimen method (slower process). For this, the used method is the second one, and each sample was tested for 30 minutes. Horn and sample are both immersed in the water container and the cavitation bubbles are generated by the horn (sonotrode) above the samples at 0.6 mm distance. At end of the experimental procedure, each sample was weighed with a digital balance to measure the material loss.

To determine the dry sliding wear resistance, the CSM Tribometer was used through the pin-on-disk method (POD). The used parameters were: load of 10 N, linear speed of 15 $cm \cdot s^{-1}$, motor speed of 475.91 rpm, distance of 750 m, working radius of 3.01 mm, temperature of 20°C and humidity of 50% (like environmental conditions). As a counter body, a steel 100Cr6 ball with a diameter of 6 mm was used. The distance of 750 m meant $39.6 \cdot 10^4$ laps respectively a time of 83 minutes.

The microhardness messurements was carried out with a ZWICK-ROELL apparatus using Vickers method [44]. A 0.3 kgf indentation force was maintained during 15 seconds. The hardness value was calculate as a average of the 5 indentations.

IOP Conf. Series: Materials Science and Engineering **416** (2018) 012025 doi:10.1088/1757-899X/416/1/012025

The corrosion resistance was achieved by potentiodynamic polarization by means of a PGP 21 potentiostat/galvanostat in a three electrode electrochemical cell, using the calomel saturated electrode as the reference electrode. The counter electrode was a platinum disc with the surface $1 cm^2$, and the X20Cr13 sample in 3.5% neutral NaCl solution as the working electrode.

Mechanical properties and chemical composition of X20Cr13 steel are presented in Table 3 and Table 4.

Table 1. Mechanical properties.

Tensile strength, Rm ($N\cdot mm^{-2}$)	Yield strength, Rp 0.2 ($N\cdot mm^{-2}$)	Elongation at yield, A (%)	Elongation at break, Z (%)
853	662	17	62.8

Table 2. Chemical composition (wt.%).

C	Si	Mn	P	S	N	Cr	Mo	Ni	Cu	Fe
0.21	0.32	0.42	0.02	0.025	0.03	12.7	0.06	0.28	0.06	85.87

3. Results and discussion

3.1. Cavitation erosion

Table 3 present the obtained results regarding the cavitation erosion behavior of the X20Cr13 steel and figure 1 and figure 2 present the graphs of cumulative mass loss and cavitation erosion rate versus time curves.

Table 3. Obtained cavitation erosion results.

Cumulated time	Period	Sample mass	Mass loss		Cavitation rate	Cumulated time	Period	Sample mass	Mass loss		Cavitation rate
			On period	Cumulative					On period	Cumulative	
t (min)	Δt (min)	m (mg)	Δm (mg)	mc (mg)	v_{ec} ($mg\cdot h^{-1}$)	t (min)	Δt (min)	m (mg)	Δm (mg)	mc (mg)	v_{ec} ($mg\cdot h^{-1}$)
0	0	15913.29	0	0	0.000	480	30	15862.05	3.31	51.24	6.560
30	30	15910.92	2.37	2.37	5.380	510	30	15858.8	3.25	54.49	6.310
60	30	15907.91	3.01	5.38	6.430	540	30	15855.74	3.06	57.55	6.160
90	30	15904.49	3.42	8.8	6.450	570	30	15852.64	3.1	60.65	6.120
120	30	15901.46	3.03	11.83	6.440	600	30	15849.62	3.02	63.67	6.130
150	30	15898.05	3.41	15.24	6.540	630	30	15846.51	3.11	66.78	6.240
180	30	15894.92	3.13	18.37	6.600	660	30	15843.38	3.13	69.91	6.280
210	30	15891.45	3.47	21.84	6.800	690	30	15840.23	3.15	73.06	6.670
240	30	15888.12	3.33	25.17	6.600	720	30	15836.71	3.52	76.58	7.140
270	30	15884.85	3.27	28.44	6.340	750	30	15833.09	3.62	80.2	6.970
300	30	15881.78	3.07	31.51	6.180	780	30	15829.74	3.35	83.55	6.750
330	30	15878.67	3.11	34.62	6.250	810	30	15826.34	3.4	86.95	6.620
360	30	15875.53	3.14	37.76	6.430	840	30	15823.12	3.22	90.17	6.370
390	30	15872.24	3.29	41.05	6.650	870	30	15819.97	3.15	93.32	6.530
420	30	15868.88	3.36	44.41	6.880	900	30	15816.59	3.38	96.7	6.990
450	30	15865.36	3.52	47.93	6.830	-	-	-	-	-	-

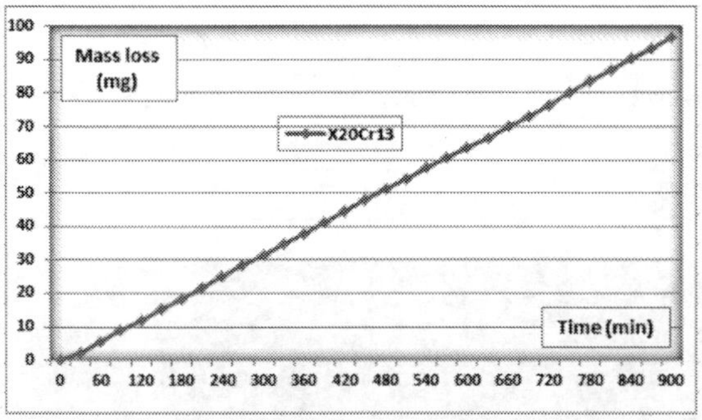

Figure 1. Cumulative mass loss vs time.

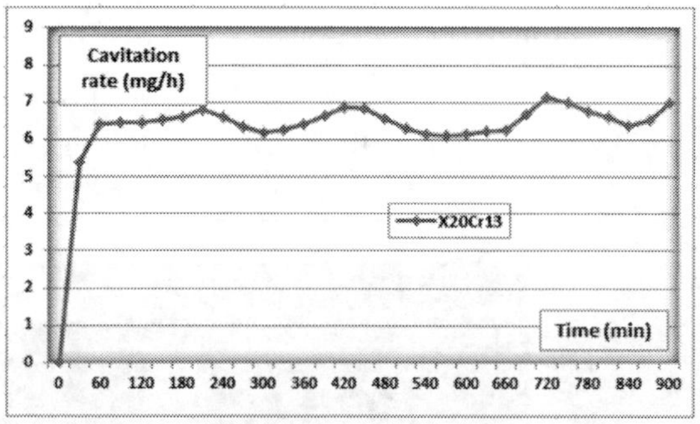

Figure 2. Cavitation erosion rate vs time.

It is observed that the mass loss process occurs linearly with an average of 3.2 mg on a period. The total mass loss was 96.7 mg, value which classifies the X20Cr13 martensitic stainless steel as having a good cavitation erosion resistance [45].

The cavitation erosion rate cross the incubation and acceleration stage for the first 60 minutes of the test and then it is maintain at the maximum rate stage [46] between the range 6 - 7 $mg \cdot h^{-1}$ until the end of the test.

3.2. *Dry sliding wear*

Evolution of friction coefficient is presented in figure 3 and as results in table 4. The measured values for wear rate, volume loss, width and depth of sample wear track are presented in table 5.

Some images regarding the wear track profiles are shown in figure 4.

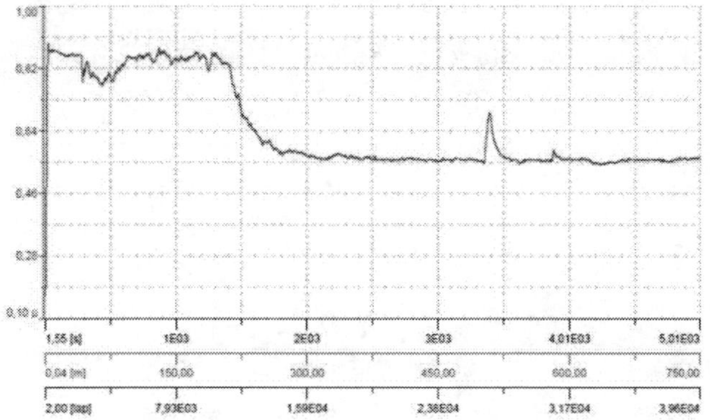

Figure 3. Evolution of friction coefficient.

The coefficient of friction evolution was starting at the 0.165 value (table 4) and was rapidly increasing until to the aprox. 0.96 value. Than vary with a decreasing trend for a distance of 75 m with the 0.82 - 0.96 value range (figure 3).

At the aprox. 225 m decreases for the next 75 m around 0.55 and is maintained around this value until around 525 m. Also, from figure 3, it can be seen that the coefficient of friction value increases suddenly to a value of 0.70, but immediately returns to the same level of 0.55 to the final distance of 750 m. For the value of 0.70, it was considered that during the test, eroded material was deposit between the ball and the sample.

Table 4. Friction coefficient results.

Start	Min	Max	Mean	Std. Dev.
0.165	0.165	0.891	0.646	0.128

Table 5. Measured values for wear rate, volume loss, width and depth of sample wear track.

Wear track depth, ***h*** (μm)	Wear track width, ***s*** (μm)	Volume loss, ***V*** (mm^3)	Wear rate, ***K*** ($mm^3N^{-1}m^{-1}$)
52.650	1397.542	0.9256	$1.23 \cdot 10^{-4}$

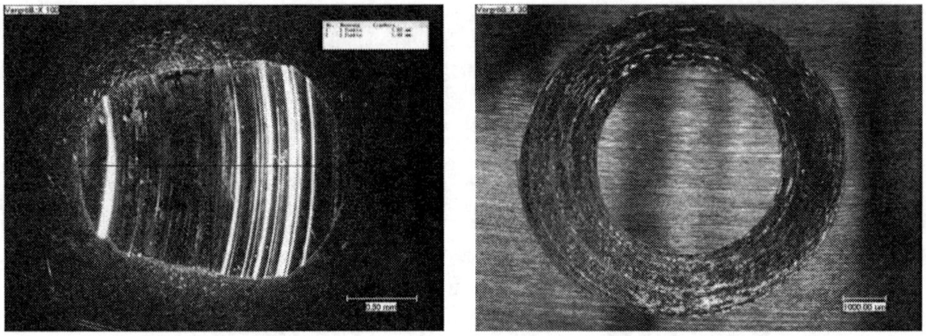

a) ball wear track *b)sample wear track*

Figure 4. Wear track profiles details.

Regarding the wear track profiles acquired with a digital microscope and with a confocal 3D laser scanning microscope, the length and the width of the ball wear was 1.92 mm respectively 1.48 mm.

3.3. Microhardness

The microhardness values (HV0.3) are presented in table 6 and in figure 5.

Table 6. Microhardness values (HV0.3).

Attempt 1	Attempt 2	Attempt 3	Attempt 4	Attempt 5	Mean value
267	261	245	252	249	255

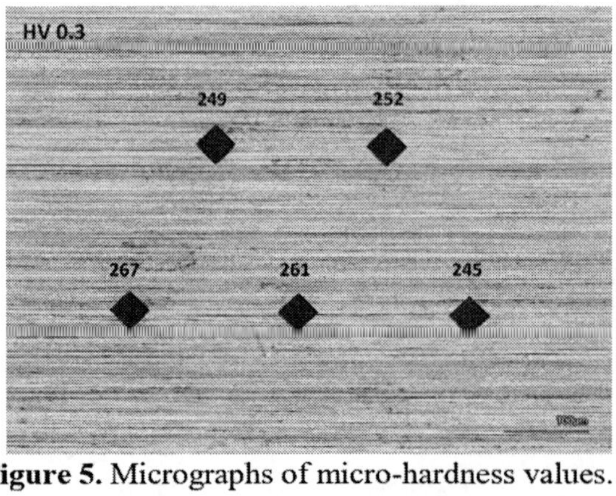

Figure 5. Micrographs of micro-hardness values.

The mean value of the microhardness is 255 HV0.3 and between the 5 indentations are small differences. The obtained values are characteristic for the martensitic stainless steels hardness [47].

3.4. Electrochemical corrosion

The obtained electrochemical corrosion values was determined by Tafel slope method. The logarithm representation of corrosion curve is shown in figure 6.

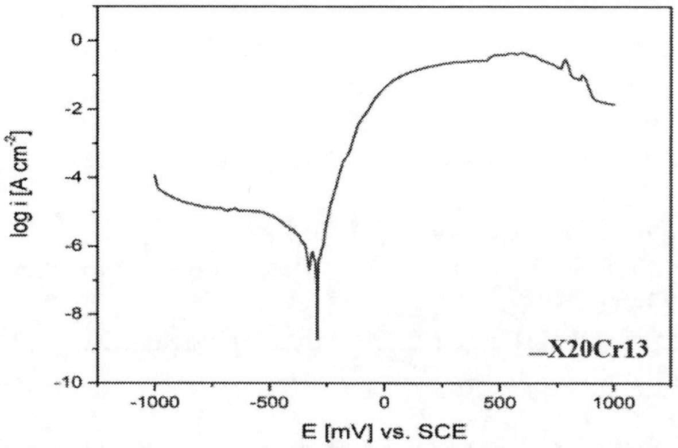

Figure 6. Logarithm representation of X20Cr13 corrosion curve.

From the logarithm representation of corrosion curve, it can be see a passivation tendency around 250 mV and that the anodic branch has a lower current densities. The current density and the potential indicate a lower reaction rate and a good / moderate corrosion resistance of X20Cr13 steel [48].

3.5. Examination of sample surfaces after tests

For the tests regarding the cavitation erosion, dry sliding wear and corrosion, figure 7 show some images with sample surfaces after this tests.

a) cavitation erosion *b) dry sliding wear* *c) corrosion*

Figure 7. Images of wear and corrosion of the sample surfaces after the tests.

It can be seen from the figure 7 that the sample surfaces showed crakes and shallow erosions generated by cavitation erosion over the entire surface; sever wear in the case of dry sliding wear test respectively an intense material destruction in the electrochemical corrosion test.

4. Conclusions

This paper has referred to the cavitation erosion that destroy rotor blades of hydraulic turbines and also to the sliding wear and corrosion that makes damege of the material surface respectively presented the experimental procedure, working parameters and the obtained results for the cavitation erosion, dry sliding wear, microhardness and corrosion of the X20Cr13 steel behaviour.

From the experimental results, it can be highlighted that the analyzed X20Cr13 steel, like a martensitic stainless steel, shown a good cavitation erosion resistance (due to chromium content), a low dry sliding wear resistance (as the coefficient of friction and wear rate results show) respectively a good / moderate corrosion resistance (that because the higher Cr content, which is favoring an easier passivation), all this compared to other materials investigated from the authors.

Regarding the Vickers microhardness obtained results, these are reliable and typical for this martensitic stainless steel.

5. References

[1] Dunca G, Isbăşoiu E C, Călinoiu C, Bucur D M and Ghergu C 2008 5th National Conference of Romanian Hydropower Engineers, Romania, May 22-23, pp 181-190
[2] Frunzaverde D, Câmpian V, Nedelcu D, Gillich G R and Mărginean G 2010 *WSEAS Transactions on Applied and Theoretical Mechanics* **5**(3) 122-31
[3] Dunca G, Bucur D M, Isbăşoiu E C, Călinoiu C and Ghergu C M 2012 *UPB Scientific Bulletin, Series D: Mechanical Engineering* **74**(1) 59-66
[4] Haţiegan C et al. 2016 IOP Conf. Series: Materials Science and Engineering, Hunedoara, Romania, May 25-26, pp 1-14
[5] Afronie E M, Manescu T S and Ianici S KOD 2012, Hungary, May 24-26, pp 515-18
[6] Ivan A, Ivan M and Both I 2010 *WSEAS Transactions on Applied and Theoretical Mechanics* **5**(3) 187-96
[7] Mesa D H, Toro A, Sinatora A and Tschiptschin A P 2003 *Wear* **55** 139-45
[8] Marta C, Suciu L and Nedelcu D 2009 *Metal Int* **14**(10) 33-38
[9] Padurean I, Trusculescu M, Arpad F, Nedelcu D and Padurean E 2008 *Metal Int* **13**(10) 60-6
[10] Tugui C A, Vizureanu P, Perju M C, Nejneru C, Istrate B and Savin A 2017 *Key Eng Mater* **750** 85-90
[11] Tugui C A, Vizureanu P, Iftimie N and Steigmann R 2016 IOP Conf. Series: Materials Science and Engineering, Romania, June 09-10, pp 1-8
[12] Bordeasu I, Popoviciu M O, Mitelea I, Balasoiu V, Ghiban B and Tucu D 2007 *Rev Chim* **58**(12) 1300-04
[13] Bordeasu I, Micu L M, Mitelea I, Utu I D, Pirvulescu L D and Sirbu N A 2016 *Mater Plast* **53**(4) 781-86
[14] Gottardi G, Tocci M, Montesano L and Pola A 2018 *Wear* **394-395** 1-10
[15] Pola A, Montesano L, Tocci M and La Vecchia G M 2017 *Materials* **10**(3) 1-13
[16] Nedeloni M D, Potoceanu N, Florea D and Chirus D 2014 *Mater Sci Forum* **782** 251-56
[17] Potoceanu N, Nedeloni M D, Chirus D and Florea D 2014 *Mater Sci Forum* **782** 257-62
[18] Nedelcu D, Nedeloni M D and Lupinca C I 2014 *Mater Sci Forum* **782** 263-68
[19] Taillon G, Pougoum F, Lavigne S, Ton-That L, Schulz R, Bousser E, Savoie S, Martinu L and Klemberg-Sapieha J E 2016 *Wear* **364-365** 201-10
[20] Candel I, Bunea F, Dunca G, Bucur D M, Ioana C, Reeb B and Ciocan G D 2014 IOP Conference Series-Earth and Environmental Science, Canada, September 22-26, pp 1-8
[21] Digulescu A, Murgan I, Candel I, Bunea F, Ciocan G, Bucur D M, Dunca G, Ioana C, Vasile G and Serbanescu A 2016 IOP Conference, France, July 04-08, pp 1-10
[22] Santa J F, Blanco J A, Giraldo J E and Toro A 2011 *Wear* **271** 1445-53
[23] Cîndea L, Haţiegan C, Pop N, Negrea R, Răduca E, Gillich G R, Dănuţ M, Nedeloni M D 2016 *Appl Therm Eng* **103** 1164-75
[24] Nedelcu D, Cojocaru V, Ghican A, Peris-Bendu F and Avasiloaie R 2015 *Mater Plast* **52**(4) 475-79
[25] Ren F, Chen F and Chen J 2014 *Adv Mater Sci Eng* **2014** 1-17
[26] Pola A, Montesano L, Gelfi M and La Vecchia G M 2012 *Metall Ital* **9** 45-49
[27] Marginean G and Utu D 2012 *Applied Surface Science* **258** 8307-11
[28] Petrogalli C, Montesano L, Pola A, Gelfi M, Ghidini A and LaVecchia G M 2015 *Procedia Engineering* **109** 154-61
[29] Hong S, Wu Y, Zhang J, Zheng Y, Zheng Y, Lin J 2016 *Ultrason Sonochem* **31** 563-69
[30] Bhagi L K, Rastogi V and Gupta P 2017 *J Mech Sci Technol* **31**(1) 17-27
[31] Štrbac N, Marković I, Mitovski A, Balanović L, Živković D and Grekulović V 2017 *Rev Metal Madrid* **53**(1) 1-7
[32] Manova D, Mandl S, Neumann H and Rauschenbach B 2005 *Surf Coat Tech* **200** 137-40
[33] Radziszewska A, Kranzmann A, Dörfel I, Feijoo M M and Solecka M 2016 *Arch Metall Mater* **61**(3) 1607-12

[34] Pascal D T, Muntean R, Kazamer N, Marginean G, Brandl W and Serban V A 2016 NANOCON 2016 Czech Republic, October 19-21, pp 775-80
[35] Amiriyan M, Blais C, Savoie S, Schulz R, Gariépy M and Alamdari H D 2017 *Metals* **17**(177) 1-12
[36] Nová I and Machuta J 2014 *Manufacturing Technology* **14**(3) 381-87
[37] Pola A, Montesano L, Gelfi M and Roberti R 2011 ESAFORM, United Kingdom, April, pp 1051-56
[38] Bala P, Krawczyk J, Lukaszek-Solek A, Pawowski B and Dziurka R 2012 The 14th international conference on Metal Forming, Poland, September 16 - 19, pp 155-58
[39] Jakopčić M, Grilec K and Lisjak D 2012 *Technical Gazette* **19**(4) 893-8
[40] Kostadin T, Cukor G and Jakovljevic S 2017 *Adv Prod Eng Manag* **12**(2) 105-14
[41] Montesano L, Gelfi M, Pola A, Colombi P and La Vecchia G M 2013 *Metall Ital* **9**(2) 3-11
[42] Rozing G, Alar V and Marušić V 2015 *Interdisciplinary Description of Complex Systems* **13**(3) 461-471
[43] Chahine G L, Franc J P and Karimi A 2014 Laboratory Testing Methods of Cavitation Erosion *Advanced Experimental and Numerical Techniques for Cavitation Erosion Prediction (Fluid Mechanics and Its Applications)* eds K.-H. Kim et al. (Springer) chapter 2 pp 21-35
[44] Kazamer N, Pascal D T, Marginean G, Serban V A, Codrean C and Utu I D 2016 *Sol St Phen* **254** 71-6
[45] Bordeasu I, Mitelea I, Popoviciu M O and Chirita C 2011 METAL 2011, Czech Republic, May 18-20, pp 1-6
[46] Bordeasu I, Popoviciu M O, Salcianu L C, Ghera C, Micu L M, Badarau R, Iosif A, Pirvulescu L D and Podoleanu C E 2016 IOP Conference Series, Romania, May 25-26, pp 1-11
[47] Marušić V, Opačak I, Nedić B, Jovanović D and Rozing G 2016 *International journal for science and innovations for the industry* **4**(2) 39-41
[48] Kazamer N, Pascal D T, Marginean G, Serban V A, Brandl W and Valean P C 2016 NANOCON pp 383-8

Acknowledgment
We are very grateful to Mrs. Dr. Gabriela Mărginean from the Westphalian University of Applied Sciences Gelsenkirchen for the laboratory equipment.

IOP Conf. Series: Materials Science and Engineering **416** (2018) 012026 doi:10.1088/1757-899X/416/1/012026

Comparative Study on Dry Sliding Wear Resistance of Carbon Steel, Alloyed Steel and Cast Iron

L Nedeloni[1,*], Z I Korka[1], D T Pascal[2], N Kazamer[2,3] and M D Nedeloni[1]

[1]Eftimie Murgu University of Reşiţa, Faculty of Engineering and Management, Piaţa Traian Vuia, No. 1-4, 320085, Reşiţa, Romania
[2]Westphalian University of Applied Sciences Gelsenkirchen, Neidenburger Str. 43, 45897 Gelsenkirchen, Germany
[3]Politehnica University of Timişoara, Department of Materials and Manufacturing Engineering, Piaţa Victoriei, No. 2, 300006, Timişoara, Romania

*Corresponding author: l.nedeloni@uem.ro

Abstract. The aim of this paper is to study and compare the dry sliding wear resistance of carbon steel, alloyed steel and cast iron, knowing that the sliding wear is a type of deterioration that appears in various practical applications. Experimental research of this study was mainly performed using a tribometer (CSM Tribometer) according to the pin-on-disk method (POD), aiming to determine the coefficient of friction (by graphic representation) and wear rate (by analytical calculation). The tested materials were some iron alloys, namely carbon steels (C25, C45), alloyed steel (34CrNiMo6) and cast iron (EN-GJS-400-15) respectively. The tests were performed complying with ASTM G99 Standard, keeping the same working conditions for all the investigated materials. The results obtained in this study were presented through tables with values, graphs and through characteristic images acquired with a digital microscope and a 3D laser scanning microscope, as well as with the aid of a specialized software (MultiFileAnalyzer) concerning the wear track profiles of the counterbody and tested specimens. The investigations revealed a lower friction coefficient for the cast iron, correlated with a reduced wear rate of the material, which leads to the fact that this type of material has a better resistance to sliding wear resistance compared with the other analysed materials.

1. Introduction

Fe-C alloys (steels and cast irons) present a great importance due to their high produced quantity and their variety of applications [1]. Steels are used in various applications like automotive industry, machinery and components subjected to wear [2]. Moreover, cast irons are used due to their relatively low costs and improved tribological performance [3].

In the case of non-alloyed and low-alloyed steels, higher carbons content in the structure (due to perlite and free secondary cementite) increases also the tensile strength, hardness and wear resistance [4]. In the case of gray cast irons, although these have good characteristics for exploitation (like vibration priming capacity), the presence of graphite in their structure causes their characteristics to decrease, but for the gray cast irons with nodular graphite, the spherical shape reduces the negative effects of its presence in the metallic mass, so that the mechanical characteristics of these castings are closer to those of steels [5].

The behaviour of steels and cast irons against wear and corrosion is usually investigated through many tests with professional laboratory equipment like: tribometer, scratch tester, vibratory apparatus or others [6-16]. With the aid of the obtained experimental results, the wear resistance of these ferrous materials can be estimated. For example, by testing the materials using a tribometer through the pin-on-disk method (POD), the mass loss, volume loss, coefficient of friction (COF) and wear rate can be calculated/estimated [17].

In this way, regarding the dry sliding wear resistance of steels and cast irons, various tests were performed to investigate the influence of carbon content of steels on the most important mechanical properties of the materials [18]. Generally, it was found that the wear resistance was affected by the applied load (in the current case the load has been varied between 5 - 25 N) and moreover, it was revealed that a lower content of carbon provides less friction [19]. The friction coefficient as well as the wear rate increases when a higher sliding rate is applied [20], but the effect of load shows greater influence on wear in comparison to sliding speed [21]. Heat treatments like laser surface hardening process (in the case of the C45 alloyed steel) has been found to have a positive influence on the dry sliding wear resistance. The significantly enhanced wear resistance is attributed to the refined microstructure and to the presence of a mechanically metastable austenite phase obtained through the laser treatment [22]. Moreover, through tempering and cryogenic treatments (in case of spring steel), the results have revealed that the material condition has a significant influence on the wear performance [23]. The addition of alloying elements like manganese [24] or titanium and chromium induced a significant effect on the behaviour of cast iron wear process [25].

In the evaluation of the dry sliding wear resistance of materials with the aid of a tribometer, a significant interest it is shown regarding the surface deposition application technologies, like thermal spraying, vapor deposition or fluoropolymer coatings [26, 27]. Working with a tribometer, the testing parameters like load, speed or distance can be successively varied and different relations or new equations between these parameters can be found [28].

Accordingly, various studies present a great interest to investigate the sliding wear behavior of materials used in gears manufacturing and of different type of coatings applied on these materials respectively [29-35]. The study of dry sliding wear behaviour of steels in comparison to other systems is essential for their performance evaluation in different applications, where the wear mechanism involves adhesion, oxidation, strain hardening, or other structural changes [36].

Therefore, this work aims to study and compare the dry sliding wear resistance of three types of iron alloys which have different carbon content and which can be used in gears manufacturing. Also, through this comparative study, the authors understand better the sliding wear process by using a tribometer according to the pin-on-disk method (POD).

2. Materials and methods

The tested materials are two carbon steels C25 (1.0406) and C45 (1.0503), an alloyed steel 34CrNiMo6 (1.6582), which was heat treated especially for the armament industry and a gray cast iron with nodular graphite EN-GJS-400-15 (0.7040).

The chemical composition of these materials is presented in table 1.

Table 1. The chemical composition of the investigated materials (wt.%).

Alloys type	C	Si	Mn	P	S	Cu	Ni	Cr	Mo	Mg	Fe
C25	0.24	0.27	0.61	0.007	0.005	0.26	0.13	0.16	0.02	-	97.91
C45	0.46	0.28	0.73	0.008	0.008	0.05	0.03	0.06	0.01	-	97.67
34CrNiMo6	0.32	0.27	0.50	0.002	0.005	0.21	1.5	1.4	0.18	-	95.2
EN-GJS-400-15	3.5	2.06	0.22	0.050	0.015	0.04	0.03	0.04	0.01	0.038	93.99

The arrangement of the tribometer was a 100Cr6 steel ball with a diameter of 6 mm for the pin holder (static friction partner) loaded against as the testing specimens. According to the instructions of the ASTM G99 Standard, all tests should be made in the same working conditions regarding the tribological environment (temperature: 20°C; atmosphere: air and humidity: 50%).

The testing parameters are presented in table 2.

Table 2. Parameters for the dry sliding wear investigations.

Load (N)	Linear speed (cm s^{-1})	Motor speed (rpm)	Radius (mm)	Distance (m)	Distance (lap)	Test duration (s)
10	15	474.3	3	750	39525	5000

The coefficient of friction was monitored and registered as a graphic representation during the tests and the wear rate was calculated according to equation 1 [17]:

$$K = \frac{2\pi r \cdot h\left(3h^2 + 4s^2\right)}{6 \cdot L \cdot d \cdot s} \quad (1)$$

where: K is the wear rate; h - wear track depth; s - wear track width; L - applied force and d - distance.

Additionally, microhardness measurements were also considered for the investigated specimens, following the ISO 6507 standard. The microhardness was measured in 5 different areas using the Vickers method on a Zwick-Roell ZHVμ-S instrument with an applied force of 0.3 kgf.

3. Results and discussions

The variation of the coefficient of friction of the four tested specimens is presented in figure 1.

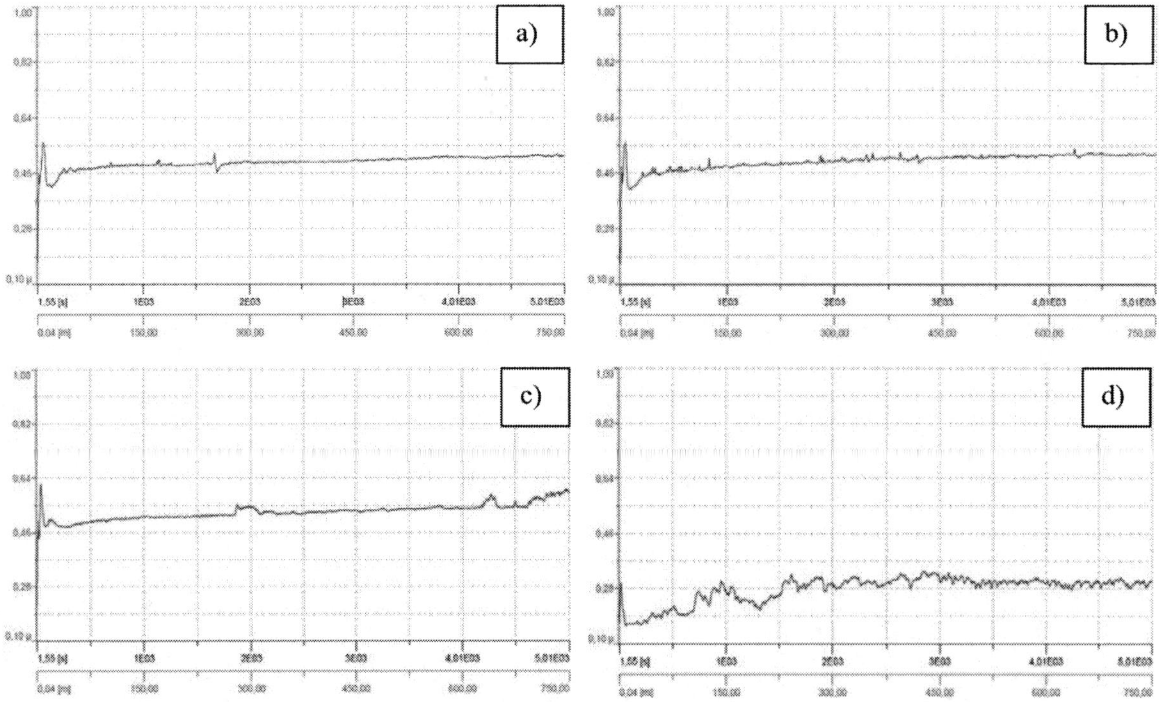

Figure 1. Variation of the coefficient of friction for a) C25; b) C45; c) 34CrNiMo6; d) EN-GJS-400-15 in time (s) and distance in meters.

The friction coefficient curves for the three steel grades are quite similar, with a maximum friction value between 0.17 - 0.22 at the beginning of the test and during the test, where the COF is fluctuate during the tests between 0.34-0.56 (here the steady state is reached [37]). For cast iron specimen (EN-GJS-400-15), COF reaches a maximum value of 0.24 at the beginning of the test and further, the value is maintained constant between 0.16 - 0.28 for a distance of 225 m and then at the end of the test (750 m) varies in the range of 0.28 - 0.34.

Table 3. Comparison of the friction coefficients values.

Alloys type	Start	Min	Max	Mean	Std. Dev.
C25	0.170	0.170	0.559	0.496	0.020
C45	0.216	0.169	0.558	0.500	0.025
34CrNiMo6	0.240	0.218	0.619	0.528	0.026
EN-GJS-400-15	0.170	0.158	0.340	0.279	0.042

Images for the specimens and ball wear track profile were acquired with a digital microscope (KEYENCE VHX-600) and with a confocal 3D laser scanning microscope (KEYENCE VKX-260K) respectively, as shown in figure 2-4.

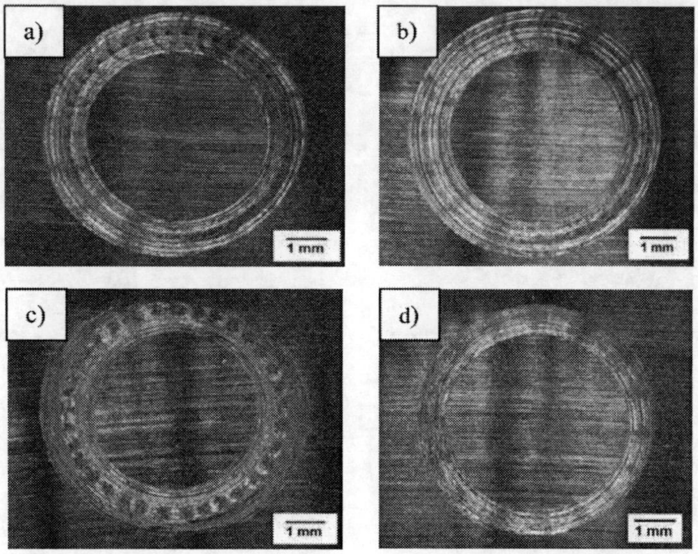

Figure 2. Specimens wear track profiles for a) C25; b) C45; c) 34CrNiMo6; d) EN-GJS-400-15.

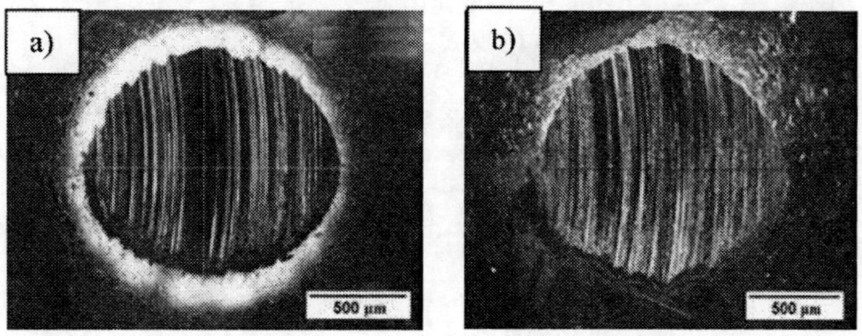

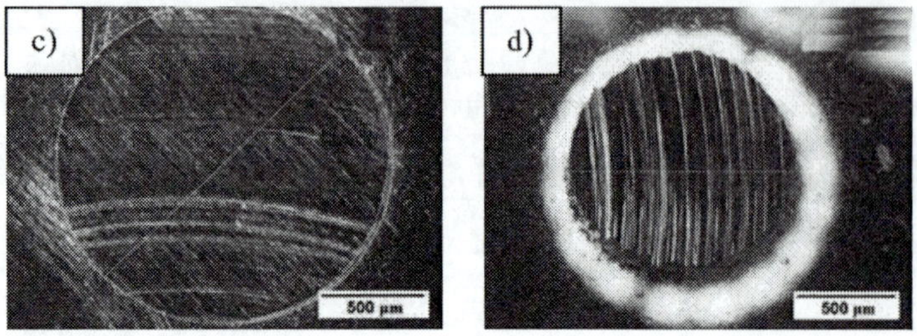

Figure 3. Ball wear track profiles for a) C25; b) C45; c) 34CrNiMo6; d) EN-GJS-400-15.

The characteristic width ***s*** and depth ***h*** of the specimens wear track profiles were measured with the MultiFileAnalyzer software and the overall calculated wear rate values are presented in table 4.

The width of the ball track profiles measured with the digital microscope is as follows: 1317 μm (C25), 1317.76 μm (C45), 1600.35 μm (34 CrNiMo6) and 1091.48 μm (EN-GJS-400-15) respectively.

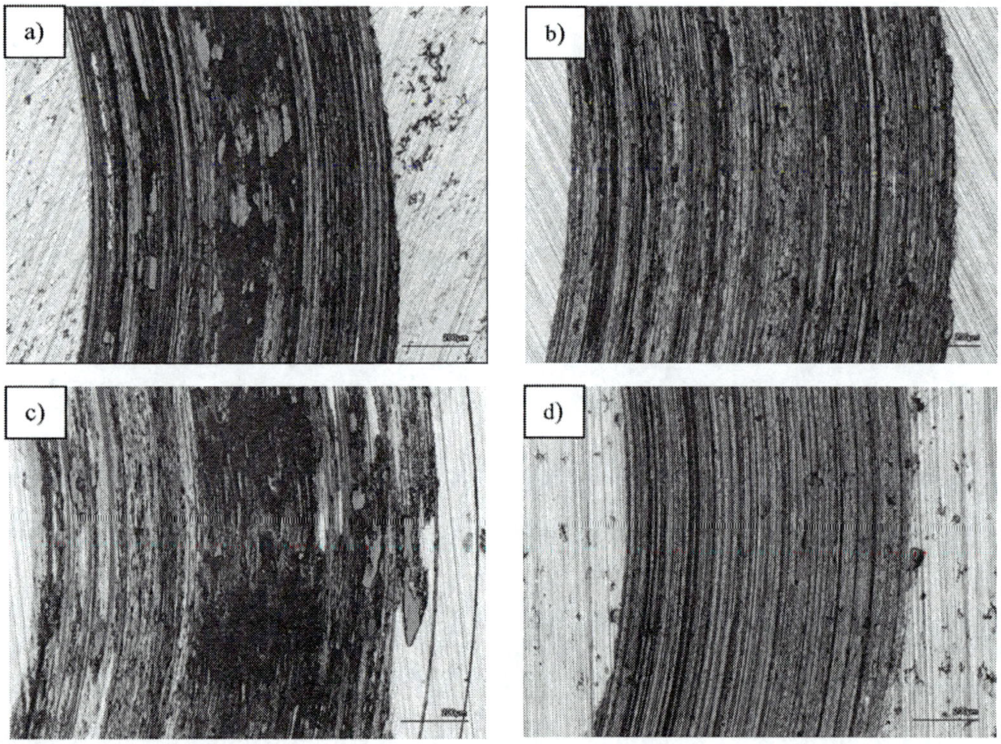

Figure 4. Area of specimens wear track profiles for a) C25; b) C45; c) 34CrNiMo6 and d) EN-GJS-400-15 (100X Magnification).

Table 4. Comparison of the wear rate values.

Alloys type	Wear track depth, ***h*** (μm)	Wear track width, ***s*** (μm)	Cross section area, ***A*** (μm^2)	Volume loss, ***V*** (mm^3)	Wear rate, ***K*** (mm^{3}N^{-1}m^{-1})
C25	9.421	941.182	5911.69	0.1114	$14.9 \cdot 10^{-6}$
C45	3.861	1161.34	2989.31	0.0563	$7.51 \cdot 10^{-6}$
34CrNiMo6	8.52	1161.55	6597.11	0.1244	$16.6 \cdot 10^{-6}$
EN-GJS-400-15	2.50	878.08	1465.24	0.0276	$3.68 \cdot 10^{-6}$

The occurrence of adhesive wear is highlighted by the removal of the material from the specimen surface, as observed in the wear track profiles presented in figure 4, particularly in the case of C25 (a) and 34CrNiMo6 (c).

The gray cast iron (EN-GJS-400-15) with nodular graphite presents a lower friction coefficient, correlated with a reduced wear rate of the material, which leads to the fact that this type of material has a better resistance to sliding wear compared with the other analyzed steels, due to the self-lubricating graphite character.

Therefore, 34CrNiMo6 suffers the most severe wear rate as compared to EN-GJS-400-15, in the presented tested conditions. The combined effect of adhesive wear is responsible for an inferior wear resistance of the C25 and C45 steels.

Correspondingly, the friction coefficient values obtained in this study for the investigated specimens are in good agreement with the dry sliding wear of steel-steel and steel-cast iron friction value range presented in literature [38-41].

Forwards, the microhardness tests results are highlighted in figure 5 and the microhardness values are presented in table 5.

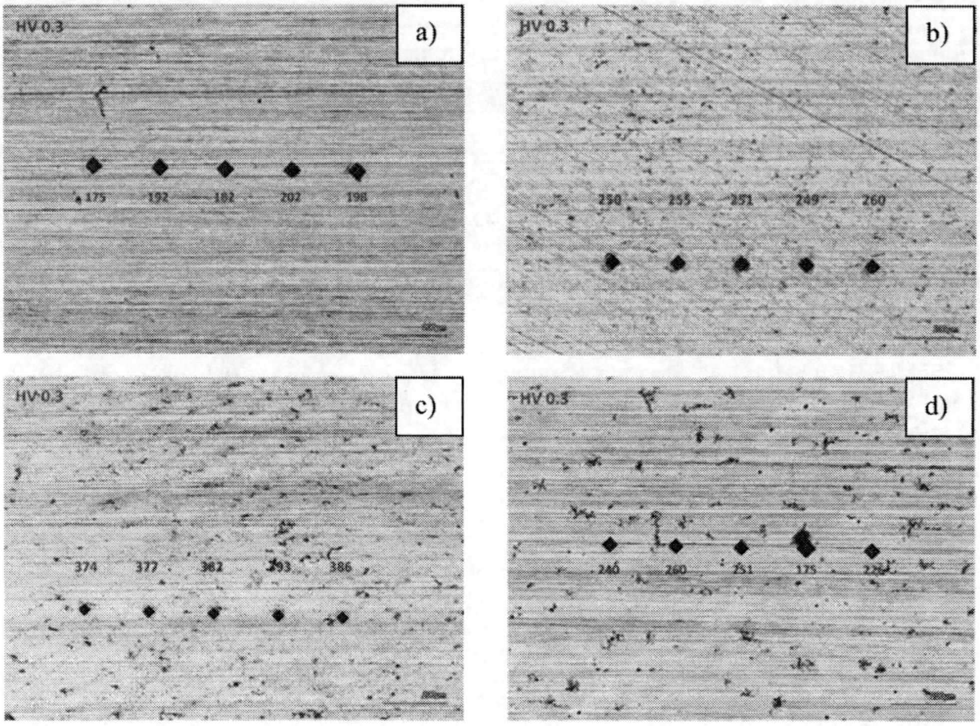

Figure 5. LM of the HV0.3 indentation of the a) C25; b) C45; c) 34CrNiMo6; d) EN-GJS-400-15.

Table 5. Microhardness values for the investigated specimens (HV0.3).

Alloys type	Attempt 1	Attempt 2	Attempt 3	Attempt 4	Attempt 5	Mean value
C25	175	192	182	202	198	190
C45	250	255	251	249	260	253
34CrNiMo6	374	377	382	393	386	382
EN-GJS-400-15	240	260	251	175	226	231

The highest microhardness value (382 HV0.3) is found for the 34CrNiMo6 steel due to the heat treatment which was subjected to. This behaviour may attributed to the ferrite-pearlite structure which is known to possess a higher microhardness [42-44].

Even if the microhardness value of the EN-GJS-400-15 cast iron is lower (231 HV0.3), in the case of dry sliding wear resistance, it displays the best behavior compared to other tested steels, which mean that the microhardness value do not influence the sliding wear resistance.

4. Conclusions

This work has presented a comparative study on dry sliding wear resistance of specimens obtained from carbon steels C25 and C45, heat treated alloyed steel 34CrNiMo6 and cast iron EN-GJS-400-15.

For the same working conditions regarding the tribological environment, the coefficient of friction (by graphic representation) and wear rate (by analytical calculation) was presented.

For the three steel grades, the COF evolution was quite similar, with a mean friction value between 0.34-0.56 during the tests, presenting a good stability. For cast iron, the COF had a mean value between 0.16 - 0.28. Moreover, a smaller wear rate was calculated for cast iron ($3.68 \cdot 10^{-6}$ mm^3 N^{-1} m^{-1}) followed by that of C45 steel ($7.51 \cdot 10^{-6}$ mm^3 N^{-1} m^{-1}). In the case of these alloys, the lowest volume loss was also estimated. Similar to other studies, the wear resistance is attributed to the carbon content.

The adhesive wear occurrence during the POD tests has been taken into consideration. The friction coefficient values are in good agreement with the dry sliding wear of steels and cast iron materials friction value range.

The current investigations revealed for the gray cast iron with nodular graphite, a lower friction coefficient, correlated with a reduced wear rate of the material, which leads to the fact that this type of material has a better resistance to sliding wear resistance compared with the other analyzed steels. Even if the microhardness value is significantly lower in the case of the analyzed cast iron, this aspect does not influence the dry sliding wear resistance.

5. References

[1] Ferhat M, Benchettara A, Amara S E and Najjar D 2014 *J Mater Environ Sci* **5**(4) 1059-68
[2] Ivan A, Ivan M and Both I 2010 *WSEAS Transactions on Applied and Theoretical Mechanics* **5**(3) 187-96
[3] Mendas M and Benayoun S 2013 *Tribol Int* **67** 124-31
[4] Lupinca C I, Frunzaverde D, Rigou I V, Cornean C E and Ciobanu I 2011 *Metal Int* **16**(4) 129-32
[5] Ruja I, Frunzaverde D, Marta C, Suciu S, Cziple F and Roşu M 2010 *Metal Int* **15**(12) 13-8
[6] Prabu S, Choudhary A, Jain A and Sharma A 2014 *Proc Mat Sci* **5** 809-16
[7] Babu P D, Buvanashekaran G and Balasubramanian K R 2013 *Proc IMechE Part J: J Engineering Tribology* **227**(10) 1138-49
[8] Alotaibi J G, Yousif B F and Yusaf T F 2014 *Proc IMechE Part J: J Engineering Tribology* **0**(0) 1-13
[9] Gracia-Escosa E, García I, de Damborenea J and Ana Conde A 2017 *J Mater Res Technol* **6**(3) 241-50
[10] Sekban D M, Aktarer S M, Yanar H, Alsaran A and Purcek G 2017 Proceedings of the 13th International Conference on Tribology, ROTRIB'16, Romania, September 22-24, pp 1-7
[11] Montesano L, Gelfi M, Pola A, Colombi P and La Vecchia G M 2013 *Metall Ital* **9**(2) 3-11
[12] Gelfi M, Gorini D, Pola A and La Vecchia G M 2016 *J Mater Eng Perform* **25**(9) 3896-903
[13] Girelli L, Pola A, Gelfi M, Masotti M N and La Vecchia G M 2017 *Metall Ital* **6** 5-10
[14] Mitelea I, Bena T, Bordeasu I and Craciunescu C M 2018 *Rev Chim* **69**(3) 611-7
[15] Janicki D 2018 *Materials* **11**(75) 1-17
[16] Lupinca C I, Nedeloni M D and Nedelcu D 2014 *Mater Sci Forum* **782** 269-74
[17] *** 2000 Standard Test Method for Wear Testing with a Pin-on-Disk Apparatus ASTM G99

[18] Gupta V K, Ray S and Pandey O P 2008 *Materials Science-Poland* **26**(3) 617-31
[19] Gupta A K and Mishra D N 2013 *International Journal of Science and Research* **2**(7) 222-4
[20] Lakshminarayana V and Balu V 2015 *J Eng Appl Sci* **10**(13) 5655-58
[21] Kumar S and Pathak V K 2016 *International Journal of Advances in Materials Science and Engineering* **5**(1) 21-7
[22] Adel K M 2014 *Proc Mat Sci* **6** 1639-43
[23] Arun K V and Swetha K V 2011 *Journal of Minerals & Materials Characterization & Engineering* **10**(4) 323-37
[24] Agunsoye J O, Ochulor E F, Talabi S I and Olatunji S 2012 *Tribology in Industry* **34**(2) 239-46
[25] Kopyciński D, Piasny S, Kawalec M and Madizhanova A 2014 *Archives of Foundry Engineering* **14**(1) 63-6
[26] Korka Z I 2009 *Research on vibration reduction in operation of cylindrical gearboxes*, Eftimie Murgu" University of Resita, Romania, PhD Thesis
[27] Korka Z I 2017 Precision Coatings *Micro and Precision Manufacturing (Engineering Materials)* ed K Gupta (Springer, Cham) chapter 8, pp 165-93
[28] Parthasarathi N L, Borah U and Albert Sh K 2013 *Computer Modelling and New Technologies* **17**(1) 51-63
[29] Kekes D, Psyllaki P and Vardavoulias M 2014 *Tribology in Industry* **36**(4) 361-74
[30] Nedeloni L, Korka Z I, Nedeloni M D and Pauliuc D 2017 *Annals of "Eftimie Murgu" University of Resita* **24**(1) 225-34
[31] Lungu M V et al. 2017 IOP Conference Series: *Materials Science and Engineering*, Romania, May 25-26, pp 1-11
[32] Lungu M V et al. 2018 Proceedings of the International Conference BALTTRIB'2017, Lithuania, November 16-17, pp 27-33
[33] Kazamer N, Pascal D T, Marginean G, Serban V A, Codrean C and Utu I D 2016 *Sol St Phen* **254** 71-6
[34] Pascal D T, Muntean R, Kazamer N, Marginean G, Brandl W and Serban V A 2016 NANOCON 2016 Czech Republic, October 19-21, pp 775-80
[35] Kazamer N, Pascal D T, Marginean G, Serban V A, Brandl W and Valean P C 2016 NANOCON 2016 Czech Republic, October 19-21, pp 383-8
[36] Banerjee A, Tungala V, Sala K, Biswas K and Maity J 2015 *J Mater Eng Perform* **24**(6) 2303-11
[37] Fox-Rabinovich G S, Gershman I, El Hakim M A, Shalaby M A, Krzanowski J E and Veldhuis S C 2014 *Lubricants* **2** 113-123
[38] Prabu S S, Prathiba S, Venkatesan N, Sharma A, Ahmed S and Shah Y A 2014 *Procedia Engineering* **97** 2110-18
[39] Prabu S S, Prathiba S, Asokan M A, Jain A, Jain N K, Ahmed S and Chourasiya P K 2014 *Procedia Engineering* **97** 2119-226
[40] Salvaro D B, Giacomelli R O, Binder R, Binder C, Klein A N and De Mello J D 2017 *Wear* **376** 803-12
[41] Giacomelli R O, Salvaro D B, Bendo T, Binder C, Klein A N and De Mello J D 2017 *Surf Coat Tech* **314** 18-27
[42] Džugan J and Konopík P 2010 Proceedings of the 1st International conference COMAT 2010 on Recent Trends in Structural Materials, Czech Republic, November 25-26, pp 256-66
[43] Nurnberger F, Grydin O, Schaper M, Bach F W, Koczurkiewicz B and Milenin A 2010 *Steel Research Int* **81**(3) 224-33
[44] Nurnberger F, Grydin O, Yu Z and Schaper M, 2011 *Metallurgical and Mining Industry* **3**(7) 79-86

IOP Conf. Series: Materials Science and Engineering **416** (2018) 012027 doi:10.1088/1757-899X/416/1/012027

Investigations on Dry Sliding Wear and Corrosion Resistance of Thermal Sprayed Molybdenum Coatings

L Păduraru[1], L Nedeloni[1,*], N Kazamer[2,3,*], R Muntean[2], D T Pascal[2], P C Vălean[2,3] and M D Nedeloni[1]

[1]Eftimie Murgu University of Reşiţa, Faculty of Engineering and Management, Piaţa Traian Vuia, No. 1-4, 320085, Reşiţa, Romania

[2]Westphalian University of Applied Sciences Gelsenkirchen, Neidenburger Str. 43, 45897 Gelsenkirchen, Germany

[3]Politehnica University of Timişoara, Department of Materials and Manufacturing Engineering, Piaţa Victoriei, No. 2, 300006, Timişoara, Romania

*Corresponding author: l.nedeloni@uem.ro; norbert.kazamer@w-hs.de

Abstract. This paper aims to investigate the characteristics of molybdenum thermal sprayed coatings deposited on 1.0570 steel substrate and compare them with the characteristics of the uncoated carbon steel (1.0570), respectively alloyed steels (1.7219 and 1.6582), which are currently used for similar applications, such as parts and structural components used at temperatures down to -100°C, aircraft, terrestrial transportation and other industries. The Mo coating deposition was performed by wire metallization and aimed to increase the wear and corrosion resistance of steels typically applied in railway transportation industry. The microhardness, wear behaviour and corrosion resistance have been investigated for all the presented materials. Molybdenum coatings deposited by wire metallization technique possess superior frictional behaviour due to the formation of an oxide tribo-film. Therefore, the friction coefficient of the presently investigated Mo coatings reached a value of 0.65, correlated with a low wear rate of about $0.65 \cdot 10^{-5}$ mm^3 N^{-1} m^{-1}. The porosity degree of the Mo coatings was found to be around 3.50%, a common value for this type of coatings. Microhardness average value of Mo coating was 480 HV0.3 and the corrosion behaviour was found to be superior in comparison to the other materials, exhibiting an inclination to passivation at around 300 mV.

1. Introduction

Various studies have been carried out lately in order to improve wear behavior of railway vehicle parts [1]. One of the main focus in improving fatigue and wear resistance of railway vehicles main components is to increase the axle size to considerably reduce the stress [2-5]. Another direction relates to the pretreatment of the raw material from which axles are manufactured by different methods or coatings, greatly improving in this way, the wear, fatigue and corrosion behavior [6-12].

The materials use for axles manufacturing are chosen according to the properties to be pursued and the conditions required by an optimal operation, in order to avoid the above-mentioned risks. Carbon steels and also the heat-treated steels are commonly used because they fulfil the resistance conditions, but in exceptional cases, molybdenum, chromium or nickel alloys may also be used [13-20].

Over the last 20 years, most researchers have attempted to replace chemical and electrochemical coatings of steel components using unconventional processes such as thermal spraying of different types of hard particles or superior materials [21-24]. Thermal spraying is an economical and efficient technological process that provides excellent tribological properties, resistance to stress, corrosion and oxidation. It has been shown that molybdenum is a promising candidate for coating various parts which are subjected to corrosion and wear conditions. Also, molybdenum as well and other refractory metals have high melting temperatures, and due to this fact, during the deposition process, the interaction between the substrate and the melted material is very high [25-29].

In this direction, the present study brings a contribution in the field of wear and corrosion resistance coatings which can be applied for railway vehicles main parts. The analyzed steels and molybdenum coating were investigated for microhardness, dry sliding wear and corrosion resistance in 3.5% NaCl solution to compare and highlight their behavior under identical test conditions. The obtained results are reliable compared to other studies [30-36] showing that the Mo coating deposition by wire metallization increases the wear performance and also the corrosion resistance.

2. Experimental procedure

The main steps of this study were deposition of the molybdenum coating through the wire metallization process, morphological and microstructural characterization of the obtained coating by scanning electron microscopy combined with energy dispersive X-ray spectroscopy (SEM and EDX), Light Microscopy (LM), Laser Scanning Confocal Microscopy (LSCM), microhardness, wear resistance and corrosion behavior for all specimens.

The chemical composition of three types of steel and the Mo coating is shown in table 1.

Table 1. The chemical composition of the specimens (%).

Specimen	C	Si	Mn	Cr	Mo	Ni	P	S
1.0570	≤0.20	≤0.55	≤1.60	-	-	-	≤0.035	≤0.035
1.7219	0.22-0.29	≤0.35	0.50-0.80	0.90-1.20	0.15-0.30	-	≤0.030	≤0.025
1.6582	0.30-0.38	≤0.30	0.50-0.80	1.30-1.70	0.15-0.30	1.30-1.70	≤0.025	≤0.025
Mo coating	-	-	-	-	99.5	-	-	-

The Mo coating was deposited by wire metallization process, using a molybdenum wire of 3.17 mm. During the deposition process, the molybdenum wire was melted by flame produced by the gas mixture combustion used in the process and subsequently the melted particles were finely pulverized on the surface of the specimen. The melting temperature of the molybdenum wire is around 2600°C and the Mo oxides are formed at approx. 700°C. The Mo particles form a fine, microporous coating on the surface of the substrate. This type of coating is considered suitable for various applications which require high wear and corrosion resistance. Figure 1 shows some characteristic images of the specimen before, during and after the Mo deposition.

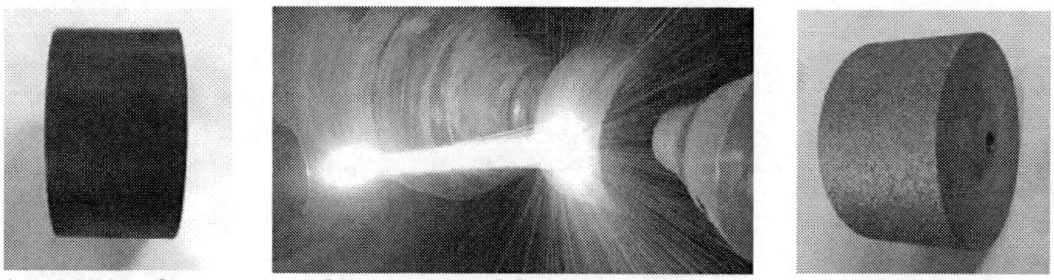

a) 1.0570 substrate *b) process of the Mo coating* *c) Mo coating*

Figure 1. Characteristic images of the Mo coating specimen and process.

The microstructure, morphology as well as thickness of the obtained Mo coating deposited onto the 1.0570 steel was investigated by SEM. Additionally, the chemical composition of the analyzed coating was determined by EDX. The images obtained using LM were further processed using ImageJ software to determine the porosity degree of the Mo coating [37,38].

The microhardness testing was performed with the aid of a Zwick-Roell ZHVμ-S instrument equipped with a known geometry indentor, following the ISO 6507 standard. The applied force was 0.3 kgf and held for 15 seconds for each indentation. The distance between two indentations was set at 0.150 mm for a number of 5 indentations for each specimen, where the final hardness was the average of the measurements.

For the investigations on dry sliding wear, tests were performed using a CSM tribometer according to the Pin-on-Disk method (POD). A 10 N load was applied using a 6 mm 100Cr6 steel ball, with a linear speed of 15 cm s^{-1}, a working radius of 3 mm and a distance of 1500 m at 20°C, in agreement to the ASTM G99-95a standard, in order to determine the friction coefficients and wear rate of the tested system.

Corrosion resistance was investigated by plotting the linear polarization curves with a potentiostat in an electrochemical cell, with three electrodes configuration, where the working electrode consisted of the material to be analyzed, a platinum disk for the counter electrode and saturated calomel electrode (SCE) as reference electrode. Linear polarization curves were recorded from a potential value of -1000 mV to 500 mV, with a scanning rate of 10 mV min^{-1} in neutral 3.5% NaCl solution, following the ISO 13129 standard.

3. Results and discussion

A good wear and corrosion behavior is directly dependent on the quality of the deposited coating, microstructure, compactness and porosity [37]. In this regard, Mo coating was examined on surface and cross-section by SEM, mainly using the BSE detector at 50X to 1000X magnification, in order to analyze the microstructure, thickness, and the quality of the coating/substrate interface (figure 2).

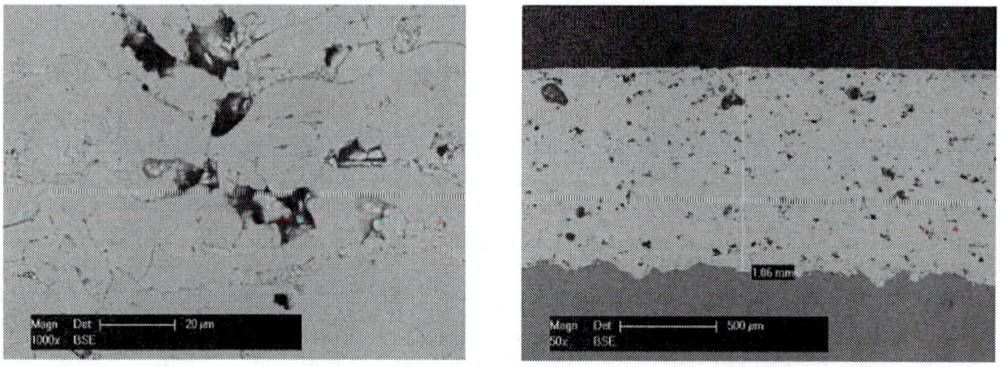

Figure 2. SEM micrographs of the Mo coating on the surface and cross-section.

Cross-section micrograph highlights a coating thickness of 1.06 mm and the interface between the coating and the substrate is clean, free from impurities, pores or cracks, which demonstrates that the adhesion of the coating to the substrate meets the essential requirements (figure 3). A good adhesion to the substrate combined with a compact, non-cracked coating leads to a qualitative coating that fulfills the protective role, especially anticorrosive.

IOP Conf. Series: Materials Science and Engineering **416** (2018) 012027 doi:10.1088/1757-899X/416/1/012027

Figure 3. SEM micrographs of Mo coating: with BSE respectively SE detector.

Additionally, EDX analysis revealed the chemical composition of the deposited coating (figure 4).

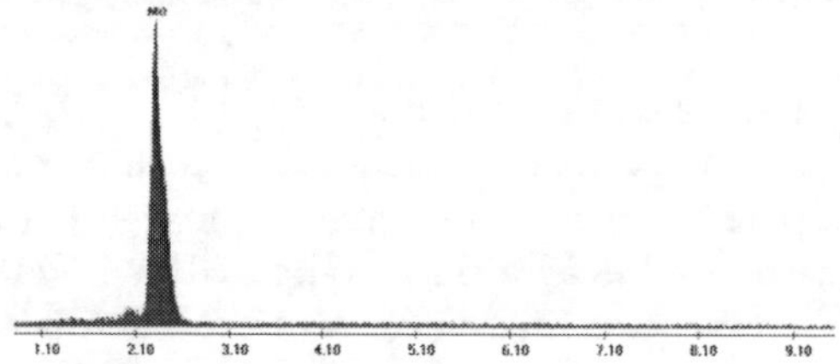

Figure 4. EDX spectrum of Mo coating.

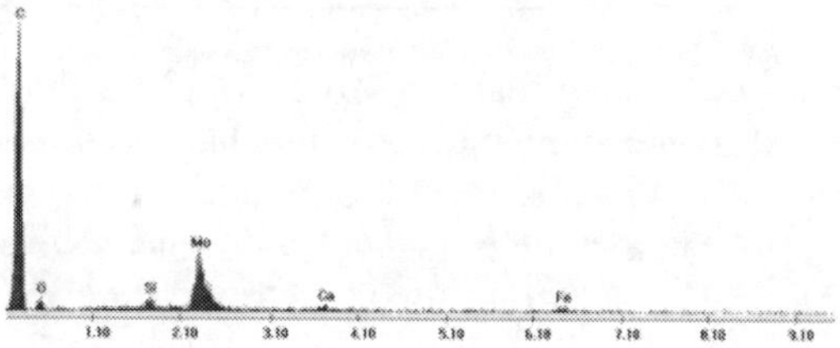

Figure 5. EDX spectrum of C inclusions.

Molybdenum is highlighted as a single element, since the wire used in the deposition process had a purity of >99.5%. The possibility of molybdenum to oxidize during the deposition process is also taken into account, since oxidation occurs very strongly at temperatures above 400°C. Once the molten molybdenum reaches the surface of the substrate, oxidation is also possible. The inclusions present in the coating were also analyzed, and the EDX spectrum shown in figure 5, highlights their chemical composition. Some of the inclusions have come from the metallographic processing of the specimen (SiC), metal oxides and other metal impurities.

The CSLM micrographs shown in figure 6 were used to determine porosity of the coating. Figure 6a reveals the micrograph of the Mo coating at 100X magnification, showing that the present pores are closed without interconnection. In figure 6b, the image was further processed using ImageJ, obtaining in this way the porosity degree. By this method, an average porosity value of 3.50% was determined. This value is specific to this type of coating [38].

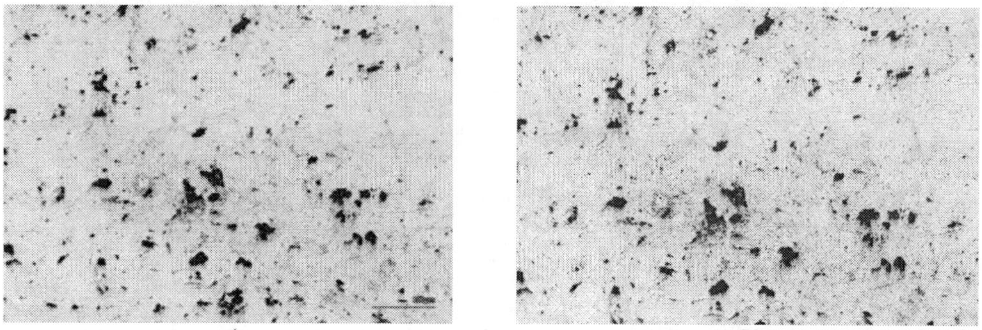

a) *b)*

Figure 6. CSLM micrographs of the Mo coating and correspondingly processed image.

The microhardness was assessed using the Vickers method [21,39] in 5 different areas, on the cross-sectional depth of the molybdenum coating. The final value of the microhardness was calculated as the average of the 5 measurements made on each type of the specimen.

Table 2 presents the minimum and maximum values obtained as well as the calculated mean value of the Mo coating in comparison to the investigated steel materials.

Table 2. Calculated values of HV0.3 microhardness.

Specimen	HV0.3 min.	HV0.3 max.	HV0.3 mean
1.0570	187	196	191
1.7219	321	326	323
1.6582	381	393	387
Mo coating	302	657	480

It can be noticed that the Mo coating has the highest value of microhardness (480 HV0.3). This is attributed to the high molybdenum hardness as property of the material as well as to the coating's compactness. The absence of large and interconnected pores leads also to a higher value of hardness. The heat-treated stainless steel 1.6582 has also high hardness (387 HV0.3), compared to 1.7219 and 1.0570 used as substrate for the Mo coating. Figure 7 shows LM micrographs of the microhardness indentations for each specimen.

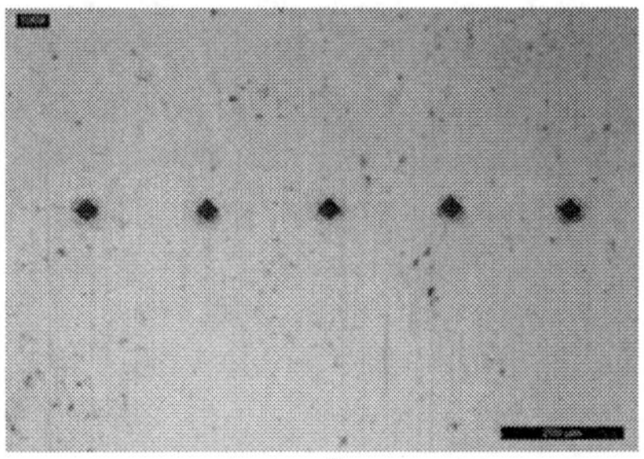
a)

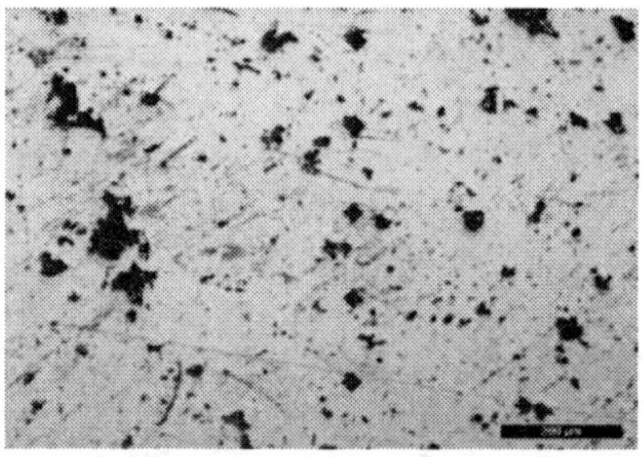
b)

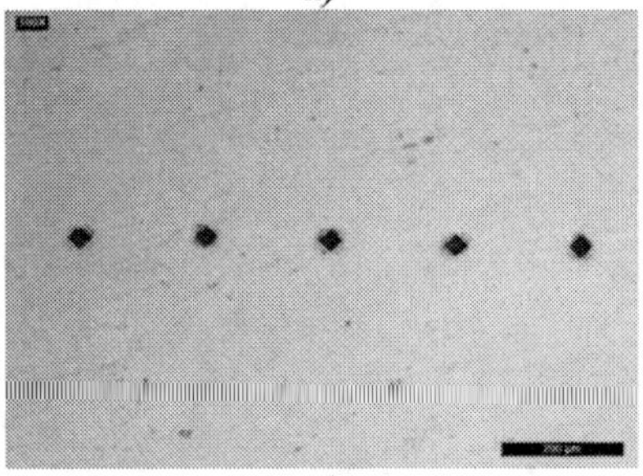
c)

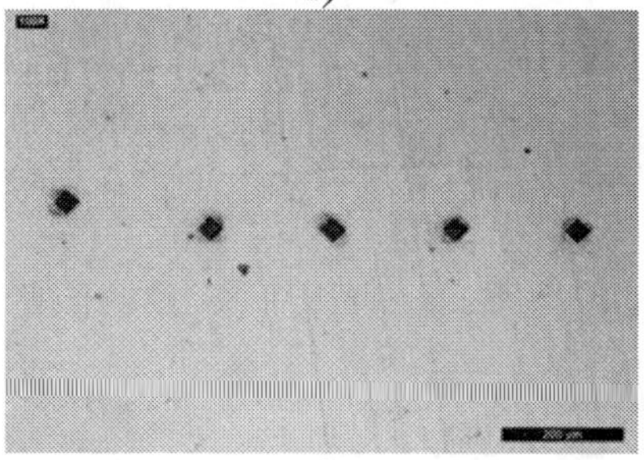
d)

Figure 7. LM micrographs of the microhardness indentations: a) 1.6582, b) Mo coating, c) 1.0570 respectively d) 1.7219.

Regarding the dry sliding wear behavior, the friction coefficient and the wear rate was obtained. Therefore, the friction coefficient was monitored with the integrated software and the wear rate was represented by volume loss depending on the counterpart, applied load and the distance on the specimen surface.

The obtained results presented in figure 8 and 9 respectively table 3 reveal a superior behavior of the Mo coating. Even if the average value of the friction coefficient is lower for the 1.0570 steel substrate compared to the Mo coating, the Mo coating show a lower value of wear rate.

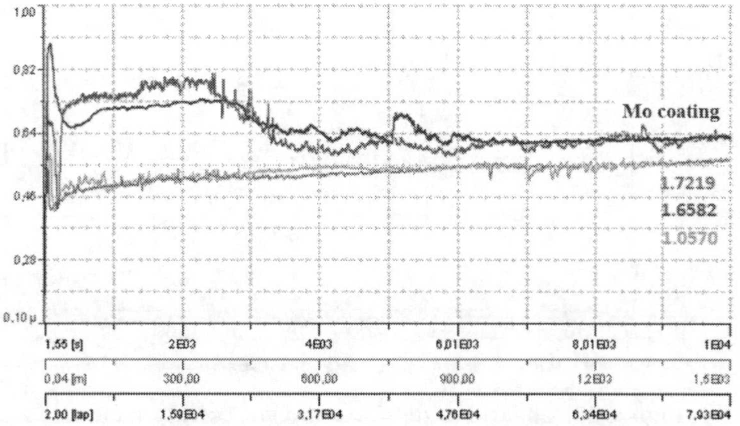

Figure 8. Evolution of friction coefficients for the investigated specimens.

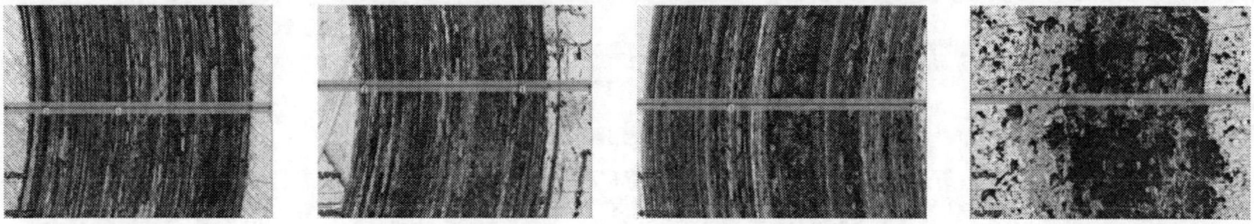

a) 1.0570 *b) 1.7219* *c) 1.6582* *d) Mo coating*

Figure 9. Wear track profiles for a) 1.0570, b) 1.7219, c) 1.6582 and d) Mo coating.

Table 3. Values of the wear rate.

Specimen	h (µm)	s (µm)	A (μm^2)	W_v (mm^3)	k ($mm^3N^{-1}m^{-1}$)
1.0570	27.15	1089.56	19730	0.3719	$2.48 \cdot 10^{-5}$
1.7219	9.41	966.44	6063	0.1143	$0.76 \cdot 10^{-5}$
1.6582	34.30	1218.42	27877	0.5255	$3.50 \cdot 10^{-5}$
Mo coating	12.02	653.50	5238	0.0987	$0.65 \cdot 10^{-5}$

The plots presented in figure 10, show the logarithmic form of the polarization curves of the investigated specimens in 3.5% NaCl solution. The values of the corrosion current as well as the corrosion potential was determined by Tafel slope method and are presented in table 4.

The corrosion currents are of the same order of magnitude for all the investigated specimens and there are no major differences, but the corrosion potential for the Mo coating is shifted to more positive values, due to the chemical composition of the coating.

The curve corresponding to the Mo coating presents a passivation tendency around 300 mV and its anodic branch is characterized by lower current densities compared to the other steels. The better corrosion behavior among the three types of steel is represented by the 1.6582 steel, which can be assigned to the higher Cr and Ni content in its chemical composition, favoring the passivation phenomenon.

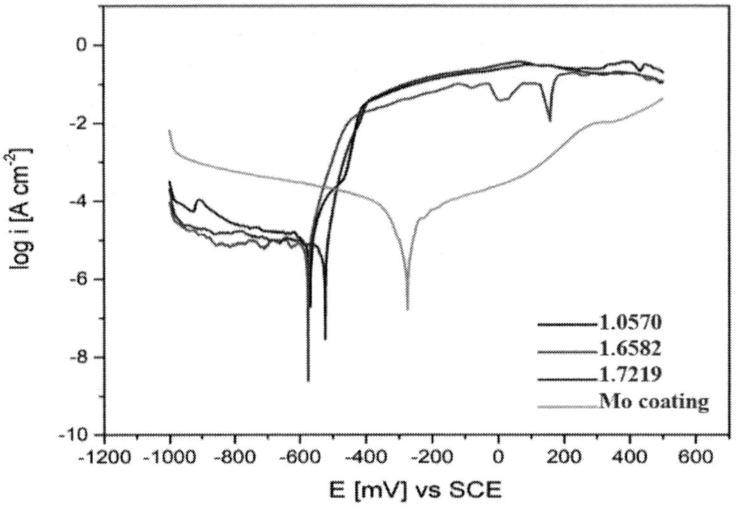

Figure 10. Logarithm representation of corrosion curves.

Table 4. Values of potentials and corrosion currents.

Specimen	E_{corr} [mV] vs. SCE	i_{corr} [A cm^{-2}]
1.0570	-577	$1{,}23 \cdot 10^{-5}$
1.7219	-534	$1{,}17 \cdot 10^{-5}$
1.6582	-587	$1{,}34 \cdot 10^{-5}$
Mo coating	-275	$3{,}38 \cdot 10^{-5}$

4. Conclusions

The investigated Mo coating possess superior characteristics, in comparison to the three analyzed steels, fact that makes it suitable for various applications. The interface between the Mo coating and the substrate was clean, without inclusions, leading to a good adhesion of the coating. CSLM micrographs reveal the presence of pores, which do not affect the structure and the resistance of the coating because they are closed and with no cracks. The average porosity was assessed to 3.50%, this value being specific to this type of wire deposition. From the other tested materials, the Mo coating showed the highest microhardness value. The friction coefficient and the wear rate were determined through POD method and the profiles were analyzed using the CSLM laser microscope.

The results showed a superior wear behavior of the molybdenum coating, with a lower friction coefficient and wear rate. The superior corrosion resistance of the Mo coating confirms the effectiveness of depositing this type of coating in order to extend the life of the railway vehicles components.

5. References

[1] Smith R A 2008 *Ironmaking and Steelmaking* **35**(7) 505-13
[2] Olofsson U and Nilsson R 2002 *J. of Rail and Rapid Transit* **216**(4) 249-64
[3] Olofsson U and Telliskivi T 2003 *Wear* **254** 80-93
[4] Sebesan I, Făinuş L and Zaharia N L 2013 *Mater Plast* **50**(4) 253-6
[5] Făinuş L, Zaharia N L, Tudose D I and Hadăr A 2016 *UPB Scientific Bulletin, Series D: Mechanical Engineering* **78**(3) 107-18
[6] Jin G, Xu B, Wang H, Li Q, Wei S 2007 *Surf Coat Tech* **201**(15) 6678-80

IOP Conf. Series: Materials Science and Engineering **416** (2018) 012027 doi:10.1088/1757-899X/416/1/012027

[7] Kazamer N, Pascal D T, Marginean G, Serban V A, Brandl W and Valean P C 2016 NANOCON pp 383-8

[8] Secosan E R, Marginean G, Ciubotariu R C, Frunzaverde D and Campian V 2015 *Annals of the „Eftimie Murgu" University of Resita* **22**(1) 385-96

[9] Davis J R 2001 Introduction to Surface Engineering for Corrosion and Wear Resistance *Surface Engineering for Corrosion and Wear Resistance* ASM International chapter 1, pp 1-10

[10] Lazar P, Bejinariu C, Sandu A V, Cazac A M, Corbu O, Sandu I G and Perju M C 2017 IOP Conference Series: Materials Science and Engineering, Romania, May 25-26, pp 1-5

[11] Lyphout C, Sato K, Houdkova S, Smazalova E, Lusvarghi L, Bolelli G and Sassatelli P 2016 *J Therm Spray Techn* **25**(1-2) 331-45

[12] Lungu M V et al. 2018 BALTTRIB'2017, pp 27-33

[13] Ivan A, Ivan M and Both I 2010 *WSEAS Transactions on Applied and Theoretical Mechanics* **5**(3) 187-96

[14] Pascal D T, Muntean R, Kazamer N, Marginean G, Brandl W and Serban V A 2016 NANOCON pp 775-80

[15] Hulka I, Şerban V A, Secoşan I, Vuoristo P and Niemi K 2012 *Surf Coat Tech* **210** 15-20

[16] Modi S C and Calla E 2001 *J Therm Spray Techn* **10**(3) 480-6

[17] Dumitrascu V, Benea L and Danaila E 2017 IOP Conference Series: Materials Science and Engineering, Iasi, Romania, May 25-26, pp 1-8

[18] Budai A M, Campian V C, Frunzaverde D, Miclosina C and Pepa D 2017 *Technical Gazette* **24**(2) 271-7

[19] Cîndea L, Haţiegan C, Pop N, Negrea R, Răduca E, Gillich G R, Dănuţ M, Nedeloni M D 2016 *Appl Therm Eng* **103** 1164-75

[20] Lungu M V et al. 2017 Conference Series: Materials Science and Engineering, Iasi, Romania, May 25-26 pp 1-11

[21] Kazamer N, Pascal D T, Marginean G, Serban V A, Codrean C and Utu I D 2016 *Sol St Phen* **254** 71-6

[22] Rigou I V, Mărginean G, Frunzăverde D and Câmpian C V 2011 TAM '11, Greece, July 14-16, pp 219-22

[23] Zajchowski P and Crapo H B 1996 *J Therm Spray Techn* **5**(4) 457-62

[24] Cojocaru V, Frunzaverde D, Campian C V, Marginean G, Ciubotariu R and Pittner A M 2010 EMESEG '10, Greece, July 22-24, pp 177-80

[25] Ozdemir I, Tekmen C, Okumus S C and Celik E 2003 *Surf Coat Tech* **174** 1964-69

[26] Wayne S F, Sampath S and Anand V 1994 *Tribol T* **37**(3) 636-40

[27] Laribi M, Vannes A B, Treheux D 2006 *Wear* **262**(11-12) 1330-36

[28] Ahn J, Hwang B and Lee S 2005 *J Therm Spray Techn* **14**(2) 251-7

[29] Ciubotariu C R, Frunzaverde D, Secosan E, Marginean G and Campian V 2016 *Stroj Vestn-J Mech E* **62**(3) 154-62

[30] Sampath S and Wayne S F 1994 *J Therm Spray Techn* **3**(3) 282-8

[31] Chi W, Sampath S and Wang H 2006 *J Therm Spray Techn* **15**(4) 773-8

[32] Wang Y, Wang D, Yan J and Sun A 2013 *Appl Surf Sci* **284** 881-8

[33] Zhao X B and Ye Z H 2013 *Surf Coat Tech* **228** 266-70

[34] Yan J, He Z, Wang Y, Qiu J and Wang Y 2016 *J Therm Spray Techn* **25**(7) 1322-29

[35] Paraschiv A, Matache G, Puscasu C and Grigorescu M 2018 MATEC Web of Conferences, Bulgaria, September 6-10, pp 1-9

[36] Yan J, Guo Y, Wang Y, Zhou P and Qiu J 2018 *Lubricants* **6**(48) 1-11

[37] Oksa M, Turunen E, Suhonen T, Varis T and Hannula S P 2011 *Coatings* **1** 17-52

[38] Tailor S, Modi A and Modi S C 2018 *J Therm Spray Techn* **27** 757-68

[39] Mendes V B and Leta F R 2003 Proceedings, XVII IMEKO World Congress, Croatia, July 22-27, pp 992-5

IOP Conf. Series: Materials Science and Engineering **416** (2018) 012028 doi:10.1088/1757-899X/416/1/012028

Cavitation Erosion and Dry Sliding Wear Research on X5CrNi18-10 Austenitic Stainless Steel

M D Nedeloni, E Birtărescu*, L Nedeloni, T Ene, A Băra and B Clavac

Eftimie Murgu University of Reşiţa, Faculty of Engineering and Management, Piaţa Traian Vuia, No. 1-4, 320085, Reşiţa, Romania

*Corresponding author: e.birtarescu@uem.ro

Abstract. In this research, the cavitation erosion and dry sliding wear behavior of X5CrNi18-10 austenitic stainless steel has been analyzed through many tests. Cavitation erosion tests have been made for 180 minutes for the two samples and then for an extended duration of 1800 minutes for the surface which exhibited the best behaviour. All this tests were made using a cavitation experimental stand with the following main parameters for the ultrasonic generator (frequency: 20 ±0.5 kHz and amplitude: 50 μm). Dry sliding wear test has been made using a tribometer with the parameters: applied force 10 N, line speed 15 cm/s and distance 750 m. The analysed X5CrNi18-10 austenitic stainless steel had showed very good cavitation erosion behaviour, but a low dry sliding wear resistance compared to other steels as can be seen in the obtained results presented as cumulative mass loss and cavitation erosion rate for the cavitation erosion tests and as coefficient of friction and wear rate for the dry sliding wear test.

1. Introduction

Electricity production by classical methods is obtained in power plants, where a major category is the hydro power plant, of which the hydraulic turbines are part [1]. For the manufacture the hydraulic turbine components, steel, cast iron, bronze as metallic materials and certain small quantities of non-metallic materials can be used [2,3]. Most components are manufactured of different types of steel, such as carbon, treated and stainless steel [4-6].

The characteristic wear mechanisms that can occur on these components are: abrasive, adhesive and sliding wear respectively erosion, which can be cavitation erosion, hydro-abrasion, sediment and sand erosion. Besides to these different types of wear, other damage can occur due to the working environment or inadequate maintenance such as: typical rotor blocking, inadequate heat treatment or erosion associated with corrosion [1]. From the hydraulic turbine components, the rotor is the most affected by cavitation erosion.

The cavitation erosion occurs as a result of the cavitation phenomenon encountered in the fluid environment where the rotors and the rotor blades operate. The phenomenon of cavitation is described as a dynamic process of forming, developing and blowing up bubbles or cavities filled with vapors and gases in the mass of a liquid [7].

In the process of implosion is an assembly of simultaneous phenomena of mechanical, chemical, electrical and thermal nature [8,9]. Different research on the cavitation erosion of materials respectively on the cavitation phenomenon has been made by several authors with regard to the negative effects of cavitation erosion, to vibrations analysis or to the cavitation vortex [10-13].

It can be said that worldwide, the research purpose through laboratory testing of the cavitation erosion of materials is to discover new materials with good behavior, materials that have high hardness, tear strength, fine grain, homogeneity etc. [14-16]. In general, materials having a high cavitation erosion resistance are of two categories: materials used in the manufacture of hydraulic turbine components and materials used for refurbishing them [17].

In order to test cavitation erosion resistance of materials, there are numerous types of laboratories apparatus such as: cavitation tunnels, vibratory cavitation apparatus and cavitating liquid jets [6]. Of the tested materials with vibratory cavitation apparatus, the less cavitation erosion resistance have Al alloys and the highest is attributed to the Stellite coating followed by stainless steel and Cr-Ni alloys [18-20]. In general, the behavior of these materials (other than Stellite) can be improved by certain treatments [21-26]. In other different technical applications, stainless steels, in particular the X5CrNi18-10 austenitic stainless steel has been tested [27-29].

At the Eftimie Murgu University of Resita the calibration of the cavitation experimental stand (as a vibratory cavitation apparatus) was performed [30,31] and by testing a series of materials, the highest cavitation erosion resistance was attributed to the stainless steel (X5CrNi18-10), then to the treated and carbon steels, followed by cast iron, bronze and other [32-34].

Also, the tribological performances of materials respectively of coatings have been extensively studied [35-37] by using tribometers according to the pin-on-disk method (POD).

In this paper, in addition to the cavitation erosion investigation, the results obtained through a professional tribometer using pin-on-disk method regarding the dry sliding wear resistance of the X5CrNi18-10 austenitic stainless steel are also presented.

2. Experimental procedure

The cavitation experimental stand consists the following components (figure 1): ultrasonic generator, piezoelectric-acoustic converter, mechanical transformer (with the role of amplifying the amplitude in the horn and in the sample), the water container (filled with distilled water) and the sample support. The water temperature is measured using a digital thermometer.

The steps for testing a sample are: cleaning and filling the water container, cleaning the sample and measuring its mass and starting the generator for tests. After the tests, using a digital balance, the sample mass is measured and the value is registered. If the mass loss of a sample is determined by weighing, the cavitation erosion rate is calculated by deriving the mass loss in relation with time.

It can be said that the generation of cavitation bubbles with the cavitation experimental stand differs from that encountered in the operation of hydraulic turbines, where some sediment, mud or sand are also encountered [1]. So, the mechanism of material erosion on this stand that simulates the cavitation bubbles is almost similar to the cavitation phenomenon that occurs in reality [38-40].

The CSM tribometer consists a sample disk and a counter body (in this research a ball 100Cr6 steel with 6 mm diameter was used). The radius on the sample is adjustable and the load and speed are predefined (figure 2). Measurements can be performed with or without lubricants and under different environmental humidity conditions. The coefficient of friction (CoF) is displayed in real time and the wear rate is achieved by measuring the wear track and determining the material volume loss.

The experimental methodology on the two stands took into account the parameters described above, as shown in G32-92 and G32-10 standards (both for cavitation erosion tests) and ASTM G99-95a standard (for dry sliding wear tests).

Regarding the X5CrNi18-10 austenitic stainless steel samples, the dimensions were Φ16x10 mm and the cavitation erosion tests were done in four periods of 180 minutes respectively on an extended duration of 1800 minute of cavitational attack. The dry sliding wear tests required only 83 minutes.

The testing parameters are shown in table 1 and table 2 and the mechanical properties respectively the chemical composition of X5CrNi18-10 are presented in table 3 and table 4.

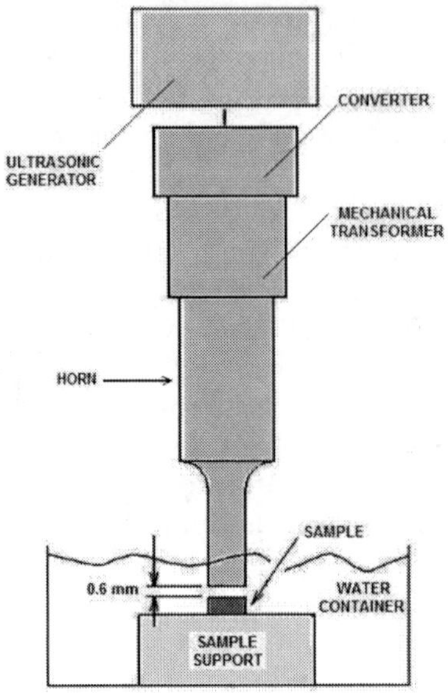

Figure 1. Scheme of cavitation experimental stand components and principle of the stationary specimen method.

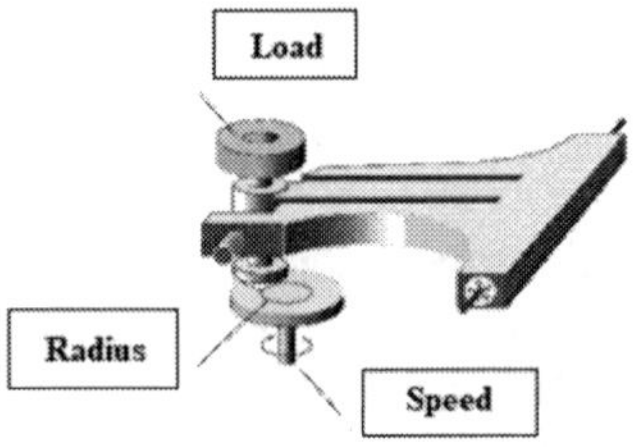

Figure 2. Principle of the pin-on-disk method.

Table 1. Parameters for cavitation erosion tests.

Frequency (kHz)	Amplitude (μm)	Water temperature (°C)	Horn and sample distance (mm)
20 ±0.5	50	25 ±2	0.6

Table 2. Parameters for dry sliding wear test.

Load (N)	Linear speed ($cm \cdot s^{-1}$)	Motor speed (rpm)	Radius (mm)	Distance (m)	Humidity (%)	Temperature (°C)
10	15	474.33	3.02	750	50	20

Table 3. X5CrNi18-10 mechanical properties.

Tensile strength Rm ($N \cdot mm^{-2}$)	Yield strength Rp 0.2 ($N \cdot mm^{-2}$)	Elongation at yield A (%)
605	294	58.4

Table 4. X5CrNi18-10 chemical composition (wt.%).

C	Cr	Mn	Co	N	Ni	P	S	Si	Cu	Fe
0.02	18.231	1.415	0.087	0.086	8.141	0.031	0.022	0.369	0.446	71.15

The high chromium and nickel content of this steel make it very resistant to corrosion [41], but also to cavitation erosion [42,43], as will be see from the research results.

3. Results and discussion

3.1. *Cavitation erosion results*

For the multiple tests of 180 minutes respectively for the extended duration of 1800 minutes, in table 5 some of the cavitation erosion results are presented comparatively.

Table 5. Cavitation erosion results.

Samples (sur)faces	Cumulated time	Period	Sample mass		Max. mass loss		Max. cavitation rate
			Before test	After test	On period	Cumulative	
	t (min)	Δt (min)	m (mg)	m (mg)	Δm (mg)	mc (mg)	v_{ec} ($mg \cdot h^{-1}$)
Sample 1-Face 1	180	15	16309.43	16308.69	0.15	0.74	1.480
Sample 1-Face 2	180	15	16308.69	16308.18	0.12	0.51	0.580
Sample 2-Face 1	180	15	16122.91	16122.1	0.11	0.81	1.220
Sample 2-Face 2	180	15	16122.11	16121.3	0.14	0.81	1.040
Sample 1-Face 2	1800	60	16308.17	16256.18	2.74	51.99	2.520

The graphs of cumulative mass loss and cavitation erosion rate versus time curves are shown in figure 3 and figure 4 for the multiple tests of 180 minutes respectively in figure 5 and figure 6 for the extended duration of 1800 minutes.

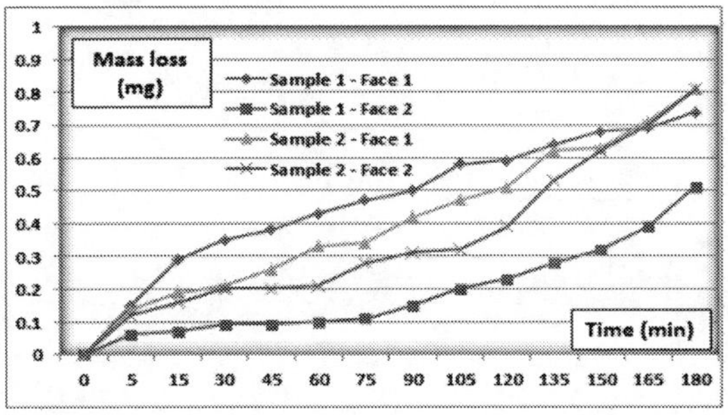

Figure 3. Cumulative mass loss vs time (180 min).

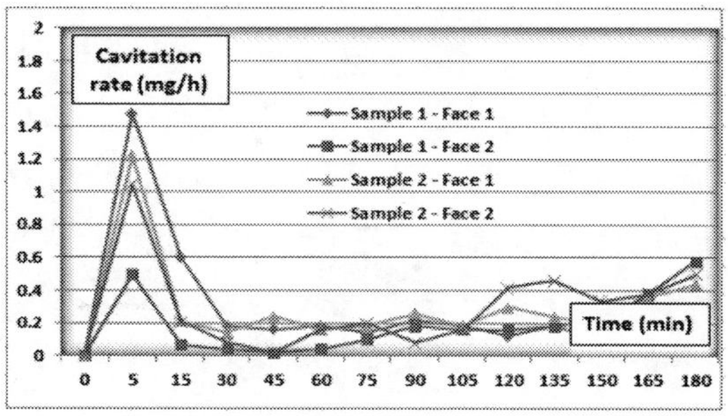

Figure 4. Cavitation erosion rate vs time (180 min).

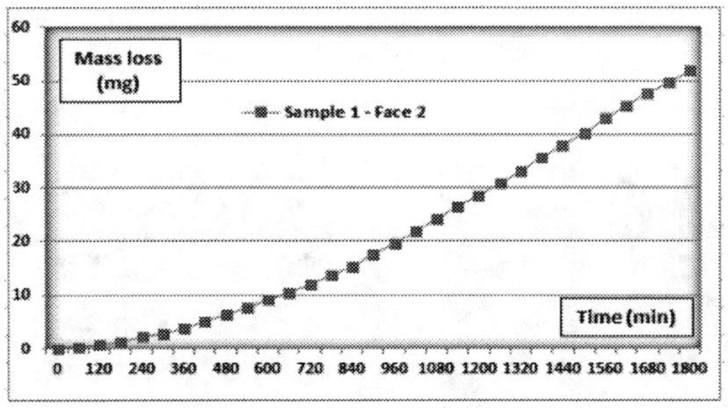

Figure 5. Cumulative mass loss vs time (1800 min).

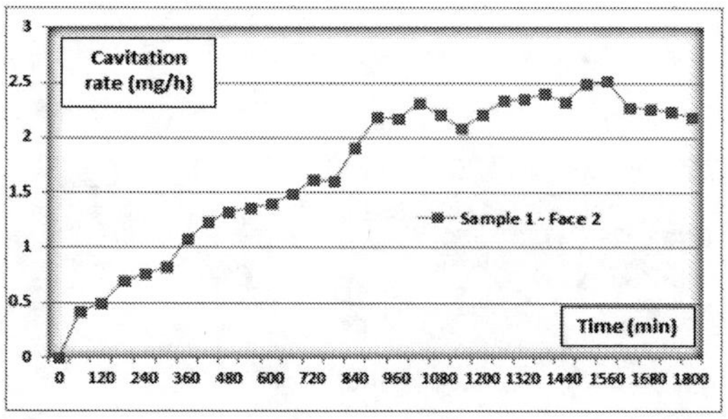

Figure 6. Cavitation erosion rate vs time (1800 min).

From the all tests of 180 minutes, regarding the mass loss, it can say that the four (sur)faces have lost eroded material as follows: 0.74 mg for Sample 1 - Face 1, 0.51 mg for Sample 1 - Face 2, 0.81 mg for Sample 2 - Face 1 and also 0.81 mg for Sample 2 - Face 2.

So the Sample 1 - Face 2 with less material loss and also the lowest value of cavitation erosion rate was subjected further for an extended duration of 1800 minutes, where after this test the total material loss was only 51.99 mg with a cavitation erosion rate not more than 2.520 $mg \cdot h^{-1}$.

Some images of the sample tested for 1800 minutes are shown in figure 7.

Figure 7. Images of Sample 1 - Face 2 after 1800 minutes of cavitational attack a) macrophotography and b) microstructure (Magn. 250X - 100 μm).

From figure 7 it can be see that the surface was heavily eroded in the center with crakes and microcrakes respectively had shallow erosions in the adjacent zones.

3.2. *Dry sliding wear results*

The research on the dry sliding wear resistance of the X5CrNi18-10 austenitic stainless steel have follow to obtain results in two directions, namely: determination of friction coefficient and wear rate. These results as shown in figures 8 and 9 respectively in tables 6 and 7.

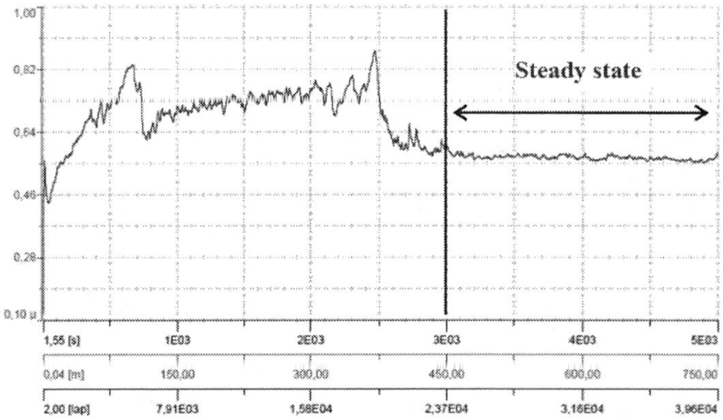

Figure 8. Coefficient of friction vs time and distance curve.

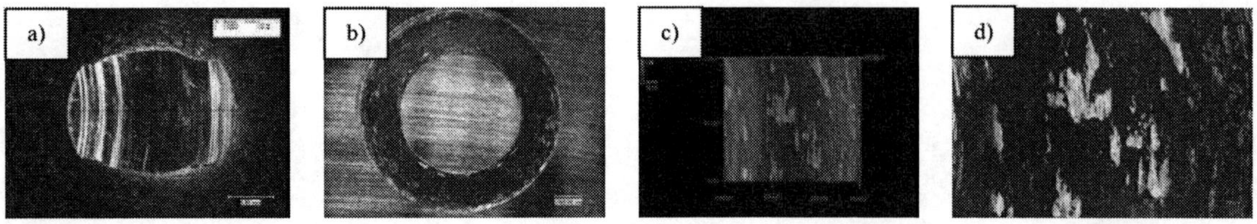

Figure 9. Wear track profiles of a) ball; b) sample; c) 3D view and d) digital micrograph.

Table 6. Friction coefficient values.

Start	Min	Max	Mean	Std. Dev.
0.188	0.124	0.873	0.642	0.088

Table 7. Wear rate results.

Wear track depth, ***h*** (μm)	Wear track width, ***s*** (μm)	Cross section area, ***A*** (μm^2)	Volume loss, ***V*** (mm^3)	Wear rate, ***K*** ($mm^3N^{-1}m^{-1}$)
44.496	1381.881	41023.99	0.7733	$1.03 \cdot 10^{-4}$

After the dry sliding wear test, it can said that the coefficient of friction evolution is starting at the 0.188 value (table 6) and increase for a distance of 100 m at the 0.82 value (figure 8); that, around 0.64 varies with an increasing trend until the maxim peak value of 0.87. From the figure 8, also can be seen that from the distance of 375 m until 450 m, CoF decreases and evolving constant with small variations around 0.57 value until the end of the distance (750 m). This means that the steady state of wear was reached [44,45] and a tribofilm has been formed [46].

Regarding the wear track profiles (which were acquired with a digital microscope and with a confocal 3D laser scanning microscope), the length and the width of the ball wear was 1.87 mm respectively 1.40 mm and regarding the sample, the volume loss of it was 0.7733 mm^3 and the wear rate value was $1.03 \cdot 10^{-4}$ mm^3 N^{-1} m^{-1} (table 7).

4. Conclusions

The paper has referred to the cavitation erosion as a result of the cavitation phenomenon that occurs in the working environment of the rotors / rotor blades of hydraulic turbines and to the cavitation erosion research purpose which is to discover materials with a very good cavitation erosion resistance and respectively regarding the tribological performance of materials by using tribometers according to the pin-on-disk method (POD).

In this way, the cavitation erosion and dry sliding wear behavior of X5CrNi18-10 austenitic stainless steel through many tests regarding the cumulative mass loss and cavitation erosion rate respectively regarding the coefficient of friction and the wear rate was investigated and it was found that the X5CrNi18-10 austenitic stainless steel possess a very good cavitation erosion resistance respectively has a low dry sliding wear resistance.

The cavitation erosion results obtained from mass loss and cavitation erosion rate curves highlighted a slowly cavitation erosion mechanism with some crakes, microcrakes respectively shallow erosions.

For the satisfactory dry sliding wear results, the values of friction coefficient are characteristic for the steel-steel friction value range (0.46 – 0.85) and correlated with the wear rate values, it is indicate that the material destruction was quite intense.

5. References

[1] Singh R, Tiwari S K and Mishra S K 2012 *J Mater Eng Perform* **21**(7) 1539-51
[2] Machuta J and Nová I 2015 *Manufacturing Technology* **15**(4) 591-6
[3] Monteiro W A, Pereira S A L and Vatavuk J 2017 *Journal of Metallurgy* **2017** 1-7
[4] Ivan A, Ivan M and Both I 2010 *WSEAS Transactions on Applied and Theoretical Mechanics* **5**(3) 187-96
[5] Tugui C A, Vizureanu P, Iftimie N and Steigmann R 2016 IOP Conf. Series: Materials Science and Engineering, Romania, June 09-10, pp 1-8
[6] Bereteu L, Bordeaşu I, Drăgănescu G, Simoiu D, Micu L and Sălceanu C 2015 *Appl Mech Mater* **801** 317-22
[7] Bordeasu I, Popoviciu M O, Salcianu L C, Ghera C, Micu L M, Badarau R, Iosif A, Pirvulescu

L D and Podoleanu C E 2016 IOP Conference Series: Materials Science and Engineering, Romania, May 25-26, pp 1-11
[8] Ghera C, Mitelea I, Bordeasu I and Craciunescu C 2016 METAL 2016, Czech Republic, April 25-27, pp 706-11
[9] Murgan I, Digulescu A, Candel I, Cornel I and Şerbănescu A 2017 Oceans 2017, Anchorage, September 17-22, pp 1-5
[10] Ghiban B, Bordeasu I, Chiban N, Badarau R, Hadar A and Serban N 2008 DAAAM, Slovakia, Octomber 22-25, pp 541-2
[11] Dunca G, Bucur D M, Isbăşoiu E C, Călinoiu C and Ghergu C M 2012 *UPB Scientific Bulletin, Series D: Mechanical Engineering* **74**(1) 59-66
[12] Candel I, Bunea F, Dunca G, Bucur D M, Ioana C, Reeb B and Ciocan G D 2014 IOP Conference Series-Earth and Environmental Science, Canada, September 22-26, pp 1-8
[13] Digulescu A, Murgan I, Candel I, Bunea F, Ciocan G, Bucur D M, Dunca G, Ioana C, Vasile G and Serbanescu A 2016 IOP Conference Series-Earth and Environmental Science, France, July 04-08, pp 1-10
[14] Kekes D, Psyllaki P, Vardavoulias M and Vekinisc G 2014 *Tribology in Industry* **36**(4) 375-83
[15] Da Cruz J R, Henke S L and Monteiro A S C 2016 *Materials Research* **19**(5) 1033-41
[16] Li D G, Chen D R and Liang P 2017 *Ultrason Sonochem* **35** 375-81
[17] Cîndea L, Haţiegan C, Pop N, Negrea R, Răduca E, Gillich G R, Dănuţ M, Nedeloni M D 2016 *Appl Therm Eng* **103** 1164-75
[18] Ciubotariu C R, Frunzaverde D, Marginean G, Şerban V A and Bîrdeanu A V 2016 *Opt Laser Technol* **77** 98-103
[19] Bordeasu I, Micu L M, Mitelea I, Utu I D, Pirvulescu L D and Sirbu N A 2016 *Mater Plast* **53**(4) 781-86
[20] Ciubotariu C R, Frunzaverde D, Secosan E, Marginean G and Campian V 2016 *Stroj Vestn-J Mech E* **62**(3) 154-62
[21] Montesano L, Gelfi M, Pola A and La Vecchia G M 2015 *Procedia Engineer* **109** 228-33
[22] Gottardi G, Tocci M, Montesano L and Pola A 2018 *Wear* **394-395** 1-10
[23] Nová I and Machuta J 2016 *Manufacturing Technology* **16**(1) 225-30
[24] Pola A, Montesano L, Tocci M and La Vecchia G M 2017 *Materials* **10**(3) 1-13
[25] Petrogalli C, Montesano L, Pola A, Gelfi M, Ghidini A and LaVecchia G M 2015 *Procedia Engineer* **109** 154-61
[26] Montesano L, Gelfi M, Pola A, Colombi P and La Vecchia G M 2013 *Metall Ital* **9**(2) 3-11
[27] Mahmoud E R I 2015 *IJERT* **4**(08) 422-7
[28] Chawla K, Saini N and Dhiman R 2013 *IOSR-JMCE* **9**(4) 18-22
[29] Ozgowicz W, Kalinowska-Ozgowicz E and Kurc A 2008 *Archives of Materials Science* **32**(1) 37-40
[30] Nedelcu D, Cojocaru V, Nedeloni M, Peris-Bendu F and Ghican A 2015 *Mechanika* **4** 272-6
[31] Nedeloni M D, Hatiegan C, Vasile O, Hamat C O, Fanica C and Gillich N 2015 *RJAV* **12**(2) 155-60
[32] Lupinca C I, Nedeloni M D and Nedelcu D 2014 *Mater Sci Forum* **782** 269-74
[33] Potoceanu N, Nedeloni M D, Chirus D and Florea D 2014 *Mater Sci Forum* **782** 257-62
[34] Nedeloni M D, Potoceanu N, Florea D and Chirus D 2014 *Mater Sci Forum* **782** 251-6
[35] Kazamer N, Pascal D T, Marginean G, Serban V A, Brandl W and Valean P C 2016 NANOCON pp 383-8
[36] Kekes D, Psyllaki P and Vardavoulias M 2014 *Tribology in Industry* **36**(4) 361-74
[37] Rozing G, Alar V and Marušić V 2015 *Interdisciplinary Description of Complex Systems* **13**(3) 461-471
[38] Bena T, Mitelea I, Bordeasu I and Craciunescu C 2016 METAL 2016, Czech Republic, May 25-27, pp 653-8
[39] Bordeasu I, Micu L M, Mitelea I, Utu I D, Pirvulescu L D and Sirbu N A 2016 *Mater Plast*

IOP Conf. Series: Materials Science and Engineering **416** (2018) 012028 doi:10.1088/1757-899X/416/1/012028

53(4) 781-86
[40] Salcianu C L, Bordeasu I, Mitelea I and Craciunescu C M 2016 METAL 2016, Czech Republic, May 25-27, pp 843-8
[41] Rozing G, Alar V, Marušić V and Samardžić I 2016 *Metalurgija* **55**(3) 441-444
[42] Padurean I, Bordeasu I, Velescu C and Oanca O 2007 *Rev Roum Sci Tech Ser Mec Appl* **52**(1) 15-26
[43] Padurean I and Nedelcu D 2008 *Rev Roum Sci Tech Ser Mec Appl* **52**(3) 203-215
[44] Bonny K, De Baets P, Vleugels J, Van der Biest O, Salehi A, Liu W and Lauwers B 2009 *Int J Refract Met H* **27** 449-457
[45] Cozza R C 2013 *Surf Coat Tech* **215** 224-233
[46] Fox-Rabinovich G S, Gershman I, El Hakim M A, Shalaby M A, Krzanowski J E and Veldhuis S C 2014 *Lubricants* **2** 113-123

Acknowledgment
We are very grateful to Mr. Prof. Viorel Câmpian from the Eftimie Murgu University of Reşiţa and to Mrs. Dr. Gabriela Mărginean from the Westphalian University of Applied Sciences Gelsenkirchen for the laboratory equipment.

IOP Conf. Series: Materials Science and Engineering **416** (2018) 012029 doi:10.1088/1757-899X/416/1/012029

An Analysis of Fracture Toughness in Concrete with Fly Ash Addition, Considering all Models of Cracking

G L Golewski[1,*]

[1]Faculty of Civil Engineering and Architecture, Lublin University of Technology, Nadbystrzycka, 40 str. 20-618 Lublin, Poland

*Corresponding author: g.golewski@pollub.pl

Abstract. This paper presents results of tests of fracture toughness of concrete modified with the addition of class F fly ash (FA), covering all three models of cracking: I, II, and III. Knowing the values of critical stress intensity factors: K^S_{Ic}, K_{IIc} and K_{IIIc} generalized fracture toughness K_c was determined. The tests were carried out on concrete composites with a different content of FA, i.e.: 0% (FA-00), 20% (FA-20) and 30% (FA-30). The tests were performed out after 28 days, i.e. the standard period of curing of the specimens. Based on the obtained results it can concluded that the class F FA additive in the amount up to 30% of mass of cement clearly changes the fracture toughness of concrete, at all models of cracking. Concrete with a 20% addition of FA – FA-20 is characterized by the most favorable mechanical parameters, whereas composite with a 30% addition of FA– FA-30 had low compressive strength and generalized fracture toughness. Moreover, the results of fracture toughness tests, are convergent qualitatively with compressive strength results.

1. Introduction

Concrete is currently the most often used and most important material in modern construction. It serves not only the role of material allowing the construction of: residential buildings, offices, public buildings and production halls, but also fulfils many other important functions in the social life of man – like for example protection of people from the effects of environmental pollution through the use of concrete structures such as: sewage disposal systems and sewage treatment plants, desulfurization installations and evacuation of combustion gases in heat plants, power plants and combined heat and power plants, acoustic screens for motorways and expressways, protection from radioactive sources in hospitals and nuclear power plants.

Due to the versatile possibilities of use, concrete composites are produced in such high quantities that it is difficult to imagine the functioning of modern civilisation without their presence in the construction industry. However, despite the fact that this materials has been widely used for decades, it has undergone frequent modifications and improvements over this time. In addition, it should be kept in mind that modern concrete, in most cases, is no longer a three-component mineral composite made of cement, aggregates and water, but is also composed of mineral additives and chemical admixtures, e.g. [1–7].

The concrete composites with mineral additives have gained more economic importance in recent years mainly due to the significant development of sustainable construction. It should be noted that 3 main postulates of sustainable construction state that:

- renewable raw materials should not be consumed faster than they can be regenerated,
- non-renewable raw materials should not be consumed faster than they can be replaced by renewable substitutes,
- pollution and waste should not be generated faster than nature can absorb them, recycle or dispose.

Unfortunately, the production of the basic binder for the production of concrete, i.e. Portland cement CEM I, contradicts the above postulates. First of all, the cement industry is one of the largest emitters of greenhouse gases (GHG), due to the fact that his branch of economy generates around 7% of total CO_2 emissions worldwide, e.g. [8]. The technological process in the cement industry, which generates the most CO_2, is the production of cement clinker in rotary kilns. Therefore, the production of pure CEM I Portland cement, containing the largest amount of clinker, causes the highest CO_2 emission. Increasing the share of mineral additives in the composition of cement and reducing the share of clinker is therefore a solution for the reduction of CO_2 emission by the cement industry.

On the other hand, the reduction of cement binder in the composition of the concrete mix has a positive effect on the reduction of CO_2 emission and also the reduction of the level of energy generated in the Portland cement production process.

Moreover, it is known that the Portland cement production process requires the consumption of not only large amounts of energy, but also a significant amount of raw materials. Therefore, an additional benefit of the presented treatments, i.e. substitution of cement with mineral additives, is the protection of natural resources such as: clay, marl, chalk and limestone – which are used to produce Portland clinker.

Taking into consideration the above reasons, it should be stated that the replacement of cement with mineral additives is an important task from the ecological point of view and is clearly environmentally friendly activity [9–11] .

Detailed classification of various materials used as substitutes for cement binder in the concrete structure is given in [12]. According to information contained therein, the most commonly used raw material from the group of mineral additives being industrial wastes is fly ash (FA).

Coal Fly ash has no elevated rates indicating its radioactivity [13] and moreover the presence of this additive affects both the concrete mix and the properties of hardened concrete. Spherical, glassy grains of FA (Fig. 1), e.g. [10, 14] cause the effect of "ball bearing", improving the workability of the mix and reducing its water demand by 5-15%. The effect of FA on the microstructure of concrete is both physical and chemical [15–17]. FA exhibit high pozzolanic activity, which leads to the formation of more C-S-H gel. This contributes to the reinforcement of the concrete microstructure both in the area of the cement matrix and in the contact zone between the aggregate and the paste. As the FA grains are smaller than cement grains, their presence in the paste improves the tightness of the cement matrix and thus the durability of the concrete. On the basis of the studies presented in [12, 18], it was shown that mature concrete with 20% FA additive had significantly smaller crack widths in the Interfacial Transition Zone (ITZ) between coarse aggregate and cement matrix than plain concrete.

Figure 1. Spherical glassy grains of fly ash.

The most common and most frequently used method of assessing the quality of concrete composites is the compressive strength test. However, this method is insufficient in the light of the research concerning the mechanism of concrete destruction. Based on this research, it is found that the concrete strength is closely connected with a phenomenon of initial cracks development. It should be added that microcracks always appear in the structure of concrete composites, and they are the result of complex processes associated with concrete curing. The inevitable defects and inhomogeneities occurring in the composite structure cause the concentration of stresses. As a consequence of these phenomena – in the place of the highest concentration of stresses – the development of microcracks begins, initially they are straight cracks and subsequently they are wing ones. During subsequent stages of material destruction, the microcracks combine into chains creating a system of spatial fractures, which consequentially leads to loosening the material structure and destruction of the structural element [19–23].

Therefore, compressive strength does not sufficiently characterize the concrete composite, the destruction of which occurs not as a result of exceeding average stresses measured during the compression strength test, but as a result of achieving sufficient energy in a place of the highest concentration of stresses (in the area of the longest, the widest or the sharpest microcrack).

Observations of the destruction process of concrete specimens during compression tests prove that frequently, an increase in force occurs even when the specimen has already cracked, that is when it has already been failured. The inadequacy of the compressive strength tests is eliminated in respect of fracture toughness tests. In these experiments, the value of critical force and critical stresses are determined at the time of rapid propagation of the initial crack.

Therefore, this paper presents results of tests of fracture toughness of concrete modified with the addition of FA, covering all three models of cracking: I, II, and III. Knowing the values of critical stress intensity factors K_{Ic}, K_{IIc} and K_{IIIc}, e.g. [24–27] generalized fracture toughness K_c was determined based on Equation (1) [28].

$$K_c^2 = K_{Ic}^2 + K_{IIc}^2 + \frac{K_{IIIc}^2}{1-\nu} \tag{1}$$

where: ν, is Poisson's ratio.

Since, in the literature there are no results of tests of fracture toughness in concrete containing FA, including all models of cracking, in this paper presents a comprehensive analysis of this problem. The tests were carried out on concrete composites with a different content of FA, i.e.: 0% (FA-00), 20% (FA-20) and 30% (FA-30). The tests were performed out after 28 days, i.e. the standard period of curing of the specimens.

2. Materials and methodology

2.1. Materials

The concrete mixtures were prepared from: Ordinary Portland Cement (OPC) CEM I 32.5 R, a pit sand with a maximum size of 2.0 mm (as fine aggregate), natural gravel with a minimum and maximum size of 2.0 and 8.0 mm (as coarse aggregate), class F fly ash (Blaine fineness – 0.364 m^2/g, specific gravity – 2.14 g/cm^3), plasticizer (Basf Liquol BV-18) and the laboratory pipeline water.

2.2. Mixture design and specimens used in the studies

The methodology used for determining the composition of the mixtures relies on replacing a suitable amount of the cement volume (the weight volume content) in the mixture with a corresponding amount of the FA to ensure that the total amount of binding material is always constant. The concrete mixture proportions are summarized in Table 1. All mixtures had the same water-binder ratio $w/b = 0.4$.

Table 1. Summary of concrete mixture proportions (kg/m³)

Mixture	OPC	FA	%FA	Sand	Gravel	Water	Plasticizer
FA-00	352	0	0				
FA-20	282	70	20	676	1205	141	2
FA-30	246	106	30				

The scope of the studies included the performance of the following tests:

- compressive strength – f_{cm}, on cubic specimens (150 mm),
- fracture toughness under mode I fracture – K_{Ic}^{S}, on beams (80 × 150 × 700 mm) with one initial crack (Fig. 2a),
- fracture toughness under mode II fracture – K_{IIc}, on cubic specimens (150 mm) with two initial cracks (Fig. 2b),
- fracture toughness under mode III fracture – K_{IIIc}, on cylindrical specimens with a diameter of 150 mm and a height of 300 mm having an initial circumferential notch (Fig. 2c).

It should be added that all tests for all types of composites were carried out on 6 specimens after 28 days of their curing.

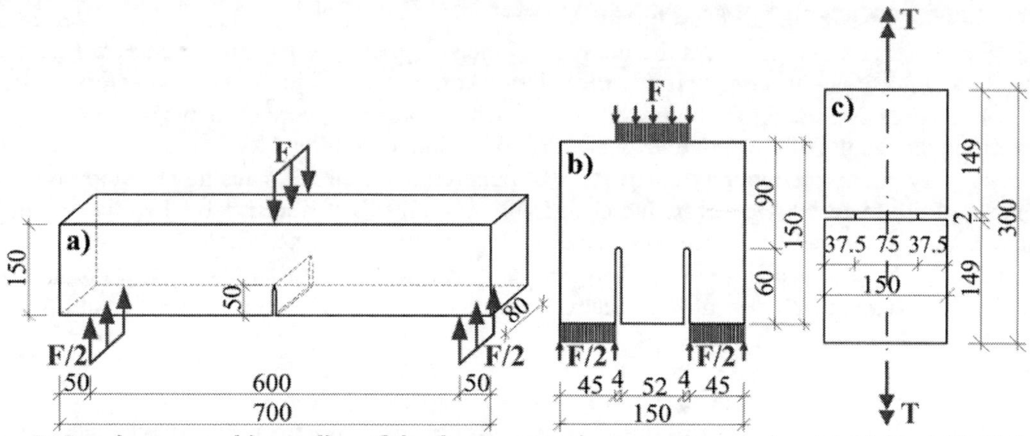

Figure 2. Specimens used in studies of the fracture toughness under mode: a) I, b) II, c) III, F-force, T-torsional moment; dimensions in mm.

2.3. Compressive strength tests and fracture toughness examination

The uniaxial compression strengths were tested using a compression machine (Walter + Bai ag) with a maximum load of 3000 kN. The compressive strengths f_{cm}, were tested with application of cubic specimens (described in section 2.2) according to the standards of series EN 12390.

Dimensions of all specimens for fracture toughness tests with loading conditions was shown in Fig 2, whereas views and descriptions of the experimental stands are presented in details in the following papers:

- [20] – for the first model of cracking,
- [16] – for the second model of cracking,
- [17,22] – for the mode III loading.

The testing of Mode I fracture toughness K_{Ic}^{S} was performed according with the draft guidelines of RILEM recommendations [29]. A detailed description of the fracture toughness test for the first model of cracking are given in [9,10,20].

Fracture toughness K_{IIc} was determined according to the equation proposed by Watkins [30]. The methodology for determining this factor was given in [16,20].

IOP Conf. Series: Materials Science and Engineering **416** (2018) 012029 doi:10.1088/1757-899X/416/1/012029

In order to determine the fracture toughness K_{IIIc}, a special innovative device was designed and manufactured [8,15, 17,22,23]. This device has also been patented. A detailed description of the fracture toughness test for the third model of cracking (anti-plane shear) are given in [8,15,17,22,23]. Fracture toughness K_{IIIc} can be calculated, according to the equation proposed by Miannay [31].

With the known values of fracture mechanics parameters for all models of cracking, it was possible to designate a generalized fracture toughness of the material – K_c from the Equation (1).

3. Results and discussion

Table 2 presents the average values of compressive strength f_{cm} and fracture toughness for all models of cracking, i.e.: K_{Ic}^{S}, K_{IIc} and K_{IIIc}. The table also includes standard deviations δ from all experiments. The last column of Table 2 gives the results of the calculated generalized fracture toughness K_c.

Table 2. Results of the research

Concrete	Analyzed parameters				
	$f_{cm} \pm \delta$	$K_{Ic}^{S} \pm \delta$	$K_{IIc} \pm \delta$	$K_{IIIc} \pm \delta$	K_c
FA-00	47.51 ± 2.55	1.06 ± 0.10	4.24 ± 0.40	2.49 ± 0.20	5.03
FA-20	48.96 ± 3.02	1.09 ± 0.11	4.39 ± 0.51	2.60 ± 0.24	5.22
FA-30	45.10 ± 3.55	0.93 ± 0.10	3.65 ± 0.42	2.14 ± 0.21	4.33

A precise analysis of the test results given in Table 2 leads to the conclusion that, after 28 days of curing, the 20% FA addition causes the increase of the analyzed mechanical parameters f_{cm} and K_c by 2% to 3%. Unfortunately, composites with a larger amount of additive are characterized by lower values of compressive strength and generalized fracture toughness by more than 5%.

Based on the results of the above tests, it can be concluded that the addition of FA in the amount of 20% has a very positive effect on the structure of concrete composites, causing an increase in fracture toughness of the material. However, increasing the amount of the additive by 10% has a significant negative effect on the values of mechanical parameters in such concrete (Table 2).

Additionally, when analyzing Table 2, it was also found that compressive strength values f_{cm} are qualitatively convergent with generalized fracture toughness K_c of concrete with the addition of FA (Table 2).

4. Conclusions

The following conclusions can be drawn from the present investigation.

1. The class F fly ash additive in the amount up to 30% of mass of cement clearly changes the fracture toughness of concrete, at all models of cracking, after 28 days of curing.
2. Concrete with a 20% addition of FA is characterized by the most favorable mechanical parameters.
3. The results of fracture toughness tests, are convergent qualitatively with compressive strength results.

5. References

[1] Horszczaruk E, Sikora P and Lukowski P 2016 Application of nanomaterials in production of self-sensing concretes: contemporary developments and prospects *Arch. Civ. Eng.* **LXII**(3) 61–72.

[2] Zhang P, Wan J, Wang K and Li Q Influence of nano-SiO_2 on properties of fresh and hardened high performance concrete: A state-of-the-art review 2017 *Constr. Build. Mater.* **148** 648–58.

[3] Alwaeli M 2013 Application of granulated lead-zinc slag in concrete as on opportunity to save natural resources *Rad. Phys. Chem.* **83** 54–50.

[4] Niewiadomski P, Hola J and Cwirzen A 2018 Study on properties of self-compacting concrete modified with nanoparticles *Arch. Civ. Mech. Eng.* **18** 877–86.

[5] Sosdean C, Marsavina L and De Schutter G 2016 Exprimental and numerical determination of the chloride penetration in cracked mortar specimen *E. J. Env. Civ. Eng.* **20**(2) 231–49.

[6] Giergiczny Z and Krol A 2008 Immobilization of heavy metals (Pb, Cu, Cr, Zn, Cd, Mn) in the mineral additions containing concrete composites *J. Hazard. Mater.* **160** 247–55.

[7] Blaszczynski T and Krol M 2015 Usage of green concrete technology in civil engineering *Proc. Eng.* **122** 296–301.

[8] Golewski G L 2018 Effect of curing time on the fracture toughness of fly ash concrete composites *Compos. Struct.* **185** 105–12.

[9] Golewski G L 2018 Green concrete composite incorporating fly ash with high strength and fracture toughness *J. Clean. Prod.* **172** 218–26.

[10] Golewski G L 2017 Improvement of fracture toughness of green concrete as a result of addition of coal fly ash. Characterization of fly ash microstructure *Mater. Charact.* **134** 335–46.

[11] Liew K M, Sojobi A O and Zhang L W 2017 Green concrete: Prospects and challenges *Constr. Build. Mater.* **156** 1063–95.

[12] Golewski G L 2018 *J. Hazard. Mater.* **357** 298–304.

[13] Golewski G L 2015 Studies of natural radioactivity of concrete with siliceous fly ash addition *Cem. Wapno Beton* **2** 106–114.

[14] Chindaprasirt P, Jaturapitakkul Ch and Sinsiri T. 2005 Effect of fly ash fineness on compressive strength and pore size of blended cement paste *Cem. Concr. Compos.* **27** 425–28.

[15] Sadowski T and Golewski G L 2018 A failure analysis of concrete composites incorporating fly ash during torsional loading *Compos. Struct.* **183** 527–35.

[16] Golewski G L and Sadowski T 2014 An analysis of shear fracture toughnesstoughness K_{IIc} and microstructure in concretes containing fly-ash *Constr. Build. Mater.* **51** 207–14.

[17] Golewski, G L and Sadowski T 2017 *Constr. Build. Mater.* **143** 444–54.

[18] Golewski G L 2018 An assessment of microcracks in the Interfacial Transition Zone of durable concrete composites with fly ash additives *Compos. Struct.* **200** 515–20.

[19] Golewski G L 2017 Generalized fracture toughness and compressive strength of sustainable concrete including low calcium fly ash *Materials* **10**(12) 1393.

[20] Golewski G L and Sadowski T 2016 Macroscopic evaluation of fracture processes in fly ash concrete *Sol. Stat. Phenom.* **254** 188–93.

[21] Golewski, G L and Sadowski T 2012 *Sol. Stat. Phenom.* **188** 158–63.

[22] Golewski G L 2017 Determination of fracture toughness in concretes containing siliceous fly ash during mode III loading *Struct. Eng. Mech.* **62** 1–9.

[23] Golewski G L 2017 Effect of fly ash addition on the fracture toughness of plain concrete at third model of fracture *J. Civ. Eng. Manag.* **23** 613–20.

[24] Berto F, Ayatollahi M and Marsavina L 2017 Mixed mode fracture *Theoret. Appl. Fract. Mech.* **91** 1.

[25] Ayatollahi M R and Saboori B 2015 T-stress effects in mixed mode I/II/III brittle frature *Eng. Fract. Mech.* **144** 32–45.

[26] Marsavina L, Constantinescu D M, Linul E, Voiconi T and Apostol D A 2015 Shear and mode II fracture of PUR foams *Eng. Fail. Anal.* **58**(2) 465–76.

[27] Marsavina L, Berto F, Negru R, Serban D A and Linul E 2017 An engineering approach to predict mixed mode fracture of PUR foams based on ASED and micromechanical modelling *Theor. Appl. Fract. Mech.* **91** 148–54.

[28] Knott J F 1976 *Fundamentals of Fracture Mechanics* Butterworths London UK.

[29] TC 89-FMT Fracture Mechanics of Concrete Test Methods 1990 Determination of fracture parameters (K_{Ic} and $CTOD_c$) of plain concrete using three-point bend tests. RILEM Draft Recommendations *Mater. Struct.* **23** 457–60.

[30] Watkins J 1983 Fracture toughness test for soil-cement samples in mode II *Inter. J. Fract.* **23**(4) R135–R138.

[31] Miannay D P 1998 *Fracture mechanics* Springer-Verlag New York USA.

IOP Conf. Series: Materials Science and Engineering **416** (2018) 012030 doi:10.1088/1757-899X/416/1/012030

Investigations on Tungsten Inert Gas Welded Magnesium Alloy

I Peter* and M Rosso

Politecnico di Torino, Department of Applied Science and Technology
Corso Duca degli Abruzzi, 24, 10129 Torino- Italia

*Corresponding author: ildiko.peter@polito.it

Abstract. Generally, magnesium alloys show interesting technological characteristics, which make them valuable for many industrial applications. Good mechanical properties, creep resistance and good castability, recyclability are only some of the most important properties remarkable for their employment in different areas. Generally, at industrial level welding technologies, like Tungsten Inert Gas technique, are used to eliminate and/or to repair the casting defects. In the present paper some details on their microstructural behaviour and mechanical performances on welded WE43 magnesium alloy are evaluated and reported. The obtained results have demonstrated that welding do not decrease the mechanical performance of the material and the heat affected zone cannot compromise the overall behaviour of the welded samples. In most of cases, the material's failure takes place nearly the welding defects.

1. Introduction

Actually, light alloys are more and more employed in different industrial applications. In order to obtain high integrity and improved performance components, the need of new alloy compositions and of innovative production techniques are required. Al-based alloys are ideal light-weighting material for automotive and aeronautical applications [1-6]. In the same time, the lowest density of Mg alloys compared to Al-based ones, allows achieving a further weight saving [7-9]. Magnesium alloys due to their extremely interesting and promising properties like high value of specific strength and stiffness, low density, remarkable damping capacity, excellent castability, high thermal conductivity and for their recyclability, are very attractive for a wide range of applications. Automotive, aerospace and electronics industries [10-12] are the most relevant areas where these alloys are employed. However, due to their hexagonal close-packed (HPC) crystal structure, magnesium alloys are characterized by reduced castability and cold workability at room temperature, which make complicate and extremely expensive the manufacturing process of components, especially with a complex shape and geometry [13]. Extended use of light alloys in automotive industry needs reliable welding technologies, because many of the produced parts need to be joined to one similar or different materials to obtain a more complex geometry. To achieve this goal, welding techniques have received particular attention. Generally, welding of Al alloys can be realized through different welding processes: Arc-welding techniques, such as metal inert gas (MIG), tungsten inert gas (TIG) and manual metal arc (MMA) [14-16]. Actually, in many industrial fields, the development of welding technologies represents a key factor to enlarge the use of Mg-based alloys too for the manufacturing of structural components. Currently, the most important welding procedures used for magnesium alloys are: TIG, Laser Beam Welding (LBW), Electron Beam Welding (EBW) and Friction Stir Welding (FSW) [17-18]. Nowadays, at industrial level, TIG and MIG fusion welding technologies are especially applied for removing and for repairing defects in Mg castings [18]. TIG is largely used at industrial scale, due to its advantages of utility and economy. However, Mg-based alloys are characterized by low melting point, high thermal conductivity, strong affinity toward oxygen and nitrogen, and for these reasons during the welding, many defects can be developed. The most relevant defects are related to the presence of porosity and cracks, wide heat affected zone (HAZ), residual stresses and distortions [19-20]. Additionally, application of TIG welding technology presents some restrictions due to its low penetration in single pass welding and its low productivity. According to the literature data, improvement of both the penetration and the quality of Mg alloys weldments can be realized through the addition of TiO2 coating [17] and the application of high frequency vibration, during the solidification of the melt [20]. According to [21-23], the increasing use of Mg-based alloys for cyclically-loaded structural applications has given rise to a growing interest for as concerns fatigue properties of cast magnesium alloys. The presence of defects, developed during the casting operations, as well as of the specific microstructural features of the alloys, such as grain size, second-phase particles and hardening precipitates, represent the elements that mainly affect the fatigue properties of these alloys [24-25]. Porosity, oxide films and intermetallic inclusions are defects that, acting as stress concentration sites, can constitute favorable nucleation sites for cracks with relevant negative consequences, decreasing both the lifetime and the fatigue strength of the alloys, mainly at elevate numbers of cycles [25]. In this paper, the microstructure, the tensile properties and the fatigue strength of ZE41 Mg alloy before and after TIG welding process carried out with double pass, are evaluated, aiming to appreciate how the welding process affects the microstructure and the mechanical strength. Fracture surface analysis completes the study, in order to identify the defects and to correlate their presence to the level of the fatigue strength.

2. Experimental

The chemical composition of the ZE41 magnesium alloy is reported in Table 1.

Table 1, Chemical composition (wt%) of the investigated ZE41 magnesium alloy

Elements	Zn	Zr	Si	Mn	Rare earths (Ce,La,Nd,Pr)	Mg
Wt%	4	0.7	0.04	0.03	1.68	Balance

Some plates have been produced by sand casting technology and have been submitted to a two-step T5 heat treatment. The thermal treatment has been carried out in the following sequence: first step performed at 330°C for 2 hours, which is useful to guarantee a stress relieving.

This has been followed by air quenching then a second ageing step carried out at 180°C for 12 hours followed by air cooling. TIG welding process has been used for the joints development, and during this step the plates have been made in contact without a rigorous control of the gap.

The following welding parameters have been adopted: 160-180 (A); 15-17 (V) and an electrode with a diameter of 3.2 mm has been used. The same ZE41 magnesium alloy has been used as filler material; finally, argon has been employed as shielding gas. The welding process has been carried out with double pass. In order to reduce the thermal shocks as much as possible, that is due to the thermal gradient existing between the base metal and the filler, the plates have been preheated to 160°C before welding. After the welding, the samples have been submitted for 12 hours to a stress relieving post heat treatment, performed at a temperature of 185°C. Samples for microstructural, mechanical and fatigue tests have been extracted from the prepared plates.

Microstructure analysis have been carried out on the samples prepared by a standard metallographic technique by mounting and polishing procedures after tensile and fatigue tests using an optical microscope, (OM, MeF4 Reichart-Jung) and Scanning Electron Microscopy (SEM, Leo 1450VP) equipped with Energy X-rays Dispersive Spectroscopy unit (EDS, Oxford microprobe) for compositional analysis. SEM has been employed for fracture surface analysis, after mechanical and fatigue tests, to determine the causes which can be associated to the failure of the samples.

On the polished surfaces Vickers micro-hardness measurements have been performed using a Volpert DU01 tester. A force of 200 gf (1.96 N) has been applied for 15 s for each measurement and a minimum of 5 indentations have been performed on each samples. Micro-hardness has been measured along a line, from the weld centre line to the base metal, with a distance of 1mm between two consecutive indentations.

3. Results and discussion

3.1. Microstructural analysis

Figure 1 reports the microstructure of base material: as it can be observed, it consists mainly of equiaxed α-Mg matrix grains and an eutectic compounds, distributed along the grain boundaries. The eutectic compound highlighted in Figure 1 and analyzed by the X-rays diffraction, reported in Figure 2 is orthorhombic T-phase (Mg7Zn3RE), which presents the following lattice parameters a = 0.99 nm, b = 1.15 nm and c = 0.98 nm.

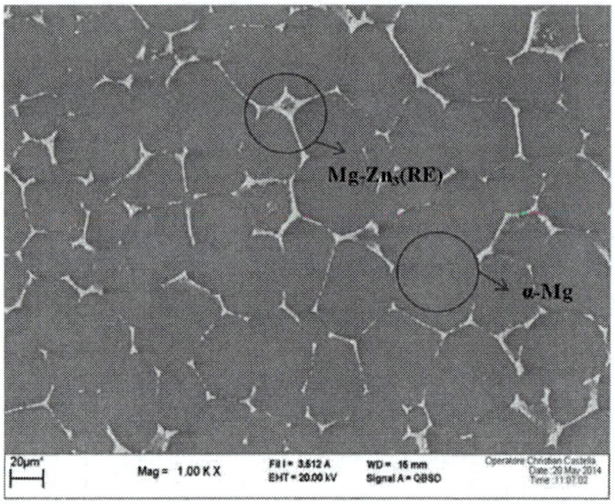

Figure 1. SEM microstructure of ZE41 magnesium alloy

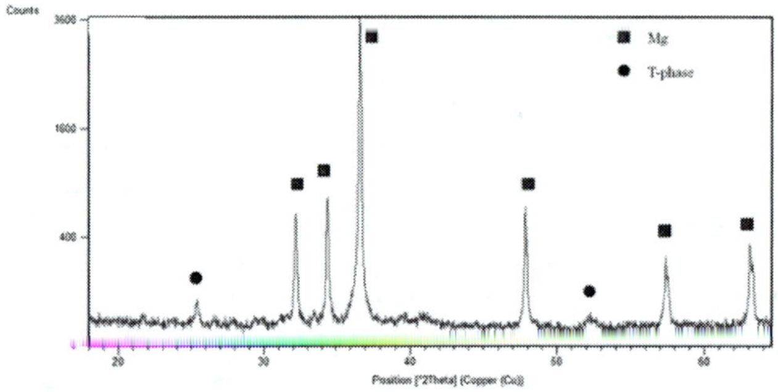

Figure 2. X-ray diffraction pattern of the ZE41 magnesium alloy

Homogeneously distributed small Zr particles as shown in Figure 3 (≤ 5μm) that have been individuated in the analysed samples. Zr represents one of the most powerful grain refinement for Mg-based alloys. Its grain refining ability is reduced by the presence of chemical elements such as Al, Mn and Si according to [27]. However, due to their low volume fraction, the Zr-rich phase has not been detected by XRD.

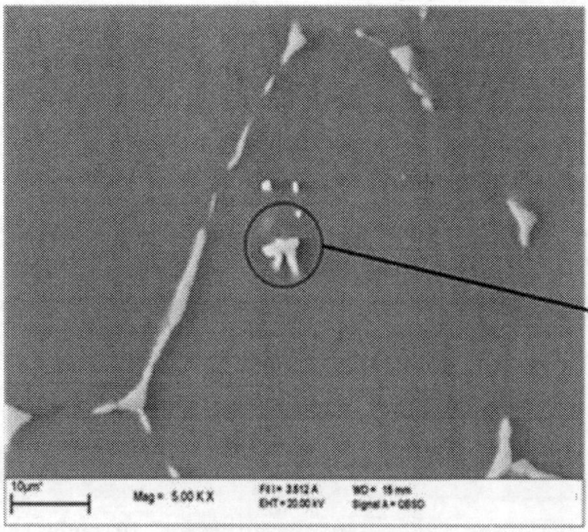

Elements	%wt	%at
Mg	11.75	32.15
Zn	11.95	12.17
Zr	76.30	55.68

Figure 3. Zr-rich particle which acts as nucleation site for α-Mg grains

The SEM micrograph reported in Figure 4, shows the transition zone between the fusion zone and the heat affected zone. Both zones are mainly composed of primary α-Mg matrix grains and eutectic compounds, T-phase (Mg7Zn3RE), along grain boundaries, as shown in Figure 1. It can be observed in Figure 5 that the three zones, namely FZ, HAZ and BM consisting of equiaxed grains have been individuated. Fine grains appear in the FZ (Figure 5). Due to the high cooling rate of the welding pool, coming from the high thermal conductivity of ZE41 Mg-based alloy, the FZ is finer with respect to the HAZ. As reported in [28], a finer grain size represents a suitable feature to reduce the susceptibility to cracking in welded castings. The conductive heat dissipated toward the side of the welding pool during the TIG welding process mainly causes the grain coarsening occurred in the HAZ. The HAZ shows a coarsened T-phase compared to the BM, as illustrated in the OM micrographs reported in Figure 5.

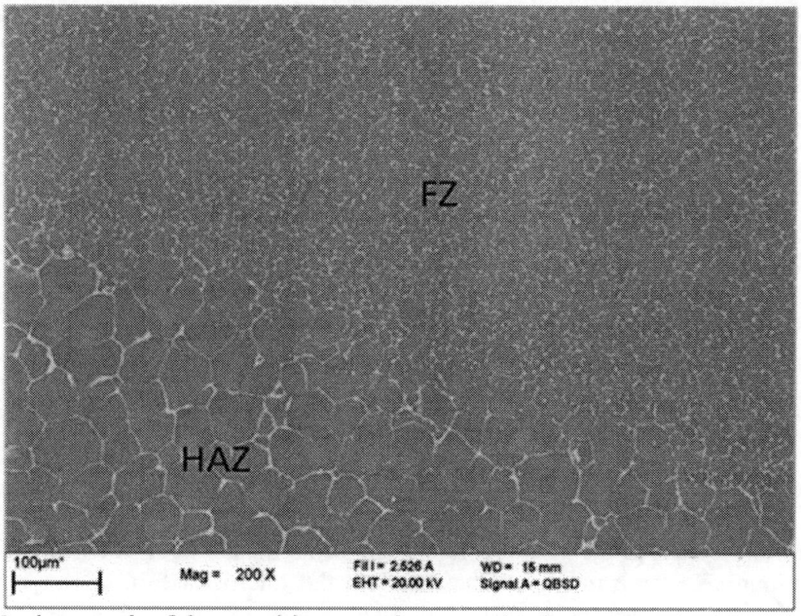

Figure 4. SEM micrograph of the transition zone between the fusion zone (FZ) and the heat-affected zone (HAZ)

IOP Conf. Series: Materials Science and Engineering **416** (2018) 012030 doi:10.1088/1757-899X/416/1/012030

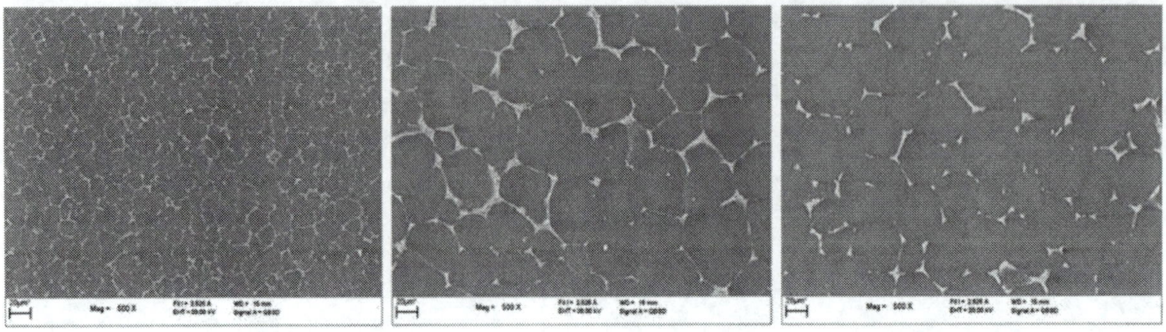

(a) (b) (c)

Figure 5. OM micrographs of: (a) fusion zone (FZ), (b) heat affected zone (HAZ) and (c) base metal (BM)

3.2. Vickers Micro-hardness

The TIG welding process causes microstructural variations, which locally influence the material properties, as reveal the Vickers micro-hardness profiles (Figure 6), of both not welded and TIG welded samples. The grain size and the presence of strengthening precipitates represent the two factors that mainly influence the micro-hardness of an alloy. In fact, according to the Hall-Petch equation [29], hardness increases when grain size decreases; however due to the Orowan hardening mechanism [30] the hardness increases if the spacing between the hardening precipitates is reduced. The highest micro-hardness values of the FZ, registered for the TIG welded samples, also compared to that of the not welded samples, are principally due to the grain refinement and the possible re-precipitation of strengthening precipitates. The micro-hardness profile of TIG welded samples shows a significant decrease in correspondence of the HAZ, because these zones are submitted to a softening effect and a grain coarsening [31]. Finally, the micro-hardness of welded samples, in the BM zones, reaches the value of not welded sample (~75 Hv).

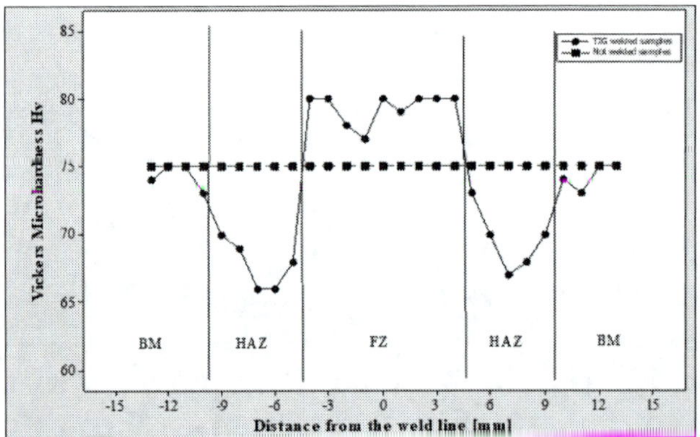

Figure 6. Vickers micro-hardness profiles on the cross section of not welded and TIG welded samples

4. Conclusions

In this research, the effects of TIG welding process on the microstructure, the tensile properties and the fatigue strength of ZE41(Mg–Zn–RE–Zr) magnesium alloy were investigated. In case of the TIG welded samples, a grain refinement in the fusion zone and a grain coarsening in the heat affected zone takes place. The distributions of the micro-hardness reflect the microstructure variations occurred in the welded samples.

5. References

[1] Guo MX, Zhang Y, Zhang XK, Zhang JS, Zhuang LZ 2016 Non-isothermal precipitation behaviours of Al-Mg-Si-Cu alloys with different Zn contents, J. Material Science Engineering A **669** 20

[2] Cavazzoni L, Miscia G, Rotondella V, Baldini 2015 Procedia Eng **109** 17
[3] Grosselle F, Timelli G, Bonollo F 2010 J. Material Science Engineering A 527: 3536–3545.
[4] Peter I, Rosso M, Bivol C 2009 Microstructure and mechanical behaviour of Al-based alloy obtained by liquid forging technique, Metal Int **2** 15
[5] Huang Y, Ye X, Hu B, Chen L 2016 Equivalent crack size model for pre-corrosion fatigue life prediction of aluminum alloy 7075-T6, Int J Fatigue **88** 217
[6] Tavares SMO, Dos Santos J, De Castro PMST 2013 Friction stir welded joints of Al–Li Alloys for aeronautical applications: butt-joints and tailor welded blanks, Theor Appl Fract Mech **65** 8
[7] Hirsch J, Al-Samman T 2013 Superior light metals by texture engineering: Optimized aluminum and magnesium alloys for automotive applications, Acta Mater **61** 818
[8] Park GH, Kim JT, Park HJ et al 2016 Development of lightweight Mg-Li-Al alloys with high specific strength, J. Alloys Compounds **680** 116
[9] Zhang W, Xiao W, Wang F, Ma C (2016) Development of heat resistant Mg-Zn-Al-based magnesium alloys by addition of La and Ca: Microstructure and tensile properties, J Alloys Compd 684:8-14
[10] Chen XH, Huang XW, Pan FS, Tang AT, Wang JF, Zhang DF (2011) Effects of heat treatment on microstructure and mechanical properties of ZK60 Mg alloy, Trans Nonferrous Met Soc China 21:754-760.
[11] Huang ML, Li HX, Bao L, Ma XB, Hao SM (2012) Partial phase relationships of Mg-Zn-Ce system at 350 °C, Trans. Nonferrous Met. Soc. China 22 681-685.
[12] Ratna Sunil B, Pradeep Kumar Reddy G, Mounika ASN, Navya Sree P, Rama Pinneswari P, Ambica I, Ajay Babu R, Amarnadh P (2015) Joining of AZ31 and AZ91 Mg alloys by friction stir welding , J. Magnes Alloy 3:330–334.
[13] Coelho R, Kostka A, Pinto H, Riekehr S, Kocak M, Pyzalla AR 2008 Mater Sci Eng A 485:20–30.
[14] Zhang L, Li X, Nie Z, Huang H, Niu L (2016) Comparison of microstructure andmechanical properties of TIG and laser welding joints of a new Al–Zn–Mg–Cu alloy, Mater Des 92:880–887.
[15] Yi J, Cao SF, Li LX, Guo PC, Liu KY (2015) Effect of welding current on morphology and microstructure of Al alloy T-joint in double-pulsed MIG welding, Trans Nonferrous Met Soc China 25 3204−3211.
[16] Peter I, Rosso M (2011) Effect of the filler metals on aluminium alloy joints, Metal Int 16:157-160.
[17] Wang L, Shen J, Xu N (2011) Mater Sci Eng A 528:7276– 7284.
[18] Min D, Shen J, Lai S, Chen J, Xu N, Liu H (2011) Effects of heat input on the low power Nd:YAG pulse laser conduction weldability of magnesium alloy AZ61, Opt Lasers Eng 49:89–96.
[19] Carlone P, Palazzo GS (2015) Characterization of TIG and FSW weldings in cast ZE41A magnesium alloy, J Mater Process Technol 215:87–94.
[20] Wen T, Liu SY, Chen S, Liu LT, Yang C (2015) Influence of high frequency vibration on microstructure and mechanical properties of TIG welding joints of AZ31 magnesium alloy, Trans Nonferrous Met Soc China 25:397−404.
[21] Li Z, Wang Q, Luo AA, Peng L, Zhang P (2015) Fatigue behavior and life prediction of cast magnesium alloys, Mater Sci Eng A 647:113–126.
[22] Luo AA, Fu PH, Peng LM, Kang XY, Li ZZ, Zhu TY (2012) Metall Mater Trans A 43:360–368.
[23] Yang F, Lv F, Yang XM, Li SX, Zhang ZF, Wang QD (2011) Enhanced very high cycle fatigue performance of extruded Mg–12Gd–3Y–0.5Zr magnesium alloy, Mater Sci Eng A 528: 2231–2238.
[24] Xu DK, Liu L, Xu YB, Han EH (2008)The fatigue behavior of I-phase containing as-cast Mg–Zn–Y–Zr alloy, Acta Mater 56:985–994.
[25] Mayer H, Papakyriacou M, Zettl B, Stanzl-Tschegg SE (2003) Influence of porosity on the fatigue limit of die cast magnesium and aluminium alloys, Int J Fatigue 25:245–256.

IOP Conf. Series: Materials Science and Engineering **416** (2018) 012031 doi:10.1088/1757-899X/416/1/012031

Influence of the Chemical Composition, Structure and Heat Treatment on the Topography and Mechanical Properties of Ti5Al4V and Ti6Al4V after Electron Beam Surface Treatment

M P Nikolova[1,*], E Yankov[1], D Nikolov[1], D Tonev[1], V Zaharieva[1], M Ormanova[2], S Valkov[2], T Hikov[1] and P Petrov[1]

[1]University of Ruse "A. Kanchev", Dept. of Material Science and Technology, 8 Studentska Str., Bulgaria
[2]Bulgarian Academy of Sciences, Institute of Electronics, 72 Tzarigradsko Chausse Blvd, Sofia, Bulgaria

*Corresponding author: mpnikolova@uni-ruse.bg

Abstract. In order to obtain a proper and accelerated osseointegration process, continuous efforts for improving the surface of the implants have been made. The physical texture, structure and mechanical properties of the implant surface are known to be an important factor in bone-implant interaction. Trying to find a solution for weak osseointegration, new modification techniques should be used and examined. In the particular study, the Ti5Al4V and Ti6Al4V alloys with different initial structure and composition were vacuum heat treated at different technological parameters in order to obtain maximum strengthening. One and the same electron beam surface modified (EBSM) by scanning electron beam was subsequently carried out for both alloys and their surface characteristics and mechanical properties were revealed by metallographic analysis, roughness, XRD and hardness measurements. The results revealed that under the selected conditions of treatment, a specific wave-like topography was obtained. The mean grains' sizes, depth of the structural changes after the surface modification and hardness and crystal lattices' changes of both alloys was assessed depending on the differences in the composition, initial structural and mechanical characteristics of the alloys in the different heat-treated conditions.

1. Introduction

Titanium is a bioinert material because it does not connect to bone or induces bone growth [1]. In recent years, there are many surface modifications methods such as plasma assisted thermochemical treatment, laser etching techniques, sandblasting, anodic chemical treatment, ion implantation that are used for improving the surface properties of implant materials [2]. These are applied to increase the surface roughness and wettability and thus to enhance effectively the mechanical performance and bone-implant anchorage. Among the surface modifications, the electron beam surface modified (EBSM) provide a new possibility to be a good candidate for surface roughening providing the opportunity to obtain reproducible isotropic textures with regular surface micro-pattern without any surface contamination, grain refinement and homogenization of the elements' distribution [3]. The higher hardness and good wear-resistance is achieved because of the high cooling rates and martensite phase precipitation.

The exclusively used α+β high-strength medical titanium alloys could be thermally processed to produce a wide range of microstructures with different properties. The thermal pre-treatment is necessary to prepare the surface of an implant for advance surface performance. The elastic modulus of

Ti tends to increase with increasing Al content and increasing the volume fracture of α [4] while the β-transus temperature increase. On the other hand, the presence of β-stabilizing elements like V, Mo, and Nb are known to increase the strength, reduce the elastic modulus and improves the passivation leading to higher resistance to corrosion [5]. This study aims at determining the response of two α+β titanium alloys with small variations in the chemical composition in as-received and heat treated conditions to the EBSM with scanning electron beam.

2. Experimental procedures

The chemical compositions of both materials used for the experiment, determined by JEOL JXCA-733 Microprobe scanning electron microscope (SEM) equipped with wavelength dispersive spectrometers (WDS), were shown in Table 1. Samples with dimensions 14×14×4 and 15×15×4 mm were cut out using the electro-erosion cutting method. The as-received (AR) samples were single solution treated (ST) for 30 min at 920 °C and 950 °C for the Ti54 and Ti64, respectively and water quenched. Half of the specimens were precipitated (P) for 4 hours at 500 °C and 540 °C for Ti54 and Ti64, respectively and air cooled. All treatments were carried out in ≤ 1 Pa vacuum. The surfaces of the samples were grounded and polished before the EBSM. To study the effect of the surface treatment, the EBSM was applied to AR, ST, and ST and precipitated (ST+P) samples.

Table 1. Chemical composition (wt. %) of the substrate material.

Element	Al	V	Fe	Mn	Co	Mo	Pd	Nb	Hf	Ti
Ti54	5.21	4.40	0.14	0.11	0.06	0.17	0.15	0.36	0.04	Bal.
Ti64	5.80	4.67	0.16	-	0.09	0.07	0.02	-	0.14	Bal.

The electron beam surface modification (EBSM) was carried out by electron beam installation Leybold Heraeus (EWS 300/15–60). The following technological parameters were applied: electron beam current – I = 30 mA, accelerated voltage – U = 52 kV, speed of the samples motion – v = 1 cm/s, electron beam frequency – f = 1 kHz, electron beam diameter - d = 0.5 mm and linear manner of scanning.

The samples were evaluated quantitatively with respect to R_a (average roughness) and R_z (maximum roughness height) after ten times scanning with a cut-off value of 0.8 mm in a direction perpendicular to grooves caused by the local melting by using a contact profilometer (Mitutoyo SJ 201 P). The light optical microscopy was performed on the top surface and cross-section of the samples using a Nikon microscope. The 14-megapixel digital camera was adapted to the microscope and used for the image acquisition after Kroll's reagent etching. Some of the micrographs were processed by Mountains Map premium v.7 software for grain size assessment.

The X-ray phase identification was performed with URD-6 diffractometer, applying Bragg-Brentano geometry and $Cu\text{-}K_{\alpha}$ radiation and working at 30 kV and 20 mA. The diffraction angle range scanned was 2θ = 30-100° with a step size of 0.05° and counting time of 2.5 s/step. The qualitative and quantitative phase analysis was held using Match!3 software.

A Vickers Hardness tester 432 SVD by Wilson-Wilpert was used for the hardness measurements of the alloy according to the standard ISO 6507-1. In order to compare the hardness changes, the researchers were made on the top of each sample with a load of 10 kg and dwell time of 10 s. Ten sequential measurements by increments of 500 μm between the imprints were made.

3. Results

The micrographs of both alloy – Ti54 and Ti64 in as-received and heat treated (ST) condition are shown in Figure 1. Because of the sub-micron size of the phases separated after the precipitation, no changes in the optical micrographs of the ST+P were detected as opposed to the ST specimens. In contrast to the coarse lamellar structure of AR Ti54, the Ti64 alloy displayed smaller lamellar colonies within the α+β grains (Figure 1a). The micrographs of the ST alloys show that the heat treatment was conducted from the α-β-range of both alloys despite the difference of 30 °C in the process temperature (Figure 1b).

Although the higher solutionizing temperature, the microstructure of the Ti64 displayed finer martensite lathes among the primary α-grains because of the fine initial structure instead of gross acicular martensite laths that were found in the ST Ti54 alloy. In order to compare quantitatively the amount of the undissolved α-grains, both ST micrographs of the bimodal structures were software processed and shown in Figure 1c. It is clearly seen that the amount of the primary α-phase is higher for the 950° C ST Ti64 than that of the 920° C water quenched Ti54 alloy.

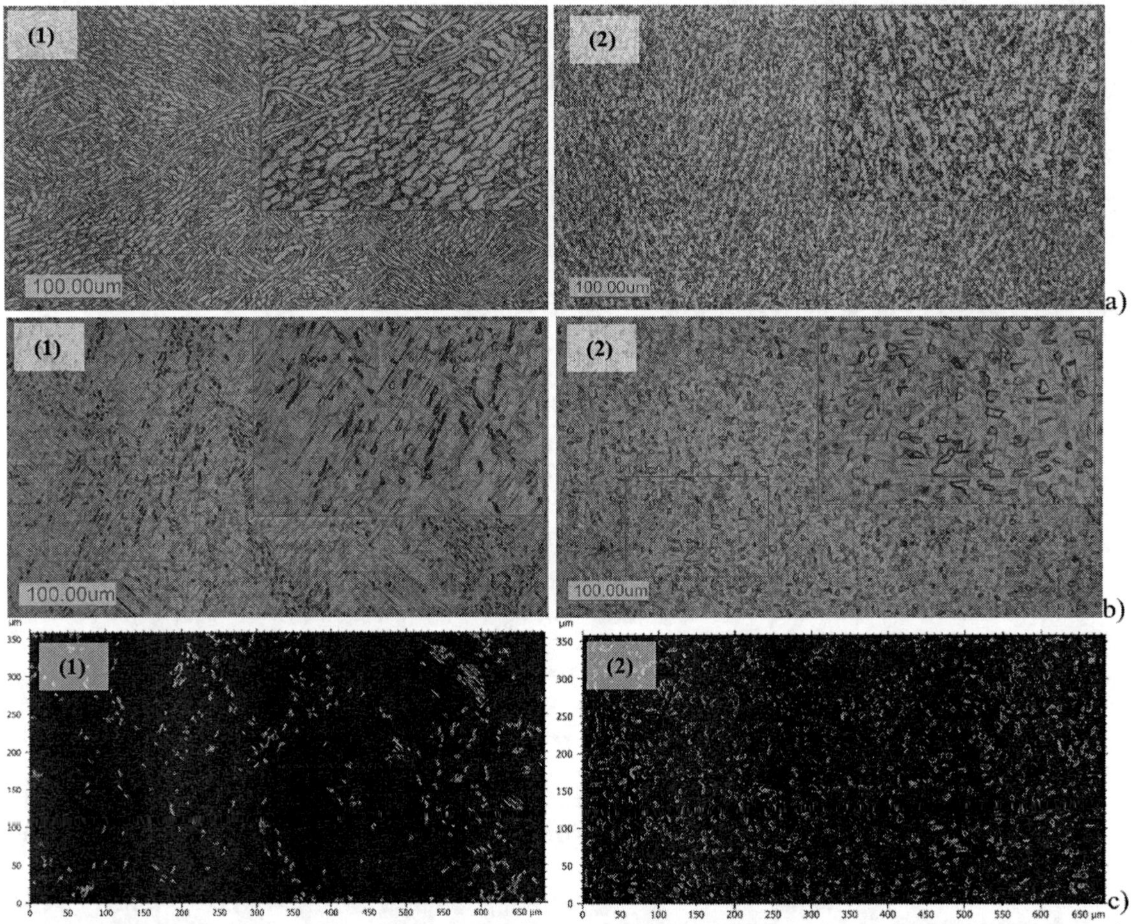

Figure 1. Optical micrographs of the AR (a), heat-treated (ST) (b) samples and software processed micrographs of the heated treated alloys (c). The numbers (1) and (2) indicates the Ti54 and Ti64 alloy respectively.

As shown in Figure 2a the higher Al content in the alloy increased not only the AS sample hardness but also straighten the material after the heat treatment process. Except for the ST+P Ti54 samples, after the EBSM treatment, the surface hardness values increased. The EBSM surface strengthening is the highest for the ST Ti64, followed by the AR Ti54 alloy. The lower temperature of α-phase precipitation hardening led to decreased surface hardness values of the ST+P Ti54 alloy. After the EBSM the depth of the structural changes was higher for the heat treated samples (Figure 2b) related to the AS materials for both alloys. The largest was the in-depth influence of the electron beam for the ST+P Ti64 allow and the smallest for the AS Ti64.

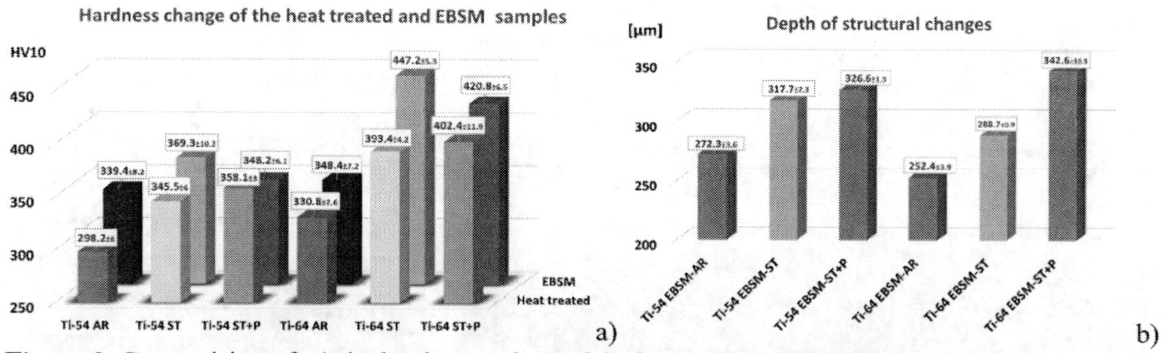

Figure 2. Comparision of: a) the hardness values of the heat treated and EBSM Ti54 and Ti64 samples and b) depth of the structural changes occurring after the EBSM.

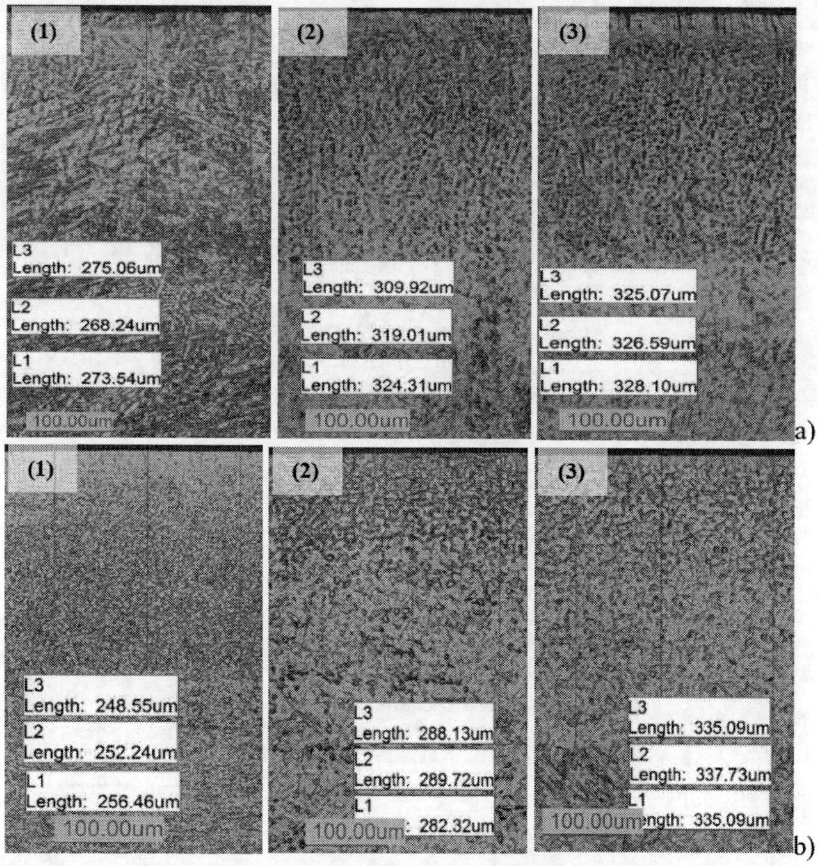

Figure 3. Micrographs of the cross-sections of the EBSM Ti54 (a) and Ti64 (b) alloys: (1) AR; (2) ST; (3) ST+P.

The structural changes occurring after the EBSM in cross-section, seen in Figure 3, indicates that the lamellar microstructure of the alloys with distorted Widmanstatten α-plates concentrated the energy of the electron beam nearer to the surface thus changing substantially the surface hardness. After the heat treatment, a tendency of an increase in the depth of the structural changes was displayed from the bimodal structure especially by the ST+P samples. Despite that similarity, the Ti64 alloy showed the greater ability for surface strengthening in ST and ST+P condition.

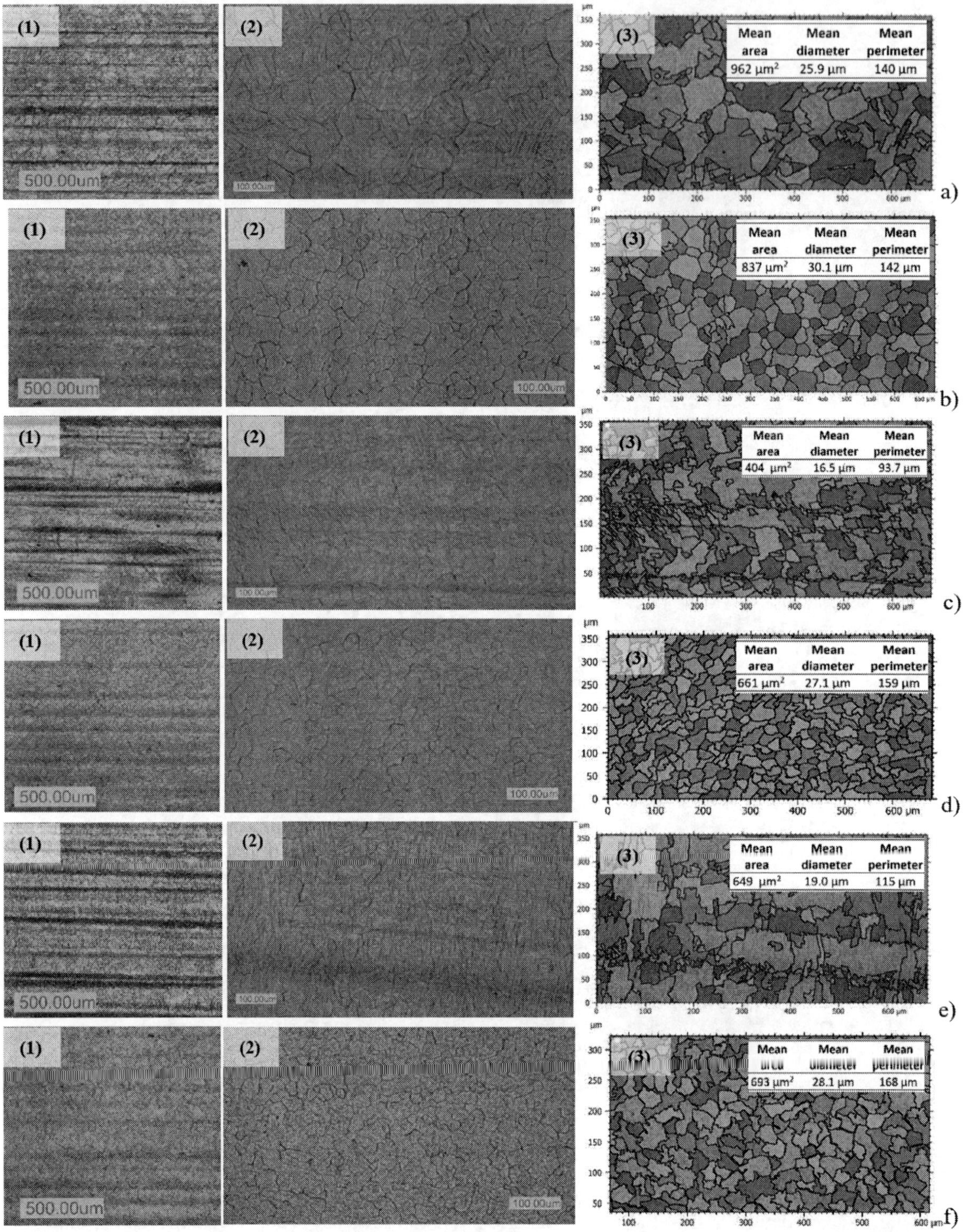

Figure 4. Topographic micrographs of the EBSM samples: a) AR Ti54; b) AR Ti64; c) ST Ti54; d) ST Ti64; e) ST+P Ti54; f) ST+P Ti64. The numbers and (3) used indicates the micro-roughening of the surface at small magnification (1), the grains in the re-melted zone (2) and software processes images determining some grains' parameters.

The topographic view of the surface response of the alloys (Figure 4) shows a morphology of wave-like patterning which is more pronounced for the Ti54 alloy and sensitive to the initial heat treatment of the material. The changes in the surface roughness values, as opposed to the polished samples with roughness values of R_a 0.11±0.02 and R_z 0.81±0.16, are presented in Table 2. The roughness values of the ST Ti54 alloy were the highest followed by AR material which tendency was also seen from the topographic micrographs (Figure 4-1). The form and size of the grains in the surface re-melted zone extensively differed for both alloys (Figure 4-2 and 4-3). Depending on the grains' size, smaller or coarser martensite laths were found at the surface of all examined samples. Despite the close mean values of the area, diameter and perimeter, the Ti54 top grains of the AR EBSM samples showed a wide disparity among the diameters compared to the AR Ti64 alloy (Figure 5-1). A tendency of grains' refinement was observed for both ST Ti54 and Ti64 alloys as opposed to the AS EBSM samples (Figure 4 c and d), but in contrast to Ti64 alloy, the variations in the mean diameter of the Ti54 grains were the highest (Figure 5-2). Similar to the ST Ti54, the microstructure of the ST+P EBSM alloy (Figure 4e) demonstrated a variety of grain sizes (Figure 5a-3) that followed an asymmetric normal distribution. The surface grains of the ST+P EBSM Ti64 alloy had slightly larger mean values of the area, perimeter and diameter (Figure 4f) than the ST sample from the same alloy (Figure 4d) and both of them showed nearly normal distributed mean diameters (Figure 5-2 and 3).

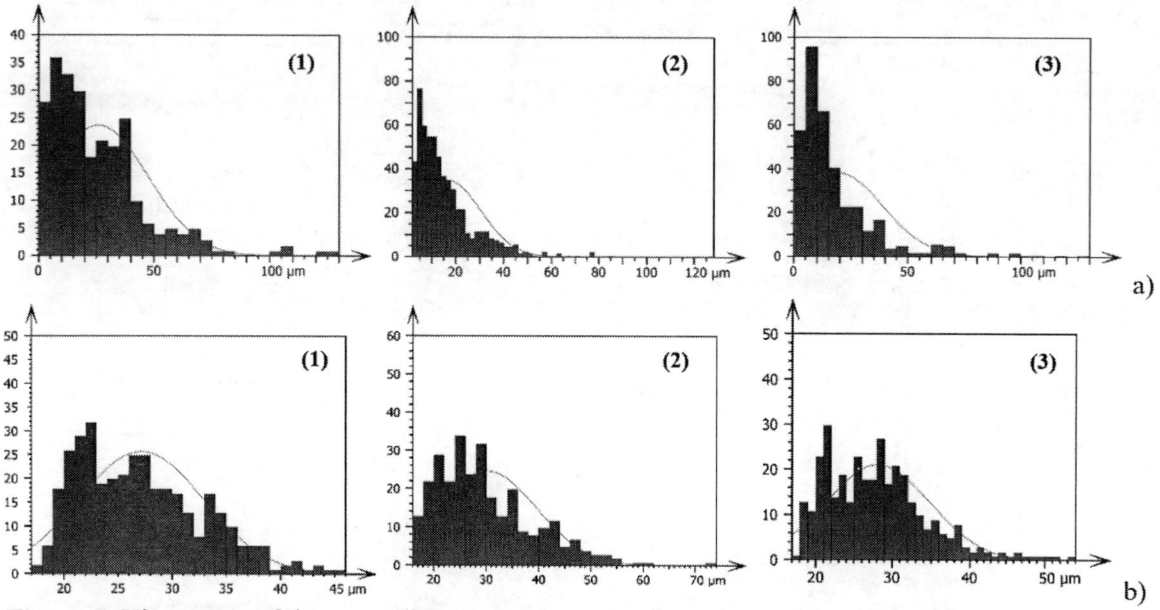

Figure 5. Histograms of the mean diameter of the grains from the motif analysis study of Ti54 (a) and Ti64 (b). The numbers indicate (1) AR; (2) ST; (3) ST+P conditions.

Table 2. Roughness values after EBSM of the alloys measured in a direction transverse to the waves. R_a stands for the average roughness values and R_z is the peak-to-valley roughness.

Sample	AR		ST		ST+P	
	R_a, [µm]	*R_z, [µm]*	*R_a, [µm]*	*R_z, [µm]*	*R_a, [µm]*	*R_z, [µm]*
Ti54 EBSM	1.21±0.2	5.97±0.9	1.28±0.3	6.89±1.5	1.07±0.3	5.32±0.9
Ti64 EBSM	0.23±0.2	1.5±0.14	0.22±0.2	1.3±0.13	0.24±0.3	1.4±0.19

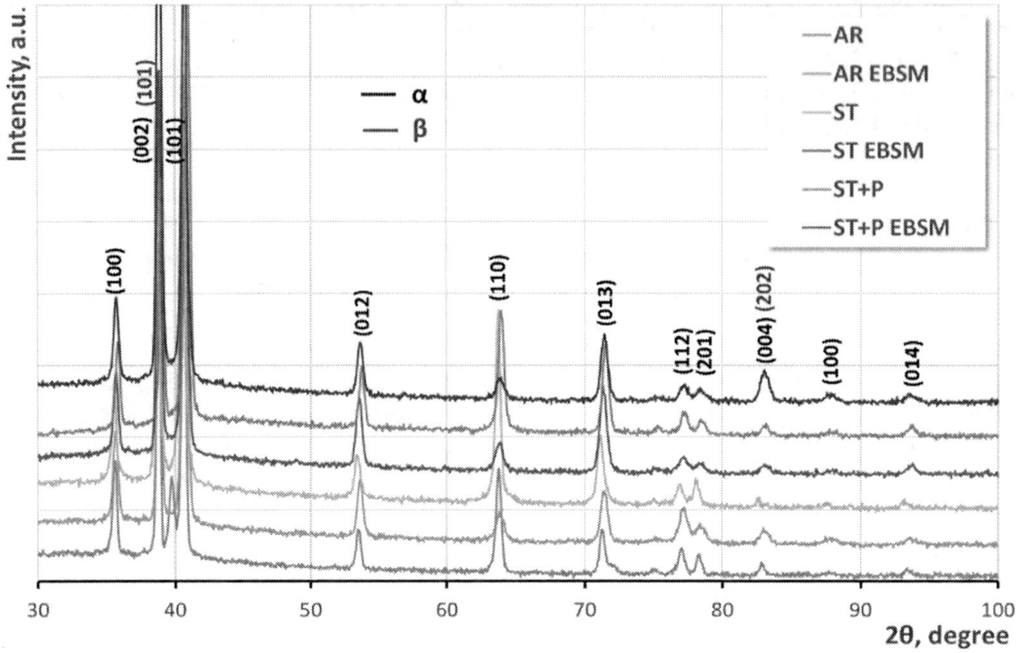

a)

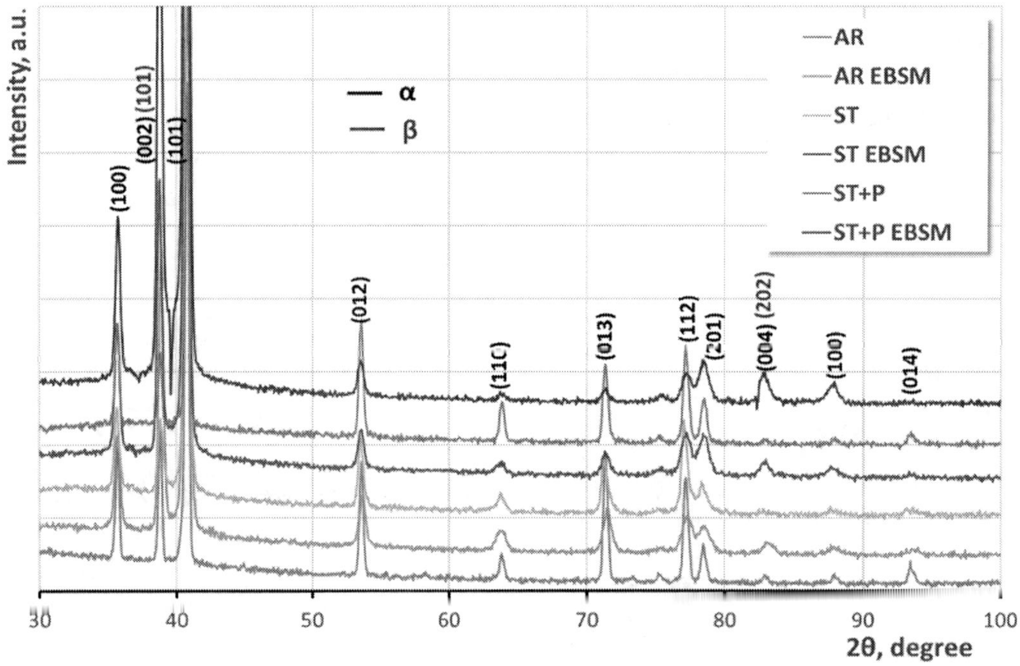

b)

Figure 6. XRD patterns of the Ti54 (a) and Ti64 (b) alloy in AR, ST, and ST+P condition before and after the EBSM.

The difference in the chemical composition of the alloy influence mainly the "c" cell parameter of the α-phase of the alloys (Tables 3 and 4). In contrast to Ti64, after the EBSM of the AR Ti54, the "c' parameter was increased for the account of "a" decrease due to the dissolution of β into α phase which explained the substation hardness change of this sample. Similar to the treated AR Ti64 alloy, the EBSM rose up the (101) and (002) intensity of the Ti64 samples. The ST samples of both alloys showed increased inter-planar spacing due to phase pre-saturation and enhanced (101) intensity as opposed to the AR samples (Figure 6). The (101) and (002) peaks' intensity was increased for the ST-EBSM Ti54 alloy at the expense of a decrease of (110) intensity. After the EBSM, both "a" and "c" parameters of the ST Ti64 alloys were simultaneously raised. The ST+P samples showed decreased parameters of the α-cells compared to the ST (Table 3 and 4) because of the alloy elements' partitioning effect. These changes tended to affect mainly the "c" cell parameter, which was afterward increased by the subsequent EBSM.

Table 3. Lattice parameters of alpha phase of Ti54 samples.

parameter	AR	AR EBSM	ST	ST EBSM	ST+P	ST+P EBSM
a, Å	2.907	2.904	2.914	2.902	2.893	2.900
c, Å	4.659	4.668	4.689	4.667	4.629	4.690

Table 4. Lattice parameters of alpha phase of Ti64 samples.

parameter	AR	AR EBSM	ST	ST EBSM	ST+P	ST+P EBSM
a, Å	2.901	2.905	2.907	2.902	2.906	2.901
c, Å	4.668	4.652	4.660	4.676	4.643	4.674

4. Conclusions

The heat treatment of the Ti64 alloy results in higher hardness values because of the obtained fine martensite laths and higher content of primary α-grains containing more Al that is a α-solution strengthening. After EBSM of both alloys, the heat treated bimodal structures are more susceptible to in-depth morphological changes than those of the AR materials, where larger surface grains are detected. The increased Al content in the α+β-alloy allows further surface hardness increase by EBSM not only in ST but in ST+P condition. The finer initial and heat treated structures of the Ti64 alloy form more homogenous in size surface grains after the EBSM while the coarse lamellar and bimodal structures of Ti54 alloy give rise to higher surface roughness values and grains that differ extensively in size. For both alloys, the heat treatment decreases the size of the re-melted grains at the surface after the modification. The EBSM have a similar effect on the crystallographic orientation of the grains predominantly increasing the (101) reflection despite the different initial grains' morphology. The higher overall content of the β-stabilizing elements enhances significantly the hardness of AR Ti54 as opposed to the AR material while the higher Al content in Ti64 contributes to the thermal stability and therefore the straightening response of the alloy to heat treatment and concentrated energy flux modification. However, these differences will have different consequences for the elastic modulus values of both alloys that is very important for the implant materials.

5. References

[1] Song H-J, Kim M-K, Jung G-C, Vang M-S and Park Y-J 2007 *Surf. Coat. Tech.* **201** 8738–8745
[2] Yetim T 2016 *J. Bionic Eng.* **13** 397-405
[3] Nikolova M P, Yankov E, Petrov P, Valkov S, Ormanova M, Zaharieva V, Tonev D and Andreeva A 2017 *Conference Proceedings Metal 2017* 1555-59
[4] Atapour M, Pilchak A L, Frankel G S and Williams J C 2011 *Mat. Sci. Eng. C-Bio. S.* **31** 885–891
[5] Manam N S, Harun W S W, Shri D N A, Ghani S A C, Kurniawan T, Ismail M H and Ibrahim M H I 2017 *J. Alloy. Compd* **701** 698-715

Acknowledgements

The financial support of this work was provided by the National Science Fund of Ministry of Education and Science, Bulgaria, under Grant project "Gradient functional nanocoatings produced by vacuum technologies for biomedical applications" with number DNº 07/3 (2016).

IOP Conf. Series: Materials Science and Engineering **416** (2018) 012032 doi:10.1088/1757-899X/416/1/012032

Solar Powder Metallurgy: Preparation of TiC – TiB_2 Ceramic Foam

J Kováčik[1,*], T Dvorák[1], Š Emmer[2], J Rodriguez[3] and I Cañadas[3]

[1] Institute of Materials & Machine Mechanics, Slovak Academy of Sciences, Dúbravská cesta 9, 84513 Bratislava, Slovakia
[2] IVMA STU, Pionierska 15, 831 02 Bratislava, Slovakia
[3] Plataforma Solar de Almeria, Ctra de Senes Km 4, 04200 Tabernas (Almeria), Spain

[*]Corresponding author: ummsjk@savba.sk

Abstract. The solar powder metallurgy was used to prepare TiC–TiB_2 ceramic foams in 40kW horizontal solar furnace at Plataforma Solar de Almeria (PSA), Spain. The foams were prepared using self-propagating high-temperature synthesis from the green compacts of powder mixture of Ti + 15 vol.% B_4C. The foaming was performed under protective argon atmosphere at various heating rates. It was shown that it is possible to prepare successfully TiC–TiB_2 ceramic foams with porosity around 45%. The foaming temperature or ignition temperature was found to be around 1380 °C. The optimal heating rate for foam preparation was found to be 220°C/min. It was confirmed that 2 pieces of precursor will join successfully to create single foam. Further, graphite and alumina moulds were tested for possible preparation of 3D shapes. The EDS investigations confirmed the creation of TiB_2, TiC and complex TiB_xC_{1-x} phases.

1. Introduction

The TiC–TiB_2 composites can be characterized not only by high melting points, high hardness, excellent wear resistance and good thermal stability, but also by superior properties such as enhanced fracture toughness and bending strength compared with constituent ceramic components [1]. For this reason they are very attractive for application as advanced structural materials in aircraft propulsion systems and space thermal protection systems.

However the application of TiC–TiB_2 composites has been limited due to the fact that they have been difficult to process. Various methods, such as reactive sintering, reaction hot pressing, spark plasma synthesis, transient plastic phase processing, and self-propagating high temperature synthesis (SHS) or termed combustion synthesis (CS) have been developed for the synthesis of TiC–TiB_2 ceramic composites with various porosity [1-4].

Recently authors used Solar Powder Metallurgy to study the preparation of fully dense titanium samples using solar furnace via pressure less sintering [5]. It was shown that significant advantage of solar furnace is simple handling and significant shortening of heating/cooling times. Therefore, the aim of the proposed work is to study experimentally a possibility to prepare TiC–TiB_2 ceramic composite foams with regard to determination of foaming temperature, dependence of final foam structure on heating rate, possibility to join various pieces of foamable precursor into the one piece of foam and finally to determine the proper mould material for TiC–TiB_2 foaming, which is necessary for preparation of complex 3D shapes.

IOP Conf. Series: Materials Science and Engineering **416** (2018) 012032 doi:10.1088/1757-899X/416/1/012032

2. Experimental

The Hydrogenated–dehydrogenated (HDH) Ti powder (purity 99.4%, supplied by Kimet Special Metal Precision Casting Co., Ltd., China) and B_4C powder (98% purity, normal grade, UK Abrasives, Inc., USA) were used for the investigations (see Figure 1). Ti powder is of typical angular shape while B_4C is of irregular shape. The powder size distributions were determined using Fritch Analysette 22 laboratory equipment. Ti powder has d_{50} = 25 μm and d_{90} = 46 μm. B_4C is standard boron carbide powder of powder size F600, i.e., 3 - 19 μm.

The powders corresponding to the composition of Ti + 15 vol.% of B_4C were dry mixed in Turbula mixer for 30 minutes. Then the powder mixture was compacted by uniaxial pressing at 280 MPa, which was optimal pressure for used pressing die. The typical geometry of green compacts before sintering was: diameter of 15.2 mm and height of 4 mm. The compacts were weighted and finally the porosity was calculated using known geometry. The observed volumetric porosity was approximately 42% for all samples.

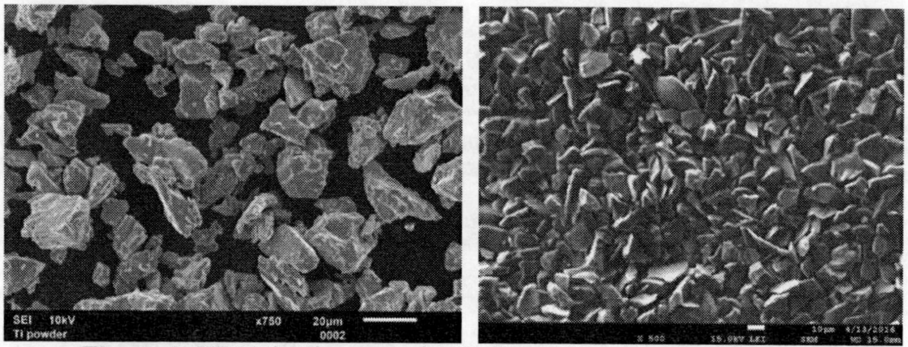

Figure 1. HDH titanium powder (left) and B_4C powder (right).

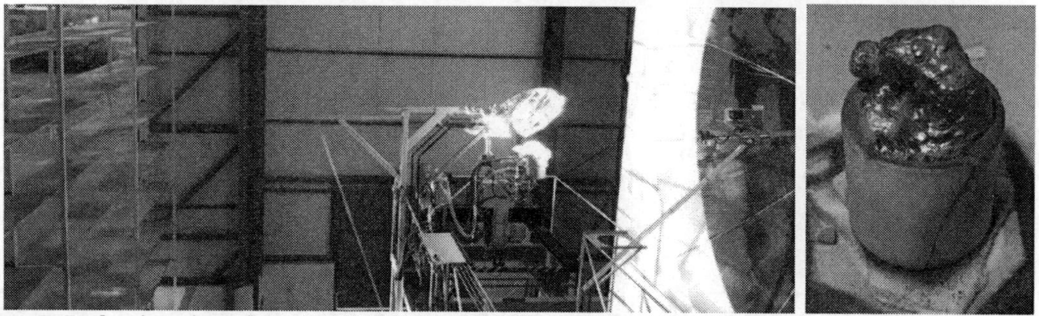

Figure 2. Left – interior of SF40 solar furnace during experiment: open shutter, vacuum/gas chamber mounted on table and concentrator; right – ceramic foam prepared in alumina tube.

The foaming experiments were performed at the horizontal solar furnace SF40 in PSA, Almeria, Spain. SF40 is able to deliver up to 40 kW power at peak concentration ratios exceeding 7000 kW/m^2. This SF40 system consists of heliostat, shutter, parabolic concentrator and test table [6]. The test table is a mobile platform on which the chamber enabling sintering of specimens in vacuum/protective atmosphere is placed and three stepping motors control the table motions (see Figure 2). The prepared sample was placed on zirconium plate (400 x 400 x 3 mm). The specimen was placed inside of chamber under the focal spot. The chamber is evacuated through a Pfeiffer turbomolecular pump and gas system is connected to it. The sample was placed to be in optimal focusing distance with regard to solar spot size either free or inside graphite/alumina tubes. For these tests graphite tube with inner diameter of 18 mm and height of 30 mm was used. The used alumina tube has inner diameter of 14 mm and height of 30 mm. The size of solar spot was approximately 90 mm. Then the chamber was evacuated (approximate value of the vacuum was around 8.10^{-1} Pa) and argon gas flow (400 l/hour under the pressure of 3 bars, technical purity argon 99.998% with maximal impurities of H_20 5 ppm mol/mol and of O_2 5 ppm mol/mol) was used during experiments.

Various heating rates were used during the experiments, the heating rate was controlled using shutter opening and was within the range of 90 – 1700 °C/min. Twenty experimental samples were foamed/sintered using SF40 solar furnace at temperatures between 1200 – 1600 °C for various dwell times between 5 to 30 minutes, most of them using heating rate of 220°C/min. The temperature was measured using pyrometer (emissivity 0.95) due to the reaction of titanium with thermocouples [5]. Remark: as the process is rapid it is not possible to use zirconium ceramic protection on thermocouples (see Figure 3). The main goal was to establish the optimal solar furnace parameters (shutter opening, temperature and time) necessary for good foaming of $TiC-TiB_2$ ceramic samples. The final porosity was determined from geometry and the weight of the samples in the case of sintering (not foamed). In the case of foamed samples that underwent foaming, the investigation of porosity was performed using X-ray micro tomography Phoenix Nanotom for non destructive analysis of the internal structure of materials.

After the experiments the samples were cut and samples for microstructure observations were prepared. Microstructure of samples was observed at PSA`s Materials Lab using Leica DMI 5000 Minverted digital microscope and at IMMS SAS using scanning electron microscope JEOL 7600F, equipped with Schottky thermal-emission cathode (thermal FEG - W-coated ZrO2) as well as Energy-dispersive X-ray energy (EDS) spectrometer from Oxford Instruments.

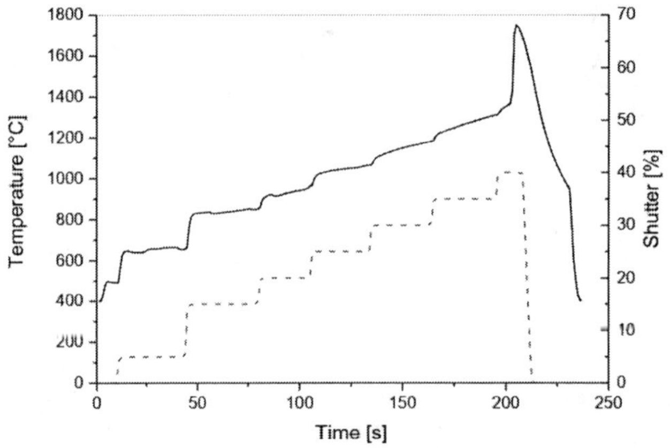

Figure 3. Typical heating/cooling curve of the solar foaming of $TiC-TiB_2$, on the secondary y axis is plotted the shutter opening (dash curve): Solar foaming, heating rate 220°C/min, direct solar radiation 825 W/m^2.

3. Results and discussion

The experimental works confirmed that when temperature of sample achieved certain foaming temperature self-propagating high-temperature synthesis took place and it was possible to create $TiC-TiB_2$ foams from Ti + 15 vol.% B_4C compacted precursor (see Figure 4). Further, it was shown that it is possible to prepare successfully $TiC-TiB_2$ ceramic foams with porosity around 45%. When the samples do not reached foaming temperature simple sintering took place with porosity varying according to sintering temperature and time.

Basically, the self-propagating high-temperature synthesis usually takes place during heating of mixtures of Ti with carbon, boron, TiC, TiN, B_4C and other additions when ignition temperature is reached [3]. In present work, at ignition temperature inside of investigated material titanium starts to react with boron carbide in a following way: $3Ti + B_4C \rightarrow TiC + 2TiB_2$. This direct reaction is highly exothermic ($\Delta H^0 = -686$ kJ/mol) and is connected with immediate increase of temperature measured by pyrometer as indicates sharp temperature increase in Figure 3. In all experimental runs the maximum observed foaming temperature was around 1750 °C. After that temperature was reached the immediate cooling of the sample was performed.

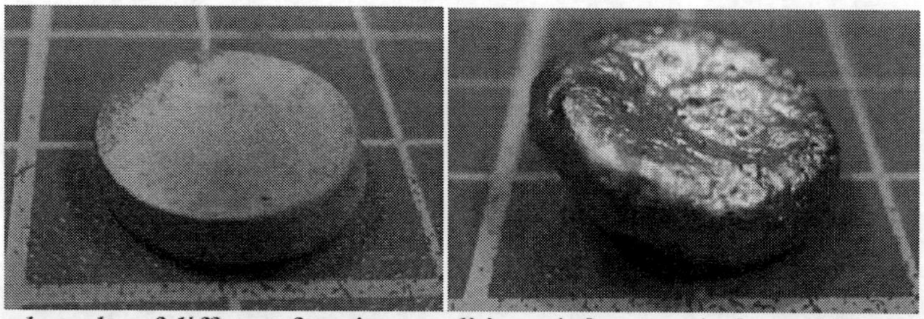

Figure 4. Typical results of different foaming conditions: left – sample foamed just below ignition temperature, heating rate 90°C/min, porosity 18%; right – sample foamed above ignition temperature, heating rate 220°C/min, porosity 44%.

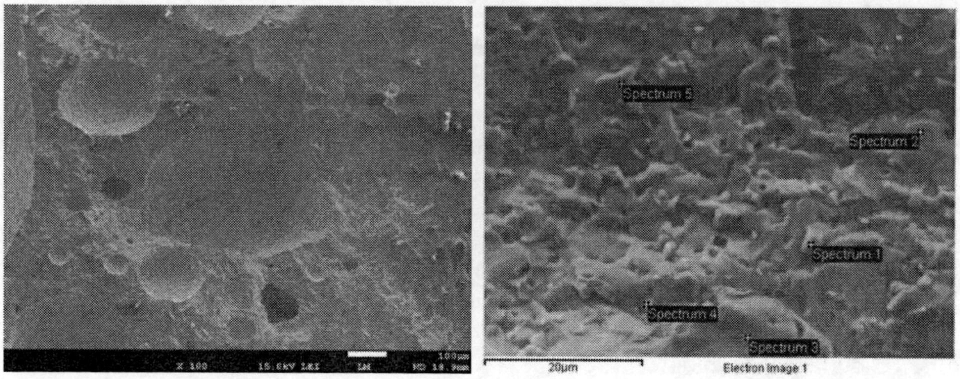

Figure 5. Microstructure of fractured foam and pore wall EDS point analyses of fractured cell wall of TiC – TiB_2 foam, solar foaming inside alumina tube at heating rate of 220°C/min: Spectra 1-3 - TiB_2, Spectrum 5 - TiC, Spectrum 4 - TiB_xC_{1-x}.

Various authors mention various ignition temperatures according to the used method of preparation: Bhaumik et al. are observed ignition temperature of 1977 °C [1]. In review work of Vallauri et al. [3] are mentioned various ignition temperatures in the range of 950°C - 1977°C according to the used technology, used materials mixtures and due to the production time and pressure. In the present work it was found that during solar foaming of the green compacts of powder mixture of Ti + 15 vol.% B_4C the foaming temperature or ignition temperature is around 1380 °C at optimal heating rate of 220°C/min. For higher hating rates it is a little bit higher due to the heterogeneous temperature distribution within the sample with regard to pyrometer spot size. Inside foam pore walls EDS investigations confirmed the creation of TiB_2, TiC and complex TiB_xC_{1-x} phases (Figure 5).

Most of foaming technologies of metals and ceramics require usually the optimization of foaming parameters [7]. It was shown, that during solar foaming of TiC–TiB_2 foam the most important parameter was heating rate of the sample. As indicates Figure 6 at low heating rates during foaming sample mostly undergoes exothermic reaction without any significant volume changes. At very high heating rates the foam not only underwent volume changes, but also certain collapse of foam is observed due to the overheating and inhomogeneous cooling during solidification. From the set of investigated heating rates of 90, 170, 220, 260, 300 and 1700 °C/min the optimal heating rate was found to be around 220°C/min.

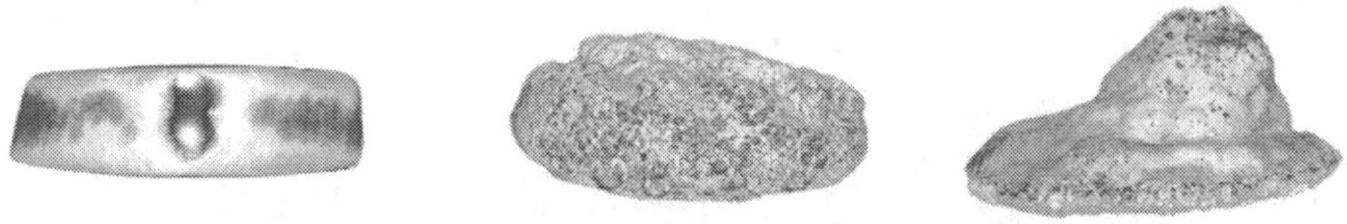

Figure 6. Effect of heating rate on solar foaming of TiC – TiB_2: left - heating rate 90°C/min; middle - heating rate 220°C/min; right - heating rate 300°C/min.

When it is necessary to foam complex 3D parts it is often required to use certain number of precursor pieces that will join during foaming into the single foam. Therefore this possibility for the investigated material was also tested. As proved in Figure 7 it is possible using solar foaming to create TiC–TiB_2 foam from more than one piece of precursor.

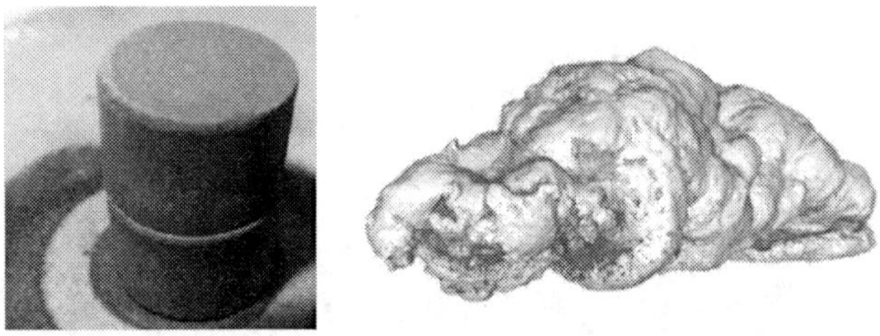

Figure 7. Joining of two precursor pieces during foaming in solar furnace: left – precursors before foaming; right – foam created from them at heating rate of 220°C/min.

Another important technological problem is a proper choice of the material for 3D moulds [8]. In this work alumina and graphite tubes were used to test the suitability of these materials for foaming of TiC–TiB_2 foam. The results are promising, but not fully satisfactory: Graphite is not suitable for due to high thermal conductivity leading to significant foam collapse during cooling process due to slower solidification (see Figure 8). On the contrary low thermally conductive alumina behaved much better. Unfortunately, the temperatures during exothermic reaction led to the reaction of foaming material with alumina. More experiments are needed in the next future. It is proposed to test zirconium or YSZ ceramic as a mould material, which will probably not react with the investigated composite mixture.

Figure 8. TiC – TiB_2 foams foamed at various moulds at heating rate 220°C/min: left – graphite mould with evident foam collapse; right - alumina mould with evident cracking of mould.

4. Conclusions

The experiments showed that it is possible to prepare TiC–TiB_2 ceramic foams with porosity around 45% using self-propagating high-temperature synthesis from the green compacts of Ti + 15 vol.% B_4C. The foaming temperature or ignition temperature is around 1380 °C and optimal heating rate is around 220°C/min. TiB_2, TiC and complex TiB_xC_{1-x} phases were observed using EDS investigations.

It was shown experimentally, that it is possible to create ceramic foam from more than one piece of precursor. Further experiments solved the problem with the proper moulding to obtain desired 3D shapes from TiC–TiB_2 ceramic foam: Graphite was not suitable due to high thermal conductivity leading to foam collapse during cooling. Low thermally conductive alumina was better, however reacted with foaming material. In the next future, it is proposed to test zirconium or YSZ ceramic as a mould material. The results are also promising for preparation of Ti based foams with certain amount of TiC–TiB_2 ceramic for biomedical applications [9].

5. References

[1] Bhaumik SK, Divakar C, Singh AK and Upadhyaya GS 2000 *Materials Science and Engineering: A* **279** 275-281
[2] Binglin Zou, Ping Shen, Zhongming Gao and Qichuan Jiang 2008 *Journal of the European Ceramic Society* **28** 2275-2279
[3] Vallauri D, Atías Adrián IC and Chrysanthou A 2008 *Journal of the European Ceramic Society* **28** 1697-1713
[4] Kobashi M, Ichioka D, Kanetake N 2010 *Materials (Basel).* **3(7)** 3939-3947
[5] Kováčik J, Emmer Š, Rodríguez J and Cañadas I 2017 *Journal of Alloys and Compounds* **695** 52-59
[6] Rodriguez J, Cañadas I and Zarza E 2016 New PSA high concentration solar furnace SF40 In: AIP Conference Proceedings 1734, 70028
[7] Lehmus D, Wichmann M and Busse M 2008 *In International Conference on Porous Metals and Metallic Foams*, L. P. Lefebvre, J. Banhart, D. C. Dunand (eds.), Montreal Canada. (Lancaster, Pa: DEStech Publications Inc) p. 51
[8] Linul E, Movahedi N and Marsavina L 2017 *Composite Structures* **180** 709-722
[9] Edalati E, Sajjadi SA and Babakhani A 2017 *Metals 2017* **7** 296

Acknowledgement

Financial support by the Access to Research Infrastructures activity in the 7th Framework Programme of the EU (SFERA Grant Agreement n. 228296) is gratefully acknowledged. Financial support of the SFERA II project – Transnational Access activities (EU 7th Framework Programme Grant Agreement n° 312643) is also gratefully acknowledged. Slovak Grant Agency VEGA grant 2/0044/17 and SRDA agency grant APVV-14-0936 financial supports are gratefully acknowledged.

IOP Conf. Series: Materials Science and Engineering **416** (2018) 012033 doi:10.1088/1757-899X/416/1/012033

Enhancement of Hydrogen Evolution Reaction on Pt-Co Layer Deposited on Copper

D A Duca[1], M L Dan[1], A Z Vidinaru[1], R Muntean[1,2] and N Vaszilcsin[1,*]

[1]Faculty of Industrial Chemistry and Environmental Engineering, Politehnica University Timisoara, Romania
[2]Department of Materials Science and Testing, Westphalian University of Applied Science, Gelsenkirchen, Germany

[*]Corresponding author: nicolae.vaszilcsin@upt.ro

Abstract. Hydrogen is still considered an excellent energy carrier and fuel of the future, since in the combustion process only water is produced, without carbon dioxide. Whereas, water splitting by electrolysis is indeed a green procedure for hydrogen production, the research on hydrogen evolution reaction (HER) is focused especially on the catalytic activity of cathodic materials. In this paper, Pt-Co alloy deposition on copper substrate as catalyst for HER is presented. Pt-Co alloy has been obtained using pulsed plating current technique in order to achieve a uniform dispersion of Pt-Co particles onto the metal surface. Morphological and structural characterization of the obtained layers has been carried out by scanning electron microscopy and XR diffractometry. Electrochemical behaviour and stability of the electrocatalytic Pt-Co layers has been emphasised by cyclic voltammetry. Further, kinetic parameters for cathodic HER have been determined by Tafel plots method in alkaline solutions. Exchange current density i_o, as well as cathodic transfer coefficient 1-α, revealed an increased catalytic activity in the temperature range of 25 – 65°C.

1. Introduction

Despite the fact that apparently Li ion batteries have just won against H_2 fuel cells in the race for automotive applications, the cathodic HER is still a challenge for researchers in order to minimize the specific power consumption [1-5]. Hydrogen still remains an excellent energy carrier and a fuel of the future, especially in the space technology and stationary power engines. On the other hand, hydrogen applications are not limited to automotive applications. It is also used as reagent in the synthesis of some large production chemicals, like ammonia and methanol, and as well as in a specific petrochemical processes: hydro-dealkylation, -desulfurization and -cracking [6,7]. Last but not least, electrochemical hydrogen production by water splitting is indeed a *green procedure*, taking into account that, in other applied industrial processes, for each ton of hydrogen an amount of 5.5 tons of CO_2 is released in the atmosphere in the steam reforming of methane, respectively 11 tons of CO_2 in the coal gasification [8].

Early researches have been focused on the electrocatalytic activity on the cathodic HER of non-noble metals having a large specific area [9,10]. Furthermore, the enhancement of HER using proton carriers in the electrolyte solution have been investigated [11,12]. Current trends in the electrochemical hydrogen production are directed towards high pressure water electrolyzers, able to operate a high current densities and elevated temperature [13,14]. In this respect, there are two

IOP Conf. Series: Materials Science and Engineering **416** (2018) 012033 doi:10.1088/1757-899X/416/1/012033

commercial technologies that use alkaline electrolysis and proton exchange membrane electrolysis. As well, under consideration is the development of solid oxide electrolyzers, operating at high temperature (800 – 900°C) [15].

In order to reduce electrolyzers costs, three kind of non-noble electrocatalysts are used: metal alloys, inorganic compounds of transition metals and doped nanostructures [9]. According to the literature data, the most active electrocatalysts for HER are platinum alloys. J. Zheng has been obtained Pt-M alloys (M = Fe, Co, Ni, Pd) supported on Ti foils, having good catalytic activity as electrodes in water electrolyzers both for hydrogen and oxygen evolution reaction [16,17]. Similar results have been reported by Y. Zang et al [18]. Small overpotential for HER (90 mV at 8 mA cm^{-2}) was registered on platinum-rhodium alloy [19]. PdNi alloy nanoparticles are highly efficient bifunctional catalysts for hydrogen and oxygen evolution reaction. Low overpotentials for HER were observed: 55 mV at a current density of 10 mA cm^{-2} in an acid solution and 187 mV in 1 mol L^{-1} KOH solution at the same current density [20]. Interesting achievement has been noted by D.M.F. Santos et al. using platinum – lanthanides (cerium, samarium and holmium) alloys in 8 mol L^{-1} KOH aqueous solution, at temperature range of 25 – 85°C. The calculated kinetic parameters showed that Pt – lanthanides alloys increase the HER rate in comparison to pure platinum [21,22]. Recently, high-performance as bifunctional electrocatalysts for HER has been reached by platinum – palladium alloy introduced in N-doped graphene [23].

In a previous paper [24], a new way to realize Pt-Co alloy as electrocatalyst has been presented, considering that it is difficult to obtain this material by cathodic deposition from a salts electrolyte solution. The Pt-Co alloy composition may be controlled just applying a pulsed plating electrochemical technique [25].

In the present paper, Pt-Co alloys deposition on copper support is described, using pulsed plating current technique in order to achieve a uniform dispersion of Pt-Co particles onto the metal surface, as well as its electrocatalytic behavior for HER.

2. Materials and methods

Experimental studies have been focused on manufacturing the Pt-Co electrodes and testing them as catalysts for HER in alkaline media.

2.1. *Electrodes manufacturing*

The electrodes have been manufactured in two steps. The first step consisted in the preparation of the non-catalytic support (Cu disks – 1 cm^2 surface), and the second one in the electrodeposition of Pt-Co alloy on this support, using pulse deposition technique.

Copper disks surface have been grinded with various grit emery papers (1,200 – 4,000 μm), polished with 3 μm polycrystalline diamond, cleaned in an ethanol ultrasonic bath for 30 minutes at 30°C and finally dried in hot air.

Pt-Co alloy deposition has been performed using a potentiostat/galvanostat Ivium Technologies Vertex, in a classic electrochemical cell, where the copper disks (0.8 cm^2 exposed area) previously prepared were used as working electrodes and a graphite bar as counter-electrode. The electrodeposition has been carried out from a weak acidic solution of K_2PtCl_4 (0.005 mol L^{-1}) and $CoCl_2{\cdot}6H_2O$ (1 mol L^{-1}) as a source of Pt^{2+}, respectively and Co^{2+} ions. The molar ratio of Pt^{2+}:Co^{2+} ions was 1:20. In the electrolyte solution, 1 mol L^{-1} KCl has been added to increase the conductivity of the solution, and 0.5 mol L^{-1} H_3BO_3 to maintain the *pH* at a constant value. During the electrodeposition, N_2 (gas) was bubbled in the solution and after, the working electrode was washed with distilled water and dried in hot air.

The composition and morphology of deposited Pt-Co layers have been characterized by scanning electron microscopy and XR diffractometry.

IOP Conf. Series: Materials Science and Engineering **416** (2018) 012033 doi:10.1088/1757-899X/416/1/012033

2.2. *Tests for HER*

Electrochemical behaviour and stability of the electrocatalytic Pt-Co layers in alkaline media has been studied by cyclic voltammetry (CV). Cyclic voltammograms (CVs) have been recorded at different concentrations of NaOH, between 1 and 6 mol L^{-1}, with multiple scan rates.

Kinetic parameters (exchange current density i_o and cathodic coefficient transfer 1-α) for HER have been calculated from linear voltammograms (LVs) by Tafel plots method. LVs have been drawn with low scan rate, 2 mV s^{-1}, in 25 – 65°C temperature range.

Electrochemical measurements have been performed using a PARSTAT 2273 potentiostat/galvanostat in a 150 mL thermostatted glass cell. Pt-Co electrodes, having 0.8 cm^2 exposed area, have been used as working electrodes, two graphite rods as counter electrodes and Ag/AgCl as reference. All further potentials from the experimental work are referred to this electrode ($E_{Ag/AgCl}$ = +0.197 V vs normal hydrogen electrode).

3. Results and discussion

3.1. *Pt-Co electrodeposition*

Pt-Co electrodeposition was made using pulse deposition method due to its advantages, as rigorous control of particle size, uniform distribution and increased adhesion.

The variables of the pulse deposition process are applied pulse time t_{on}, pause time t_{off}, current density i and number of deposition cycle. This study focuses on the variation of these parameters in order to achieve the optimal characteristics. Nucleation process and the size of the deposited alloy particles are influenced by t_{on} and i, whilst during t_{off}, the metal ions concentration from the vicinity of interface is completed and desorption of the previously adsorbed ions occurs. In order to determine the optimal parameters, electrodeposition tests have been done varying the current density, deposition and pause time of a pulse and number of cycles. Thus, current densities ranging between 250 and 1,000 A m^{-2} were applied for pulses whose t_{on} varied from 10 to 20 ms, and t_{off} from 5 to 100 ms. The number of cycles varied between 10,000 and 20,000. During t_{off}, the density of applied current was equal to 0.

3.1.1. *Current density influence on Pt-Co layer morphology.*

To study the current density influence on the morphology of the deposited layers, tests have been done applying three current densities: 250, 750 and 1,000 A m^{-2}, while the other parameters were maintained constant (t_{on} = 10 ms, t_{off} = 100 ms and 10,000 deposition cycles).

Figure 1 presents SEM images obtained for the Pt-Co layers deposited on the Cu support at the specified current densities.

a)

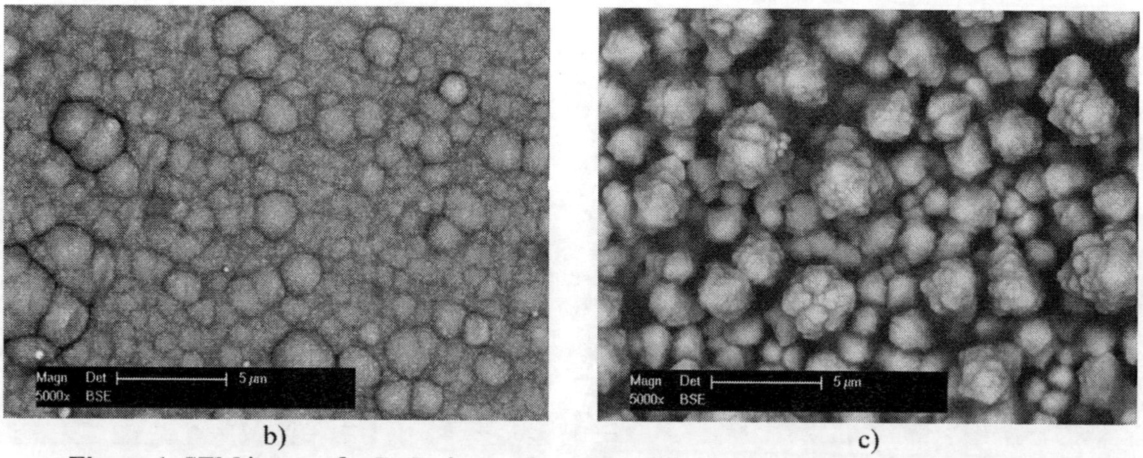

Figure 1. SEM images for Pt-Co layers deposited with: a) 250, b) 750 and c) 1000 A m^{-2}.

From the analysis of the SEM micrographs and knowing that the current density at which the electrodeposition process is carried out influences significantly the quality and structure of the deposited layer, it can be observed that for 250 A m^{-2} current density (the lowest used in experimental studies) Pt-Co microcrystalline structure is obtained, with evenly distributed particles over the surface of the copper substrate. Increasing the current density has as effect the increase of the deposited particles size, as at 1,000 A m^{-2} current density dendrites appear. Consequently, layers obtained at this current density are inappropriate to be used as catalyst for HER.

In order to determine the thickness of the Pt-Co alloy layers deposited on the copper support by pulse deposition, SEM analysis were performed in a section of the layer. Current density influences also the thickness of the deposited layers; at a value of 250 A m^{-2} a Pt-Co layer of 1.5 μm thickness is obtained, and this value increases up to 5 μm at 1,000 A m^{-2}.

SEM micrographs presented above revealed the significant influence of current density on the quality and crystalline structure of Pt-Co layers deposited on copper support. It was found that depositions obtained at 250 A m^{-2} are optimal for use in the enhancement of HER because of its finest microcrystalline structure, evenly distributed on the support. Despite that thicker layers are obtained at higher current densities, they do not qualitatively correspond because the particle size is large or even in the form of dendrite.

3.1.2. Influence of pulse deposition time on Pt-Co layer morphology

Specific pulsation parameters, t_{on} and t_{off} influence both the chemical composition of the alloy and quality of the galvanic deposit obtained (microcrystalline structure, nucleation and growth of crystallites). In the case of a binary alloy, such as Pt-Co, during t_{off} (i = 0 A m^{-2}), the less noble metal cobalt can be dissolved on the surface of the cathode by the noble metal, platinum, according with a cementation reaction:

$$Pt^{2+} + 2Co \rightarrow Co^{2+} + Pt \qquad (1)$$

Also, a longer t_{off} between pulses is recommended for the metal ions concentration at interface to return to its initial value, homogenizing the solution due to the migration/diffusion of the Pt^{2+} and Co^{2+} ions from the solution bulk to the emergent zone of the cathode [24].

In preliminary experimental studies, t_{off} value ranged from 5 to 100 ms. Based on the theoretical aspects and on the obtained Pt-Co layers, t_{off} = 100 ms was chosen to be the optimal value. A more obvious influence is the deposition time t_{on} specific of a pulsation.

Figure 2 presents SEM images of Pt-Co layers obtained after 10,000 deposition cycles, at 250 A m^{-2} current density and t_{off} = 100 ms , varying the pulse deposition time t_{on} from 10 to 20 ms.

IOP Conf. Series: Materials Science and Engineering **416** (2018) 012033 doi:10.1088/1757-899X/416/1/012033

a) b)

Figure 2. SEM images for Pt-Co layers deposited at t_{on} of a) 10 and b) 20 ms.

From the analysis of SEM micrographs, it can be observed that increasing the deposition time for a pulse from 10 to 20 ms leads to a finer microcrystalline structure with a higher density of crystals. The optimal value chosen was $t_{on} = 20$ ms.

3.1.3. Number of deposition cycles influence on Pt-Co layer morphology

The number of cycles or the number of applied pulses has also a considerable influence on the deposition of the Pt-Co layers obtained on copper substrate from weak acid electrolyte solution. Figure 3 presents SEM images of the Pt-Co layers obtained by pulse deposition technique at 250 A m^{-2} current density, t_{on} – 20 ms, t_{off} – 100 ms, varying the number of deposition cycles from 10,000 (Figure 3a) to 20,000 (Figure 3b).

a) b)

Figure 3. SEM images for Pt-Co layers obtained after a) 10,000 and b) 20,000 deposition cycles.

Analysing the SEM micrographs, it can be seen that increasing the number of deposition cycles, as the particle size remains almost constant, increases the nucleation rate and so, the number of crystals deposited on the surface of the metal support. This leads to obtaining an electrode with microcrystalline structure and a large specific surface. Also, the thickness of the deposited alloy layer increases, implicit increasing the corrosion resistance, respectively the duration of use of these electrodes for HER in alkaline media.

3.1.4. Structural characterization of deposited Pt-Co layer

XRD measurements have been performed in order to determine the structural characteristics, and so the alloy phase composition of the Pt-Co layer. Beneath, XRD patterns of Pt-Co alloy deposited on Cu are shown at two values of current density: 250 (Figure 4a) and 750 (Figure 4b) A m^{-2}.

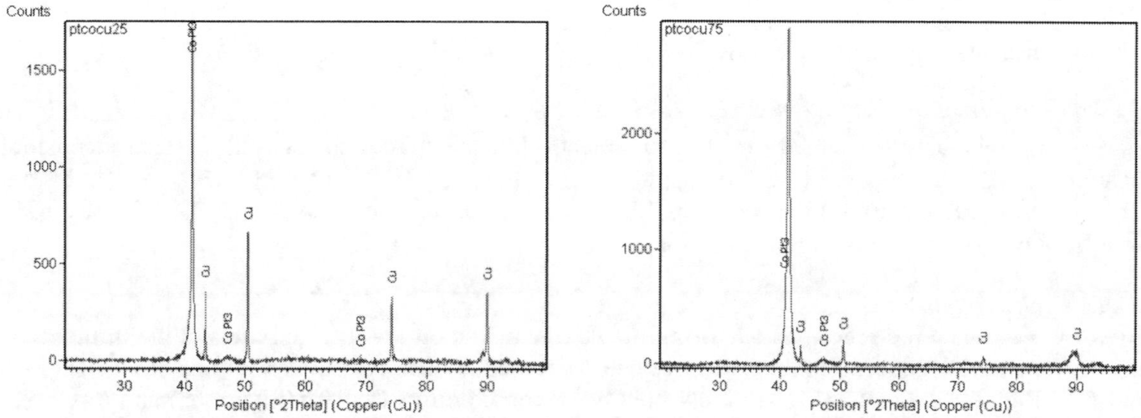

a) b)

Figure 4. XRD patterns for Pt-Co layers deposited on Cu substrate at a) 250 and b) 750 A m^{-2}.

For both Pt-Co samples, a single alloy phase $CoPt_3$ was detected. Unalloyed metallic Pt is not found in the electrodes structure. The peaks observed at 2θ values of 43°, 51°, 74° and 90°, characteristic for copper correspond to the Cu material support used in the electrodeposition process.

Figure 5 shows comparatively the XRD patterns for the working electrode, pure Pt and also pure Co. It can be observed the diffraction patterns of $CoPt_3$ alloy phase and pure Pt are similar, no additional peaks being observed. Also, no additional peaks for pure Co can be observed.

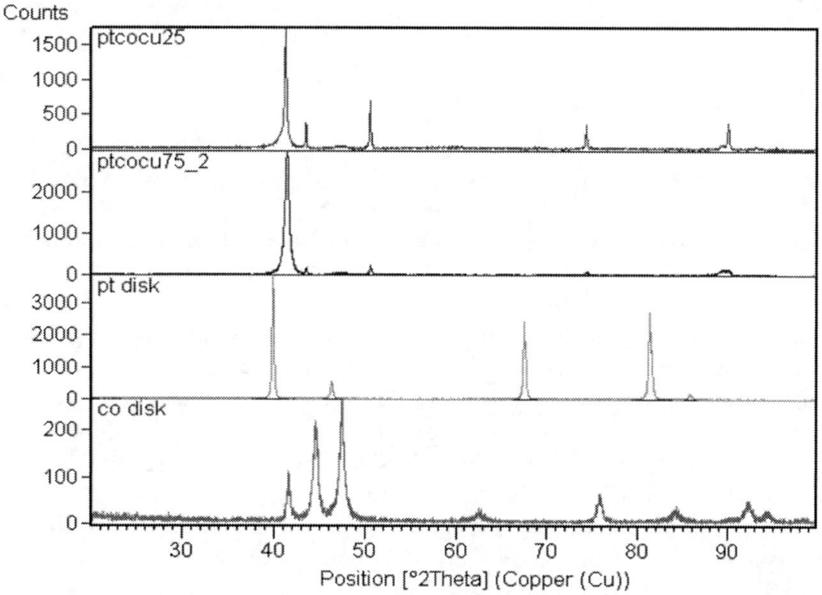

Figure 5. XRD patterns for working electrode, pure Pt and pure Co.

The Pt-Co characteristic peaks are slightly shifted at higher angles compared with pure Pt peaks, aspect indicating a contraction of the lattice constant and also a high alloying degree due to incorporation of Co atoms into the Pt-Co alloy structure [24].

Previous studies revealed that decreasing Pt-Pt bond length, the binding energy is also reduced and so the catalytic activity increase considerable [24].

Pure Co could not be identified in Pt-Co structure, meaning all cobalt atoms have been incorporated in the Pt structure, forming Co-Pt alloy.

3.2. Tests for HER

Based on previously presented results, the electrodes manufactured by pulse electrodeposition technique at 250 A m^{-2} current density, applied pulse time t_{on} = 20 ms, paused time t_{off} = 100 ms and 10,000 (Pt-Co-10k electrode), respectively 20,000 number of deposition cycles (Pt-Co-20k electrode) have been tested for HER.

3.2.1. Cyclic voltammetry

In order to evaluate the processes occurring in alkaline media on Pt-Co electrode and the influence of electrolyte solution concentration, CVs have been recorded at different concentrations of NaOH (1 – 6 mol L^{-1}). Figure 6 presents comparatively the CVs recorded on Pt-Co-10k (Figure 6a) and Pt-Co-20k (Figure 6b) with 10 mV s^{-1} scan rate.

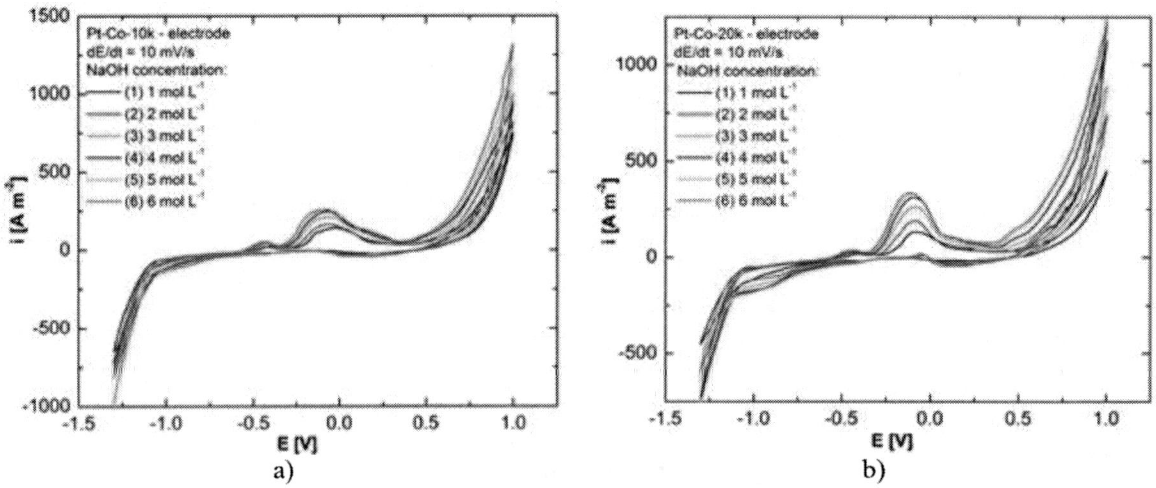

Figure 6. CVs recorded on a) Pt-Co-10k and b) Pt-Co-20k electrode in different concentrations of NaOH, at 10 mV s^{-1} scan rate.

Analysing the CVs, starting from open circuit potential towards anodic polarization, on both electrodes, two oxidation peaks, around -0.45 V and -0.15 V are observed. They correspond to the oxidation of Co in two steps (Co → Co^{2+} and Co^{2+} → Co^{3+}). Further, around + 0.6 V, a less pronounced wave, better visible on Pt-Co-20k electrode, attributed to platinum oxides formation is noticeable. At more positive potential values, oxygen evolution reaction begins. Sweeping the potential in opposite direction, in the cathodic domain it can be seen the reduction wave of Co^{3+} and Co^{2+} from the proximity of the interface to metallic Co (at -0.75 V), followed by HER.

One may observe that HER depends on the NaOH concentration. A concentration up to 3 mol L^{-1} NaOH enhances HER, its potential value being shifted towards positive values with the concentration increase. At higher concentrations, between 4 and 6 mol L^{-1}, the characteristic potential for HER becomes more negative, so the overpotential increases.

IOP Conf. Series: Materials Science and Engineering **416** (2018) 012033 doi:10.1088/1757-899X/416/1/012033

3.2.2. Linear voltammetry

To study the effect of the electrolyte solution concentration on HER and to determine the kinetic parameters of this process, LVs on both Pt-Co-10k and Pt-Co-20k electrodes have been recorded at concentrations between 1 and 6 mol L^{-1} NaOH in 25 – 65°C temperature range. Figure 7 presents comparatively the LVs drawn on the two electrodes with 2 mV s^{-1} scan rate at 25°C.

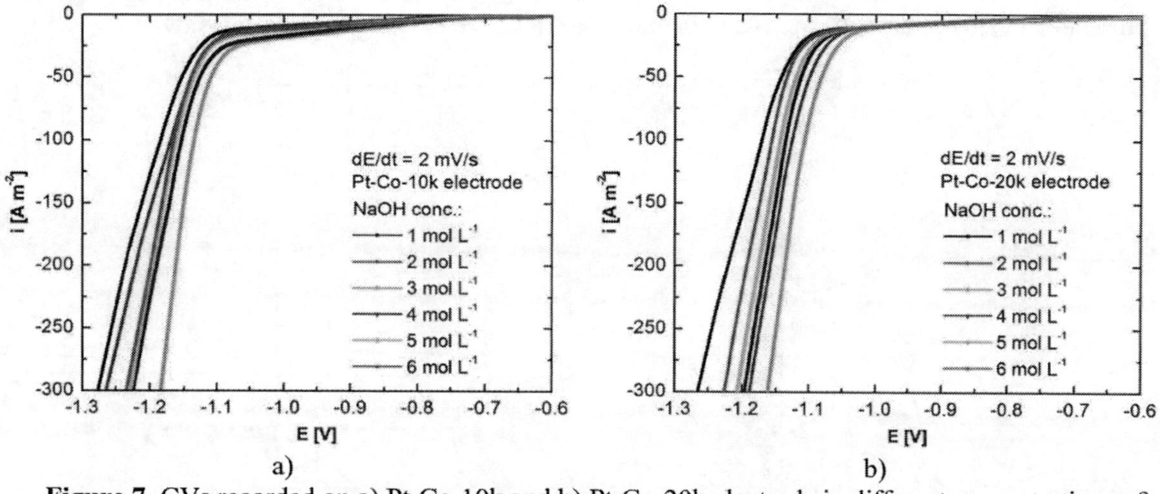

Figure 7. CVs recorded on a) Pt-Co-10k and b) Pt-Co-20k electrode in different concentrations of NaOH, at 25°C; 2 mV s^{-1} scan rate.

In both cases, a decrease of HER overpotential is noticed with the increase of NaOH concentration, most obvious for 3 mol L^{-1}, when a shift of approximately 100 mV towards more positive values is observed as against 1 mol L^{-1}, the lowest concentration of NaOH used in the experimental study. At higher concentrations this catalytic effect is diminished. One can conclude this is the optimum concentration to be used for HER on Pt-Co electrodes.

Knowing from the literature data the favourable effect of temperature rise on HER, LVs have been recorded at different temperatures, between 25 and 65°C. Figure 8 shows the LVs drawn in 3 mol L^{-1} NaOH.

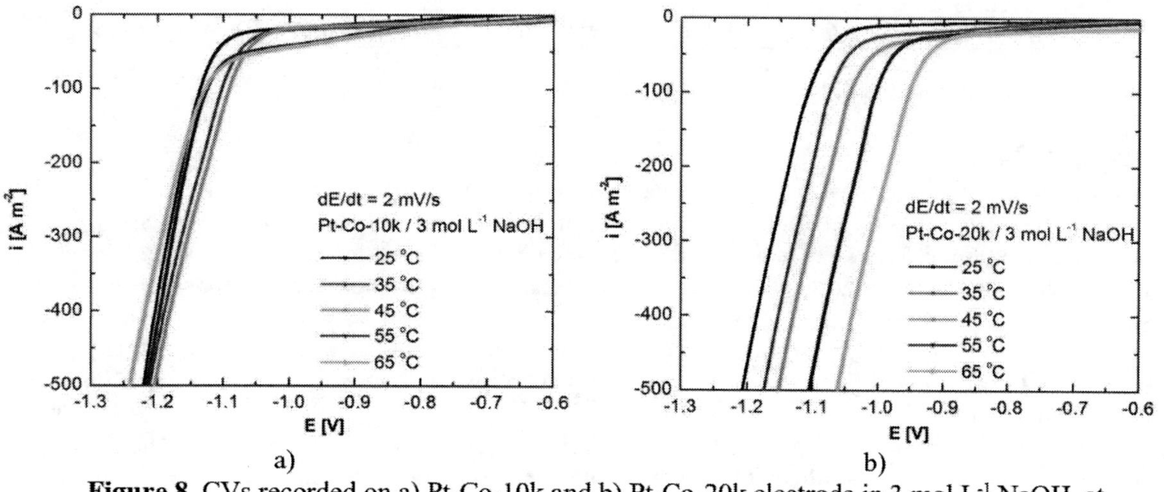

Figure 8. CVs recorded on a) Pt-Co-10k and b) Pt-Co-20k electrode in 3 mol L^{-1} NaOH, at different temperatures; 2 mV s^{-1} scan rate.

Analysing the LVs, it is observed HER depends on the thickness of the Pt-Co layer. As expected, Pt-Co-20k is a better fit for the enhancement of HER, the deposited layer being thicker, as previously presented. For this electrode, an overpotential decrease of about 150 mV is seen for the temperature increase from 25 to 65°C.

For a better understanding of HER, from the linear dependence η = f (lg|i|), kinetic parameters (exchange current density i_o and cathodic coefficient transfer 1-α) have been calculated. Figure 9 presents Tafel plots for HER on Pt-Co-20k electrode in different concentrations of NaOH.

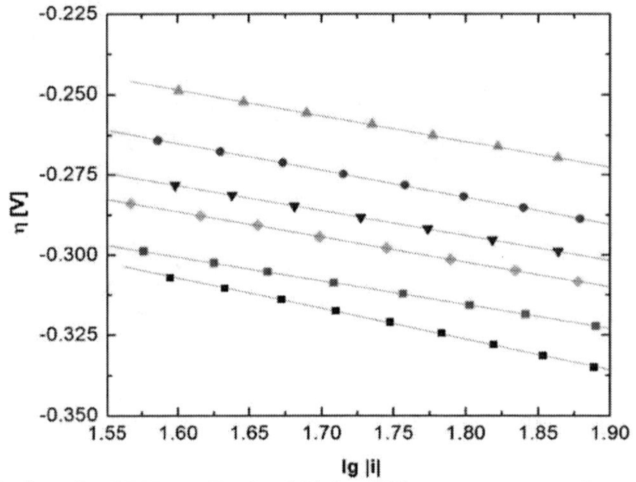

Figure 9. Tafel plots for HER on Pt-Co-20k in different concentrations of NaOH at 25°C.

The calculated values of exchange current density i_o and cathodic coefficient transfer 1-α are presented in table 1.

Table 1. Kinetic for HER on Pt-Co-20k electrode in 3 mol L^{-1} NaOH at 25°C.

Pt-Co-20k	NaOH conc. [mol L^{-1}]					
electrode	1	2	3	4	5	6
1-α	0.44	0.54	0.59	0.55	0.53	0.5
i_o [A m^{-2}]	0.008	0.021	0.025	0.011	0.010	0.009

Similar results have been reported by other authors [10,26,27].

4. Conclusions

In this study, new type of Pt-Co electrodes have been manufactured and tested for HER.

First, the characteristic parameters for pulse electrodeposition technique have been varied in order to find the optimal values to achieve the desired characteristics for Pt-Co deposited layer. The influence of current density i, pulse time t_{on}, and t_{off}, and also the number of deposition cycles was studied.

It was found that a high current density leads to large particles or even dendrite formation. It is preferably to carry out the electrodeposition at 250 A m^{-2} when compact, even distributed and small particles are formed.

As regards the applied and paused time of a cycle, it was concluded the longer t_{on}, and t_{off} are, the finer microcrystalline structure with a higher density of crystals and optimal platinum amount is obtained. The optimal values chosen were t_{off} = 100 ms and t_{on} = 20 ms.

The number of electrodeposition cycles does not influence significant the microcrystalline structure but have a strong influence on the layer thickness, the higher the number of cycles is, the thicker the layer is.

XRD patterns demonstrate that using the above parameters by pulse electrodeposition $CoPt_3$ homogenous alloy is obtained.

Electrochemical behaviour of $CoPt_3$ alloy in alkaline solutions has been studied by CV. Furthermore, kinetic parameters for HER on $CoPt_3$ electrode were determined. The results revealed the best catalytic effect is obtained in 3 mol L^{-1} NaOH.

5. References

[1] Opitz A, Badami B, Shen L, Vignarooban K and Kannan AM 2017 *Renew. Sust. Energ. Rev.* **68** 685-92
[2] Budde-Meiwes H, Drillkens J, Lunz B, Muennix J, Rothgang S, Kowal J and Sauer DU 2013 *J. Automob. Eng*, **227** 761-76
[3] Schipper F and Aurbach D 2016 *Russ. J. Electrochem.* **52** 1229-58
[4] Wang H and Gao L 2018 *Current Opinion in Electrochemistry* **7** 7-14
[5] Safizadeh F, Ghali E and Houlachi G 2015 *Int. J. Hydrogen Energy* **40** 256-74
[6] Zuttel A, Borgschulte A and Schlapbach L 2008 *Hydrogen as a Future Energy Carrier* (Weinheim: Wiley-VCH Verlag)
[7] Press RJ, Santhanam KSV, Miri MJ, Bailey AV and Takacs GA 2008 *Introduction to Hydrogen Technology* (New York: Wiley and Sons Publication)
[8] Vaszilcsin N, Kellenberger A and Medeleanu M 2016 *Kimya Problemleri* **4** 366-71
[9] Eftekhari A 2017 *Int. J. Hydrogen Energy* **42** 11053-77
[10] Kellenberger A, Vaszilcsin N, Brandl W and Duteanu N 2007 *Int. J. Hydrogen Energy* **32 (15)** 3258-65
[11] Cretu R, Kellenberger A, Medeleanu M and Vaszilcsin N 2014 *Int. J. Electrochem. Sci.* **9(8)** 4465-77
[12] Jakab A, Vaszilcsin N, Manea F and Dan M 2015 *Studia UBB Chemia* **60 (3)** 63-76
[13] Bensmann B, Hanke-Rauschenbach R, Pena Arias IK and Sundmacher K 2013 *Electrochem. Acta* **110** 570-80
[14] Marangio F, Santarelli M and Cali M 2009 *Int. J. Hydrogen Energy* **34** 1143-58
[15] Marangio F, Pagani M, Santarelli M and Cali M 2011 *Int. J. Hydrogen Energy* **36** 7807-15
[16] Zheng J 2017 *Appl. Surf. Sci.* **413** 72-82
[17] Zheng J 2017 *Appl. Surf. Sci.* **413** 360-5
[18] Zhang Y, Li P, Yang X, Fa W and Ge S 2018 *J. Comp. Alloys 732* 248-56
[19] Lu J, Zhang L, JingS, Luo L and Yin S 2017 *Int. J. Hydrogen Energy* **42** 5993-9
[20] Chen Jw, Chen Jd, Yu D, Zhang M, Zhu H and Du M 2017 *Electrochem. Acta* **246** 17-26
[21] Santos DMF, Sequeira CAC, Maccio D, Saccone A and Figueiredo JL 2013 *Int. J. Hydrogen Energy* **38** 3137-45
[22] Santos DMF, Sljukic B, Sequeira CAC, Maccio D, Saccone A and Figueiredo JL 2013 *Energy* **50** 486-92
[23] Zhong X, Qin Y, Chen X, Xu W, Zhuang G, Li X and Wang J 2017 *Carbon* **114** 740-8
[24] Muntean R, Pascal DT, Marginean G and Vaszilcsin N 2017 *Int. J. Electrochem. Sci.* **12** 4597-609
[25] Woo S, Kim I, Lee JK, Bong S, Lee J and Kim H 2011 *Electrochem. Acta* **56** 3036-41
[26] Lasia A and Rami A 1990 *J. Electroanal. Chem.* **294(1-2)** 123-41
[27] Krstajic N, Popovic M, Grgur B, Vojnovic M and Sepa D 2001 *J. Electroanal. Chem.* **512** 16-26

Acknowledgement
This work was partially supported by Politehnica University Timisoara in the frame of PhD studies.

IOP Conf. Series: Materials Science and Engineering **416** (2018) 012034 doi:10.1088/1757-899X/416/1/012034

Comparison of the Phase Composition and Nanohardness of Gradient TiN/TiO_2 Coatings on Ti5Al4V Alloy Deposited by Different PVD Methods

S Valkov[1*], M P Nikolova[2] E Yankov[2], T Hikov[1], R. Bezdushnyi[3], D Dechev[1], N Ivanov[1] and P Petrov[1]

[1]Bulgarian Academy of Sciences, Institute of Electronics, 72 Tzarigradsko Chausse Blvd, Sofia, Bulgaria
[2]University of Ruse "A. Kanchev", Dept. of Material Science and Technology, 8 Studentska Str., Bulgaria
[3]Sofia University "St. Kliment Ohridski", Faculty of Physics, 5 James Bourchier Blvd., Sofia, Bulgaria

*Corresponding author: stsvalkov@gmail.com

Abstract. The effect of the deposition technology of gradient TiN/TiO_2 coatings, applied on Ti5Al4V substrate, on the phase composition and mechanical properties has been studied. The films have been applied by reactive magnetron sputtering and cathodic arc and glow-discharge techniques. The coatings' properties have been studied by X-ray diffraction (XRD) and nanoindentation tests. The results in the present study show that the application of both methods are capable to form polycrystalline TiN and TiO_2. The coatings deposited by magnetron sputtering are monophasic and polycrystalline as TiO_2 is in the form of anatase. Those, applied by cathodic arc and glow-discharge technique, TiO_2 layer is in the form of double-phase structure of rutile and anatase. The measured hardness is similar for the coatings deposited via both methods. The hardness of magnetron sputtered coatings is 6.3±1.1 GPa. The values for coatings obtained by glow-discharge is 6.1±1.4 GPa.

1. Introduction

Titanium alloys have been widely used as hard tissue replacement materials. However, poor tribological properties and possible release of metal products from the alloy in the physiological environment could compromise the biocompatibility of the alloy. The deposition of a stable titanium dioxide films on hard and wear resistant TiN film could overcome the major drawbacks of the Ti alloys whilst promote positive interaction between the metallic implant and the existing bone tissue. These materials are applicable in many branches, such as fabrication of different tools which work in abrasive environments, food processing industries, where surfaces must be cleanable and resistant to microbial contamination, for manufacturing of implants due to their biocompatibility [1-3]. They can also be used as self-cleaning and bio-sensor coating, biomedical materials, etc. [4,5].

The discussed coatings can be obtained by a number of different techniques. The ones used by us are different physical vapor deposition methods. It should be noted that both magnetron sputtering and cathodic arc and glow-discharge provide advantages for controlling the deposition process. This guarantees the reproducibility of the film formation and control of its properties [6-9].

The aim of our present study is to investigate the influence of different deposition methods on the mechanical properties and phase composition of the as-deposited multilayer TiN/TiO_2.

2. Experimental procedures

The chemical compositions of the bare Ti substrate and the target material used for the experiment, determined by JEOL JXCA-733 Microprobe scanning electron microscope (SEM) equipped with wavelength dispersive spectrometers (WDS), were shown in table 1. Samples with dimensions 14×14×4 mm were cut out of 16 mm thick sheet material using the electro-erosion cutting method. The as-received samples were single solution treated (ST) for 30 min at 920 °C and water quenched. Half of the specimens were precipitated (P) for 4 hours at 500 °C and air cooled. All treatments were carried out in ≤ 1 Pa vacuum. The surfaces of the samples were grounded and polished before the EBSM. To study the effect of the surface treatment, the EBSM was applied to as-received, ST and ST and precipitated (ST+P) samples.

Table 1. Chemical composition (wt. %) of the substrate material.

Element	Al	V	Fe	Mn	Co	Cr	Mo	Pd	Nb	Hf	Ti
Substrate	5.21	4.40	0.14	0.11	0.06	-	0.17	0.15	0.36	0.04	Bal.

The electron beam surface modification (EBSM) was carried out by electron beam installation Leybold Heraeus (EWS 300/15–60). The following technological parameters were applied: electron beam current – $I = 20$ mA, accelerated voltage – $U = 52$ kV, speed of the samples motion – $V = 0.5$ cm/s, electron beam frequency – $f = 1$ kHz, scanning electron beam with sinusoidal dithering. Afterwards, all samples were ultrasonically washed with absolute ethanol, acetone, isopropanol, rinsed with distilled water for 5 min and dried.

The TiN/TiO_2 coating was applied by two different methods: reactive magnetron sputtering and cathodic arc and glow-discharge method using Ti targets with purity of 99.8%. The deposition of the magnetron deposited TiN layer took a place in the Ar-N_2 atmosphere, as the working pressure was 1.2×10^{-1} Pa, during the deposition process, the substrate had been heated to 350° C. The TiO_2 film was realized in the pure O_2 environment, the working pressure was 7×10^{-2} Pa, as the substrate temperature was decreased down to 180° C. The thickness of each coating (i.e. TiN and TiO_2, respectively) was about 1 μm assessed by the resonant frequency of quartz plate (Intermetalix IL800) coated within the same process conditions. In order to minimize the residual stresses and the oxidation, the samples with deposited coating were retrieved from the vacuum chamber after the achievement of room temperature.

For the cathodic arc deposition of the TiN, a sidewall positioned evaporating system in a cubic vacuum chamber with water-cooled walls was used. The samples were hanged near the center of a clockwise-rotating with a frequency of 0.5 Hz turntable. To ensure the coating stress relaxation and necessary adhesion a very thin pure layer from the target (at 2.5×10^{-1} Pa in Ar atmosphere for 5 min., bias 600 V, 110 A arc current) was applied. The TiN film was made in a pure N_2 atmosphere at 340 °C substrate temperature for a time of 60 min., 110 A arc current, bias 250 V and 7.5×10^{-1} Pa pressure in the working chamber. The TiO_2 film was made by glow plasma discharge using the uppermost located sputtering system in the same chamber. A bias voltage of 1340 V (720 mA current) in a pure O_2 atmosphere at a pressure of 6×10^{0} Pa were applied for a deposition time of 240 min. Both layers' thickness was attained by means of calotest measurements: TiN ~ 1.6 μm, TiO_2 ~ 0.6 μm.

The coatings were characterized by X-ray diffraction analysis using URD6 Seiferd&Co diffractometer with CuKα radiation. The registration of the patterns was within 20° to 70° at 2θ scale

IOP Conf. Series: Materials Science and Engineering **416** (2018) 012034 doi:10.1088/1757-899X/416/1/012034

with 0.1° step and counting time of 10 s per step. The experiments were conducted in symmetrical Bragg-Brentano (B-B) mode.

The mechanical properties of the coatings were characterized by Nanomechanical Tester (Bruker, USA). The software program prepared for this experiment consisting of 4 lines with 12 indentations each (total of 48 indentations) and 80 μm spacing was utilized. The applied force for each indentation was 50 mN.

3. Results and discussions

Figure 1 and 2 presents the XRD patterns of the multilayer coatings, deposited on Ti5Al4V alloy. Peaks corresponding to α-Ti from the substrate, as well as TiN and a-TiO_2 can be seen in the diffractogram of the sample obtained by magnetron sputtering (figure 1). Diffraction maxima corresponding to r-TiO_2, β-TiO_2, etc. were not observed, which means that the coatings are mono-phase (polycrystalline TiN and TiO_2).

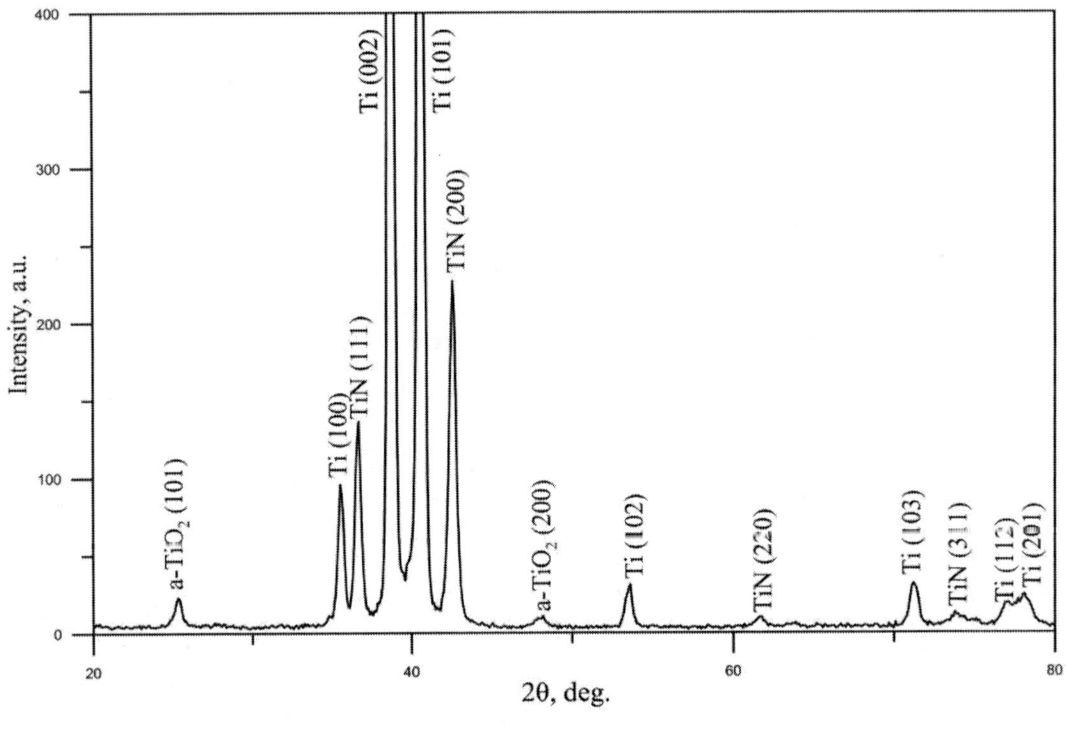

Figure 1. XRD of multilayer TiN/TiO_2 deposited by magnetron sputtering

The diffractogram of the sample obtained by cathodic arc and glow-discharge (figure 2) shows also peaks belonging to α-Ti from the substrate, as well as TiN and a-TiO_2. Also, diffraction maxima corresponding to r-TiO_2 are observed, which means that the obtained coatings are double-phase.

These results are in agreement with those published in other studies, where the formation of TiO_2 coatings has been studied. It should be noted that in our case, during the deposition process via magnetron sputtering, no bias voltage was applied. This is the reason for the TiO_2 to be in single phase with absence of rutile. This result correlates with the results published in ref. [10]. The presence of rutile phase in the glow-discharge deposited coatings is due to the applied high bias voltage. As stated in other studies [10, 11] the rutile phase can be induced in the TiO_2 coating by applying bias voltage.

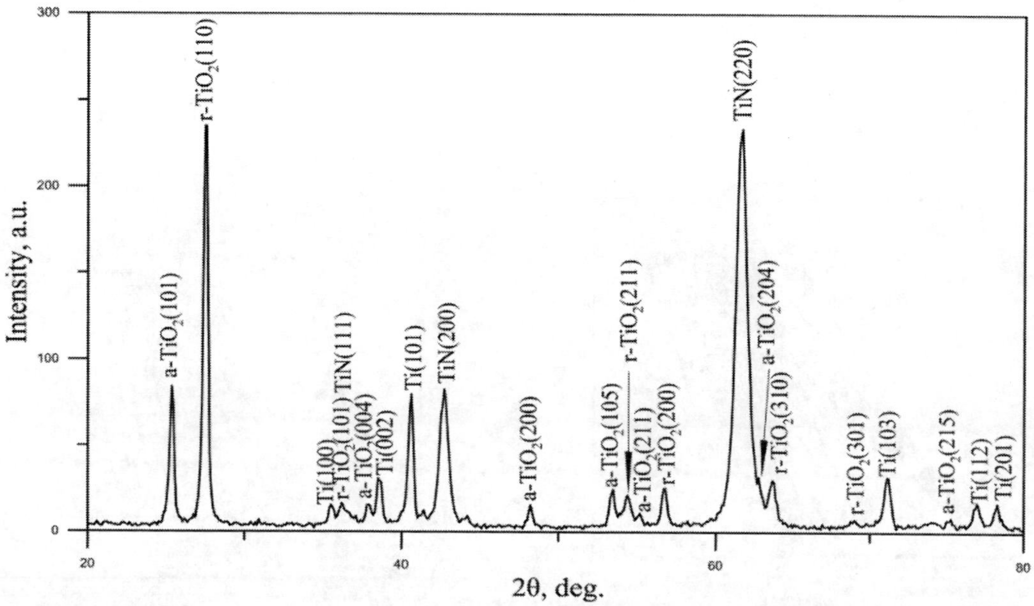

Figure 2. XRD of multilayer TiN/TiO_2 deposited by cathodic arc and glow-discharge

The mechanical properties of the coatings have been studied by nanoindentation tests and the results are summarized in table 2. Both coatings were tested by a load of 50 *mN*. The hardness values were evaluated from the load displacement curves [12, 13], which are shown in figure 3 and 4. From the figures it can be seen that the penetration depth of the indentor after each load is different. This scattering of the penetration depths can be explained by the very different surface topography. The treatment of the substrates by scanning electron beam is capable to increase the surface roughness leading to a larger number of peaks and valleys with higher amplitude. The penetration of different surface formation strongly reflects on the penetration level during the measurement.

Table 2. Nanohardness of the deposited multilayer TiN/TiO_2 coating

Deposition method	Hardness, GPa
Magnetron sputtering	6.3 ± 1.1
Cathodic arc and glow-discharge	6.1 ± 1.4

The measured hardness of the TiN/TiO_2 coatings deposited on Ti5Al4V substrates by the two different methods show similar values, namely 6.3 *GPa* for the specimen deposited by magnetron sputtering and 6.1 *GPa* for the one deposited by glow-discharge.

The authors [10] have reported that the application of bias voltage and presence of double phase structure of rutile and anatase when considering TiO_2 film results in increase the hardness. However, the results obtained in the present study can be explained by the nature of the two deposition methods. Using magnetron sputtering yields denser coatings than cathodic arc [14]. The denser coatings leads greater hardness. In our case, the application of bias voltage during the cathodic arc and glow discharge method leads to an increase in density of the atoms near the substrate, hence increase the density and change the phase composition (figure 2) of TiO_2 film in the coating. The results obtained in present study show that the mechanical properties of the deposited TiN/TiO_2 coatings depend on the phase composition and on the deposition method. This is the reason that both methods are capable to produce coatings with similar hardness.

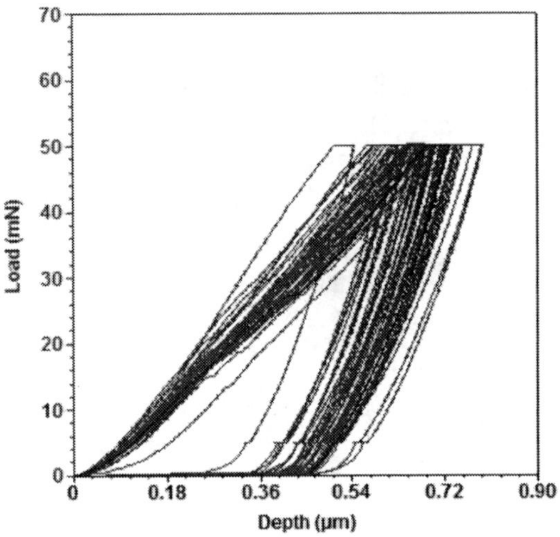

Figure 3. Load displacement curves of multilayer TiN/TiO_2 deposited by magnetron sputtering

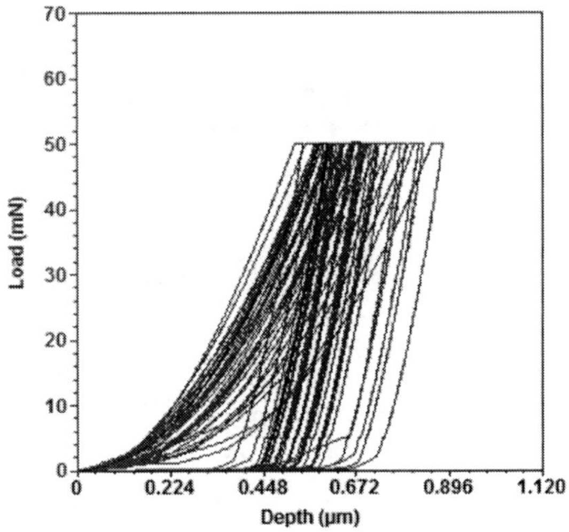

Figure 4. Load displacement curves of multilayer TiN/TiO_2 deposited by cathodic arc and glow-discharge

4. Conclusions

The results in the present study show the difference in phase composition of multilayer TiN/TiO_2 coatings on Ti-5Al-4V substrate obtained by two PVD methods, magnetron sputtering and cathodic arc and glow-discharge. Magnetron sputtered coatings are monophasic and polycrystalline with presence of anatase TiO_2, whereas the samples obtained by the cathodic arc and glow-discharge method are double-phase and polycrystalline with the presence of anatase and rutile TiO_2.

It is shown that the samples have similar hardness. The coatings obtained by the magnetron sputtering method have hardness about 6.3 *GPa*, while those obtained by the cathodic arc and glow-discharge method have 6.1 *GPa*.

5. References

[1] Babinova R, Smirnov V, Useenov A, Kravchuk K, Gladkikh E, Shapovalov V, Mylnikov I, Mechanical properties of titanium nitride films obtained by reactively sputtering with hot target 2017 *J. Phys.: Conf. S.* **872** 012035

[2] Jang B, Lee C, Choi J, Kwon J, Kim H, Park C, Kim H, The Mechanical Properties of TiN Films Grown by a Pure N_2 Plasma Sputtering 2017 *Mat. Sc. F.* **878** 89-95

[3] Jabbari Y S A, Fehrman J, Barnes A C, Zapf A M, Zinelis S, Berzins D W, Titanium nitride and nitrogen ion implanted coated dental materials 2012 *Coatings* **2** 160–78

[4] Pisarek M, Holdynski M, Roguska A, Kudelski A, Czachor M J, TiO_2 and Al_2O_3 nanoporous oxide layers decorated with silver nanoparticles — active substrates for SERS measurements 2014 *J. S. St. Electrochem.* **18** 3099–109

[5] Pan D, Fan H, Li Z, Wang S, Huang Y, Jiao Y, Yao H, Influence of substrate on structural properties and photocatalytic activity of TiO_2 films 2017 *M. & N. Letters* **12** 82-6

[6] Brunckova H, Medvecky L, Kovalcikova A, Fides M, Mudra E, Durisin J, Skvarla J, Kanuchova M, Structural and mechanical properties of lanthanide doped $La_{1/3}Nb_{0.8}Ta_{0.2}O_3$ thin films prepared by sol–gel method 2017 *Smart Mat. Str.* **26** 045009
[7] Park S J, Choi D J, Synthesis of porous Al_2O_3/ZrO_2 nanocomposites by chemical vapour deposition 2017 *Adv. Appl. Cer.* **116** 236-41
[8] Han J G, Recent progress in thin film processing by magnetron sputtering with plasma diagnostics 2009 *J. Phys* D*: Appl Phys* **42** 043001
[9] Johnson M, Cote P, Modeling Magnetron Sputter Deposition 2006 *Mat. Manuf. Proc.* **21** 628–33
[10] Bait L, Azzouz L, Madaoui N, Saoula N, Influence of substrate bias voltage on the properties of TiO_2 deposited by radio-frequency magnetron sputtering on 304L for biomaterials applications 2017 *Appl. Surf. Sc.* **395** 72-7
[11] Yumoto H, Matsudo S, Akashi K, Photocatalytic decomposition of NO_2 on TiO_2 films prepared by arc ion plaing 2002 *Vacuum* **65** 509-14
[12] Oliver W C, Pharr G M, An improved technique for determining hardness and elastic modulus using load and displacement sensing indentation experiments 1992 *J. Mat. Res.* **7** 1564-83
[13] Oliver W C, Pharr G M, Measurement of hardness and elastic modulus by instrumented indentation: Advances in understanding and refinements to methodology 2004 *J. Mat. Res.* **19** 3-20
[14] Pauleau Y 1992 Physical vapour deposition techniques I: evaporation and sputtering *Advanced Techniques for surface engineering* ed W Gissler and H A Jenh (Springer) 135-179

Acknowledgments
The present research was supported by a contract with Bulgarian National Science Fund, contract №: DN 07/3 2016

IOP Conf. Series: Materials Science and Engineering **416** (2018) 012035 doi:10.1088/1757-899X/416/1/012035

Effect of Coffee Grounds Addition on Efflorescence in Fly Ash-based Geopolymer

D Mierzwiński[1,*], K Korniejenko[1], M Łach[1], J Mikuła[1] and J Krzywda[1]

[1] Institute of Materials Engineering, Faculty of Mechanical Engineering, Cracow University of Technology, Jana Pawła II 37, 31-864 Cracow, Poland.

*Corresponding author: dariusz.mierzwinski@mech.pk.edu.pl

Abstract. The article aim is to analyse the influence of addition of the cheap and environmental friendly material that can reduce – coffee grounds on the efflorescence effect. The additives should reduce this effect trough decreasing pH. The geopolymer matrix was prepared from fly ash from Skawina CHP plant and sand in ratio 1:1. The samples were prepared using sodium promoter - 12M sodium hydroxide solution combined with the sodium silicate solution and coffee grounds (1, 3 and 5% by mass of the composite). The samples were prepared according to the methodology described in the standard EN 12390-1. The empirical part of the research is based on: visual analyse for efflorescence, absorptivity test, pH test and compressive strength test. The results show the influence of coffee grounds on the pH reduction from about 11 to 9.21, the elimination of efflorescence and the reduction of the compressive strength.

1. Introduction

Nowadays, Portland cement is commonly used material in construction industry. This technology has a lot of disadvantages among others: high energy consumption, high level of CO_2 emissions and questionable durability. The most promising alternative is the technology of alkali-activation. The material is created by reaction between alkaline solutions and solid aluminiosilicates. Unfortunately, the activation process required high pH [1,2]. The one of the limitation of wide using this materials is common problem with efflorescence caused by excess sodium oxide remaining unreacted [1] The unreacted sodium oxide creates white efflorescence on the surface – it is sodium carbonate heptahydrate Na2CO3 · 7H2O. It can be the significant problem, because it decreases not only aesthetic value for the final products, but also mechanical properties of the material [3,4]. The research on additives shows that certain admixtures can eliminated or limited efflorescence effect, exemplary alumna-rich additives such as natural pozzolan or zeolites [4,5].

Coffee grounds are an organic waste from the food and beverage industry. There are usually destined for landfill [6,7]. They have barely acidic pH (reported usually about 5.0) [6,8].

2. Experimental procedure

Specimens based on geopolymer matrix and additive of coffee grounds in amount: 1 (samples marked as P1), 3 (P3) and 5% (P5) by mass were investigated. The variable includes different amount of reinforcement and reference samples based on matrix material without addiives (samples marked as P0).

2.1. *Materials*

2.1.1. Geopolymer matrix. The geopolymer matrix was made from fly ash from the CHP plant in Skawina (Poland) and sand in ratio 1:1. This kind of fly ash is rich in oxides such as SiO_2 and Al_2O_3. The oxides content is presented in table 1. The fly ash particles morphology was typical for such by-products (from coal combustion) and suitable for the process of alkali-activation.

Table 1. Chemical composition of the fly ash.

	SiO_2	Al_2O_3	Na_2O	CaO	Fe_2O_3	MgO	K_2O	SO_3	TiO_2	P_2O_5	BaO
[% mass]	55.89	23.49	0.59	2.72	5.92	2.61	3.55	0.16	1.09	0.82	0.2

The process of alkali-activation has been made by 12M sodium hydroxide solution combined with the sodium silicate solution. The ratio of liquid parts to dry parts was 0.4 by mass.

2.1.2. Coffee grounds. The coffee grounds came from coffee brewed traditionally. There were mixture of different species of coffee, collected and dried. Next, there were added to the composites in amount: 1, 3 and 5% by mass.

2.2. *Specimens*

The alkaline solution was prepared by means of pouring the aqueous solution of sodium silicate over the solid sodium hydroxide. The tap water replaced the distilled one. The solution was mixed and left until its temperature became stable and the concentrations equalized. The fly ash, sand, coffee grounds and alkaline solution were mixed about 15 minutes to receive the homogeneous paste by using low speed mixing machine. Then, the mixture was poured into moulds on cylindrical shape with dimensions: Ø = 50 mm and h = 100 mm.. The samples were hand-formed and then subjected to vibratory removal of air bubbles. Tightly closed moulds were heated in the laboratory dryer for 24h at 75°C. Then, the samples were unmoulded.

3. Analytical procedure

3.1. *Microstructure research*

Scanning electron microscope (SEM) type JEOL JSM 820 with EDS has been used for microstructure research. The research has been made for the samples previously broken while compressive strength test. The samples were covered with a thin layer of gold with JEOL JEE-4X vacuum sputter. The investigations were made at various magnifications.

3.2. *Efflorescence - visual examination*

The cured samples - after 7 days, there were put into distilled water. The water level was on the half of each sample. The specimens were left there 7 days. Then, the efflorescence process has been initiated. It became more intense during the next few week. The visual examination were made after 28 days.

3.3. *Compressive strength test*

The compressive strength test was carried out on cylindrical shape with dimensions: Ø = 50 mm and h = 100 mm. Test was performed on a testing machine MATEST. The specimens were cured in humidity 50% and temperature 21°C and investigated after 28 days and 3 months. The compressive strength test was carried out according to the methodology described in the standard PN-EN 13791. The characteristic compressive strength is the lowest value from (equation 1):

$$f_{ck,is} = min\begin{Bmatrix} f_{cm(n),\,is} - k \\ f_{is,lowest} + 4 \end{Bmatrix} \tag{1}$$

where:

$f_{ck,is}$ – characteristic compressive strength of concreate 'in-situ'
$f_{cm(n),is}$ – average compressive strength of concreate 'in-situ', for 'n' samples
$f_{is,lowest}$ – the lowest result for compressive strength of concreate 'in-situ',
k – coefficient, it is depended on the number of results 'n' and is equal: for n = 10 ÷ 14 → k = 4; for n = 7 ÷ 9 → k = 5; for n = 3 ÷ 6 → k = 6,

After taking under consideration the correction coefficient γ = 0,85 the equation 2 is applied:

$$f_{ck,is} = min \begin{Bmatrix} f_{cm(n),\,is} - k \\ f_{is,lowest} + 4 \end{Bmatrix} \cdot \frac{1}{0{,}85} \tag{2}$$

3.4. *Water absorption tests*

Firstly, the cured samples - after 7 days, were weighted and next immersed completely in distilled water. Then, after next 7 days they were weighted again. The water absorption was calculated according the equation (3).

$$n_w = \frac{G_2 - G_1}{G_1} \cdot 100\% \tag{3}$$

where:
G_1 – average mass of dry samples,
G_2 – average mass of samples after 7 days in distilled water.

In the second step, the samples were again dried during the 7 days and after this time they were weighted.

3.5. *pH test*

The cured samples - after 14 days, were tested according to pH. The first 7 days after de-moulded, specimens were cured in humidity 50% and temperature 21°C. Next 7 days there were immersed in distilled water. The pH of the leachate was measured using a pH meter (CP-105 type).

4. Experimental results and discussion

4.1. *Microstructure research*

The SEM observations show differences between the samples reinforced by coffer grounds and samples without reinforcement (figure 1.).

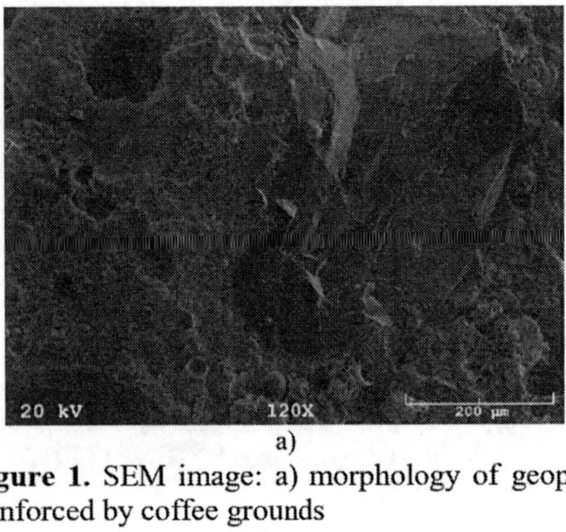

a)

b)

Figure 1. SEM image: a) morphology of geopolymer matrix b) morphology of geopolymer matrix reinforced by coffee grounds

The EDS studies show the coffer grounds were impregnated by the alkaline solution used in the geopolymerization process. The weight ratio of Al/Si in sample without reinforcement was 0.5 and the weight ratio of Na/(Al+Si) was 1.0. The samples with coffee grounds have following ratios:

- for geopolymer matrix: weigh ratio of Al/Si – 0.4, and weigh ratio of Na/(Al+Si) -0.1.
- for reinforcement: weigh ratio of Al/Si – 0.2, and weigh ratio of Na/(Al+Si) - 0.4.

4.2. Efflorescence - visual examination

The results of the visual examination of efflorescence are shown in the figure 2. The efflorescence occurred on all samples, however it had different character depending on amount of additive. On the samples with 5% additive, the efflorescence was rather directly up the line of water and it was reduced on the upper surface. It is probably caused by "closing" capillary pore system by coffee grounds.

Figure 2. Efflorescence – the samples with coffee grounds: a) 1% b) 3% c) 5%

Carbonate efflorescence on concrete surfaces is formed as a result of the carbonation of calcium hydroxide - $Ca(OH)_2$. Calcium hydroxide, also called portlandite, is one of the cement hydration products (phases of silicate Portland clinker). Carbonate efflorescence occurs when the calcium hydroxide transported through the capillary pore system gets to the surface of the concrete element.

The rate and intensity of carbonation is influenced by CO_2 concentration and ambient humidity. In concrete saturated with water, the carbonation process is practically non-existent, and in the case of dried concrete, carbonation occurs very slowly. Generally, the carbonation occurs when there is a water film on the surface of the capillaries, in which the rapidly diffusing CO_2 dissolves and reacts with the calcium ions. The fastest carbonation process is in the range of 50% to 70% relative humidity [9].

A similar mechanism occurs in the case of geopolymers and geopolymer concretes. The difference in the chemical composition of the geopolymers comparison with traditional concretes results in the possibility of the appearance of sodium carbonate heptahydrate on the surface, which is revealed quite quickly in the right hydrothermal conditions [3].

The intensity of carbonation of concrete exposed to the external environment increased with the porosity of the concrete, it is susceptibility to cracks and scratches and the w/c ratio in the mixture [10,11]. It also depend on density of the concrete mixture (the lower density favours the carbonization), the concrete care and the concentration of CO_2 in the ambient air (the higher amount of CO_2 favours the carbonization) [12, 13].

For conventional concretes, efflorescence is generally harmless except for discoloration [14]. The standards for concrete products (eg PN-EN 1338: 2005) allow the occurrence of the efflorescence as a natural element of concrete maturation. In the case of geopolymers the efflorescence could influence on the mechanical properties.

4.3. Compressive strength test

The results of the compressive strength test are shown in the figure 3.

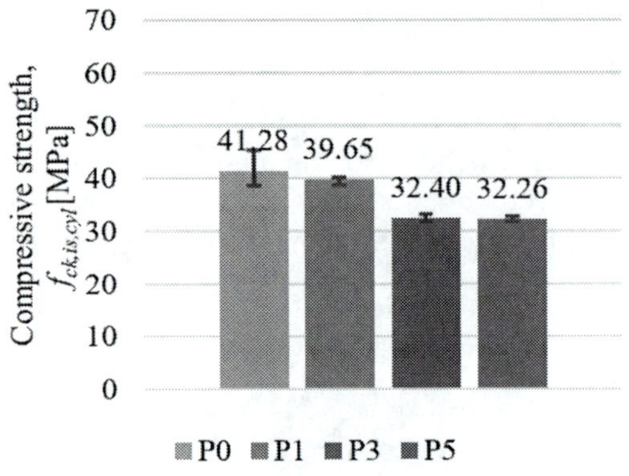

a)

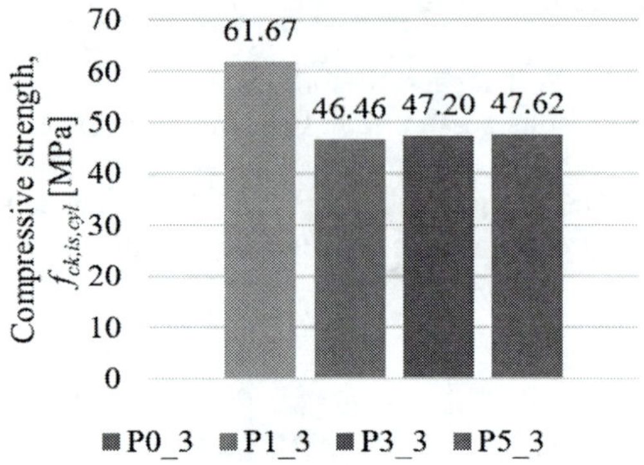

b)

Figure 3. The results of the compressive strength test: a) after 28 days b) after 3 months.

The results of the research show the alignment of strength values for samples with the coffee grounds in relation to plane samples after 28 days.

4.4. Water absorption tests

The results of the water absorption test are presented on the figure 4. The blue columns are for first stage and the red one for second stage of the process.

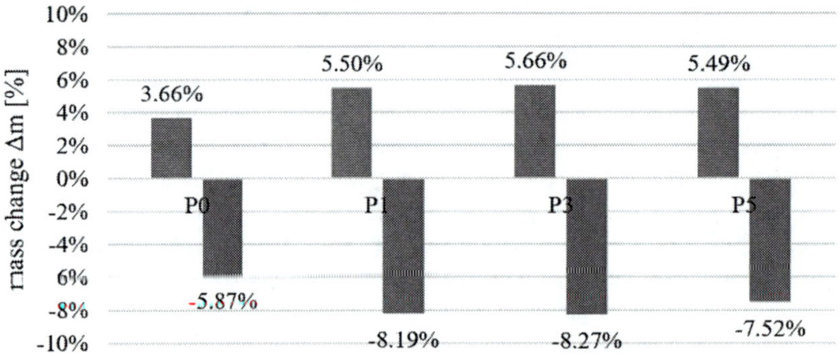

Figure 4. The results of the water absorption test (upper: after 7 days of soaking; red: after re-drying)

The results of the research show that the samples containing the additive of coffee grounds have better absorbent properties compared to the plane samples (without reinforcement). Regardless of the amount of the coffee grounds, the difference in absorbability is relatively small. The additive of coffee grounds largely allows the moisture to evaporate as a result of increased porosity.

4.5. pH test

The results of the pH test are presented in the table 2.

Table 2. The pH values of the leachate.

	P0	P1	P3	P5
pH	10,99	9.61	9.33	9.21

The results of the research confirm the decrease of pH of the leachate for the samples containing coffee grounds. The change in the weight content of coffee grounds in geopolymer samples did not significantly change the pH. Coffee grounds significantly reduce the pH of the leachate in comparison with plane samples (without reinforcement).

5. Conclusions

Geopolymer composites reinforced with coffee grounds in amount 1, 3 and 5% by weight as well as plain (without reinforcment) specimens have been produced and characterized. The results of the research and literature review show that the additive of the coffee grounds to geopolymers could prevent the efflorescence. In the same time, the composites have reasonable properties for many application, exemplary in construction industry.

The results of the research show the influence of coffee grounds on the composite properties. The additive of coffee grounds reduced the pH from about 11 to 9.21. At the same time, the amount of efflorescence on the surface decreased, but also the mechanical properties came down. The results of the compressive strength test show 20% reduction of the compressive strength for samples containing 5% of coffee grounds compared to the reference samples (plane samples). Also, the results of the water absorption for the samples with reinforcement are worse. However, the mechanical properties for the samples with reinforcement are lower, these samples have more repeatable results. The additive regulates the process of curing and seems to protect the material against changeable conditions.

6. References

[1] Kani EN, Allahverdi A, Provis JL 2012 *Cement & Concrete Composites* **34** 25

[2] Davidovits J 2015 *Geopolymer Chemistry and Applications*, 4th ed. (Saint-Quentin, France: Geopolymer Institute) p 644

[3] Zhang Z, Wang H, Provis JL, Reid A 2013 *Prec. Concrete Institute of Australia's Biennial National Conference (Concrete 2013): Understanding Concrete* (Gold Coast, Australia) p 1

[4] Zhang Z, Wang H, Provis JL, Reid A, Wang H 2014 *Cement & Concrete Research* **64** 30

[5] Kua TA, Arulrajah A, Horpibulsuk S, Du YJ, Suksiripattanapong C 2017 *Clean Technologies and Environmental Policy* **19 (1)** 63

[6] Wang JB, Li FS, Zhou ZH, Du P, Xu D, Xie N, Cheng X, Liu Y 2018 *Construction and Building Materials* **158** 683

[7] Kondamudi N, Mohapatra SK, Misra M 2008 *Journal of Agricultural and Food Chemistry* **56 (24)** 11757

[8] Sanchez-Hernandez JC, Domínguez J, Vermicompost derived from spent coffee grounds: assessing the potential for enzymatic bioremediation, *Handbook of Coffee Processing By-Products. Sustainable Applications* ed, Galanakis CM (Elsevier Inc., London- San Diego-Cambridge-Oxford) pp. 369–398

[9] Omosebi O, Maheshwari H, Ahmed R, Shah S, Osisanya S, Hassani S, De Bruijn G, Cornell W, Simon D 2016 *Cement & Concrete Composites* **74** 54

[10] Šefflová M, Pavlů T 2016 *Key Engineering Materials* **722** 228

[11] Zhang L, Sun L 2017 *Advances in Structural Engineering* **20 (3)** 228

[12] Shen Q, Pan G 2017 *Materials Science Forum* **902** 74

[13] Cui H, Tang W, Liu W, Dong Z, Xing F 2015 *Construction & Building Materials* **93** 522

[14] Nowak-Michta A 2012 *Czasopismo Techniczne* **109 (2)** 77

Acknowledgments

This work was supported by the ERANet-LAC 2nd Joint Call (http://www.eranet-lac.eu) and funded by National Centre for Research and Development, Poland, under grant: "Development of eco-friendly composite materials based on geopolymer matrix and reinforced with waste fibers".

IOP Conf. Series: Materials Science and Engineering **416** (2018) 012036 doi:10.1088/1757-899X/416/1/012036

Influence of pH on the Corrosion Behaviour of Laser-Processed Co-Cr Dental Alloys in Artificial Saliva

M L Dan[1], L-V Costea[1,*], C E Savencu[2], L Porojan[2]

[1] Faculty of Industrial Chemistry and Environmental Engineering, University Politehnica Timişoara, Romania
[2] Department of Dental Prostheses Technology, University of Medicine and Pharmacy V. Babeş, Timişoara, Romania

*Corresponding author: liviu.costea@upt.ro

Abstract. The materials used in dental restorations must present, aside from a mechanically stable and fracture resistant structure and a high level of biocompatibility, a strong resistance towards corrosion over a wide range of pH values. This research focuses on the relation between two different CAD-manufacturing procedures of base Co-Cr alloys used in dental restorations regarding the assessment of their corrosion resistance in artificial saliva at various pH levels. We comparatively appraised the corrosion behaviour of two cobalt-chromium dental alloys manufactured by two alternative computer assisted technologies namely selective laser melting (SLM) and selective laser sintering (SLS). Several electroanalytical techniques like linear polarization, cyclic voltammetry and electrochemical impedance spectroscopy have been employed to study the processes taking place at the metal/electrolyte interface. Preliminary results reveal that computer assisted processing technologies could represent an alternative when considering the manufacture of complex restorations.

1. Introduction

The oral cavity represents an extremely complex and ever changing environment implying moisture, large temperature variations that can range from about 5^{0}C while consuming cold fluids or ice cream to more than 55^{0}C during the intake of hot meals, coffee or tea, relatively high forces during mastication as well as a substantial pH variation. In healthy individuals, saliva has a complex composition based mainly on water soluble electrolytes, proteins, hormones, as well as bacteria while it's pH is very close to neutral, having values of about 6.7 to 7.4 [1, 2]. These values can dramatically change due to various physical conditions like gastroesophageal reflux disease, drug intake, consumption of citric fruits or juices. All these factors require that the materials used in dental restorations must present, aside from a mechanically stable and fracture resistant structure and a high level of biocompatibility, a strong resistance towards corrosion over a wide range of pH values.

The most important categories of materials used in dentistry are represented by ceramics, polymers, composites and metals/alloys [3]. The aforementioned material clusters find their use mainly for dental prostheses as well as restorations. Many of the new computer aided technologies used for the fabrication of metallic structures implemented in the last decade can be divided into two major classes based on the treatment of the base material. While milling of pre-processed starting materials falls in the category of subtractive manufacturing, the more recently presented selective laser melting (SLM) and selective laser sintering (SLS) techniques are seen as additive processing technologies [4]. The present research focuses on the investigation of the corrosion behaviour of two Co-Cr alloys produced using the latter fabrication methods, within Fusayama artificial saliva by means of specific electroanalytical methods. The pH values used in this investigation ranged from 2.5 to 6.5 with the aim of closely simulating the various physiological conditions to which the restoration material could be exposed.

Base Co-Cr alloys are widely used for manufacturing removable partial dentures, dental crowns, fixed as well as removable prostheses, due to their high strength, excellent resistance to wear, and nonetheless their reasonable price [5]. The traditional casting technology used for over a century for obtaining structures used in dental restorations, has been proven to be extremely time-consuming and labour-intensive. It has proven its limitations while machining advanced designed parts [6]. That is the reason why new additive manufacturing technologies made their way in prosthetic dentistry, gaining popularity due to higher precision and speed, compared to traditional casting methods. New additive manufacturing techniques are mainly processes that conjunct computer aided design to a high intensity laser beam with the purpose to gradually melt (or sinter) the metallic powder used as raw material, resulting in complex and relatively dense final products. The technology not only delivers very accurate manufacturing, but could significantly reduce production costs [7]. The mentioned Laser based processing technologies known as Selective Laser Sintering (SLS) and Selective Laser Melting (SLM) are currently employed to produce dental restoration devices. A high power laser is used to partially (SLS) or fully (SLM) melt a metal powder to form specific complex shaped objects. Even though comprehensive studies of the biocompatibility of Co-Cr dental alloys have been conducted, it is well known that the engaged processing technology can highly influence the surface microstructure and in consequence corrosion behaviour and thus biocompatibility [8].

The complex electrolyte environment found in the oral cavity is subjected to multiple changes regarding parameters such as temperature, humidity, pH and composition, all of these can induce corrosion and thus damage to dental alloys, hence affecting their biocompatibility [9, 10]. The existence of narrow interproximal spaces, where food and plaque generated by microorganisms can accumulate in the vicinity of dental restorations may create more acidic conditions that, in turn, can facilitate corrosive reactions [11]. Corrosion of these medical devices may lead to metal ions originating from the base alloy being released into the body. These ions can travel to different body regions through the digestive passage, lungs and oral tissues exerting toxic as well as allergic reactions [12]. This is why the in-depth knowledge of the mechanisms governing the corrosion processes affecting the above mentioned alloys is of paramount importance.

The present study aims to elucidate the corrosion mechanisms of the studied alloys at different acidity levels of the medium in order to further improve the employed manufacturing techniques towards reducing potential harmful agents that could be released during long term use, thus making them less prone to locally or systemically occurring adverse reactions.

2. Experimental

Samples of cylindrical shape with the diameter of 10 mm and the height of 2 mm have been obtained through two distinct machining processes, namely: the CAD selective laser sintering technique using a commercial Co-Cr base material and the CAD selective laser melting technology using Co-Cr powder. The chemical composition of the investigated probes, as given by the manufacturer, is detailed in Table 1. After cooling down, the probes have been air-blasted with alumina particles and mechanically finished and polished using specific techniques. Oxide-firing (at 950 – 980°C) was undertaken, as required for porcelain veneering for both samples, the metal surface was finally blasted with aluminium oxide (75 μm).

All electrodes obtained as above were finished using silicon carbide and 1μm grit diamond paste respectively, degreased and treated ultrasonically in distilled water, rinsed and dried prior to be subjected to electrochemical investigations. Electrochemical investigations have been undertaken using a BioLogic SP150 potentiostat/ galvanostat along with a conventional three-electrode cell system. All electrochemical measurements have been undertaken using Fusayama artificial saliva as electrolyte solution [13]. Electrochemical impedance spectroscopy (EIS) investigations have been conducted using a BioLogic SP150 potentiostat/galvanostat equipped with an EIS module, within the frequency between 100 kHz and 10 MHz. The amplitude of the alternating voltage was of 10 mV. A number of 60 points have been recorded for each spectrum with a logarithmic distribution of 10 points per decade. The experimental data has been fitted using the ZView – Scribner Associates Inc. software and equivalent electrical circuits by applying the Levenberg – Marquardt least squares complex non-linear fitting algorithm [14].

Table 1. Composition of the analysed probes (expressed as percentage by weight)

Probe	Co	Cr	Mo	Mn	W	Si	Fe	C
SLS	59	25	3.5	<1.5	9.5	<1	<1.5	<1.5
SLM	59	25	3.5	<1.5	9.5	<1	<1.5	<1.5

3. Results and Discussion

The protecting layer coating the majority of the base alloys used in dental restorations is a key factor in inhibiting the release of toxic metal ions in the human body. In order to closely re-create the various situations found in the oral cavity in terms of acidity, we have comparatively studied the corrosion behaviour of two samples fabricated using new CAD assisted technologies, on the basis of a Co-Cr alloy.

The linear polarization curves (Tafel plots) of the samples obtained through selective laser melting (SLM) and selective laser sintering (SLS) in Fusayama artificial saliva electrolyte at acidity levels varying from pH 2.5 to 6.5 are depicted in Figure 1.

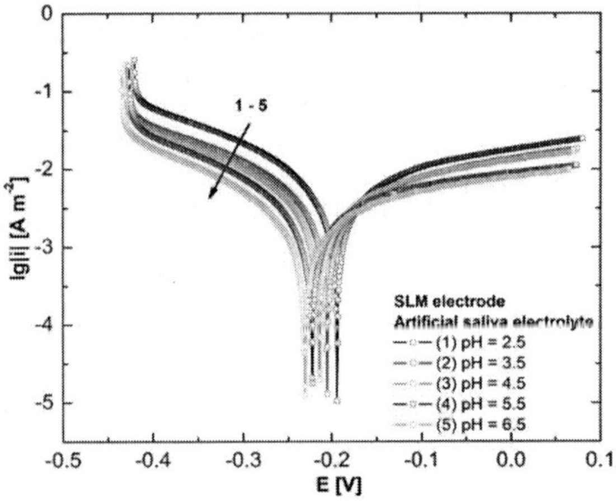

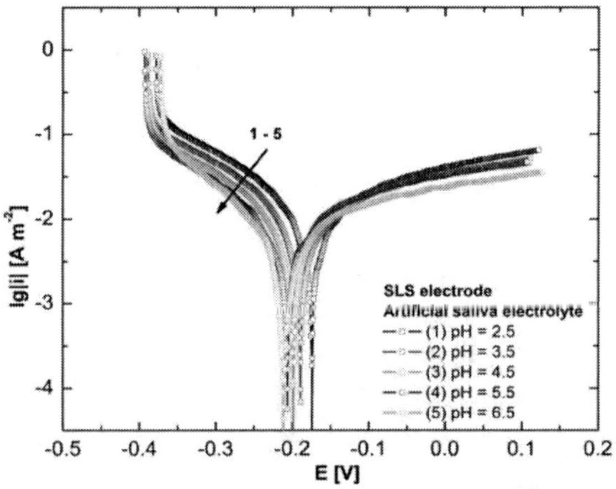

Figure 1. Tafel polarization curves recorded for the SLM (a) and SLS (b) samples in artificial saliva electrolyte at 25°C and various pH values at a scan rate of 1 mV·s^{-1}

The investigated samples exhibit, as expected, very similar corrosion behaviours in the studied media, but nevertheless, some important differences do occur. In the case of the laser melted sample (SLM), the Tafel plot (Fig. 1a) exhibits a strong current increase at potentials lower than 45 mV, most probably due to the reduction of physically dissolved oxygen or the reduction of H^+ ions, taking into account that the artificial saliva solution was in permanent contact to atmospheric oxygen. Scanning towards more positive potentials reveals the transition region between the cathodic and the anodic field characterized by a sudden decrease of the current density at potentials around -0.2 V. It is to be pointed out that the corrosion potential of the studied sample (E_{corr}) shifted towards more negative values with increasing pH, while the corrosion current density (i_{corr}) increased significantly with increasing acidity of the medium. At potentials beyond E_{corr} a relatively large passive region is revealed, with slowly growing currents over no a potential range up to about 1.0 V. No transpassive behaviour has been recorded under these circumstances within the studied electrochemical window, revealing that the metal oxides formed during passivation are adherent and have a highly protective character, as confirmed by other studies in the field [15].

The potentiodynamic characteristics of the sintered probe (Fig. 1b) are highly similar in shape, but closer analysis reveals that in the latter case the magnitude of the corrosion currents is slightly higher. The corrosion potentials shift towards more negative values with acidity increase, suggesting a direct influence of the medium on the corrosion kinetics. However all potentials are situated at more anodic levels compared to the SLM sample witch could be an indication for a slightly higher corrosion affinity. The transpassive region, characteristic for the breakdown of the passive coating, does not come up, leading to the assumption that the formed film is kinetically stable on the studied potential domain.

Specific Tafel parameters like corrosion rate (v_{corr}), and corrosion potential (E_{corr}), polarization resistance (R_p), corrosion current density (i_{corr}), computed for each sample at different values of solution pH are listed in Table 2. The data shows that, the corrosion potential in the case of the SLM sample varies negatively with increasing pH, while the laser sintered (SLS) probe exhibits quite the opposite behaviour, with the most negative corrosion potential recorded at the most acidic pH. The corrosion rate however, is of the same order of magnitude in both samples and remains the highest in the medium with the lowest pH.

Table 2. Polarization parameters for the corrosion process of the studied samples in artificial saliva at different pH values

Sample	*pH*	i_{corr} [μA cm^{-2}]	E_{corr} [mV]	$-b_c$ [mV dec^{-1}]	b_a [mV dec^{-1}]	R_p [kΩ]	$v_{corr} \cdot 10^3$ [mm year^{-1}]
SLM	2.5	0.63	-194	458	232	26.2	8.56
	3.5	0.50	-206	466	225	38.2	6.79
	4.5	0.41	-214	476	212	41.7	5.58
	5.5	0.33	-222	481	198	51.5	4.52
	6.5	0.24	-231	430	188	62.3	3.24
SLS	2.5	0.68	-219	385	187	23.4	9.17
	3.5	0.57	-210	379	182	32.2	7.69
	4.5	0.45	-198	374	171	38.7	6.09
	5.5	0.36	-187	343	165	49.1	4.91
	6.5	0.26	-175	309	145	60.8	3.49

A more in-depth investigation to further clarify the corrosion processes taking place at the alloy-solution interface for both probes implies cyclic voltammetry experiments in artificial saliva at two pH values at a scan speed of 100 mV s^{-1}. The potential window covered a relatively wide potential domain, reaching from a cathodic value of -1.5 V, up to an anodic switching potential of +1.5 V. Figure 2a) depicts the cyclic voltammograms recorded on an electrode obtained through the SLS technique, at pH values of 2.5 and 6.5 respectively. Scanning from the open circuit potential towards positive values reveals a single, relatively broad anodic peak. The peak height recorded at pH 2.5 is about 4.8 A m^{-2} whereas the pH increase leads to a pronounced lowering of the current signal, the value dropping to about only 2.5 A m^{-2}. The increase of the peak height at more acidic values can be an indication for a rate increase of the studied corrosion process, possibly due to the breakdown of the passive film formed at the electrode surface. Finally, at more positive values, the oxygen evolution process triggers a sharp increase of the current density.

Switching the scan direction towards negative potentials reveals a weak cathodic signal at potentials of about +0.25 V in both media, probably due to the reduction of adsorbed oxygen at the electrode surface, followed by a quite large passive region in both media, with no detectable signals up to the potential at which the hydrogen evolution reaction causes a swift current increase at potentials starting with -0.75 V. As expected, the oxygen and hydrogen evolution processes are highly influenced by the pH of the medium, both shifting towards more negative values with increasing acidity.

The superimposed voltammograms of the two samples, namely SLM and SLS, recorded in highly acidic artificial saliva of pH 2.5 at the same sweep rate of 100 mV s^{-1} are suggestively depicted in Fig.2b. Although the shapes of the current-voltage dependencies exhibit very resembling characteristics, indicating that the corrosion mechanisms on the two investigated electrode materials might follow similar patterns, there is one difference that strikes the eye. While the peak height corresponding to the sintered Co-Cr alloy (SLS) is about 4.8 A m^{-2}, the same peak recorded under the same conditions on the laser melted (SLM) electrode is with about 0.4 A m^{-2} shorter.

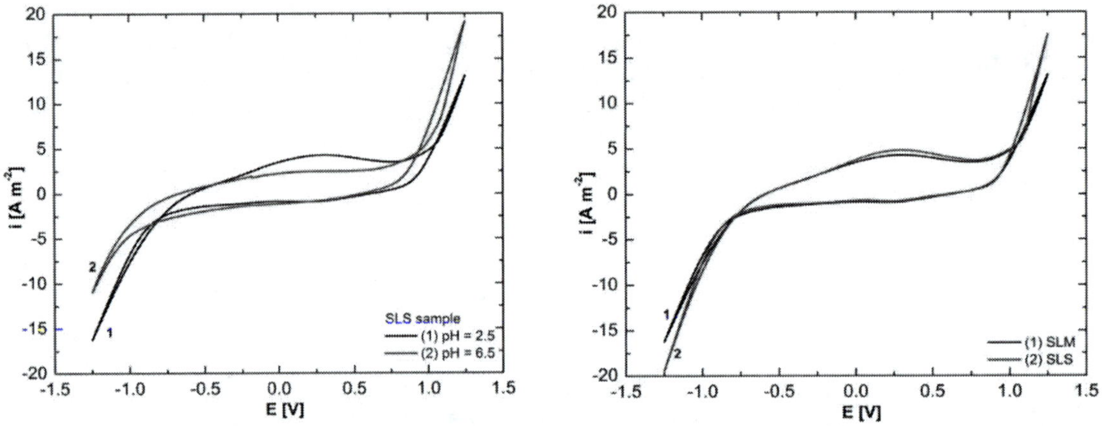

a) b)

Figure 2. Cyclic voltammograms recorded in Fusayama artificial saliva electrolyte at a scan rate of 100 mV s^{-1} on (a) SLS sample at pH 2.5 and 6.5 and (b) SLS and SLM samples at pH 2.5

This may be an indication for a slight increase of the corrosion processes taking place on the sintered sample. On the SLS electrode, the potential window is characterized by a cathodic shift of the potential at witch oxygen evolution occurs. The hydrogen evolution reaction occurs at more positive potentials on the SLS electrode as compared to the SLM probe. Both aforementioned differences may denote a weak catalytic influence of the electrode material over the two processes.

Chronoamperometric investigations on both SLS and SLM samples have been recorded in artificial saliva at various pH, as presented in Fig.3, with the aim of an intrinsic elucidation of the corrosion properties of the studied samples.

Applying a similar constant potential of +250 mV vs. E_{corr} for both, SLS as well as SLM manufactured electrodes and recording the variation of the current density over a timeframe of 15 minutes on the whole array of pH values reveals some new insights regarding the investigated processes. After an initial swift decrease of the corresponding current densities, both samples exhibit an almost constant throughput but there are some mentionable differences. While the SLS sample shows an expectable corrosion behaviour in terms of higher current densities at lower pH, indicating that the more aggressive the medium, the faster the corrosion rate becomes, in the case of the laser melted probe, the transition towards lower current in the case of the pH 5.5 solution occurs only after about 300 seconds after applying the potential step. In the latter case, the different slopes of the constant current region at pH 4.5 and 5.5 respectively, suggest a kinetic hindrance of the investigated processes. The data listed in Table 3 display distinctively higher corrosion current densities recorded for the samples machined by laser sintering compared to those obtained by the laser melting technique, suggesting that the base alloys obtained by the former technology might not efficiently develop a protective and compact passive coating.

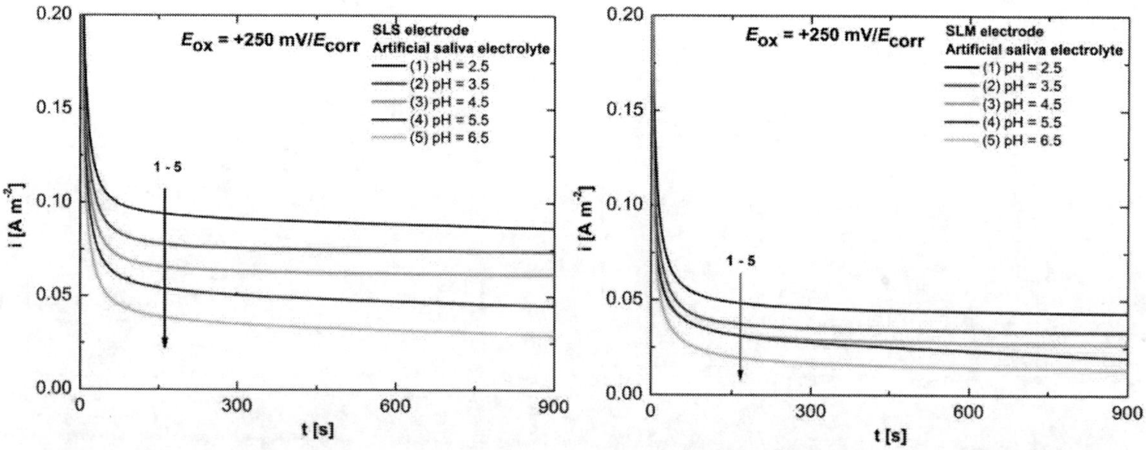

a) b)

Figure 3. Chronoamperometric measurements recorded at +250 mV vs. E_{corr}, for the anodic oxidation of SLS (a) and SLM (b) electrodes in artificial saliva electrolyte at various pH for 15 min

Table 3. Chronoamperometric data for SLS and SLM samples at +250 mV vs. E_{corr} in artificial saliva electrolyte at different pH values

pH	Sample	$i_{oxidation}$ [mA m^{-2}]	Sample	$i_{oxidation}$ [mA m^{-2}]
2.5		86		43
3.5		73		33
4.5	SLS	62	SLM	28
5.5		46		20
6.5		30		13

Electrochemical impedance spectroscopy (EIS), has been employed to get further information regarding the passive film formation as well as the corrosion behaviour of the Co-Cr specimens obtained through CAD aided laser meting (SLM) and laser sintering (SLS) respectively, at the pH values of interest. The obtained results are depicted in Figs. 4 and 5 as Nyquist as well as Bode plots for SLM and SLS specimens respectively.

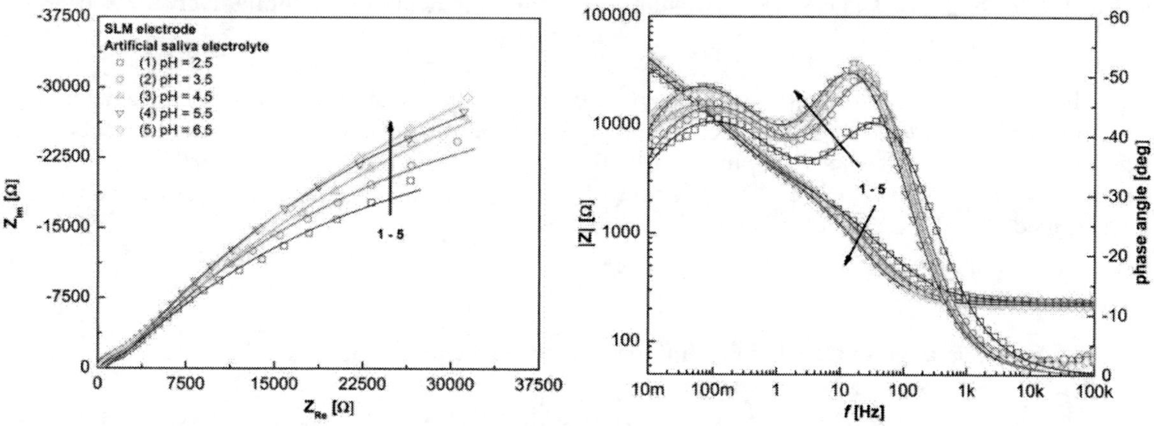

a) b)

Figure 4. Nyquist (a) and Bode plots (b) recorded at corrosion potentials on SLM samples in artificial saliva electrolyte at different pH values

IOP Conf. Series: Materials Science and Engineering **416** (2018) 012036 doi:10.1088/1757-899X/416/1/012036

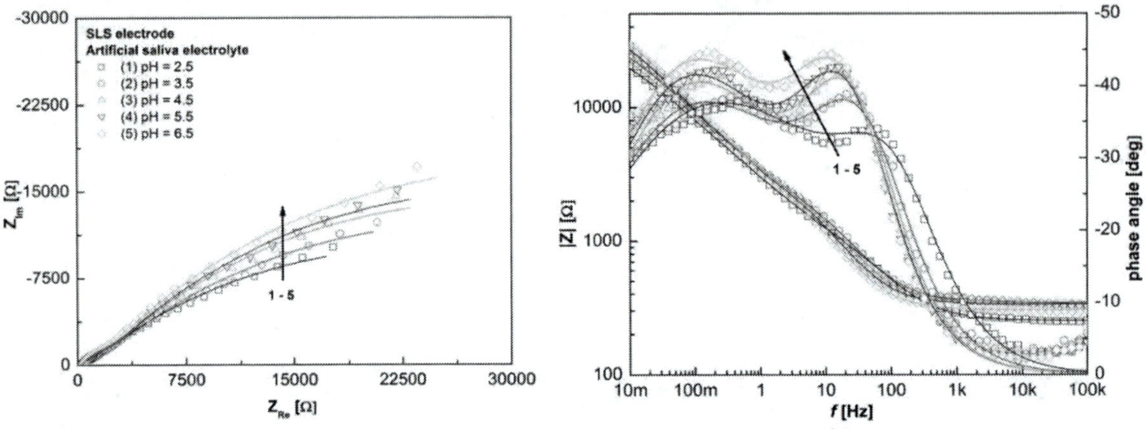

a) b)

Figure 5. Nyquist (a) and Bode plots (b) recorded at corrosion potentials on SLS samples in artificial saliva at different pH values

The metal/electrolyte boundary can be considered as corresponding to an electrical circuit, as a combination of electronic components. The estimation of the obtained EIS data is generally based on an equivalent electrical circuit (EEC), used in order to characterize the processes occurring at the outer layer of the investigated dental alloys. The results have been evaluated and fitted using an EEC as presented in Fig. 6. The presented circuit uses a series combination of the uncompensated solution resistance, R_s, combined with two (*CPE-R*) parallel combinations, each one composed of a constant phase element (*CPE*) and a charge transfer resistance, *R* [16, 17].

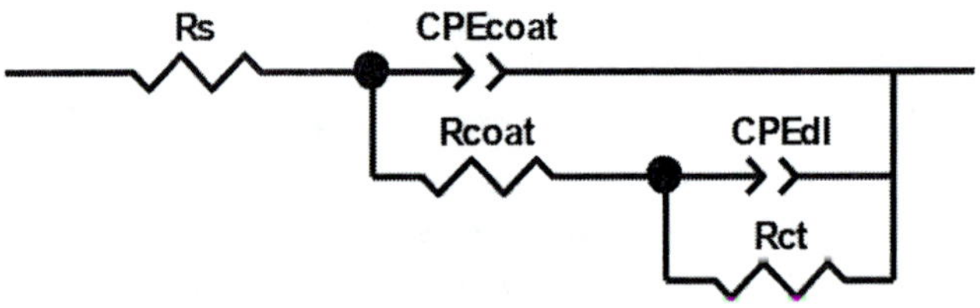

Figure 6. Equivalent electrical circuit model used for generating the fitting data of the recorded EIS spectra

The constant phase element (*CPE*) was placed into the circuit instead of a pure capacitance. It is a non-ideal capacitor used for the characterization of real electrochemical systems taking into account that the passive films at the electrode surface usually shows specific defects and imperfections [18]. This element accounts more precisely for the processes that take place at the electrode/electrolyte interface. The impedance of the *CPE* element is given by Equation (1):

$$Z_{CPE} = 1/T(j\omega)^n \quad (1)$$

in which *T* is a parameter that is contingent to the capacity of the double layer, *n* is an exponent ranged between 0 and 1 which describes the *CPE* angle.

It has been shown that the corrosion processes taking place at the interface between the studied Co-Cr dental alloys and artificial saliva, imply the formation and growth of an adhering passive coating consisting of several types of metal oxides [19]. The physical correspondence of the EEC displayed in Fig. 6 consists of the association of the alloy/electrolyte interface ($CPE_{dl}R_{ct}$) and the passive coating/electrolyte interface ($CPE_{coat}R_{coat}$) [17].

The computed results obtained after fitting are depicted as solid lines in Figs. 5 and 6 while the individual points represent the measured data. Calculated parameters for the elements comprising the EEC are detailed in Table 4 corresponding to the corrosion process of dental alloys obtained through the SLM and SLS technologies respectively, in artificial saliva. The values of the double layer capacity have also been calculated.

Table 4. Calculated values of the EEC elements for the corrosion process of SLM and SLS samples in artificial saliva

Sample	*pH*	R_s [Ω cm²]	coating/electrolyte interface			alloy/electrolyte interface			$C_{dl} \cdot 10^4$ [F cm⁻²]	$Chi^2 \cdot 10^3$
			$T \cdot 10^5$ [F cm⁻² sⁿ⁻¹]	*n*	R_{coat} [kΩ cm⁻²]	$T \cdot 10^4$ [F cm⁻² sⁿ⁻¹]	*n*	R_{ct} [kΩ cm⁻²]		
SLM	6.5	21.7	2.49	0.85	3.42	0.87	0.68	147	2.77	1.36
	5.5	22.1	2.20	0.83	3.27	0.99	0.65	130	3.85	1.17
	4.5	23.1	1.90	0.81	2.88	1.05	0.62	117	4.66	1.56
	3.5	23.3	1.61	0.79	2.51	1.10	0.60	101	5.50	1.64
	2.5	24.0	1.44	0.78	2.25	1.19	0.58	87.6	6.38	0.75
SLS	6.5	30.3	3.02	0.82	2.70	1.20	0.62	61.7	4.03	3.37
	5.5	33.4	2.57	0.81	2.35	1.24	0.61	56.5	4.42	3.78
	4.5	34.6	2.12	0.80	2.05	1.30	0.58	53.4	5.40	4.86
	3.5	34.2	2.10	0.79	1.71	1.32	0.54	50.9	6.56	4.01
	2.5	31.2	1.70	0.77	1.39	1.42	0.52	47.1	7.88	1.76

Highly similar shapes of the electrochemical impedance spectra displayed in Figs. 4 and 5, as well as the utilization of the same equivalent circuit shown in Fig. 6, used in order to accurately fit the experimental data obtained for the Co-Cr samples finished through laser melting (SLM) and laser sintering (SLS) in artificial saliva, is suggestive evidence that the processes taking place at the metal/electrolyte interface follow the same pattern, namely the coating of the metal surface with a passive oxide layer [20].

The values of for chi-square at an order of magnitude of 10^{-3} are an indication for a highly accurate correlation between the experimental data and the EEC model parameters [21]. Both specimens show R_{ct} values of the order of 10^6 kΩ cm⁻² over the whole array of studied pH values, suggesting that the high corrosion resistances [17] are a result of a compact layer coating the whole exposed surface and thus kinetically hindering oxidation processes that could cause corrosion of the base alloy.

As expected, corrosion resistance of both specimens follows a descending pattern along with the rise of the solution acidity. The charge transfer resistance, R_{ct}, that characterizes the dental alloy/electrolyte interface is inversely related to the corrosion rate, *i.e.* the higher the resistance, the slower the process of corrosion gets, hence the base material is more resistant towards corrosion in the given medium. Twice as big R_{ct} values exhibited by the laser melted samples (SLM) compared to SLS suggest a higher corrosion resistance of the former obtained alloy. High values, of about 10^3 Ω, of the charge transfer resistance R_{coat}, that quantifies the formation of the oxide layer on the alloy surface are an indication of a strong adhesion of the formed oxide film, while protecting efficiently against the ongoing corrosion processes. These values, being significantly higher for the SLM specimen, come to confirm the better corrosion resistance of the latter. All of the above, corroborated to the greater values of the double layer capacity C_{dl} calculated in the case of the laser melted probe come to further validate the higher degree of protection conferred by the passive coating in the latter case.

4. Conclusion

The studied Co-Cr dental alloys manufactured by CAD-aided laser melting and laser sintering exhibit, as expected, fairly similar electrochemical responses while studied in artificial saliva with variable acidity. The attempted electrochemical studies revealed satisfying corrosion inhibition offered by the passive coating formed on the alloy surface. A noteworthy difference lies in the higher degree of resistance towards corrosion exhibited by the Co-Cr alloy manufactured using the selective laser melting technology in Fusayama artificial saliva as compared to the laser sintered specimen. The corrosion rate varied towards higher values with increasing acidity in both cases. The higher corrosion resistance exhibited by the SLM specimen over the entire range of pH values between 2.5 and 6.5 comes to confirm our previous results obtained on the afore-mentioned alloys [22].

The coating of the alloy surface with a compact passive oxide layer, kinetically hindering further corrosion phenomena was outlined and elucidated by electrochemical impedance spectroscopy. The growth of the protecting layer being favoured on the surface of the SLM sample results in a greater protection thus a higher resistance towards corrosion of the probe crafted by laser melting.

5. References

[1] Naumova E A, Sandulescu T, Bochnig C, Al Khatib P and Lee W K 2014 *Sci. Rep.* **4** 1-9
[2] Hans R, Thomas S, Garla B, Dagli R J and Hans M K 2016 *Scientifica* **2016** 1-6
[3] Anusavice K J, Shen C and Rawls H R 2013 Phillips' science of dental materials 12th ed., Saunders, St. Louis
[4] Al Jabbari Y S, Koutsoukis T, Barmpagadaki X and Zinelis S 2014 *Dent. Mat.* **30** e79-e88
[5] Mengucci P, Barucca G, Gatto A, Bassoli E, Denti L, Fiori F, Girardin E, Bastinoni P, Rutkowski B and Czyrska-Filemonowicz A, 2016 *J. Mech. Beh. Biomed. Mater.* **60** 106-17
[6] Béreš M, Silva C C, Sarvezuk P W C, Wu L, Antunes L H M, Jardini A L, Feitosa A L M, Žilková J, de Abreu H F G and Filho R M 2018 *Mat. Sci. Eng.* **A 714** 36–42
[7] Alageel O, Abdallah M N, Alsheghri A, Song J, Caron E and Tamimi F, 2017 *J. Biomed. Mat. Res. B: Appl. Biomat.* **00B** 1-12
[8] Puskar T, Lapcevic A, Arandjelovic S, Radulovic S, Budak I, Vukelic D J and Jevremovic D 2015 *Metalurgija* **54** 481-84
[9] Kim E-C, Kim M-K, Leesungbok R, Lee S-W and Ahn S-J 2016 *Dent. Mater.* **32** 1394-1405
[10] Pan Y, Jiang L, Lin H and Cheng H 2017 *Dent. Mater. J.* **36** 82-87
[11] Kim H R, Kim Y K, Son J S, Min B K, Kim K H and Kwon T Y 2016 *Mater. Lett.* **178** 300-03
[12] Pang A M, Shetty M, Prasad D K and Kanathila H 2016 *J. Clin. Diagn. Res.* **10** ZC23-ZC27
[13] Geis Gerstorfer J and Weber H 1985 *Dtsch. Zahnärztl.* **40** 87–91
[14] Porojan L, Savencu C E, Costea L V, Dan M L and Porojan S D 2018 *Int. J. Electrochem. Sci.* **13** 410-23
[15] Hsu R W W, Yang C-C, Huang C-A and Chen Y-S 2005 *Mat. Chem. Phys.* **93** 531-38
[16] Fojt J, Joska L, Malek J and Sefl V 2015 *Mater. Sci. Eng.* C **56** 532-37
[17] Furko M, Jiang Y, Wilkins T A and Balazsi C 2016 *Mater. Sci. Eng.* C **62** 249-59
[18] Iacoban S, Mareci D, Bolat G, Munteanu C and Souto R 2015 *Metall. Mat. Trans.* B **46** 1011-21
[19] Metikoš-Huković M, Pilić Z and Babić R and Omanović D 2006 *Acta Biomat.* **2** 693-700
[20] Brett C M A and Trandafir F 2004 *J. Electroanal. Chem.* **572** 347–54
[21] Saji V S and Choe H C 2009 *Trans. Nonferrous Met. Soc. China.* **19** 785-90
[22] Savencu C E, Costea L V, Dan M L and Porojan L 2018 *Int. J. Electrochem. Sci.* **13** 3588-3600

IOP Conf. Series: Materials Science and Engineering **416** (2018) 012037 doi:10.1088/1757-899X/416/1/012037

The Anisotropy Effect of Closed-Cell Polyisocyanurate (PIR) Rigid Foam under Quasi-Static Compression Loads

E Linul[1,*], L Marsavina[1] and M Georgescu[2]

[1]Department of Mechanics and Strength of Materials, Politehnica University of Timisoara, 1 Mihai Viteazu Avenue, 300 222 Timisoara, Romania
[2]Department of Metal Construction and Mechanics of Construction, Politehnica University of Timisoara, 1 Ioan Curea str., 300 224 Timişoara, Romania

[*]Corresponding author: emanoil.linul@upt.ro

Abstract. Lightweight polymeric cellular materials are increasingly used in many industrial applications because of their height mechanical and energy absorption properties. This paper investigates the effect of foam anisotropy under quasi-static compression tests. A 35 kg/m^3 closed-cell rigid Polyisocyanurate (PIR) structural foam was tested in the experimental programme with a loading speed of 10 mm/min. Experimental results show that the investigated foam presents a highlighted anisotropic behavior and the main mechanical properties such as Young's modulus, yield stress, plateau stress and onset strain of densification varies with changing the loading direction. In addition, during compression, the energy absorption performance of PIR foam is strongly affected by the anisotropy.

1. Introduction

In comparison with fully dense solid materials, cellular materials such as polymeric [1-3] and metallic [4-8] foams exhibit low density with good shear and fracture strength, excellent stiffness-to-weight ratio, high porosity and exceptional ability to absorb impact energy at almost constant pressure. Therefore, due to their comprehensive physical and mechanical properties, cellular materials have expanded their application in different fields from packaging to aerospace industries [4]. In the last three decades, an extensively number of analytical [9-11], numerical [12-15] and experimental [16, 17] investigations were used in order to determine tensile [18, 19], compression [20], shear [21, 22], bending [23-25] and fracture [26, 27] properties of PU foams, under static and impact [28-30] loading conditions. In these studies the influence of density, anisotropy, loading speed, temperature, solidification rate, microstructure, size effect, constituent material and different fracture criteria were investigated.

On the other hand, in addition to the PU foams, Polyisocyanurate (PIR) rigid foams are a relatively new class of foam materials with low densities and high mechanical, physical, thermal and acoustic properties. PIR foams offer potential significant energy absorption performances in lightweight and stiff structures starting from the automotive, aerospace and especially building industry, up to many other specialized engineering applications. Only few research studies are related to manufacturing [3, 31, 32] and mechanical characterization [11, 15] of PIR foams. Studies on understanding the shear flow in PIR foam core-glass/epoxy face sheet sandwich composites by varying geometry (core thicknesses) and skin to core weight ratios were performed by Reddy [33]. Mode I fracture toughness in the foam rise direction has been experimentally characterized by Andersons et al. [34] for polyisocyanurate foams produced using polyols derived from rapeseed oil and filled with a montmorillonite nanoclay.

Most of these studies are focused on the determination of physical, thermal and acoustic properties and less on mechanical behavior. As a complement to recent studies, this paper investigate the effect of anisotropy of a rigid PIR foam on the main mechanical properties, under quasi-static compression tests.

2. Materials and experimental tests setup

2.1. Materials

A large closed-cell rigid polyisocyanurate (PIR) foam plate of 320 mm × 320 mm × 54 mm, having density of 35 kg/m^3 was produced (see Fig. 1a). In order to investigate the anisotropy effect, the resulted foam samples were tested under three different loading directions as follows: an "out-of-plane" loading direction-named as *direction (1)*, and two "in-plane" loading directions-named *direction (2)* and *direction (3)*, according to Fig. 1b. Five cubic samples (50 mm × 50 mm × 50 mm) were provided for each loading condition.

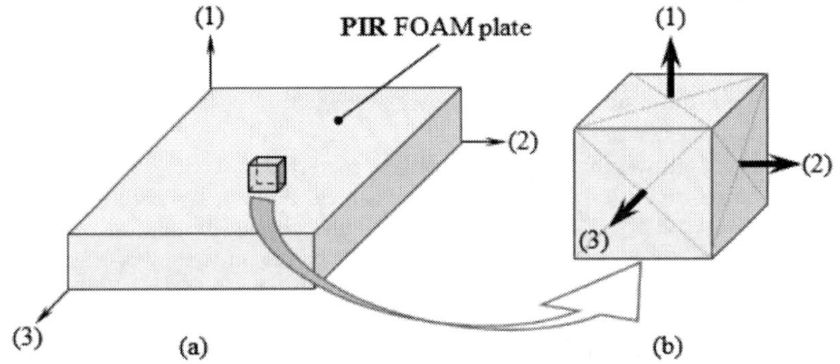

Figure 1. Produced PIR foam plate (a) and different loading directions (b)

2.2. Experimental program

The experimental tests were carried out on a universal Zwick Proline Z005 quasi-static testing machine having the maximum load cell of 5 kN. The tests were performed at room temperature (25°C), with a loading speed of 10 mm/min, according to ASTM D 1621-00 standard [35]. The loading was applied slowly, continuously and without shocks.

3. Results and discussion

Figure 2 presents the typical compressive stress (σ)-strain (ε) curves for three specified loading directions (one out-of-plane and two in-plane), up to approximately 90% engineering strain.

Regardless of the loading direction, all investigated PIR specimens show the classic stress-strain behavior of polymeric foams, which consist of three different regions [36-39]: (i) a linear-elastic region (from which the compressive modulus-E is determined) where the stress rises sharply up to a specified yield point; followed by (ii) a plateau region (where the plateau stress-σ_p is determined), over a large deformation; and ends with (iii) an abrupt increase in stress named onset strain of densification (OSD).

From Fig. 2 it is observed that only in the direction (1) a peak yield stress is obtained, located immediately after the linear-elastic region. On the other two directions (2) and (3) there is a smooth transition from the elastic region to the plateau region. Also, an extended and approximately constant plateau with some small stress drops (when plastic collapse of foam occur) is highlighted in direction (1), while a slight strengthening/hardening for the other two loading directions is obtained. The OSD of PIR foam represents the start of the cell-wall interactions, which enhance the compressive resistance of the solid material of which the foam is made. It was observed that the OSD depends on several factors: foam material, loading speed, testing temperature [40] etc. According to the Fig. 2 and data summarized in Table 1 it seems that OSD also depends on the foam anisotropy, this property decreasing from a value of 66.43% (for direction (1)) to a value of 58.81% (for direction (3)).

IOP Conf. Series: Materials Science and Engineering **416** (2018) 012037 doi:10.1088/1757-899X/416/1/012037

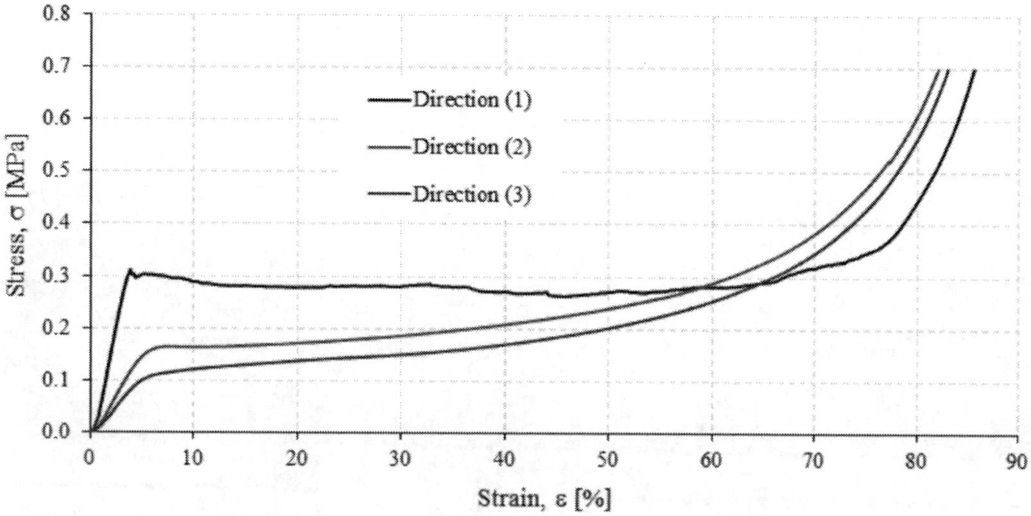

Figure 2. Stress-strain curves of investigated PIR foam

The anisotropic behavior of the PIR foam is also highlighted by the average values of the compression characteristics presented in Table 1. The values of the mechanical characteristics E, σ_y, σ_p are superior in direction (1) compared to the directions (2) and (3) about 3 times for the modulus of elasticity, respectively 2 times the values of the yield and plateau stresses.

Table 1. The mechanical properties of investigated PIR foams under different loading directions.

Loading direction	Geometrical parameters			Mechanical properties			
	b_1 [mm]	b_1 [mm]	h [mm]	E [MPa]	σ_y [MPa]	σ_p [MPa]	ε_D [%]
(1)				10.42	0.32	0.28	66.43
(2)	50	50	50	3.71	0.16	0.18	60.06
(3)				2.88	0.12	0.15	58.81

According to Fig. 3, different collapse mechanisms were observed during quasi-static compression tests. Under loading direction (1) the tested specimen present only one straight deformation band, approximately located at the same level on each side (near the middle of the specimen height) and perpendicular to the loading direction (see Fig. 3a), while the specimen loaded on direction (3) shows two deformation bands on each side, arbitrarily located on the specimen height. Otherwise, the direction (2) is deformed similarly to (1), but two of the sample faces presents curvilinear deformation surfaces.

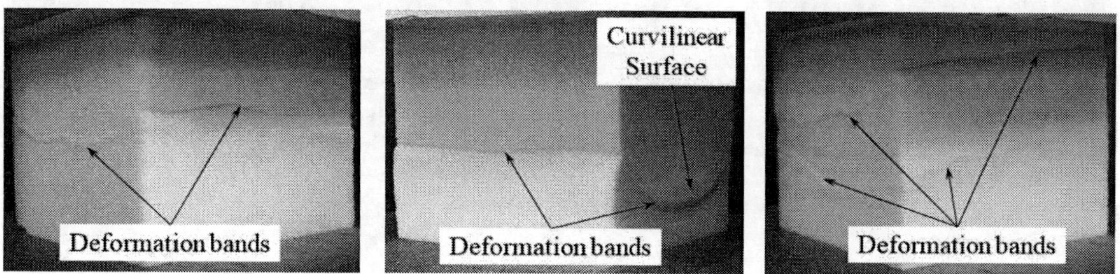

a) Direction (1) b) Direction (2) c) Direction (3)

Figure 3. Deformed specimens under different loading directions

The energy absorption (W) was calculated using Eq. (1) [41-46], and a variation of W with strain for different loading directions is shown in Fig. 4a.

$$W = \int_0^{\varepsilon} \sigma d\varepsilon \tag{1}$$

As it can be seen from Figure 4b, the energy absorption at onset strain of densification (W_D) decreases by up to 50% for direction (3) compared to direction (1).

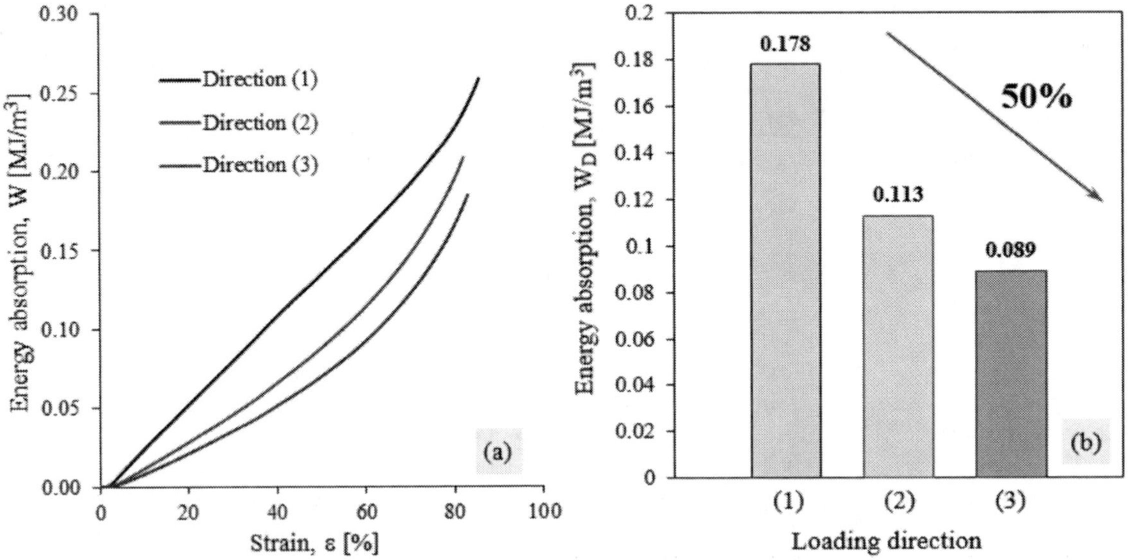

Figure 4. Energy absorption-strain curves (a) and W_D values (b) for different loading directions

The biggest difference (of 67%) in terms of energy absorption is obtained in the area of small strains, where the plateau region begins (around 10% strain); this difference decreasing significantly up to 27% at large strains (80% strain). The W difference, between different loading directions, can be observed most easily by analyzing the data from Table 2 and Fig. 5.

Table 2. The energy absorption values of investigated PIR foams for different loading directions.

Loading direction	Energy absorption at different strains, W [MJ/m³]							
	10%	20%	30%	40%	50%	60%	70%	80%
(1)	0.024	0.052	0.080	0.108	0.135	0.162	0.192	0.227
(2)	0.012	0.028	0.046	0.066	0.088	0.114	0.147	0.195
(3)	0.008	0.021	0.036	0.051	0.070	0.093	0.122	0.166

It is very important to correctly position the closed-cell PIR foam in practical applications because it presents a strong anisotropic behavior [47]. Therefore, by simply placing the foam in the correct position in the designed resistance element, the mechanical properties will increase by over 50%.

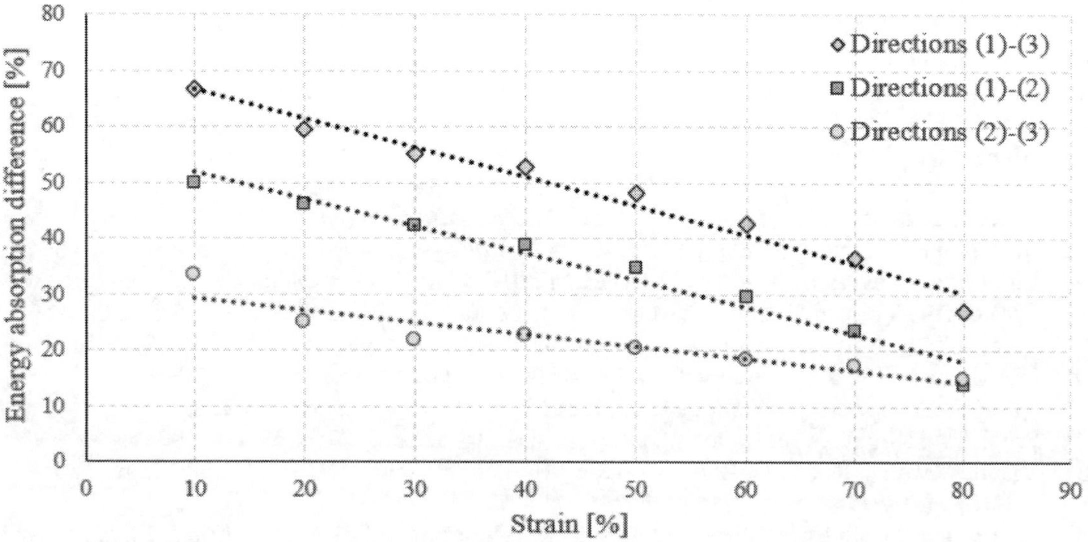

Figure 5. W difference-strain variations for different loading directions

4. Conclusions

In order to determine the effect of anisotropy, the quasi-static compression properties of a closed-cell Polyisocyanurate (PIR) rigid foam (35 kg/m^3 density) was investigated under different loading directions (directions (1), (2) and (3)).

It was obtained that the PIR foam shows higher mechanical properties in out-of-plane loading direction (direction (1)) than in-plane (directions (2) and (3)), about 3 times for E, respectively 2 times the values of σ_y and σ_p (see Table 1).

Furthermore, the energy absorption capacity of PIR foam is strongly affected by the anisotropy. The biggest energy absorption difference (67% for low strains and 27% for high strains) was obtained between directions (1) and (3), and the smallest difference (34% for low strains and 15% for high strains) between directions (2) and (3)-see Table 2 and Fig. 5.

5. References

[1] Marsavina L, Constantinescu DM, Linul E et al. 2016 *Eng. Fract. Mech.* **167** 68-83
[2] Andersons J, Kirpluks M, Stiebra L, Cabulis U 2015 *Theor. Appl. Fract. Mech.* **78** 8-14
[3] Hejna A, Kirpluks M, Kosmela P et al. 2017 *Ind. Crop. Prod.* **95** 113-125
[4] Ashby MF, Evans AG, Fleck NA, Gibson LJ, Hutchinson JW, Wadley HNG, 2000 *Metal Foams: A Design Guide*, Butterworth-Heinemann
[5] Linul E, Marsavina L, Kováčik J 2017 *Mat. Sci. Eng.-A Struct.* **690** 214-224
[6] Marsavina L, Kovacik J, Linul E 2016 *Theor. Appl. Fract. Mech.* **83** 11-18
[7] Linul E, Movahedi N, Marsavina L 2017 *Compos. Struct.* **180** 709-722
[8] Movahedi N and Linul E 2017 *Mater. Lett.* **206** 182-184
[9] Birsan M, Sadowski T, Marsavina L et al. 2013 *Int. J. Solids. Struct.* **50** 519-530
[10] Marsavina L, Berto F, Negru R et al. 2017 *Theor. Appl. Fract. Mech.* **91** 148-154
[11] Andersons J, Kirpluks M, Stiebra L, Cabulis U 2016 *Mater. Design* **92** 836-845
[12] Şerban DA, Linul E, Voiconi T, Marsavina L, Modler N 2015 *Iran. Polym. J.* **24** 515-529
[13] Linul E and Marsavina L 2011 *Iran. Polym. J.* **20(9)** 736-746
[14] Marsavina L, Linul E, Voiconi T, Negru R 2016 *IOP Conf. Ser: Mater. Sci. Eng.* **123(1)** 012060
[15] Wang YC and Foster A 2017 *Fire Saf. J* **90** 1-14
[16] Negru R, Marsavina L, Voiconi T et al. 2015 *Theor. Appl. Fract. Mech.* **80** 87-95
[17] Voiconi T, Negru R, Linul E et al. 2014 *Frattura ed Integrità Strutturale* **30** 101-108
[18] Marsavina L, Constantinescu DM, Linul E et al. 2014 *Eng. Fract. Mech.* **129** 54-66

[19] Aliha MRM, Linul E, Bahmani A, Marsavina L 2018 *Polym. Test.* **67** 75-83
[20] Linul E, Şerban DA, Marsavina L, SadowskiT 2017 *Arch. Civ. Mech. Eng.* **17(3)** 457-466
[21] Marsavina L, Constantinescu DM, Linul E et al. 2015 *Eng. Fail. Anal.* **58** 465-476
[22] Apostol DA, Stuparu F, Constantinescu DM et al. 2016 *Materiale Plastice* **53(4)** 685-688
[23] Apostol DA, Stuparu F, Constantinescu DM et al. 2016 *Materiale Plastice* **53(2)** 280-282
[24] Linul E and Marsavina L 2015 *P. Romanian Acad. A* **16(4)** 522-530
[25] Voiconi T, Linul E, Marsavina L et al. 2014 *Solid State Phenomena* **216** 116-121
[26] Marsavina L, Linul E, Voiconi T et al. 2015 *Frattura ed Integrita Strutturale* **34** 444-453
[27] Linul E, Marsavina L, Sadowski T, Kneć M 2012 *Solid State Phenomena* **188** 205-210
[28] Şerban DA, Voiconi T, Linul E, Marsavina L, Modler N 2015 *Materiale Plastice* **52(4)** 537-541
[29] Ciecierska E, Jurczyk-Kowalska M, Bazarnik P et al. 2016 *Compos. Struct.* **140** 67-76
[30] Marsavina L, Linul E, Voiconi T, Sadowski T 2013 *Polym. Test.* **32** 673-680
[31] Kurańska M, Prociak A, Cabulis U et al. 2017 *Ind. Crop. Prod.* **95** 316-323
[32] Prociak A, Kurańska M, Cabulis U, Kirpluks M 2017 *Polym. Test.* **59** 478-486
[33] Vishnu Vardhan Reddy G, Ramya Suresh ME, Padmanabhan K 2017 *Mater Today: Proceedings* **4(8)** 9170-9178
[34] Andersons J, Cabulis U, Stiebra L, Kirpluks M, Sparnins E 2017 *Int. J. Fract.* **205(1)** 111-118
[35] ASTM D 1621-00: *Standard test method for compressive properties of rigid cellular plastics*
[36] Linul E, Şerban DA, Marsavina L, Kovacik J 2017 *Fatigue & Fracture of Engineering Materials & Structures* **40(4)** 597-604
[37] Linul E, Marsavina L, Kovacik J, Sadowski T 2017 *P. Romanian Acad. A* **18(4)** 361-369
[38] Kovácik J, Jerz J, Mináriková N et al. 2016 *Frattura ed Integrita Strutturale* **36** 55-62
[39] Voiconi T, Linul E, Marsavina L et al. 2014 *Key Engineering Materials* **601** 254-257
[40] Linul E, Voiconi T, Marsavina L, Sadowski T 2013 *Journal of Physics: Conference Series* **451** 012002
[41] Linul E, Serban DA, Voiconi T et al. 2014 *Key Engineering Materials* **601** 246-249
[42] Linul E, Movahedi N, Marsavina L 2018 *J. Alloy. Compd.* **740** 1172-1179
[43] Taherishargh M, Linul E, Broxtermann S, Fiedler T 2018 *J. Alloy. Compd.* **737** 590-596
[44] Movahedi N, Linul E, Marsavina L 2018 *J. Mater. Eng. Perform.* **27(1)** 99-108
[45] Marsavina L, Cernescu A, Linul E, Scurtu D, Chirita C 2010 *Materiale Plastice* **47(1)** 85-89
[46] Linul E, Movahedi N, Marsavina L 2018 *Materials* **11(4)** 554
[47] Linul E, Serban DA, Marsavina L 2018 *Phys. Mesomech.* **21(2)** 178-186

Acknowledgment
This work was supported by a grant of the Romanian Ministery of Research and Innovation, CCCDI – UEFISCDI, project number PN-III-P1-1.2-PCCDI-2017-0391 / CIA_CLIM - Smart buildings adaptable to the climate change effects, within PNCDI III". Dr. E. Linul work was supported (partial supported) by research grants PCD-TC-2017.

IOP Conf. Series: Materials Science and Engineering **416** (2018) 012038 doi:10.1088/1757-899X/416/1/012038

Evaluation of the mechanical properties and failure mechanism of fibres formed in municipal wastewater systems

D Micota[1], S V Galatanu[1,*], L Marsavina[1] and S Muntean[1,2]

[1] Politehnica University of Timisoara, 1 Mihai Viteazu Blvd., 300222, Timisoara, Romania
[2] Romanian Academy – Timisoara Branch, 24 Mihai Viteazul Blvd., 300223, Timisoara, Romania

[*]Corresponding author: sergiu.galatanu@upt.ro

Abstract. The rainwater axial pumps are specially designed to operate in municipal wastewater systems. The fibres and wastes are discharged during storms with this type of pumps. Several problems have occurred in service due to the fibres clogging up the impeller. Therefore, the chemical compositions of the waste, microscopic analyses, tensile and shearing tests were performed on the debris collected into the rainwater pumps in order to determine it structure and mechanical properties. The failure mechanism of fibres was also determined. It was confirmed that the tensile and shear strength of wet fibres are higher than the dry ones. Also, the value obtained by shearing tests is higher than the values obtained by tensile tests for both wet and dry fibres. This approach represents a useful engineering tool for determination of mechanical properties and failure mechanism of fibres formed in the municipal wastewater systems that lead to clogging up the impeller.

1. Introduction

The wastewater treatment plant (WTP) of Timisoara was rehabilitated during 2009 – 2011 and it was commissioned in March 2011. The Timisoara's WTP was designed to provide services for 440000 inhabitants. Today, more than 350000 inhabitants (97% from the total number of Timisoara's inhabitants), are connected to the city's sewerage system. The annual average volume of the wastewater treated in Timisoara's WTP is more than 42.1 million m^3. Up to 10% of this volume is rainwater. The maximum capacity of Timisoara's WTP is 3 m^3/s. The additional flow over the maximum capacity of the WTP from rains and storms is discharged in Bega River using seven axial rainwater pumps with maximum flow rate of 4 m^3/s on each pump. Two rainwater pumps have had catastrophic failure that took them out of operation for several weeks, Figure 1.

Nowadays, the wastewaters contain a higher amount of synthetic cloth and artificial fibres such as tissues, wipes, and dishcloths. The most common solid parts found in municipal wastewater are organic and often consist of long and stringy shapes, such as fibres, Figure 2. Also, the infiltration sand is trained by the wastewater stream [1].

The major problem is the catastrophic failure of the rainwater pump impeller. The failure is caused by clogging up the pump impeller, Figure 3. For this type of axial rainwater pumps, the gap between impeller blades and casing is clogged leading to the radial deflection of the impeller. The impeller deflection is followed by mechanical friction between it blades and casing producing the catastrophic failure of the rainwater pumps.

IOP Conf. Series: Materials Science and Engineering **416** (2018) 012038 doi:10.1088/1757-899X/416/1/012038

Isono et al. [2] have conducted an experimental study on the pump clogging phenomena showing that the length of the string significantly affects the probability of passing. The reason for clogging is versatile [3-5]. Any object that fits into the municipal wastewater pipe systems can be involved. Gerlach et al. [6] have shown that the hydraulic performance of the wastewater pumping station discovered that clogging not only leads to pump failures but also reduces the attainable flow rates greatly. Additionally, clogging results is found in a loss of energy-efficiency.

Figure 1. Fibres accumulation on the blades of the impeller of the rainwater pumps. One impeller blade is missing due to the catastrophic failure.

Figure 2. Typical rag ball removed from the impellers of the rainwater pumps.

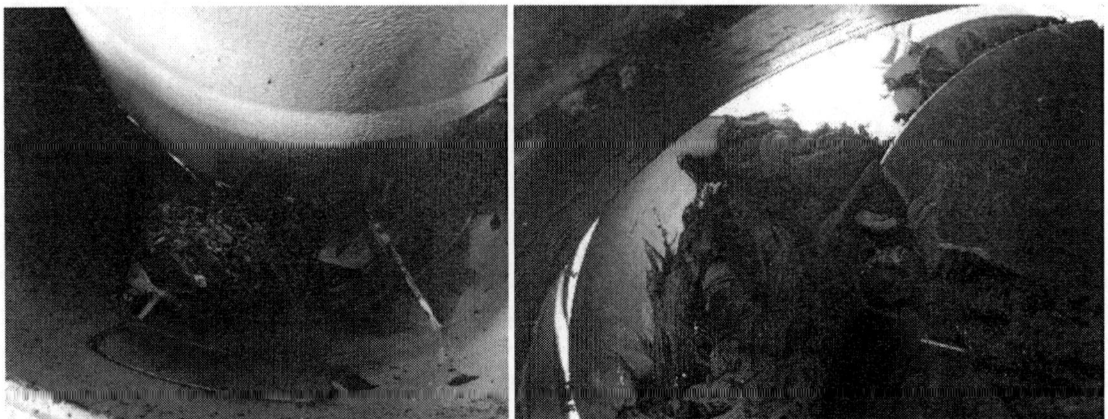

Figure 3. Fibres accumulation in the blade regions of pluvial pumps: stay vanes (left) and impeller (right).

The purpose of this paper is to determine the properties of the fibres accumulated in the rainwater pumps that lead to clogging up the impeller. The paper includes laboratory investigation and test results concerning wastes collected from the municipal wastewater systems. This part is divided into three subchapters. The first two subchapters present the procedures and results for determining the chemical composition and microscopic examination of the waste and next subchapter present the tensile and shear tests performed on the fibres and the results of investigations. The main conclusions are summarized in the last section.

2. Laboratory investigations and test results

The chemical compositions, microscopic analyses, tensile and shear tests were performed in order to determine the mechanical properties of the fibres collected into the rainwater pumps in order to investigate the causes that lead to catastrophic failure of the rainwater pumps. The mechanism of breaking the fibres collected from the rainwater pumps is proposed based on the results obtained from the microscopic analysis of the breaking zones. It was noted that our results are supported by Hearles et al. [7].

2.1. Chemical compositions

The chemical composition of the municipal wastes from the rainwater pump was determined in two steps. The initial weight of investigated municipal waste was 16.5 g. First, the waste was immersed into 100 ml distilled water during 24 hours to determine the amount of water-soluble substances. Next, the waste was sunk into 100 ml petroleum ether during 24 hours to determine fat-soluble substances. At the end of the procedure, it was recorded that the municipal waste did not change its appearance.

The results obtained for the chemical composition of the municipal waste is presented in Figure 4.

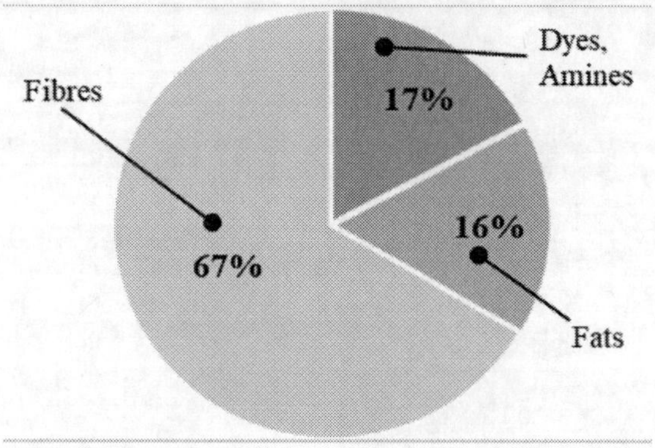

Figure 4. Chemical composition of the waste.

Following the procedure, it was noticed that 17% of the initial mass of the waste is composed by dyes and amines dissolved into the distilled water, 16% are fats selected into petroleum ether and 67% are fibres.

2.2. Microscopic examination

The microscopic examination was performed with the optical microscope. The microscopic analysis was considered in the breaking zones of the fibres (wastes) on transversal and longitudinal directions, respectively. The breaking mechanism of the fibres includes that the interaction between fibrils will be very weak and they will be separated independently. After breaking of all fibrils the two ends are separated, followed by the break. Figure 5 shows the schematic representation of independent fibrillary break.

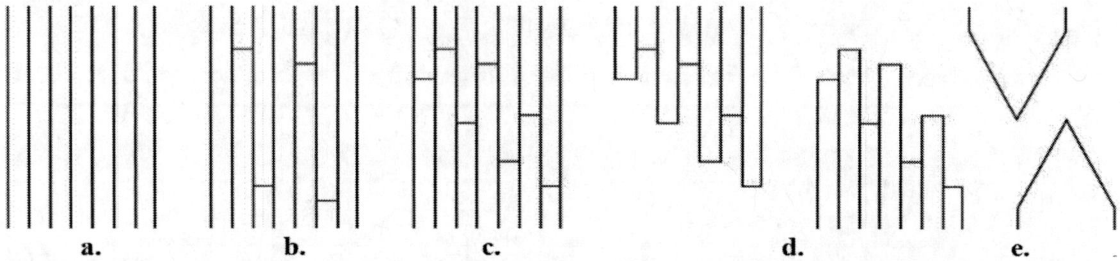

Figure 5. Schematic representation of independent fibrillary break. (a) Structure of separate fibrils, only weakly linked, (b) Under sufficient tension, fibrils begin to break, (c) Finally all fibrils have broken and the ends can separate, (d) Two broken ends, (e) Possible collapse to tapering ends, [7].

Figure 6 shows a microscopic examination of the waste, in which the fibres breaking mechanism can be seen. The fibres are formed from assembly of crystalline fibrils. Fibrils break down progressively after which the remaining fibres cannot longer supports the load, and the failure occurs in the direction of the main stresses at 45 degrees. The fibres breaking area are presented in Figure 7 while a detail of the fats from the fibres is evidenced in Figure 8.

The microscopic examinations performed on the fibres in transversal direction, respectively in longitudinal direction indicate the structure of a composite material consisting from fibres, sand, stones and fats, Figure 9.

Figure 6. Microscopic analysis of the waste: fibres breaking mechanism for shear tests

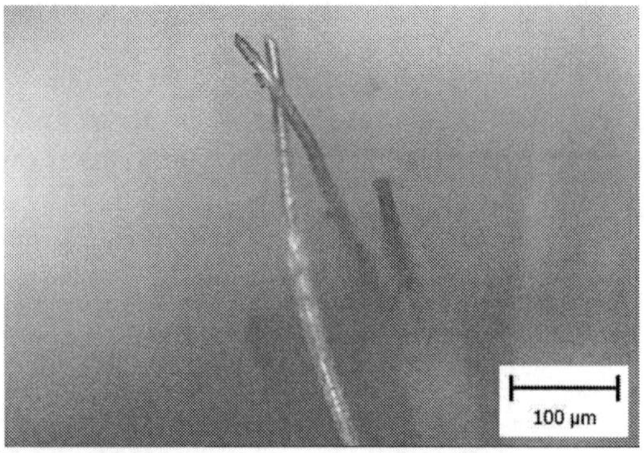

Figure 7. Breaking area of the fibres

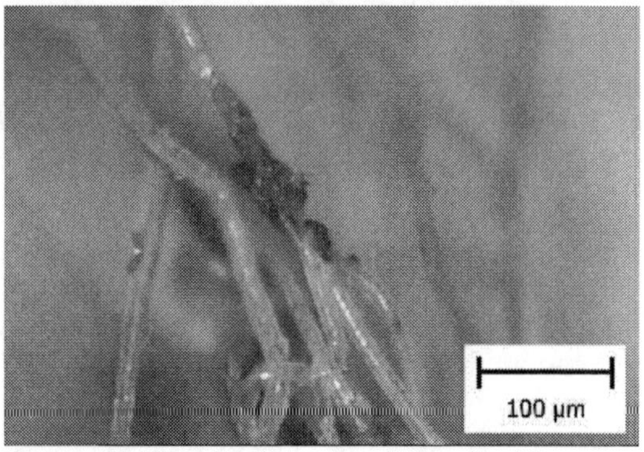

Figure 8. Detail of the fats from the fibres

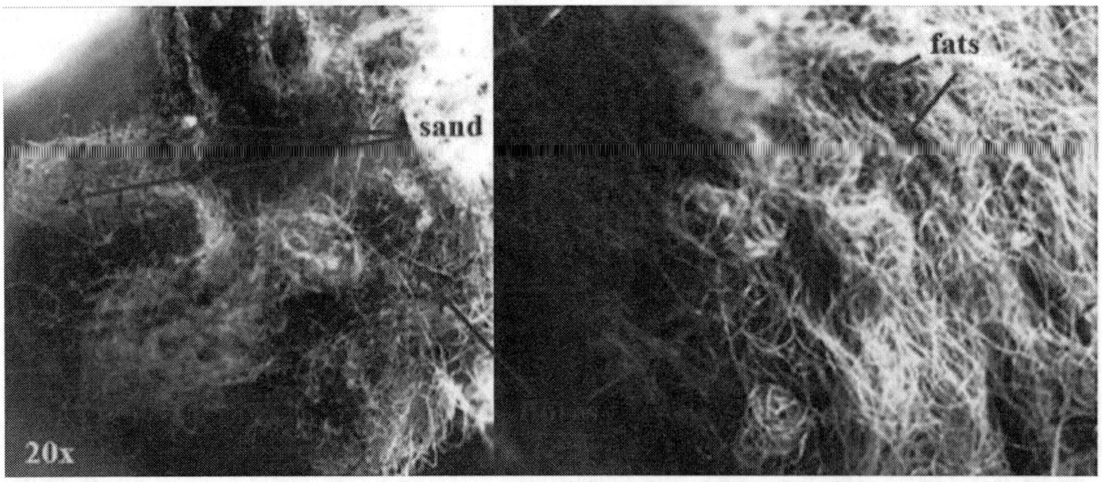

Figure 9. Microscopic analysis of the waste material: transversal section (left) and longitudinal section (right).

2.3. Tensile and shear tests

Tensile and shear tests have been conducted in the Strength of Materials Laboratory from Politehnica University Timisoara. The tests were performed on a Zwick Proline Z005 universal test machine with a maximum force of 5 kN at the ambient temperature. Figure 10 shows the universal test machine and the grip for the tensile tests is detailed in Figure 11.

The tests were performed using a control displacement with a loading speed of 10 mm/min both for dry and wet fibres. The wet fibres were kept in water during 24 hours before to start the test. Several cross sections were measured along to the fibre in order to estimate it average diameter.

Figure 10. Zwick Proline Z005 universal test machine

The shear tests were performed on a special device designed and built to determine the shearing strength of the metal bars. Figure 12 presents the shearing device installed on the universal testing machine.

A blank test was only performed on the shearing device itself to determine the characteristic force-displacement curve of the device. The shearing force of the fibre is determined by eliminating the drive force of the shearing device from the shearing force recorded at the fibres breakage. This force value was used to determine the shearing strength of the fibres.

The specimens were cut from the same cords that were also tested for tensile strength.

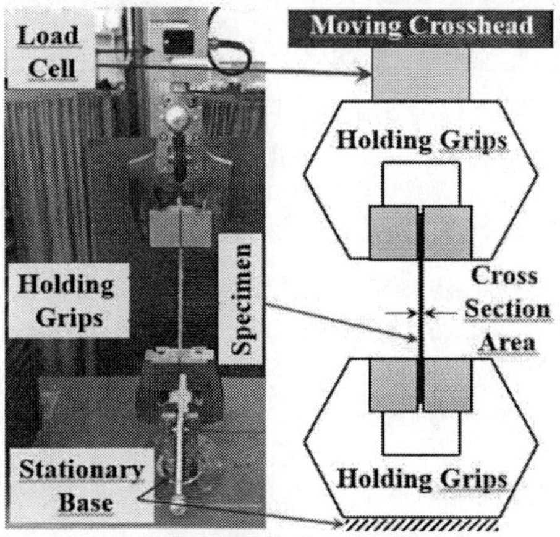

Figure 11. Grip detail for tensile test

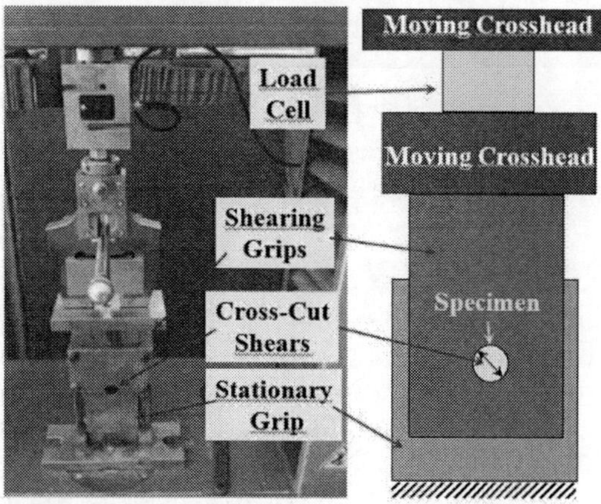

Figure 12. Grip detail for shearing device

The force – displacement curves obtained on several specimens of the dry/wet fibres under the tensile tests are given in Figure 13. One can observe a large spreading of the result due to the anisotropic structure of the fibres and the variation of the number of fibres in the tested specimens. Also, one can distinguish a different behaviour of the dry and wet fibres under the tensile tests because the water acts like a plasticizer and it strongly modifies the mobility of the amorphous part of macromolecules and shifts, similar as temperature [8 – 9]. The moisture drastically change the properties of fibres. Figure 14 presents the force – displacement curves obtained on several specimens of the dry/wet fibres under

IOP Conf. Series: Materials Science and Engineering **416** (2018) 012038 doi:10.1088/1757-899X/416/1/012038

shearing tests. The effect of the spring from the shear fixture (long initial linear part) was subtracted in order to calculate maximum shear force. The force – displacement curves have a linear-elastic area up to maximum force after which some of them are breaking immediately and others have a progressive degradation mechanism.

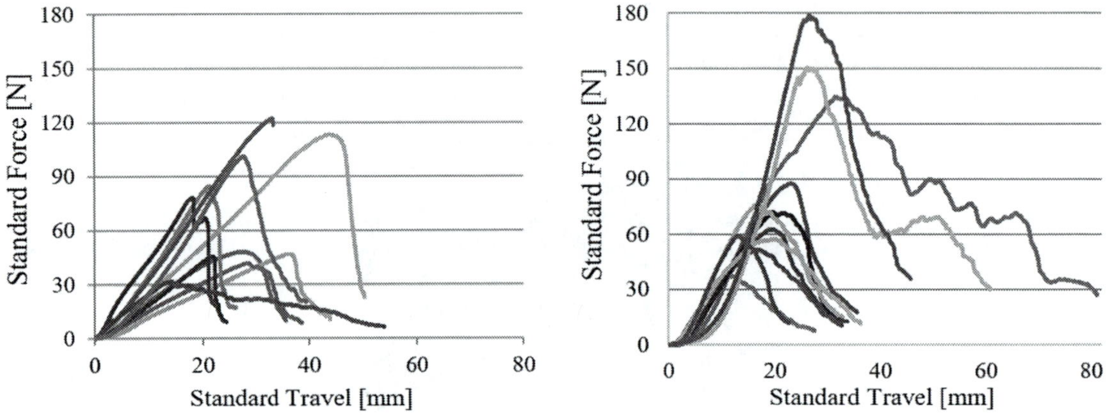

Figure 13. Force – displacement curves obtained on several specimens of fibres under the tensile tests: dry fibres (left) and wet fibres (right)

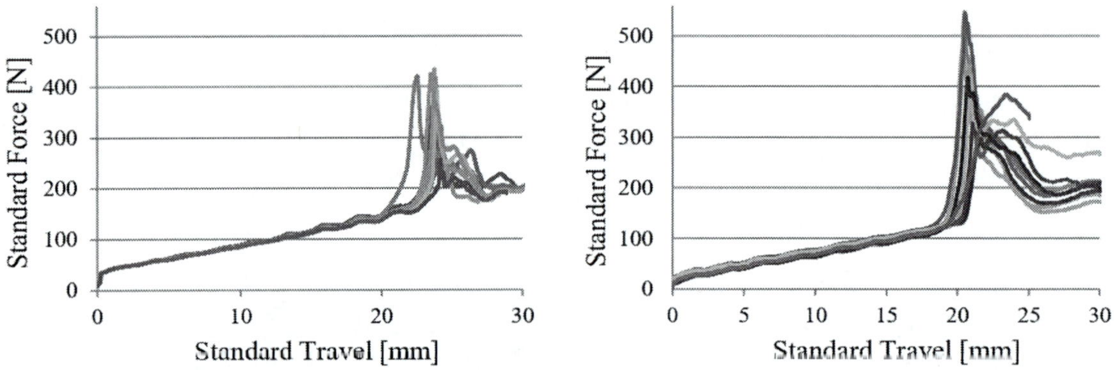

Figure 14. Force – displacement curves obtained on several specimens of fibres under the shearing tests: dry fibres (left) and wet fibres (right)

In Figure 15 are presented several broken fibres specimens following the shearing tests. The fibres specimens tested have different thicknesses and structures because of them are formed as irregular strings during storms discharge. A total of 40 specimens were used for the tests: 10 specimens for tensile tests for dry fibres, 10 specimens for tensile tests for wet fibres, 10 specimens for shearing tests for dry fibres and 10 specimens for shearing tests for wet fibres.

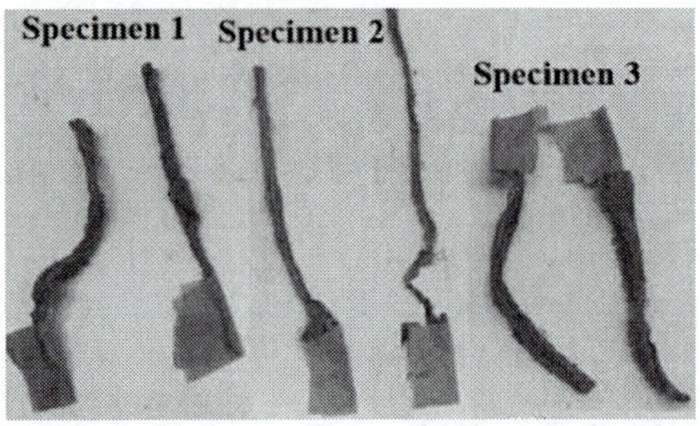

Figure 15. Three fibres specimens broken under shearing tests.

The average values for tensile and shearing strengths of the dry and wet fibres are presented in Figure 16. One can observe that the average value of the tensile strength for the dry fibres is 5.73 MPa and 27.01 MPa for the wet fibres, respectively. It is noted that the tensile strength of the wet fibres is 4.7 times higher than that of the dry fibres. Niconov et al. [9] have arrived at the same conclusion testing climbing cords because moisture has an effect analogous to the temperature effect on the cords and fibres.

The average value for shearing strength of the dry fibres is 16.55 MPa, respectively 55.13 MPa for wet fibres. It can be seen that the average shearing strength is 2.9 times higher than the tensile strength for the dry fibres. The difference between shearing and tensile strengths is given by fact that in case of the tensile tests breakage is achieved progressively by tearing, the fibres being non-homogeneous in their length while in the case of shearing tests all area of fibres is loaded directly. Also, the composite structure of the waste/specimens (fibres, sand, and so on) plays an important role during the shearing tests in order to define it strength. The average shearing strength is 3.3 times higher for the wet fibres than the dry ones.

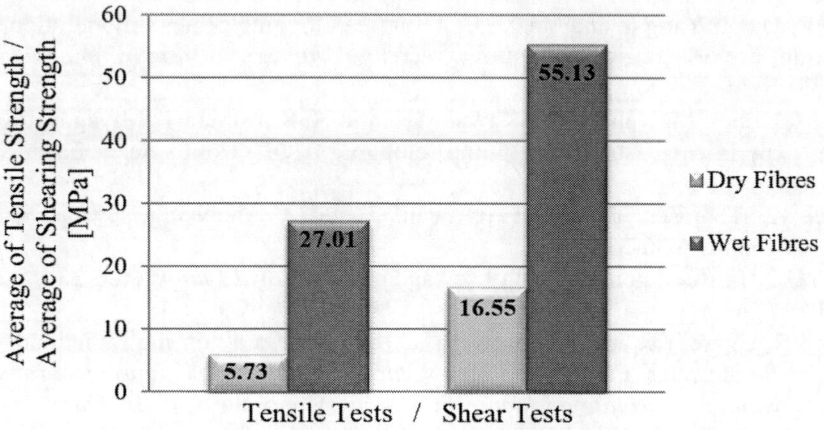

Figure 16. Average of tensile and shearing strength for the dry and wet fibres.

A correlation between the maximum force and the fibres cross section area is obtained for tensile/shear data in Figure 17 with lower coefficients (R^2=0.320/0.581 for tensile/shear tests) due to non-homogeneity (different composition and structure) of the tested wet fibres. These correlations will be used to estimate the force required to break by stretching or shearing the fibres which clog the gap between impeller blade and casing.

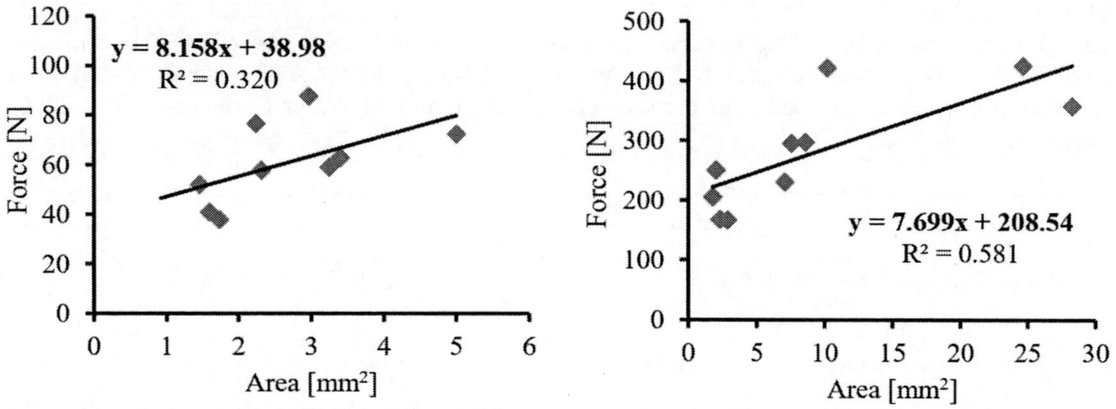

Figure 17. The experimental data (♦) and it linear regression function (—) for the wet fibres under tension (left) and shearing (right) tests.

3. Conclusion

The paper presents our experimental investigations performed on waste collected from the municipal wastewater system. The chemical composition and microscopic analysis of the waste material is performed. As a result, the waste composition determined in our case is: 17% dyes and amines, 16% fats and 67% fibres. The mechanical properties of the dry and wet fibres under tensile and shear conditions are obtained. The breaking mechanism and tensile and shearing strength of the fibres was identified. It was noted that the results for tensile strength of wet fibres is higher with 21.2% than the results of dry fibres. The results for shearing strength of the wet fibres are higher with 28.2% than the results of dry fibres. The results show that the shearing strength of the fibres is higher with 51% than the tensile strength. The results lead to the conclusion that wet fibres require a higher force for breaking both in tensile and shearing. This approach represents a useful engineering tool for determination of mechanical properties and failure mechanism of fibres formed in the municipal wastewater systems that lead to clogging up the impeller.

4. References

[1] Wang Z, Du X, Yang Y and Ye X 2012 Surface clogging process modeling of suspended solids during urban stormwater aquifer recharge *Journal of Environmental Sciences* **24** (8): 1418–1424

[2] Isono M, Nohmi M, Uchidal H, Kawai M, Kudo H, Kawahara T, Miyagawa K and Saito S 2014 An experimental study on pump clogging, IOP Conf. Ser.: Earth Environ. Sci. **22** 012009:1-11

[3] Höchel K, Thamsen P U and Rauwald H 2013 Analysing sewage handling problems, *World Pumps* 6:30-33

[4] Moore G 2011 Intelligent control of sewage pumps *World Pumps* **6**:26-33

[5] *** 2012 Clogging: not just throughlet size *World Pumps* **9**:34-38

[6] Gerlach S, Ugarelli R and Thamsen P U 2016 Case Study on the Functional Performance of a large Wastewater Pumping Station *International Symposium on Transport Phenomena and Dynamics of Rotating Machinery* Hawaii, Honolulu April 10-15

[7] Hearles J W S, Lomas B and Cooke W D, *Atlas of fibre fracture and damage to textile*, Secound Edition, Department of Textiles, University of Manchester Institute of Science and Technology

[8] Lawrence S St, Willett J L and Carriere C J 2001 Effect of moisture on the tensile properties of poly(hydroxy ester ether) *Polymer* **42** 5643 – 5650

[9] Nikonov A, Saprunov I, Zupančič B and Emri I 2011 Influence of moisture on functional properties of climbing ropes *International Journal of Impact Engineering* **38** (11):900-909.

Acknowledgment

The authors are gratefully for the financial support of Bridge Grant TANAGRA/63BG2016 project code PN-III-P2-2.1-BG-2016-0082 entitled "Transfer of knowledge to the economic operator" - financed by the Romanian Government and also for the support of Municipal Water Company AQUATIM SA Timişoara.

IOP Conf. Series: Materials Science and Engineering **416** (2018) 012039 doi:10.1088/1757-899X/416/1/012039

CoCr Based Alloys in Current Dental Prosthetic Applications

L Reclaru[1] and L C Ardelean[2,*]

[1]Corrosion & Biocompatibility Department, VVSA, branch of Richemont International SA Varinor Innovation, Switzerland

[2]"Victor Babes" University of Medicine and Pharmacy Timisoara, Faculty of Dental Medicine, Dept 1, Technology of Materials and Devices in Dental Medicine, 9 Revolutiei 1989 Bd., Timisoara, Romania

[*]Corresponding author: lavinia_ardelean@umft.ro

Abstract. CoCr alloys are traditionally manufactured by casting, but nowadays alternatives such as CAD/CAM milling or 3D Printing Technologies (Direct metal laser sintering (DMLS), Selective laser sintering (SLS), Selective laser melting (SLM), Electron beam melting (EBM)) are available. CoCr blanks to be milled in the green body state (unsintered metal powder held together by a binder) and then densely sintered try to overcome the hardness of the alloy and to ease the milling process.

The aim of this paper is to compare the quality of prosthetic pieces manufactured by CAD/CAM milling, CAD/CAM sintering and Selective laser sintering.

Because the chemical composition of the CoCr alloys used is comparable, differences in quality of the prosthetic pieces are the result of the manufacturing technology. In terms of structure and corrosion resistance, the CAD/CAM milling sample proved to be the best choice and the SLS sample the worst.

1. Introduction

CoCr alloys, which are more often used in dentistry, represent an alternative to NiCr alloys, known as potential allergens. [1,2] Because attempts to produce better dental alloys, e.g. CoCr based alloys dopped with precious metals, were quite a disappointment [3,4,5], nowadays manufacturers are trying to alternatively improve the technological process.

It is well known that CoCr alloys have excellent corrosion resistance, being mainly used for manufacturing removable partial dentures and metal ceramic fixed partial dentures, which involve accurate frameworks. [3,6] Unfortunately, the high hardness makes them difficult to handle both for the dental technician and the dentist, when traditionally manufactured by casting. Casting uses bulk metal, it is laborious and time consuming. When specific flowing parameters of the casting device are not properly set, castability problems and porosities may frequently appear.

New technologies which ease the manufacturing of CoCr alloys are now available:

1. Milling (origin: bulk metal)

CAD/CAM milling technology is based on a virtual impression or a 3D image. The CAD software virtually designs the prosthetic piece and the unit connected to the computer actually mills it. This is

IOP Conf. Series: Materials Science and Engineering **416** (2018) 012039 doi:10.1088/1757-899X/416/1/012039

quite a hard task in case of CoCr blanks, high demands such as coolant delivery, rigidity of the machine, being placed on the milling device. [6]

A great improvement was made by using units with 4 or 5 axes which permit milling and dry or wet grinding of very good quality prosthetic pieces. Milled elements do not show structure defects such as porosities or cracks.

2. CAD/CAM sintering (origin: powder)

This new technology uses metal blanks with a wax-like texture which allows easy dry milling using system's units. These CoCr blanks are similar to partially sintered zirconia blanks and can be easily milled in this green body state (unsintered metal powder held together by a binder). [7] Afterwards they are debinded and densely sintered in a special furnace. The result apparently has a homogeneous structure. Distortion-free frameworks without contraction cavities are usually obtained. [8]

3. 3D Printing Technologies (origin: powder)

CoCr alloys may be manufactured by using the following 3D Printing technologies:

- Selective laser sintering (SLS) which uses a high power laser to fuse small particles of metal powder into a desired 3D object. Based on a virtual image, the CoCr alloy powder is slowly added, layer by layer, as the 3D CAD software measures thousands of cross-sections to determine exactly how each layer of the prosthetic piece is to be constructed. [9]

The adjustable platform the pieces are built upon moves itself to a height equal to the thickness of each layer that is being built. After scanning each cross-section, the powder bed is lowered by one layer thickness. Additional powder is than added on top of previous solidified layer and sintered. The powder is maintained at an elevated temperature so that it fuses easily upon exposure to the laser. [10]. The SLS technique enables obtaining accurate prosthetic pieces with proper mechanical properties. [9]

- Direct metal laser sintering (DMLS)

In this case the metal powder, with no binder or fluxing agent, is melted by a high power laser beam. The absence of the polymer binder avoids the burn-off and infiltration steps, the result being a piece with higher density and detail resolution compared to SLS. [10]

- Selective laser melting (SLM) uses a high powered laser to melt metallic powders together, followed by heat treatment and post processing. [10]

- Electron beam melting (EBM) uses an electron beam in a high vacuum to melt the metal powder layer by layer. The pieces obtained using EBM are dense, extremely strong and void-free. [10]

2. Experimental part

CoCr alloys for milling, CAD/CAM sintering and SLS were analyzed. Their composition (as given by the manufacturer) shows no great difference, being comparable to that of CoCr alloys for casting. CAD/CAM sintering alloy blanks contain an additional organic binder.

The CoCr samples were embedded in a cold-curing resin on a methyl methacrylate basis, then polished. Electrolytic etching was performed. A metallographic microscope was used for observing the microstructures of the alloys and a scanning electron microscope equipped with an EDX system for local phase analysis and microanalysis was also used.

2.1 Milling (origin: bulk metal)

In figures 1-3 the micrographic structures of the CoCr alloy without and with chemical attack are shown. There are no abnormalities to be noticed, the images being specific for CoCr alloy bulk structures.

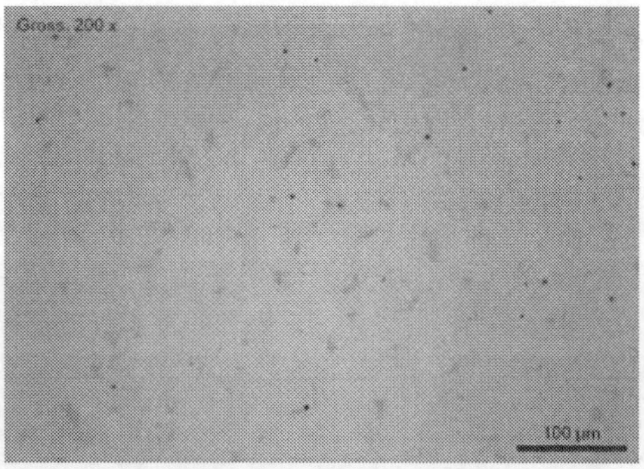

Figure 1. Micrographic structure of the CoCr alloy without chemical attack.

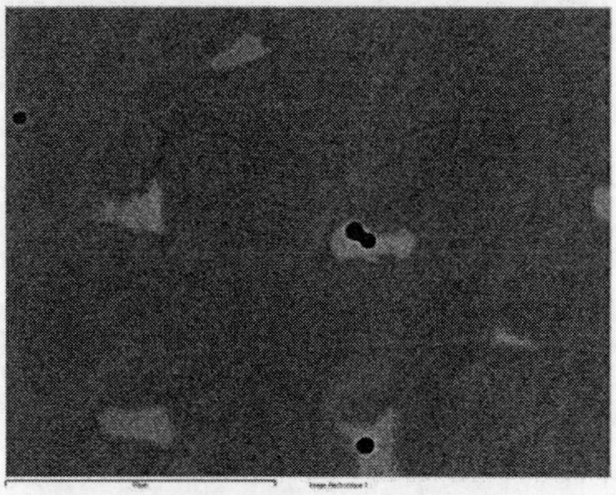

Figure 2. SEM observation of the CoCr alloy without chemical attack.

The chemical composition of the phases is shown in Table 1.

Table 1. The chemical composition of the phases.

Spectrum	C	O	Al	Si	Cr	Mn	Fe	Co	Mo	Total
Spectrum 1: matrix				0.85	28.45		0.85	64.46	5.38	100.00
Spectrum 2: white phase				1.07	38.61		0.38	35.48	24.46	100.00
Spectrum 3: inclusion	14.37	37.13	0.47	17.74	12.26	1.18		14.72	2.13	100.00

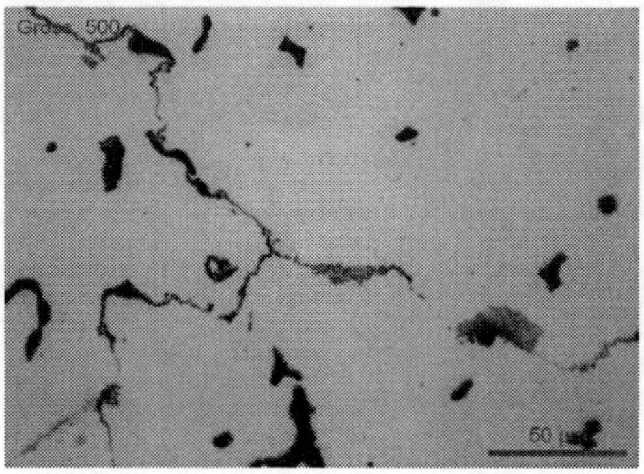

Figure 3. Micrographic structure of the CoCr alloy after chemical attack.

IOP Conf. Series: Materials Science and Engineering **416** (2018) 012039 doi:10.1088/1757-899X/416/1/012039

2.2 CAD/CAM sintering (origin: powder)

The micrographic structure without chemical attack shows micrometric porosities, homogeneously distributed, with no noticeable significant defects. (Figure 4)

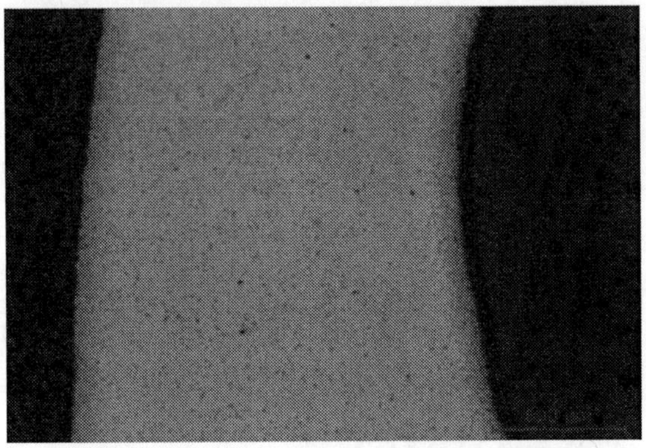

Figure 4. Micrographic structure without chemical attack.

After sintering, the fine and homogeneous microstructure reveals micrometric porosities, homogeneously distributed, typical for sintered pieces. (Figure 5)

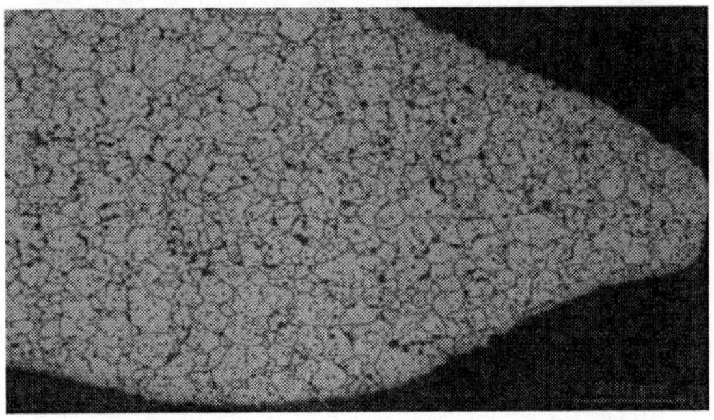

Figure 5. Microstructure after sintering.

2. 3 Selective laser sintering (origin: powder)

The structures examined in vertical and horizontal samples are shown in figures 6-7.

Figure 6. Metallographic observation of a vertical sample.

Figure 6 (vertical sample) shows CrCo piles which are the result of the laser action when building the prosthetic piece.

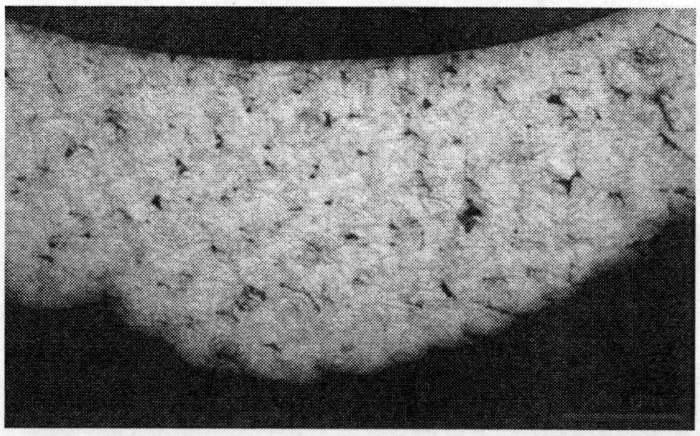

Figure 7. Metallographic observation of a horizontal sample.

The horizontal samples show porosity in the structure of CoCr layers (Figure 7) so further corrosion tests were made, namely electrochemical evaluation by potentiodynamic polarisation in artificial Fusayama saliva (composition: NaCl 0.4 g/l; KCl 0.4g/l; NaH_2PO_4* H_2O 0.69 g/l; $CaCl_2$*H_2O 0.79 g/l and urea 1.0 g/l.) at aconstant 37°C temperature and pH 5.

Scanning electron microscopy (SEM) examination of the sample surface after the corrosion test shows cracks. (Figure 8)

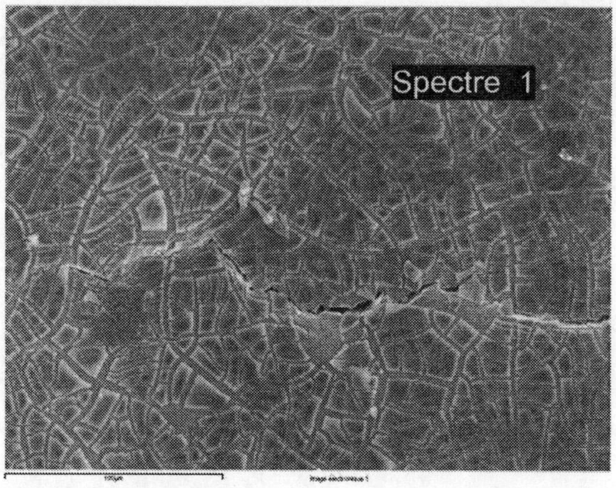

Figure 8. SEM micrograph of the corroded surface after the polarization test.

The spectral composition analysis by EDX (spectrum 1) shows sodium and chlorine oxygen, which are clearly corrosion products. (Table 2)

Table 2. The spectral composition of the SLS sample.

Spectrum	O	Na	Si	Cl	Cr	Co	Mo	Total
Spectrum 1	24.53	1.19	1.12	0.59	28.76	34.30	9.52	100.00

3. Conclusions

In conclusion, the CAD/CAM milling sample proved to be the best in terms of structure and corrosion resistance, the worst being the SLS sample. The structure and corrosion behavior of CoCr alloys highly depends on the manufacturing process. In case of milling the alloy is in a bulk form, with a compact structure, with no noticeable impurities. No heating/sintering is involved in the manufacturing process, which might alter the alloy's properties. In case of CAD/CAM sintering, corrosion problems may occur due to the presence of the binder and because of the final sintering stage. In case of SLS, the powder nature of the alloy and the manufacturing process may cause corrosion. 3D Printing is constantly developing, so CoCr alloys manufactured using these technology will probably become serious competitors for prosthetic dental applications. The low price and the ease of execution using virtual impressions and design are some of the advantages. Traditional casting may be successfully replaced by 3D Printing and CAD/CAM milling, which represent a logical step towards the future.

4. References

[1] Reclaru L, Unger RE, Kirkpatrick CJ, Susz C, Eschler PY, Zuercher MH, Antoniac I and Luthy H 2012 *Mater. Sci. and Eng.* C **32** 1452

[2] Rusu LC, Bortun CM, Tanasie G, Podariu AC, Baderca F, Solovan C and Ardelean L 2014 *Rom. J. of Morf. and Embr.* **55** 3

[3] Ardelean L, Reclaru L, Bortun C and Rusu LC 2015 *Superalloys* ed M Aliofkhazraei (Rjeka: InTech) **7** 141

[4] Ardelean L, Reclaru L, Bortun C, Ghiban B and Rusu LC 2010 *Met. Int.* **15** 31

[5] Ardelean L, Reclaru L, Bortun C and Rusu LC 2016 *Micro and Nanotechnologies for Biotechnology* ed SG Stanciu (Rjeka: InTech) **6** 123

[6] Ardelean L, Reclaru L, Bortun C and Rusu LC 2016 *Solid State Phenomena, Adv. Mater. and Struct. VI* **254** 132

[7] http://www.amanngirrbach.com

[8] Geis-Gerstorfer J, Schille C and Schweizer E 2013 *Dent. Dial.* **14** 20

[9] Reclaru L, Ardelean L, Rusu LC and Sinescu C 2012 *Solid State Phenomena, Adv. Mater. and Struct. IV* 412

[10] http://www.popular3dprinters.com

IOP Conf. Series: Materials Science and Engineering **416** (2018) 012040 doi:10.1088/1757-899X/416/1/012040

The Effects of Screen Sizes on the Surface Properties of Tepered Steel Treated by Active Screen Plasma Nitriding

D Kovács[1], A Kemény[1], A Bonyár[2] and J Dobránszky[3]

[1] Budapest University of Technology and Economics, Faculty of Mechanical Engineering, Department of Material Science and Engineering,
1111, Budapest, Műegyetem rakpart 3, Hungary.
[2]Budapest University of Technology and Economics, Department of Electronics Technology,
1111, Budapest, Egry József street 18, Hungary
[3] MTA–BME Research Group for Composite Science and Technology,
1111, Budapest, Műegyetem rakpart 3, Hungary

[*]Corresponding author dorina@eik.bme.hu

Abstract. Active screen plasma nitriding (ASPN) is a novel thermochemical surface treatment which has many specialities compared to the conventional direct current plasma nitriding (DCPN). The edge effect is eliminated, the hardness and layer thickness on the surface are homogeneously distributed. In this study, 42CrMo4 alloyed steel was treated with different screen sizes (screen diameter, hole size) active screen plasma nitriding. The parameters of treatment are similar in all cases (4 hours at 490°C and 2,2 torr). The nitrided samples were characterized by atomic force microscope (AFM) to analyse the roughness of the surface. The cross-section hardness of the samples and the thickness of the nitrided layers were also measured.

1. Introduction

The nitriding process is widely used for improving the hardness and the wear resistance of steels. DCPN has become a standard industrial technology, but it has some technological issues such as edge effect and hollow cathode effect [1–3]. Over the years, ASPN appeared in the plasma nitriding treatment processes. The main advantages are based on the replacement of the glow discharge region. This means, that the plasma forms on the screen, instead of the workpiece, which causes homogeneous hardness and layer thickness on the surface of the specimen, therefore the issues of the DCPN are eliminated. Considering that the plasma is formed only on the screen, and the worktable is isolated from the voltage source, the samples are heated by radiation. ASPN also generates nitrogen mass transfer to the surfaces of the specimens [4, 5]. Several researchers have reported the effects of the nitriding parameters, such as the distance between the screen and the sample [6–8], the effect of different screen sizes [9–11] on the surface roughness [12–16] and also wear and corrosion resistance improvement [17–20].

Surface roughness parameters, which can be quantified with AFM, are frequently used as a typical measurement of mechanical surface properties. In our particular case, the interaction between ions and the sample surface during the sputtering corresponds to the variation of the roughness [21–23]. The wear and corrosion resistance are depending on the base material, but these material properties are not evaluated in this research.

The aim of this research is to investigate the influence of the active screen hole sizes on the layer thickness, cross-section hardness profile and most importantly the surface roughness of the active screen plasma nitrided samples.

2. Materials and methods

In this study, tempered 42CrMo4 low alloy steel was used with the following chemical compositions shown in Table 1. The sample disks were 20 mm in diameter and 6 mm in thickness. The surface of the samples was mechanically ground with 80 to 2500-grit SiC paper and polished with 3 μm diamond suspension. The samples were cleaned with acetone and dried before placed in the vacuum chamber.

Table 1. Chemical composition of materials in wt.%

	C	Si	Mn	P	S	Cr	Mo
42CrMo4	0.38 – 0.45	≤0.4	0.6 – 0.9	≤0.025	≤0.035	0.9 – 1.2	1.5 – 0.3

The active screen was made of 1.0330 type steel. The screen dimensions were ∅100×85 mm with ∅5 mm, ∅18 mm and ∅45 mm holes and the wall thickness was 0.8 mm. Figure 1 shows the schematic workspace of the chamber and the screens with the different hole diameters and the distance of their centers. All treatments were carried out using the same parameters as shown in Table 2 and following the same procedures. First, the chamber was pumped to a base pressure of 2×10^{-1} torr, then the chamber was flushed with argon. After the flushing, the pressure was set to 2,2 torr with the nitriding gas. The plasma was produced on a negatively polarized screen (cathode) and the anode (base of the chamber), which was held at ground potential. The temperature was monitored using an isolated K-type thermocouple under the workpiece. After the nitriding process, the workpiece was cooled down from the treatment temperature to the room temperature.

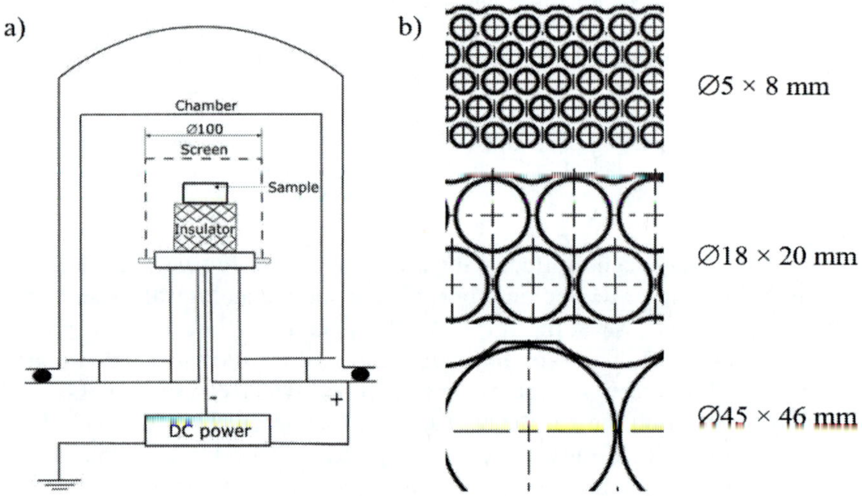

Figure 1.a) Schematic diagram of the ASPN chamber
b) Hole diameters of active screens

Table 2. Parameters of plasma nitriding

Voltage (V)	Current (A)	Gas mixture N_2:H_2(%)	Temperature (°C)	Pressure (torr)	Time (h)
490 – 540	0.9 – 1.5	25:75	490	2.2	4

IOP Conf. Series: Materials Science and Engineering **416** (2018) 012040 doi:10.1088/1757-899X/416/1/012040

Scanning electron microscope (SEM) was used to examine the morphology of the samples and measure the thickness of the nitrided layer. For the microhardness measurement Buehler IndentaMet 1105 equipment was used with Vickers head. The measuring load was 10 g for 11 s time. Contact mode AFM images were taken with a Veeco (lately Bruker) diInnova scanning probe microscope (SPM) with Bruker DNP-10 probes, using the longest cantilever with the smallest spring constant (k = 0.03 N/m). The sampling rate of the image acquisition was 512x512 with 1 Hz scan rate. The obtained images were post-processed with the Gwyddion 2.36 software [24]. Only standard background correction was applied on the images to remove piezo movement and sample tilt effects.

3. Results and discussion

The results of the different hole sized ASPN processes are compared with DCPN on the same nitriding parameters.

3.1. Thickness and hardness measurements

Scanning electron microscope was used to determine the depth of the compound zone. Figure 2 shows the cross-sections of the samples. The layer thickness was 5.8 µm with DCPN, 3.2 µm with ∅5 mm holes of AS (active screen), 5.5 µm with ∅18 mm holes of AS and 5. µm with ∅45 mm holes of AS. The layer thickness doesn't increase significantly when the hole size is increased over ∅18 mm. It means that there is a critical transition hole size between ∅5 mm and ∅18 mm which needs further investigation. The highest thickness is at DCPN treatment because of the direct heating with current. However, in this case the layer is not uneven, and nitride networks were founded at some place of the diffusion zone along the grain boundaries.

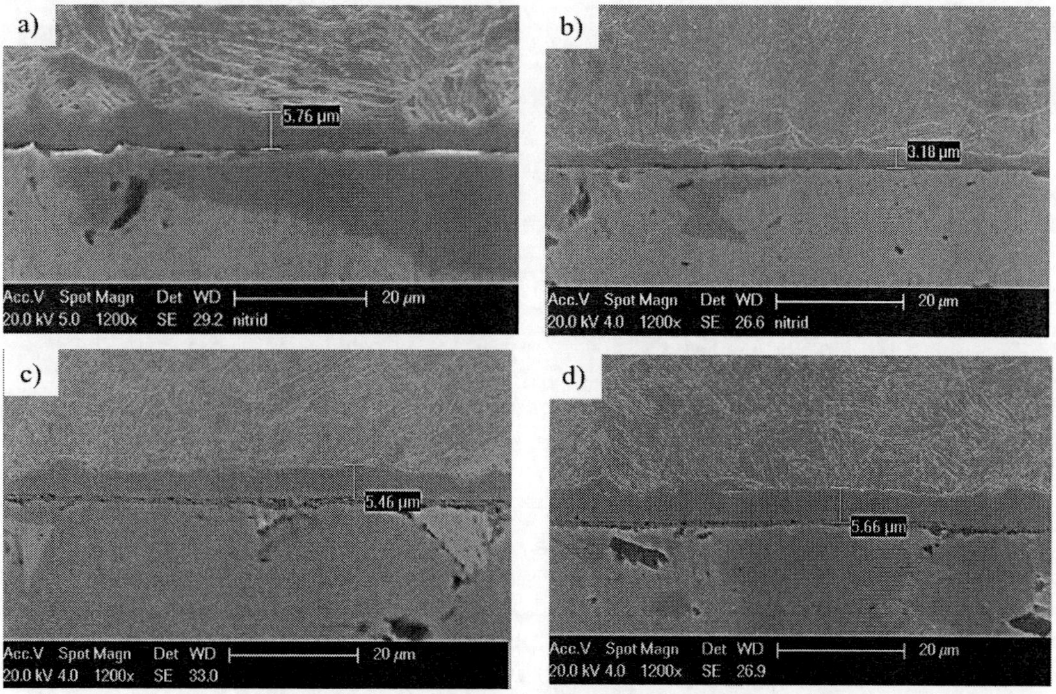

Figure 2. Compound layer thickness of steel 42CrMo4 a) DCPN, b) ∅5 mm holes of AS, c) ∅18 mm holes of AS, d) ∅45 mm holes of AS treated samples

Figure 3 shows the cross-section hardness profiles of the investigated samples and the maximum hardnesses are found in Table 3. Each curve is made from 3 different measurements on different places

IOP Conf. Series: Materials Science and Engineering **416** (2018) 012040 doi:10.1088/1757-899X/416/1/012040

of the cross-section, but all started from the layer, through the diffusion zone, towards the base material. The standard deviation is also shown on the diagram in every point.

Table 3. Maximum hardnesses of nitrided samples

	DCPN	∅5 mm holes of ASPN	∅18 mm holes of ASPN	∅45 mm holes of ASPN
HV0.01	1050	787	912	698

From the diagram it can be read that according to the layer thickness values, the DCPN sample has the highest hardness. This seems good, but the DCPN method always creates an edge effect, where the layer is not uniform with the other parts of the sample and the hardness values near the edge effect are differ from the common hardness of the nitrided sample [6, 8, 20]. It is also interesting, that the ∅18 mm hole size screen effects the highest hardness values among the ASPN treated samples. It can be declared as an optimum among the other screens. This statement is also backed by the layer thickness measurements. The layer thickness depends on the quantity of formed nitrides. Nitrides have significantly higher hardness than the raw material, so theoretically the thicker layers should have higher hardness. However, the hardness values of the ∅45 mm hole size ASPN treated sample should be considered. A possible explanation for the lower values, is that the holes are too large in the screen, so it is not able to create a fully even layer where there is and open area of the screen. Although this possibility should be examined further.

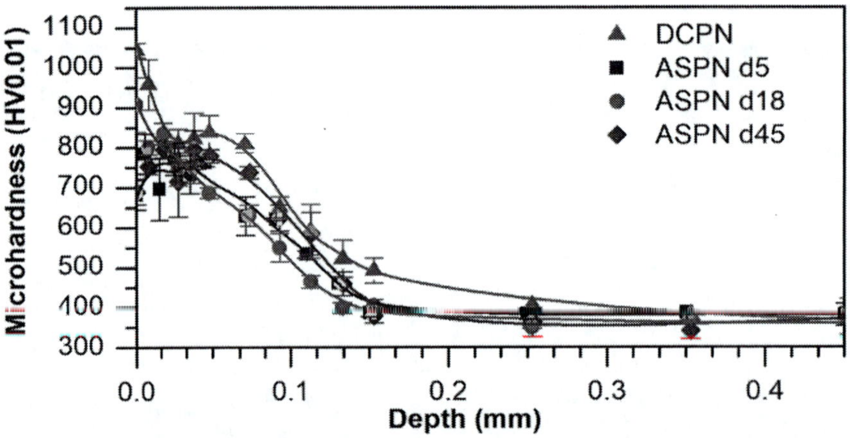

Figure 3. Hardness profile of plasma nitrided samples

3.2. Surface roughness measurements with AFM

The surface topography of the samples was investigated by AFM, as shown in figure 4. It can be clearly seen that the surface roughness of the nitrided samples can even be 30-40 times higher than the polished reference sample (S_a is 0.75±0.05 nm for the reference, 15±0.05 nm for the DCPN sample and is between 35-40 nm for the ASPN samples, measured on 2×2 μm images). The ∅18 mm hole size screen has the highest roughness. If the dimension of the hole size is increasing, the roughness is also increasing until the critical transition size. Each results of previous measurements confirm the ∅18 mm hole size screen is a transition size among the hole size dimensions. It has to be noted that the surface roughness is not sufficient alone to conveniently characterize the effect of nitriding. As can be seen in figure 4, the surface after the treatment is structured in different levels. Besides the grains caused by the process, we can observe surface waviness with hills and valleys (especially for the DCPN treated samples in figure 4 c)-d)), and also larger, complex structures (which resemble aggregated grains, most visible in the 10×10 μm images of figure 4 e)-j) for the ASPN samples with larger hole diameters). This means, that the treated surfaces have different characteristic features depending on the technological parameters, and since these features have different spatial frequency, based on the selected scan-size, the measured surface roughness values will be affected differently. By using smaller scan sizes it is possible to focus on the grainy nitride structures. It is clear that the ASPN samples have significantly larger grain sizes compared to the DCPN sample, moreover, the treatment with ∅18 mm hole size resulted in the largest average grains. At larger scan sizes the mentioned complex structures define the surface roughness, which increases with a factor of 1.4-2.3 by increasing the scan size to 10×10 μm from 2×2 μm, depending on the samples. These structures with smaller spatial frequency could also be attributed to aspects of the treatment, but their detailed characterization is out of the scope of the current paper.

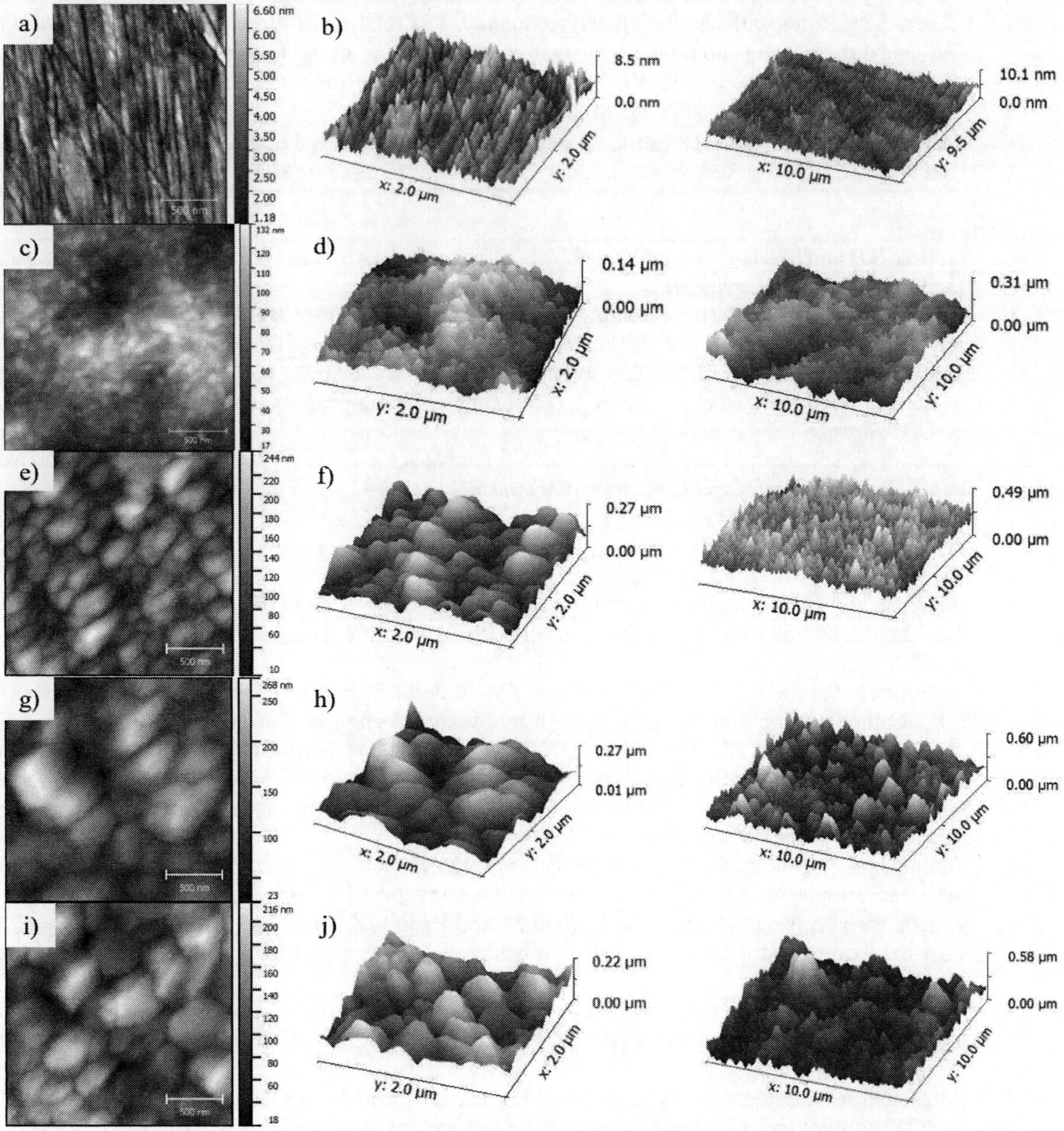

Figure 4. Contact-mode 2D and 3D AFM topography images made on a)-b) polished, c)-d) DCPN treated, e)-f) ∅5 mm holes of AS treated, g)-h) ∅18 mm holes of AS treated, i)-j) ∅45 mm holes of AS treated samples.

4. Conclusions

- The highest thickness is obtained by using DCPN treatment, but the edge effect causes an uneven layer thickness. It seems that the ASPN process has a critical transition from the point of view of the screens hole sizes, which treatment influences the layer thickness.
- Also, the DCPN treatment resulted the highest hardness in the nitrided layer. The ASPN treated samples have similar hardness, but their maximum hardnesses are lower than the DCPN treated sample. After all, the highest hardness formed when the screen with ∅18 mm holes was used.
- The surface roughness of the nitrided samples can even be 30-40 times higher than the polished originals, depending on the process, but besides the nitride grains the ASPN samples show characteristic features with smaller spatial frequency. The relation of these structures with the technological parameters should be investigated in more detail in the future.
- The AFM investigation of the surfaces showed that the sample treated by using an ∅18 mm hole size screen has the highest average grain size.
- All things considered, ∅18 mm hole could be considered as the optimal size of the screen for the ASPN process, based on the investigated samples in this research.

5. References

[1] Andrea S B and Mária K B 2012 *Nitridálás – korszerű eljárások és vizsgálati módszerek Miskolc* (Miskolc: Miskolci Egyetem)
[2] Asm International 1991 Heat Treating *ASM Int. Mater. Park. OH* **4** 860
[3] Pye D 2003 *Practical NITRIDING and Ferritic Nitrocarburizing* (Ohio: ASM International)
[4] Hamann S, Börner K, Burlacov I, Hübner M, Spies H J and Röpcke J 2013 Spectroscopic studies of conventional and active screen N 2-H2 plasma nitriding processes with admixtures of CH 4 or CO2 *Plasma Sci. Technol.* **22**
[5] Yazdani A, Soltanieh M and Aghajani H 2015 Active screen plasma nitriding of Al using an iron cage: Characterization and evaluation *Vacuum* **122** 127–34
[6] Nishimoto A, Nagatsuka K, Narita R, Nii H and Akamatsu K 2010 Effect of the distance between screen and sample on active screen plasma nitriding properties *Surf. Coatings Technol.* **205** 8–11
[7] de Sousa R R M, de Araújo F O, da Costa J A P, Dumelow T, de Oliveira R S and Alves C 2009 Nitriding in cathodic cage of stainless steel AISI 316: Influence of sample position *Vacuum* **83** 1402–5
[8] de Sousa R R M, de Araújo F O, da Costa J A P, de S. Brandim A, de Brito R A and Alves C 2012 Cathodic Cage Plasma Nitriding: An Innovative Technique *J. Metall.* **2012** 1–6
[9] Nishimoto A, Matsukawa T and Nii H 2014 Effect of Screen Open Area on Active Screen Plasma Nitriding of Austenitic Stainless Steel *ISIJ Int.* **54** 916–9
[10] Nishimoto A, Tokuda A and Akamatsu K 2009 Effect of Through Cage on Active Screen Plasma Nitriding Properties *Mater. Trans.* **50** 1169–73
[11] Ahangarani S, Mahboubi F and Sabour A R 2006 Effects of various nitriding parameters on active screen plasma nitriding behavior of a low-alloy steel *Vacuum* **80** 1032–7
[12] Karimzadeh N, Moghaddam E G, Mirjani M and Raeissi K 2013 The effect of gas mixture of post-oxidation on structure and corrosion behavior of plasma nitrided AISI 316 stainless steel *Appl. Surf. Sci.* **283** 584–9
[13] Ganesh Sundara Raman S and Jayaprakash M 2007 Influence of plasma nitriding on plain fatigue and fretting fatigue behaviour of AISI 304 austenitic stainless steel *Surf. Coatings Technol.* **201** 5906–11
[14] Allenstein A N, Lepienski C M, Buschinelli A J A and Brunatto S F 2013 Plasma nitriding using high H2 content gas mixtures for a cavitation erosion resistant steel *Appl. Surf. Sci.* **277** 15–24
[15] Borgioli F, Fossati A, Matassini G, Galvanetto E and Bacci T 2010 Low temperature glow-discharge nitriding of a low nickel austenitic stainless steel *Surf. Coatings Technol.* **204** 3410–7
[16] Hirsch T, Clarke T G R and da Silva Rocha A 2007 An in-situ study of plasma nitriding *Surf.*

Coatings Technol. **201** 6380–6
[17] Köster K, Kaestner P, Bräuer G, Hoche H, Troßmann T and Oechsner M 2013 Material condition tailored to plasma nitriding process for ensuring corrosion and wear resistance of austenitic stainless steel *Surf. Coatings Technol.* **228** S615–8
[18] Li Y, Wang L, Xu J and Zhang D 2012 Plasma nitriding of AISI 316L austenitic stainless steels at anodic potential *Surf. Coatings Technol.* **206** 2430–7
[19] Li Y, Wang L, Shen L, Zhang D and Wang C 2010 Plasma nitriding of 42CrMo low alloy steels at anodic or cathodic potentials *Surf. Coatings Technol.* **204** 2337–42
[20] Alves C, de Araújo F O, Ribeiro K J B, da Costa J A P, Sousa R R M and de Sousa R S 2006 Use of cathodic cage in plasma nitriding *Surf. Coatings Technol.* **201** 2450–4
[21] Li Y, Wang Z and Wang L 2014 Surface properties of nitrided layer on AISI 316L austenitic stainless steel produced by high temperature plasma nitriding in short time *Appl. Surf. Sci.* **298** 243–50
[22] de Sousa R R M, de Araújo F O, Ribeiro K J B, Mendes M W D, da Costa J A P and Alves C 2007 Cathodic cage nitriding of samples with different dimensions *Mater. Sci. Eng. A* **465** 223–7
[23] Öztürk O, Okur S and Riviere J P 2009 Structural and magnetic characterization of plasma ion nitrided layer on 316L stainless steel alloy *Nucl. Instruments Methods Phys. Res. Sect. B Beam Interact. with Mater. Atoms* **267** 1540–5
[24] Hossein A and Behrangi S 1988 *Plasma nitriding of steels* **17**
[25] Hubbard P, Dowey S J, Partridge J G, Doyle E D and McCulloch D G 2010 Investigation of nitrogen mass transfer within an industrial plasma nitriding system II: Application of a biased screen *Surf. Coatings Technol.* **204** 1151–7
[26] Hubbard P, Partridge J G, Doyle E D, McCulloch D G, Taylor M B and Dowey S J 2010 Investigation of nitrogen mass transfer within an industrial plasma nitriding system I: The role of surface deposits *Surf. Coatings Technol.* **204** 1145–50

Acknowledgments
The research reported in this paper was supported by the Higher Education Excellence Program of the Ministry of Human Capacities in the frame of Nanotechnology research area of Budapest University of Technology and Economics (BME FIKP-NANO). This work is partially supported by Richter Gedeon Plc. This research has been partially supported by the ''ÚNKP'' program of the Hungarian Government.

IOP Conf. Series: Materials Science and Engineering **416** (2018) 012041 doi:10.1088/1757-899X/416/1/012041

Experimental Investigation of Mechanically Coupled Composite Specimens in the ENF Test Configuration

J Rzeczkowski[1], S Samborski[1] and J Paśnik[1]

[1]Lublin University of Technology, Department of Applied Mechanics, Nadbystrzycka 36 St., Lublin 20-618, Poland

E-mail: kubarzeczkowski@op.pl

Abstract. This paper reports an experimental study on the mode II interlaminar fracture of carbon/epoxy multidirectional (MD) laminates as well as on composite laminates with specific stacking sequence exhibited mechanical couplings. The aim of the analysis was to determine the mode II critical strain energy release rate (c-SERR, G_{IIC}) and an applicability of the data reduction schemes. Specimens were subjected to the end notched flexure (ENF) test. During the experiment the applied load and the load point displacement were registered. Additionally, the crack onset and all propagation values were observed and marked on the specimens edges. Moreover, the acoustic emission (AE) data were registered. The values of the mode II c-SERR were obtained with assumptions of the Direct Beam Theory (DBT) and Corrected Beam Theory (CBT). The experiment proved that the fracture toughness depends on the specimen delamination interface. For the asymmetric laminates, the mode II c-SERR reached significantly greater values for those specimens where the bending-extension coupling (BE) occurred.

1. Introduction

Fiber-Reinforced Laminated Composites (FRLC) are widely used for responsible parts for example in the aircraft. Their physical and mechanical properties have made them a superior candidate in many applications. Meanwhile, some sorts of defects and damage modes frequently appear in structures made of these materials [1]. Delamination is one of the most dangerous forms of damage modes in laminated composites therefore characterization of delamination resistance is thus of great relevance in design of composite parts. The basic modes of delamination include mode I (tensile opening), mode II (in-plane shear) and mode III (transverse shear). Among them the end-notched flexure (ENF) test is currently standardized for the measurement of the mode II interlaminar critical strain energy release rate of unidirectional (UD) laminates [2]. Most of the present design approaches are based on macromechanics, and many structures are subjected to bending loads, which naturally give rise to significant mode II in $[0°]_n$ composites. Owing to their high stiffness and ability to sustain self-similar crack growth conditions, UD specimens are particularly convenient for interlaminar fracture tests. However, most applications involve multidirectional (MD) laminates and it is observed that delamination generally occur between plies with different fiber directions, and the interface angles (defined as the ply angles of the two layers that forming the interface) are thought to be main factor among various factors that may influence the delamination in multidirectional laminates. It is, therefore, essential to study the interlaminar fracture of multidirectional specimens if critical strain energy release rates are to be applied in design. Delamination has a high tendency for intra-ply cracking between off-axis plies, which induces crack branching and/or delamination migration from the orginal interface [3,4]. Various studies were already reported on the mode II fracture of MD laminates [5-9]. For the mode II critical strain energy release rate (mode II c-SERR) its calculation formulae come from the beam theory [10]. During the test, the applied load and traverse displacement are recorded. The crack propagation is also observed and registered. To determine the maximum load point and propagation onset at the load-displacement plot, the Acoustic Emission (AE) signal can be registered as additional data [11]. It allows to increase the accuracy of determining the crack initiation moment. The results of the ENF tests showed that mode II critical strain energy release rate distribution was changing with the delamination interface ply angle and the type of mechanical coupling [12].

2. Experiment

The ENF test specimens beam were subjected to three-point bending in accordance with ASTM D7509 Standard [2]. The samples consisted of multidirectional composite laminates containing a PTFE insert at the mid-plane that served as delamination initiator. During the test samples with delamination interfaces [0/ α], [α/ α] and [α/- α] were used. Additionally, bending-twisting (BT) coupled laminate with the respective ply sequence: [α/0/ α/ α/0/- α/0/- α/- α/- α/- α/0/- α/ α/0/0/ α/ α] and bending extension (BE) coupled laminate [α/- α/0/- α/0/ α/90/ α/- α] were examined. The fiber orientation angle α took the following values {30°, 45°, 60°, 90°}. The specimens dimensions were width b=20 mm, total length $2L$=150 mm and thickness $2h$=5.5mm.

All test were carried out with the Shimadzu universal testing machine at 1 mm/min crosshead speed according to the ASTM D 7905 indications. All specimens were loaded through loading roller acting in half-length of samples. During the ENF test the applied load and the load point displacement were registered. Also the crack propagation along the edges of the specimen was observed and recorded. Figure 1 shows the deformed ENF specimen with the test parameters, including the applied load P, the load point displacement δ and the crack length a.

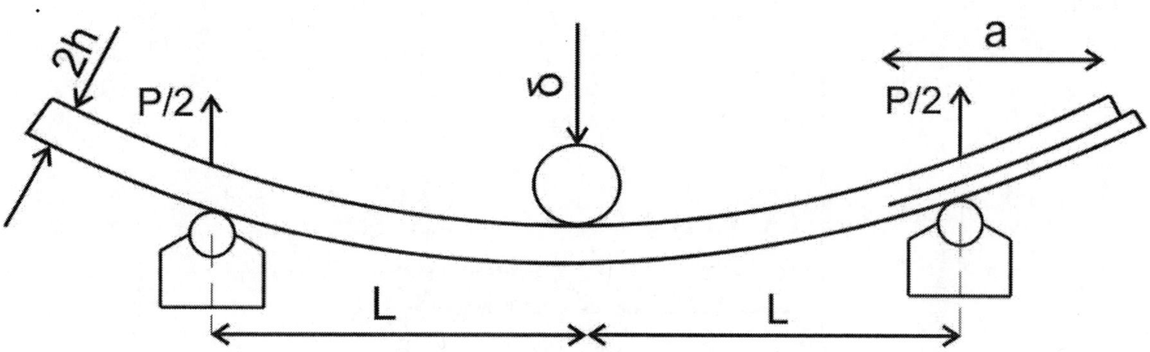

Figure 1. Deformed ENF Specimen

To determine mode II critical strain energy release rate two calculation methods were used: the Direct Beam Theory (DBT) and Corrected Beam Theory (CBT).

For the first method, G_{IIC} was determined from the following formula:

$$G_{IIC} = \frac{9a^2P\delta}{2b(2L^3 + 3a^3)} \tag{1}$$

For corrected beam theory CBT, G_{IIC} is determined as follows:

$$G_{IIC} = \frac{9a^2P^2}{16b^2h^3E_f} \tag{2}$$

Here E_f is the axial modulus and is calculated according to equation (3) and C_0 is the initial compliance at $a{=}a_0$ as determined from the experimental compliance procedure.

$$E_f = \frac{L^3}{4bh^3C_0} \tag{3}$$

In order to increase of the accuracy of detect initiation of the peak force values related to the fracture process initiation, the acoustic emission (AE) recordings were made. The delamination onset point was indicated by the first AE phenomenon. Figure 2 shows load-displacement curves and the energy of the acoustic emission signal. The two initiation failure envelopes P_{max} and AE are very close to each other. This results suggests that the peak force of the load displacement curve and the first acoustic signal both correspond to the failure initiation.

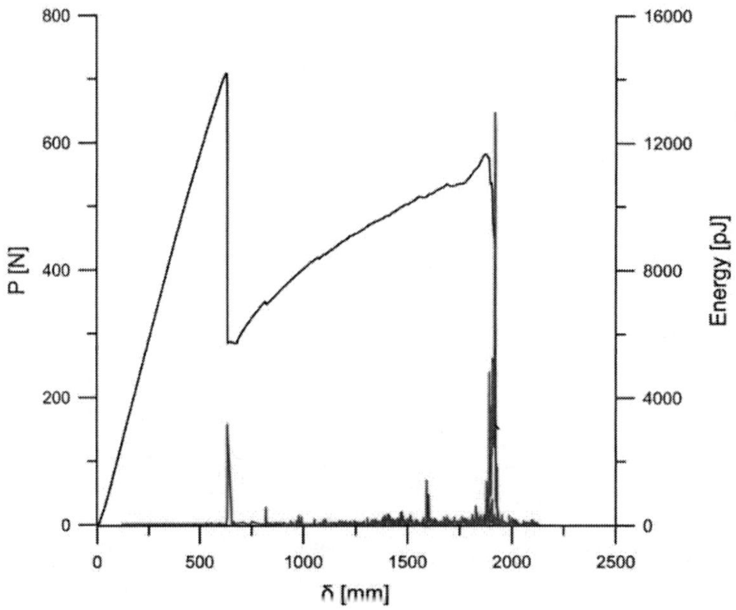

Figure 2. Load/displacement curves and energy of acoustic signal

3. Results and discussion

Figure 3 shows the experimental values of the initial G_{IIC} for specimens with the interface α/α. The greatest value of mode II c-SERR was equal 1.5 N/mm for samples with the interface 0°/0°. In configuration 90°/90° this value was on average level of 1.1 N/mm. Specimens with the interface 30°/30°, 45°/45° and 60°/60° exhibited lower values of the G_{IIC}. This effect can be induced by the couplings, as explained in [9,12,13]. For these interfaces it could be observed, that along with ply angle growth, the value of critical strain energy release rate also increased.

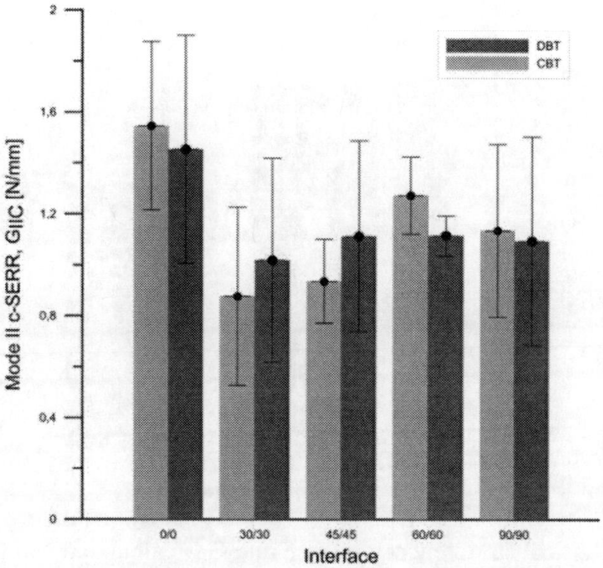

Figure 3. Mode II critical strain energy release rate G_{IIC} calculated using the DBT and the CBT method for the α/α interface

For the interface 30°/-30° the mode II c-SERR was on average level 1.3 N/mm and in comparison to the interface 30°/30° this value was about 0.3 N/mm greater. The G_{IIC} both for the interface 45°/-45° and 60°/-60° reached similar average values as in the interfaces 45°/45° and 60°/60°. The experimentally obtained results of the c-SERR for specimens with the interface α/-α were shown in Figure 4.

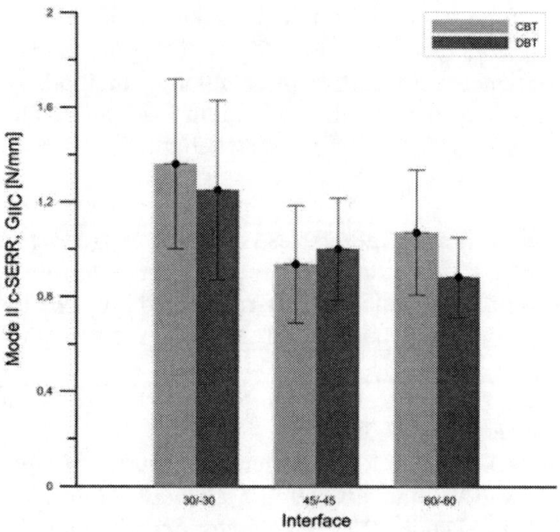

Figure 4. Mode II critical strain energy release rate G_{IIC} calculated using the DBT and the CBT method for the α/-α interface

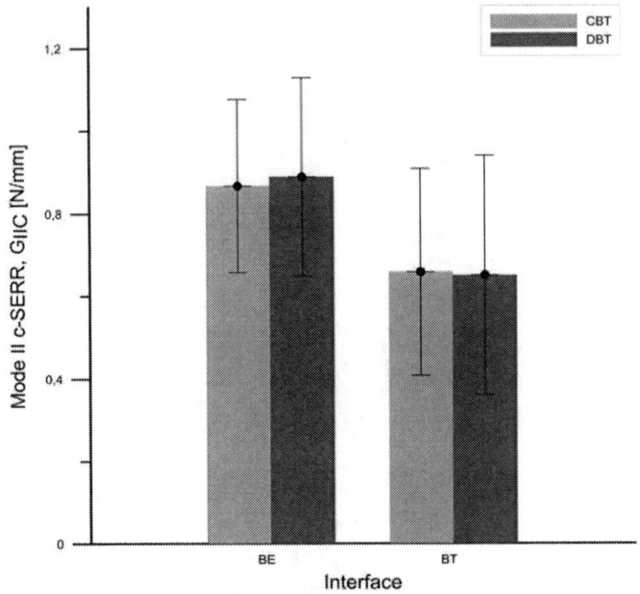

Figure 5. Mode II critical strain energy release rate G_{IIC} calculated using the DBT and the CBT method for specimens with mechanical couplings

Mechanical coupling laminates exhibited similar results of calculation of the mode II critical strain energy release for the corrected beam theory and the direct beam theory. For specimens with the bending-extension (BE) coupling the G_{IIC} was about 0.87 N/mm and for the bending-twisting (BT) couplings was equal 0.65 N/mm (Figure 5).

For the CBT and DBT methods the differences between the results of calculations of the mode II c-SERR for all specimens were minimal.

4. Conclusions

The experimental analysis of the standard ENF test for the multidirectional and coupled laminate beam specimens was performed with respect to the ASTM D7905 Standard. The critical strain energy release rate (GIIC) was obtained using different calculation methods. Acoustic emission signal was used as a additional data in order to recognize maximum load point. The results show that mode II c-SERR depends on specimens interface and mechanical coupling.

5. References

[1] Shokrieh MM, Zeinedini A and Ghoreishi SM 2017 *Comp. Struct.* **171** 19-31
[2] ASTM D 7905/7905M-14. Standard test method for determination of mode II interlaminar fracture toughness of unidirectional fiber-reinforced polymer matrix composites 2001.
[3] Pereira AB, de Morais AB, Marques PT and de Castro 2004 *Compos. Sci. Technol.* **64** 1653-59
[4] Zao L, Wang Y, Zhang J, Gong Y, Lu Z, Hu N and Xu J 2017 *Compos. Part. B.* **131** 196-208
[5] Chai H 1990 *Int. J. Fract.* **43** 117-31
[6] Polaha JJ, Davidson BD, Hudson RC and Pieracci A 1996 J. *Reinf. Plast. Compos.* **15** 141-73
[7] Choi NS, Kinloch AJ and Williams JG 1999 *J. Compos. Mater.* **33** 73-100
[8] Ozdil F, Carlsson LA and Davies P 1998 *Compos. Sci. Technol.* **58** 1929-38
[9] Samborski S and Rzeczkowski J 2018 *AIP Confernece Proceedings* **1922** 080010
[10] Ducept F, Davies P and Gamby D 2000 *Intern. Journ of Adhes & Adhesiv.* **20** 233-44
[11] Kubiak T, Samborski S and Teter A 2015 *Comp. Struct.* **133** 921-29

[12] Samborski S 2017 *Comp. Struct.* **163** 342-49
[13] Samborski S 2016 *Comp. Struct.* **152** 477-487

Acknowledgments: The paper was financially supported by the Ministerial Research Project No. DEC-2016/21/B/ST8/03160 financed by the Polish National Science Centre.

IOP Conf. Series: Materials Science and Engineering **416** (2018) 012042 doi:10.1088/1757-899X/416/1/012042

Assessment of Mass Variation by Vibration-based Methods

L Barboni[1], GR Gillich[1,*] and ZI Korka[1]

[1]"Eftimie Murgu" University of Resita

*Corresponding author: gr.gillich@uem.ro

Abstract. We present in this paper a study regarding the effect of mass variation on the vibration response of a beam-like structure. During operation, the structures can be exposed to the action of supplementary masses which additionally load them and change their dynamic behavior. These loads can be therefore observed and assessed from the changes in the natural frequencies since the mass increase leads to a frequency decrease. To find the effect of increasing the mass of a beam slice, we used the law of conservation of mechanical energy. The total stored energy is unaltered from the mass change and equals the total kinetic energy. Hence, increasing a slice's mass has as a consequence the velocity decreases, thus the frequency decreases. We succeed to demonstrate that the slice position in the beam is crucial for the frequency change in each bending mode and found a relation for the calculus of a coefficient that can be used to predict the frequency changes due to a local mass alteration. The relation was successfully tested against experiments.

1. Introduction

During operation, structural elements are subject to designed or accidental loads which can act temporary or permanently. Undesired quasi-permanent loads are, for example, those produced by adherent particles or, especially for wood or similar elements, by moisture. These loads, since not considered in the design process, can affect the safety of the entire structure. Because it is hard to evaluate such accidental loads by direct measurements, engineers face the important task of developing indirect methods to assess supplementary masses on structural elements.

The literature is rich in papers presenting vibration-based methods used nowadays to assess changes in structures [1]-[5]. Most of such methods, dedicated to identifying the structure region where mass and stiffness are affected, are based on measuring the changes in the natural frequencies [6]. A technique involving sensor clusters which simultaneously detect stiffness and mass changes is presented in [7]. How the occurrence of mass changes in satellites can be performed, is shown in [8], while the effect of a moving mass represented by a vehicle on the measured modal parameters of a bridge from traffic-induced excitation is described in [9]. The last two mentioned papers consider just mass changes because the structural stiffness is not affected by the additional mass. In [10], a different approach is presented, the link between the mass and the structure being done by a spring system.

Our previous research was dedicated to contrive relations that precisely predict the frequency changes due to mass or stiffness alteration in a slim beam element [11]-[14]. It was found that the global stiffness decrease [15] or the total energy stored in the beam [16] has to be considered if prediction of natural frequency changes is approached. The research presented in this paper takes into account exclusively the influence of local mass increase on the natural frequencies and make use of the principle of conservation of the mechanical energy. The finding is used to define patterns usable in the process of assessing mass variation.

2. Materials and methods

As the changes of the natural frequencies are used to evaluate the condition of beam-like structures, the paper aims to present a relationships elaborated by the authors, which is able to assess any mass change. For exemplification, a cantilever beam which is supporting an additional mass m_a placed at a certain distance is presented. The analysed beam (Figure 1.a) is made of steel and has a prismatic shape defined by length L = 1000 mm, height H = 5 mm and width B = 50 mm. As such, the main elements defining the stiffness of the steel bar are: the area of the cross section $A = B{\cdot}H$ and the moment of inertia of plain area $I = (B{\cdot}H^3)/12$. The other factors which are characterizing the mechanical properties of the beam are: the longitudinal elasticity modulus E, the Poisson's ratio ν and the volumetric mass density ρ. Those values, taken from the software library, are indicated in Table 1.

a.

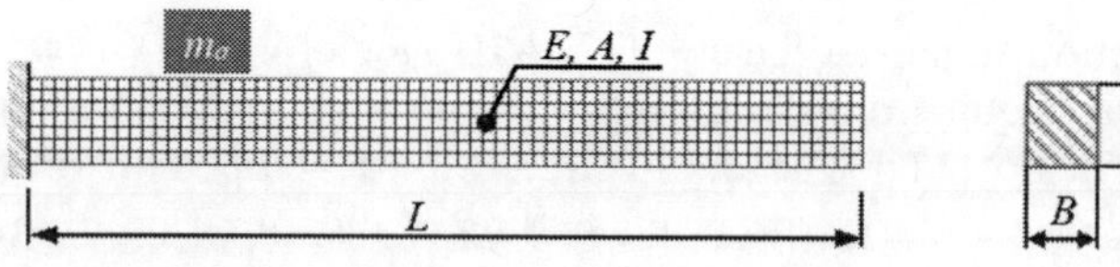

b.

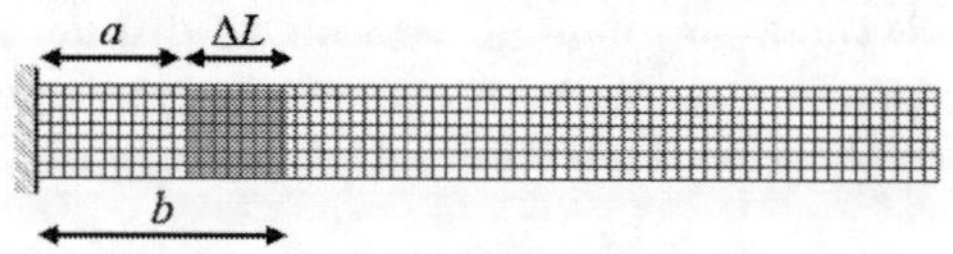

Figure 1. Cantilever beam with additional mass (a) and its FE model (b).

The analytic approach is based on conservation of the mechanical energy applied to the Euler-Bernoulli beam. How the frequency dependency on the non-uniform distributed beam mass was derived is presented in next section.

Table 1. Beam dimensions and material's physical-mechanical properties.

Mass density original beam ρ [kg/mm³]	Equiv. mass density (one additional mass) ρ_1 [kg/mm³]	Equiv. mass density (two additional masses) ρ_2 [kg/mm³]	Young modulus E [N/m²]	Poisson ratio υ [-]
7850	12637.5	17425	$2{\cdot}10^{11}$	0.3

The research presented in this paper took place in three stages. In the first stage, we calculated the natural frequencies of the beam and, based on the relationship we propose in the next section, the modified natural frequencies of the beam after placing one additional mass m_a = 47.875 g and two additional masses $2{\cdot}m_a$= 95.75 g respectively.

In the second stage, to test the contrived relation, the natural frequencies were found using the modal analysis module of ANSYS. A finite element (FE) model meshed by using hexahedral elements with maximum edge size 2 mm was created, resulting in about 37500 elements. Instead of the additional masses, the density of the corresponding beam segment, located between a = 300 mm and b = 340 mm, was increased (see Figure 1.b and equivalent mass densities in Table 1). The simulation was performed for the same two cases of additional masses as above described. To test the proposed relation, the coefficients considering mass position and weight were applied to the frequencies of the original beam and the estimated frequencies were compared with those attained by FEM.

In the last stage, in order to calibrate the FEM model, the natural frequencies were measured on the test stand shown in Figure 2. The clamping of the beam (1) was done with a universal vice (2), similar to those used on machine- tools. The additional masses were simulated by placing two annular magnets 3 on each bar side at the required distance from the fixed end. The use of a magnetic weight has the advantage of not hampering the deformation of the bar in the contact area.

Figure 2. Beam with additional mass prepared for experimental tests.

For collecting the vibration signals, a Kistler 8772A10 accelerometer (4) was employed, which was connected through a signal acquisition module NI 9234 to a computer running a special application designed in LabView software [17]. The application was developed in order to ensure and exact processing of the collected data and for a precise labelling of the vibration frequencies [18].

3. Deriving the natural frequencies of the beam loaded with an additional mass

For a cantilever beam, the natural frequencies f_i for the bending vibration modes are calculated with the relationship:

$$f_i = \frac{\alpha_i^2}{2\pi}\left(\frac{EI}{mL^3}\right)^{1/2}, \tag{1}$$

where: m is the mass of the beam and α_i are the wave numbers for the beam, which are calculated from the characteristic vibration equation:

$$\cos\alpha\cosh\alpha + 1 = 0\,. \tag{2}$$

Now, let's consider an additional mass m_a placed on the beam at a distance x from the fixed end (Figure 1a). Reducing the additional mass m_a to the width B and the height H of the beam, on a segment of length ΔL, the volumetric mass density of the beam is considered to be increased, while maintaining the stiffness constant (Figure 1b). On the rest of the beam, the mass density remains unchanged. The model shown in Figure 1b is adopted to derive the consequence on the natural frequencies of the additional mass placed on the beam.

First, the own mass of the cantilever beam is neglected. Considering a mass m_P placed at a distance x from the fixed end of the beam, it covers a distance $w_i(x)$ in a quarter-period $T_i/4$. Another mass m_E placed at the free end of the beam covers in the same period of time the distance $w_i(L)$. These distances are connected on the bending vibration mode shape of the cantilever beam, being calculated as:

$$w_i(x) = \cosh\alpha_i x - \cos\alpha_i x - \frac{\cos\alpha_i L + \cosh\alpha_i L}{\sin\alpha_i L + \sinh\alpha_i L}(\sin\alpha_i x - \sinh\alpha_i x)\,. \tag{3}$$

Figure 3 shows the distances $w_i(x)$ and $w_i(L)$ computed for the vibration modes 1 and 3. The kinetic energy of the mass m_P may be calculated as:

$$U_{Ki}(x) = \frac{m_P}{2}\left[\frac{w_i(x)}{T_i}\right]^2, \tag{4}$$

with T_i the period of the oscillation. The mass m_E, the kinetic energy can be expressed as:

$$U_{Ki}(L) = \frac{m_E}{2}\left[\frac{w_i(L)}{T_i}\right]^2, \tag{5}$$

IOP Conf. Series: Materials Science and Engineering **416** (2018) 012042 doi:10.1088/1757-899X/416/1/012042

In order to obtain in both cases the same natural frequencies, the kinetic energies expressed by Eq. 4 and 5 have to be equal. Thus, the dependence between the two masses, for the ith vibration mode, can be expressed as:

$$m_{Ei} = m_{Pi}\left[\frac{w_i(x)}{w_i(L)}\right]^2 = m_{Pi}[\overline{w}_i(x)]^2, \tag{6}$$

where $\overline{w}_i(x)$ stands for the dimensionless transverse displacement of a point situated at the length x from the fixed end. This proportion takes values between zero and one.

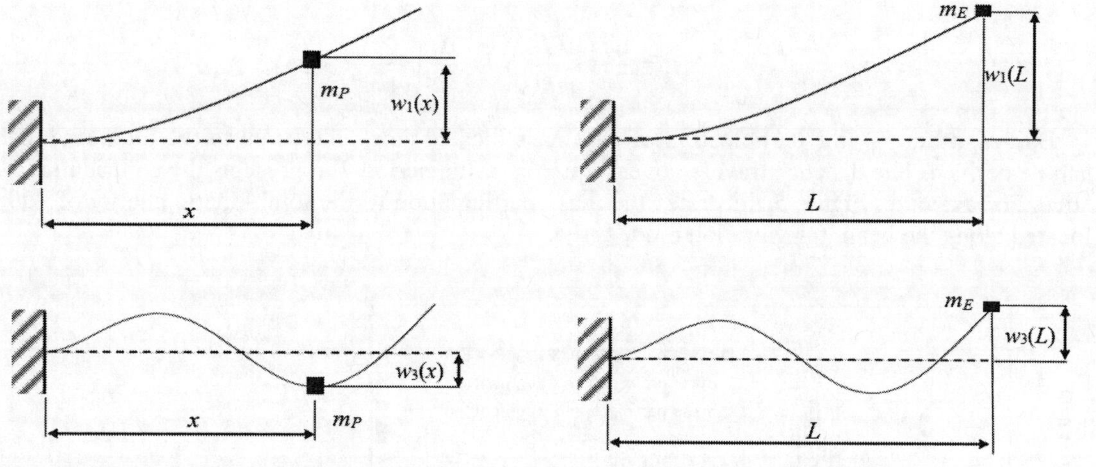

Figure 3. Displacements computed for vibration modes 1 and 3 at two dynamic systems with masses located in different positions.

Analysing Eq. 6, it can be concluded that for any mass located on a cantilever beam at a distance x from the fixed end, an equivalent mass situated on the free end may be assumed. The relation between these two masses is given by the square of the dimensionless transverse displacement $[\overline{w}_i(x)]^2$. That's why, for dynamic reasons, the equivalent mass of a beam may be obtained by adding the equivalent masses of all beam slices. Figure 4 illustrates this approach for bending vibration mode 3 of the beam.

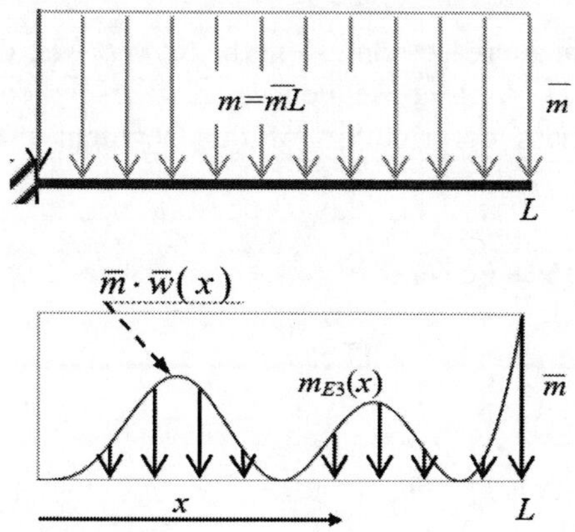

Figure 4. Cantilever beam statically load by a distributed inertial and the mass participation to the kinetic energy for vibration mode three.

Genuinely, the total equivalent specific mass m_i^{eq}, which corresponds to the ith bending vibration mode is acquired by multiplying the specific mass of the beam $\bar{m}$ with the area located below the graph of the function $[\bar{w}_i(x)]^2$. Therefore, we have:

$$m_i^{eq} = \bar{m}L\int_0^L [\bar{w}_i(x)]^2 dx = 0.25\bar{m}L. \tag{7}$$

As the curves are having the same nature, the area located below the graph of the function $[\bar{w}_i(x)]^2$ is always 0.25 for each bending vibration mode. Consequently, the natural frequencies of the beam may be formulated by using the equivalent mass as:

$$f_i = \frac{\alpha_i^2}{2\pi}\left(\frac{EI}{\bar{m}L^4}\right)^{1/2} = \frac{\alpha_i^2}{2\pi}\left(\frac{EI}{4m_i^{eq}L^3}\right)^{1/2}. \tag{8}$$

Based on Eq. (7), the participation of any mass located along the beam to the total kinetic energy can be derived. The distance traveled over a quarter of the period T_i represents the contribution of each slice. For example, Figure 5 illustrates the mass participation to the total kinetic energy of 100 slices located along the beam for vibration mode three.

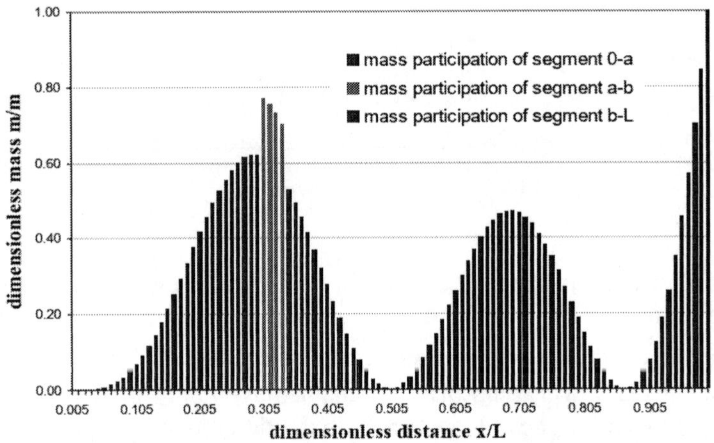

Figure 5. Mass participation to the total kinetic energy of 100 slices located along the beam for vibration mode three.

Next, let's consider that the specific mass of the beam increases on a length $\Delta L = b - a$, due to placing on the beam an additional mass m_a (see Figure 1). As a consequence, on that segment, the specific mass increases to $\bar{m}_a$. Thus, the equivalent mass contribution of this segment may be expressed as:

$$m_{ai}^{eq} = \bar{m}_a L\int_a^b [\bar{w}_i(x)]^2 dx = k_i^{a-b}\bar{m}_a L. \tag{9}$$

Similarly, the equivalent specific masses of the two segments located between *0* and *a*, respective *b* and *L*, are:

$$m_{i-1}^{eq} = \bar{m}L\int_0^a [\bar{w}_i(x)]^2 dx = k_i^{0-a}\bar{m}L, \tag{10a}$$

$$m_{i-3}^{eq} = \bar{m}L\int_b^L [\bar{w}_i(x)]^2 dx = k_i^{b-L}\bar{m}L. \tag{10b}$$

In case that the additional mass would be distributed along the whole bar length results $k_i^{0-L} = 0.25$. Presuming the beam maintains its stiffness constant, the increase of the mass density of the beam on the portion located between *a* and *b*, changes the natural frequencies of the bar as follows:

$$f_{Mi} = \frac{\alpha_i^2}{2\pi}\left(\frac{EI}{4(m_{i-1}^{eq} + m_{ai}^{eq} + m_{i-3}^{eq})L^3}\right)^{1/2}, \tag{11}$$

which may be also expressed as:

$$f_{Mi} = \frac{\alpha_i^2}{2\pi}\left(\frac{EI}{4(k_i^{0-a}\overline{m} + k_i^{a-b}\overline{m}_a + k_i^{b-L}\overline{m})L^4}\right)^{1/2}, \tag{12}$$

With regard to Eq. (8), one can write Eq. (12) as:

$$f_{Mi} = f_i\left(\frac{\overline{m}}{4(k_i^{0-a}\overline{m} + k_i^{a-b}\overline{m}_a + k_i^{b-L}\overline{m})}\right)^{1/2} = c_{Mi}f_i. \tag{13}$$

were c_{Mi} takes into account the existence of an additional mass m_a, and can be used to estimate the frequency for FEM and measurements. Forwards, the frequency decrease can be written as:

$$\Delta f_{Mi} = f_i\left[1 - \left(\frac{\overline{m}}{4(k_i^{0-a}\overline{m} + k_i^{a-b}\overline{m}_a + k_i^{b-L}\overline{m})}\right)^{1/2}\right] = (1 - c_{Mi})f_i. \tag{14}$$

Consequentially, the relative frequency shift (RFS), which is obtained by dividing the frequency decrease to the initial frequency of the beam, in the corresponding vibration mode, becomes:

$$\Delta \overset{\circ}{f}_{Mi} = \frac{\Delta f_{Mi}}{f_i} = 1 - \left(\frac{\overline{m}}{4(k_i^{0-a}\overline{m} + k_i^{a-b}\overline{m}_a + k_i^{b-L}\overline{m})}\right)^{1/2} = 1 - c_{Mi}. \tag{15}$$

RFS can be found for several bending vibration modes; these define patterns able to characterize the occurrence of an additional mass of given dimension and position.

4. Results and discussion

The natural frequencies of the first six vibration modes for the original beam and for the beam on which was placed one (m_a) and two additional masses ($2\cdot m_a$), respectively, are shown in Table 2. The natural frequencies were obtained by EM and experimental measurements.

Table 2. Natural frequencies obtained by FEM and experimental measurements.

Mode No.	FEM simulations results: Frequency [Hz] Original beam	Beam with m_a	Beam with $2\cdot m_a$	RFS^{FEM} for beam with m_a [%]	RFS^{FEM} for beam with $2\cdot m_a$ [%]	Measurement results: Frequency [Hz] Original beam	Beam with m_a	Beam with $2\cdot m_a$	RFS^{meas} for beam with m_a [%]	RFS^{meas} for beam with $2\cdot m_a$ [%]
1	4.0901	4.0852	4.0805	0.117	0.232	4.0806	4.0757	4.0711	0.105	0.221
2	25.627	25.236	24.859	1.525	2.996	24.701	24.326	24.027	1.514	2.725
3	71.757	69.963	68.397	2.500	4.682	70.61	68.897	67.498	2.426	4.406
4	140.63	140.02	139.5	0.433	0.803	138.22	137.67	137.14	0.398	0.781
5	232.53	230.88	229.43	0.709	1.333	232.39	230.87	229.53	0.654	1.231
6	347.46	340.28	334.64	2.066	3.689	341.52	335.17	330.04	1.859	3.361

Table 3. Frequencies estimated for the beam subjected to additional masses and the subsequent RFSs.

Mode No.	Orig. beam Freq. [Hz]	Estimation - beam with m_a c_{Mi} [-]	Freq. [Hz]	Error [%]	Estimation - beam with $2 \cdot m_a$ c_{Mi} [-]	Freq. [Hz]	Error [%]	Estimated RFS^{FEM} Beam with m_a [%]	Beam with $2 \cdot m_a$ [%]
1	4.09	0.9988	4.0852	-0.0017	0.9976	4.0805	-0.0013	0.115	0.230
2	25.627	0.9847	25.236	-0.0024	0.9702	24.859	-0.0182	1.523	2.979
3	71.757	0.9743	69.963	0.0620	0.9506	68.397	0.2639	2.560	4.934
4	140.63	0.9955	140.02	0.0155	0.9910	139.50	0.0897	0.449	0.892
5	232.53	0.9923	230.88	0.0549	0.9848	229.43	0.1802	0.764	1.511
6	347.46	0.9774	340.28	0.1945	0.9563	334.64	0.7040	2.256	4.367

The values presented in Table 2 show that the RFS values obtained by FEM and experiments are very similar, which confirms the validity of the Eq. (15).

Table 3 exemplifies how the natural frequencies of a beam subjected to one or two additional masses can be estimated, starting from the values of the natural frequencies obtained by FEM on the original bar and applying the coefficient c_{Mi} and Eq. (13). One can see that the percentage differences between the estimated frequencies and those obtained by FEM are less than 1%. The table is completed with the RFS values obtained by estimation (calculus).

In order to highlight the differences between RFS values obtained by FEM and the estimated ones, Figure 6 provides two charts, each plotted for one, respectively two additional masses.

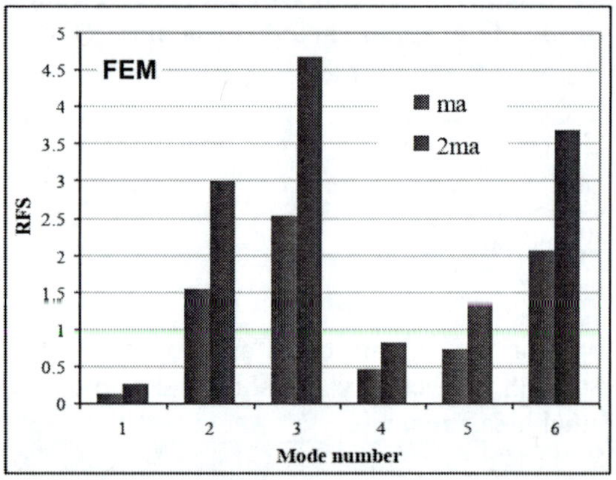

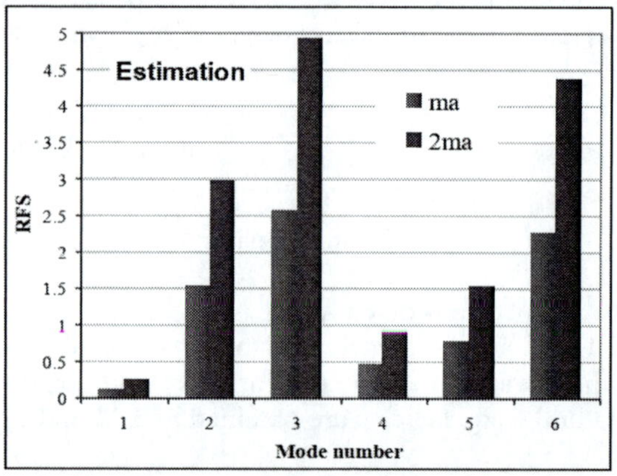

Figure 6. Relative frequency shifts for the case that one respectively two additional masses are placed on the beam segment between $a = 300$ mm and $b = 340$ mm; left graph represents the values obtained from FEM simulations while the right graph stay for the values obtained by applying Eq. (13).

A good equal was found by comparing the data directly resulted from FEM with the estimations. This further confirms the validity of the analytical formulas for the calculation of the frequencies.

5. Conclusion

A mathematical relation that allows the estimation of the natural frequencies of a beam which supports an additional mass was proposed herein. The relation contains a coefficient that can be calculated in respect to the position and size of the additional mass. It has practical utility in analytically determining of patterns if the natural frequencies of the continuous bar are known. The use of these patterns turns the position and size identification of the additional mass into an inverse problem.

The RFS values reflect changes in natural frequencies caused by the existence of an additional mass on the beam. The distribution of these values in vibration modes reflects the position of the additional mass on the beam. By normalization, it results a string of values ranging from 0 to 1, identical for any mass increase, just like in the case of beams subjected to stiffness loss [14].

6. References

[1] Zhou Y L, Qian X, Cao H and Wahab M A 2017 Damage detection in structures using frequency response functions ensemble with extended cosine based indicator *J. Phys. Conf. Ser.* **843** 012021

[2] Teodorescu-Draghicescu H, Stanciu A, Vlase S, Scutaru L, Calin M R and Serbina L 2011 Finite element method analysis of some fibre-reinforced composite laminates *Optoelectron. Adv. Mat.* **5**(7) 782-785

[3] Gillich G R, Birdeanu E D, Gillich N, Amariei D, Iancu V and Jurcau C S 2009 Detection of damages in simple elements *Annals of DAAM & Proc. of 20th Intern. Symp.* **20**(1) 623-625

[4] Bighamian R and Mirdamadi H R 2013 Input/output system identification of simultaneous mass/stiffness damage assessment using discrete-time pulse responses, differential evolution algorithm, and equivalent virtual damped SDOF *Struct. Control Hlth.* **20**(4) 576-592

[5] Vlase S 2012 Dynamical response of a multibody system with flexible elements with a general three-dimensional motion *Rom J. Phys.* **57**(3-4) 676-693

[6] Doebling S W, Farrar C R, Prime M B and Shevitz D W 1996 *Damage Identification and Health Monitoring of Structural Systems from Changes in Their Vibration Characteristics: A Literature Review* (Report LA-12767-MS, Los Alamos National Laboratory, NM: USA)

[7] Mei Q and Gül M 2014 Novel Sensor Clustering–Based Approach for Simultaneous Detection of Stiffness and Mass Changes Using Output-Only Data *J. Struct. Eng.* **141** (10) 04014237

[8] Kim C Y, Kim N S, Yoon J G and Jung D S 1999 Effect of vehicle mass on the measured dynamic characteristics of bridges from traffic-induced vibration test. *SPIE Proceedings* 1106-1111

[9] Maji A and Vernon B 2012 Vibration-based monitoring to detect mass changes in satellites, *Proc. SPIE* **8348** Health Monitoring of Structural and Biological Systems 83481G

[10] Wu J J 2006 Use of equivalent mass method for free vibration analyses of a beam carrying multiple two-dof spring-mass systems with inertia effect of the helical springs considered *Int. J. Numer. Meth. Eng.* **65**(5) 653-678

[11] Negru I, Praisach Z I, Gillich G R and Vasile O 2015 About the neutral axis distortion due to cracks and its influence upon the beams natural frequencies *Rom. J. Acoust. Vib.* **12**(1) 35-38

[12] Gillich G R, Wahab M A, Praisach Z I and Ntakpe J L 2014 The influence of transversal crack geometry on the frequency changes of beams *Proc. of Int. Conf. on Noise and Vibration Engineering ISMA2014* (Leuven, Belgium) 485-498

[13] Gillich G R, Praisach Z I, Wahab M A, Gillich N, Mituletu I C and Nitescu C 2016 Free Vibration of a Perfectly Clamped-Free Beam with Stepwise Eccentric Distributed Masses *Shock Vib.* **2016** 2086274

[14] Gillich G R, Praisach Z I and Negru I 2012 Damages Influence on Dynamic Behaviour of Composite Structures Reinforced with Continuous Fibers *Mater. Plast.* **49**(3) 186-191

[15] Nitescu C, Gillich G R, Wahab M A, Manescu T and Korka Z I 2017 Damage severity estimation from the global stiffness decrease *J. Phys. Conf. Ser.* **842** 012034

[16] Gillich G R, Tufoi M, Korka Z I, Stanciu E and Petrica A 2016 The Relations between Deflection, Stored Energy and Natural Frequencies, with Application in Damage Detection *Rom. J. Acoust. Vib.* **13**(2) 87-93

[17] Gillich G R, Frunzaverde D, Gillich N and Amariei D 2010 The use of virtual instruments in engineering education *Procedia Soc. Behav. Sci.* 2(2) 3806-3810

[18] Gillich G R, Mitletu I C, Praisach Z I, Negru I and Tufoi M 2016 A method to enhance frequency readability for early damage detection *J. Vibr. Eng. Technol.* **3**(5) 637- 652

Expired Domestic Drug - Paracetamol - as Corrosion Inhibitor for Carbon Steel in Acid Media

D A Duca[1], M L Dan[1,*] and N Vaszilcsin[1]

[1]Faculty of Industrial Chemistry and Environmental Engineering, Politehnica University Timişoara, Romania

*Corresponding author: mircea.dan@upt.ro

Abstract. In this paper, the possibility to use expired drug paracetamol as corrosion inhibitor for carbon steel in acid solutions has been investigated. 0.5 mol L^{-1} sulfuric acid and 1 mol L^{-1} hydrochloric acid have been used as corrosive media in the experimental studies. Electrochemical behaviour of paracetamol and its stability in test solutions have been examined by cyclic voltammetry. Further, the inhibitory effect has been studied by several methods: linear polarization, electrochemical impedance spectroscopy and chronoamperometry. As well, Tafel plots method was used in order to determine kinetic parameters of the corrosion process.

1. Introduction

In recent years, the use of expired drugs as corrosion inhibitors for different metals and alloys represents a new challenge in terms of recovering the active compounds from pharmaceutical drugs existing in increased amounts, both in households and pharmacies, hospitals, etc. The properties that recommend the use of active substances from expired drugs in corrosion prevention are high molecular size and the presence of hetero atoms like nitrogen, sulphur, oxygen and π bonds in their structures [1]. Also, most of these compounds are high soluble in aqueous solutions and stable in different aggressive environments [2]. It is reported that after the expiration date, for more than 90% of the drugs, the active constituents do not degrade and maintain their stability for a long period [3,4]. Based on presented considerations, expired drugs are safe to be used and because of their non-toxic characteristics and negligible negative impact on the environment, these commercial pharmaceutical products seem to be potential candidates to replace traditional toxic chemical corrosion inhibitors [1]. Moreover, the expired drugs were included in *green corrosion inhibitors* family.

In the context of an increasing amount of expired drugs, their use as corrosion inhibitors can not only reduce environmental pollution, but can also lead to a reduction of disposal and degradation costs of these expired drugs.

The use of expired paracetamol as a corrosion inhibitor for metals and alloys in acidic media, was mentioned for the first time in 2012, but until now no complete experimental studies have been performed [5].

In this paper, the possibility to use paracetamol (PCTM) drug as corrosion inhibitor for carbon steel in acid solution was investigated. 0.5 mol L^{-1} sulfuric acid and 1 mol L^{-1} hydrochloric acid have been used as corrosive media in the experimental tests. The electrochemical behaviour of paracetamol and its stability in test solutions have been examined by cyclic voltammetry. Further, the inhibitory effect

IOP Conf. Series: Materials Science and Engineering **416** (2018) 012043 doi:10.1088/1757-899X/416/1/012043

has been studied by linear polarization, electrochemical impedance spectroscopy and chronoamperometry. Tafel plots method was used in order to determine kinetic parameters.

2. Experimental part

2.1. Materials

In order to study the corrosion behaviour of carbon steel in acid media (H_2SO_4 and HCl) and the inhibitory effect of commercial paracetamol expired drug, different concentrations of organic compound between 10^{-6} and 10^{-3} mol L^{-1} have been used. The corrosive media was prepared from sulfuric acid (Merck, p.a. 95-97%) and hydrochloric acid (Merck, p.a. 37%). Paracetamol ($C_8H_9NO_2$) IUPAC name is N-(4-hydroxyphenyl)acetamide and its chemical structure is presented in figure 1.

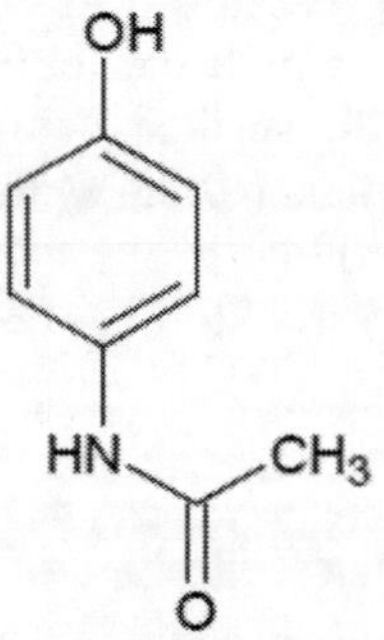

Figure 1. Paracetamol chemical structure.

The corrosion test electrode was a cylindrical disc cut from a carbon steel sample type OL52-3k with the elemental composition presented in table 1.

Table 1. Elemental composition of the carbon steel samples.

Element	Fe	C	Si	Mn	P	S	Al	Cu
wt%	98.58	0.196	0.034	0.896	0.013	0.037	0.023	0.125

2.2. Methods

Electrochemical behaviour of paracetamol expired drug in both acidic media has been studied by cyclic voltammetry (CV) on platinum electrode. Linear polarization (LV) (Tafel method), chronoamperometry and electrochemical impedance spectroscopy (EIS) were carried out in order to emphasize the inhibitive properties of paracetamol on carbon steel corrosion process. The experimental set-up consisted of a conventional three-electrode typical glass cell of 100 mL and a Biologic SP150 potentiostat/galvanostat. The potentiostat was connected with two graphite counter electrodes, a carbon steel working electrode with 0.8 cm^2 surface area and the saturated Ag/AgCl reference electrode. All potentials are referred to this electrode ($E_{Ag/AgCl}$ = 0.197 V/NHE).The working electrode surface was polished before each experiment with different grit abrasive paper, cleaned in ultrasonic bath and rinsed with distilled water. The electrode potential was stabilized for 60 minutes before each measurement.

Cyclic voltammograms (CVs) have been recorded at different scan rates, between 5 and 500 mV s^{-1} while linear polarization (LVs) only at low scan rate (1 mV s^{-1}) in order to ensure quasi-stationary conditions.

EIS studies were performed using the SP-150 impedance module, in frequency range 0.1 Hz - 100 kHz and alternative voltage amplitude of 10 mV. For each spectrum, 60 points were collected, with a logarithmic distribution of 10 points per decade. The experimental EIS data have been fitted to the electrical equivalent circuits (EEC) by CNLS Levenberg – Marquardt method using ZView – Scribner Associates Inc. software.

3. Results and discussion

3.1. Cyclic voltammetry

In order to notice the electrochemical behaviour of paracetamol and obtain preliminary information about how this drug can influence the corrosion process of carbon steel, CVs were recorded with different scan rates on platinum electrode in acid solutions without and with all paracetamol concentrations used in experimental studies.

In figure 2, the CVs recorded at 10 mV s^{-1} scan rate are depicted. The base curve obtained in 0.5 mol L^{-1} H_2SO_4 and 1 mol L^{-1} HCl solutions presents the characteristics of polarization curves drawn on platinum in acid solutions. On the CVs recorded in 0.5 mol L^{-1} H_2SO_4 solution without paracetamol (Figure 2a) at anodic polarization starting from open circuit potential (OCP) only the curve corresponding to oxygen evolution reaction (OER) can be observed. On the backward scan, a cathodic peak appears at about +0.50 V, which can be associated with the reduction of superficial platinum oxides formed on the electrode surface concomitant with the reduction of superficial remanent oxygen. At more negative potentials (-0.25 V) the hydrogen evolution reaction (HER) can be observed. Supplementary, on CVs recorded in 1 mol L^{-1} HCl solution (Figure 2b) on anodic branch at +1.35 V a pronounced peak corresponding to chlorine evolution reaction on platinum electrode can be distinguished.

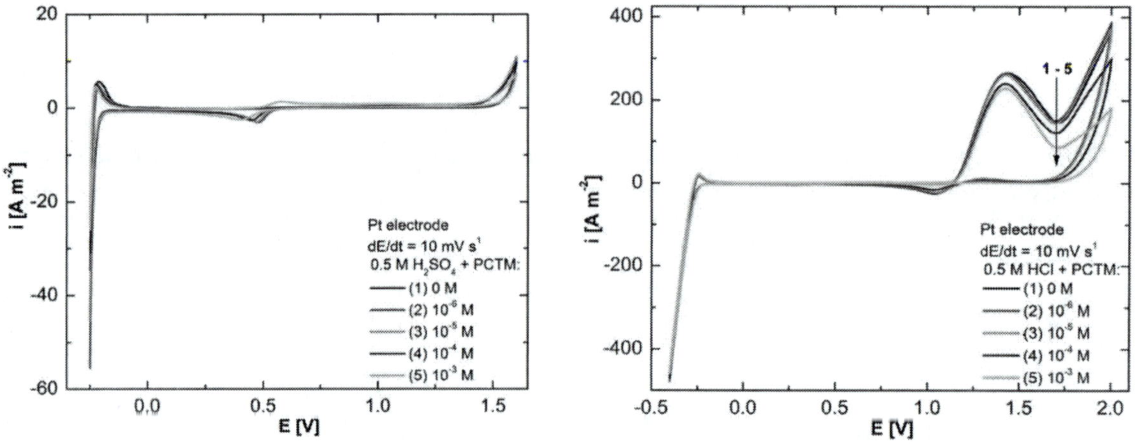

a) b)

Figure 2. CVs recorded on Pt electrode in 0.5 M H_2SO_4 (a) and in 1 M HCl (b) in the absence/presence of paracetamol at 10 mV s^{-1}scan rate.

The effect of adding expired paracetamol in the 1 mol L^{-1} HCl electrolyte solution leads to a current decrease of chlorine evolution reaction and a shift of OER potential towards more positive values. No supplementary anodic or cathodic peaks characteristic for expired paracetamol oxidation or reduction processes was recorded on CVs. One can conclude this compound is stable for all added concentrations in 1 mol L^{-1} HCl solution.

On CVs plotted in 0.5 mol L^{-1} H_2SO_4 with paracetamol, on anodic branch at approximate +0.6 V a small intensity oxidation peak assigned with the organic compound oxidation can be seen. This is most pronounced for 10^{-3} mol L^{-1} paracetamol (the higher concentration used in the corrosion tests) added in the solution. It can be assumed that paracetamol oxidation process is reversible because in the cathodic domain the reduction peak is enhanced.

In order to identify how paracetamol concentration influences the cathodic HER and anodic OER, polarization curves have been recorded separately in both acids at higher sensitivity (5 mV s^{-1}), in cathodic domain as well as in anodic one (Figures 3 and 4).

IOP Conf. Series: Materials Science and Engineering **416** (2018) 012043 doi:10.1088/1757-899X/416/1/012043

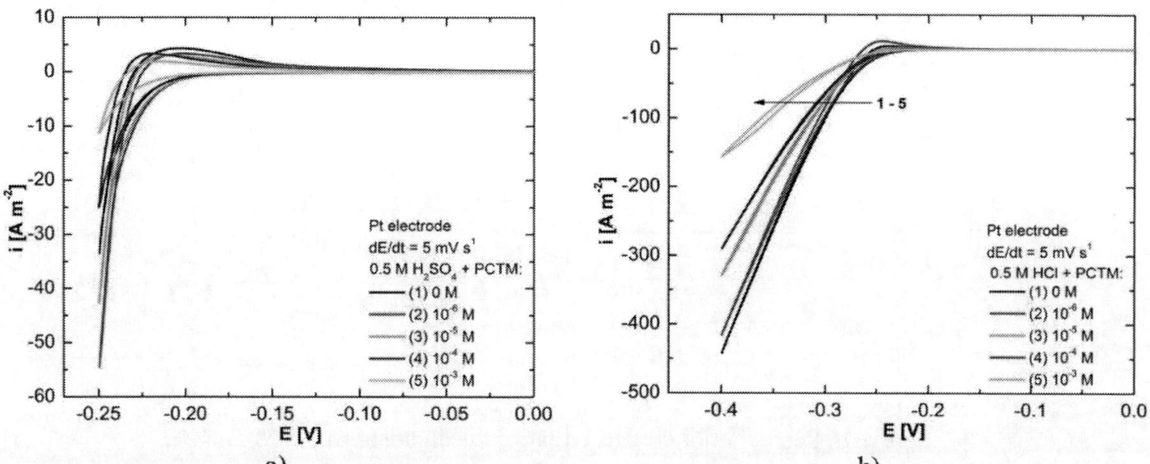

a) b)

Figure 3. CVs for cathodic polarization recorded on Pt electrode in 0.5 M H_2SO_4 (a) and in 1 M HCl (b) in the absence/presence of paracetamol at 5 mV s^{-1}scan rate.

According to figure 3, HER is inhibited due to the adsorption on platinum electrode of paracetamol molecules; the effect is more evident when 1 mol L^{-1} HCl is used as a corrosive agent.

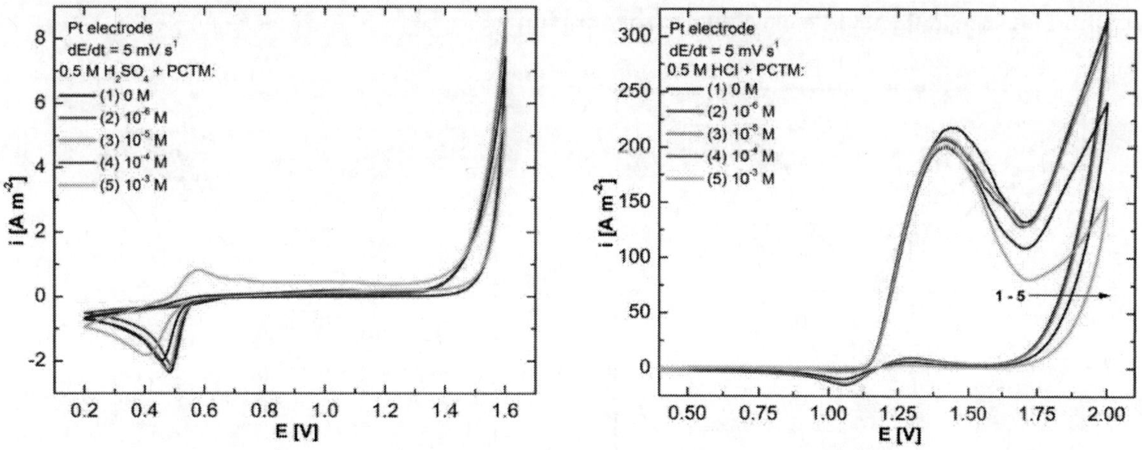

a) b)

Figure 4. CVs for anodic polarization recorded on Pt electrode in 0.5 M H_2SO_4 (a) and in 1 M HCl (b) in the absence/presence of paracetamol at 5 mV s^{-1}scan rate.

The anodic CVs recorded on Pt electrode in 1 mol L^{-1} HCl solution (Figure 4b) highlight an inhibition of both anodic OER and chlorine evolution reaction processes, stronger effect is noticeable as paracetamol concentration increases in the electrolyte solution. OER process is not influenced by the presences of the expired drug in 0.5 mol L^{-1} H_2SO_4 solution (Figure 4a).

However, as previously noted, in 0.5 mol L^{-1} H_2SO_4 solution paracetamol is oxidized, the peak corresponding to this process being recorded on the CVs anodic plateau at +0.55 V. Paracetamol electrochemical oxidation in acidic media proceeds in two steps according to the mechanism shown in figure 5 [6,7]. In first step, by releasing of two electrons and two protons paracetamol (N-(4-hydroxyphenyl)acetamide) is oxidized to N-acetyl-p-benzoquinone-imine and in the second step, this compound will hydrolyze. Although it was initially assumed that paracetamol oxidation process is irreversible, the shape of the corresponding cathode peak recorded on CVs at +0.45 V and literature data shows that this process is cvasi-reversible [7].

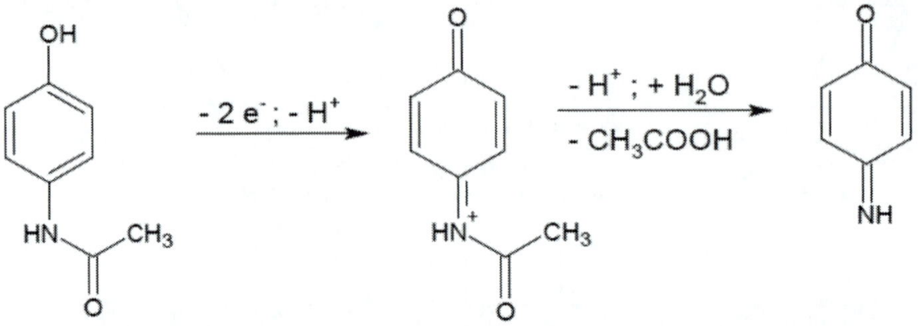

Figure 5. Paracetamol electrooxidation mechanism in 0.5 M H_2SO_4.

3.2. *Linear voltammetry*

LVs shown in figure 6 were recorded at low scan rate, (1 mV s^{-1}) in the range ±250mV/OCP, at 25°C on OL52-3k electrodes in aggressive solutions (0.5 mol L^{-1} H_2SO_4 and in 1 mol L^{-1} HCl) in the absence and presence of paracetamol different concentrations (10^{-6}-10^{-3} mol L^{-1}). LVs were plotted after removing dissolved oxygen from the acid solutions to avoid secondary reactions that may occur on the electrode surface and after 60 minutes conditioning at OCP in order to reach a quasi-equilibrium state at the interface metal/electrolyte solution.

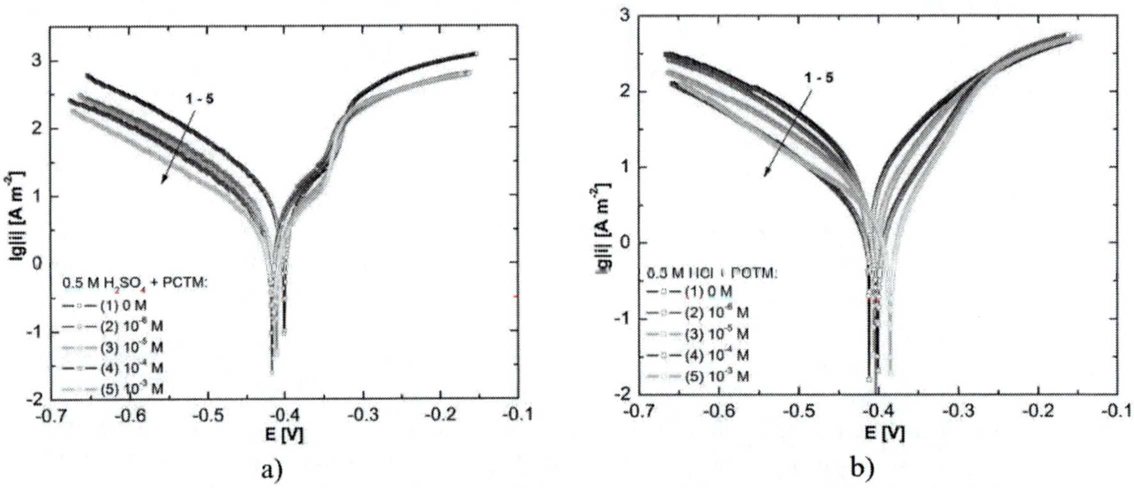

Figure 6. Tafel curves on OL52-3k electrodes in 0.5 M H_2SO_4 (a) and in 1 M HCl (b) in the absence/presence of paracetamol expired drug, scan rate: 1 mV s^{-1}.

Polarization parameters for OL52-3k corrosion in both acid media, 0.5 mol L^{-1} H_2SO_4 and 1 mol L^{-1} HCl, i.e., corrosion potential (E_{corr}), corrosion current density (i_{corr}), cathodic and anodic Tafel slopes (b_c and b_a), polarization resistance (R_p) and corrosion rate (v_{corr}) were determined extrapolating the potentiodynamic curves. The inhibition efficiency (*IE*) (Equation 1) and surface coverage degree (θ) with the paracetamol molecules have been calculated and the results are gathered in table 1.

$$IE(\%) = \left(\frac{i^{o}_{corr} - i^{inh}_{corr}}{i^{o}_{corr}}\right) \cdot 100 \quad (1)$$

where i^{o}_{corr} and i^{inh}_{corr} are the uninhibited, respectively inhibited corrosion current densities.

Table 2. Polarization parameters for OL52-3k corrosion process in acid solutions.

Solution	PCTM conc (M)	E_{corr} (mV)	i_{corr} (mA cm^{-2})	$-b_c$ (mV)	b_a (mV)	R_p (Ω)	v_{corr} (mm/year)	IE (%)	θ
	0	-398	3.31	197	126	6.72	51.0	-	-
	10^{-6}	-419	1.91	181	114	16.9	29.4	42.3	0.42
H_2SO_4	10^{-5}	-413	0.72	164	102	19.4	11.5	78.3	0.78
	10^{-4}	-417	0.30	154	89	21.9	4.76	90.9	0.91
	10^{-3}	-414	0.21	146	77	28.2	6.38	93.7	0.94
	0	-410	1.30	216	165	21.7	27.1	-	-
	10^{-6}	-404	0.84	190	130	29.8	17.1	35.4	0.35
HCl	10^{-5}	405	0.43	182	112	33.0	9.43	67.1	0.67
	10^{-4}	-400	0.31	148	77	50.8	3.16	76.5	0.77
	10^{-3}	-384	0.19	138	63	54.6	2.10	85.8	0.86

From the results presented in table 2, one can see that OL52-3k corrosion rate is diminished by increasing the paracetamol concentration added in aggressive solution, maximum inhibition efficiency values of 86% in HCl and 94% in H_2SO_4 solutions were obtained for 10^{-3} mol L^{-1} expired drug added in acid solution. The presence of paracetamol and its oxidation products cause significant changes of the anodic and cathodic Tafel slopes. This suggest paracetamol can be classified as a mixed-type inhibitor.

OL52-3k corrosion process mitigation in the presence of paracetamol can be attributed to its adsorption on the metal surface blocking thus the active sites. In the acid aqueous solutions, the adsorption of expired drug molecules can be considered as a quasi-substitution process between the paracetamol from the aqueous phase and water molecules at the metal surface, according to equation 2 [8,9]:

$$PCTM_{(sol)} + nH_2O_{(ads)} \rightarrow PCTM_{(ads)} + nH_2O_{(sol)} \quad (2)$$

where $PCTM_{(sol)}$ and $PCTM_{(ads)}$ are paracetamol molecules in the acid aqueous solutions, respectively in adsorbed state and n is the water molecules number replaced by PCTM.

3.3. Electrochemical impedance spectroscopy.

Figures 7 and 8 illustrat EIS spectra expressed as Nyquist and Bode plots recorded for OL52-3k corossion in 0.5 mol L^{-1} H_2SO_4 and 1 mol L^{-1} HCl solutions in the absence and presence of paracetamol different concentrations used in experimental studies.

On Nyquist diagrams, OL52-3k corrosion process is described by a single semi-circle corresponding to a single charge transfer process during carbon steel dissolution in acid media. As can be seen, each paracetamol concentration added in aggressive solutions has a distinctive semicircular graph in Nyquist spectra. The semicircle diameter can be associated with charge transfer resistance and also, with the corrosion rate. The semicircle diameter increases substantially with increase the paracetamol concentration added in test solutions, which means that the OL52-3k corrosion rate decreases. In other words, the inhibition efficiency of paracetamol is increasing in the same way [10,11].

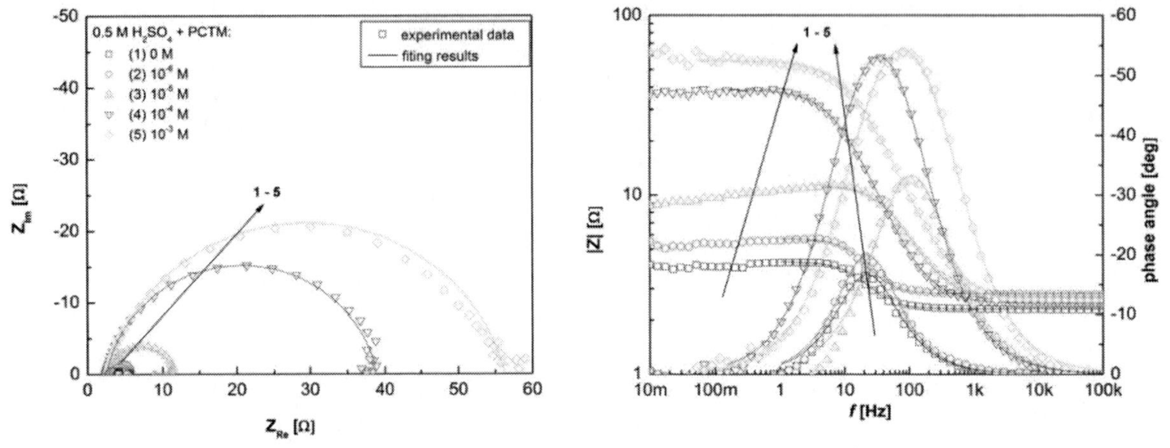

Figure 7. Nyquist (a) and Bode (b) plots on OL 52-3k in 0.5 M H_2SO_4 without and with different concentrations of paracetamol expired drug.

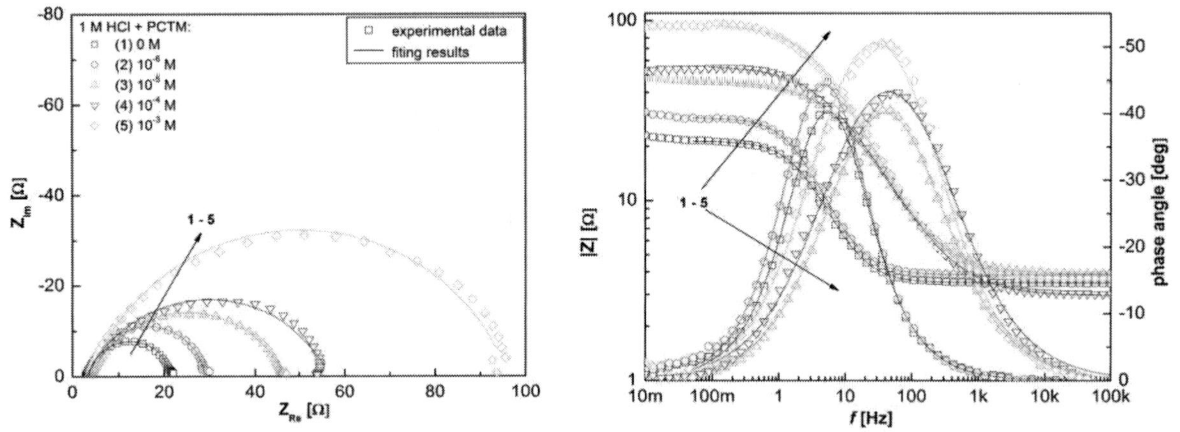

Figure 8. Nyquist (a) and Bode (b) plots on OL 52-3k in 1 M HCl without and with different concentrations of paracetamol expired drug.

For a precise corrosion process description, the experimental EIS data were fitted using the electrical equivalent circuit (EEC) shown in figure 9, characteristic for carbon steel corrosion in acid media, using a complex non-linear least squares (CNLS) procedure [10-12].

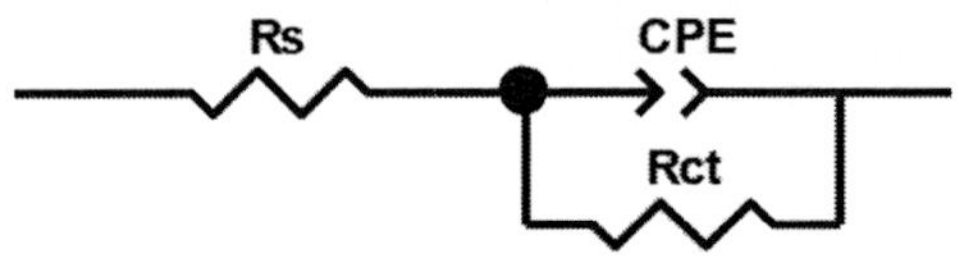

Figure 9. EEC for modeling OL52-3k corrosion in acid solution.

The EEC consists of a solution resistance R_s in series with a parallel connection of a constant phase element (*CPE*) accounting for the double layer capacity and a charge transfer resistance R_{ct}. The resistance R_s represents the uncompensated solution resistance. In EEC used for modelling OL52-3k corrosion process the ideal capacitor characterized by double layer capacity (C_{dl}) is replaced by a constant phase element (*CPE*) because it represents more precisely the real electrochemical systems behaviour (depressed loops in EIS diagrams). The impedance of *CPE* element is described by the Equation 3 [13]:

$$Z_{CPE} = 1/T(j\omega)^n \quad (3)$$

where *T* is a parameter proportional with double layer capacity, n is an exponent ranged between 0 and 1 which describes CPE angle.

The fitting results are depicted as continuous line in EIS spectra from figures 7 and 8 and the obtained values of the EEC elements are presented in table 3 for OL52-3k samples corrosion in both acid media. Also, double layer capacity values (C_{dl}) were calculated using Equation 4 [13]:

$$C_{dl} = T^{1/n}\left(\frac{1}{R_s} - \frac{1}{R_{ct}}\right)^{\frac{n-1}{n}} \quad (4)$$

Expired paracetamol inhibition efficiency values were calculated from the charge transfer resistance values (R_{ct}) from table 3 according to the Equation 5 [11,14]:

$$IE(\%) = \frac{(1/R_{ct}^0) - (1/R_{ct}^i)}{(1/R_{ct}^0)} \cdot 100 \quad (5)$$

where R_{ct}^0 and R_{ct}^i are the charge transfer resistance in acid solutions without respectively with different paracetamol concentrations.

Table 3. Calculated values of the EEC elements for OL52-3k corrosion in acid solutions.

Solution	PCTM conc (M)	R_s (Ω cm^2)	$T\cdot 10^4$ (F cm^{-2} s^{n-1})	n	R_{ct} (Ω cm^2)	$C_{dl}\cdot 10^5$ (F cm^{-2})	Chi^2 $\cdot 10^3$	IE (%)	θ
H_2SO_4	0	2.32	5.23	0.94	1.93	33.1	2.4	-	-
	10^{-6}	2.82	3.37	0.91	3.30	17.2	1.5	41.6	0.42
	10^{-5}	2.69	1.64	0.91	8.80	8.27	4.2	78.1	0.78
	10^{-4}	2.48	0.91	0.89	35.4	4.45	1.4	94.6	0.95
	10^{-3}	2.52	0.45	0.86	53.8	1.70	3.9	96.4	0.96
HCl	0	3.48	5.51	0.92	14.6	35.7	2.3	-	-
	10^{-6}	3.85	4.72	0.88	23.2	25.9	1.9	37.1	0.37
	10^{-5}	3.88	2.34	0.85	44.5	10.2	0.5	67.3	0.67
	10^{-4}	2.94	1.44	0.80	55.7	4.20	2.2	73.9	0.74
	10^{-3}	3.59	0.99	0.78	95.6	2.59	2.6	84.8	0.85

For a one time constant process, described by EEC from figure 9, R_{ct} value is directly proportional to the corrosion resistance of OL52-3k samples in acid solution. The fitting data from table 3 clearly shows that R_{ct} value increases when increased paracetamol drug amounts in acid solutions, confirming the results obtained in linear polarization experiments. Consecutively with increase of R_{ct} values the C_{dl} values are reduces. The decrease of C_{dl} values is according to the Helmholtz model in which the double layer capacitance is inversely proportional to the surface charges and can result from a decrease in local dielectric constant or an increase in the electric double layer thickness, suggesting that paracetamol molecules are adsorbed at the carbon steel electrode/acid solution interface [11,14]. This means that the paracetamol molecules were adsorbed by the gradual replacement of water on the electrode surface and form an adherent protective film on OL52-3k samples, decreasing the electrochemical dissolution rate for carbon steel in acid media [11,12]. The order of 10^{-3} for chi-square values indicate an excellent correlation between the experimental EIS data and the chosen EEC model values. Also, the inhibition efficiency (*IE*) increases with the increasing of expired drug concentration added in acid solutions. The inhibitory efficiency values calculated from EIS data (Table 3) are comparable to those obtained in linear polarization studies (Table 2).

3.4. Chronoamperometric studies

Chronoamperometric measurements were carried out in order to further elucidate the comparative corrosion behaviour of the OL52-3k samples in acid solutions without and with different concentrations of paracetamol drug at anodic polarization [15,16]. Current-time (*i-t*) curves have been recorded for a period of 15 minutes in the case of two different values applied anodic potential, +25 and +250 mV/E_{corr} for carbon steel samples in all test solution. Figure 10 shows the *i-t* curves recorded on OL52-3k electrode in 1 mol L^{-1} HCl solution in the absence and presence of paracetamol.

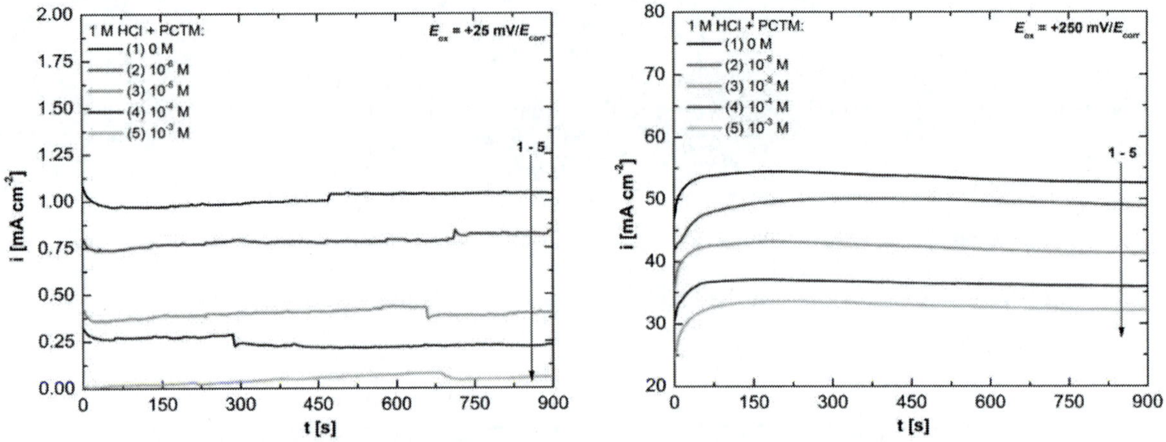

a) b)

Figure 10. Chronoamperometric measurements recorded at +25 mV/ E_{corr} (a) and +250 mV/E_{corr} (b) for the anodic oxidation of OL52-3k electrodes in 0.5 M HCl without and with different paracetamol concentrations for 15 min.

Analyzing the graphical *i-t* results, a common feature of carbon steel electrooxidation at two potential values is the rapid stabilization of current density after about 5 minutes, probably due to the formation and growth of a passive oxide film uniformly distributed on the metal surface [15]. The shape of presented curves indicates a continuous carbon steel oxidation and the current density characteristic of this process is depending on the paracetamol concentration. The current density values recorded after 15 minutes for all chronoamperometric measurements are shown in table 4.

Table 4. Chronoamperometric data for OL52-3k oxidation in test solutions.

PCTM conc (M)	0.5 M H_2SO_4		1 M HCl	
	E_{ox} = +25 mV/E_{corr}	E_{ox} = +250 mV/E_{corr}	E_{ox} = +25 mV/E_{corr}	E_{ox} = +250 mV/E_{corr}
	i (mA cm^{2})			
0	6.47	78.2	1.04	52.6
10^{-6}	5.41	69.1	0.82	48.8
10^{-5}	3.62	50.3	0.41	41.4
10^{-4}	2.95	47.7	0.22	36.1
10^{-3}	2.07	43.7	0.06	31.9

In acid solution without expired drug, on the electrode surface, a protective layer based on corrosion products is formed, which provides a partial anticorrosive protection [17]. The current density lower values recorded for the oxidized samples in the acid media with paracetamol are due to the formation of a protective film of organic molecules on the OL52-3k active surface, according to the mechanism described by equation 2 and explained based on EIS data, which prevents or decreases the rate of the carbon steel corrosion.

The current densities values specific for the OL52-3k samples anodic oxidation in 0.5 mol L^{-1} H_2SO_4 in the absence, respectively presence of paracetamol expired drug, are higher than those recorded in 1 mol L^{-1} HCl, indicating that the corrosion rate of this carbon steel type is higher in sulfuric acid according to the experimental results obtained in linear polarization method and EIS studies.

4. Conclusions

The presented studies shows excellent inhibition properties of paracetamol expired drug for the corrosion of carbon steel type OL52-3k in 0.5 mol L^{-1} H_2SO_4 and 1 mol L^{-1} HCl. The paracetamol concentration added in acid media influences significantly the inhibition efficiency, calculated from Tafel polarization method and confirmed by EIS data. The maximum inhibition efficiency values of 94% in 0.5 mol L^{-1} H_2SO_4 and 86% in 1 mol L^{-1} HCl were obtained for 10^{-3} mol L^{-1} paracetamol concentration added in both acid solutions. Based on the linear polarization results, the expired drug acts predominantly as a mixed-type inhibitor (anodic and cathodic). Cyclic voltammetry studies concluded that the expired paracetamol are electrochemical stable in HCl solution. In H_2SO_4 solution it is oxidized in two steps mechanism according to a quasi-reversible process, the oxidation products formed being adsorbed on the active surface of the samples, contributing to the increase of the inhibitory efficiency. The results show that paracetamol molecules or its oxidation products form a barrier layer on the OL52-3k samples surface by adsorption, thus inhibiting the corrosion process.

5. References

[1] Singh P, Chauhan DS, Srivastava K, Srivastava V and Quraishi MA 2017 *Int. J. Ind. Chem.* **8** 363-72
[2] Karthikeyan S, 2016 *Int. J. Chemtech. Res.* **9** 251-9
[3] Stark G, Fawcett JP and Tucker IG *Pharm. J.* **258** 637-40
[4] Lyon RC, Taylor JS, Porter DA, Prasanna HR and Hussain AS 2006 *J. Pharm. Sci.* **95** 1549-60
[5] Vaszilcsin N, Ordodi V and Borza A 2012 *Int. J. Pharm.* **15** 241-4.
[6] Zidan M, Tee TW, Abdullah AH, Zainal Z and Kheng GJ 2011 *Int. J. Electrochem. Sci.* **6** 279-88
[7] Nematollahi D, Shayani-Jam H, Alimoradi M and Niroomand S 2009 *Electrochim. Acta* **54** 7407-15
[8] Dan ML, Vaszilcsin N, Labosel MA and Pancan B, 2014 *Chem. Bull. "POLITEHNICA" Univ. Timisoara* **73** 13-18
[9] Andreani S, Znini M, Paolini J, Majidi L, Hammouti B, Costa J and Muselli A 2016 *J. Mater. Environ. Sci.* 7 187-95
[10] Al-Amiery AA, Kadhum AAH, Kadihum A, Mohamad AB, How CK and Junaedi S 2014 *Materials* **7** 787-804
[11] Galai M, El Gouri M, Dagdag, El Kacimi Y, Elharfi A and Ebn Touhami M 2016 *J. Mater. Environ. Sci.* 7 1562-75
[12] Kadhum AAH, Mohamad AB, Hammed LA, Al-Amiery AA, San NH and Musa AY 2014 *Materials* **7** 4335-48
[13] Savencu CE, Costea LV, Dan ML and Porojan L 2018 *Int. J. Electrochem. Sci.* **13** 3588-600
[14] El-Taib Heakal F, Fouda AS and Zahran SS 2015 *Int. J. Electrochem. Sci.* **10** 1595-615
[15] Porojan L, Savencu CE, Costea LV, Dan ML and Porojan SD 2018 *Int. J. Electrochem. Sci.* **13** 410-23
[16] Sherif E-SM, Abdo HS and El Abedin SZ 2017 *Int. J. Electrochem. Sci.* **12** 1600-11
[17] Sherif E-SM 2012 *Int. J. Electrochem. Sci.* **7** 1482-95

Acknowledgment
This work was partially supported by *Politehnica* University Timisoara in the frame of PhD studies.

Eco-friendly Improvement of Water Erosion Resistance of Unstable Soils with Biodegradable Polymers

A Donayre[1], L F Sanchez[1], S Kim[2], R Aguilar[2] and J Nakamatsu[1,*]

[1] Science Department, Pontificia Universidad Catolica del Peru, Lima 32, Peru
[2] Engineering Department, Pontificia Universidad Catolica del Peru, Lima 32, Peru

[*]Corresponding author: javier.nakamatsu@pucp.pe

Abstract. The improvement of water erosion resistance of soils is gaining the attention of the scientific community. Erosion represents a threat to agricultural productivity because of the loss of valuable superficial soil and nutrients. Cultivation soils erosion is one of the main contributors to the desertification process, which is, itself, a global problem. Recently, polysaccharides have been tested as additives to improve earth construction materials with respect to their mechanical properties and water erosion resistance. This paper studies the use of biodegradable polymers to improve soil stabilization with respect to erosion by water. Aqueous solutions of the polysaccharides chitosan and carrageenan were used to improve wettability and runoff resistance of sandy and loose soils. Chitosan becomes positively charged when dissolved in dilute acidic solutions, while carrageenan becomes negatively charged. Their capacity to form polyelectrolytes in aqueous solutions may be a contributor in retaining cohesiveness in loose soils.

1. Introduction

The soil has been an important component in human development, both for its use as a construction material and because of its fundamental role in agriculture. Some of the problematics concerning these two areas are related to its erosion. The propensity to erosion is originated by the presence of weakly-bonded small particles in the surface of the material [1]. Erosion occurs when an external agent such as wind, flowing water or mechanical friction grounds down particles causing the loosening of the already weak bonds, affecting the integrity of the material. As a consequence, erosion facilitates the loss of soil components either in the form of fugitive dust, as a result of loose particles being blown off and suspended in the air, or as sediments being carried away. Dust suspension is one of the causes of air pollution and can have negative effects: blocking sunlight affecting vegetation and crops development, altering near water bodies' characteristics such as pH and turbidity [2], and creating potential respiration health issues. On the other hand, loose soil particles displacement menaces the agricultural productivity due to the loss of valuable surface soil and its nutrients [3], which is one of the causes of desertification [4].

Synthetic and natural chemical products have been used as additives to modify loose soil properties and improve its strength and stability, which improves the resistance to water and wind erosion [4] through techniques like cementation, chemical linkage and waterproofing [5]. The effects of numerous synthetic additives, like a mixture of sulfonate-urea-melamine-formaldehyde resins and poly(vinyl alcohol) [1] and polyacrylamides, xanthates and cellulose charged microfibrils [3], over runoff resistance tests have been studied.

Biodegradable polymers have also been used as additives, carrageenan and chitosan have been studied as an additive in adobe, evaluating their water erosion resistance through drip tests [5, 6]. Biodegradable additives have the advantage of avoiding both soil pollution by chemical accumulation and the formation of toxic intermediates during their degradation process [3]. Therefore, application of these additives can help fight soil erosion for vegetation growth and control fugitive dust generated in gravel and dirt roads.

2. Experimental part

For this study, biodegradable polymers were used as additives, specifically carrageenan and chitosan, spraying their aqueous solutions over sand and soil surfaces. Three tests were conducted to evaluate impermeability and water-erosion resistance of treated soils: Measurement of the contact angle of a sessile water-drop and drip resistance test of treated sand and soil; and water runoff resistance over a sand surface.

2.1. Preparation of the additives and loose soil samples

Carrageenan was extracted from algae *Chondracanthus chamissoi* while chitosan was purchased from Aldrich. Additives consisted of solutions of carrageenan in deionized water and chitosan in a 1% (v/v) aqueous acetic acid solution. For contact angle measurements and drip tests, solutions of 0.5 % (w/v) were used, while for the runoff tests 0.5 % (w/v) as well as 0.2 % (w/v) solutions were used. For each test, a reference sample (control) consisting of the material without additive was also evaluated. The soil was sieved in order to remove gravel and larger particles. Both the soil and sand were slightly compacted by manual pressure for contact angle and drip tests.

2.2. Contact angle of a sessile water-drop

Two sets, one for sand and the other for soil, of five samples each were prepared: A petri dish (6 cm of diameter) was filled with the respective material, sand or soil. Samples were weighed before and immediately after spraying the additive solution. Table 1 shows the additives and quantities used to cover each sample, as well as the average contact angles measured.

Table 1. Samples and contact angles measured.

Sample*	Additive quantity** ($mg{\cdot}cm^{-2}$)		Contact angle (°)
	Carrageenan	Chitosan	
Control-Sand	0	0	n.m.
Sand-Car	0.503	0	n.m.
Sand-Chi	0	0.220	140.7 ± 5.5
Sand-Car/Chi	0503	0.470	131.0 ± 12.2
Sand-Chi/Car	0370	0.220	140.3 ± 46.0
Control-Soil	0	0	n.m.
Soil-Car	0.816	0	n.m.
Soil-Chi	0	0.644	106.8 ± 5.9
Soil-Car/Chi	0.715	0.513	133.7 ± 4.1
Soil-Chi/Car	0.323	0.724	138.6 ± 3.8

* Car/Chi = First layer of carrageenan, second (top) layer of chitosan.

** Additive quantity corresponds to dry polymer added.

n.m. = Could not be measured (water drop was immediately absorbed).

Samples were left to dry at room temperature for 1 day after additive application. Once dry, 20 μL deionized water drops were deposited gently on their surface with a microsyringe. For each sample, five drops were randomly distributed and their contact angles were measured. Ten measurements were performed for each type of sample, when possible.

2.3. Water drip resistance test

For this test, four sets, two of sand and two of soil (each consisting of 3 samples) were prepared. Samples were similar to the ones used for contact angle measurements, a petri dish filled with sand or soil with a section of the container's wall cut and replaced by a filter to avoid water stagnation. Table 2 shows the additives and their amount. Control samples were evaluated but omitted from the table since they contained no additive.

Water erosion was evaluated by adaptation of a Spanish standard test for compressed soil blocks [7]. According to this test, water drops were allowed to fall vertically from a height of 1 meter at a flow rate of 50 $mL \cdot min^{-1}$ directly over the sample, placed at a slope of 27° in relation to the horizontal. Erosion damage was evaluated at 1 minute and changes were registered by video and photos. The test was continued for the samples that showed no damage after 1 minute of dripping until significant erosion was observed.

Table 2. Samples and drip resistance test results.

Sample	Additive quantity* ($mg \cdot cm^{-2}$)		Time of damage observed (s)
	Carrageenan	Chitosan	
Control-Sand-1	0	0	0
Sand-Chi/Car-1	0.636	0.596	15
Sand-Car/Chi-1	0.665	0.569	600
Control-Sand-2	0	0	0
Sand-Chi/Car-2	0.594	0.523	10
Sand-Car/Chi-2	0.607	0.542	300
Control-Soil-1	0	0	0
Soil-Chi/Car-1	0.640	0.922	13
Soil-Car/Chi-1	0.858	0.587	220
Control-Soil-2	0	0	0
Soil-Chi/Car-2	0.527	0.697	12
Soil-Car/Chi-2	0.779	0.498	170

* Additive quantity corresponds to dry polymer added.

2.4. Water runoff resistance test

A tray of dimensions 36.5 x 23.4 x 0.8 cm (length, width and depth), was filled with sand. This tray was tilted 15° from the horizontal. A continuous stream of water (234 $mL \cdot min^{-1}$ flow rate) fell from a height of 5 cm and, after impact, it flowed over the sand surface for 1.5 minutes. Two tests were performed, in the first one, half of the sample's surface was sprayed with a chitosan 0.5% solution, while the remaining half was left without additive (control). In the second test, half was coated with carrageenan 0.5% as a bottom layer and chitosan 0.5% on top, while the other half of the surface was coated with the same additives (same order of application) but with a concentration of 0.2% each. Videos and pictures were taken.

3. Results and Discussion

3.1. Contact angle of a sessile water-drop

Water drops were not formed in control samples nor in the ones treated with carrageenan only, water applied was immediately absorbed. In contrast, water drops formed in samples Sand-Chi, Sand-Car/Chi and Sand-Chi/Car and their soil homologs, as shown in figures 1 and 2. Angles were obtuse, as table 1 shows, which reveal a hydrophobic surface and suggests improvement in the impermeability of the material. On sand, all angles fall within a range. Angles for soil samples showed similar values as the additives used were the same, except for the sample Soil-Chi, which shows a lower value. An important difference between soil and sand is that water drops on sand samples (Sand-Chi, Sand-Car/Chi and Sand-Chi/Car) remained on the surface for more than one day, while the ones formed on top of the soil were absorbed more rapidly. In the case of Soil-Chi, the drop was absorbed in about 3 seconds. For Soil-Chi/Car, the drop lasted around 128 seconds, while for Soil-Car/Chi sample, the absorption time was 67 seconds. The difference in contact angle results may be due not only to soil and sand chemical composition but also to their particle size, porosity and affinity to water. Even more, surface texture is also important when evaluating contact angles, in this respect, the lotus effect is explained by surface rugosity at the micro and nanoscale. On the other hand, the viscosity of the polymer solution might affect its deposition and distribution with soil or sand particles on the surface, carrageenan forms more viscous solutions than chitosan. And finally, once samples are dry, chitosan's insolubility in neutral water (from sessile water-drop) should form a less permeable film than carrageenan, which is water soluble.

Figure 1. Water drops formed on samples a) Sand-Chi, b) Sand-Car/Chi and c) Sand-Chi/Car.

Figure 2. Water drops formed on samples a) Soil-Chi, b) Soil-Car/Chi and c) Soil-Chi/Car.

3.2. Water drip resistance test

Results for sand and soil samples after 1 minute of drip test were similar and are shown in figure 3. Control samples of both materials showed instant erosion from dripping and after 30 s, a wide and deep hole was formed. Samples covered by a layer of chitosan and a top layer of carrageenan revealed the formation of a narrow hole after 10-15 seconds, exactly where the drop impacted the surface. The best resistance to water drop erosion was exhibited by samples covered by a layer of carrageenan and a top layer of chitosan in both samples, soil and sand, showed no damage at all after 1 minute. The time at which these samples showed damaged differed and can be observed in table 2.

IOP Conf. Series: Materials Science and Engineering **416** (2018) 012044 doi:10.1088/1757-899X/416/1/012044

Figure 3. Pictures of drip test results for soil and sand samples, after 1 minute.

3.3. Water runoff test

Sand with no additives was completely carried away by the flow of water as figure 4(B) illustrates. In comparison, all the samples treated with chitosan or a combination of carrageenan and chitosan solutions, as shown in figures 4(A), 4(C) and 4(D), resisted water erosion and avoided sand sliding, and water came out clear at the end of the tray after flowing over them. Lower concentrations of additives were also observed to successfully protect the samples.

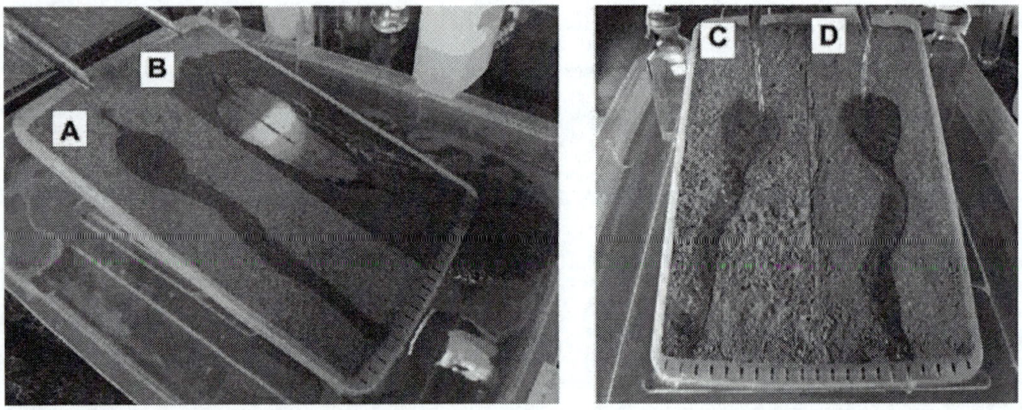

Figure 4. Water runoff tests over sand: (A)-Chitosan 0.5 % against (B)-Control; (C)-Carrageenan 0.5 % and chitosan 0.5 % (top) against (D)-Carrageenan 0.2 % and chitosan 0.2 % (top)

Results and observations of all the tests suggest that chitosan was effective in changing the properties of the surface of loose soil and sand. Chitosan efficacy can be due to its cationic nature in solution (polycation) which can bind strongly to soil and sand particles, which are known to hold negatively charged surfaces [8]. On the other hand, carrageenan itself could not protect loose soil nor sand against water erosion, which may be due to its solubility in water. However, electrostatic interactions between carrageenan and chitosan (polyelectrolytes of opposite charges) may help stabilize the particles even more. Moreover, the higher viscosity of polymer solutions, which allows the formation of 3-dimensional bridges between the polymeric solution and the particles [8], together with their capacity to form films when the solvent is eliminated, may also contribute to hold particles attached to each other and in place.

4. Conclusions

Improvement of water erosion resistance and impermeability of soil and sand surfaces by application of biopolymers chitosan and carrageenan was demonstrated. As contact angle measurements of water drops suggest, waterproofing was achieved for all samples in which chitosan was applied (only momentarily in the case of soil), either as a sole additive or as part of a mixture, and also, when carrageenan was applied on top of a chitosan layer. Drip test's results revealed that a combination of layers of chitosan and carrageenan improved water erosion resistance, being the sample with chitosan on top the one that showed the best performance. A combination of 0.2% chitosan solution on top and 0.2% carrageenan on the bottom was also able to avoid sand from being carried away by water flow, as the runoff test results evidenced. Altogether, an additive consisting of chitosan on top of a carrageenan layer seems to be an effective means of soil and sand stabilization.

5. References

[1] Lahalih S M and Hovakeemian G 1988 *Ind. Eng. Chem.* **27** 1806-10
[2] Anderson J and Gesford A 2007 Environmentally sensitive maintenance for dirt and gravel roads. https://www.epa.gov/sites/production/files/2015-10/documents/environmentallysensitive maintenance_dirtgravelroads.pdf. (accessed: February 2018)
[3] Orts W J, Sojka R E, Glenn G M and Gross R A 2001 Biopolymer additives for the reduction of soil erosion losses during irrigation *Biopolymers from Polysaccharides and Agroproteins, ACS Symposium Series* eds R A Gross and C Scholz (Washington DC: American Chemical Society) pp. 102-16
[4] Lahalih S M 1998 *Ind. Eng. Chem.* **37** 420-6
[5] Nakamatsu J, Kim S, Ayarza J, Ramirez E, Elgegren M and Aguilar R 2017 *Constr. Build. Mater.* **139** 193-202
[6] Aguilar R, Nakamatsu J, Ramirez E, Elgegren M, Ayarza J L, Kim S, Ortega L and Pando M A 2016 *Constr. Build. Mater.* **114** 625-37
[7] AENOR, UNE 41410:2008: Bloques de tierra comprimida para muros y tabiques: Definiciones, especificaciones y métodos de ensayo 2008 Madrid
[8] Hataf N, Ghadir P and Ranjbar N 2018 *J. Clean. Prod.* **170** 1493-500

Acknowledgments
This work was supported by the Direccion de Gestion de la Investigacion at Pontificia Universidad Catolica del Peru through the program PAIN-2017 (awarded to A Donayre) and by the Chemistry Department.

IOP Conf. Series: Materials Science and Engineering **416** (2018) 012045 doi:10.1088/1757-899X/416/1/012045

Identification of Delamination in Multilayered Composites

C Tufisi[1], G R Gillich[1,*], O Vasile[2], Z I Korka[1] and C O Hamat[1]

[1]"Eftimie Murgu" University of Resita
[2]Politehnica University of Bucuresti

[*]Corresponding author: gr.gillich@uem.ro

Abstract. The paper presents the study of the vibrational behavior of laminated composites whose sheets are debonded but still in contact. The aim is to show that the delamination occurring in a given region of the structure produces different frequency shifts for the different bending vibration modes, dependent on the position and extent of the delamination. The vibrational response that comprises the frequency shifts of several modes constitute the delamination signature and can be therefore be used as a patter in damage detection procedures. The paper presents a finite element analysis comprising composite structures with transversal cracks followed by different debondings which highlight the dynamic structural response individualized by damage position and severity.

1. Introduction

Damages influence the vibrational behavior of beam-like structures by diminishing the natural frequencies as a result of stiffness loss [1]. The stiffness loss, as well as the frequency decrease, directly depends on the damage shape and depth. Actual literature is rich in studies concerning detection of transverse cracks or delamination [2]-[6]. However, few studies deal with cracks of a more complex configuration as L or inverted T cracks that usually occur in composites [7] are, because of the difficulty in approaching such cases.

In our prior research we found a relation, applicable for isotropic or laminated beams, that permits to predict the frequency changes if the size and location of a transverse crack are known [8]-[11]. This relation leads to results similar to those obtained if a torsional massless spring is used to simulate the crack, but has the advantage of simplicity. In this paper, we compare the effect of transverse cracks with those of cracks having an L shape resulted as the longitudinal extension of a transversal crack.

2. Frequency shift due to transverse cracks

The mathematical relation predicting the frequency of the bending modes for a beam with a transverse crack, prior developed by the authors [12], imply the frequency of the healthy beam f_{i-U}, a term $\bar{\phi}_i''(x)$ representing the value of the normalized modal curvature that defines the crack location and a term $\gamma(0,a)$ representing the damage severity, which depends on the damage depth a. In a concentrated form, the relation of the damaged beam's frequency $f_{i-D}(x,a)$ takes the form:

$$f_{i-D}(x,a) = f_{i-U}\left[1-\gamma(0,a)\cdot(\bar{\phi}_i''(x))^2\right] \tag{1}$$

The square of the normalized modal curvature (or bending moment) for a given crack position x indicates the energy stored by the healthy beam in an infinitesimal slice dx in which the crack is located. The square of the normalized modal curvature is different for the different vibration modes, and, due to normalization, obviously takes values between 0 and 1 for all modes [13]. Distance x is taken from the fixed end. Figure 1 shows the energy distribution for the weak-axis bending vibration mode three. One can observe that there are locations where the crack produces no energy decrease and consequently it has no effect on the natural frequencies; this happens when it is located on inflection points. On the other hand, it produces main frequency alteration when it is positioned on local maxima of the modal curvature [14].

The damage severity $\gamma(0,a)$ is the function representing the highest stiffness decrease due to a crack of depth a, which for the cantilever beams occurs for the crack located at the fixed end (i.e. $x = 0$). We demonstrated that for an isotropic beam the function representing the damage severity versus damage depth is continuous and can be derived using the beam deflection in healthy and damaged state [15]. A similar frequency evolution is found as using the fracture mechanics method, but the severity estimation using beam deflection is a simpler approach [16]. For multilayered beams, the function has disruptions, occurring when the crack totally separates the different layers [17]. An example of severity curves for a steel beam is presented in figure 2 with continuous line with square bullets, while the case of a composite beam of same thickness with five layers is plotted with simple continuous line.

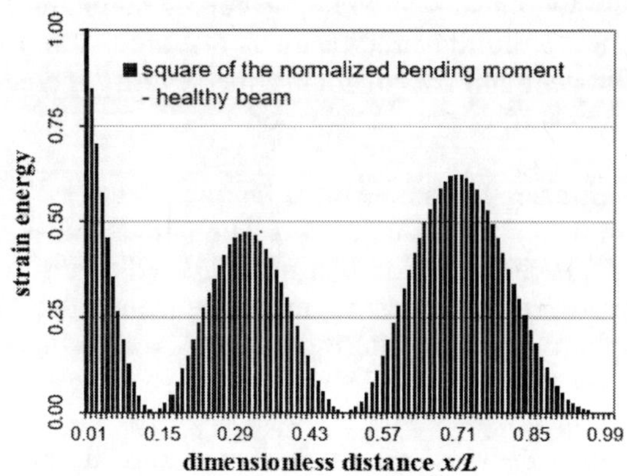

Figure 1. Stored energy distribution for the bending vibration mode three.

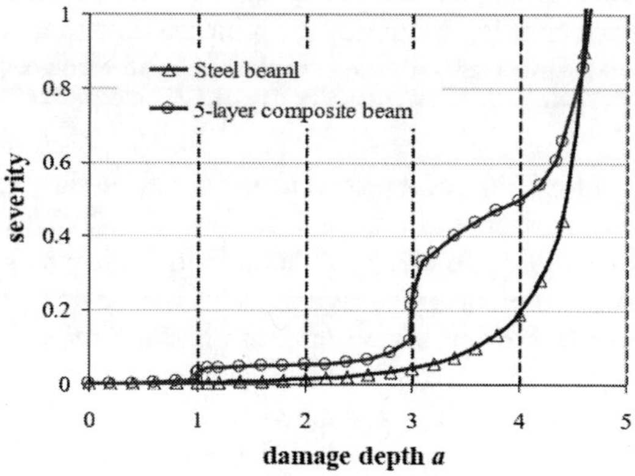

Figure 2. Damage severity for a steel beam and a 5-layered composite beam, respectively.

The frequency decrease is found from the mathematical relation (1) as:

$$\Delta f_i = f_{i-U} - f_{i-D}(x,a) = f_{i-U} \cdot \gamma(0,a) \cdot (\bar{\phi}_i''(x))^2 \quad (2)$$

As a consequence, the relative frequency shift for any bending vibration mode $i = 1...n$, denoted $\Delta \bar{f}_i(x,a)$, is the ratio between the frequency shift of that mode and the corresponding frequency of the undamaged beam [18]-[19]:

$$\Delta \bar{f}_i(x,a) = \frac{f_{i-U} - f_{i-D}(x,a)}{f_{i-U}} = \gamma(0,a) \cdot (\bar{\phi}_i''(x))^2 \quad (3)$$

From relations (2) and (3) clearly results that the frequency shift curves, i.e. the evolution curves of the cracked beam's frequency for different locations along the beam for a given mode, are similar for the isotropic and laminated beams, respectively. The single difference consists in the amplitude of the shift $\gamma(0,a)$. From figure 2 one can observe that a crack with depth 2 mm has the severity 0.005 for the layered beam, while the severity for the steel beam is obviously lower (approximately 0.0015). On the other hand, one can observe that the severity 0.05 is achieved in the isotropic beam for the crack depth 3.1 mm; this crack will provide the same effect as the 2 mm crack in the composite beam.

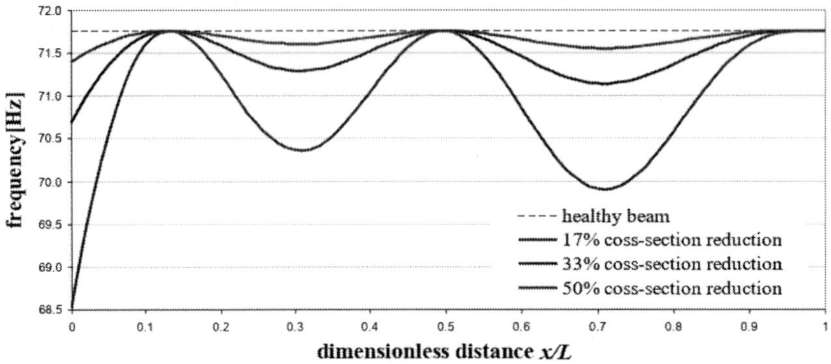

Figure 3. Frequency shift curves for mode three for three damage depths.

Figure 3 illustrates the frequency shift curves belonging to bending vibration mode three for a steel beam for cracks reducing the cross-section area with 17%, 33% and 50% respectively. Here one can observe that, by increasing the damage depth iteratively with a step of 17%, the curves maintain their shape but the frequency drop increases exponentially. For a multilayered beam as that studied in [17], consisting of five layers with 1 mm thickness (three of steel and two of foam), the severity increases faster with the crack depth, without altering the frequency shift curves in any mode.

3. The effect of transverse crack extension by delamination on the natural frequencies

This section is dedicated to the study of a crack with complex shape, consisting in a transverse extension followed by a delamination. It was shown in previous section that multilayered beams with transverse cracks behave as isotropic beams, the only difference consisting in the frequency decrease rule. Based on this observation and for reason of simplicity, without altering the achieved results, we employ in the study presented herein a beam having all layers of steel. Actually, the specimen is a steel beam with an L-shaped crack.

It is expected to attain significantly bigger frequency decreases in if the crack spreads also in longitudinal direction [20]. Thus, to compare the effect of a transverse crack with an L-shaped crack, normalization of results is necessary.

3.1. Materials and method

The modal analysis was performed by means of the finite element method (FEM) by involving the ANSYS simulation software. The specimen was a prismatic beam of steel, having the dimensions and material properties presented in Table 1. Boundary conditions corresponding to a cantilever beam were taken into consideration, meaning that the left beam end was constrained and the right beam end free. The model was meshed by using hexahedral elements having the maximum edge size 2 mm, resulting in a total of 37500 elements.

Table 1. Beam dimensions and material's physical-mechanical properties

Length L [mm]	Width B [mm]	Thickness H [mm]	Mass density ρ [kg/mm^3]	Young modulus E [N/m^2]	Poisson ratio υ [-]
1000	50	5	7850	$2\cdot10^{11}$	0.3

First, the natural frequencies for the first six weak-axis bending vibration modes of the healthy beam were targeted. The attained results are presented in Table 2.

In the second stage, the modal analysis comprised damaged beams of the same dimensions and material as the healthy one. The damage consists of a transverse crack with depth 2.5 mm, which is extended to the right in the longitudinal direction with a 50 mm "delamination", as shown in Figure 4.

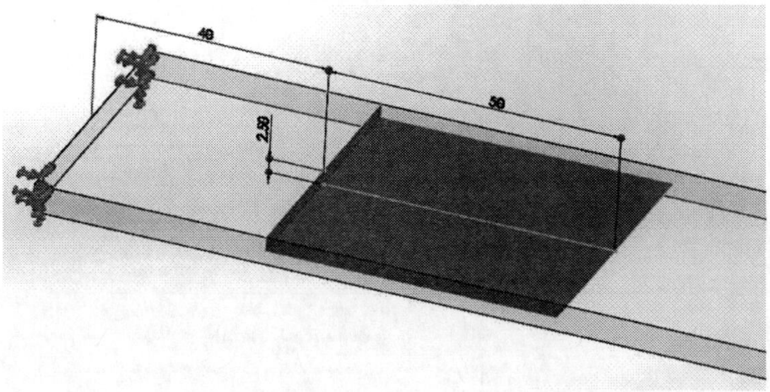

Figure 4. Fixed beam end with an L-shaped crack starting at distance 40 mm.

The damage is positioned with its left end at 20 mm from the fixed beam end, and is iteratively removed to the right with a step of 20 mm; last simulated damage has the left end at 940 mm from the fixed end, thus 47 damage cases being simulated.

The aim of this paper is to compare the dynamic behavior of beams with a transverse crack with those of the damaged beam an L-shaped crack. Therefore, the relative frequency shift curves for both cracks are plotted. For the transverse crack this is made by involving relation (1), while for the L-shaped crack the results obtained from the FEM supported analysis are taken. For better comparing the results, the severity $\gamma(a)$ is artificially increased by a scaling factor. This factor is individually chosen for each mode, in order to achieve in both curves the same value for the biggest shift, as indicated by point A for vibration mode 2 in Figure 5.

3.2. Results and discussion

Modal analysis was first performed for the healthy beam. The results, in terms of the frequencies for the first six bending vibration modes, are presented in Table 2. These fit the values obtained with the analytical method, the errors being less than 0.5%. Afterwards, the frequencies for the beam with an L-shaped crack were extracted. Examples of values obtained for the crack position at 240 mm, 500 mm and 760 mm from the fixed beam end are presented in Table 2, along with the values obtained for the transverse crack by calculus by employing relation (1). One can easily observe that, as expected, the effect of the L-shaped crack effect is much greater than that of the transverse crack.

Table 2. Frequencies obtained by simulation for the healthy beam, the beam with a transverse crack and the beam with an L-shaped crack placed at different locations on the beam.

Mode	Frequency [Hz]						
	Healthy beam	$x = 240$ mm		$x = 500$ mm		$x = 760$mm	
		Transverse crack	L-shaped crack	Transverse crack	L-shaped crack	Transverse crack	L-shaped crack
1	4.090	3.997	3.265	4.066	3.835	4.088	4.074
2	25.627	25.614	25.147	24.973	19.973	25.427	23.629
3	71.757	70.659	61.452	71.753	70.728	69.961	56.824
4	140.631	137.715	127.344	137.111	119.671	137.037	117.195
5	232.533	231.150	228.712	232.521	224.884	230.871	213.681
6	347.462	346.995	314.753	338.762	306.723	347.077	337.822

To deeply understand the phenomena of frequency shift due to different type of damage, we plotted the frequency shift curves for both types of crack. For comparison reasons, the severity of the transverse crack is artificially increased, as explained in previous sub-section.

IOP Conf. Series: Materials Science and Engineering **416** (2018) 012045 doi:10.1088/1757-899X/416/1/012045

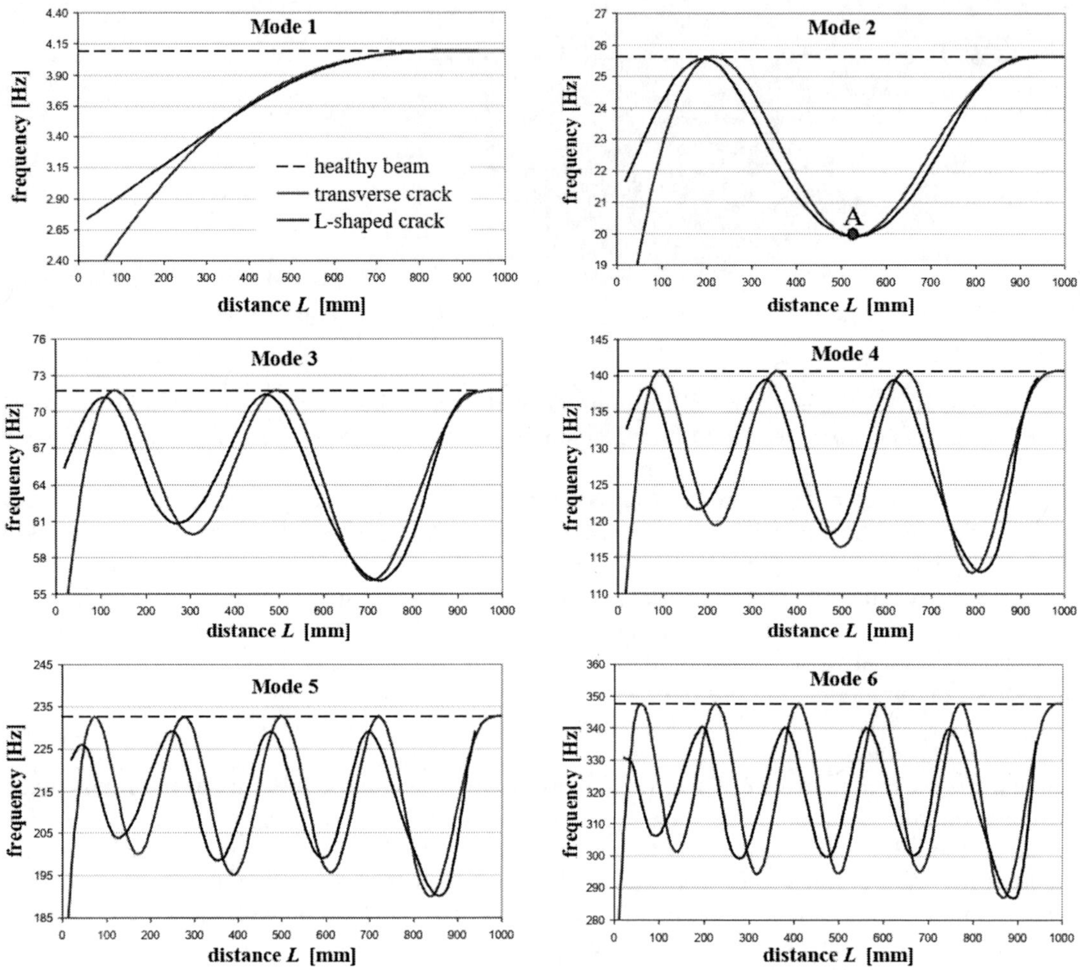

Figure 5. Frequency versus crack location for both types of damages for six vibration modes.

The way how an L-shaped crack diminishes the natural frequencies of the beam is shown by the evolution of the relative frequency shift with the crack position. One can observe that, in comparison with the transverse crack, the shifts due to an L-shaped crack are significantly greater in amplitude.

4. Conclusion

The frequency shift curves for a beam with an L-shaped crack were determined by simulation and compared with the values deduced for the transverse crack with an original mathematical relation. It was found that the two crack types produce different shifts, but the curves representing these shifts are somehow similar.

The key finding of the study refers to the largest frequency shift of the particular modes, which for the L-shaped crack does not longer manifest at the fixed end. An exception is the vibration mode one, which has a single maximum, right at the fixed end. This shows that achieving a unique severity for all modes is no longer possible.

Analyzing the frequency shift curves plotted for the L-shaped crack, it has been observed that the points of extrema do not match those found for the transverse crack. It was also noticed that the frequency of the beam with an L-shaped crack never reaches the frequency of the healthy beam. The phenomenon is increasingly pronounced for higher-order modes and the cracks closer to the fixed end. This happens differently to the beam with a transverse crack. In this case, no frequency drop is remarked if the transverse crack is located at a point of inflection because here bending does not manifest and thus stress and strain are null. In the case of L-shaped cracks, a larger beam segment loses its rigidity due to the crack configuration. Thus, along with the damage on the beam slice located at the inflection point, which does not affect the frequency, also the neighbor slices are affected. This further damage produces a frequency drop that is visible in the frequency shift curves of the beams with L-shaped cracks.

In fact, in the case of L-shaped cracks, the beam manifests as having multiple transverse cracks of various types along the segment that includes the delamination, or as a beam with a segment of lower rigidity but without loss of mass. It is the challenge that faces the authors to find the mathematical relation expressing the frequency shift curves for the L-shaped crack starting from this conclusion. However, for damage detection purpose, the damage patterns can be contrived using the FEM simulation results.

5. References

[1] Gillich G R, Birdeanu E D, Gillich N, Amariei D, Iancu V and Jurcau C S 2009 Detection of damages in simple elements *Annals of DAAM & Proc. of 20th Intern. Symp.* **20**(1) 623-625

[2] Yang Z B, Radzienski M, Kudela P an Ostachowicz W 2016 Scale-wavenumber domain filtering method for curvature modal damage detection *Compos. Struct.* **154** 396-404

[3] Vosoughi A R 2015 A developed hybrid method for crack identification of beams *Smart Struct. Syst.* **16**(3) 401-414

[4] Mazanoglu A 2015 A novel methodology using simplified approaches for identification of cracks in beams *Lat. Am. J. Solids Stru.* **12**(13) 2460-2479

[5] Zhang K and Yan Z 2017 Multi-cracks identification method for cantilever beam structure with variable cross-sections based on measured natural frequency changes *J. Sound Vib.* **387** 53-65

[6] Khnaijar A and Benamar R 2017 A new model for beam crack detection and localization using a discrete model *Eng. Struct.* **150** 221-230

[7] Ravi J T, Nidhan S, Muthu N and Maiti S K 2018 Analytical and experimental studies on detection of longitudinal, L and inverted T cracks in isotropic and bi-material beams based on changes in natural frequencies *Mech. Syst. Signal Pr.* **101** 67-96

[8] Gillich G R, Mituletu I C, Praisach Z I, Negru I and Tufoi M 2017 Method to enhance the frequency readability for detecting incipient structural damage *IJST-Trans. Mech. Eng.* **41**(3) 233-242

[9] Ntakpe J L, Gillich G R, Mituletu I C, Praisach Z I and Gillich N 2016 An Accurate Frequency Estimation Algorithm with Application in Modal Analysis *Rom. J. Acoust. Vib.* **13**(2) 98-103

[10] Gillich G R, Praisach Z I and Negru I 2012 Damages influence on dynamic behaviour of composite structures reinforced with continuous fibers *Mater. Plast.* **49**(3) 186-191

[11] Ntakpe J L, Praisach Z I Mituletu I C Gillich G R and Muntean F 2017 Natural frequency changes of two-span beams due to transverse cracks *J. Vibr. Eng. Technol.* **5**(3) 229-238

[12] Gillich G R and Praisach Z I 2014 Modal identification and damage detection in beam-like structures using the power spectrum and time-frequency analysis *Signal Process.* **96**(A) 29-44

[13] Gillich G R Praisach Z I and Iavornic C M 2012 Reliable method to detect and assess damages in beams based on frequency changes *Proc. of the ASME 2012 International Design Engineering Technical Conf.& Computer and Information in Engineering Conf. IDETC/CIE 2012* (Chicago, Illinois, USA) DETC2012-70094

[14] Gillich G R, Wahab M A, Praisach Z I and Ntakpe J L 2014 The influence of transversal crack geometry on the frequency changes of beams *Proc. of Int. Conf. on Noise and Vibration Engineering ISMA2014* (Leuven, Belgium) 485-498

[15] Nitescu C, Gillich G R, Wahab M A, Manescu T and Korka Z I 2017 Damage severity estimation from the global stiffness decrease *J. Phy. Conf. Ser.* **842** 012034

[16] Gillich G R, Minda A A, Praisach Z I and Protocsil C 2014 Stress intensity factor vs. transversal deflection - How to determine the damage severity? *Rom. J. Acoust. Vib.* **11**(1) 25-30

[17] Gillich G R, Maia N, Mituletu I C, Tufoi M, Iancu V and Korka Z I 2016 A new approach for severity estimation of transversal cracks in multi-layered beams *Lat. Am. J. Solids Stru.* **13**(8) 1526-1544

[18] Gillich G R and Praisach Z I 2013 Detection and quantitative assessment of damages in beam structures using frequency and stiffness changes *Key Eng. Mater.* **569-570** 1013-1020

[19] Gillich G R, Tufoi M, Korka Z I, Stanciu E and Petrica A 2016 The Relations between Deflection, Stored Energy and Natural Frequencies, with Application in Damage Detection *Rom. J. Acoust. Vib.* **13**(2) 87-93

[20] Manescu T, Gillich G R, Manescu T S, Suciu C 2014 Optical Evaluation of Surface Deffects Influence upon the Contact Area between a Metallic Spherical Punch and a Flat Surface *Mater. Plast.* **51**(2) 157-161.

IOP Conf. Series: Materials Science and Engineering **416** (2018) 012046 doi:10.1088/1757-899X/416/1/012046

W – TiB_2 Composite Material for Electro-Spark Deposition

J Kováčik[1,2,*], Š Emmer[2], J Kulasa[3], M Lis[3], S Malara[3], J Rodriguez[4] and I Cañadas[4]

[1] Institute of Materials & Machine Mechanics, Slovak Academy of Sciences, Dúbravská cesta 9, 84513 Bratislava, Slovakia
[2] IVMA STU, Pionierska 15, 831 02 Bratislava, Slovakia
[3] Institute of Non-Ferrous Metals, ul. Sowińskiego 5, 44-100 Gliwice, Poland
[4] Plataforma Solar de Almeria, Ctra de Senes Km 4, 04200 Tabernas (Almeria), Spain

[*]Corresponding author's e-mail: ummsjk@savba.sk

Abstract. The W – TiB_2 composite materials were prepared by spark plasma sintering at INM Gliwice, Poland and by solar powder metallurgy at Plataforma Solar de Almeria (PSA), Spain. Composites were prepared by sintering of green compacts of W powder with 60 vol.% of TiB_2 powder. During the spark plasma sintering the sintering temperature was 1800 °C, pressure of 22kN, time of 300 s and heating rate of 1.67 °C/s. In solar furnace the sintering temperature was 1820 °C, pressure less, time 600 s and heating rate of 2.33 °C/s. The resulting composites prepared by spark plasma sintering possess 95 - 97 % of theoretical density. The composites prepared in solar furnace have porosity of 49%. Both composites are brittle and prone to cracking. The following phases were observed by EDS and XRD analyses: After solar sintering only W and TiB_2 phases were observed. After spark plasma sintering WB and TiB_2 phases were observed. The reason is different method of preparation. Finally it was proved experimentally that both composites can be successfully used as electrodes for electro spark deposition of hard wear resistant layers on metallic substrates.

1. Introduction

Titanium diboride (TiB_2) based cermets have recently been studied for their potential applications to cutting tools and wear resistant materials because of their superior thermo - mechanical properties under various conditions [1 -2].

Resistance spot welding [3] is a process in which contacting metal surface points are joined by the heat obtained from resistance to electric current. It is widely used in the automobile manufacturing industry to weld the sheet metal to form a car and is completely automated. A limiting factor in industrial manufacturing using resistance spot welding is the life of the welding electrodes. A short life necessitates frequent electrode dressing or changing, resulting in lower production rates and higher costs. Also, weld quality is more variable towards the end of the life of an electrode, particularly when welding coated steels. Therefore it was proposed to use titanium diboride based coatings prepared by electro spark deposition process [4] on these electrodes containing tungsten insert [3]. However the situation during resistance spot welding is complex and single titanium diboride layers are not fully suitable in this case. Therefore it was propose to use composite electrodes for electro spark deposition. It is expected that tungsten with 60 vol.% of TiB_2 will be proper choice to protect the working surface of resistance spot welding electrodes.

IOP Conf. Series: Materials Science and Engineering **416** (2018) 012046 doi:10.1088/1757-899X/416/1/012046

The question is how to prepare desired composition simply and cost effectively. Usually this can be done using powder metallurgy methods. One possibility to prepare dense titanium diboride and tungsten is spark plasma sintering (SPS) [1, 2]. However up to now no experimental work concerning direct SPS of mixtures of tungsten with TiB_2 can be found in the literature.

Recently authors used Solar Powder Metallurgy (SolPM) to study the preparation of fully dense titanium goods using solar furnace via pressure less sintering [5]. It was shown that significant advantage of solar furnace is simple handling and significant shortening of heating/cooling times. Therefore it was proposed to use also this method of preparation of desired composite material. This approach is also brand new from materials science point of view.

Therefore, the aim of the proposed work is to study experimentally a possibility to prepare dense W + 60 vol.% TiB_2 powder compacts using Spark plasma sintering (SPS) and Solar powder metallurgy (SolPM) for electro spark deposition.

2. Experimental

The TiB_2 powder (supplier 3M Technical Ceramics, Germany) of angular shape with grain size < 32μm and d_{50} = 15.5μm (see Fig. 1) was used in experiments as filling phase. Composition of used powder according to supplier is following: Ti > 66 %, B > 29 %, C < 2.0 %, B_2O_3 < 2.5%, O < 2.5 %, N < 0.8 % and Fe < 0.3 %. Tungsten powder (unknown origin) of granular shape with grain size < 10 μm was used as matrix powder. The powder mixture corresponding to the composition of W + 60 vol. % TiB_2 was dry mixed in Turbula mixture for 30 minutes. The densities of used powders considered prior mixing were: W 19.25 kg.m^{-3}, TiB_2 4.52 kg.m^{-3}.

Figure 1. SEM of TiB_2 powder (left) and the mixture of W + 60 vol. % TiB_2 powders (right).

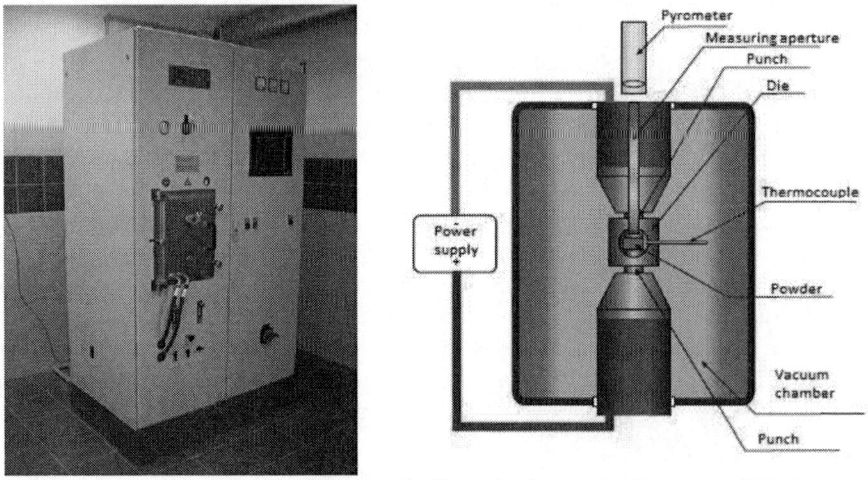

Figure 2. Used FCT-HP D5/2 SPS machine (left) and schematic diagram of SPS process (right).

IOP Conf. Series: Materials Science and Engineering **416** (2018) 012046 doi:10.1088/1757-899X/416/1/012046

Spark plasma sintering was performed directly on green powder mixture using FCT-HP D5/2 SPS machine (FCT Systeme GmbH, Germany) at IMN Gliwice. Into the graphite die of inner diameter of 30 mm 45 g of powder mixture was put prior experiments. Then the die was closed, the inner space was evacuated and the following SPS parameters were used: sintering temperature of 1800 °C, pressure of 22kN, time of 300 s, heating rate of 1.67 °C/s and initial air pressure at process start less than 3 hPa.

After SPS the samples were sandblasted to remove graphite envelope, polished and used directly for metallographic (SEM microscope Zeiss Evo MA10 equipped with EDS spectrometer BrukerXFlash®5010) and XRD observations using XRD7 difractometer (Seifert-FPM, Germany). The density of samples was determined using Archimedes method. It was confirmed that during spark plasma sintering it was possible to achieve 95 - 97 % of theoretical density (10.41 $g.cm^{-3}$). Finally prismatic electrodes □ 5x5 mm with the length of 25 mm were cut from the disks using electric discharge machining for ESD deposition tests.

For Solar powder metallurgy the powder mixture was compacted by uniaxial pressing at 280 MPa, which was optimal pressure for used pressing die. The typical geometry of green compacts before sintering was: diameter of 15.2 mm and height of 9 mm. The compacts were weighted and finally the porosity was calculated using known geometry. The observed porosity prior solar sintering was in the porosity range of 43 - 47%.

The sintering experiments were performed at the horizontal solar furnace SF40 in PSA, Almeria, Spain. SF40 is able to deliver up to 40 kW power at peak concentration ratios exceeding 7000 kW/m^2. This SF40 system consists of heliostat, shutter, parabolic concentrator and test table [6]. The prepared sample was placed on zirconium plate (400 x 400 x 3 mm). The specimen was placed inside of chamber under the focal spot (see Figure 3). The size of solar spot was approximately 90 mm. Then the chamber was evacuated (approximate value of the vacuum was around 8.10^{-1} Pa) and argon gas flow was used during experiments.

Experimental samples were sintered using SF40 solar furnace at temperatures between 1770 – 1820 °C for dwell time of 10 minutes under the heating rate of 2.33 °C/s. The temperature was measured using pyrometer (emissivity 0.95) [5]. The final porosity was determined from geometry and the weight of the samples and was in the range of 48 - 50%.

After the experiments the samples were cut and samples for microstructure observations were prepared. Microstructure of samples was observed at PSA`s Materials Lab using Leica DMI 5000 Minverted digital microscope and at IMMS SAS using scanning electron microscope JEOL 7600F, equipped with Schottky thermal-emission cathode (thermal FEG - W-coated ZrO2) as well as energy EDS spectrometer from Oxford Instruments. Finally prismatic electrodes □ 4x4 mm with the length of 8 mm were cut from the disks using electric discharge machining for ESD deposition tests.

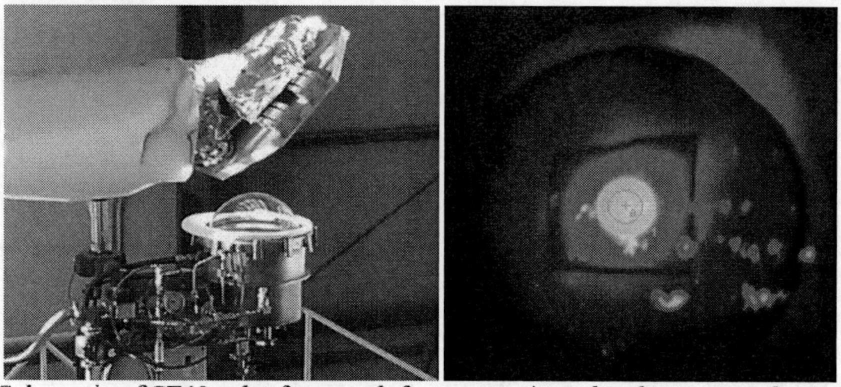

Figure 3. Schematic of SF40 solar furnace: left - vacuum/gas chamber mounted on test table; right - solar spot size during solar test (IR camera picture).

3. Results and discussion

The experimental works confirmed that it is possible to prepare electrodes from both composite materials via SPS and SolPM. These electrodes can be successfully used for electro spark deposition of hard wear resistant layers on metallic substrates, see for example Figure 4. While both materials are mechanically stable to be used as electrode for ESD process, significant differences in porosity and composition were observed according to used powder metallurgy densification process.

Figure 4. ESD deposition of prepared composite material on AW7075 aluminium alloy substrate: left – SPS prepared electrode; right - SolPM prepared electrode. Substrate size of ϕ 30 mm x 10 mm.

Primary difference was observed in final porosity of the sample. The composite material prepared by spark plasma sintering possesses 3 – 5 % porosity. On the contrary the composites prepared via solar powder metallurgy in solar furnace have porosity in the range of 43 - 47%. It was observed that both composites are brittle and prone to cracking during cooling process.

The most important difference was observed in microstructure (see Figure 5). The main reasons of microstructure differences are control process parameters such as temperature and applied pressure: Spark plasma sintering at 1800 °C under the pressure of 22kN led to more or less homogeneous microstructure (one dominant phase) with certain amount of pores. It ought to be kept in mind that the outer surface of SPS sample is observed where more voids, pores and defects are concentrated. On the contrary inner microstructure of SolPM prepared composite (sintering temperature 1820 °C and no pressure) is fully porous with two different phases inside of composite material.

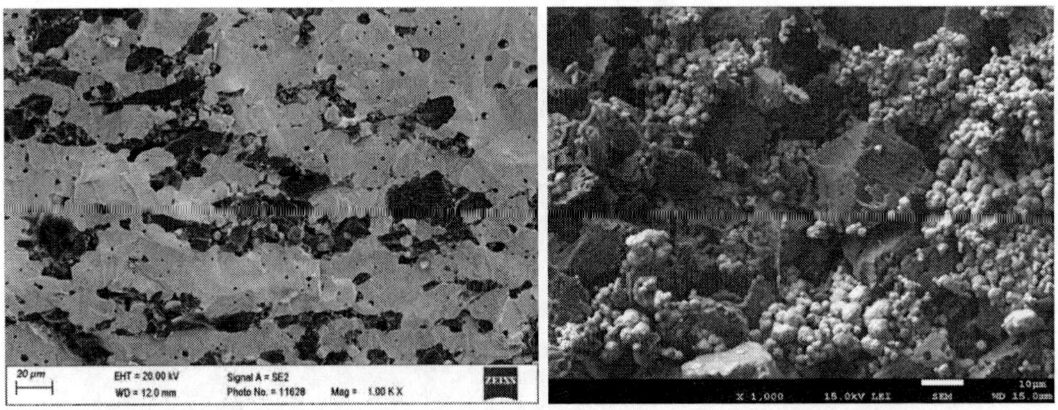

Figure 5. Microstructure of composites prepared by SPS (left) and SolPM (right).

To distinguish the created phases EDS and XRD analyses were performed. Inside the composite prepared at solar furnace only W and TiB_2 phases were observed using ESD analyses (Figure 6). Moreover slight expansion of the sample in both axial and radial direction was observed after solar sintering. This 2 – 4% expansion can be connected with rearranging of the structure due to relaxation of internal stresses created during compacting of powders. The most probably sintering process led to the slight reaction between tungsten and TiB_2 powders as solid composite that can be carefully machined was created. Anyway the reaction zone is so small, that it was not possible to detect these phases using XRD analysis. The strength of such material is relatively low and it was quite difficult to avoid cracking of this material during machining of ESD electrodes.

Figure 6. Microstructure of SolPM composite sample and EDS area analyses: Spectrum 1 - TiB_2, Spectrum 2 - tungsten.

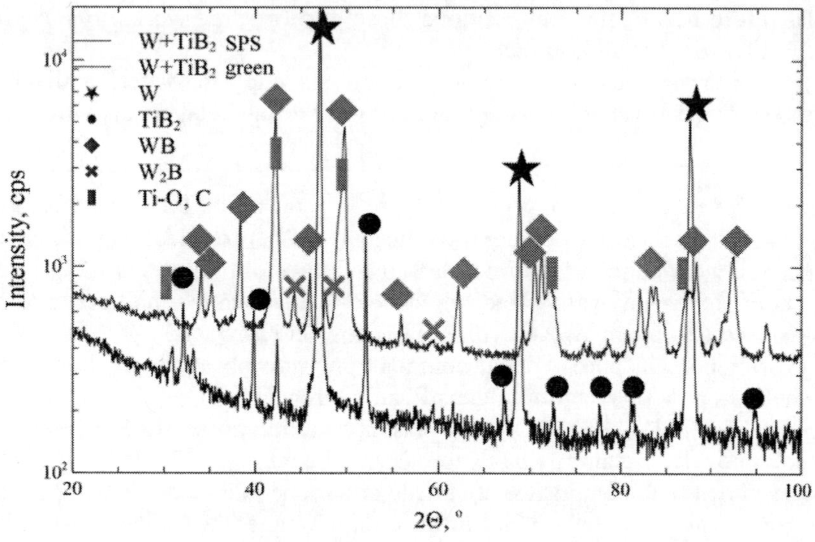

Figure 7. XRD analyses of original green powder mixture and polished SPS composite sample.

Ahn et al. [7] have studied titanium diboride based cermets creation via HIP sintering of green compacts of TiB_2-W with compositions between 5 and 40 vol.% of W at a pressure of 200MPa and temperatures above 1750 °C. They have formed cermets with relative density of more than 99%. Analysis of these samples by XRD, EPMA and Auger spectra, showed the presence of TiB_2, β-WB and $WTiB_2$, However no metallic tungsten was detected between 5 and 30 vol.% of W. This more or less coincides with the observed phases for both methods of preparation. Moreover it indicates that during solar sintering the real temperature was probably below 1750 °C. According to the literature no reaction

IOP Conf. Series: Materials Science and Engineering **416** (2018) 012046 doi:10.1088/1757-899X/416/1/012046

between W and TiB_2 occurs up to 1100°C [8]. Strashinskaya [9] observed that in 1 to 5 hour tests the reaction of a powdered mixture of W+TiB_2 started at < 1800°C. Compact tungsten reacted with TiB_2 powder at 1900°C and with compact TiB_2 at 2100°C. All experiments were performed at about 5 x 10^{-5} Torr and tungsten boride phases were formed. According to Gorbacheva et al. [10], titanium borides are stable in contact with W in the temperature range from 1500 to 2000°C. There was neither diffusion of B into W nor adherence of both components to each other; only traces of W_2B were be detected. Leitnaker et al. [11] observed that tungsten reacted with TiB_2 in a mole ratio TiB_2 :W=1:1 to form a possibly ternary compound with a lattice close to that of TaB. The formation of $WTiB_2$ as ternary compound was also noted when powder mixtures (grain size < 40 μm) of W and TiB_2 (containing - 0.3 wt% C) in the range 0.5 to 100 mol% TiB_2 were pressed and sintered at 1800 to 2400°C and pressure of 3.33 Pa. Eutectics appeared on each side of this compound in the quasi-binary system W-TiB_2, one at about 12 mol% TiB_2, the other at about 88 mol% TiB_2 [12].

After spark plasma sintering WB and TiB_2 phases were observed in final composite using XRD analysis. It was observed that the main phase is WB phase. WB phase is rhombic with space group of Cmcm # 63 and corresponding space group constants of a = 3.19 Å, b = 8.4 Å, c = 3.07 Å. Further small amount of W_2B and TiB_2 phases were observed. Due to the rests of graphite envelope on the outer surface of samples also Ti - O,C phase was observed (see Figure 7).

There are following reasons for small reaction of used tungsten and TiB_2 powders during SolPM: No applied pressure. Argon protective atmosphere present in solar furnace that is cooling a bit the sample during sintering. The pyrometer readings of the temperature are not precise. For example overestimation of temperature for 6% was observed during solar sintering of Ti [5]. Therefore the real sintering temperature was probably lower for the investigated composite in solar furnace. Moreover, the solar radiation was scarcely 870 W.m^{-2} during October 2017 in PSA, Spain. Therefore at 100% shutter opening it was not possible to achieve higher temperature during sintering of the investigated samples at 40 kW solar furnace. The experiments were done at maximal reachable temperature at given weather conditions. Therefore it is planned to continue these experimental works in solar furnaces with high power flux (SF60, Spain, Odeillo, France).

Summarising, both methods led to the preparation of composite materials that can be successfully used as electrodes for electro spar deposition of hard wear resistant layers on various metallic substrates.

4. Conclusions

The experiments proved that it is possible to prepare W + 60 vol.% TiB_2 based composites using Spark plasma sintering (SPS) and Solar powder metallurgy (SolPM) for electro spark deposition electrodes. As no experimental work concerning direct SPS of mixtures of tungsten with TiB_2 can be found in the literature this approach is innovative and brand new. Moreover for the first time also Solar Powder Metallurgy was used to prepare W + 60 vol.% TiB_2 powder compacts.

Following differences in porosity and composition were observed for both prepared composites according to the used powder metallurgy densification process:

The composite material prepared by spark plasma sintering possesses 3 – 5 % porosity. Composites prepared via solar powder metallurgy have porosity in the range of 13 – 17%.

It was observed that both composites are brittle and prone to cracking during cooling process.

Inside the composite prepared at solar furnace only W and TiB_2 phases were observed. Moreover slight expansion (2 – 4%) of the sample in both axial and radial direction was observed. The most probably sintering process led to only slight reaction between tungsten and TiB_2 powders. Composite were carefully machined to avoid cracking during preparation of ESD electrodes.

After spark plasma sintering rhombic dominant WB cermet phase was observed inside composite. Further traces of W_2B, TiB_2 and Ti - O,C phases were also observed in final composite using XRD analysis. The easy machining via electric discharge machining was used to prepare electrodes for ESD deposition.

There are following reasons for small reaction of used tungsten and TiB_2 powders during SolPM: No applied pressure, cooling argon protective atmosphere, overestimation of temperature (lower real temperature inside of material), and limiting power flux at 100% shutter opening at 40 kW solar furnace - the experiments were done at power flux at very good weather conditions. Therefore it is planned to continue these experimental works in solar furnaces with high power flux.

References

[1] Zhang Z H, Shen X B, Wang F C, Lee S K, Fan Q B, Cao M S 2012 *Scripta Mater.* **66** 167, doi.org/10.1016/j.scriptamat.2011.10.030.

[2] Demirskyi D, Sakka Y 2015 *J. European Ceramic Society* **35** 405, doi.org/10.1016/j.jeurceramsoc.2014.08.007

[3] Emmer Š, Gábrišová Z, Kováčik J, Sejč P, Lis M, Kulasa J 2017 *Kovove mater* **55** 385 DOI: 10.4149/km2017_6_385

[4] Kováčik J, Baksa P, Emmer Š 2016 Acta Metallurgica Slovaca **22** 52-9 dx.doi.org/10.12776/ams.v22i1.628

[5] Kováčik J, Emmer Š, Rodríguez J, Cañadas I 2017 *Journal of Alloys and Compounds* **695** 52-59, doi.org/10.1016/j.jallcom.2016.10.147

[6] Rodriguez J, Cañadas I, Zarza E 2016 New PSA high concentration solar furnace SF40 In*: AIP Conference Proceedings* **1734**, 70028, doi.org/10.1063/1.4949175

[7] Ahn D-G, Kawasaki A, Watanabe R 1996 *Materials Transactions*, JIM, **37** 1078-83 doi.org/10.2320/matertrans1989.37.1078

[8] Best E, Kuhn P, Kurtz W, List H 1986 *W Tungsten: Metal, Chemical Reactions with Inorganic and Organic Compounds*, Bergmann, H, Gras, D, Hinz, I, Ernst Koch, Kurtz, W, Vetter U, (eds), W Wolfram Tungsten (System-Nr 54) W / A-B / A / 7, Springer Berlin Heidelberg

[9] Strashinskaya L V 1976 *Fiz Khim Mekh Mater* **12** 117/9 ; *Soviet Mater Sci* **12** 330/2

[10] Gorbacheva T B, Krylov Yu I , Mikova N M 1973 *Nauchn Tr Vses Nauchn Issled Proektn Inst Tugoplavkikh Metal Tverd Splavov* **14** 239/43; CA 82 (1975) No 159443

[11] Leitnaker J M , Krikorian N H, Krupka M C 1962 *J Electrochem Soc* **109** 66

[12] Yasinskaya GA and Groisberg MS 1963 Interaction of titanium boride with niobium and tungsten, *Powder Metall Met Ceram* **2** 457 doi.org/10.1007/BF00774190

Acknowledgement

Financial support by the Access to Research Infrastructures activity in the 7th Framework Programme of the EU (SFERA Grant Agreement n. 228296) is gratefully acknowledged. Financial support of the SFERA II project – Transnational Access activities (EU 7th Framework Programme Grant Agreement n° 312643) is also gratefully acknowledged. Slovak Grant Agency VEGA grant 2/0044/17 and SRDA agency grant APVV SK-PL-2015-0003 financial supports are gratefully acknowledged.

Comparative Study on the Corrosion Resistance of CAD/CAM and Conventional Processed Co-Cr Dental Alloys in NaCl Solutions

L-V Costea[1], M L Dan[1,*], C E Savencu[2], L Porojan[2]

[1] Faculty of Industrial Chemistry and Environmental Engineering, University Politehnica Timişoara, Romania
[2] Department of Dental Prostheses Technology, University of Medicine and Pharmacy V. Babeş, Timişoara, Romania

*Corresponding author: mircea.dan@upt.ro

Abstract. The present study emphasizes on the structural implications of the machining procedure on base Co-Cr dental alloys in terms of corrosion resistance. We comparatively assessed the corrosion behaviour of two cobalt-chromium dental alloys: one manufactured by alternative computer assisted technologies obtained by CAD/CAM milling (MIL) and the other one by conventional melting- casting technology (CAS) in NaCl solutions. The samples have been evaluated using several electroanalytical techniques like: linear polarization, cyclic voltammetry (CV), electrochemical impedance spectroscopy (EIS). Preliminary testing is showing that the new computer assisted processing technologies are comprised as a viable alternative towards traditional manufacturing methods.

1. Introduction

The most common categories of dental materials are composite materials, ceramics, polymers and various metal alloys [1]. Many of these compounds are used in modern dentistry as parts of various medical devices with the aim of re-establishing the physical as well as the aesthetical condition of the patient [2]. Dental materials based on Co-Cr alloys are widely employed in several restoration techniques because of their resistance towards corrosive agents and their remarkable mechanical characteristics [3]. The obtained structures mounted in the oral cavity, are constantly exposed to an environment with a plethora of dynamically changing physical and chemical parameters. Temperature variations, fluctuations of the oral pH, various electrolytes originating from food or fluid intake as well as from specific medical treatments and physical conditions of the patient as well as bacteria, exert a high level of strain on the used materials [4] leading to corrosion phenomena taking place at the surface of dental alloys in direct contact with the oral tissue. The recently introduced computer aided technology (CAD/CAM) differs fundamentally from the classical manufacturing methods in terms of time efficiency as well as final restoration quality [5] this makes the final products to behave contrastingly towards corrosion compared to the classically obtained samples.

The casting technique of cobalt-chromium (Co-Cr) dental alloys has been used for decades to produce dental restorations [6]. Alternatively, dental alloys can be processed using CAD/CAM technology, as hard or soft milling [7]. Co-Cr alloys exhibit elevated melting points, a greater hardness and relatively lower ductility compared to other noble alloys, which makes them more difficult to process [8]. The hard milling technique offers the advantage of machining industrial prefabricated

IOP Conf. Series: Materials Science and Engineering **416** (2018) 012047 doi:10.1088/1757-899X/416/1/012047

blanks without the characteristic casting defects and/or porosities [8]. The microstructure morphology of the Co-Cr alloy varies depending on the used manufacturing technique hence interfacial, chemical as well as mechanical properties are associated as well with the employed processing technology of the base material.

Although various studies were conducted over the years regarding the biocompatibility of various dental alloys, no consensus has yet been reached on this issue. Corrosion behavior of the metal compositions represents a key factor concerning the immunological response towards dental alloys. Metal ions released in surrounding tissues can induce cytotoxicity with apoptosis (cell death), allergic reactions, inflammation and discoloration [9]. Dental restorations residing in the oral cavity are permanently exposed to an ever changing complex environment with fluctuating physical and chemical parameters like temperature variations, shifting of the oral pH value, various electrolytes originating from food or fluid intake as well as from certain medications. All of the above together with the naturally forming of bacterial plaque that could come in contact with the base material, provide ideal conditions for the occurrence of degradation phenomena at the metal surface, thus its partial biodegradation [10-12].

The corrosion resistance of Co-Cr dental alloys is influenced by numerous factors, the most significant being: chemical composition, microstructure and manufacturing technique [10-12]. Considering the relatively short time span in which cobalt-chromium base alloys processed by means of CAD/CAM technologies have been employed on a large scale in dentistry, the assessment of their corrosion behavior by closely simulating the conditions within the oral cavity necessitate further investigations.

2. Experimental

Two cylindrical samples with the base diameter of 10 mm and 2 mm height were employed in two distinct manufacturing processes, namely: the melting-casting technique using a commercial Co-Cr base material and the CAD/CAM milling technology using prefabricated Co-Cr dental alloy blocks. The chemical composition of the investigated probes, as given by the manufacturer, is detailed in Table 2. The samples employed in the CAS technique were firstly made of wax, invested and casted using a vacuum pressure casting machine. The samples were divested after cooling down, air-blasted with alumina particles and mechanically finished and polished using specific instruments. The samples manufactured by CAD milling were obtained using a 5 axis milling machine from a prefabricated block of type 4 Co-Cr dental alloy. Oxide-firing (at 950 – 980°C) was performed, as required for porcelain veneering for all samples, the metal surface was blasted with fresh aluminum oxide (75 μm).

All samples were finally finished using silicon carbide sheets and 1μm grit diamond paste respectively, degreased and sonicated in distilled water, rinsed and dried prior to undergoing electrochemical investigations. Electrochemical investigations have been undertaken using a BioLogic SP150 potentiostat/ galvanostat equipped with an EIS module along with a conventional three-electrode cell system. All electrochemical measurements have been undertaken using Fusayama artificial saliva as electrolyte solution [13]. Electrochemical impedance spectroscopy (EIS) investigations have been conducted within frequencies ranging from 100 kHz to 10 MHz the amplitude of the alternating voltage was of 10 mV. A number of 60 points have been recorded for each spectrum with a logarithmic distribution of 10 points per decade. The experimental data has been fitted using the ZView – Scribner Associates Inc. software and equivalent electrical circuits by applying the Levenberg – Marquardt least squares complex non-linear fitting algorithm [14]

3. Results and Discussion

Dental restorations are being constantly exposed to the ever-changing conditions within the oral cavity and, since the content of sodium chloride in the human saliva is very close to that of seawater [15], corrosion of the base metal alloys becomes a key factor. In order to investigate the phenomena taking place at the metal - electrolyte interface, we've conducted several comparative electrochemical measurements of the two types of samples in saline solutions at different temperatures, closely resembling the conditions in the oral cavity.

The corrosion behavior of the cast sample immersed in 3.5% aqueous saline solution was investigated by recording the Tafel polarization curves at four different temperatures (Figure 1). The manufacturing technique seems to exert a notable influence on the voltammetric response of the studied samples. While the corrosion potential of the casted sample (CAS) shifts towards more positive values with increasing temperature (Figure 1a), the sample obtained by CAD milling (MIL) shows only a slight deviation of the corrosion potential with respect to the increase of the solution temperature.

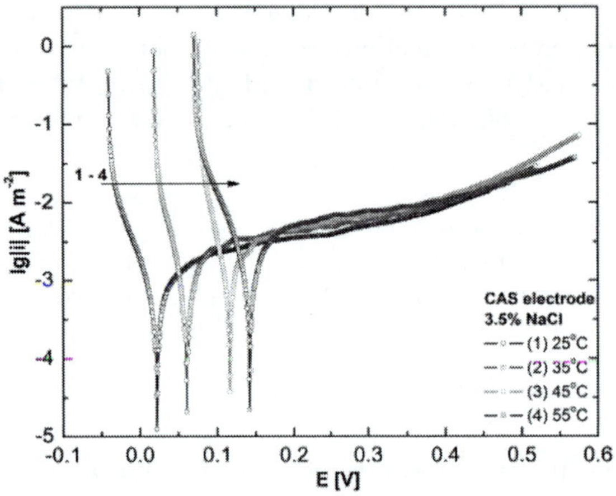

(a)

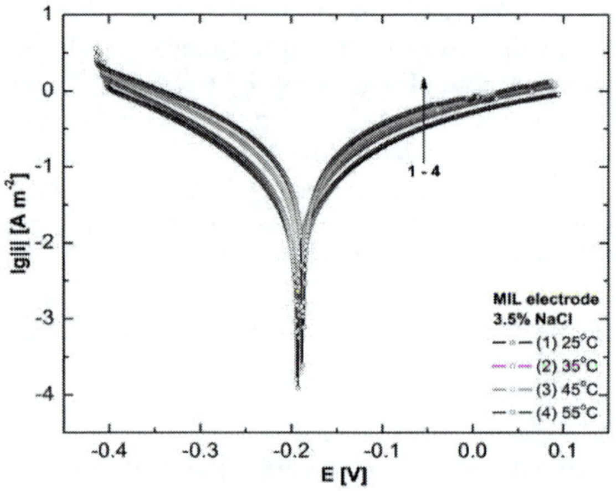

(b)

Figure 1. Tafel polarization curves recorded for the cast sample (a) and for the CAD milled sample (b) in 3.5% NaCl solution, at different temperatures and a scan rate of 1 mV·s^{-1}

The classically manufactured CAS sample exhibits a sharp current density increase at potentials below 50 mV (at 25°C), most probably due to the reduction of physically dissolved oxygen or the reduction of H^+ ions. The first process being more likely to occur noting that deaeration of the electrolyte solution has not been undertaken prior to the voltammetric measurements. Scanning towards more positive potentials reveals the active-passive region, characterized by a swift decrease of the current density, followed by a relatively large passivation region extending from about 100 mV to about 250 mV (at 25 °C) where the current density remains almost constant. Passivity seems to partially break down at potentials higher than 250 mV (at 25°C), probably due to the presence of Cl^- anions [16] suggesting a transpassive region with a slight current increase at potentials up to 550 mV, where the passivating film is most probably dissolved and water is oxidized. This behavior, is in accordance with other literature data [17]. Temperature increase of the electrolyte solution leads to a shift of the corrosion potential (E_{corr}) towards more positive regions, altogether with the passivation plateau and the transpassive zone.

The data for the CAD-manufactured (MIL) sample differ significantly in terms of electrochemical corrosion response (Figure 1 b). In this case, the corrosion potential remains relatively stable at about -190 mV over the whole temperature array, suggesting a higher susceptibility towards electrochemical corrosion compared to the CAS sample. The potentiodynamic analysis reveals a highly symmetrical behaviour of the cathodic region compared to that of the anodic field, both sides showing transpassive regions with a sharp current increase at about -400 mV probably due to oxygen reduction, with no passivating plateau.

Tafel parameters, corrosion rate (v_{corr}), as well as other significant electrochemical data like corrosion current density (i_{corr}), corrosion potential, polarization resistance (R_p), recorded for each of the two samples at various electrolyte temperatures are detailed in Table 1. The data reveals that, the more negative corrosion potential in the case of the CAD milled (MIL) sample is in accordance to its corrosion rate which is significantly higher compared to the casted (CAS) sample, probably due to the lack of a compact and nonconductive passivating coating.

Table 1. Polarization parameters for the corrosion behaviour of the investigated samples in 3.5% NaCl electrolyte at various temperatures

Sample	T [°C]	i_{corr} [μA cm^{-2}]	E_{corr} [mV]	$-b_c$ [mV dec^{-1}]	b_a [mV dec^{-1}]	R_p [kΩ]	$v_{corr} \cdot 10^3$ [mm year^{-1}]
MIL	25	5.85	-197	195	160	10.7	73.2
	35	7.77	-194	203	163	7.91	106
	45	12.57	-190	239	187	5.92	171
	55	17.83	-186	265	208	4.73	243
CAS	25	0.054	22.1	304	29.7	163	1.42
	35	0.180	71.1	333	25.5	128	2.45
	45	0.244	103	362	21.9	101	3.32
	55	0.290	117	401	27.2	86.7	4.06

In order to elucidate the corrosion behaviour of the investigated probes, we recorded the corresponding cyclic voltammograms in 3.5% saline solution at four different temperatures at a scan speed of 100 mV s^{-1}. The recorded voltammetric data covered a wide potential domain, reaching from a negative value of about -1.25 V, towards the Hydrogen evolution reaction, to a positive polarization of 1.25 V at which the metal ion release from the base alloy is likely to take place. A relative outcome of the obtained cyclic voltammograms for the two studied samples is shown in Figure 2b.

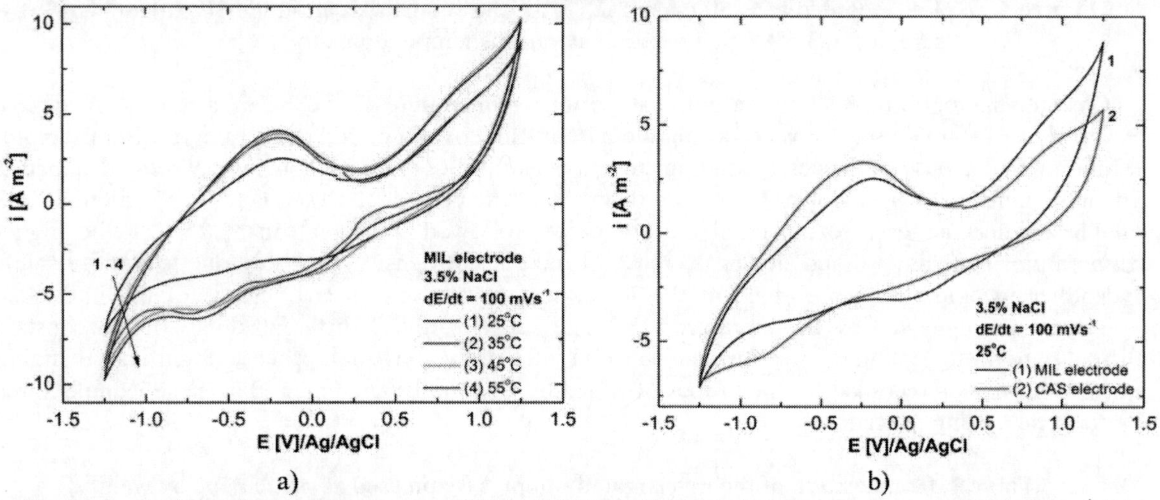

Figure 2. Cyclic voltammograms recorded in 3.5% NaCl electrolyte at a scan rate of 100 mV s^{-1} for the MIL sample at different temperatures (a) and superimposed for MIL and CAS samples at 25°C (b)

Varying the potential from the open circuit potential towards more positive values on the MIL sample, reveals a steady increase of the current density recorded for all measured temperatures, however, a gap is observed between the recordings for 25-35°C and 45-55°C respectively. A weakly defined peak at about 0.8 V is revealed before the current increases due to the oxygen evolution reaction. The reverse scan shows a weakly defined signal at about -0.1 V and a sharper peak at -0.6 V. The latter maximum does not occur at a temperature of 25°C and seems to be linked to a relatively broad signal occurring at approximately -0.25 V within a quasireversible process. The cathodic scan is limited negatively by the hydrogen evolution reaction at potentials around -1.25 V.

The comparative view of the voltammograms recorded for the two samples at the same temperature reveals a significant difference in their electrochemical behaviour namely the CAS sample exhibits a wider passivating region thus being less susceptible towards corrosion, in accordance to the data obtained from the Tafel analysis, while the MIL sample does not reveal any passivating zone at all.

Both samples have been subjected to chronoamperometric studies in 3.5% saline solution at the whole array of temperatures, as shown in Figure 3, with the aim of additionally clearing up the corrosion conduct of the investigated probes.

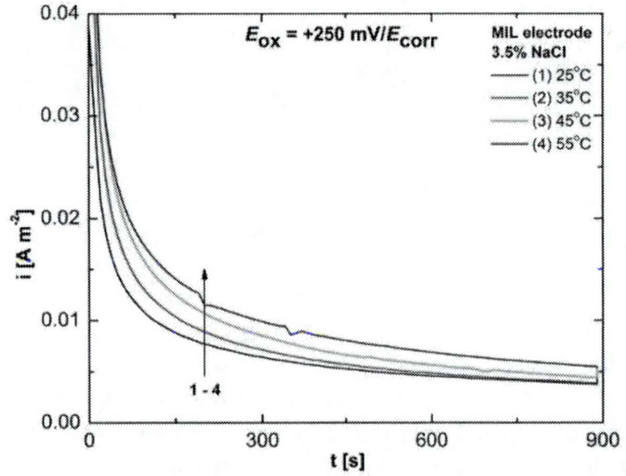

a)

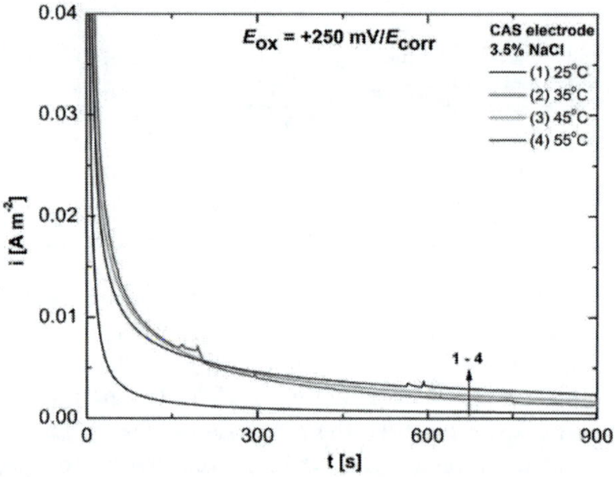

b)

Figure 3. i-t plots recorded at +250 mV vs. E_{corr}, for the anodic oxidation of MIL (a) and CAS (b) samples in 3.5% NaCl solution at various temperatures for 15 min.

Chronoamperograms for both samples at various temperatures have been raised at an applied potential of +250 mV vs. E_{corr} by recording the current flow over a period of 15 minutes. Both samples exhibit rapid decrease of the corresponding current densities but there are some noteworthy diffeneces. The MIL sample shows a slower but steady decrease of the current at all temperatures while the CAS sample exhibits a nearly exponential current decay followed by a plateau at 25°C while higher temperatures generate a much slower decrease of the current density, probably due to a more stable and adhering oxide coating formed at the metal surface as well as an increase of the interfacial resistance, as suggested by the literature [18]. The samples reveal different slopes of the constant current region, suggesting distinct mechanisms and rates of the corrosion process. Significantly higher currents have been recorded for the probes obtained by CAD milling compared to those manufactured by classical casting procedures.

Table 2. Composition of the investigated samples (expressed as percentage by weight)

Sample	Co	Cr	Mo	Mn	W	Si	Fe	C
CAS	61.5	26	6	<1	5	<1	<1	0
MIL	61	28	0	0.25	8.5	1.65	<0.5	<0.1

In order to further investigate the mechanism of the corrosion processes taking place at the surface of the studied samples, electrochemical impedance spectra (EIS) have been recorded. Impedance studies have been conducted at the corresponding corrosion potentials (E_{corr}) in aqueous 3.5% NaCl solution within a temperature range between 25 and 55°C. The obtained results are depicted as

IOP Conf. Series: Materials Science and Engineering **416** (2018) 012047 doi:10.1088/1757-899X/416/1/012047

Nyquist and Bode plots for the samples obtained by CAD milling (Fig.4) and casting (Fig.5) respectively.

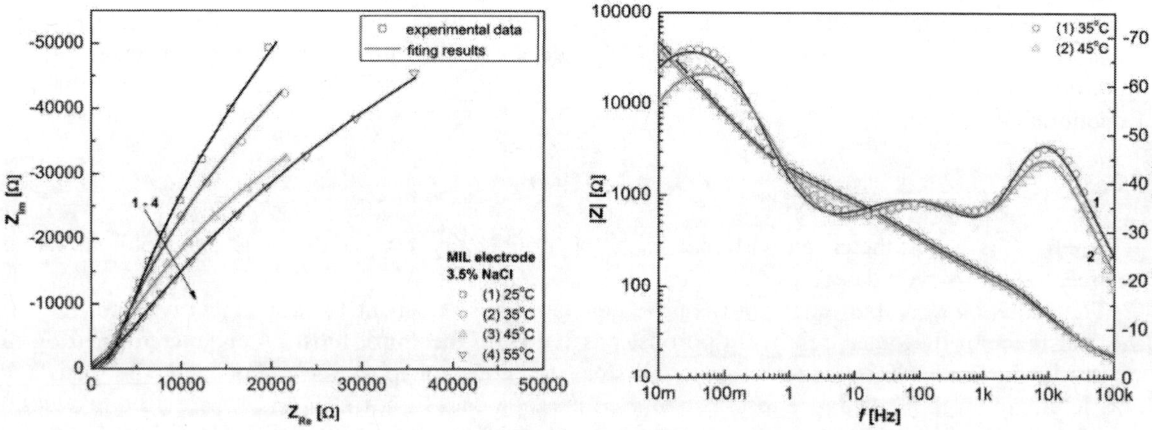

a) b)

Figure 4. Nyquist (a) and Bode plots (b) recorded at the corresponding corrosion potentials on MIL samples in 3.5% NaCl solution at various temperatures

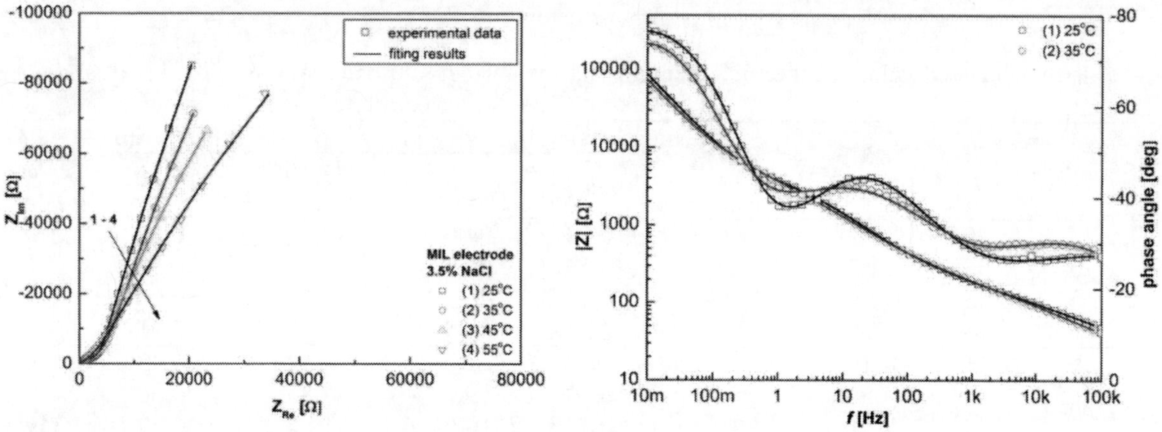

a) b)

Figure 5. Nyquist (a) and Bode plots (b) recorded at the corresponding corrosion potentials on CAS samples in 3.5% NaCl solution at various temperatures

The obtained results on the investigated dental alloys lead us to propose an equivalent electrical circuit (EEC) of the type $R_s(CPE_{pore}R_{pore})$ $(CPE_{film}R_{film})$ $(CPE_{dl}R_{dl})$ based on a three layer structure to fit the EIS data. The layers are most probably composed of an inner compact film and an outer relatively porous passive film, as shown in Figure 6, in order.

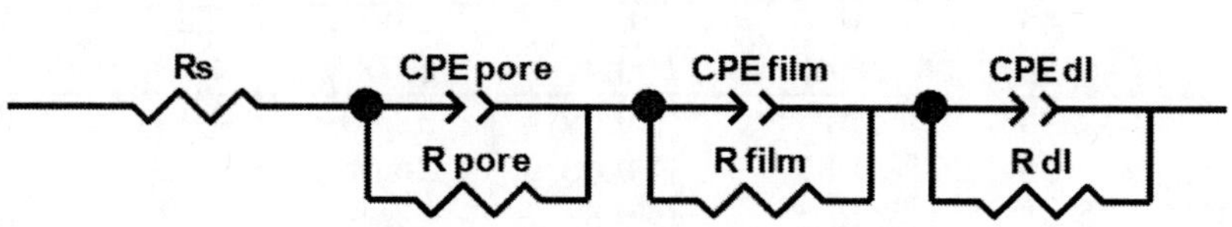

Figure 6. Equivalent electrical circuit model used for fitting the recorded EIS spectra

IOP Conf. Series: Materials Science and Engineering **416** (2018) 012047 doi:10.1088/1757-899X/416/1/012047

The EEC consists of the solution resistance R_s in series with three parallel connections consisting of the charge transfer resistance R and a constant phase element (*CPE*), each of the connections being specific to the layers mentioned above, while R_s represents the uncompensated solution resistance. In order to closely simulate a real electrochemical system, the ideal capacitor (*C*) is usually replaced by a constant phase element (*CPE*) because it represents the processes that take place at the electrode/electrolyte interface more accurately. The impedance of the *CPE* element is given by Equation (1):

$$Z_{CPE} = 1/T(j\omega)^n \tag{1}$$

where T is a parameter proportional with the double layer capacity, n is an exponent ranged between 0 and 1 which describes *CPE* angle.

The terms CPE_{pore} and R_{pore} represent the constant phase element for the capacitive behavior and the resistance of the outer, relatively porous, passive film, the third, most outer layer formed on the electrode surface. CPE_{film} and R_{film} stand for the constant phase element and the resistance characteristic for the impedance response of the inner compact film. CPE_{dl} and R_{dl} are used to describe the charge transfer process occurring at the interface of the investigated dental alloys. The fitting results are depicted as continuous lines in Figures 5 and 6 whereas the hereby obtained values for the circuit elements corresponding to the corrosion process of the CAD milled (MIL) and the conventionally casted (CAS) samples determined in 3.5% NaCl electrolyte are given in Table 3. The values for the double layer capacity have also been computed.

Table 3 Calculated values of the EEC elements for the corrosion process of the studied MIL and CAS samples in 3.5% NaCl electrolyte

Samples	T [°C]	R_s [Ω cm^{-2}]	porous passive film			compact film		
			$T \cdot 10^5$ [F cm^{-2} s^{n-1}]	n	R [Ω cm^{-2}]	$T \cdot 10^5$ [F cm^{-2} s^{n-1}]	n	R [Ω cm^{-2}]
MIL	25	14.6	10.6	0.58	2406	0.17	0.87	118
	35	11.2	9.26	0.57	1833	0.20	0.88	91
	45	12.8	8.99	0.56	1511	0.35	0.90	71
	55	14.8	8.72	0.55	1234	0.33	0.91	47
CAS	25	4.74	6.2	0.52	225	4.52	0.47	4048
	35	4.56	5.0	0.50	190	7.83	0.50	3704
	45	7.51	4.1	0.47	167	10.4	0.52	3496
	55	8.33	3.4	0.45	139	12.2	0.56	3011

Samples	T [°C]	charge transfer process			$C_{dl} \cdot 10^4$ [F cm^{-2}]	$Chi^2 \cdot 10^4$
		$T \cdot 10^5$ [F cm^{-2} s^{n-1}]	n	$R_{dl} \cdot 10^5$ [Ω cm^2]		
MIL	25	20.2	0.87	3.51	3.82	5.56
	35	15.1	0.83	2.68	3.22	8.47
	45	10.6	0.79	1.60	2.25	4.27
	55	6.48	0.73	1.07	1.33	1.85
CAS	25	16.7	0.89	32.0	4.15	0.56
	35	14.2	0.86	26.6	3.73	0.84
	45	12.5	0.83	17.1	3.61	0.91
	55	10.7	0.80	10.9	3.52	0.38

The similar shape of the impedance spectra shown in Fig. 3 and 4 as well as the use of the same equivalent electrical circuit (EEC) shown in Figure 5 for the fitting of the experimental impedance spectroscopy data obtained for the corrosion study of the investigated Co-Cr alloys obtained through CAD milling and by conventional casting in 3.5% aqueous NaCl, suggests highly similar mechanisms governing the processes taking place at the dental alloy/electrolyte interface for the two samples. Although the corrosion most probably implies the formation of two layers of metal oxides on the surface of the alloy, one being porous whereas the second being a more compact passivating layer, there are notable differences concerning the corrosion rate of the two probes. Considering the values of the charge transfer resistance for each of the two layers, we can assume that the forming of the porous oxide layer is favoured on the MIL sample whereas on the CAS probe the compact passivating oxide layer seems to be predominant, conferring a higher corrosion resistance of the base alloy. The charge transfer resistance R_{dl} is proportional to the corrosion rate, *i.e.* the smaller the resistance, the faster the corrosion process will occur. Taking into account that the values of R_{dl} corresponding to the CAS sample are up to an order of magnitude greater than those obtained for the MIL probe is an indication for a significantly higher corrosion resistance of the first within the investigated medium, compared to the latter. As is to be expected, the rate of corrosion in saline solution rises along with the temperature increase within the extent between 25-55 °C for the samples obtained through CAD milling (MIL) as well as for the casted ones (CAS). With the aim of undertaking an in-depth analysis of the studied corrosion phenomena, we determined the capacity of the double layer (C_{dl}) for each sample at temperature values within the investigated array. The relatively high values for C_{dl} exhibited by the CAS sample stand for a greater resistance towards corrosion.

4. Conclusion

The presented electrochemical studies conducted in 3.5% NaCl solution reveal a greater resistance towards corrosion associated to the Co-Cr dental alloys obtained through casting compared to the ones manufactured by CAD milling, whereas the corrosion rate varied ascending, proportional to temperature, as expected.

The investigated dental alloys reflect notable differences regarding the electrochemical behaviour towards corrosion based on two completely different manufacturing technologies. The Tafel polarization curves as well as the impedance analyses acknowledge a relatively higher corrosion rate as well more negative corrosion potentials exhibited by the CAD processed sample. The corrosion potential of the casted sample varies strongly with temperature, whereas the CAD milled probe shows no such behaviour. More thorough investigations are necessary in order to determine the corrosion mechanisms of the two samples. The corrosion behaviour of the studied samples in the above mentioned medium resembles the one determined in artificial saliva [19].

The electrochemical impedance data revealed that two layers are formed on the surface of the investigated probes in aqueous 3.5% NaCl electrolyte, one being more porous and the second being compact with a passivating effect. The growth of the second protecting layer being favoured on the surface of the CAS sample results in a greater protection thus a higher resistance towards corrosion of the probe crafted by conventional casting.

5. References

[1] Anusavice K J, Shen C and Rawls H R 2013 Phillips' science of dental materials 12th ed., Saunders, St. Louis
[2] Duraccio D, Mussano F and Faga M G 2015 *J. Mater. Sci.* **50** 4779–812
[3] Reclaru L, Lüthy H, Eschler P-Y, Blatter A and Susz C 2005 *Biomat.* **26** 4358-65
[4] Upadhyay D, Panchal M A, Dubey R S, and Srivastava V K 2006 *Mat. Sci. Eng.* A **432** 1-11.
[5] Krieger M *Sidekick Dental Mag.* http://sidekickmag.com/dental-technology/cadcam technology-you-cant-afford-not-to-have-it/
[6] von Fraunhofer J A, 1994, *Corrosion* vol. 1, ed L L Shreir, R A Jarmon, and G T Burstein 3rd

Ed., (Oxford: Butterworth Heinmann) p 155
[7] Kim H R, Kim Y K, Son J S, Min B K, Kim K H and Kwon T Y 2016 *Mat. Lett.* **178** 300-3
[8] Kim H R, Jang S H, Kim Y K, Son J S, B K Min, Kim K H and Kwon T Y 2016 *Materials* **9** 596
[9] Pan Y, Jiang L, Lin H and Cheng H 2017 *Dent. Mat. J.* **36** 82–87
[10] Mutlu-Sagesen L, Ergun G and Karabulut E 2011 *Dent Mat. J.* **30** 598–610
[11] Bhaskar V and Subba Reddy V V 2010 *J. Indian Soc. Pedod. Prev. Dent.* **28** 6-12
[12] Lu C, Zheng Y, Zhong Q 2017 PLoS ONE 12 e0174440
[13] Geis Gerstorfer J, Weber H 1985 *Dtsch Zahnärztl* **40** 87–91
[14] Porojan L, Savencu C E, Costea L V, Dan M L and Porojan S D 2018 *Int. J. Electrochem. Sci.* **13** 410-23
[15] Sharma M, Ramesh Kumar A V, Singh N, Adya N, Saluja B 2008 *J. Mat. Eng. and Perf.* **17** 695-701
[16] Burstein G T 2010 Passivity and Localized Corrosion *Shreir's Corrosion* vol 2 ed. T J A Richardson (Elsevier Science) chap 2.02.3 p 735
[17] Galateanu B, Golgovici F, Hudita A, Stan M, Dinescu S, Costache M, Demetrescu I and Popescu A 2016 *Mat. Corr.* **67** 1096
[18] Diaz I, Martinez-Lerma J F, Montoya R, Llorente I, Escudero M L and García-Alonso M C 2017 *Bioelectrochem.* **115** 1-10
[19] Savencu C E, Costea L V, Dan M L and Porojan L 2018 *Int. J. Electrochem. Sci.* **13** 3588-3600

IOP Conf. Series: Materials Science and Engineering **416** (2018) 012048 doi:10.1088/1757-899X/416/1/012048

Assessment of the Distribution of Nanofibers in Membranes Based on Microstructure Photos

P Wolszczak[1,*] **and C Garcia**[2,3]

[1] Department of Automation, Faculty of Mechanical Engineering, Lublin University of Technology, Nadbystrzycka 36 St, Lublin, 20-618, Poland
[2] Mechanical and Aerospace Department, University of Strathclyde, 75 Montrose Street, Glasgow, G1 1XJ, UK
[3] IMDEA Materials Institute, C/ Eric Kandel 2, Getafe, 28906, Spain

[*]Corresponding author: p.wolszczak@pollub.pl

Abstract. This paper presents a methodology for estimation of fibers distribution in fibrous mat. Distribution of fibers is important for the mechanical and electrical properties of created new materials. For this study, membranes of polyvinylidene fluoride and polyvinyl pyrrolidone nanofibers with different sizes and random orientation are prepared via electrospinning. The fiber distribution evaluation of the membranes was based on microstructure photos. The developed method can be used as a practical tool to estimate the distribution of fibers.

1. Introduction

Electrospinning is a simple method to produce polymer fibers by charging high voltage to polymer solutions or melts. Over the past two decades, this technique has attracted considerable attention among the research community due to its low cost, versatility and simplicity for the fabrication of fibers with a wide variety of materials. Additionally, electrospinning allows the preparation of fibrous mats with different morphologies by the modification of the polymer solutions and processing parameters used during the electrospinning.

Due to the high surface area and porosity, electrospun nanofibers are excellent materials for several applications as filtration membranes [1], triboelectric sensors [2, 3] and advanced composite materials [4, 5]. For these applications, the morphology and the distribution of the fibers are important for the performance and properties of the created nanomaterials. Therefore, it is urgent to develop methods to measure accurately the diameter and distribution of the fibers.

The fiber diameters and its distribution are usually measured from scanning electron microscope (SEM) images using a ruler (manual analysis). The mentioned method is very time-consuming and can only be used for relative small size samples because the area of analysis is very small as compared to the total size of the fibrous material. Furthermore, manual analysis is not suitable for automated measurements during higher production rate of nanofibers [6]. In this work, we suggest an alternative method to estimate the distribution of the fibers based on digital image analysis. The developed method is automatized, versatile and more time-efficient method than the manual analysis used for most of researchers.

In this paper, membranes of polyvinylide fluoride (PVDF) and polyvinyl pyrrolidone (PVP) fibers are prepared using the technique of electrospinning. The aim of this work is to investigate the distribution of the fibers in the fabricated membranes using image analysis processing. The main advantages of digital image analysis respect to conventional manual analysis is that it can be used for large size materials and saves considerable time. To the best of our knowledge, very little research has been published about this method and the majority the studies utilized manual analysis to measure the distribution of the fibers in electrospun membranes. Therefore, there is a need to propose alternative measurement methods as digital image analysis.

2. Preparation of nanofiber membranes

The aim of the section is to describe the procedure used to fabricate the membranes of PVDF and PVP nanofibers using the technique of electrospinning.

In the preparation of polyvinylidene fluoride nanofibers, PVDF with a molecular weight of 275,000 g mol^{-1}, N, N- dimethylformamide (DMF) and acetone solvents are purchased from Sigma Aldrich. Initially, PVDF in form of pellets is dissolved in a solvent mixture of DMF and acetone (40/60) to prepare chemical solutions at three different polymer concentrations of 16 %, 20 % and 24 % w/v respectively. Subsequently, the homogeneous PVDF solutions are transferred to an electrospinning machine (Nanon-01A) to be spun under the following conditions: applied voltage 15 kV, feed rate 1 ml/h, distance between the collector and needle tip 15 cm, static collector and room temperature. As a results, membranes of random PVDF nanofibers with three different sizes are obtained as illustrated in section 4.

During the preparation of polyvinyl pyrrolidone electrospun fibers, PVP with a molecular weight of 360,000 g mol^{-1} and ethanol solvent are acquired from Sigma Aldrich as well. First, PVP in form of powder is dissolved in ethanol to prepare a 10% w/v polymer solution. Then, the same laboratory electrospinning apparatus is used to spun the polymer solution using the following conditions: applied voltage 18 kV, feed rate 0.5 ml/h, distance between the collector and needle tip 12 cm, static collector and room temperature. Finally, membranes of PVP nanofibers with random morphology and microscale size are prepared as reported above.

3. Analysis of the fibre diameter distribution

This section introduces the methodology used to determinate the distribution of the fibers. The investigation is organized as follows: First, membranes of PVDF and PVP submicron fibers are produced via electrospinnning (section 2). Second, photos of the nanofibers are obtained using a scanning electron microscope (SEM). Finally, the SEM images are processed by manual and automatic analysis with the aim to measure the distribution of the fibers.

3.1. Manual analysis

The software Image J 1.45s is used to measure the fibers diameters from the SEM images. The average fiber diameter value and standard deviation is calculated based on the measurements of 100 fibers.

3.2. Automatic analysis

Figure 1 shows the processing algorithm used to analyze the nanostructured membranes. The scanning electron microscope images show fibers with disordered weave, variable diameter and uneven exposure conditions. Different SEM image areas are characterized by different brightness and sharpness. Therefore, image processing and measurement was conducted on square slices with a side of 200 pixels as shown in Figure 1a. Firstly, the image slices were processed with a Canny filter to emphasize the edges of the fibers (Figure1b). The edge detection in the supplied binary image contained errors due to uneven illumination of the cylindrical surface of wide fibers. As a result,

IOP Conf. Series: Materials Science and Engineering **416** (2018) 012048 doi:10.1088/1757-899X/416/1/012048

additional edges were formed on the areas of the fiber surface. Figure 1 (c) shows a detailed image of the multiple edges. Secondly, geometric edge detection was performed using a Hough filter as indicated in Figure 1d. The direction of the gradient was defined perpendicular to the line as observed in the figure. Next, the lines were joined in pairs surrounding the fibers. In this way, a geometric image of the SEM view was obtained (Figure 1e).

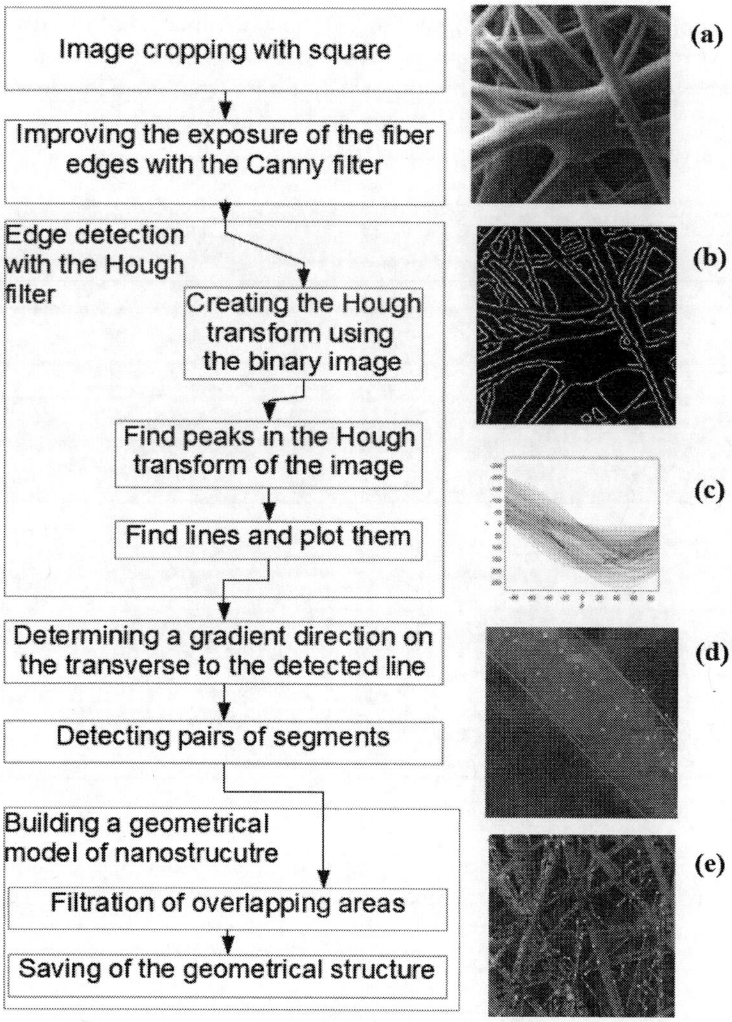

Figure 1. Stages of the SEM automatic image processing algorithm using Canny and Hough filters.

Some of the methods to characterize the microstructure require the identification of fiber edges on microscopic images of sample cross-sections. Others are based on quantitative measurements of the areas occupied by the components of the composite [7-11]. They consist on counting the cases where the characteristics of lower level, such as the values of color components of the pixel in fig.1a have a value lower than the threshold. The advantage of using a geometrical model of the microstructure is that allows the calculation of almost all kinds of measures and indicators. In this work, the fiber diameters were obtained using the geometric images of the studied structures as indicated in fig.1e. It consisted of measuring the distance between the paired lines. During the research, the efficiency of the algorithm was improved by adjusting the processing parameters to the images of various structures. The accuracy of the results was proved by measuring the degree of coverage of the SEM image with measurements. Areas defined by overlapping pairs of fibers with similar angles and spacing were discarded in the analysis.

4. Results and discussion

In this study, membranes of PVDF and PVP electrospun fibers with different sizes are produced using the electrospinning procedure detailed in Section 2. The SEM images of the fibrous mats are shown in figure 2. The SEM images show good quality nanofibers distributed randomly in the membrane with different fiber diameter sizes. The distribution of nanofibers from the SEM images was determined by manual and digital image analysis with the aim to test the accuracy of the automatic method. The experimental results of the analysis are given in figure 3.

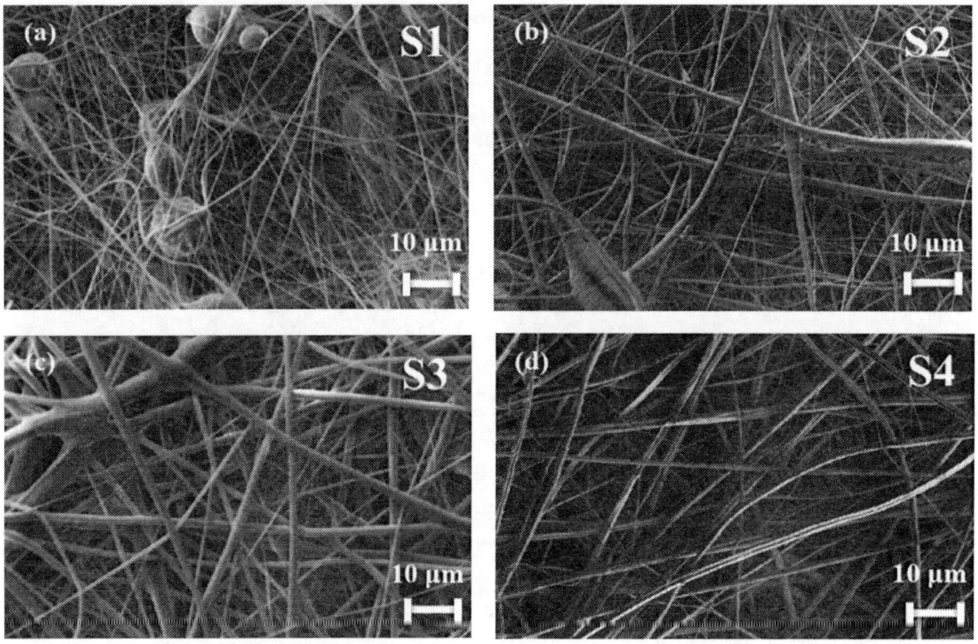

Figure 2. SEM images of nanofibers with different fiber diameters and random orientations used in the digital image analysis: (a) PVDF spun at 16% w/v, (b) PVDF spun at 20% w/v, (c) PVDF spun at 24 % w/v and (d) PVP spun at 10 % w/v.

4.1. Comparison between manual analysis and automatic analysis

Figure 3 shows the results of measurements made manually and automatically. The manual measurements were made using the ruler tool in the software for image analysis. The location of the measured sections and measuring points were decided by the operator. The authomatic measurements were made using the algorithm presented in the article. The Processing parameters were chosen adjusted to a series of representative images. It can be assumed that the geometric images obtained using the automated method are burdened with defects consisting in detecting ostensible edges. These defects lead to the formation of systematic errors of measurements of fiber diameters.

The trend of average fiber diameters in measurements made by hand and automatically are similar as shown in figure 3a. The fiber diameters increase with increasing polymer concentration. The results are characterized by a high value of the standard deviation (figure 3.b), which is caused by the occurrence of high outliers values. This is also evidenced by average values that are higher than median values.

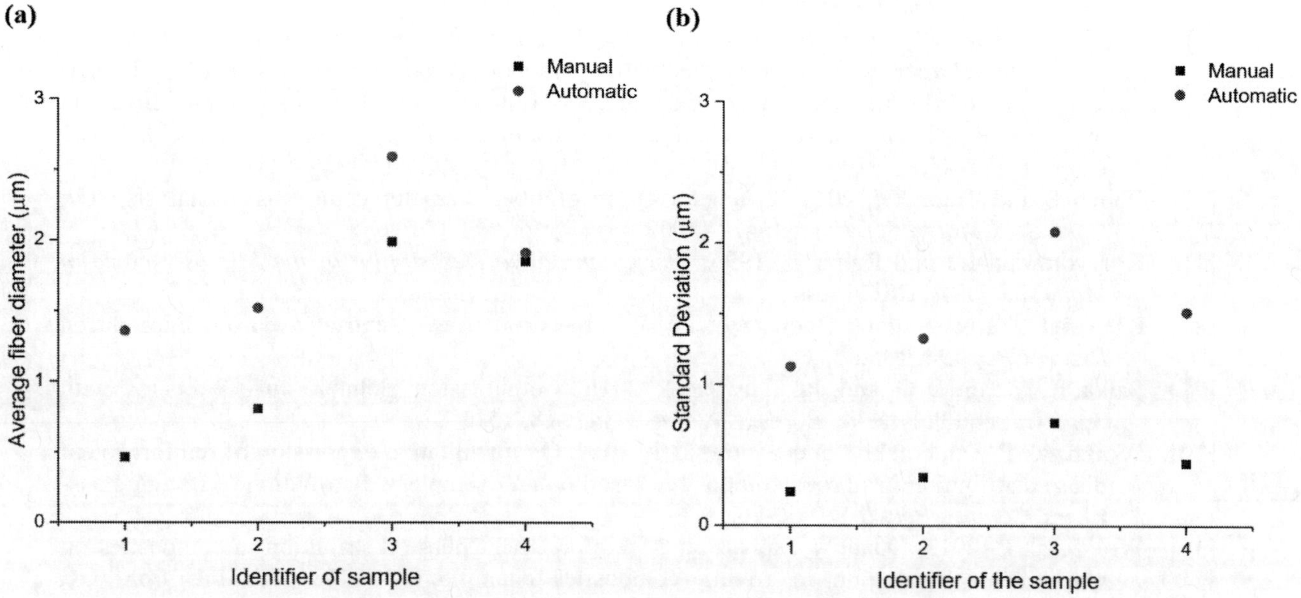

Figure 3. Results of fiber diameters measurements made manually and automatically using digital image analysis: (a) average fiber diameters and (b) standard deviations.

5. Conclusions

The processing of SEM images with random nanofibers and multiple fibre diameters, is a complicated task. By the digital image analysis (automatic measurement), edges and positions of fibers are recorded, while manual analysis allow for measurement of diameters or quantitative measurements with the use of image elements. Digital image analysis provides a geometric model of the structure and multiple measurements, including randomly or systematically arranged testing elements like lines or cycloids.

The article compares manual and automatic measurements carried out by digital image analysis. Manual measurements are subjective, because an operator chooses the number and localization of measurements. However, in the automatic method systematic errors are present. Due to the inability of the possibility of direct measurement it is difficult to estimate measurement error. The accuracy of measurements in the automatic method consist in verification by the operator, based on the distinguished geometrical features.

During the manual and automatic measurements observed similar differences between the average values characterizing the different types of samples. Assuming that the values of the systematic errors caused by the automated method are constant (having its source in the parameters of the algorithm of automatic process), it can be used in comparative studies of various structures.

6. References

[1] Zhu M, Han J, Wang F, Shao W, Xiong R, Zhang Q, Pan H, Yang Y, Samal SK, Zhang F and Huang C, 2017, Electrospun nanofibers membranes for effective air filtration, *Macromol. Mater. Eng.*, **302**, 1-27

[2] Garcia C, Trendafilova I, Guzman de Villoria R and Sanchez del Rio J, 2018, Triboelectric nanogenerator as self-powered impact sensor, *MATEC Web Conf.,* **148**, 14005

[3] Garcia C, Trendafilova I, Guzman de Villoria R and Sanchez del Rio J, 2018, Self-powered

pressure sensor based on the triboelectric effect and its analysis using dynamic mechanical analysis, *Nano Energy*, **50**, 401-9

[4] Garcia C, Wilson J, Trendafilova I and Yang L, 2017, Vibratory behaviour of glass fibre reinforced polymer (GFRP) interleaved with nylon nanofibers, Compos Struct, **176**, 923-32

[5] Garcia C, Trendafilova I, Zucchelli A and Contreras J, 2018, The effect of nylon nanofibers on the dynamic behaviour and the delamination resistance of GFRP composites, *MATEC Web Conf.*, **148**, 14001

[6] Baheti S and Tunak M, 2017, Characterization of fiber diameter using image analysis, *IOP Conf. Ser., Mater. Sci. Eng.*, **254**, 142002

[7] Kurzydłowski KJ and Ralph B, 1995, *The Quantitative Description of the Microstructure of Materials*, USA, CRC Press

[8] Klepka T, Garbacz T and Bociaga E, 2014, Some properties of extruded porous microducts, *Przem Chem*, **94**, 1352-5

[9] Pałka K, Adamek G and Jakubowicz J, 2016, Compression Behavior of Ti Foams with Spherical and Polyhedral Pores, *Adv Eng Mater*, **18**, 1511-8

[10] Wolszczak P, Samborski S and Sadowski T, 2017, On quantitative expression of reinforcement phase distribution in fibrous composites based on an exemplary distribution of roving glass-fibers, *Compos Part B* **129** 66-76

[11] Wolszczak P, 2013, Quantity expression of reinforcement phase distribution in composites on glass-fibers distribution in roving composites example, ICCS17: 17th Int. Conf. on Composite Structures 17-21 June (Porto, Portugal), p 119.

Acknowledgments
The authors wish to thank IMDEA Materials Institute for its valuable support in terms of materials and equipment

IOP Conf. Series: Materials Science and Engineering **416** (2018) 012049 doi:10.1088/1757-899X/416/1/012049

The Experimental Assessment of the Deformability of Cement and Cement-asphalt Matrices with Rubber Powder Additive by Application of DIC System

J Kukiełka[1,*] and T Sadowski[2]

[1] Department of Roads and Bridges, Lublin University of Technology, Nadbystrzycka 40 str., 20-618 Lublin, Poland

[2] Department of Solid Mechanics, Lublin University of Technology, Nadbystrzycka 40 str., 20-618 Lublin, Poland

Abstract. The paper presents an experimental study of deformability of beams made of cement and asphalt-cement matrices with addition of rubber powder 25÷40% in relation to binders. 4-Point Bending (4-PB) tests were performed with series of specimens after 28 days of hardening. Tests were done at various temperatures, i.e. 5°C, 23°C and 50°C. Monitoring of deformation process was fully recorded by Digital Image Correlation (DIC) system ARAMIS up to the final failure.The cracks in the asphalt-cement matrices with rubber powder initiate when stress state exceeded about 50% of their flexural strength (fct,flex2). The obtained experimental results lead to the general conclusion that addition of rubber powder to different matrices causes increase of deformability, from 2 to 3 times in relation to the pure cement matrix. Moreover, it can be stated that composite materials such as asphalt-cement concrete (ACC), mineral-cement-emulsion mixes (MCEM) with rubber powder additive are more flexible in comparison to the pure cement concrete (CC),

1. Introduction

Composite materials, for instance cement concretes (CC), asphalt-cement concretes (ACC), mineral-cement-emulsion mixes (MCEM), as well as those with added rubber powder, are considered to be heterogeneous. Different types of matrices, as a continuous phase of such complex materials and inclusions reinforcing a matrix, are essential components of composite structures. Therefore, the overall properties of considered composites depend on mechanical properties of matrices and inclusions.

In Western Europe and the USA asphalt-cement matrices made by adding asphalt emulsion (AEm) to cement paste are most frequently used in roads recycling; in such matrices the amount of cement in recycled mineral mix usually does not exceed 2%. Cement accelerates asphalt emulsion setting and the initial strength of the mix [1, 2, 3, 4, 5, 6]. The MCEM containing more than 2% of cement have properties similar to those of cement-stabilised mineral mixes [5].

The proposed addition of rubber powder to the MCEM should contribute to improved durability and reduced resilient modulus, as may be concluded from preliminary research [7].

The use of waste tires in road construction was described in the monograph based on the review of national and foreign research [8].

Chippings and granules were used in numerous lab tests of cement concretes intended for the construction of yards and playing fields while crumb rubber was used for instance as an addition to hot asphalt in “dry process” or “wet process” technology [9]. Crumb rubber in “dry process” is added to mixing plant in the amount of 3÷4% of the mix mass. Hot mix-asphalt with added crumb rubber shows increased resistance to temperature changes, for instance to thermal cracks in road pavement layer. Asphalt-rubber binders used for surface treatment and crack relief layers are known from a few industrial technologies, e.g. McDonald [10] or SAMI (Stress Absorbing Membrane Interlayer) [11]. The properties and usability of crumb rubber depend on the rubber type, manner of grinding, shape of particles, specific surface and modification manner [9, 12, 13].

Asphalt pavements are generally designed based on two criteria:

(1) asphalt concrete fatigue,

(2) subgrade compressive strain.

The asphalt concrete fatigue equation contains calculation of tensile strain in the bottom of asphalt layer [14]. According to this criterion the destructive tensile strain of asphalt mixture should be specified. Different fatigue criteria should be applied to pavement with the MCEM. For instance in publication [15] fatigue equation of pavement with the MCEM was proposed and the minimal level of damage strain which guarantees one million load cycles was established $\varepsilon_6 = 170*10^{-6}$ m/m – for fatigue damage parameter equal to $D_f = 30\%$.

In this paper we present results of the 4-PB deformation tests on the above composites with different matrices, starting from purely elastic response, cracks initiation and their further propagation up to the final failure.

In order to test deformation stages we used beams of dimensions 250×50×50 mm made of cement or asphalt-cement matrices with addition of rubber powder. The whole deformation process was recorded using the 3D optical DIC system ARAMIS, which is very useful in case of many different construction materials, e.g. [16-26]. Deformation analysis makes it possible to determine the deformation value in the tested specimen plane from 0.05% to >100% with accuracy of 0.01 % [27]. The system allows to locate cracks and trace their propagation from the width of 0.01 mm [16, 17, 28] and was successfully used in works concerning concrete resistance to cracks for the analysis of initial cracks propagation [28, 29, 30, 31].

2. Materials used for matrix preparation

In accordance with Polish requirements concerning the MCEM [3] Portland cement CEM I or Portland composite cement CEM II class 32.5 or 42.5, which complies with the requirements of PN-EN 197-1 standard "Cement. Part 1. Composition, requirements and conformity criteria for common-use cements" should be used. Portland cement CEM I 42.5 R was adopted to make matrices.

Currently, in deep cold recycling slow-setting overstable asphalt emulsion C60 B10 ZM/R as per PN-EN 13808:2013-10 is used and replaces previous C60 B5 R emulsion as per PN-EN 13808:2010. Asphalt emulsion C60 B5 R was adopted to make matrices.

The rubber powder used in tests was made by milling rubber at room temperature to obtain 0/1 mm grading.

3. Specimen preparation and testing

The subject of research were the following compositions of matrices (W – water, CEM – cement, AEm – asphalt emulsion, RP – rubber powder) :

A – W:CEM = 1:2 – cement matrix with known properties in the case of concretes [32] is also used in lean concretes and aggregates stabilised with hydraulic binders,

B – CEM:RP = 1.5:1 – cement matrix with added rubber powder used in standard and lean concretes [33, 34, 35],

C – AEm:CEM = 3:5 – asphalt-cement matrix with dominant content of cement which occurs in most mixes used in cold recycling in Poland [3],

D – AEm:CEM:RP = 1:1:1 – asphalt-cement matrix with added 0/1 mm rubber powder with the same content of components (m/m),

E – AEm:CEM:RP = 1:1.5:1 – asphalt-cement matrix with added 0/1 mm rubber powder with increased content of cement (m/m),

F – AEm:CEM:PR = 1:2:1 – asphalt-cement matrix with added 0/1 mm rubber powder with the cement content corresponding to the content of the remaining components (m/m).

From each matrix composition 9 beams of 250×50×50 mm in dimensions were made. All the matrix compositions had the W/CEM ratio within the range of 0.5÷0.6 in order to obtain plastic consistence and to prevent segregation of components in the moulds. The specimens were consolidated by gravitational compaction and were left in moulds for 24 hours; after this time they were demoulded and stored for 28 days in room temperature, protected from water evaporation.

To obtain strain fields using the ARAMIS system it is necessary to introduce points pattern on the specimen surface, which is scanned by the DIC system and then its digital image is created. In the experimental tests it was assumed that vertical displacement rate of the loading head during 4-PB process was equal to 0.2 mm/min. The DIC system enables saving the entire deformation test as a video, which makes thee analysis of the whole specimen destruction process possible from the beginning of loading until its destruction [27].

The horizontal tensile strain measurements ε_t were concentrated in a central bottom part of the specimen, between the points at which loading was applied. The analysed fragment of the specimen has length and height equal to approximately 50 mm and 10 mm, respectively.

4. The results of asphalt cement matrices tests

The flexural strength in the 4-PB tests were performed in temperatures 5°C, 23°C and 50°C. For each matrix composition we prepared 3 beams, which were stored for at least 4 hours in a climate chamber prior to the test. The experimental tests were performed immediately after the specimens were removed from the chamber. The sample deformations ε_t were measured by application of the ARAMIS and were presented in Fig. 1.

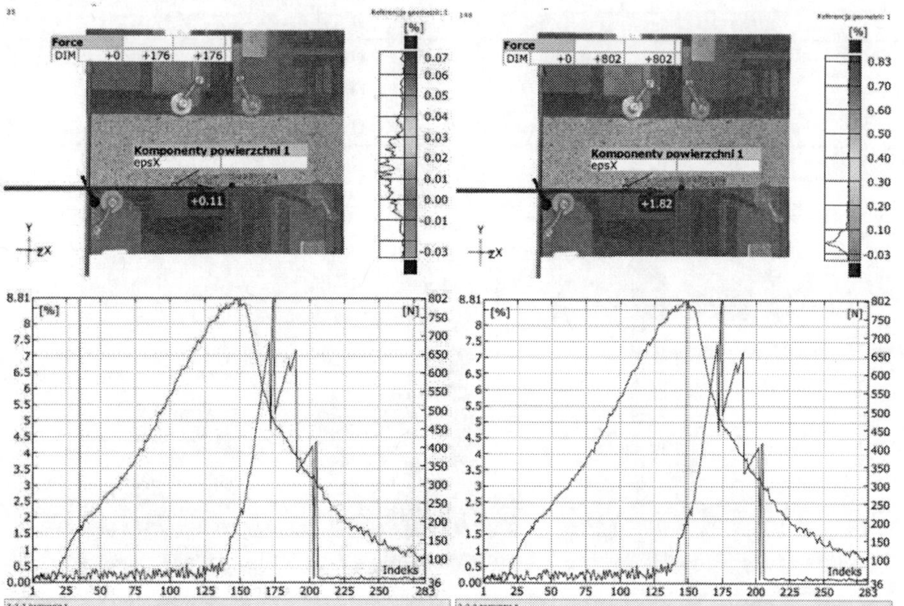

Initiation of crack Destruction of specimen

Figure 1. The 4-PB flexural strength of the beam made of the F matrix measured in 5°C

The results of experimental measurements and calculations are summarised in Table 1 and Fig. 2.

Table 1. The average values of tensile strains ε_t and flexural stresses for characteristic points of deformation processes of specimens made of cement and asphalt-cement matrices with added rubber powder.

Matrix	Temperature of tested beams [^{0}C]	Tensile strain initiating a crack ε_t [%]	Flexural stress at the moment of cracking σ_{flex2} [MPa]	Flexural strength $f_{ct,flex2}$ [MPa]	$\frac{\sigma_{flex2}}{f_{ct,flex2}}$
	5	0.05	0.71	1.29	0.55
A	23	0.07	0.37	0.67	0.54
	50	0.06	0.21	0.33	0.63
	5	0.08	0.84	1.26	0.67
B	23	0.23	0.52	1.03	0.51
	50	0.17	0.20	0.70	0.28
	5	0.12	2.50	2.67	0.94
C	23	0.10	0.88	0.97	0.91
	50	0.17	0.52	0.85	0.61
	5	0.31	0.51	0.89	0.57
D	23	0.16	0.28	0.64	0.44
	50	0.18	0.14	0.34	0.42
	5	0.15	0.80	1.28	0.62
E	23	0.13	0.56	0.78	0.72
	50	0.14	0.19	0.40	0.47
	5	0.15	0.66	1.12	0.66
F	23	0.17	0.29	0.51	0.56
	50	0.13	0.18	0.32	0.55

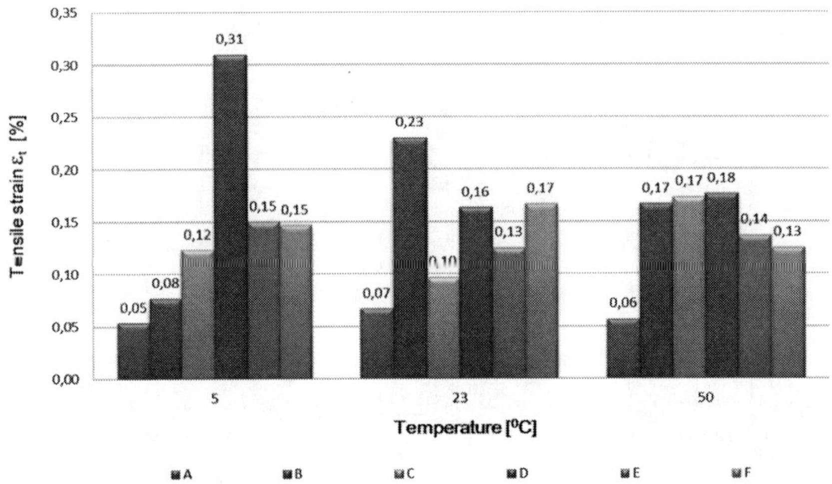

Figure 2. The values of the tensile strain ε_t initiating cracks during 4-PB strength tests for beams made of matrices: A, B, C, D, E, F after their maturing during 28 days and tested in 5°C, 23°C, 50°C.

5. Conclusion

IOP Conf. Series: Materials Science and Engineering **416** (2018) 012049 doi:10.1088/1757-899X/416/1/012049

(1) The performed tests of mechanical properties of specimens taken from cement and asphalt-cement matrices with addition of rubber powder reveal that deformability of the considered composites increased from 2 to 3 times in relation to cement matrix while influencing flexural strength 0,7 to 1,0 times.
(2) The cracks initiation in the AEm:CEM:RP matrices appears when stress exceeded about 50% of their flexural strength ($f_{ct,flex2}$). In case of the AEm:CEM matrix this stress was about 90% in temperature less than 23^{0}C. In temperature 50^{0}C cracks initiation took place respectively when stress exceeded 40% and 60% .
(3) The tensile strain corresponding to cracks initiation a in the cement matrix W:CEM = 1:2 was equal to ε_t=0.05% regardless of the test temperature. In case of the AEm:CEM:RP = 1:2:1 and AEm:CEM:RP = 1:1.5:1 matrices the appropriate values of strains were ε_t= 0.13÷0.15%. The largest tensile strain ε_t was obtained for matrices with high content of the rubber powder, i.e. for CEM:RP = 1.5:1 and AEm:CEM:RP = 1:1:1. The experimentally estimated tensile strain were equal to ε_t = 0.23% and ε_t = 0.31%, respectively.
(4) The influence of temperature on the threshold of critical deformations causing cracks initiation is not significant.
(5) One can conclude, summarizing, that the most important factor which influences the deformability of matrices is their composition and the amount of added rubber powder.
(6) It is necessary to formulate a numerical model describing the mechanical response of the beams including formation of the process zone in the crack tip and its propagation, e.g. [36 – 48].

6. References

[1] Asphalt Recycling and Reclaiming Association (ARRA) 2001 *Basic asphalt recycling manual*, Publication No. NHI01-022 (Annapolis, Maryland, USA)
[2] Merrill D, Nunn M E and Carswell I 2004 *A guide to the use and specification of cold recycled materials for the maintenance of road pavements*. TRL Report 611, Transport Research Laboratory, (Crowthorne, UK).
[3] General Directorate for National Roads and Motorways 2014 *Manual design and embedding of mineral-cement-emulsion (MCE) mixtures*. Warsaw, Poland
[4] Xu J, Huang S, Qin Y and Li F 2011 *Int. J. of Pavement Research and Technology* **4** 48–55
[5] *Pavement Recycling Guidelines* 2003 PIARC Committee C7/8 – Road Pavements.
[6] *Road Pavement Rehabilitation Techniques Using Enhanced Asphalt Mixtures* 2004 Final Technical Report PARAMIX.
[7] Kukiełka J 2017 *Civil Engineering and Architecture* (*Budownictwo i architektura)* **16** 53-63
[8] Horodecka R, Kalabińska M, Piłat J, Radziszewski P and Sybilski D 2002 Waste tires management in road construction. *Book of Road and Bridge Research Institute (Zeszyt Instytutu Badawczego Dróg i Mostów)* **54**
[9] Epps J A 1994 *Uses of Recycled Rubber Tire in Highways Practice*. NHRP Synthesis, TRB National Research Council Press (Washington D.C.)
[10] McDonald C H 1969 Asphalt Rubber Compounds and Their Applications For Pavement *California Streets and Highway Conference*
[11] Hardy T A, Brock L G 1985 *Stress Absorbing Membrane Interlayer. Experimental Feature* Final Report. Project OR 77-03. Federal Highway Administration.
[12] Monismith C L 1985 Improved Asphalt Mix Design. *Proceedings Association of Asphalt Pavenig Technology*. **54**
[13] Takellou H B 1991 *Benefits of Recycling Waste Tires in Rubber Asphalt Paving*. Transportation Research Board, (Washington D.C.)
[14] Asphalt Institute 1982 *Research and Development of the Asphalt Institute's Thickness Design Manual (MS-1)*, 9th edition. Research Report 82-2.
[15] Kuźniewski J, Skotnicki Ł, Szydło A 2015 *Bulletin of the Polish Academy of Sciences.*

Technical Sciences **63**, DOI: 10.1515/bpasts-2015-0012

[16] Sadowski T, Marsavina L, Peride N, Craciun E-M 2009 *Comput. Mat. Sci.* **46** 687-693

[17] Marsavina L, Linul E, Constantinescu DM, Apostol D, Voiconi T, Sadowski T 2014 *Eng. Fract. Mech.* **129** 54-66

[18] Sadowski T, Golewski P 2012 *Comp. Mater. Sci.* **52** 293-297

[19] Nakonieczny K, Sadowski T 2009 *Comp. Mater. Sci.* **44** 1307-1311

[20] Sadowski T, Nakonieczny K 2008 *Comput. Mat. Sci.* **43** 171-178

[21] Marsavina L, Sadowski T 2007 *Int. J. Frac.* **145** 237-243

[22] Sadowski T, Samborski S 2008 *Comput. Mat. Sci.* **43** 75-81

[23] Postek E, Sadowski T 2011 *Compos. Interf.* **18** 57-76

[24] Sadowski T, Golewski P 2012 *Comp. Mater. Sci.* **64** 285-288

[25] Marsavina L, Linul E, Voiconi T, Sadowski T 2013 *Polymer Testing* **32** 673-680

[26] Sadowski T, Marsavina L 2011 *Comput. Mat. Sci.* **50** 1336-1346

[27] ARAMIS v6 User Manual, GOM mbH, Mittelweg 7-8 D-38106 Braunschweig

[28] Ajdukiewicz C, Gajewski M and Mossakowski P 2010 *Logistyka* **6** 28–35

[29] Goszczyńska B and Tworzewska J 2014 *Przegląd Budowlany* **12** 44–49

[30] Golewski G, Sadowski T 2014 *Constr. Build. Mat.* **51** 207-214

[31] Golewski G, Golewski P, Sadowski T 2012 *Comput. Mat. Sci.* **62** 75-78

[32] Brandt A M 2010 Concrete diagnosis based on structure testing. *56 Konferencja Naukowa Komitetu Inżynierii Lądowej i Wodnej PAN oraz Komitetu Nauki PZITB*. Kielce – Krynica. p 57–71

[33] Ołdakowska E 2015 *Ecological Engineering*. **43** 49–54. DOI: 10.12912/23920629/58902

[34] Sadowski T and Pietras D 2014 *Solid State Phenomena* **216** 67-72

[35] Sgobba S, Marano G P, Borsa M and Molfetta M 2010 Use of Rubber Particles from Recycled Tires as Concrete Aggregate for Engineering Applications. *Second International Conference on Sustainable Construction Materials and Technologies*. Università Politecnicadelle Marche, Ancona, Italy. Vol. 1 **465** p 47–58

[36] Gajewski J, Sadowski T 2014 *Comp. Mater. Sci.* **82** 114-117

[37] Dębski H, Sadowski T 2014 *Comput. Mat. Sci.* **83** 403-411

[38] Ivanov IV, Sadowski T, Pietras D 2013 *Europ. Phys. J. Special Topics* **222** 1587-1595

[39] Marsvina L, Sadowski T, Kneć M 2013 *Eng. Fract. Mech.* **108** 139-151

[40] Birsan M, Sadowski T, Marsavina L, Linul E, Pietras D 2013 *Int. J. Solids and Struct.* **50** 519-530

[41] Balawender T, Sadowski T, Knec M 2011 *Arch. Metal. Mat.* **56** 439-446

[42] Balawender T, Sadowski T, Golewski P 2011 *J. Adhesive Sci. Technol.* **25** 2391-2407

[43] Sadowski T, Bec J 2011 *Comput. Mat. Sci.* **50** 1269-1275

[44] Burlayenko V, Sadowski T 2012 *Comput. Mat. Sci.* **52** 212-216

[45] Burlayenko V, Sadowski T 2012 *Finite Elements in Analysis and Design* **62** 49-64

[46] Marsavina L, Sadowski, T 2008 *Polymer Testing* **27** 941-944

[47] Samborski S, Sadowski T, 2010 *Jour. Amer. Cer. Soc.* **93**, 3607-3609

[48] Sadowski T, Golewski P, Kneć M 2014 *Comp. Struct.* **112**, 66-77

IOP Conf. Series: Materials Science and Engineering **416** (2018) 012050 doi:10.1088/1757-899X/416/1/012050

The Effect of Manufacturing on the Press fit Insertion Force

D I Maxim[1,*], L Marsavina[1], L Rusu[1]
[1]University Politehnica Timisoara, Romania

[*]Corresponding author: maxim_daniel@yahoo.com

Abstract. In the automotive industry and not only, it is useful to have a very fast and simple assembly process between parts, so it is used the press fit process between parts. Usually, the press fit force depends on part geometry, radial interference between parts and material behavior. In this study was used one shaft (only one dimension) and thin sheet metal with different size of hole in order to have more interference values between parts. This work will present a comparison of the press fit force between the two types of manufacturing process (machining and blanking) of the hole in sheet metal. The numerical analysis is calibrated with the experimental measurement for machined hole in sheet metal.

1. Introduction

Because of the high productivity requirement, the press fits are used more and more in the assembly processes. Assembling by press fit belongs to the wider group of mechanical joining processes in the industry with high productivity.

The assembling by press fits is created by the differences in geometry of two mating parts, driven to the high contact pressure between the two surfaces in contact. This contact pressure holds the two parts in position after mounting and in service. The strength of this mounting is mainly depending by friction coefficient, interference value and contact surface.

This study is focused to find the difference between two geometries of the hole, one obtained by machining process and the other one obtained by blanking process. Because of the simplicity to obtain the machined part, the test is done for machined part.

In the press fit process are important the length of the contact, the Poisson effect and friction coefficient. The ratio between length and diameter of press fit has influence on Poisson effect, which mean for small ratio, the Poisson effects can be neglected but the friction coefficient is more important [1]. According to [1] [2], we can use the Lamé formula to calculate the pressure from contact:

$$p=\frac{1}{2}\left(\frac{\delta}{b}\right)E_e \qquad (1)$$

where:

$$E_e=\frac{2}{\frac{1}{E_o}\left(\frac{c^2+b^2}{c^2-b^2}-\vartheta_o\right)+\frac{1}{E_i}\left(\frac{b^2+a^2}{b^2-a^2}-\vartheta_i\right)} \qquad (2)$$

$$F_i=\mu pA \tag{3}$$

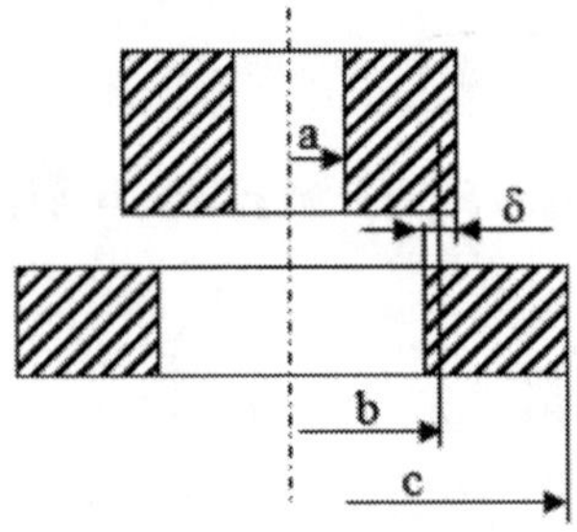

Figure 1. Geometry of the press fit assembly parts

with:
p – contact pressure,
δ – radial interference value between parts, Figure 1,
b – radius of interference, Figure 1,
a – inner radius of inner cylinder, Figure 1,
c – outer radius of outer cylinder, Figure 1,
E_o – elastic modulus of outer cylinder,
E_i – elastic modulus of inner cylinder,
ϑ_o – Poisson ratio of outer cylinder material,
ϑ_i – Poisson ratio of inner cylinder material,
F_i – insertion force or friction force,
μ – friction coefficient,
A – contact area.

According to the DIN 7190 [3] standard in the calculation of the interference must be taken in account the roughness of the surfaces in contact. The formula to calculate the interference is:

$$U_w=U-0.8*(R_{zA}+R_{zI}) \tag{4}$$

where:
U_w – calculated interference
U – interference
R_{zA} – roughness of the hub
R_{zI} – roughness of the shaft

In the [4], is modeled the roughness of the parts and consider in simulation by homogenized finite macro-element. For small interferences is useful to consider the roughness of the parts but for bigger interferences the asperities quickly became plastic and their behavior has relatively little influence on the average pressure.

The thermal stress and thermal expansion added over the assembly tightness can drive to the failure of the assembly. The alternates of the loads can have to be take care when is used the press fit assembly process [5].

2. Characterization of material

A tensile test using a "dog-bone" specimen was performed from cold-rolled electrical steel. The material is tested on 3 directions (Rolling Direction (RD), 45° from RD, Transversal Direction (TD)), at loading speed 5 mm/min which mean a strain rate around 0.0033 s^{-1} on Instron 8874 testing machine at room temperature. The specimen tested has thickness of 0.65 mm with complete geometry presented in Figure 2.

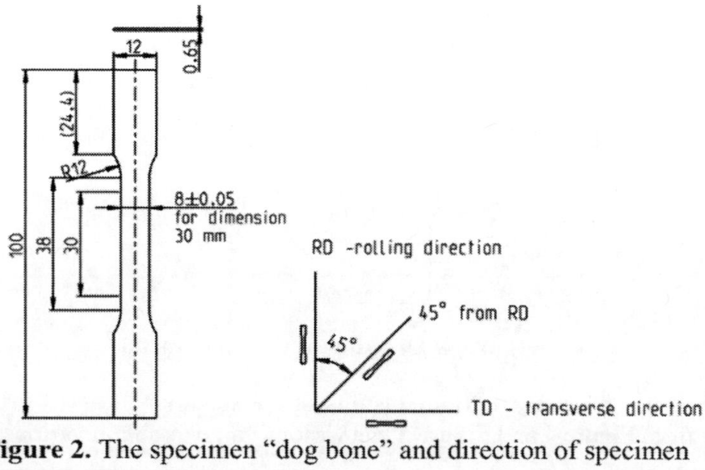

Figure 2. The specimen "dog bone" and direction of specimen

The engineering stress –strain curves are presented in Figure 3. The data is used to calculate de the press fit force, in numerical simulations.

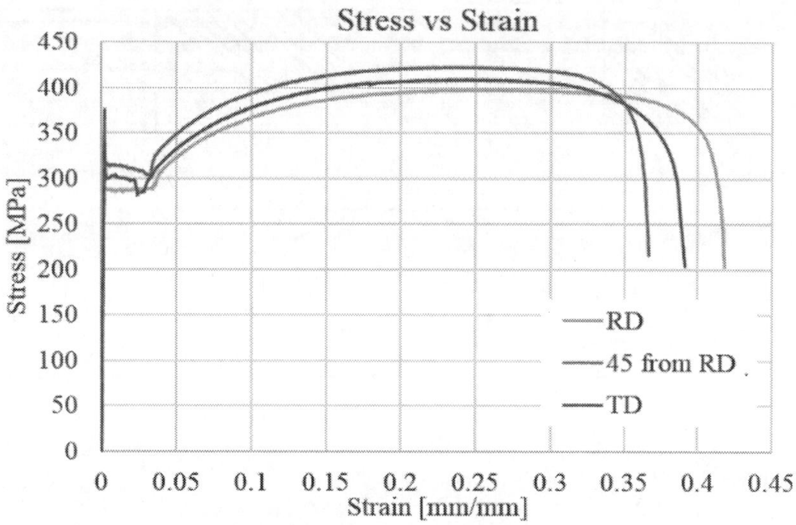

Figure 3. Stress – strain curve

The material used for sheet metal is anisotropic and has different properties on the rolling direction (RD) and transversal direction (TD). The characterization of the material for 45° from RD show a big value for tensile stress but less tensile strain at break. Yield stress for TD is with 5% bigger than the yield stress for RD and ultimate tensile stress for TD is with 3% bigger than ultimate tensile stress for RD.

In the numerical simulation an isotropic material was used, with the properties averaged from the 3 curves to have faster simulations results. In the further work the anisotropy of material will be considered.

3. Geometry for press fits

Below is presented the geometry used in the numerical analysis and in the test. The sheet metal has thickness 0.65 mm. The shaft material is stainless steel and has special geometry for one end to do more easily the press fit.

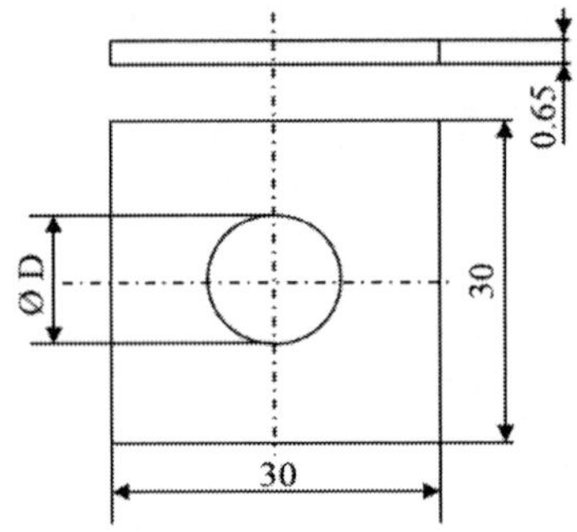

Figure 4. Drawing of the sheet metal

Figure 4 presents the geometry and dimensions of the sheet metal, used for tests and simulations. The dimension ØD from Figure 4 and Figure 5 has various value to obtain various tightness and the shaft has ØD=8 mm±0.005 mm.

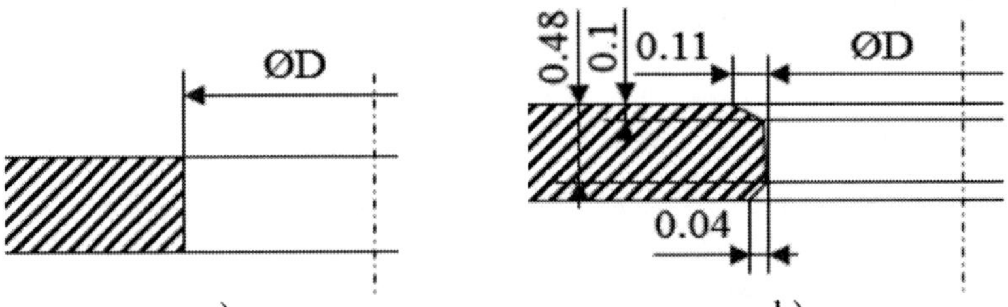

Figure 5. Profile of the sheet metal

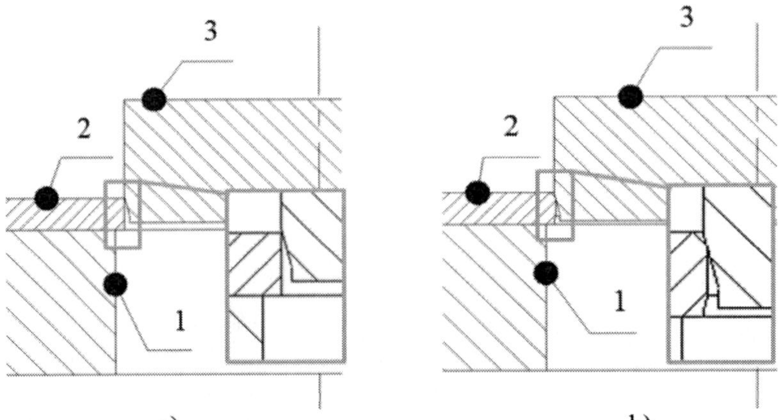

Figure 6. Scheme of the press in process

In Figure 6 it is represented the scheme of the press in process for assembly. We have the support (1), the sheet metal (2) and the shaft (3). The Figure 6a represent the press in process with the sheet metal profile machined on CNC machine (see Figure 5a) and the Figure 6b represent the press in process with the sheet metal profile made by blanking process (see Figure 5b).

4. Numerical simulations

Transient structural simulation was performed using Ansys Workbench 17.1 package. The 3D model was used in the simulation with higher order element type SOLID186 with 20 nodes/element, for all the parts.

It was compared the press force between the two profiles of the hole from the sheet metal, one machined hole (see Figure 5a) and the other one obtained by blanking (see Figure 5b). The diameter of the shaft has a constant value, but for the sheet metal hole more values were considered to obtain more values of the interference between parts.

The material models were linear for the shaft and support (E=210 GPa) and elastic-plastic bilinear for the sheet metal (yield strength =300 MPa and tangent modulus 864.4 MPa). The geometric parts were meshed in 15920 elements, around 71760 nodes for the sheet metal; in 27135 elements, around 114183 nodes for shaft and in 1944 elements, around 10512 nodes for support.

The support (1) from Figure 6 was considered fixed support like boundary condition. The symmetric behavior of the frictional contacts was considered between parts. The load was applied as displacement of shaft (1 mm) in the direction from the top surface of the sheet metal.

Our goal was to compare the profiles of the hole from sheet metal to understand the difference between them. In Figure 7 presents difference between machined profile and blanking profile hole. It is observed that the insertion force, in case of blanking profile, is lower than in case of machined profile hole.

The insertion force for profile machined has almost the same value when the interference is bigger than 0.010 mm. In case of blanking profile, the insertion force is increasing continuously until 0.030 mm interference.

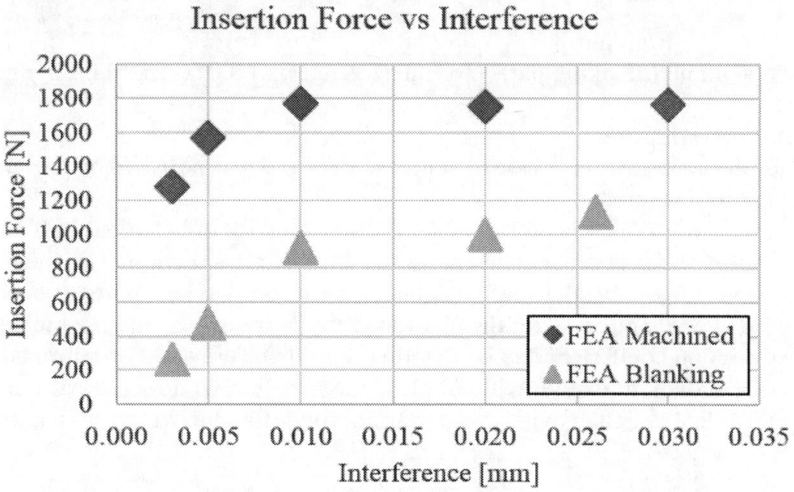

Figure 7. Insertion Force vs. Interference (machined vs blanking)

Figure 8 presents the strain for machined profile with interference 0.01 mm /0.02 mm /0.03 mm on radius and in Figure 9 is shown the strain for blanking profile with interference 0.01 mm /0.02 mm /0.026 mm on radius. For blanking profile was used the maximum interference 0.026 mm because of the convergence issue. It can be observed the higher maximum strain in case of blanking profile. The strain for blanking sheet metal (see Figure 9) is 4 times bigger than the machined sheet metal (see Figure 8). The high strain appears do to the profile of the hole of sheet metal. The small amount of material is in "front of shaft" and high friction occurs on blanking hole of sheet metal on small surfaces.

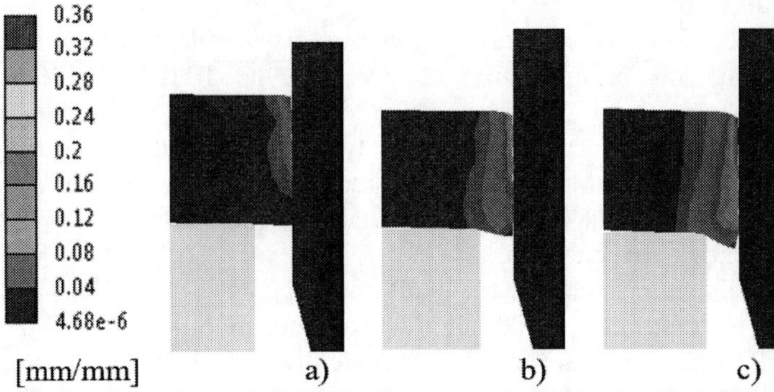

Figure 8. Total strain (machined parts) for interferences: a) 0.01 mm, b) 0.02 mm, c) 0.03 mm

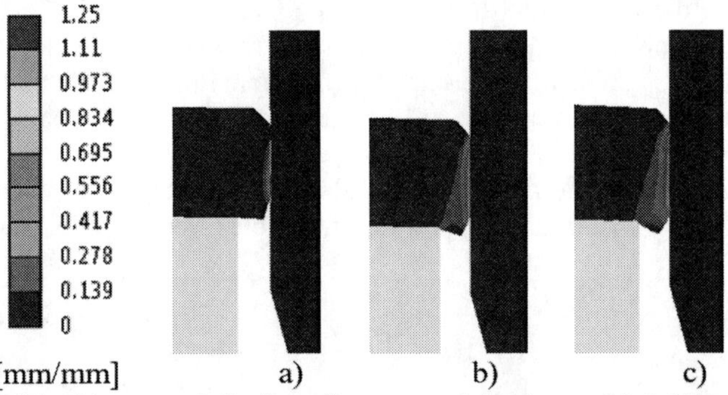

Figure 9. Total strain (blanking parts) for interferences: a) 0.01 mm, b) 0.02 mm, c) 0.026 mm

5. Experimental investigation

The shaft used for the experiment is made of stainless steel and was polished on the surface which should be in contact with the sheet metal.

First measurement was done to determine the friction coefficient between shaft and sheet metal part. The most important is the friction coefficient on the axial shaft direction. The measurement was done with the system from Figure 10, the shaft was clamp between two half parts of sheet metal. The shaft was linked by wire over the pulley with the mass load. By knowing the mass which act on shaft and the loading mass the friction coefficient can be determinate. There were tests performed, one with cleaned parts (cleaned with alcohol) and other without clean the parts but without special lubricants. The results are presented in Table 1 together with the dry friction coefficient which was used in the numerical simulations.

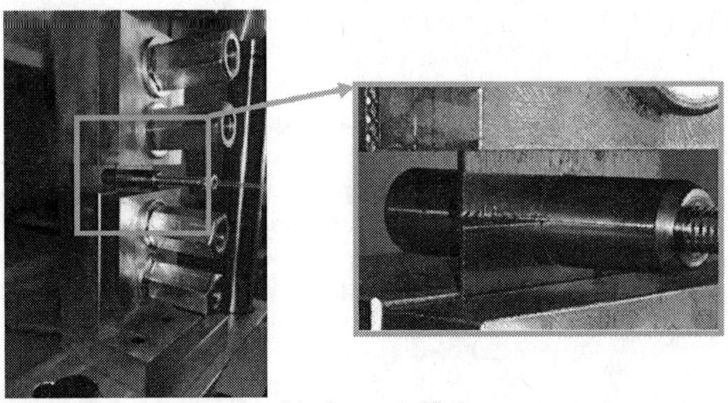

Figure 10. System for friction coefficient measurements

IOP Conf. Series: Materials Science and Engineering **416** (2018) 012050 doi:10.1088/1757-899X/416/1/012050

Table 1. The friction coefficient.

	Partially Dry	Dry
Friction coefficient	0.4467	0.4975

The insertion tests were performed on machine Zwick Z005 at room temperature and constant speed of insertion 6 mm/min.

The maximum force obtained at 1 mm displacement was used to compare with the numerical simulation. The displacement of shaft was approximately 2 mm to insert the shaft until the cylindrical zone of the shaft (see Figure 12).

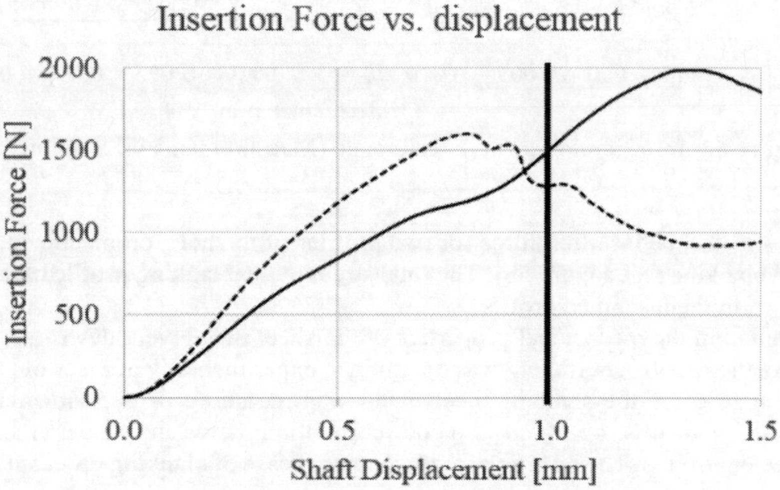

Figure 11. Typical insertion force vs. displacement of shaft

After 1 mm insertion either an increase or decreased of force was observed (see Figure 11). However, for comparison with simulation results only the load corresponding to 1 mm insertion was considered.

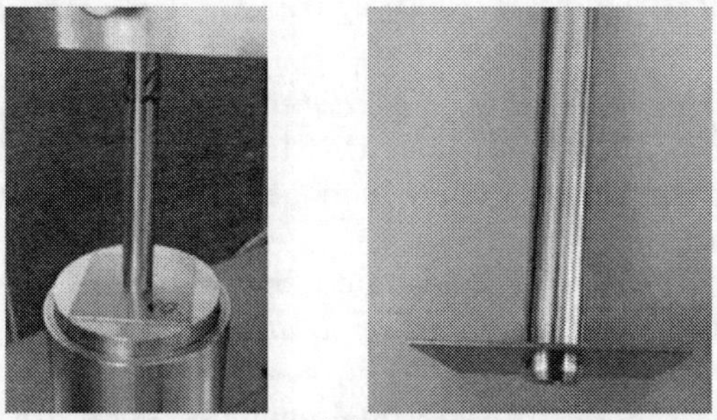

Figure 12. Insertion device and the shaft inserted in sheet metal

In Figure 13 can be observed a comparison of insertion force between measurements and simulations in case of machined profile of the hole from sheet metal. A very quickly increase of force at small interference has been observed, which means the material of sheet metal has a fast transition from elastic domain to the plastic domain.

IOP Conf. Series: Materials Science and Engineering **416** (2018) 012050 doi:10.1088/1757-899X/416/1/012050

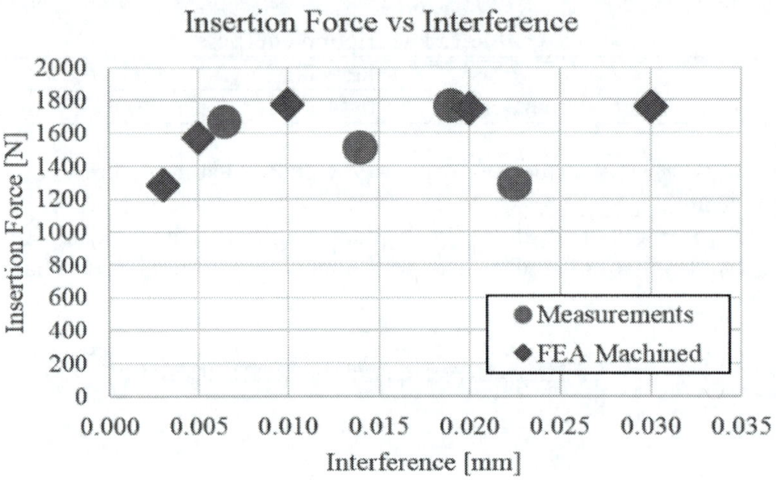

Figure 13. Insertion Force vs. Interference (Measurements vs machined profile)

6. Conclusion

The paper investigates the insertion force for two profiles of the hole, on machined and (see Figure 5a) and one named blanking (see Figure 5b). The total strain in the blanking profile case is approximately 4 times bigger than in the machined profile.

Prior to simulation the mechanical properties of the sheet metal were determined on three different directions. Also, the friction coefficient was determined experimentally in cases of partially dry and dry.

The insertion force of the shaft in sheet metal was measured in experimentally for the case of machined profile and was observed a good correlation between measurements and numerical simulations (see Figure 13). The insertion force is lower in case of blanking process than in the machined process.

7. References

[1] Selvage CC 1979 Assembly of interference fits by impact and constant force methodes, *Cambrige: Massachusetts Institute of Technology*

[2] Schmid SR, Hamrock BJ and Jacobson BO 2014 *Fundamentals of Machine Elements*, New York: CRC Press

[3] *ISO 7190:2001-02, Interference fits - Calculation and design rules*

[4] Boutoutaou H, Bouaziz M and FontaineJ F 2013 Modelling of interference fits with taking into account surfaces roughness with homogenization technique *International Journal of Mechanical Sciences* 21-31

[5] Hao D and Wang D 2013 Finite-element modeling of the failure of interference-fit planet carrier and shaft assembly *Engineering Failure Analysis* 184-196

[6] *ISO 6892-1:2009, Metallic materials - Tensile testing - Method of test at room temperature*

[7] ***, Instron Loadcell Test Certificate, Catalogue No. 2527-201, Serial No. 53930, Capacity 25 kN, United Kingdom, 30 Oct 2008

[8] Lodygowski T, Rakowski J and Litewka P 2013 *Recent Advances in Computational Mechanics* Poznan: CRC Press

[9] *Ansys Workbench 17.1 Documentation,* 2017

IOP Conf. Series: Materials Science and Engineering **416** (2018) 012051 doi:10.1088/1757-899X/416/1/012051

Metal Foam-Filled Tubes as Plastic Dissipaters in Earthquake-Resistant Steel Buildings

C Ghiani[1], E Linul[2,*], MC Porcu[1], L Marsavina[2], N Movahedi[3] and F Aymerich[1]

[1]Department of Mechanical, Chemical and Materials Engineering, University of Cagliari, via Marengo 2, Cagliari, Italy
[2]Department of Mechanics and Strength of Materials, Politechnica University of Timisoara, 1 Mihai Viteazu Avenue, 300 222 Timisoara, Romania
[3]Independent Researcher (Graduated from Semnan University), Isfahan, Iran

[*]Corresponding author: emanoil.linul@upt.ro

Abstract. The aim of this paper is to investigate whether metal-foam-made devices can be effective to dissipate seismic energy in buildings during strong earthquakes. To this purpose, non-linear numerical analyses of concentrically braced steel buildings under recorded ground motions have been carried out, while some experimental tests on metal-foam specimens and metal-foam-filled tubes have been performed. Foam-based devices are assumed to be inserted within the diagonal braces of the considered steel frame to dissipate energy by plastic deformation during strong earthquakes. To apply the experimental data, a scaled numerical model of the prototype building has been implemented by means of the similitude theory and the Buckingham Π theorem. The results of the study provide a preliminary assessment of the potential of metal foam-based dissipaters to reduce the seismic effects in civil structures.

1. Introduction

Metal foams are a relatively novel class of materials used for several applications mostly in the mechanical engineering field [1-3]. Among their high-performance properties are weight-to-stiffness ratio, damping and energy dissipation, thermal resistivity and acoustic absorption [4, 5]. In particular, owing to their excellent energy absorption characteristics, metal foams are widely used in the mechanical, aerospace and automotive applications, as, for instance, shock mitigation in vehicles, cf. e.g. [6]. On the contrary, they are still scarcely used in civil engineering, as addressed in [7]. However, the capability to dissipate energy that metal foams typically possess could be exploited to control the seismic effects on buildings.

Different strategies are actually adopted to reduce the seismic stress on civil structures during exceptional events. Typically they are based on (i) the ductile behavior of structural elements or connections, [8], (ii) base-isolation [9] or mass reduction [10] techniques, (iii) dampers or dissipaters introduced in the structure [11]. The feasibility of using metallic foams to dissipate seismic energy in concentrically-braced (CB) steel buildings was firstly explored in [12] while the effectiveness of filling tubular steel elements with metallic foams to increase their ductility and bucking resistance was shown in [13]. Although CB frames may perform very well under seismic actions, cf. [14], diagonal braces were found to experience perform severe concentration of plastic-deformation in their mid-length section, which may lead to premature brace fracture and frame failure [15]. For this reason modified or hybrid diagonal braces were proposed [12, 15].

The present paper aims to further investigate on the possible application of metallic foams (MFs) and foam-filled tubes (FFTs) within the diagonal braces of CB steel frames to dissipate large amount of energy during severe earthquakes. The results of some experimental tests were exploited to obtain an idealized elastic-plastic behavior of foam-based devices to be implemented in the numerical model. Non-linear dynamic numerical analyses were carried out on full-scale and scaled models of a case-study steel building with and without metal-foam-based devices. The results leaded to evaluate the percentage of stress reduction achievable under real ground motions. This study is just a preliminary phase of a wider investigation involving cyclic tests on foams and foam filled tubes, as well as several parametric non-linear analyses on different steel structures under many severe earthquakes.

2. Experimental tests on metallic foams and foam-filled tubes

Closed-cell metallic (aluminum alloy) foams (Fig. 1a) with two different densities (420 kg/m^3 and 750 kg/m^3) were manufactured by casting technique. All details about aluminum alloy foams preparation has been described in Ref. [16]. The cylindrical specimens (height = 20 mm, diameter = 20 mm) were cut from large foam blocks using Electric Discharge Machining. The obtained specimens were tested both separately (only foam) [17] and as core material in cylindrical thin-walled (1 mm thickness) 304 stainless steel tubes (see Fig. 1b) [18, 19]. Quasi-static uniaxial compression tests were carried out at room temperature on a 100 kN LBG testing machine (see Fig. 1c), with a constant crosshead speed of 10 mm/min, according to the ISO13314-11 standard [20].

The stress-strain diagrams obtained for 420-MF (420 kg/m^3 foam density), 750-MF (750 kg/m^3 foam density) and 420-FFT (750 kg/m^3 foam-filled tube) specimens under quasi-static compression tests are provided in Fig. 2. The behavior obtained in this case is typical to other closed-cell metallic foams and advanced composite foam structures, highlighting three very important regions in the field of energy absorption: linear-elastic, plateau and densification regions [21, 22].

(a) MF specimen (b) FFT specimen (c) The 100 kN LBG quasi-static testing machine

Figure 1. The specimens before and after the loading (a, b), and experimental test set-up (c)

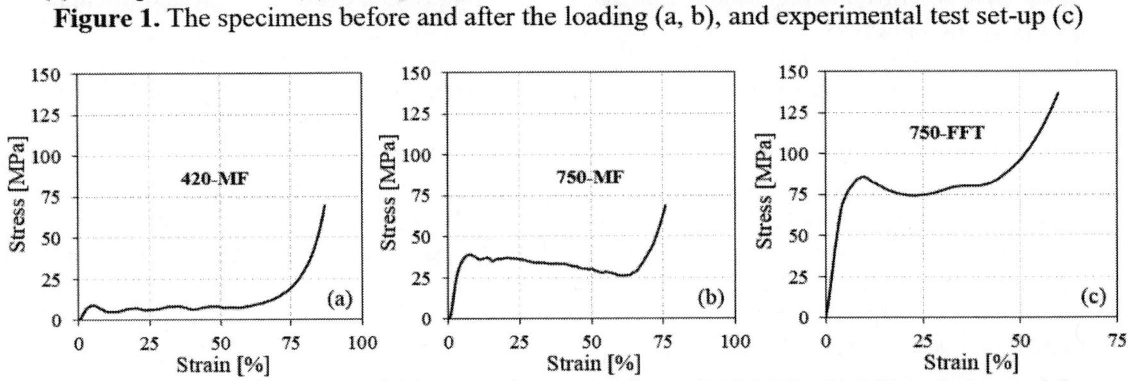

Figure 2. Stress-strain curves of (a) 420 kg/m3 metal foam (420-MF); (b) 750 kg/m3 metal foam (750-MF) and (c) foam-filled-tube (750-FFT) under quasi-static compression loading.

The results of the experimental tests evidenced a remarkable ductile and dissipative behavior of both MFs and FFTs under compression loads. A comparison between the diagrams in Fig. 2 evidences that the yield stress may range from values lower than 10 MPa to values greater than 100 MPa depending on the foam density and on the properties of tubes. In particular, changing the thickness of the tube or lightening the tube wall may result in lowering the yield stress of FFTs.

It is to note, however, that the experimental results presented in this section may give just a coarse idea of the behavior of MFs and of FFTs under seismic actions. Further tests should, in fact, be performed to obtain the hysteretic behavior of MFs and FFTs (accounting for stiffness and/or strength degradation) under cyclic loads.

3. Numerical model of a braced steel frame with metal-foam-based dissipaters

A four-story inverted V bracing framed steel building was considered, making reference to [14]. Some geometrical details of the building are provided in Fig.3. In Table 1 are given the static loads calculated according to Eurocode 1 [23], while Table 2 lists the cross section types of the steel elements. A 3D model of the building was implemented with the finite element program SAP2000 [24]. Columns and beams were modelled through elastic straight frame elements. Fixed beam to column connections and fixed constraints to the ground were assumed. Rigid diaphragms were considered at the floor slabs. Diagonal braces were modelled as capable to resist only axial actions (truss elements). A cylindrical MF dissipater was supposed to be inserted in the mid of each diagonal brace, as drafted in Fig. 3. The MF plastic dissipater was modelled through a 1-degree of freedom (longitudinal) plastic hinge with kinematic hardening unloading. Calculated according to EC8 [25], lumped masses are considered acting at the building floors. A damping ratio of 5%, which is a common value for steel structures was assumed. Time-history non-linear analyses were carried out under a strong recorded earthquake, see Fig. 3d, supposed to be applied both in the x and y directions.

Table 1. Static loads

Dead loads	Live loads	Wind load
4.44 kN/ m^2	2.00 kN/m^2	0.49 kN/m^2

Table 2. Cross section types

Columns	Beams	Diagonal braces
W14x211	W14x48	PIPE8XXS

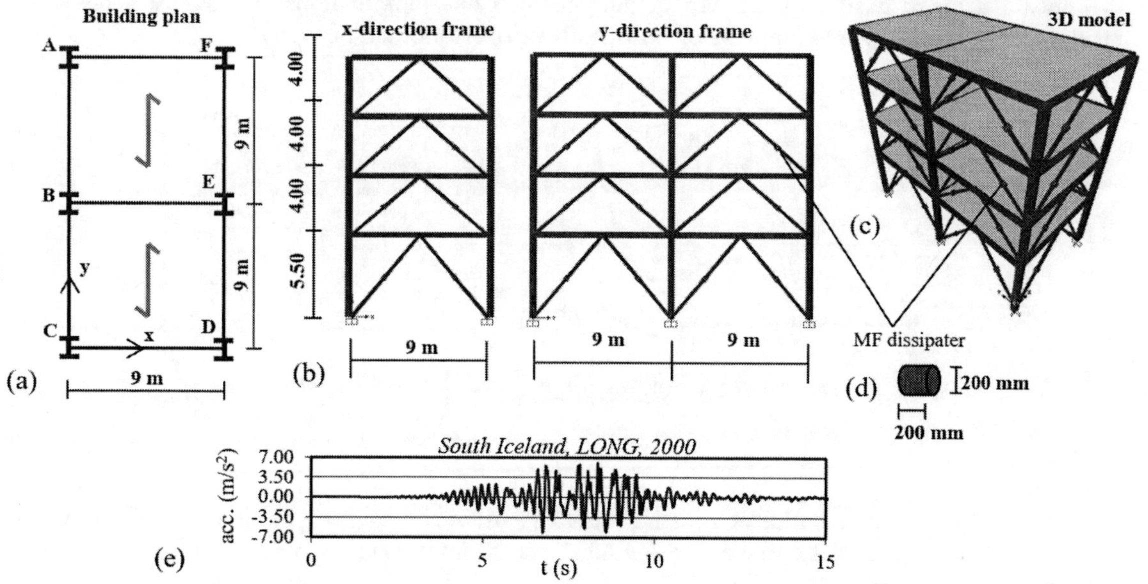

Figure 3. Case-study building. (a) Building plan; (b) frames in the x and y direction; (c) 3D model; (d) size of the MF dissipater; (e) South Iceland 2000 earthquake.

IOP Conf. Series: Materials Science and Engineering **416** (2018) 012051 doi:10.1088/1757-899X/416/1/012051

The experimental data provided in Section 2 can be exploited to calibrate the behavior of the foam-based devices to be introduced within the numerical model, as discussed in Section 3.1. Due to the small-size of the cylindrical specimens tested (20 mm diameter and height), however, the numerical investigation should be performed on a scaled model of the prototype building of Fig.1. Based on a dimensional analysis, the procedure to obtain a scaled model is presented in Section 3.2.

3.1. MF plastic dissipaters

Although exhibiting a remarkable ductile behavior under compressive loading, foams typically have a very low resistance under tension loading. Of course, the tension strength may be strongly increased if foam-filled tubes are considered instead of bare foams. For a preliminary concept analysis, however, MF dissipaters are here assumed to be made by bare foam and to resist only to compression.

To size the MF device, the following equation can be used:

$$\sigma_y = \frac{4F_y}{\pi D_f^2} \tag{1}$$

where F_y denotes the axial yielding load value for the MF dissipater. By assuming that any building member (even the MF dissipaters) should resist static and wind loads by deforming in the elastic range, the value of F_y can be derived from a linear static analysis. As derived from it, the value $F_y = 400\ kN$ is thus set for the present case-study. By assuming that the diameter of the cylindrical MF device is almost equal to the outer diameter of the tubular braces (PIPE8XXS), that is $D_f = 200\ mm$, the following yield stress for the foam device should be considered: $\sigma_y \cong 12.50\ MPa$.

The idealized bilinear stress-strain curve plotted in black in Fig. 4a was thus assumed in the present investigation. As can be inferred form Fig.4a, the idealized curve is slightly higher than the experimental one relevant to the 420-MF specimen. The force-displacement curve implemented in SAP2000 to represent the behavior of the MF device is given in Fig. 4b. Conventionally, compression stress and strain have negative sign for SAP2000. Since MF devices can only withstand compression loads, the diagonal braces are expected to work alternately during an earthquake. This assumption is rather strong and imply a detailed design of the MF devices for practical applications. Should FFT devices be considered instead, a tension-compression constitutive behavior could be considered.

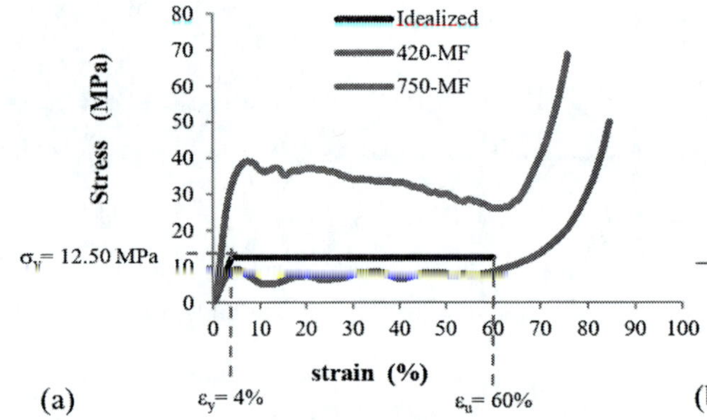

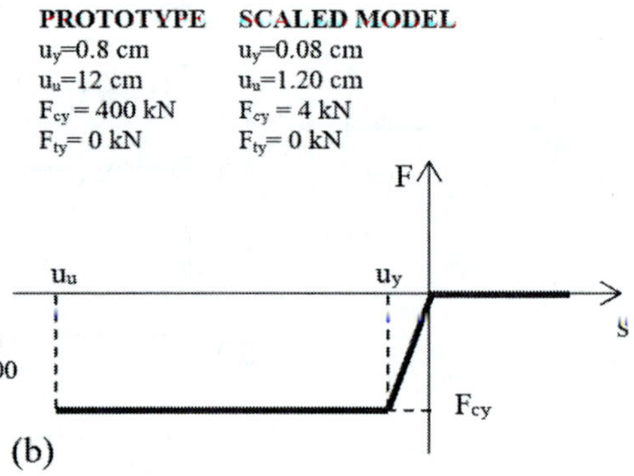

Figure 4. (a) Idealized bilinear constitutive stress-strain curve of MF dissipaters; (b) force-displacement curve of the MF dissipater as implemented in SAP2000.

The yield and failure values of force and displacement adopted for the MF devices are provided in Fig.4b. To avoid buckling of the hybrid brace (consisting of two equal steel tubes with a central cylindrical MF device) the critical load P_{cr} should be larger than the yield compression load F_{cy}. By considering the hybrid brace as a stepped column, the value of P_{cr} can be obtained by solving the following transcendental equation with a trial-and-error method [26]:

$$tan\left[L_s\left(\frac{P_{cr}}{E_sI_s}\right)^{1/2}\right]tan\left[L_f\left(\frac{P_{cr}}{E_fI_f}\right)^{1/2}\right]=\left(\frac{E_fI_f}{E_sI_s}\right)^2 \quad (2)$$

Here E_s and E_f are the elastic moduli while I_s and I_f are the inertia moments of steel and foam cross sections, respectively. In the present case, $E_s = 199948\ MPa$, $E_f = 312.50\ MPa$, $I_s = 1.227\ mm^4$ and $I_f = 0.991\ mm^4$. Due to symmetry with respect to the mid-section, it can be put $L_s = L/2$ and $L_f = a/2$, being L the length of the hybrid element and a the length of the MF device (here $L_s = 3569\ mm$ and $L_f = 100\ mm$). The value $P_{cr} = 806.19\ kN > F_{cy}$ was thus finally obtained.

Since, however, the diameter of the MF dissipaters ($D_f = 200\ mm$) to be introduced in the full-scale building (prototype) is about ten times greater than the diameter of the specimens tested in laboratory ($20\ mm$) [27], a scaled model should be introduced in the numerical investigation, as discussed in the next section.

3.2. Scaled model of the steel building

To exploit experimental results in the numerical investigation, a scaled model was considered in the present study. This is a rather usual approach in Civil Engineering, due to the very large dimensions of the full-scale models, see e.g. [28, 29]. The stress σ in a point of interest is chosen as the dependent variable giving the seismic response of the steel building and assumed to be function of 6 independent parameters: time (t), density of materials (ρ), elastic properties of materials (E), ground acceleration (a), gravity acceleration (g) and all geometric dimensions (L). That is:

$$\sigma = f(\ t,\ \rho,\ E,\ a_g,\ g,\ L\ ) \quad (3)$$

The fundamental dimensions of the mechanical problem being three, according to Buckingham's Π theorem, the number of independent dimensionless π-parameters is (6-3)=3. Eq. (1) then becomes:

$$\frac{\sigma}{E} = g\left(\frac{t}{L}\sqrt{\frac{E}{\rho}},\ \frac{a}{g},\ \frac{gL\rho}{E}\right) \quad (4)$$

with

$$\Pi_1 = \frac{\sigma}{E}\ ,\quad \Pi_2 = \frac{t}{L}\sqrt{\frac{E}{\rho}}\ ,\ \Pi_3 = \frac{a}{g},\quad \Pi_4 = \frac{gL\rho}{E}\ , \quad (5)$$

Owing to the fact that the gravity acceleration cannot be scaled (except for very specific test conditions), from equality $(\Pi_3)_p = (\Pi_3)_m$ (subscript p and m denoting quantity related to prototype and to scaled model, respectively) it is found that the scale factor for accelerations is $S_g = S_a = 1$. As a consequence, from equality $(\Pi_4)_p = (\Pi_4)_m$ follows that $S_E/S_\rho = S_L$. As usually done in structural dimensional analysis, the same elastic modulus of the materials is assumed in the prototype and in the model, i.e. $S_E = 1$, which implies $S_\sigma = 1$ and $S_\rho = 1/S_L$. Table 3 summarizes the scale factors for the main physical variables of the considered problem. Aside from those equal to 1, all scale factors in Table 3 are function of the length scale factor S_L. In the present investigation we put $S_L = 10$. This value is given in fact by the ratio between the diameter of the foam devices to be inserted in the prototype (200 mm) and the diameter of the tested specimens (20 mm). When modelling the constitutive behavior of the MF device in the scaled model the values given in Fig.4b should be considered, as derived by the scaling procedure.

Table 3. Scale factors

physical variable	unit	Scale factor
Length (L)	mm	$1{:}\,S_L$
Area (A)	mm^2	$1{:}\,S_L^2$
Volume (V)	mm^3	$1{:}\,S_L^3$
Density (ρ)	kg/ mm^3	$S_L{:}1$
Mass (M= ρV)	kg	$1{:}\,S_L^2$
Time (t)	s	$1{:}(S_L)^{1/2}$
Frequency (f=1/t)	Hz	$(S_L)^{1/2}{:}1$
Acceleration (a)	mm/s^2	1:1
Force (F=Ma)	kN	$1{:}\,S_L^2$
Young's modulus (E)	kN/mm^2	1:1
Stress (σ)	kN/mm^2	1:1

4. Results of the numerical investigation

A full-scale model (prototype) and a scaled model of the case-study building were modelled with SAP2000. First, the case of a linear elastic model of the building without MF dissipaters was studied. Secondly, the non-linear MF dissipaters were introduced in both the prototype and the scaled models. The first five frequencies of the two models are compared in Table 4 for the two considered cases (with and without MF dissipaters). As expected, the frequencies of the non-linear model (with the MF dissipaters), which is a more flexible system, are smaller than those of the linear model. It can be also noted that the scale factor for frequency $S_f = \sqrt{S_L} \cong 3.16$ (see Table 3) is always fulfilled both for linear and nonlinear models. Therefore, a good accuracy of the prototype response prediction through the scaled model is expected to be found.

Table 4. First five numerical eigen-frequencies

	without MF dissipaters			with MF dissipaters		
Freq.	Prototype	Scaled-model	Scale factor	Prototype	Scaled-model	Scale factor
1	5.46 Hz	17.27 Hz	3.16:1	2.11 Hz	6.69 Hz	3.16:1
2	6.53 Hz	20.67 Hz	3.16:1	2.36 Hz	7.47 Hz	3.16:1
3	14.90 Hz	47.14 Hz	3.16:1	2.62 Hz	8.28 Hz	3.16:1
4	17.12Hz	54.16 Hz	3.16:1	6.41 Hz	20.30 Hz	3.16:1
5	23.80 Hz	75.27 Hz	3.16:1	7.03 Hz	22.24 Hz	3.16:1

Some of the results of the numerical investigation are provided in Figs. 5-7. The shear values $V_x(t)$ and $V_y(t)$ at the base of column B are compared in Fig.5 for the scaled model with and without MF devices. It is found that a stress reduction up to 50 % can be achieved. Very similar results were found for the other columns of the scaled model. The same reduction (in percent) was found also for the prototype. However, a reduction in stress can often be paid in terms of displacement demand. The diagrams of Fig. 6 highlight, in fact, that a larger interstorey drift is found in the building with MF dissipaters. However, the EC8 [25] limit value $d_{LIM} = 0.01h/v$ is not exceeded. For the prototype it is $d_{LIM} = 100\ mm$ ($h = 4\ m$ and the importance factor $v = 0.4$), while $d_{LIM} = 10\ mm$ for the scaled model. A residual (plastic) displacement in both directions is kept after the seismic excitation.

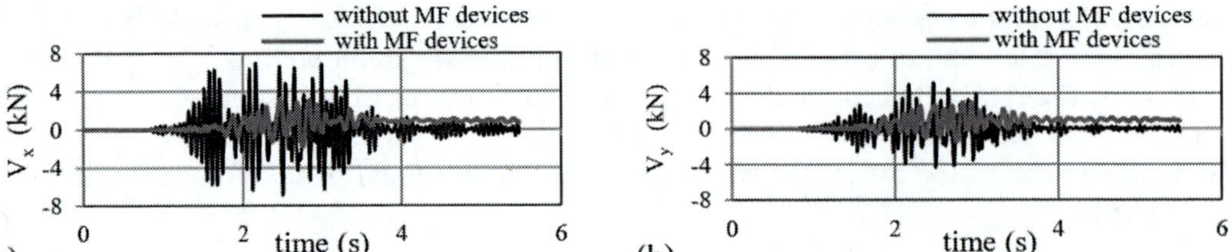

(a) (b)

Figure 5. Scaled model. Column B base shear in (a) x-direction and (b) y-direction.

The hysteresis cycles of the MF dissipaters placed within two first story concentric braces are given in Fig. 7a while the corresponding axial loads are compared in Fig. 7b, showing that the expected alternate behavior is rightly exhibited by any couple of concentric diagonal braces.

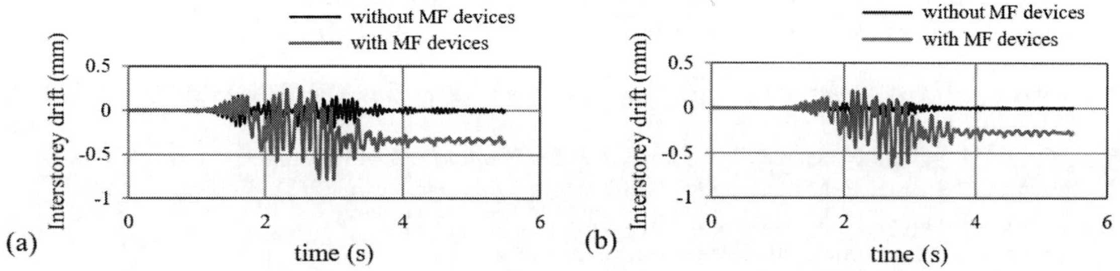

Figure 6. Interstorey drift at the upper storey in (a) x-direction and (b) y-direction

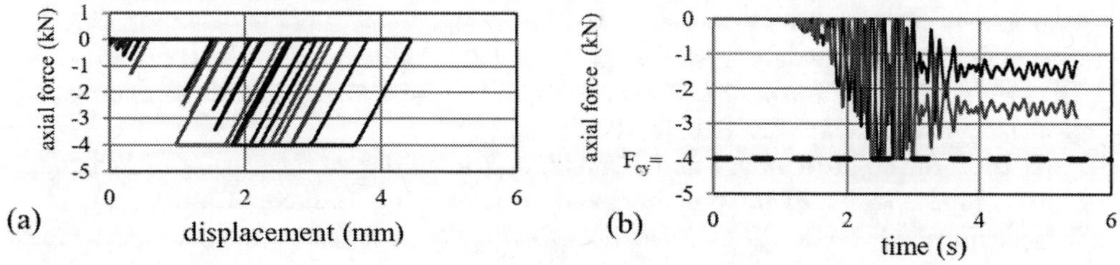

Figure 7. (a) Hysteresis cycles and (b) axial loads in the MF dissipaters of two opposite braces.

5. Conclusions

Similarly to other anti-seismic devices, also a foam-made dissipater is expected to behave like a mechanical fuse where damage concentrates under strong horizontal actions, one of the main advantages being that it can be dismounted and replaced after the seismic event. Of course, the practical design of foam-made devices entails different requirements also depending on whether bare foam devices or foam-filled tubes are considered and should be tuned according to the objectives of the seismic performance to be achieved by the building. One of the main concern of this work was to carry out a preliminary study on the effectiveness of MF dissipaters to reduce the seismic effects in buildings during severe earthquakes. To this purpose, a concept analysis was developed, based on some experimental data and involving linear and non-linear dynamic analyses of a 3D case-study steel building under a real ground motion. The experimental results and the numerical simulations on which the paper is based are just a preliminary stage of a wider investigation that will involve cyclic loading tests, full-scale specimens, suites of significant earthquakes and different building types.

6. References

[1] Simancik F, Degischer HP, Körner C, Singer et al. 2002 *Handbook of cellular metals: production, processing, applications*, Wiley, Weinheim
[2] Marsavina L, Kovacik J, Linul E 2016 *Theor. Appl. Fract. Mech.* **83** 11-18
[3] Linul E, Marsavina L, Kováčik J 2017 *Mat. Sci. Eng.-A Struct.* **690** 214-224
[4] Taherishargh M, Linul E, Broxtermann S, Fiedler T 2018 *J. Alloy. Compd.* **737** 590-596
[5] Kovácik J, Jerz J, Mináriková N et al. 2016 *Frattura ed Integrita Strutturale* **36** 55-62
[6] Duarte I, Vesenjak M, Krstulović-Opara L 2014 *Compos. Struct.* **109** 48-56
[7] Smith BH, Szyniszewski S et al. 2012 *J. Constr. Steel Res.* **71** 1-10
[8] Porcu MC 2017 Ductile Behavior of Timber Structures under Strong Dynamic Loads *Wood in Civil Engineering* Giovanna Concu (Ed.) InTech DOI: 10.5772/65894
[9] Kelly JM 1986 *Soil Dyn. Earthq. Eng.* **5(4)** 202-216

IOP Conf. Series: Materials Science and Engineering **416** (2018) 012051 doi:10.1088/1757-899X/416/1/012051

[10] Porcu MC 2013 *WIT Transactions on the Built Environment* **132**, 333-344
[11] Symans MD, Charney F A et al. 2008 *J. Struct. Eng.* **134(1)** 3-21
[12] Moradi M 2011 *Structural applications of metal foams considering material and geometrical uncertainty*, 37-88, PhD Thesis, University of Massachusetts Amherst.
[13] Moradi M and Arwade SR 2014 *Struct. Eng. Mech* **51(6)** 1017-1036
[14] Sabelli R, Mahin S, Chang C 2003 *Eng. Struct.* **25(5)** 655-666
[15] Vayas I (Ed) 2017 *Innovative anti-seismic devices and systems* The INNOSEIS Project RFCS-02-2015, pp. 1-423.
[16] Linul E, Movahedi N, Marsavina L 2018 *J. Alloy. Compd.* **740** 1172-1179
[17] Movahedi N, Linul E, Marsavina L 2018 *J. Mater. Eng. Perform.* **27(1)** 99-108
[18] Linul E, Movahedi N, Marsavina L 2017 *Compos. Struct.* **180** 709-722
[19] Movahedi N and Linul E 2017 *Mater. Lett.* **206** 182-184
[20] ISO13314-2011 Mechanical testing of metals-Ductility testing-Compression test for porous and cellular metals
[21] Linul E, Şerban DA, Marsavina L, Kovacik J 2017 *Fatigue & Fracture of Engineering Materials & Structures* **40(4)** 597-604
[22] Linul E, Marsavina L, Kovacik J, Sadowski T 2017 *P. Romanian Acad. A* **18(4)** 361-369
[23] EN, B. (1991). 1-2: 2002 *Eurocode 1: Actions on structures—Part 1-2: General actions—Actions on structures exposed to fire* British Standards
[24] CSI, SAP2000 rel.17 2014 *Integrated Finite Element Analysis and Design of Structures Basic Analysis Reference Manual* Computers and Structures Inc. Berkeley California USA
[25] CEN. 2004 *Eurocode 8: Design of Structures for Earthquake Resistance*, European Comitee for Standardization, Brussels, 1-232
[26] Timoshenko SP and Gere JM 1961 *Theory of elastic stability* McGrawHill-Kogakusha Ltd Tokyo
[27] Linul E, Movahedi N, Marsavina L 2018 *Materials* **11(4)** 554
[28] Stavridis A, Shing B, Conte J 2010 Design, Scaling, Similitude and Modeling of Shake-Table Test Structure, *Shake Table Training Workshop* San Diego
[29] Moncarz P and Krawinkler H 1981 *Theory and Application of Experimental Model Analysis in Earthquake Engineering* Rpt. No. 50 John Blume Earthquake Eng. Ctr. Stanford Univ.

Acknowledgments
This work was partially supported by research grants PCD-TC-2017 and by *Fondazione Sardegna* (University of Cagliari, Italy). The Visiting Professor Program of Regione Sardegna, the ERASMUS+STUDIO Program (ROTIMISOA04) and the MOSTA (Staff Mobility) Program of the University of Cagliari are also acknowledged.

IOP Conf. Series: Materials Science and Engineering **416** (2018) 012052 doi:10.1088/1757-899X/416/1/012052

Application of Topological Measures in Estimation of Reinforcing Fibers Distribution in Composites

P Wolszczak[1], **T Sadowski**[2] **and S Samborski**[3]

[1] Lublin University of Technology, Faculty of Mechanical Engineering, Department of Automation, Nadbystrzycka 36, 20-618 Lublin, Poland

[2] Department of Solid Mechanics, Faculty of Civil Engineering and Architecture, Lublin University of Technology, Nadbystrzycka 40, 20-618 Lublin, Poland

[3] Department of Applied Mechanics, Lublin University of Technology, Nadbystrzycka 36, 20-618 Lublin, Poland

Abstract. Roving composites are used in the production process of critical supporting structures. The mechanical properties of the composites depend on the even distribution of the fibers. Monitoring of geometrical measures of composites microstructure can be used for prediction of material strength and durability. In the paper results from the research performed on the production samples of girders of helicopter blades are presented. Some geometrical properties, like: the average fiber diameter D, matrix film thickness average G_{AB} and its standard deviation were used for specimens classification by the value of the flexural elastic modulus. The topological measure G_{AB} is intendent to evaluate of the fibers distribution. Applicability of the measure has been checked based on strength tests results. The uniformity of the fiber distribution expressed by the averaged G_{AB} measurement allows for estimation of the composite strength. Additionally the quantitative character of the measure G_{AB} take in defects occuring in the internal structure of composites.

1. Introduction

Composite materials are used in the production of machines to achieve special strength properties while maintaining their light weight. Composites containing fibers allows obtaining anisotropic properties of composites, which depend on the orientation of reinforcing fibers. The composites made of continuous fibers (filaments) achieve high resistance to transverse load to the direction of fibers.

An important property that affects the strength of unidirectional composites is the uniformity of the distribution of reinforcement in the composite matrix. This property is related to other qualities that affect the strength and impact strength of the composite. The durability of the composite depends to a large extent on the surface of the interface connecting the reinforcement phase with the warp [1-5]. Considering the thickness of the interface, the significance of the relative volume of the interface in the composite can also be considered. When the thickness of the interface reaches the minimum value, this means the presence of contacting fibers, which promote crack propagation. Cracks lead to permanent weakening of the composite. Therefore, it seems relevant to the authors of the article to assess the uniformity of the distribution of the composite reinforcing fibers, to develop a uniformity measure of the distribution allowing for the assessment of the quality of the structure and to assess the strength of the composite [6-10].

For this purpose, an experiment was conducted consisting in the preparation of composite samples builds of glass fibers and epoxy resin, in which all design parameters and quality indicators resulting from the used technology were similar and samples differed only in the distribution of fibers in the transverse plane to the fiber direction. A comparison of the overall strength and microstructural structure of the samples was then made. Based on these results, the analysis of the correlation of geometrical properties and results of the strength test was carried out. The results of the analysis allowed for the development of a measure of distribution in the form of an average warp thickness in a plane perpendicular to the fiber direction, surrounding individual fibers [11-15].

On the basis of the literature analysis, it was initially estimated that the impact of the reinforcement fiber distribution, although significant, would be lower than the influence of defects in the microstructural structure, such as the presence of air bubbles and contacting fibers [16-17]. Therefore, it was decided to obtain samples from the production process of responsible load-bearing structures, in which demanding quality procedures appropriate for the aviation industry are applied. With this choice, the impact of defects on the overall strength of the samples has been minimized.

2. Experiment

The study used glass fibers with a diameter of 10 μm (made at the Krosglass plant in Krosno, Poland) and epoxy resin (Epidian 5). The fibers were covered to create an interface with an epoxy resin. The fibers were rewound in strands so that the maximum glass content (relative volume) could be obtained in the composite material. Supersaturation of the fibers was carried out manually using instruments, therefore the average relative volume of glass was 51 %. Higher values of relative volume can be obtained using machine supersaturation. From such material was prepared responsible aerial supporting (load-bearing) structures and witness samples, which were used in further studies.

The samples-witness were tested on three-bending strength and short beam shear method. The following mechanical strength characteristics were determined: the force at yield point F_1, and maximum elastic deformation L_1, the maximum force F_{max} and deflection in shear test L_2, and the beam deflection angle φ. The analysis was limited to the elastic response of the composites and the value of modulus of elasticity E_f can be calculated with the following equation:

$$E_f = \frac{F_{max} L^3}{4BH^3 f} \tag{1}$$

where F_{max} is the maximal force of elastic deflection in the short beam shear test (SBST), B, H, L are width, thickness, and length of the specimen, and f the beam deflection.

Then the metallographic section from samples after strength tests were prepared and microscopic observations were made. Observations in the plane parallel to the direction of the fiber axis allowed for assumption that the distribution of fibers in the plane of transverse defects to the axis of the fibers is identical to that in the plane destroyed during the strength test. The fibers were laid parallely. On the sections of a few millimeters between both sections, the parallelism of the fibers was not observed, because it was less than measurement resolution. The cross-sections were used to develop geometric images of microstructures consisting of the position of fiber centers X, Y and diameter D.

3. Definitions of applied topological measures for estimation of fibers distribution

Voronoi tessellations were used during microscopic measurements. There were measured: the surface area of the matrix surrounding any individual fiber A_m, fiber diameter D_f, the distance between neighboring fibers G_{1i}, where i is a number of neighboring fibers, $G_{2j}(\varphi)$ is the radial thickness of voronoi polygon (figure 1) measured with angular intervals ($\Delta\varphi = 5$ deg), where j varies from 1 to 360 with 5 degree angle, G_{AD} is the average film thickness G_2 calculated from the matrix area A_m and the fiber diameter D_f.

$$A_m = \frac{\pi}{4}\left[\left(D_f + 2G_{xx}\right)^2 - D_f^2\right] \tag{2}$$

IOP Conf. Series: Materials Science and Engineering **416** (2018) 012052 doi:10.1088/1757-899X/416/1/012052

$$G_2 \approx G_{AD} = \frac{1}{2}\left(\sqrt{D_f^2 - 4\frac{A_m}{\pi}} - D_f\right) \tag{3}$$

G_{AB} is the fractional of the Voronoi polygon with matrix area A_m by the circuit fiber.

$$G_{AB} = \frac{A_m}{\pi D_f} \tag{4}$$

4. Analysis of results

Figure 2 presents the distributions describing the dispersion of thickness G_{1i}, G_{2j}, and figure 3 presents distributions describing the dispersion of averaged G_{AB}. The presented exemplary results were calculated for a sample comprising 187 composite fibers.

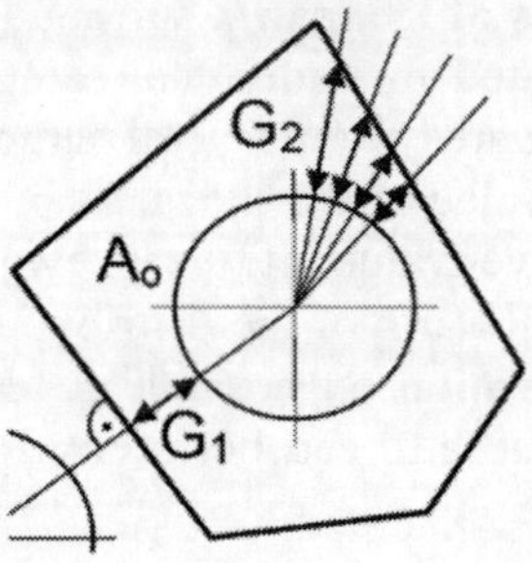

Figure 1. The local thickness of the matrix separated around a single fiber with Voronoi polygon.

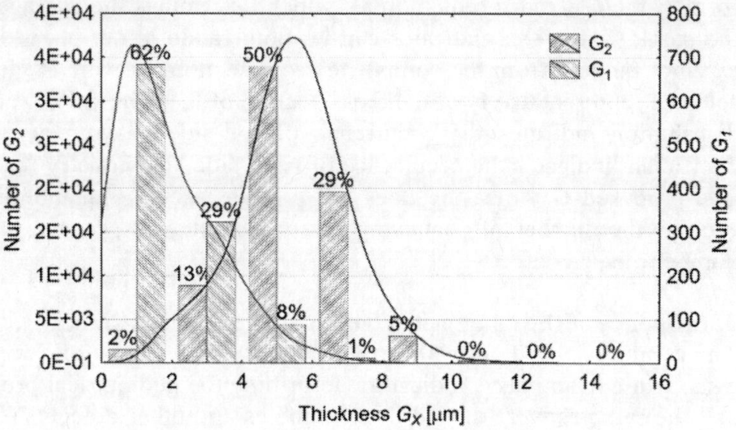

Figure 2. Example of distributions describing the local thickness spreads G_{1i} and G_{2j} measured for a sample image of a composite cross-section comprising 187 fibers.

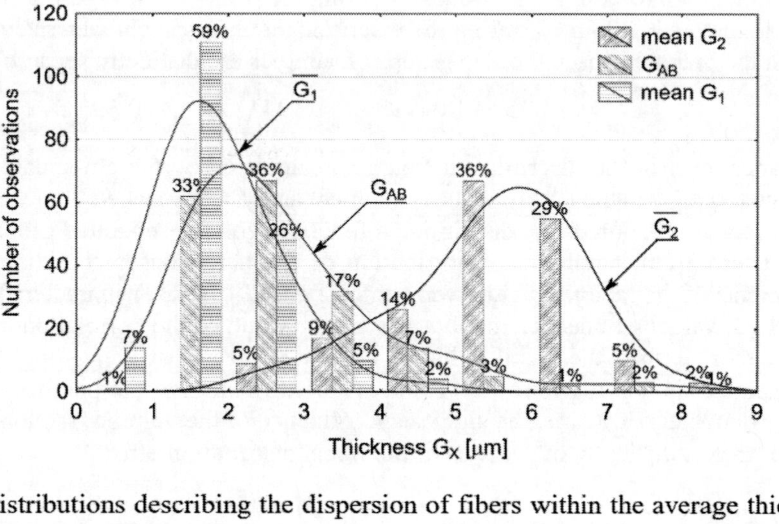

Figure 3. The distributions describing the dispersion of fibers within the average thickness $\overline{G}_1$, $\overline{G}_2$ and $\overline{G}_{AB}$ calculated for 187 fibers.

The quoted results make it possible estimation of the actual micro constructions of the tested composite. This model is presented in the form of randomly spaced parallel reinforcing fibers having random value of diameters, surrounded by a matrix of annular shape and the random thickness G_X. On the basis of figure 2 and figure 3, various types of deviations from a homogeneous model structure of the composite may be determined by a global mean and medians values. For example, from figure 3 follows that in the sample considered, about 7 % of the fibers were surrounded by a layer of resin with an average thickness of less than 1 μm, which practically means the contact between the fibers. Left skewed distribution of the thickness G_1 (figure 2) proves the advantage of smaller thickness than the mean value calculated for this set. This conclusion confirms the thickness distribution G_{2j}, which in turn is characterized by right-side skewness.

Averaging the local film thicknesses within individual fibers leads to the expected effect in the form of reduced skewness, making the dispersion size of G_1, G_2 and G_{AB} more closely related to the normal distribution.

In order to evaluate the homogeneity of the geometrical structure, the measure is proposed in the form of the G_{AB} index (equation 5), which determines the average thickness of the matrix surrounding the single fiber. This indicator can be both random, i.e. it can be calculated for individual samples taken at random from the composite cross-section, as well as global calculated from several samples selected representatively of the cross-section of the composite. A negative value of the skewness of the distribution indicates the occurrence of left-sided asymmetry. A positive value of the skewness coefficient indicates the occurrence of right-sided asymmetry (extended right arm of the distribution). The proposed G_{AB} measure does not provide accurate quantitative information on possible defects, such as the number of contacting fibers, nevertheless, on the basis of this ratio can be inferred the number of such defects.

5. Influence of the composite structure on its strength

The results of SBST tests allowed to choose two groups of samples, which differ with results of the tests, while quantitative indicators describing the geometrical properties of the samples were identical and the samples were devoid of air bubbles and other defects. The factor differentiating the structure of the samples was fiber arrangement. A discriminant analysis was conducted, which distinguished three main features of the microstructure of the samples that influenced the diversification of strength properties.

In the figure 4, the difference between groups of samples-witnesses was characterized according to the value of the maximum force F_{max} [N] obtained during the SBST tests. Figure 5 shows the values of the geometric parameters highlighted during the discriminant analysis, characterized by two groups of samples. Thus in the prepared space the two groups of samples are distinctly separable.

$$D_1 = 14.357 + 1.649\, G_{AB} - 16{,}944\, G_{1\min} - 0.396\, V_f \quad (5)$$

The significance level of the discriminant function, calculated by the chi-square test used to gave the p-value below 0.0001 and the coefficient of canonical analysis R reached high value (0.915). These values proved strong correlation of the discriminant function with the separated groups of specimens.

In order to determine the individual contribution of characteristics in building a discriminatory function, coefficients of the factor structure were calculated. G_{AB} (0.837) characteristics and V_f (0.810) have a similar, high values of coefficients, but the higher value of the correlation coefficient reaches G_{AB}. The third factor reached the small value of G_{1min} = 0.188, what means the lack of significant effect of fiber diameter in our analysis. This result is a confirmation of the plan for obtaining samples of the same fiber diameter. However, the significant influence of the volume fraction of V_f glass in this group of samples makes difficulty in analysis of the fibers distribution effect.

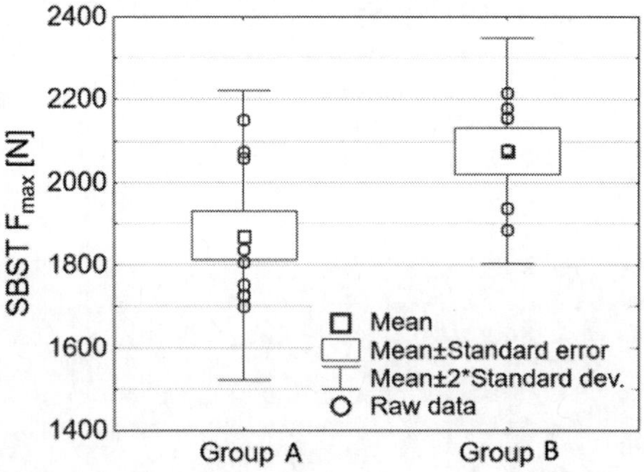

Figure 4. Statistics of maximum force values F_{max} [N] during short-beam strenght test in group A and B [18].

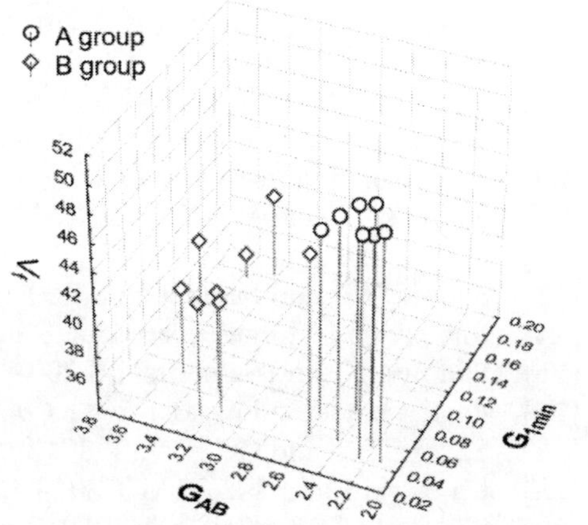

Figure 5. Average values of V_f, G_{AB} and G_{1min} for individuals specimens in groups A and B [19].

6. Summary and conclusions

In this paper we proposed several topological measures to estimate fibers distribution in the polymer matrix composites. In particular, we analysed defects occuring in the internal structure of composites. The topological measure G_{AB} allows to evaluate the distribution of fibers. The uniformity of the fiber distribution expressed by the averaged G_{AB} measurement allows for estimation of the composite strength. The proposed measures find application in monitoring of composites quality. In further studies, it is advisable to collect the results allowing for determination of the regression function of all features characterizing the uniformity of fiber distribution with respect to strength characteristics. The model can be extended by inclusions of damage or cracking processes, as it was done for other composite materials, e.g. [20-30].

7. References

[1] Pyrz R 2000 *Morphological characterization of microstructures. Comprehensive Composite Materials* ed Kelly A, Zweben C (Oxford: Elsevier Science) 465-78
[2] Bochenek B and Pyrz R 2004 *Comput Mater Sci* **31** 93-112
[3] Bieniaś J, Dębski H, Surowska B and Sadowski T 2012 *Comput Mater Sci* **64** 168-72
[4] Ostapiuk M, Surowska B and Bieniaś J 2014 *Compos Interface* 21 **4** 309-18
[5] Chang S-H, Parinov I and Topolov V Y 2014 *Advanced Materials Physics Mechanics and Applications* (Cham, Heidelberg, New York, Dordrecht, London: Springer)
[6] Kurzydłowski K J and Ralph B 1995 *The Quantitative Description of the Microstructure of Materials* (London: CRC)
[7] Torquato S 1997 *Phys Rev Lett* **79** 681
[8] Werwer M, Cornec A and Schwalbe K-H 1998 *Comput Mater Sci* **12** 124-36
[9] Jones R M 1999 *Mechanics of Composite Materials* (Philadelphia/Levittown/London: Taylor and Francis Inc)
[10] Wada A and Fukuda H 1999 *Compos Sci Technol* **59** 89–95
[11] Sadowski T and Samborski S 2005 *Key Eng Mater* **290** 86-93
[12] Samborski S and Sadowski T 2010 *J Am Ceram Soc* 93 **11** 3607-9
[13] Dziubińska A, Gontarz A, Horzelska K and Pieśko P 2015 *Procedia Manufacturing* **2** 337-41
[14] Kubiak T, Samborski S and Teter A 2015 *Compos Struct* **133** 921-9
[15] Dziubińska A, Majerski K and Winiarski G 2017 *Adv Sci Technol Res J* 11 **4** 147-58
[16] Wolszczak P and Cechowicz R 2011 *Examination of the influence of shear micro-geometrical properties on transverse elasticity the modulus of roving composite materials used in critical constructions*. 32nd Risø International Symposium on Materials Science, Roskilde, Denmark
[17] Wolszczak P 2013 *Quantity expression of reinforcement phase distribution in composites on*

glass-fibers distribution in roving composites example, ICCS17 Porto, Portugal
[18] Wolszczak P, Samborski S and Sadowski T 2018 *Micromechanics and Nanomechanics of Composite Solids* ed S Meguid and G Weng (Switzerland: Springer, Cham) **15** 425-36
[19] Wolszczak P, Sadowski T and Samborski S 2017 *Compos Part* B **129** 66-76
[20] Marsavina L and Sadowski T 2007 *Int. J. Frac.* 145 237
[21] Sadowski T and Marsavina L 2011 *Comput. Mat. Sci.* **50** 1336
[22] Golewski G and Sadowski T 2014 *Const. Build. Mat.* **51** 207
[23] Sadowski T and Samborski S 2003 *Comp. Mat. Sci.* **28** 512
[24] Marsavina L, Linul E, Voiconi T and Sadowski T 2013 *Polym. Test.* **32** 673
[25] Sadowski T and Samborski S 2003 *J. Am. Cer. Soc.* **86** 2218
[26] Postek E and Sadowski T 2011 *Comp. Interf.* **18** 57
[27] Sadowski T and Golewski P 2012 *Comput. Mat. Sci.* **64** 285
[28] Gajewski J and Sadowski T 2014 *Comp. Mater. Sci.* **82** 114
[29] Sadowski T and Golewski P 2012 *Comput. Mat. Sci.* **52** 293
[30] Dębski H and Sadowski T 2014 *Comput. Mat. Sci.* **83** 403

IOP Conf. Series: Materials Science and Engineering **416** (2018) 012053 doi:10.1088/1757-899X/416/1/012053

Manufacturing and Compressive Mechanical Behavior of Reinforced Polyurethane Flexible (PUF) Foams

E Linul[1,*], PA Linul[2,3], C Valean[1], L Marsavina[1] and D Silaghi-Perju[1]

[1]Department of Mechanics and Strength of Materials, Politehnica University of Timisoara, 1 Mihai Viteazu Avenue, 300 222 Timisoara, Romania
[2]Politehnica University of Timisoara, Faculty of Industrial Chemistry and Environmental Engineering, 6 Vasile Parvan Blvd., 300 223, Timisoara, Romania
[3]National Institute of Research for Electrochemistry and Condensed Matter, Aurel Paunescu Podeanu Street 144, 300 569 Timisoara, Romania

[*]Corresponding author: emanoil.linul@upt.ro

Abstract. This paper presents a manufacturing method and mechanical behavior of unreinforced (U-PUF) and reinforced (R-PUF) polyurethane flexible foams. In order to obtain a new type of reinforced polyurethane composite foams, an amount of 0.5% (of the foam liquid mass) aluminum microfibers were used. The mechanical characterization of both U-PUF and R-PUF foams was carried out at room temperature through quasi-static compression tests. The compression tests were performed on cubic specimens with a loading speed of 10 mm/min. The obtained and investigated foam showed a density of 150 kg/m^3. It was found that the new obtained R-PUF foams crushes progressively during compression and shows higher mechanical properties (in terms of compressive modulus, yield stress, plateau stress and densification strain) compared to U-PUF foam. Particular attention was given to the foam crush performances. Experimental results indicated that aluminum microfibers significantly improve the foam energy absorption capabilities in compression.

1. Introduction

Polyurethanes (PU) are polymers resulting from the polyaddition reaction between isocyanates and polyalcohols. They are used as foams, elastomers, plastic polyurethanes and mixtures [1]. Polyurethane foams can be rigid (PUR) or flexible (PUF) with open, closed and partly open-partly closed cellular structure [2-4]. Owing to the unique combination of excellent physical, thermal, acoustic and mechanical properties [5-8], PUR foams are used in various applications, but especially as core material for sandwich panels [9], while flexible ones are widely used in applications of different industries such as automobiles, furniture, transport, textiles and fibers [10-12].

Many analytical [13-15], numerical [16, 17] and experimental [18-20] results are reported on PUR foams, while literature shows far fewer results for PUF foams. In recent years, emphasis has been placed on the manufacturing of flexible PU foams with superior mechanical properties. For this purpose, different materials have been used as reinforcement to obtain PUF composite foams [10-12, 21, 22]. Usman et al. [21] investigates the effect of $CaCO_3$ filler in PUF foam matrix for achieving high mechanical properties at a low production cost. The effect of post-consumer PET (polyethylene terephthalate) as reinforcement filler in flexible PU foams was studied by Mello and coworkers [22]. They evaluated the physical, mechanical and morphological characteristics of the PUF foams. Also,

IOP Conf. Series: Materials Science and Engineering **416** (2018) 012053 doi:10.1088/1757-899X/416/1/012053

indentation hardness, tensile strength and elongation at break of the flexible PU foam composites were examined by Latinwo et al. [23] as a function of the filler content (calcite and dolomite) in the PU matrix.

Compared to reinforced materials used by other researchers, in the present paper we used aluminum fibers resulting from the aluminum processing process. These reinforcements are very cheap, obtained without further processing and the foam manufacturing acting as recycling process. Aluminum fibers with lengths between 5-10 mm were collected and introduced into the foam matrix material by a well-established procedure. Their use has led to a considerable increase in mechanical properties, while maintaining low foam density (due to low fiber density). Also, the energy absorption crush performances presents values to be considered for future possible applications that contain flexible foams

2. Materials and methods

In order to obtain reinforced polyurethane flexible (R-PUF) foams, aluminum microfibers were added into both component A (isocyanate) and component B (polyol) separately, and, in order to get a good dispersion, each component has been stirred with the aluminum fibers for about 3 minutes. After individual stirring, component A was mixed with component B and stirred together for 30 seconds; afterwards, the R-PUF foam has been allowed to dry at room temperature for 24 hours. Before being added to the polyurethane components, the aluminum fibers were dried at 80°C for about 1 hour. The same procedure was followed for obtaining unreinforced polyurethane flexible (U-PUF) foams.

After the drying and hardening process, large foam blocks were obtained. The density (150 kg/m^3) of the foam material under consideration was measured using mass and specimens dimensions. The U-PUF and R-PUF foams microstructures are presented in Fig. 1.

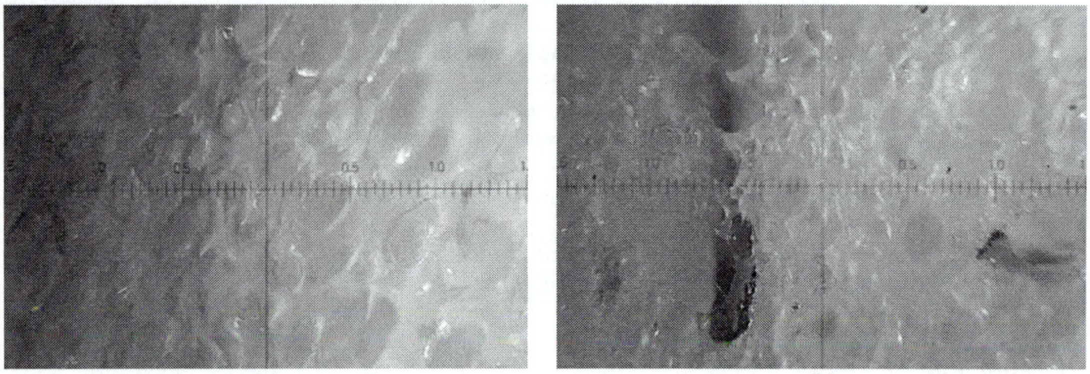

(a) (b)

Figure 1. Microstructure of unreinforced (a) and reinforced (b) PUF foam specimens

To establish the material properties, uniaxial quasi-static compression tests on the foam cubic specimens were carried out. A 5 kN Zwick Roell testing machine has been used for this purpose. The experimental tests were performed with a loading speed of 10 mm/min. The foam material properties were established in the controlled room temperature and humidity conditions, according to [24].

3. Results and discussion

Typical load (F)-displacement (Δ) curves derived from the compression tests were converted to representative compressive stress (σ)-strain (ε) curves using geometrical parameters of the foam specimens. The σ-ε compression behavior is presented in Fig. 2a for both U-PUF and R-PUF foams. In Fig. 2b are shown the compressive damage stages of specimen in strain range from 0 to 80%. Both U-PUF and R-PUF composite foams presents a progressive elasto-plastic collapse behavior during compression test.

IOP Conf. Series: Materials Science and Engineering **416** (2018) 012053 doi:10.1088/1757-899X/416/1/012053

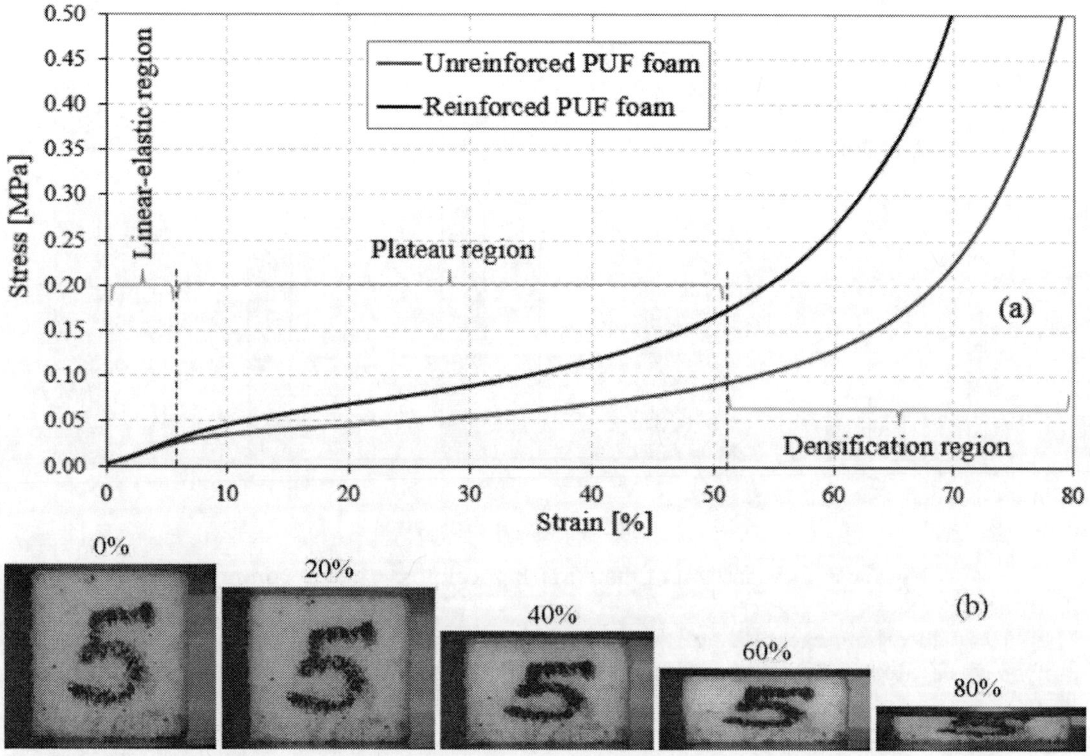

Figure 2. Stress-strain curves (a) and deformation sequences (b) of investigated PUF foam

The U-PUF and R-PUF foams show stress-strain profiles similar to other foams [25-28], which consists of a linear elastic region followed by a strain hardening (plateau) region. By comparing the reinforced samples results to the unreinforced ones, an increase in compressive modulus of elasticity by 12.6% is observed, as observed from Table 1.

Table 1. The main mechanical properties of unreinforced and reinforced PUF foams.

PUF foam type	Mechanical properties						
	E [MPa]	σ_y [MPa]	ε_y [%]	σ_p [MPa]	ε_D [%]	σ_D [MPa]	W_D [MJ/m^3]
Unreinforced	0.4510	0.0220	4.7900	0.0563	54.0900	0.1029	0.0293
Reinforced	0.5160	0.0295	5.8100	0.0960	51.0900	0.1735	0.0422

During strain hardening the strength of the foam material is increased and in the plateau area there is a large absorption energy. In this case, the major difference from other works [29-31] is that in this case the transition from the elastic region to the plateau region is smooth [32-34] without the presence of a peak stress drop. Upon further loading the specimens in compression, the stress starts rising in foam samples up to around 50% strain value after which there is a sudden increase in stress at a small increase in the strain. This last region is named densification region and starts from onset strain of densification (OSD) up to final failure. At the moment of OSD the R-PUF foam shows an energy absorption value about 0.0422 MJ/m^3 with 31% more than U-PUF foam.

All the reinforced foams show an improvement in compressive strength compared to the unreinforced PUF foams, as shown in Fig. 3. Long aluminum microfibers induce a change in foam matrix strength and stiffness making them more stiff and strong as they are reinforced.

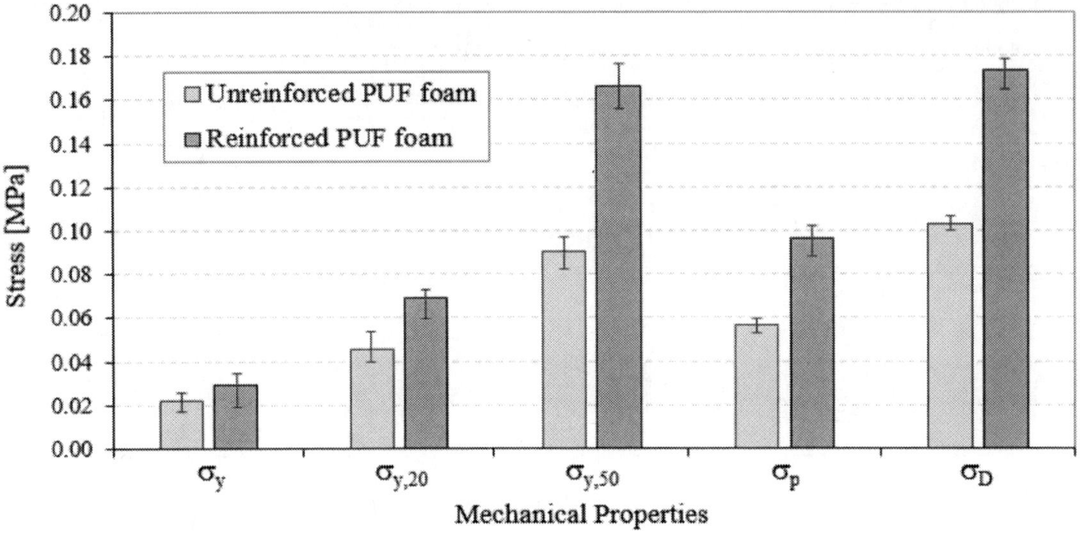

Figure 3. Comparison of main mechanical properties in compression

The energy-absorbing capability of these unreinforced and reinforced polyurethane flexible foams specimens was evaluated according to Refs. [35-42] and presented in Fig. 4 as a function of strain.

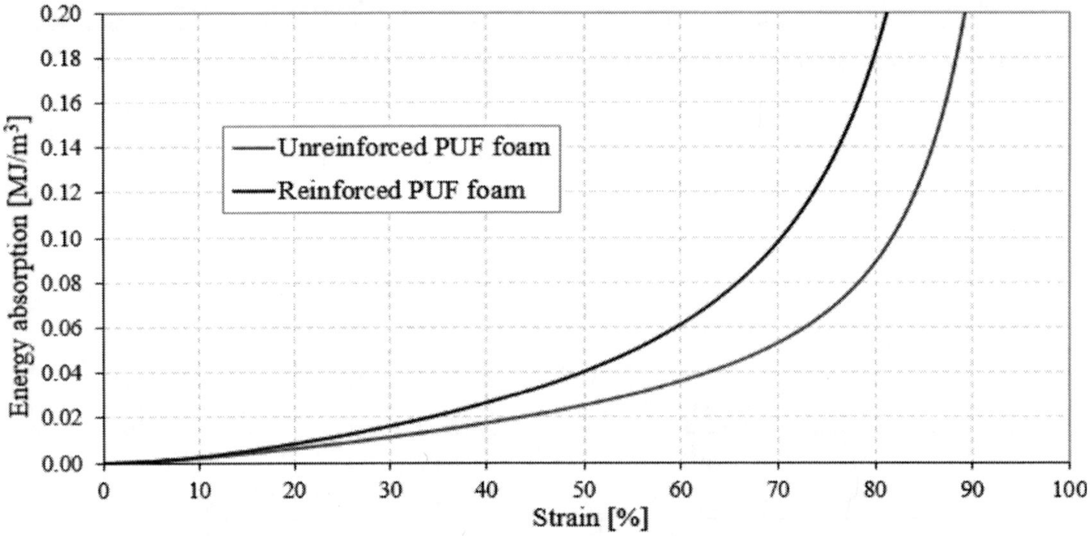

Figure 4. Energy absorption variation with strain of PUF foams

The absorbed energy by the R-PUF foam specimen is larger compared to the U-PUF foam to the fact that the long aluminum microfibers take over an important part of the quasi-static load. From Fig. 4 it is to be noted that in both cases (U-PUF and R-PUF foams), the energy absorption curves shows the same behavior with approximately the same W values up to a strain of 15%. The biggest difference (in the terms of W) being to a deformation of 80%. This energy absorption difference (W_{diff}) is presented in Fig. 5 as a function of ε.

Table 2 presents the energy absorption values corresponding to different strains.

Table 2. The energy absorption values of investigated PUF foams.

PUF foam type	Energy absorption at different strains, W [MJ/m^3]							
	10%	20%	30%	40%	50%	60%	70%	80%
Unreinforced	0.0021	0.0062	0.0112	0.0174	0.0253	0.0361	0.0529	0.0884
Reinforced	0.0025	0.0083	0.0162	0.0264	0.0403	0.0613	0.098	0.1824

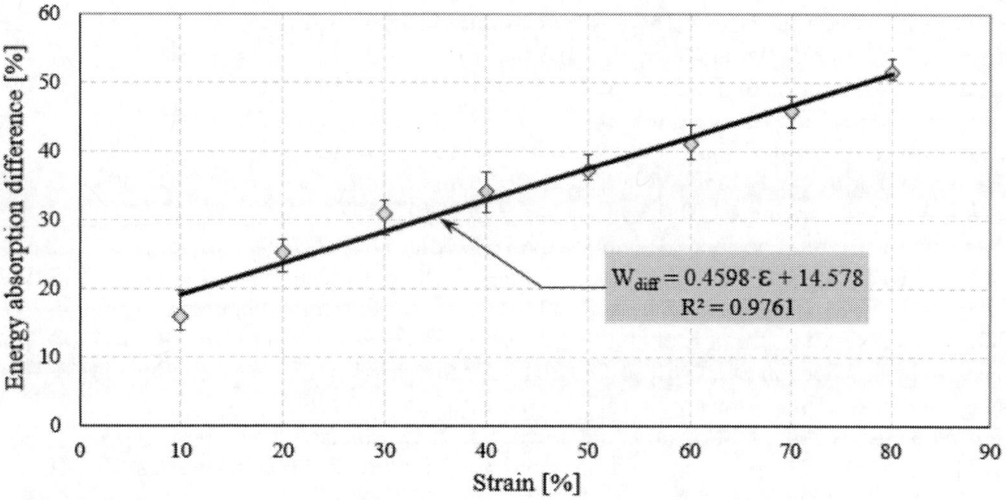

Figure 5. Energy absorption-strain variation of investigated foams

In Fig. 5 the energy absorption difference rises linearly from a value of 16% for low strains to a maximum value of 51.54 %, associated to high strains.

It is therefore necessary to explore new foam materials that during crash events would provide enough energy absorption capabilities together with high mechanical properties and stable collapse mechanisms [43, 44]. Reinforced polyurethane composite foams may provide a good solution to this problem, especially if they are used as core material in advanced composite structures (sandwich structures, foam-filled tubes etc.). From the current investigation can be argued that instead of regular/standard polyurethane foams the reinforced polyurethane foams should be studied in more detail.

4. Conclusions

This paper discloses both a method for manufacturing and mechanical characterization of closed-cell polyurethane foams reinforced with aluminum microfibers. Compressive properties of unreinforced and reinforced composite flexible foams are investigated and compared in the present work.

In the present study it has been found that the new produced R-PUF foam (150 kg/m^3 density) exhibit high mechanical properties per mass compared to U-PUF foam, which could be very promising composite foam material for the engineering applications.

A comparison of energy absorbing capabilities of U-PUF and R-PUF foams was presented. It was found that reinforced polyurethane composite foams have better crashworthiness performances (almost double) than standard foams used currently.

Moreover, using this foam manufacturing process a beneficial waste management has been achieved.

5. References

[1] Suleman S, Khan SM, Gull N et al. 2014 *Int. J. Innovation Scientific Research* **12** 165-169
[2] Marsavina L, Constantinescu DM, Linul E, Stuparu FA, Apostol DA 2016 *Eng. Fract. Mech.* **167** 68-83

[3] Apostol DA, Stuparu F, Constantinescu DM et al. 2016 *Materiale Plastice* 53(2) 280-282
[4] Gibson LJ, Ashby MF 1997 *Cellular Solids: Structure and Properties.* 2nd Edition, Cambridge University Press, Cambridge
[5] Aliha MRM, Linul E, Bahmani A, Marsavina L 2018 *Polym. Test.* **67** 75-83
[6] Şerban DA, Voiconi T, Linul E, Marsavina L, Modler N 2015 *Materiale Plastice* **52(4)** 537-541
[7] Voiconi T, Linul E, Marsavina L et al. 2014 *Solid State Phenomena* **216** 116-121
[8] Linul E, Marsavina L, Sadowski T, Kneć M 2012 *Solid State Phenomena* **188** 205-210
[9] Marsavina L, Constantinescu DM, Linul E et al. 2015 *Eng. Fail. Anal.* **58** 465-476
[10] Zhang B, Liu H, Han J 2018 *J Phys Chem. Solids* **115** 199-207
[11] Chen MJ, Shao ZB, Wang XL et al. 2012 *Ind. Eng. Chem. Res.* **51** 9769-9776
[12] Lefebvre J, Bastin B, Bras ML et al. 2004 *Polym. Test.* **23** 281-290
[13] Negru R, Marsavina L, Voiconi T et al. 2015 *Theor. Appl. Fract. Mech.* **80** 87-95
[14] Marsavina L, Berto F, Negru R et al. 2017 *Theor. Appl. Fract. Mech.* **91** 148-154
[15] Birsan M, Sadowski T, Marsavina L et al. 2013 *Int. J. Solids Struct.* **50** 519-530
[16] Linul E and Marsavina L 2011 *Iran Polym. J.* **20(9)** 736-746
[17] Şerban DA, Linul E, Voiconi T, Marsavina L, Modler N 2015 *Iran. Polym. J.* **24** 515-529
[18] Apostol DA, Stuparu F, Constantinescu DM et al. 2016 *Materiale Plastice* **53(4)** 685-688
[19] Marsavina L, Linul E, Voiconi T et al. 2015 *Frattura ed Integrita Strutturale* **34** 444-453
[20] Voiconi T, Negru R, Linul E et al. 2014 *Frattura ed Integrità Strutturale* **30** 101-108
[21] Usman MA, Adeosun SO, Osifeso GO 2012 *J. Minerals Mater. Charact. Eng.* **11(3)** 311-320
[22] Mello D, Pezzin SH, Amico SC 2009 *Polym. Test.* **28** 702-708
[23] Latinwo GK, Aribike DS, Susu AA, Kareem SA 2010 *Nat. Sci.* **8(6)** 23-31
[24] ASTM D3574 - 17 Standard Test Methods for Flexible Cellular Materials-Slab, Bonded and Molded Urethane Foams
[25] Marsavina L, Kovacik J, Linul E 2016 *Theor. Appl. Fract. Mech.* **83** 11-18
[26] Linul E, Şerban DA, Marsavina L, Kovacik J 2017 *Fatig. Fract. Eng. Mat. Struct.* **40(4)** 597-604
[27] Taherishargh M, Linul E, Broxtermann S, Fiedler T 2018 *J. Alloy. Compd.* **737** 590-596
[28] Kovácik J, Jerz J, Mináriková N et al. 2016 *Frattura ed Integrita Strutturale* **36** 55-62
[29] Linul E, Şerban DA, Marsavina L, SadowskiT 2017 *Arch. Civ. Mech. Eng.* **17(3)** 457-466
[30] Linul E, Serban DA, Voiconi T et al. 2014 *Key Engineering Materials* **601** 246-249
[31] Linul E, Voiconi T, Marsavina L, Sadowski T 2013 *J. Phys. Conf. Ser.* **451** 012002
[32] Linul E and Marsavina L 2015 *P. Romanian Acad. A* **16(4)** 522-530
[33] Marsavina L, Linul E, Voiconi T, Negru R 2016 *IOP Conf Ser: Mater. Sci. Eng.* **123(1)** 012060
[34] Voiconi T, Linul E, Marsavina L et al. 2014 *Key Engineering Materials* **601** 254-257
[35] Linul E, Movahedi N, Marsavina L 2018 *J. Alloy. Compd.* **740** 1172-1179
[36] Movahedi N, Linul E 2018 *Mater. Lett.* **217** 194-197
[37] Linul E, Marsavina L, Kováčik J 2017 *Mat. Sci. Eng.-A Struct.* **690** 214-224
[38] Movahedi N, Linul E, Marsavina L 2018 *J. Mater. Eng. Perform.* **27(1)** 99-108
[39] Linul E, Movahedi N, Marsavina L 2017 *Compos. Struct.* **180** 709-722
[40] Marsavina L, Cernescu A, Linul E, Scurtu D, Chirita C 2010 *Materiale Plastice* **47(1)** 85-89
[41] Movahedi N, Linul E 2017 *Mater. Lett. 206* 182-184
[42] Linul E, Marsavina L, Kovacik J, Sadowski T 2017 *P. Romanian Acad. A* **18(4)** 361-369
[43] Linul E, Movahedi N, Marsavina L 2018 *Materials* **11(4)** 554
[44] Linul E, Serban DA, Marsavina L 2018 *Phys. Mesomech.* **21(2)** 178-186

Acknowledgment
This work was supported (partial supported) by research grants PCD-TC-2017.

IOP Conf. Series: Materials Science and Engineering **416** (2018) 012054 doi:10.1088/1757-899X/416/1/012054

Micromechanical Behaviour of Asphalt Concrete Based on X-ray Computed Tomography Images and Random Generation of Heterogeneous Aggregates Skeleton

Y El Haloui[1,*], F F Tehrani[2,3], J Absi[4], F Courreges[5], M Elomari[1], F Allou[2] and Ch Petit[2]

[1] Université Cadi Ayyad, UCA, FSSM LAEPT, Boulevard Prince My Abdellah, P.B. 2390, 40000 Marrakech, Maroc.
[2] Université de Limoges, GC2D, boulevard Jacques Derche, 19300 Egletons, France.
[3] Conservatoire national des arts et métiers Paris, 292 Rue Saint-Martin, 75003 Paris, France.
[4] Université de Limoges, IRCER, UMR CNRS 7315, 12 Rue Atlantis, 87068 Limoges, France.
[5] Institut de Recherche Xlim, UMR CNRS 7252 - 123, avenue Albert Thomas - 87060 Limoges Cedex, France.

[*]Corresponding author: yassine.elhaloui@edu.uca.ac.ma

Abstract. The main objective of this work is to investigate the influence of the microstructure characteristics on the mechanical behaviour of asphalt concrete. The heterogeneous microstructure of an asphalt concrete is influenced by the aggregate content, shape, orientation and contacts. In previous works, the asphalt concrete is treated as homogeneous material and its heterogeneous microstructures are ignored. While the distribution of its heterogeneous microstructure can strongly influence the intensity of deformation at local scale. Therefore, it is important to create more precisely a micromechanical model containing information on heterogeneous microstructure. A comparison study has been done between the microstructure issues from X-ray computed tomography (CT) and digital models of random aggregate generation. The dynamic modulus of the viscoelastic matrix and elastic properties of aggregates are used as input parameters in the numerical simulations. The obtained results at global scale (complex modulus and phase angle) and at local scale (local strain intensity) of each model are compared. This comparison leads to confirm the feasibility of the proposed heterogeneous random generation of aggregate skeleton at global and local scales. These results provide the basis for an interesting discussion on material design optimization in order to increase the durability of the road pavement materials.

1. Introduction

An asphalt mixture is a composite material that comprises aggregates, binder and air voids [1]. Aggregates usually take up about 85% of the volume of asphalt concrete with different sizes, irregular shapes and random locations. Bituminous materials can be modelled at different length scales, i.e. from bitumen scale to asphalt mixture scale [2]. Its mechanical behaviour is considered as viscoelastic and depend on temperature and loading frequency. The associated characteristics are usually described through use of the dynamic modulus and phase angle [3]. In the most previous studies, the asphalt concrete is treated as homogeneous material and its microstructures are ignored [4,5]. However, it has been reported that the deformation and strength of asphalt concrete are not only influenced by volume fraction of its components but also affected by the spatial distribution of its microstructures [6]. Thus, to investigate the mechanical behaviours of asphalt concrete, it is imperative to notice more accurately a micromechanical model containing information of components and microstructures.

The X-ray computed tomography (CT) method is a digital image processing technology which can be used to capture the internal microstructure of composite materials like asphalt mixture. Two dimensional (2D) or three dimensional (3D) numerical specimens are generated based on slices capture. In general, X-ray CT involves three steps as following: 1) Images acquisition. 2) Images processing. 3) Images analysis. X-ray CT has been used to paid attention for asphalt mixture in the past 15 years. For example, Dai in [7] constructed 2D and 3D micromechanical models using X-ray CT to predict the dynamic modulus and phase angle of asphalt mixture. It was concluded that the 3D model offered a high prediction of mechanical behavior of asphalt mixture than the 2D model.

2. Scope and objective

In this paper, asphalt mixture is modelled as heterogeneous material contains irregular aggregate skeleton localised in random position and bonded with asphalt mastic. The X-ray computed technology and random aggregate generation have been respectively applied for acquisition and generation of the internal microstructure of material. In this work, the aggregate images were taken from an X-ray tomography of a real asphalt mixture samples (see figure 1). In parallel the virtual microstructure of the material was generated from a custom software (called "MOA", French acronym for Random Object Modeler) developed in our laboratory. The finite element modelling is performed with the commercial finite element code ABAQUS [8]. The viscoelastic properties of the bitumen and mastic are calibrated with the laboratory experimental measurement and introduced into a user material subroutine within ABAQUS code. Thereafter the viscoelastic mastic is combined with elastic aggregates to predict the properties of asphalt mixture.

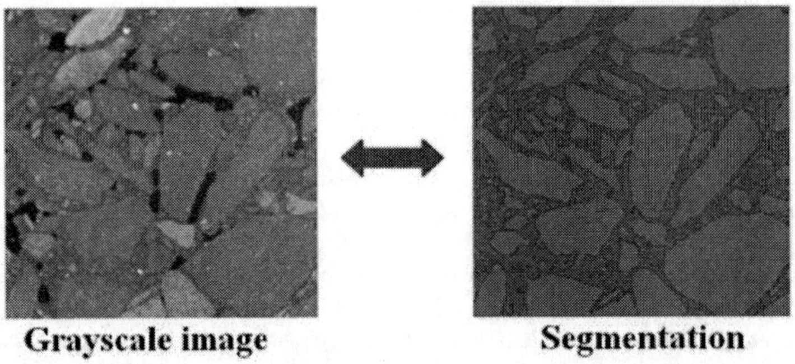

Figure 1. Illustration of gray image acquired by X-ray CT and segmentation process

3. Materials and experimental results

The asphalt mastic is defined as the combination of a 50/70 penetration grade asphalt bitumen and limestone filler sieved between 63 and 100 μm. A laboratory test was conducted by Delaporte *et al* [9]. Bitumen, mastic and asphalt mixture exhibit a thermo-rheological simple behaviour: the data (φ, |E*|) have a unique curve in Black space. The dynamic modulus and phase angle master curves obtained at a reference temperature of 10°C are shown in figure 2. In this study, the asphalt mixture is composed of aggregates as inclusions and mastic as matrix. To be specified in numerical simulation, the viscoelastic properties of mastic should be calibrated based on figure 2 and then defined in ABAQUS like Prony parameters. The step of calibration can be performed by Viscoanalyse code [10]. First, the complex modulus |E*| and phase angle φ are input in a normalized file recognizable by *Viscoanalyse* code; then, the Prony parameters can be adjusted by specifying the number of Maxwell elements.

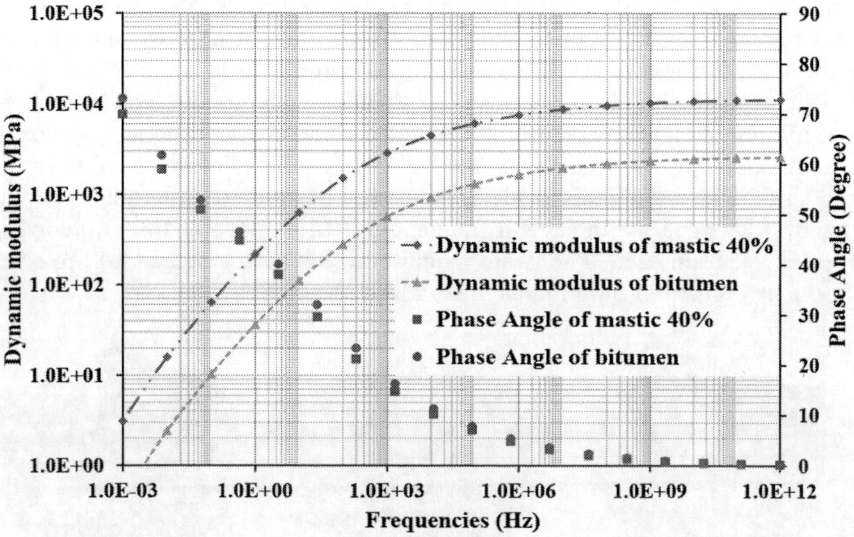

Figure 2. Modulus and phase angle master curves (Tr = 10°C) of bitumen 50/70, mastic 40% and asphalt mixture [9]

The type of aggregate used in this study was limestone. In general, aggregates are assumed to behave elastically. In this study, an estimated aggregates elastic modulus of 70 GPa and Poison's ratio of 0.15 were considered. These properties will be used as input parameters of simulation to define the mechanical behaviour of aggregates. In the other hand, the aggregate size distribution in asphalt mixture concrete is given in Table 1. This gradation was used in the formulation of the considered asphalt mixture specimen. It will be used also to generate the internal microstructure of this materials through use the heterogenous generation method by MOA code. This part is detailed in the following section.

Table 1. Particle size distribution in an asphalt mixture

Sieve size (mm)	12.5	10	8	6.3	4	2	1	0.5
Passing percentage (%)	100	92	68.9	52.7	42.2	32.1	21.7	15.4

4. 2D and 3D micromechanical modelling of bituminous composites

The MOA program, developed in our laboratory in C++ language, is an original software capable of automatically generating samples derived from a matrix with a given percentage of air voids and containing a known number of inclusions featuring different shapes, extending from spherical to polyhedral and randomly placed in the volume defined by the matrix in the form of a rectangular prism or cylinder.

To avoid a situation in which the distance between two randomly created objects becomes less than the minimum resolution of the FEM code, the generator imposes a minimum inter-object distance (equal to one-tenth of the smallest acceptable radius). Moreover, the inclusions might intersect the external surface of the container, which could provide an accurate representation of the final sample obtained by cutting from a larger fabricated sample. Then, a Boolean operation serves to differentiate the inclusion part from the matrix part. Once such geometries have been imported into the ABAQUS code, a fully cohesive state at the matrix/inclusion interface is imposed on the entire inclusion surface by use of the "tied contact" function, which has been preloaded in the FEM code. [11,12]

As shown in figure 3, 2D and 3D microstructure of numerical models, that includes aggregates and matrix, were generated from MOA software as a rectangular and a rectangular prism specimen, respecting the gradation curve of the materials.

It is noteworthy that direct aggregate to aggregate contact is neglected in the current study. However, the existence of a thin bitumen and mastic film is assumed between each aggregate and the other, which simulates to some extent the adsorption of aggregate to some of the bitumen phase and mastic phase. This asphalt film allows to simulate a viscoelastic contact condition between different aggregates. The minimum element size in the generated finite element model was adapted in function of the minimum thickness of the bitumen and mastic between the adjacent separated aggregates.

On the top surface of the digitals models, a uniaxial tension-compression cyclic load is applied. cyclic strain (sinusoidal signal) with several frequencies (0.1, 1, 10 and 100 Hz) corresponding to the experimental tests have been used. The strain magnitude should not exceed 10^{-4} in order to remain in the linear viscoelastic domain of bituminous materials [13].

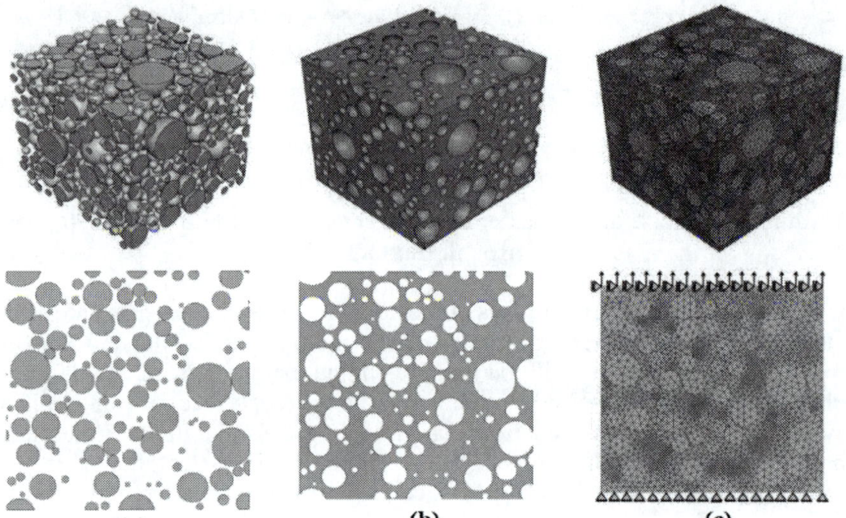

Figure 3. Two and three-dimensional micromechanical model a) Inclusions, b) Matrix, c) Finite element meshed model

5. Numerical simulation results

5.1. Global scale

The dynamic modulus $|E^*|$ and phase angle φ are the most required parameters to describe viscoelastic materials like bituminous composites. To obtain these parameters, four methods can be used: i) Experimental tests, ii) rheological models, iii) analytical formulations and iv) numerical simulations. In this research, numerical simulation method has been considered to determine $|E^*|$ and φ. The magnitude of the dynamic modulus was calculated using the last two cycles for each frequency, with the equation:

$$|E^*| = \frac{|\sigma_{max} - \sigma_{min}|}{|\varepsilon_{max} - \varepsilon_{min}|} \quad (1)$$

The phase angle was calculated by tracking the computation points of maximum/minimum stress and strain. For each cycle, 100-time increments (computation points) were used to achieve a high resolution.

Figure 4a and figure 4b present the master curves of the complex modulus and phase angle of asphalt mixture obtained respectively by the random aggregate generation and X-ray computed tomography methods. It can be observed that the results obtained by both approaches agree reasonably. The average difference between simulation results at global scale is less than 4 % for complex modulus and 3 % for phase angle. In order to evaluate the influence of microstructure on the mechanical response of material at local scale, the intensity of strain at different zones of numerical models were obtained and compared to global imposed strain (ε = 10^{-4}).

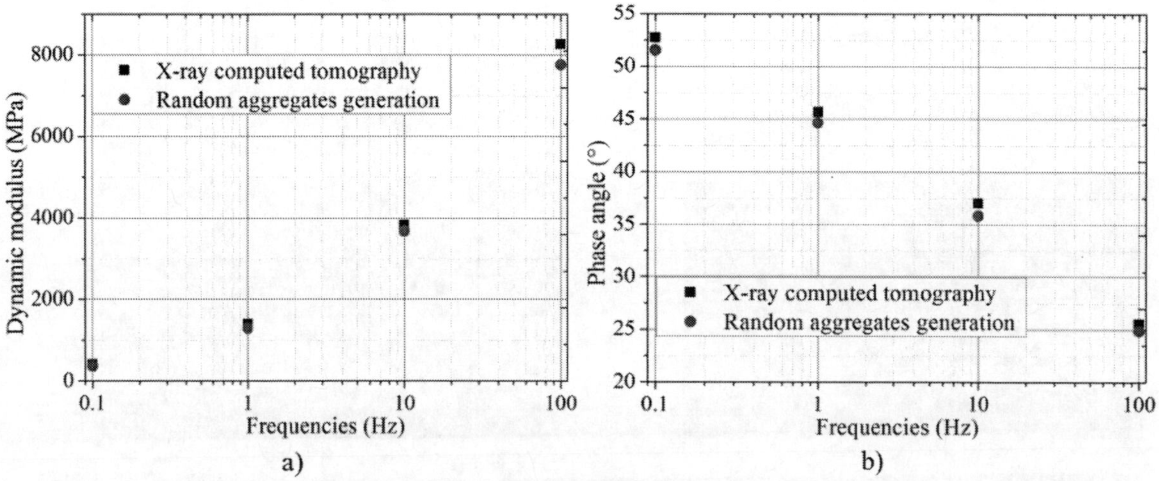

Figure 4. Comparison between random generation and X-ray tomography methods, a) complex modulus, b) phase angle

5.2. Local scale

The distributions of local vertical strain obtained respectively by X-ray computed tomography and random aggregates generation methods are presented in Figure 5a and figure 5b. It can be observed that the deformation intensity ratio depends generally on the position and the thickness of the thin viscoelastic layer of matrix. As it is shown, the intensity of strain at local scale can be reached up to 10 times greater than imposed global strain.

5.3. Final discussion

In this study, numerical investigations of the mechanical behaviour, at global and local scale, of asphalt mixture through the use of X-ray CT and heterogenous generation were conducted. The effect of inclusions shape irregularity was shown in our previous study [6]. Two categories of aggregates shape were used: polygonal inclusions and spherical inclusions. At global scale, the corresponding complex modulus of each type of aggregates were determined. The values corresponding to the spherical inclusions were situated just below those of the polygonal inclusions. In addition, both master curves display a regular variation versus the frequency. Therefore, the shape of aggregates does not affects the overall mechanical behaviour. At local scale, stress values of polygonal inclusions were found greater than that of spherical inclusions due to the high angularity of polygonal aggregates in compression in comparison to the case of spherical aggregates. Thus, considering heterogeneous models rather than homogenous ones can makes sense. In fact, by ignoring the microstructure of asphalt mixture, the intensity of strain within the equivalent media of this material will be close to the global applied strain (10-4). However, it was found that, taking into account biphasic model, the level of strain is 10 times greater than the initial imposed strain ($\varepsilon=10^{-4}$). This large variation affects various local phenomena within asphalt mixture such as self-heating.

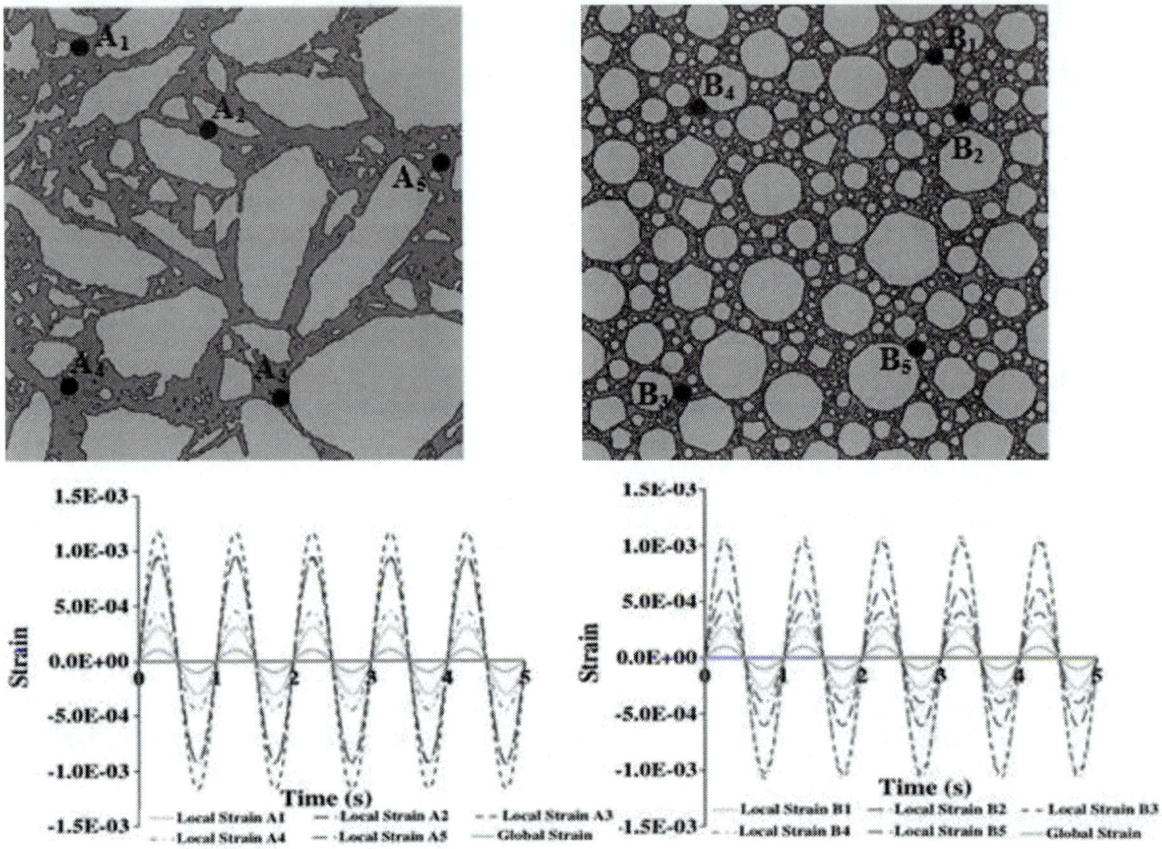

Figure 5. Strain intensity obtained at different zones of (a) X-ray tomography, (b) random aggregate generation models

6. Conclusions

In this paper, we developed a heterogeneous numerical approach able to calculate the mechanical properties of hot mix asphalt from the properties of its components. The microstructures of the material were obtained by two methods: random aggregates generation and X-ray computed tomography. In order to investigate the influence of microstructure into mechanical response of asphalt mixes, the complex modulus and phase angle issue from the two methods under study, were compared and a good agreement was observed. Moreover, the intensity of deformation within the viscoelastic matrix at local scale was investigated and again an interesting and converged results were obtained. In summary, the proposed heterogeneous numerical approach, yields very acceptable results for highly heterogeneous media of asphalt mixes at global and local scales in comparison to those issues from X-ray method.

7. References

[1] Masad E, Muhunthan B, Shashidhar N and Harman T 1999 Quantifying laboratory compaction effects on the internal structure of asphalt concrete *Transportation Research Record: Journal of the Transportation Research Board* **1681** 179-185

[2] Fakhari Tehrani F, Absi J, Allou F and Petit C 2016 Micromechanical modeling of bituminous materials complex modulus at different length scale *International Journal of Pavement Engineering* **19** 1-12

[3] Di Benedetto H, Olard F, Sauzéat C and Delaporte B 2004 Linear viscoelastic behaviour of bituminous materials. From binders to mixes *Road Materials and Pavement Design* **5 (1)** 163-202

[4] Huang B, Shu X, Li G and Chen L 2007 Analytical modeling of three-layered HMA mixtures *International Journal of Geomechanics* **7 (2)** 140-148

[5] Shu X and Huang B 2008 Micromechanics-based dynamic modulus prediction of polymeric asphalt concrete mixtures *Composites Part B: Engineering* **39 (4)** 704-713

[6] Fakhari Tehrani F, Absi J, Allou F and Petit C 2013 Heterogeneous numerical modeling of asphalt concrete through use of a biphasic approach: Porous matrix/inclusions *Computational Materials Science* **69** 186-196

[7] Dai Q 2011 Two-and three-dimensional micromechanical viscoelastic finite element modeling of stone-based materials with x-ray computed tomography images *Construction and building materials* **25 (2)** 1102-1114

[8] Hibbitt K 2007. ABAQUS version 6. 7: theory manual, users' manual. *verification manual and example problems manual*

[9] Delaporte B, Di Benedetto H, Chaverot P and Gauthier G 2007 Linear viscoelastic properties of bituminous materials: From binders to mastics *Proc of the Technical Sessions. Asphalt Paving Technology: Association of Asphalt Paving Technologists* **76** 455-494

[10] Chailleux E, Ramond G, Such, C and de La Roche C 2006 A mathematical-based master curve construction method applied to complex modulus of bituminous materials *Road Materials and Pavement Design* **7 (1)** 75-92

[11] Fakhari Tehrani F, Absi J, Allou F and Petit C 2016 Viscoelastic properties of bituminous composites using multiscale heterogeneous numerical simulation and micromechanical analytical self-consistent model *Rilem bookseries* **13** 133-138

[12] Fakhari Tehrani F, Absi J, Allou F and Petit C 2013 Heterogeneous numerical modeling of asphalt concrete through use of a biphasic approach: Porous matrix/inclusions *Computational Materials Science* **69** 186-196

[13] Airey G D and Rahimzadeh B 2004 Combined bituminous binder and mixture linear rheological properties *Construction and Building Materials* **18 (7)** 535-548

IOP Conf. Series: Materials Science and Engineering **416** (2018) 012055 doi:10.1088/1757-899X/416/1/012055

Experimental Verification of the DCB Test Configuration Applicability to Mechanically Coupled Composite Laminates

J Rzeczkowski[1,*], S Samborski[1] and J Paśnik[1]

[1] Lublin University of Technology, Department of Applied Mechanics, Nadbystrzycka 36 St., Lublin 20-618, Poland

*Corresponding author: kubarzeczkowski@op.pl

Abstract. This article presents an experimental determination on mode I critical strain energy release rate (c-SERR, G_{IC}). Multidirectional (MD) composite beams were subjected to the double cantilever beam (DCB) test during which the applied load and the load point displacement were registered. The maximum peak force was taken into account in calculation of the G_{IC}. It was computed using three different methods. On the basis of the performed test it was found that the critical strain energy release rate depends on the specimen interface.

1. Introduction

The Fiber-Reinforced Laminate Composites (FRLC) are commonly used in many fields of engineering. For example, these materials are used in contemporary load-carrying structures. Therefore, it demands from the designers thorough recognition of mechanical and strength properties. Delamination is particularly dangerous damage form in laminates. From the engineering and design point of view there is a vital need to assess the risk of a composite structure failure in result of delamination propagation. The fracture toughness (G_{IC}) values can be determined for mode I fracture in the double cantilever beam (DCB) test. This parameter is today counted to basic mechanical characteristics of composite materials beside the Young moduli and the Poisson coefficient [1,2]. The respective ASTM standard [3] covers the measurement of G_{IC} in the form of the critical strain energy release rate (c-SERR) of unidirectional (UD) laminates. In fact, UD specimens are generally used for the DCB test due to the high flexural stiffness and ability to maintain self-similar crack propagation. However, most applicationsin composite structures involve multidirectional (MD) laminates in which delaminations tend to occur between layers of different orientations [4]. Several studies were conducted of mode I fracture of MD laminates with delamination in α/α, $\alpha/-\alpha$ and $0/90$ interfaces [5-9]. The major problem is the development of interply damage, often associated with extensive fibre bridging and non-linearity. Moreover, MD laminates usually presents elastic couplings. This pehenomena can be written in a block-matrix form as a constitutive equations [10] describing all possible couplings in a laminate.

$$\begin{Bmatrix} \mathbf{N} \\ \mathbf{M} \end{Bmatrix} = \begin{bmatrix} \mathbf{A} & \mathbf{B} \\ \mathbf{B} & \mathbf{D} \end{bmatrix} \begin{Bmatrix} \boldsymbol{\varepsilon}^0 \\ \boldsymbol{\kappa}^0 \end{Bmatrix} \tag{1}$$

The key constituents of block matrix in square brackets are respectively; the extensional stiffness matrix (**A**), the coupling stiffness matrix (**B**) and the bending stiffness matrix (**D**). The presence of coupling can be the source of significant errors in G_{IC} measurements [11-12]. For example, from the practical point of view, in case of the DCB configuration the bending-twisting couplings may cause variability of the initial SERR along delamination front in coupled laminated beam. It is thus important to study mode I fracture of mechanically coupled composite materials. In this article the values of the mode I c-SERR were obtained by using of three calculation methods: the Modified Beam Theory (MBT), Compliance Calibration Method (CCM) and Modified Compliance Calibration (MCC). As expected the experimental test showed that result of calculations for mode I interlaminar fracture toughness in mechanically coupled laminated composites depends on the specimens delamination interface as well as the couplings induced by specific stacking sequences.

2. Experiment

Experimental DCB test were performed on multidirectional composite laminates with delamination interfaces [0/ α], [α/ α] and [α/- α]. Additionally, bending-twisting (BT) coupled laminate with the respective ply sequence: [α/0 α/ α/0/- α/0/- α/- α/- α/- α/0/- α/ α/0/0/ α/ α] and bending extension (BE) coupled laminates [α/- α/0/- α/0/ α/90/ α/- α] were examined. The fiber orientation angle α set was {30°, 45°, 60°, 90°}. The specimens dimensions was width b=20mm, total length 150mm and thickness $2h$=2,75mm.

The tests were carried out in Shimadzu machine at 1 mm/min crosshead speed according to the ASTM D 5528 procedure. All specimens were loaded through piano hinges bonded to the top and bottom legs of a specimen. During the DCB test the applied load P and the load point displacement δ were registered. Also the crack onset and all propagation values along the edges was visually observed and marked on the opposite sides of the specimens. In figure 1was shown the DCB experimental set-up. In the mode I tests, the fiber bridging phenomena which causes growing resistance during delamination process was observed as illustrated figure 2.

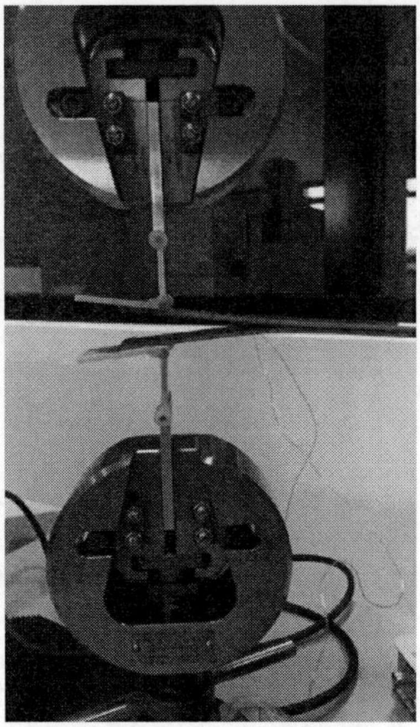

Figure 1. DCB experimental setups

Figure 2. Deformed DCB specimens

To determine mode I critical strain energy release rate three different calculation methods were used: the MBT, the CCM and the MCC methods, respectively.

IOP Conf. Series: Materials Science and Engineering **416** (2018) 012055 doi:10.1088/1757-899X/416/1/012055

MBT method uses correction parameter Δ which may be determined experimentally by generating a least squares plot of the cube root of compliance $C^{1/3}$ as a function of delamination length a. The load and displacements corresponding to the visually observed delamination onset on the edge and all the propagation values were used to generate this plot. Mode I interlaminar fracture toughness is expressed following equation:

$$G_{IC} = \frac{3P\delta}{2b(a+|\Delta|)} \tag{2}$$

The CCM method determines additional parameter n which is the slope of $ln(C)$ versus $ln(a)$ curve. Therefore, the critical SERR is calculating as follows:

$$G_{IC} = \frac{nP\delta}{2ab} \tag{3}$$

In the MCC method, the first critical SERR can be expressed the following equation:

$$G_{IC} = \frac{3P^2C^{2/3}}{2A_1bh} \tag{4}$$

where A_1 is the slope of a least squares plot of delamination length normalized by specimen thickness a/h, as a function of the cubic root of compliance $C^{1/3}$.

3. Results and discussion

Table 1 collects the results of calculation mode I critical strain energy release rate calculated using the three different methods.

Table 1. Values of mode I critical strain energy release rate for different delamination interfaces and calculation methods

Specimens Interface	G_{IC} [N/mm]		
	MBT	CMM	MCC
0°/30°	0.327	0.319	0.374
0°/45°	0.807	0.830	0.872
0°/60°	0.957	0.894	0.947
0°/90°	0.867	0.874	0.847
30°/30°	0.746	0.749	0.686
30°/-30°	0.892	0.865	0.730
45°/45°	0.719	0.757	0.738
45°/-45°	0.527	0.516	0.546
60°/60°	1.113	1.039	1.039
60°/-60°	0.821	0.816	0.712
90°/90°	1.095	1.040	1.029

The results show influence of delamination interface on G_{IC} both for samples with interface 0/α, α/α as well α/-α. In case of the 0/α interface differences can be observed between mode I c-SERR value. For the interface 0°/60°, G_{IC} increases significantly above the one determined for the 0°/30° configuration. Its value was 0.957 N/mm, while for the interface 0°/30° only 0.327 N/mm. For the angle α {45°, 60° 90°} the c-SERR values are on similar level. Also close results

IOP Conf. Series: Materials Science and Engineering **416** (2018) 012055 doi:10.1088/1757-899X/416/1/012055

of the critical strain energy release rate calculation were both for the interface 30°/30° and 45°/45° as well as 60°/60°and 90°/90°but along with angle growth also a growth of G_{IC} values was observed. For α {30°, 45°} this value was on average level of 0.732 N/mm and for α {60°, 90°} reached about 1.104 N/mm. Comparing 45°/45° and 45°/-45° configuration the difference between the critical strain energy release rate is about 0.250 N/mm. Similar situation was observed in case of the 60°/60° and 60°/-60° interface. The smallest difference of G_{IC} occurred for the 30°/30° and 30°/-30° layups. The biggest value of mode I critical strain energy release rate was for the 60°/60° interface and reached 1.113 N/mm. Also the 90°/90° interface characterized itself with a high level of the G_{IC} value i.e. 1.095 N/mm. At most, the rest of specimen reached c-SERR value on the level of 0.800 N/mm except 45°/-45° interface (0.527 N/mm) and the 0°/30° (3.327 N/mm), which is the smallest value in the whole experiment.

In figure 3 was presented the values of the mode I critical strain energy release rate for specimens with specific ply sequences induces the bending-extension (BE) coupling and having the superior meaning in the case of mode I experimental test - the bending-twisting (BT) coupling. For the BT samples the value of mode I c-SERR was on the level of 0.800 N/mm and for the BE specimens it reached about 0.450N/mm.

For the mechanically coupled specimens, three different data reduction schemes were used as it was mentioned before. The values of the mode I critical strain energy release rate were on similar level for all methods.

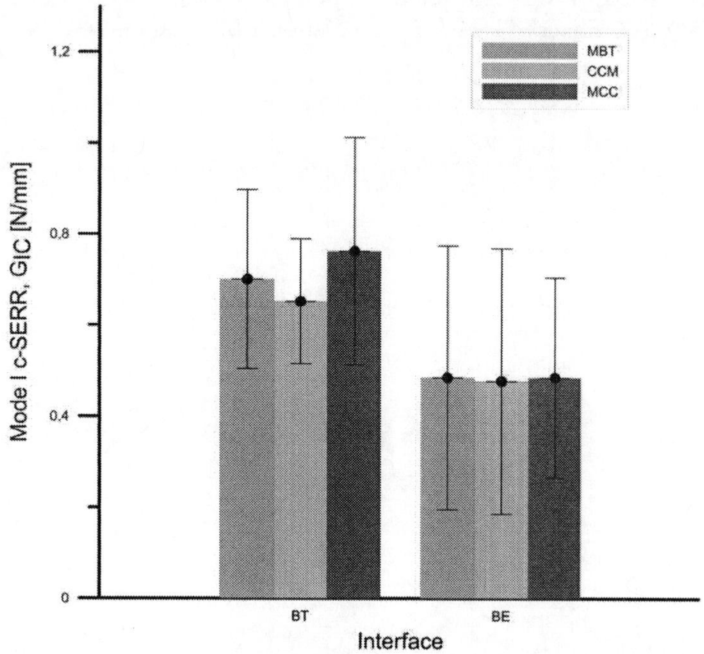

Figure 3. Mode I critical strain energy release rate for specimens with bending-twisting and bending-extension couplings

4. Conclusions

The experimental analysis of the DCB test configuration according to the ASTM 5528 Standard in case of the mechanically coupled laminated composite beams was performed. In particular, the effect of different interfaces of plies neighboring the delamination plane was discussed. The critical strain energy release rate (G_{IC}) was obtained using different calculation methods. The results shows that mode I c-SERR depends on specimens interface. Also, as it was expected, the bending-twisting coupling had strongest effect on the c-SERR values in the DCB configuration applicability to mechanically coupled specimens.

5. References

[1] Samborski S 2016 *Comp. Struct.* **152** 477-487
[2] Samborski S and Rzeczkowski J 2018 *AIP Confernece Proceedings* **1922** 080010
[3] ASTM D 5528-01. Standard test method for mode I interlaminar fracture toughness of unidirectional fiber-reinforced polymer matrix composites 2001
[4] Pereira A B, de Morais A B, de Moura M F S F and Magalhaes A G 2005 *Elsevier.* **36** 1119-27
[5] Chai H 1984 *Composites.* **15** 277-90
[6] Laksimi A, Benzeggagh M L, Jing G, Hecini M and Roelandt J M 1991 *Compos Sci. Technol.* **41** 147-64
[7] Robinson P and Song D Q 1992 *Compos. Mater.* **33** 73-100
[8] Polaha J J, Davidson B D, Hudson R C and Pieracci A J 1996 *Reinf. Plast. Compos.* **15** 141-73
[9] Choi N S, kinloch A J and Williams J G 1999 *J. Compos Mater.* **33** 73-100
[10] York C B 2013 *Compos A* **44** 140-8
[11] de Morais A B, de Moura M F, Marques A T and de Castro P T 2002 *Compos. Sci. Technol* **62** 679-86
[12] Davidson B D, Kruger R and Konig M 1996 *Eng. Fract. Mech.* **55** 557-69

Acknowledgments: The paper was financially supported by the Ministerial Research Project No. DEC-2016/21/B/ST8/03160 financed by the Polish National Science Centre.

Issues of Direct Application of Fracture Toughness Determination Procedures to Coupled Composite Laminates

S Samborski[1], J Rzeczkowski[1] and J Paśnik[1]

[1]Department of Applied Mechanics, Lublin University of Technology, Nadbystrzycka 36, 20-618 Lublin, Poland

Abstract. The article presents examples of difficulties related to direct application of the standardized DCB and ENF test methods to composite laminates with mechanical couplings. It reveals delamination as one of the most common damage forms in the considered group of engineering materials. The paper is a part of the study presented in detail in other papers published elsewhere [2-4,11]. The Finite Element Method simulations of the interlaminar cleavage formation and growth prior to the physical experiments is shown as necessary. The advantages of the numerical methods of delamination modeling, such as the virtual crack closure technique in planning the experiments and introducing essential modifications in data reduction schemes are presented. The presence of unexpected mode-mixity in the standard fracture-toughness tests applied to the mechanically coupled composite laminates is also shown.

1. Introduction

Vast field of applications of composite laminates in contemporary engineering, for example in structural parts of the aircraft, ships or automobiles could still be broadened, if the non-symmetric laminates, exhibiting mechanical couplings came to play. As shown by York [1], there is a poorly investigated group of multidirectional continuous fiber reinforced polymeric (CFRP) laminates where different coupling effects call forth hundreds of coupled layups. In the effect, the multidirectional laminates exhibit distortions in the Strain Energy Release Rate (SERR) distributions along delamination front and different mode mixities [2-4], such that direct application of the standardized test methods is difficult or uncertain. Nevertheless, basic mechanical properties and material constants must be determined experimentally, so any improvement of the existing test procedures (ASTM D5528, ISO 15024, ISO 15114, ASTM D7905) are welcome, the more that the laminated composites are prone to occurrence of various defects. Delamination is one of the most common damage forms in the CFRP laminates. The finite element (FE) simulations of the fracture toughness test configurations - prior to the experiments - are invaluable source of knowledge and, as such, are necessary. The numerical methods of delamination modeling, for example the virtual crack closure technique (VCCT) enable proper planning of the experiments and introducing essential modifications in the data reduction schemes - by testing many FE-model configurations with various mechanical couplings. The experimental data obtained in the double cantilever beam (DCB), the end-notched flexure (ENF) or other tests confirm the numerical predictions.

2. Fundamentals of the mechanical coupling phenomena

According to the Classical Lamination Theory (CLT), the coupling phenomena are commonly expressed in the form of a block-matrix relations [2-4]:

$$\begin{Bmatrix} \mathbf{N} \\ \mathbf{M} \end{Bmatrix} = \begin{bmatrix} \mathbf{A} & \mathbf{B} \\ \mathbf{B} & \mathbf{D} \end{bmatrix} \begin{Bmatrix} \boldsymbol{\varepsilon}^0 \\ \boldsymbol{\kappa}^0 \end{Bmatrix} \tag{1}$$

where **N** and **M** represent the laminate's internal forces and moments, respectively; $\boldsymbol{\varepsilon}^0$ is the vector of strains in the mid-plane and $\boldsymbol{\kappa}^0$ – the vector of curvatures. There are also stiffness matrices: **A** - the extensional stiffness matrix, **B** - the coupling stiffness matrix and **D** - the bending stiffness matrix.
The coupling matrix **B** may take six different forms [1]: $\mathbf{B}_0$ with all constituents equal zero, $\mathbf{B}_S$ for a simple laminate, $\mathbf{B}_F$, when all elements are finite; $\mathbf{B}_L$ with the B_{11} and B_{22} terms different than zero, $\mathbf{B}_T$ when the B_{16} and B_{26} constituents are non-zero and the $\mathbf{B}_{LT}$ being a superposition of the former two. Combinations of the above listed forms of the matrices **A**, **B** and **D** give numerous possible couplings at different total numbers of plies [1-4].

3. Methodology of the FEM simulations

The finite element analyses were performed with the ABAQUS/CAE code, ver. 16.4.2 with its virtual crack closure technique (VCCT) facility [5-7]. The propagation of an interlaminar crack – delamination was simulated by a separation of node pairs. The FE mesh was constructed of the S4 and the S4R elements [2-4] of the nominal size 2.5 mm, densified along the front to 0.5 mm. The DCB/ENF beam model was composed of two laminated parts bonded together [2,3]. A total of 11280 quadrilateral elements was used, having 82 026 degrees of freedom. The simulations were performed with the DELL's T3500 workstation equipped with the Intel's Xeon W3540 8-core processor clocked at 2.93GHz with the RAM memory capacity of 12GB and the RAID-5 SATA disk matrix. Composite material characteristics accepted in the model definition are collected in Table 1; the data were taken from the literature [8] for the SEAL's Texipreg HS160RM carbon-epoxy prepreg laminate.

As stated by Zou [9], one should always expect mixed fracture modes. Thus, for the sake of generality, the Reeder Law has been chosen as the fracture criterion [6,10]:

$$G_{\text{eq-c}} = G_{\text{Ic}} + \left(G_{\text{IIc}} - G_{\text{Ic}}\right)\left(\frac{G_{\text{II}} + G_{\text{III}}}{G_{\text{T}}}\right)^{\eta} + \left(G_{\text{IIIc}} - G_{\text{IIc}}\right)\left(\frac{G_{\text{III}}}{G_{\text{II}} + G_{\text{III}}}\right)\left(\frac{G_{\text{II}} + G_{\text{III}}}{G_{\text{T}}}\right)^{\eta}, \tag{2}$$

where $G_{\text{eq-c}}$ stood for the equivalent critical SERR and $G_{\text{T}} = G_{\text{I}} + G_{\text{II}} + G_{\text{III}}$.

The criterion for delamination initiation was:

$$G_{\text{T}} \geq G_{\text{eq-c}} \tag{3}$$

4. Results and discussion

In this section, the most interesting cases of the DCB/ENF simulation results are presented. In figure 1 the results for the DCB beam model configuration are presented. Note, that the "BT" abbreviation stands for the bending-twisting coupling, expected to be the most problematic in the practical realization, from the point of view of conforming the regulations of the relevant ASTM D5528 Standard. Indeed, when compared to those obtained for the unidirectional (UD) layup, the results for the directional coupled configurations exhibit strong skewness.

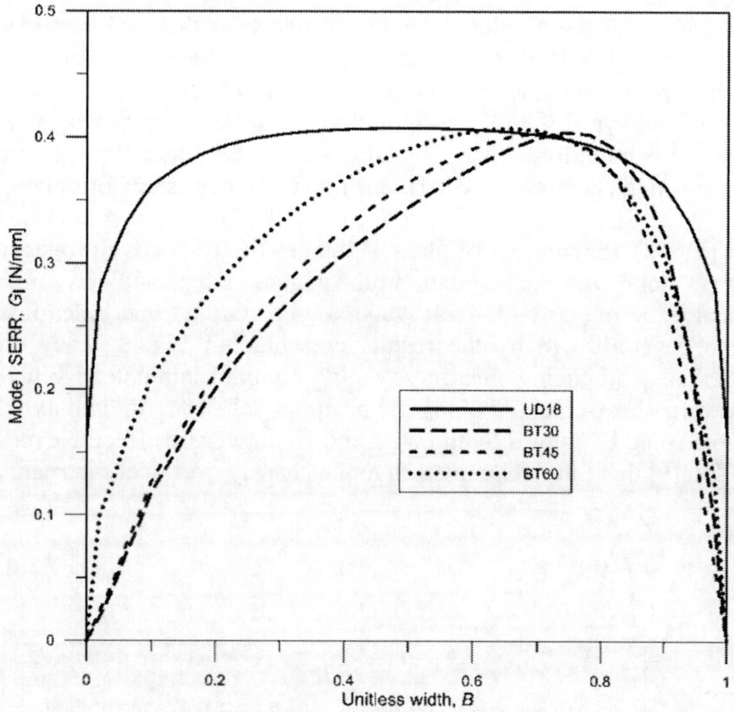

Figure 1. Skewness of the SERR plots induced by coupling

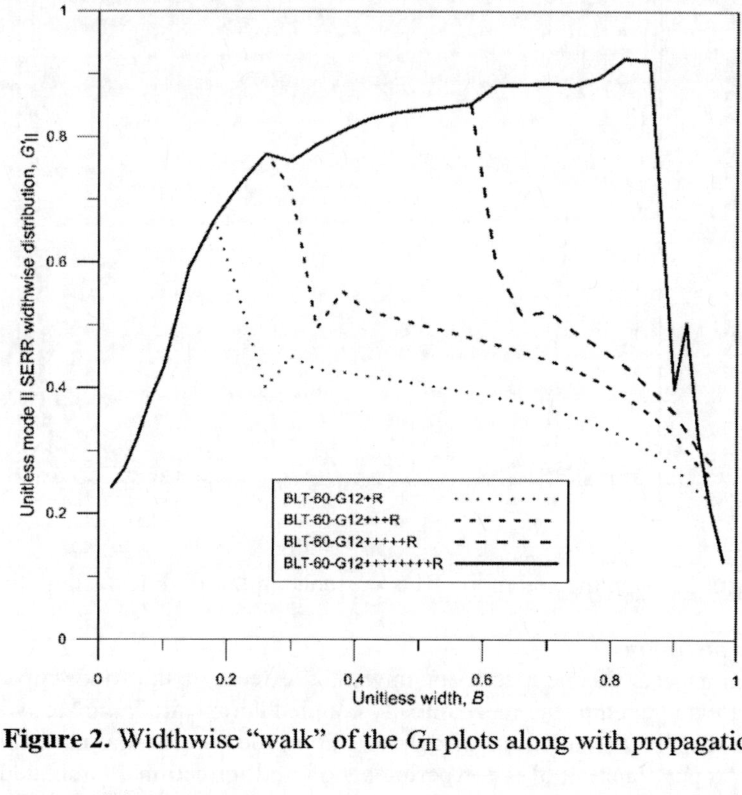

Figure 2. Widthwise "walk" of the G_{II} plots along with propagation

The considered BT sequence was composed of 18 plies in each leg of the DCB beam, as follows: [θ/0/θ/θ/0/-θ/0/-θ/-θ/-θ/-θ/0/-θ/θ/0/0/θ/θ], with the fiber orientation angle θ = 30, 45 or 60 degrees,

IOP Conf. Series: Materials Science and Engineering **416** (2018) 012056 doi:10.1088/1757-899X/416/1/012056

respectively. The skewness of the G_I plots was the strongest for the moderate fiber angles within the (0, 90°) interval and it disappeared for θ close to 90 degrees. These observations induced some doubts about the self-similarity of the mode I delamination propagation in case of the coupled laminates.

Concerning the second type of test configuration – the ENF, another interesting effects were observed in the FEM simulations. Namely, for the "BLT" coupling (simultaneous bending-extension/extension-twisting/shearing-bending), during subsequent steps of delamination propagation, maximum of the mode II SERR distribution moved ("walks") from one side of the beam model to another. In the plot (Fig. 2), the number of plusses denotes the respective propagation steps. The shift of the maximum of G_{II} along with the simulation time suggested a possible asymmetry in delamination front in the physical experiments – the cleavage might start at one side of the specimen. This conclusion seems to be confirmed by the results presented in Fig. 3. They come from the ENF simulation for the extension-bending/shearing-twisting coupled laminate. Even though the assumed test configuration extorts shear-cracking along the beam model's longitudinal axis, the strong mode III effects came out. This could be caused by the well-known anticlastic effect. Nevertheless, it confirmed the observation of Zou [9] that pure delamination modes hardly ever occur in reality.

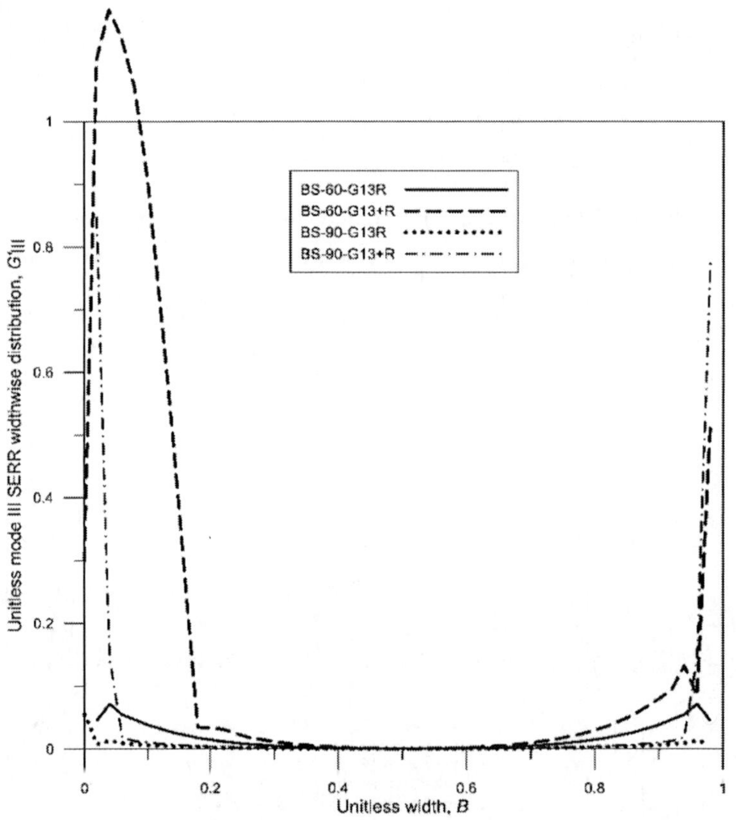

Figure 3. Unexpected mode III occurrence in the ENF test simulation

5. Summary and conclusions

In the paper several aspects of unexpected or unwanted effects of the mode-mixity in the standard fracture-toughness tests applied to the mechanically coupled composite laminates were displayed. The presented results of the FEM simulations of delamination processes in laminated beams are a part of the study aimed at proper planning of the experiments on multidirectional laminate layups. Additional parameters describing the skewness of the SERR plots are necessary in order to modify the standardized test methods and data reduction schemes [2-4]. The numerical outcomes are now being

IOP Conf. Series: Materials Science and Engineering **416** (2018) 012056 doi:10.1088/1757-899X/416/1/012056

verified in laboratory tests with specimens exhibiting different couplings, as well as various delamination interfaces [11].

6. References

[1] York CB 2010 *J. Aerosp. Eng.* **23** 219
[2] Samborski S 2016 *Comp. Struct.* **152** 477
[3] Samborski S 2017 *Comp. Struct.* **163** 342
[4] Samborski S 2018 *Comp. Struct.* (in press)
[5] Rybicki EF and Kanninen MF 1997 *Eng. Fract. Mech.* **9** 931
[6] Simulia 2014
[7] Burlayenko V and Sadowski T 2008 *Budownictwo i Architektura* **2** 95
[8] de Moura MF, Campilho R and Goncalves J 2008 *Compos. Sci. Technol.* **68** 2224
[9] Zou Z, Reid S, Soden P and Li S 2001 *Int. J. Solids Struct.* **38** 2597
[10] Reeder JR 1992 *NASA TM-1992-104210*
[11] Samborski S and Rzeczkowski J 2018*AIP Conf. Proc.* **1922**

Acknowledgement: This paper was financially supported by the Ministerial Research Project No. DEC-2016/21/B/ST8/03160 financed by the Polish National Science Centre.

An Influence of UV Ageing Process on Tensile Strength and Young's Modulus of Polymeric Fiber Composite Materials

P Golewski[1,*] and T Sadowski[1]

[1]Lublin University of Technology,
Faculty of Civil Engineering and Architecture,
Nadbystrzycka 40, 20-618 Lublin, Poland

[*]Corresponding author: pgolewski@gmail.com

Abstract. The aim of this study was to estimate the influence of the UV ageing process of the polymer matrix composites (PMC). Samples were prepared using 2 types of prepregs in the form of: (1) fabrics and (2) unidirectional fibers. The samples were aged in an accelerated weathering chamber for various times. Then the uniaxial tension tests were conducted using the DIC (digital image correlation) method to monitor deformation process of the PMC in time. Changes of the tensile strength and the Young modulus were experimentally estimated and compared to reference samples without ageing.

1. Introduction

The PMC, apart from static, dynamic and thermal loads [10-11] are also exposed to the natural environment.

The UV radiation for epoxy matrix composites without any coatings e.g. in the form of paint, is destructive by erosion. During the UV radiation the polymer chains may break down and CO, CO_2 and H_2 may be released. This ensures that the matrix becomes more brittle especially at low temperatures [1].

The PMC are often used for the production of wind turbine blades, building structures, ship hulls that are exposed both to: (a) the UV radiation and (b) moisture penetration. With the simultaneous influence of both factors, [6], a significant increase of the Young modulus was obtained for the composite with the isophthalic polyester matrix and fiberglass reinforcement. This is related to combination of water and the formation of longer chains of the polyester matrix.

The PMCs are also used in aerospace. In this case, ageing can be carried out as synergy of:

(1) high vacuum,
(2) the UV radiation,
(3) cyclic changes of temperature,
(4) atomic oxygen occurring in the atmosphere.

These are factors occurring on a low Earth's orbit, and research requires a special laboratory chamber, which was used in [9].

Natural fibers with polymer matrixes are increasingly used currently. In [2] the authors investigated composite based on natural fibers of flax. The tests were carried out both using UV radiation and water spray. Samples were aged for the following times: 500 h, 1000 h and 1500 h. The strength reduction after 1500 h was 29.9% and the Young's modulus reduction was 34.9%. However, it should be mentioned, that the samples were made by hand lay up method, which could have a significant impact on the results. In the paper [8] the bio-epoxy jute-basalt hybrid composite material was investigated. The ageing tests were carried out for the following steps: 14, 28, 56 and 84 days. Addition of basalt fiber layers allowed for more that 2-fold increase of the Young's modulus and bending strength. Generally, it can be concluded that composites with natural fibers such as flax, hemp, jute and sisal are more sensitive to aging.

Among the analyzed works, uniaxial tensile tests [4], three point bending tests [8] and Charpy impact tests [12] are most often used. However, there is no reports in the literature concerning the strain fields distributions in the analyzed aged samples. Therefore, the aim of this paper is to fill this gap.

2. Samples ageing

The tests were carried out using the QUV LU-8047-TM chamber. It allows to restore atmospheric conditions:

(1) solar radiation,
(2) water condensation,
(3) rain-induced erosion.

In this study only the UV radiation system was used. Ageing was carried out using UVA-340 lamps, which provide the best possible simulation of sunlight in the critical short wavelength region from 365 nm down to the solar cutoff of 295 nm. Its peak emission was at 340 nm. The radiation power was 0.89 W/m^2 and this was the value consistent with ASTM G154. The temperature that the samples reached was 50°C.

The tests were carried out for two types of PMC composites reinforced by:

- a fabric - Gurit prepreg EP121-C20-45, 0.23mm thick elementary layer,
- unidirectional fibers - Gurit prepreg EP137-CR527_100-35, having 0.1mm thick elementary layer.

For samples with the EP121 material, layering system 0_4 was used, and for material the EP137 it was $[0,90]_4$. Aging tests were carried out for 4 times: 144 h, 288 h, 432 h and 576 h. In the middle of each cycle (every 72 h), the chamber was opened and the samples were rotated to the other side, so that the ageing process was uniformly done for both sides. Each series consisted of 3 samples, and the reference samples were not subjected to the UV radiation. A total of 30 samples were tested (15 for EP121 and 15 for EP137).

Just after the first ageing cycle, yellow discolorations appeared on the surface. Therefore, after ageing tests, samples were subjected to microscopic observations (figure 1).

a)

b)

c)

d) e)
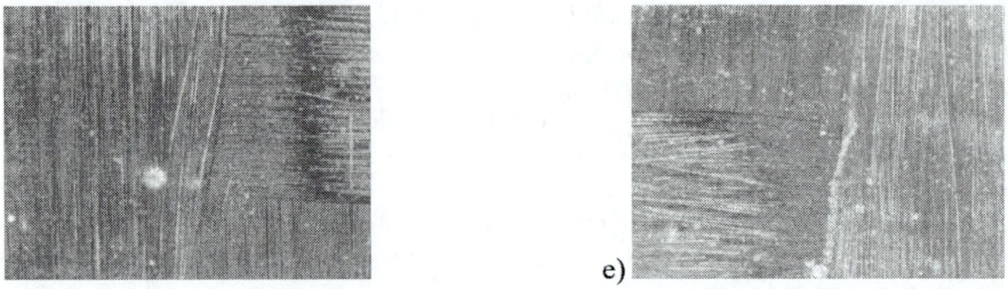

Figure 1. Microscopic observations: a) 0 h, b) 144 h, c) 288 h, d) 432 h, e) 576 h.

For the subsequent ageing cycles, the intensity of discoloration did not change, however, after wiping the sample surface with a white and wet fabric, the fabric became colored. This demonstrates the chemical destruction of the PMC matrix outer layer and the formation of microparticles [3]. Therefore, degradation may be accelerated if there will be a movement of the microparticles caused by water flow under gravity, which happens when condensation or rain occurs.

In [7] in addition to yellow discolorations, the authors also observed crack formations for the PMC materials with reinforcement by fibers made of nylon and polypropylene. Tests were carried out both for: (1) the UV radiation and (2) the moisture absorption. Photo degradation of the PMCs causes formation of ester, aldehyde, formate and propyl end groups. The UV exposure may bring positive or negative changes concerning the values of elastic modulus and strength.

3. Laboratory tests

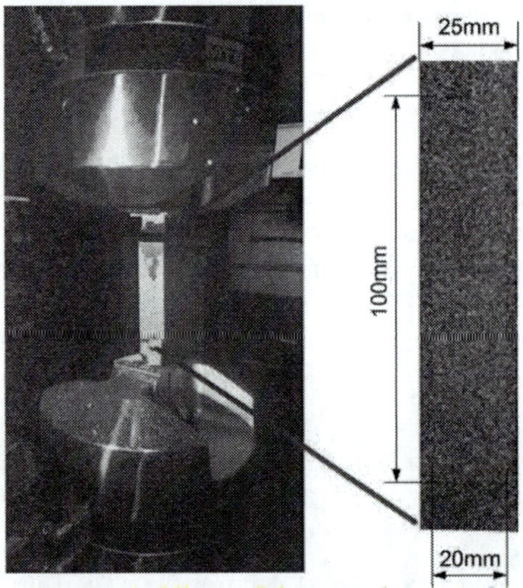

Figure 2. View of the sample central part after test and DIC observation areas

Laboratory tests were carried out for samples in accordance with the ASTM-D3039. The dimensions of the samples were 25 x 250 mm. The thickness for the EP121 material was 0.96 mm and for the EP137 0.84 mm. Both the wide and thickness of individual samples were measured with a micrometer with an accuracy of 0.001 mm.

Uniaxial tensile tests were carried out using the MTS 100 kN testing machine (figure 2) at a speed of 2 mm/min. The DIC system ARAMIS was coupled with testing machine, what allowed to determine displacement fields in the samples and further calculation of the elastic properties and strains of the considered PMC materials. The images were saved with frequency equal to 0.5 s.

4. Analysis of experimental results

The analysed PMCs behave as linearly elastic materials and their basic parameters like: the Young's modulus (E), the Poisson ratio (ν), tensile strengths (σ_{yy}^{f}), strains at failure (ε_{yy}^{f}) were collected in table 1.

Table 1. Results summary

	E [MPa]		ν [-]		σ_{yy}^{f} [MPa]		ε_{yy}^{f} [%]	
	EP121	EP137	EP121	EP137	EP121	EP137	EP121	EP137
0 h	52733	74807	0.050	0.035	732	1288	1.36	1.64
144 h	51232	74104	0.055	0.036	766	1343	1.43	1.72
288 h	53971	74871	0.054	0.035	800	1424	1.43	1.81
432 h	52761	74265	0.059	0.038	765	1307	1.41	1.66
576 h	53996	74378	0.066	0.039	748	1358	1.35	1.72

On the side edges of the samples we observed free fibers that did not carry load or they were deformed in an accidental directions. Therefore, we limited the measurements area to the central part of the sample, i.e. to the field 20 mm x 100 mm to omit introduction of errors (figure 2).

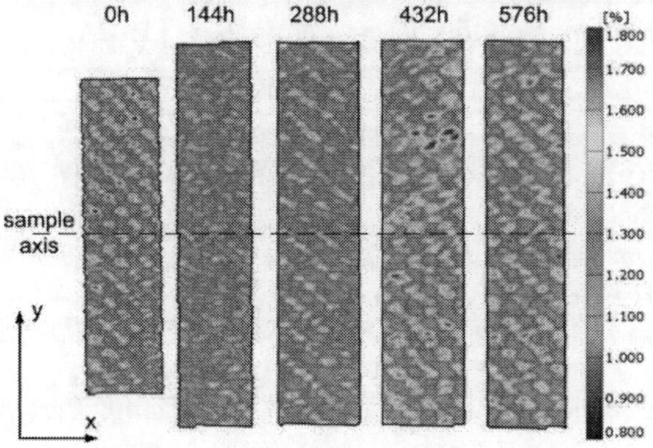

Figure 3. ε_{yy} strains in the EP121 for the axial force equal to 18,050 N

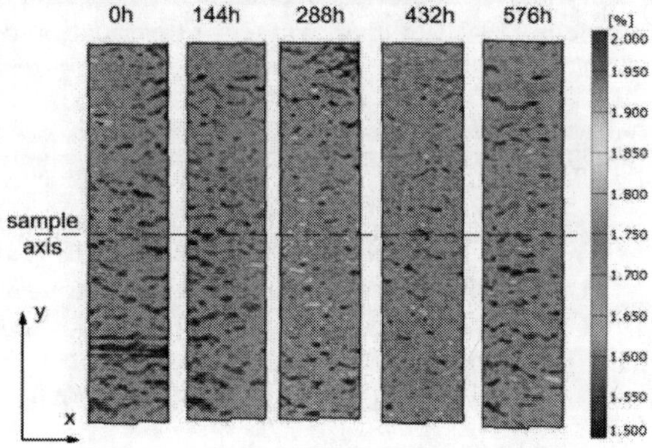

Figure 4. ε_{yy} strains in the EP137 for the axial force equal to 27,500 N

Figure 3. shows the normal strain fields ε_{yy} for the PMC reinforced by the EP121 for all ageing periods and the load level 18,050 N. For a better comparison, the scales of strain fields were unified to the range of 0.8% - 1.8%. Due to the fact, that reinforcement material is a twill fabric, we obtained characteristic strands at an angle 45°. Uniform values along the strands are distorted in particular for the 432 h sample. For the sample aged 576 h, the lowest strain value was observed. This indicates a disturbance during load transfer between the fibers and the matrix.

Figure 4 shows the normal strain fields ε_{yy} for the EP137 reinforcement at the load equal to 27,500 N. This force corresponds to the maximum effort in samples. The legend for each sample was unified to the range of 1.5% - 2%. The EP137 prepreg is in the form of unidirectional fibers, hence the deformation strands are arranged perpendicular to the sample axis. Only for the 288 h sample increased strain values can be observed close to the sample axis and equal to 2%.

5. Conclusions

In this paper, the results of UV ageing tests were presented for the two types of PMCs with reinforcements in the form of fabric and unidirectional fibers. The DIC system ARAMIS was used to observe in details changes of elastic properties during uniaxial tensile tests. The following conclusions can be formulated:

- just after the first ageing cycle in the EP121 material, discolorations appeared at strands interlacing area, i.e. in places rich in resin,

- discoloration are the results of matrix UV decomposition and are easy to remove. Therefore, if the additional condensation or rain-induced erosion occurs, subsequent layers of the matrix would be exposed and the ageing effect would be more visible. The occurrence of above mentioned phenomena will be included in next studies.
- taking into account the σ^{f}_{yy} tensile strength for both EP121 and EP137, there is an increase of 9.2% and 10.5%, respectively for 288 h, while for the 576 h the increase of strength is smaller and is around of 5.4% for both materials. As compared to the reference samples, strain at failure are also increased.
- it should be noted, that there is no UV influence when analyzing values of the Young's modulus for both materials and all ageing times.
- during the tensile tests for aged samples, characteristic cracking sound was heard much earlier than for reference samples. This indicates that such tests should be carried out using acoustic emission.

The paper will be extended to include degradation processes during deformation and formulation of analytical or numerical models. One can apply a continuum damage approach as it was done for example in [12-18]. Other manner of degradation description is multiscale modeling, e.g. [19-34].

References

[1] Pochiraju K, Tandon G and Schoeppner G 2012 *Long-Term Durability of Polymeric Matrix Composites* (Springer)

[2] Yan L, Chouw N and Jayaraman K 2015 *Materials and Design* **71** 17

[3] Lu T, Solis-Ramos E, Yi Y B and Kumosa M 2017 *Comp. Sci. Technol.* **153** 273

[4] Cysne B A, Fulco A, Guerra E, Arakaki F, Tosatto M, Costa M and Melo J 2017 *Comp. Part B* **110** 298

[5] Nicholas J, Mohamed M, Dhaliwal G, Anandan S and Chandrashekhara K 2016 *Comp. Part B* **94** 370

[6] Mouzakis D, Zoga H and Haliotis C 2008 *Comp. Part B* **39** 467

[7] Chevali V, Dean D and Janowski G 2010 *Polym. Degrad. Stabil.* **95** 2628

[8] Fiore V, Scalici T, Badagliacco D, Enea D, Alaimo G and Valenza A 2017 *Comp. Struct.* **160** 1319

[9] Han J-H and Kim Ch-G 2006 *Comp. Struct.* **72** 218

[10] Golewski P and Sadowski T 2018 *J. Europ. Cer. Soc.* **38** 2920

[11] Golewski P and Sadowski T 2017 *Arch. Metall. Mater.* **62** 2295

[12] Sadowski T and Samborski S 2003 *Comp. Mat. Sci.* **28** 512

[13] Sadowski T and Samborski S 2003 *J. Am. Cer. Soc.* **86** 2218

[14] Sadowski T and Samborski S 2008 *Comp. Mater. Sci.* **43** 75

[15] Marsavina L, Linul E, Voiconi T and Sadowski T 2013 *Polym. Test.* **32** 673

[16] Marsavina L, Linul E, Constantinescu DM, Apostol D, Voiconi T and Sadowski T 2014 *Eng Fract. Mech.* **129** 54

[17] Sadowski T, Marsavina L, Peride N and Craciun E M 2009 *Comput. Mat. Sci.* **46** 687

[18] Birsan M, Sadowski T, Marsavina L, Linul E and Pietras D 2013 *Int. J. Solids Struct.* **50** 519

[19] Nemat-Nasser S and Horii M 1999 *Micromechanics: overall properties of the heterogeneous materials* (Elsevier, Amsterdam – New York – Oxford – Tokyo)

[20] Sadowski T and Marsavina L 2011 *Comput. Mat. Sci.* **50** 1336

[21] Sadowski T 2012 *Comput. Mat. Sci.* **64** 209

[22] Postek E and Sadowski T 2011 *Comp. Interf.* **18** 57

[23] Sadowski T and Pankowski B 2016 *Comp. Struct.* **143** 388

[24] Marsavina L and Sadowski T 2007 *Int. J. Frac.* **145** 237

[25] Sadowski T and Golewski P 2012 *Comput. Mat. Sci.* **52** 293

IOP Conf. Series: Materials Science and Engineering **416** (2018) 012057 doi:10.1088/1757-899X/416/1/012057

[26] Dębski H and Sadowski T 2014 *Comput. Mat. Sci.* **83** 403
[27] Sadowski T and Golewski P 2012 *Comput. Mat. Sci.* **64** 285
[28] Nakonieczny K and Sadowski T 2009 *Comput. Mat. Sci.* **44** 1307
[29] Sadowski T and Nakonieczny K 2008 *Comput. Mat. Sci.* **43** 171
[30] Marsavina L and Sadowski T 2009 *Comput. Mat. Sci.* **45** 693
[31] Marsavina L and Sadowski T 2009 *Comput. Mat. Sci.* **44** 941
[32] Golewski G and Sadowski T 2014 *Const. Build. Mat.* **51** 207
[33] Sadowski T and Postek E, Denis C 2007 *Comput. Mat. Sci.* **39** 230
[34] Gajewski J and Sadowski T 2014 *Comp. Mater. Sci.* **82** 114

Acknowledgement

The authors would like to acknowledge the Wit-Composites (Poland) for samples preparation and dr M. Kneć for help in samples testing.

This work was financially supported by Ministry of Science and Higher Education within the statutory research number S/20/2018.

IOP Conf. Series: Materials Science and Engineering **416** (2018) 012058 doi:10.1088/1757-899X/416/1/012058

Comparison of the Corrosion Behaviour of Differently Deposited PVD TiN/TiO$_2$ Coatings on Electron Beam Modified Ti5Al4V Alloy

M P Nikolova[1,*], E Yankov[1], T Hikov[2], D Dechev[2], N Ivanov[2], S Parshorov[3], V Zaharieva[1] and P Petrov[2]

[1]University of Ruse "A. Kanchev", Dept. of Material Science and Technology, 8 Studentska Str., Bulgaria
[2]Bulgarian Academy of Sciences, Institute of Electronics, 72 Tzarigradsko Chausse Blvd, Sofia, Bulgaria
[3]Bulgarian Academy of Sciences, Institute of Metal Science, Equipment and Technologies with Hydro- and Aerodynamics Centre "Akad. A. Balevski", 67 Shipchenski Prohod Blvd, Sofia, Bulgaria

[*]Corresponding author: mpnikolova@uni-ruse.bg

Abstract. In the particular study, the micropatterning of the Ti5Al4V alloy in as-received, solution treated and solution and precipitated condition was done using electron beam surface modification (EBSM) by scanning electron beam. The nanostructured TiN/TiO$_2$ films were deposited over the EBSM Ti5Al4V substrates using dc magnetron sputtering and cathodic arc and glow-discharge methods. The electrochemical potentiodynamic tests were performed in Ringer saline at 37±0.05 °C and a pH of 5.7 in order to determine the corrosion resistance behavior of the coated and uncoated EBSM substrates as compared with the bare as-received, solution treated and solution and precipitated samples. The presented results gave an idea of the corrosion behavior of the examined implant systems in aggressive environment and external conditions close to the body right.

1. Introduction

The constituting materials for hard endogenous prosthesis should ensure the mechanical stability of the bone-implant system. The former should also lower the friction forces and enhance the anchoring thus permitting fast bone to implant fixation. The exclusively used α+β titanium alloys as supporting materials suffer from toxicity, low fatigue and wear at sliding resistance [1]. When using a functional coating that lowers the friction and stimulates osseointegration, some drawbacks of the titanium alloys could be overcome. Hard and thin TiN coatings generally present high tribological properties while the TiO$_2$ is among the mostly used biocompatible materials showing good cell adhesion [2]. The combination of nanoscale coating deposited on microscale roughened surface (known as the hierarchical modification) is thought to be favorable for the cell adhesion and formation of components of the extracellular matrix [3]. The electron beam surface modification (EBSM) gives the opportunity of obtaining reproducible isotropic textures with regular surface pattern without any surface contamination of the titanium alloy [4]. Simultaneously, the nanotexture of a PVD coating and the micropatterning of the surface could significantly change the corrosion resistance of the titanium alloys. The TiN and TiO$_2$ films are generally thought to be chemically inert. As the PVD deposition process usually generates many small defects such as pinholes which size and quantity are highly dependent on the preparation conditions, the corrosion behavior of a coated implant system will depend on the extent of these defects. Yilbas B S et al. [5] reported that an adequate defect-free and dense structure of a TiN coating could improve considerably the corrosion resistance of a material. Other authors [6] stated that the TiO_2 coatings improve the corrosion resistance by acting as protective layer on steel surface. According to Perillo P M [7], the charge carriers' density of a TiO_2 film on TiN one is significantly higher than that of TiO_2 formed on pure Ti. The titanium alloys themselves are able to form amorphous titanium dioxide that is stable over a wide range of potentials and pH [8]. In case of α+β titanium alloy, the β-phase preferentially dissolves during the long-term exposure in aggressive corrosion media [9]. Hence, treatments that change the phase composition of the alloy such as quenching, precipitation as well as electron-beam modification should affect the corrosion properties of the material. In this scenario, the intention of the particular work is to compare the electrochemical behavior of Ti alloy before and after the heat-treatment process and EBSM with those coated by magnetron sputtering and cathodic arc and glow-discharge methods.

2. Experimental procedures

The chemical compositions of the bare Ti substrate and the target material used for the experiment, determined by JEOL JXCA-733 Microprobe scanning electron microscope (SEM) equipped with wavelength dispersive spectrometers (WDS), were shown in Table 1. Samples with dimensions 14×14×4 mm were cut using the electro-erosion cutting method. The as-received (AR) samples were single solution treated (ST) for 30 min at 920 °C and water quenched. Half of the specimens were precipitated (P) for 4 hours at 500 °C and air cooled. All treatments were carried out in $\leq$ 1 Pa vacuum. The surfaces of the samples were grounded and polished before the EBSM. To study the effect of the surface treatment, the EBSM was applied to AR, ST, and ST and precipitated (ST+P) samples.

Table 1. Chemical composition (wt. %) of the substrate material.

Element	Al	V	Fe	Mn	Co	Cr	Mo	Pd	Nb	Hf	Ti
Substrate	5.21	4.40	0.14	0.11	0.06	-	0.17	0.15	0.36	0.04	Bal.

The EBSM was carried out by electron beam installation Leybold Heraeus (EWS 300/15–60). The following technological parameters were applied electron beam current – I = 20 mA, accelerated voltage – U = 52 kV, speed of the samples motion – v = 0.5 cm/s, electron beam frequency – f = 1 kHz, electron beam diameter - d = 0.5 mm and linear manner of scanning. Afterward, all samples were ultrasonically washed with absolute ethanol, acetone, isopropanol, rinsed with distilled water for 5 min and dried.

The TiN/TiO_2 coating was applied by two different methods - reactive magnetron sputtering (MS) and cathodic arc and glow-discharge method using Ti targets with purity of 99.8%. The deposition of the magnetron TiN layer took a place in the Ar-N_2 atmosphere, as the working pressure was 1.2×10^{-1} Pa, during the deposition process, the substrate had been heated to 350° C. The TiO_2 film was realized in the pure O_2 environment, the working pressure was 7×10^{-2} Pa, as the substrate temperature was decreased down to 180° C. The thickness of each coating (i.e. TiN and TiO_2, respectively) was about 1 μm assessed by the resonant frequency of quartz plate (Intermetalix IL800) coated within the same process conditions. In order to minimize the residual stresses and the oxidation, the samples with deposited coating were retrieved from the vacuum chamber after the achievement of room temperature.

For the cathodic arc deposition of the TiN, a sidewall positioned evaporating system in a cubic vacuum chamber with water-cooled walls was used. The samples were hanged near the center of a clockwise-rotating with a frequency of 0.5 Hz turntable. To ensure the coating stress relaxation and necessary adhesion, a very thin pure layer from the target (at 2.5×10^{-1} Pa in Ar atmosphere for 5 min., bias 600 V, 110 A arc current) was applied. The TiN film was made in a pure N_2 atmosphere at 340 °C substrate temperature for a time of 60 min., 110 A arc current, bias 250 V and 7.5×10^{-1} Pa pressure in the working chamber. The TiO_2 film was made by glow plasma discharge using the uppermost located sputtering pure Ti system in the same chamber. A bias voltage of 1340 V (720 mA current) in a pure O_2 atmosphere at a pressure of 6×10^{0} Pa were applied for a deposition time of 240 min. Both layers' thickness (TiN ~ 2.7 μm, TiO_2 ~ 0.8 μm) was attained by calotest measurements.

The X-ray phase identification was performed with URD-6 diffractometer, applying Bragg-Brentano geometry and Cu-Kα radiation and working at 30 kV and 20 mA. The diffraction angle range scanned was 2θ = 20-100° with a step size of 0.05° and counting time of 2.5 s/step. The light optical microscopy was performed on the top surface and cross-section of the samples after Kroll's reagent etching using a Nikon microscope equipped with a 14-megapixel digital camera.

The electrochemical study was held in experimental set-up comprised the electrochemical cell with a platinum counter electrode and a saturated Ag/AgCl reference electrode. Both electrodes together with the working one with 0.2 cm^2 exposed area were immersed in 80 ml naturally aerated Ringer solution (RS) for infusion (8,60 g/L NaCl, 0.30 g/L KCl, and 0.33 g/L $CaCl_2.2H_2O$) at 37±0.05 °C to simulate the in-body environment. The pH of the solution equal to 5.70±0.02 at 37° C was measured using a portable HI 8424 pH meter (Hanna instruments) that was fitted with a pH glass electrode (HI 1230B). The open circuit potentials (OPC) were measured at 0.25 s intervals as a function of the immersion time for 1000 s. After the OPC, potentiodynamic polarization curves were recorded by sweeping the potential starting at a cathodic potential about 250 mV below the E_{ss} (steady state potential) up to +2000 mV vs. Ag/AgCl at a scan rate of 1 mV.s^{-1} using a potentiostat 263A (EG&G Princeton Applied Research) coupled to a PC by a controller. All OPC and potentiodynamic measurements were repeated at least twice in order to ensure repeatability of the experimental results. The corrosion parameters E_{corr} (corrosion potential) and I_{corr} (corrosion current density) were obtained from the polarization curves by Tafel extrapolation method. I_{pass} (passivation current density) was attained by measuring the current density values around the middle of the passivation region of the polarization plot. The polarization resistance (R_p) was determined from the Stern-Geary equation.

3. Results and discussions

The micrographs of the AR, ST, and ST+P Ti54 together with the corresponding topographic images of the re-melted grains obtained on the surface after the EBSM of the material are shown in Figure 1.

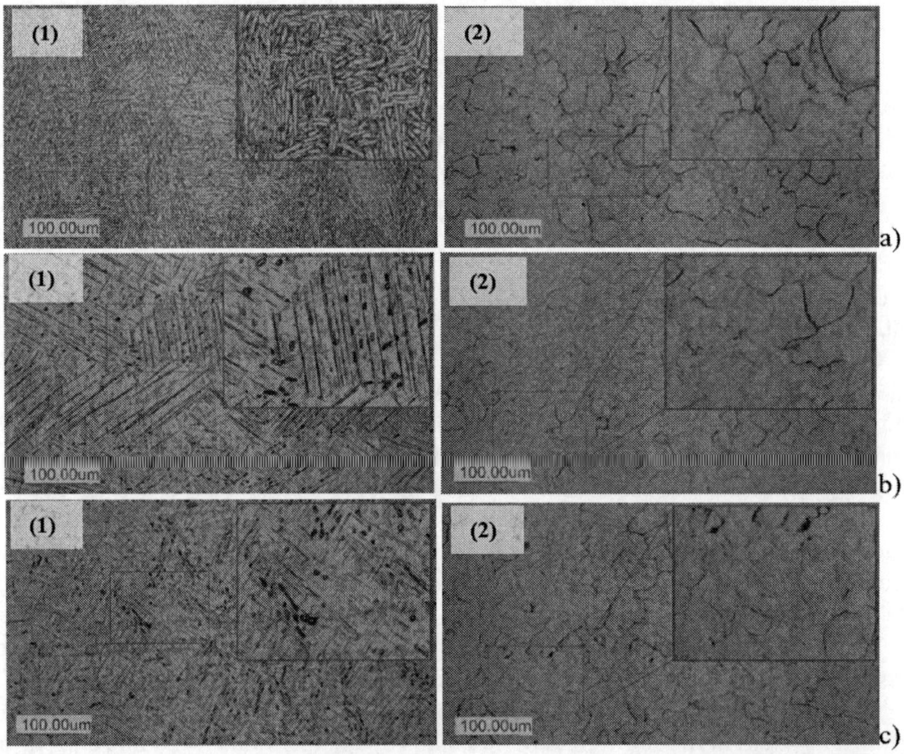

Figure 1. Micrographs of the AS (a), ST (b) and ST+P (c) samples before (1) and after the EBSM (2). The images marked with "1" are cross-sections of the AR and heat-treated samples, while the micrographs "2" represent topographic views after the EBSM.

IOP Conf. Series: Materials Science and Engineering **416** (2018) 012058 doi:10.1088/1757-899X/416/1/012058

The XRD spectra of all coated and non-coated materials are displayed in Figure 2. The coarse α+β lamellar structure of the AR material (Figure 1 a-1) after the ST from α+β range contained predominantly gross α'-martensite (Figure 1 b-1) formed due to the diffusionless β→α' transformation and some undissolved globular α-phases in between them. As the primary α phase and α'-martensite had the same inter-planar spacing of the HCP crystal structure, their distinction by XRD analysis is difficult. The precipitation after the ST led to element partitioning effect. As known, α'-martensite decomposes to the ordered Ti_3Al phase that was not detected by optical microscopy observations (Figure 1 c-1) but the shift of the ST+P α-reflections towards higher angle, as opposed to the ST sample (Figure 2), confirmed the decreased inter-planer spacing of the phase. The EBSM changed the XRD patterns of the material and the top surface morphology where well-defined boundaries of the smaller-sized grains were distinguished (Figure 1-(2)). Within these grains, fine martensite laths together with small pits at the grains' boundaries were present. The MS coatings deposited on the surface of the EBSM samples indicated only small anatase peaks of the surface oxide layer as well as strong (111) and (200) nitride reflections (Figure 2). In contrast, the glow-discharge deposited oxide on the arc-PVD nitride showed two distinctive TiO_2 phases – anatase and rutile, dominated by the rutile phase. The arc deposited nitride sub-layer mainly consisted of (200) and (220) textured crystal lattices.

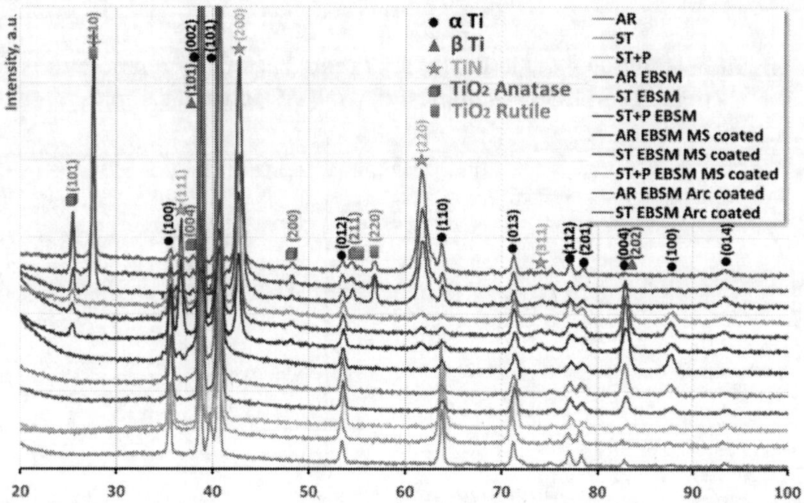

Figure 2. XRD patterns of the AS, heat treated, EBSM and coated samples by arc and MS methods.

The steady state potentials (E_{ss}) in time of the AR, heat-treated and EBSM samples (Figure 3 a) displayed a positive shift of the E_{ss} after precipitation of the alloy (from -475 mV to -320 mV) suggesting an increase in the corrosion resistance of the alloy in ST+P condition. Simultaneously, the ST+P EBSM curve was located at the top with the ST EBSM curve just below it. Hence, the re-melted and homogenized surface structures obtained after EBSM showed better corrosion resistance especially when the alloy was in ST and ST+P conditions probably because of the thicker passive layer. The E_{ss} of the coated samples (Figure 3 b) outperformed the potentials of the non-coated samples with the exception of the EBSM ST and ST+P specimens. The spontaneous passivation process of the AR EBSM Arc coated sample was interrupted by fluctuations probably because of short terms activation and passivation changes. In general, all OCPs of the coated samples showed close values varying between -16 and -92 mV vs. Ag/AgCl. However, the E_{ss} shift to more positive or negative values is not a certain criterion of low or high corrosion rate.

The coated and uncoated alloy displayed similar passive properties with low current density values suggesting low corrosion ratio. The corrosion potential of the ST (E_{corr} = -0.447 V) sample was slightly higher than that of the AR (E_{corr} = -0.486 V) while that of the ST+P was better (E_{corr} = -0.286 V) than that of the ST one (Figure 4 a). These findings could be attributed to the heterogeneous α+β composition of the AR alloy [4] resulting in lower corrosion resistance of the material because of galvanic cell formation [10]. The results suggest that both α and Ti_3Al phases that had similar crystallographic structures decreased the crystallographic mismatch of the phases existing within the lamellar structure of the AR and outperformed the corrosion resistance of the unstable acicular structure of the ST sample. Excluding the small peak in the current density of the ST+P sample at ~1.75V, the passivation behavior of all three samples represented lack of potential breakdowns in the anodic part of the potentiodynamic curve indicating high pitting corrosion resistance during the electrochemical measurements.

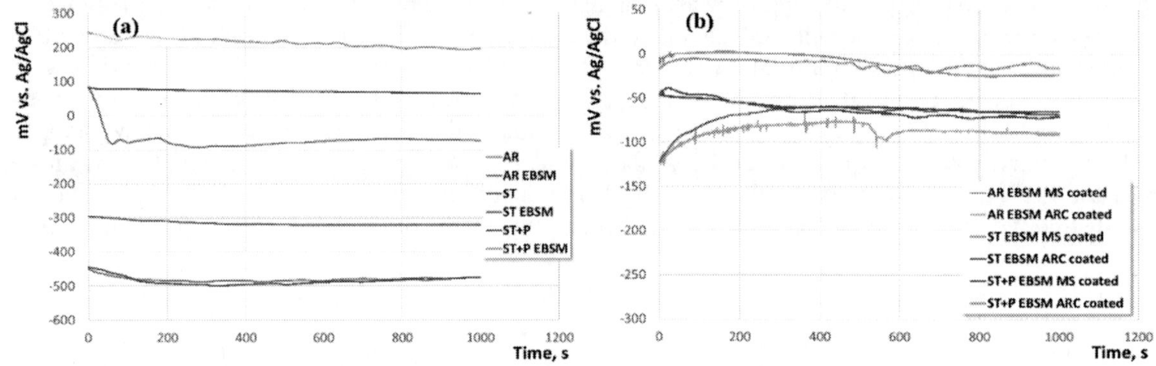

Figure 3. Open circuit potential $E = f(t)$ in RS (pH 5.7) and 37±0.05°C temperature of the: a) AR, heat treated and EBSM samples; c) TiN/TiO_2 coated samples

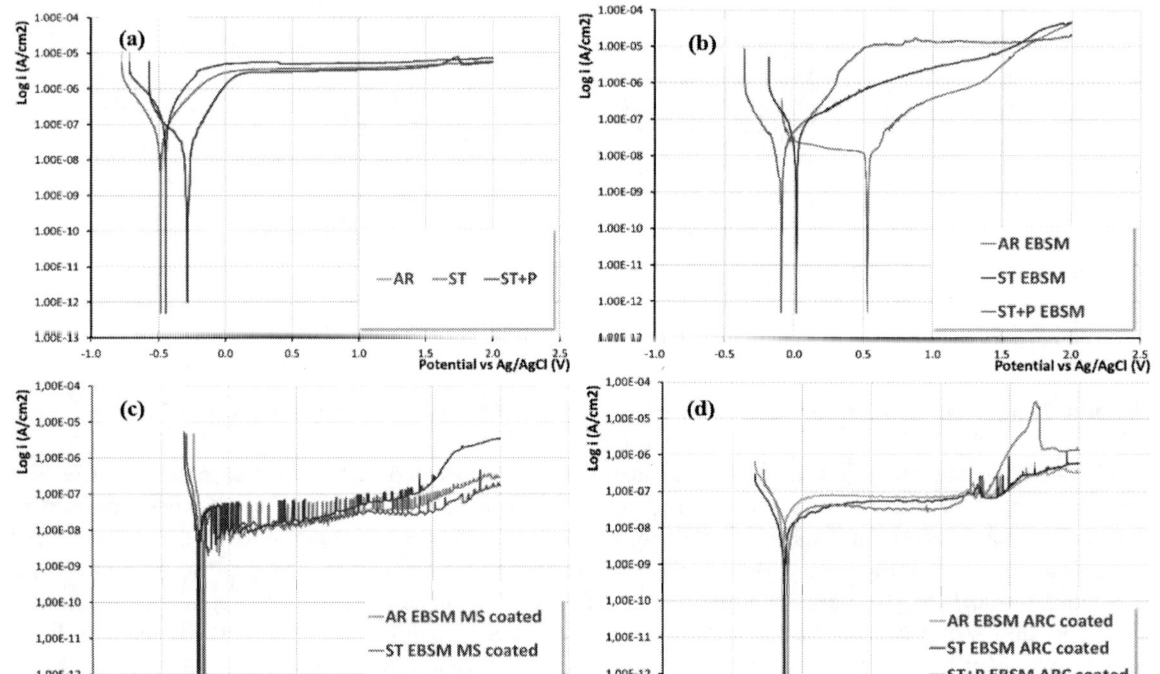

Figure 4. Potentiodynamic polarization curves $log/i/ = f(E)$ of the samples in a RS (pH 5.7) at 37±0.05 °C: a) AR and heat-treated alloy; b) EBSM alloy in AR, ST, and ST+P condition; c) MS coated EBSM alloy; d) Arc coated EBSM alloy.

All corrosion potentials of the EBSM samples indicated a tendency of more positive shift (Figure 4 b). Therefore, the evenly distributed alloying elements in all phases form uniformly distributed oxide clusters. The current densities of the ST and ST+P EBSM samples increased with the potential. In contrast, after breaking down of the initial film of the AR EBSM specimen, another stable state was formed but at higher passive current density (~13.3 μA/cm^2) as opposed to the AR (~3.9 μA/cm^2), ST (~5.3 μA/cm^2) and ST+P (~3.4 μA/cm^2). For the heat-treated and EBSM samples with smaller grains detected on the re-melted surface [4], the positive slope in the anodic current density indicated that the protective film was becoming less protective and more defective [7]. The intensification of the interaction of chloride ions with the surface could be associated with the increase in roughness of the EBSM samples while the lack of flat passive area could be due to the surface grains' morphology.

The ceramic structures of the MS and Arc deposited coatings (Figure 4 c and d) increased the R_p, charge transport and improve the electrochemical barrier properties of the surface. In contrast to the previous samples, the cathodic polarization curves of the coated samples did not show active dissolution areas. The oscillation in the current density of the MS coated samples in the passive region (Figure 4 c) could suggest the anatase broke down by micro-sized pitting nucleations and re-passivations (metastable pits formation [8]). These localized breakdowns were observed throughout the whole anodic parts of the curves. Hence, the electrochemical behavior of the MS deposited coating conforms to the morphology of the surface porosity. However, the passive corrosion currents of all MS coated samples had extremely low values indicating that a very protective film was formed.

The preliminarily formed anatase- and rutile-containing layers deposited on the surface of the Arc-obtained TiN (Figure 4 d) decreased the accessibility of the solution to the TiN that contains many diffusion paths around the droplets. The current density in the cathodic domain was almost one order of magnitude lower than the MS coated samples indicating decreased reduction reactions. The passive current density varied among ~0.076 μA.cm^{-2} and ~0.036 μA.cm^{-2} for the AR and St+P EBSM and Arc coated samples, respectively, indicating a higher corrosion resistance compared to the non-coated alloy. At potentials above ~1.2 V vs. Ag/AgCl, the current density increased forming peaks that some authors [11, 12] associated this with the formation of Ti compounds or structural changes inside the film [13]. In this area, localized breakdowns of the passive oxide by the anions were also detected in the potentiodynamic curve. In contrast to the MS coated samples, the microscopic breakdown events occurred after the passivation region (~ 0 - 1.25 V). The rapid increase of the current indicated the transition into transpassive state until stable pits were formed. At potentials higher than 1.5 – 1.8 V second passive region was observed at higher passivation current density. It is obvious that the oxide layers inhibit a process of anodic dissolution.

Table 2. Values of the electrochemical parameters: E_{corr} - corrosion potential; I_{corr} - corrosion current density; I_{pass} - passivation current density; R_p - polarization resistance; E_{ss} – steady state potential

Ti54 material	**E_{corr} (mV)** vs. Ag/AgCl	**I_{corr} (μAcm^{-2})**	**I_{pass} (μAcm^{-2})**	**R_p (MΩ)**	**E_{ss}, (mV)** vs. Ag/AgCl
AR	-486	0.04	3.86	0.20	-475
ST	-447	0.22	5.25	0.22	-475
ST+P	-286	0.017	3.33	0.24	-320
AR+EBSM	-104	0.02	13.30	-	-73
ST+EBSM	19	0.04	2.77*	-	64
ST+P+EBSM	552	0.015	0.39*	-	194
AR+EBSM MS coated	-190	0.034	0.034	1.8	-16
ST+EBSM MS coated	-219	0.030	0.029	2.7	-65
ST+P EBSM MS coated	-230	0.042	0.052	6.7	-71
AR+EBSM Arc coated	-140	0.084	0.072	6.6	-89
ST+EBSM Arc coated	-135	0.014	0.053	24.7	-68
ST+P EBSM Arc coated	-101	0.014	0.036	11.9	-25

* I_{pass} for ST+EBM and ST+P+EBM was determined at 1V.

The increase of E_{corr} seen in the coated and especially in the EBSM samples indicated that the material was less prone to corrosion (Table 2). With regard to the corrosion current density that is proportional to the corrosion rate, the ST+P, ST+P EBSM, and ST+P EBSM Arc coated samples showed slightly lower values compared to the ST sample, although of the same order of magnitude (Table 2). A tendency of decreasing the corrosion current density after heat treatment of the EBSM alloy was seen for the Arc coated materials while the corrosion current densities of the MS coated specimens displayed very close values. On increasing the potential, most samples formed protective anodic oxide films but the passive current of the coated specimens showed significantly lower values than the AR, heat treated and AR+EBSM samples. The R_p values which represents the resistance to corrosion in the particular solution indicated the highest values for the Arc coated samples and the lowest for the AR specimen. This implies that the coated samples have a very good corrosion resistance and low rate of ion release from the surface.

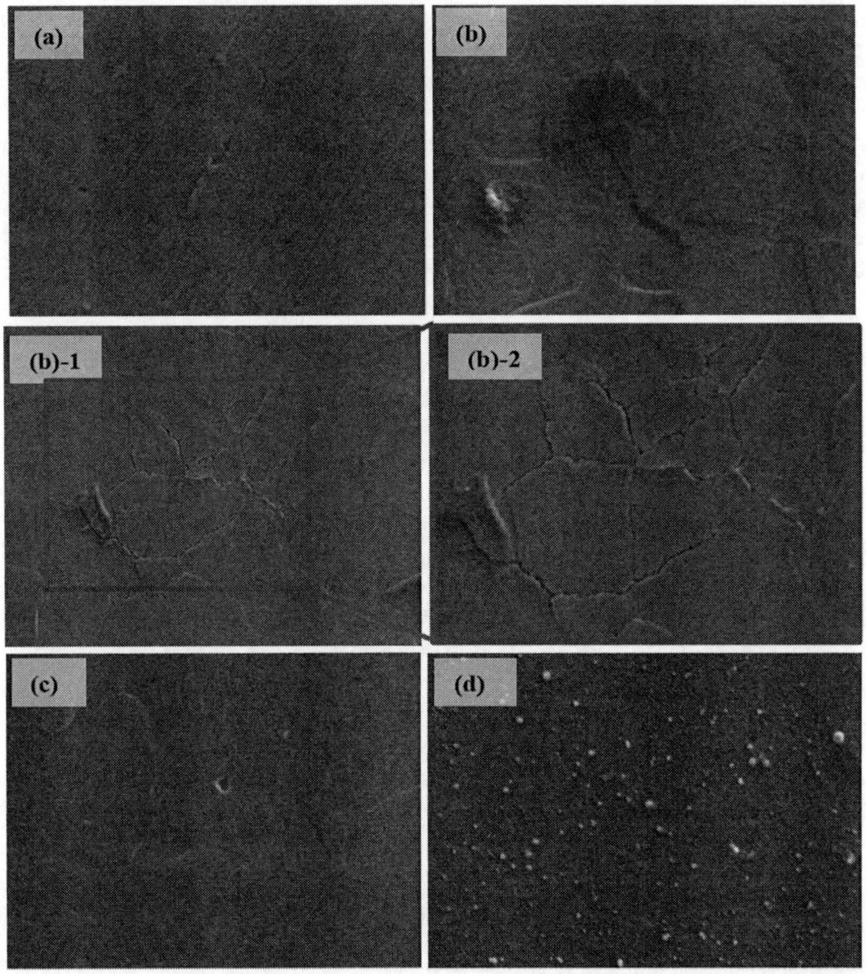

Figure 5. SEM micrographs (SE mode) of the surface of the samples (1000×) after the corrosion tests: (a) ST; (b, b-1 (600×) and b-2) different areas of the AR+EBSM surface; (c) AR+EBSM MS coated; (d) AR+EBSM Arc coated sample.

Figure 5 showing the SEM micrographs after the electrochemical tests, indicated the presence of localized defect areas. The detrimental aggressive attack of the Cl^- ions on the materials' surfaces caused the formation of lots of small pits over the entire surface of the ST sample. As seen in Figure 5 b, corrosion process triggered surface changes spread over larger areas of the AR and EBSM sample, predominantly located around the grain boundaries. Rarely, grain boundary pits surrounding and separating certain grains and some located inter-granularly (Figure 5b-1) were also noticed. Few pits could be observed on the surface of the MS coated sample (Figure 5 c) while those on the AR EBSM Arc coated specimen appeared to be smaller but with the higher amount (Figure 5 d), which explained its slightly higher I_{corr} values.

4. Conclusions

As the passivity of the titanium alloy occurs by the formation of a stable oxide film, the presence of chlorides in the solution enhances the stability and passivation and thus the innate resistance of the alloy in AR and heat-treated (ST and ST+P) condition. The ST+P treatment and EBSM strongly influence the OPC values of the Ti54 alloy in RS but the treatment with high energy fluxes makes the surface more susceptible to pitting corrosion especially for the finer grained ST and EBSM samples that show lack of passive area in the particular conditions. The improved values of the electrochemical parameters of the EBSM alloy could be attributed to the higher passivity of the homogenized electron-beam re-melted surface. However, these layers could not be perfect isolating barriers to ions release but they could act as an intermediate barrier that reduces the corrosion rate. This is so because as seen from the electrochemical parameters, the corrosion resistance is much dependent on the passivation ability of the substrate and the micro-defects in the coating. The RS stimulates pits formation on the coated samples but at different time periods – metastable pits in the whole anodic part of the MS coated samples and stable pits after 1.25 V anodic potential for the ARC and glow discharged oxidized samples. Further examination of the exact mechanism of the growing metastable pits in physiological solution should be performed. The results from the particular study show that the deposited coatings could be allocated to the noble metal surface treatments that could be used for facilitating biocorrosion prevention. Both complex TiN/TiO_2 coatings offer improved corrosion resistance and capability to withstand the aggressive conditions of the RS at 37 °C.

5. References

[1] Manam N S, Harun W S W, Shri D N A, Ghani S A C, Kurniawan T, Ismail M H and Ibrahim M H I 2017 *J. Alloys Compd* **701** 698-715

[2] Kommireddy D S, Patel A A, Shutava T G, Mills D K and Lvov Y M 2005 *J. Nanosci. Nanotechnol.* **5** (7) 1081-1087

[3] Misra S K and Boccaccini A R 2007 Biodegradable and bioactive polymer/ceramic composite scaffolds, *Tissue Engineering Using Ceramics and Polymers*, 72-92, ISBN: 9781845691769

[4] Nikolova M P, Yankov E, Petrov P, Valkov S, Ormanova M, Zaharieva V, Tonev D and Andreeva A 2017 Electron Beam Surface Modification Of Ti5Al4V Alloy For Biomedical Applications *Conf. Proc. Metal 2017*, pp 1555-59

[5] Yilbas B S, Sahin A Z, £oban A 1995 Wear and corrosion properties of PVD TiN coated Ti-6Al-4V materials, Transactions on Engineering Sciences **8**, ISSN 1743-3533

[6] Burnat B, Dercz G, Blaszczyk T 2014 J Mater Sci: Mater Med **25** 623–63

[7] Perillo P M 2015 Corrosion Behaviour of Titanium Nitride Coating on Titanium and Zircaloy-4, American Journal of Materials Science and Application **3** (2) 18-25

[8] Heakal F E-T, Shehata O S and Tantawy N S 2014 *Int. J. Electrochem. Sci.* **9** 1986-2004

[9] Atapour M., Pilchak A L, Frankel G S, Williams J C 2011 Corrosion behavior of β titanium alloys for biomedical applications, Mater Sci & Eng C **31** 885–891

[10] Song H-J, Kim M-K, Jung G-C, Vang M-S and Park Y-J 2007 *Surf & Coat Techn* **201** 8738–8745

[11] Assis S L, Wolynec S and Costa I 2006 Corrosion characterization of titanium alloys by electrochemical techniques *Electrochim Acta* **51** 1815–1819

[12] Pohrelyuk I M, Fedirko V M, Tkachuk O V and Proskurnyak R V 2014 Corrosion Resistance of Titanium Alloys with Oxynitride Coatings in Concentrated Inorganic Acids *Mater. Sci.* **50** (2) 269-276

[13] Zhu M, Wang R, Chen C, Zhang H and Zhang G 2017 Electrochemical study on the corrosion behavior of Ti_3SiC_2 in 3.5% NaCl solution, *RSC Advances* 7, 12534–12540

Acknowledgments

The present research was supported by a contract with National Science Fund (NSF) of Bulgaria, (D№ 07/3 (2016), Gradient functional nanocoatings produced by vacuum technologies for biomedical applications).

IOP Conf. Series: Materials Science and Engineering **416** (2018) 012059 doi:10.1088/1757-899X/416/1/012059

Description of Brittle Behaviour of Alumina/Zirconia 2 Phase Ceramics

T Sadowski[1], L Marsavina[2], E Craciun[3] and D Pietras[1]

[1] Department of Solid Mechanics, Lublin University of Technology, Nadbystrzycka 40 str., 20-618 Lublin, Poland
[2]Polytechnika University Timisoara, Department Strength of Materials, Blvd. M.Viteazu 1, Timisoara, 300222, Romania
[3] Ovidius University of Constanta, Faculty of Mechanical, Industrial and Maritime Engineering, Bd. Mamaia 124, 90052, Consatanta, Romania

Abstract. Alumina/zirconia 2 phase ceramics is innovative composite material applied for thermal barrier coatings of critical elements of jet engines. Its internal microstructure is complex as it consists of: (1) 2 phases of monolithic ceramics, (2) initial porosity and (3) other technological defects. Moreover, during loading process one can observe gradual microcracking, which is strictly connected with the loading path. The actual microstructure becomes much more complex. The crucial point in description of the behaviour of 2 phase ceramics is correlation of microcracking processes with the type of applied loading, for example: uniaxial tension, uniaxial compression or 2D and 3D states of stress. This different gradual composite degradation in tension or compression is typical for brittle materials. Here we describe both microcracking models of 2 phase ceramics associated with types of loading.

1. Introduction

Modern polycrystalline materials are created as mixture of different phases with various technological processes. Usually they have very complex internal microstructures, which include grains created from monolithic ceramic phases, system of interfaces or interphases and technological defects, e.g. [1, 2]. Examples of analytical or numerical modelling are presented in [3-14]. Modern nanoceramic composites have more complicated internal structures, e.g. [15, 16]. The similar situation is observed in case of so called functionally graded materials where the thermo-mechanical properties have continuous changes in 1 or 2 directions, e.g. [17-28]. Typically, ceramic matrix composite (CMC) consists of elastic phases. However, in case of advanced ceramic materials like cermets, interphases or interfaces between elastic grains are plastic e.g. [29-40].

Estimation of the overall properties of heterogeneous materials were described in [41-47]. Postulation existence of the Representative Volume Element (RVE) in microscale and assumption of statistical homogeneity of the CMC allows for formulation of general results relating to bounds of micropotentials and strain as well as complementary energies functionals. Assessment of the complex composites response containing: (1) different phases, (2) impurities and (3) different kinds of structural defects (dislocations, pores, cracks etc.) inside the material is possible by application of averaging procedure over the RVE.

The analysed alumina/zirconia ceramics can have different volume contents of both phases. However, the experimental investigations were done for small amount of zirconia up to 20%. To describe brittleness of the ceramic composite we considered different states of loading and cracks

IOP Conf. Series: Materials Science and Engineering **416** (2018) 012059 doi:10.1088/1757-899X/416/1/012059

propagation processes under pure tensile or compression. The matrix material is Al_2O_3 (phase I), whereas the phase II, i.e. ZrO_2 has different thermal expansion coefficient what causes initiation of small amount of microcracks and micropores during technological cooling. These initial microdefects cause stress concentrations, when material is subjected to loading. Microdefects can develop to longer mesocracks and finally can create dominant macrocracks leading to polycrystalline ceramics failure. We applied mesoscale model, using averaging procedure over the Representative Surface Element (RSE), for description of different cracking phenomena at micro- and meso- scales, which developed during different deformation stages.

2. Micromechanical model for 2 phase ceramic polycrystalline composites

Figure 1 presents advanced stages of deformations of the Al_2O_3/ZrO_2 polycrystalline ceramics with system of initial defects (e.g. pores, microcracks etc.) and developed complex state of different types of cracks. The whole deformation processes in analysed ceramics associated with nucleation and growth of different defects can be described as sequences of the following stages:

- creation of initial defects during technological cooling – micropores and microcracks are distributed inside grains or grain boundaries due to mismatch of the thermal expansion coefficient of 2 phases creating the composite polycrystal,
- development of initiated microcracks by their growth along the nearest grain boundaries due to the fact, that grain boundaries have lower value of fracture resistance in comparison to grains. This process ends in creation of mesocracks, which occupy the whole straight segment of grain boundaries.
- creation of the wing cracks when mesocracks start to kink and spread along the sequent segment of the polycrystalline grain boundaries system. In advanced stages of pure tensile or compressive deformations the polycrystalline material has complex internal structure, figure 1a and 1b.

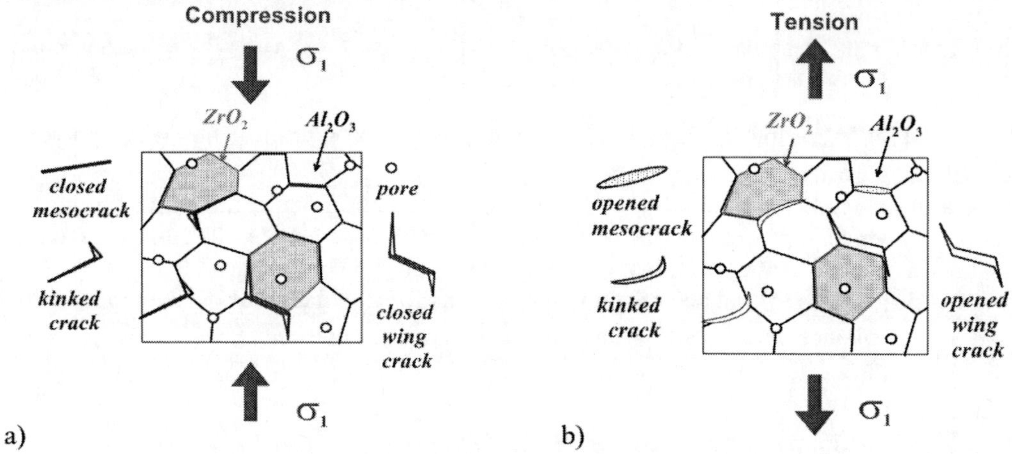

Figure 1. Advanced stage of deformation of the Al_2O_3/ZrO_2 polycrystalline ceramics - cracked microstructure due to a) compression, b) tension

All the above described staged of deformation can be expressed by the following constitutive equations using Voigt notation:

$$\{\bar{\varepsilon}_i\} = \{\bar{S}_{ik}^{\mathrm{CMC}}(\sigma_m, p_{\mathrm{ZrO_2}}, p_o, N^{(s)})\}\ \{\sigma_k\} \tag{1}$$

for $i,k = 1,...,6$. In (1) $\{\bar{\varepsilon}_i\}$ denotes the strain vector, $\{\bar{S}_{ik}^{\mathrm{CMC}}\}$ is the averaged elastic compliance matrix, $\{\sigma_k\}$ is the stress vector. Moreover, $p_{\mathrm{ZrO_2}}$ is the volume content of the second phase (ZrO_2), p_o is the

IOP Conf. Series: Materials Science and Engineering **416** (2018) 012059 doi:10.1088/1757-899X/416/1/012059

initial porosity parameter and $N^{(s)}$ are sets of numbers defining different kinds of discontinuities "s" developing inside the material, figure 1a and 1b.
In the advanced state of deformation the averaged elastic compliance tensor $\left\{\bar{S}_{ik}^{\mathrm{CMC}}\right\}$ can be decomposed to:

$$\left\{\bar{S}_{ik}^{\mathrm{CMC}}(\sigma_r, p_{\mathrm{ZrO_2}}, p_o, N_c^{(s)})\right\} = \begin{cases} (1 - p_{\mathrm{ZrO_2}} - p_o)\left\{S_{ik}^{\mathrm{Al_2O_3}}\right\} + \\ + p_{\mathrm{ZrO_2}}\left\{S_{ik}^{\mathrm{ZrO_2}}\right\} + \\ + \left\{S_{ik}^{p_o}(p_o)\right\} + \\ + \sum_{(s)}\left\{S_{ik}^{(s)}(\sigma_r, p_{\mathrm{ZrO_2}}, p_o, N^{(s)})\right\} \end{cases} . \quad (2)$$

$\left\{S_{ik}^{\mathrm{Al_2O_3}}\right\}$ is the compliance elastic matrix of the matrix material Al_2O_3, $\left\{S_{ik}^{\mathrm{ZrO_2}}\right\}$ describes the elastic properties of the ZrO_2. $\left\{S_{ik}^{p_o}(p_o)\right\}$ is the part of the compliance elastic matrix related to the existence of the initial porosity. The last part of (2) includes all different types of defects (micro-, meso- and macrocracks) inside the RSE.

The third part of the averaged elastic compliance tensor (2) $\left\{S_{ik}^{p_o}(p_o)\right\}$ describes initial porosity of the ceramic matrix composite. This part was estimates in [47] as:

$$\left\{S_{ik}^{p_o}\right\} = \begin{Bmatrix} 3/4 & -1/4 & -1/4 & 0 & 0 & 0 \\ -1/4 & 3/4 & -1/4 & 0 & 0 & 0 \\ -1/4 & -1/4 & 3/4 & 0 & 0 & 0 \\ 0 & 0 & 0 & 1 & 0 & 0 \\ 0 & 0 & 0 & 0 & 1 & 0 \\ 0 & 0 & 0 & 0 & 0 & 1 \end{Bmatrix} \frac{4[1-(v_o^{\mathrm{CMC}})^2]}{E_o^{\mathrm{CMC}}} \left[\frac{p_o}{1-p_o}\right]. \quad (3)$$

Here E_o^{CMC} and v_o^{CMC} are the Young's modulus and the Poisson's ratio of the initial polycrystalline ceramics without porosity.

As it is seen in figure 1, in the advanced stages of deformation one can distinguish the following types of defects, which appear inside the material: (1) opened mesocracks, (2) closed mesocracks, (3) wing cracks with opened central part, (4) wing cracks with closed central part.
Following [48-50] we can formulate constitutive equations for calculation of the averaged strain vector $\left\{\bar{\varepsilon}_i^{\mathrm{mc}}\right\}$ and the compliance matrix $\left\{S_{ik}^{\mathrm{mc}(s)}\right\}$ for single active opened mesocracks "s" developed inside the RSE:

$$\left\{\bar{\varepsilon}_i^{\mathrm{mc}}\right\} = \sum_{(s)}\left\{S_{ik}^{\mathrm{mc}(s)}\right\}\left\{\sigma_k\right\} = \left[\int_{D_m}^{D_M}\int_{\phi_l^{\mathrm{mc}}}^{\phi_2^{\mathrm{mc}}} M_{ij}^{\mathrm{mc}}(\phi^{(s)})\, \sigma_j(\phi^{(s)}) p_1^{\mathrm{mc}}(\phi^{(s)}) \left[\frac{D}{4}\right]^2 p_2^{(\mathrm{gr})}(D)\, \mathrm{d}\phi^{(s)}\, \mathrm{d}D\right] \frac{2\pi}{AE_o^{\mathrm{CMC}}} N_o^{\mathrm{mc}}, \quad (4)$$

where N_o^{mc} is a number of active mesocracks in the RSE of the area A. $D \in (D_m, D_M)$ and is the size of the grains inside the RSE. D_m and D_M are maximum and minimum value of grains diameters in the RSE. $M_{ij}^{\mathrm{mc}}(\phi^{(s)})$ is the matrix describing functions of inclination $\phi^{(s)}$ of the mesocracks "s" in relation to horizontal axis, figure 2. $p_1^{\mathrm{mc}}(\phi^{(s)})$ is the distribution function of the mesocracks inclination angle $\phi^{(s)}$. $p_2^{(\mathrm{gr})}(D)$ denotes the distribution function of grains length.

In case of closed mesocracks it is necessary to include friction coefficient μ along the cracks length, i.e. function $M_{ij}^{(\mathrm{mc})}(\phi^{(s)}, \mu)$ depends on 2 variables, e.g. [48-50].

$$\left\{\bar{\varepsilon}_i^{\mathrm{mc}}\right\}=\left[\int_{D_{\mathrm{m}}}^{D_{\mathrm{M}}}\int_{\phi_1^{\mathrm{mc}}}^{\phi_2^{\mathrm{mc}}} M_{ij}^{\mathrm{mc}}\left(\phi^{(s)},\mu\right)\,\sigma_j(\phi^{(s)})p_1^{\mathrm{mc}}(\phi^{(s)})\left[\frac{D}{4}\right]^2 p_2^{(\mathrm{gr})}(D)\,\mathrm{d}\phi^{(s)}\,\mathrm{d}D\right]\frac{2\pi}{AE_{\mathrm{o}}^{\mathrm{CMC}}}N_{\mathrm{o}}^{\mathrm{mc}}\,. \tag{5}$$

The wing cracks are presented in figure 2. They consist of 1 central mesocrack and 2 wings. For example, the mechanical response of mesocracks creating wings and additional strains due to wings opening for the tension (figure 2b) can be expressed by:

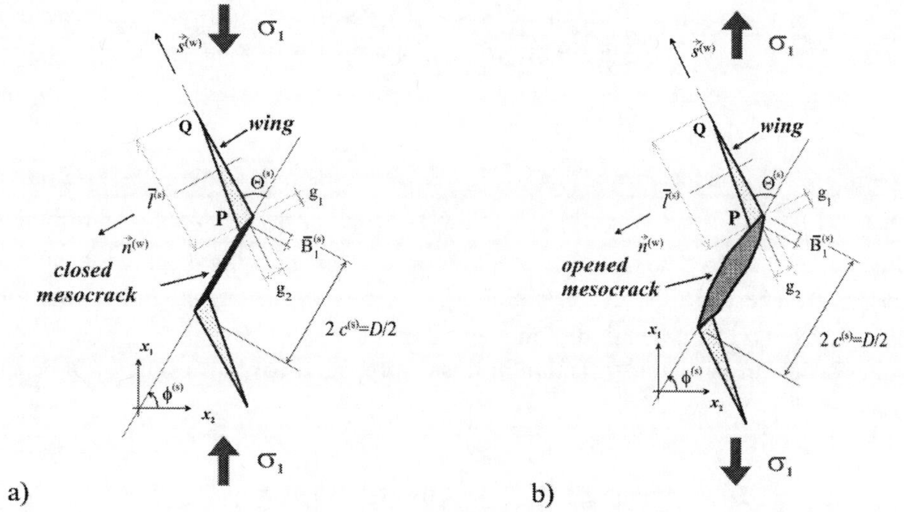

Figure 2. Advanced stage of degradation of the Al_2O_3/ZrO_2 described by wing crack models a) compression state, b) tension state

$$\left\{\bar{\varepsilon}_i^{\mathrm{wc}}\right\}=\left\{\begin{array}{l}\dfrac{2}{A}\displaystyle\int_{D_{\mathrm{m}}}^{D_{\mathrm{M}}}[\int_{\phi_1^{\mathrm{wc}}}^{\phi_2^{\mathrm{wc}}}\bar{M}_{ij}^{\mathrm{mc}}\left(\phi^{(s)}\right)\,\sigma_j(\phi^{(s)})\,p_3^{\mathrm{mc}}(\phi^{(s)})\left(\frac{D}{4}\right)^2\bar{p}_2^{(\mathrm{gr})}(D)\,\mathrm{d}\phi^{(s)}]\,\mathrm{d}D+\\ +\displaystyle\int_{\bar{l}_1}^{\bar{l}_2}p_{\mathrm{k}_1}(\bar{l}^{(s)})[\int_{\theta_1^{\mathrm{wc}}+\phi_1^{\mathrm{wc}}}^{\theta_2^{\mathrm{wc}}+\phi_2^{\mathrm{wc}}}M_{ij}^{\mathrm{wc}}\left(\phi^{(s)},\theta^{(s)}\right)\,\sigma_j(\phi^{(s)})\,p_{\mathrm{k}_2}(\theta^{(s)}+\phi^{(s)})\,\mathrm{d}(\theta^{(s)}+\phi^{(s)})]\bar{l}^{(s)}\mathrm{d}\bar{l}^{(s)}+\\ +\dfrac{c^{(s)2}}{A}\displaystyle\int_{\bar{l}_1}^{\bar{l}_2}p_{\mathrm{k}_1}(\bar{l}^{(s)})[\int_{\theta_1^{\mathrm{wc}}+\phi_1^{\mathrm{wc}}}^{\theta_2^{\mathrm{wc}}+\phi_2^{\mathrm{wc}}}M_{ij}^{\mathrm{wc}}\left(\phi^{(s)},\theta^{(s)}\right)\sigma_j(\phi^{(s)})\,p_{\mathrm{k}_2}(\theta^{(s)}+\phi^{(s)})\,\mathrm{d}(\theta^{(s)}+\phi^{(s)})]\,\bar{l}^{(s)2}\mathrm{d}\bar{l}^{(s)}\end{array}\right\}\frac{\pi N_c^{\mathrm{wc}}}{E_{\mathrm{o}}^{\mathrm{CMC}}}\,. \tag{6}$$

Here, $\left\{\bar{\varepsilon}_i^{\mathrm{wc}}\right\}$ is the averaged strain vector, N_c^{wc} denotes a number of the wing cracks inside the RSE. $\bar{M}_{ij}^{\mathrm{mc}}\left(\phi^{(s)}\right)$ is the matrix describing mesocracks which are able to create 2 wings at their tips, $M_{ij}^{\mathrm{wc}}\left(\phi^{(s)},\theta^{(s)}\right)$ - the matrix of spatial placement of the wing "s". $\phi_1^{\mathrm{wc}}\le\phi^{(s)}\le\phi_2^{\mathrm{wc}}$ is the fan of all mesocracks creating wings, $p_3^{\mathrm{mc}}(\phi^{(s)})$ - is the distribution function of the active mesocracks creating wings. $D/2=2c^{(s)}$ is the length of the straight grain boundary segment. $\bar{p}_2^{(\mathrm{gr})}(D)$ is the distribution function of D. $\bar{l}^{(s)}=l^{(s)}/c^{(s)}$ - is the nondimensional wing length, $\bar{l}_1\le\bar{l}^{(s)}\le\bar{l}_2$ - is the fan of the wings lengths inside the RSE. $\theta_1^{\mathrm{wc}}+\phi_1^{\mathrm{wc}}\le\theta^{(s)}+\phi^{(s)}\le\theta_2^{\mathrm{wc}}+\phi_2^{\mathrm{wc}}$ is the fan of inclination angles of the wings. $p_{\mathrm{k}_1}(\bar{l}^{(s)})$ and $p_{\mathrm{k}_2}(\theta^{(s)}+\phi^{(s)})$ are the distribution functions of the wings lengths and the inclination angles.

For the wing crack models with closed mesocrack (figure 2a) it is necessary to include friction coefficient μ and the functions: $M_{ij}^{(\mathrm{mc})}(\phi^{(s)})$ and $M_{ij}^{\mathrm{wc}}\left(\phi^{(s)},\theta^{(s)}\right)$ should be replaced by $M_{ij}^{(\mathrm{mc})}(\phi^{(s)},\mu)$ and $M_{ij}^{\mathrm{wc}}\left(\phi^{(s)},\theta^{(s)},\mu\right)$, respectively.

In order to describe the set of cracks development inside the RSE we used the strain energy release rate criterion. For one closed mesocrack or the wing crack we have:

$$G^{(s)}=\begin{Bmatrix} G^{\mathrm{mc}}(\sigma_r,c^{(s)},\phi^{(s)}) \\ G^{\mathrm{mc}}(\sigma_r,c^{(s)},\phi^{(s)},\mu) \\ G^{\mathrm{wc}}(\sigma_r,c^{(s)},\phi^{(s)},\bar{l}^{(s)},\theta^{(s)}) \\ G^{\mathrm{wc}}(\sigma_r,c^{(s)},\phi^{(s)},\bar{l}^{(s)},\theta^{(s)},\mu) \end{Bmatrix}=\frac{1-[\nu_{\mathrm{o}}^{\mathrm{CMC}}]^2}{E_{\mathrm{o}}^{\mathrm{CMC}}}\left(K_I^2+K_{II}^2\right)\geq 2\gamma_{\mathrm{gb}}^{\mathrm{CMC}}, \tag{7}$$

where, $G^{\mathrm{mc}}(\sigma_r,c^{(s)},\phi^{(s)},\mu)$ and $G^{\mathrm{wc}}(\sigma_r,c^{(s)},\phi^{(s)},\bar{l}^{(s)},\theta^{(s)},\mu)$ are the strain energy release rate function for the mesocrack "s" or the wing crack. $\gamma_{\mathrm{gb}}^{\mathrm{CMC}}$ is the surface energy between grain boundaries of the polycrystalline composite. In case of the opened mesocracks or the wing crack we have functions G^{mc} and G^{wc} with $\mu=0$. The analysis of crack kinking to the nearest grain boundary was presented for example in [51-62], but with application of numerical models.
K_I and K_{II} in (7) are stress intensity factors in tips of propagating cracks. They can be expressed for the single crack as:

$$K_I=\begin{Bmatrix} -D\left[\sin\phi^{(s)}(\cos\phi^{(s)}-\begin{pmatrix}\mu=0\\ \mu\sin\phi^{(s)}\end{pmatrix})\sin\theta^{(s)}\right]\Big/\left[\pi\left(l^{(s)}+l^{*}\right)\right]^{(1/2)}+ \\ +(\pi l^{(s)})^{(1/2)}\frac{1}{2}\left[1-\cos 2(\theta^{(s)}+\phi^{(s)})\right] \end{Bmatrix}\sigma_1 \tag{8}$$

$$K_{II}=\begin{Bmatrix} D\left[\sin\phi^{(s)}(\cos\phi^{(s)}-\begin{pmatrix}\mu=0\\ \mu\sin\phi^{(s)}\end{pmatrix})\right]\cos\theta^{(s)}\Big/\left[\pi\left(l^{(s)}+l^{*}\right)\right]^{(1/2)}+ \\ +(\pi l^{(s)})^{(1/2)}\frac{1}{2}\sin 2(\theta^{(s)}+\phi^{(s)}) \end{Bmatrix}\sigma_1. \tag{9}$$

In case of the opened mesocracks or wing cracks with opened central part $\mu=0$. Moreover, $l^{(s)}+l^{*}$ is so called equivalent length of wings, e.g. [41, 63].

3. Numerical results

To illustrate level of brittleness of the considered ceramic composite Al_2O_3/ZrO_2 we discussed tension and compression deformation processes. Both composite components: Al_2O_3 and ZrO_2 are characterized by the following initial properties.

- Al_2O_3 – the Young's modulus $E_0 = 400$ GPa and the Poisson's ratio $\nu_0 = 0.22$ and the fracture toughness $K_{Ic} = 4.5$ MPam$^{1/2}$,
- ZrO_2 - the Young's modulus $E_0 = 200$ GPa and the Poisson's ratio $\nu_0 = 0.25$ and the fracture toughness $K_{Ic} = 10$ MPam$^{1/2}$.

The mean diameter of the CMC structure was 10^{-5} m, whereas $D_{\mathrm{m}}=8\cdot 10^{-6}$m and $D_{\mathrm{M}}=12\cdot 10^{-6}$m

.

Figure 3 presents deformation processes under compression and tension for composite including 10% and 20 % ZrO_2. Characteristic points corresponding to the mesocracks and the wing cracks are specified. Application of the second phase (ZrO_2)leads to decrease of the load capacity of the Al_2O_3, however, increases its deformability and the fracture toughness. The compressive strength is approximately 10 times higher in comparison to tension. Additionally, a comparison with performed experiment confirms applicability of the model to modelling of the 2-phae CMCs. The experiments were done with sample of diameter 6 mm and height 20 mm with testing machine MTS 250 kN.

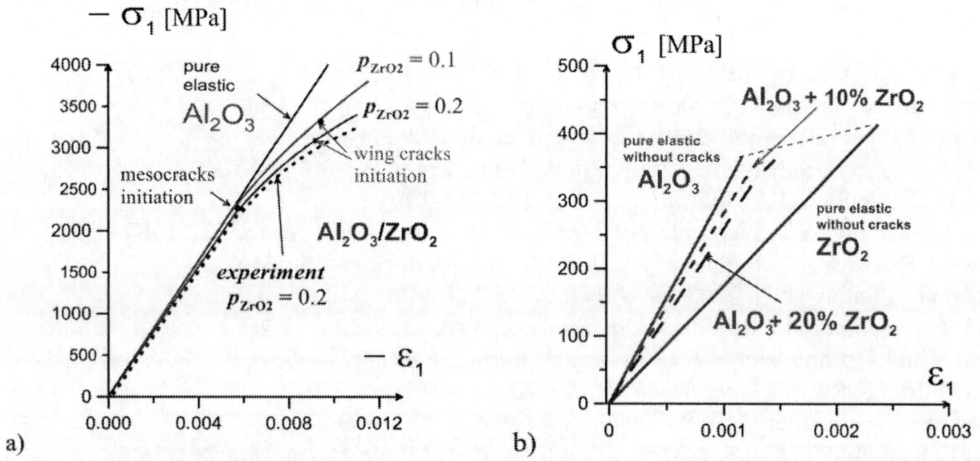

Figure 3. Deformation processes of the ceramic composite for: a) compression and b) tension

4. Final remarks

Performed numerical analysis and experimental tests with Al_2O_3/ZrO_2 ceramic composite lead to the following conclusions:

- The brittle behaviour of the composite is strong.
- Level of the volume contents of: (1) ZrO_2 and (2) porosity strongly influence the material behaviour under mechanical loading.
- Microcracks develop under mixed mode of fracture along straight segment of grain boundaries to form mesocracks,
- Futher growth of the mesocracks is by intergranular kinking to form the wing cracks.
- The dominant wing cracks can develop by intergranul or transgranular manner leading to the composite failure.

5. References

[1] Rice R W 1998 *Porosity of ceramics* (New York: Marcel Dekker Inc.)
[2] Fett T and Munz D 1999 *Mechanical Properties, Failure Behaviour, Materials Selection, Ceramics* (Berlin Heidelberg New York: Springer)
[3] Gömze L A and Gömze L N 2009 *Epitöanyag: JSBCM* **61** 38-42
[4] Sadowski T and Samborski S 2003 *J. Am. Cer. Soc.* **86** 2218-21
[5] Gömze L A and Gömze L N *IOP Conf. Ser.: Mater. Sci. Eng.* **47** 012033
[6] Sadowski T and Samborski S 2003 *Comp. Mat. Sci.* **28** 512-17
[7] Espinoza H D and Zavattieri P D 2003 *Mech. Mater.* **35** 333-64
[8] Espinoza H D and Zavattieri P D 2003 , *Mech. Mater.* **35** 365-94
[9] Sadowski T and Marsavina L 2011 *Comput. Mat. Sci.* **50** 1336-46
[10] Sadowski T 2012 *Comput. Mat. Sci.* **64** 209-11
[11] Sadowski T and Samborski S 2008 *Comp. Mater. Sci.* **43** 75-81
[12] Ghosh D, Banda M, Akurati S, Kang H and Fakharizadeh O 2017 *Scripta Mater.* **138** 139-44
[13] Sadowski T and Pankowski B 2016 *Comp. Struct.* **143** 388-94
[14] Sadowski T and Golewski G 2008 *Comp. Mat. Sci.* **43** 119-26
[15] Winter M 2002 *Nanocrystalline ceramics, synthesis and structure* (Berlin: Springer)
[16] Koch C C, Ovid'ko I A, Seal S and Veprek S 2007 *Structural nanocrystalline materials:*

fundamentals and applications (Cambridge: Cambridge Univ. Press)
[17] Suresh S and Mortensen A 1998 *Fundamentals of functionally graded materials* (Cambridge: The University Press)
[18] Birsan M, Sadowski T, Marsavina L, Linul E, Pietras D 2013 Int. J. Solids Struct. **50** 519-530
[19] Marsavina L, Linul E, Voiconi T, Sadowski T 2013 *Polymer Testing* **32** 673-680
[20] Sadowski T and Golewski P 2012 *Comput. Mat. Sci.* **52** 293-97
[21] Sadowski T and Golewski P 2012 *Comput. Mat. Sci.* **64** 285-88
[22] Nakonieczny K, Sadowski T 2009 *Comput. Mat. Sci.* **44** 1307-1311
[23] Sadowski T, Nakonieczny K 2008 *Comput. Mat. Sci.* **43** 171-178
[24] Sadowski T, Ataya S, Nakonieczny K 2009 *Comp. Mater. Sci.* **45** 624-632
[25] Sadowski T, Neubrand A 2004 *Int. J. Fract.* **127** 135-140
[26] Ivanov IV, Sadowski T, Pietras D 2013 *Europ. Phys. J. Special Topics* **222** 1587-1595
[27] Sadowski T, Ataya S, Nakonieczny K 2009 *Comp. Mater. Sci.* **45** 624-632
[28] Marsavina L, Sadowski T 2007 *Int. J. Frac.* **145** 237-243
[29] Siegl L S and Exner H E 1987 *Metall. Trans. A.* **18A** 1299-308
[30] Siegl L S and Fischmeister H F 1988 *Acta Metall.* **136** 887-97
[31] Ravichandran K S 1994, *Acta Metal. Mater.* **42** 143–50
[32] Hönle S and Schmauder S 1998 *Comput. Mater. Sci.* **13** 56-60
[33] Felten F, Schneider G and Sadowski T 2008 *Int. J. Ref. Mat. Hard. Mat.* **26** 55-60
[34] Li W, Wang H, Wang L, Hou C, Song X, Liu X and Han X 2017 *Mat. Res. Letters* **5** 55-60
[35] Postek E and Sadowski T 2011 *Comp. Interf.* **18** 57-76
[36] Sadowski T, Postek E, Denis C 2007 *Comput. Mat. Sci.* **39** 230-236
[37] Sadowski T, Hardy S, Postek 2005 *Comput. Mat. Sci.* **34** 46-63
[38] Sadowski T, Hardy S, Postek E 2006 *Mat. Sci. Eng. A* **424** 230-238
[39] Dębski H, Sadowski T 2014 *Comput. Mat. Sci.* **83** 403-411
[40] Gajewski J, Sadowski T 2014 *Comp. Mater. Sci.* **82** 114-117
[41] Nemat-Nasser S, Horii M 1999 *Micromechanics: overall properties of the heterogeneous materials* (Elsevier, Amsterdam – New York – Oxford – Tokyo)
[42] Gross D, Seelig T 2006 *Fracture mechanics with an introduction to micromechanics* (Springer, Berlin- Heidelberg)
[43] de Borst R, Sadowski T (Eds) 2008 *Lecture notes on composite materials* (Springer-Verlag, Wien-New York)
[44] Schmauder S, Mishnayevsky L 2009 *Micromechanical and nanosimulation of metals and composites* (Springer-Verlag, Berlin Heidelberg)
[45] Sadowski T, Trovalusci P (Eds) 2014 *Multiscale Modelling of Complex Materials* (Springer, Wien NewYork)
[46] Altenbach H, Sadowski T (Eds) 2015, *Failure and Damage Analysis of Advanced Materials* (Springer, Wien NewYork)
[47] Kachanov M 1993 *Advances in Appl. Mech.* **30** 259-445
[48] Sadowski T 1994 *Mech. Materials* **18** 1-16
[49] Sadowski T 1994 *Int. J. Damage Mech.* **3** 212-233
[50] Sadowski T 1995 *Int. J. Damage Mech.* **4** 293-318
[51] Marsavina L, Sadowski T 2009 *Comput. Mat. Sci.* **45** 693-697
[52] Marsavina L, Sadowski T 2009 *Comput. Mat. Sci.* **44** 941-950
[53] Golewski G, Sadowski T 2014 *Const. Build. Mat.* **51** 207-14
[54] Marsavina L, Linul E, Constantinescu DM, Apostol D, Voiconi T, Sadowski T 2014 *Eng. Fract. Mech.* **129**, 54-66
[55] Bieniaś J, Dębski H, Surowska B, Sadowski T 2012 *Comput. Mat. Sci.* **64** 168-172
[56] Golewski G, Golewski P, Sadowski T 2012 *Comput. Mat. Sci.* **62**, 75-78
[57] Dębski H and Sadowski T 2017 *Comp. Struct.* **159** 121-27
[58] Sadowski T, Neubrand A 2004 *Int. J. Fract.* **127** 135-140

IOP Conf. Series: Materials Science and Engineering **416** (2018) 012059 doi:10.1088/1757-899X/416/1/012059

[59] Sadowski T and Nowicki T 2008 *Comput. Mat. Sci.* **43** 235-41
[60] Sadowski T, Bec J 2011 *Comput. Mat. Sci.* **50** 1269-1275
[61] Burlaynko VN, Sadowski T 2012 *Comput. Mat. Sci.* **52** 212-216
[62] Sadowski T, Marsavina L, Peride N, Craciun E M 2009 *Comput. Mat. Sci.* **46** 687-693
[63] Horii H and Nemat-Nasser S 1985 **90** 3105-3112.

Acknowledgements
The results presented in this paper were obtained within the framework of research grant No. UMO/2016/21/B/ST8/01027 financed by the National Science Centre, Poland.

IOP Conf. Series: Materials Science and Engineering **416** (2018) 012060 doi:10.1088/1757-899X/416/1/012060

Computational Advantages of the Local Strain Energy Density for Fracture and Fatigue Design

F Berto[*] and L Sandnes

Department of Mechanical and Industrial Engineering, Norwegian University of Science and Technology, Trondheim, Norway

[*]Corresponding author: filippo.berto@ntnu.no

Abstract. This field of brittle fracture has engaged researchers from various fields of engineering from the early days until today. There are nowadays different approaches for the fracture and fatigue design. A high potential has been found in the strain energy density (SED) approach recently applied under different fracture and fatigue loading conditions. This approach has strong numerical advantages especially if applied to advanced materials and complex structures. Different problems and applications are presented in this paper, considering some recent outcomes at different scale levels, showing the possibility of considering a coarse mesh without any definition of a control volume. The potential of the method fits well with the application to connections of any type and in particular with welded joints. The main aim of this work is to compare the SED when applied to traditional welding processes as friction stir welding, and advanced technologies as hybrid bonding.

1. Introduction

In steel structures welding is a widely used and accepted joining technique. A weld line represents a geometrical discontinuity (a notch) and both weld toes and weld roots act as stress concentrators resulting in severe localized stresses, which strongly affect the static and the fatigue strength of a structure. The structural integrity of welded joints is further worsened by the manufacturing process, which may introduce defects, distortions and residual stresses that can be very detrimental for the fatigue and fracture behavior as shown by Fricke [1]. In this complex scenario, another challenge is the multiaxial stress state that usually is present in complex structures [2–4]. The ability to perform an accurate fatigue assessment of welded connections is an issue which must be properly addressed in order to avoid in-service unexpected failures. The criteria, which can be found in the current force standards and in the literature, vary depending on the parameters to be considered meaningful for the fatigue resistance. These criteria can be divided into different categories depending on the complexity of the stress analysis performed on structural details for the final calculation, as widely discussed by Radaj [5]. It is possible to distinguish criteria based on nominal stress, structural stress, local stress or other well-established methodologies concerning residual life assessments carried out on the basis of real defects/cracks/flaws or assumed crack-like defects, according to linear elastic fracture mechanics, as shown by Maddox [6] and Gurney [7]. Basically, according to Radaj et al. [8], it is possible to distinguish between global and local approaches. The former ones are less time consuming methods, whereas the latter are more advanced and powerful methods which require a large computational effort and high professional skills and competence. Despite this scenario, current standards are lacking in giving a real guidance on how to perform a reliable fatigue assessment. Still, most of them do not refer to any local concept and lead to use the nominal stress method. Even so, no recommendations exist on how to derive the nominal stress from a finite element model, so that it is left to the engineering assessment of a designer to establish which nominal stress is the suitable for performing a fatigue check, as recently pointed out by Hobbacher [9]: "At this point no clear definition of nominal stress does exist, nor a recommendation, how to derive it from the finite element stress plot. It is more than astonishing that several new design codes do not refer to any local concepts, leaving the designer alone with his modern computational equipment."

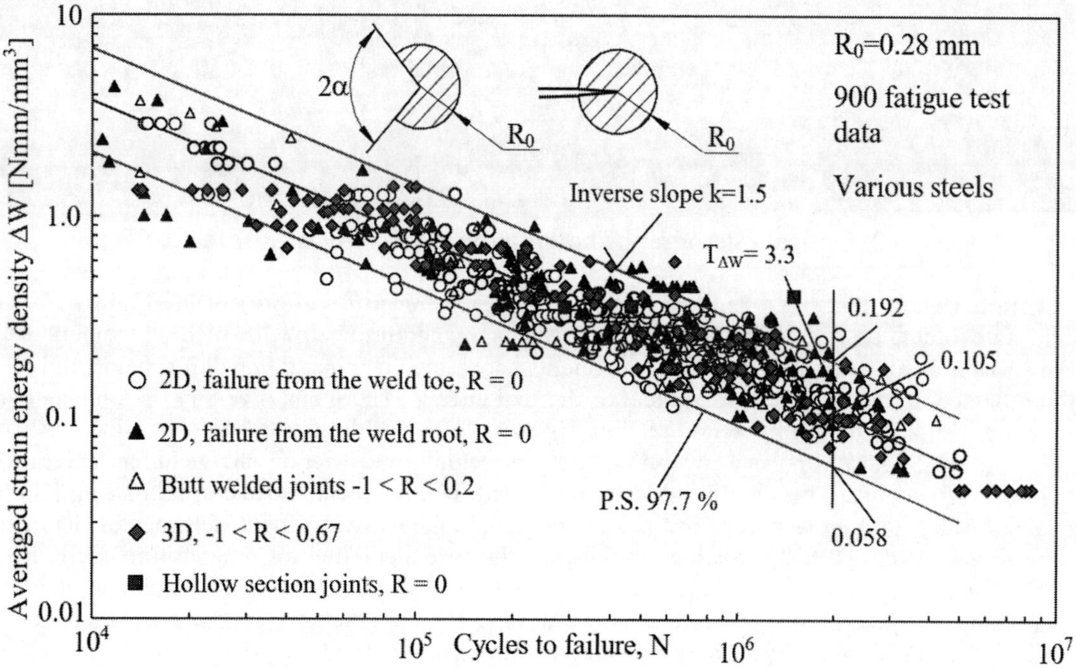

Figure 1. Fatigue strength of fillet welded joints made of steels as a function of the averaged local strain energy density; scatter band defined by mean value ± 2 standard deviations; fatigue crack initiation at weld toe or weld root; diagram taken from [15].

The strain energy density (SED) approach has been formulated in [10,11]. According to this criterion, the total strain energy density, averaged over a control volume surrounding the weld root or weld toe, is the parameter which controls the fatigue resistance of a joint. Failure occurs when the SED reaches a critical value, dependent on the material but independent of the weld geometry, which corresponds to the Beltrami criterion [12]. The underlying approach to which this idea can be related is Neuber's concept of elementary structural volume [13,14]. The critical volume is represented by a portion of a circular sector at the weld toe and a full circular sector at the weld root. In case of steel welded joints the radius of the control volume is equal to 0.28 mm (see Figure 1). Thereafter, it has been proved that SED can successfully predict brittle and high cycle fatigue failures of pre-cracked U- or V-notched specimens made out of several materials, including metals and ceramics, trying to propose SED as an all-around method [15-17]. This method keeps the robustness of the notch stress intensity approach, as shown by Livieri and Lazzarin [18], because closed form relations between the strain energy density and the relevant NSIFs exist and at the same time it does not require extremely refined meshes which is an impressive advantage dealing with complex structures [19, 20]. The advantage of the coarse mesh is surely amplified if the control volume has not been modelled in the real structure and this could play a fundamental role also in the automatic on time topology optimization of a complex component or structure subjected to in service loadings. Aim of this paper is to briefly discuss some advantages of the SED approach considering some recent forefront applications and emerging technologies: simulation based design, fracture and fatigue property assessment. All fields together will enable novel advanced design of complex parts, their direct conversion into products as well as the guarantee of their compliance through properly defined and tailored failure criterions. This interdisciplinary and integrative research will streamline the manufacturing workflow all the way from the idea to the final product and will enable designers and engineers to better transform their ideas into reality.

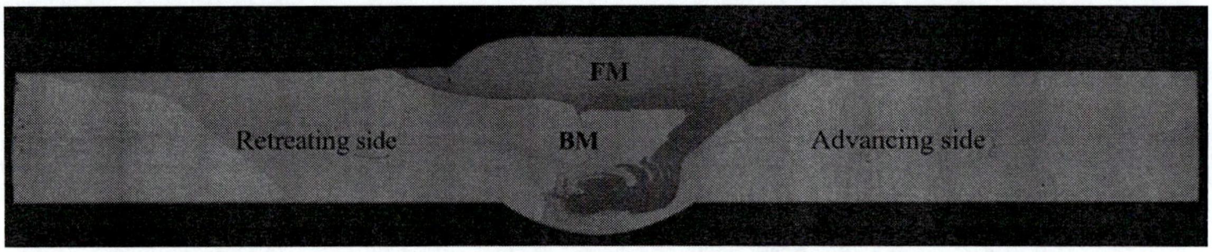

Figure 2. Optical micrograph showing the macrostructure of the HYB joint.

2. Application of the Strain Energy Density to a new generation of butt welded joints

Solid state joining processes produce sound joints at a temperature below the melting temperature of the parent material. Bonding between the components is obtained using extreme deformation and diffusion under the action of mechanical, electrical or thermal energy. There is a diversity of solid state joining processes available for commercial use. For instance, forge welding, ultrasonic welding, cut welding, explosion welding, cold pressure welding, friction welding and friction stir welding. The process of choice is dependent on the final product to be joined. Hybrid metal extrusion &bonding (HYB) is a novel solid state joining technique that has the potential to drastically reduce detrimental characteristics of traditional welding [20, 21]. By operating below the base metal melting temperature, it eliminates the emergence of a fusion zone (FZ), reduces residual stresses as well as the chances of intermetallic compounds (IMCs) compared to traditional welding. A typical geometry is shown in Figure 2.

Here, the HYB geometry has been reanalyzed in terms of the volume-free SED approach, under plane strain condition. Two different finite element software have been used: Ansys® and Straus7®. Linear plane finite elements have been adopted, namely Plane182 and Quad4 in Ansys® and Straus7®, respectively. In both software default element formulation and default free meshing criteria have been set. The finite element models are shown in Figure 2, where only a half of the geometry has been considered taking advantage of the symmetry conditions. It has been applied a nominal stress equal to 102 MPa, the Young's modulus and the Poisson's ratio equal to 70000 MPa and 0.3, respectively.

The stress singularity, as well as the mode 1 NSIF, have been computed in Ansys® by using a very refined mesh with the smallest element having size of the order of 10^{-5} mm and are shown in Figure 3. Afterwards, the volume-free SED approach has been used adopting free-generated meshes. The notch opening angle 2α is equal to 135°. The values of the strain energy density and of the NSIFs obtained by means of the volume-free approach (in both Ansys® and Straus7®) are reported in Table 1 and are compared to those obtained by means of the direct approach. The maximum error between the K_1 value accurately determined from the whole local stress field and by using the volume-free approach is 0.8% and 1.0% in Ansys® and in Straus7®, respectively (see Table 1). Obtained results show a robustness of the volume-free approach with respect to different software. This will open the doors to the SED application for topology optimization of complex structures.

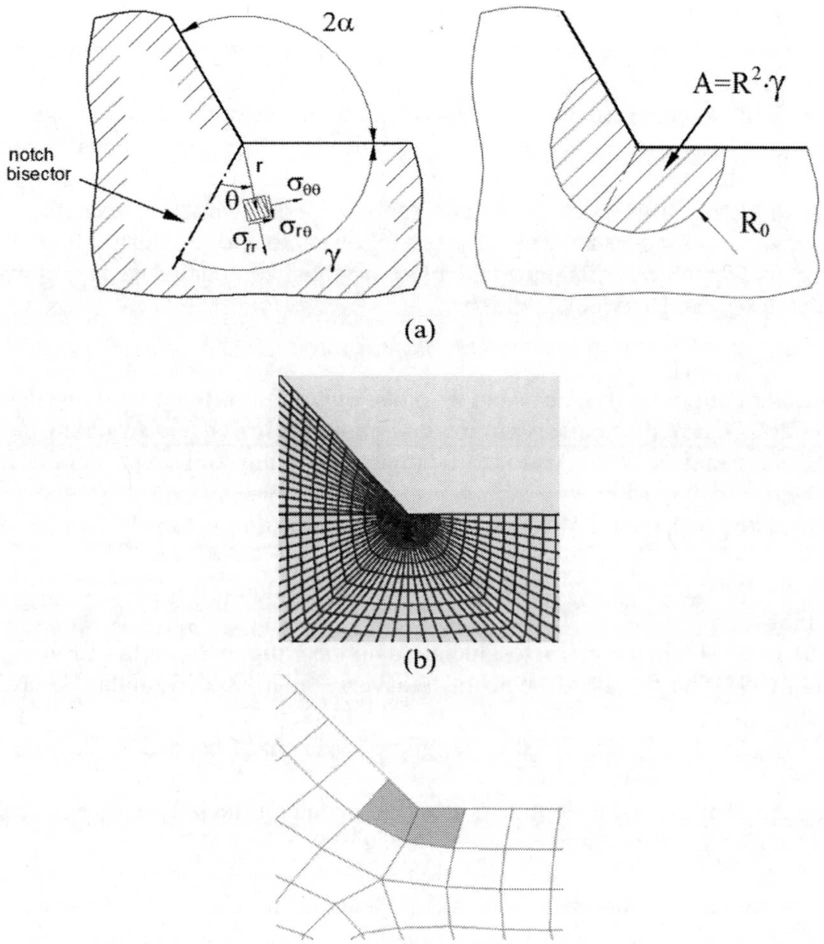

(c)

Figure 3. Butt welded joint. Very refined mesh generated in Ansys® to determine K_I from local stress field (b). Coarse mesh free-generated in Ansys® to determine K_I from the volume-free SED and Coarse mesh free-generated in Straus7® to determine K_I from the volume-free SED (c)

Table 1. Comparison between the values of the NSIFs calculated: from local stress, from the direct SED and from the volume-free SED, with different software. R_0=0.28 mm is considered.

	Software	K_1 [MPa mm$^{0.326}$]	Δ%
Local stress	Ansys®	151.2	
Direct SED	Ansys®	150.3	-0.6
SED Volume-free	Ansys®	152.3	0.8
	Straus7®	152.6	1.0

3. Conclusions

The fatigue assessment and service life of welded joints is an ambitious task, for which global and local approaches are available. Global approaches, which comprise the nominal stress and the hot spot stress methods, are generally recognized by the national and international design codes and are under permanent amendment. Local approaches, which include the strain energy density based approaches, are very powerful and if well applied very advantageous. In fact, they are useful for the design development in industry in respect of satisfactory reliability, higher strength and longer life. Additionally, they are important as a means of assessing strength and life of structural elements without reference to the global approach. This must take place if global approach data are not available in cases of unconventional design and service conditions.

4. References

[1] Fricke W 2003 Fatigue analysis of welded joints: State of development. *Mar Struct* **16** 185–200
[2] Marquis G 2007 Current Trends in Multiaxial Fatigue Research and Assessment. ICMFF9
[3] Fatemi A, Shamsaei N 2010 Multiaxial fatigue modeling and some simple approximations. ICMFF9
[4] Sakane M, Zhang S, Kim T 2011 Notch effect on multiaxial low cycle fatigue. *Int J Fatigue* **33** 959–68
[5] Radaj D 1990 Design and analysis of fatigue resistant welded structures. Cambridge: Woodhead Publishing
[6] Maddox SJ 1987 The Effect of Plate Thickness on the Fatigue Strength of Fillet Welded Joints
[7] Gurney TR 1991 The fatigue strength of transverse fillet welded joints. Cambridge: Abington Publishing
[8] Radaj D, Sonsino CM, Fricke W 2006 Fatigue assessment of welded joints by local approaches. Cambridge: Woodhead Publishing
[9] Hobbacher AF 2010 New developments at recent update of the IIW recommendations for fatigue of welded joints and components. *Steel Constr* 3
[10] Lazzarin P, Zambardi R 2001 A finite-volume-energy based approach to predict the static and fatigue behavior of components with sharp V-shaped notches. *Int J Fract* **112** 275–98
[11] Lazzarin P, Berto F 2005 Some expressions for the strain energy in a finite volume surrounding the root of blunt V-notches *Int J Fract* **135 (1-4)** 161-185
[12] Beltrami E 1885 Sulle condizioni di resistenza dei corpi elastici. *Rend Del Reg Ist Lomb* **XVIII** 704–14.
[13] Neuber H. Kerbspannungslehre (in German), 2nd Edition. Berlin: Springer Verlag; 1958.
[14] Neuber H. Kerbspannungslehre (in German), 3rd Edition. Berlin: Springer Verlag; 1985.
[15] Berto F, Lazzarin P 2009 A review of the volume-based strain energy density approach applied to V-notches and welded structures. *Theor Appl Fract Mech* **52** 183–94
[16] Berto F, Lazzarin P 2014 Recent developments in brittle and quasi-brittle failure assessment of engineering materials by means of local approaches. *Mater Sci Eng* **75** 1–48
[17] Livieri P, Lazzarin P 2005 Fatigue strength of steel and aluminium welded joints based on generalised stress intensity factors and local strain energy values. *Int J Fract* **133** 247–76
[18] Lazzarin P, Berto F, Gomez FJ, Zappalorto M 2008 *Int J Fatigue* **30** 1345–57
[19] Lazzarin P, Berto F, Zappalorto M 2010 Rapid calculations of notch stress intensity factors based on averaged strain energy density from coarse meshes: theoretical bases and applications. *Int J Fatigue* **32** 1559–67
[20] Ø. Grong, Method and device for joining of metal components, particularly light metal components. 2006, Google Patents.
[21] Aakenes UR, Grong Ø, Austigard T 2014 Application of the hybrid metal extrusion & bonding (HYB) method for joining of AA6082-T6 base material, *Materials Science Forum* **794** 339-344

IOP Conf. Series: Materials Science and Engineering **416** (2018) 012061 doi:10.1088/1757-899X/416/1/012061

Numerical Analysis of Two Types Polymeric Fibre Composite Materials with Different Reinforcement Architecture for Creation of Innovative Snap-fit Joints

P Golewski[1,*] and T Sadowski[1]

[1]Department of Solid Mechanics,
Faculty of Civil Engineering and Architecture,
Lublin University of Technology,
Nadbystrzycka 40, 20-618 Lublin, Poland

[*]Corresponding author: pgolewski@gmail.com

Abstract. In this study we analyzed application of two types reinforcements of the polymer matrix composites (PMC): (1) fabric prepreg, (2) unidirectional prepreg and their hybrids for production of modern snap–fit joints. The numerical calculations were made in Abaqus program. Moreover, different layer systems and geometries were proposed to get the most effective joints. For each kind of joints the opening forces were assessed and Tsai-Hill strength criterion for estimation of a damage parameter inside of the PMC was applied. The proposed methodology leads to designing of more optimal joints for various engineering structures.

1. Introduction

Snap-fit joints were used for many years in various fields of technology starting from everyday objects (e.g. battery cover of remote control unit) and ending with joints in building constructions [2]. Despite this, they have not been subjected to scientific research so far, in contrast to conventional joints: rivets, adhesive, whose disadvantage is the presence of a significant number of parameters affecting their strength [10, 12, 14, 15]. The first study on the construction and design of snap-fit joints is P. R. Bonenberger's textbook, but it only concerns components made of plastic. Despite the simplicity of their construction (the joint consist of only two pieces), they often have advantage shapes and can be manufactured not only from plastics but also from metals and fiber composites. The problem, which occurs during their manufacturing is the dimensional tolerance, which is difficult to achieve, e.g. due to plastic shrinkage. This causes appearance of excessive interference or clearance. In both cases, the joint will be poor quality. This issue was analyzed by the authors at work [3] and they proposed an optimized shape in which the joint has a constant force during its closing and opening.

In many cases, the snap-fit joints are inseparable, because they are situated inside the structure. Therefore, the current works are also focused on the use of shape memory materials [4, 6-7] for joint of beams. After heating the entire structure to appropriate temperature, the beams bend and thus two parts can be disconnected.

From the point of view of closing process it is necessary to consider a force and tactile feedback (FTF) which is important problem [5]. The force feedback refers to the force that must be overcome to close the joint, whereas the tactile feedback refers to vibrations that arise with an energy release during closing. Currently, joined parts are also manufactured using a 3D printing technology, what can increase popularity of the snap-fit joints. However, during printing, it is necessary to comply with the relevant rules [8].

The other material used for manufacturing of the snap-fit joints are aluminum alloys used in aviation [11, 13]. To prepare the proper shapes of the joints an incremental methods and a loss methods can also be applied. However up to now there is no laboratory tests in this area.

The other innovative materials used for manufacturing of the snap-fit joints are different types of polymer matrix composites (PMC) obtained for example with application of prepregs and cured in an autoclave. In this area, there are also no reports in the literature. Therefore, in this paper we presented results of numerical simulations for joints:

- made of different types materials and different reinforcement architecture (layer systems),
- having 3 shape geometries: a flat, a concave and a convex.

As a result of the simulations, the forces that needed to open the joints in stretching or shearing processes were determined. The obtained simulation results can be useful for designers of the snap-fit joints.

2. Analyzed models

The snap-fit joints are composed of two parts. Most often they have a beam shapes with a gripping section, which is fixed in a socket [2]. To modeling of the snap-fit joints behavior one can apply also a cantilever beam. The closing force of the joint depends on the beam stiffness and in order to guarantee uniform stress distribution in the cross section it is necessary to introduce variable rigidity.

This study focused on modeling of the gripping section, which is usually a flat surface, figure 1a. Depending on the desired value of the opening force, the beam's surface can be inclined to appropriate angle in relation to the socket. However, up to now, there is no solutions for including curvature in the gripping section of the snap-fit joint. Therefore, in this studies we analyzed 3 different shapes of the gripping part (figure 1): the flat, the convex and the concave.

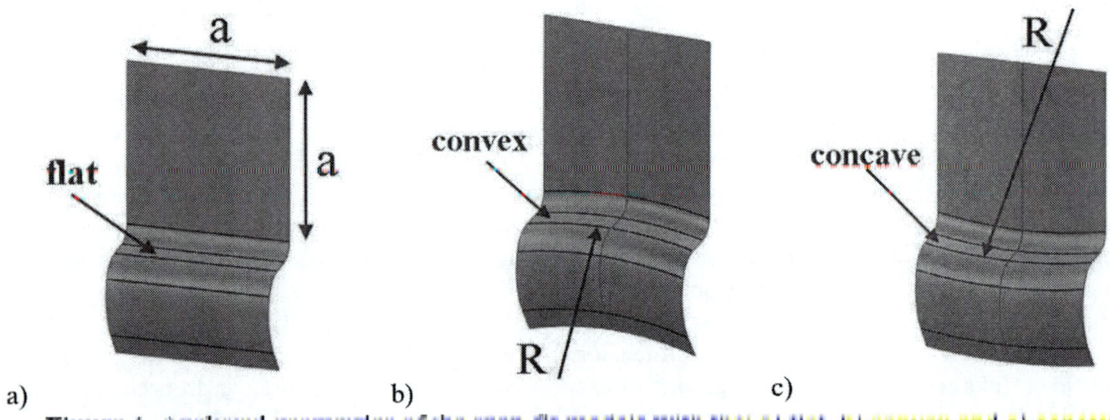

Figure 1. Analyzed geometries of the snap-fit models with the: a) flat, b) convex and c) concave shapes

Dimension "a" in all models was equal to 20 mm. In our models we used 3 values of radius "R" – 75 mm, 62,5 mm, 50 mm for creation of the concave and convex models.

3. Numerical model

Numerical tests were made in the Abaqus program. Totally of 17 simulations were carried out:

- 5 for the model with the flat gripping part (for tension),
- 6 for the model with the convex gripping part (3 for tensile and 3 for shearing tests),
- 6 for the model with the concave gripping part (3 for tensile and 3 for shearing tests).

Tensile and shearing tests were modeled as incremental deformation process in direction of displacement u1 or u2 applied to the upper edge of the snap-fit joints as shown in figure 2.

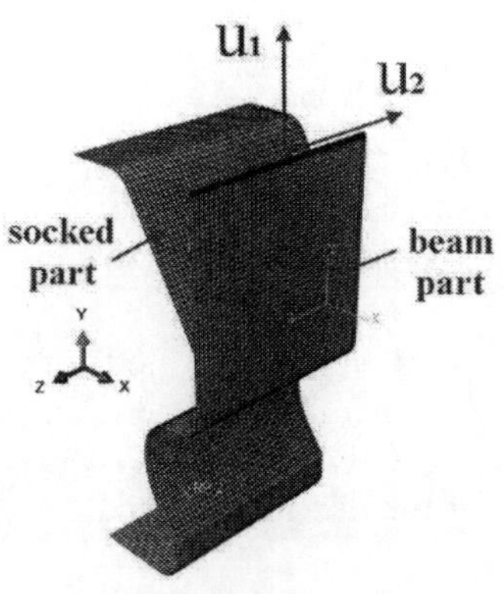

Figure 2. Mesh of finite elements

Due to the fact that both parts have a relatively small thickness, shell elements were used in an amount: 4960 S4R for the socket and 3200 S4R for the beam with a gripping part.

In simulations, the socket was treated as an infinitely rigid element which was realized by removing all degrees of freedom.

Two types of reinforcing materials in the PMC were used:

- a fabric - Gurit prepreg EP121-C20-45 - 0.23 mm thickness of the single elementary layer,
- unidirectional fibers - Gurit prepreg EP137-CR527_100-35 - 0.1mm thickness of the single elementary layer.

Five different layer configuration (systems) were considered for joints with the flat gripping part:

1 system - 2xEP121 0° - this is a reference PMC model,
2 system - EP137 0° / EP121 0° / EP137 0°,
3 system - EP137 0° / 2xEP137 90° / EP137 0°,
4 system - EP121 0° / EP137 0° / EP121 0°,
5 system - EP137 0° / EP137 45° / EP137 -45° / EP137 0°.

0° direction refers to a direction along the y axis. In simulations used general contact with friction coefficient equal to 0.1 between contacting surfaces of the socked and beam parts.

4. Numerical results of the analysis

The numerical analysis was divided into 3 stages:

(1) in the first stage, the snap-fit joint with the flat gripping part was analyzed with different arrangement of the composite layers. The joint was subjected to uniaxial tensile deformation until the force dropped completely. The reference model (model 1) had a 2xEP121 0° layer system. The EP121 material is made of a twill fabric and has the Young's modulus equal to 55 GPa, i.e. smaller value than in case of aluminum. In the system 2, to make the joint more rigid, two layers of unidirectional fibers (Young's modulus 148 GPa) were used in place of one layer twill fabric. The stiffness increased by about 50% and the weight decreased by about 11,5%. On the other hand, the stiffer joint causes an increase of stresses and for system 2 the permissible limit resulting from the Tsai-Hill criterion has been exceeded.

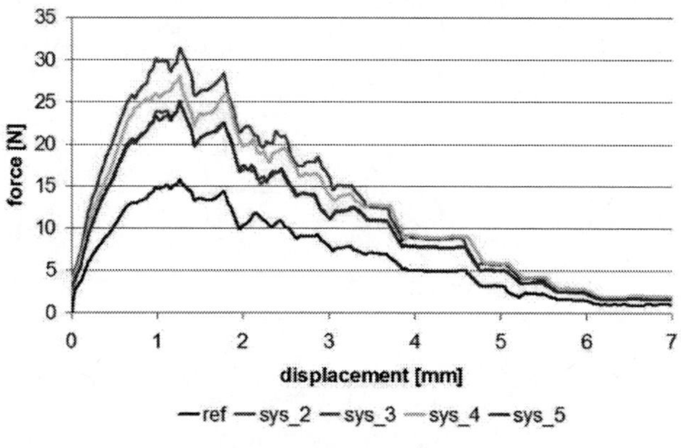

Figure 3. Force-displacement diagrams for the snap-fit joins with the flat gripping part

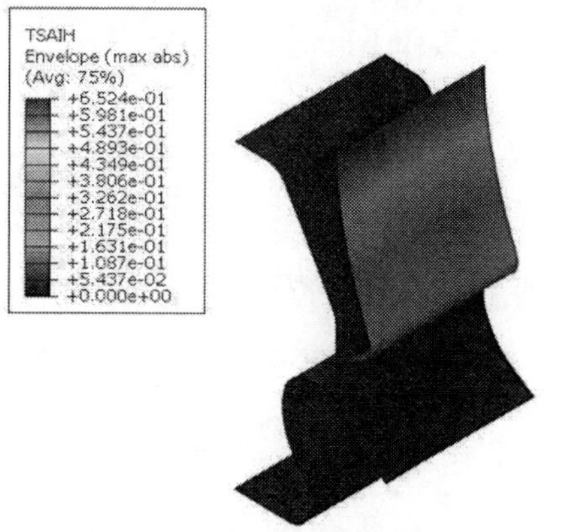

Figure 4. Tsai-Hill criterion (model 1)

IOP Conf. Series: Materials Science and Engineering **416** (2018) 012061 doi:10.1088/1757-899X/416/1/012061

A summary of the results for all systems is shown in figure 3. The fields for the Tsai-Hill criterion and thus the level of material effort are shown, for the system 1 in figure 4. In table 1, all results for 5 analyzed layer systems were collected.

Table 1. Summary of results for models with flat gripping part

	1 (ref) system	2 system	3 system	4 system	5 system
max Tsai-Hill	0,65	1,1	0,78	0,78	0,79
max force [N]	15,7	31,3	24,9	28,1	25,1
mass [g]	1,48	1,31	1,14	1,77	1,14

(2) in the second stage of the numerical studies, tensile deformation tests were performed for the concave and the convex joints. It should be emphasized, that the reference layer system for the convex and the concave model are shown in figure 5 and figure 6 respectively. However, in the both cases there is no clear effect of geometry change on the value of the maximum force.

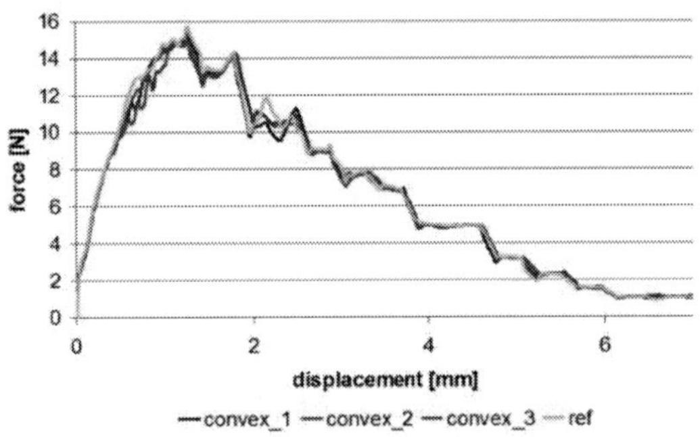

Figure 5. Force-displacement diagrams for the snap-fit joins with the convex gripping part

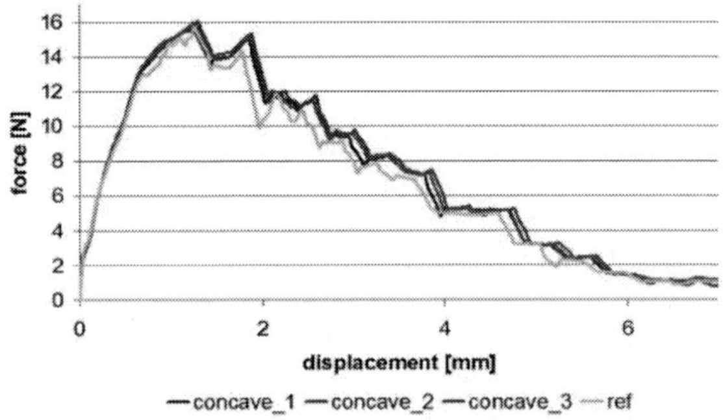

Figure 6. Force-displacement diagrams for the snap-fit joins with the concave gripping part

(3) the third stage of the numerical investigations the concave and convex joints were subjected to shear deformation tests along the axis "z". The joint with flat gripping part was not tested, because in this case the movement along the "z" axis is not limited, i.e. the force force is equal to zero. For the convex and the concave joints, we noticed change of the shear force when the radius "R" changes from 75 mm to 50 mm (figure 6 and 7). The maximum shear force is an order of magnitude smaller in comparison to the maximum tensile force. However, in many cases such strength reserve may be sufficient to protect the joint from slipping of the one parts.

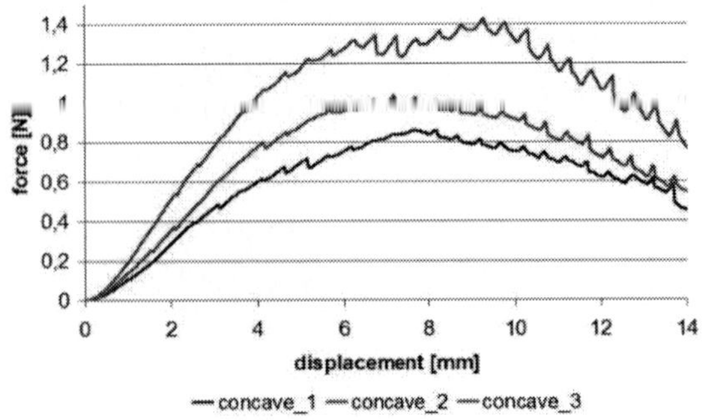

Figure 7. Concave models

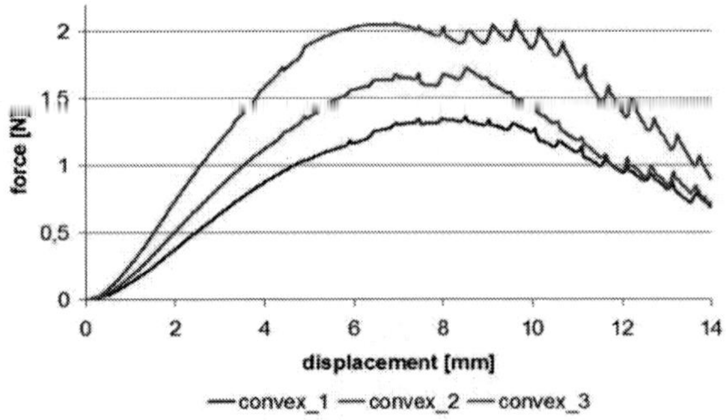

Figure 8. Convex models

Testing of the snap-fit joints in various load states is also justified if we consider connectors application simplifying the real shape of the joints. For large structural assemblies with complex shapes and connected by system of snap-fit joints, it is not possible precise modeling of this structural element due to complicated geometrical shape. But knowing the characteristics of the single joint it is possible to reduce this system of snap-fit joints to the new one simplified by application set of connectors.

5. Conclusions

The paper provides information about influence of different reinforcement architecture of the PMC and geometric shapes on the strength of the snap-fit joints. As distinguished from plastic and metals the PMC provide significantly wider design possibilities. From the simulations carried out the following conclusions can be formulated for models with flat gripping part:

- replacement on one fabric layer (model 1) into two layers of unidirectional fibers (model 2) doubles the opening force,
- except for model 2, the material effort estimated by the Tsai-Hill hypothesis not exceeded limiting value and was in the range of 0.65 – 0.79.

In case of the tensile deformation tests, the shape change of the gripping part does not influence substantially the carrying force. However, if the joint would work in a complex load state, e.g. shear state due to twisting of the whole structure, the following conclusions were drawn:

- it is better to use models with the convex gripping part. Then the increase of the joint opening force is about 45% higher with respect to the concave models,
- the radius "R" (figure 1b and 1c) have a substantial influence on the shear joint strength. The radius change from 75 mm to 50 mm results in increase of the joint strength by about 67% for the concave models and about 50% for the convex models.

The numerical model can be extended to include damage or cracking process, as it was done for other composite materials, e.g. [16 – 26].

6. References

[1] Bonenberger P R 2005 *The first Snap-Fit Handbook, Creating and Managing Attachments for Plastic* (Hanser Publisher, Munich)
[2] Patent N0.: US008397461B2
[3] Chen Y H and Lan C C 2012 *Mechanism and Machine Theory* **55** 34
[4] Li H, Jin K, He B and Chen Y 2012 *CIRP Annals - Manufacturing Technology* **61** 31
[5] Rusli L, Luscher A and Sommerach C 2010 *Int. Journal of Industrial Ergonomics* **40** 618
[6] He B, Li H and Jin K 2013 *Materials and Design* **47** 539
[7] Carrell J, Tate D, Wang S and Hang H-Ch 2011 Journal of Cleaner Production **19** 2066
[8] Klahn C, Singer D and Meboldt M 2016 *Procedia CIRP* **50** 264
[9] Roos M, Horn B, Gramlich S, Ulbrich S and Kloberdanz H 2016 *Procedia CIRP* **50** 123
[10] Sadowski T and Golewski P 2017 *Arch. Metall. Mat.***62** 1577
[11] Sadowski T, Golewski P and Radoiu V 2016 *Solid State Phen.* **254** 1
[12] Golewski P and Sadowski T 2017 *Int. J. Adhesion and Adhesives* **77** 174
[13] Sadowski T and Golewski P 2015 *Arch. Metall. Mat.* **60** 2743
[14] Sadowski T and Golewski P 2014 *Key Eng. Mat.* **607** 49
[15] Sadowski T and Golewski P 2013 *Arch. Metall. Mat.* **58** 579
[16] Marsavina L and Sadowski T 2007 *Int. J. Frac.* **145** 237
[17] Sadowski T and Marsavina L 2011 *Comput. Mat. Sci.* **50** 1336
[18] Golewski G and Sadowski T 2014 *Const. Build. Mat.* **51** 207
[19] Sadowski T and Samborski S 2003 *Comp. Mat. Sci.* **28** 512
[20] Marsavina L, Linul E, Voiconi T and Sadowski T 2013 *Polym. Test.* **32** 673
[21] Sadowski T and Samborski S 2003 *J. Am. Cer. Soc.* **86** 2218
[22] Postek E and Sadowski T 2011 *Comp. Interf.* **18** 57
[23] Sadowski T and Golewski P 2012 *Comput. Mat. Sci.* **64** 285

[24] Gajewski J and Sadowski T 2014 *Comp. Mater. Sci.* **82** 114
[25] Sadowski T and Golewski P 2012 *Comput. Mat. Sci.* **52** 293
[26] Dębski H and Sadowski T 2014 *Comput. Mat. Sci.* **83** 403

Acknowledgement

Financial support of Structural Funds – European Regional Development Funds (ERDF) Project No: INNOLOT/I/5/NCBiR/2013; program "INNOLOT – Innovative Aviation" coordinated by the National Center for Research and Development (NCBiR – Poland); Title: "Block Structures – Mechanical joining innovations to replace conventional fasteners in aerostructures" Period: 1.12.2013 – 31. 11.2018 is gratefully acknowledged.

This work was financially supported by Ministry of Science and Higher Education within the statutory research number S/20/2018.

IOP Conf. Series: Materials Science and Engineering **416** (2018) 012062 doi:10.1088/1757-899X/416/1/012062

Changes in the Mechanical Properties of Ti Samples with TiN and TiN/TiO_2 Coatings Deposited by Different PVD Methods

E Yankov [1*], M P Nikolova[1], D Dechev[2], N Ivanov[2], T Hikov[2], S Valkov[2], V Dimitrova[3], M Yordanov[3], and P Petrov[2]
[1]University of Ruse "A. Kanchev", Dept. of Material Science and Technology, 8 Studentska Str., Bulgaria
[2]Bulgarian Academy of Sciences, Institute of Electronics, 72 Tzarigradsko Chausse Blvd, Sofia, Bulgaria
[3]Technical University of Sofia, Faculty of Engineering and Pedagogy, Sliven, Bulgaria

*Corresponding author: eyankov@uni-ruse.bg

Abstract. The metallic implants represent a major class of the hard tissue replacement materials. In order to enhance the surface properties of the titanium alloys multicomponent coatings containing ceramic phases are employed to improve the tribological performance, corrosion resistance and biocompatibility. The mechanical properties (elastic modulus, necking region, loss of adhesion) of arc PVD TiN and TiN/TiO_2 coatings deposited on unalloyed Ti (99 wt. %) foil were examined by nanoindentation, uniaxial tensile test while applying a new approach - using thermal imaging camera during the test. The hardness of the TiN coating reached values of 9649 MPa and 67 GPa, while that of the TiN/TiO_2 was lowered down to 8774 MPa and 58.5 GPa which low values are closer to that of the cortical bone. The mechanical behavior of TiN/TiO_2 coated material characterized it as a more plastic system indicating a good deformability while the TiN displayed more fragile behavior. There were no signs of loss of adhesion or loss of coating integrity up to maximum load for all tested TiN and TiN/TiO_2 coated samples. The thermal analysis proves that the coated samples show lower thermal conductivity, which is very important for the performance of an endosseous dental implant for example.

1. Introduction

Identifying the mechanical properties of a material is an issue of importance for its specific application. Various technologies are being used in order to enhance the surface characteristics of different materials [1-3]. As technologies develop, different methods are effectively implemented of surface modifications for metals and alloys used in implantology. All of them aim at increasing the biocompatibility and addressing the shortcomings of a particular implant material. Titanium and its alloys are in the scope of many scientific works applying a different kind of coatings [4-8]. The major drawbacks of these orthopedical materials are that they are relatively soft and have poor wear resistance [9]. Up to the present day, the scientists continue to seek for surface modifications that increase the mechanical biocompatibility and in vivo performance of the exogenous material. The ceramic coatings of biocompatible or bioactive materials could successfully enhance both osseointegration and mechanical properties of the surface. Unfortunately, such coatings are fragile and special conditions of deposition and implantation are required. If the coating is fragmented, corrosion and inflammatory processes leading to early explantation may occur. In this respect, the study focuses on the examination of the change of the mechanical properties of PVD deposited TiN after subsequent glow discharge oxidation

IOP Conf. Series: Materials Science and Engineering **416** (2018) 012062 doi:10.1088/1757-899X/416/1/012062

that produces the superficial TiO_2 film. The coatings were deposited on 0.1 mm thick Ti foil enabling to assess the change in the mechanical properties by uniaxial tensile test and nanoindentation.

2. Experimental procedures

The substrate material used was Ti foil received in the form of sheets with the following dimensions: 100×300 mm and thickness of t_0 = 0,1 mm. The chemical composition shown in Table 1 was determined by JEOL JXCA-733 Microprobe scanning electron microscope (SEM) equipped with wavelength dispersive spectrometers (WDS).

Table 1. Chemical composition (wt. %) of the substrate material.

Element	Al	V	Mo	Sb	Hf	Pd	Cd	Fe	Ti
Bare Ti	0.11	0.26	0.33	0.12	0.076	0.055	0.019	0.002	Bal.

Samples were cut out from the sheet material in one direction with respect to the rolling one as shown in Figure 1 c. The sheet samples were with the following dimensions: L_o = 50 mm, L_c = 90 mm, L_t = 150 mm and b_0 = 12.5 ± 0.2 mm.

For the cathodic arc deposition of the TiN, a sidewall positioned evaporating system in a cubic vacuum chamber with water-cooled walls was used. The samples were hanged near the center of a clockwise-rotating with a frequency of 0.5 Hz turntable. To ensure the coating stress relaxation and necessary adhesion a very thin pure layer from the target (at 2.5×10^{-1} Pa in Ar atmosphere for 5 min., bias 600 V, 110 A arc current) was applied. The TiN film was made in a pure N_2 atmosphere at 280-290 °C substrate temperature for a time of 60 min., 108 A arc current, bias 250 V and 7.5×10^{-1} Pa pressure in the working chamber. The TiO_2 film was made by glow plasma discharge using the uppermost located sputtering system in the same chamber. A bias voltage of 1340 V (720 mA current) in a pure O_2 atmosphere at a pressure of 6×10^{0} Pa were applied for a deposition time of 240 min. Both layers' thickness (TiN - 2.7 µm, TiO_2 - 0.8 µm) was attained by means of calotest measurements.

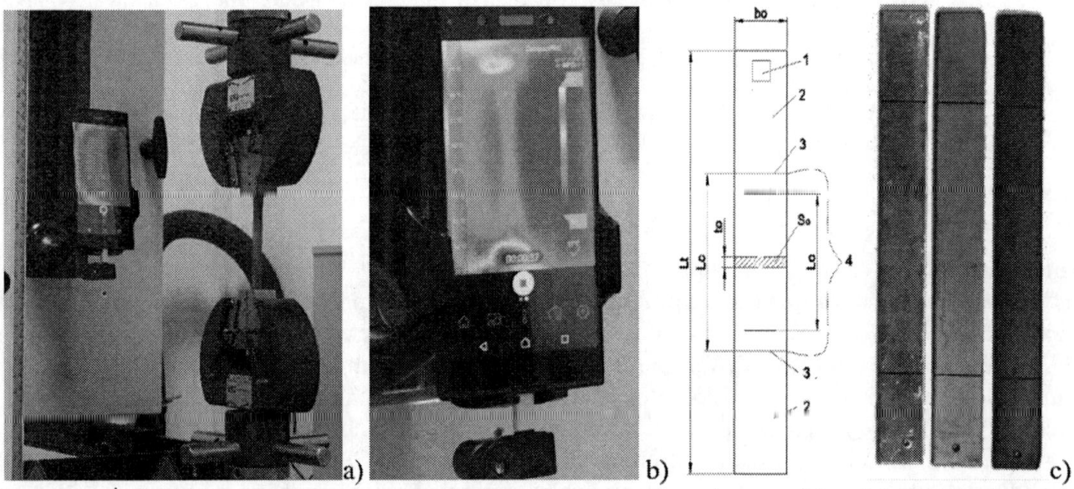

Figure 1. Tensile test equipment and samples: a) universal testing machine; b) SEEK Thermal Compact Pro camera together with the screen and clamp; c) test samples for the uniaxial tensile test (ISO-6892-1:2016) – scheme and appearnce of the uncoated, TiN and TiN/TiO_2 coated samples.

The static uniaxial tensile test was performed on a universal testing machine Instron-3384 (Figure 1a) on the base of the same standard, supplied with a software Bluehill-3. Five of each sample – uncoated, TiN-coated and TiN/TiO_2 coated were examined. For each test control measurements of the working length of testing, height and width were held and the values were set in the used test method via the software program. For the purpose of the testing method, a constant strain rate of 1 mm/min until the load fell down to 300 N and capturing intervals 10 ms of was used. The data were saved in an appropriate digital form for subsequent analysis. Simultaneously, SEEK Thermal Compact Pro camera with the following characteristics: temperature range – -30 ÷ +330 °C; the size of the thermal sensor - 320x240 pixels, thermal response below 70 mK, spectral range - 7.5÷14 μm, capturing rate - 15 frames per second was used to obtain thermographic images. The thermal camera was equipped with a mobile device with workable software Seek Thermal v.2.1.1.3 (Figure 1 b). The set was adjusted in position to the front of the thermal chamber of the testing machine by a stalk. The camera and machine synchronization were done by joining their software. The real-time monitoring and registration of the thermal fields at the surface of the sample and in the room were obtained during the tensile test. The data collected from the set of thermographic images were used for assessment of the degree of non-uniformity of the thermal field that leads to stresses and deformations occurring in the samples.

The nanohardness values of the TiN and TiN/TiO_2 coating were measured using fine polished Ti foil in order to minimize the surface roughness values. A Vickers nanoindentation tester FISCHERSCOPE H100 using a load force of 20 mN was utilized to determine the Martens hardness HU, plastic hardness H_{pl} and Young's modulus E of the coating. The coating failures were evaluated by JEOL JXCA-733 (Japan) employing accelerating voltage of 19.8 kV.

3. Results and discussions

The values of the coating hardness calculated from the force-depth data of the nanoindentation test are plotted in Table 2. Taking into account the thickness of the coatings, it could be expected that the measured values were influenced by the substrate material used.

Table 2. Hardness values of the TiN and TiN/TiO_2 coating measured with 20 mN load. The abbreviations used stands for HU – universal hardness; H_{pl} - plastic hardness; h – penetration depth, E - Young's modulus, W_{total} – total energy for elastic and plastic deformation, W_r – plastic deformation part of W_{total}

Coating	HU, MPa	H_{pl}, MPa	h, μm	E, GPa	W_{total}, nJ	W_r, %
TiN	5667±799	9649±4226	0.365±0.06	67±5.1	2.1±1.2	24.7±10.3
TiN/TiO_2	4708±546.8	8774±816.	0.401±0.03	58.5±4.8	6.7±1.8	41±14.8

The oxide deposition on the surface of the nitride resulted in a decrease of hardness and elastic modulus thus closing the mismatch between the bone (E = 9-28.4 GPa [10]) and implant modulus. The nanohardness of arc PVD TiN measured by Hernández L C et al. [11] grown at -300 V bias substrate was higher (17.6 MPa) than the values assessed in the present work. This difference could be attributed to the higher nitrogen content in the gas atmosphere used in the particular study, higher load, the pure Ti substrate material, surface roughness and the resultant coating texture. The universal hardness of the TiN/TiO_2 coating decreased by about 16.92% as opposed to the TiN coated sample. The TiN/TiO_2 plastic hardness reduced by 9.07% on average while the elastic modulus lessened by 12.78 % compared to the TiN coating. These facts imply that the oxidation of the nitride has led to internal stresses' reduction and improvement of plasticity. That is why the total energy for elastic and plastic deformation was increased by 3.2 % after the TiO_2 deposition and simultaneously the plastic deformation part of W_{total} rose up by 66% as opposed to the nitride coating.

After averaging the uniaxial tensile tests results of four tested samples of each kind (uncoated Ti, TiN and TiN/TiO_2 coated), the gathered values were processed and the diagram considered as a representative was the one with the closest mechanical properties to the average results from each indicator diagram. The representative indicator diagrams for the bare Ti, TiN and TiN/TiO_2 coated samples are shown in Figure 2.

For all three samples, because of the smooth transition from elastic to plastic area, the tensile strength at yield had not been established. The absolute extension of the Ti after the TiN deposition significantly reduced its values for the account of an increase in the tensile strength (Figure 2 a). After oxidizing the nitride, a stable downward trend of reducing the tensile strength together with a slight increase in the absolute elongation was observed. Because the applied load and the extension are not additive values, it is not appropriate to be compared. In order to compare the values from the tensile tests, a tensile stress

IOP Conf. Series: Materials Science and Engineering **416** (2018) 012062 doi:10.1088/1757-899X/416/1/012062

(σ) – tensile strain (ε) diagram is presented in Figure 2 b. This diagram allows to identify the strength values of the samples - Young's modulus (E), load at the elastic limit (R_e), tensile stress at maximum load (R_m) and tensile stress at tensile strength (σ_u). Additionally, the plasticity characteristics - elongation at tensile strength (ε_m), elongation at break (ε_u) and relative contraction after the break (Z) for the substrate and the coatings were also determined. The calculated values are presented in Table 3.

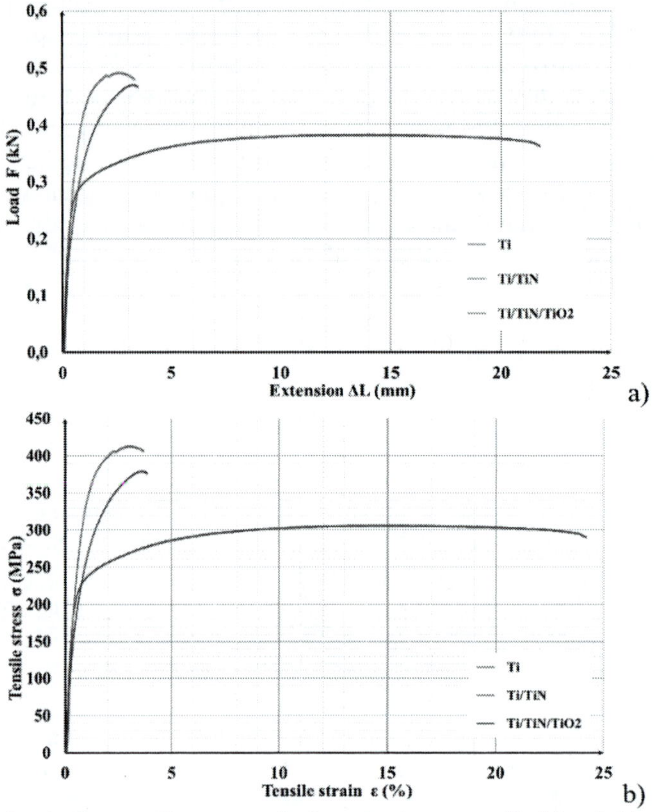

Figure 2. Representative indicator diagrams of: a) extension – applied load, b) tensile stress – tensile strain.

Table 3. Tensile test values of the tested samples: R_e - load at the elastic limit; E - Young's modulus, R_m - tensile stress at maximum load; σ_u - tensile stress at tensile strength; ε_m - elongation at tensile strength; ε_u - elongation at break; Z - relative contraction after the break.

Coating	R_e , MPa	E, GPa	R_m, MPa	σ_u, MPa	ε_m, %	ε_u, %	Z,%
Ti	184.7±5.4	64.1±7.7	306.7±10.6	305.6±10.6	14.6±1.7	14.9±1.9	17.6±3.5
Ti/TiN	227.2±22.1	70.8±12.8	398.5±10.9	398.5±12.2	1.9±0.1.1	1.9±0.4	3.8±0.9
Ti/TiN/TiO_2	154.4±42.2	64.3±17.7	380±13.4	365.5±	2.4±0.5	4.7±0.8	6.1±1.8

The load at the elastic limit for the bare Ti samples showed average values of 184.7 MPa while those coated with TiN indicated higher average values - 227.2 MPa, which were 18.72 % higher than those of the uncoated sample. For the TiN/TiO_2 coated samples, the average load at elastic limit decreased to 154.4 MPa, which means 32.03 % reduction, as opposed to the only nitride coated sample. The results from the load at elastic limit of the coated samples indicate that the elastic strength of the materials lies with the surface film, especially when it has enhanced mechanical properties. The elastic modulus of the Ti foil amounted to about 64.1 GPa and after the TiN deposition it rose up to the average value of 70.8 GPa that means an increase of 9.55 %. After the oxidation, the modulus decreased to 64.3 GPa which was 9.13% less than that of the TiN coating. The decrease in Young's modulus for the TiN/TiO_2

coating points that the oxidized material would have higher toughness against brittle fracture because of the lower internal stresses. The tensile stress at maximum load for the nitride coated sample (R_m - 398,5 MPa) was about 23.04% higher than that of the bare substrate (R_m - 306.9 MPa) for the account of a decrease of plasticity of the TiN (ε_m – 1.86 %) of over 7 times as opposed to the Ti substrate (ε_m – 14.61 %). After oxidation of the coating, the tensile stress at a maximum load slightly decreased (R_m - 380 MPa) with 4.66 % on average. This had a beneficial effect on the plasticity of the TiN/TiO_2 that rose up with 29 % compared to the TiN coated sample. There were no signs of loss of adhesion (de-bonding from the substrate or in between the layers) or loss of coating integrity up to maximum load for all TiN and TiN/TiO_2 coated samples.

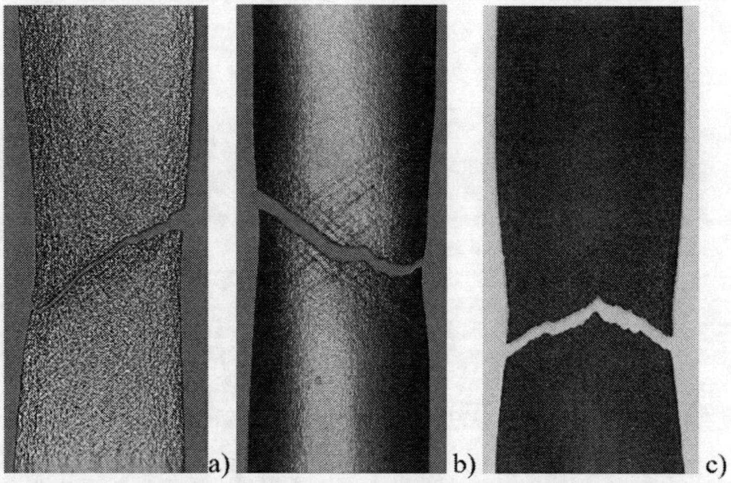

Figure 3. Macrographs of the tested samples in the ruptured area after the uniaxial tensile test: a) bare Ti foil, b) TiN coated sample, c) TiN/TiO_2 coated sample.

The macrographs of the tested samples in the place of the necking region are presented in Figure 3. Local reorientations and fractures were identified only near the necking region and fractured surfaces. For the coated samples Lüders bands also appeared at the surface. By analogy with the steels, their appearance could be associated with the presence of interstitial N in the Ti substrate because of the diffusion of the element during the deposition process that was characterized by heavy ion bombardment and increased temperature of the substrate material. The strength in the fractures area after the TiN deposition (σ_u – 398.5 MPa) indicated 23.3% higher values as opposed to the bare Ti foil (σ_u – 305.6 MPa). This change was accompanied by lowering the elongation at break with about 8 times (Table 2). After the TiN deposition, the relative contraction after the break (Z – 3,8%) also lessened by 4 times compared to the bare Ti (Z – 17,6%) indicating the ability of the surface to retain the initial geometrical form of the implant if fractures or local effects occur. The oxidized coated samples indicated Z values close to 6.1 %, which were 1.58 times lower than the values of the TiN-coated sample. The mechanical behavior for the TiN/TiO_2 coated material characterized it as a more plastic system (Figure 3.c) compared to the TiN coated samples (Figure 3.b). These results complied with the measured nanoindentation values. The $\varepsilon_u/\varepsilon_m$ for the TiN/TiO_2 coating was equal to 1.94 indicating a good deformability while that ratio for the TiN coated sample amounted 0.998 suggesting more fragile behavior.

The SEM micrographs of the bare Ti sample shown in Figure 4 a, clearly demonstrates the roughening of the surface near the necking zone due to the extensive grains' reorientation. At higher magnification, some sets of twins and slip lines formed by the higher stress in the hexagonal-close packed (hcp) lattice were also seen. The observed grains re-orientation and the inter-crystal changes conform to the higher value of the relative contraction after the break (Z = 17.6%) of the bare Ti. The surface around the necking zone of the coated samples (Figure 4 b and c) looked flatter because of the blocked movements of the crystal lattices by the hard uppermost film and the close interlocking between their surfaces. Near to the raptured zone, the nitride and oxynitride coatings showed a similar way of cross-section fracturing with respect to the direction of the tensile axis of the PVD films without forming areas with missing parts of the coating. It is clear that the critical load of fracture of the TiN coating is higher than that of the Ti foil. When increasing the applied load after this moment the Ti was being deformed which lead to local contractions mainly in depth than in width (Z = 3.8 %) of the TiN coated sample. The oxidation formed more favorable gains' boundary zones for plastic deformation of the TiN/TiO_2 coating. This fact is evidenced by the increased value of the relative contraction after the break (Z = 6.1 %) suggesting higher plasticity of the coating and more brittle substrate behavior probably because of the higher N diffusion in depth due to the longer deposition process contributing to the very good adhesion. A modern approach evaluating and analyzing the mechanical behavior by thermographic analysis was applied during the uniaxial tensile test.

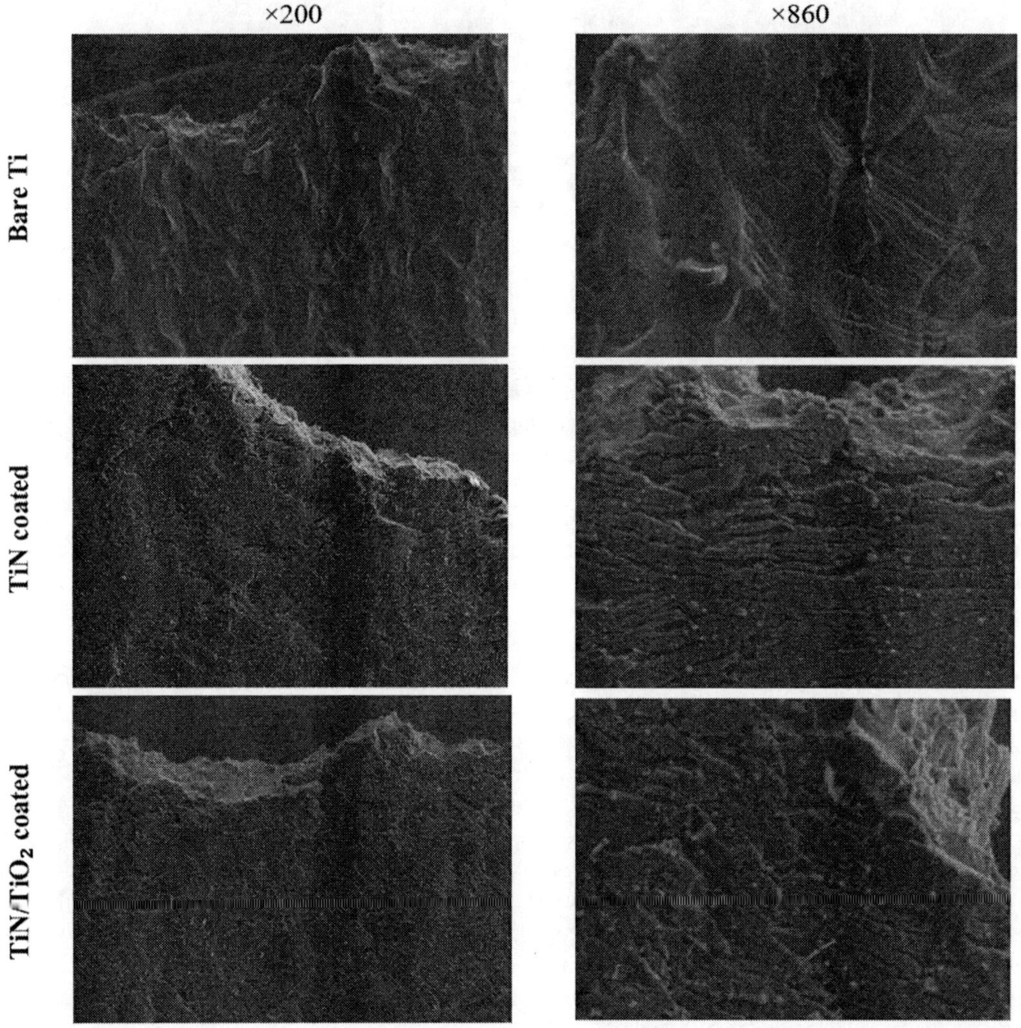

Figure 4. SEM micrographs of the fractures areas after the tensile test at different magnifications.

The thermographic images captured in the beginning (elastic region) and at the end (plastic region and necking area) of the tensile tests of each sample (bare Ti, TiN, and TiN/TiO_2 coated) are shown in Figure 5. The highest was the emitted temperature (27.8 °C) from the bare Ti samples while the lowest one (25.4 °C) was registered for the TiN coated specimen in the elastic area. This fact suggests a hindered heat transfer of the TiN coating and its low plasticity. The emitted temperature in the elastic area from the TiN/TiO_2 coated sample was equal to 26.2 °C implying the enhanced heat transfer and plasticity as opposed to the TiN coated specimen. All tested samples demonstrated the highest values of emitted temperatures in their elastic regions. After moving into the plastic region, the temperature gradually decreased and that behavior differed from that established for steels. Additionally, the necking zones in all samples occurred in the colder areas of the samples, not in the hotter ones.

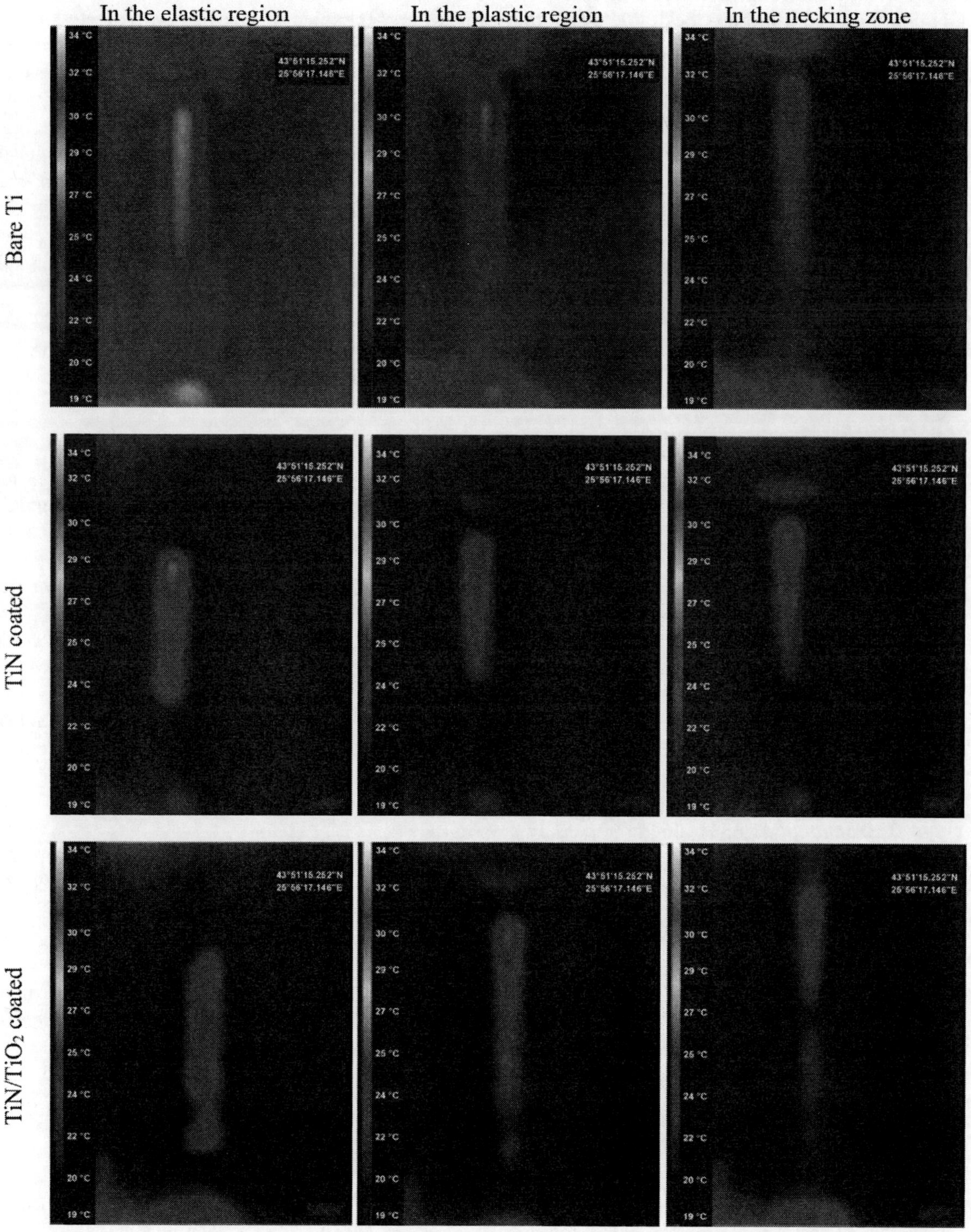

Figure 5. Thermographic images captured during the tensile tests of the bare and coated samples.

4. Conclusions

The deposition of coatings on Ti foil allows for determining different mechanical characteristics of coated systems by using uniaxial tensile test. The method proves to be sensitive to the changes occurring in the mechanical properties of the arc deposited TiN after the glow discharge oxidation when compared to single nitride coating. The decrease in the elastic modulus values after oxidation of the arc deposited TiN coating is an important desirable effect that brings the mechanical properties of the tested system closer to that of the cortical bone. Simultaneously, the plasticity of the TiN/TiO_2 coated sample increases its values with 29% compared to the TiN one at the expense of a small decrease in the hardness and tensile stress at a maximum load. The demonstrated high adhesion strength of the coated systems would prevent the intense fragmentation and separation of parts of the surface film that could trigger inflammatory processes in vivo. The observed differences in the values of the mechanical properties determined by the nanoindentation and tensile tests are dictated by the differences in the tested volumes. The tensile tests results provide an overall outline of the mechanical properties of the tested systems while the nano-hardness results describe the local elastic and plastic behavior of a certain surface area. The thermal analysis proves that the coated samples show lower thermal conductivity, which is a very important fact for the performance of an endosseous dental implant for example.

Additional experiments for determining the changes in the chemical and phase compositions of the coated systems will be needed to explain the differences in the mechanical properties of the tested materials.

5. References

[1] Chawla V, R Jayaganthana and Chandra R 2008 *Mater*. Character. **59** 1015 – 1020
[2] Yan C, Hao L, Hussein A, Wei Q and Shi Y 2017 Microstructural and surface modifications and hydroxyapatite coating of Ti-6Al-4V triply periodic minimal surface lattices fabricated by selective laser melting, *Mater Sci Eng C Mater Biol.* **75** 1515–1524, doi: 10.1016/j.msec.2017.03.066.
[3] Sidambe A T 2014 Biocompatibility of Advanced Manufactured Titanium Implants - A Review *Materials* **7** 8168-8188, doi:10.3390/ma7128168.
[4] Kaluđerović M R, Schreckenbach J P and Graf H-L 2016 Titanium dental implant surfaces obtained by anodic spark deposition – From the past to the future *Materials Science and Engineering C* **69** 1429–1441
[5] Zhang F, Lin L-X, Wang G-W, R Hu, Lin C-J and Chen Y 2012 A high-throughput electrochemical impedance spectroscopy evaluation of bioresponsibility of the titanium microelectrode array integrated with hydroxyapatite and silver, Electrochim. Acta **85** 152– 161
[6] P Mazón, García-Bernal D, Meseguer-Olmoc L, F Cragnolini and De Azad P N 2015 *Ceramics International* **41** 6631–6644
[7] J Kim, W J Lee and Park H W 2016 Mechanical properties and corrosion behavior of the nitriding surface layer of Ti-6Al-7Nb using large pulsed electron beam (LPEB) J Alloys Compd **679** 138-148
[8] Maho A, Detriche S, Delhalle J and Mekhalif Z 2013 *Materials Science and Engineering C* **33** 2686–2697
[9] El-Hadad S, Waly M and Khalifa W 2015 Microstructure evolution and mechanical properties of alpha-beta heat treated Ti-6Al-4V alloy, *22nd Intern. Conf "Advances and Trends in Engineering Materials and their Applications"*, Montreal
[10] Zysset P K 2009 Indentation of bone tissue: a short review, *Osteoporos Int.* **20** 1049–1055
[11] Hernández L, Ponce L, Fundora A, López E and Pérez E 2011 Nanohardness and Residual Stress in TiN Coatings *Materials* **4** 929-940

Acknowledgement

The financial support of this work was provided by the National Science Fund of Ministry of Education and Science, Bulgaria, under Grant project "Gradient functional nanocoatings produced by vacuum technologies for biomedical applications" with number DN№ 07/3 (2016).

IOP Conf. Series: Materials Science and Engineering **416** (2018) 012063 doi:10.1088/1757-899X/416/1/012063

A Method to Precise Determine the Young's Modulus from Dynamic Measurements

L Barboni[1], G R Gillich[1,*], C P Chioncel[1], C O Hamat[1] and I C Mituletu[1]

[1]"Eftimie Murgu" University of Resita

*Corresponding author: gr.gillich@uem.ro

Abstract. Knowing the mechanical characteristics of materials is crucial for properly interpreting the results attained from experiments. In this paper, we present a method to experimentally determine the Young modulus of various materials by employing multi-modal analysis. The method, which implies finding the structures' natural frequencies, is based on the resonance test. If the natural frequencies are known, the dynamic Young's modulus can be extracted from the relation that expresses these frequencies in respect to the stiffness. An advanced signal processing algorithm is proposed for precise frequency evaluation. The algorithm constitutes the core of the method; herein we demonstrate how it is involved to extract the frequencies of several bending vibration modes. Tests were made by involving non-contact and impulsive excitation, respectively, the fist one providing best repeatability.

1. Introduction

The elastic properties for linearly elastic isotropic solids are described by two independent constants, namely Young's modulus E and the shear modulus G. These elastic constants of materials have great importance for the design of engineering structures. For that reason, numerous experimental techniques to find moduli E and G were developed. These techniques can be divided into two groups: (*i*) static methods, and (*ii*) dynamic methods.

Attempts reported in [1]-[3] are among the first made to find the elastic properties of isotropic materials basing on nondestructive vibration tests. The equations established in the first paper for computing moduli E and G from the fundamental flexural and torsional frequencies of prisms and cylinders were found from the Timoshenko beam theory. Advances in calculating the Young modulus from the proper fundamental frequencies along with an improvement of the experimental techniques is proposed in [2] and [3], respectively. In fact, these three papers have formed the base of the ASTM standard test method [4] for characterization of dynamic elastics properties based on the impulse excitation method [5].

Nowadays, there are also numerous methods to estimate the elastic properties of structures made of non-isotropic and/or inhomogeneous materials as concrete [6], timber [7]-[8] or laminated composites [9]-[14] involving vibration-based methods.

Some aspects of our prior research concerned with the effect of mass variation on the frequencies of beams with constant or variable cross-section [15]-[17]. We also studied the behavior of beam-like structures, to find out how the transition from the behavior of a beam to that of a plate is influenced by the width of the structural element [18]. For all this research precise frequency estimation was necessary [19]. In the present investigation, we show how signal processing algorithm applied for determination of the Young's modulus of laminated composite plates.

2. The principle of deriving Yong's modulus from vibration measurements

The dynamic method involved to calculate Young's modulus base on the univocal relationship between it and the dimensions, weight and natural frequencies of the beam for a given support configuration. This method presents following advantages: it is rapid, nondestructive, implies the application of low loads and can be performed at extreme temperatures. Because the method is nondestructive, it enables monitoring the materials properties and makes possible observing their variation in time.

From the point of view of performing the experiments, the method consists of two stages: in the first stage, the excitation and signal acquisition and processing are performed to detect the natural frequencies; in the second one, mathematical relations and computational procedures are performed to obtain E from the natural frequencies.

Basically, there are three vibration mode types that we can use to find Young's modulus of a prismatic beam with an asymmetric cross-section around one axis, namely out-of-plane and in-plane bending modes, respectively the longitudinal vibration mode. Unlike the other two modes of vibration, the out-of-plane modes can be easier excited, and the risk of occurring combined loads is minimal. Current methods used to excite the test specimen are impulsive-based [4]. From our research, we found that sweep-sine excitation or an excitation applied for short-time in a controlled frequency bandwidth provides better results [20]. This aspect is treated in next section.

In most cases of using bending vibration modes, a slender beam that respects the Euler-Bernoulli theory is chosen as the test specimen. To this aim, the length L per thickness h ratio must be greater than 200 [7]. Obviously, for in-plane vibration, the ratio considers the beam's width b and length L. The specimen has, usually, free-free or fixed-free end conditions. These supports are preferred because almost perfect boundary condition can be ensured and therefore uncertainties due to supports are avoided. Henceforth fixed-free end conditions are considered.

Knowing the end conditions, we can express the mathematical relation between Young's modulus E and the beam dimensions L, B respectively H, weight m and natural frequencies f_i, as:

$$E = \frac{mL^3}{I}\left(\frac{2\pi f_i}{\alpha_i^2}\right)^2, \qquad (1)$$

where I is the moment of inertia of plain area, derived as $I = (B \cdot H^3)/12$, and α_i are coefficients related to the beam's support condition and the mode numbers. Table 1 presents the first six coefficients derived for the beam with fixed-free end conditions (cantilever beam).

Table 1 Dimensionless coefficients for calculating the frequencies of the cantilever beam.

Mode No.	1	2	3	4	5	6
α_i [-]	1.87510	4.69409	7.85476	10.99554	14.13717	17.27876

All parameter values on the right side of relation (1) can be measured with accuracy, except for the natural frequency, for which an advanced estimation algorithm is required. The following sections present an algorithm developed by the authors, which allows high-precision natural frequency evaluation and how it is used in the experimental procedure to find Young's modulus through the herein proposed dynamic method.

3. Precise frequency evaluation by post-processing the vibration signal

As already shown, vibration-based assessment of E modulus is supported by the precision of the frequency estimation method. Unfortunately, the vibration response to impulsive excitation is rapidly damped and offers, in standard frequency evaluation is performed, a spectrum with rare spectral bins [21]. This happens because the frequency resolution Δf depends on the acquisition or analysis time t_S in accordance to the relation $\Delta f = 1/t_S$. Therefore, the estimated frequency f_E can differ from the true frequency f_R with an amount ε, this error being limited to $\varepsilon = \Delta f / 2$.

IOP Conf. Series: Materials Science and Engineering **416** (2018) 012063 doi:10.1088/1757-899X/416/1/012063

We propose, in this section, an algorithm to improve the frequency readability by increasing the number of the spectral bins, without being necessary to lengthen the acquisition time. This is made by superposing more spectra, each of them found from the original signal cropped with different time lengths. The original signal has the time length t_S, the number of samples N_S and the constant sampling time Δt, nominated also as time resolution. In this example, the individual spectra were obtained by calculating the Discrete Fourier Transform (DFT), but the algorithm also works well if the spectra are accomplished from the Power Spectrum (PS) or Power Spectral Density (PSD).

Due to leakage, the amplitude value indicated at any other position than the actual frequency f_R is less than the true amplitude value. Finding the real frequency-amplitude pair consists in shortening the original signal to which a DFT is afterward applied, until the amplitude achieves a clear maximum. This process is repeated for any targeted vibration mode. Applying the algorithm involves performing the following steps:

Step 1. Establish the coarse value of the targeted frequency, for example from a standard DFT or from numerical simulation, resulting the evaluated frequency f_E :

Step 2. Find the coarse signal period T_E from the relation:

$$T_E = \frac{1}{f_E}; \tag{2}$$

Step 3. Establish the time length $t_{S\text{-prim}}$ for the original signal. It must exceed a number of n complete cycles, each having the length of period T_E . A good option is to choose:

$$t_{S\text{-prim}} = (n+0.45)\cdot T_E \quad \text{for} \quad 4<n<6; \tag{3}$$

Step 4. Find the length of the last signal used in the analysis from the relation:

$$t_{S\text{-fin}} = (n-0.45)\cdot T_E; \tag{4}$$

Step 5. Calculate the number of samples that are contained in the original respectively last analyzed signal from the relations:

$$N_{S-prim} = \frac{t_{S-prim}}{\Delta t} \quad \text{respectively} \quad N_{S-fin} = \frac{t_{S-fin}}{\Delta t}; \tag{5}$$

Step 6. Calculate with the following relation the total number of samples to be reduced:

$$N_{S-red} = N_{S-prim} - N_{S-fin}; \tag{6}$$

Step 7. Establish the number of iterations k;

Step 8. Calculate the number of samples to be reduced by iteration, from the relation:

$$N_{S-it} = \frac{N_{S-red}}{k}; \tag{7}$$

Step 9. Round down the result of relation (7) to attain N_{S-it} as an integer;

Step 10. Perform a DFT for the original signal and the k signals attained after iterative cropping with N_{S-it} samples.

Step 11. Superpose the k+1 spectra to achieve an overlapped spectrum;

Step 12. Extract the frequency that corresponds to the biggest amplitude in the bandwidth of interest from the overlapped spectrum. This frequency is very close or identical with the true frequency component in the analyzed signal.

An overlapped spectrum for a large $N_{S\text{-}it}$ is presented in figure 1. Reducing the number of samples by iteration, i.e. small $N_{S\text{-}it}$, results in a better frequency estimation. A detailed image of the top amplitudes for denser spectral lines in the overlapped spectrum is presented in figure 2. The total number of samples N_S does not influence the frequency resolution in one spectrum, but if large enough, it allows a fine tuning of the iteration process. The fines resolution in the overlapped spectrum is achieved if one sample is reduced by iteration, i.e. $N_{S-it} = 1$.

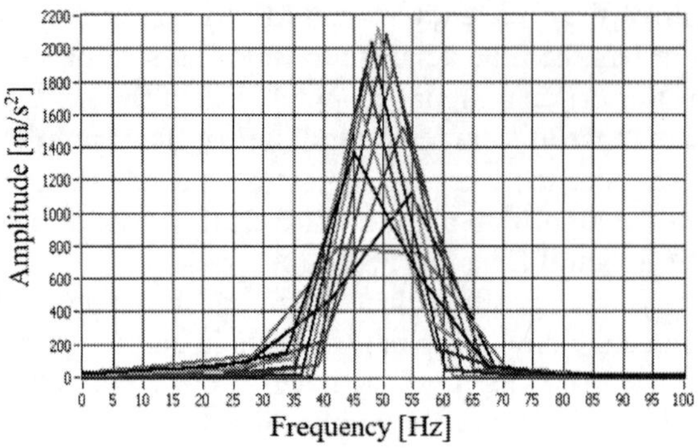

Figure 1. Overlapped spectrum with large number of samples extracted by iteration.

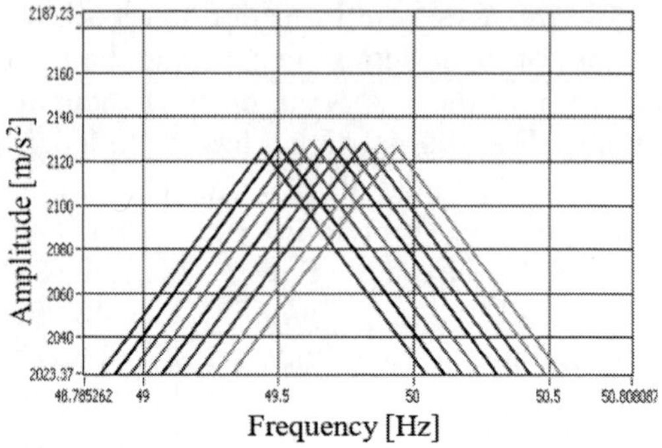

Figure 2. Top of an overlapped spectrum with few samples extracted by iteration.

Note that this procedure permits estimating the frequency for a selected vibration mode. If more frequencies are targeted, following steps 1 to 12 must be repeated for all selected modes independently.

4. Experimental setup and results

For the experimental tests, a stand developed in the "Eftimie Murgu" University of Resita was used. The stand allows an accurate measurement of the natural frequencies for beam-like structures. The test stand includes a rigid frame, on which a vise is mounted, allowing simulating cantilever beam conditions. The beam position is chosen to have the weak axis in horizontal plane and to avoid curving under the own weight. So, no transversal forces act in out-of-plane vibration. In figure 3, in order to highlight all the elements of the stand, the beam was drawn rotated by 90°.

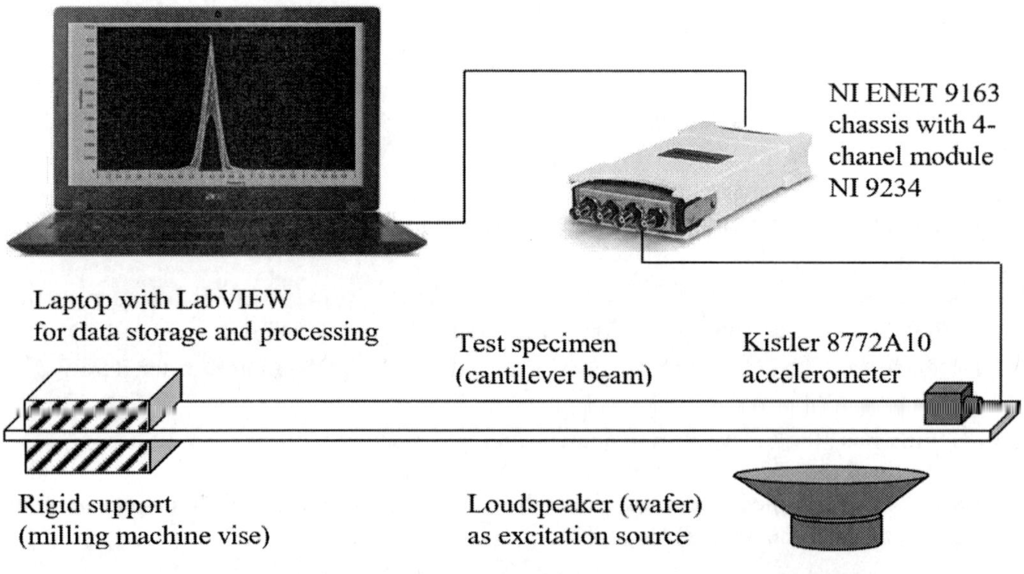

Figure 3. Schematic representation of the experimental stand.

For measuring the natural frequencies of the beam, the stand was equipped with a data acquisition and transmission system, which comprises of a Kistler 8772A10 accelerometer, a NI9234 acquisition module and a ENET-9163 chassis. The accelerometer was fixed close to the beam end by wax bonding. The effect of the mass of the accelerometer can be eliminated by considering a concentrated mass placed at a known distance from the fixed end of the beam [15]. In this paper we didn't reveal

IOP Conf. Series: Materials Science and Engineering **416** (2018) 012063 doi:10.1088/1757-899X/416/1/012063

this aspect, the focus being set on the accuracy of the frequency determination. Due to the oscillations of the beam, the accelerometer generates electrical impulses proportional to the sensed acceleration. For reading and storing the natural frequencies, a computer connected to the data transmission system was used. The exact frequency extraction was done using a special developed application in LabView software. The free segment of the tested bar has the dimensions: *L*= 1000 mm, *B*= 50 mm, *H*= 5 mm and the weight *m*= 1.9625 kg.

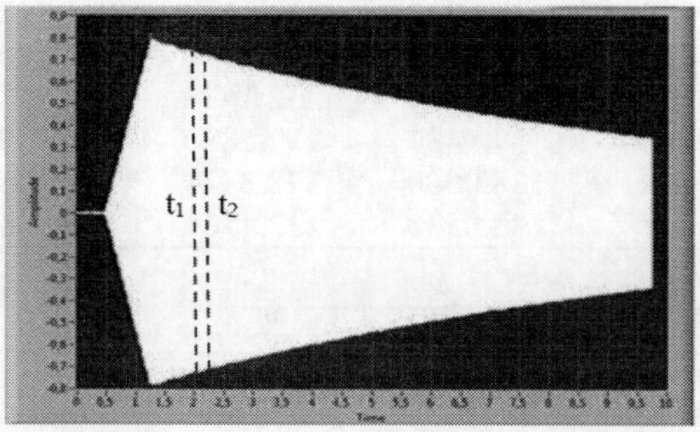

Figure 4. Response of a cantilever beam excited with acoustic pressure at 71Hz for one second.

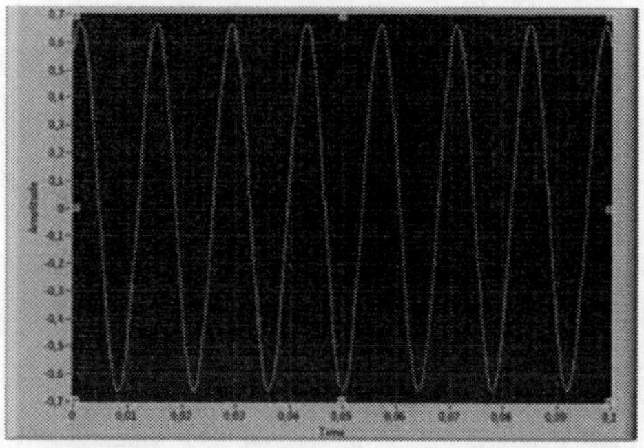

Figure 5. Detail of the portion extracted from the free vibration response of the bar.

Excitation of the beam was made by two alternative methods: with an impulse hammer respectively by acoustic waves with harmonic frequency for a short duration (1 second). The acoustic waves were transmitted by a frequency generator, then amplified and emitted by a loudspeaker. The acoustic (non-contact) excitation has the advantage that energy can be transmitted to the beam in a controlled manner only in the frequency domain of interest, avoiding vibrations in other modes. Figure 4 exemplifies the beam response to an excitation produced by an acoustic wave with a frequency of 71 Hz for one second. As it can be observed, at the end of the excitation period, a significant amplitude increase of the response signal was noticed, while after stopping the excitation signal, the beam is damped vibrating, the amplitude of the response signal decreasing continuously. Figure 5 shows a detail for a very short time (0.1 seconds) extracted from vibration response of the beam.

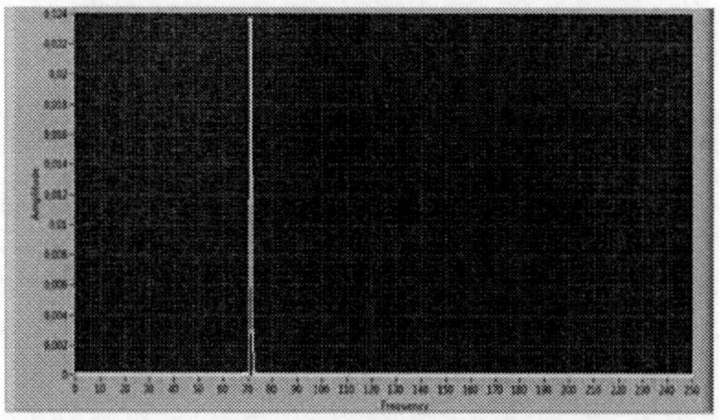

Figure 6. Spectrum of signal extracted from the free response of the acoustically excited beam.

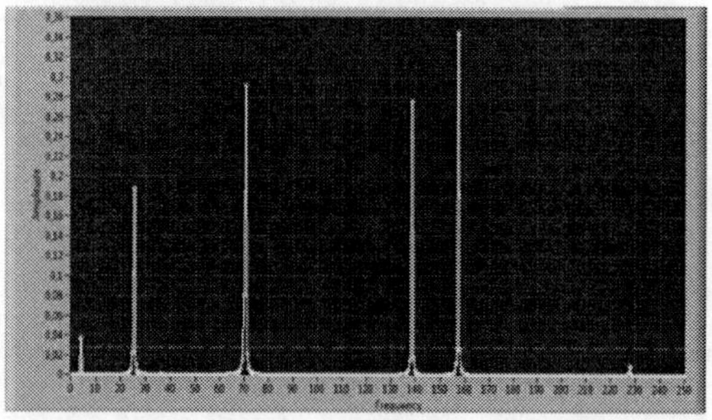

Figure 7. Spectrum of signal extracted from the free response of the impulsively excited beam.

Figures 6 and 7 illustrate comparatively the spectrum of the signals extracted from the free response of the beam excited with a concentrated energy, respectively by impulse. One can observe that, in the case of acoustic excitation, only the frequency corresponding to the analyzed mode is identified (figure 6). Therefore, excitation with low amplitude is sufficient and so nonlinearities are avoided. In the case of impulse excitation, several vibration modes can be observed (figure 7) and an important signal-to-noise ratio is reported.

Tables 2 and 3 present the measured frequencies, achieved by the two methods, for the first six out-of-plane vibration modes. For each vibration mode, the average was calculated for 5 measurements.

Table 2. Natural frequencies measured by impulse excitation.

Measurement	Natural frequency [Hz]					
no.	f_1	f_2	f_3	f_4	f_5	f_6
1	3.9995	24.9249	70.6922	138.0000	232.3350	341.5270
2	4.1099	24.6944	70.6964	138.3170	232.8470	341.8800
3	4.1099	24.8192	70.6469	138.0000	232.1430	341.5680
4	4.1199	24.6945	70.5982	137.9490	232.5880	341.9890
5	4.0960	24.9972	70.6469	138.0000	232.3370	341.5680
Average	4.0871	24.8260	70.6561	138.0530	232.4500	341.7060

Table 3. Natural frequencies measured by acoustic excitation.

Measurement	Natural frequency [Hz]					
no.	f_1	f_2	f_3	f_4	f_5	f_6
1	4.0598	26.0621	72.0168	141.2410	234.0830	348.7590
2	4.0598	26.0531	72.0168	141.2410	234.0820	348.7590
3	4.0623	26.0729	72.0185	141.2410	234.0820	348.7590
4	4.0623	26.0919	72.0185	141.2410	234.0220	348.7590
5	4.0623	26.0984	72.0168	141.2410	234.0530	348.7590
Average	4.0613	26.0757	72.0175	141.2410	234.0644	348.7590

From the tables above, one can see that the frequencies measurements have a better repeatability in the case of acoustic excitation. This may be explained by the fact that the acoustic excitation of the beam is done in the area of the natural frequency which will be measured. Furthermore, the differences between the two methods are in average at about 2%, the biggest difference (about 5%) being experienced at the frequencies of vibration mode no. 2.

Table 4. Determination of Young's modulus from dynamic measurements

Excitation	Young's modulus E [GPa]					
method	f_1	f_2	f_3	f_4	f_5	f_6
Impulse	201.0	188.8	195.1	194.0	201.2	194.9
Acoustic	198.5	208.3	202.7	203.0	204.0	203.0

Based on the average frequencies experimentally determined by the two methods and the beam parameters, Young's modulus E was calculated using relation (1). The results are shown in table 4, the differences to the known value of Young's modulus for structural steel, E= 200 GPa, being small.

5. Conclusion

We propose in this paper a method to identify Young's modulus by involving the natural frequencies of the out-of-plane vibration modes, supported by a non-contact excitation method and an advanced frequency estimation algorithm, developed to this aim. The results achieved by involving these techniques ensure better repeatability as the case of using impulsive excitation, due to low signal-to-noise ratio and avoidance of nonlinearities. Also, it was found that higher modes provide more stable results if acoustic excitation is involved. As expected, more dispersed results for the impulsive excitation tests were observed. This qualify the acoustic excitation and the signal post-processing algorithm, introduced in this paper, to be used for experimental determination of Young's modulus, with the remark that higher modes provide more stable results.

IOP Conf. Series: Materials Science and Engineering **416** (2018) 012063 doi:10.1088/1757-899X/416/1/012063

6. References

[1] Pickett G 1945 Equations for computing elastic constants from flexural and torsional resonant frequencies of vibrating prisms and cylinders *ASTM Proc.* **45** 846–865.

[2] Spinner S, Reichard T W and Tefft W E 1960 A Comparison of Experimental and Theoretical Relations Between Young's Modulus and the Flexural and Longitudinal Resonance Frequencies of Uniform Bars, *J. Res. NBS A Phys. Ch.* **64A**(2)

[3] Spinner S and Tefft W E 1961 A Method for Determining Mechanical Resonance Frequencies and for Calculating Elastic Moduli from These Frequencies *ASTM Proc.* **61** 1221–1238

[4] ASTM Standard E 1876 – 07, Standard Test Method for Dynamic Young's Modulus, Shear Modulus, and Poisson's Ratio by Impulse Excitation of Vibration

[5] Wesolowski M, Barkanov E, Rucevskis S, Chate A and La Delfa G 2009 Characterisation of elastic properties of laminated composites by non-destructive techniques *17th Int. Conf. on Composite Materials: Book of Abstracts* (United Kingdom: Edinburgh, 27-31 July, 2009)

[6] Lee K M, Kim D S and Kim J S 1997 Determination of Dynamic Young's Modulus of Concrete at Early Ages by Impact Resonance Test, *KSCE J. Civ. Eng.* 1(1) 11–18

[7] Roohnia M 2014 An Estimation of Dynamic Modulus of Elasticity in Cantilever Flexural Timber Beams *Drvna Industrija* **65**(1) 3-10

[8] Kubojima Y, Tonosaki M and Yoshihara H 2006 Young's modulus obtained by flexural vibration test of a wooden beam with inhomogeneity of density, *J. Wood Sci.* **52** 20–24

[9] Yu L, Wang J and Xia D 2015 Vibration Method for Elastic Modulus of Glued Laminated Beams *Proc. of 3rd Int. Conf. on Mat., Mech. and Manufact. Eng. (IC3ME 2015)* 1603-1606

[10] Giaccu G F, Meloni D, Valdès M and Fragiacomo M 2017 Dynamic determination of the modulus of elasticity of maritime pine cross-laminated panels using vibration methods, *WIT Trans. on Ecology and the Envir.* **226** 571 – 579

[11] Ivanova Y, Partalin T and Georgiev I 2016 Characterisation of elastic properties of laminated composites by ultrasound and vibration, *Scientific Proc. of Scienticic Techn. Union of Mech. Eng.* **1**(187) 418-425

[12] Gillich G R, Samoilescu G, Berinde F and Chioncel C P 2007 Experimental determination of the rubber dynamic rigidity and elasticity module by time-frequency measurements *Mater. Plast.* **44**(1) 18-21

[13] Iancu V, Vasile O and Gillich G R 2012 Modelling and Characterization of Hybrid Rubber-Based Earthquake Isolation Systems *Mater. Plast.* **49**(4) 237-241

[14] Nedelcu D, Gillich G R, Cziple F, Ciuca I and Padurean I 2008 Considerations about using polymers in adaptive guardrails construction *Mater. Plast.* **45**(1) 47-52

[15] Gillich G R, Praisach Z I, Wahab M A, Gillich N, Mituletu I C and Nitescu C 2016 Free Vibration of a Perfectly Clamped-Free Beam with Stepwise Eccentric Distributed Masses *Shock Vib.* **2016** 2086274

[16] Gillich G R, Praisach Z I and Negru I 2012 Damages Influence on Dynamic Behaviour of Composite Structures Reinforced with Continuous Fibers *Mater. Plast.* **49**(3) 186-191

[17] Gillich G R, Tufoi M, Korka Z I, Stanciu E and Petrica A 2016 The Relations between Deflection, Stored Energy and Natural Frequencies, with Application in Damage Detection *Rom. J. Acoust. Vib.* **13**(2) 87-93

[18] Tufoi M, Gillich G R, Hatiegan C, Gillich N and Lorenz P 2015 Some Aspects Regarding the Transition from Beam to Plate Behavior of Vibrating Structures *Rom. J. Acoust. Vib.* **12**(1) 62-68

[19] Ntakpe J L, Gillich G R, Mituletu I C, Praisach Z I and Gillich N 2016 An Accurate Frequency Estimation Algorithm with Application in Modal Analysis *Rom. J. Acoust. Vib.* **13**(2) 98-103

[20] Ntakpe J L, Gillich G R, Minda A A, Mituletu I C and Gillich N 2017 Two-Stage Interpolation Algorithm for Precise Frequency Estimation of Vibration Signals *Rom. J. Acoust. Vib.* **14**(2) 125-130

IOP Conf. Series: Materials Science and Engineering **416** (2018) 012064 doi:10.1088/1757-899X/416/1/012064

Cobalt Layered Perovskite Type $Y_{0.5}Ca_{0.5}BaCo_4O_7$ as Anodic Material for Sulphite Oxidation in Neutral Media

AF Enache[1], ML Dan[1,*], MA Lăboşel[1] and N Vaszilcsin[1]

[1]Faculty of Industrial Chemistry and Environmental Engineering, Politehnica University Timişoara, Romania

[*]Corresponding author: mircea.dan@upt.ro

Abstract. In this paper, the behaviour of an anodic catalyst made by calcium doped cobalt layered perovskite type 114 electrode ($Y_{0.5}Ca_{0.5}BaCo_4O_7$) has been investigated for the electrooxidation process of sulphite ions in neutral media (1 mol L^{-1} Na_2SO_4). This research is necessary to understand the oxidation mechanism on the surface of this type of electrode. Voltammetric techniques (cyclic and linear) have been applied in order to describe the electrooxidation mechanism and to find the optimum parameters for sulphite anodic oxidation. Kinetic parameters for the process occurring on the working electrode surface have been determined by Tafel plots method. Further, electrochemical impedance spectroscopy was performed to confirm the sulphite oxidation mechanism on $Y_{0.5}Ca_{0.5}BaCo_4O_7$ electrode. Optimum characteristic parameters, such as current density, oxidation potential range for sulphite electrooxidation and SO_3^{2-} ions transformation degree in test solutions, have been obtained by chrono-electrochemical methods (chrono -amperometry, -potentiometry, -coulometry).

1. Introduction

In the last years, the applications of metal and metal oxide nanostructures have attracted the interest of researchers. The modified electrodes have become the materials more and more used in the studies of the electrocatalytic oxidation and reduction kinetics of many important redox systems [1, 2].

$Y_{0.5}Ca_{0.5}BaCo_4O_7$ layered perovskite was defined as semiconductor material [3]. One of the most investigated transitional metals mixed oxides is 114 cobalt perovskites family due to their structural, magnetic and electrochemical properties. Experimental researches have shown that there is a correlation between compound structure and his properties, especially due to the variable cobalt ions oxidation number [4, 5].

Modified electrodes have a wide range of applications, especially in electrochemical technology, energy conversion and storage systems, information storage, electrochromic and display devices, as well in electroanalysis [6-9].

Electrochemical behaviour of $Y_{0.5}Ca_{0.5}BaCo_4O_7$ has been studied both in alkaline and neutral solutions and had as starting point the electrochemical intercalation of oxygen inside of transitional metal oxide network as well as the experimental results regarding the electrochemical behaviour of cobalt studied by voltammetric techniques and electrochemical impedance spectroscopy [10-15]. Based on the electrochemical properties, especially its uncommon oxygen intake-release capability, these layered cobalt perovskite type 114 may be proposed for use as anodic materials with electrocatalytic properties in different fuel cells with alkaline or neutral electrolyte.

IOP Conf. Series: Materials Science and Engineering **416** (2018) 012064 doi:10.1088/1757-899X/416/1/012064

Anodic oxidation of SO_2 is a component of the hybrid sulphur cycle for hydrogen production, being also used for fuel gas desulfurization [16]. The composition and concentrations of the compounds with S(IV) in aqueous solutions depend on pH value. In solutions having pH < 1.8 sulphur dioxide predominates and when pH > 7.2 mainly sulphite ions are present in solution [17, 18]. It was found that oxidation rate is strongly dependent on the nature of electrode materials, porosity and active surface of electrodes. These effects are explained by the influence of adsorption of Sulphur chemical species [17]. Also, the potential range for sulphite anodic oxidation is depending on the chosen anodic material, the type of electrolyte (acid, neutral or alkaline) and the sulfite ions concentration added in the solution.

The general aim of this study is to identify the suitable conditions for the recycling of sulphites resulting as byproducts in many industrial processes. This research on electrochemical behaviour of sulphite ions will allow its use in a Na_2SO_3/O_2 fuel cell operable in order to produce green energy from waste materials.

2. Experimental part

The $Y_{0.5}Ca_{0.5}BaCo_4O_7$ perovskite has been prepared using solid state reaction, by mixing Y_2O_3 (Aldrich 99,99%), $CaCO_3$ (Aldrich 99,99%), $BaCO_3$ (Aldrich 99,99%) and $CoO_{1.38}$ (99,99% Normapur) according to the stoichiometric cation ratio. The powder has been reground after decarbonation at 600°C, fired in air for 48 hours at 1100°C and then removed rapidly from furnace. The mixture was then reground and pressed into discs of 1 cm^2 and sintered in air for 24 hours at 1100°C. The structure of obtained $Y_{0.5}Ca_{0.5}BaCo_4O_7$ perovskite has been checked by X-Ray powder diffraction.

A three-electrode undivided cell connected to SP 150 Bio-Logic potentiostat/galvanostat was used during all measurements. Two graphite roads were used as counter electrode and Ag/AgCl as reference electrode. Test solution was deaerate before each measurement using a nitrogen purging system.

Electrochemical experiments were carried out in 1 mol L^{-1} NaOH solution in the absence and presence of different sulphite concentrations (10^{-3}, 10^{-2}, 10^{-1}, $5\cdot10^{-1}$ and 1 mol L^{-1}). Test solutions were prepared from Na_2SO_4 and Na_2SO_3 reagents (p.a. ≥ 98%, Merck).

Cyclic voltammograms have been recorded at different scan rates (5 - 500 mV s^{-1}) and linear polarization at low scan rate (1 mV s^{-1}) in order to ensure quasi-stationary conditions.

Electrochemical impedance spectroscopy (EIS) has been performed using the SP-150 impedance module, in frequency range between 100 kHz and 10 mHz. An alternative voltage amplitude of 10 mV was applied. For each spectrum, 60 points were collected, with a logarithmic distribution of 10 points per decade. The experimental EIS data have been fitted to the electrical equivalent circuit (EEC) by CNLS Levenberg – Marquardt method using ZView – Scribner Associates Inc. software.

Chrono-amperometry, -coulometry and -potentiometry techniques were applied in order to determine the transformation degree of sulphite to sulphate at six potential values betweend -0.25 and 1.00 V.

3. Results and discussion

3.1. Cyclic voltammetry

Cyclic voltammograms (CVs) recorded at 50 mV s^{-1} scan rate, on $Y_{0.5}Ca_{0.5}BaCo_4O_7$ compound in neutral solution without and with different sodium sulphite concentrations, are presented in Figure 1a. It can be observed that in the absence of sulphite ions at anodic polarization a distinctive peak appears at approximate +1 V potential value, which is associated with Co(II) to Co(III) oxidation inside of perovskite structure. At more positive potential oxygen evolution reaction (OER) is unfolding. In test solutions containing different concentration of sulphite ions, the characteristic potential for OER on the electrode surface is shifted to more positive values, respectively the potential of hydrogen evolution reaction is shifted to more negative ones.

IOP Conf. Series: Materials Science and Engineering **416** (2018) 012064 doi:10.1088/1757-899X/416/1/012064

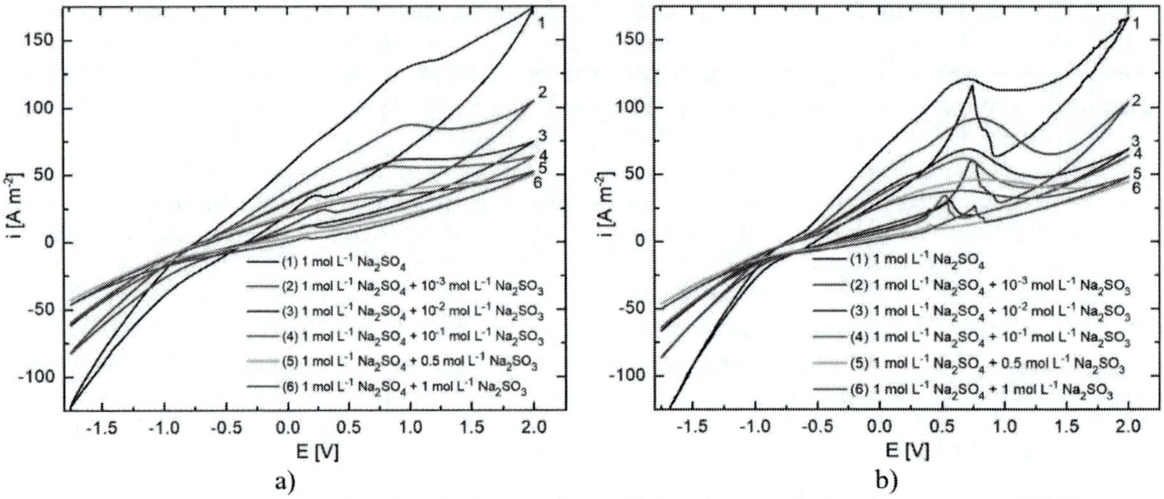

Figure 1. CVs recorded on $Y_{0.5}Ca_{0.5}BaCo_4O_7$ electrode in neutral solution in presence of different concentrations of Na_2SO_3, scan rate 50 mV s^{-1} (a) and 5 mV s^{-1} (b).

As well, on the voltammograms anodic branch at more positive potentials than the open circuit potential (OCP) no additional oxidation peak is registered, the only one present being the peak specified for the solution without sulphite ions. This is the first important information about the mechanism of sulphite electrochemical oxidation on layered cobalt perovskite electrodes in neutral media. The intensity of the anodic peak that appears in the range of +0.50 ÷ +1.25 V decreases more and more with the addition of a higher sulphite concentration.

Figure 1b presents the cyclic curves plotted for same test solutions at low scan rate (5 mV s^{-1}). Decreasing the scan rate provides the possibility to accurately identify the peaks corresponding to the oxidation processes that take place on $Y_{0.5}Ca_{0.5}BaCo_4O_7$ electrode, Co(II) to Co(III) inside to perovskite and SO_3^{2-} to SO_4^{2-} on electrode surface.

In neutral media, electrochemical behaviour of $Y_{0.5}Ca_{0.5}BaCo_4O_7$ perovskite can be described by the reversible reaction (Eq. 1).

$$Y_{0.5}Ca_{0.5}BaCo_4O_7 + \delta H_2O \leftrightarrow Y_{0.5}Ca_{0.5}BaCo_4O_{7+\delta} + 2\delta H^+ + 2\delta e^- \quad (1)$$

Due to the structure flexibility, $Y_{0.5}Ca_{0.5}BaCo_4O_7$ can realise an excess or deficit of oxygen ions (δ) into its structure, property common to 114 layered cobalt perovskite family [10]. After the electrode preparation step, using the method described previous, in the structure of $Y_{0.5}Ca_{0.5}BaCo_4O_{7\pm\delta}$ value of δ is 0 and cobalt average oxidation number is +2.375, being present both Co(II) and Co(III) ions.

The sulphite oxidation process are mediated by Co(II)/Co(III) redox couple, the diminution of oxygen content by electrochemical reduction in neutral media would favour the processes occurring on the perovkite electrode surface by decreasing the cobalt average oxidation number.

3.2. *Linear voltammetry*

Linear voltammograms (LVs) have been drawn at very low scan rate (1 mV s^{-1}) in order to provide quasi-stationary conditions at the electrode.

The analysis of LVs shown in Figure 2, specific potential ranges of both oxidation processes that occur at the interface perovskite electrode/neutral electrolyte were identified in all test solutions with different sulphite ions concentrations. One can be distinguished two oxidation domains:

I - potential values between -0.25 and +1.25 V characteristic for Co(II) to Co(III) oxidation inside to perovskite structure for all test solutions and SO_3^{2-} to SO_4^{2-} oxidation on perovskite electrode surface for solution with sulphite ions;

II - OER on anode surface at potential values higher than +1.25 V described by Eq. (2).

$$2H_2O \leftrightarrow O_2 + 4H^+ + 4e^- \quad (2)$$

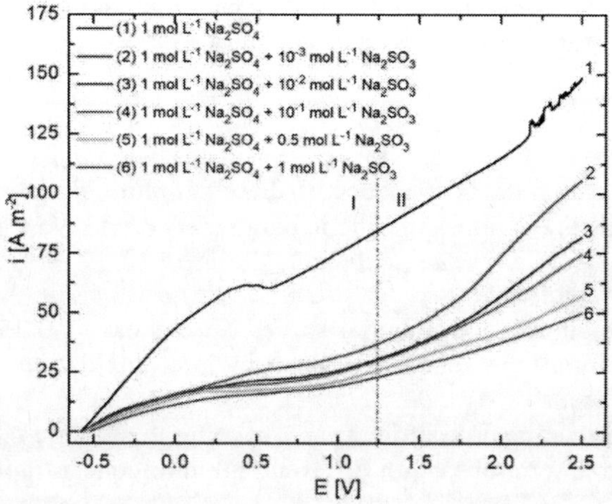

Figure 2. LVs recorded on $Y_{0.5}Ca_{0.5}BaCo_4O_7$ electrode in neutral solution in absence and presence of different concentrations of Na_2SO_3, scan rate 1 mV s^{-1}.

Based on LVs, kinetic parameters (anodic transfer coefficient α and exchange current density i_o) for electrochemical oxidation of sulphite to sulphate in neutral solution on $Y_{0.5}Ca_{0.5}BaCo_4O_7$ electrode have been calculated for each electrolyte solution using Tafel plots method (Figure 3). The kinetic parameters values are presented in Table 1.

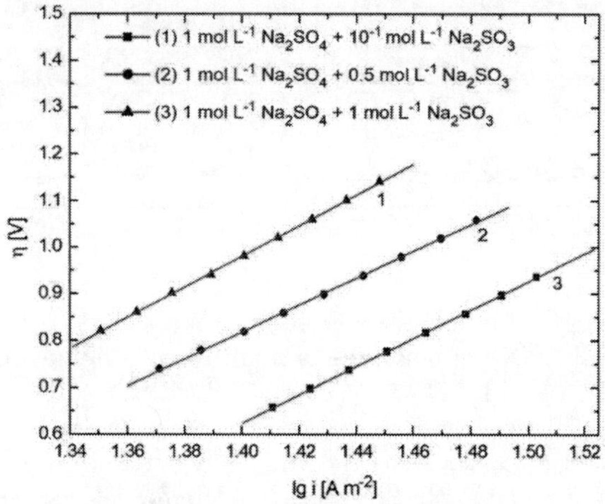

Figure 3. Tafel plots for sulphite oxidation on $Y_{0.5}Ca_{0.5}BaCo_4O_7$ electrode in neutral solution with different concentrations of Na_2SO_3.

Table 1. Kinetic parameters values for sulphite oxidation in neutral solution.

Na_2SO_3 conc. (mol L^{-1})	α	i_o (A m^{-2})
0.1	0.021	3.05
0.5	0.020	3.30
1	0.018	3.01

Transfer coefficient values prove the complexity of the anodic mechanisms for sulphite ions in neutral solution on this layered cobalt perovskite electrode.

High values obtained for exchange current density are characteristic for fast charge transfer processes. In these situation, it can be appreciated that the rate determining step is the charge transfer given by Eq. (3) similarly for the direct oxidation of sulphite [19].

$$SO^{2-}_{3\,(ads)} \rightarrow SO^{\bullet-}_{3\,(ads)} + e^- \quad (3)$$

Taking into account that on the electrode surface, in the potential range of sulphite oxidation at higher than +1.00 V, two parallel processes occur (direct sulphite oxidation and atomic or molecular oxygen generation), the values obtained for kinetic parameters α and i_o are just apparent.

3.3. *Chronoamperometric studies*

Analyzing the LVs from Figure 2, six potential values were chosen (-0.25, 0.00, +0.25, +0.50, +0.75 and +1.00 V) to highlight the sulphite direct/indirect oxidation processes that occur at electrode/electrolyte interface.

Figure 4a shows the chronoamperometric curves, relating to sulphite oxidation, for 60 minutes, on working electrode in neutral solutions, with different sulphite concentrations, at a potential value of +0.75 V.

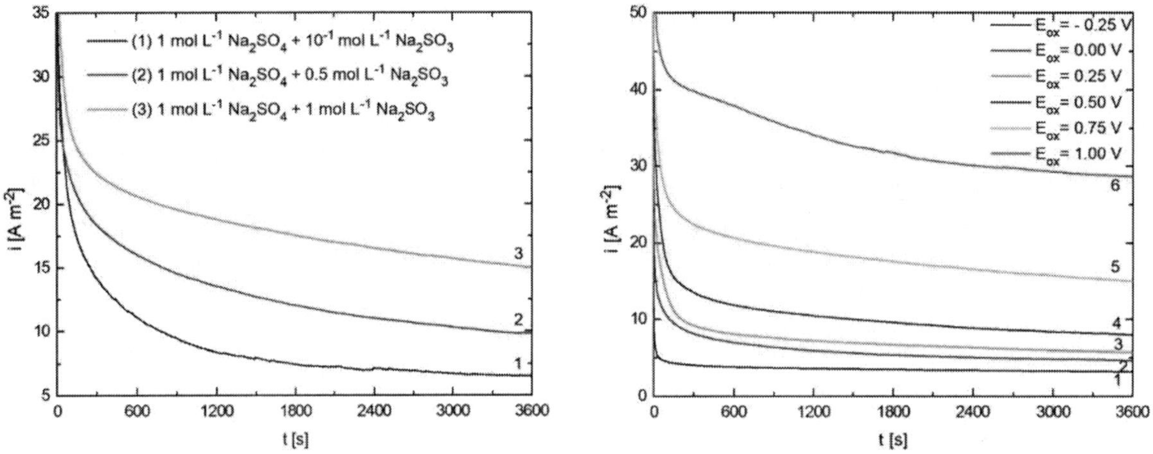

Figure 4. Current-time curves for sulphite electrooxidation process on $Y_{0.5}Ca_{0.5}BaCo_4O_7$ layered perovskite electrode in 1 mol L^{-1} Na_2SO_4 solution with different sulphite concentrations at +0.75 V (a) and in 1 mol L^{-1} Na_2SO_4 + 1 mol L^{-1} Na_2SO_3 solution (b).

In Figure 4b, the variation of current density in time registered on $Y_{0.5}Ca_{0.5}BaCo_4O_7$ electrode in neutral solution containing 1 mol L^{-1} Na_2SO_3 is presented for all six potential values at which the experiments have been carried out.

It can be observed that in the characteristic potential range for sulphite electrooxidation in neutral solution, current densities are strongly dependent on anodic potential and SO_3^{2-} ions concentration in electrolyte. The increase of sulphite concentration from 10^{-1} to 1 mol L^{-1} has a result in an increase of about 150% of the current density when oxidation process is carried out at E_{ox} = +0.75 V, Also, at same SO_3^{2-} concentration (1 mol L^{-1}) the current density increases from 2.5 A m^{-2} for E_{ox} = -0.25 V to 27.5 A m^{-2} for E_{ox} = 1.00 V. The approximate constant value of current densities in time during chronoamperometric tests in neutral solution on $Y_{0.5}Ca_{0.5}BaCo_4O_7$ electrode confirms that sulphite ions concentration used in experimental studies is high enough, its slightly decrease is depending on the decrease of SO_3^{2-} ions concentration near the interface.

IOP Conf. Series: Materials Science and Engineering **416** (2018) 012064 doi:10.1088/1757-899X/416/1/012064

3.4. Chronocoulometric studies

The amount of electricity consumed for electrooxidation of sulphite ions in the characteristic potential range for each test solution has been determinated by chronocoulometry.

In Figure 5a the sulphite transformation degree during anodic oxidation as a function of time for different sulphite concentration is presented, when the electrochemical process is carried out at +0.75 V. The highest value of the transformation degree has been obtained for the test solution containing the lowest concentration of sulphite (10^{-1} mol L^{-1}).

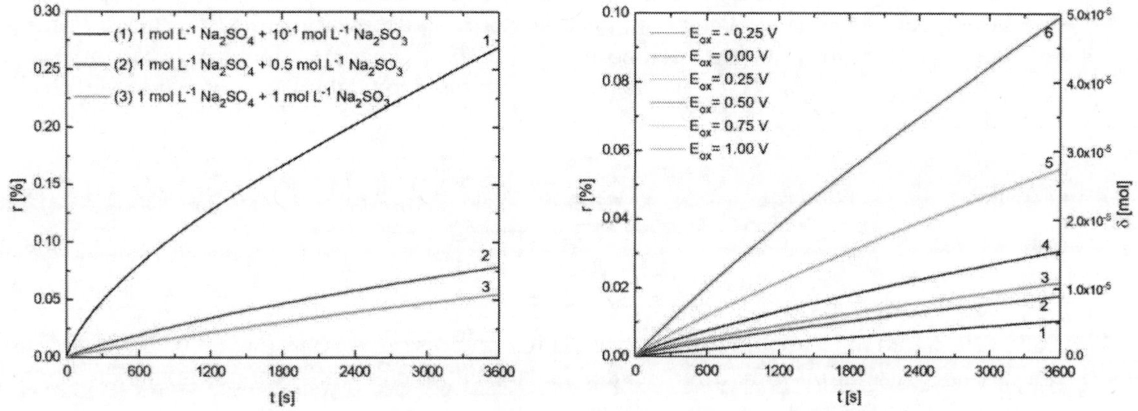

a) b)

Figure 5. Chronocoulometric curves for sulphite electrooxidation process on $Y_{0.5}Ca_{0.5}BaCo_4O_7$ electrode in neutral solution with different sulphite concentrations at +0.75 V (a) and in 1 mol L^{-1} Na_2SO_4 + 1 mol L^{-1} Na_2SO_3 solution (b).

Based on these data, the evaluation of number of sulphite moles changed in the anodic reaction (δ) and transformation degree (r) of sulphite to sulphate has been possible using Faraday's laws [20]. The results for 1 mol L^{-1} sulphite added in neutral solution at all six potentials are presented in Figure 5b.

3.5. Electrochemical impedance spectroscopy

Based on the studies presented above cyclic and linear voltammograms, Tafel plots method and chrono-electrochemical data, EIS was used to obtain specific information for sulphite ions oxidation on $Y_{0.5}Ca_{0.5}BaCo_4O_7$ electrodes in Na_2SO_4 solution with different Na_2SO_3 concentrations between 10^{-3} and 1 mol L^{-1} at 6 oxidation potential values in the range -0.25 and + 1.00 V.

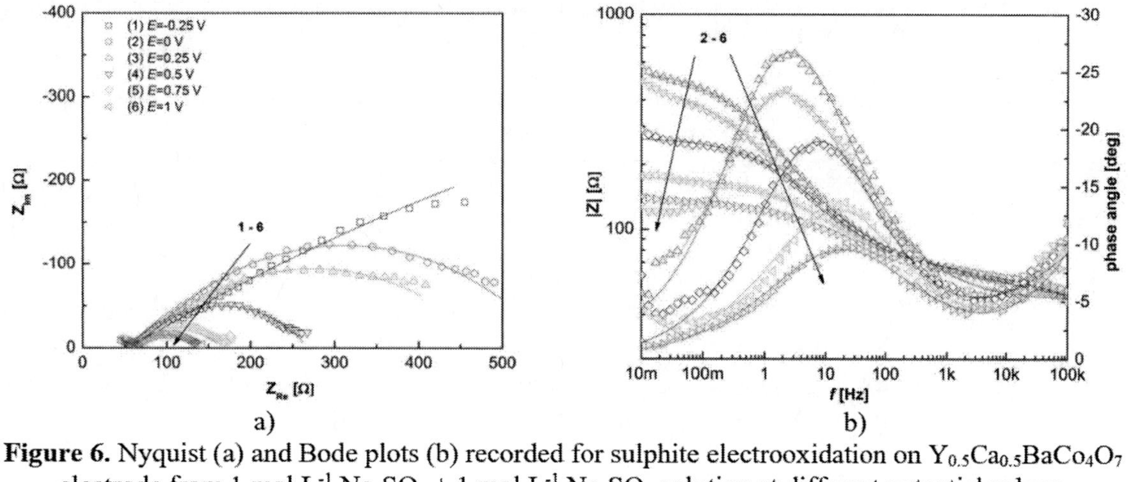

a) b)

Figure 6. Nyquist (a) and Bode plots (b) recorded for sulphite electrooxidation on $Y_{0.5}Ca_{0.5}BaCo_4O_7$ electrode from 1 mol L^{-1} Na_2SO_4 + 1 mol L^{-1} Na_2SO_3 solution at different potential values.

The obtained results presented in form of Nyquist and Bode complex plane representation are shown in Figure 6 for 1 mol L^{-1} Na_2SO_3 added in neutral solution.

The equivalent electrical circuit (EEC) shown in Figure 7 was used to fit the impedance data obtained for sulphite electrooxidation on layered cobalt perovskite electrode in 1 mol L^{-1} Na_2SO_4 with all Na_2SO_3 concentrations used in electrochemical studies in the potential range between -0.25 and +1.00 V. The EEC consists in a solution resistance *Rs* in series with two parallel connections with a constant-phase element and a resistor (*CPE* - *R*). First one (CPE_{el} - R_{el}) was used to fit impedance data recorded at the high frequency values, which are attributed to a charge-transfer process at the layered cobalt perovskite electrode surface. The second connection was necessary to fit EIS data specific to the lower frequency values being composed of CPE_{ox} element for characterization of $Y_{0.5}Ca_{0.5}BaCo_4O_7$ electric double-layer capacitance in sulphite oxidation process and the charge transfer resistance (R_{ct}).

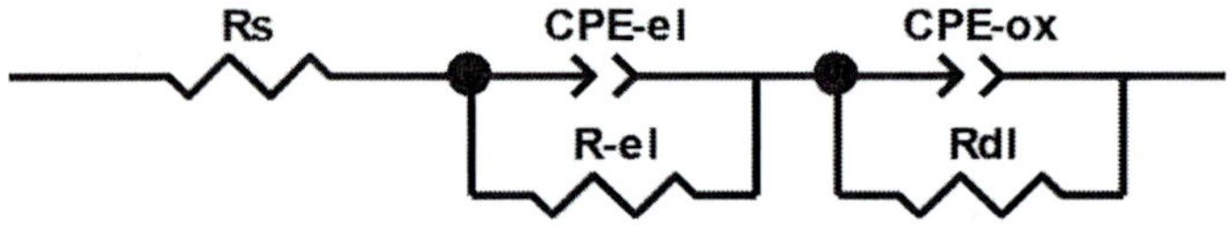

Figure 7. EECs used for modelling sulphite oxidation process on $Y_{0.5}Ca_{0.5}BaCo_4O_7$ electrode from 1 mol L^{-1} Na_2SO_4 solution with different sulphite concentration at different potential values.

The Bode diagrams shape confirm the chosen EEC for fitting EIS data obtained for sulphite electrooxidation process on $Y_{0.5}Ca_{0.5}BaCo_4O_7$ electrode in neutral solution. The fitting results for sulphite oxidation on this type of electrode are plotted as continuous line in graphical EIS sprectra and the corresponding values of EEC elements are shown in Tables 2. On the other side, the double layer capacity (C_{dl}) values have been calculated for each studied layered cobalt perovskite electrode/neutral electrolyte interface.

Analyzing the results from Table 2, it can be observed that transfer resistance values (R_{ct}) decrease significantly with the increase of sulphite concentration in neutral solution, especially if the process is led to an oxidation potential value higher than + 0.50 V, indicating that the sulphite anodic oxidation occurs with higher rate, confirming chrono -amperometric and -coulometric data. For the same sulphite concentration used in experimental studies, the charge transfer resistance decreases with the polarization increasing, the minimum values being recorded for an oxidation potential of +1.00 V. As well, the double layer capacity (C_{dl}) have been calculated and, as expected, values of this parameter increase with the oxidation potential at which the electrolysis process is carried out. The order of 10^{-3} ÷ 10^{-4} for chi-square values indicate an excellent correlation between the experimental electrochemical impedance data and the chosen EEC model.

Table 2. Calculated parameters of EEC elements for sulphite electrooxidation in neutral solution on layered cobalt perovskite electrode

SO_3^{2-} conc. (M)	*E* (V)	R_s (Ω cm²)	CPE_{el}		R_e (Ω cm²)	CPE_{ox}		R_{ct} (Ω cm²)	$C_{dl}\cdot10^5$ (F cm⁻²)	$Chi^2\cdot10^3$
			$T\cdot10^4$ (F cm⁻² s^(n-1))	n		$T\cdot10^4$ (F cm⁻² s^(n-1))	n			
10^{-3}	-0.25	3.78	6.90	0.25	44.7	1.37	0.52	1609	3.48	1.18
	0.00	3.88	5.38	0.27	46.5	1.58	0.54	1170	3.69	0.84
	0.25	3.50	4.02	0.30	49.3	1.88	0.56	680	3.75	0.50
	0.50	3.48	1.81	0.32	55.9	2.32	0.57	393	3.81	0.88
	0.75	4.09	1.00	0.34	59.9	2.57	0.59	286	4.19	0.53
	1.00	4.44	0.75	0.36	61.6	2.66	0.60	252	4.40	0.28

SO_3^{2-} conc. (M)	E (V)	R_s (Ω cm^2)	CPE_{el} $T\cdot 10^4$ (F cm^{-2} s^{n-1})	CPE_{el} n	R_e (Ω cm^2)	CPE_{ox} $T\cdot 10^4$ (F cm^{-2} s^{n-1})	CPE_{ox} n	R_{ct} (Ω cm^2)	$C_{dl}\cdot 10^5$ (F cm^{-2})	$Chi^2 \cdot 10^3$
10^{-2}	-0.25	3.19	8.33	0.26	61.9	1.44	0.54	1488	3.89	4.50
	0.00	3.23	6.26	0.29	64.5	1.64	0.55	1094	4.07	0.42
	0.25	3.59	6.44	0.31	65.6	1.96	0.57	654	4.24	1.70
	0.50	3.75	4.47	0.34	67.9	2.35	0.59	376	4.47	1.06
	0.75	3.87	2.49	0.37	74.4	2.67	0.60	269	4.54	0.32
	1.00	3.62	2.25	0.41	81.2	2.87	0.61	228	4.91	1.42
10^{-1}	-0.25	3.59	9.42	0.34	85.3	1.61	0.54	1336	4.26	0.66
	0.00	4.26	9.10	0.36	94.5	1.83	0.55	956	4.48	0.63
	0.25	4.30	8.36	0.39	101	2.04	0.58	635	4.74	1.00
	0.50	4.01	6.47	0.40	105	2.76	0.59	305	4.97	4.50
	0.75	4.01	4.84	0.43	108	3.00	0.61	235	5.43	3.56
	1.00	3.78	4.01	0.47	112	3.15	0.63	183	5.70	4.63
0.5	-0.25	3.10	11.32	0.37	104	1.63	0.55	1250	4.48	1.48
	0.00	3.52	10.91	0.38	107	1.99	0.56	817	4.69	2.15
	0.25	3.48	10.38	0.41	113	2.24	0.58	582	5.14	4.97
	0.50	3.91	8.63	0.43	117	3.28	0.59	243	5.55	2.80
	0.75	3.58	6.53	0.47	120	3.24	0.62	181	5.85	3.61
	1.00	3.79	5.99	0.50	124	3.28	0.65	140	6.20	2.81
1	-0.25	3.64	11.63	0.40	107	1.78	0.56	1085	4.86	0.49
	0.00	3.82	11.16	0.42	111	2.08	0.57	747	5.17	0.22
	0.25	4.22	10.26	0.44	114	2.34	0.59	521	5.41	0.43
	0.50	3.88	9.64	0.47	117	3.26	0.61	200	5.85	2.78
	0.75	4.28	7.51	0.51	122	3.72	0.64	115	6.14	1.94
	1.00	4.16	7.05	0.54	130	3.21	0.69	89	6.52	0.45

4. Conclusions

Experimental data presented in this paper have confirmed the possibility to oxidize SO_3^{2-} to SO_4^{2-} on an electrode made of cobalt layer perovskite type 114.

The electrocatalytic effect of this type of electrode has been studied for sulphite oxidation in neutral solution applying cyclic and linear voltammetry, chronoamperometry and electrochemical impedance spectroscopy.

The characteristic range of the optimal potential has been identified and current densities specific for sulphite oxidation and transformation degrees depending on the electrolysis time have been determined by electrochemical methods for each sulphite concentration in the neutral electrolyte solution. The obtained kinetic parameters (α and i_o) show that the overall process is controlled by the charge transfer step. This fact was confirmed by EIS studies, based on which charge transfer resistance (R_{ct}) and double-layer capacity (C_{dl}) were determined for different potential of the chosen range and for each sulphite concentration added in the test solution. The significant value of exchange current density i_o (3.30 A m^{-2}) emphasizes an important catalytic effect of the working electrode for the studied process.

The electrochemical behavior of sulphite ions on $Y_{0.5}Ca_{0.5}BaCo_4O_7$ electrode, especially their facile oxidability in neutral media, suggests that this type of electrode could be used in a Na_2SO_3 (aq) / O_2 fuel cell.

5. References

[1] Fujihira M, Bard AJ and Stratmann M 2007 *Encyclopedia of Electrochemistry* vol 10 (New York: Wiley)
[2] Heli H, Eskandari I, Sattarahmady N and Moosavi-Movahedi AA 2012 *Electrochim. Acta* **77** 294-301
[3] Valldor M 2006 *Solid State Sci.* **8** 1272-80
[4] Wang S, Hao H, Zhu B, Jia J and Hu X 2008 *J. Mater. Sci.* **43** 5385-89
[5] Zhang Y, Wu X and Han R 2012 *J. Chin. Ceram. Soc.* **40** 289-93
[6] Sattarahmady N and Heli H 2011 *Anal. Biochem.* **409** 74-80
[7] Heli H, Majdi S and Sattarahmady N 2010 *Sens. Actuators B* **145** 185-93
[8] Redepenning JG 1987 *Trends Analyt. Chem.* **6** 18-22
[9] Mortimer RJ 1997 *Chem. Soc. Rev.* **26** 147-56
[10] Dan ML, Pralong V, Vaszilcsin N, Kellenberger A and Duteanu N 2011 *J. Solid State Electrochem.* **15** 1227-33
[11] Enache A, Vaszilcsin N and Dan ML 2015 *Annals Univ.Oradea, Fascicle Environ. Prot.* **25** 186-92
[12] Grenier JC, A. Wattiaux A, Doumerc JP, Dordor P, Fournes L, Chaminade JP and Pouchard M 1992 *J. Solid State Chem.* **96** 20-30
[13] Grenier JC, Bassat JM, Doumerc JP, Etourneau J, Fang Z, Fournes L, Petit S, Pouchard M and Wattiaux A 1999 *J. Mater. Chem.* **9** 25-33
[14] Ismail KM and Badawy WA 2000 *Applied Electrochem.* **30** 1303-11
[15] Dan ML, Vaszilcsin N, Kellenberger A and Duteanu N 2011 *Studia UBB Chemia* **56** 119-27
[16] Allen JA, Rowe G, Hinkley JT and Donne SW 2014 *Int. J. Hydrogen Energy* **39** 11376-89
[17] O'Brien JA, Hinkley JT, Donne SW and Lindquist SE 2010 *Electrochim. Acta* **55** 573-91
[18] Zelinsky AG 2016 *Electrochim. Acta* **188** 727-33
[19] Skavas E and Hemmingsen T 2007 *Electrochim. Acta* **52** 3510-17
[20] Dan ML, Vaszilcsin N and Enache AF 2016 *Chem. Bull. "POLITEHNICA" Univ. Timisoara* **75** 58-65

Acknowledgement
This work was partially supported by Politehnica University Timisoara in the frame of PhD studies.

Effect of Silica Fume and Siliceous Fly Ash Addition on the Fracture Toughness of Plain Concrete in Mode I

D M Gil[1]and G L Golewski[1,*]

[1]Faculty of Civil Engineering and Architecture, Lublin University of Technology, Nadbystrzycka, 40 str. 20-618 Lublin, Poland

[*]Corresponding author: g.golewski@pollub.pl

Abstract. Nowadays, structural concretes should meet strict requirements such as: strength, durability, and resistance to adverse weather conditions and corrosion. Therefore, research on modifying concrete with mineral additives and chemical admixtures is conducted. Siliceous fly ash (FA) and silica fume (SF) are by-products of industrial processes and their application to the concrete mixtures can improve both parameters of composites, as well as greatly contributing to environmental protection. This paper presents the results of fracture toughness tests, specified at the Mode I fracture (tension at bending; opening mode), based on RILEM Draft Recommendations. The experiments were carried out after 28 days of curing of plain concrete, which consisting the following additives: 10% SF + 0% FA, 10% SF + 10% FA, 10% SF + 20%FA. To assess mechanics parameters compressive strength tests and fracture toughness tests were conducted. Fracture toughness tests were carried out on MTS 809 Axial/Torsional Test System.In our studies, the fracture toughness of the composites based on the experimental results of critical stress intensity factor $- K_{Ic}^{s}$. Detailed analysis of the obtained results lead to conclusion that the combined addition of FA and SF clearly changes the fracture toughness according to the Mode I fracture.

1. Introduction

The 21st century is a huge challenge for construction engineers. It is marked by the search for new technological solutions and modern construction materials. Are used, among others hybrid innovative techniques based on the synergy of the combined components together. An example of such materials are binary, ternary or even quaternary binders with the use of mineral additives, e.g. [1-4].These include, among others, fly ash cements with the addition of silica fume [5-8]. Currently, many research centers conduct research both on the separate use of SF [9,10] and FA for concrete, as well as tests assessing the properties of composites, to which a combined addition of both materials was used.

Siliceous fly ash (FA) is a by-product obtained in the process of hard coal combustion performed in electric power stations and in thermal-electric power stations [11-13]. It is removed by mechanical collectors or electrostatic precipitators as a fine particulate residue from the combustion gases before they are discharged to the atmosphere. Furthermore, FA from coal combustion processes is not hazardous from the radiological point of view [14].

Silica fume (SF) is a byproduct of producing silicon metal or ferrosilicon alloys. SF consists primarily of amorphous (non-crystalline) silicon dioxide (SiO_2). The individual particles are small, approximately 100 times smaller than the size of the average cement particle. Because of its chemical and physical properties, i.e.: fine particles, large surface area, and the high SiO_2 content, SF is a very reactive pozzolan when used in concrete. Moreover it should be added that, modification of the concrete structure with active mineral additives, such as FA or SF, is justified by ecological as well as economic reasons, e.g. [15].

Damages to concrete in the form of microcracks and cracks are one of the main reasons for deterioration of its physical and mechanical parameters and finally destruction of the material. Therefore, in order to increase the durability of concrete and other materials, it is necessary to know not only the basic strength parameters of these composites, but also the parameters of fracture mechanics, e.g. [8,9,16-29].

2. Experimental details

The paper presents the results of research on the macroscopic assessment of fracture toughness of concrete according to the first model of cracking. Own research allowed to define the basic parameter of fracture mechanics that is the critical stress intensity factor –(K^{S}_{Ic}) [30-35]. Additional studies were compressive strength (f_{cm}) and the splitting tensile strength (f_{ct}).

The mixtures were made with the following composition of additives:
0% FA+10% SF (FA-00+SF-10),
10%FA+10%SF (FA-10+SF-10),
20% FA+10% SF (FA-20+SF-10).

The type of materials used for making the mixtures and the mix proportion of the concretes composition are shown in Table 1. The compositions of concrete mixtures were selected in such a way that one can compare the effect of both pozzolanic additives, used in different amounts, on the analyzed concrete parameters. In all series of concrete, the constant amount of the binder substitute at 10% was SF.A variable parameter was the addition of FA, which replaced the Portland cement in the amount of: 0, 10 and 20%.

Table 1. Details of concrete mixtures (kg/m³)

Component of the mixture \ Type of concrete	FA-00+SF-10	FA-10+SF-10	FA-20+SF-10
Portland cement CEM I 32.5R	317	282	246
Siliceous fly ash	0	35	70
Silica fume - concentrated	35	35	35
Water	141	141	141
Sand 0–2 mm	676	676	676
Gravel 2–8 mm	1205	1205	1205
Plasticizer – Stacheplast 125	2	2	2

For each type of concrete mixture were made:
6 cubic samples (150 mm) for compressive strength tests according to PN-EN12390-3,
6 cubic samples (150 mm) for splitting tensile strength tests according to PN-EN12390-6,
6 beams (80 x 150 x 700 mm) for fracture toughness tests at mode I fracture.

After demoulding the samples cured for 14 days in water and then until the 28th day in laboratory conditions. After 28 days, the strength tests of concrete were made on the Controls Advantest 9 hydraulic press. The obtained average results are presented in Table 2. In this Table also gives the coefficients of variation v_{fcm} and v_{fct} for the tests carried out.

Table 2. Compressive strength and spliting tensile strength results

Type of concrete	f_{cm} (MPa)	v_{fcm} (%)	f_{ct} (MPa)	v_{fct} (%)
FA-00+SF-10	59.40	4.38	4.22	8.63
FA-10+SF-10	59.50	3.73	3.52	6.99
FA-20+SF-10	56.33	7.34	3.30	3.44

3. Mode I fracture test

Fracture toughness tests were carried out on MTS 809 Axial/Torsional Test System. The study was based on the RILEM draft recommendation [36] in a stand similar to that presented in [24,34]. The experiments were performed for the loading scheme presented in Fig. 1. Sample dimensions in detail were also shown in Figure 1.

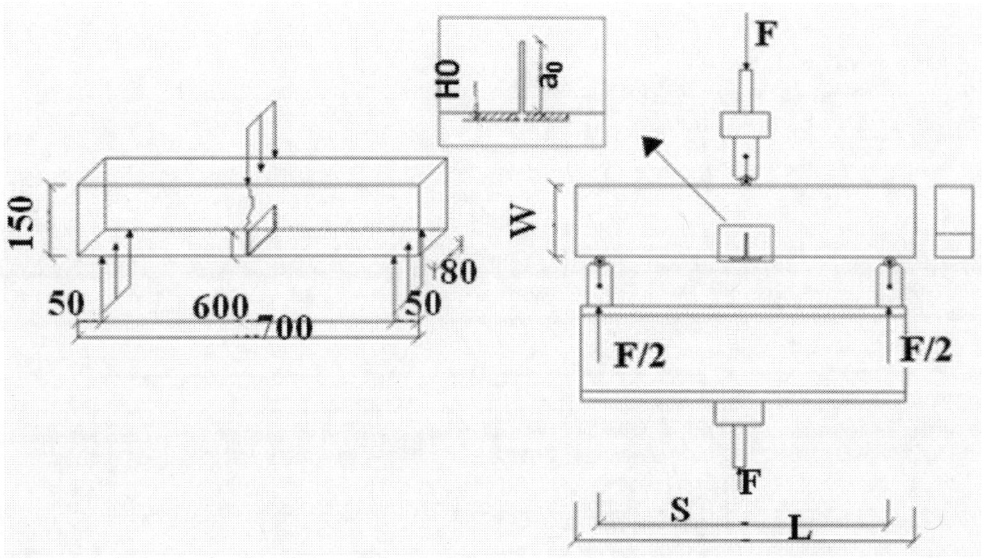

Figure 1. Schematic drawings of the specimen used in the fracture toughness examination according to mode I: HO – clamp gauge holder thickness, a_0 – height of initial crack, CMOD – crack mouth opening displacement.

To assess the fracture toughness of concrete, in the Mode I fracture, we used beams with the dimensions 700 x 150 x 80 mm which had one initial centrally crack. The beams were subjected to 3-point bending test. They were made in demountable bolted wooden forms (Fig. 2). The assumed size of the initial crack in the beams was achieved by actually concreting flat steel plates, having a thickness of 3 mm. A special experimental stand was prepared as shown in Fig. 4. All necessary results, needed to determine the critical stress intensity factors for concretes, were obtained with application MTS 809 testing machine. The width of the initial crack opening during the tests was measured using a crack opening sensor that is the MTS clip gage axial extensometer, which was placed on the clamping test grips.

Figure 2. A sample for testing after formation

The specimens placed in the experimental stand were subjected to cyclic loading process performed quasi statically. Loading rate was selected so that the maximum load was reached in approximately 5 min. The applied load was reduced at approximately 95% post-peak load. After reducing the load to zero, the

IOP Conf. Series: Materials Science and Engineering **416** (2018) 012065 doi:10.1088/1757-899X/416/1/012065

test specimen was loaded again. After that the cycles were repeated until the beams were broken into 2 parts (Fig. 4). The whole cyclic deformation processes were described by the following curves: Force (*F*) – crack mouth opening displacement (CMOD) (Fig. 3) and *F* – deflection (*f*).
The obtained results allowed for determination of the fracture toughness K_{Ic}^{S} using the formula given in RILEM recommendations [36].

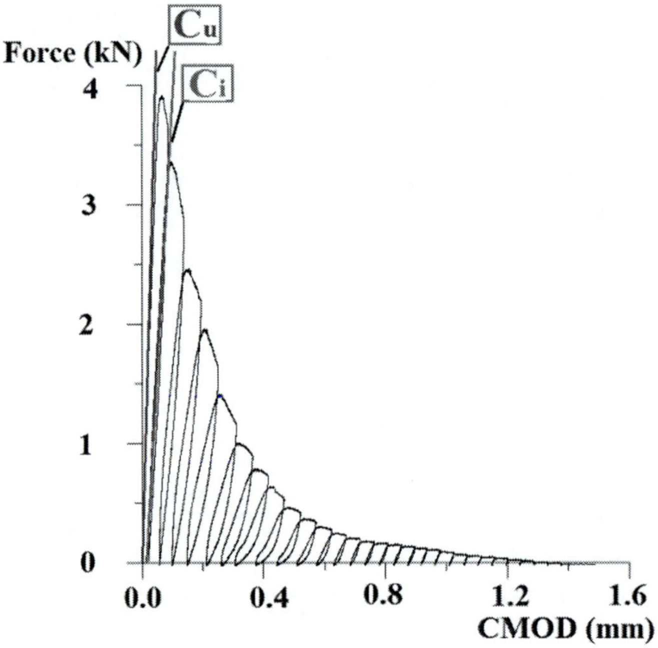

Figure 3. An exemplary failure curve for FA10 + SF10 series concrete (Force-CMOD): C_i – initial compliance, C_u – unloading compliance.

Table 3. Results of fracture toughness tests.

Type of concrete	K_{Ic}^{S} (MN/m$^{3/2}$)	v (%)
FA00+SF10	1.34	6.99
FA10+SF10	1.40	15.23
FA20+SF10	1.33	5.57

Based on the results obtained from the study, it can be observed that after 28 days the highest values were obtained using a combined addition of 10% SF and 10% FA. Based on the analysis of the results obtained in own research (Table 3), it can be concluded that the addition of FA to concrete caused an increase in the basic parameter of fracture mechanics - K_{Ic}^{S}. In the case of FA00+SF10 and FA20+SF10 by 5.0%.

The use of 10% FA to the mixture did not reduce the compressive strength. Slightly compressive strength decreased with 20% FA.

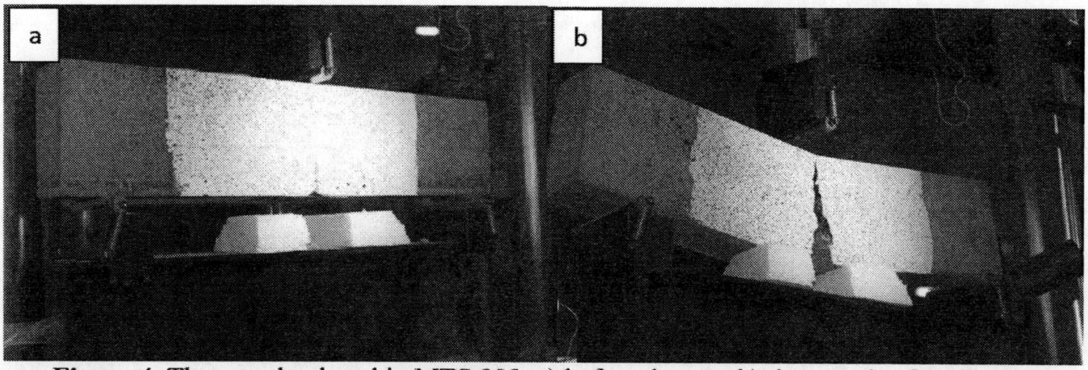

Figure 4. The sample placed in MTS 809. a) before the test, b) the sample after the test.

4. Conclusions

From this research the following conclusions can be formulated:

1. The FA and SF additives change the fracture toughness K_{Ic}^{S} in relation to the ash concretes.
2. The concretes with the FA and SF has a higher from 20% to 30% are fracture toughness compared to reference concrete and concrete with the addition of only FA.
3. The addition of SF to ash concrete results in a significant increase in their compressive and tensile strength.
4. The most favorable compressive strength and fracture toughness results are obtained by replacing 20% of the cement mass in the concrete mix with the total addition of FA and SF in the same amount.

5. References

[1] Bernal J, Reyes E, Massana J, León N and Sànchez E 2018 Fresh and mechanical behawior of a self-compacting concret with additions of nano-silica, silica fume and ternary mixtures *Constr. Build. Mater.* **160** 196–210.
[2] Dave N, Mistra AK, Srivastava A and Kaushik SK 2016 Experimental analysis of strength and durability properties of quaternary cement binder and mortar *Constr. Build. Mater.* **107** 117– 124.
[3] Vivek SS and Dhinakaren G 2017 Durability characteristics of binary blend high strength SCC *Constr. Build. Mater.* **146** 1–8.
[4] Niewiadomski P, Hola J and Cwirzen A 2018 Study on properties of self-compacting concrete modified with nanoparticles *Arch. Civ. Mech. Eng.* **18** 877–886.
[5] Nochaiya T, Wongkeo W and Chaipanich A 2010 Utilization of fly ash with silica fume and properties of Portland cement–fly ash–silica fume concrete *Fuel* **89** 768–774.
[6] Giergiczny E and Giergiczny Z 2010 Influence of siliceous fly ash variable quality on the propeties of cement-fly ash composties *Cem. Wapno Beton* **3** 157–163.
[7] Rathish Kumar P, Sumanth Reddy C and Saleem Baig M D 2014 Compressive strenght performance of high strength concretes using binary supplementary cementitious materials *Cem. Wapno Beton* **1** p 8–16.
[8] Zhang P, Qiao-Yan Guan Q Y and Zhang T H2016 Fracture behavior of fly ash concrete containing silica fume *Struct. Eng. Mech.* **59**(2) 261–275
[9] Prokopski G and Langier B 2000 Effect of water/cement ratio and silica fume addition on the fracture toughness and morphology of fractured surfaces of gravel concretes *Cem. Concr. Res.* **30** 1427–1433.
[10] Sosdean C, Marsavina L and De Schutter G 2016 Experimental and numerical determination of the chloride penetration in cracked mortar specimens *Eur. J. Env. Civ. Eng.* **20**(2) 231–249.
[11] Ahmaruzzaman M 2010 A review on the utilization of fly ash *Prog. Ener. Comb. Sci.* **36** 327–363.
[12] Golewski G L 2018 An assessment of microcracks in the Interfacial Transition Zone of durable concrete composites with fly ash additives *Compos. Struct.* **200** 515–520.
[13] Golewski G L 2018 Evaluation of morphology and size of cracks of the Interfacial Transition Zone (ITZ) in concrete containing fly ash (FA) *J. Hazard. Mater.* **357** 298–304.

[14] Golewski G L 2015 Studies of natural radioactivity of concrete with siliceous fly ash addition *Cem. Wapno Beton* **2** 106–114.

[15] Alwaeli M 2011 Economic calcus of the effectiveness of waste utilization processed as substitutes of primary materials *Env. Prot. Eng.* **37** 51–58.

[16] Peride N, Craciun E M, Carabineanu A and Marsavina L 2008 Mixed mode crack propagation in advanced materials *J. Optoelectr. Advan. Mater.* **10**(11) 2870–2875.

[17] Craciun E M, Carabineanu A and Peride N 2008 Antiplane interface crack in a prestresssed fiber-reinforced elastic composite *Comput. Mater. Sci.* **43**(1) 184–189.

[18] Golewski G L 2017 Generalized fracture toughness and compressive strength of sustainable concrete including low calcium fly ash *Materials* **10**(12) 1393.

[19] Golewski G L 2017 Effect of fly ash addition on the fracture toughness of plain concrete at third model of fracture *J. Civ. Eng. Manag.* **23** 613–620.

[20] Golewski G L 2017 Determination of fracture toughness in concretes containing siliceous fly ash during torsional loading *Struct. Eng. Mech.* **62**(1) 1–9.

[21] Golewski G L and Sadowski T 2017 The fracture toughness the K_{IIIc} of concretes with fly ash (FA) additives *Constr. Build. Mater.* **143** 444–454.

[22] Golewski G L and Sadowski T 2014 An analysis of shear fracture toughness K_{IIc} and microstructure in concretes containing fly-ash *Constr. Build. Mater.* **51** 207–214.

[23] Golewski G L and Sadowski T 2012 Experimental investigation and numerical modeling fracture processes under Model II in concrete composites containing fly-ash additive at early age *Sol. Stat. Phenom.* **188** 158–163.

[24] Golewski G L 2018 Green concrete composite incorporating fly ash with high strength and fracture toughness *J. Clean. Prod.* **172** 218–226.

[25] Golewski G L 2017 Improvement of fracture toughness of green concreto as a result of addition of coal fly ash. Characterization of fly ash microstructure *Mater. Charact.* **134** 335–346.

[26] Golewski G L 2018 Effect of curing time on the fracture toughness of fly ash concrete composites *Compos. Struct.* **185** 105–112.

[27] Golewski G and Sadowski, T 2006 Fracture toughness at shear (mode II) of concretes made of natural and broken aggregates *Brittle Matrix Compos.* **8** 537–546.

[28] Marsavina L Constantinescu D M, Linul E, Voiconi T and Apostol D A 2015 Shear and mode II fracture of PUR foams *Eng. Fail. Anal.* **58**(2) 465–476.

[29] Berto F, Ayatollahi M and Marsavina L 2017 Mixed Mode Fracture: Recent Developments *Theor. Appl. Fract. Mech.* **91** 1.

[30] Cristescu N D, Craciun E M and Soos E 2004 *Mechanics of elastic composites* Boca Raton: Chapman&Hall/CRC Press.

[31] Konkol J and Prokopski G 2016 Fracture toughness and fracture surfaces morphology of metakaolinite-modified concrete *Constr. Build. Mater.* **123** 638–648.

[32] Linul E, Movahedi N and Marsavina L 2017 The temperature effect on the axial quasi-static compressive behawior of ex-situ aluminum foam-filled tubes *Compos. Struct.* **180** 709–722.

[33] Sadowski T and Golewski G L 2018 A failure analysis of concrete composites incorporating fly ash torsional loading *Compos. Struct.* **183** 527–535.

[34] Golewski G L and Sadowski T 2016 Macroscopic evaluation of fracture processes in fly ash concrete *Sol. Stat. Phenom.* **254** 188–193.

[35] Marsavina L, Berto F, Negru R, Serban D A and Linul E 2017 An engineering approach to predict mixed mode fracture of PUR foams based on ASED and micromechanical modelling *Theor. Appl. Fract. Mech.* **91** 148–154.

[36] TC 89-FMT Fracture Mechanics of Concrete Test Methods 1990 Determination of fracture parameters (K_{Ic} and $CTOD_c$) of plain concrete using three-point bend tests. RILEM Draft Recommendations *Mater. Struct.* **23** 457–460.

IOP Conf. Series: Materials Science and Engineering **416** (2018) 012066 doi:10.1088/1757-899X/416/1/012066

The Degrading Phenomenon of the Cylindrical Bearing Rollers Due to Multiple Collisions During the Manufacturing Process

T Manescu[1,*], G Balan[2] and R Avram[2]

[1]"Aurel Vlaicu" University of Arad, Bv. Revoluţiei, no. 77, 310130 Arad, România
[2]"Eftimie Murgu" University of Reşiţa, Sq. Traian Vuia, no. 1-4, 320085 Reşiţa, România

[*]Correspoding author: manescu.tiberiu@gmail.com

Abstract. In the various production phases of the bearing rollers, multiple collisions occur, which lead to defects in the finite product. These small defects (dents) have a negative effect during exploitation: they can be the initiation point in fissure spreads which lead to vibrations and a high level of noise. Thus, the bearing is degrading. For the current study various analytical calculations have been performed in order to highlight the energy transfer between the two elements. They have been validated through the finite element method (FEM).The results obtained emphasize that critical areas may appear in the test piece. In those areas the equivalent stress is close to the yield point of the material used in bearing rollers manufacturing.

1. Introduction

During the manufacturing process of the bearing rollers, one of the causes which lead to defects is the collision of the rollers at various working stations. These collisions lead to dents which can later produce bearing failures.[1] The defect analysis can be approached under many different angles [2-4]. If rollers with significant permanent deformations are part of the bearing assembly, vibrations and noises are to be expected during its functioning, see [5],[6]. The vibration phenomenon is studied by consecrated researchers related to this topic [7],[8]. The basic rating life which is the life associated with 90 % reliability, with commonly used high quality material, good manufacturing quality and with conventional operating conditions is treated under the ISO 281:2007 norm [9]. However, the norm is only a guideline for the lifetime assessment and it cannot introduce parameters for a large variety of factors that influence the bearing life rating: rolling elements with/without defects, corrosion, electrical erosion, etc.

Thus, it is important to analyse the multiple collisions during the workflow in order to evaluate the critical manufacturing process where the defects occur. The contact problems of bodies are approached by Hertz [10], Hartnett [11], de Mul [12] and other researchers specialised in this engineering field [13-15]. Due to the complexity of the manufacturing processes and regarding the fact that the mass production of the rollers is not able to handle them in a one piece flow, the defects cannot be totally eliminated, but with the proper approach they can significantly be reduced.

This paper contains an extensive study of the 16x24 mm cylindrical rollers using two distinctive methods: analytic and FEM for the rollers at the classification working station. The aim of the current article is to study the energy transfer during the impact of the rollers and to conclude if it could lead to permanent deformations.

2. Analytic study

Analytic calculations have been performed for a roller bearing 16x24 mm. For the chosen type the preconditions have been determined as presented in table 1: A-section surface; V - volume; M - mass; G – weight.. In addition, the Earth gravity is considered g=9,8 [m/s^2].

Table 1. Geometrical and mechanical parameters of the roller

d [mm]	L [mm]	A [mm^2]	V [mm^3]	M [kg]	G [N]
16	24	201,06190	4825,4856	0,037880	0,3712

The path of a roller bearing from the manufacturing machine (CP) to the collecting recipient (MB) is presented in figure 1. Following notations occur: the sliding surface at a 30 degree angle (J), the orifice that connects the sliding surface with the collecting recipient (O), (G)- the roller weight, I and II represent the roller position (horizontal and vertical) and h1, h2, h3 are the height of the sliding surface, the orifice and the height of the collecting recipient. h1+h2+h3= h (total height) 430mm. V is the notation for the velocity of the roller.

The analysis has been performed for two cases: A - Collision head to head, and B - Collision head to generatrix, with an angle between 0 and 60 degrees. The two cases are specific for the analysed manufacturing line. For other geometry of the collecting recipient and different velocity of the rollers the collision angles are different, thus another analysis should be performed.

In order to calculate the velocity of the rollers when they reach the bottom of the collecting recipient, the law of mechanical energy conservation is applied.

$$Ep = Ec \tag{1}$$

where Ep represents the potential energy of the roller at CP and Ec is the kinetic energy when the roller reaches the collecting recipient in case I or II.

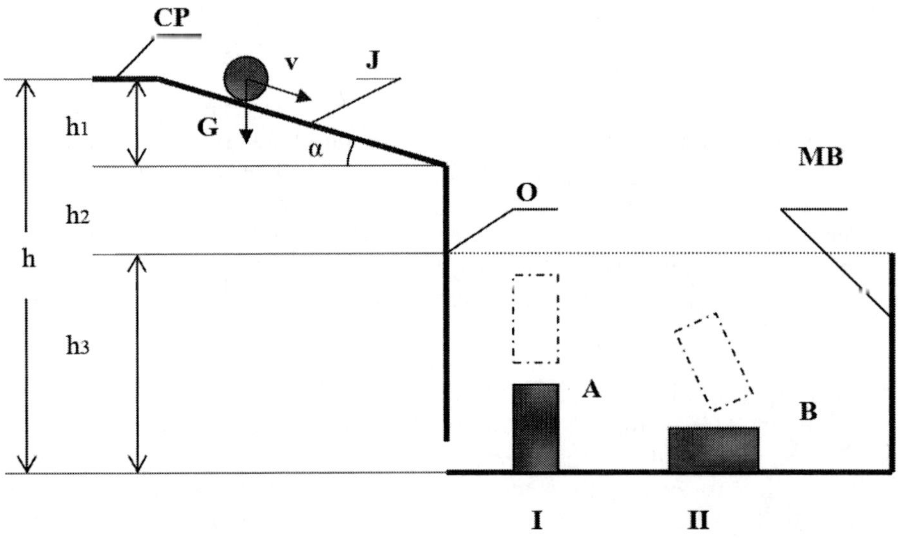

Figure 1. The collision of bearing rollers during production.

The maximum velocity at the impact moment can be calculated:

$$v_{max} = \sqrt{2gh} \tag{2}$$

where h is the maximum height before fall = 0,43 [m]

For the above mentioned values: v_{max}= 2,903 ≈ 3,0 [ms^{-1}].

The approximate analytical solution regarding shock between solids can be approached using the law of mechanical energy conservation.

It is defined as an impact coefficient or dynamic multiplier (Ψ) the ratio between the instantaneous force at the impact moment P and the weight G of the moving body. In fig.1 h is the height and the solid body deformation arrow is measured in the impact direction.

$$\Psi = \frac{P}{G} \tag{3}$$

Due to the small deformations values, a simplified equation will be used to determine the impact coefficient:

$$\Psi = \sqrt{\frac{2h}{\Delta s}} \tag{4}$$

Δs (static deformation) will be calculated for the roller chosen in table 1 and the two impact situations A and B (fig. 4)

Between the static and dynamic deformation, static and dynamic stress, following relations occur:

$$\Delta din = \Psi \Delta s \tag{5}$$

$$\sigma din = \Psi \Delta s \tag{6}$$

For the analytical calculation case I-A (fig.2) the static stress and deformation are calculated using the relation (7) and (8):

$$\sigma = \frac{Gi}{Ai} \tag{7}$$

$$\Delta_{si} = \frac{G_i L_i}{E A_i} \tag{8}$$

where E isYoung's Modulus of the roller material.

The formulas for the analytical calculation of case II B are as following: static stress (9), maximum bending moment (10), axial stress (11), static deformation (12), axial inertia moment (13) as listed below:

$$\sigma_{si} = \frac{M_{imax}}{W_i} \tag{9}$$

$$M_{imax} = \frac{G_i l_i}{4} \tag{10}$$

$$Wi = \frac{\pi d_i^3}{32} \tag{11}$$

$$\Delta_{si} = \frac{G_i L_i^3}{48 E I_i} \tag{12}$$

$$I_i = \frac{\pi d_i^4}{64} \tag{13}$$

It is necessary to calculate the value of the dynamic force (instantaneous force at the impact moment) with the relation (14.a), the static force is the weight of the roller (14.b).

$$Fdin_i = \Psi Fst_i \tag{14.a}$$

$$Fst_i = G_i \tag{14.b}$$

IOP Conf. Series: Materials Science and Engineering **416** (2018) 012066 doi:10.1088/1757-899X/416/1/012066

The analytical calculations performed for the cases A and B are centralized in Table 2 and Table 3, with the initial data E=2,1*10^5 N/mm^2 and h=0,43m

Table 2. Head-head contact

σ_{s1}	MPa	0,0018463170	Δ_{s1}	mm	0,0000002110
Ψ_1	(2h/Δs)^0,5	63841,0645468651	F_{din1}	N	23699,33535
σ_{din1}	MPa	117,8708414938	Δ_{din1}	mm	0,0134709533

Table 3. Head-generatrix contact (impact angle 0)

F_1=G_1	N	4,154	M_{max1}	Nmm	2,22734400
Iz_1	mm^4	3216,990877	W_{z1}	mm^3	402,1238597
σ_{s1}	MPa	0,00184632	Δ_{s1}	mm	0,00000016
Ψ_1	(2h/Δs)^0,5	73717,31707	F_{din1}	N	27365,63731
σ_{din1}	MPa	136,1055342	Δ_{din1}	mm	0,01166619

The study for the rollers collision was conducted through FEM for two cases: A head to head and B head to generatrix, angle 0 and 45 degrees.

Dynamic analysis head to head impact (Case A). The FEM model contains two bearing rollers: the first is in motion with a velocity of 3m/s and the second is static (fig.2). The mesh has a total of 70329 nodes and 65150 elements (fig. 3). The maximum von Misses stress is 151.8 [MPa] as shown in figure 4., located at the lower part of the roller (fig. 5). The maximum deformation is 0,22697 [mm] (fig. 6 and 7).

Figure 2. Rollers impact

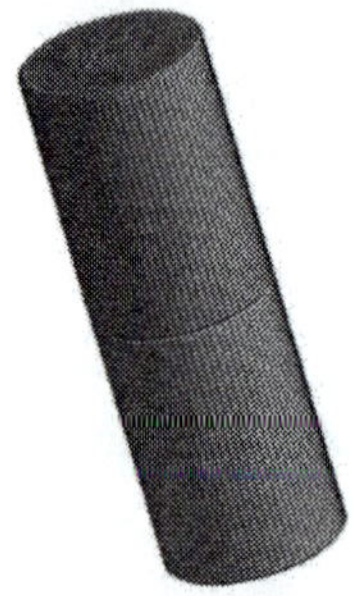

Figure 3. Grid

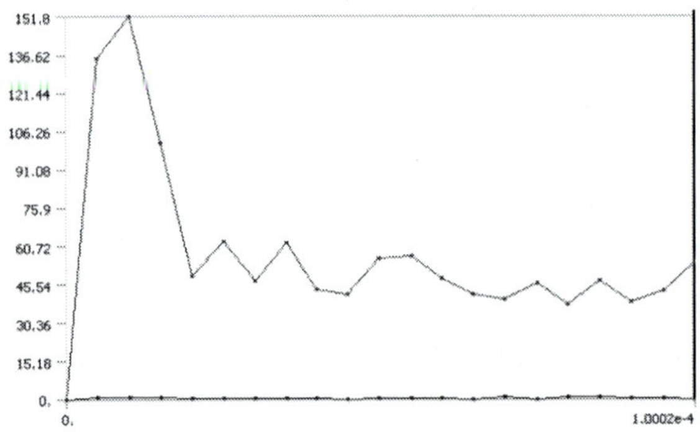

Figure 4. Von Misses Equivalent Stress

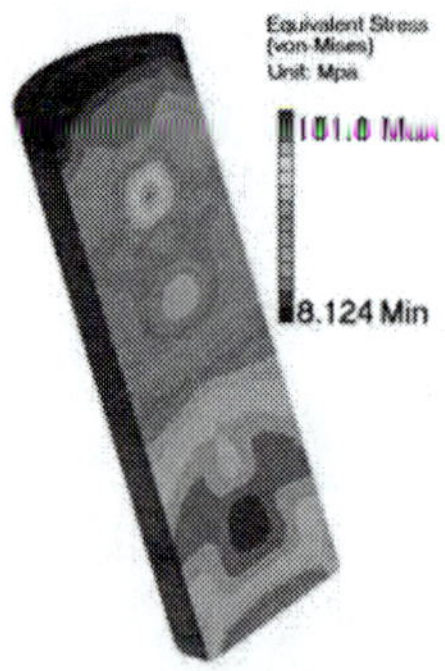

Figure 5. Equivalent Stress

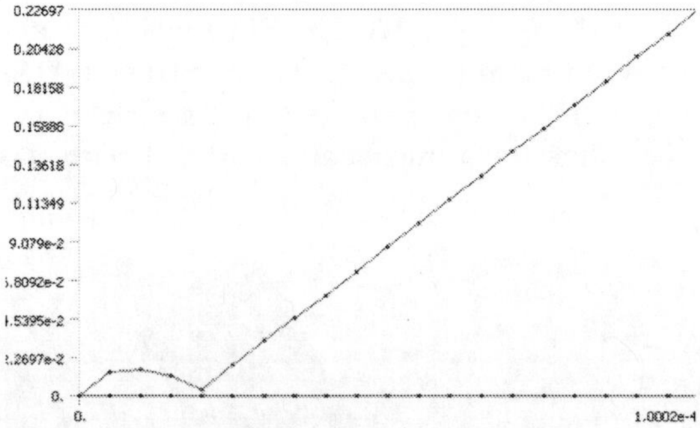

Figure 6. Total deformation graphic

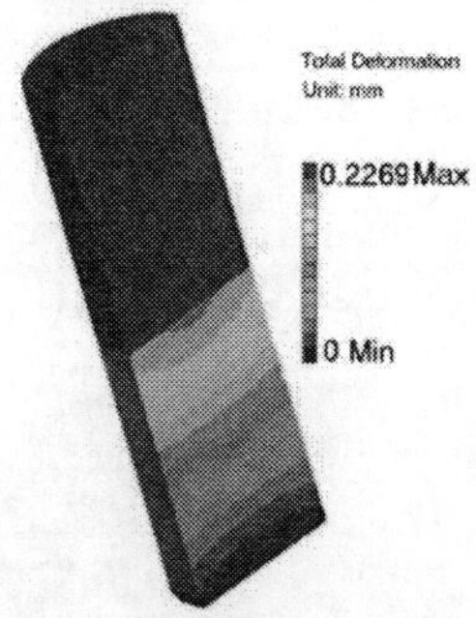

Figure 7. Total deformation

Dynamic analysis of the head-generatrix impact (angle 0°, case B). The FEM model contains two bearing rollers: the first is in motion with a velocity of 3m/s and the second is static (fig.8). The mesh has a total of 79380 nodes and 73617 elements (fig. 9). The maximum von Misses stress is 266.86 [MPa] as shown in fig 10., located at the impact area of the two rollers (fig. 11). The maximum deformations are shown in fig. 12 and 13.

Figure 8. Rollers impact

Figure 9. Grid

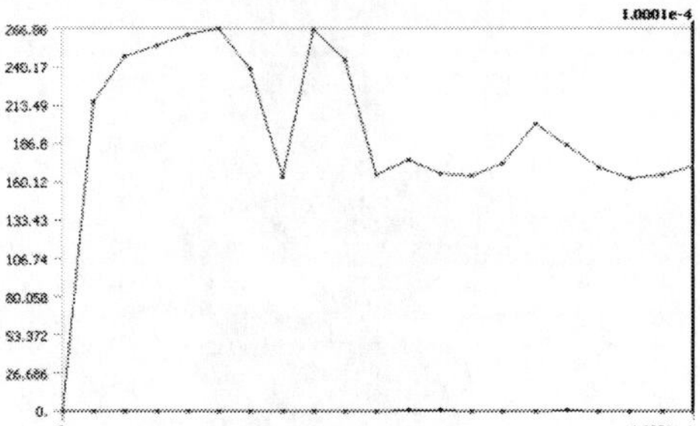

Figure 10. Von Misses Equivalent Stress

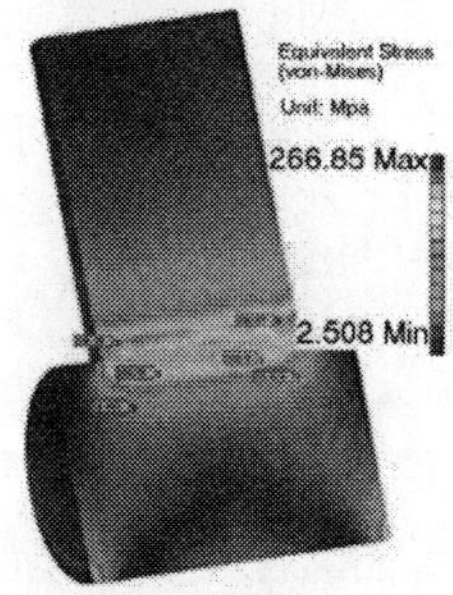

Figure 11. Equivalent Stress

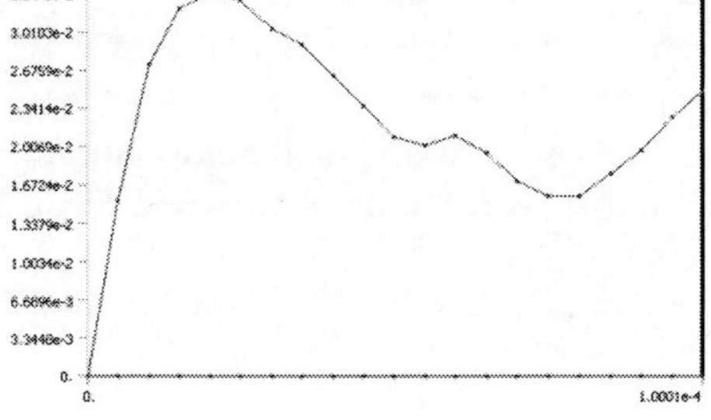

Figure 12. Total deformation graphic

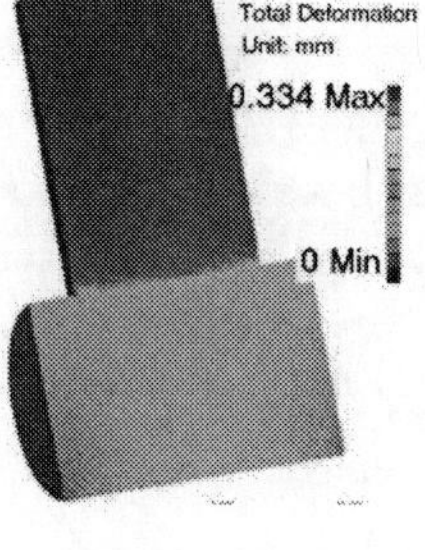

Figure 13. Total deformation

IOP Conf. Series: Materials Science and Engineering **416** (2018) 012066 doi:10.1088/1757-899X/416/1/012066

Dynamic analysis of the head-generatrix impact (angle 45°, case B). The FEM model contains two bearing rollers: the first is in motion with a velocity of 3m/s and the second is static (fig.14). The mesh has a total of 62500 nodes and 57720 elements (fig. 15). The maximum von Misses stress is 428.33 [MPa] as shown in fig 16., located at the edge of the roller at the impact area (fig. 17). The maximum deformation is 0,27885 [mm] (fig. 18 and 19).

Figure 14. Rollers impact

Figure 15. Grid

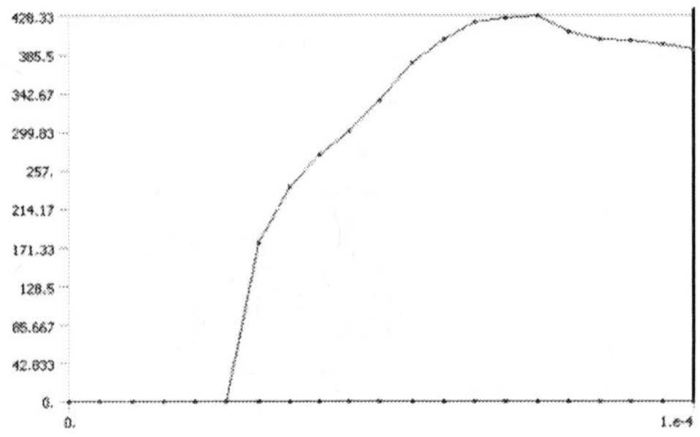

Figure 16. Von Misses Equivalent Stress

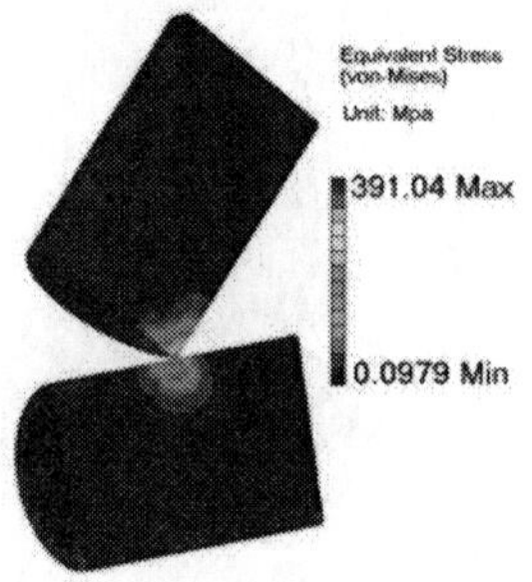

Figure 17. Equivalent Stress

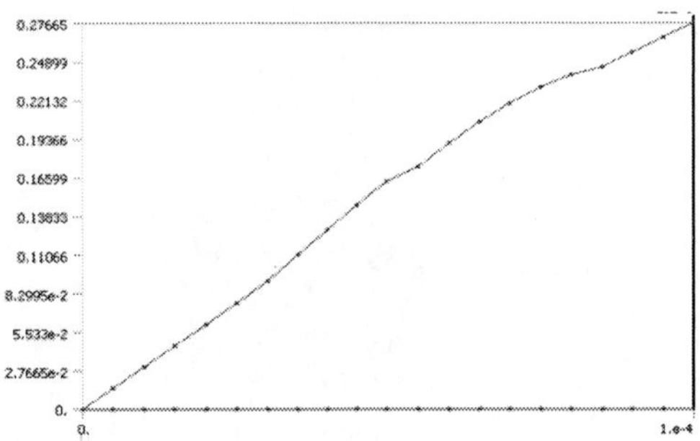

Figure 18. Total deformation graphic

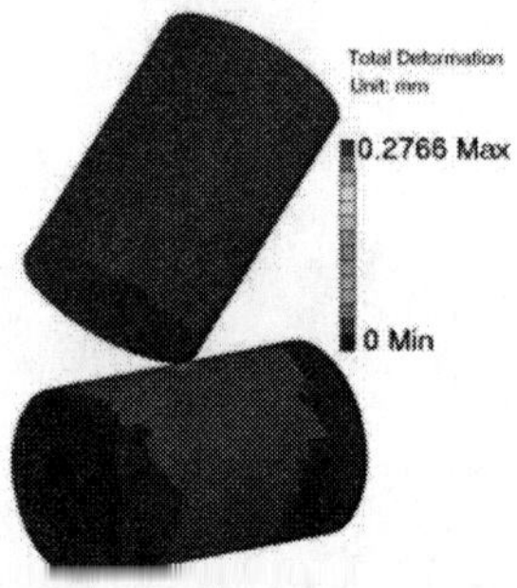

Figure 19. Total deformation

3. Result comparison

The comparison between the maximum values of the von Misses stress and total deformations calculated analytical and FEM results are centralized in Table 4 (von Misses stress) and Table 5(total deformations).

Table 4. Maximum equivalent stress [MPa]

A head - head		B head - generatrix	
Analytic	FEM	Analytic	FEM
117,87	151,8	136,1	266,86

Table 5. Total maximum deformations [mm]

A head - head		B head - generatrix	
Analytic	FEM	Analytic	FEM
0,01347	0,22697	0,01167	0,03448

4. Conclusions
Collisions are phenomenons during which the bodies exchange between themselves energy. The duration is short enough so that the external loads have are not able to modify the system impulse.

The studies conducted present the contact in two distinct cases: rollers hit each other head- head and head-generatrix with the angle 0-60 degrees. These situations occur frequently during the manufacturing process of the small bearing rollers. The worst case is when the collision angle is 30 degrees.

The von Misses equivalent stress distribution show that the critical points (where the yield point is exceeded) appear inside the rollers body and can cause permanent deformations on the rollers surface.

5. References
[1] Upadhyay R K, Kumaraswamidhas L A and Sikandar A M, 2013 *Case Studies in Engineering Failure Analysis* **1** 15–17

[2] Nélias D, Boucly V and Brunet M, 2005 Elastic-Plastic Contact Between Rough Surfaces: Proposal for a Wear or Running-In Model, **128**(2), 236-244

[3] Sukumar N, Moës N, Moran B and Belytschko T, 2000 *Int J Numer Meth Eng*, Volume **48**, Issue 11, 20 August

[4] Selim F A, Wells D P, Harmon F, Kwofie J, Lancaster G and Jones J L, 2003 *AIP Conference Proceedings* 680, 499

[5] Mănescu T, Mănescu T Şt and Iavornic C, 2011 28th Danubia-Adria Symposium, *Society for Mechanical Engineering*, Siofok, Hungary, pag. 186-188
Manescu T, Gillich G R, Manescu T S and Suciu C, 2014 *Mater. Plast.*, **51**(2), pp. 157-161

[6] Gillich G R, Bratu P, Frunzaverde D, Amariei D and Iancu V, 2010 *4th WSEAS International Conference on Computer Engineering and Applications*, Harvard Univ, Cambridge, MA, USA, **27-29**, pp. 96-103

[7] Gillich G R, Gillich N, Chioncel C P and Cziple F, 2008 *J Environ Prot Ecol*, **9**(2), pp. 465-473

[8] ISO 281:2007 Rolling bearings -- Dynamic load ratings and rating life

[9] Hertz H, 1982 Über die Berührung fester elastischer Körper und über die Härte, *Verhandlungen des Vereins zur Beförderung des Gewerbefleißes*, pp. 449-463

[10] Hartnett M J, 1980 A General Numerical Solution for Elastic Body Contact Problems, ASME, *Applied Mechanics Division*, **39**, pp. 51-66

[11] de Mul J M, Kalker J J and Fredriksson B, 1986 The Contact Between Arbitrarily Curved Bodies of Finite Dimensions, Transactions of the ASME, *J. Tribol*, **108**, pp. 140-148

[12] Bhushan B, 1998 Contact mechanics of rough surfaces in tribology: multiple asperity contact, *Tribology Letters*, Volume **4**, Issue 1, pp 1–35

[13] Polonsky I A and Keer L M, 1999 Fast Methods for Solving Rough Contact Problems: A Comparative Study, *J. Tribol* **122**(1), 36-41

[14] Vijayakar S, 1991 A combined surface integral and finite element solution for a three-dimensional contact problem, *INT J NUMER METH ENG* ,Volume **31**, Issue3

IOP Conf. Series: Materials Science and Engineering **416** (2018) 012067 doi:10.1088/1757-899X/416/1/012067

Structural and Optical Properties of Perovskite–Type Compounds Obtained by Ultrasonic Method

P Sfirloaga[1*], M Stoia[2], M Poienar[1] and P Vlazan[1]

[1]National Institute for Research and Development in Electrochemistry and Condensed Matter, Timisoara, Condensed Matter Department, P. Andronescu no.1, 300254, Romania
[2]"Politehnica" University of Timisoara, P-ta Victoriei no. 2, Timisoara 300006, Romania

*Corresponding author: paulasfirloaga@gmail.com

Abstract. Perovskite–type $NaTaO_3$ (NaT) and $NaNbO_3$ (NaN) materials have been successfully obtained by ultrasonic method with immersed sonotrode in the reaction medium followed by annealing at 600°C, for 6 h. The influence of thermal treatment on phase composition, particle size, and morphology of the obtained materials have been investigated by X-ray diffraction (XRD), Diffuse reflectance ultraviolet-visible spectroscopy (UV-Vis), Fourier transform infrared spectroscopy (FT-IR), Scanning electron microscopy (SEM), and semiquantitative elemental analysis – EDX. The average size of the crystallites of NaT materials was found to be 37.7 nm while that of NaN materials was found to be 30.5 nm. The optical band gap was determined using Kubelka–Munk equation from UV-Vis spectroscopic measurements and is estimated to be 4.05 eV for $NaTaO_3$ and 3.87 eV for $NaNbO_3$ perovskite-type samples. EDX semiquantitative analysis has highlighted the purity of the obtained materials, and quantification confirmed the stoichiometry of the compounds (ABO_3).

1. Introduction

In recent years, continuous efforts are being made worldwide to improve the performance characteristics of ABO_3-type materials suitable for practical applications [1]. Ceramics with ABO_3 type structure have been widely used in various electronic and microelectronic devices such as capacitors, piezoelectric transducers, pyroelectric detectors/sensors, memory devices, microwave tunable devices, etc, $NaNbO_3$ and $NaTaO_3$ being well-known pervoskite oxides which possess attractive physical properties [2].

It has been shown that the preparation method has a great influence on the morpho - structural properties of the perovskite materials [3-5]. The conventional techniques for synthesizing tantalates and niobates powders are based on the solid state reaction, which requires long periods of heating (800 °C) and that leads to inhomogeneity in composition [6-8].

Sol–gel using alkoxides, hydrothermal and colloid emulsions are time consuming and involve highly unstable alkoxides and difficult to maintain in reaction conditions [9]. The soft-chemical route allows a better control of physico-chemical properties through manipulation of synthesis parameters. However, this method implies long reaction times and thermal treatment [5]. In this context, the use of new techniques

IOP Conf. Series: Materials Science and Engineering **416** (2018) 012067 doi:10.1088/1757-899X/416/1/012067

working at room temperature, capable to reduce the reaction time, as well as to manipulate the final product properties must be taken into account.

The aim of this study was to prepare nanoparticles of $NaTaO_3$ and $NaNbO_3$ using the ultrasonic method with immersed sonotrode in the reaction medium and to investigate their properties. Physico-chemical properties were investigated using X-ray diffraction (XRD), ultraviolet (UV) diffuse spectroscopy, FT-IR studies, scanning electron microscopy (SEM), and semiquantitative elemental analysis - EDX.

2. Experimental procedure

The synthesis of the materials with perovskite structure by ultrasonic method with immersed sonotrode in the reaction medium was performed. All chemicals were of analytical grade and used without further purification. In order to obtain $NaTaO_3$ and $NaNbO_3$ materials, tantalum (V) ethoxide ($Ta(OC_2H_5)_5$), ammonium niobate (V) oxalate ($C_4H_4NNbO_9$ x H_2O), sodium citrate ($C_6H_5Na_3O_7$x H_2O) and sodium hydroxide (NaOH) were used.

In a first time, the solution for obtaining NaT material was prepared as follows: tantalum ethoxide was added dropwise in 50 mL bidistilled water and the precipitation with 7 M NaOH solution up to pH 11 was achieved afterwards. In case of NaN material, ammonium niobate oxalate in $5x10^{-3}$ M concentration was dissolved in 50 mL bidistilled water, to which $5x10^{-3}$ M sodium citrate solution was added. For better precipitation 7 M NaOH solution was added until the pH 9. The solutions were subjected to ultrasonic process with an ultrasonic processor Sonics Vibra Cell type at 80 °C, 60 min, with 80% amplitude and pulsations 10 seconds On / Off 5 seconds.

The obtained precipitates were filtered and washed with bidistilled water and alcohol for the removal of soluble products, and then were dried at 110 °C, for 2 hours. The resultant powders were a mixture of phases and to obtain single phase perovskite structure, they were thermally treated for 6 hours at 600 °C, at a heating rate of 5°C/min.

The structural properties of the as-obtained materials were determined by X-ray diffraction (PANalytical X'Pert PRO MPD Diffractometer with Cu tube). The light absorption properties of the materials were studied by UV–Vis diffuse reflectance spectroscopy (UV-VIS) using a Lambda 950 Perkin Elmer device and FT-IR studies were performed with a Shimadzu Prestige FT-IR spectrometer, in KBr pellets, in the range 500 - 4000 cm^{-1}. The morphology and quantitative analysis of the powders were investigated by SEM using an Inspect S PANalytical SEM / EDX.

3. Results and discussions

X-ray diffraction patterns and crystalline structures of NaT and NaN materials are presented in Figure 1, Figure 2 respectively. The XRD patterns were recorded from 20° to 80° (2θ) using CuKα radiation with λ=1.54060Å. From Figure 1, it can be seen that the materials are well crystallized, the diffraction peaks being all indexed by the perovskite structure. Also, the diffraction patterns were refined with X'Pert HighScore Plus program, using CIF 1521386 (for NaT) and CIF 5910011 (for NaN). Following the Rietveld refinement analysis of diffraction data (for NaT material) the following crystallographic data were obtained: tetragonal symmetry, space group P4/mbm, unit cell parameters a=b=5.5552 and c= 3.9338 Å, $\alpha = \beta = \gamma = 90$ °, V=121.40 Å (according to JCPDS No. 01-073-0881). For the NaN sample the following parameters are obtained: cubic symmetry, space group Pm-3m, unit cell parameters a=b=c= 3.9 (2) Å, $\alpha = \beta = \gamma = 90$ °, V= 59.48 Å (according to JCPDS No. 00-019-1221).

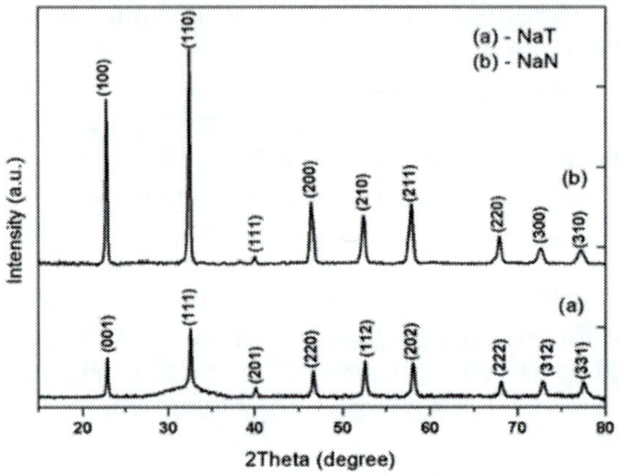

Figure 1. XRD patterns for NaT (a) and NaN (b) samples.

The crystallite size was estimated from the XRD patterns (with Scherrer's equation) and the average size of the crystallites was 37.7 nm for NaT sample and 30.5 nm for NaN sample.

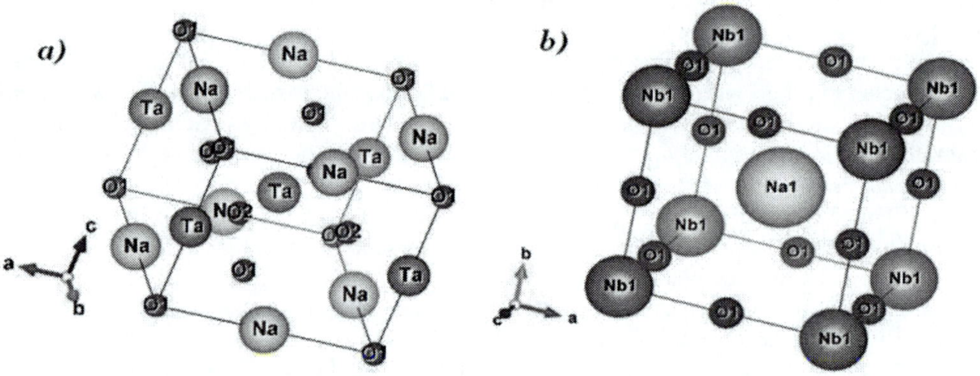

Figure 2. Crystalline structures of NaT (a) - space group P4/mbm and NaN (b) - space group Pm-3m materials.

Figure 3 shows the absorption spectra of tantalates and niobates materials recorded between 300 and 800 nm using diffuse reflectance module. The optical band gap was determined using Kubelka–Munk equation and is estimated to be 4.05 eV for $NaTaO_3$ perovskite-type sample and 3.87 eV for $NaNbO_3$. As already shown in [10], the band gap of niobate is smaller than those of tantalate and this can be due to the levels of the conduction bands consisting of Nb 4d that are more positive than those of Ta 5d [10].

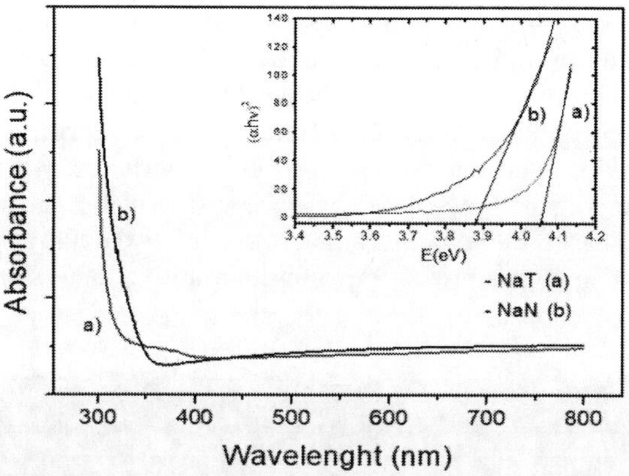

Figure 3. Diffuse reflectance spectra of the NaT (a) and NaN (b) materials. Insert: optical band gap calculated from the UV-Vis spectroscopic measurements by using the Tauc plots of $(\alpha h\nu)^{1/2}$ versus E(eV).

On the other hand, it should be mentioned that different values are reported in the literature for the $NaTaO_3$ samples obtained by different synthesis methods (sol-gel, hydrothermal and/or solid state) [5, 11] and this can be related to the deviation from 180^0 of the Ta-O-Ta bond angle in the perovskite type structure [12, 13].

FT-IR studies of the tantalate and niobate (NaT and NaN) synthesized by ultrasonic method with immersed sonotrode in the reaction medium are presented in Figure 4. The FT-IR spectrum of the NaT powder exhibits beside the wide band in the range 400-1000 cm^{-1}, with maximum around 600 cm^{-1} characteristic to the vibrations of Ta – O bonds [1], some bands characteristic to alkoxide groups (1352 and 1438 cm^{-1} corresponding to the δ(C—H) bending vibrations [14]. More, a large band at 3385 cm^{-1} together with a weak band located at 1631 cm^{-1}, characteristic to ν(OH) and δ(HOH) vibrations of water molecules associated by hydrogen bonds [15] are observed for NaT material.

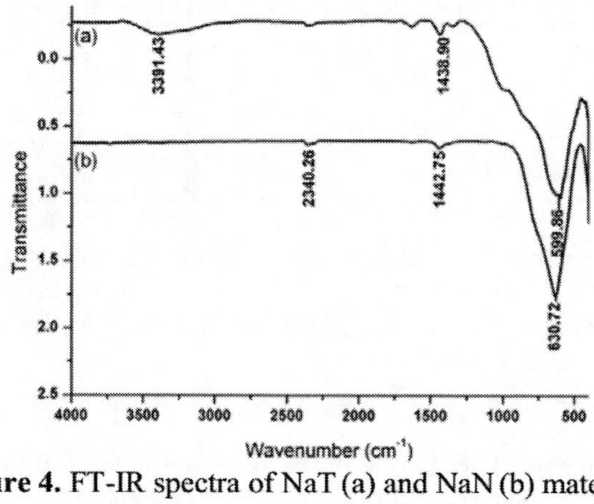

Figure 4. FT-IR spectra of NaT (a) and NaN (b) materials.

The FT-IR spectrum of the NaN powder exhibits only the wide band with maximum located at 630 cm^{-1}, characteristic according to the literature to the streching vibrations of Nb – O bonds [16] or to the vibration mode of niobium - dioxo bridge [17].

IOP Conf. Series: Materials Science and Engineering **416** (2018) 012067 doi:10.1088/1757-899X/416/1/012067

The morphology of the NaT and NaN samples synthesized by ultrasonic method with immersed sonotrode in the reaction medium and annealed at 600°C was studied by scanning electron microscopy (SEM) as shown in Figure 5. From Figure 5.a) agglomerated particles can be observed. It is important to mention that the observed morphology is quite different to that reported by other authors. Using the hydrothermal method, it was observed the formation of cubic particles, with the grain size about 100 nm [18]. Regarding NaN sample (Figure 5.b), it can be observed that the particles are agglomerated in rod-like formations with various sizes (a few micrometers), due to the ultrafine nature of the sample. To highlight the purity of the obtained materials, the semiquantitative analysis EDX was performed (Figure 6).

Figure 5 . SEM morphology for NaT (a) and NaN (b) materials.

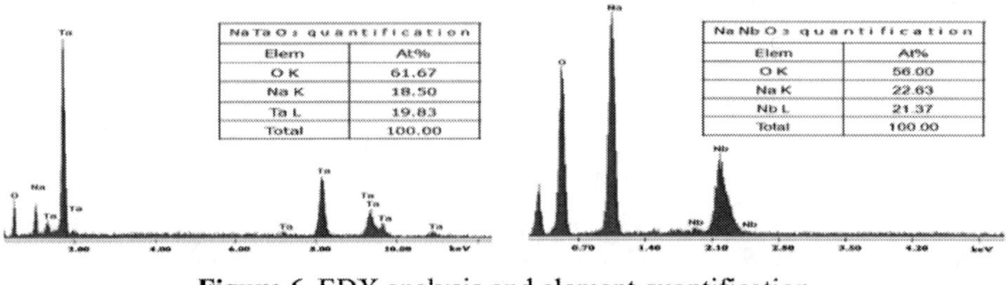

Figure 6. EDX analysis and element quantification.

Thus, one can observe the obtained materials are without any impurities, and quantification confirmed the stoichiometry of the compounds (ABO_3).

4. Conclusions

By ultrasonic method with immersed sonotrode in the reaction medium followed by thermal treatment at 600°C at, $NaTaO_3$ and $NaNbO_3$ materials with cubic symmetry has been successfully synthesized. The crystallite size of $NaTaO_3$ and $NaNbO_3$ was estimated to be 37.3 nm and 30.5 nm, respectively. The calculated value of optical band gap indicated that band gaps of niobates are smaller than those of tantalates. SEM images showed rod-like formations of $NaNbO_3$ comparative with agglomerated particles for $NaTaO_3$ with the expected stoichiometry. Therefore, one can conclude that the ultrasonic method with immersed sonotrode in the reaction medium is a fast and effective technique to obtain single phase perovskite type materials of $NaTaO_3$ and $NaNbO_3$.

References

[1] Roy S K, Singh S N, Kumar K and Prasad K 2013 *Adv. Mater. Res.* **3** 173
[2] Seo I-T, Choi C H, Song D, Jang MS, Kim B Y, Nahm S, Kim Y S, Sung T H, Song H C 2013 *J. Am. Ceram. Soc.* **96** 1024
[3] Paul M, Tabuchi M, West A R 1997 *Chem. Mater.* **9** 3206
[4] Osterloh F E 2008 *Chem. Mater.* **20** 35
[5] Vázquez-Cuchillo O, Manzo-Robledo A, Zanella R, Elizondo-Villareal N, Cruz-López A, 2013 *Ultrason. Sonochem.* **20** 498
[6] Shanker V, Samal S L, Pradhan G K, Narayana C and Ganguli A K 2009 *Solid State Sci.* **11** 562
[7] An C, Tang K, Wang C, Shen G, Jin Y, Qian Y 2002 *Mater. Res. Bull.* **37** 1791
[8] Maciel G S, Rakov N, de Araujo C B, Lipovskii A A, Tagantsev D K 2001 *Appl. Phys. Lett.* **79** 584
[9] Samuel V, Gaikwad A B and Ravi V 2006 *Bull. Mater. Sci.* **29** 123
[10] Liua J W, Chen G, Li Z H and Zhang Z G 2007 *Int. J. Hydrogen Energy* **32** 2269
[11] Hu C-C and Teng H 2007 *Appl. Catal. A:* Gen. **331** 44
[12] Hamada N, Sawada H, Solovyev I, Terakura K 1997 *Physica B* **237–238** 11
[13] Mizoguchi H, Eng H W, Woodward P M 2004 *Inorg. Chem.* **43** 1667
[14] Yang S-H, Cai Y-N, Yang H-P and Jin S-M 2009 *Trans. Nonferrous Met. Soc. China* **19** 1504
[15] Titantah J T and Karttunen M 2013 *Scientific Reports* **3** 2991
[16] Li P, Ouyang S, Zhang Y, Kako T and Ye J 2013 *J. Mater. Chem.* A **1** 1185
[17] Weihong L, Hong W, Kecheng L 2010 *J Sol-Gel Sci. Technol.* **55** 229
[18] Li X and Zang J 2011 *Catal. Commun.* **12** 1380

Acknowledgements

This work was supported by the Experimental Demonstrative Project 48PED/2017. The authors thank the X-ray Diffraction Laboratory from National Institute for Research and Development in Electrochemistry and Condensed Matter.

IOP Conf. Series: Materials Science and Engineering **416** (2018) 012068 doi:10.1088/1757-899X/416/1/012068

Thin Wall AW3103 Pipe – Dynamic Load Concerns

RS Maruta[1], R Nagy[1] and K Menyhardt[1,*]

[1]Department of Mechanics and Strength of materials, Politehnica University Timisoara, Romania

[*]Corresponding author: karoly.menyhardt@upt.ro

Abstract. Surveys from all over the automotive industry are showing an increase in the usage of aluminium year after year. Compressed air systems are part of a modern car's assembly and are constantly under load from vibrations. Because of this, the life span of the material (product) is significantly shortened. Throughout this paper we will present a series of tests and results, which shed some light on the viability of aluminium pipes in dynamic conditions. These tests were made on raw and custom shaped samples, to study the effects of form and tempering.

1. Introduction

In the last few years, there has been an increase in aluminium usage among automobile manufacturers. As predicted, due to its fairly good mechanical properties and 100% recyclability, aluminium remains "the fastest growing automotive material over competing materials and is entering its most unprecedented growth phase"[1].

The European automotive industry is considered technically the most advanced and highly innovative [2]. Because of political pressure to reduce fuel consumption and CO_2 emissions, the efforts for automotive companies to reduce total mass has shifted from using steel to lightweight materials, one solutions being aluminium alloys. During the last decade the average amount of aluminium used in cars has doubled, and based on this trend, its usage will keep growing.

In recent years, aluminium piping has become a solid alternative to other piping materials for compressed air systems. Conventional compressed air systems equipped with moisture trap will still have some moisture in the system and corrosion will occur. Additionally, loose scale deposits collect over time and create pressure drops.

Also, aluminium's high thermal conductivity maximises heat transfer in applications such as heat exchangers. Thanks to its low cost and easy formability into complex shapes, aluminium is increasingly utilised in these applications. Heat exchangers also take advantage of aluminium's: high strength, low density and corresponding light weight, machinability, mechanical strength and shape retention after machining.

Aluminium alloy EN AW-3103 or AA 3003 is a good candidate for extruded round tubes because of its long life. It is primarily used in various types of 'mechanical expanded' type heat exchangers (condensers and evaporators). AW3103 is manganese alloy with a high corrosion resistance. The alloy has an average strength and is suitable for welding and bending.

2. Static load premises

Thin walled AW3103 aluminium pipe has been brought to our attention due to its erratic act in live conditions. Pipes of various shapes and sizes had the same symptoms.

The two main type of tubes were sorted into: inner diameter (i.d.) 9.5 mm – outer diameter (o.d.) 12 mm and i.d. 15 mm – o.d.18 mm (Figure 1). In order to assess their dynamic response, some static tests were performed to acquire their elastic modulus.

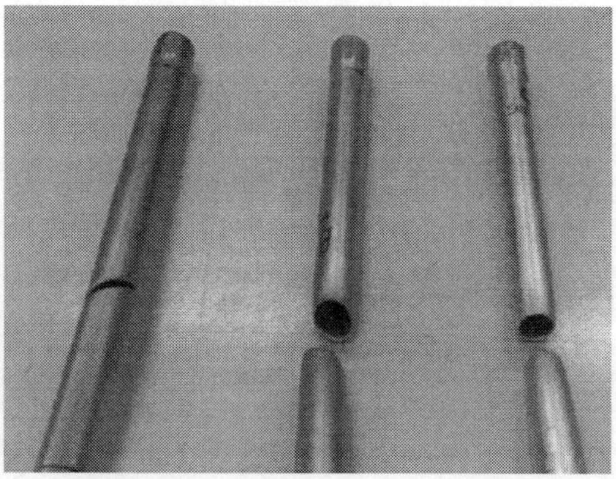

Figure 1. Untested 9.5-12 mm tubes and tested 15-18 mm tubes

Using a universal tensile testing machine, the specimens were subjected to a controlled tension until failure. This way we directly measured ultimate tensile strength, breaking strength, maximum elongation determining Young's modulus. Considering the number of specimens tested, 18-24 elements, the results (Table 1) were pretty inconsistent and sparse thus raising first questions about the material [3, 4].

Table 1. Extreme and average values for test specimens

Parameters	i.d.9.5 mm – o.d. 12 mm			i.d.15 mm – o.d. 18 mm		
	min	max	avg	min	max	avg
E [N/mm2]	38375	67575	65132	17964	73248	63080
Fm [kN]	0.73	5.21	5	8.39	15.53	11
Rp0.1 [N/mm2]	11.96	104.66	99	72.56	105.62	97
Rp0.2 [N/mm2]	12.61	109.93	104	87.73	109.75	99
Rm [N/mm2]	17.02	121.31	114	108.89	204.568	123

In order to be able to plot a fatigue (Wohler) curve, we needed stress values for the pipes (Figure 2), and with the aid of previously determined elastic modulus we did a comparison of strain resulted from load cell and stress resulted from strain gauges applied on the pipes in the elastic region.

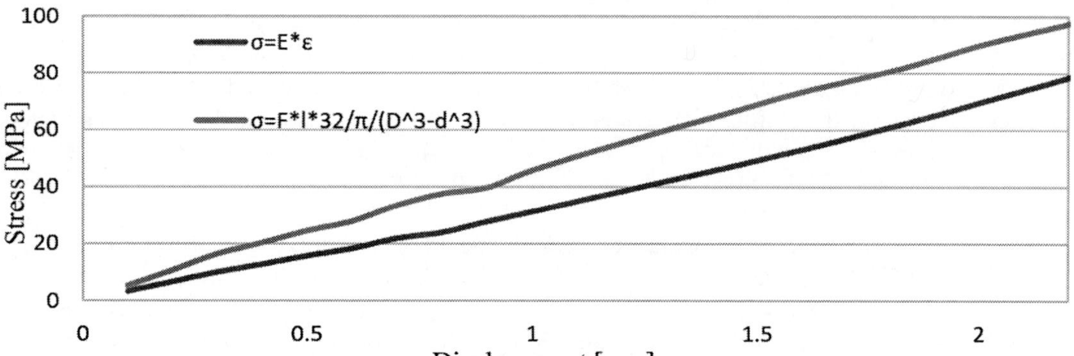

Figure 2. Stress comparison of average values for 6 specimens (i.d 15 mm – o.d. 18 mm)

The test specimen was precisely tensioned by a high precision stepper motor (1/1440 mm) and the stress was measured with both through strain gauges attached on the specimen and load cell attached to the tensioning cable (Figure 3).

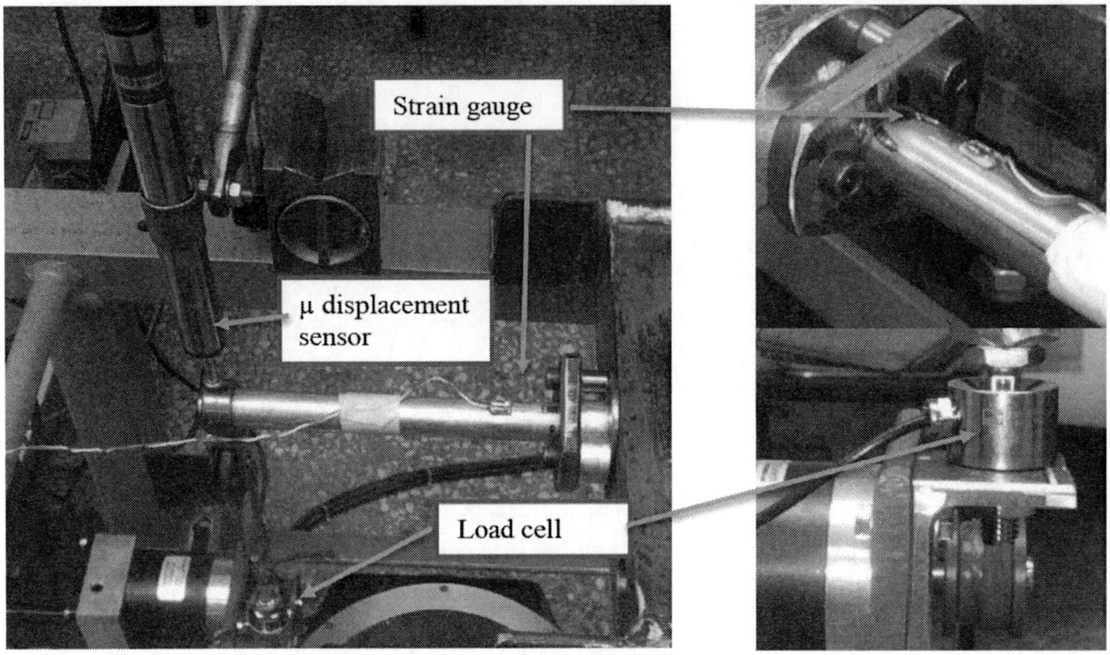

Figure 3.Stress validation set-up

3. Dynamic load testing

Having an idea about the stress values at different amplitudes / displacements, we pursued to test the life limit of the two type of tube in two separate architectures. The fatigue experiments have been carried out with custom specimen arrangements depending on the diameter of the tubes.

3.1. Forced vibration at resonance

For the i.d.9.5 – o.d.12 mm pipes there was a special curvature requested, and the specimens were oscillated at resonance (around 37-38Hz) with discrete acceleration on an electrodynamic shaker in order to maintain a specific displacement/strain, ergo stress.

On a group of 18 test specimens vibrated till fracture, there was again a considerable gap between the numbers of cycles for a specific stress value (Figure 4). The values for the Y-axis was specifically omitted not to be in breach of contract. Overall, the fatigue curve of the tested pipes do resemble the AW3103 Wohler curve, but a little bit reduced in number of cycles. Throughout the fatigue test, for different accelerations, a specific strain was maintained with the aid of inductive sensors which monitored the maximum displacement of the pipe's free end. The setup rig for this scenario is shown in Figure 5.

Special care was needed to monitor the acceleration on all 3 axes due to the peculiar nature of the test: the grips needed to be sturdy enough not to overdamp the specimen's vibration and at the same time, due to resonance the acceleration on all 3 axes would act out due to the mass grips and its center of gravity. During these test Bruel&Kjaer Type 4506 and 4513 micro accelerometers were used and they were harshly taxed for hundreds of hours at acceleration exceeding even 160 m/s^2.

Although the pipes were machined formed and basically had the same form, curvature and mass, their initial resonance frequency was not the same, indicating a difference in them, and this frequency decreased inversely proportional with the duration of the tests.

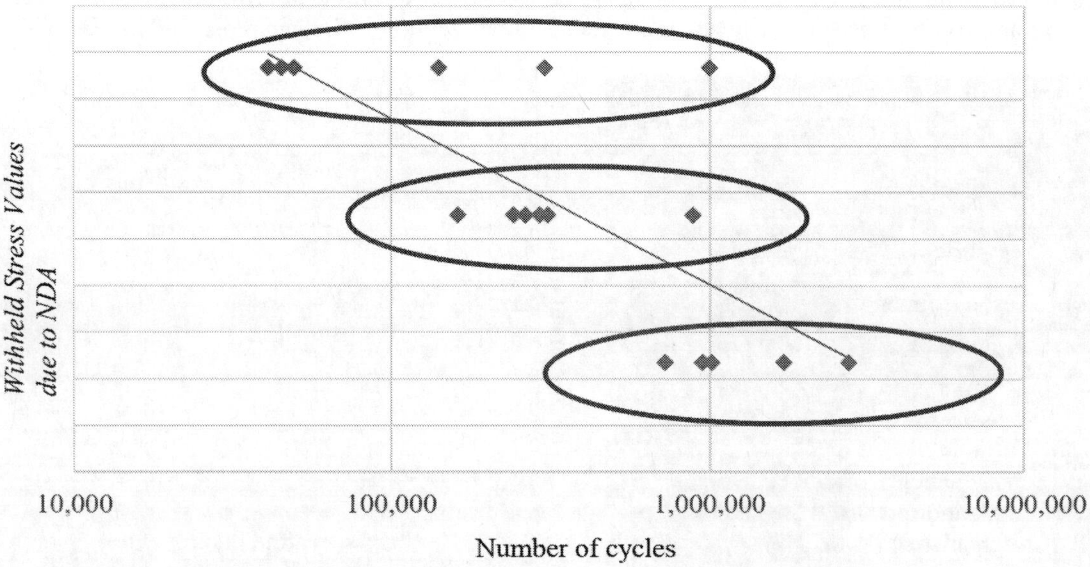

Figure 4. Sparse values for stresses based on number of cycles

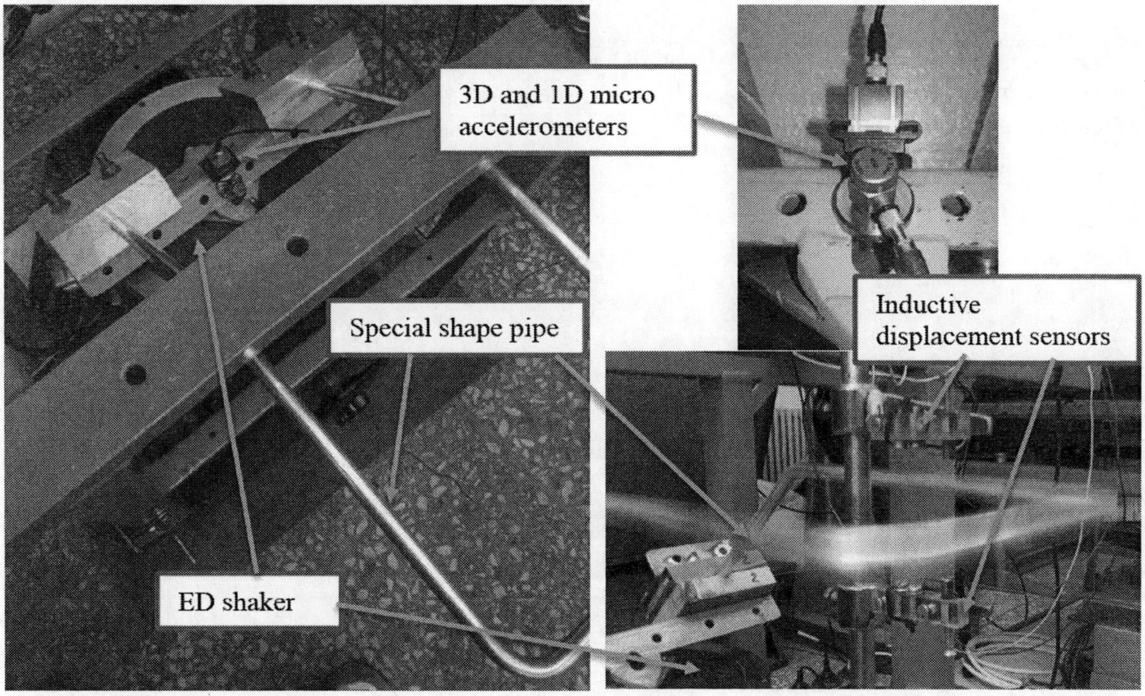

Figure 5. ED shaker with pipe for resonance fatigue set-up, static and during resonance

IOP Conf. Series: Materials Science and Engineering **416** (2018) 012068 doi:10.1088/1757-899X/416/1/012068

3.2. Forced vibration at constant displacement

For the i.d.15 mm – o.d.18 mm pipes there was a different test setup, and the specimens were mechanically oscillated with aid of a specially developed crank and connecting rod mechanism able to obtain and maintain symmetric displacements between 0.4 and 2 mm. Several attempts were made to achieve and control this small displacement, and the final version is the one presented in Figure 6.

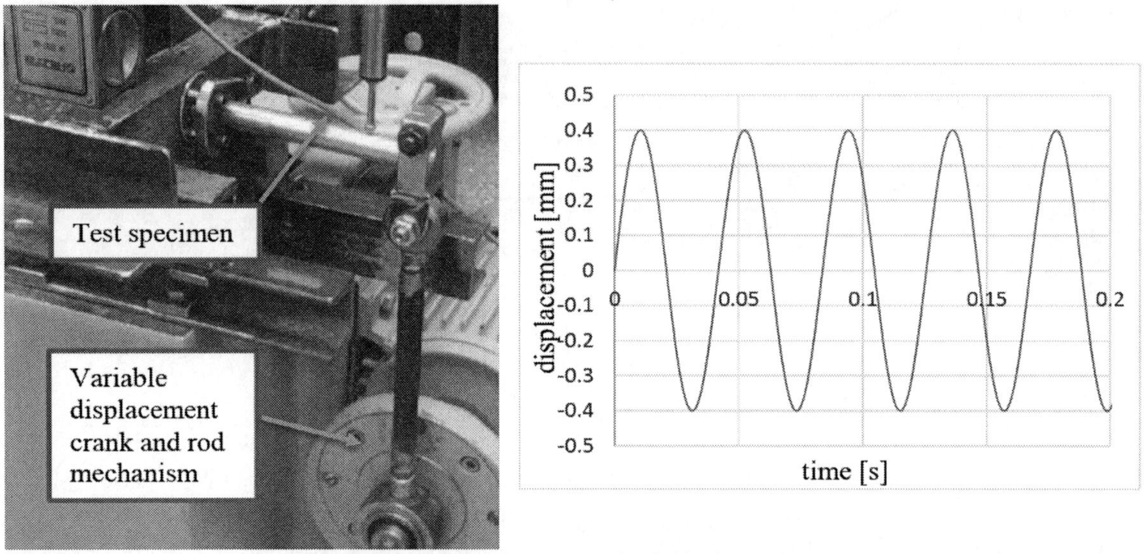

Figure 6. 0.4mm symmetrical displacement

On a number of 12 test specimens ruptured so far, there is a significant gap between values on the Wohler curve. Again, in order to avoid breach of contract, values for the stresses have been omitted (Figure 7).

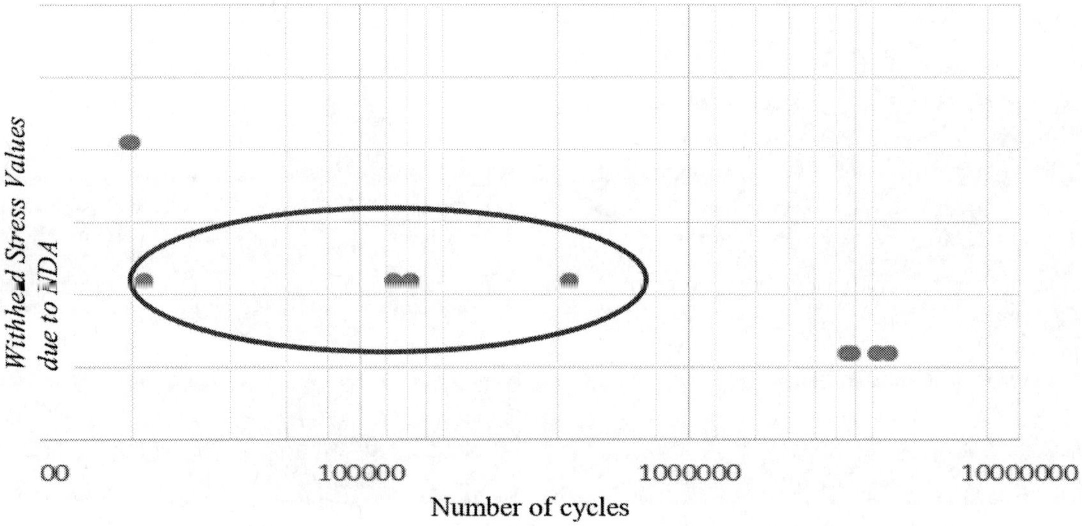

Figure 7. Sparse values for stresses based on number of cycles

These tubes, had connection couplings welded at their end and one may mitigate that this procedure would act as tempering. None the less, the welding procedure is an automated one, so this would affect all the specimens the same way. From the previous data obtained from static tests, there wasn't a specific convergence in values.

From the same material, with same dimensions, there were three type of coupling attachment methods: laser welding, tinning and snug fit. Although there is obvious variations between the life cycles of these assemblies, they all show a common treat, in the way that there is no specific point of convergence for a specific value of stress as a function of number of cycles.

4. Conclusion and results

The goal of this paper is to point out the premature failure in AW3103 tubes that are used in the automotive industry, caused by material flaws, mainly due to material heterogeneity. To this end, several samples from two types of tubes of same material where tested statically to determine their elastic modulus and dynamically to determine their life cycle.

From a statics point of view they failed due to the lack of a common convergence point or a narrow dispersion range for the elastic module.

From a dynamics point of view their "stress-number of cycles" curves did not present the same converging point for each stress value, but rather a pretty wide range.

Based on these findings and the specifics of the tests performed we can rule out automated mechanical forming of the tubes as a main cause for premature failure of finite products made from AW3103 and suggest looking into the quality of the raw material from which these tubes are made or their handling before machining (previous/multiple plastic deformation).

Although the geometry of the first type of tubes, i.d.9.5 - o.d.12 mm, would clearly make them fracture at the base of the fixture, during the fatigue tests there was a case in which one of the tubes cracked in a bended area where the stress values were lower (Figure 8), once again proving material problems and early onset of fracture!

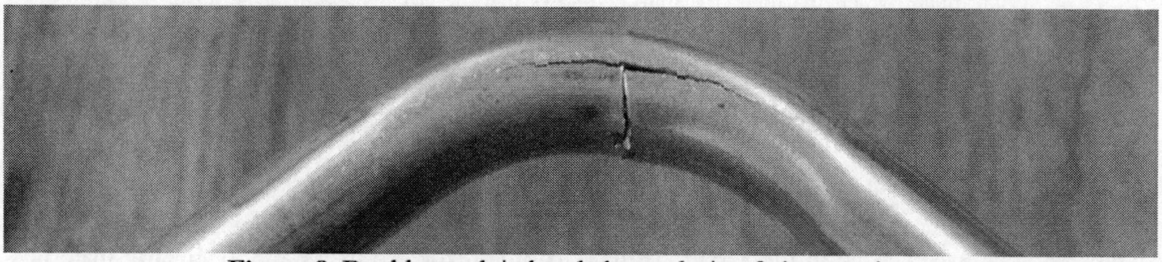

Figure 8. Double crack in bended area during fatigue testing

5. References

[1] http://www.autonews.com/article/20180205/OEM01/180209972/aluminum-trend-is-only-growing, last access date 2018-06-12

[2] Hirsch J, Recent development in aluminium for automotive applications 2014 *Transactions of Nonferrous Metals Society of China* **24** 1995-2002

[3] Ochi H et al., Evaluation of Tensile Strength and Fatigue Strength of Commercial Pure Aluminum/Tough Pitch Copper Friction-Welded Joints by Deformation Heat Input 2008 *Materials Transactions* **49** 2786-91

[4] Campbell FC *Elements of Metallurgy and Engineering Alloys* 2008 Chapter 14 ASM International

IOP Conf. Series: Materials Science and Engineering **416** (2018) 012069 doi:10.1088/1757-899X/416/1/012069

Study on the Obtaining of $MnSi_{1.75}$ Intermetallic Compound by Mechanical Alloying

V Cebotari[1], T F Marinca[1], F Popa[1], B V Neamţu[1], N A Sechel[1], R Hirian[2] and I Chicinaş[1,*]

[1]Materials Science and Engineering Department, Technical University of Cluj-Napoca, 103-105, Muncii Avenue, 400641 Cluj-Napoca, Romania
[2] Babeş-Bolyai University, Faculty of Physics, 400084, Cluj-Napoca, Romania

[*]Corresponding author: ionel.chicinas@stm.utcluj.ro

Abstract. Thermoelectric higher manganese silicide (HMS) having the composition $MnSi_{1.75}$ material was obtained by mechanical alloying after 18 hours of milling. As starting materials, manganese and silicon elemental powders were used. After 18 hours of milling beside the HMS in the powders, MnSi phase was obtained as well. After only 4 hours of milling, both binary MnSi phases are found in material besides a significant amount of non-reacted elements. Milling up to 10 hours leads to the silicon characteristic peaks disappearance and the HMS and MnSi phases amount increase. Subsequent annealing of the milled powders was performed in order to promote internal stresses removal. The 4 hours annealing of the powder at the temperature of 350 °C under high vacuum (10^{-6} Torr) leads to stress release and an increase in the MnSi phase quantity. The particle size analysis shows that the powder diameter is in 0.1-125 μm range. A large amount of powder is in the range of 15-125 μm, but there is also a small amount that is in the 0.1-15μm range. Scanning electron microscopy (SEM) was used to characterize the powders morphology.

1. Introduction

The thermoelectric materials can convert a temperature difference into electrical energy. This conversion is possible due to the Seebeck effect. The performance of the thermoelectric materials is characterized by the Seebeck coefficient alongside with the figure of merit (ZT). The figure of merit of a thermoelectric material depends on the Seebeck coefficient, electrical conductivity, temperature and respectively thermal conductivity of the material [1, 2].

Higher manganese silicide (HMS) is one of many thermoelectrical materials intensively studied today. The thermoelectric materials in the Mn-Si system are attractive because the constituent elements have a widespread in the earth's crust, are not toxic and have good chemical stability [3]. HMS is a thermoelectric compound with *p*-type conduction [4]. HMS have the general formula $MnSi_y$ where (y=1.67-1.77) and crystallized in tetragonal structure with manganese in corners and silicon inside of tetragon in the form of a spiral. The currently studied HMS compounds from the Mn-Si system are $Mn_{11}Si_{19}$, $Mn_{26}Si_{45}$, $Mn_{15}Si_{26}$, $Mn_{27}Si_{47}$, Mn_4Si_7 [5-7].

The preparation of HMS by conventional arc melting is difficult due to the high melting points of the elements and the high time required to obtain the structural homogeneity [4, 8]. Mechanical alloying represents an alternative technique to obtain HMS. The obtaining of the HMS compounds by mechanical alloying is studied in several articles, where different milling parameters (ball to powder ration, vials and/or main disk speed etc.) are varied, leading to the complete formation reaction being spanned in the range of 3 to 24 hours [9-11].

In the present study, using a planetary ball mill, the formation of the HMS compound was studied for milling times up to 18 hours. The phase evolution, the morphology, and the particle size distribution as the milling time increase are discussed.

2. Experimental

For obtaining the HMS compound manganese powder (purity 99.3% and -325 mesh particle class) and silicon powder (purity 99.9% and -100 mesh particle class) from Alfa Aesar was used. The amount of each component was weighed with an analytical balance in order to obtain the following chemical composition: $MnSi_{1.75}$. The homogenization of powders was performed using a Turbula spatial homogenizer for 15 minutes. The blended powder was loaded in a stainless steel vial with 89 stainless steel balls (14 mm). The milling was performed in a Fritsch Pulverisette 6 planetary ball mill under argon atmosphere. The ball to powder ratio was 10:1 and the main disk rotation speed was 400 rpm. The process was carried out in an argon atmosphere. Sampling was made inside of glove bag with argon atmosphere to avoid the oxidation. A thermal treatment for stresses release was performed at 350 °C for 4 hours in high vacuum.

X-ray diffraction (XRD) studies were carried out with a Buker D8 Advance diffractometer using Cu Kα (λ = 1.54183 Å) radiation in the angular range: 2θ=20-110^0. To record high-resolution data, the acquisition time for the X-ray patterns was 12 hours. The particle size analyzer Analysette 22 Nano Tec was used to record the particle size distribution of the powders mechanically milled. From the particle size distribution the D10, D50, and D90 parameters were determined. D10, D50, and D90 represent the mean powder particle diameter of the 10, 50 and 90 % of the total powder volume [12]. The JEOL-JSM 5600 LV scanning electronic microscope (SEM) was used to study the morphology of powders.

3. Results and discussion

The XRD patterns for $MnSi_{1.75}$ are presented in figure 1 at different milling time and for the initial mixture of powders. In the diffraction pattern of initial mixture, only the characteristic peaks for silicon and manganese elements are identified. After 4 hours of mechanical alloying, the corresponding peaks of silicon decreases in intensity and widens due to the reduction of the crystallite size [13]. On the other hand, the manganese peaks increase in intensity and width. At the same milling time, the HMS compound and an undesired MnSi phase appear as a consequence of solid state reaction initiation.

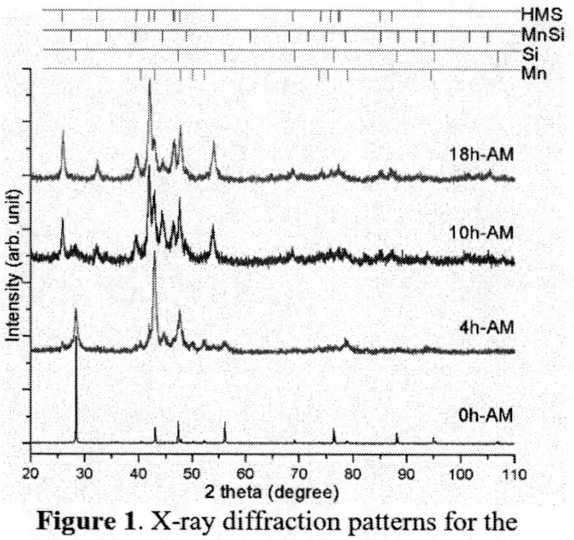

Figure 1. X-ray diffraction patterns for the initial mixture of powders and for 4, 10 and 18 hours milled powders.

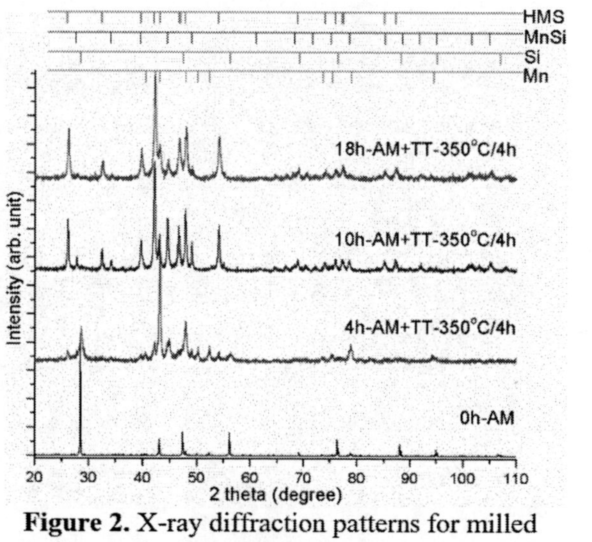

Figure 2. X-ray diffraction patterns for milled powders and subsequently annealed at 350 °C for 4 hours.

The diffraction patterns for 10 hours milled powder shows the HMS, MnSi and initial elements (Mn and Si) peaks. The unreacted amount of initial elements quantity is small. The majority phase in the material is the HMS compound. The complete reaction of elements was achieved after 18 hours of milling. However the 18 hours diffraction patterns contains both Mn-Si phases (HMS and MnSi). The amount of MnSi phase at this milling time is smaller compared to the samples milled for 10 hours. It is not clear what process leads to decreasing the MnSi phase quantity, but it can be posible that this phase transforms under high stress induced by milling process into the HMS phase. The analysis of the X-ray diffraction patterns, did not evidenced any oxidation of the powders, as reported in [8], since the milling and the sampling was performed in high purity argon atmosphere.

Shin et all [7], using a planetary ball mill (Fritsch, Pulveristte5) with ball to powder ration 20:1 and balls with 5 mm in diameter obtain a complete reaction of elements after 12 hours at 400 rpm. Comparing the results obtained in this paper and the results presented by Shin it is concluded that the use of more energetic conditions (as compared to our milling conditions) can reduce the time needed for alloying. It can also be seen that the steps through which the powder passes are comparable only at slightly higher times. Therefore, it can be said that the use of a higher ball to powder ratio conduct to reducing the time needed for alloying. Figure 2 shows XRD patterns for the samples mechanically milled for different duration and subsequently subjected to a heat treatment. In comparison with the peaks for as-milled powder, after heat treatment, the peaks become narrow because the internal stresses removal. Heat treatment at 350 °C for 4 hours has also the effect of enhancing the compound formation. In the case of 10 hours milled samples and heat treated, the disappearance of the silicon maximum ($2\theta \approx 28.4°$) and the maximum of manganese ($2\theta \approx 43°$) is observed. This can be assigned to the complete reaction of those phases. In the diffraction patterns corresponding to powders milled 18 hours and heat treated there are not significantive changes.

In figure 3 the SEM images for mechanical milled powders and for elemental powders (as reference sample) are presented. In the figure 3.a, the image for manganese elemental powders is shown. The irregular shape and sharp edges of manganese powder are evidenced. Manganese powder contains particle sizes ranging from a few micrometers to several tens of micrometers. In contrast, the silicon powders presented in figure 3.b, contains only large particles. The morphology of silicon powders is approximately the same as the manganese powder. In the image presented in figure 3.c) it can be seen that smaller manganese particles are well mixed with larger silicon particles. After 4 hours of mechanical milling, the morphology of particles is completely modified, sharp edges disappears and small, agglomerated particles are noticed.

Figure 3. SEM images: a) Mn powder, b) Si powder, c) 0h-MA milled powder, d) 4h-MA milled powder, e) 10h-MA milled powder and f) 18h-MA milled powder

In figures 3.d, 3.e and 3.f, it can be seen that the milled powders consists in agglomerates of smaller, fine particles. The dimension of agglomerates increases with milling time. At the milling time investigated, it can be noticed that the particle size is below 10 μm, but the size of the agglomerates exceeds a few tens of micrometers. Particle size are comparable to those reported in [9], with the difference that no agglomerated particles were reported in the reference.

In the image 4 is shown the particle size distribution for powders milled for 4, 10 and 18 hours, respectively. At all of these milling times, it's possible to identify that in the milled powders exist a very small amount of fine powders and a large amount of large particles. This result is in concordance with SEM images. The curve corresponding to 4 hours of milling is Gaussian type distribution. The curves for 10 and 18 hours of milling are slight asymmetrical.

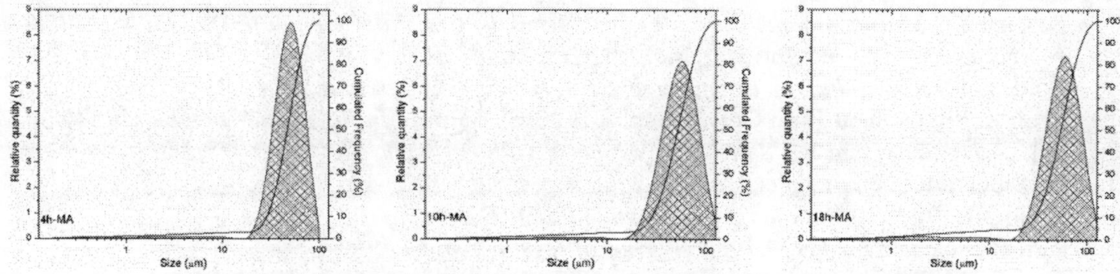

Fig. 4 Particle size distribution for powders milled 4, 10 and 18 hours.

Using the particle size distribution data, the average of particle size was calculated and this is presented in figure 5. The D10 parameter and remains constant up to 18 hours of milling, with a value of about 29 μm. The decrease of parameter D10 in the range of 4 to 10 hours is attributed to the fragmentation processes, especially in the silicon large particles.

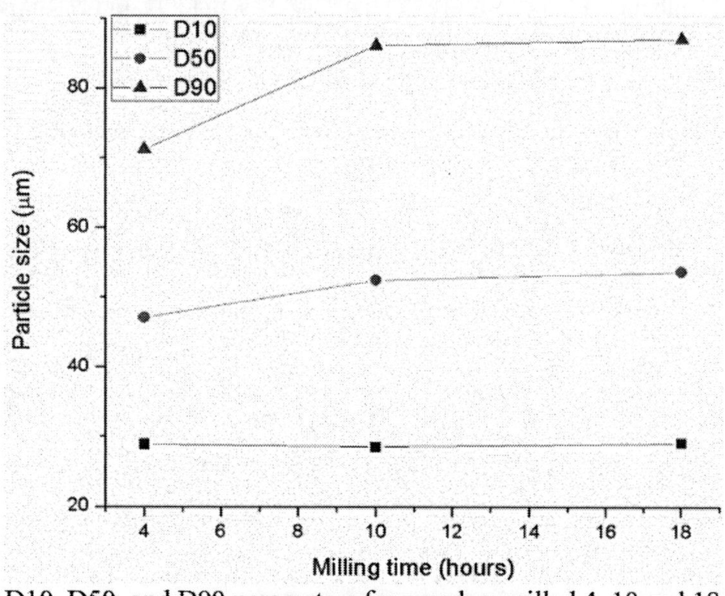

Fig. 5. D10, D50, and D90 parameters for powders milled 4, 10 and 18 hours.

In the case of a D50 parameter, the powder particle size increases more visible in the 4 to 10 hours interval and slightly in 10 to 18 hours interval. The D50 parameter is ranging from 47 μm corresponding to 4 hours of milling to 54 μm at 18 hours of milling. This trend is also observed for D90 parameter. Increased growth in the range of 4 to 10 may be due to the fact in this milling time range a large powder quantity reacts and the welding process are slightly higher than the fracture ones. In the range 10 – 18 hours, agglomeration of particles containing the HMS phase is more significant.

4. Conclusions

HMS having the composition MnSi1.75 was obtained after mechanical alloying up to 18 hours of milling. The prepared powders are not single-phase powders, the HMS phase being accompanied by a minor MnSi phase. The quantity of MnSi impurity phase is reduced via increasing of the mechanical milling duration. Heat treatments at 350 °C for 4 hours are sufficient to conduct to stresses release. After 18 hours of mechanical alloying material shows agglomerates of very fine powders.

5. References

[1] Tedenac J C *Multicomponent silicides for thermoelectric materials phase stabilities, synthesis, and device tailoring* SpringerBriefs in Materials 2017.
[2] Goldsmid H J *Introduction to Thermoelectricity* Springer Series in materials science 121.
[3] Chen X, Shi L, Zhou J and Goodenough J B 2015 *Journal of Alloys and Compounds* **641** 30–36
[4] Itoh T and Uebayashi S *J. Jpn. Soc. Powder metallurgy* **63** 7
[5] De Riddera R and Amelinckx S 1971 *Mat. Res. Bull.* **6** 1223-1234
[6] Kojima T 1979 *Journal of crystal growth* **47** 589-592
[7] Kawasumi L and Sakata M 1980 *Journal of crystal growth* **49** 651-589.
[8] Truong D Y N, Kleinke H and Gascoin F 2015 *Intermetallics* **66** 127-132
[9] Shin D K, Jang K W, Ur S C and Kim I H 2013 *Journal of electronic materials* **42**(7)
[10] Famengo A, Battiston S, Saleemi M, Boldrini S, Fiameni S, Agresti F, Toprak M S, Barison S and Fabrizio M 2013 *Journal of Electronic Materials* **42**(7)
[11] Lee Y G, Choi M K, Kim I H and Ur S C 2012 *Journal of Ceramic Processing Research.* **13**(6) 816-819
[12] Marinca T F, Chicinaş H F, Neamţu B V, Chicinaş I, Isnard O, Popa F and Păşcuţă P 2016 *Advanced Powder Technology* **27** 1588–1596
[13] Marinca T F, Chicinaş H F, Neamţu B V, Isnard O, Pascuta P, Lupu N, Stoian G and Chicinas I 2016 *Materials Chemistry and Physics* **171** 336-345

Acknowledgment
This work was supported by the Romanian National Authority for Scientific Research, CNCS – UEFISCDI [grant number PN-III-P2-2.1-PED-2016-1816].

Prediction of the Biomechanical Behavior of All-Ceramic Dental Restorations after Cyclic Loading using FEA

L Porojan[1,*], S Porojan[2] and F Topală[3]

[1]Department of Dental Prostheses Technology (Dental Technique), School of Dentistry, "V. Babeş" University of Medicine and Pharmacy, 9 Revolutiei 1989 Blv., 300041 Timişoara, Romania
[2]Department of Oral Rehabilitation (Dental Technique), School of Dentistry, "V. Babeş" University of Medicine and Pharmacy, 9 Revolutiei 1989 Blv., 300041 Timişoara, Romania
[3]Department of Dental Prosthodontics, School of Dentistry, "V. Babeş" University of Medicine and Pharmacy, 9 Revolutiei 1989 Blv., 300041 Timişoara, Romania

[*]Corresponding author: sliliana@umft.ro

Abstract. The mechanical performance of dental ceramic materials is commonly approached experimental by measuring the fracture strength, which reflects its static behavior. In normal service dental restorations are loaded far below their critical load, either continuously or under repetitive conditions. Unlike static stress, which is analyzed with calculations for a single stress state, fatigue damage occurs when stress at a point changes over time. The objective of the present study is to compare alternative loading areas of molar zirconia crowns with different designs, veneered with various overpressing ceramics and to predict the biomechanical behavior of the prosthetic restored teeth after cyclic loading, using a FEA. The fatigue behavior of various all-ceramic restorations will be investigated using a wide range of load levels (400-3200 N) till 10^8 cycles, during axial and para-axial loading. The Fatigue Safety Factor was calculated, and values less than one indicated failure before the design life is reached. The failure behavior was interpreted relative to the load levels, areas, and restoration designs. Axial loading led to fatigue failure at lower loads (500 N) compared to para-axial loading (2000 N). Related to the crown design, the failure behavior didn't change significant.

1. Introduction

Understanding mechanical properties is the first step to estimate clinical performance of biomaterials. Moreover, the data are always required for finite element analysis (FEA) which has become exceptionally popular in modern biomaterials science research [1]. Fracture toughness (KIC) which is a prominent factor to fracture of brittle materials, reveals resistance of material against crack extension [2]. For all-ceramic restorations zirconia framework design and the veneering material may influence the biomechanical behavior of the restoration [3]. In current practice, the framework of all-ceramic bilayered crowns is obtained by milling an even thickness copies, rather than using a scientific-based design. The continuously developing possibilities of actual CAD softwares allow achieving digital improved designs of the frameworks, regarding the control of the veneer thickness in case of the cutback designs [4]. Among all mechanical parameters, the most crucial one for dental restorations is fatigue behavior as it is the major cause of failures [5,6]. The stress-life (Wöhler) curve is the oldest way to characterize the fatigue resistance against cyclic loading [7]. It shows the relationship between the stress

IOP Conf. Series: Materials Science and Engineering **416** (2018) 012070 doi:10.1088/1757-899X/416/1/012070

amplitude (sa) and the number of cycles to failure (N) [8]. This method not only shows the fatigue limit, but also predicts the number of cycles in the given fatigue load.

In the past, it was thought that brittle materials like ceramics do not suffer fracture due to fatigue in ambient temperature. This belief came from the fact that plastic deformation causes fatigue and as the plastic deformation in brittle materials is not evident, they cannot undergo fatigue. Fatigue circumstances have now been found in ceramics, although the mechanism seems to be due to degradation of toughening rather than plastic deformation [9].

Different studies investigated the fatigue behavior and longevity of ceramic materials, but only for specific load levels and a limited total number of cycles of 10^4 [10-13]. However, no research has been conducted on fatigue behavior of various CAD/CAM ceramics using a wide range of load levels. Therefore, with the rapid development of the CAD/CAM ceramics in dentistry, there is a need to understand, predict and compare the fatigue behavior of those ceramics [14].

2. Aim of the research

The objective of the fatigue investigations was to compare alternative loading areas of molar zirconia crowns veneered with various overpressing ceramics and to predict the biomechanical behavior of the prosthetic restored teeth based after cyclic loading.

3. Materials and methods

For the experimental analyses, a maxillary first molar was chosen. A nonparametric modeling software (Blender 2.57b) was used to obtain the 3D tooth shape. The collected data were used to construct three dimensional models using Rhinoceros (McNeel North America) NURBS (Nonuniform Rational B-Splines) modeling program. Beginning from the anatomic contour of the crown, the tooth prepared for all-ceramic bilayered crowns with a 1 mm chamfer margin. The occlusal surface was then reduced by 1.5–2 mm. The axial walls were prepared with 6° convergence.

Two geometric models with differential coping designs were developed, framework designs suggested by the soft of the CAD/CAM system Cercon Art 3.2 (Degudent, Hanau, Germany). First, a uniform thickness of 0.5 mm was chosen for the framework. Second, a cutback design was prepared as same as for metal-ceramic crowns in order to obtain uniform, adequate thickness and support for the veneering ceramics. It was assumed that prosthetic crowns were made of zirconia-based ceramic veneered with overpressing lithium disilicate reinforced ceramics. For all cases a digital model of the bilayer crown was designed to occupy the space between the original tooth form and the prepared tooth form.

The geometric models were imported in the finite element analysis software ANSYS and meshed. To perform calculations, each tooth model was divided into a structural solid elements joined in nodes. The nodes on the bottom surface of the abutment tooth were constrained in all directions. The crown materials were assigned using literature data. The interfaces between models where modelled automatically by ANSYS (the type was bonded, and the behaviour symmetric). In order to simulate the stress distribution, the Young's module and Poisson's ratios were introduced: Young's modulus (GPa) 18 for dentin, 61 for glass ceramic, and 205 for zirconia and Poisson's ratio 0.27 for dentin, 0.21 for glass ceramic, and 0.31 for zirconia.

Different loading values were assumed. A distributed load between 400 N and 3200 N was applied in the central fossa area of the crown models (axial) and on the cusp slope (para-axial), to the direction of insertion by a 3D finite element ball model (6 mm in diameter). Characterizing the capability of a material to survive the many cycles a component may experience during its lifetime is the aim of fatigue analysis.

The ANSYS Fatigue Module supports a wide variety of fatigue analysis. The comprehensive calculations and results enable engineers to evaluate their designs for avoiding failures under real world conditions. Stress Life is based on S-N curves (Stress – Cycle curves) and has traditionally dealt with relatively high numbers of cycles and therefore addresses High Cycle Fatigue (HCF), greater than 10^5 cycles inclusive of infinite life.

IOP Conf. Series: Materials Science and Engineering **416** (2018) 012070 doi:10.1088/1757-899X/416/1/012070

4. Results and discussions

Unlike static stress, which is analyzed with calculations for a single stress state, fatigue damage occurs when stress at a point changes over time. Constant amplitude, proportional loading was chosen. A total number of cycles between 10^6 and 10^8 was investigated.

Once the type of fatigue analysis and the loading type were chosen, the next decision is whether to apply a mean stress correction.

For Stress Life mean stress can be accounted for directly through interpolation between material curves. Fatigue Life can be determined over the whole model. The contour plot shows the available life for the given fatigue analysis. If loading is of constant amplitude, this represents the number of cycles until the part will fail due to fatigue.

Fatigue Safety Factor is a contour plot of the factor of safety with respect to a fatigue failure at a given design life. The maximum Factor of Safety displayed is 15. For Fatigue Safety Factor, values less than one indicate failure before the design life is reached [15].

For 400 N cyclic loading the safety factor was over 1 in all cases, regardless of design and loading area (Figures 1, 2).

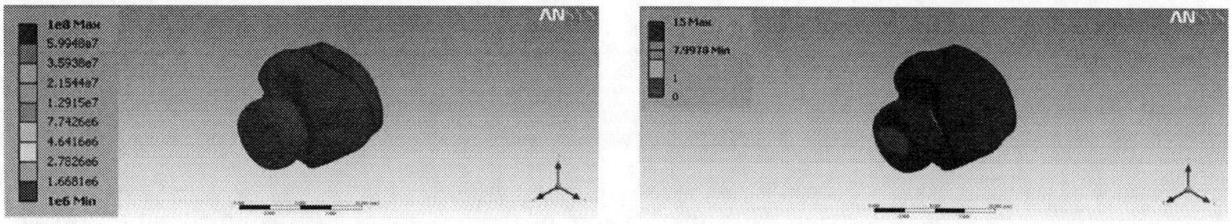

Figure 1. Fatigue Life and Fatigue Safety Factor in the bilayer ceramic molar crown with an uniform thickness of the veneer for a 400 N axial cyclic loading.

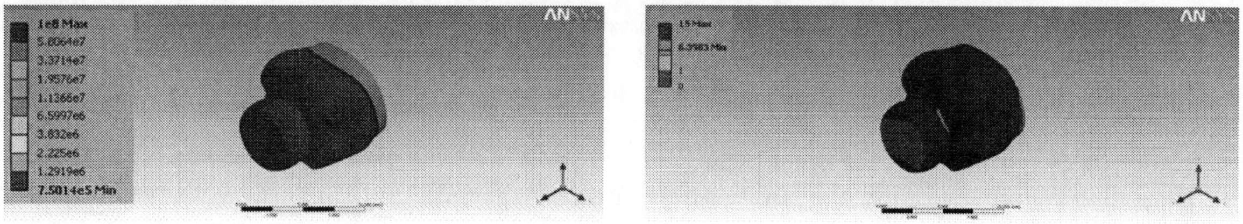

Figure 2. Fatigue Life and Fatigue Safety Factor in the bilayer ceramic molar crown with an uniform thickness of the veneer for a 500 N axial cyclic loading.

For 500 N cyclic loading the safety factor was under 1 in cases of para-axial loading, leading to failure at 7.5 x 10^5 cycles (for an uniform thickness of the veneer) and 3.44 x 10^5 cycles (for an uniform thickness of the framework). For axial loading the safety factor becomes smaller than 1 for cyclic loading with 3200 N, at 5.52 x 10^4 cycles (for an uniform thickness of the veneer) and 5.63 x 10^4 cycles (for an uniform thickness of the framework) (Figure 3).

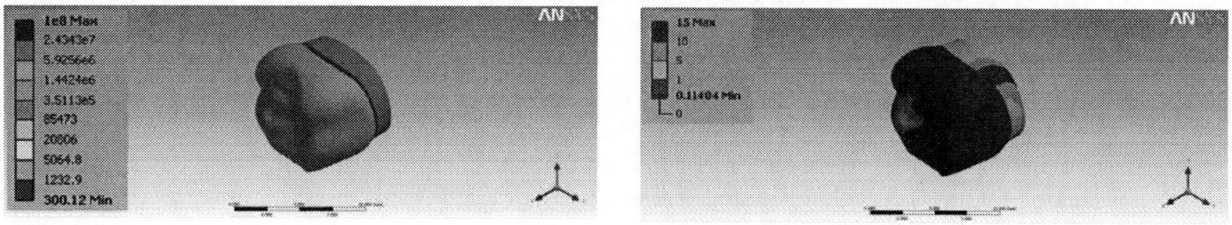

Figure 3. Fatigue Life and Fatigue Safety Factor in the bilayer ceramic molar crown with an uniform thickness of the framework for a 3200 N para-axial cyclic loading.

Registered Fatigue Safety Factor values for samples with different designs (uniform thickness of the veneer and uniform thickness of the framework) and loading areas (for axial and para-axial loading) are presented in Table 1.

Table 1. Fatigue Safety Factor values for all loaded samples.

Force [N]	Axial loading		Para-axial loading	
	uniform thickness of the veneer	uniform thickness of the framework	uniform thickness of the veneer	uniform thickness of the framework
400	7.9978	8.0029	1.0326	1.0265
500	6.3983	6.4023	0.74442	0.7984
1000	3.1991	3.2012	0.37221	0.36492
1500	2.1328	2.1341	0.24814	0.24328
2000	1.5996	1.6006	0.1861	0.18246
2500	1.2797	1.2805	0.14888	0.14597
3000	1.0664	1.0671	0.12407	0.12164
3200	0.99973	1.0004	0.11632	0.11404

There are no differences between the two types of crown designs in the axial load in terms of Fatigue Safety Factor. In case of para-axial loads, the lifetime of the restorations is extremely short. For this reason an occlusal balancing is needed as close as possible to ideal - which leads to a longer life of restorations.

5. Conclusions

Within the limitations of the studies, the following conclusions can be drawn:

- The fatigue behavior of various all-ceramic restorations was investigated using a wide range of load levels (400-3200 N) till 10^8 cycles.
- Axial loading leads to fatigue failure at higher loads (3200 N) compared to the para-axial (500 N).
- Related to the crown design, the failure behavior didn't change significant.
- The occlusal balancing is the most important factor for the lifetime of the restorations.

References

[1] Della Bona A, Matinlinna J P 2014 *Handbook of Oral Biomaterials* Pan Stanford USA.
[2] Della Bona A, Corazza P H, Zhang Y 2014 *Dent Mater* Characterization of a polymer-infiltrated ceramic-network material. 30 pp 564–569.
[3] Porojan L, Topala F, Porojan S, Savencu C 2017 *Dental Materials Journal* Effect of frame design and veneering material on biomechanical behavior of zirconia dental crowns veneered with overpressing ceramics 36(3) pp 275-281
[4] Porojan S, Porojan L 2016 *ACM Proceedings of the 8th International Conference on Information Management and Engineering* A Digital Approach for Anterior All-ceramic Bilayered Crowns Design Istanbul, TURKEY 99 pp75-79
[5] Sandu L, Faur N, Bortun, C 2007 *Journal of Prosthetic Dentistry* Finite element stress analysis and fatigue behavior of cast circumferential clasps 97(1) pp 37-44.
[6] Keulemans F, Palav P, Aboushelib M M N, vanDalen A, Kleverlaan C J, Feilzer A J 2009 *Dent Mater* Fracture strength and fatigue resistance of dental resin-based composites 25 pp 1433–1441.
[7] Cameron D W, Hoeppner D W 1996. *ASM Handbook*.19.ASM International, USA.
[8] Schijve J 2009 Fatigue of Structures and Materials 2nd ed. Springer, Netherlands.

[9] Watchman J B, Cannon W R, Matthewson M J, 2009 In: Mechanical Properties of Ceramics 2nd ed. JohnWileyand Sons, USA.
[10] Studart A R, Filser F, Kocher P, Gauckler L J, 2007 *Biomaterials* In vitro lifetime of dental ceramics under cyclic loading in water 28 pp 2695–2705.
[11] Belli R, Geinzer E, Muschweck A, Petschelt A, Lohbauer U, 2014 *Dent Mater* Mechanical fatigue degradation of ceramics versus resin composites for dental restorations 30 pp 424–432.
[12] Guess P C, Zavanelli R A, Silva N R F A, Bonfante E A, Coelho P G, Thompson V P 2010 *Int Prosthodont* Monolithic CAD/CAM lithium disilicate versus veneered Y-TZP crowns: comparison of failure modes and reliability after fatigue 23 pp 434–442.
[13] Magne P, Schlichting L H, Maia H P, Baratieri L N, 2010 *J Prosthet Dent* In vitro fatigue resistance of CAD/CAM composite resin and ceramic posterior occlusal veneers 104 pp 149–157.
[14] Homaei E, Farhangdoost K, Tsoi J K H, Matinlinna J P, Pow E H N 2016 *J Mech Behav Biomed Mater* Static and fatigue mechanical behavior of three dental CAD/CAM ceramics 59 pp 304-313.
[15] Browell R 2006 Calculating and Displaying Fatigue Results, ANSYS, Inc.

Stress and Strain Determination Occurring in Contact Area of the 450/75 ACSR Conductor

D Achiriloaiei[1,*], S V Galatanu[1] and I Vetres[1]

[1] Politehnica University of Timisoara, 1 Mihai Viteazu Blvd., 300222, Timisoara, Romania

[*]Corresponding author: daniel.achiriloaiei@yahoo.com

Abstract. The ACSR conductors are the metallic ropes stretched freely between poles and fixed onto them by supporting elements, which ensure the circulation of electrical energy. The special importance of this electrical conductors, coupled with the very high costs for repairs and replacements have imposed that the lifetime of power lines conductors to be between 50 and 80 years. Energy interruptions due to broken conductors have catastrophic impact from an economic, social and national security point of view. The present paper aims to determine stress and strain which are developing into contact area of the wires in the high voltage conductors. The study was performed using new conductors and 42 years aged conductors. Therefore an analytical calculation was performed in order to understand how the contact types are developed in conductor. The local stress and strain were determined experimental and a series of polynomial equations have been proposed. Theoretical and experimental researches have shown that the shape and size of imprints occurring at the contact points between wires have a significant influence on stress and strain of the conductors. An assessment of contact stresses under plastic deformations cannot be made without taking into consideration the characteristics of the imprints.

1. Introduction

To determine the life of conductor in the literature numerous attempts have been made, but considering the small number of vibration and their effects records applied, they are not relevant enough to the conductor. Response produced by the galloping vibrations, as well as wind vibrations is a major issue in assessing the durability of electrical conductors, under the circumstances that experimental studies have highlighted the devastating role of combined action of contact and fatigue on the durability of aluminium [1].

Even if regular conductor inspection is recommended, hidden defects (especially those generated at the wires contact or between the conductor and the clamps) may cause premature breaking of them. Layout of the high voltage lines significantly influence the occurrence and development of vibrations. Thus, it has been demonstrated that on a section although some common features are encountered, there are several particular situations, depending on the position in the section [2, 3]. Breaking of the component elements of the conductor (the wires) is a first step in deficient operation and a first step to final break. It has been noticed that the breaking of conductors is one of the serious electrical shock loads which a line section may have, through power interruptions and high voltage imbalances which can lead to cascade breaks of the conductors. Under normal operating conditions, the conductors are almost constantly subject to low amplitude wind vibrations. At the same time, they have to sustain their own weight, and in the fastening areas mechanical stresses arise due to increased local pressures around the clamps and within the many devices they come into contact with.

The problem of normal contact in case of plastic deformations is particularly difficult because they are far superior to the elastic ones. In addition, the convex profile of the wire causes additional complications as shown by Achiriloaiei et al. [4] which is a first step towards understanding the phenomena that appear at the contact between wires. Determination of the plastic deformed area is impossible in this case, consequently, the contact pressure can only be approximated. For such applications the use of numerical methods is recommended and experimental validation of the results.

Critical contact areas of high voltage lines with multilayer conductors remain in the fasteners regardless of aerodynamics, and are manifested by degradation of the conductors. To simulate the contact types between wires and to identifying the effects created have been kept the same angles of imprinting that were used in the study of Achiriloaiei et al. [4]. The wires were subjected to an imprint test using three different angles 30°, 60°, 90°. Many of the points where contact stresses develop are located in the area of the clamps, between different layers or between the last layer and the body of the clamping clamps, as can be seen in Figure 1. In Figure 1 is marked with arrows the areas where stress has maximum values and the directions of application of the contact pressures.

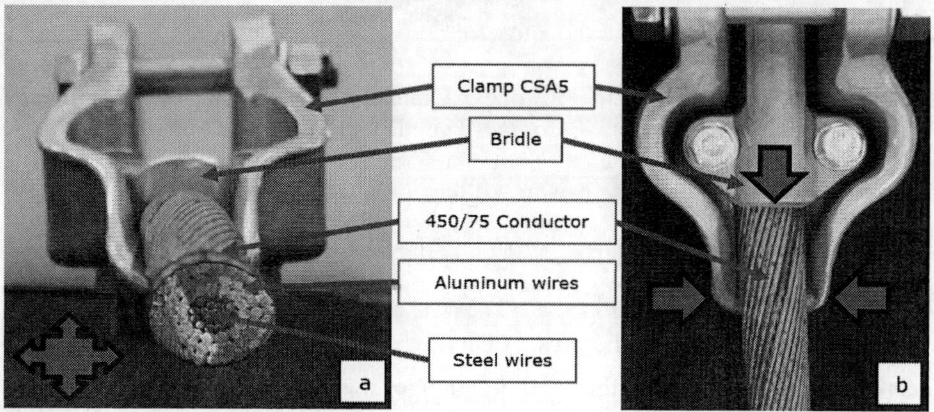

Figure 1. Fixing the conductor to the clamp; a) the section of the conductor - directions and meanings of movement; b) conductor-clamp contact area

Conductor behaviour for multiaxial load creates, beyond the complex contact stress, also gradients of a physical magnitude of stresses at the edge of the contact surface, where cracks can be created and propagations thereof. It must also be taken into account that local contact stress, under the action of friction, is distributed differently depending on the geometry of the contact, pressures, materials in contact and tribology of the contact surface (science and technology of interaction between two contact surfaces in relative motion). The contacts surface, in the case of conductor wires, is in the form of non-conforming couplings or Hertzien and is changing with the type of contact variation and material behaviour.

2. Analytical method

In the case of wires the calculations that are made are similar to those of the cylindrical bodies. Aluminium wires used in electric conductors are considered as cylinders wrapped helically whose radius depends on their position within the layer they belong to. When a normal contact force acts on the bodies local deformations occur and the geometrical configuration of the contact ellipse may change. For a Hertzian contact, the contact pressure is given by the below expression, considering that p_0 is mean pressure on the contact surface:

$$p = p_0 \sqrt{1 - \left(\frac{x}{dy}\right)^2 - \left(\frac{y}{dx}\right)^2} \tag{1}$$

Figure 2 shows the layout of two wires in contact, a specific case of interlayer contact.

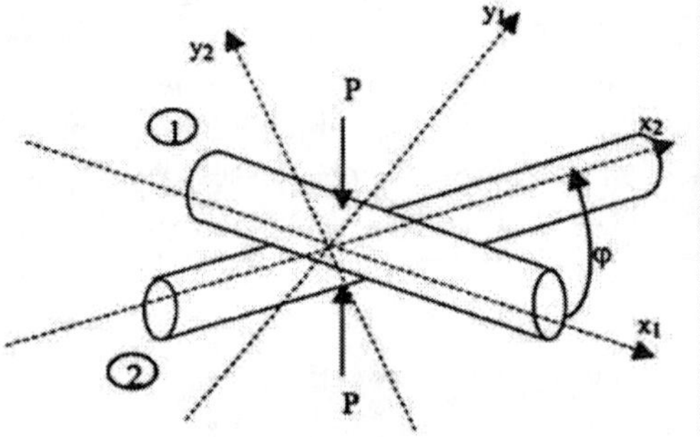

Figure 2. The contact between two wires with the same diameter

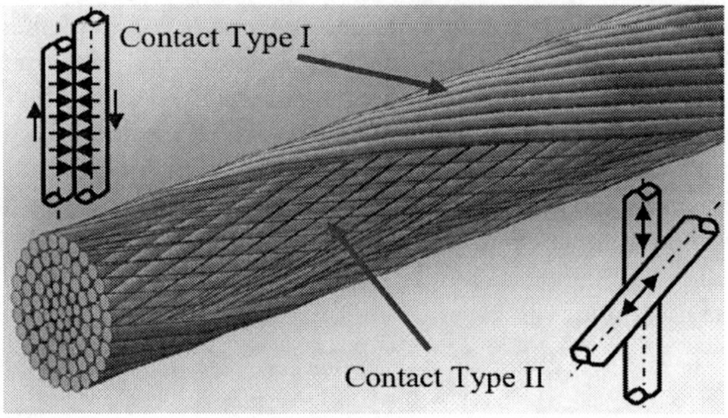

Figure 3. Contact types: Contact Type I – linear contact; Contact Type II – point-to-point contact

Under these conditions, the equations can be written:

$$R_1 = \frac{\rho_1}{\sin^2 \alpha_1}; \qquad R_2 = -\frac{\rho_2}{\sin^2 \alpha_2} \tag{2}$$

Where $\varphi = |\alpha_1 - \alpha_2|$ and ρ_1, ρ_2 are the rays of curvature of the two wires in contact, and α_1 and α_2 represent their directions in relation to the axial direction of the conductor.

Ouaki et al. [6] in his work considers two geometric profiles whose contact surfaces can be approximated by two polynomial functions δ (x, y). Under conditions of a common axis system (x, y) for the two bodies of the conductor wires having the same radius $R_1 = R_2 = r$, their proximity becomes:

$$\delta(x, y) = Ax^2 - By^2 \tag{3}$$

The coefficient A and B depends on the curves of the two bodies and their orientation towards each other.

$$A = \frac{1}{2r}(1 - \cos\varphi); \qquad B = \frac{1}{2r}(1 + \cos\varphi) \tag{4}$$

In the literature it is considered dx - higher axis of the ellipse and dy - the smallest axis of the ellipse, and it is indicated that eccentricity of the contact ellipse, e_c, it does not depend on applied force and can be approximated in the form of:

$$e_c = \frac{dy}{dx} \approx \left(\frac{B}{A}\right)^{-\frac{2}{3}} \tag{5}$$

It can be seen that when φ = π/2, dx = dy, the contact area is reduced to a circle, and when φ = π, the ellipse becomes a rectangle of width 2dy and length $e_c = \infty$ as shown in Figure 3.

Due to the linear-elastic character shown at the moderate tensile loading and the helical layout of the wires, in contact areas is being exercised continuous load such as bonding condition (reaction).

Under these forces, the conductor is in balance and it creates a certain state of loading in the form of the two types of contacts. These contacts have shape and dimensions of crystalline interstices changes at the structural level, developing continuous efforts in the mass of wires that lead to the formation of new crystals. Thus, in the mass of the conductors there is a series of efforts, stresses and strains, which influence the lifetime of them and is summed up for tasks that occur in operation

3. Laboratory investigations and test results

3.1. *Material, equipment and methods*

Aluminum Conductor Steel Reinforcement (ACSR) conductor used for laboratory determinations is 450/75. The new wires were taken from a conductor produced by IPROEB Bistrita according to Stas 35/1999. The aged wires are of the same type and were taken from the active conductors on the LEA network, Porţile de Fier-Arad from the Timişoara-Arad section and which have been in service for 42 years. The materials used in the tests are aluminium wire with a purity of 99.5% and a diameter of d = 2.95 mm, which are found in the three layers of windings.

Tensile tests and imprints have been conducted in the Strength of Materials Laboratory from Politehnica University Timisoara. The tests were performed using a tensile-compression testing machine, model Zwick/Roell of 5 kN at the 20°C and the relative humidity around 50 ± 5 %. The microscopic examination on the surface of the wires was performed with the optical microscope. The microscopic analysis was considered in the imprinted area of the wires in order to obtain 3D deformations. Taking into account the imprint angle φ which influences the appearance and size of the imprints, the dependence between the force of pressing and the development of imprints on the surfaces of the new wires has been established.

The indentation of wires was done by a contact loading, using a fixing device which renders the real contact between the conductor wires, figure 4.

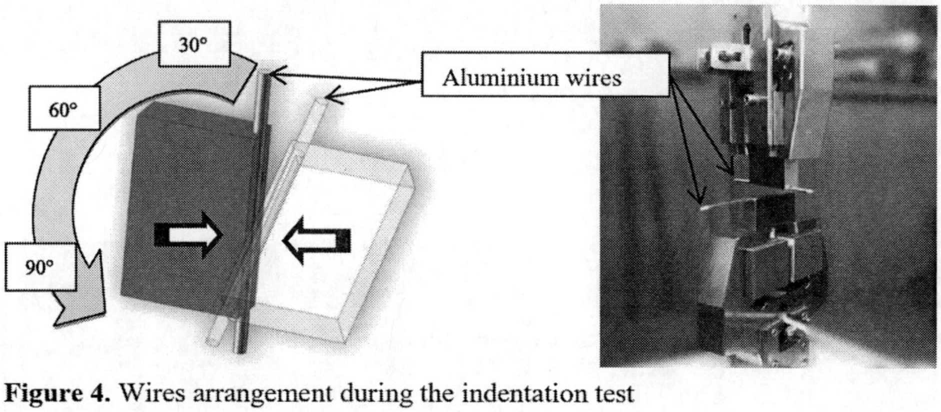

Figure 4. Wires arrangement during the indentation test

The values of the imprints φ have been set to be consistent with those most commonly encountered in reality, with values of 30°, 60°, 90°. At the same time, the normal pressure values were divided into 6 loading steps of 0,54N/mm^2, 1,08N/mm^2, 1,62N/mm^2, 2,16N/mm^2, 2,7N/mm^2 and 5,4N/mm^2.

3.2. *Tensile and imprint tests*

The tests revealed the mechanical behavior of the Al wires from a new and old conductor. The imprints size have been measured and the wires was subjected to tensile tests. Tensile tests were done on wires taken from each three types of imprints. Using the measurements made on the plastic deformed areas, the imprints variation diagrams were drawn according to the force of the press. With the help of the imprint variation diagrams, a series of polynomial equations have been developed to calculate the area of the deformed surface by the contact pressure, depending on the force of the

IOP Conf. Series: Materials Science and Engineering **416** (2018) 012071 doi:10.1088/1757-899X/416/1/012071

pressing force. These equations can also be used in the case of existing imprints when it is desired to determine the pressing forces. There are two sets of equations for new and aged wires.

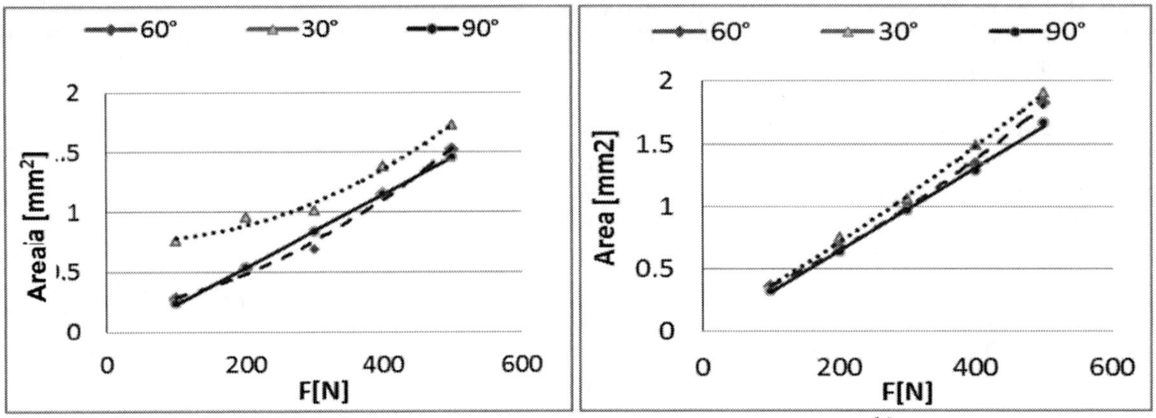

a) b)

Figure 5. Comparative variation of the inprinted surface by force of pressure, for variable φ: a) new wires; b) aged wire

$$A_{n30} = \left(4 \cdot 10^{-6}\right) F^2 - 0{,}0003 \cdot F + 0{,}7691 \tag{6}$$
$$A_{n60} = \left(4 \cdot 10^{-6}\right) F^2 - 0{,}0008 \cdot F + 0{,}1712 \tag{7}$$
$$A_{n90} = 0{,}0031 \cdot F + 0{,}0714 \tag{8}$$

$$A_{i30} = 10^{-6} \cdot F^2 + 0{,}0031 \cdot F + 0{,}0467 \tag{9}$$
$$A_{i60} = \left(2 \cdot 10^{-6}\right) F^2 + 0{,}0022 \cdot F + 0{,}1253 \tag{10}$$
$$A_{i90} = 0{,}0033 \cdot F - 0{,}0198 \tag{11}$$

Where: A_{n30} – Area of imprint for new wire and φ=30°, 60°, 90°
A_{i30} – Area of imprint for aged wire and φ=30°, 60°, 90°

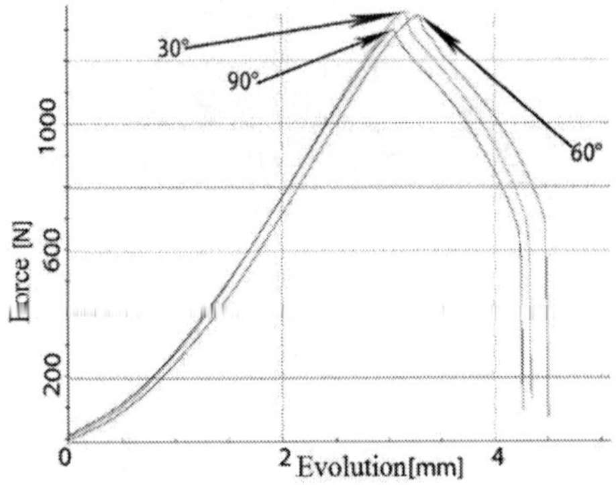

Figure 6. Characteristic diagram of imprinted wires

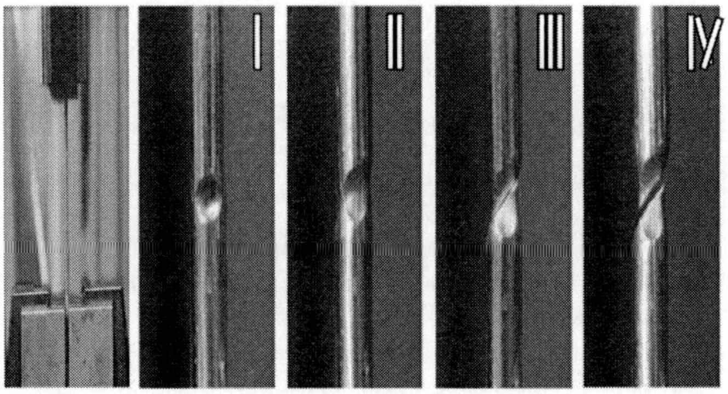

Figure 7. Evolution of traction of the imprinted wire

The traction characteristic pattern was plotted for the imprinted Al wires using 1kN constant force at different imprint angles. In case of imprints with different angles, the size of the imprint varies greatly and influences the tensile breaking strength. Local deformations are deeper for 90° imprint angles and tensile strength is lower than for 30 ° and 60 ° imprint angles. Another very important aspect is related to the section where the breakage occurs and the shape it has.

3.3. Stresses and strains estimation

The purpose of static analyses performed on aluminium wires is to determine deformations that occur in the contact area, resulting in the appearance of imprints. The data from the laboratory measurements of new wires were introduced into the SigmaPlot graphics program, with which the graphs of the plastic deformations were plotted. In the case of tensions, it was chosen to present tensions calculated according to the von Mises criterion. They present the three-dimensional variations of the imprints, with the emphasis on the dependence between the imprint surface and the depth of penetration. Was followed, the distribution of the equivalent stresses, σ_{ech} von Mises at the element level and the distribution of the equivalent plastic deformation.

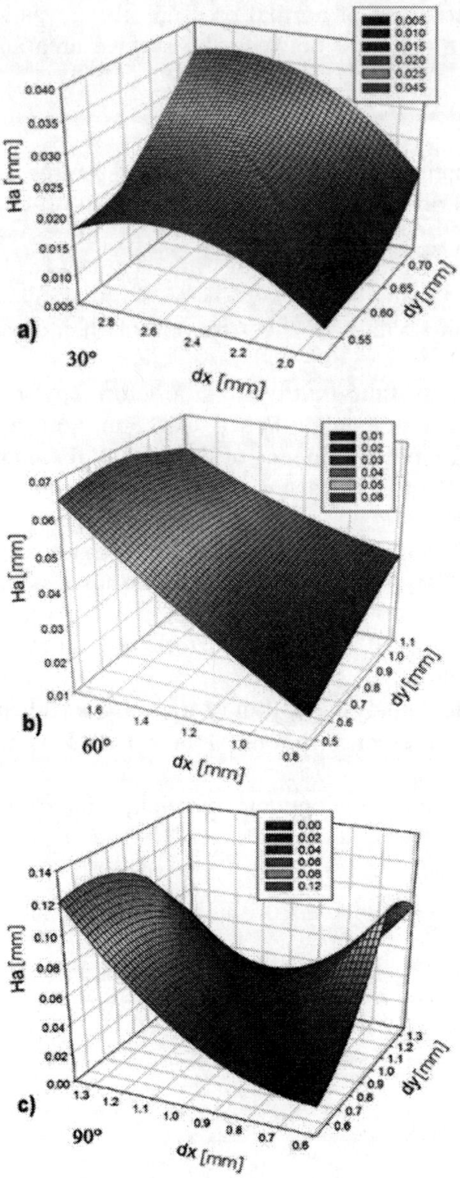

Figure 8. Three-dimensional variation of inprints for the inprint angles φ

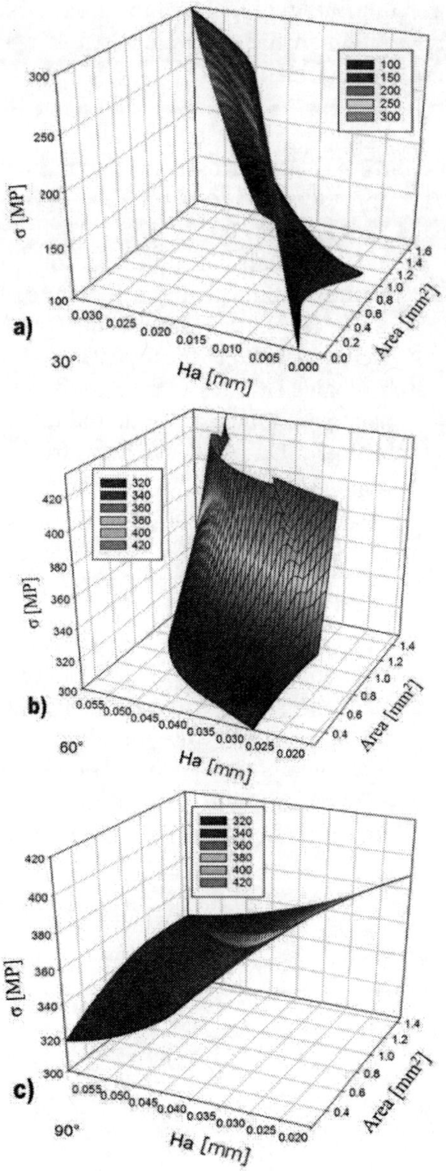

Figure 9. Variance of von Mises stresses for the three inprint angles

4. Conclusions

Solutions provided by elastic analyses have proven to be inadequate for this type of contacts. The contact area is too large compared to the size of the contact in the bodies to make the analyses on half of the plane. Comparison of tensions with the 2D solution for cylinder contact showed the importance of free wire surface and pressure applied to imprints. The elasto-plastic results are similar to those for cylinder contact but with particularities of pre-existing plastic deformations.

The existence of imprints, of any kind, causes the traction fracture to develop from the imprint section, as in Figure 7. This is not confirmed for small imprint forces due to structural inhomogeneity. In contrast to the breakage of unapplied samples, where the fissure development plane is normal to the axis, in the case of imprints the fissure develops in a plane inclined at 45° to sample axis. This fissure passes through the imprint area, joining the marginal micro-portions, and cuts in the thinned area.

As the size of the imprints increases, the tensile strength is more heavily reduced (dropping with 10% for imprints with a 2.5 mm^2 area). Using linear variations of normal pressure, there was a non-linear variation in the equivalent stress and a change in the ratio between the surface area and the depth of the imprints.

5. References

[1] Martins L.H.L., Rossino L.S., Filho W.W.B., Araujo J.A., Detailed design of fretting fatigue apparatus and tests on 7050–T7451 Al alloy, Tribology - Materials Surfaces and Interfaces, Vol. 2, 2008, pp. 39–49

[2] Kim H.S., Mall S., Investigation into three-dimensional effects of finite contact width on fretting fatigue, Finite Element in Analysis and Design, Vol. 41, 2005, pp. 1140–1159

[3] Suzanne Guerard, Power Line Conductors, a Contribution to the Analysis of their Dynamic Behaviour, Ph.D thesis, 2011

[4] Achiriloaiei D., Neş C.S., Dumitru I., Influenţa contactului dintre fire şi a deformaţiilor locale asupra conductoarelor de înaltă tensiune, Revista Universităţii Petrol–Gaze din Ploieşti, 2015

[5] Achiriloaiei D., Neş C.S., Dumitru I., Cernescu A., On the influence of the contact force and the angle between the wires on ACSR conductors' mechanical properties and durability, Key Engineering Materials, Vol. 665, pp.33-36, 2016

[6] Ouaki B., Analyse expérimentale et théorique de l'endommagement par fatigue à pettits débattements des fils de conducteurs électriques au voisinage d'une pince de suspension, Ph-D These, 1998

[7] Ramey G.E., Townsend J.S., Effect of clamp on fatigue of ACSR conductors, Journal of Energy Division, ASCE (EYl) 1981, Vol. 107, pp.103–119

[8] Zhu Hui-ling, Zhu Xin-yin, Wang Yu-lin, Numerical analysis of sliding friction coefficient and the normal pressure between the cable wire, International journal on smart sensing and intelligent systems vol. 7, no. 4, 2014

[9] EPRI Transmission Line Reference Book: wind induced conductor motion, Electric Power Research Institute, Palo Alto, California, United States, 2006.

IOP Conf. Series: Materials Science and Engineering **416** (2018) 012072 doi:10.1088/1757-899X/416/1/012072

New Electrodes Based on Pt-Co Alloys Used as Anodic Materials for Sulphite Oxidation in Alkaline Fuel Cells

A F Enache[1], M L Dan[1], R Muntean[2] and N Vaszilcsin[1,*]

[1]Faculty of Industrial Chemistry and Environmental Engineering, Politehnica University Timişoara, Romania
[2] Department of Materials Science and Testing, Westphalian University of Applied Sciences, Gelsenkirchen, Germany

*Corresponding author: nicolae.vaszilcsin@upt.ro

Abstract. In this paper, electrooxidation of sulphite ions on Cu/Pt-Co electrodes in alkaline media (1 mol L^{-1} NaOH) has been investigated in order to establish the mechanism and to determine optimal parameters of the anodic process. Electrochemical behavior of sulphite ions on Cu/Pt-Co electrodes in alkaline solutions has been studied by voltammetric techniques (cyclic and linear), and kinetic parameters (current density i_0 and anodic transfer coefficient α) have been determined by Tafel plots method. Furthermore, electrochemical impedance spectroscopy was carried out to confirm the sulphite oxidation mechanism on this type of electrode. Chrono-electrochemical techniques have been applied in order to determine the optimum characteristic parameters: current density, oxidation potential for sulphite electrooxidation and transformation degree of SO_3^{2-} ions in the anodic process at different electrolysis times.

1. Introduction

Among noble metals tested in order to obtain appropriate anodes for both alkaline and acid media, platinum exhibits the highest catalytic activity [1]. Limited resources and high cost of Pt are considered significant impediments for a wide commercialization of fuel cells. Pt-Co alloys represent an attractive challenge in the development of advanced electrocatalysts, which would considerable decrease not only the content of Pt, but would also enhance the catalytic activity and durability [2]. Alloying Pt in the form of nanoparticles with 3d transition metals (Fe, Co, and Ni) has shown much more potential in tailoring the atomic and electronic structures of catalytically active materials than any other approach [3].

Several anodes made of metal oxides, as well as electrodes with modified surface have been used throughout time in order to study the electrochemical oxidation of sulphite ions (SO_3^{2-}) and sulphur dioxide dissolved in water ($SO_{2(aq)}$) [4-6]. Much attention was paid to the oxidation of sulphite and sulphur dioxide on the electrodes based of noble metals (Pt, Au, Pd) due to their high catalytic activity [7-11]. The oxidation rate of tetravalent sulphur compounds was noticed to be strongly dependent on the electrode material nature, electrocatalytic properties of electrode, active electrode surface, state of metal surface and specific potential range for electrooxidation process. For noble metals these effects are related mainly to the influence of adsorbed sulphur species [12, 13] on the kinetics of the electrode process. However, the oxidation rate of sulphur dioxide and sulphite ions might also be influenced by oxidation processes on the surface of noble metals [14, 15].

The general purpose of this study is to obtain an alloy catalyst with increased selectivity and activity, low energy consumption and long lifetime, operable in a SO_3^{2-} / O_2 fuel cell. As well, its aim is to reduce the amount of noble metal by obtaining nanometric particles, thus increasing the active electrodes surface, improving electrochemical reactivity and reducing process costs.

2. Experimental part

Working electrode has been obtained by electrodeposition of Pt-Co on a smooth copper disc with 1.2 cm^2 surface. The electrodeposition of Pt-Co alloy catalyst by the pulsed current plating method was done into an acidic solution, platinum ions coming from K_2PtCl_4 solution and cobalt ions from $CoCl_2 \cdot 6H_2O$ solution. The molar ration between Pt^{2+} and Co^{2+} ions in electrolyte solution was 1:20. 1 M KCl and 0.5 M H_3BO_3 were added into solution in order to achieve optimum conductivity and pH stabilization. The pulsed current deposition parameters were t_{on} = 20 ms, t_{off} = 100 ms, i = 250 A m^{-2} and 20000 cycles. These are the optimum values in order to obtain Pt-Co alloy catalyst particles with size distribution in the nanometric range on copper support.

For sulphite anodic oxidation in alkaline media, a three-electrode undivided cell connected to SP 150 BioLogic potentiostat/galvanostat was used during all measurements. Pt-Co alloys prepared catalysts were used as working electrode with 0.8 cm^2 active surface. Also, two graphite roads - counter electrodes and Ag/AgCl - reference electrode, equipped the electrolysis cell. A nitrogen purging system was used to de-aerate the solution before each electrochemical measurement.

Electrochemical experiments were carried out in 1 mol L^{-1} NaOH solution in the absence and presence of different sulphite concentrations (10^{-3}, 10^{-2}, 10^{-1}, $5 \cdot 10^{-1}$ and 1 mol L^{-1}). Reagents NaOH and Na_2SO_3 (p.a., ≥ 98%, Merck) were used in order to prepare the test solutions.

Cyclic voltammograms have been recorded at different scan rates, between 5 - 500 mV s^{-1} and linear polarization at low scan rate (1 mV s^{-1}) in order to ensure quasi-stationary conditions.

Electrochemical impedance spectroscopy (EIS) studies were performed using the SP-150 impedance module, in frequency range 0.1 Hz - 100 kHz and alternative voltage amplitude of 10 mV. For each spectrum, 60 points were collected, with a logarithmic distribution of 10 points per decade. The experimental EIS data have been fitted to the electrical equivalent circuits (EECs) by CNLS Levenberg – Marquardt method using ZView – Scribner Associates Inc. software.

Chronoamperometry, chronocoulometry and chronopotentiometry methods were applied in order to determine the sulphite electrooxidation process parameters as: current density, potential range, and transformation degree, at five oxidation potential values.

3. Results and discussion

3.1. Cyclic voltammetry. Cyclic voltamograms (CVs) have been recorded in alkaline solution in order to identify the processes occurring at the Pt-Co alloys electrode/alkaline electrolyte interface.

In Figure 1 the CVs registered for three types of electrodes (Pt, Cu, Cu/Pt-Co) in presence of 10^{-1} mol L^{-1} sulphite are shown. On the curves drawn for the electrodes containing copper (Cu and Cu/Pt-Co), the appearance of a pick can be observed at approximately -0.25 ÷ +0.25 V corresponding to the copper oxidation process. At advanced anodic polarization, on the three electrodes the peak specific to sulphite ions oxidation appears at +0.50 ÷ +0.75 V.

In Figure 2 the CVs ploted on Cu/Pt-Co electrode in alkaline solution with and without sulphite ions are presented. CVs recorded on Pt electrode are represented at a higher resolution than for Cu and Cu/Pt-Co electrodes in order to highlight the electrochemical reactions occuring at electrode/electrolyte interfaces. It can be observed that, at cathodic polarization, the peak assigned to the oxygen reduction reaction is revealed only in the absence of sulphite ions. That means, the oxygen anodically formed has been consumed for sulphite ions oxidation in a chemical reaction.

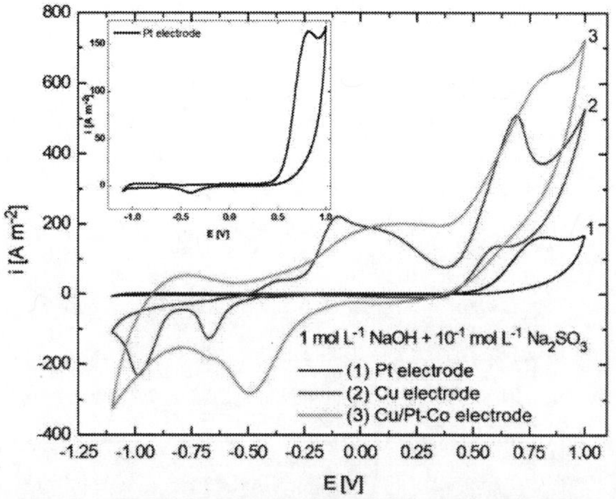

Figure 1. CVs recorded on different electrodes in alkaline solution in presence of 10^{-1} mol L^{-1} Na_2SO_3, scan rate 500 mV s^{-1}.

At advanced cathodic polarization, it can be seen that only hydrogen evolution reaction (HER) takes place.

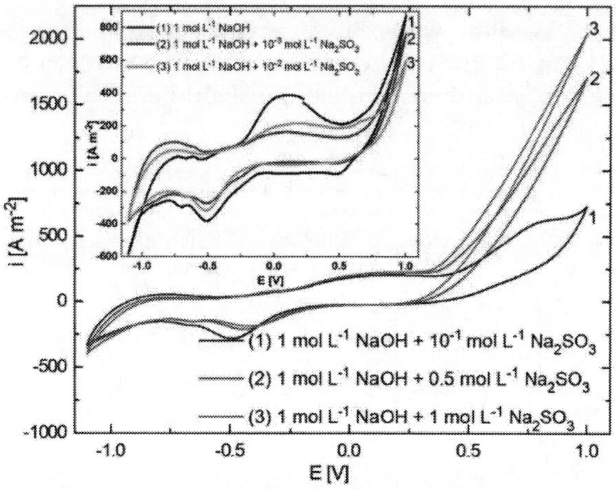

Figure 2. CVs recorded on Cu/Pt-Co electrode in alkaline solution with and without Na_2SO_3, scan rate 500 mV s^{-1}.

As well, at anodic polarization, the sulphite oxidation (Eq. 1) occurs simultaneously with atomic oxygen formation (Eq. 2), as it has been shown in previous papers [16-18].

$$SO_3^{2-} + 2HO^- \rightarrow SO_4^{2-} + H_2O + 2e^- \quad (1)$$

$$2HO^- \rightarrow O + H_2O + 2e^- \quad (2)$$

3.2. Linear voltammetry. Linear voltammograms (LVs) have been drawn at low scan rate (1 mV s^{-1}) in order to provide quasi-stationary conditions at the electrode.

In Figure 3 the LVs plotted for 3 different electrodes containing (Pt, Cu, Cu/Pt-Co) in presence of 10^{-1} mol L^{-1} Na_2SO_3 are presented. The electrochemical processes which take place on all 3 electrodes have been highlighted by the occurrence of peaks at different potential values. Thus, a peak at about -0.30 V appears as a result of copper oxidation and at approximately +0.40 V the sulphite oxidation occurs followed by oxygen evolution reaction (OER).

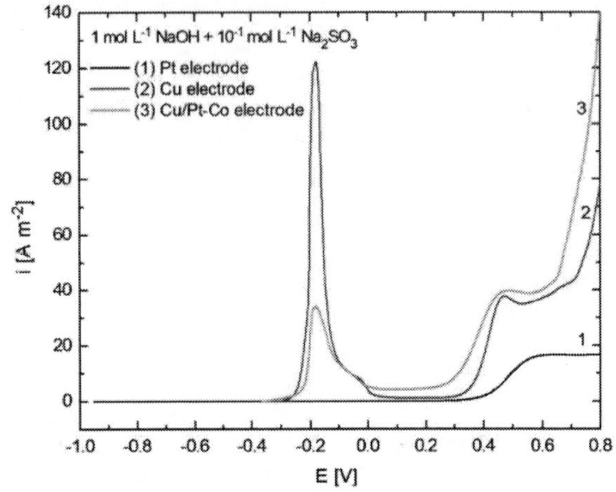

Figure 3. LVs recorded on different electrodes in alkaline solution in presence of 10^{-1} mol L^{-1} Na_2SO_3, scan rate 1 mV s^{-1}.

Analyzing the LVs recorded in alkaline media in presence of different concentrations of sulphite (Figure 4), it can be seen that sulphite oxidation current increases with the sulphite concentration added in electrolyte. That proves that at the anode a chemical reaction between atomic oxygen and sulphite occurs simultaneously with electrochemical ones. The oxidation current for low concentration of sulphite (10^{-3} and 10^{-2} mol L^{-1}) is almost indistinguishable due to an insignificant amount of SO_3^{2-} ions in the solution at electrode/electrolyte interface.

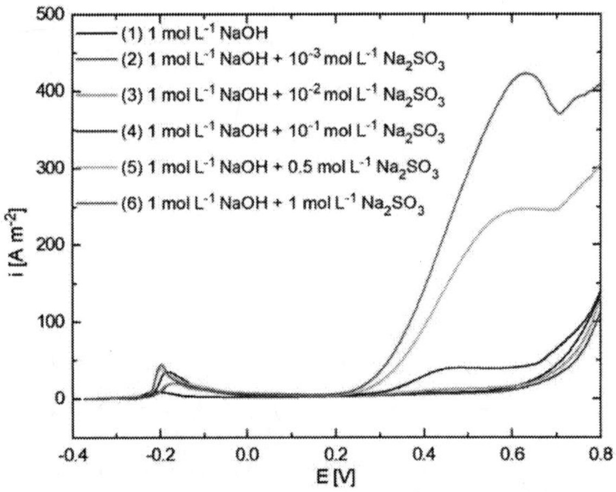

Figure 4. LVs recorded on Cu/Pt-Co in alkaline solution with and without Na_2SO_3, scan rate 1 mV s^{-1}

In order to determine the kinetic parameters (transfer coefficient α and exchange current density i_o) for anodic oxidation of sulphite Tafel plots method presented in Figure 5 has been applied.

The calculated kinetic parameters are presented in Table 1. Obtained high values of exchange current density are characteristic for fast charge transfer processes. In these circumstances, it can be appreciated that the rate determining step for the atomic oxygen generation (Eq. 2) is controlled by charge transfer step.

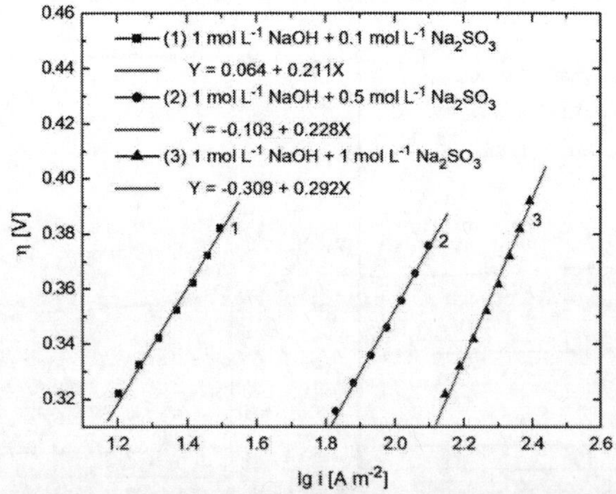

Figure 5. Tafel plots for sulphite oxidation on Cu/Pt-Co electrode in alkaline solution with different concentrations of Na_2SO_3.

Table 1. Kinetic parameters for sulphite oxidation in alkaline solution.

Na_2SO_3 conc. (mol L^{-1})	*b* (mV dec^{-1})	α	i_o (A m^{-2})
0.1	211	0.28	0.74
0.5	228	0.26	1.57
1	292	0.20	2.88

Similarly, for the direct oxidation of sulphite, the rate determining step is the charge transfer given by Eq. (3) [19].

$$SO_{3\,(ads)}^{2-} \rightarrow SO_{3\,(ads)}^{\bullet-} + e^{-} \quad (3)$$

In case of reduced sulphite concentrations (10^{-3} and 10^{-2} mol L^{-1}), the diffusion of sulphite ions from the bulk of solution to the electrode becomes the determining step.

Taking into account that on the electrode/electrolyte interface, in the potential range of sulphite oxidation, two parallel processes occur (direct sulphite oxidation and atomic oxygen generation), the values determined for α and i_o are just apparent.

3.3. Chronoamperometric studies. Analyzing the LVs from Figure 4, five potential steps were chosen (+0.35, +0.45, +0.55, +0.65 and +0.75 V) to highlight the anodic processes that occur at electrode/electrolyte interface. The first four values correspond to the sulphite direct/indirect oxidation in test solutions and the last one is assigned to OER.

In Figure 5a are presented the chronoamperometric curves, relating to sulphite oxidation, for 60 minutes, on working electrode in alkaline solutions, with different sulphite concentrations, at a potential value of +0.55 V.

In Figure 6b, the variation of current density in time registered on Cu/Pt-Co electrode in alkaline solution containing 1 mol L^{-1} Na_2SO_3 is presented for all five potential values at which the experiments have been carried out.

Based on chronoamperometric data it can be observed that the characteristic potential range for SO_3^{2-} oxidation in alkaline solution, between +0.35 and +0.65 V, does not depend on the sulphite concentration, but the current density increase with the increasing of sulphite concentration. At more positive potential values, SO_3^{2-} oxidation and OER occur simultaneously.

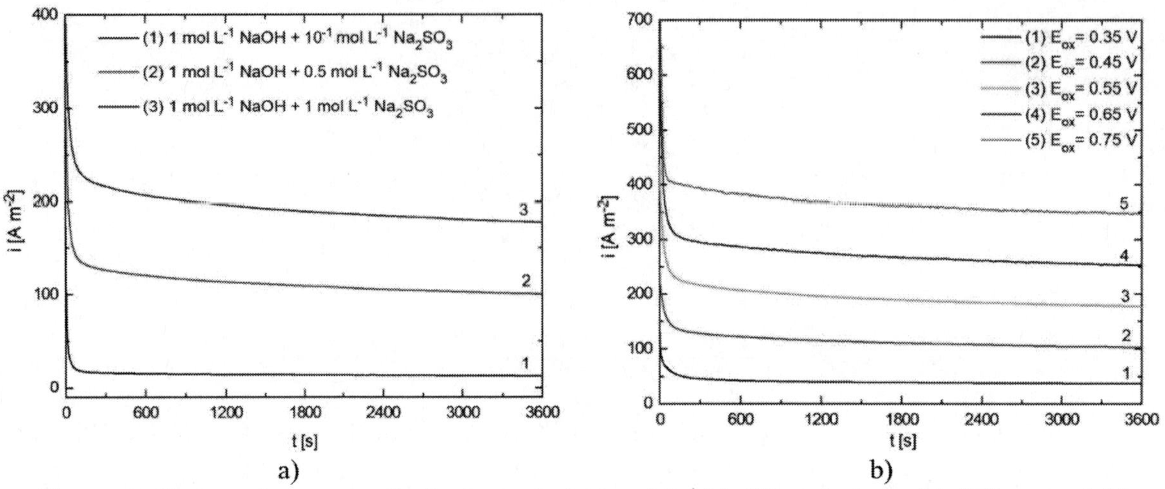

Figure 6. Current-time curves for Pt electrode in 1 mol L^{-1} Na_2SO_4 solution with different sulphite concentrations at +1 V (a) and in 1 mol L^{-1} Na_2SO_4 + 1 mol L^{-1} Na_2SO_3 solution (b).

Correlating these values with cyclic and linear voltammetric data, it can be established that the amount of sulphite added in alkaline solutions stimulates the first stage of OER.

3.4. Chronocoulometric studies. Chronocoulometric technique has been applied in order to determine the amount of electricity consumed for electrooxidation of sulphite ions in the characteristic potential range for each test solution.

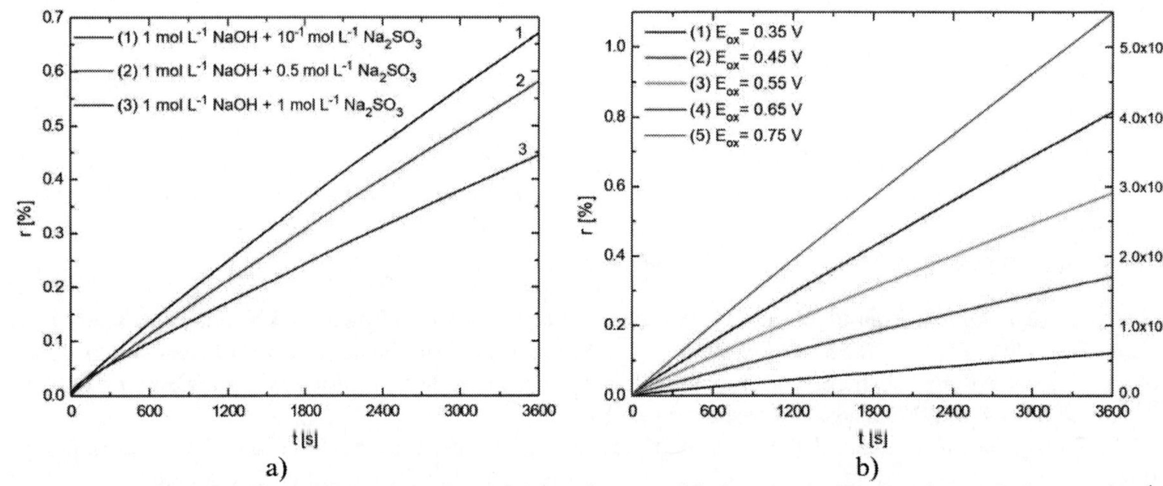

Figure 7. Chronocoulometric curves for sulphite electrooxidation on Cu/Pt-Co electrode in 1 mol L^{-1} Na_2SO_4 solution with different sulphite concentrations at 1 V (a) and in 1 mol L^{-1} Na_2SO_4 + 1 mol L^{-1} Na_2SO_3 solution (b).

Figure 6a presents the sulphite transformation degree during anodic oxidation as a function of time and sulphite concentration, when the electrochemical process is carried out at +0.55 V. Sulphite concentration is inversely proportional to the transformation degree, so the highest value of the transformation degree has been obtained for the test solution containing the lowest concentration of sulphite (0.1 mol L^{-1}).

IOP Conf. Series: Materials Science and Engineering **416** (2018) 012072 doi:10.1088/1757-899X/416/1/012072

Based on chronocoulometric data, the evaluation of number of sulphite moles changed in the anodic reaction (δ) and transformation degree (r) of sulphite to sulphate has been possible using Faraday's laws [20]. The results for 1 mol L^{-1} sulphite added in alkaline solution at all five potentials are presented in Figure 6b.

3.5. Electrochemical impedance spectroscopy. Based on cyclic and linear voltammetry confirmed by chronoamperometric data, the non-destructive experimental method electrochemical impedance spectroscopy (EIS) was used to obtain specific information for sulphite oxidation in alkaline media on Cu/Pt-Co electrode. EIS data weere recorded in test solutions with sulphite concentration between 10^{-3} and 1 mol L^{-1} at 4 oxidation potential values in the range +0.35 and + 0.65 V. The obtained results are shown in Nyquist and Bode complex plane representation given in Figure 8 for 0.5 mol L^{-1} Na_2SO_3 added in alkaline solution and Figure 9 for all test solution at same oxidation potential value, +0.45 V.

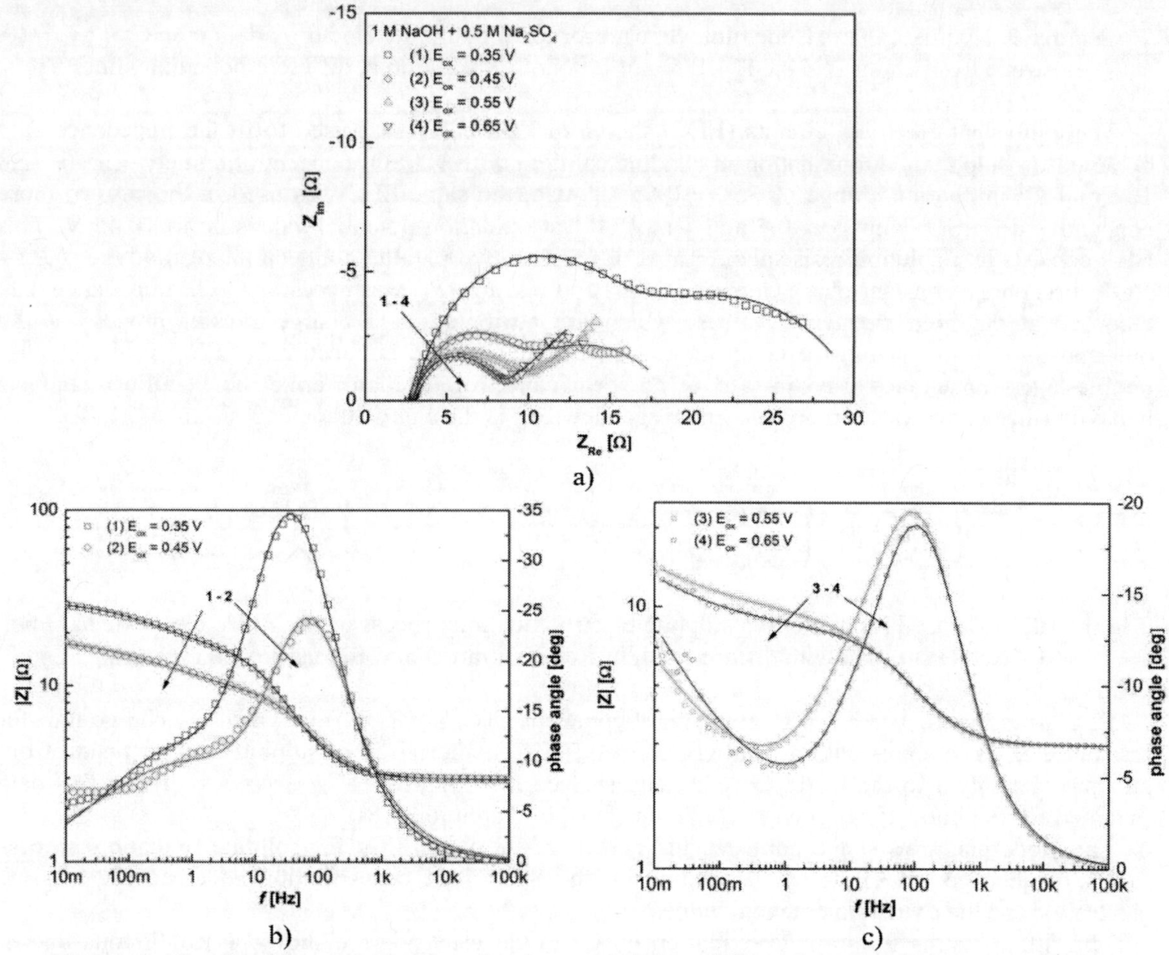

Figure 8. Nyquist (a) and Bode plots (b, c) recorded for sulphite electrooxidation on Cu/Pt-Co electrode from 1 mol L^{-1} NaOH + 0.5 mol L^{-1} Na_2SO_3 solution at different potential values.

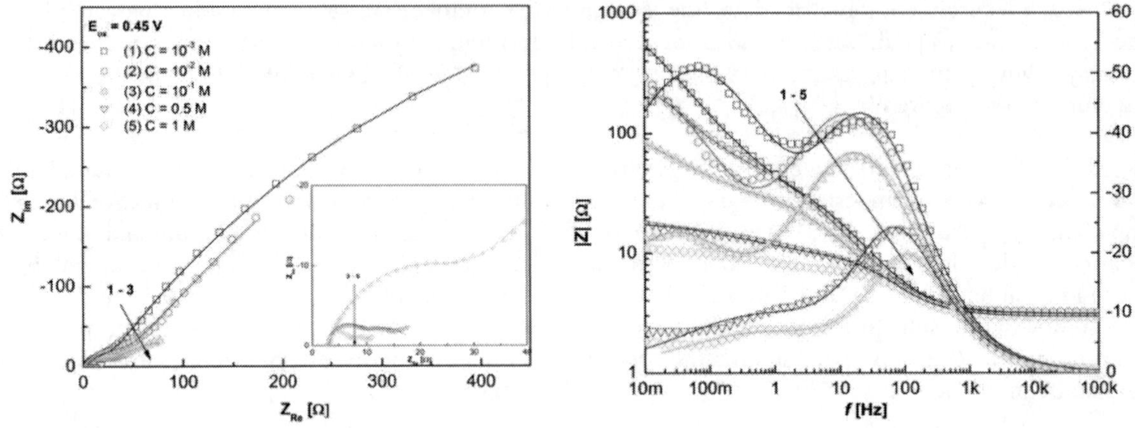

a) b)

Figure 9. Nyquist (a) and Bode plots (b, c) recorded for sulphite electrooxidation on Cu/Pt-Co electrode from 1 mol L^{-1} NaOH + 0.5 mol L^{-1} Na_2SO_3 solution at different potential values.

The equivalent electrical circuits (EEC) shown in Figure 10a was used to fit the impedance data obtained for sulphite eletrooxidation in alkaline solution with Na_2SO_3 concentration between 10^{-3} and 10^{-1} mol L^{-1} in potential range +0.35 ÷ +0.65 V. Also, the same EEC was used in the case of more concentrated Na_2SO_3 solutions, 0.5 and 1 mol L^{-1}, at oxidation potential values up to +0.45 V. This EEC consists in a solution resistance *Rs* in series with a two parallel connections of the type (*CPE* - *R*). In first one, a constant-phase element (CPE_e) and resistor (R_e) was necessary to fit impedance data recorded at the high frequency values, which are attributed to a charge-transfer process at the outermost surface of Cu/Pt-Co electrode [21,22]. In the second connection, (CPE_{dl}) is the electric double-layer capacitance used instead of an ideal capacitor to characterize the Pt alloys catalysts behaviour in electrooxidation process and resistance (R_{ct}) is the charge transfer resistance [21,22].

a) b)

Figure 10. EECs used for modelling sulphite electrooxidation process on Cu/Pt-Co electrode in 1 mol L^{-1} NaOH solution with different sulphite concentration at different potential values.

EEC from Figure 10b has a single parallel connection ($CPE_{dl}(R_{ct}W_s)$) in which the charge transfer resistance R_{ct} is in series with a Warburg element (W_s) characteristic of sulphite ions diffusion from alkaline electrolyte to the Cu/Pt-Co electrode surface. A similar EEC was used to fit the EIS data obtained for sulphite oxidation on Pt electrodes in neutral solution [18].

The Bode diagrams shape confirms the choice of the two EECs for sulphite oxidation process fitting on this type of electrode, depending on the SO_3^{2-} ions concentration added to the alkaline electrolyte and the oxidation potential value.

The fitting results for sulphite oxidation on Cu/Pt-Co electrode are shown as continuous line in graphical EIS sprectra and the corresponding values of EEC elements are shown in Tables 2 and 3. Also, the double layer capacity (C_{dl}) values have been calculated for each studied electrode/electrolyte interface.

Table 2. Calculated parameters of EEC elements from figure 9a for sulphite oxidation in alkaline media

SO_3^{2-} conc. (M)	*E* (V)	R_s (Ω cm^2)	CPE_e $T\cdot10^5$ (F cm^{-2} s^{n-1})	CPE_e n	R_e (Ω cm^2)	CPE_{dl} $T\cdot10^4$ (F cm^{-2} s^{n-1})	CPE_{dl} n	R_{ct} (Ω cm^2)	$C_{dl}\cdot10^6$ (F cm^{-2})	$Chi^2\cdot10^3$
10^{-3}	0.35	3.09	3.17	0.77	24.8	1.05	0.65	1300	1.30	1.81
	0.45	3.09	2.75	0.78	22.7	0.91	0.67	1164	1.77	1.38
	0.55	3.10	2.46	0.78	21.3	0.90	0.71	837	3.16	1.59
	0.65	3.15	1.93	0.88	20.4	0.86	0.72	56.4	3.80	1.76
10^{-2}	0.35	3.11	4.46	0.70	45.4	0.74	0.61	1179	2.88	1.76
	0.45	3.10	4.05	0.72	37.2	2.42	0.62	997	3.02	1.72
	0.55	3.13	3.27	0.74	32.7	2.39	0.63	480	3.60	1.83
	0.65	3.18	1.19	0.84	28.5	2.16	0.65	26.5	4.37	1.33
10^{-1}	0.35	2.94	5.50	0.69	38.3	2.00	0.66	209	5.97	0.59
	0.45	2.92	4.66	0.70	24.3	6.93	0.58	124	7.92	0.51
	0.55	2.94	3.15	0.72	21.6	6.70	0.60	85.3	1.08	0.35
	0.65	3.00	2.20	0.77	15.2	6.64	0.61	19.5	15.1	2.53
0.5	0.35	2.94	2.41	0.80	12.0	9.88	0.48	15.7	2.36	0.61
	0.45	2.91	2.05	0.83	5.43	9.21	0.56	10.7	10.8	1.72
1	0.35	2.88	1.80	0.85	8.05	15.50	0.47	8.28	5.13	0.21
	0.45	2.88	1.43	0.87	3.52	13.0	0.54	5.23	23.4	0.35

Table 3. Calculated parameters of EEC elements from figure 9b for sulphite oxidation in alkaline media

SO_3^{2-} conc. (M)	*E* (V)	R_s (Ω cm^2)	CPE_{dl} $T\cdot10^4$ (F cm^{-2} s^{n-1})	CPE_{dl} n	R_{ct} (Ω cm^2)	W_R (Ω cm^2)	W_T (H cm^2)	W_P	$C_{dl}\cdot10^5$ (F cm^{-2})	$Chi^2\cdot10^3$
0.5	0.55	2.93	3.05	0.73	5.44	24.1	186	0.17	3.16	0.47
	0.65	2.95	2.68	0.75	5.06	8.58	48.9	0.37	3.44	0.67
1	0.55	2.88	3.46	0.76	3.29	22.2	245	0.26	7.84	0.70
	0.65	2.85	3.03	0.78	2.97	5.19	89.8	0.37	9.64	0.55

Analyzing the results from Tables 2 and 3, it can be observed that transfer resistance values (R_{ct}) decrease significantly with the increase of sulphite concentration added in alkaline solution, indicating that the sulphite anodic oxidation occurs with higher rate, confirming voltammetric and kinetic data. For the same sulphite concentration used in experimental studies, the value of the charge transfer resistance decrease with increasing polarization, the minimum values being recorded for an oxidation potential value of +0.65 V. Also, the double layer capacity (C_{dl}) have been calculated and, as expected, values of this parameter increases with the increasing of anodic potential value at which the sulphite ion oxidation process is carried out. The order of 10^{-3} for chi-square values indicate very good correlation between the experimental EIS data and the chosen EEC model values.

4. Conclusions

Experimental results have shown that SO_3^{2-} ions can be easily oxidized on Cu/Pt-Co electrode, in alkaline media, either directly on the electrode or indirectly with the atomic oxygen electrochemically generated.

The obtained data showed that the potential range at which the electrode processes are carried out does not depend on the sulphite concentration and the increase of the sulphite concentration stimulates the atomic and molecular oxygen generation reactions. As expected, the degree of sulphite transformation depends on the electrolysis time and sulphite concentration in the electrolyte solution.

The electrochemical behavior of SO_3^{2-} suggest that type of electrode can be used as anodic material for sulphite electrooxidation, especially in a SO_3^{2-} / O_2 alkaline fuel cell.

5. References

[1] Scialdone O 2009 *Electrochim. Acta* **54** 6140-7
[2] Mukerjee S, Srinivasan S, Soriaga MP and McBreen J 1995 *J. Electrochem. Soc.* **142** 1409-22
[3] Haifeng L, Dongguo L, Arvydas P, Nenad M and Stamenkovic VR 2016 *Nano Energy* **29** 149-65
[4] Brevett CAS and Jonson DC 1992 *J. Electrochem. Soc* **139** 1314-9
[5]Quijada C, Morallon E, Vazquez JL and Berlouis LEA 2000 *Electrochim. Acta* **46** 651-9
[6] Allen JA, Rowe G, Hinkley JT and Donne SW 2014 *Int. J. Hydrogen Energy* **39** 11376-89
[7] O'Brien JO, Hinkley JT, Donne SW and Lindquist SE 2010 *Electrochim. Acta* **55** 573-91
[8] Seo ET and Sawyer DT 1965 *Electrochim. Acta* **10** 239-52
[9] Samec Z and Weber J 1975 *Electrochim. Acta* **20** 403-12
[10] Varga K, Baradlai P and Vertes A 1997 *Electrochim. Acta* **42** 1143-55
[11] Westbroek P, De Strycker J, Van Uytfanghe K and Temmerman E 2001 *J. Electroanal. Chem.* **516** 83-8
[12] Quijada C, Vázquez JL and Aldaz A 1996 *J. Electroanal. Chem.* **414** 229-33
[13] Zelinsky AG 2016 *Electrochim. Acta* **188** 727-33
[14] Tolmachev YV and Scherson DA 2004 *Electrochim. Acta* **49** 1315-9
[15] Zelinsky AG and Pirogov BY 2017 *Electrochim. Acta* **231** 371-8
[16] Enache A, Dan M, Kellenberger A and Vaszilcsin N 2017 *Bulg. Chem. Commun.* **49** 139-46
[17] Enache AF, Dan ML and Vaszilcsin N 2017 *Chem. J. Moldova* **12** 102-9
[18] Enache AF, Dan ML, Vaszilcsin N, 2018 *Int. J. Electrochem. Sci.* **13** in press
[19] Skavas E and Hemmingsen T 2007 *Electrochim. Acta* **52** 3510-7
[20] Dan ML, Vaszilcsin N and Enache AF 2016 *Chem. Bull. "POLITEHNICA" Univ. Timisoara* 2016 **75** 58-65
[21] Yuan H, Guo D, Qiu X, Zhu W and Chen L 2009 *J. Power Sources* **188** 8-13
[22] Zhou W, Du Y, Ren F, Wang C, Xu J and Yang P 2010 *Int. J. Hydrogen Energy* **35** 3270-9

Acknowledgment
University *Politehnica* Timisoara supported partially this work in the frame of PhD studies.

IOP Conf. Series: Materials Science and Engineering **416** (2018) 012073 doi:10.1088/1757-899X/416/1/012073

Modelling the Bending Behaviour of Novel Fibre-Reinforced Sandwich Structures with Polyurethane Foam Core

O Weißenborn[1,*], S Geller[1], J Jaschinski[1] and M Gude[1]

[1]Institute of Lightweight Engineering and Polymer Technology, Technische Universität Dresden, Holbeinstraße 3, 01307 Dresden, Germany

[*]Corresponding author: oliver.weissenborn@tu-dresden.de

Abstract. In contrast to conventional manufacturing methods for sandwich structures, in this paper a foamable polyurethane system is used to impregnate textile layers and to create the sandwich foam core in one single process step. Given this simultaneous process, a high adaptability of the resulting mechanical properties can be achieved by varying the textile reinforcement of the sandwich top layers, the foam core density and the sandwich thickness, respectively. Within this paper, a design of experiment approach is implemented by using statistical analysis of variance (ANOVA) to investigate the influence of selected geometrical and material parameters on the resulting bending properties of these novel sandwich structures. A partial factorial design helps to identify the sensitivity of individual parameters and their interaction, as well as the correlation between sandwich morphology and bending performance. Based on the gained results, statistical models will help to identify parameters with statistical significance. Therefore, optimized sandwich structures according to stress-related requirements and geometrical constraints can be designed.

1. Introduction

Sandwich structures are considered being excellent weight-optimized structural components for the use in lightweight applications [1]. In this context, fibre-reinforced plastics with their excellent adjustable mechanical properties are established materials for the use as sandwich face sheets. In many cases, low-density polymer foams, such as polyurethane, serve as core material by stabilizing the sandwich structure and contributing with an improved bending and buckling resistance as well as high energy absorption abilities [2, 3]. Conventional manufacturing technologies for sandwich structures usually include hand lay-up methods (e.g. VARI or VARTM) by draping of textiles on a prefabricated polymer foam core [4, 5, 6] or the reaction injection moulding of the foam material into a predefined cavity made of fibre-reinforced face sheets [7, 8]. Other processing technologies such as the long-fibre injection method involve the spraying of a foamable polyurethane resin with cut glass fibres on a paper honeycomb core [9]. In contrast to these processing technologies, a foamable polyurethane resin was used in the current work to impregnate textile structures and create the sandwich foam core simultaneously.

Although sandwich structures are already well established in numerous applications, many research studies focus on design optimization. In this context, the limitations of local failure modes, e.g. face-core debonding [10] or foam core collapse [11] and the improvement of mechanical properties [12, 13] are among the main motivation. Comprehensive studies were performed to predict the failure behaviour of sandwich panels by developing analytical and numerical models taking different face-sheet–core material combinations into account [14, 15, 16]. However, most of these works did not focus on a sound statistical investigation of the gained experimental data using ANOVA.

In the present work the bending performance of these novel sandwich structures is analysed using a statistical design of experiments approach. Based on the experimental results, geometrical parameters with statistical significance are selected and statistical models are developed to predict the mechanical properties bending stiffness and maximum bending moment. The main motivation for using these models is anticipated for the load adapted design of sandwich components with geometrical constraints, such as limited wall thickness or limitations in total component mass.

2. Materials and manufacturing of sandwich structures

The process of sandwich manufacture is divided into several steps, which include the preparation of textile preforms (1), the mixing and spraying of the polyurethane foam inside an open mould (2), the insert of a second textile layer (3) as well as chemical cross-linking and curing (4, see Figure 1) [17]. Initially, glass-fibre plain weave textiles (Type 92105), provided by P-D Interglas Technologies GmbH, with a grammage of 163 g/m² were inserted into the heated open mould (~ 80 °C).

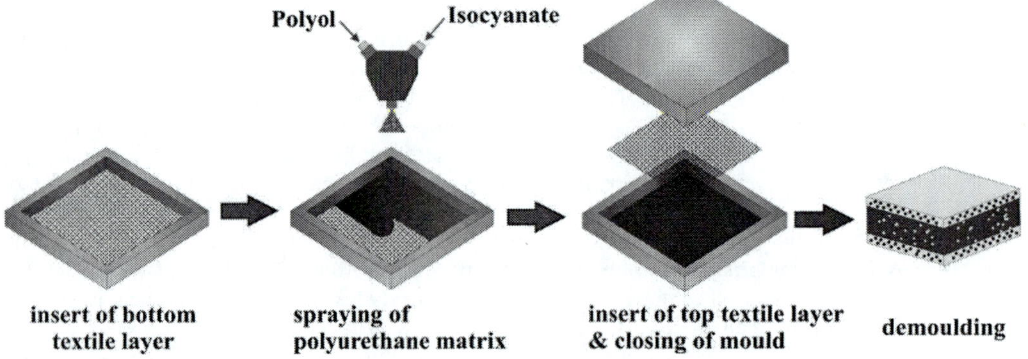

Figure 1. Manufacturing process of sandwich structures using a polyurethane spray coat method

The processing of the reactive polymer was performed using a polyurethane spray coat method (KraussMaffei Technologies GmbH). As matrix material serves the polyurethane system Elastoflex E3851/102 (BASF Polyurethanes GmbH) [18]. Mould height and foam matrix density were adapted according to the statistical design plan between 8 to 12 mm and 650 to 900 kg/m³, respectively. After a reaction time of 300 seconds, the sandwich plates were demoulded and stored at room temperature for post-curing.

3. Statistical experimental design, plan of experiments, 4-Point-bending-tests

Experimental design using statistical approaches is a common method to improve and optimize scientific and engineering issues. In the current work, a 3^2-factorial design was implemented, involving three identified factors, each with at least two levels of interest. For the factor *amount of layers*, all three levels were included. Beyond that, one additional experimental run is included, being the central point of the design. Preliminary processing studies helped to estimate the magnitude of the response change of each factor and to identify predominant processing restrictions [17]. In the current work, sandwich structures with three different thicknesses (8.0 mm, 10.0 mm and 12.0 mm), three foam core densities (650 kg/m³, 775 kg/m³ and 900 kg/m³) and sandwich top layers with each up to 3 layers of fibre-reinforcement were manufactured. Both factors foam core density and sandwich thickness are considered being numerical and continuous. In contrast, the number of textile layers have numerical and nominal characteristics.

After final curing of the polyurethane foam, specimens for mechanical testing with dimensions according to the German standard DIN 53 293 were produced using water jet cutting. The 4-point-bending test was performed using a ZWICK universal testing machine. The applied force was measured with a 10 kN load cell. At the given load speed of 5 mm/min, the deformation of the sandwich was measured with an ARAMIS 5M optical system (GOM GmbH) and local deflections at the maximum point (fm) and opposite of the compression die (fs) were determined. Analysis and evaluation of the experimental data including calculation of bending stiffness and bending moment were performed using ORIGINPro 2016G (OriginLab Corporation). For statistical analysis, evaluation and modelling the software JMP Pro 12.1.0 (SAS Institute GmbH) was used.

4. Results and discussion

Bending stiffness was calculated using data in the linear elastic region of the force-deflection curve, bending moment at the point of maximum force measured. Three replicates were tested for each configuration.

Table 1 presents the average value of all replicates including statistical scatter and their specific density-related value. However, for statistical modelling the individual results were used instead of the average values.

Table 1. Average results of mechanical investigations.

Run	Amount of layers [1]	Foam density [kg/m³]	Sandwich thickness [mm]	Bending stiffness [kNmm²]	Specific bending stiffness	Bending moment [Nmm]	Specific bending moment
1	1	650	8	3,614 ± 1.5 %	5,405	18,039 ± 8.5 %	26.97
2	1	650	12	13,064 ± 4.1 %	20,156	33,330 ± 1.2 %	51.42
3	1	775	10	9,263 ± 5.7 %	11,779	28,895 ± 3.9 %	36.74
4	1	900	8	7,545 ± 6.7 %	8,484	23,686 ± 5.2 %	26.64
5	1	900	12	15,072 ± 3.5 %	17,451	37,953 ± 6.4 %	43.94
6	2	650	8	7,226 ± 1.4 %	10,167	25,908 ± 4.8 %	36.45
7	2	650	12	17,916 ± 4.4 %	26,753	42,337 ± 6.2 %	63.22
8	2	900	8	8,991 ± 5.4 %	9,392	32,343 ± 3.3 %	33,79
9	2	900	12	20,136 ± 9.1 %	22,006	51,232 ± 5.3 %	55.99
10	3	650	8	8,869 ± 8.6 %	11,776	29,314 ± 3.2 %	38.92
11	3	650	12	24,562 ± 9.0 %	34,801	50,901 ± 4.8 %	72.12
12	3	900	8	10,708 ± 9.3 %	10,500	36,992 ± 5.4 %	36.27
13	3	900	12	26,637 ± 4.4 %	28,775	51,579 ± 3.5 %	55.72

Table 2. Analysis of variance for bending stiffness.

Source	SDQ	Mean square	df	Test F	p-values
D	4,787E13	4,787E13	1	24.035	0.0001
L	4,256E14	2,128E14	2	106.848	0.0001
T	1,240E15	1,240E15	1	622.730	0.0001
D*L	1,979E12	0,989E12	2	0.4969	0.6139
D*T	3,788E11	3,788E11	1	0.1902	0.6662
L*T	8,347E13	4,173E13	2	20.9558	0.0001
D*L*T	2,590E12	1,295E12	2	0.6502	0.5299
Error	5,377E13	1,991E12	27	-	-
Total	1,856E15	-	38	-	-

Initially, the influence of sandwich thickness (T), foam core density (D) and number of layers (L) was studied through statistical design for the bending stiffness of the sandwich structures. The analysis of variance was performed for a level of confidence of 95 %. In

Table 2, the parameters T, D and L were evaluated considering their F-distribution and p-value. Critical F-distribution was determined being $F_c \approx 2.2$ for $df_1 = 11$ (parameter degrees of freedom) and $df_2 = 27$ (error degrees of freedom). Factors with "Test F" $< F_c$ and $p > 0.005$ were rejected because the effects have a level of significance below 95 %. Thus, the factors D*L, D*T and D*L*T were rejected and a new analysis of variance was performed (**Error! Not a valid bookmark self-reference.**).

Table 3. Analysis of variance for stiffness with significance effects.

Source	SDQ	Mean square	df	Test F	p-values
D	4,787E13	4,787E13	1	26.0864	0.0001
L	4,256E14	2,128E14	2	115.9650	0.0001
T	1,240E15	1,240E15	1	675.8645	0.0001
L*T	8,347E13	4,174E13	2	22.7439	0.0001
Lack of fit	5,702E12	0,950E13	6	0.4660	0.8270
Error	5,872E13	1,835E12	32	-	-
Total	1,856E15	-	38	-	-

The corresponding model describing the bending stiffness for the studied range in relation to the geometrical factors with statistical significance is given by:

$$\mathbf{E\,(D,L,T) = 13{,}658{,}126 + 1{,}153{,}124 \cdot \left(\frac{(D-775)}{125}\right) + \begin{Bmatrix} -3{,}946{,}071 \text{ (for L = 1)} \\ -90{,}423 \quad \text{(for L = 2)} \\ 4{,}036{,}494 \text{ (for L = 3)} \end{Bmatrix} + 5{,}869{,}471 \cdot \left(\frac{(T-10)}{2}\right) +}$$
$$\left(\frac{(T-10)}{2}\right) \cdot \begin{Bmatrix} \mathbf{-1{,}625{,}222 \text{ (for L = 1)}} \\ \mathbf{-410{,}885 \quad \text{(for L = 2)}} \\ \mathbf{2{,}036{,}108 \quad \text{(for L = 3)}} \end{Bmatrix} \qquad \text{(Eq.1)}$$

Comparing the experimental and predicted results, the statistical model can be described with $r^2 = 0.9714$ and a normalized RMSE = 0.104. The statistical model indicates that the bending stiffness is depending on all studied parameters, but each having a different statistical significance. In particular, the predominant factor is the sandwich thickness followed by the number of textile layers. Since the increase of all parameters contributes to a rising bending stiffness, concurrent effects of the factors T and L with an almost equal statistical significance to the foam density were detected. The corresponding response surface for the bending stiffness (left) and their relation to the structure density (right) are given in Figure 2.

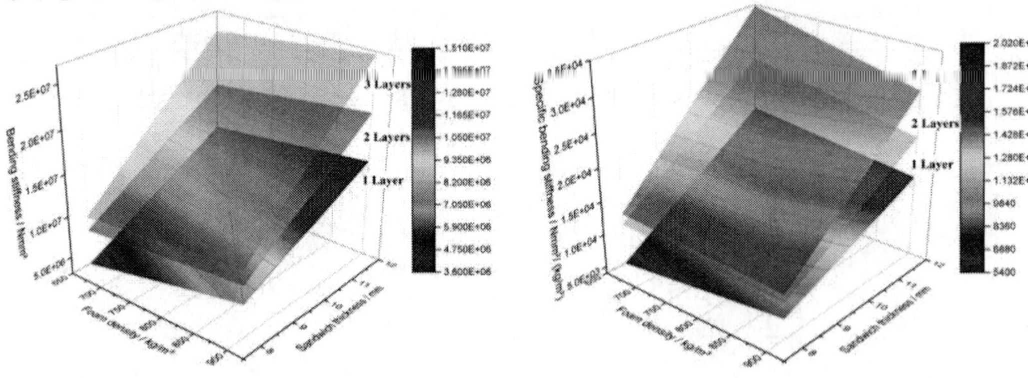

IOP Conf. Series: Materials Science and Engineering **416** (2018) 012073 doi:10.1088/1757-899X/416/1/012073

Figure 2. Response curve surfaces for the variation of bending stiffness as a function of foam density, sandwich thickness and amount of textile layers (left) and the bending stiffness related to the sandwich structure density (specific bending stiffness, right)

Since the parameter "L" (number of textile layers) is assumed to be a nominal factor, each diagram shows three planes, each resembling sandwich structures with the particular amount of textile layers. It is evident from the gained results, that the increase of foam density causes only a slight increase in bending stiffness. In contrast, the sandwich thickness has a major impact on the bending stiffness. Also, each additional layer of textile reinforcement contributes to an improved stiffness significantly. However, the gain in stiffness depends on the interaction of several factors. For instance, the increase in stiffness from a single layer to three layers has a larger impact at a higher sandwich thickness of 12 mm compared to 8 mm. By taking the structural density into account, the specific bending stiffness in relation to the studied factors is analysed (see Figure 2, right). Here, the reduction of the foam core density from 900 kg/m³ to 650 kg/m³ overcompensates the contribution of the foam core density to the bending stiffness. Thus, optimized configurations for lightweight sandwich structures can be found at lower foam core density.

The maximum bending moment is the second studied mechanical property with high relevance for sandwich structures. An analysis of variance of the factors D, L, T and their interaction was conducted in analogy to the bending stiffness (Table 4). Again, factors with "Test F" $< F_c$ and $p > 0.005$ were rejected because of insufficient levels of significance below 95 %.

Table 4. Analysis of variance for bending moment.

Source	SDQ	Mean square	df	Test F	p-values
D	288,257,011	288,257,011	1	49.6178	0.0001
L	1,370,538,453	685,269,226	2	117.9557	0.0001
T	2,552,768,888	2,552,768,888	1	439.4091	0.0001
D*L	19,478,985	9,739,492	2	1.6765	0.2059
D*T	7,741,007	7,741,007	1	1.3325	0.2585
L*T	19,421,701	9,710,850	2	1.6715	0.2068
D*L*T	34,336,189	17,168,094	2	2.9552	0.0691
Error	156,857,840	5,809,549	27	-	-
Total	4,449,400,077	-	38	-	-

Also for the bending moment, all studied main parameters D, L and T are assumed to be statistically significant. However, the remaining interacting factors appearing in table 4 (D*L, D*T, L*T and D*L*T) are rejected and a new analysis of variance was performed (Table 5).

Table 5. Analysis of variance for bending moment with significance effects.

Source	SDQ	Mean square	df	Test F	p-values
D	288,257,011	288,257,011	1	41.2080	0.0001
L	1,370,538,453	685,269,226	2	97.9632	0.0001
T	2,552,768,888	2,552,768,888	1	364.9332	0.0001
Lack of fit	81,969,339	10,246,167	8	1.7092	0.1436
Error	237,835,725	9,147,527	26	-	-
Total	4,449,400,077	-	38	-	-

Based on the experimental data, a statistical model considering the significant factors D, L and T with the following equation was determined:

$$\mathbf{E(D,L,T)} = \mathbf{36{,}177.4} + \mathbf{2{,}829.7} * \left(\frac{(\mathbf{D}-\mathbf{775})}{\mathbf{125}}\right) + \begin{Bmatrix} \mathbf{-7{,}796.9\ (for\ L=1)} \\ \mathbf{1{,}777.7\ \ (for\ L=2)} \\ \mathbf{6{,}019.2\ \ (for\ L=3)} \end{Bmatrix} + \mathbf{8{,}420.8} * \left(\frac{(\mathbf{T}-\mathbf{10})}{\mathbf{2}}\right) \qquad \text{(Eq. 2)}$$

The retrieved model is described by $r^2 = 0.9465$ and a normalized RMSE = 0.074. The response surface of the bending moment as a function of the considered geometrical parameters is plotted in Figure 3 (left). Again, the results indicate that the sandwich thickness has the main impact on the bending performance, followed by the number of textile layers and lastly the foam density. The results also show a significant increase in maximum bending moment as the number of textile layers increases from a single to double layer at a low sandwich thickness. The gain in bending moment by further increase on up to three textile layers is considerably lower, especially at a sandwich thickness of 12 mm.

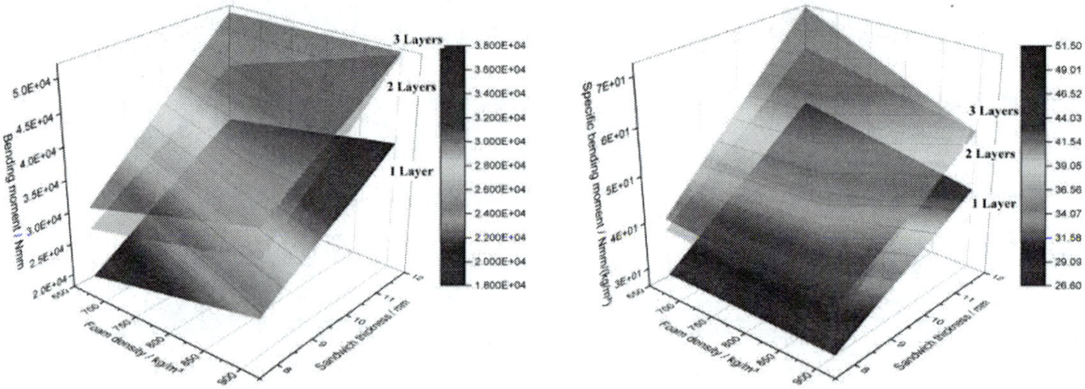

Figure 3. Response curve surfaces for the variation of bending moment as a function of foam density, sandwich thickness and amount of textile layers (left) and the bending moment related to the sandwich structure density (specific bending moment, right)

The response surface of the density-related bending moment is plotted in Figure 3 (right). Here, the results indicate a significant increase of the specific bending moment by reducing the foam core density, especially at a higher sandwich thickness. In analogy to the results of the specific bending stiffness, the reduction of the foam core density reduces the component density significantly and therefore overcompensates the contribution of the foam core to the bending stiffness.

5. Conclusions

In this paper, the influence of thickness, amount of reinforcing textile layers and foam core density on the bending performance of novel sandwich structures was examined using a design of experiments approach. Evaluation of the experimental data revealed a statistical significance of the parameters T, L and D as well as of the interacting parameter L*T on the bending stiffness and the bending moment, respectively. With the gained statistical models, reliable predictions of the bending performance can be obtained. The plotted response surfaces provided for bending stiffness, bending moment and their density-related values could help to optimize the design of these novel sandwich structures.

6. References

[1] Schaedler T A and Carter W B 2016 *Architected cellular materials* Annual Review of Materials Research **46:1** 187-210

[2] Gibson L J and Ashby M F 1997 *Cellular Solids. Structure and properties* Cambridge University Pres

[3] Dogan A and Arikan V 2017 *Low-velocity impact response of E-glass reinforced thermoset and*

thermoplastic based sandwich composites Composites Part B **127** 63-69

[4] Mohamed M, Anandan S, Huo Z, Birman V, Volz J and Chandrashekhara K 2015 *Manufacturing and characterization of polyurthane based sandwich composite structures* Composite Structures **123** 169-1

[5] Nunes J P and Silva J F 2016 *Sandwich composites in aerospace engineering in Advanced composite materials for aerospace engineering* (Cambridge: Woodhead Publishing) p 129-174

[6] Calabrese L, Bella G D and Fiore V 2016 *Manufacture of marine composite sandwich structures in Marine applications of advanced fibre-reinforced composites* ed J Graham-Jones and J Summerscales (Cambridge: Woodhead Publishing) p 57-78

[7] Rohleder M and Jakob F 2016 *Foam injection molding in Specialized injection molding techniques* ed P Heim (Oxford: William Andrew Publishing) p 53-106

[8] Karlsson K F and Åström B T 1997 *Manufacturing and applications of structural sandwich componen* Composites Part A: Applied Science and Manufacturing **28:2** 97-111

[9] Hufenbach W, Gude M and Geller S 2012 *Function integrated fibre-reinforced polyurethane composi with cellular matrix for intelligent lightweight structures* Conference on Cellular Materials (Dresden)

[10] Sun Z, Li D, Zhang W, Shi S and Guo X 2017 *Topological optimization of biomimetic sandwich structures with hybrid core and CFRP face sheets* Composites Science and Technology **142** 79-90

[11] Sun Z, Shi S, Guo X, Hu X and Chen H 2016 *On compressive properties of composite sandwich structures with grid reinforced honeycomb core* Composites Part B: Engineering **94** 245-52

[12] Zhang J, Supernak P, Mueller-Alander S and Wang C 2013 *Improving the bending strength and energ absorption of corrugated sandwich composite structure* Materials & Design **52** 767-73

[13] Padmanabhan K 2014 *Strength-based design optimization studies on rigid polyurethane foam core-gl and carbon-glass fabric face sheet/epoxy matrix sandwich composites* Mechanics of Advanced Materi and Structures **21:3** 191-6

[14] Vitale J P, Francucci G, Xiong J and Stocci A 2017 *Failure mode maps of natural and synthetic fiber reinforced composite sandwich panels* Composites Part A: Applied Science and Manufacturing **94** 21 25

[15] Martins R, Reis L and Marat-Mendes R 2016 *Finite element prediction of stress-strain fields on sandwich composites* Procedia Structural Integrity **1** 66-73

[16] Steeves C A and Fleck N A 2004 *Collapse mechanisms of sandwich beams with composite faces and foam core loaded in three-point bending. Part II: Experimental investigation and numerical modelling* International Journal of Mechanical Sciences **46:4** 585-08

[17] Gude M, Geller S and Weissenborn O 2014 *Integral manufacture of fiber-reinforced sandwich structu with cellular core using a polyurethane spray-coat method* Conference on Cellular Materials (Dresder

[18] Elastoflex E 3851/102 Version 15.02.2007 Material Data Sheet BASF Polyurethanes GmbH

Acknowledgments
The authors would like to express their gratitude towards the Federal Ministry for Economic Affairs and Energy of the Federal Republic of Germany and the AiF Projekt GmbH for funding the project SHAPE (project number: ZF4024707LP7).

IOP Conf. Series: Materials Science and Engineering **416** (2018) 012074 doi:10.1088/1757-899X/416/1/012074

Determination of Material Characteristics of Bearing Balls Through Static and Dynamical Actuation

T Manescu[1,*], G Balan[2] and R Avram[2]

[1]"Aurel Vlaicu" University of Arad, Bv. Revoluţiei, no. 77, 310130 Arad, România
[2]"Eftimie Murgu" University of Reşiţa, Sq. Traian Vuia, no. 1-4, 320085 Reşiţa, România

[*]Corresponding author: manescu.tiberiu@gmail.com

Abstract. In order to achieve a normal behavior of the bearing during exploitation, it is necessary to study the characteristics of the rolling elements material. The experiments are meant to determine the mechanical characteristics of the material used in ball bearing manufacturing (e.g. tensile strength, fatigue). The trials conducted have been split into two different categories: in a first stage statically tests have been performed, in which the bearing ball has been tested for compression until breakage using a special device developed for this purpose. The second stage has been dedicated to dynamical tests, using a specially designed fatigue test machine for bearing balls. The experimental results have been statistically worked out, acquiring conclusive results for the trials performed.

1. Introduction

Bearing analysis concerns a wide number of specialists from various engineering fields. Several approaches can be identified: vibration analysis [1],[2], fatigue tests, lifetime assessments with different lubricants, etc. Most of the scientific research is related to the bearing assembly, without highlighting the importance of the bearing components (inner ring, outer ring, cage, rolling elements). If the bearing components have defects, a shorter lifetime of the bearing is to be expected. The authors study in the current paper the bearing balls mechanical characteristics in order to evaluate if they can have a significant impact in the bearing

The analysis performed in the current paper is orientated in two directions: the first one is related to mechanical compression trials performed with a specialized device executed for this task and the second one is related to fatigue issues of the bearing balls.

2. Materials and methods

2.1. Mechanical compression device

The trials have been performed on a universal testing machine, [3]. Thus, there have been six sets of bearing balls tested at compression (with five balls each set) with the diameters: D_w = 6,8; 8.8; 12,7; 15,0; 17,8 and 20 mm.

For the mechanical compression trials, a specialised device was designed and executed (figure 1). The bearing balls or semi balls are fixed in the device as shown in figure 2. The devices maneuverability is facile and it can be mounted in the universal testing machine with ease. The force until breakage is recorded by the universal machine. The whole ball can be tested for static trials and the semi ball can be tested for the stress and deformation study with the electric resistive tensiometry method, using the adequate tensiometric stamps with a big measurement base [4]. However, in the current paper only the necessary force until breakage was evaluated by the authors.

Figure 1. Specialised device for bearing balls compression test

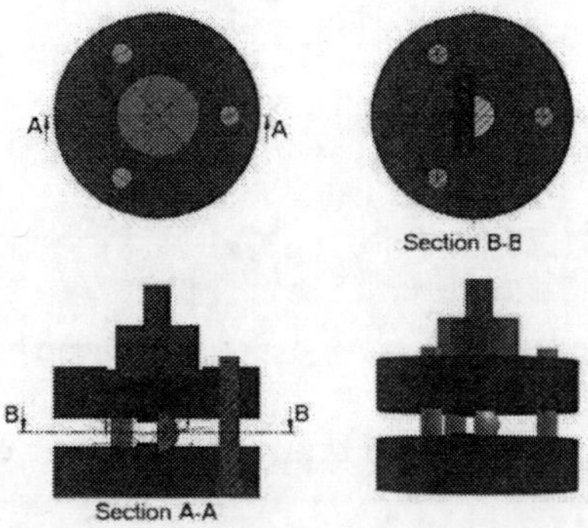

Figure 2. Schematic of the specialised device for bearing balls compression tests

2.2. Fatigue trial machine

The bearing balls where tested on the machine presented in figure 3. In order to obtain similar results during all the tests performed, several parameters had to be carefully monitored and maintained in a narrow range of values. The tests performed where under constant load (F=3600kN/test chamber) and the test chamber was permanently greased, using Mobil SHC Cibus 68. The oil flow was at around 1500ml/min. In every test chamber there were 6 balls tested in each trial, using the sudden death method (until the failure of one of the balls). The machine stopped the test every time the vibration sensor detected higher vibrations in the assembly, meaning one of the bearing balls had a material pitting.

The balls are located in a cage with 12 holes (figure. 4c), working in a vertical position (figure. 4.b.). The horizontal load F_a is shown in figure 4.a. The test chamber has two cages, allowing up to 24 simultaneous ball trials. The machine, described in detail in [5]-[9], is presented in figure 3.

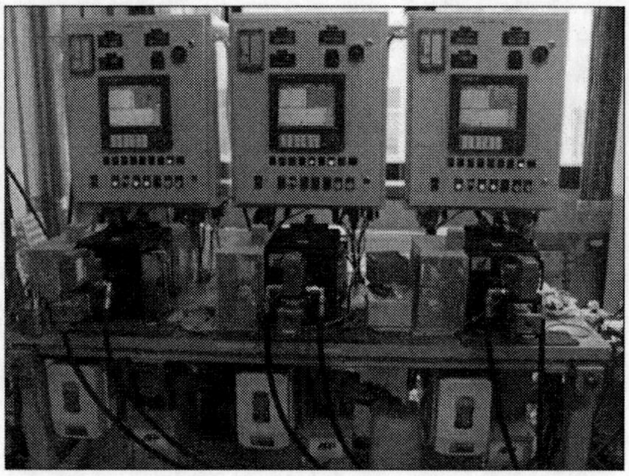

Figure 3. Fatigue testing machine for bearing balls

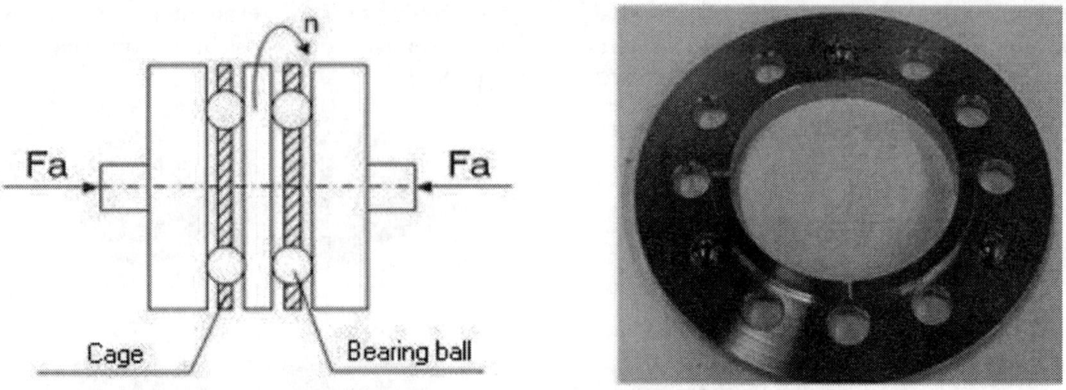

Figure 4. Schematic of the test chamber and the cage

The rolling elements of a bearing are made of chrome alloy steel: 100Cr6 for smaller bearings and 100CrMn6 for large bearings. The material used to conduct the trials is 100Cr6 with $E=2\cdot10^5$ MPa, Poisson coefficient $\mu = 0{,}29$ and admissible stress $\sigma_a = 6000$ MP, also known as the admissible contact pressure p_a.

A total of 120 bearing balls with the diameter $D_w = 12{,}7$ mm, without production defects have been tested for fatigue at $n = 200\cdot10^6$ rotations.

3. Results and discussions

3.1. Mechanical compression trials

The balls have been tested until breakage (figure.5). The balls had no manufacturing defects and no wear. The force at the breaking moment Fr [N] has been validated as the arithmetic mean of the individual values obtained for each ball in the respective set.

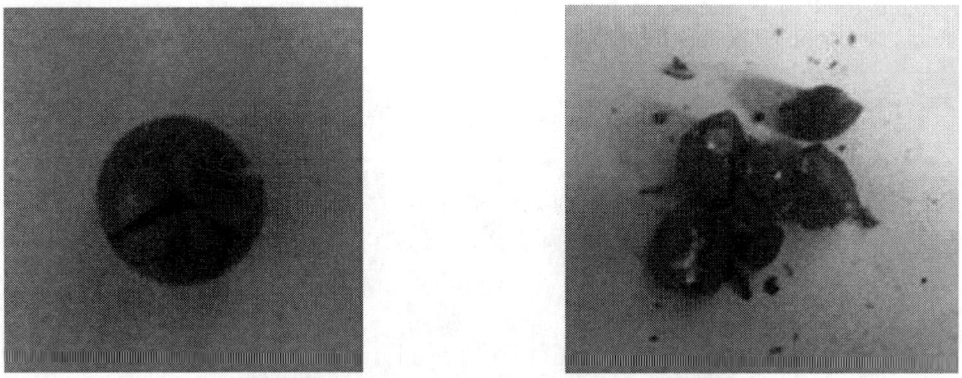

Figure 5. Ball tested at compression

It can be observed that the intensity of the necessary force in order to break the ball F_r is dependent to the ball diameter D_w and the value is increasing linear, as shown in figure 6, where the tendency line is correspondent to the ball diameter/breakage force.

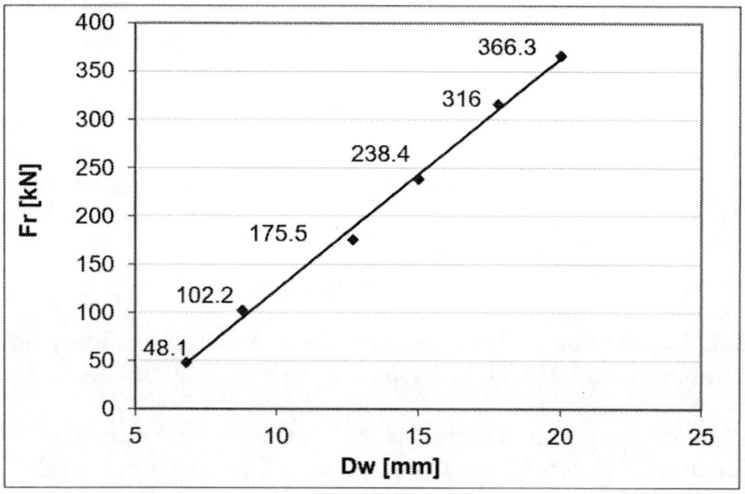

Figure 6. Breakage force variation depending on the ball diameter

3.2. Fatigue trials of bearing balls

After the testing of 120 bearing balls [10]-[12], a total of 60 balls have been randomly selected for the analysis, shown in figure 7.

Figure 7. Bearing balls D_w =12,7 mm after n = $200 \cdot 10^6$ rotations

Initial parametrical data for the bearing balls:

- Initial diameter: $D_w = D_o = 12{,}7$ mm
- Maximum section area: $A_0 = 506{,}7074791$ mm^2
- Volume: $V_0 = 1072{,}530831$ mm^3
- Material density: $\rho = 7{,}80$ g/mm^3
- Initial weight: $m_0 = 8{,}36570$ g

The data has been measured individually for each ball tested after n = $200 \cdot 10^6$ rotations. The results have been obtained by calculating the mean of the measured parameters:

- Final diameter $D_{med} = 12{,}686567$ mm
- Final weight $m_{med} = 8{,}35238$ g

The results after the weight measuring sequence of the balls highlight the loss of weight due to fatigue:

$D_{med} = D_o - D_{med} = 0{,}013433$ mm

$m_{med} = m_0 - m_{med} = 0{,}01332$ g

The fatigue becomes relevant only after a certain number of rotations (n> 100.000.000 rotations)[13]

IOP Conf. Series: Materials Science and Engineering **416** (2018) 012074 doi:10.1088/1757-899X/416/1/012074

4. Conclusions

The current paper presents a study regarding the static and dynamic behavior of the bearing balls of different sizes. The results obtained through the two test methods (mechanical and fatigue) emphasise that the bearing balls have a high liftetime expectance, exceeding the lifetime calculated for the bearing assembly according to ISO 281:2007 norm. [14] Thus, the breakdown of a bearing is not caused by the rolling elements as long as they do not have production defects.

5. References

[1] Ntakpe J L, Gillich G R , Mituletu I C, Praisach Z I and Gillich N, 2016 An Accurate Frequency Estimation Algorithm with Application in Modal Analysis, *Romanian Journal Of Acoustics And Vibration*, **13**(2), pp. 98-103

[2] Nitescu C, Gillich G R, Wahab M A, Manescu T and Korka Z I, 2017 Damage severity estimation from the global stiffness decrease, 12th International Conference On Damage Assessment Of Structures, in: *ROM J PHYS*, Vol. **842**

[3] Gafiţanu M, Năstase D, Cretu S and Olaru D, 1985 *Rulmenţi, Proiectare şi Tehnologie*, vol.1 şi 2, Editura Tehnică Bucureşti

[4] Mgharbel A, Delanoë-Ayari H and Rieu J-P, 2009 Measuring accurately liquid and tissue surface tension with a compression plate tensiometer, *HFSP J.* (3) 213–221

[5] Manescu T Jr, 2014 Analiza duratei de viaţă a bilelor de rulment de dimensiuni mici, Ştiinţă şi Inginerie, Sebeş, vol. **25**, pag. 575-580, ISSN 2067-7138

[6] Manescu T Jr, 2014 Analiza dimensională şi calitativă a suprafeţelor bilelor de rulment încercate la oboseală, *Ştiinţă şi Inginerie*, Sebeş, vol. **25**, pag. 581-588, ISSN 2067-7138.

[7] Manescu T Jr, 2014 Uzura bilelor de rulment încercate la oboseală, *Ştiinţă şi Inginerie*, Sebeş, vol. **25**, pag. 589-598.

[8] Manescu T jr, 2014 Contribuţii privind analiza stării de tensiuni şi deformaţii în bilele rulmenţilor radiali, de mici dimensiuni. cu si fara defecte de fabricatie, Phd. Thesis, UEM Resita

[9] Mănescu T Jr, Iancu V and Pomoja F, 2015 Incercarea de compresiune a bilelor de rulment, *Ştiinţă şi Inginerie*, Sebeş, vol. **27**, pag. 575-580

[10] Gillich G R, Bratu P, Frunzaverde D, Amariei D and Iancu V, 2010 Identifying Mechanical Characteristics of Materials with Non-linear Behavior using Statistical Methods, *4th WSEAS International Conference on Computer Engineering and Applications*, Harvard Univ, Cambridge, MA, USA, pp. 96-103

[11] Sadeghi F, Jalalahmadi B, Trevor S.S., Raje N and Nagaraj K A, 2009 A Review of Rolling Contact Fatigue, J. Tribol **131**(4)

[12] SKF General Catalogue 6000 EN, 2008

[13] SKF, Catalogue 5000E, 2003

[14] ISO 281:2007 Rolling bearings - Dynamic load ratings and rating life

IOP Conf. Series: Materials Science and Engineering **416** (2018) 012075 doi:10.1088/1757-899X/416/1/012075

Application of the CZM Technique to Delamination Analysis of Coupled Laminate Beams

J Paśnik[1], S Samborski[1,*] and J Rzeczkowski[1]

[1]Lublin University of Technology, Department of Applied Mechanics, Nadbystrzycka 36 St., Lublin 20-618, Poland

[*]Corresponding author: s.samborski@pollub.pl

Abstract. These days composite materials are widely used as elements of various machinery. They are applied as reliable machinery parts e.g. in aviation or automotive. This paper is referred to coupled laminates, which are one of composite materials types. Most common damage form in laminates is delamination. The aim of this paper is to show how to use Cohesive Zone Model technique (CZM) to model fracture propagation in laminate specimens. Finite element models were constructed using shell type elements in ABAQUS, which allows to analyze fracture propagation involving cohesive zone effect [5]. This operation needs many material constants which obtaining experimentally is problematic . The FE analysis is helpful in reasonable planning the stand tests and shows the accuracy of experimental studies. It also provides a lot of relevant information on laminate properties. Boundary conditions, loads and the obtained fracture toughness distributions of the numerical models of coupled laminate beams correspond to the mode I and mode II delamination propagation: the double-cantilever beam (DCB) and the end-notched flexure (ENF) [1, 3].

1. Numerical modeling technique

Cohesive zone theory was applied to fracture modeling by Dugdale in 1960 and Barenblatt in1962 separately. Since then there were many extension of this concept. For example it was recognized that cohesive elements are attractive in modeling composite laminates.

Abaqus offered two types of cohesive behavior: element-based and surface-based cohesive behavior. Cohesive elements can be used for modeling very detailed adhesive connections or for delamination in composites. Where comes to composites it must create a “zero” thickness adhesive layer – in point of fact an intermediate glue material, which is actually replaced in finite element model by cohesive element, is very thin. Modeling such materials is based on traction-separation law. In this paper FE model was created using surface-based cohesive behavior which is simplified and easier way to establish cohesive connections with traction-separation interface behavior. As distinct from cohesive elements it is defined as a surface interaction property in Abaqus. Damage is also an interaction property, material is defined only by Engineering Constants (Young’s modulus, Poisson’s coefficient and shear modulus). Surface-based modeling is mostly applied when interface thickness is so small it can be neglected. Surface-based behavior can be restricted to portion of surface that are initially in contact. It also allows define specific data like fracture energy as a function of the ratio of normal to shear displacement (mode mix) at the interface. Finite sliding, surface-to-surface option in contact is not allowed in Abaqus/Standard, so in this model small sliding was used.

2. Model loading and boundary conditions

One of the main aspects of this paper is an impact of different boundary conditions on load – displacement curve. Fig. 1 represents boundary conditions of End Notched Flexure test [4-7] model in Abaqus. It can be seen that both ends of specimen cannot move – all displacements are prevented. The vertical displacement of 10 mm, along z-direction, is applied to the line in the middle of top surface. Displacement of this line in the x- and y-directions were also prevented. With those BCs, specimen is supposed to conform mode II during fracture propagation.

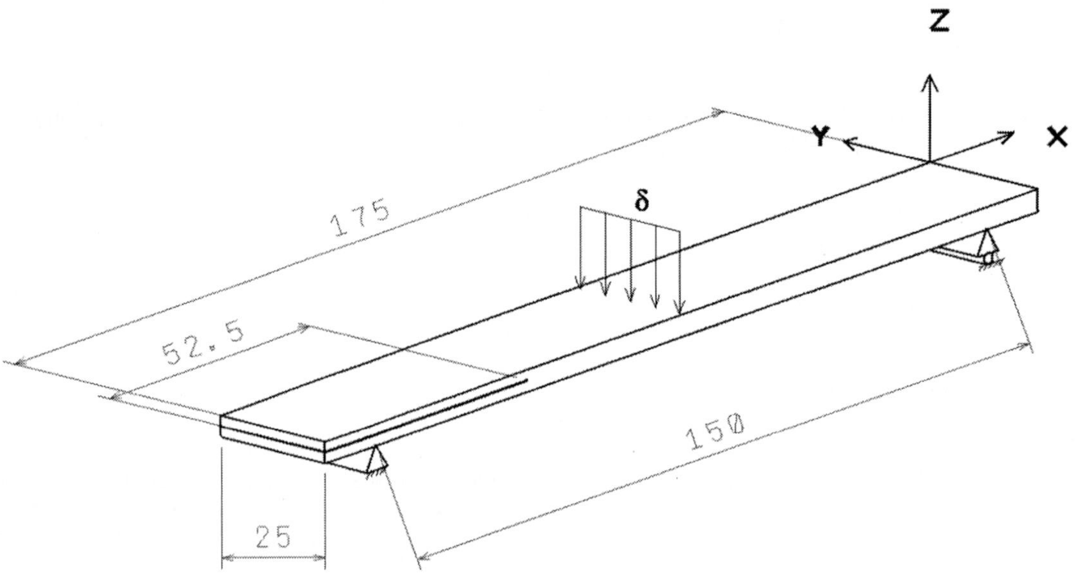

Figure 1. ENF specimen dimensions and boundary conditions

At this point it could be considered how will model respond when boundary conditions will be different. Numerical analyses were made also for other BCs, which are similar to that mentioned before. In addition to true ENF boundary conditions there were also prevented rotation along x axis of the specimenin order to avoid detachment of the specimen from support. This is due to observation of experimental tests of those laminates which tend to twist when bent.

2.1. Numerical models

Finite element method analyses were made using ABAQUS/CAE environment, ver. 6.14.5. Models of specimen were made as shell elements, which are the element with zero thickness. Material of the specimens is SEAL Texipreg HS160RM . The moduli were $E_1 = 109000$ MPa, $E_2 = E_3 = 8819$ MPa, $\nu_{12} = \nu_{13} = 0.342$, $\nu_{23} = 0.38$, $G_{12} = G_{13} = 4315$ MPa, $G_{23} = 3200$ MPa and ply thickness was 0.2 mm. In order to acquire accurate result it is logical to densify finite element mesh near delamination front and also at beam model's egdes, as figure 2 shows.

IOP Conf. Series: Materials Science and Engineering **416** (2018) 012075 doi:10.1088/1757-899X/416/1/012075

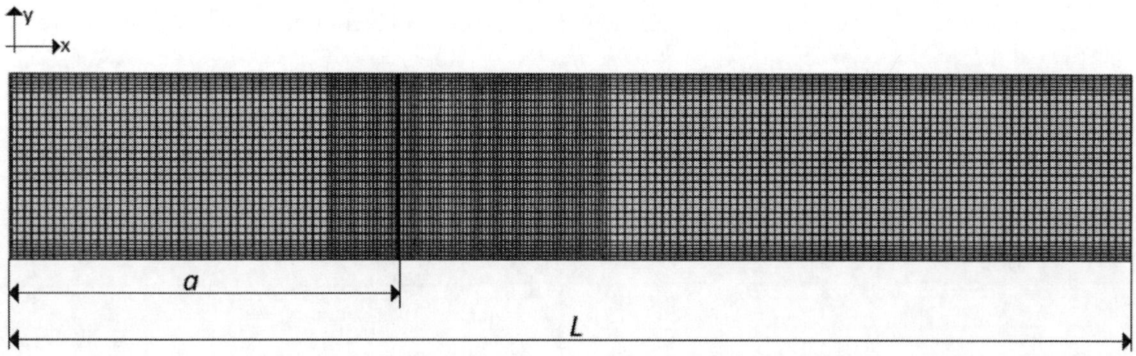

Figure 2. Finite element mesh

Fig. 3 shows force – displacement plots with two different boundary conditions – in first case (BC1) boundary conditions are the same as in true ENF test, in the other one (BC2) there are also prevented rotation along x-direction in points of support.

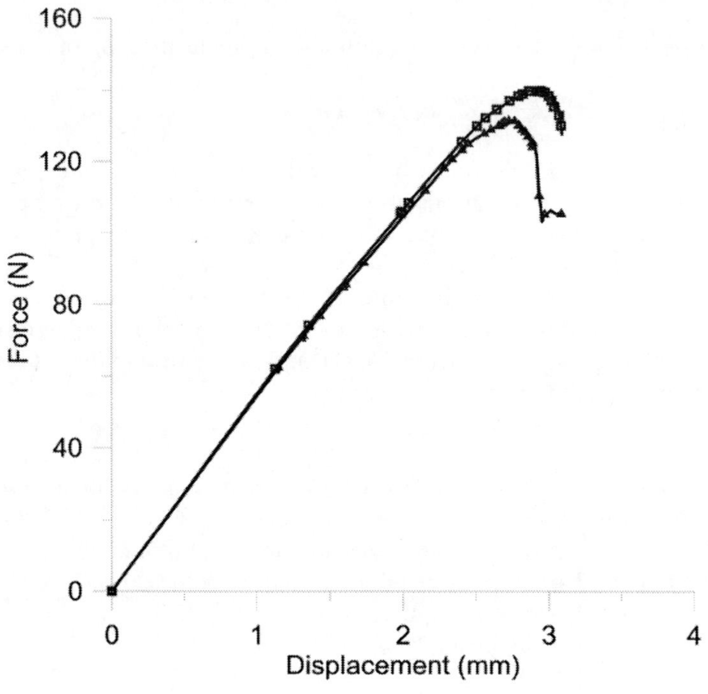

Figure 3. Load displacement curve for bending-twisting specimen

3. Results and discussion

Finite element model was made in Abaqus/Standard. It was modeled using shell elements which are easy to apply and shorten significantly the time of numerical calculations. Cohesive zone model technique was applied here to model delamination in coupled laminates instead of Virtual Crack Closure

Technique (VCCT). As fig. 4 shows, CZM technique is relevant technique to model fracture propagation [6].

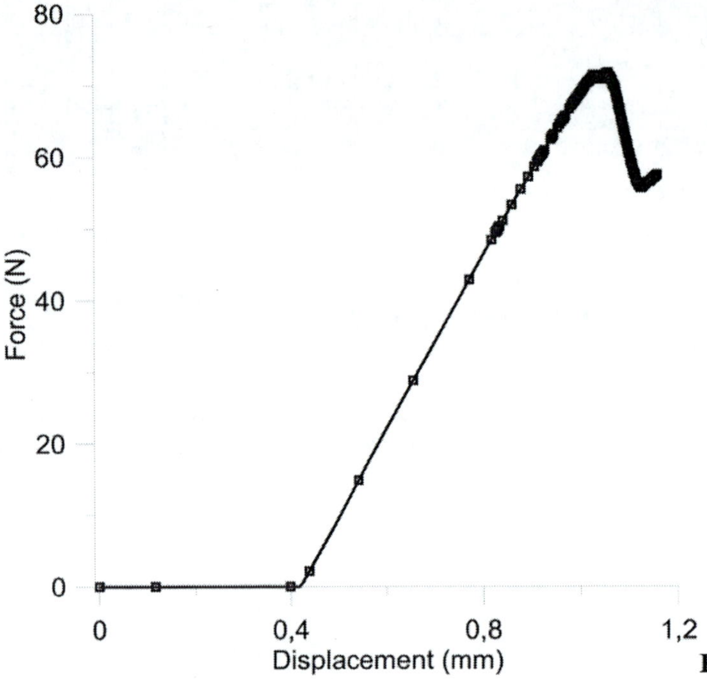

Figure 4. Load displacement curve for CZM Simple [+45/-45]$_9$ specimen

Force increase begins at displacement of 0,4 mm because of numerical reasons. For Abaqus, it is easier to start analysis when there are some straight calculations at the very beginning. In the first part of process force grow linearly, then first fracture occur and force drop sharply. Such force application is compatible with experimental force distribution as well as numerical analysis with VCCT [6]. The main difference between CZM and VCCT techniques is the time of calculations. Analysis using cohesive behavior needs much more time to succeed than model with VCCT. There must be further tests whether CZM reflects real effects in coupled laminates better than VCCT and, in relation to that, if it is reasonable to expend much time to get more valuable results.

4. Conclusions

As proved above, cohesive zone model can be widely applied not only for adhesive bonding but also for modeling delamination in composites. It has numerous advantages: it easier to use with FEM than VCCT, nonlinearities are not negligible with CZM in contrast to other methods and that is why it can be alternative to virtual crack closure technique. However there is one concern about using CZM – it needs much more computing power than VCCT because it requires higher mesh density to provide a more reliable results which cause much more time to do the analysis.

5. References

[1] S. Samborski, Analysis of the end-notched flexure test configuration applicability for mechanically coupled fiber reinforced composite laminates, Composite Structures 163 (2017), 342–349.

[2] S. Samborski, Numerical analysis of the DCB test configuration applicability to mechanically coupled Fiber Reinforced Laminated Composite beams, Composite Structures 152 (2016), 477–487.

[3] K. Dadej, Interlaminar cracking resistance of nonhomogeneous composite beams, Composites Theory And Practice 17:3(2017) 144-148.
[4] .B. de Morais, Analysis of crack propagation in double cantilever beam tests of multidirectional laminates, Mechanics of Materials 35 (2003) 641-652.
[5] R. D. S. G. Campilho, Interlaminar Fracture Characterization of a Carbon-Epoxy Composite in Pure Mode II, Materials Science Forum 636-637 (2010) 1518-1524.
[6] V. N. Burlayenko, T. Sadowski, FE modeling of delamination growth in interlaminar fracture specimens, Budownictwo i Architektura 2 (2008) 95-109.
[7] ABAQUS Online Documentation, Version 6.14,© DassaultSystèmes, 2014

Acknowledgment
This paper was financially supported by the Ministerial Research Project No. DEC-2016/21/B/ST8/03160 financed by the Polish National Science Centre.

IOP Conf. Series: Materials Science and Engineering **416** (2018) 012076 doi:10.1088/1757-899X/416/1/012076

Influence of Synthesis Method on the Morphology and Properties of ABO_3 Materials

P Vlazan[1], M Stoia[2], P Svera[1], M Poienar[1], P Sfirloaga[1,*]

[1]National Institute for Research and Development in Electrochemistry and Condensed Matter, Timisoara, Condensed Matter Department, P. Andronescu no.1, 300224, Romania,
[2]Faculty of Industrial Chemistry and Environmental Engineering, "Politehnica" University of Timisoara, 6 Vasile Parvan Avenue, 300223 Timisoara, Romania

[*]Corresponding author: paulasfirloaga@gmail.com

Abstract. Materials of potassium niobate ($KNbO_3$) with orthorhombic structure were successfully synthesized using two methods: solvothermal at 220 °C for 12 hours and sol-gel technique at 140 °C for 4h; both obtained samples were heat treated at 600 °C for 4 hours, and the influence of the synthesis methods on their properties was studied. $KNbO_3$ materials were prepared using stoichiometric amounts of $NbCl_5$, KOH as reactants and 1,2 propanediol as reaction medium. The structure and morphology of the $KNbO_3$ powders were investigated by X-ray diffraction (XRD), scanning electron microscopy (SEM), energy dispersive spectroscopy (EDAX), Fourier transform infrared (FTIR) and Raman spectroscopy.

1. Introduction

Ferroelectric and piezoelectric materials without lead with perovskite ABO_3-type structure have attracted considerable interest among scientists due to their interesting properties.

Perovskite oxides were reported to be promising gas sensing materials due to their attractive properties such as excellent gas sensitivity, mixed ionic-electronic conductivity, great doping flexibility, ability to accommodate to large amount of dopants and defects, high melting and decomposition temperature [1,2]. Ceramics materials have been widely used in electric devices, such as ultrasonic motors, piezoelectric transformers, and acceleration sensors [3].

Potassium niobate ($KNbO_3$) is one of the most important perovskite-type materials with electro-optic, nonlinear-optic, piezoelectric, luminous, and photocatalytic properties [4]. The properties of perovskite materials are influenced by the preparation method, reaction time, or applied thermal treatment. Many synthetic methods have been reported up to now, such as solid state reaction [5], sol–gel method [6], chemical co-precipitation method [7], citrate complex precursor method [8, 9], and combustion method [10]. Also, there are many reports of powders preparation of unique shape by means of hydrothermal method which has the advantage that could not only produce fine powder but also control the morphology to form one-dimensional nanostructure or thin film [11], solvothermal method - it is very similar to the hydrothermal route the only difference being that the reaction medium consist of organic solvents. Thus the solvothermal method gains one the benefits of both the sol-gel and hydrothermal processes [12] and microwave –assisted process [13]. Use of microwave energy as the heating source of the aqueous solution, brings positive in the hydrothermal synthesis because the volumetric heating is responsible for a more homogeneous crystallization, and the fast kinetic process lowers the synthesis costs [14]. In the case

of hydrothermal synthesis, the characteristics of materials can be altered by changing certain experimental reaction parameters: temperature, time, solvent type, and other.
In this paper, we investigated the $KNbO_3$ perovskite materials synthesized by solvothermal and sol-gel method in order to study the effect of the synthesis method on the structural and morphological properties. The obtained powders were characterized by XRD techniques, SEM microscopy, FT-IR and Raman spectroscopy.

2. Experimental procedure

The nanostructures of $KNbO_3$ were synthesized by sol-gel and solvothermal method using niobium chloride ($NbCl_5$, Aldrich Products), potassium hydroxide (KOH, Merck, Germany), citric acid and the organic solvent 1,2 propanediol as reaction medium. All of these starting materials were used without any further purification.

For solvothermal synthesis a mixture of $5x10^{-3}$ $NbCl_5$ was dissolved in 1,2 propanediol and KOH 7M solution was added drop by drop under continuous stirring on a magnetic stirrer. The solvothermal reaction was carried out in 50 ml autoclave Teflon-lined stainless - steel. The autoclave was heated at 220ºC in air oven, reaction duration being 12 h. The precipitate obtained was separated by filtration and washed with distilled water and ethyl alcohol and then dried in the air oven at 80°C for 2 hours.

In the case of $KNbO_3$ powders preparation by sol-gel method, additional precursors: citric acid sol.1% as chelating agent and ethyl alcohol as gelling agent were used. Sol-gel method allows obtaining single phase compound at low temperatures.

The preparation of $KNbO_3$ nanostructures was achieved as follows: stoichiometric amounts of niobium (V) chloride was dissolved in 1,2 propanediol to which sol.1% citric acid was added as chelating agent and ethyl alcohol as gelling agent. Precipitation of the $KNbO_3$ phase was carried out with 7M sol. KOH at room temperature. After this, the temperature was raised to 140°C and was maintained at this value until a black resin was formed.

The resin obtained was then thermally treated at 600ºC for 4 hours. After the thermal treatment, the compound was grinded in an agate mortar in order to obtain a powder sample which was then studied morpho-structurally.

The perovskite phase confirmation and crystal structure of the compounds were determined by the X-ray diffractometer PANalyticalX'Pert Pro MPD with Cu Kα radiation (λ = 1.54060Å, at room temperature in the range 20–80° with scan rate of 2° per minute and step size of 0.013°. The lattice parameters and the average crystallite size were calculated using HighScore Plus-software.

The morphological analysis and energy dispersive x ray analysis were achieved using scanning electron microscope (Model Inspect S).

The presence of water and other chemical bonds in the powders was determined by FT-IR spectrometry with a Shimadzu Prestige FT-IR spectrometer, in KBr pellets, in the range 400 - 4000 cm^{-1}. Raman measurements were performed using Shamrock 500i Spectrograph (ANDOR, United Kingdom) coupled with a CCD (cooled charge-coupled device) detector. The 514.5 nm laser was used as excitation source with 2 mW power. Olympus BX51 microscope equiped with a x50 long focus objective with NA 0.40 was used to focus the laser light on the sample and collect the scattered light into the spectrometer.

3. Results and discussions

X-ray diffraction analysis was carried out to investigate the structure of the prepared powders. The XRD diffraction spectra of the samples synthesized by solvothermal (Sample (a)) and sol-gel (Sample (b)) methods are almost similar to each other, except for a minor variation (0.1°) of the peaks which are shifted to lower 2θ angle values in the case of the sample (a). The structural parameters for the $KNbO_3$ samples were obtained by Rietveld refinement of the powders diffraction data using X'Pert HighScore Plus software. Figure 1. shows that the samples are well-crystallized and single phase, indexed with the card JCPDS card no. 01-071-0946 and all diffraction peaks for both samples can be attributed to orthorhombic structure.

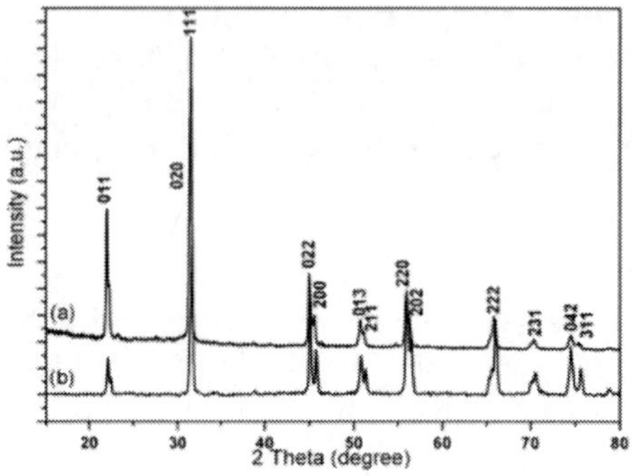

Figure 1. X-ray diffraction for $KNbO_3$ material obtained by solvothermal (a) and sol-gel (b) methods

The diffraction patterns for both samples have been refined with Amm2 space group. The unit cell parameters for Sample (a) are: a = 3.97(1) b= 5.67(1) and c = 5.79(1) Å, and for Sample (b) are: a = 3.982(4), b = 5.71(1), and c = 5.69(1) Å. It can be observed a slight increase for *a* and *b* parameter, while for *c* parameter a smaller value is obtained the case of Sample (b).

The crystallite size for $KNbO_3$ samples, synthesized by solvothermal and sol-gel methods were calculated with Scherrer's formula, as following [15]: the nanoparticle dimensions calculated were around 53.3 nm for sample (a) and 273.4 nm for sample (b).

Figure 2 shows the SEM images and EDX elemental analysis spectrum of $KNbO_3$ nanoparticles (a) and cubes (b) synthesized by means of solvothermal and sol-gel method, respectively.

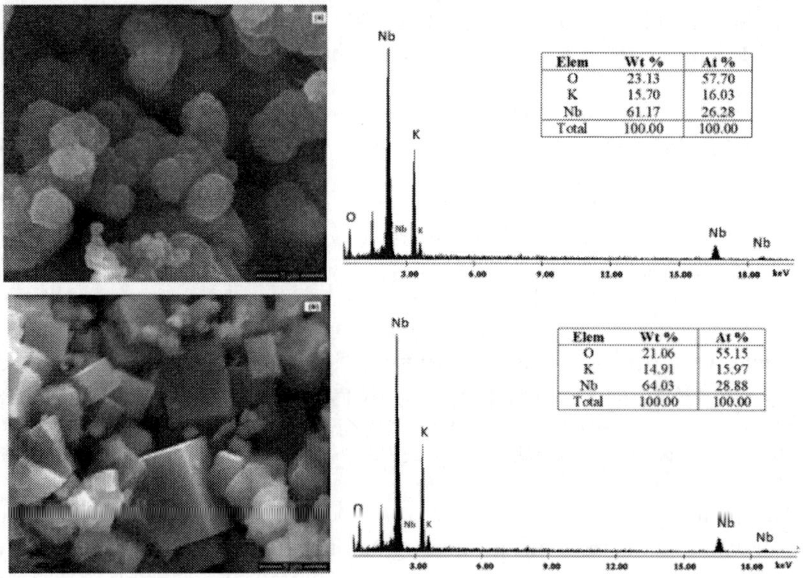

Figure 2. SEM images and EDX analysis spectrum of $KNbO_3$ synthesized through solvothermal (a) and sol-gel (b) method

The morphology of the particles is different among the two structure types depending on the reaction medium and obtaining method. Thus, by solvothermal method the particles have spherical shapes, agglomerated in asymmetrical formations (Figure 2a) while by sol-gel method cube-shaped particles have been obtained (Figure 2b).

Energy Dispersive X-ray analysis (EDX) was carried out to identify the purity of the $KNbO_3$ nanoparticles synthesized by the two methods. From the EDX analysis it was determined that the samples did not contain any impurities. High intensity peaks corresponding to Nb and K elements were clearly observed in the EDX patterns of both the samples.
Results of the FT-IR analysis show that the organic compounds are almost completely combusted during the heat treatment (see Fig. 3).

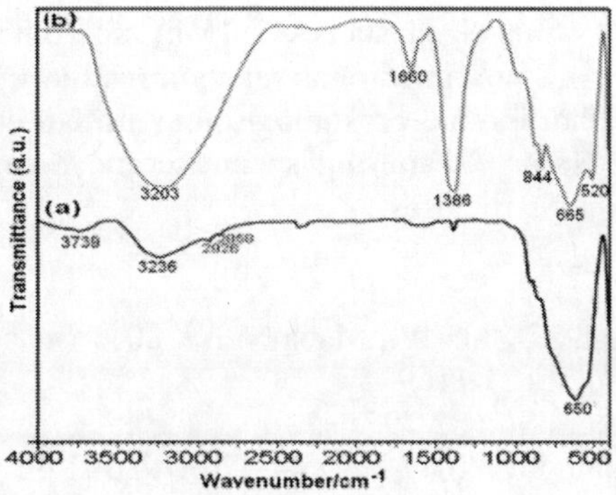

Figure 3. FT-IR spectrum of $KNbO_3$ synthesized through solvothermal (a) and sol-gel (b) method

Both FT-IR spectra exhibit a large band in the range 450-1000 cm^{-1}, with a maximum around 660 cm^{-1}, characteristic for the vibrations of the Nb-O bonds [16].

The FT-IR spectra of $KNbO_3$ powder obtained by sol-gel method presents a wide band in the 3000-3600 cm^{-1}range, with a maximum at 3203 cm^{-1}, probably for the superposition of the bands characteristic to hydrogen bonded O – H groups [17] with the bands characteristic to the vibrations of C-H bonds from residual citric acid, used in synthesis [18]. The bands located at 1660 cm^{-1} and 1386 cm^{-1} can be assigned to the antisymmetric stretching and symmetric vibrations of carbon-oxygen bonds from COO- groups of the residual citric acid [19].

The Raman spectra of $KNbO_3$ materials obtained through different method are shown in the Figure 4. The present peaks are attributed to the $KNbO_3$ in crystalline form [20], more exactly in orthorhombic crystalline form [21], which correspond to transverse and longitudinal phonon modes, $B_1(TO_4)$, $B_1(TO_3)$, $A_1(TO_3)$ and $A_1(LO_3)$, observed at 281.2, 530.0, 596.2 and 834.6 cm^{-1} for $KNbO_3$ obtained by solvothermal method, and $B_1(TO_4)$, $A_1(TO_1)$, $B_1(TO_3)$, $A_1(TO_3)$ and $A_1(LO_3)$, observed at 281.2, 294.3, 530.0, 596.2 and 834.6 cm^{-1} for $KNbO_3$ obtained by sol-gel method, respectively [22].

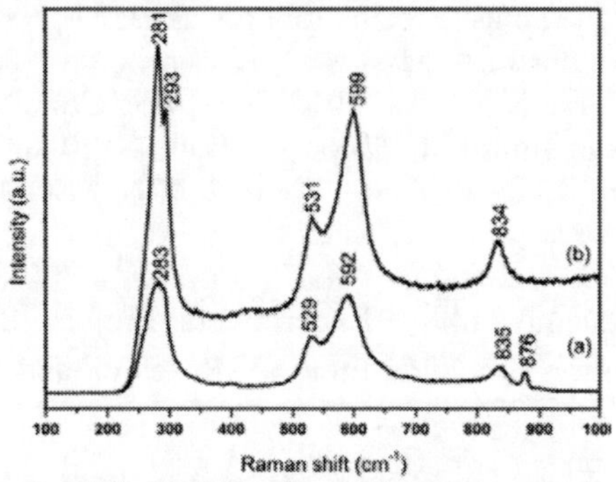

Figure 4. Raman spectra of orthorhombic $KNbO_3$ (Bmm2 space group) obtained by solvothermal (a) and sol-gel (b) methods

At 530.0 and 596.2 cm^{-1} appear Raman bonds that represent a double-degenerate symmetric O–Nb–O stretching vibration, which are sensitive to the change of phase structure or presence of other compounds [23].

Increase in intensity of the peak in the Raman spectra indicates increase of grain size of materials which was obtained by sol-gel method.

4. Conclusions

$KNbO_3$ nanoparticles have been synthesized successfully by solvothermal and sol–gel method, using citric acid and the organic solvent 1,2 propanediol as reaction medium. The obtained materials consist of micro and nanoparticles, with orthorhombic crystallographic structure. Similar results are obtained for both samples from infrared and Raman scattering spectroscopy. Small modifications occurred due to some difference in the corresponding structures.

5. References

[1] Moos R., Izu N., Rettig F., Reiß S., Shin W., Matsubara I. 2011 Sensors 11 3439
[2] Fergus J. W. 2007 Sensor Actuat. 123 1169
[3] Haertling G.H. 1999 J Am Ceram Soc 82 797
[4] Wang Y., Chen Z., Ye Z., Huang J.Y. 2012 J. Cryst. Growth 341 42
[5] Kinoshita T., Senna M., Doshida Y. and Kishi H. 2012 Ceram. Int. 38 1897
[6] Cao Y., Xiao L., Wang W., Choi D., Nie Z., Yu J., Saraf L. V., Yang Z., Liu J. 2012 Solid State Sci. 14 655
[7] Simon-Seveyrat L, Hajjaji A, Emziane Y, Guiffard B., Guyomar D. 2007 Ceram. Int. 33 35
[8] Kakimoto K-I, Ito T., Ohsato H. 2009 Ferroelectrics 380 196
[9] Kokubo T., Kakimoto K., Ohsato H. 2008 Key Eng. Mater. 368-372 1883
[10] Kingsley J.J., Pederson L.R. 1993 Materials Letters 18 89
[11] Kumada N., Kyoda T., Yonesaki Y., Takei T. and Kinomura N. 2007 Mater. Res. Bull. 42 1856
[12] Nakashima K., Ueno S. and Wada S. 2014 J. Ceram. Soc. Jpn. 122 547
[13] Prado-Gonjal J., Arevalo-Lopez A. M. and Moran E. 2011 Mater. Res. Bull. 46 222
[14] Paula A.J., Parra R., Zaghete M.A., Varela J.A. 2008 Mater. Lett. 62 2581
[15] Cullity BD 1978 Element of X-ray Diffraction, 2nd edition, Addison-Wesley publishing company, Reading, MA, 87
[16] Wang Y., Yi Z., Li Y., Yang Q. and Wang D. 2007 Ceram. Int. 33 1611
[17] Nohman A.K.H., Ismail H.M., Hussein G.A.M.1995 J Anal Appl Pyrol 34 265
[18] Franklin D.S. and Guhanathan S. 2014 Polym. Bull. 71 93
[19] Coste S., Lecomte A., Thomas P., Merle-Mejean T., Champarnaud-Mesjard J.C. 2007 J. Sol-Gel Sci Techn 41 79
[20] De Andrade J.S, Pinheiro A. G., Vasconcelos I.F., Sasaki J. M., Paiva J.A.C., Valente M.A., Sombra A.S.B. 1999 J. Phys. Condens. Matter 11 4451
[21] Wan S., Zhang B., Sun Y., Tang X., You J. 2015 Cryst. Eng.Comm. 17 2636
[22] Raja S., Ramesh Babu R., Ramamurthi K., Moorthy Babu S. 2018 Ceram. Int. 44 3297
[23] Sun S. Ge Y., Zhao Y., Yuan X., Zhao Y. and Zhou H. 2016 RSC Adv. 6 113038

Acknowledgement:
This work was supported by the Demonstrative Experimental Project **48PED/2017**. The authors thank the X-ray Diffraction Laboratory from National Institute for Research and Development in Electrochemistry and Condensed Matter, Timisoara.

Structure Characteristics of Silicon Alloyed Ductile Cast Iron

S C Stan[1], M Chisamera[1], I Riposan[1,*], E L Neacsu[1] and A M Cojocaru[1]

[1]Materials Processing & Ecometallurgy, Politehnica University of Bucharest, 313, Independentei Spl., 6th District, 060042 Bucharest, ROMANIA

*Corresponding author: iulian.riposan@upb.ro

Abstract. Experiments deal with the solidification pattern of FeSiCaMgRE-treated and un-inoculated and inoculated [Ca,Ce,S,O-FeSi] three ductile iron compositions [2.5%Si; 4%Si and 4%Si-1.6%Mo (Si:Mo=2.5)]. Structure analysis is performed in 20mm diameter bar, resin sand mould solidification. Increasing of the Si content leads to increasing the graphite amount. Mo addition in high Si-iron limits the graphitizing effect of Si, but the graphite amount remains higher compared with conventional iron. A similar evolution is registered for graphite nodule count, while Si-Mo iron has lower nodule count compared with referred iron. For the same production conditions, graphite nodularity [according to ISO 945 evaluation] is negatively affected by Si and Si-Mo: it decreases from 95% up to 80% value, but remains at the accepted level, which means that the selected inoculant seems to be effective in these irons. For high Si-ductile irons, the nodularity evaluation must be more carefully considered, by including a minimum limit of the accepted sphericity shape factor. Increasing of the Si content leads to a prevalent ferritic matrix. Supplementary addition of Mo at the upper limit in Si-Mo affects the ferritic effect of Si, resulting an intermediary ferrite and pearlite mixture. It is found that inoculation is important for high-Si, and particularly for Si-Mo alloyed irons, requiring a high efficiency inoculation procedure.

1. Introduction

Ductile cast iron [nodular or spheroidal graphite morphology] is a ferrous material, subjected to casting process, typically including [wt.%] 3.4 – 3.8 C, 1.8 – 2.8 Si, 0.1 – 0.6 Mn, < 0.05 P, < 0.02 S, 0.03 – 0.06 Mg_{res}, with or without rare earth elements [RE] contribution. The structure includes graphite particles at nodular [spheroidal] morphology and ferrous matrix, mainly as ferrite, pearlite or ferrite and pearlite mixture, without free carbides and limited other phases amount. The mechanical properties of ductile iron are significantly superior to compact graphite cast irons and especially to grey [lamellar graphite] cast irons.

Ductile cast iron is obtained by a double metallurgical treatment: (a) modification and (b) inoculation. Modification treatment of the iron melt mainly influences graphite nucleants growth by increasing their compactness degree, from lamellar morphology [the highest capacity for stress concentration] through intermediate graphite morphologies (vermicular/compacted) up to the highest level, for nodular/spheroidal morphology [the lowest stress concentration]. Inoculation is a graphitizing treatment applied to the molten iron immediately prior to casting, with direct effects on the primary structure (austenite, carbides, eutectic cells, graphite characteristics). Indirectly it influences the eutectoid structure, especially pearlite/ferrite ratio. Generally, inoculation is applied to forestall solidification at excessive eutectic undercooling degree, favourable for carbides occurrence or/and undesired graphite morphologies [1].

Increasing silicon in cast irons leads to complex effects, basically in the following sequence: 1) free carbides prevention; 2) reduce the pearlite / ferrite ratio up to a fully ferritic structure; 3) alloying of ferrite; 4) excessive ferrite alloying forming hard Fe_xSi_y compounds; 5) lowers carbon content and corresponding decrease in amount of graphite [drastically at more than 10%Si]. The first effect is beneficial in all cases for mechanical properties and machinability, while decreasing the pearlite to ferrite ratio is most important for improved ductility / toughness behavior (Figure 1). But in many cases, the graphitizing role of silicon is critical, and in conjunction with a higher pearlite / ferrite ratio, a higher silicon content at more than 2.5%Si requires the use of other pearlite forming elements (Cu, Ni, Sn etc) [2].

The literature review reveals that three groups of silicon alloyed ferritic ductile cast irons [3 – 20%Si] are very attractive for their specific applications: (a) Si-strengthened ferritic materials [3.2 – 4.3%Si], to replace conventional ferrite-pearlite grades [Si < 3%], with drastically reduced hardness variation [4HB], increased cutting tool life, and consistently better mechanical properties [Rm = 450 - 650MPa; $Rp_{0.2}$ = 350 – 500MPa; A=10 - 20%]; (b) Si-Mo ferritic grades (3.5 – 5.5%Si, 0.2 – 2.0%Mo) with superior mechanical properties [Rm=450 - 550MPa; $Rp_{0.2}$= 275 – 440MPa; A=4 – 10%)] and improved resistance to oxidation at high temperatures; (c) corrosion resistance cast irons, especially for more than 14%Si [2]. The actual standardized ductile cast iron grades, as conventional irons [ferrite/ferrite + pearlite/pearlite matrix] and ferritic Si/Si-Mo alloyed irons are illustrated by Figure 2.

Ongoing research programs were reported in recent years, revealing important information on the behaviour of high silicon and silicon – molybdenum cast irons at room temperature and up to 1000°C temperature, mainly used in the automotive industry. It is found that silicon highly segregated around nodules, which is the origin of the initiation and propagation of cracks fracture. As result, the conventional ferritic ductile iron, at less than 2.5%Si has the best impact strength, while higher silicon content favours brittle cleavage. Generally, it is about a higher Si content, and a lower impact strength. Supplementary alloying (Ni, Al) has beneficial effects in mechanical properties.

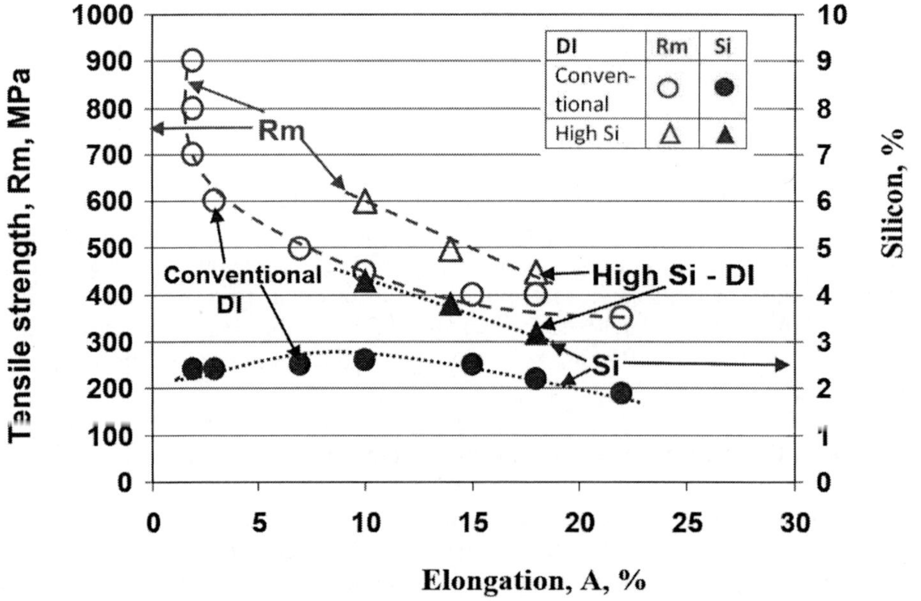

Figure 1. Tensile Strength [Rm] – Elongation [A] – Silicon [Si] relationship in Conventional Ductile Irons and High Si Ductile Irons

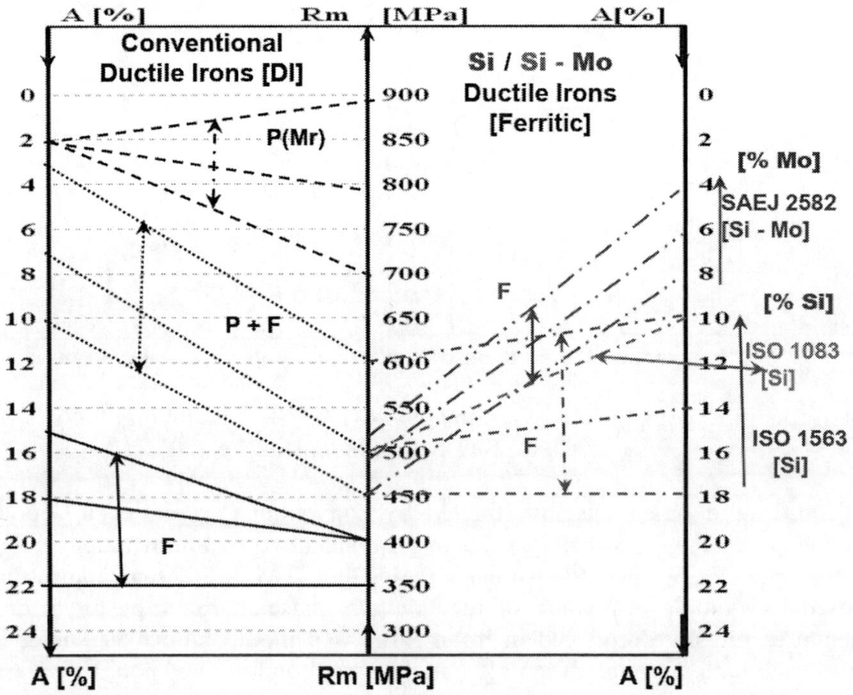

Figure 2. Standardized Conventional Ductile Iron [DI - EN 1563:2011; ISO 1083:2004], Si [DI-Si - ISO 1083:2004; EN 1563:2012] and Si–Mo [DI-(Si-Mo) - SAE J2582: 2004] alloyed Ductile Irons

Silicon alloyed ductile irons show tolerance of higher amounts of pearlite and carbide stabilizing elements such as 0.6-1.0%Mn, 0.3-0.6%Cr, 0.26%V.

In Si-Mo alloyed ductile irons, maximum working temperature increases, from 700-750°C up to 900-1000°C by supplementary alloying, with Al and contribution of Ni, Co, Cr, V, Ti, Zr [SiMoAl - X DI]. Silicon and aluminium have specific effects on the transformation temperature, the oxide layer thickness and new formed phases. So, silicon and aluminium association improves the durability of the protective oxide layer but with a negative effect on the spheroidal graphite morphology, increasing the ductile iron embrittlement. For this reason, compacted graphite CGI –SiMoAl application seems to be attractive. New data are obtained as the specific corrosion and fatigue behaviour in dry and humid air and diesel exhaust gas and also as the crack growth mechanisms in different oxidation and corrosion media [2 - 17].

The main objective of the present paper is to evaluate the structure characteristics of three representative materials, as conventional (2.5%Si), Si-alloyed (4%Si) and Si-Mo alloyed (4%Si – 1.6%Mo, at lower Si : Mo=2.5 ratio) ductile cast irons, mainly for the influence of Si and Si-Mo alloying and inoculation treatment.

2. Materials and methods

Three heats are produced in a 10kg coreless induction furnace, 8000Hz frequency: the first one at a normal silicon content (I, at 2.5%Si), as a reference, and the last two heats at 4.0%Si level (II) and 4%Si + 1.6%Mo (III), respectively. A selection of these experimental alloying grades is more detailed in a recent work [2]. Figure 3 shows the specific of the chemistry of the materials used in these experiments. For silicon alloying, a high purity ferrosilicon alloy [70.9%Si, 0.038%Ca, 0.06%Al, Fe bal] is added in the furnace; ferromolybdenum is used for molybdenum correction and 2.0 wt.% FeSiCaMgRE alloy [43.03%Si, 1.87%Ca, 1.35%Al, 10.35%Mg, 1.1%TRE, Fe bal] is used for nodularising treatment by tundish-treatment method. Inoculation in the ladle was applied, after Mg-treatment, with 0.3wt.% Ca,Ce,S,O-FeSi alloy [70-76%Si, 0.75-1.25%Ca, 0.75-1.25%Al, 1.5-2.0%TRE, S and O bearing compounds totalling less than 1%, Fe bal].

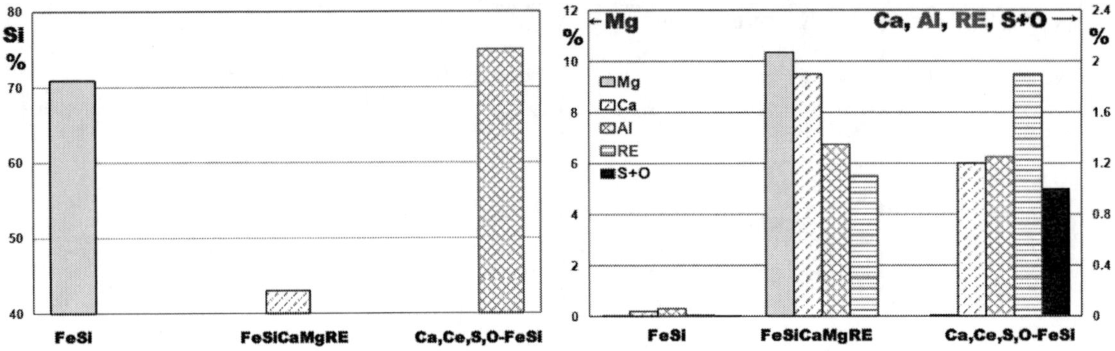

Figure 3. Chemical composition of treatment alloys [FeSi - Si alloying; FeSiCaMgRE - nodulizer; Ca,Ce,S,O-FeSi – Inoculant; Fe-Bal]

The solidification process is investigated by conventional ceramic cup cooling curve analysis having a modulus of approximately 0.75cm [equivalent to a 30mm diameter bar], high cooling / solidification rate W_1, W_2 and W_3 wedge samples – ASTM A 367 and 20mm diameter bar sample, usually used for quickly evaluation of the graphite phase morphology by macro-fracture analysis. Experimental castings are produced in furan resin sand mould (about 89 Dieter Hardness and 1400 N/cm^2 tensile strength on standard samples) in the same conditions as pouring parameters.

Details on the cooling curves parameters during solidification, mainly as representative temperatures and undercooling during eutectic reaction and the end of solidification have been presented in a previous paper [2], which also evaluated the formation of dark coloured porous region, including chunky-graphite, in the thermal centre of wedge castings. The present paper focuses on the structure characteristics in the 20mm diameter bar samples.

3. Results and discussion

The final chemical composition of all inoculated ductile irons is within a very narrow range, for 0.035 – 0.055% Mg_{res}, as Figure 4 shows. According to 4.0 – 4.4% carbon equivalent [CE= %C+0.3 (%Si + %P) – 0.03 %Mn + 0.4%S] level, the experimental cast irons occupies eutectic range.

Silicon, an important influencing factor on carbon equivalent level, acts to decrease the solubility of carbon in molten iron, and supports graphite formation during the cooling process up to the end of solidification, instead of the free carbide [cementite], despite that the carbide formation route is more likely from an energy point of view. On the other hand, silicon decreases the solubility of carbon in solid iron, favouring carbon diffusion from austenite to the existent graphite particles during the eutectoid reaction, encouraging ferrite formation, thereby decreasing pearlite content. In this way, more than 2.0%Si content could protect the ferrite formation, decreasing the pearlitic formation factor Px [18], in the presence of the most representative pearlite forming elements (Equation 1):

$$P_x = 3.0\ (\%Mn) - 2.65\ (\%Si - 2.0) + 7.75\ (\%Cu) + 90\ (\%Sn) + 357\ (\%Pb) + 333(\%Bi) + 20.1\ (\%As) + 9.60\ (\%Cr) + 71.7\ (\%Sb) \quad (1)$$

As the radii of the Si and Fe atoms are different, a stress in the lattice is developed, resulting in a solid-solution hardening, in the body-centred cubic lattice of ferrite. Consequently silicon alloyed ferrite has superior characteristics of strength and hardness, but lower ductility and toughness properties. Manganese content is typically for ferrite-pearlite matrix ductile irons, at 0.35 – 0.45% level. Minor elements are controlled at 0.0.04 – 0.06% content Cu, Cr, Ni, 0.01 - 0.02% Al, Ti and less than 0.01%

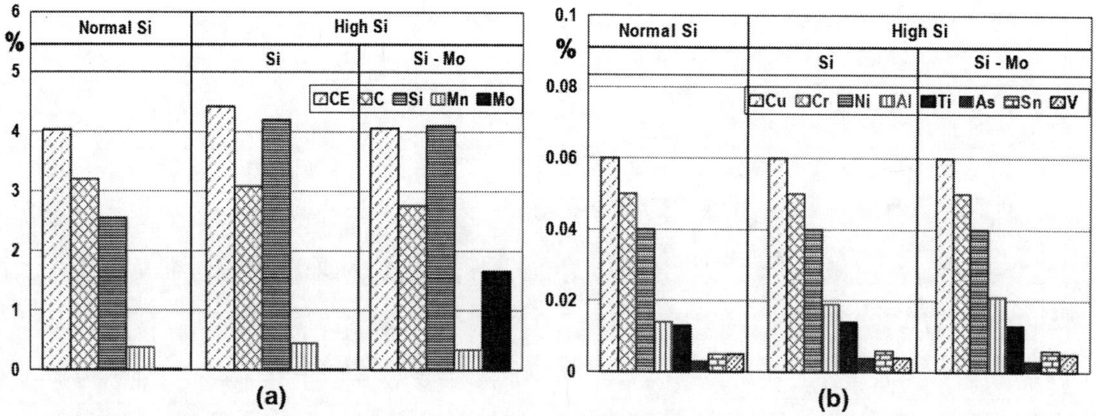

Figure 4. Chemical composition of tested ductile cast irons, as the base (a) and minor (b) representative elements [CE-carbon equivalent]

As, Sn, V. Considering the mentioned content of Si, Mn and minor elements, Px = 1.34 for conventional/normal ductile iron is obtained, typically for ferrite and pearlite mixture metal matrix. This parameter decreases at negative values [Px = -2.8] for high silicon content. According to Thielemann [18], the complex action of chemical composition for anti-nodularising effects is illustrated by Factor K, defined by Equation 2:

$$K= 4.4\ (\%Ti) + 2.0\ (\%As) + 2.4\ (\%Sn) + 5.0\ (\%Sb) + 290\ (\%Pb) + 370\ (\%Bi) + 1.6\ (\%Al) \quad (2)$$

The three experimental ductile cast irons are defined by K = 0.14 – 0.18 factor. For K < 0.2 level, all of the test cast irons are characterized by a low anti-nodularising action. So, from an anti-nodularising elements effect point of view no degenerated graphite morphologies are expected, for a 0.035 – 0.055% Mg residual level.

The structure characteristics are evaluated with Automatic Image Analysis [analySIS® FIVE Digital Imaging Solutions software], according to ISO 945 specifications. Influence of silicon and silicon-molybdenum alloying on the nodule count, graphite amount and graphite nodularity in un-etched samples, inoculated ductile cast irons, is illustrated by Figure 5.

Increasing silicon content leads to an increasing graphite amount, according to the graphitizing effect of this element in cast iron, for less than 0.02%Mo content. Addition of a high content of molybdenum in high silicon level ductile iron limits the graphitizing effect of silicon, but the graphite amount remains higher compared with conventional, un-alloyed iron. A similar evolution is registered for graphite nodule count: a favourable effect of silicon alloying, while molybdenum contributes to decreasing the nodule count of high silicon ductile iron. It is noted that Si-Mo alloyed iron has lower graphite nodule count compared with referred conventional, un-alloyed iron.

As a simple expression, graphite nodularity included in Figure 5b represents the ratio between the amount of graphite particles defined as nodular (spheroidal) graphite [V and VI form, ISO 945] and the total graphite particles identified in the structure, in specific analysis conditions. In actual foundry practice, graphite nodularity is typically used as a true measure of ductile iron quality. For the same irons chemistries and Mg-treatment and inoculation conditions, graphite nodularity is quite negative for being affected by silicon alloying. Nodularity decreases from 95% up to 80% level, but remains at the accepted parameter for ductile irons, which means that the selected inoculant seems to be effective and recommended for high silicon ductile cast irons production. It is confirmed that silicon acts as an element which affects the morphology of spheroidal graphite.

A recent paper [19] shows those important differences in graphite nodularity and nodule count can arise depending on the shape factors of graphite particles that are selected, such as the Sphericity

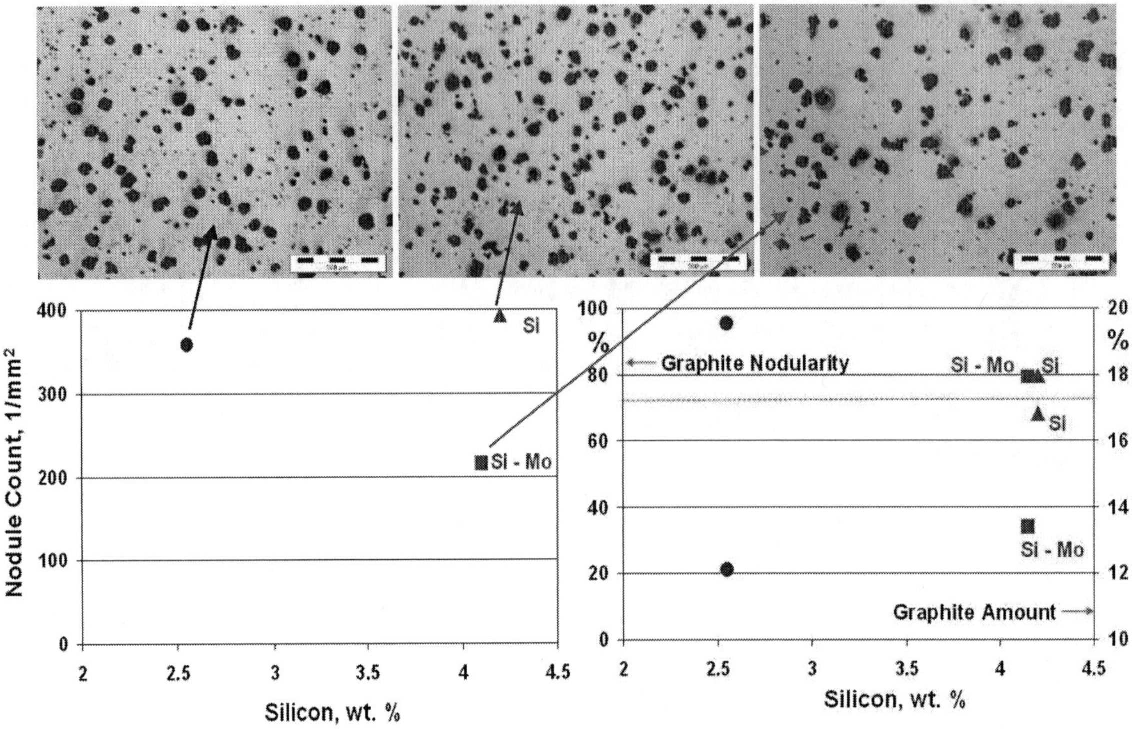

Figure 5. Graphite phase general view and characteristics in inoculated ductile cast irons [a) nodule count; b) amount and nodularity]

Shape Factor - SSF. This parameter considers area (A_G) and real perimeter (P_G) [the sum of the pixel distances along the closed boundary] of the graphite particles [$SSF = 4\pi A_G / P_G^2$]. Using a minimum SSF value of 0.5, at least 90% nodularity is observed for all of the tested irons. If the minimum limit of the shape factor is increased to a SSF value of 0.625, the nodularity appears to fall in the range of 85 to 95%, which would be typical of most ductile iron castings. Using the highest minimum shape factor SSF value of 0.8, corresponding to the highest compactness degree level for the graphite particles, the nodularity range fell to a 68 to 90% range. In many cases, this would be insufficient for some of the inoculated irons to be accepted as high performance ductile iron.

As microstructures presented in Figure 5 point out, not only the ratio of nodular graphite is affected by silicon alloying (nodularity decreases), but also (and especially) the graphite particles morphologies, as compactness degree also decreases. For this reason, it seems that for high silicon ductile cast irons, the graphite nodularity evaluation must be more carefully considered, by including a minimum limit of the accepted SSF.

Metal matrix view and the ferrite amount of these materials in Nital-etched samples are shown in Figure 6. Increasing silicon in cast irons leads to complex effects: no free carbides presence and reduced pearlite / ferrite ratio up to a mainly fully ferritic structure.

In the present research program, it is found that the conventional, un-alloyed ductile cast iron [2.5%Si] presents a normal pearlite and ferrite mixture metal matrix. Increasing of the silicon content [4%Si] leads to a prevalent ferritic matrix. Supplementary addition of molybdenum at the upper limit in Si-Mo ductile irons affects the ferritic effect of silicon, resulting an intermediary ferrite and pearlite mixture.

IOP Conf. Series: Materials Science and Engineering **416** (2018) 012077 doi:10.1088/1757-899X/416/1/012077

4. Conclusions

Experiments deal with the solidification pattern of FeSiCaMgRE-treated and un-inoculated and inoculated [Ca,Ce,S,O-FeSi] three ductile iron compositions [2.5%Si; 4%Si and 4%Si-1.6%Mo (Si:Mo=2.5)].

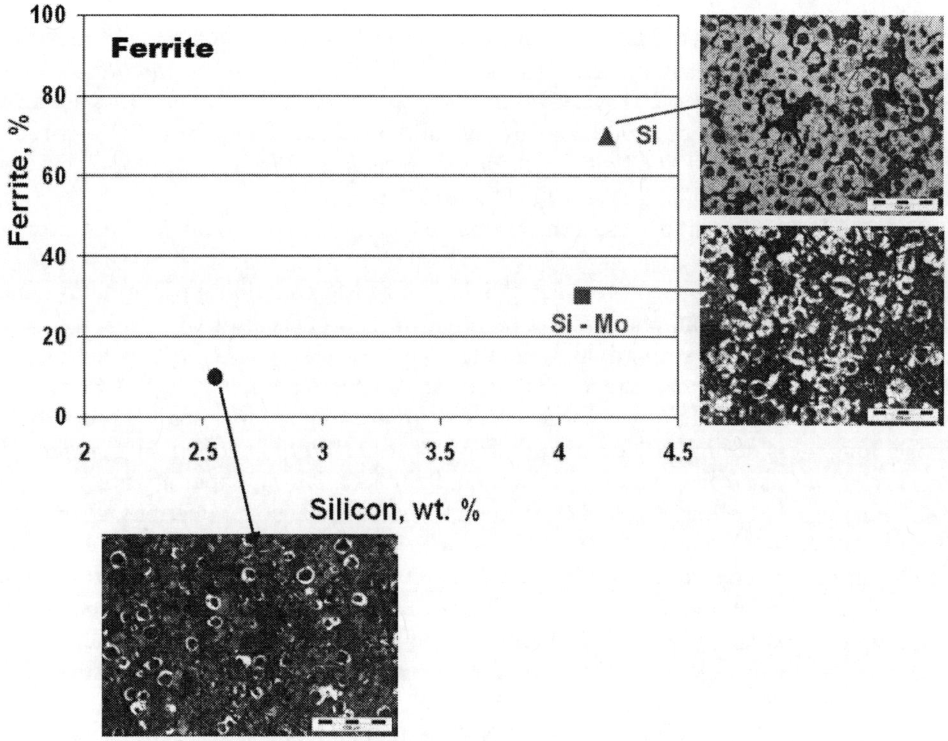

Figure 6. Metal matrix general view and ferrite amount in inoculated ductile cast irons

In completed solidification conditions [20 mm bar diameter, resin sand mould] the evidence presented has led to the following conclusions referring to the structure characteristics:

- As it was expected, the increasing of the Si content leads to the increasing of the graphite amount, but high Mo addition in high Si-iron limits the graphitizing effect of Si, with graphite amount remaining higher compared with conventional iron.
- A similar evolution is registered for graphite nodule count, with the highest level for Si-alloying, while Si-Mo iron has lower nodule count compared with referred iron.
- For the same production conditions, graphite nodularity [according to ISO 945 evaluation] is negatively affected by Si and Si-Mo alloying: it decreases from 95% up to 80% value, but remains at the accepted level, which means that the selected inoculant appears to be effective in these irons.
- For high Si-ductile irons, the nodularity evaluation must be more carefully considered, by including a minimum limit of the accepted sphericity shape factor.
- Increasing of the Si content leads to a prevalent ferritic matrix. Supplementary addition of Mo at the upper limit in Si-Mo affects the ferritic effect of Si, resulting an intermediary ferrite and pearlite mixture.

5. References

[1] Riposan I and Skaland T 2017 Modification and inoculation of cast iron. *ASM Handbook on the Cast Iron Science and Technology* vol 1A, ed D.M. Stefanescu (American Society of Materials), pp160-176.

[2] Stan S, Riposan I, Chisamera M and Barstow M 2018 Solidification pattern of silicon alloyed ductile cast irons. *Proc. 122nd AFS Congress*, Fort Worth, TX, USA, Paper 18-022.

[3] Stets W, Loblich H, Gassner G and Schumacher P 2013 Solution strengthened ferritic ductile iron according DIN EN 1563:2012. *Proc. 2013 Keith Millis Symp. on Ductile Cast Iron* (Nashville, TN, USA, Oct. 2013) 283-292.

[4] Dommaaschk C 2017 Chances and limits of high silicon ductile iron. *Proc. 2nd South African Metal Casting Conf. & WFO Technical Forum,* (Emperors Palace, Kempton Park, Gauteng, South Africa, March 2017), Session 11, Plenary, Paper No. 2.

[5] Menk W 2014 Development of cast iron for exhaust applications. *European Cast Iron Meeting*, (Nancy, France, 2014).

[6] Dekker L and Tonn B 2016 Occurrence and behavior of Mo containing precipitates in nodular cast iron at high temperatures. *Int. J. Cast Met. Res.* **29** (1-2) 85-91.

[7] Menk W 2017 A new high strength high ductile nodular iron. *Proc.11th Int. Symp. on the Science and Processing of Cast Iron* (Jonkoping, Sweden, Sept. 4-7, 2017), Paper no. 90.

[8] Weiss P, Riebisch M and Buhrig-Polaczek A 2017 Improving the mechanical properties and impact toughness of high silicon ductile iron. *Proc. 11th Int. Symp. on the Science and Processing of Cast Iron* (Jonkoping, Sweden, Sept. 4-7, 2017), Paper no. 56.

[9] Papi K, Tunzini S and Menk W 2014 Cast iron alloys for exhaust applications. *Proc. 10th Int. Symp.on the Science and Processing of Cast Iron* (Mar del Plata, Argentina, 2014), Paper 07.

[10] Scheidhauer N, Dommaschk C and Wolf G 2017 Oxidation resistant cast iron for high-temperature application. *Proc. 11th Int. Symp. on the Science and Processing of Cast Iron* (Jonkoping, Sweden, Sept. 4-7, 2017), Paper no. 55.

[11] Ekström M and Jonsson S 2014 *Proc. 10th Int. Symp. on the Science and Processing of Cast Iron* (Mar del Plata, Argentina, 2014), Paper 06.

[12] Ebel A, Yapi Brou S, Malard B, Lacaze J., Monceau D and Vaissiere L 2017 *Proc. 11th Int. Symp. on the Science and Processing of Cast Iron* (Jonkoping, Sweden, Sept. 4-7, 2017), Paper no. 08.

[13] Huyan F, Larker R, Rubin P and Hedstrom P 2014 Effect of solute silicon on the lattice parameter of ferrite in ductile irons. *ISIJ Int.* **54** (1) 248–250.

[14] Alhussein A, Risbet M, Bastien A, Chobaut JP, Balloy D and Favergeon J 2014 Influence of silicon and addition elements on the mechanical behavior of ferritic ductile cast iron. *Mater. Sci. Eng. A*, **605**, 222-228.

[15] Larker, R, 2008 *Proc. 2008 Keith Millis Symp. on Ductile Cast Iron* (Las Vegas, NV, USA, 2008)

[16] Glavas Z, Strkalj A and Stojakovic A 2016 The properties of silicon alloyed ferritic ductile irons. *Metalurgija*, **55** (3), 293-296.

[17] de la Torre U, Loizaga A, Lacaze J and Sertucha J 2014. *Mater. Sci. Techn.*, **30** (12), 1425-1431.

[18] Thielemann T 1970 Zur wirkung van spurenelementen in gusseisen mit kugelgraphit, *Giessereitechnik* (1), 16-24.

[19] Riposan I, Uta V and Stan S 2017 *Metall. Res. Technol.* **114** (4) Article Number: 416, 9 pages.

Acknowledgments

This work was partially financed by the grant of the Romanian National Authority for Scientific Research and Innovation, CNCS/CCCDI - UEFISCDI, project number PN-III-P2-2.1-PED-2016-1793, within PNCDI III and the mobility grant of the Romanian Ministry of Research and Innovation, CNCS – UEFISCDI, project number PN-III-P1-1.1- MC-2017-1151, within PNCDI III.

IOP Conf. Series: Materials Science and Engineering **416** (2018) 012078 doi:10.1088/1757-899X/416/1/012078

Fracture Behavior of Monolithic Hot-Pressed Posterior Ceramic Dental Crowns

L Porojan[1,*], L Rusu[2], A Boloş[3], C Savencu[1] and S Porojan[3]

[1]Department of Dental Prostheses Technology (Dental Technique), School of Dentistry, "V. Babeş" University of Medicine and Pharmacy, 9 Revolutiei 1989 Blv., 300041 Timişoara, Romania
[2]Department Mechanics and Vibrations, Faculty of Mechanical Engineering, Politehnica University, 1 Mihai Viteazu Street, 300222 Timisoara, Romania
[3]Department of Oral Rehabilitation (Dental Technique), School of Dentistry, "V. Babeş" University of Medicine and Pharmacy, 9 Revolutiei 1989 Blv., 300041 Timişoara, Romania

*Corresponding author: sliliana@umft.ro

Abstract. Glass ceramic restorations are wide used in dental practice due to their advantages as life-like appearance, excellent biocompatibility, chemical inertness, close abrasion resistance to enamel, but their most common mechanical complications are chipping and catastrophic fracture. The purpose of the study was to evaluate monolithic ceramic posterior crowns with anatomical design, achieved by hot-pressed glass ceramics, without filler Cergo (Degudent) and leucite reinforced Vita PM9 (Vita) by experimental analyses, in order to study their fracture behavior under compressive load. The modes of fracture will be correlated with the type of material, fracture load and displacement magnitude. For the experiments complete crown preparation were made on a first upper molar. The specimens were subjected to failure testing. Digital images of the fractured crowns fragments were used to verify the pieces in all situations. The failure occurred by cracks, splitting into two or more parts, with or without fracture of the die. The correlation of the displacement depending on the load highlights the peaks, which can be correlated with cracks produced in time. Within the limitations of this study, the tested hot-pressed monolithic ceramic crowns show similar fracture strength and fracture behavior.

1. Introduction

Ceramic crowns are developed as aesthetic restorations because of their life-like aspect, and are characterized by excellent biocompatibility, chemical inertness, close abrasion resistance to enamel, but they have different strength values and structural characteristics related to the composition, and chosen design [2-4]. Glass ceramic restorations are wide used in dental practice due to their advantages mentioned above, but their most common mechanical complications are chipping and catastrophic fracture.

Monolithic silicate ceramic is highly aesthetic and has been used over time for inlays, onlays and veneers with success. However, for bigger restorations, like crowns, more complete fractures were registered [5,6]. Silicate ceramic restorations are brittle and fragile if they are designed very thin. Therefore it is important to chose the optimal thickness and design [7,8].

Monolithic all-ceramic crowns are more and more chosen in practice, compared to the traditional bilayered restorations, because of the disadvantages, related to the complex fabrication, resulted residual stresses, and delamination of the veneers [9-10]. Zirconia and glass-ceramics are aesthetic materials of choice, due to their relatively high modulus and fracture strength [11-15]. All-ceramic crowns absorb occlusal stresses due to their stiffness, but brittleness is still a concern, because different failures of monolithic crowns are described in dental practice [16-18]. These failures could be "abrupt" or catastrophic, with a complete cracking of the crown and sometimes of the dental structure. Other types of fracture involve with chipping of veneering ceramics. There are not many studies which describe the cracks related to the material properties and crown design [19,20].

An important disadvantage of all-ceramic restorations is the manufacturing procedure, associated with a high thermal expansion coefficient and a low thermal conductivity. Therefore the susceptibility to the thermal shock is important for the further behavior. The exposure of all-ceramic restorations to rapid temperature changes during the manufacturing process (pressing, sintering, grinding, polishing) leads to thermal stresses in the material. The pressure of sandblasting, grinding and polishing during processing, and external loading in use, such as occlusal forces during mastication could provide an associated mechanical stress. The combined effect of both thermal and mechanical stress can lead to failure of all-ceramic dental restorations in time [21,22].

Most glass ceramic restorations are indicated to be glazed. Glazing can increase the fracture strength, the wear resistance, produce a smoother surface, and can prevent nucleation and crack propagation [23-25]. Experimental analyses should reproduce the clinical behavior, including the occlusal load and lead to failures similar to those observed in restorations after clinical use [26-28]. Chipping is frequently shown on functional cusps and marginal ridges. These failures are repairable or not, and depend on the location and extension. Catastrophic failure, or fracture throughout the total thickness of the restoration divides it into two or more pieces.

The purpose of the study was to evaluate monolithic ceramic posterior crowns with anatomical design achieved by hot-pressed glass ceramics, without filler Cergo (Degudent) and leucite reinforced Vita PM9 (Vita) by experimental analyses, in order to explain their fracture behavior under compressive load. The modes of fracture will be analyzed, correlated with the type of material, fracture load and displacement magnitude.

2. Materials and methods

Two restorative materials for monolithic dental crowns were selected for the study, both hot pressed glass ceramics. Experimental analyses were made on a full crown preparation made on a first upper molar from a typodont model. The axial reduction was 1.5 mm with a convergence angle of 6 degree, and the occlusal one was 1.5-2 mm. For the finish line a 1 mm deep chamfer was chosen. All sharp angles of the preparation were rounded in order to avoid stress accumulation. The preparation was replicated with a composite material Structur 3 (Voco, Cuxhaven, Germany), which has properties similar to the tooth structure. The dies were scanned using Cercon Eye and

dedicated CAD (computer aided design) software Cercon Art (Degudent, Hanau, Germany) was used for the design of crowns with full anatomic design. In order to obtain identical samples for the experiments, sixteen wax-ups were milled from a Cercon base cast material. These were invested, and specific steps were covered according to the manufacturer's instructions in order to obtain heat pressed ceramic crowns using Cergo (Degudent, Hanau, Germany) (samples C1-C8) and Vita PM 9 (Vita, Bad Säckingen, Germany) (samples V1-V8) ceramics. The restorations thickness was verified by periodic measurements at several locations. The crowns were glazed and adhesive luted with Maxcem Elite (Kerr, Biberach, Germany) on the composite dies.

A digital load was applied to the occlusal surface of the crowns during adhesive setting and the the samples were maintained 24 hours before loading. An Instron 8874 (Instron, Norwood, MA, USA) testing device was used for failure testing. A stainless steel ball with a diameter of 6 mm was fixed to the upper crosshead of a universal testing machine and positioned upon the tested crown. The vertical load was aligned with the tooth's axis. Due to the radius of the sphere, the contact points were

IOP Conf. Series: Materials Science and Engineering **416** (2018) 012078 doi:10.1088/1757-899X/416/1/012078

located adjacent to the central fossa. A 0.2 mm thick rubber foil was positioned between the crown and the sphere in order to reduce peak stresses at the contact points. A compressive load was applied at a crosshead speed of 2 mm/min, and failure was recorded.

3. Results

The crack was observed in the monolithic crowns after the ultimate failure of the crown. Therefore the crack propagation could not be determined visually, but in all cases the origin was on the occlusal surface.

Digital images of the crowns fragments were used to verify the pieces in all cases. The final fracture occurred after cracks, splitting the crown into two and or more parts, with or without fracture of the composite die. In the majority of cases (samples V1, V4, V7, C2, C3, C4, C5, C7, C8) final fracture occurred by splitting through the crown and composite die into two or more parts. The splitting line was centric or eccentric. The splits deviated slightly from the fissure plane, suggesting that fracture may find a path through the cusps (in case of centric split) (Figure 1) or along the fissure (in case of eccentric split) (Figure 2). In these cases the crown failure appeared to be abrupt at the critical load, with no visible sign of precursor cracking prior to splitting. Other cases are characterized by fractures of the crown associated with fissure in the composite die (samples, V5 and C1) (Figure 3), crown fractures (samples V2, V3, V6, C6) (Figure 4) or only cracks in the crowns, mostly radiar (sample V8) (Figure 5). For Cergo samples fractures were more critical.

The graphically correlation of the displacement with the load highlights a series of peaks which could be associated with crown cracks (Figure 6). An increased number of peaks are associated with multiple fragments of the fractured samples.

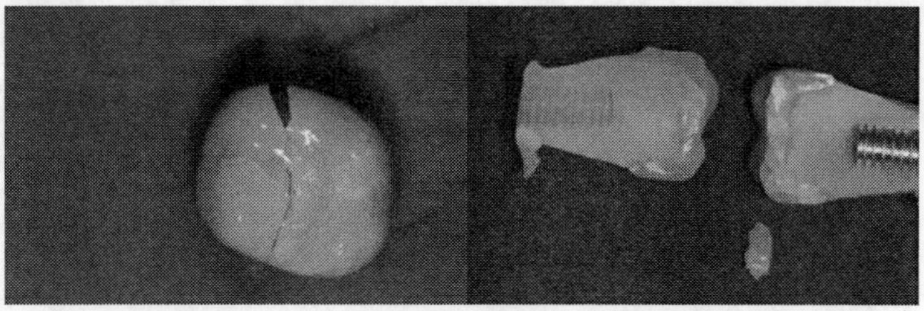

Figure 1. Centric splitting of the crown and underlying die (sample V1).

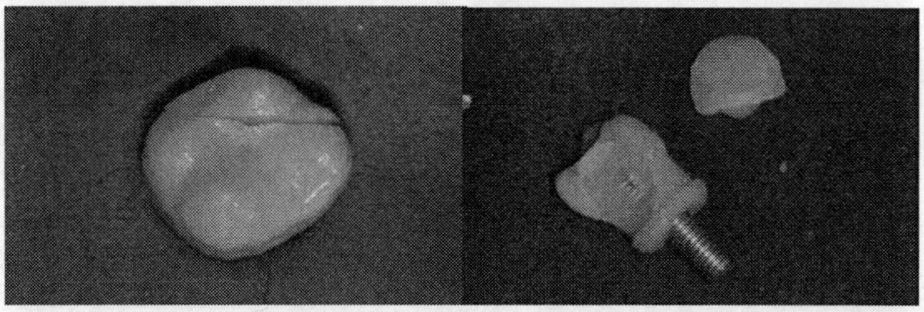

Figure 2. Eccentric splitting of the crown and underlying die (sample C3).

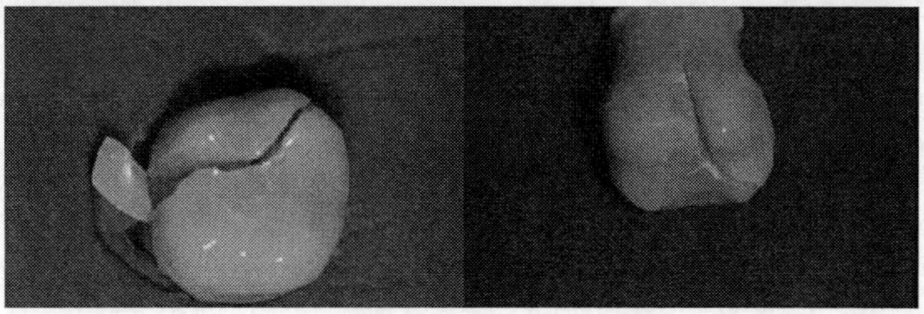

Figure 3. Eccentric splitting of the crown and cracking of the die (sample V5).

IOP Conf. Series: Materials Science and Engineering **416** (2018) 012078 doi:10.1088/1757-899X/416/1/012078

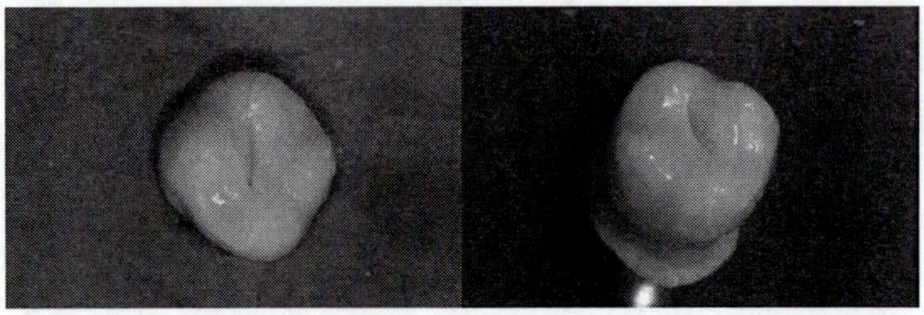

Figure 4. Fracture of the crown (sample C6).

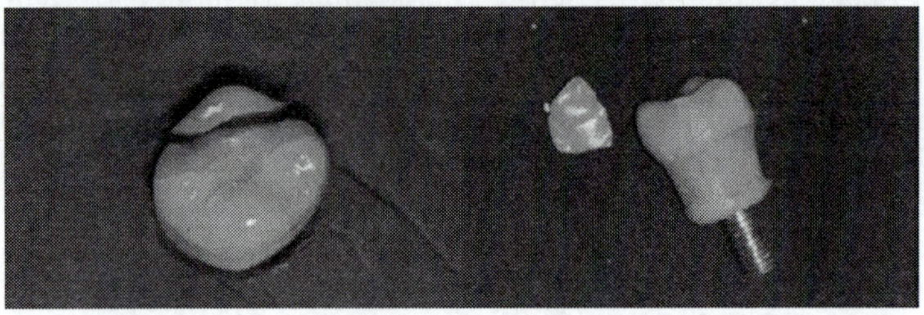

Figure 5. Chipping of the crown (sample V8).

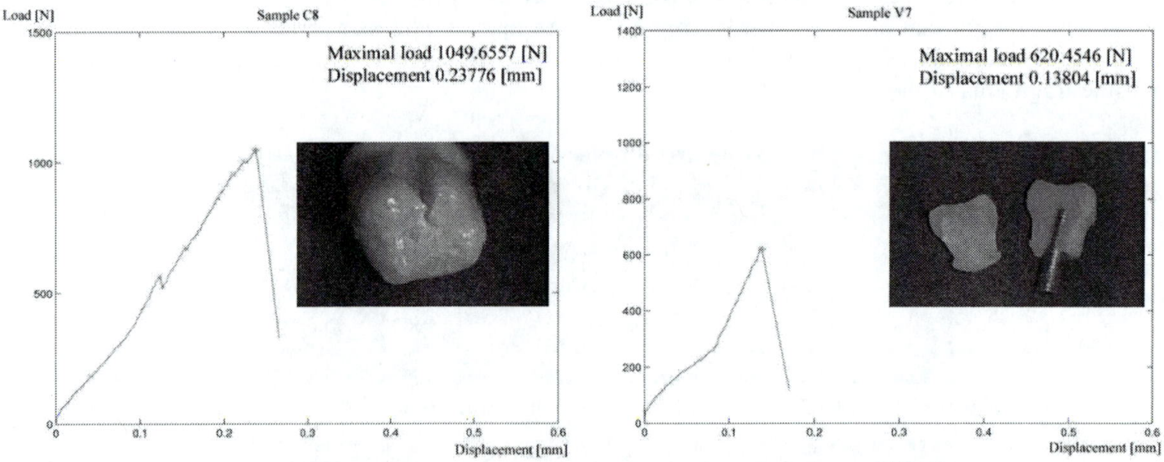

a) b)

Figure 6. Graphical representation recorded during mechanical compression tests: a) multiple peaks for sample C8, correlated with cracks in the ceramic crown, and fracture of the die, b) single peak for sample V7, correlated with a fracture line in the ceramic crown, and fracture of the die.

The maximal loads registered for each sample can be correlated with the displacements. The mean values were 918.16 N, and 0.21 mm for Cergo samples, 666.20 N, and 0.22 mm for Vita samples.

For Vita samples there were greater differences between the studied samples, while at Cergo the fracture values were closer to each other. Between the studied groups, the difference is not significant ($p > 0.05$), their fracture behavior is similar.

4. Discussions

The study examined splitting of anatomically molar crowns, made of hot pressed ceramics, without filler and leucite reinforced. Splitting was induced by axial loading with a sphere placed in the central fossa. Experiments were made on manufactured hot pressed ceramic crowns and critical loads to failure were measured. To simulate clinical conditions, the study used anatomic crowns, instead of disks or bars, bonded to a composite material which presents a mechanical behavior similar to human teeth.

The fracture loads found in this study (620.45 to 1270.35 N) were higher than those of normal human mastication (60 to 250 N) and in most cases even higher than the occlusal forces related to bruxism, tooth clenching, and parafunctions (approximately 800 N) [29,30].

Fatigue behavior could result in slow crack growth from the surface defects over time and contribute to the failure. Therefore, the possibility of chipping or fractures in clinical service is not excluded and can be attributed to the fatigue [31,32]. The main challenge for the design of ceramic restorations includes the estimation of their failure possibility. Other studies show that the strength of a ceramic material is inversely proportional to the square root of the critical size of crack-like defects that are randomly distributed in the material [33]. Due to the position, length and orientation of critical cracks, the strength of ceramics vary unpredictably from component to component, even if identical specimens are tested.

5. Conclusions

Glass ceramics are brittle and fragile materials, and crack initiation and propagation can result in failures of the restorations during functions.

The estimation of the failure probabilities of monolithic ceramic restorations under expected loading conditions is a challenge in this field.

The modes of failure have to be correlated with the type of the material. Within the limitations of this study, the tested hot-pressed monolithic ceramic crowns show similar fracture strength and fracture behavior.

References

[1] Zhao T, Qin Y, Zhang P, Wang B, Yang J 2014 *Ceram Int* High-performance, reaction sintered lithium disilicate glass-ceramics 40 pp 12449–12457.

[2] Heintze S D, Cavalleri A, Forjanic M, Zellweger G, Rousson V 2008 *Dent Mater* Wear of ceramic and antagonist- a systematic evaluation of influencing factors in vitro 24 pp 433–449.

[3] Porojan L, Porojan S, Rusu L, Bolos A, Savencu C 2017 *MATEC Web of Conferences* Edited by: Amir H F A; Khiew P S Experimental Analyses for The Mechanical Behavior of Pressed All-Ceramic Molar Crowns with Anatomical Design, 108, International Conference on Mechanical, Aeronautical and Automotive Engineering (ICMAA), Malacca, Malaysia, UNSP 02001.

[4] Porojan L, Savencu C, Porojan S 2016 *Revista de Chimie* Applications of Heat-pressed Ceramics for Single Tooth Restorations 67 pp 123-126

[5] Malament K A, Socransky S S, Thompson V, Rekow D 2003 Survival of glassceramic materials and involved clinical risk: variables affecting long-term survival *Pract Proced Aesthet Dent* (Suppl), S5-S11.

[6] Skouridou N, Pollington S, Rosentritt M, Tsitrou E, 2013 *Dent Mater* Fracture strength of minimally prepared all-ceramic CEREC crowns after simulating 5 years of service 29 e70-e77.

[7] Guess P C, Schultheis S, Wolkewitz M, Zhang Y, Strub J R 2013 *J Prosthet Dent* Influence of preparation design and ceramic thicknesses on fracture resistance and failure modes of premolar partial coverage restorations 110 pp 264–273.

[8] Sieper K, Wille S, Kern M 2017 *J Mech Behav Biomed Mater* Fracture strength of lithium disilicate crowns compared to polymer-infiltrated ceramic-network and zirconia reinforced lithium silicate crowns 74 pp 342-348.

[9] Christensen G J 1999 *J Am Dent Assoc* Porcelain-fused-to-metal vs nonmetal crowns 130 pp 409-511.

[10] Zhang Y, Lee JJ-W, Srikanth R, Lawn BR 2013 *Dent Mater* Edge chipping and flexural resistance of monolithic ceramics 29 pp 1201-1208.

[11] Gehrt M, Wolfart S, Rafal N, Reich S, Edelhoff D 2013 *Clin Oral Invest* Clinical results of

lithium-disilicate crowns after up to 9 years of service. Ig 17 pp 275-284.

[12] Pieger S, Salman A, Bidra AS 2014 *J Prosthet Dent* Clinical outcomes of lithium disilicate single crowns and partial fixed dental prostheses: a systematic review 112 pp 22-30.

[13] Zesewitz TF, Knauber AW, Northdurft F P 2014 *Int J Prosthodont* Fracture resistance of a selection of full-contour all-ceramic crowns: an in vitro study 27 pp 264-266.

[14] Johansson C, Cmet G, Rivera J, Larsson C, Vuit Von Steyern P 2014 *Acta Odontol Scand* Fracture strength of monolithic all-ceramic crowns made ofhigh-translucent yyttrium-oxide-stabilized zirconium dioxide compared to porcelain-veneered crowns and lithium disilicate crowns 72 pp 145-153.

[15] Magne P, Carvalho AO, Bruzi G, Giannini M 2015 *J Prosthet Dent* Fatigue resistance of ultrathin CAD/CAM complete crowns with a simplified cementation process 114 p 574-579.

[16] Fasbinder DJ, Dennison J B, Heys D, Nelva G 2010 *J Am Dent Assoc* A clinical evaluation of chairside lithium disilicate cad/cam crowns: a two-year report 141 10S–4

[17] Kern SM, Sasse M, Wolfart S 2012 *J Am Dent Assoc* Ten-year outcome of three-unit fixed dental prostheses made from monolithic lithium disilicate ceramic 143 pp 234-240.

[18] Reich S, Schierz O 2013 *Clin Oral Investig* Chair-side generated posterior lithium disilicate crowns after 4 years 17 pp 1765-1772.

[19] Zhang Y, Mai Z, Barani A, Bush M, Lawn B 2016 *Dent Mater* Fracture-resistant monolithic dental crowns 32 p 442-449.

[20] Mores R T, Borba M, Corazza P H, Della Bona Á, Benetti P 2017 *J Prosthet Dent* Influence of surfacefinishing on fracture load and failure mode of glass ceramic crowns 118 pp 511-516.

[21] Mora G P, O'Brien W J 1994 *J. Biomed. Mater. Res.* Thermal shock resistance of core reinforced all-ceramic crown systems 28 pp 189-194.

[22] Dimitrijević M, Pošarac M, Jančić-Heinemman R, Majstorović J, Volkov- Husović T, Matović B 2008 *Appl. Ceram.* Thermal shock resistance of ceramic fibre composites characterized by non-destructive methods, process 2 pp 115-119.

[23] Silva TM, Salvia AC, Carvalho RF, Pagani C, Rocha DM, Silva EG2014*J Prosthodont Res*Polishing for glass ceramics: which protocol? 58 p 160-170.

[24] Benetti P, Kelly J R, Della Bona A 2013 *Dent Mater* Analysis of thermal distributions in veneered zirconia and metal restorations during firing 29 p 1166-72.

[25] Swain M V 2009 *Acta Biomater* Unstable cracking (chipping) of veneering porcelain on all-ceramic dental crowns and fixed partial dentures 5 pp 1668-1677.

[26] Sailer I, Makarov N A, Thoma D S, Zwahlen M, Pjetursson B E 2015 *Dent Mater* All-ceramic or metal-ceramic tooth-supported fixed dental prostheses (FDPs)? A systematic review of the survival and complication rates. Part I: single crowns (SCs) 31 pp 603-623.

[27] Brunot-Gohin C, Duval J L, Azogui E E, Jannetta R, Pezron I, Laurent-Maquin D, Gangloff SC, Egles C. 2013 *Dent Mater* Soft tissue adhesion of polished versus glazed lithium disilicate ceramic for dental applications 29 e205-212.

[28] Scherrer SS, Quinn JB, Quinn GD, Kelly JR 2006 *Int J Prosthodont* Failure analysis of ceramic clinical cases using qualitative fractography19 pp 185-92

[29] Kelly J R, Benetti P Ceramic materials in dentistry: historical evolution and current practice. Aust Dent J 2011;56 (Suppl 1) pp 84-96.

[30] Brunot-Gohin C, Duval JL, Azogui EE, Jannetta R, Pezron I, Laurent-Maquin D, et al. 2013 *Dent Mater* Soft tissue adhesion of polished versus glazed lithium disilicate ceramic for dental applications 29 e205-12.

[31] Mores R T, Borba M, Corazza P H, Della Bona Á, Benetti P 2017 *J Prosthet Dent* Influence of surface finishing on fracture load and failure mode of glass ceramic crowns 118 p 511-516.

[32] Scherrer S S, Quinn J B, Quinn G D, Wiskott H W 2007 *Dent Mater* Fractographic ceramic failure analysis using the replica technique 23 p 1397-404.

[33] Serkan N, Chunsheng L 2012 *Ceram Int* Fracture statistics of dental ceramics: Discrimination of strength distributions 38 pp 4979-4990.

Compressive Behavior and Energy Absorption Capability of Reinforced Closed-Cell Aluminum Alloy Foams

E Linul[1,*], L Marsavina[1] and J Kovacik[2]

[1]Department of Mechanics and Strength of Materials, Politehnica University of Timisoara, 1 Mihai Viteazu Avenue, 300 222 Timisoara, Romania
[2]Institute of Materials and Machine Mechanics, Slovak Academy of Sciences, Dúbravská cesta 9, 845 13 Bratislava, Slovak Republic

[*]Corresponding author: emanoil.linul@upt.ro

Abstract. In recent years aluminum foams and lightweight composite structures with foam core became popular in aerospace as well as in the automotive industry, because of their unique mechanical properties combined with a very low density. This paper investigates and compares the mechanical behavior, the collapse modes and energy absorption capability of unreinforced and reinforced (with 2 different reinforcement positions) aluminum foam composite structures. Quasi-static compressive tests have been undertaken under three different directions and the reinforcement effect on the main mechanical properties (Young's modulus, yield stress, plateau stress and densification) was investigated. The tested closed-cell aluminum alloy foams with a density of 325 kg/m^3 were reinforced with stainless steel mesh. It has been shown that the compression strength and energy absorption performances of the composite foams can be optimized through the correct positioning of the reinforcements.

1. Introduction

The current interest in every field of today's engineering world is to increase the use of lightweight materials and integrate them into vehicle designs in order to improve the economy efficiency and reduce the structure weight, without affecting other attributes such as cost, performance, comfort, safety etc. [1-4]. During the last two decades, the fully dense solid materials (e.g. steel, Al, Cu, polymers etc.) have lost slowly their applicability, while the use of porous materials (metallic foams, polymeric foams etc.) has experienced a rapid growth mainly due to their applications as core material in advanced composite structures [5-8]. Porous materials are a relatively new class of structural and functional materials exhibiting unique mechanical, physical and other properties that make them attractive in a many wide range of critical structural engineering applications [9-12]. Moreover, these cellular materials have high stiffness-to-weight ratio, high ability to absorb impact energy, exceptional heat transfer ability, good formability, good corrosion resistance and recycling potential that makes them ideal candidates for replacing high density materials [13-16].

Different types of foam materials [17-21] have been investigated experimentally under static [22-24] and impact [25-27] loading conditions. Of the most commonly used metal foams are those produced from aluminum alloys and in particular closed-cell aluminum alloy foams [28, 29]. Most studies have been carried out on conventional aluminum foam [30-32] while a small number of investigations have focused on metal matrix syntactic foams or composite foams [33, 34]. Thermal properties of closed-cell aluminum foams prepared by melt foaming technology was evaluated by Wang et al. [35], while Xia and coworkers [36] investigated the acoustic properties of closed-cell aluminum foams with different macrostructures. On the other hand, the mechanical behavior of aluminum foams is investigated in Refs. [1, 5, 17, 25, 28-34]. These studies have focused on the effect of temperature, density, microstructure, anisotropy, strain rate, reinforcements and foaming parameters on main mechanical properties of aluminum foams.

Therefore, this paper investigates the compressive behavior and energy absorption crush performances of different types of closed-cell aluminum alloys foams (unreinforced and reinforced foams) under quasi-static loads. Two types of composite foams with both longitudinal and transversal reinforcements were evaluated, and the optimal foam design is highlighted.

2. Materials and methods

Stainless steel mesh with diamond shape of 6 mm × 3 mm and 1 mm (width) × 0.5 mm (thickness) was used as reinforcement materials (Fig. 1a). The austenitic Cr-Ni stainless steel reinforcements (X5CrNi 18-10) with a composition of Cr 17-19.5 %, Ni 8-10.5%, C<0.07% presents a square weight of 3.4 kg/m^2. Powder metallurgical route was used for manufacturing of metal foam matrix composites [37].

The stainless steel mesh were inserted into the mold together with precursor and during the foaming process the reinforcements were infiltrated with molten material, thus forming a metallurgical bond between foam and steel (Fig. 1b). Foamable AlSi10 precursor material containing 0.4 wt.% of TiH_2 as foaming agent was foamed, kept at high temperatures (620°C) and solidified. The chemical composition of obtained closed-cell aluminum alloy foam is 10 wt.% Si and aluminum balanced.

a) b) c) d)

Figure 1. Used materials: stainless steel mesh as reinforcement (a), formed metallurgical bond (b), obtained aluminum foam plate (c) and processed cubic specimens (d)

Large composite foam plate was foamed (Fig. 1c), and processed as cubic specimens (25 mm × 25 mm × 25 mm) (Fig. 1d). In order to prevent any damage to the cellular structure, the cubic foam samples were cut by Electric Discharge Machining (EDM).

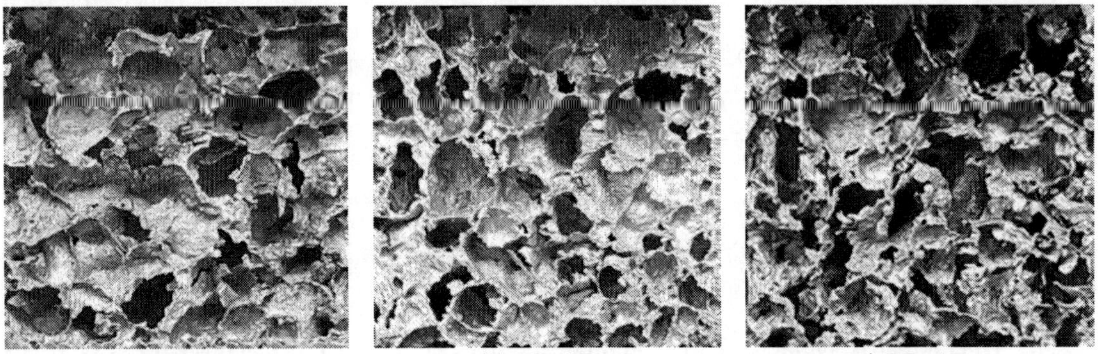

a) UR-AFs b) TR-AFs c) LR-AFs

Figure 2. The macrostructure morphology of the investigated composite foams

IOP Conf. Series: Materials Science and Engineering **416** (2018) 012079 doi:10.1088/1757-899X/416/1/012079

All cubic specimens were cut from the same reinforced plate under three different loading directions, as follow: direction (1) – unreinforced aluminum foams (**UR-AFs**); direction (2) – foams with longitudinal reinforcements (**LR-AFs**) and direction (3) – foams with transversal reinforcements (**TR-AFs**). The macrostructure morphology of the investigated composite foams is presented in Fig. 2 for each direction. The obtained average foam density was 325 kg/m^3.

Quasi-static compression tests were carried out at room temperature (~25°C) on a 100 kN LBG testing machine, with a constant crosshead speed of 10 mm/min. Five specimens were provided for each loading direction according to the ISO13314-11 standard [38]. Figure 3 shows the initial and deformed specimens after mentioned three different loading directions.

Figure 3. Initial and deformed specimens

3. Results and discussion

Figure 4a presents the typical stress-strain curves up to approximately 80% engineering strain, while Fig. 4b presents different deformation sequences (0%, 20%, 40%, 60% and 80%) of unreinforced and reinforced aluminum foams.

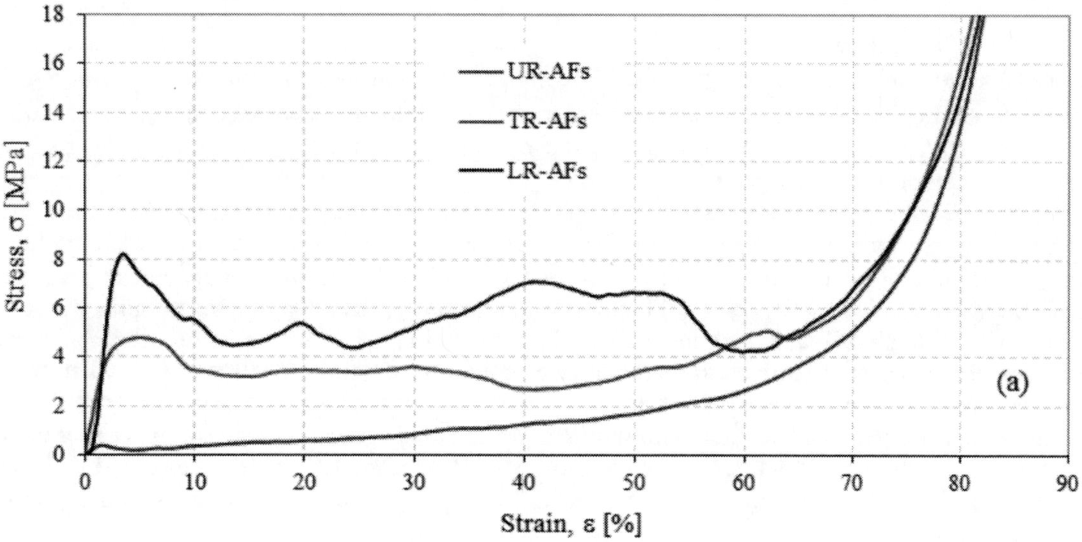

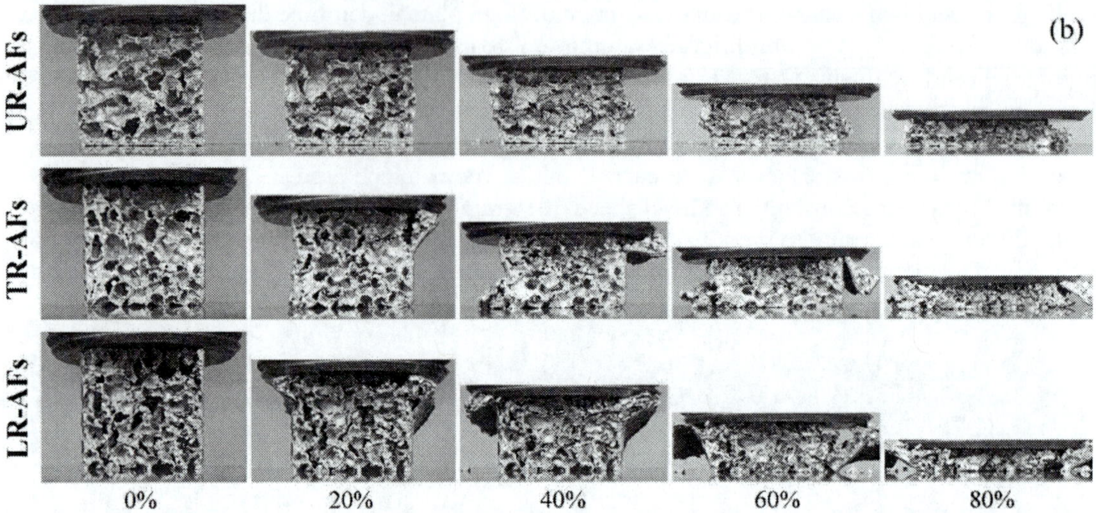

Figure 4. Stress-strain curves (a) and deformation sequences (b) for different investigated AFs

The main mechanical properties of investigated closed-cell aluminum alloy foam specimens were determined based on [38] and are presented in Table 1, depending on the loading directions.

Table 1. The main mechanical properties of investigated closed-cell aluminum foams.

Reinforcement type	Linear-elastic region			Plateau region			Densification region		
	E [MPa]	σ_y [MPa]	ε_y [%]	$\sigma_{20\%}$ [MPa]	$\sigma_{50\%}$ [MPa]	σ_p [MPa]	σ_D [MPa]	ε_D [%]	W_D [MJ/m^3]
UR-AFs	34.17	0.39	1.49	0.57	1.68	0.88	1.14	38.25	0.23
TR-AFs	235.69	4.78	5.06	3.46	3.38	3.33	4.33	57.87	1.98
LR-AFs	476.80	8.20	3.44	5.32	6.63	5.40	7.02	70.46	3.93

In Table 1, the following notations were used: E is the compressive modulus determined from this linear-elastic region; σ_y - compressive yield stress (the first point on the σ-ε curves from which an increase in strain occurs without an increase in stress); ε_y - yield strain corresponding to σ_y; $\sigma_{20\%}$, and $\sigma_{50\%}$ compressive strength at 20% and 50% strain; σ_p - plateau stress; ε_D - onset strain of densification (OSD); σ_D - compressive strength corresponding to OSD.

The obtained stress - strain curves of investigated AFs are commonly to those studied by other groups of researcher both for unreinforced and reinforced cellular structures or porous structures [39-42]. The σ-ε curve exhibits three different definite regions: (i) an initial linear-elastic behavior at low stresses (<5% strain), controlled by the cell-wall bending and stretching; (ii) a long plateau region where foam cells collapse through cell-wall buckling, cell-wall plastic yielding, cell-wall brittle crushing and/or cell-wall fracture; and (iii) a densification region at large compressive strains (>40% strain), when the cells have almost completely collapsed [43, 44].

It has been observed that UR-AFs foam shows approximately a linear increase in stress after the yield limit (on the plateau region), while the both reinforced (TR-AFs and LR-AFs) foams exhibit oscillating character during plateau region from the very early stages of compression up to onset strain of densification.

By using reinforcements, the compressive modulus increase by 7 times for TR-AFs and 14 times for LR-AFs, compared to unreinforced foams. On the other hand, as it can be observed from Table 1, the compressive strength increases much more (up to 21 times for LR-AFs and 12 times for TR-AFs) for reinforced foams than those unreinforced. Also, reinforced foams show a much longer plateau region, with an onset strain of densification with 33.90% (for TR-AFs) and 45.71% (for LR-AFs) higher, compared to UR-AFs which presents an OSD of 38.25%. At the same time, the energy absorption values at OSD are up to 17 times higher for reinforced than for the unreinforced ones.

According to the obtained results (Fig. 4a and Table 1) it was found that the drop stress (which occurs after the first maximum compressive strength) reaches up to 67% of the yield stress for reinforced foams, while this value is about 49% for unreinforced foams. Figure 5 presents the drop stress amplitudes ($\Delta\sigma$) of investigated foams under different loading directions. The difference of amplitudes between the unreinforced and reinforced foams, in terms of drop stress, increases linearly.

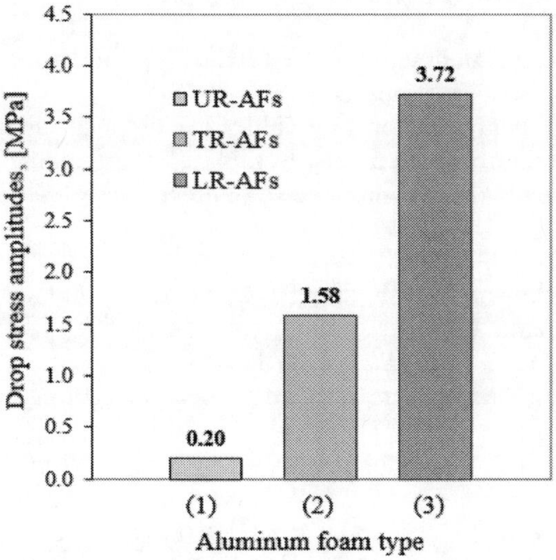

Figure 5. $\Delta\sigma$ amplitudes of AFs.

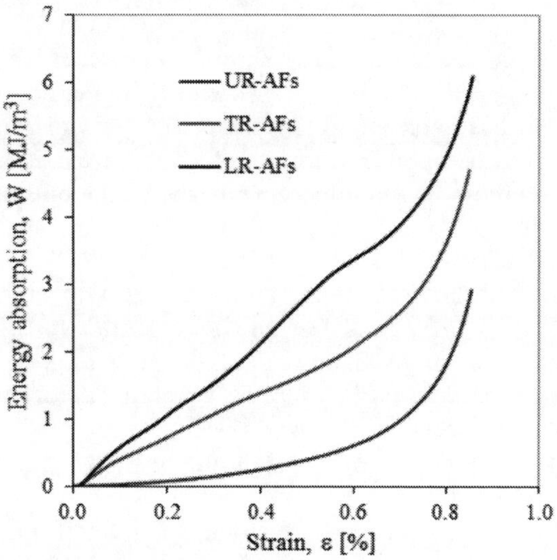

Figure 6. W - ε variation of AFs.

Due to their cellular structure, the closed-cell foam structures presents high energy absorption capability than the solid materials from which they are made [45, 46]. The energy absorption-strain variation of closed-cell aluminum alloy foams is presented in Figure 6 for all loading directions. The area under true stress - strain curve, from its origin up to a given value of strain, represents the energy required or stored in the material before its failure [47]. Most of the energy (around 80% of the total amount of energy) is absorbed in the plateau region of the σ-ε curve because the collapse mechanisms occur [48].

Table 2. The energy absorption values of investigated AFs for different loading directions.

Reinforcement type	Energy absorption at different strains, W [MJ/m^3]							
	10%	20%	30%	40%	50%	60%	70%	80%
UR-AFs	0.03	0.07	0.14	0.25	0.39	0.61	0.98	1.81
TR-AFs	0.4	0.73	1.08	1.4	1.69	2.08	2.6	3.62
LR-AFs	0.58	1.07	1.54	2.14	2.82	3.38	3.9	4.91

As can easily be seen from Figure 6, but especially from Table 2, the use of reinforcements significantly increases the ability of the aluminum foams to absorb energy. Major differences have been achieved in the area of small deformations (around 10%) where LR-AFs foam has up to 20 times better energy-absorption capacities than UR-AFs foam, while TR-AFs up to 14 times compared to UR-AFs foam. In the case of large deformations (over 60% strain), this difference considerably decreases approximately linearly to about 4 times for LR-AFs and about 3 times for TR-AFs, compared to unreinforced aluminum foams. Moreover, remarkable W differences can be seen also between the two directions where reinforcements are positioned differently. The foams with longitudinal reinforcements shows energy absorption capacities up to 40% better than the foams with transversal reinforcements. Therefore, a correct positioning of reinforcements leads to a significant increase of foam crush performances, which can sometimes be very important.

4. Conclusions

The uniaxial quasi-static compressive behavior and energy absorption capabilities of unreinforced and reinforced closed-cell aluminum alloy foams are experimentally investigated.

The following conclusions can be drawn:

- The reinforced foams show higher mechanical properties than unreinforced foams (up to 21 times);
- The LR-AFs shows significantly higher crush performances compared to TR-AFs;
- It has been found that the compression strength and energy absorption values of composite foams can be optimized through the correct positioning of the reinforcements. In this case the foams with longitudinal reinforcements shows the optimal design of foam composite structures.

5. References

[1] Ashby MF, Evans A, Fleck NA, Gibson LJ, Hutchinson JW, Wadley HNG, Delale F 2000 *Metal Foams: A Design Guide.* Butterworth-Heinemann, USA
[2] Birsan M, Sadowski T, Marsavina L et al. 2013 *Int. J. Solids Struct.* **50** 519-530
[3] Marsavina L, Linul E, Voiconi T, Constantinescu DM, Apostol DA 2015 *Frattura ed Integrita Strutturale* **34** 444-453
[4] Marsavina L, Constantinescu DM, Linul E, Stuparu FA, Apostol DA 2016 *Eng. Fract. Mech.* **167** 68-83
[5] Taherishargh M, Katona B, Fiedler T, Orbulov IN 2016 *J. Compos. Mater.* **51(6)** 773-781
[6] Negru R, Marsavina L, Voiconi T et al. 2015 *Theor. Appl. Fract. Mech.* **80** 87-95
[7] Apostol DA, Stuparu F, Constantinescu DM et al. 2016 *Materiale Plastice* **53(4)** 685-688
[8] Marsavina L, Constantinescu DM, Linul E et al. 2015 *Eng. Fail. Anal.* **58** 465-476
[9] Mansoor B, Nassar H, Shunmugasamy VC, Khraisheh MK 2015 *Mat. Sci. Eng. A-Struct.* **628** 433-441
[10] Linul E and Marsavina L 2011 *Iran. Polym. J.* **20(9)** 736-746
[11] Aliha MRM, Linul E, Bahmani A, Marsavina L 2018 *Polym. Test.* **67** 75-83
[12] Marsavina L, Constantinescu DM, Linul E, Apostol DA, Voiconi T 2014 *Eng. Fract. Mech.* **129** 54-66
[13] Apostol DA, Stuparu F, Constantinescu DM et al. 2016 *Materiale Plastice* **53(2)** 280-282
[14] Şerban DA, Linul E, Voiconi T, Marsavina L, Modler N 2015 *Iran. Polym. J.* **24** 515-529
[15] Linul E and Marsavina L 2015 *P. Romanian Acad. A* **16(4)** 522-530
[16] Marsavina L, Cernescu A, Linul E, Scurtu D, Chirita C 2010 *Materiale Plastice* **47(1)** 85-89
[17] Katona B, Szebényi G, Orbulov IN 2017 *Mat. Sci. Eng. A-Struct.* **679** 350-357
[18] Movahedi N and Linul E 2018 *Mater. Lett.* **217** 194-197
[19] Voiconi T, Negru R, Linul E et al. 2014 *Frattura ed Integrità Strutturale* **30** 101-108
[20] Marsavina L, Berto F, Negru R et al. 2017 *Theor. Appl. Fract. Mech.* **91** 148-154
[21] Linul E, Serban DA, Marsavina L 2018 *Phys. Mesomech.* **21(2)** 178-186
[22] Marsavina L, Linul E, Voiconi T, Sadowski T 2013 *Polym. Test.* **32** 673-680
[23] Linul E, Marsavina L, Sadowski T, Kneć M 2012 *Solid State Phenomena* **188** 205-210
[24] Voiconi T, Linul E, Marsavina L, Sadowski T, Knec M 2014 *Solid State Phenomena* **216** 116-121
[25] Myers K, Katona B, Cortes P, Orbulov IN 2015 *Compos. Part A-Appl. S.* **79** 82-91
[26] Linul E, Voiconi T, Marsavina L, Sadowski T 2013 *J. Phys. Conf. Ser.* **451** 012002
[27] Şerban DA, Voiconi T, Linul E, Marsavina L, Modler N 2015 *Materiale Plastice* **52(4)** 537-541

[28] Movahedi N and Mirbagheri SMH 2016 *Strength Mater.* **48(3)** 444-449
[29] Voiconi T, Linul E, Marsavina L et al. 2014 *Key Engineering Materials* **601** 254-257
[30] Kovácik J, Jerz J, Mináriková N et al. 2016 *Frattura ed Integrita Strutturale* **36** 55-62
[31] Marsavina L, Kovacik J, Linul E 2016 *Theor. Appl. Fract. Mech.* **83** 11-18
[32] Movahedi N, Linul E, Marsavina L 2018 *J. Mater. Eng. Perform.* **27(1)** 99-108
[33] Kádár K, Máthis K, Orbulov IN, Chmelík F 2016 *Mater. Lett.* **173** 31-34
[34] Taherishargh M, Linul E, Broxtermann S, Fiedler T 2018 *J. Alloy. Compd.* **737** 590-596
[35] Wang H, Zhou XY, Long B, Yang J, Liu HZ 2016 *Trans. Nonferrous Met. Soc. China* **26** 3147-3153
[36] Xia X, Zhang Z, Zhao W, Li C, Ding J, Liu C, Liu Y 2017 *J. Mater. Sci. Techn.* **33** 1227-1234
[37] Linul E, Marsavina L, Kováčik J 2017 *Mat. Sci. Eng.-A Struct.* **690** 214-224
[38] ISO13314-2011 Mechanical testing of metals-Ductility testing-Compression test for porous and cellular metals
[39] Szlancsik A, Katona B, Bobor K, Májlinger K, Orbulov IN 2015 *Mater. Des.* **83** 230-237
[40] Linul E, Movahedi N, Marsavina L 2017 *Compos. Struct.* **180** 709-722
[41] Marsavina L, Linul E, Voiconi T, Negru R 2016 *IOP Conf. Ser: Mater. Sci. Eng.* **123(1)** 012060
[42] Linul E, Şerban DA, Marsavina L, Kovacik J 2017 *Fatigue & Fracture of Engineering Materials & Structures* **40(4)** 597-604
[43] Linul E, Movahedi N, Marsavina L 2018 *J. Alloy. Compd.* **740** 1172-1179
[44] Movahedi N and Linul E 2017 *Mater. Lett.* **206** 182-184
[45] Linul E, Serban DA, Voiconi T, Marsavina L, Sadowski T 2014 *Key Engineering Materials* **601** 246-249
[46] Linul E, Marsavina L, Kovacik J, Sadowski T 2017 *P. Romanian Acad. A* **18(4)** 361-369
[47] Linul E, Şerban DA, Marsavina L, SadowskiT 2017 *Archives of Civil and Mechanical Engineering* **17(3)** 457-466
[48] Linul E, Movahedi N, Marsavina L 2018 *Materials* **11(4)** 554

Acknowledgment
This work was supported (partial supported) by research grants PCD-TC-2017.

IOP Conf. Series: Materials Science and Engineering **416** (2018) 012080 doi:10.1088/1757-899X/416/1/012080

Mechanics of Active Mechano-Chemical Responsive Polymers

R Brighenti[1,*], F Artoni[1] and M P Cosma[1]

[1]Department of Engineering & Architecture, University of Parma
Parco Area delle Scienze 181/A, 43124 Parma, Italy

[*]Corresponding author: brigh@unipr.it

Abstract. The need for smart materials in advanced applications, has promoted significantly the research in the field of responsive (or active) materials. They are capable of some physical response to external stimuli, such as temperature, pH, light, mechanical stress, etc. so they can adapt themselves to the surrounding environment. In the present paper, we propose a mechanical model to describe the response of polymers containing mechanically responsive molecules (mechanophores), having the ability to assume two distinct geometrically stable configurations under mechanical actions or due to a change of the environment's pH. When linked to the polymer's chains, under a proper mechanical stress, they switch from one state to the other, triggering a macroscopic deformation of the material. Such an internal deformation, driven by the mechanophores' activation, can be exploited to get a desired functionality of the polymer. Our model starts from the mechanical behavior at the microscopic level, and through the mechanics of a single chain and chain statistics, it is scaled-up to the (mesoscopic) continuum level. An energy-based approach is adopted: the free energy W is obtained by adding up the contribution of the network deformation (W_{net}), of the mixing term related to the solvent uptake (W_{mix}) and of the switching of the hosted responsive molecules (W_{sw}). Numerical simulations demonstrate that the material response under chemical or mechanical stimuli, shows a permanent deformation due to the activation of the switchable molecules, enabling the design of smart self-adaptive polymers.

1. Introduction

Polymeric materials have been subjected to a wide diffusion in engineering applications, thanks to their affordability and to their wide range of relevant mechanical properties, such as lightweight, tunable stiffness, fracture toughness, etc.; for these reasons a growing demand of proper mechanical predicting models is steadily increasing [1].

By harnessing their special microstructure, made of a network of long entangled linear chains connected together in several junction points, various macromolecular-based theories have been proposed [2, 3]. Models based on some simplifying assumptions of the spatial chains arrangements have also been successfully proposed [4, 5] as well as models based on phenomenological aspects of their observed mechanical behavior [6].

Thanks to the network-like molecular structure, in recent years molecules having particular responsiveness have been inserted into the polymer backbone, leading to materials with unusual properties, usually referred to as 'smart'. Smart materials are capable to respond to various external stimuli such as temperature change, pH [7], mechanical stress [8], light, electric and magnetic fields [9], etc.

Responsive polymer-based materials are able to modify their chemical and/or physical properties upon external stimuli, and they open the way to the solution of complex problems where the stimuli come from living systems, such as controlled drug delivery, tissue repair, artificial muscles, biosensing, etc. [10, 11].

Among the above listed responsive materials, the sensibility to mechanical stress play a relevant role. The well-known term '*mechanophore*' has been introduced for molecules that undergo chemical modification – usually because of the presence of weak bonds – when triggered by mechanical forces transmitted by the polymer chain to which the mechanophores are connected.

Furthermore, pH sensible molecules [7] have also attracted attention because of the possibility to develop materials suitable to face complex problems, such as controlled drug delivery, development of smart membranes, viscosity control, etc.

We present, here, a micromechanical-based model for polymers whose cross-links have been replaced by mechanophores, which are constituted by molecules able to change their shape and size through the switching mechanism between two stable conformation states. The conformation change is triggered by mechanical stress or chemical stimuli and is captured based on the deformation features taking place at the molecular level. Through the statistics of the chain network, the micromechanical model is scaled-up to the mesoscopic level, and it can be readily adopted to model smart responsive polymers at the continuum scale. The last part of the paper is devoted to the simulation, through the developed theoretical framework, of the response of polymers with embedded chemo-mechanical molecules under various stress history, with or without the presence of a solvent responsible for pH change.

2. Mechanics of a polymer chain linked to a switchable molecule

We consider here the kinematics and the mechanics of a single polymeric chain linked to a switchable molecule.

2.1. Deformation of a polymer chain linked to a switchable molecule

Let us consider a single chain with an initial end-to-end distance $r_0 = l_{0p} = b \cdot \sqrt{N}$ (N, b are the number of Kuhn's segments in the chain and their length, respectively), linked to a switchable molecule (Fig. 1a) with an initial size l_{0m}. Under a force whose intensity is below the molecule's switching threshold value f'_m, the chain only deforms and stretches to λ_p (Fig. 1b). If the threshold force is overcome, $f_p > f'_m$, the switchable molecule opens an amount $2\delta s_m$ and the stretch in the chain decreases to:

$$\lambda'_p = \lambda_p - (\lambda_m - 1) \cdot \alpha \tag{1}$$

being $\lambda_m = \delta s_m / l_{0m}$ and $\alpha = l_{0m}/l_{0p}$.

In a real polydisperse polymer, the chains of the network have an end-to-end distance that can be described through a statistical distribution of the number of their Kuhn's segments, $p(N)$.

Let us assume that the average chain has $\overline{N}$ segments and that the stretch referred to such a

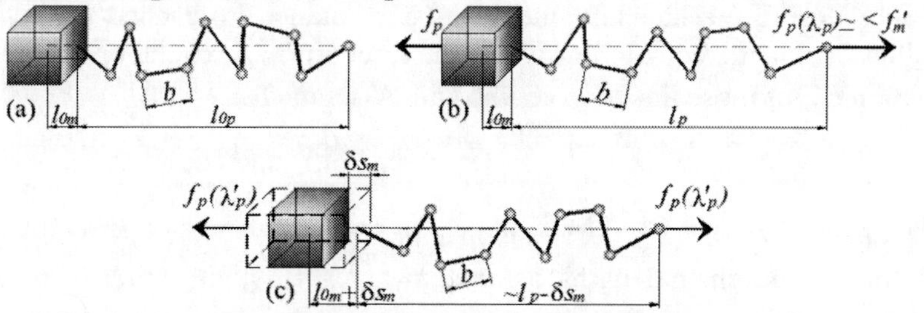

Figure 1. Scheme of a single polymer chain linked to a switchable molecule (a). Deformation before (b) and after (c) the molecule switches.

chain is $\lambda_0 = l_p/(b\sqrt{\overline{N}})$; by adopting the affine deformation hypothesis, the average stretch of the network $\bar{\lambda}$, is:

$$\tilde{\lambda} = 1 + (\lambda_0 - 1)\sqrt{\bar{N}} \int_{N_{min}}^{+\infty} \frac{p(N)}{\sqrt{N}} dN \tag{2}$$

Finally, the stretch in a chain with N segments can be referred to the above mesoscopic deformation:

$$\lambda_p(N) = 1 + (\tilde{\lambda} - 1)\sqrt{\bar{N}/N} \tag{3}$$

The effective stretch is

$$\lambda'_p(N) = \frac{l_p - \delta s_m}{\sqrt{N} \cdot b} = 1 + \frac{(\tilde{\lambda} - 1)\sqrt{\bar{N}}\, b - \delta s_m \cdot \mathcal{H}(\tilde{\lambda}, N)}{\sqrt{N} \cdot b} \tag{4}$$

being $\mathcal{H}(\tilde{\lambda}, N)$ the step function that defines the close ($\mathcal{H}(\tilde{\lambda}, N) = 0$) or open ($\mathcal{H}(\tilde{\lambda}, N) = 1$) state of the molecule. Its value depends on the average stretch $\tilde{\lambda}$ and on the number N of the chain, being the force induced in the chain by the average stretch $\tilde{\lambda}$ dependent on the number N, i.e. $f_p = f_p(\tilde{\lambda}, N)$.

In the following step, a standard Gaussian probability function will be adopted for N, i.e. $p(N) = \frac{1}{\delta N \cdot \sqrt{2\pi}} \exp\left[-\frac{(N-\bar{N})^2}{2 \cdot \delta N^2}\right]$, where δN^2 is the variance and $\bar{N}$ the mean value of the distribution.

2.2. Mechanics of the chain-switchable molecule system

The mechanics of rubberlike polymers is dominated by entropic effects, and their response is mainly governed by the entropy of the polymeric chains, that depends on the chain's length and on the statistical distribution of the chains' end-to-end distance within the network. The force $f_p(N)$ in a chain with N segments can be determined through the strain energy function ψ as follows:

$$f_p = \frac{\partial \psi}{\partial r} = \frac{k_B T}{b} \beta, \quad \psi = N k_B T \cdot \left(\frac{\beta}{bN} r + \ln \frac{\beta}{\sinh \beta}\right), \quad \beta = \mathcal{L}^{-1}\left(\frac{r}{bN}\right) = \mathcal{L}^{-1}\left(\frac{\lambda_p}{\sqrt{N}}\right) \tag{5}$$

where the length-dependent Langevin-chain model, suitable for polydisperse polymers stretched close to their maximum extension ($\lambda_{pmax}(N) = \sqrt{N}$), has been used ($\mathcal{L}^{-1}(x)$ is the inverse of the Langevin function defined as: $\mathcal{L}(x) = \coth(x) - x^{-1}$); k_B and T are the Boltzmann constant and the absolute temperature, respectively, and $r = \lambda_p \cdot r_0$, $r_0 = b\sqrt{N}$ are the end-to-end chain distance in the deformed and in the stress-free state, respectively. If a switchable molecule is linked to the polymer chain, the force f_p transmitted to the molecule (Fig. 1b), can trigger the switch to the open state, leading to a reduction (chain relaxation) of the force itself to the value $f_p = \frac{k_B T}{b} \beta'$, with $\beta' = \mathcal{L}^{-1}(\lambda'_p/\sqrt{N})$. For the sake of simplicity, all chains with N segments are assumed to have the same initial end-to-end distance r_0, so that $p(r, N) = \delta(r, r_0)$, where δ is the Dirac delta function.

3. Molecule switching in the presence of a pH change

The volume fraction, at a given time, of the open molecules embedded in the polymer, must be determined according to the kinetic equilibrium concept, since the thermal fluctuations of the network can induce opening/closing mechanisms until a proper balance is reached [12]. For a given chain length N, the following kinetic equation provides the fraction $h_{N,op}$ of the open molecules among the total volume fraction ϕ_N of those linked to chains with N segments:

$$\frac{d}{dt} h_{N,op}(\boldsymbol{F}, C_s) = k_A - (k_A + k_D) \cdot h_{N,op}(\boldsymbol{F}, C_s) \tag{6}$$

where $\phi_{N,op} = h_{N,op}\phi_N$, $\phi_{N,cl} = (1 - h_{N,op})\phi_N$, ($\phi_N = \phi_{N,op} + \phi_{N,cl}$); an equivalent value of h_{op} can be defined for all the chain lengths as: $\tilde{h}_{op} = \int_{N_{min}}^{+\infty} h_{N,op} \cdot p(N) dN$. In (6), reaction rates constants k_A, k_D (where the subscripts 'A' and 'D' stand for activation, i.e. opening, and deactivation, i.e. closing) depend on the chemical nature of the switching phenomenon, and are influenced by both the mechanical stress and by the pH (the standard Arrhenius equation can be used for quantifying such influences). The dependence on the deformation and the solvent concentration has been indicated by writing $\boldsymbol{F}, C_s$ (deformation gradient tensor and solvent concentration, respectively) as arguments of $h_{N,op}$. Since the switchable molecules are randomly distributed in the 3D space filled by the polymer, their effect on the network is to induce an isotropic expansion. By considering all the chain lengths, the total volume fraction of open molecules is $\phi_{op}(F, C_s) = \tilde{h}_{op}(F, C_s) \cdot \phi$, where $\tilde{h}_{op} = \phi_{op}(F, C_s)/\phi = \int_{N_{min}}^{+\infty} h_{N,op}(N) \cdot p(N) dN$ is the fraction of open molecules over the total number of switchable molecules, irrespectively of the chain length. The variation in volume induced by the molecule opening is provided by:

$$J_{op} = 1 + (\langle J_m \rangle - 1) \cdot \tilde{h}_{op}(\boldsymbol{F}, C_s) \cdot \phi \tag{7}$$

where $\langle J_m \rangle = \det\langle \boldsymbol{F}_m \rangle = \det\left(\int_\Omega \boldsymbol{A}^T \boldsymbol{F}_{op} \boldsymbol{A} d\Omega\right) = [(\lambda_{m1} + \lambda_{m2} + \lambda_{m3})/3]^3$ is the variation of volume of an ideal material made only by open switchable molecules characterized by the principal stretches $\lambda_{m1}, \lambda_{m2}, \lambda_{m3}$. The mean isotropic stretch due to the opening mechanisms of the involved molecules is $\lambda_{op} = J_{op}^{1/3}$. The presence of the solvent induces also swelling of the polymer with a further volume increase (polymer-solvent mixing). By solving the equilibrium equation $\partial W_{mix}/\partial J_{sw} = 0$ (see below) the steady-state volume increase, J_{sw}, due to swelling can be obtained. Being the polymer network uncompressible, the total volume change J_{tot} is induced only by swelling and by the opening mechanism:

$$J_{tot} = J_{sw}(1 - \phi) + \phi \cdot J_{op} \tag{8}$$

Finally, the concentration of the solvent C_s results to be proportional to $(J_{sw} - 1)$, i.e. to the volume increment of the polymer due to swelling and the related average chain stretch is $\lambda_{sw} = J_{sw}^{1/3}$.

4. Continuum model of a polymer containing switchable molecules

For a polymer whose chains are linked together only through switchable molecules, the mechanical energy per unit volume is given by:

$$W = W_{net} + W_{mix} + W_{op} \tag{9}$$

where

$$W_{net} = (1 - \phi) \cdot n \cdot \bar{N} \cdot k_B T \left(\frac{\beta'}{bN} r + \ln \frac{\beta'}{\sinh \beta'} \right) - \pi \cdot (J_e - 1) - \sum_{i=1}^{3} P_{ei} \cdot \lambda_{ei} \tag{10a}$$

$$W_{mix} = \frac{k_B T}{v_s} \cdot \left[(J_{sw} - 1) \ln\left(1 - \frac{1}{J_{sw}}\right) - \frac{\chi}{J_{sw}} \right], \quad J_{sw} = (J_{tot} - \phi \cdot J_{op})(1 - \phi)^{-1} \tag{10b}$$

and W_{op} is the energy associated to the switchable molecules open per unit volume. In the above relations, $\bar{N}$ is the number of segments of the representative chains depending on the distribution function $p(N)$, π is the hydrostatic pressure constraint, P_{ei} are the components of the applied external stresses, v_s is the volume of a single solvent molecule and χ is Flory's parameter that governs swelling. In (10a) $J_e = [1 + J_{tot} - J_{sw}(1 - \phi) - \phi(1 + J_{op})](1 - \phi)^{-1} \cong 1$ is the volume change of the polymer network due to external mechanical actions and n is the number of chains per unit volume; the chain stretch λ_p can be evaluated according to the 8-chain model [4] as $\lambda_p = \sqrt{(\lambda_1^2 + \lambda_2^2 + \lambda_3^2)/3}$, where $\lambda_1, \lambda_2, \lambda_3$ are the mesoscopic stretches of the polymer, while λ'_p is evaluated as:

$$\lambda'_p = J_{tot}^{1/3} \tag{11}$$

by using the total volume change J_{tot} of the polymer. Finally, λ'_p is used to calculate β', see Eq. (5).

The above stated problem, involving non-trivial mathematical functions, can be solved numerically by linearizing the stationary condition $\partial W_{net}/\partial \boldsymbol{F} - \boldsymbol{P}_e = 0$.

5. Numerical examples

The above presented model is used to simulate the mechanical response of a polymer with switchable molecules under the action of a pH change and under a mechanical stress history. The chains distribution $p(N)$ has been assumed to follow a Gaussian law, with an average value $\bar{N} = 100$

and deviation $\delta N = 5$; the problem's parameters adopted for all the simulations are as follows: shear modulus $\mu = nk_BT = 2$ MPa, $\chi = 0.6$, reaction rates without chemical or mechanical stress $k_{A0} = 10^{-6}$Hz, $k_{D0} = 10^{-4}$Hz.

5.1. Polymer with switchable molecules exposed to an acid environment

In the first example, the polymer with switchable molecules ($\langle J_m \rangle = 2.5$) is exposed to an acidic solvent. We assume that the stoichiometric ratio is equal to 1 (i.e. 1 mol of solvent is required to open 1 mol of switchable molecules), and that the acid concentration in the solvent is 0.1%; the volume fraction of the molecules has been assumed equal to be 2% (Fig. 2a) and 10% (Fig. 2b). The volume expansions of the polymer, J_{sw}, J_{op}, J_{tot} *vs* time are illustrated; it appears that the molecule expansion enhances the volume increase of the polymer. The fraction of open molecules $\tilde{h}_{op}$ increases in time more rapidly for $\phi = 2\%$ than for $\phi = 10\%$ because a higher quantity of acid is required at a given time to open the switchable molecules.

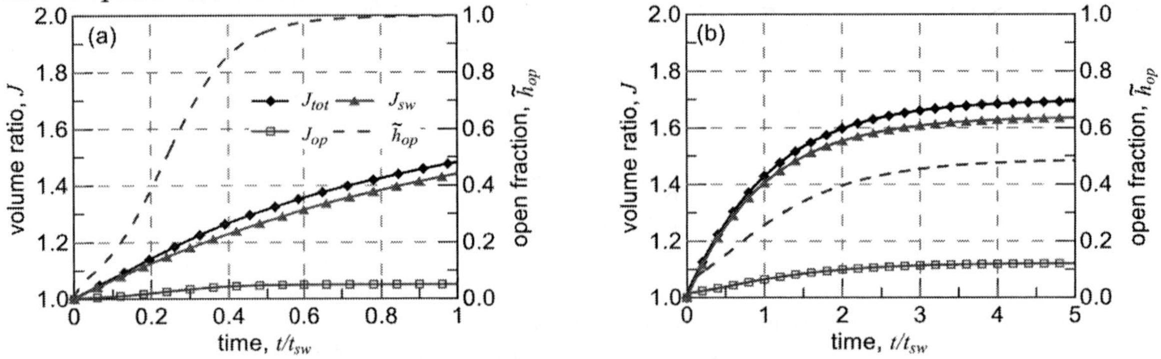

Figure 2. Evolution in time of the fraction of open switchable molecules for a polymer under the action of pH changing for $\phi = 2\%$ (a) and $\phi = 10\%$ (b).

5.2. Polymer with switchable molecules under a cyclic mechanical stress

In the second example, the polymer with switchable molecules (two cases are considered: $\langle J_m \rangle = 2.5$, and $\langle J_m \rangle = 8.0$) is subjected to a mechanical stress history; the volume fraction of the molecules has

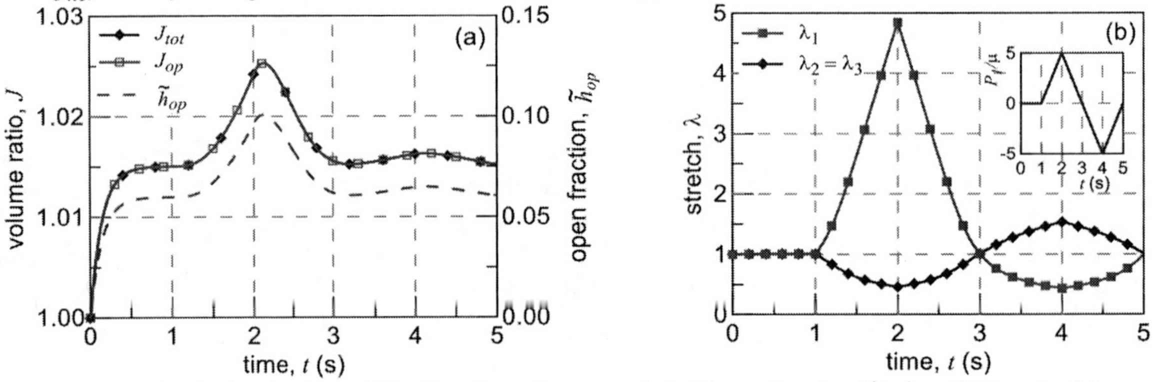

Figure 3. Evolution in time of the fraction of open switchable molecules ($\langle J_m \rangle = 2.5$) and of the volumetric deformation under the action of a cyclic mechanical stress (a). Longitudinal (λ_1) and transversal stretches (λ_2, λ_3) vs time (b).

been assumed equal to 2% and the resulting expansion and stretches are shown in Fig. 3 and in Fig. 4. Molecule expansion, triggered by the mechanical force, promotes a greater deformation of the material with respect to the case without switchable molecules.

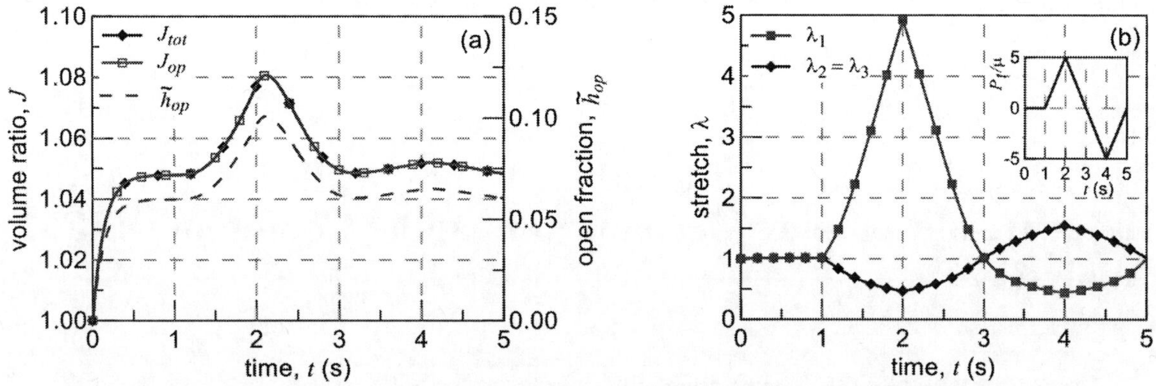

Figure 4. Evolution in time of the fraction of open molecules ($\langle J_m \rangle = 8.0$) and of the volume deformation under a cyclic stress (a). Longitudinal (λ_1) and transversal stretches (λ_2, λ_3) (b).

6. Conclusions

In the present paper we have considered the mechanics of a polymer network containing switchable molecules (mechanophores) that can assume only two different conformation states, close and open. The switching mechanism can be induced by a pH change or by mechanical stress. The theoretical framework, developed at the microscale level, accounts for the deformation induced in the polymer by both the switching of the molecules and by solvent uptake (swelling). The model has been readily upscaled to the continuum, and has been used to simulate the mechanical response of a polymeric material under the influence of a solvent which induces a pH change, and also under a mechanical stress history. The model is physically-based and is suitable to be straightforwardly implemented in a FE code for real simulations of elements made of smart responsive materials.

7. References

[1] Peplow, M. 2016 *Nature*, 536(7616), 266-267.
[2] Flory, P. J., & Rehner Jr, J. (1943). Statistical mechanics of cross-linked polymer networks I. Rubberlike elasticity. *J. Chem. Phys.*, 11(11), 512-520.
[3] Treloar, L. R. G. (1943). The elasticity of a network of long-chain molecules. I. Trans. Faraday Soc., 39, 36-41.
[4] Arruda, E. M., & Boyce, M. C. (1993). A three-dimensional constitutive model for the large stretch behavior of rubber elastic materials. *J. Mech. Phys. Sol.*, 41(2), 389-412.
[5] Boyce, M. C., & Arruda, E. M. (2000). Constitutive models of rubber elasticity: a review. Rubber Chem. Tech., 73(3), 504-523.
[6] Ogden, R. W. (1972, February). Large deformation isotropic elasticity-on the correlation of theory and experiment for incompressible rubberlike solids. In Proc. Royal Soc. London A: Math., *Phys. Eng. Sci.*, 326(1567), 565-584.
[7] Dai, S., Ravi, P., & Tam, K. C. (2008). pH-Responsive polymers: synthesis, properties and applications. *Soft Matter*, 4(3), 435-449.
[8] Li, J., Nagamani, C., & Moore, J. S. (2015). Polymer mechanochemistry: from destructive to productive. *Accounts Chem. Res.*, 48(8), 2181-2190.
[9] Manouras, T., & Vamvakaki, M. (2017). Field responsive materials: photo-, electro-, magnetic- and ultrasound-sensitive polymers. *Polym. Chem.*, 8(1), 74-96.
[10] Wei, M., Gao, Y., Li, X., & Serpe, M. J. (2017). Stimuli-responsive polymers and their applications. *Pol. Chem.*, 8(1), 127-143.
[11] Roy, D., Cambre, J. N., & Sumerlin, B. S. (2010). Future perspectives and recent advances in stimuli-responsive materials. *Prog. Polym. Sci.*, 35(1), 278-301.
[12] Hänggi, P., Talkner, P., & Borkovec, M. (1990). Reaction-rate theory: fifty years after Kramers. *Rev. Modern Phys.*, 62(2), 251-341.

IOP Conf. Series: Materials Science and Engineering **416** (2018) 012081 doi:10.1088/1757-899X/416/1/012081

Mechanical Response of Combined Thin-Walled Boxes under Influence of Complex Mechanical Loading Conditions - FEM Analysis

T Sadowski[1,*] and M Nowicki[1]

[1]Department of Solid Mechanics, Lublin University of Technology, Nadbystrzycka 40 str., 20-618 Lublin, Poland

[*]Corresponding author: t.sadowski@pollub.pl

Abstract. The subject of the analysis are thin-walled aluminum boxes connected by mechanical fasteners only or with the addition of cohesive layer. The analyzed elements have a great potential in the construction of aircraft engines as well as in other applications where thin-walled components are used. Cohesive connections, are now increasingly used in construction of structural components. Therefore, this paper presents results and comparison of selected thin-walled structure connected with mechanical connectors and cohesive layers. The main objective in this paper was to perform FEM analysis and to determine mechanical response of the selected thin-walled structure with different types of connections under quasi-static load. Aluminum boxes were subjected to a complex load, such as bending and twisting. The numerical model includes 2 damage processes in the: (1) aluminum, and (2) fasteners. This allowed us to determine locations of the most stressed, and most exposed to damage areas inside of the whole thin-walled box. The obtained results also allow to determine degree of usefulness of cohesive joints in the construction of the thin-walled structures under influence of quasi-static loads.

1. Introduction

Thin-walled box sections are commonly used in many fields of technology, such as aerospace industry, automotive industry, naval industry, civil engineering etc. These types of local structural elements connection and high level of energy absorption during load is very important for the whole engineering structures [1]. There are many joining techniques of thin-walled components, such as adhesion [2, 3], welding and spot welding [4-6], clinching [7-10], riveting [11], and combinations of the aforementioned [12-19], are used most frequently. The interesting types of connections are thin-walled beams and columns, e.g. [20-23].

This paper is focused on the influence of an adhesive layer on the equivalent stiffness and strength of the analyzed thin-walled aluminum boxes (figure 1) subjected to complex mechanical loading. The thin-walled FEM model is made of two aluminum parts. We considered 2 types of the boxes connected by:

- mechanical connectors only - the first case,
- hybrid connection consisting of the mechanical connectors and adhesive layers - the second case.

After analyzing several different numerical models it turned out that the most effective is to simplify 3-D exact geometry of the aluminium box with mechanical clips by replacing them with mechanical connectors, i.e. the "*connector*" type of joining available in the ABAQUS program, figures 1-3. Both boxes were subjected to complex mechanical loading i.e. bending and twisting, figure 4.

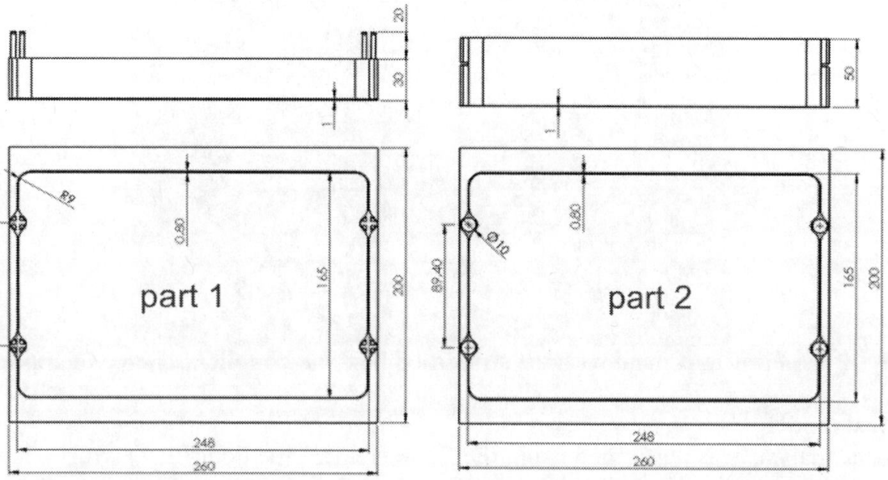

Figure 1. Geometries of the 2 parts forming of the thin-walled box

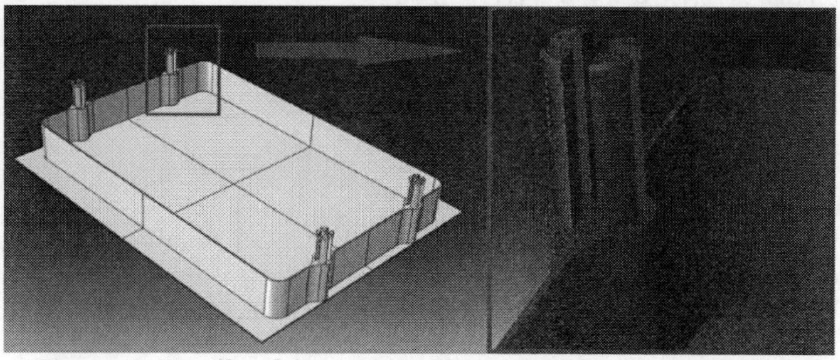

Figure 2. Details of the exact 3-D model of the thin-walled box

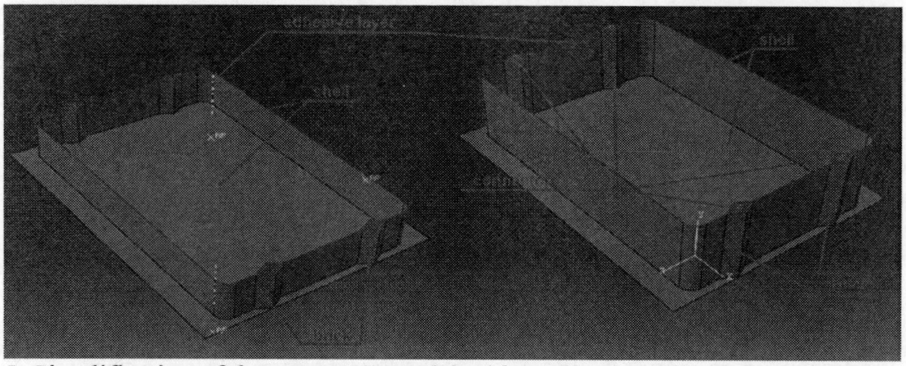

Figure 3. Simplification of the exact 3-D model with application of "*connector*" type of joining

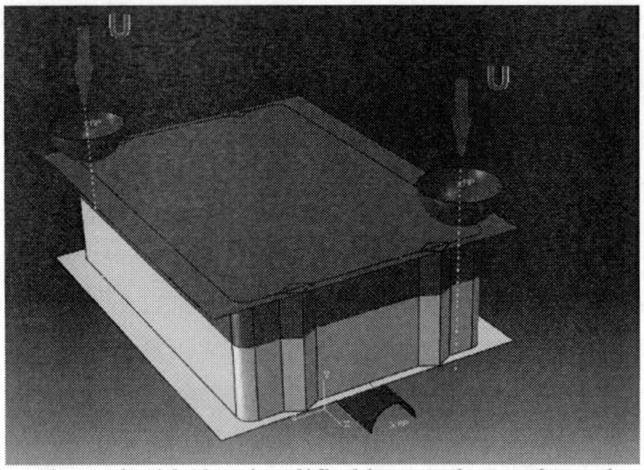

Figure 4. Experimental stand with the simplified box and complex scheme of applied loads

2. Numerical model

The numerical analysis was performed using the finite element method (FEA) with ABAQUS Explicit code. The complex numerical model consists of three types of elements:

- the 2 basic parts of the thin-walled made of aluminum 7075,
- 4 mechanical clips (figure 2) or their simplified version by 4 "*connectors*" (figure 3),
- cohesive layer created by adhesive Loctite 9514.

All of above components were modeled by various type of materials with different damage degradation processes. Geometry simplification of the 4 mechanical clips by corresponding mechanical connectors significantly accelerated calculation time and allowed getting the similar results with high level of accuracy.

2.1. Ductile damage model – aluminum

To describe aluminum parts of the boxes an elastic-plastic model with damage was used, e.g. [12-19, 25], figure 5. The damage parameter D starts to gradually increase from initial value 0, for equivalent fracture strain $\bar{\varepsilon}_0^{pl}$, to ended value 1, for equivalent fracture strain $\bar{\varepsilon}_f^{pl}$ at the time of complete damage.

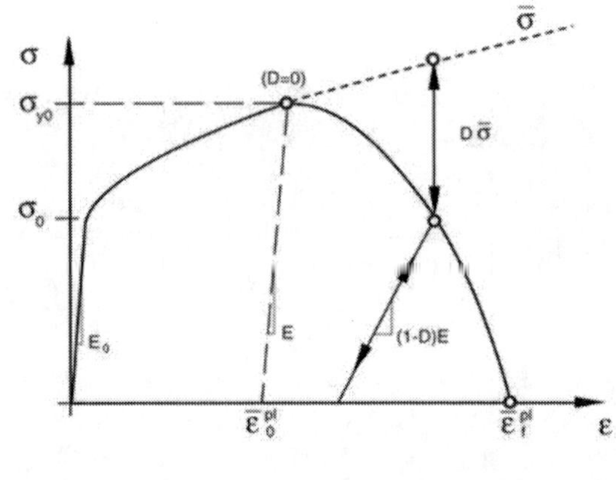

Figure 5. Ductile damage model (DDM) of the elasto-plastic material

The model assumes that the equivalent plastic strain $\bar{\varepsilon}_0^{pl}$ at the onset of damage is a function of stress triaxiality η and an equivalent plastic strain rate $\dot{\bar{\varepsilon}}_0^{pl}$. Stress triaxiality parameter is equal to $\eta = -p / q$, where p is pressure stress, and q is equivalent von Mises stress.

The criterion for damage initiation is met when:

$$\omega_D = \int \frac{d\,\bar{\varepsilon}^{pl}}{\bar{\varepsilon}_0^{pl}\left(\eta,\dot{\bar{\varepsilon}}_0^{pl}\right)} = 1, \tag{1}$$

where ω_D is a state variable that increases monotonically with the plastic deformations. When the $\bar{\varepsilon}_0^{pl}$ point is exceeded, damage develops. State of stress σ is a function of the damage parameter D, and stress due to undamaged response $\bar{\sigma}$, and it is equal to:

$$\sigma = (1-D)\bar{\sigma}\,. \tag{2}$$

The current value of the Young's modulus E is:

$$E = (1-D)E_0\,, \tag{3}$$

where E_0 is the initial value of the Young's modulus.

Material parameters of aluminum used in FEM analysis are shown in Table 1.

Table 1. Material parameters of aluminum

	Mass density [kg/m³]	Young`s modulus [GPa]	Poisson ratio [-]	Yield point [MPa]	Fracture energy [mJ]
Aluminum 7075	2700	69.6	0.34	470	22

2.2. Model of mechanical connector [25]

In order to simplify the 3-D thin-walled box the mechanical connector constitutive model (force-displacement characteristics) were determined with the FEA for the exact model subjected to different load cases. The FEA analyzes for: pulling, shearing and bending were made with the 3-D solid model, shown in figure 2. The basic strength characteristics of the connector model are collected in Table 2.

Table 2. Strength parameters of the connector

	P_0 [N]	P_{max} [N]	M_0 [Nm]	M_{max} [Nm]
pulling	423	1920	-	-
shearing	920	2250	-	-
bending	-	-	645	5740

To describe the constitutive behavior of the connector we adopted the *Cartesian* + *Cardan* type of joining. This combination allows determination of the joint properties for shifts in each direction and rotation for each axis. There is also the possibility of a detailed description of the connector in the strength context with regard to the elastic part, the plastic part and the damage process (figure 6). The description of the connectors work is limited to determining of the force-displacement relation.

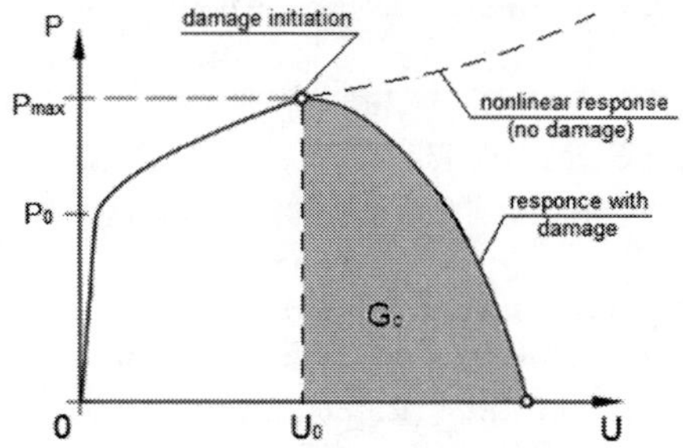

Figure 6. Constitutive behavior of the connectors

After elaboration the new connector model we created the simplified thin-walled box presented in figure 3, which was subsequently subjected to the complex state loading, figure 4.

2.3. Adhesive interface connection along edges of the parts 1 and 2
To describe the adhesive connection (figure 3), the surface-based cohesive behavior was used, e.g. [12-19, 25-35]. Negligibly small interface thickness using the traction-separation constitutive model (figure 7) was introduced between the 2 parts of the thin-walled box.

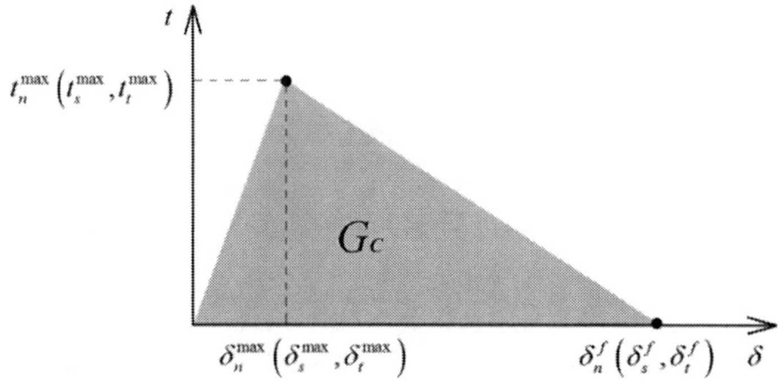

Figure 7. Traction-separation model for adhesive interface connection.

The cohesive behavior was modeled as a linear elastic traction-separation with damage. Figure 7 shown this model, were: δ is contact separation, t is contact stress (traction), and G_c is fracture energy. Traction is the function of contact force F and current area at each contact point A.

$$t = \frac{F}{A} \tag{4}$$

The lower indices in δ_n , δ_s and δ_t determine the direction: n - normal, s - shear 1, t - shear 2.
Damage initiation occurs when the maximum stress criterion is satisfied:

$$MAX\left\{\frac{\langle t_n \rangle}{t_n^{max}}, \frac{t_s}{t_s^{max}}, \frac{t_t}{t_t^{max}}\right\} = 1 \tag{5}$$

In this cohesive model damage evolution describes the degradation of the adhesive stiffness, and it is based on fracture energy G_c. Table 3 contains adhesive parameters adopted in the numerical model.

Table 3. Strength parameters of adhesive layer

	Mass density [kg/m³]	Young`s modulus [GPa]	Tensile strength [MPa]	Shear strength [MPa]	Fracture energy [mJ]
Loctite 9514	1440	1.46	44	45	0.9

3. Results
The numerical analyzes carried out allow to assess assumptions correctness in formulation of the simplified model. Figure 8 shows the distribution of forces versus displacement for the exact and simplified models. The force distributions show that the convergence is very good and a small difference occurs only in the maximum force values. However, the difference in time of calculations is huge, because the process of counting in the exact 3-D model is more than 40 times longer than in case of the simplified model.

By comparing the force distributions in figure 9, the influence of the adhesive layer on the quality of the joint can be clearly seen. A greater increase in stiffness in the hybrid connection is clearly visible in the initial loading stage. After exceeding the critical failure stresses in the cohesive connection model, the adhesive layer is totally damaged. Then the stiffness of the entire box is reduced and the system connected only by the mechanical connectors. For the analyzed model, the damage to the adhesive layer occurred at the displacement level of 6 mm.

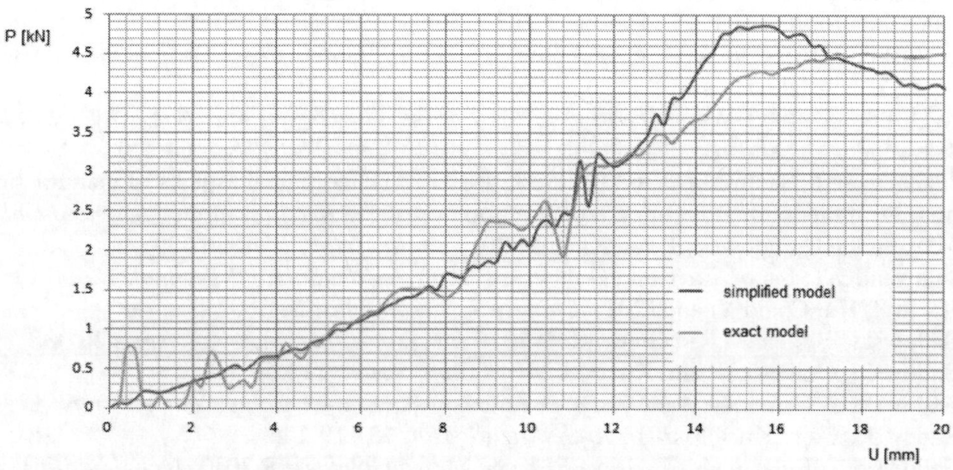

Figure 8. Force-displacement diagram. The simplified and the 3-D exact solid model.

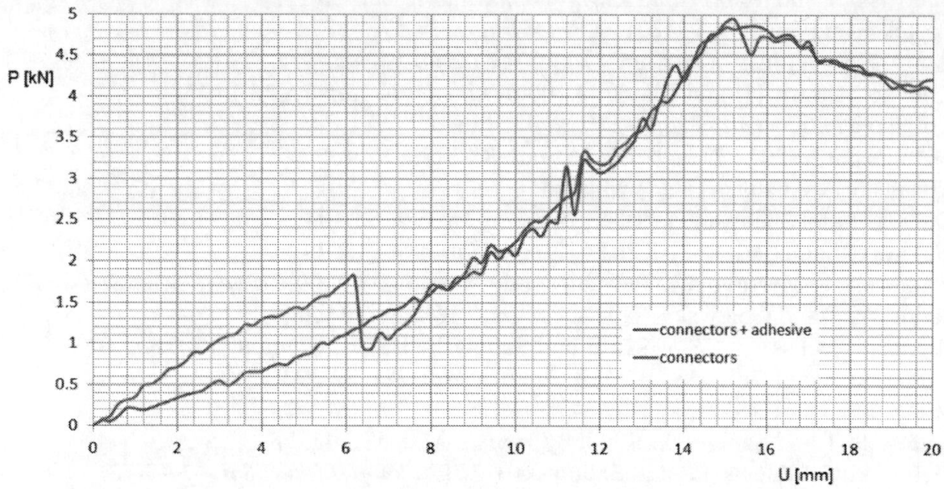

Figure 9. Force-displacement diagram. Two kinds of connection.

4. Conclusions

Summarizing, the following conclusion can be formulated:

- A simplified model of the complex shaped clips and boxes strongly accelerate the calculation time.
- The obtained results for the simplified model do not differ qualitatively from the results of the exact 3-D model.
- Simplified solutions give the opportunity to analyze the work of more complex and larger structures with low equipment costs.
- The application of the additional adhesive layer increases the stiffness of the element in the initial phase of the work of the thin-wall structure. But after failure of the adhesive interface, the structure response is the same for both cases. The use of the adhesive joint is justified in the initial deformation stage of the entire element.

IOP Conf. Series: Materials Science and Engineering **416** (2018) 012081 doi:10.1088/1757-899X/416/1/012081

5. References

[1] Alghamdi A 2001 *Thin-Walled Struct.* **39** 189
[2] Belingardi G and Scattina A 2013 *Int. J. Adhesion & Adhesives* **40** 31
[3] Yamashita M, Kenmotsu H, Hattori T 2013 *Thin-Walled Struct.* **69** 45
[4] Rusińskia E, Kopczyński A, Czmochowski J 2004 *J. Mater. Proc. Technol.* **157–158** 405
[5] Liu C, Zhang J X and Xue C B 2011 *Fusion Eng. Design* **86** 288
[6] Fu D, Zhou C, Li C, Wang G and Li L 2014 *Trans. Nonferrous Met. Soc. China* **24** 657
[7] Balawender T, Sadowski T and Golewski P 2012 *Comp. Mater. Sci.* **64** 270
[8] Balawender T, Sadowski T and Golewski P 2015 Technological aspects of manufacturing and numerical model ling of clinch-adhesive joint *Springer Brief in Applied Sciences and Technlogy*
[9] Balawender T, Sadowski T and Kneć M 2011 *Arch. Metall. Mat.* **56** 439
[10] Balawender T, Sadowski T and Golewski P 2011 *J. Adh. Sci. and Technol.* **25** 2391
[11] Kaifu Z, Hui C and Yuan L 2011 *Chinese J. Aeronautics* **24** 369
[12] Sadowski T, Kneć M and Golewski P 2010 *Int. J. of Adhesion & Adhesives* **30** 338
[13] Sadowski, Golewski P and Kneć M 2014 *Comp. Struct.* **112** 66
[14] Sadowski T, Golewski P and Zarzeka-Raczkowska E 2011 *Comp. Mater. Sci.* **50** 1256
[15] Sadowski T and Kneć M 2013 *Arch. Metall. Mat.* **58** 119-125
[16] Sadowski T, Balawender T, Śliwa R, Kneć M and Golewski P 2013 *Arch. Metall. Mat.* **58** 163
[17] Sadowski T, Kneć M and Golewski P 2014 *J. Adhesion* **90** 346
[18] Sadowski T and Marsavina L 2011 *Comput. Mat. Sci.* **50** 1336
[19] Postek E and Sadowski T 2011 *Comp. Interf.* **18** 57
[20] Liu Q, Xing H, Ju Y, Ou Z and Li Q 2014 *Comp. Struct.* **117** 1
[21] Kotelko M, Lim TH and Rhodes J 2000 *Thin-Walled Struct.* **38** 179
[22] Kotełko M 2004 *Thin-Walled Struct.* 42, 153
[23] Elchalakani M 2014 *Thin-Walled Struct.* **82** 278
[25] Abaqus 6.11 Analysis User`s Manual.
[26] Dębski H and Sadowski T 2014 *Comput. Mat. Sci.* **83** 403
[27] Gajewski J and Sadowski T 2014 *Comp. Mater. Sci.* **82** 114
[28] Sadowski T, Marsavina L, Peride N and Craciun EM 2009 *Comput. Mat. Sci.* **46** 687
[29] Sadowski T and Golewski P 2012 *Comput. Mat. Sci.* **64** 285
[30] Bieniaś J, Dębski H, Surowska B and Sadowski T 2012 *Comput. Mat. Sci.* **64** 168
[31] Sadowski T and Golewski P 2012 *Comput. Mat. Sci.* **52** 293
[32] Golewski G and Sadowski T 2014 *Const. Build. Mat.* **51** 207
[32] Sadowski T and Samborski S, 2008 *Comput. Mat. Sci.* **43**, 75
[33] Golewski G, Golewski P and Sadowski T 2012 *Comp. Mater. Sci.* **62** 75-78.
[34] Marsavina L and Sadowski T 2007 *Int. J. Frac.* **145** 237
[35] Sadowski T and Samborski S 2003 *J. Am. Cer. Soc.* **86** 2218

Acknowledgments

Financial support of Structural Funds – European Regional Development Funds (ERDF) Project No: INNOLOT/I/5/NCBiR/2013; program “INNOLOT – Innovative Aviation” coordinated by the National Center for Research and Development (NCBiR – Poland); Title: “Block Structures – Mechanical joining innovations to replace conventional fasteners in aerostructures” Period: 1.12.2013 – 31. 11.2018 is gratefully acknowledged.

This work was financially supported by Ministry of Science and Higher Education within the statutory research number S/20/2018.

Influence of the Process Parameters on the Shape and Dimensions of the Parts Obtained by Selective Laser Sintering (SLS)

D I Stoia[1,*], L Marsavina[1] and C Vigaru[1]

[1] Politehnica University Timisoara, Mihai Viteazul no.1, Timisoara, Romania

[*]Corresponding athor: dan.stoia@upt.ro

Abstract. Selective laser sintering is an additive technology that uses the energy of a laser beam together with an additional heating source in order to generate the high enough temperatures for particle bonding. Due to the multitude of process parameters that have to be controlled during the process, inconsistencies in shape, size and mechanical properties of the parts occurs in all additive manufacturing technologies. The purpose of the paper was to quantify the influence of the part position along the front-back (Y) axis of the machine, on the thickness and cross section tapper angle of parts. The material used in study was Alumide and the sinterization process was accomplished on Formiga P 100 machine. The idea of the study relay on experimental observations conducted along the time. These indicate that the geometry of the part is significantly influenced by the positioning of the models in the building plane. The statistic data processing prove that the parts are significantly different from the geometrical point of view when grows in different locations of the building plane.

1. Introduction

The additive technologies became wide spread both in industry and among hobbyist. Independently on the type of technology that is used, the accuracy of the parts represents a challenging research issue. The inconsistencies in shape, size and mechanical properties of the parts are associated with a series of technological parameters that are specific to each technology [1-7].

Galantucci et al approaches the analysis of dimensional performance of the parts obtained by mean of two additive systems that both use fused deposition modeling technique. Using a digital microscope they measure three orthogonal dimensions of the parts in order to establish the relationship between the technological factors and the parts dimensional error. The differences between the real and nominal specimen dimensions were in the range of 2-3 tenths of millimeters [8].

P.J. Nuñez et al analyze several elements of parts obtained by fused deposition modelling. They focus on dimensional precision, flatness error and surface texture. The purpose of the study was to establish the exact tolerance and surface finish that this process can offer. The poor dimensional control of the part was associated with physical contractions during curing. These lead to deviations in order of hundreds of millimeters for a certain layer thickness. An important observation was that decreasing the layer thickness to minimum, the deviations grows then times, in order of tenths of millimeters [9].

Tobias Lieneke et al proposes a methodical procedure designed to determine the dimensional tolerances of the parts obtained by FDM process. The tolerances were derived from measurements acquired using a micrometer screw. By comparing with ISO 286-1 standard, the tolerances of the manufactured parts reaches IT classes between 09 and 14. All the deviations are in the range of 2-3 tenths of millimeters [10].

Giovanni Moroni et al propose a new approach in establishing the tolerances of additive manufactured parts. The approach combines the current tolerancing practices with an X-ray scanning method. The volumetric representation is still based on voxels but it is enriched by the product and

IOP Conf. Series: Materials Science and Engineering **416** (2018) 012082 doi:10.1088/1757-899X/416/1/012082

manufacturing information. The volumetric representation of the measured points versus nominal geometric coordinates produces a confirmation of the correct tolerance field [11].

Goodridge R. D. et al concatenate some studies on the factors that influence the selection and selective laser sintering of the polymers. Factors like layer thickness, energy density, part orientation and placement, powder degradation and reuse, post build cooling rate, temperature distribution inside the machine were considered to influence the mechanical properties and geometry of the parts. One of the conclusions was the absence of commercial available system (machine) that solves the inconsistencies in dimensions and properties, caused by the uneven temperature distribution [12].

G. Berti et al conduct mechanical testing in order to characterize the composite PA- Al2O3, manufactured by selective laser sintering. After building a series of samples in different orientation respecting to the transversal axis of the machine they subject them to the mechanical testing. The samples present evident anisotropy in the growing direction (z-axis) and seem to be not sensitive to the sintering x and y directions at the room temperature. In addition, they conclude that the effect of sintering direction becomes more evident at higher temperatures [13].

This study underlines the geometrical differences of the parts according to the front-back (Y axis) direction of the machine. The idea of the study relay on the experimental results observed along the time, and which indicates that the parts that grows closer to the machine's door, have better dimensional stability.

2. Materials and methods

Selective laser sintering is an additive technology that uses the energy of a laser beam in order to generate the high enough temperatures for particle bonding (figure 1). The parts were build on EOS Formiga P 100 machine, all in the same conditions, the single process variable being the location of the parts in the building environment (figure 2). By keeping all other process parameters constant (energy density, layer thickness, material, scaling factors, type of laser scanning, building and removal chambers temperatures, cooling conditions and so on), the differences in geometrical characteristics of the parts can be put on the part positioning in the building envelope.

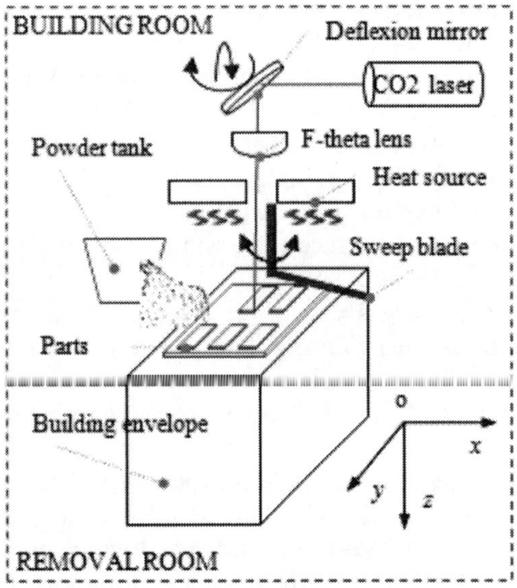

Figure 1. SLS Machine structure [14].

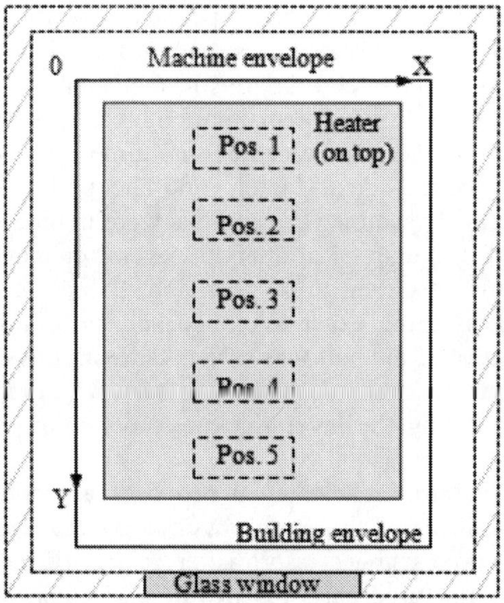

Figure 2. Part positioning in building envelope.

The material used was *Alumide*. This is an EOS product that consists of homogeneous mechanical mixture of polyamide 12 (PA12) particles and aluminium whiskers. During sinterization process, the polyamide part of the mixture fuses together gripping inside the aluminium whiskers. The result of this process is a still porous part but with a better density than the polyethylene alone. The presence of aluminium gives to the object a better rigidity and facilitates coatings and polishing. Beside these, better machinability, superior mechanical properties and more reliable dimensional accuracy can be encounter in alumide.

The sample part was designed in Solidworks 2013 according to the ASTM D638-14 standard. In the figure 3 and 4 the sample shape can be observed, together with the nominal rectangle cross section A-A and real cross section obtained on manufactured parts. Having different widths on button and top surfaces, the tapper angle can be calculated for the section. The full tapper angle (2α) was computed (equations 1 and 2) based on W1 and W2 widths of the section and the height T of the section (figure 4), all measured in the cross sectional plane of the sample part.

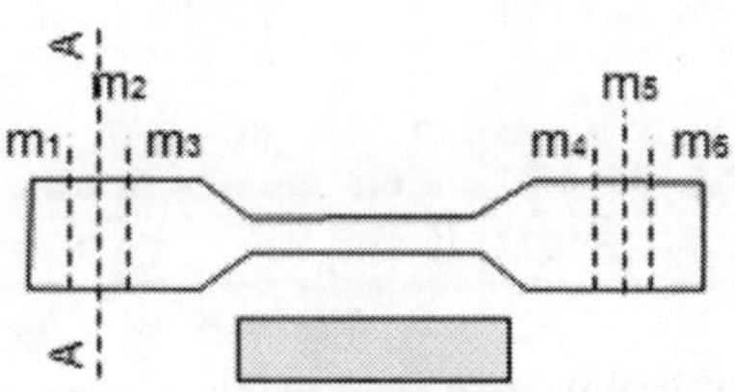

Figure 3. Measurement points and cross section of the sample part.

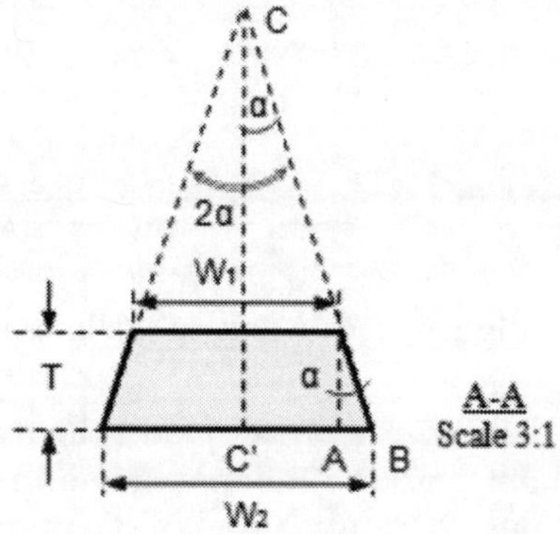

Figure 4. Real A-A section of the sample part. Scale 3:1.

$$tan\ \alpha = \frac{AB}{T} = \frac{(W_2 - W_1)}{2} \cdot \frac{1}{T} \tag{1}$$

$$2\alpha = 2\text{atan}\left(\frac{W_2 - W_1}{2 \cdot T}\right) \tag{2}$$

A total number of 15 parts were manufactured. The part distribution was on three layers. In the first building layer the parts were oriented along the X axis of the machine. In the second layer (5 mm on the top) the parts orientation was 45 degrees to the X axis while in the third layer the parts were oriented along the Y axis (90 deg to the X axis). Every layer contains 5 samples which run from position 1 to 5. The first position was always closer to the machine center while the fifth position was closed to the machine's access door. The option of growing parts at different orientation was decide in order to eliminate the relation between the measurements and a certain part orientation.

The measurements were conducted using a digital caliper that has a 0.01 mm resolution. Nine sections were measured on each of the samples (figure 3), resulting 27 measurement values for every dimension (T, W1 and W2) at each sample position.

3. Results and discussion

Raw data of thickness and full tapper angle were presented in the figures 5 and 6. The thickness is a representation of measured data while the tapper angle is a computational value that takes into account both widths variables. Observing the data distribution on both graphs, for every part positioning, no evident tendency can be identified. Therefore, statistic data processing is required.

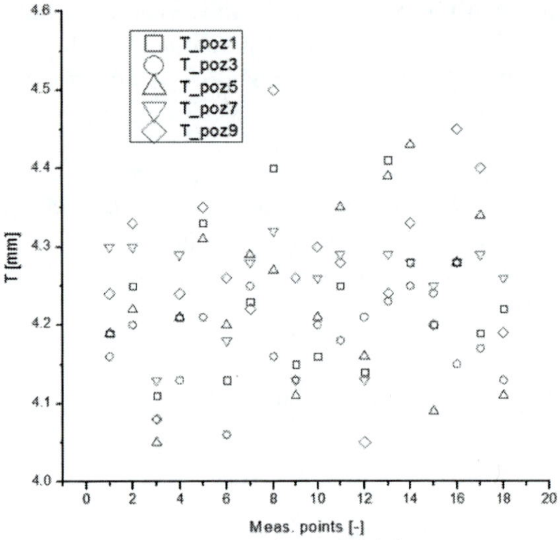

Figure 5. Raw data of thickness.

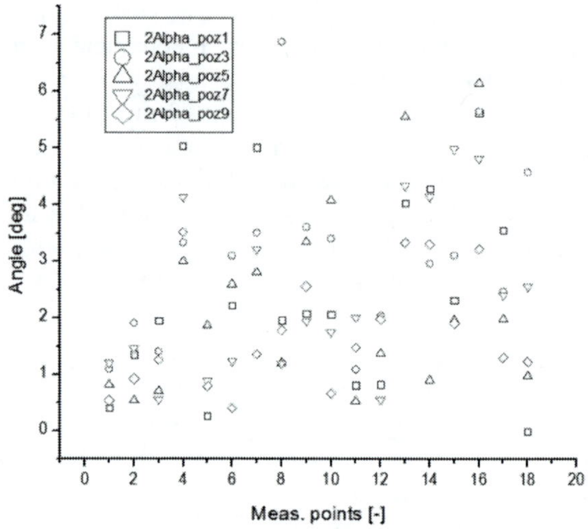

Figure 6. Raw data of full tapper angle.

For the 3D geometrical model the nominal values were: 4.20 mm for thickness and 0° for tapper angle, both constant. Before sending the 3D model to the machine this was scaled by the following coefficients: 2.3% for X and Y directions and 1.3% for Z. The scaling coefficients were applied for compensating the shrinkage that occurs during cooling phase, in order to obtain results as close as possible to the nominal dimensions.

The normal percentiles were computed using the thickness and tapper angle data of all sample parts. These graphs (figure 7 and 8) show a good data grouping around the mean (small standard deviation) and also a normal distribution. The percentiles of thickness indicate that 40% of the values are smaller than the nominal value 4.2 mm while all the other 60 % are larger. This may lead to the conclusion that the scaling factors used (X2.3%, Y2.3% and Z1.3%) are overestimating the shrinkage. On the other hand, there is no percentage of parts that possessed zero degree of tapper angle.

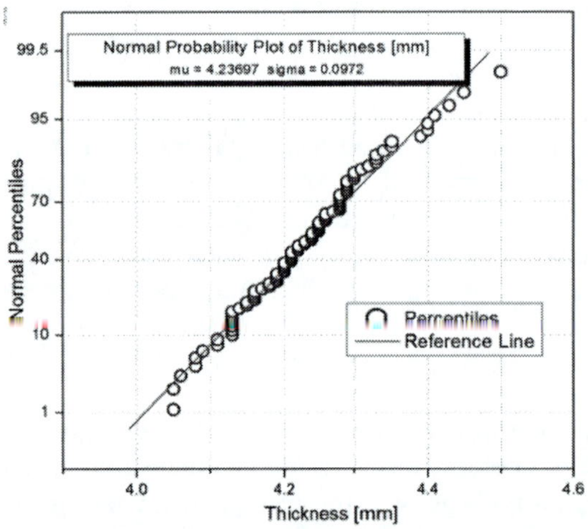

Figure 7. Normal percentiles of thickness.

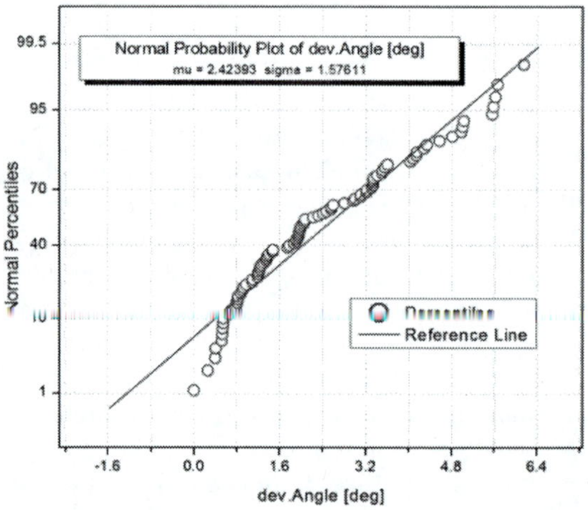

Figure 8. Normal percentiles of tapper angle.

In order to establish the variation of the thickness and tapper angle with the position in the building plane, the data were sorted on five series: Pos. 1 to 5. Under the *position 1* data series the nine measurements of three parts can be found. The results were averaged and the standard deviation was computed. By following the same protocol for all the five positions, the data were processed. The graphic representation of these can be found in the figures 9 and 10.

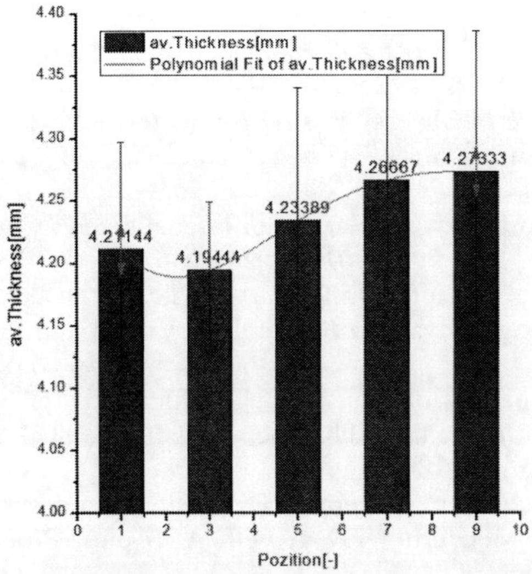

Figure 9. Bar chart and regression for thickness.

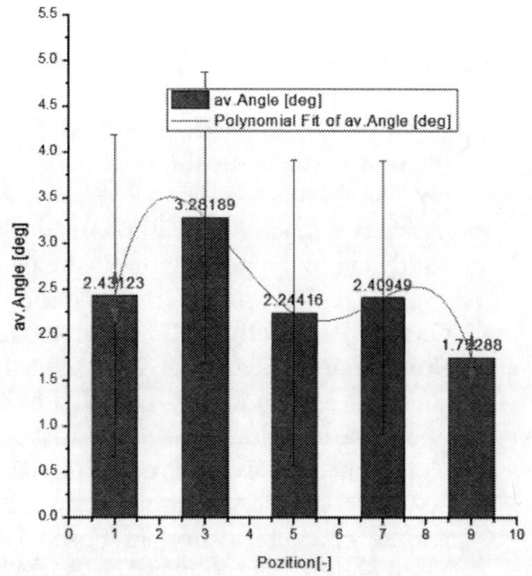

Figure 10. Bar chart and regression for tapper angle.

The polynomial 4th order regression lines are fitting the data with an $R^2 = 1$. The polynomial regression is accepted in this case because the edge of the intervals corresponds to parts that grown close to the physical border of the building envelope.

By analyzing the data distribution and the standardized residuals (not presented in the paper) for W1, W2 and T dimensions, a Gaussian data distribution was confirmed.

The statistical difference of thickness and tapper angle between the first and the fifth positions of the parts was put in evidence by the probability test. The null hypothesis H0 was defined: there is no significant difference in parts thickness/tapper angle recorded for the positions 1 and 5. The alternative hypothesis Ha was defined: the thickness/tapper angle difference is significant.

The results of p values: 0.198866 for thickness and 0.166779 for tapper angle are both higher than 0.05 and therefore H0 is rejected and the alternative hypothesis Ha is accepted.

4. Conclusions

This study underlines the geometrical differences of the alumide parts obtained by selective laser sintering according to the front-back (Y axis) direction of the P100 Formiga machine. The results represent a case study and the generalization to all types of SLS machines have to be done with cautious.

The 40% of thickness are smaller than the nominal 4.2 mm value while all the other 60 % are larger.

Both thickness and tapper angle of the parts are significantly influenced by the front-back (Y axis) positioning of the models in the building plane, as statistics indicate. Also, the two parameters exhibit a reverse effect.

5. References

[1] Salmoria G.V, Ahrens C.H, Klauss P, Paggi R.A, Oliveira R.G., Lago A 2007 Rapid manufacturing of polyethylene parts with controlled pore size gradients using Selective Laser Sintering, Materials Research, Vol. **10**, No. 2, 211-214.

[2] Pham D.T, Gault R.S 1998 A comparison of rapid prototyping technologies, Int. J. Mach. Tool.

Manu. **38** 1257–1287.

[3] Es-Said O. S, Foyos J, Noorani R, Mendelson M, R. Marloth, BA Pregger, Effect of Layer Orientation on Mechanical Properties of Rapid Prototyped Samples, Mater. Manuf. Process. **15**(2000) 107-122.

[4] Mousa A. A, 2016, Experimental investigations of curling phenomenon in selective laser sintering process, *Rapid Prototyping Journal*, Vol. **22** Iss 2 pp. 405 – 415

[5] Pilipović A, Valentan B, Šercer M 2016 Influence of SLS processing parameters according to the new mathematical model on flexural properties, *Rapid Prototyping Journal*, Vol. **22** Iss 2 pp. 258 – 268.

[6] Griessbach S, Lach R,. Grellmann W, Structure–property correlations of laser sintered nylon 12 for dynamic dye testing of plastic parts, Polym. Test. **29**:8 (2010) 1026-1030.

[7] Yeong W Y, Chua C K, Implementing Additive Manufacturing for Medical Devices: A Quality Perspective, High Value Manufacturing: Advanced Research in Virtual and Rapid Prototyping, CRC Press, 2014, 115-120

[8] Galantucci M, Bodi I, Kacani J, Lavecchi F Analysis of dimensional performance for a 3D open-source printer based on fused deposition modeling technique, 3rd CIRP Global Web Conference, Elsevier, Procedia CIRP **28** (2015) 82 – 87

[9] Nuñez P J, Rivas A, García-Plaza E, Beamud E, Sanz-Lobera E Dimensional and surface texture characterization in Fused Deposition Modelling (FDM) with ABS plus, Procedia Engineering **132** (2015) 856 – 863, Elsevier

[10] Lieneke T, Denzer V, Guido A, Adam O, Zimmer D Dimensional tolerances for additive manufacturing: Experimental investigation for Fused Deposition Modeling, Elsevier, Procedia CIRP **43** (2016) 286 – 291

[11] Moroni G, Petro S, Polini W Geometrical product specification and verification in additive manufacturing, CIRP Annals - Manufacturing Technology **66** (2017) 157–160

[12] Goodridge R D, Tuck C J, Hague R J M, Laser sintering of polyamides and other polymers, Progress in Materials Science **57** (2012) 229–267

[13] Berti G, D'Angelo L, Gatto A, Iuliano L 2010 Mechanical characterisation of PA-Al2O3 composites obtained by selective laser sintering, *Rapid Prototyping Journal* **16**(2010) 124 – 129

[14] Toth-Taşcău M et al., Influence of the Energy Density on the Porosity of Polyamide Parts in SLS Process, Solid State Phenomena, Vol. **188**, pp. 400-405, 2012

IOP Conf. Series: Materials Science and Engineering **416** (2018) 012083 doi:10.1088/1757-899X/416/1/012083

The Effect of Zinc Stearate Modification on Aging Characteristics of Asphalt Binder

P Hajikarimi[1,*], H Shahsavari[2], M Rahi[3], F Sharif[4] and F M Nejad[1]

[1]Hafez Avenue, Department of Civil & Environmental Engineering, Amirkabir University of Technology, Tehran, Iran
[2]Narmak, Department of Civil Engineering, Iran University of Science & Technology, Tehran, Iran
[3]Shahid Tondgooyan Blvd., R&D Department, Pasargad Oil Company, Tehran, Iran
[4]Hafez Avenue, Department of Polymer & Color Engineering, Amirkabir University of Technology, Tehran, Iran

[*]Corresponding author: phajikarimi@aut.ac.ir

Abstract. Different types of modifiers have been investigated to improve aging characteristics of asphalt binders. In this research, the effect of zinc stearate modification has been studied on aging resistance of asphalt binder. For this purpose, viscous flow properties and other rheological characteristics of neat and modified asphalt binders have been evaluated by using conventional experimental tests and superpave protocol test. In order to evaluate rheological behavior of asphalt binder in presence of zinc stearate, a neat asphalt binder sample with three modified samples containing 0.5, 1 and 2 wt% of zinc stearate were aged by means of the Rolling Thin Film Oven Test (RTFOT). The Brookfield rotational viscosity (RV) was implemented at six different temperatures from 100 to 200 degree of centigrade. Furthermore, the dynamic shear rheometer (DSR) was used for evaluating complex modulus (G*) and phase angle (δ) of all neat and modified samples at 10, 20, 30, 40, 50 and 60°C in frequency sweep mode. To evaluate effect of adding zinc stearate by using viscosity results, proportion of asphalt binder's viscosity after aging to its viscosity before aging was defined as viscosity aging index (VAI). Results showed that adding zinc stearate to neat asphalt binder reduces the viscosity aging index. Results of frequency sweep test indicated that zinc stearate causes the values of complex shear modulus in aging asphalt binder approach to neat state which means the effect of zinc stearate modification is positive for improving aging resistance of asphalt binder.

1. Introduction

Aging of asphalt binder during its life cycle from mixing with aggregates to fully destruction has always been one of the main concerns of researchers in the field of pavement's materials and design [1,2]. With respect to such a phenomenon, the initial rheological characteristics of asphalt binder will be changed which in some cases improves and in some other cases weakens the performance of the asphalt pavement [3]. For example, loss of volatile and light oils at high service temperatures leads to hardening asphalt binder due to aging and increases the rutting resistance of asphalt binders. However, at low service temperature the asphalt binder becomes harder and more brittle and the risk of cracking increases [4]. Generally, controlling and decreasing the effect of aging on rheological characteristics of unaged asphalt binder were topics of numerous researches in the field of asphalt binder modification.

Numerous modifiers have been carried out to increase aging resistance of asphalt binder. Some of these modifiers have been used as a rejuvenator of recycled asphalt mixture and some others as an anti-aging to enhance asphalt binder's rheological and mechanical performance. Zinc dialkyldithiophosphate (ZDDP), carbon black, octabenzone (UV531) and bumetrizole (UV326) are some well-known substances that have been used as an anti-aging agent. However, the effect of these anti-aging agents is not significant on the properties of asphalt binders [5]. Nanotechnology has been gradually penetrated into the field of asphalt modification [6]. Fini et al. proved that the addition of bio-binder which was derived from waste swine manure can reduce the viscosity of the asphalt binder. Furthermore, they indicated that bio-binder has the potential to improve thermal cracking performance of conventional asphalt binders [7].

In this research, zinc stearate was used as an additive to enhance anti-aging properties of original asphalt binder. For this purpose, one neat asphalt binder sample and three modified ones containing 0.5, 1 and 2 wt% of zinc stearate were aged by means of the rolling thin film oven test (RTFOT) and then two classical tests were performed including penetration test and softening point. The Brookfield rotational viscosity (RV) was also carried out on these samples. Generally, the viscosity of aged asphalt binder is greater than un-aged asphalt binder. In this paper, the viscosity index has been calculated by dividing viscosity of aged asphalt on viscosity of un-aged asphalt which obviously increases due to the loss of volatile and light oils. This index will be somewhat larger than one. Therefore, observing the viscosity index can be a suitable method for assessing the effect of zinc stearate on the aging characteristics of asphalt binder. In addition, investigation of rheological behavior of modified asphalt binder can provide useful information on the behavior of asphalt binder in presence of zinc stearate. For this purpose, the dynamic shear rheometer (DSR) was used at test temperatures between 10 and 60°C.

2. Experimental program

2.1. Materials

In this study samples were prepared using asphalt binder with penetration grade of 60/70 (the common asphalt binder in Iran) acquired from the Tehran Petroleum Refinery of Pasargad Oil Company located in Tehran, Iran. The Zinc Stearates (ZS) is a micronized and hydrophilic white powder that used in its formal sense is soap. This industrial-grade material is derived from the reaction of fatty acids with a combination of zinc metal. This material widely used in industrial, which derived from the reaction of fatty acids with a combination of zinc. Stearates are used as a lubricant for the polymer industries. Polymers are long chain molecules that are viscous and sticky in their high melting points. The friction force between polymer-polymer, polymer-metal, polymer-fillers, fillers-fillers and fillers-metal causes difficulty for polymers flow. The appropriate solution for aforementioned problem is reducing the coefficient of surfaces friction. Among several types of lubricants, zinc stearate is an important additive in the process of polymers production. It is used as an internal lubricant for molding. Figure 1 shows appearance shape at the room temperature and the molecular structure of zinc stearates.

2.2. Preparation of the samples

In this research, there are four different asphalt samples including BASE, ZS05 (0.5%), ZS1 (1%) and ZS2 (2%) which are fabricated at 140°C by mixing neat asphalt binder and specified percentage of zinc stearate for 45 min with a low shear mixer.

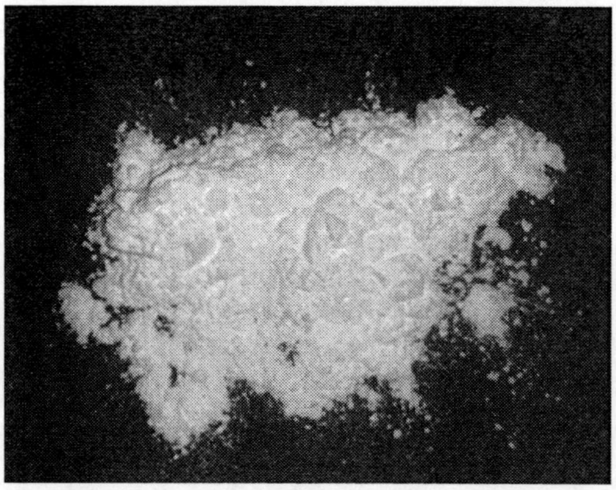

(a)

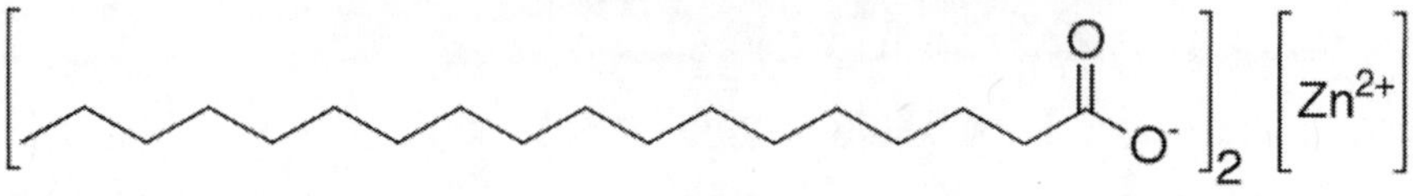

(b)

Figure 1. (a) Appearance and (b) chemical structure of zinc stearate

To simulate the short-term aging (hardening or oxidation) characteristics of the asphalt binders' samples, the RTFO test was used. Therefore, two samples of asphalt binder, including neat and RTFO-aged asphalt were obtained. In the RTFO test, the asphalt binder specimens should be heated inside the main container so that the temperature does not exceed 150°C. Then Pour 35 g of asphalt with a precision of 0.001 into eight sample bottles. The asphalt binder samples are heated up in a 163°C oven until it is completely fluid and pourable. The samples are kept in the oven, on temperature, with air flow of 400 min/ml and the carriage rotating for 85 min. When mass-change bottles have cooled to a safe handling temperature, their weights are measured to the nearest 0.001 g and recorded.

2.3. Test methods

In this study, the most important classical tests, including the ASTM D5 test (penetration point) and softening point Test (ASTM D36), which are well-known and used in the industrial literature and consultants and contractors have been performed on all samples. These experiments were carried out on unaged samples. In addition to the conventional tests, the superpave protocol experiments included rolling thin film oven test (RTFOT) and rotational viscosity (RV) were implemented at 100, 120, 135, 160, 180 and 200°C. Furthermore, the dynamic shear rheometer (DSR) was used for evaluating complex modulus (G*) and phase angle (δ) of all neat and modified samples in frequency sweep mode from 0.1 to 100 Hz. The diameter of the test sample was 25 mm with thickness of 1 mm. Both experiments were carried out in two cases of aged and unaged neat and modified samples.

3. Results and Discussion

3.1. Conventional experimental tests

Table 1 indicates results of conventional experimental tests. As it can be seen adding zinc stearate has no significant effect on physically characteristics of asphalt binder including its penetration grade and softening point. Therefore, it can be concluded that addition of stearates does not change physical properties of asphalt binder including consistency and heat sensitivity.

IOP Conf. Series: Materials Science and Engineering **416** (2018) 012083 doi:10.1088/1757-899X/416/1/012083

Table 1. Penetration and softening point of neat and modified asphalt binders

Sample Code	Penetration (0.1 mm)	Softening Point (°C)
BASE	61	48.4
ZS05	60	48.5
ZS1	64	48.7
ZS2	59	49.5

3.2. *Superpave protocol test*

Figure 2 depicts results of rotational viscosity test based on the viscosity aging index (VAI). This index can be calculated according to the following equation:

$$\text{Viscosity Aging Index} = \frac{\text{Viscosity}_{Aged\ Binder}}{\text{Viscosity}_{Unaged Binder}} \quad (1)$$

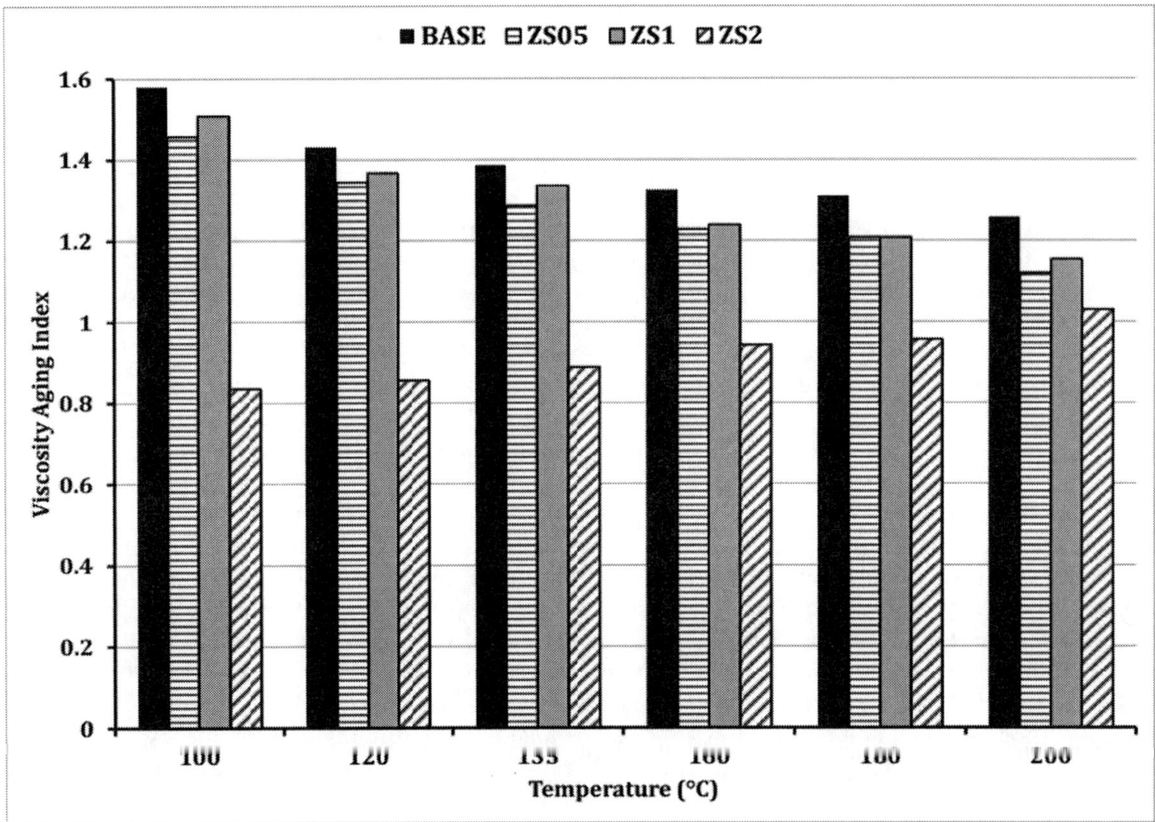

Figure 2. Viscosity versus temperature for neat and zinc stearate modified asphalt binder

Figure 2 illustrates that the addition of zinc stearate can reduce the viscosity aging index. It is apparent that increasing the percentage of ZS causes further reduction of VAI. For 2% of ZS this index becomes less than 1. These results proved that introducing ZS to neat asphalt binder can control and reduce the effects of aging and cracking. On this basis, it can be concluded that ZS as a modifier can enhance the aging characteristics of asphalt binder.

The dynamic shear rheometer (DSR) was used for evaluating complex modulus (G*) and phase angle (δ) of all neat and modified samples at 10, 20, 30, 40, 50 and 60°C in the frequency sweep mode. The results clearly showed in figure 3, 4, 5 and 6. These figures illustrate that increasing of temperature led to decreasing of complex modulus (G*). As it can be seen in these figures, the graph of unaged binder and RTFO aged modified samples converge to each other that means the positive effect of this modifier on the hardening control of asphalt binder.

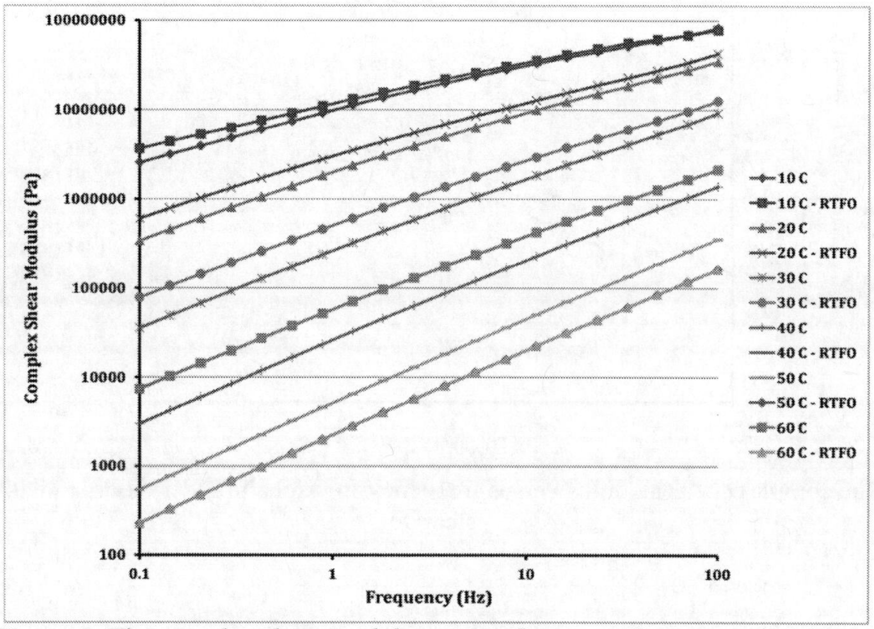

Figure 3. The complex shear modulus versus frequency for neat asphalt binder

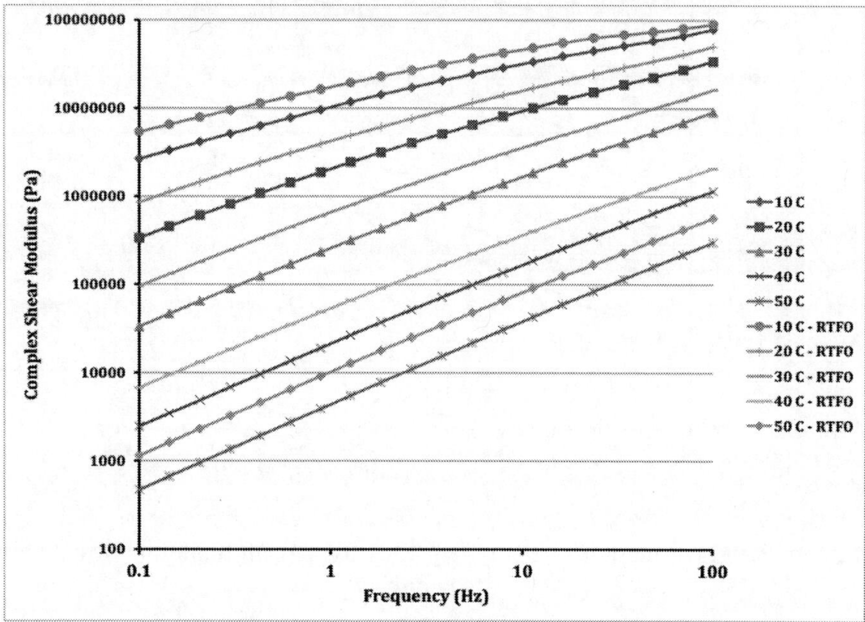

Figure 4. The complex shear modulus versus frequency for asphalt binder modified with 0.5% of zinc stearate

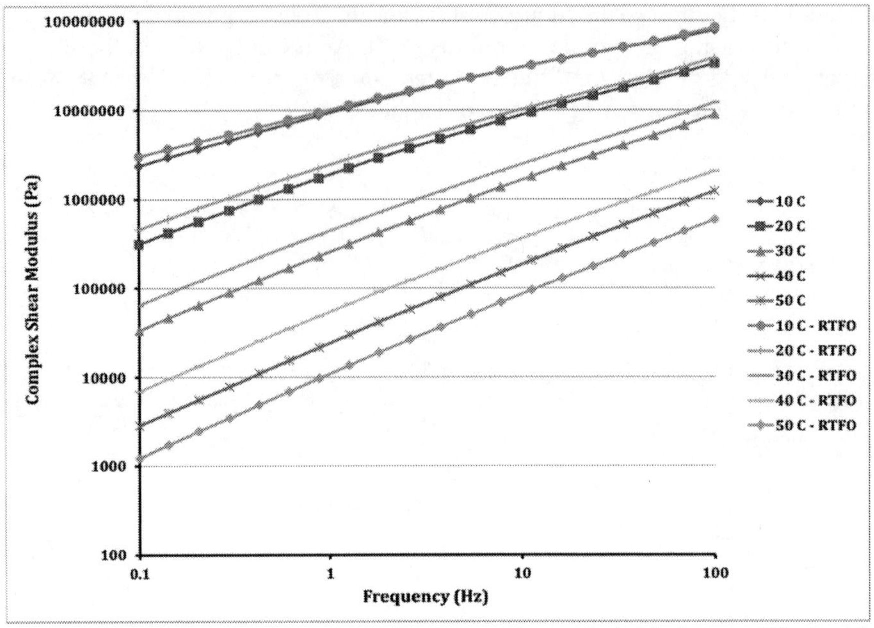

Figure 5. The complex shear modulus versus frequency for asphalt binder modified with 1% of zinc stearate

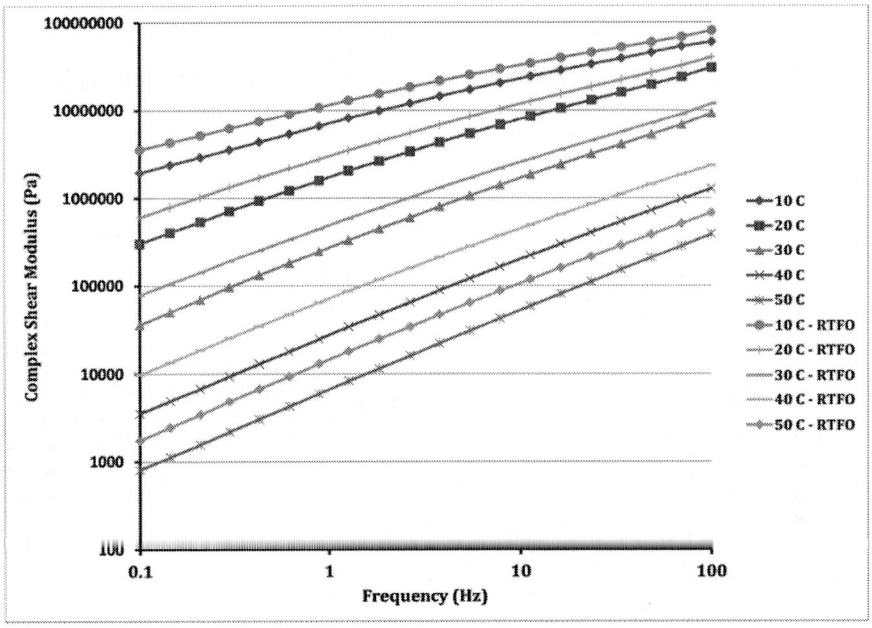

Figure 6. The complex shear modulus versus frequency for asphalt binder modified with 2% of zinc stearate

For example in figure 5 which is related to the addition of 1% by weight of stearates to neat asphalt binder, it is clearly visible that for three temperatures of 10, 20 and 30°C, the complex shear modulus of neat sample and modified samples containing 0.5% of ZS approach to each other that confirms the results of the rotational viscosity test. Furthermore, according to the set of figures it is clear that the addition of stearates, particularly at 1% and 2%, causes the graphs to shift upwards which means increasing the stiffness and the complex shear modulus of the modified sample relative to neat asphalt binder.

4. Conclusion

The effect of zinc stearate as a modifier on aging resistance of asphalt binder was discussed in this paper. The following conclusions can be drawn:

- Zinc stearate as a modifier can be concluded that addition of stearates does not change physical properties of asphalt binder including consistency and heat sensitivity.
- Addition of zinc stearate can reduce the viscosity aging index. It means that positive effect of this modifier on aging characteristics of asphalt binders.
- Results of frequency sweep test indicated that zinc stearate causes the values of complex shear modulus in aging asphalt binder approach to neat state which means the effect of zinc stearate modification is positive for improving aging resistance of asphalt binder.
- the results of a complex shear modulus illustrate that adding zinc stearates, especially in 1 and 2% by weight increases the asphalt binder stiffness

5. References

[1] Martin KL, Davison RR, Glover CJ and Bullin JA 1990 Asphalt aging in Texas roads and test sections. *Transportation Research Record* (1269)
[2] Zhang ZQ, LIANG XL and Li P 2005 Evaluation method of asphalt aging properties [J]. *Journal of Traffic and Transportation Engineering*
[3] Petersen, J. C., Robertson, R. E., Branthaver, J. F., Harnsberger, P. M., Duvall, J. J., Kim, S. S., Anderson, D. A., Christiansen, D. W. and Bahia, H. U. 1994 "Binder Characterization and Evaluation. Volume 1: Binder characterization and evaluation." *Strategic Highway Research Program*
[4] Bahia, H. U., Hanson, D. I., Zeng, M., Zhai, H., Khatri, M. A. and Anderson, R. M. (2001). "Characterization of modified asphalt binders in superpave mix design." *National Cooperative Highway Research Program*
[5] Cong P, Wang J, Li K, Chen S. 2012 *Fuel* 97:678-84
[6] Fang C, Yu R, Liu S and Li Y. 2013 *Journal of Materials Science & Technology*. **29**(7):589-94
[7] Oldham DJ, Fini EH and Chailleux E 2015 *Construction and Building Materials* **86**:75-84

IOP Conf. Series: Materials Science and Engineering **416** (2018) 012084 doi:10.1088/1757-899X/416/1/012084

Simultaneous hot isostatic pressing and solution annealing of aluminum cast alloys followed by instantaneous aging at elevated temperatures

S. Hafenstein[1] **and E. Werner**[1]

[1] Technical University of Munich, TUM Department of Mechanical Engineering, Institute of Materials Science and Mechanics of Materials, Garching b. München, Germany

E-mail: `hafenstein@wkm.mw.tum.de`

Abstract. Age-hardenable aluminum - silicon - magnesium cast alloys are widely applied in automotive and aircraft industries. Hot isostatic pressing is a common method to increase the materials fatigue resistance by reducing the inner porosity of aluminum cast materials. In the presently followed industrial production route hot isostatic pressing is performed preceding the standard heat treatment consisting of solution annealing and aging. In previous work Hafenstein et. al. demonstrated the possibility to achieve an oversaturated condition with a sufficient amount of magnesium and silicon atoms dissolved in the aluminum matrix by performing rapid temperature changes in advanced hot isostatic presses. Hence, it becomes possible to combine hot isostatic pressing and solution annealing within a single process step and to perform aging at elevated temperatures immediately after hot isostatic pressing without the necessity of separate solution annealing. The present study evidences that the design of a fully integrated process which comprises not only hot isostatic pressing and solution annealing but also aging at elevated temperatures is possible thereby shortening further the overall processing time. It is shown that the benefit with respect to strength in such a combined process is due to the shortening of the time at room temperature between solution annealing and aging at elevated temperatures.

1. Introduction

Pores are major sites for the initiation of fatigue cracks in aluminum cast material. Due to thermally activated dislocation creep and diffusion creep processes, inner porosity of aluminum cast material can be reduced by hot isostatic pressing (HIP) [1–4]. Hot isostatic pressing is conventionally done preceding to the regular heat treatment, which is composed of solution annealing, quenching and aging. This ensures high and reproducible component quality and meets the high demands on fatigue resistance for applications in the automotive and aircraft industries [5]. The high strength of age-hardened aluminum cast alloys is achieved by solution annealing, followed by quenching and aging at elevated temperatures which is performed within two separate process steps following hot isostatic pressing [6]. An oversaturated condition of magnesium and silicon atoms dissolved in the aluminum matrix is required in order to form a disperse distribution of various precipitates which increase the materials strength during artificial aging [7–10]. Magnesium and silicon atoms will remain as substitutional alloying elements within the aluminum matrix and form an oversaturated condition if the quenching rate after solution annealing is high enough to avoid diffusion during quenching. Due to the limited capability to perform rapid temperature changes within a standard hot isostatic press, densification, homogenization and quenching cannot be performed within a single step [11, 12]. Conducting three separate process steps is a significant cost factor opposing a broad industrial exploitation of such production routes. It has been shown that an increase of the cooling rate after hot isostatic pressing can ensure a sufficient amount of magnesium and silicon atoms dissolved in the aluminum matrix to achieve age-hardenability of the aluminum cast material [13,14]. Therefore, the costs of a production route which includes hot isostatic pressing could be possibly reduced by combining hot isostatic pressing and solution annealing in a single process step [13]. It is shown that instantaneous aging after hot isostatic pressing offers benefits with respect to strength by avoiding pre-aging at room temperature as is the case in standard heat treatment routes. The aim of the present study is to investigate the possibility of a combined process comprising hot isostatic pressing, solution annealing and aging. The findings of this work include results of mechanical testings and microstructural analysis performed on hot isostatically pressed samples of the aluminum cast alloy AlSi7Mg0.3 (A356).

2. Experimental

2.1. Casting

Samples of the alloy AlSi7Mg0.3 were produced via sand casting by Georg Fischer AG, Schaffhausen. Figure 1 shows a schematic illustration of the cast samples. The melt was degassed and refined by the addition of strontium. All alloying elements were added to the melt at a temperature of 730 °C. The chemical composition of the Alloy A356 is defined in the standard *EN 1706:2010* [15]. The measured composition is given in Table 1 and was analyzed by spark emission spectroscopy using a SPECTROMAXx LMM16 spark emission spectrometer (Spectro, Germany). The melt was exposed to a gassing procedure in order to achieve an enhanced casting porosity of $V_P = 0{,}17\,\%$. Heat treatment and hot isostatic pressing was performed on cylinders possessing a diameter of 16 mm and a length of 70 mm which were machined from the cast material.

Table 1. Chemical analysis of the aluminum cast alloy AlSi7Mg0.3

Element	wt.-%		Deviation wt.-%
Si	7.323	±	0.008
Fe	0.125	±	0.002
Cu	0.022	±	0.001
Mn	0.029	±	0.001
Mg	0.345	±	0.001
Ni	0.005	±	0.001
Zn	0.088	±	0.072
Ti	0.125	±	0.003
Al	balance		

2.2. HIP process and heat treatment

The mechanical properties of the alloy AlSi7Mg0.3 were determined in as-cast and various hot isostatically pressed conditions. Samples of the series *HIP 540°C 2h - 0.2K/s* and *HIP 540°C 2h - 0.2K/s - A* were subjected to conventional hot isostatic pressing at 540 °C for 120 min without subsequent solution annealing. The quenching rate within the conventional HIP process was

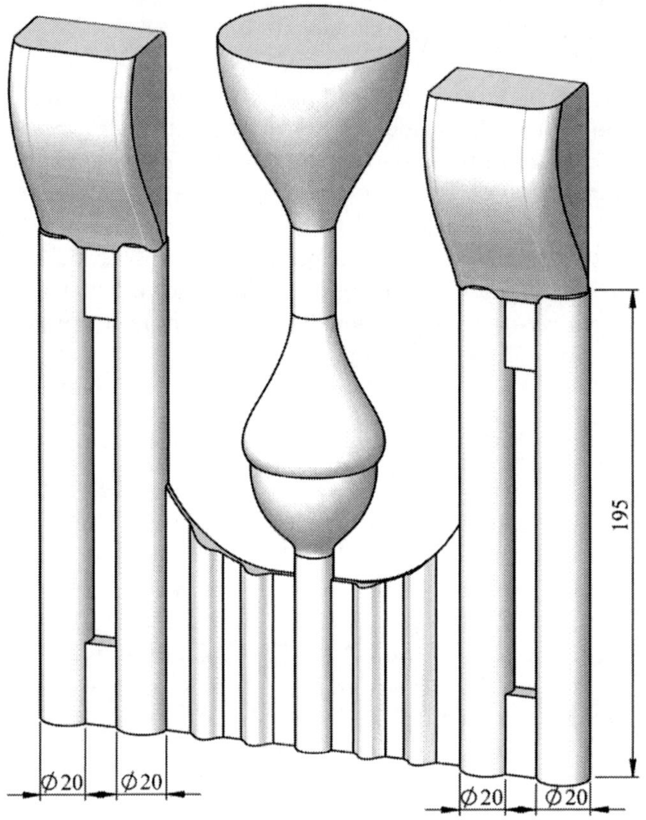

Figure 1. Schematic illustration of the samples produced by sand casting; units in mm

about 0,2 K/s within the temperature range between 540 and 200 °C. Samples of the conditions *HIP 540°C 2h - 0.2K/s* were hot isostatically pressed and aged for more than seven days at room temperature. Samples of the hot isostatically pressed condition *HIP 540°C 2h - 0.2K/s - A* were also pre-aged at room temperature and then aged at 165 °C for 150 min. The temperature over time profile of the two heat treatment conditions is shown in Figure 2. Samples of the heat treatment conditions *HIP 540°C 2h - 7K/s* and *HIP 540°C 2h - 7K/s - A* were hot isostatically pressed at 540 °C for 120 min and quenched with a rate of 7 K/s. Samples of the conditions *HIP 540°C 2h - 7K/s* were hot isostatically pressed and aged for more than seven days at room temperature. Samples of the hot isostatically pressed condition *HIP 540°C 2h - 7K/s - A* were additionally aged at 165 °C for 150 min. The temperature over time profile of the two heat treatment conditions *HIP 540°C 2h - 7K/s* and *HIP 540°C 2h - 7K/s - A* is shown in Figure 3. Samples of the heat treatment conditions *HIP 540°C 2h - 7K/s - DA* were hot isostatically pressed at 540 °C for 120 min, quenched with a rate of 7 K/s after hot isostatic pressing and then instantaneously aged at 165 °C for 150 min. The temperature over time profile of this direct-age (DA) condition is shown in Figure 4.

IOP Conf. Series: Materials Science and Engineering **416** (2018) 012084 doi:10.1088/1757-899X/416/1/012084

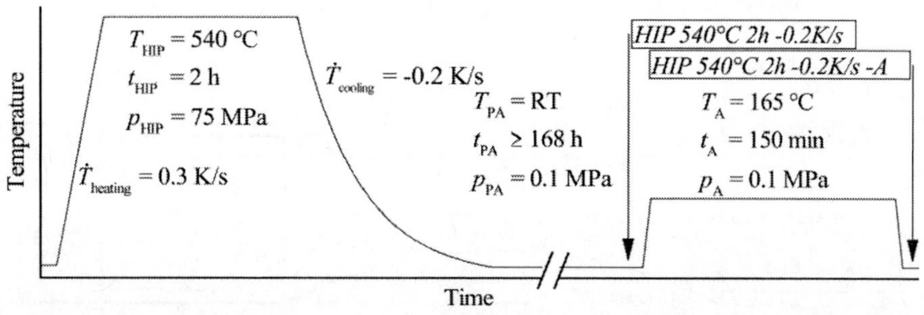

Figure 2. Temperature over time profile for the heat treatment conditions *HIP 540°C 2h - 0.2K/s* and *HIP 540°C 2h -0.2K/s -A*.

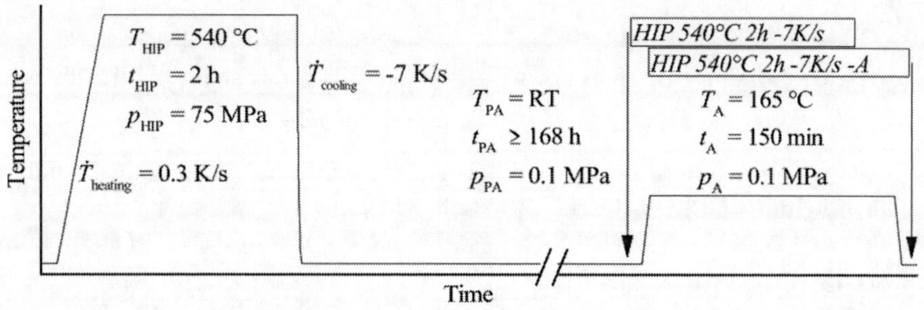

Figure 3. Temperature over time profile for the direct-age heat treatment condition *HIP 540°C- 2h -7K/s* and *HIP 540°C 2h -7K/s -A*.

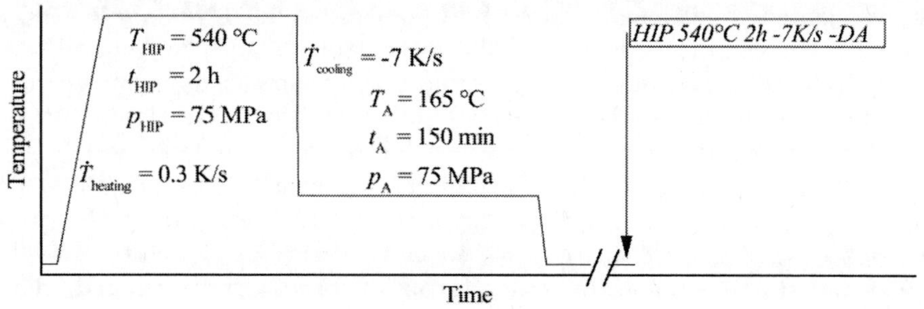

Figure 4. Temperature over time profile of the heat treatment conditions *HIP 540°C 2h -7K/s - DA*.

2.3. Mechanical testing

The sample shape was chosen according to DIN 50125:2004-01, type B, size M10 [16]. Tensile testing was carried out on an Instron 4505 tensile testing machine connected to a computer using the operating software Bluehill 2. An extensometer was used to measure the strain parallel to the cylinder axis. The tests were performed at a strain rate of $\dot{\varepsilon} = 10^{-4}\,\mathrm{s}^{-1}$. Round fatigue test samples, with dimensions as shown in Figure 5, were machined from the center of the heat treated cylinders. Fatigue resistance testing was carried out on a RUMUL high frequency resonant testing machine (Russenberger Prüfmaschinen AG, Switzerland). Due to the limited capacity of the HIP vessel, the fatigue resistance of the hot isostatic pressed conditions was compared at a single alternating stress level (R = -1, $\sigma_{\mathrm{m}} = 0\,\mathrm{MPa}$). The maximum amplitude was set at $\sigma_{\mathrm{a}} = 75\,\mathrm{MPa}$.

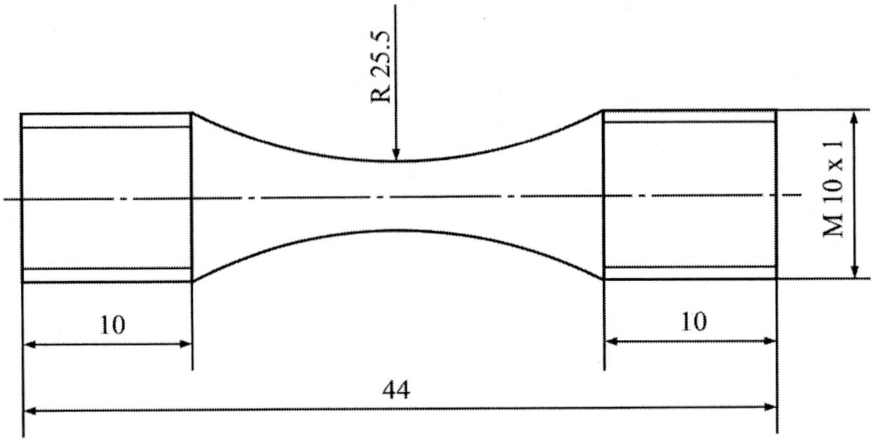

Figure 5. Drawing of the samples used for fatigue testing. Units in mm.

Hardness measurements where performed according to DIN EN ISO 6506-2 using an M4U-025 hardness testing machine (EMCO-TEST, Austria) [17].

3. Results and Discussion

Porosity affects different material properties with various intensities [18]. It is known that the porosity has a minor effect on the material's yield stress and hardness when compared to its effect on ultimate tensile strength, elongation at fracture and fatigue resistance [19, 20]. The fatigue resistance and the elongation at fracture of the *as-cast* sample series were found to be the lowest of all investigated conditions, whereas hardness and yield stress where at least as high as higher level then the ones of the conventionally hot isostatically pressed conditions *HIP 540°C-2h-0.2K/s* and *HIP 540°C 2h-0.2K/s-A*. This indicates that the quenching rate during sand casting is the same or even higher when compared to the standard HIP treatment. The preferred starting point for crack initiation is the interface between the matrix and large precipitates or pores inside the material [2, 21–24]. The number of cycles for crack initiation in fatigue resistance tests is significantly larger than the number of cycles during the growth of these cracks. The low fatigue resistance of the *as-cast* sample series is therefore a result of the high number of preferred starting points for crack initialization. It can be noted, that the fatigue resistance and the elongation at fracture is significantly increased by hot isostatic pressing even thought the yield stress, the hardness and the ultimate tensile strength of the conventionally hot isostatically pressed conditions *HIP 540°C 2h-0.2K/s* and *HIP 540°C 2h-0.2K/s-A* are as low as for the *as-cast* condition. The reason for the low yield stress, hardness and ultimate tensile strength of the conventionally hot isostatically pressed conditions can be found in the microstructural properties of these conditions, which are influenced by the low cooling rate of 0.2 K/s in the conventionally hot isostatic pressing process and the resulting distribution of alloying elements. A low cooling rate goes along with long exposure times at high temperatures, which results in stable and coarse precipitates of the alloying elements silicon and magnesium formed at elevated temperatures [25]. Therefore there are not enough silicon and magnesium atoms in solid solution which could form needle-like β'' precipitates during artificial aging that are known to be the main cause of strength improvement [26, 27]. These results indicate that the quenching rate of 0.2 K/s, following densification in the conventional hot isostatic pressing process is not high enough to produce a suitable oversaturated solid solution. The application of a second heat treatment step dedicated to artificial aging at 165 °C for 150 min leads to a slight increase of the properties yield stress, hardness, ultimate tensile strength, but causes lower values

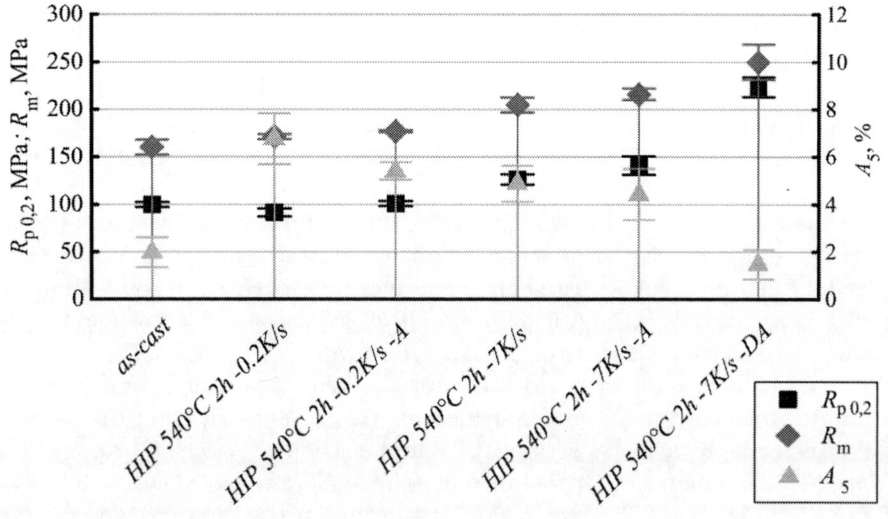

Figure 6. Tensile test results ($R_{p\,0,2}$, R_m and A_5) of all investigated samples corresponding to the performed thermal treatment; the error bar represents the standard deviation between three conducted measurements per heart treatment condition.

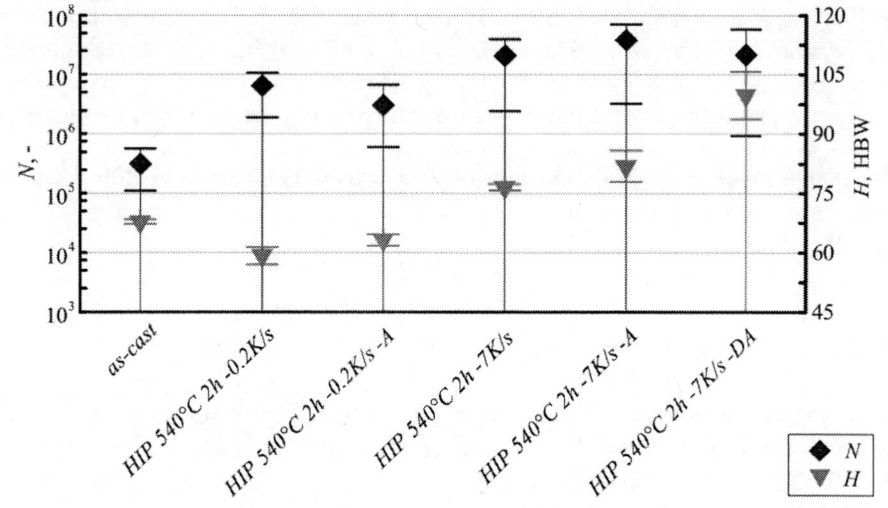

Figure 7. Cycles until failure (N) and Brinell Hardness (H) of all investigated samples corresponding to the performed thermal treatment; the error bar represents the standard deviation between three conducted measurements per heart treatment condition.

for the elongation at fracture.

With an increase of the cooing rate after hot isostatic pressing from 0,2 to 7 K/s, the yield stress of the conditions aged at room temperate (*HIP 540°C 2h -0.2K/s* and *HIP 540°C 2h -7K/s*) is increased by 38 % from 91 to 126 MPa. Concurrently the ultimate tensile strength increases by 19 % from 172 to 204 MPa and hardness by 31 % from 59 to 77 HBW. The high strength of the *HIP 540°C 2h -7K/s* heat treatment condition causes a significant higher fatigue resistance and a lower elongation at fracture when compared to the conventional condition *HIP 540°C 2h -0.2K/s*. After artificial aging at 165 °C for 150 min the heat treatment condition *HIP 540°C 2h -7K/s - A* exhibits a yield stress of 140 MPa, an ultimate tensile strength of 116 MPa and a hardness of

82 HBW. Therefore the strength level as well as the fatigue resistance of the artificially aged condition is higher than they are after room temperature aging, whereas elongation at fracture is slightly lower. It has been demonstrated previously that a quenching rate of 7 K/s after hot isostatic pressing is high enough to archive an oversaturated solid solution and enables aging to be performed after hot isostatic pressing without the necessity of separate solution annealing [13,14]. Artificial aging at 165 °C for 150 min was applied directly after hot isostatic pressing for the heat treatment condition *HIP 540°C 2h - 7K/s - DA* such samples exhibit the highest mechanical properties. When compared to the heat treatment condition *HIP 540°C 2h - 7K/s - A* which was artificially aged after one week of pre-aging at room temperature but with the same aging parameters, the heat treatment condition *HIP 540°C 2h - 7K/s - DA* shows superior mechanical properties. The yield stress of the directly aged condition *HIP 540°C 2h - 7K/s - DA* is 224 MPa and therefore 58 % higher than the yield stress of the heat treatment condition *HIP 540°C 2h - 7K/s - A*. Ultimate tensile strength and hardness of this direct-age condition are 250 MPa and 100 HBW, and respectively, are also significantly higher than that of the pre-aged samples series *HIP 540°C 2h - 7K/s - A*. The fatigue resistance of *HIP 540°C 2h - 7K/s - DA* is at the same level as that of *HIP 540°C 2h - 7K/s - A*. The elongation at fracture of the directly aged condition is found to be the lowest of all investigated sample series.

The influence of pre-aging at room temperature on the resulting mechanical properties of an aluminum alloy has been subject of previous investigations [28–30]. Short instantaneous aging of 5 min at 150 °C can reduce the detrimental effect of room temperature pre-aging on the resulting mechanical properties [31]. It has been shown hat the influence of pre-aging depends on the alloying composition as well as on the duration and the temperature [32–36]. The detrimental effect is based on the clustering of silicon and magnesium atoms during pre-aging [37]. The composition of the clusters formed in the early stage of the precipitation sequence depends on the temperature during their formation. The composition of the clusters affects the resulting mechanical properties of the alloy by promoting or restraining the formation of strength improving β'' precipitates during artificial aging which are the main source of strength improvement [38, 39].

4. Summary and Outlook

Quenching rates of 7 K/s within the temperature range between 540 and 200 °C are high enough to archive a suitable oversaturated solid solution after hot isostatic pressing. A fully integrated process which comprises not only hot isostatic pressing and solution annealing but also aging at elevated temperatures shortens the overall process time. Due to the clustering of silicon and magnesium atoms during room temperature pre-aging, a combined process with immediate aging at elevated temperatures possesses benefits with respect to strength. Taking into account the outstanding mechanical properties achieved by direct aging at elevated temperatures, even lower cooling rates than 7 K/s could be high enough to archive adequate material strength.

References

[1] Brummer M, Hoffmann H and Werner E 2010 *Proceedings of the 12th International Conference on Aluminium Alloys, September 5-9, Yokohama, Japan* ed Kumai S, Umezawa O, Takayama Y, Tsuchida T and Sato T pp 1095–1100 ISBN ISBN978-4-905829-11-9
[2] Wang Q, Apelian D and Lados D 2001 *Journal of Light Metals* **1** 73–84 ISSN 14715317
[3] Wang Q G and Davidson C J 2001 *Journal of Materials Science* **36** 739–750
[4] Graf W 2008 *Zeitschrift für Werkstoffe, Wärmebehandlung, Fertigung* **63**
[5] Lei C S C, Frazier W E and Lee E W 1997 *The Journal of The Minerals, Metals & Materials Society* **49** 38–39

[6] Altenpohl D 2013 *Aluminium und Aluminiumlegierungen* (*Reine und angewandte Metallkunde in Einzeldarstellungen* vol 19) (Berlin and New York: Springer-Verlag) ISBN 9783662302453
[7] Dutta I and Allen S M 1991 *Journal of Material Science Letters* **10** 323–326
[8] Andersen S J, Zandbergen H W, Jansen C, Tundal U and Reiso O 1998 *Acta Materialia* **46** 3283–3298
[9] Jacobs M H 1972 *Philosophical Magazine* **26** 1–13 ISSN 0031-8086
[10] Edwards G A, Stiller K, Dunlop G L and Couper M J 1998 *Acta Materialia* **46** 3893–3904
[11] Brummer M 2013 *Wärmebehandelndes Heißisostatisches Pressen von Aluminiumgusslegierungen* Dissertation Technische Universität München München
[12] Ostermeier M 2009 *Heißisostatisches Pressen von Aluminium- und Magnesiumguß* Dissertation Technische Universität München München
[13] Hafenstein S, Brummer M, Ahlfors M and Werner E 2016 *HTM Journal of Heat Treatment and Materials* **71** 117–124 ISSN 1867-2493
[14] Hafenstein S, Brummer M, Ahlfors M and Werner E 2016 *Giesserei Praxis* **7-8** 316–321
[15] DIN EN 1706:2010 Dezember 2013 Aluminium und Aluminiumlegierungen – Gussstücke – Chemische Zusammensetzung und mechanische Eigenschaften
[16] DIN 50125 Juli 2009 Prüfung metallischer Werkstoffe – Zugproben
[17] DIN EN ISO 6506-2 April 2016 Metallische Werkstoffe - Härteprüfung nach Brinell - Teil 2: Überprüfung und Kalibrierung der Prüfmaschinen
[18] Houria I M, Nadot Y, Fathallah R, Roy M and Maijer D M 2015 *International Journal of Fatigue* **80** 90–102 ISSN 01421123
[19] Schindelbacher G 1993 *Giesserei Praxis* **19** 381–392
[20] Skrinsky Y 2002 *Einfluß von heiß- und kaltisostatischem Pressen auf die statischen mechanischen Werkstoffkennwerte von Gußteilen aus Aluminiumlegierungen* Dissertation Otto-von-Guericke-Universität Magdeburg
[21] Ran G, Zhou J E and Wang Q G 2008 *Journal of Materials Processing Technology* **207** 46–52 ISSN 09240136
[22] Ceschini L, Morri A and Sambogna G 2008 *Journal of Materials Processing Technology* **204** 231–238 ISSN 09240136
[23] Ceschini L, Morri A, Toschi S and Seifeddine S 2016 *Materials Science and Engineering A* **653** 129–138
[24] Yi J Z, Gao Y X, Lee P D and Lindley T C 2004 *Materials Science and Engineering A* **386** 396–407
[25] Milkereit B, Wanderka N, Schick C and Kessler O 2012 *Materials Science and Engineering: A* **550** 87–96 ISSN 09215093
[26] Miao W and Laughlin D 1999 *Scripta Materialia* **40** 873–878 ISSN 13596462
[27] Thomas G 1961 *Journal of the Institute of Metals* **90** 57–63
[28] Pashley D W, Jacobs M H and Vietz J T 1967 *Philosophical Magazine* **16** 51–76 ISSN 0031-8086
[29] Colley L J, Wells M A and Poole W J 2014 *Canadian Metallurgical Quarterly* **53** 138–150 ISSN 0008-4433
[30] Miao W F and Laughlin D E 2000 *Metallurgical and Materials Transactions A* **31** 361–371
[31] Zhen L and Kang S B 1997 *Scripta Materialia* **36** 1089–1094 ISSN 13596462
[32] Dorward R C 1973 *Metallurgical Transactions* **4** 507–512
[33] Sjölander E and Seifeddine S 2010 *Journal of Materials Processing Technology* **210** 1249–1259 ISSN 09240136
[34] Murali S, Arunkumar Y, Chetty P V J, Raman K S and Murthy K S S 1997 *JOM* **49** 29–33 ISSN 1047-4838
[35] Carrera E, Alejandro G J, Talamantes S J and Colás R 2011 *Metallurgical and Materials Transactions B* **42** 1023–1030 ISSN 1073-5615
[36] Ceschini L, Morri A and Morri A 2013 *Journal of Materials Engineering and Performance* **22** 200–205
[37] Geuser F D, Lefebvre W and Blavette D 2006 *Philosophical Magazine Letters* **86** 227–234 ISSN 0950-0839
[38] Murayama M and Hono K 1999 *Acta Materialia* **47** 1537–1548
[39] Serizawa A, Hirosawa S and Sato T 2008 *Metallurgical and Materials Transactions A* **39** 243–251

Impact of Exploitation Flaws on Load Capacity of Tempered Glass Stairs Assessed by Numerical Analysis with XFEM

D Pietras[1,*], T Sadowski[1] and L Wójcik[2]

[1]Department of Solid Mechanics, Lublin University of Technology,
40 Nadbystrzycka Street, Lublin,20 – 618,Poland
[2] Industrial Partner: "*Filip i Maciejewski Architekci S.C.*" (Architectural office),
10b/13 Wojsławicka Street, Chełm, 22 - 100, Poland

*Corresponding author: d.pietras@pollub.pl

Abstract. Nowadays the structures made of tempered glass are widespread in civil engineering. The most popular are building facades and self-supporting stairs, where the glass is applied to carry the service loadings. In structures such as stairs the exploitation process leads to surface damage. The flaws emerged on the surface cause significant reduction of the structure load capacity. Therefore, it is purposeful to acquire the knowledge about the impact of flaws on glass strength to increase the safety of civil structures. The main aim of this paper is to assess the load capacity of the tempered glass slab with a flaw on the surface which is the major element of stairs made of glass. In order to do this a new numerical model of the one-step stair was built in ABAQUS. The residual stress was calculated by modelling of the quenching process. An initial flaw was introduced to the stair model after the tempering process. The conducted calculations in bending state of considered slab indicated that the flaw 30 x 2 x 0.3 mm in cantilever beam 1000 x 300 x 10 mm could be treated as critical.

1 Introduction

Each part of the building construction is exposed to wear due to environmental influence or attrition. The glass is sensitive to small cracks which are the exploitation result. When the part made of glass is carrying the loadings, the knowledge of response to applied loadings is important to increase the safety. This paper aims to calculate the real strength of tempered glass construction parts which contain the flaws. The glass fracture toughness was analysed in [1]. The method to improve mechanical properties of soda-lime-silica glass introduces the residual stress by tempering or ion exchange. Since the desired surface compressive stress generated in chemical methods has relatively small depth [2], in the civil engineering the most popular is tempering process. The phenomena occurring in the material during the tempering process were presented in [3-5].

The tempered glass stairs usually are constructed as laminar. Analytical approach to analyse laminated glass elements was presented in [6, 7]. The finite elements to analyse the layered glass plates were presented in [8]. The surface tolerance of tempered glass is worthless, due to this fact the development of methods to assess the strength of construction parts with a scratched surfaces are desired by construction industry. The analyses in this paper were performed in macro-scale, example of analyses in another scales are presented in [9-11]. The crack sensitivity of building materials may be analysed by using of artificial neural networks [12]. The analyses of behaviour of popular building materials in shear mode is the subject of test of many researchers [13].

2 Materials and methods

The soda-silica-lime glass which is the most popular in civil engineering applications was considered. The numerical analyses were conducted in ABAQUS finite element method. The dimensions of analysed glass stair were 1000 mm x 300 mm x 10 mm (figure 1). Homogenous slab was considered.

The residual stress was introduced by modelling of the tempering process. The user subroutine was programmed to consider the viscous and temperature dependent properties of glass in cooling process. The detailed data about calculation of residual stress due to tempering process is shown in [3-5]. The boundary conditions during the tempering process are presented in figure 1.

The modelling of the tempering process was divided into two parts:

- analysis of the heat flow during the cooling process of considered slab,
- stress analysis for viscous material.

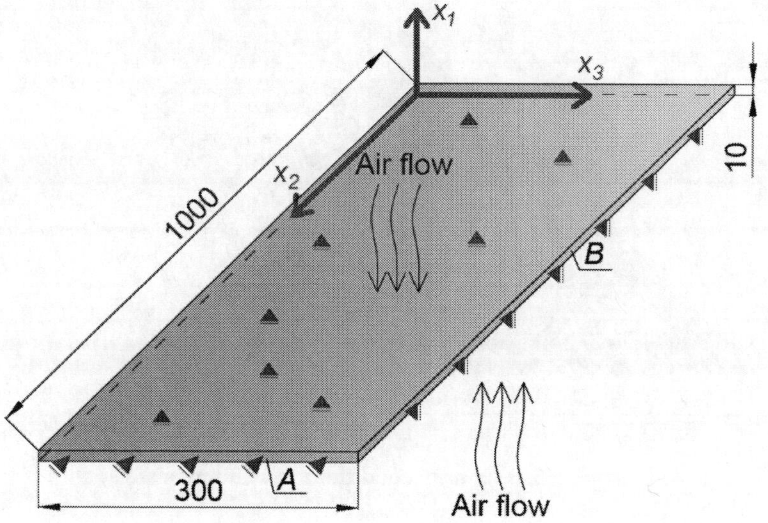

Figure 1. Boundary conditions during the tempering process.

The initial state was modelled as a stress free body at 670°C. The analysis was carried out using finite elements of the DC3D8 type in the HEAT TRANSFER step type with the option TRANSIENT. The cooling process of the slab by air with temperature 20°C was analysed. A heat transfer coefficient was assumed to be equal to h_1= 200 W /m^{2}K for surfaces perpendicular to the direction of heat flow and 100 W/m^{2}K for parallel surfaces. It was assumed that the cooling device was blowing the air from top and bottom (along x_1 axis, figure 1). As a signal to complete the analysis, the temperature change was chosen by less than 0.02°C in the next time step. The maximum temperature change in the next step was set to 40°C. The initial time was set to 10^{-6} s. The cooling time that equalled 583.659 s was obtained. The solution was found in 71 increments.

The previously obtained temperature fields in time domain were imported to the analysis of the stress state made with the VISCO step type. During tempering the considered slab was supported by locking the displacements in x_1 direction in bottom surface as well as for displacements on edges A and B in x_2 and x_3 directions, respectively (figure 1). The tolerance parameter of deformation resulting from creep parameter was assumed to be 0.005. The obtained residual stress was imported in a scratched stair analysis.

To calculate the stress intensity factors the Extended Finite Element Method [XFEM] was used. The considered flaw had a rectangular cross-section and dimensions of 30 mm x 0.3 mm, which was based on the observation of the acceptable flaws declared by many manufacturers. The dimensions of considered flaw were much smaller than the size of the analysed stair, due to this fact the submodelling method was applied. The element being analysed had a flaw in the centre of the panel perpendicular to the length of the element (figure 2). A single flaw was considered to obtain a clear result of the defect in the glass slab. The size of the mesh of finite elements was equal to 3 mm for general model and 0.3 mm for submodel was assessed in mesh study analysis made for simple cases and compared to reference values [14]. It is well known that considered crack position is not critical, but these the crack position was choosen due to high probability occurrence of the exploitation flaws in this area. The mesh of the submodel was 3 times densified in the crack vicinity.

The boundary conditions of considered bending state are shown in figure 2. The mechanical load *F* was applied as the SURFACE TRACTION to avoid the undesirable concentrations.

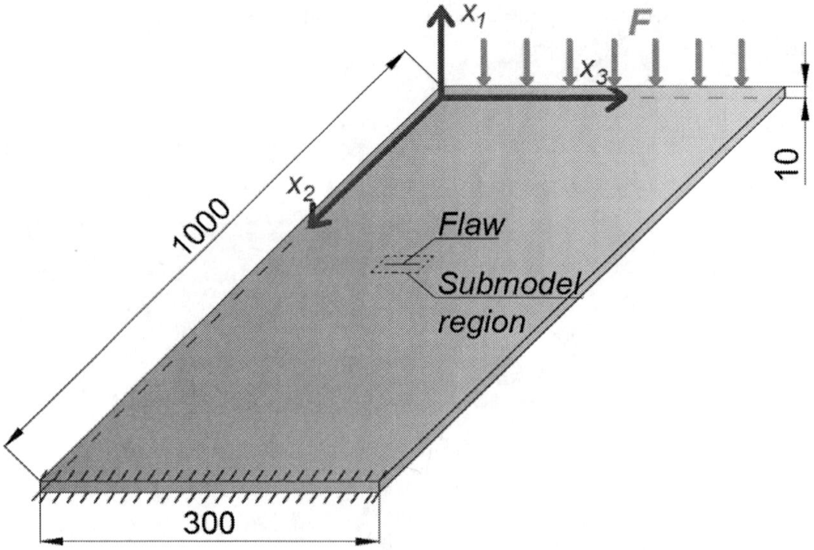

Figure 2. Boundary conditions in bending state.

The residual stress state in volume element of the considered stair is shown in figure 3. The residual stress tensor component σ_{11} has small values, due to this fact it can be considered as plane stress. After the superposition of the stress state caused by bending in upper surface stress state may be represented by figure 4. When the stress tensor component σ_{22} takes positive values the tension state in direction x_2 emerges. In tensioned surface the crack is able to grow until the slab failure. As shown in figure 4, when σ_{22} reaches positive values the element begins to work in complex stress state. Therefore, the fracture analyses have to take into account the mix-mode model of fracture.

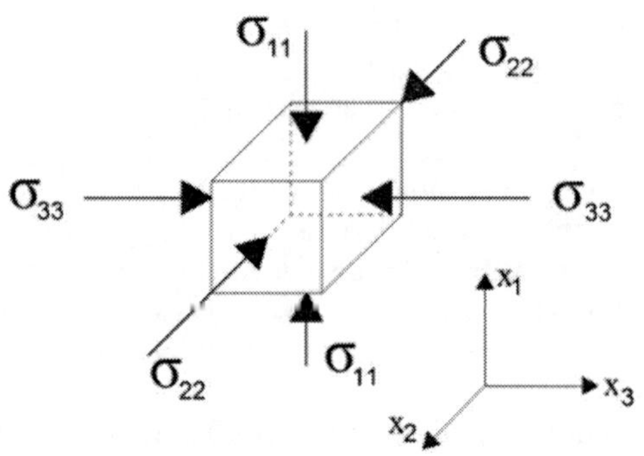

Figure 3. Residual stress state near the surface of the slab.

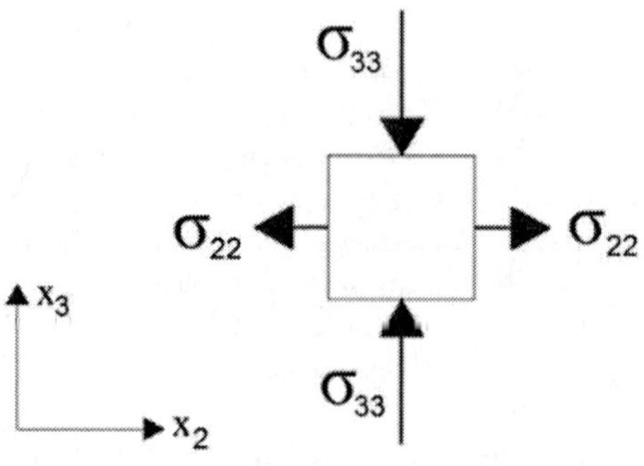

Figure 4. Stress state on the surface of the slab in bending conditions.

To calculate the stress intensity factors the linear fracture mechanics theory was applied. The contact between the flaw surfaces was omitted because of its high thickness (2 mm). The mix-mode fracture criterion for soda-lime-silica glass was taken from [15].

$$\frac{K_{II}}{K_c}\left(\frac{K_I}{K_c}\right) = -7.371\left(\frac{K_I}{K_c}\right)^5 + 12.980\left(\frac{K_I}{K_c}\right)^4 - \\ +7.927\left(\frac{K_I}{K_c}\right)^3 + 1.231\left(\frac{K_I}{K_c}\right)^2 - 0.073\left(\frac{K_I}{K_c}\right) + 1.160 \quad (1)$$

3 Results and discussion

3.1 Residual stresses

The temperature timing during the tempering process is shown in figure 5. The cooling of the glass slab is faster on the surface than inside, which is in accordance with the practice. The cooling process duration to room temperature is approximately 300 s.

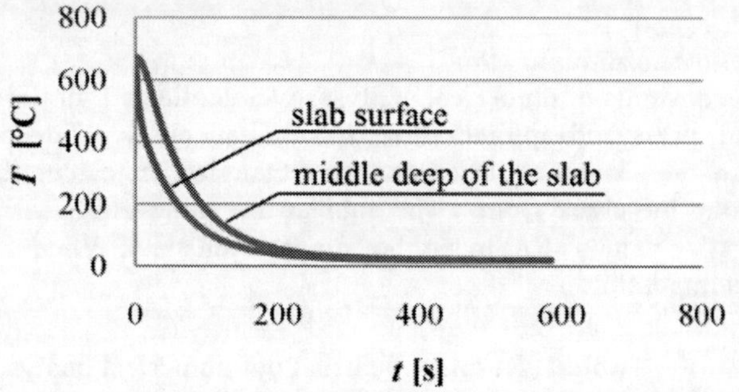

Figure 5. Temperature timing during tempering process.

The development of residual stress is shown in figure 6. The stress is increasing nonlinearly during the whole cooling process.

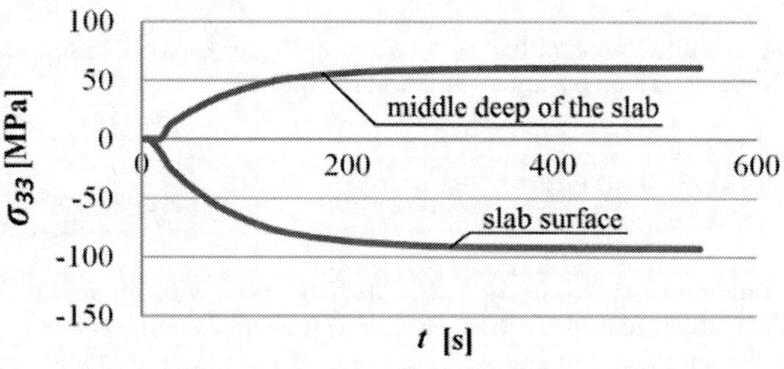

Figure 6. The residual stress growth during the tempering process.

Derived stress tensor components σ_{11} and σ_{22} distribution across the considered slab is shown in figure 7.

IOP Conf. Series: Materials Science and Engineering **416** (2018) 012085 doi:10.1088/1757-899X/416/1/012085

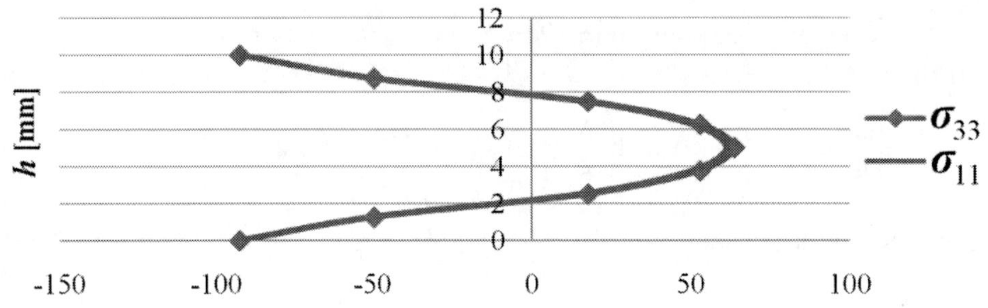

Figure 7. Stress distribution in the cross-section of the element obtained in tempering process.

3.2 Bending state of the considered stair

In table 1 the achieved results of numerical analyses were collected. In the calculations were taken into account the residual stress and applied force F simultaneously. The considered crack shape is a cuboid 30x2x0.3 mm, therefore the crack front is a rectangle. The calculated stress intensity factors are the mean values along the crack front. The sign in the value of K_{II} and K_{III} means the direction of shear state. The negative values of K_I mean the compression state. When the K_I has negative values the crack propagation is impossible.

Table 1. Results of carried out numerical analyses

F [kN]	K_I $[\mathrm{MPa}\sqrt{\mathrm{m}}]$	K_{II} $[\mathrm{MPa}\sqrt{\mathrm{m}}]$	K_{III} $[\mathrm{MPa}\sqrt{\mathrm{m}}]$
1.35	3.845	-0.034	-0.002
1.65	7.816	0.435	0.002

On the basis of table 1 the ratios K_I to K_c and K_{II} to K_c were calculated. The relation of ratio $K_{II}/K_c(K_I/K_c)$ can be described by linear function as follows:

$$\frac{K_{II}}{K_c}\left(\frac{K_I}{K_c}\right)=0.118\left(\frac{K_I}{K_c}\right)-0.642. \tag{2}$$

The above relation is plotted in figure 8. The blue curve with squares is the fracture law described by formula (1).The slight red line is the response of the considered structure to applied loadings which is described by formula (2). The intersection point of the blue curve and red straight line means the crack growth

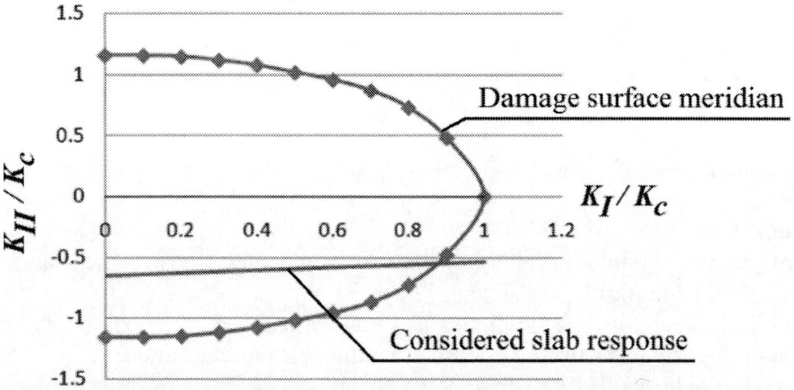

Figure 8. Comparison of response of considered slab and fracture criterion.

The intersection point calculated by comparison of the formulas (1) and (2) is:

$$\left(\frac{K_1}{K_c}\right) = 0.881. \tag{3}$$

The maximal value of K_I after consideration the impact of shearing mode is equal to 0.881 K_{II}. The critical value of stress intensity factor K_{IC} for glass used in civil engineering was estimated in [16] as equal to $0.76\ \text{MPa}\cdot\sqrt{\text{m}}$. The relation between K_I and applied force F based on table 1 is presented in figure 9.

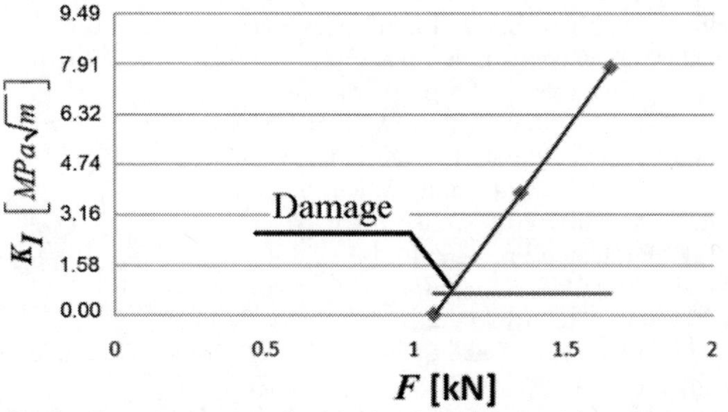

Figure 9. Assessment of critical value of force F(see figure 2)

Dependence between F *and* K_I can be interpolated by using linear regression:

$$K_I(F) = 418.5F - 443.4. \tag{4}$$

When the K_I reaches the critical value $K_c = 0.881 \cdot 0.76\ \text{MPa}\cdot\sqrt{\text{m}}$, the force F takes the critical value $F_{c\_cracked}$ = 1.11kN. The critical value F of uncracked slab in the middle span calculated by using simple Timoshenko's beam theory in the middle span of the beam can be found on the base of the following equation:

IOP Conf. Series: Materials Science and Engineering **416** (2018) 012085 doi:10.1088/1757-899X/416/1/012085

$$\frac{0.5 \cdot 6}{bh^2} LF_{c\_uncracked} - \sigma_{res}^{\max} \leq 65MPa\,. \tag{5}$$

where $\sigma_{res}^{\max} = 98.4MPa$ are the residual stress in the surface of the slab. $F_{c\_uncracked}$ = 1.634 kN. The reduction of maximal load after the slab stretching is 32.1%.

4 Conclusions

On the basis of performed numerical analyses the following conclusions can be drawn:

- The impact of complex stress state in tempered glass to reduce the fracture toughness is 12%. (formula 3). Therefore the predictions of the critical forces in tempered glass constructions carried out by means of simplified numerical models are inaccurate.
- The scratch with depth that equals 0.3 mm and length 30 mm reduces the load capacity of considered glass slab by 32.1%.

The above formulated model can be adopted to similar problems solution for other engineering materials used in aerospace, mechanical and civil engineering, e.g. [17 – 23].

5 References

[1] Nielsen JH, · Olesen JF and Stang H 2009 *Exp. Mech.* **49** 855
[2] Abrams MB, Green DJ, Jill S 2003 J. Non-Cryst. Solids **321** 10
[3] Carre H, Daudeville L 1996 J.Phys IV **06** C1-175
[4] Daudeville L and Carre H 1998 J. Therm Stresses **21** 667
[5] Nielsen JH, Olesen JF and Poulsen PN 2010 Comput Struct **88** 96
[6] Galuppi L and Royer-Carfagni GF 2012 Eng Struct **38** 53
[7] Birsan M, Sadowski T, Marsavina L, Linul E and Pietras D 2013 Int. J. Solids Struct. **50** 519
[8] Ivanov IV, Velchev DS, Georgiev NG, Ivanov ID and Sadowski T 2016 Meccanica **51** 341
[9] Sadowski T and Samborski S 2003 Comp. Mat. Sci. **28** 512
[10] Dębski H and Sadowski T 2014 Comput. Mat. Sci. **83** 403
[11] Sadowski T and Marsavina L 2011 Comput. Mat. Sci. **50** 1336
[12] Gajewski J and Sadowski T 2014 Comp. Mater. Sci. **82** 114
[13] Golewski G and Sadowski T 2014 Const. Build. Mat. **51** 207
[14] Murakami Y 1987 Pergamon Press, Oxford,
[15] Ayatollahi MR and Aliha MRM 2009 Int. J. Solids Struct. **46** 311
[16] Gong J, Chen Y and Li Ch 2001 J. Non-Cryst. Solids **279** 219
[17] Marsavina L, Linul E, Voiconi T and Sadowski T 2013 *Polym. Test.* **32** 673
[18] Sadowski T and Samborski S 2003 *J. Am. Cer. Soc.* **86** 2218
[19] Ivanov IV, Sadowski T and Pietras D 2013 *Europ. Phys. J. Spec. Topics* **222** 1587
[20] Golewski P and Sadowski T 2018 *J. Europ. Cer. Soc.* (on-line) https://doi.org/10.1016/j.jeurceramsoc.2018.01.008
[21] Postek E and Sadowski T 2011 *Comp. Interf.* **18** 57
[22] Sadowski T and Golewski P 2012 *Comput. Mat. Sci.* **64** 285
[23] Sadowski T and Golewski P 2012 *Comput. Mat. Sci.* **52** 293

Acknowledgement
This work was financially supported by Ministry of Science and Higher Education within the statutory research number S/20/2018.

IOP Conf. Series: Materials Science and Engineering **416** (2018) 012086 doi:10.1088/1757-899X/416/1/012086

Investigation of the Mechanical Properties of Acrylonitrile Butadiene Styrene (ABS)-Nanosilica Reinforced Nanocomposites for Fused Filament Fabrication 3D Printing

P Kyratsis[1] and D Tzetzis[2,*]

[1]Western Macedonia University of Applied Sciences, Department of Mechanical Engineering and Industrial Design, Kila Kozani GR50100, Greece
[2] International Hellenic University School of Science and Technology, 14km Thessaloniki - N. Moudania, Thermi GR57001, Greece

[*]Corresponding author: d.tzetzis@ihu.edu.gr

Abstract. To expand the capabilities of 3D printed structures generated from commercial thermoplastic printers, this paper presents the fabrication of polymer filaments that contain inorganic nanoparticles. Nanosilica was dispersed ultrasonically into acrylonitrile butadiene styrene (ABS) and extruded into filaments for fused filament fabrication (FFF) 3D printing. The produced filaments had nanosilica compositions of 5% and 10%wt. A dynamic ultra-micro hardness tester was used in order to characterize the mechanical behavior of the filaments. Also, 3D printed tensile specimens were used to obtain the strain-strain curves and monitor the tensile strength of such structures. The morphology of the nanosilica/ABS composite was studied with a scanning electron microscope (SEM). The results from the dynamic ultra-micro hardness tester reveal that the incorporation of nanosilica particles increases the stiffness and hardness of the ABS material especially in the case of 10%wt concentration. However, for such concentration 3D printing was not possible due to the severe embrittlement of the material. 3D printed structures produced in the form of tensile specimens from the 5%wt nanocomposites display a higher breaking point stress and stiffness than those printed from the pure ABS polymer.

1. Introduction

3D printing processes have had a tremendous impact on the field of design, in the form of rapid prototyping and toolmaking and, more recently, part production. It has been widely used in various manufacturing fields such as aerospace, automobile, biomedical, building and many others [1-4]. Novel materials are now fabricated incorporating additives with unique characteristics. Fused filament fabrication takes advantage of such novel materials produced by combining the base polymers with additives in specific concentrations and structures. The vast majority of the materials used in FFF machines are ABS, PLA and Nylon, with bulk strengths between 30–100 MPa and elastic moduli in the 1.3–3.6 GPa range with those numbers greatly reduced in printed components [5]. However, recently composite materials have been used in 3D printing processes and novel materials are now fabricated incorporating additives with unique characteristics [6-9]. Such studies aim towards the understanding of how well these filaments perform when used in 3D printing and what enhancements are realized when additives are added to the unfilled polymer. It is critical though that such 3D printed materials might result in a range of unidentified mechanical properties of the final part due to the additive nature of 3D printing. Also, the effects of an array of processing parameters, such as raster angle, printing speed, layer thickness on the mechanical properties of FFF 3D printed parts have been well researched as documented in the literature [10-15].

An alternative technique for the determination of the mechanical properties of polymers is the instrumented indentation technique. This is a simple but powerful testing technique, which can provide useful information about the mechanical properties of materials [16]. Various studies have compared the results obtained from indentation instrumented test with the results obtained from the traditional tensile tests especially for the elastic modulus calculation [17–18].

The current paper aims to characterize the material behavior of 3D printing materials using the instrumented indentation technique. Filaments were produced with ABS reinforced with 5% and 10% nanosilica. The morphology of the nanocomposite systems was characterised by scanning electron microscope. 3D printed constructs were also realised in the form of tensile specimen in order to assess the effect of the nanoadditive in the 3D printed state.

2. Materials and fabrication of 3D printed specimens

The main material that has been used in the current paper is Acrylonitrile Butadiene Styrene (ABS) which is a terpolymer made by polymerizing styrene and acrylonitrile in the presence of polybutadiene. Such material was provided in pellets form ColorFabb (Netherlands). Nanosilica particles of 14nm in diameter with concentrations of 5 and 10%wt were dispersed in diluted in acetone ABS over 30 minutes at room temperature in a probe sonicator (UP200S by Hielscher Ultrasonics GmbH). The 5%wt nanosilica reinforced ABS filaments provided good 3D prints, however the concentration of 10%wt was not possible to be printed due to the overall embrittlement of the filament and fracture during handling and processing. Figure 1 shows the ABS and 5%wt nanoilica/ABS filaments as taken from an optical microscope.

Tensile specimens were 3D printed using a BCN3D Sigma printer with a 3 mm extrusion nozzle insert. Each specimen was printed individually at the centre of the printing bed in order to produce all specimens as similarly as possible. Only system integrated (default) variations of 3D printing parameters were employed in the comparative analysis. The printing parameters employed were: nozzle extrusion temperature of 210°C, heat bed temperature of 55°C, deposition line (layer) height 0.2 mm, deposition line width 0.4 mm and printing speed of 50 mm/sec. Deposition speed was therefore not considered a variable in this study and a constant extrusion velocity was selected for all specimens, based on device parameters (e.g. effective printing range). Also, printing was carried out in a standard laboratory conditions without temperature or humidity control. The specimens employed were rectangular with no rounded edges. The specimens were printed in vertical orientation a monolithic shell.

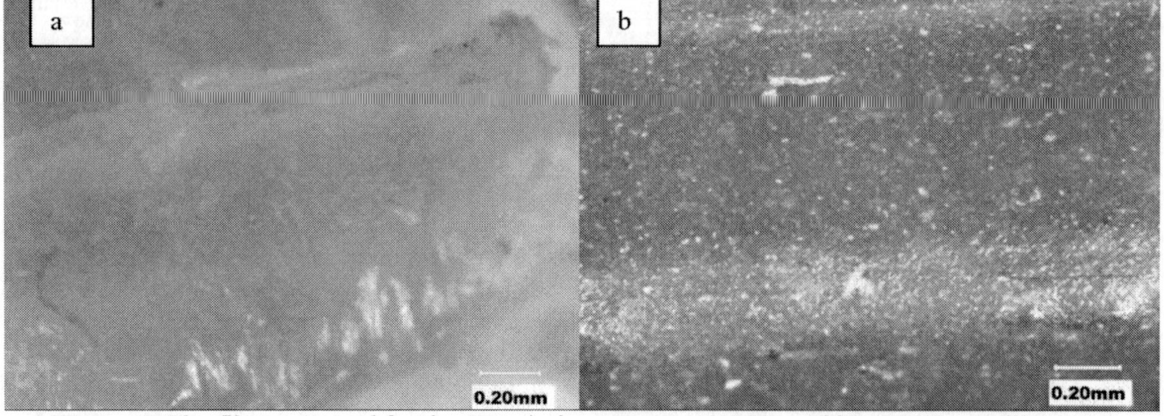

Figure 1. The filaments used for the 3D printing process: a) ABS and b) 5%wt nanosilica/ABS

3. Experimental tests

3.1. Dynamic ultra-micro hardness test

Investigation of the mechanical behavior of the filaments was performed using a dynamic ultra-micro-hardness tester (DUH-211; Shimadzu Co., Kyoto, Japan) fitted with a triangular pyramid indenter tip (Berkovich indender). The specimens were fixed onto an attached holder. Dynamic micro-indentation mainly involves applying a controlled load (P) through a diamond tip that is in contact with a surface. The penetration depth (h) of indentation is continuously recorded as a function of load. Figure 2 shows a schematic illustration of the typical micro-indentation load–penetration depth curve obtained from dynamic micro-indentation tests. During indenter loading and unloading, the specimen is subjected to both plastic deformation (hp) and elastic deformation (he). The total deformation (ht) is the sum of hp and he in the micro-indentation load–penetration depth curve. Moreover, the indentation hardness and elastic modulus can be obtained from the indentation load and penetration depth data. The indentation hardness (Hit) of the samples is a measure of the resistance to permanent deformation damage and it was calculated from the following equation

$$H_{it} = \frac{P_{max}}{A_p} \quad (1)$$

where P_{max} is the maximum force and A_p is the projected area of contact between the indenter and the sample. A_p is calculated from the following equation:

$$A_p = 23{,}96\, h_c^2 \quad (2)$$

where h_c is the depth of contact of the indenter with the sample calculated as follows:

$$h_c = h_t - \frac{3}{4}(h_t - h_r) \quad (3)$$

where h_r is derived from the load-displacement curve and is the intercept of the tangent to the unloading cycle at P_{max} with the displacement axis.

The elastic modulus (E) of the sample was calculated from the following equation:

$$\frac{1}{E_r} = \frac{1-v_s^2}{E_s} + \frac{1-v_i^2}{E_i} \quad (4)$$

where $E_{i,s}$, and $v_{i,s}$ are the elastic modulus and Poisson's ratio, respectively, for the indenter and the specimen. For a diamond indenter, E_i is 1140 GPa and v_i is 0.07.

In this study, the dynamic micro-indentation test was carried out with peak loads (P_{max}) of 500 mN. The load rate was kept constant at 13.324 mN/s, and the hold time at the maximum load was set to 3 s. The dynamic micro-indentation results, such as indentation hardness and elastic modulus, were obtained as the average values of five measurements.

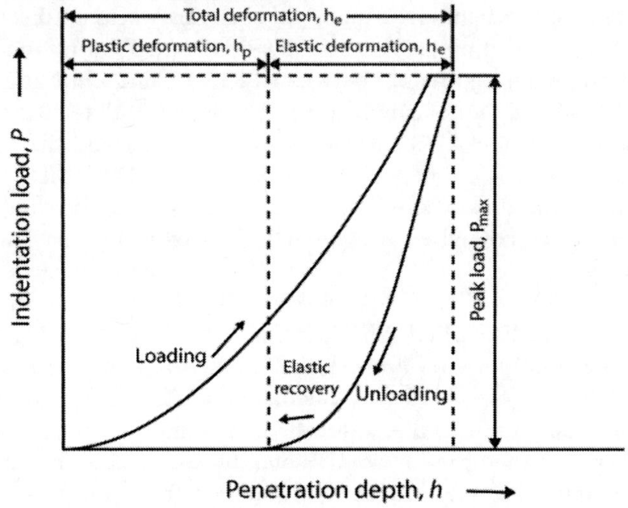

Figure 2. Schematic illustration of the micro-indentation load versus penetration depth curve obtained from the testing.

3.2. Tensile Test

Tensile tests were performed at room temperature (23°C) on a Testometric (UK) universal testing machine equipped with a 50KN load cell at a constant crosshead speed of 2 mm/ min. The dimensions of the 3D printed specimens shown in Figure 3. The modulus was calculated within the linear part of the stress–strain curves. All presented data corresponds to the average of at least five measurements.

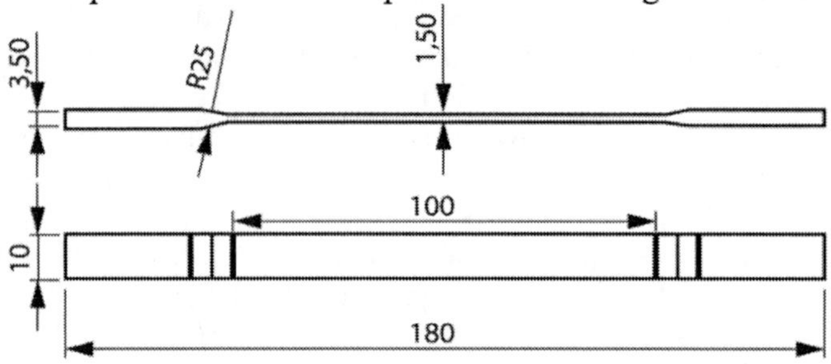

Figure 3. Tensile test specimen dimensions.

4. Results and discussion

The morphology of the ABS reinforced with nanosilica particles waw studied using Scanning Electron Microscope (SEM) Phenom ProX and the results are shown in Figure 4. Clearly, the insertion of nanosilica had a profound effect on the overall microscopic appearance and surface texture of the filament which has almost perfect cylindrical form following the extrusion procedure as seen from Figure 4(a). The surface texture is rough with characteristic local fracture areas as illustrated in Figure 4(b,c). At much higher magnification as shown in Figure 4(d), although the image is a bit out of focus due to the limit of the SEM capabilities, it becomes obvious that the dispersion of the nanosilica is not very good judging from the agglomerations that are formed as large circular patterns. These nanosilica agglomerated regions could behave as stress concentrators during testing, which would probably contribute to lowering the durability of 3D printed parts. Typical indentation force–penetration depth curves of the materials under study in a filament form are shown in Figure 5(a). The indentation force-penetration depth curves were obtained during indenter loading and unloading. For all samples the indentation force-penetration depth curves for various indentation peak loads correspond very well with one another. This indicates the high accuracy and reproducibility of the indentation method for the studied materials under mechanical load. The indentation force–penetration depth curves for all materials under test indicated creep phenomenon of the specimen at the peak force of 500 mN. There were no large differences in creep behavior among the samples while no discontinuities or steps were found on the loading curves, indicating that no cracks were formed during indentation The indentation depths at the peak load range approximately between 14 to 17 μm while the ABS/nanosilica 10wt% curve showed characteristically higher elastic recovery (high elastic deformation).

The indentation modulus for the ABS samples was 1310 MPa and the hardness 117,1MPa. The addition of 5%wt silica nanoparticles increased the modulus to 2015 MPa, which is approximately 54% increase, while the addition of 10%wt nanosilica increased the modulus to 2242MPa which is approximately 71% increase as compared to neat ABS. As expected the hardness also increased at around 10% and 20,2% by the addition of 5%wt and 10%wt silica nanoparticles with values of 128.9 MPa and 140.89 MPa respectively. There is a marked difference of stress-strain curves under tension between the unfilled and nanosilica filled ABS polymer of as shown in Figure 5(b). The maximum stress is 45,6 MPa for the pure polymer and 51,8MPa for the nanocomposite with lower strain to failure values. The modulus of pure ABS was calculated as 2288 MPa and it was improved by the addition of the nanosilica up to 2751MPa. Overall, this trend can be profoundly attributed to the softness of the polymer chain structure of neat ABS and the stiffness of the nanosilica particles. These values are not directly comparable with the results obtained from the dynamic ultra-micro hardness test since such test measures the performance of the bulk material, while the tensile tests account for the layering structure of the 3D printing operation.

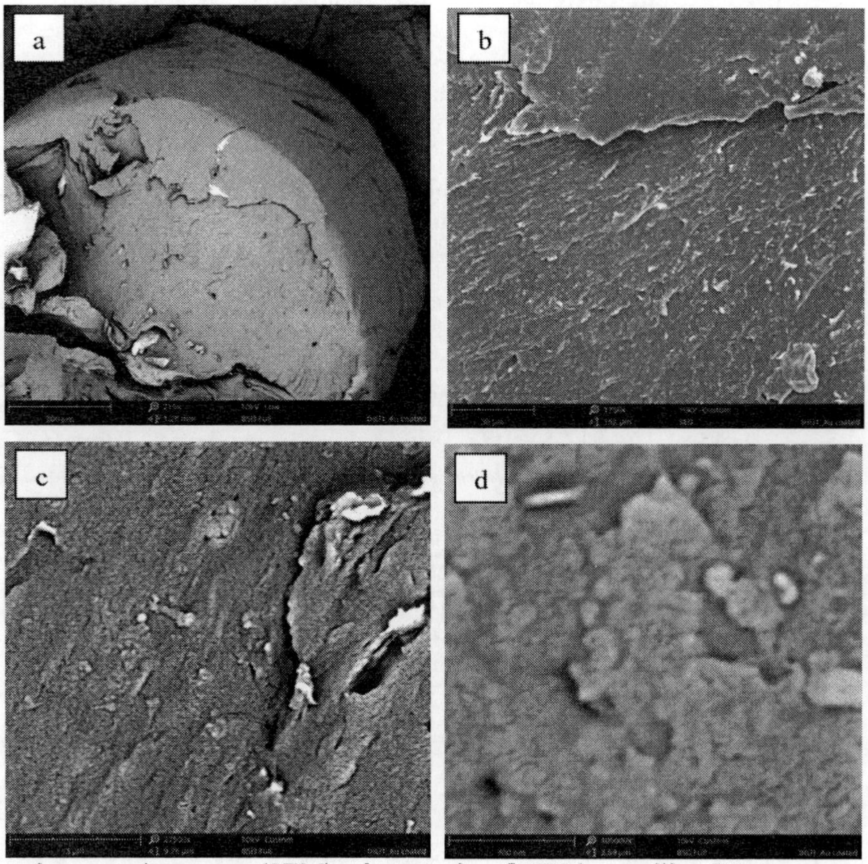

Figure 4. Scanning electron microscope (SEM) micrographs of ABS/ nanosilica 5%wt nanocomposite showing a)low magnification of the filament with the characteristic cylindrical shape, b,c) higher magnification of the nanocomposite showing a relatively rough surface probably from the nanosilica addition, and d) very high magnification and low focused image, depicting rounded- shaped agglomerations.

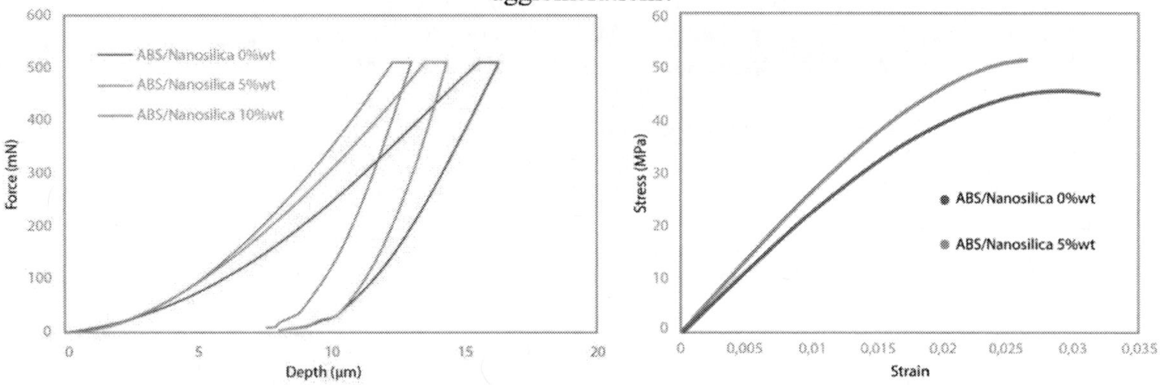

Figure 5. a)Force–depth indentation profiles of ABS and its nanocomposites as a function of nanosilica loading, b) Stress-strain curves of ABS and nanosilica/ABS 3D printed specimens

5. Conclusion

The current paper presents results for mechanical testing of generated filaments for 3D printing by the addition on nanosilica particles in ABS matrix. Nanocomposites of 5% and 10%wt were formed and tested using a dynamic ultra-micro hardness test. The results reveal that the stiffness and hardness of the ABS material especially in the case of 10%wt concentration is increased. However, for such concentration 3D printing was not possible due to the severe embrittlement of the material. Morphological examination of the filament have shown characteristic roughness due to nanosilica and agglomerated nanomaterial areas from relatively pure dispersion. Tensile specimens were 3D printed using the same comparative printing parameters for both pure ABS and 5%wt nanocomposite. Deposition speed was not considered a variable and all specimens were 3D printed vertically only with a shell. As expected, the 3D printed structures produced in the form of tensile specimens from the 5%wt nanocomposites have shown a higher breaking point stress than those printed from the pure polymer.

6. References

[1] Kumar LJ and Krishnadas Nair CG 2017 Current Trends of Additive Manufacturing in the Aerospace Industry: in Advances in 3D Printing & Additive Manufacturing Technologies (ISBN 978-981-10-0811-5) ed DI Wimpenny, PM Pandey, LJ Kumars (Springer), pp 39-54
[2] Zadpoor AA and Malda J 2017 *Annals of Biomedical Engineering* **45**(1) p 1
[3] Biswas K, Rose J, Eikevik L, Guerguis M, Enquist P, Lee B, Love L, Green J and Jackson R 2016 J Sol Energy Eng **139**(1)
[4] Guo N and Leu MC 2013 *Front. Mech. Engineer* **8**(3) 215
[5] Tymrak B, Kreiger M, and Pearce J 2014 *Mater Design* **58** 242
[6] Campbell TA and Ivanova OS 2013 *Nano Today* **8**(2) 119
[7] Compton BG and Lewis JA 2014 *Adv Mater* **26**(34) 5930
[8] Chizari K, Arjmand M, Liu Z, Sundararaj U and Therriault D 2017 *Mater Today Commun* **11** 112
[9] Yang Y, Chen Z, Song X, Zhang Z, Zhang J, Shung KK, Zhou Q and Chen Y 2017 *Adv Mater* **29**(11) 1
[10] Tandon GP, Whitney TJ, Gerzeski R, Koerner H and Baur J 2017 *Mechanics of Composite and Multi-functional Materials* vol 7, eds W Ralph R. Singh, G Tandon, P Thakre, P Zavattieri and Y Zhu (Springer) p 63.
[11] Türk DA, Brenni F, Zogg M and Meboldt M 2017 *Mater Des* **118** 256.
[12] Zaldivar RJ, Witkin DB, McLouth T, Patel DN, Schmitt K and Nokes JP 2017 *Addit. Manuf.* **13** 71.
[13] Huang B and Singamneni S 2015 *J Compos Mater* **49**(3) 363.
[14] Tian X, Liu T, Yang C, Wang Q and Li D 2016 *Composites Part A* 88 198.
[15] Mansour G, Tzetzis D and Bouzakis KD 2013 *Tribol Ind* **35** 190
[16] Tzetzis D, Tsongas K and Mansour G 2017 *Mater Res* **20**(6) 1571
[17] Tzetzis D, Mansour G, Tsiafis I and Pavlidou E 2013 *J Reinf Plast Comp* **32** 160
[18] Mansour G and Tzetzis D Nanomechanical 2013 *Polym-Plast Technol* **52** 1054

IOP Conf. Series: Materials Science and Engineering **416** (2018) 012087 doi:10.1088/1757-899X/416/1/012087

Topologic Study of a Novel Periodic Cellular Core for Sandwich Panels

M A Iftimiciuc[1,*], S Lache[1] and M N Velea[1]

[1]Department of Mechanical Engineering, Transilvania University of Brasov, 29 Eroilor Blvd, 500036 Braşov, Romania

[*]Corresponding author: mihaela.iftimiciuc@unitbv.ro

Abstract. The use of sandwich panels in industries such as aerospace, naval and automotive is extending nowadays, due to their high mechanical performances with respect to low-weight ratios. In this regard, the concept of 'lightweight structures' has been already acknowledged as the alternative solution in addressing one of the major issues when it comes to the future vehicle design and construction. This paper aims to present a novel cellular structure used as core in sandwich structures, obtained through a mechanical expansion process. A topological study is carried out, proving the competitive relative density of this new structure compared to the already existing ones.

1. Introduction

The sandwich structures have been intensively studied within the last decades, being considered as alternative solution to traditional materials in various applications that require minimal mass while maintaining high stiffness and strength. Nowadays a considerable number of solutions is offered when it comes to material use and architecture of the core [1].

This has powered the use of sandwich panels not only in the aircraft industry, but as well in the naval, railway, civil and also automotive industries [2]. The latter has started to consider using lightweight structures when the vehicle's weight begun to be an issue (i.e. in order to comply to the strict environmental regulations and standards, which require manufacturers to decrease the levels of fuel consumption and gas emissions) [3]. Moreover, the attempts to develop and extensively use of the electric vehicles has constantly stimulated researchers to identify new alternatives to replace the traditional materials.

The essential part of any sandwich assembly is represented by the cellular core, which has to meet two main requirements: first of all, from a structural point of view, the role of the core is to maintain constant the distance between the lateral sheets and to stabilize them in such a manner to prevent local buckling; secondly, a low density is required together with a high shearing and bending stiffness and strength on different directions [2].

Although the honeycomb represents the best solution for the cellular core due to its good mechanical properties, its high production costs is still a big disadvantage that prevents its use at a very large scale

Consequently, significant research in the area of lightweight cellular structures has been carried out resulting different periodical topologies such as: sinusoidal, triangular, trapezoidal, corrugated etc. These newly developed cellular topologies are intended to offer multi-functionality to the sandwich structure due to a set of essential mechanical properties such as: low relative densities, high compressive, bending and shearing stiffness together with a significant reduction of manufacturing costs [4].

Research conducted at the Transilvania University of Braşov resulted in designing a periodic cellular structure, ExpaAsym, easy to manufacture and at a significantly reduced price comparing to the honeycomb. Although it has proven to have a lower relative density and comparable mechanical properties [5], the low bonding surface raises some problems when sandwich panels are constructed using ExpaAsym as a core [6].

This paper aims to present a novel cellular structure obtained through the same manufacturing process as ExpaAsym (i.e. mechanical expansion) but offering a considerably higher bonding surface. A topologic study is carried out, proving the competitive relative density of this new structure compared to the already existing ones.

2. The novel cellular core

The two core structures designed as alternatives to other corrugated cores used for sandwich panels are presented in figure 1: a) – ExpaAsym, which has been already validated through analytical and experimental studies [5, 6] and b) – the novel core structure, still under development but considered an interesting alternative for overpassing the bonding issues identified at ExpaAsym. The novel cellular structure has been developed taking into consideration the advantages offered by the lattice truss architectures (load-bearing abilities, energy absorption capacities etc. [4, 7]) and its topological study is to be presented in the following paragraphs.

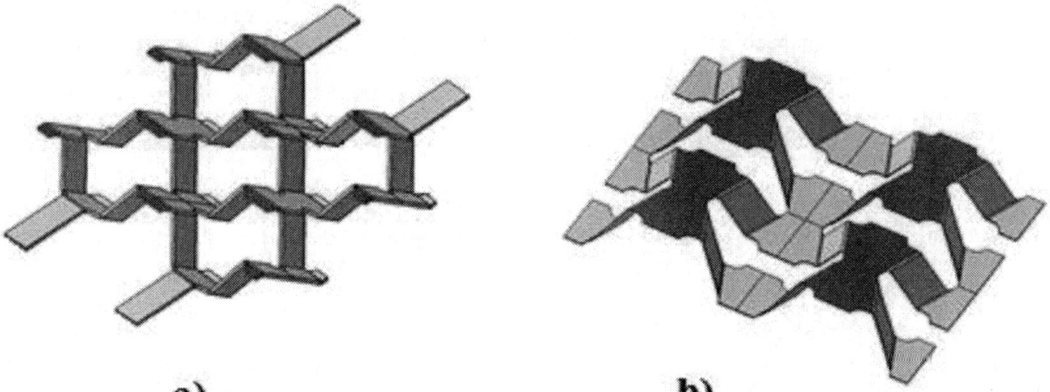

a) **b)**

Figure 1. Cellular core structures proposed for sandwich panels: a) ExpaAsym b) novel cellular core.

The manufacturing technology for this type of corrugated core is relatively simple. On a single metallic sheet, a pattern of cuts and perforations is performed. A series of bend lines is subsequently applied, at an internal angle (*A*), in order to have a full control on the mechanical expansion process. The resulting metal sheet is afterwards expanded at a specific angle (*B*), by applying a force along a direction perpendicular to the pattern of cuts while the opposite edges are kept fixed along the direction of the applied force, thus resulting the expanded unit cell, figure 2.

IOP Conf. Series: Materials Science and Engineering **416** (2018) 012087 doi:10.1088/1757-899X/416/1/012087

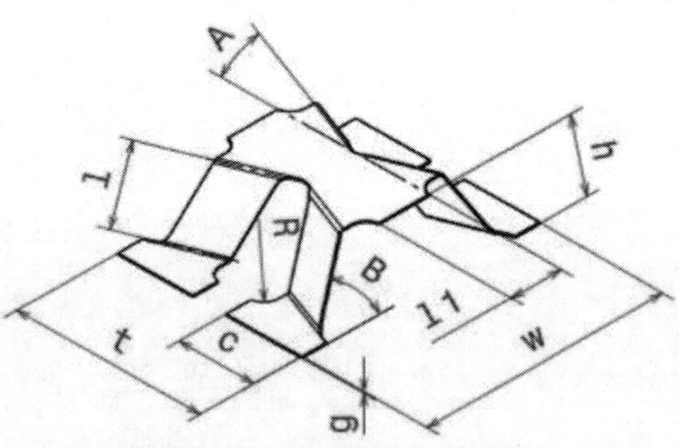

Figure 2. The novel cellular core.

The geometry of the expanded cellular structure is defined by the following parameters: internal angle (A); expansion angle (B); length of the arm's base (c); length of the cell's arm (l), radius of the perforation (r), distance between two perforations (l_1), thickness of the base material (g), length of the unit cell (w); width of the unit cell (t), height of the unit cell (h), figure 2.

In regard to the parameters afore mentioned, the gauge dimensions (h, t, w) can be defined with the help of the equations below:

$$w = 2l_1 + 2l\cos(B) + c\tan(A) + 2R \tag{1}$$

$$t = 2c + z + 2\left[l^2\cos(B)^2 - \left(l\cos(B) - R + 2g + \frac{c\tan(A)}{2}\right)^2\right]^{1/2} \tag{2}$$

$$h = g + l\sin(B) \tag{3}$$

3. Relative density of periodic cellular structures

The parameter that offers the possibility to evaluate the lightweight characteristic of corrugated cores is the relative density (ρ_r), which is represented by the ratio between the volume of base material from which the cellular core is manufactured, and the resulting volume of the generated structure [8].

$$\rho_r = \frac{v_m}{v_s} \tag{4}$$

3.1. Study upon the relative density of existing topologies

In order to determine the performance of the newly developed cellular structure, a study between the most common topologies used in design and construction of sandwich panels has been carried out. In this regard, four topologies have been taken into consideration, for which the relative density has already been determined in the literature [9, 10]: triangular, trapezoidal, honeycomb and ExpaAsym.

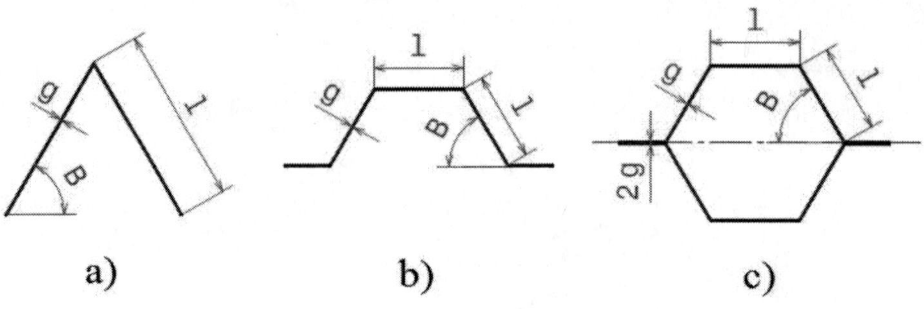

Figure 3. Main core topologies for sandwich panel construction: a) triangularcores, b) trapezoidal cores, c) honeycomb cores.

For triangular cores, figure 3a), the relative density can be defined as [9]:

$$\rho_r = \frac{2}{\sin(2B)} \frac{g}{l} \quad (5)$$

In the case of trapezoidal cores, Fig. 3b), the relative density can be determined as [10]:

$$\rho_r = \frac{g(2l + 2l_1)}{(2l_1 + 2l\cos(B))(g + l\sin(B))} \quad (6)$$

For honeycomb cores, Fig. 3c), the relative density can be defined as [10]:

$$\rho_r = \frac{2g(2l + 2l_1)}{(2l_1 + 2l\cos(B))(2g + 2l\sin(A))} \quad (7)$$

For the ExpaAsym structure, the relative density can be determined using the following formula [2, 6]:

$$\rho_r = \frac{g}{(1 + \cos(A))l\sin(A)} \quad (8)$$

Although the ExpAsym structure represents a competitive solution from the relative density point of view, there is still an aspect which requires further attention – the relatively low bonding surface still rises some problems when sandwich panels are constructed using this topology as a core, due to the complexity of assembling the lateral faces [6].

3.2. The relative density of the novel cellular core

By analysing the topology of the newly developed structure we can conclude that the value of the internal angle (A) depends on the value of the radius of the perforation (R) and the length of the arm's base, using the formula below:

$$A = \arctan\left(\frac{2R}{c}\right) \quad (9)$$

Using the formulae for the gauge dimensions, the relative density of the novel cellular core can be determined as follows:

$$\rho_r = \frac{4cg(l + l_1 + 2R) - bgl_1 - 8glr - 4\pi gR^2}{wth} \quad (10)$$

For a better understanding of the structure's behaviour, tri-dimensional cases were taken into consideration. Each of them contains a set of fixed parameters (l1=10 mm, g=0.3 mm and c=15 mm) and a set of variables: the radii of the perforations (3,4 and 5 mm) and the length of the cell's arm, in the form of three ratios, ***l=c, l=2c, l=3c***.

As previously shown in eq. (6), (7) and (8), the gauge dimensions are depending on the value of the expansion angle (*B*). Taking this into consideration, an essential aspect is the influence the latter has on the relative density of the core, the results being shown in the graph in figure 4.

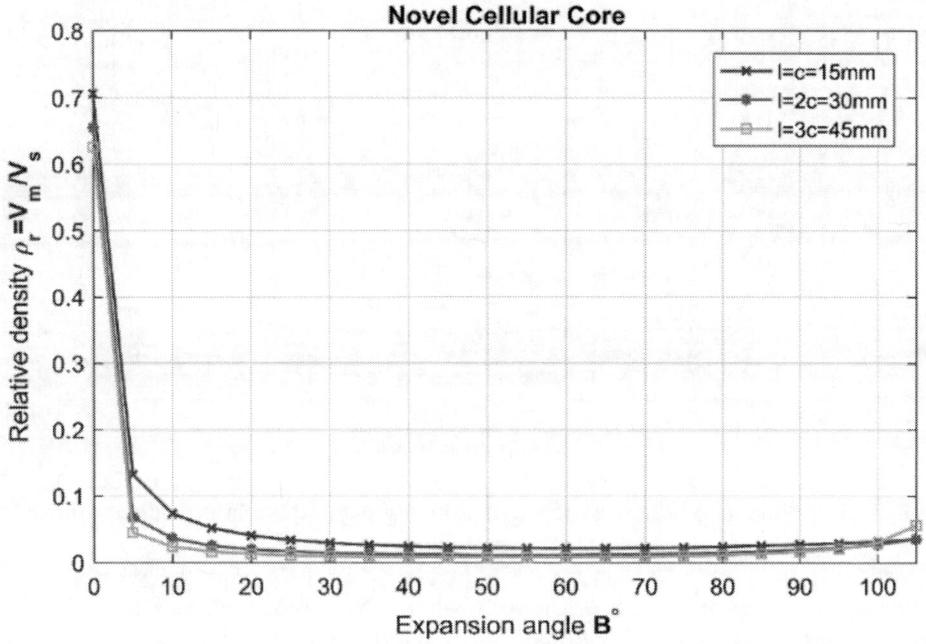

Figure 4. The dependency between the expansion angle and the relative densityof a unit cell

This graph shows the dependency between the expansion angle and the relative density of a single unit cell. The value for the relative density decreases when increasing the expansion angle.

In order to highlight the advantages of the newly developed cellular structure, compared to the ones mentioned above (triangular, trapezoidal, honeycomb, ExpaAsym), the relative density has been determined for a single unit cell, while maintaining several relations between the established parameters. The length of the cell's arm (*l*) has been kept the same for all the types of structures mentioned above (***l=15 mm, 30 mm and 45 mm***). The value for the expansion angle has varied in the interval [0°-120°] according to how permissive the topology is, while the width (*t*) and the height (*h*) have been determined according to the length (*w*) as follows: ***t=w/2*** for triangular and trapezoidal cores and ***h=w/2*** for the honeycomb.

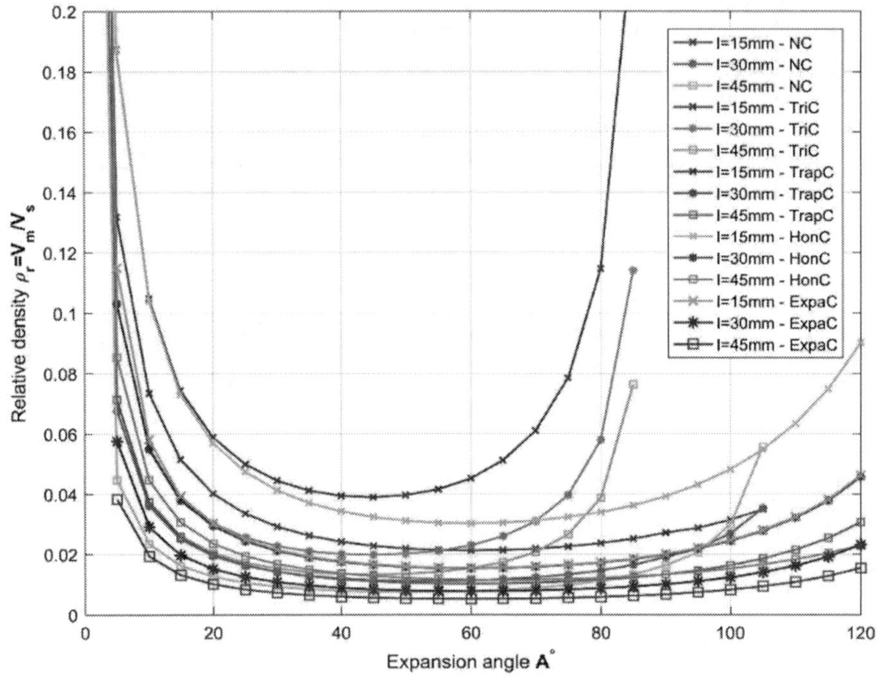

Figure 5. The dependency between the expansion angle and the relative density for the studied topologies.

The results are shown in the graph in figure 5, where: ***NC*** - novel cellular core, ***TriC*** – triangular cellular core, ***TrapC*** – trapezoidal cellular core, ***HonC*** – honeycomb cellular core, ***ExpaC*** – ExpaAsym cellular core.

4. Discussions

The values for the relative density for the cellular cores mentioned above have been obtained with the use of eq. (4). The common parameter for all the geometries taken into consideration within this study was considered the length of the cell's lever arm (*l*), which has resulted in three dimensional cases: ***l=15 mm, 30 mm and 45 mm***.

As it can be easily observed in the graph in figure 5, the relative density has a similar flow for all the studied topologies, in all of the established geometric cases; it decreases when increasing the expansion angle; reaching minimal values in the range [30°-70°], figure 6. Another important aspect worth outlined is that the relative density decreases when increasing the length of the cell's lever arm, due to its direct proportionality with the volume of void in the cellular structure. Thus, the maximum values have been obtained when ***l=15 mm***, while the minimum ones were encountered when ***l=45 mm***.

For the triangular topology, the variation of the relative density has values in the range [0.01-0.06], while for the trapezoidal structure, the relative densities are set [0.01-0.04], with the advantage that the latter offers a significant bonding surface, which could deliver a cost efficient manufacturing process.

Compared to the other two structures mentioned above, the relative density for the honeycomb cellular core has the same evolution, with identical values as for the trapezoidal, in the given range of [0.01-0.04]. It represents closed topology consisting in two trapezoidal structures bonded to each other, but it also requires a double quantity of base material for creating a single unit cell than the latter. Together with a complex tri-dimensional geometry, a time consuming manufacturing technology and a bonding surface equal to the thickness of the base material, the resulting production costs are still extremely high even with nowadays technology. Despite this main disadvantage, it is successfully used in a large number of applications due to its high mechanical properties (strength and stiffness).

The ExpaAsym cellular core was designed to account for this specific disadvantage offered by the honeycomb and as shown in figure 6, its relative density has been reduced by half, due to its complex, yet easy obtainable topology, with values between 0.005-0.002. In the case of an expansion angle equal to 60°, the result is a honeycomb-like structure, which offers the advantages of reducing the material use by 50%, while providing a consistent bonding surface as opposed to the honeycomb. Besides these two main benefits, it is also significantly easier to manufacture by mechanical expansion, and thus it represents a potentially less expensive solution.

In the case of the newly developed cellular structure, the relative density is comparable with the one of the ExpaAsym, and varies in the range of 0.008-0.03. Having a similar manufacturing process, the main advantage offered by this topology is the consistent bonding surface it provides. This is an important issue to take into consideration when manufacturing sandwich panels, since it can simplify the assembling process and significantly reduce its production costs.

The mechanical expansion of the structure is being exerted on all three dimensions (directed by a set of bending lines at an internal angle *A*) thus; it provides a stable structure, which could lead in producing a substantially wider area of sandwich core using the same amount of base material.

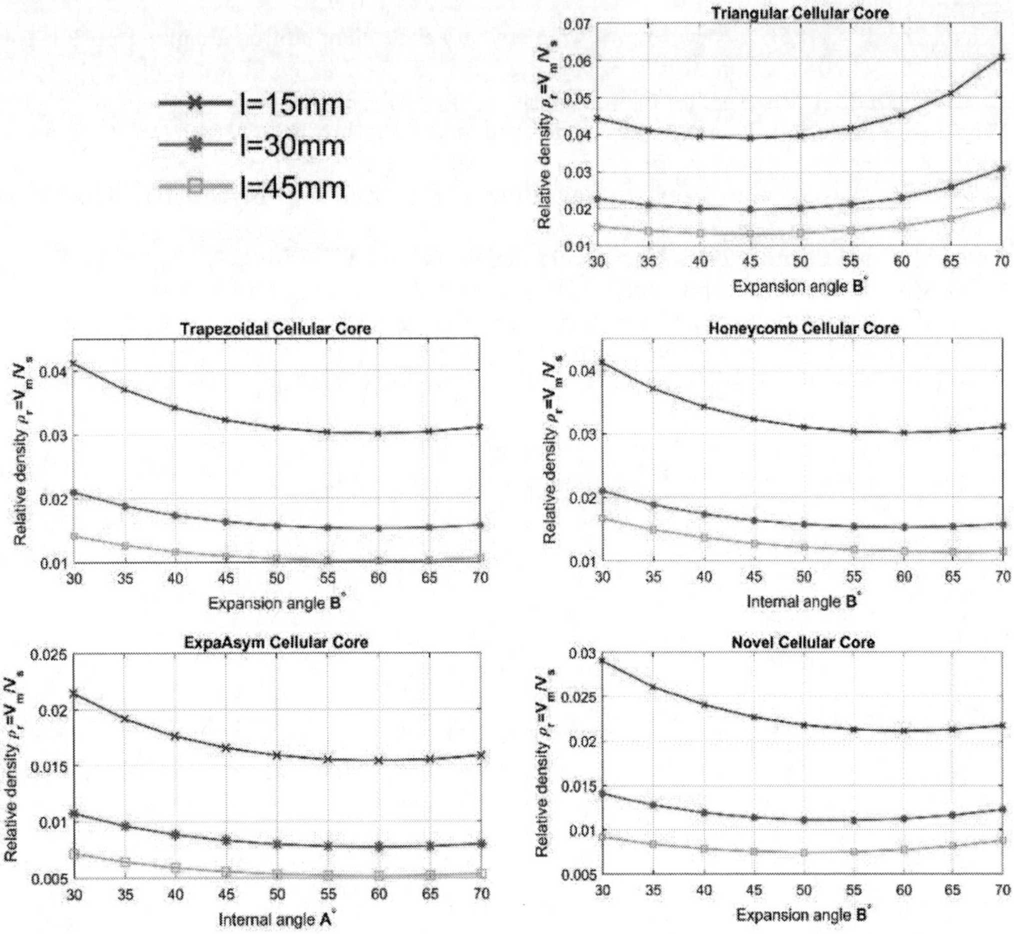

Figure 6. Relative densities for the studied cellular structures in the interval [30°-70°].

5. Conclusions

The presented novel cellular core has been developed in order to keep up with the demands on the nowadays market and the main purpose of this paper was to perform an initial evaluation of its performance.

In order to achieve this goal, relative density has been taken into consideration as being the most important parameter when it comes to evaluating the lightweight capacity of cellular cores used in sandwich panel construction.

In this regard, a comparison between the relative densities of the novel core has been performed, as opposed to the ones of four existing cellular structures from the literature, only to conclude that it represents a promising solution for this industry.

The main advantages offered by this new design are consisting in a low relative density of the core, a significant bonding surface and a reduction in usage of base material. Together with a simple manufacturing method, they represent an important set of assets in the attempt in making a difference in designing lightweight structures.

A further objective of this research is to analyse the novel cellular structure's strength and stiffness in different mechanical load scenarios and to optimize its topology in order to obtain the highest mechanical properties.

6. References

[1] Zenkert 2009 *An introduction to sandwich structures* Student edition chapter 1 pp 1.1-1.4
[2] Velea M N 2011 *Lightweight cellular structures – Design, Modeling and Analysis* ed Universității Transilvania din Brașov (Brașov) pp 12-13
[3] Worldwide Emission Standards Passenger cars and light duty 2016-2017 *Delphi Innovation for the new world*
[4] Zhunli Tan, Lishuo Bai, Bingzhe Bai, Bing Zhao, Zhiqiang Li and Hongliang Hou 2016 *Materials and Design* **92** 724–730
[5] Velea M N and Lache S 2011 *Mechanics of Materials* **43** 377-388
[6] Velea M N, Wennhage P and Lache S 2012 *Materials and Design* **36** 679-686
[7] Li-Jia Feng, Jian Xiong, Li-Hong, Yang Guo-Cai, Yu Wen Yang and Lin-Zhi Wu 2017 *Int. Journal of Mechanical Sciences* **134** 589–598
[8] Wentao H, Jingxi Liu, Bo Tao, De Xie, Jiayi Liu and Min Zhang 2016 *Composite Structures* **158** 30–43
[9] Biagi R and Bart-Smith H 2012 *Int. Journal of Solids and Structures* **49** 3901–14
[10] Velea M N and Lache S 2009 *Proc. Engineering* **10** 287–292

IOP Conf. Series: Materials Science and Engineering **416** (2018) 012088 doi:10.1088/1757-899X/416/1/012088

Impact Modelling of Cermet Composites

E Postek[1,*], T Sadowski[2]

[1] Department of Information and Computational Science, Institute of Fundamental Technological Research Polish Academy of Sciences, Pawinskiego 5B, 02-106 Warsaw, Poland

[2] Department of Solid Mechanics, Lublin University of Technology, Nadbystrzycka 40 str., 20-618 Lublin, Poland

*Corresponding author: epostek@ippt.pan.pl

Abstract. The WC/Co is one of broad class of cermet materials (CM) that are applied in fabrication of machining tools. It means that they are subjected to different kind of dynamic loadings. They have very good mechanical properties. However, the degradation of the CM material under dynamic load has not been enough thoroughly investigated. Experimental results yield that the dissipation of the fracture energy in WC/Co samples is due to Co ductile failure at the WC and Co interface [1] and/or dimple rupture mechanism [2]. Stress concentrations about stress raisers such as grain bounds cause microcracking that is further propagated by dynamic loading. However, there are no such predictions for impact conditions. The main goals of the presentation are to explore the formerly created models of the two-phase composite [3, 4] towards impact conditions impacts and give qualitative predictions of the crack and plastic strains initiation for such cases. It has been found that microcracks development process and stress fields depend on impact velocity and the existence of discontinuities: (1) inside the Co binder and (2) the interface between the binder and the grains. Estimation of the microcracks distribution by means of damage parameter is given.

1. Introduction

New technological processes very often need up-to-date materials of polycrystalline structure. Ceramic Matrix Composites (CMC) and monolithic ceramics are presented in, for example [1, 2]. Different models of the materials are given in, for example, [3-14]. Further on, nanoceramics is of more complicated structure [15, 16] and the same concerns functionally graded materials [17-21]. The other CMs materials, for example, titanium/molybdenum carbides and tungsten carbide with cobalt binders (WC/Co) are described in [22-28]. Methodologies of assessment the properties of complex materials are presented in [29-36].

Behaviour of novel composites in the context of the multiscale modelling is commonly investigated, [37-40]. For example, two-scale approach that decomposes the case into levels of the respective fine and coarse scale is derived by de Borst [37]. The multiscale modelling encounters the fundamental problem of coupling the different scales taking into account different discretization paradigms, i.e. finite elements, finite differences, meshless methodologies, etc. A variational multiscale approach pretends to be the successful methodology for scales coupling, [39]. The partition-of-unity approach for multiscale models of CM and CMC appear to be successful [37].

CM behaviour due to impact loading is more complex than one phase polycrystalline ceramics. It concerns the modelling of the CMs as well.

For example, the WC/Co properties were examined experimentally, [22, 23, 41]. Concerning our model of impact of CMs, we apply meso-mechanical way of modelling for the RVE containing grains and thin continuous intergranular layers. Experiments on thin metallic intergranular layers in CMs have been described, for instance, in: [22, 23] for WC/Co, [42] for a Pt layer on Al_2O_3, [43] for Al_2O_3 bonded with pure Al and Al with addition of 4% Mg, [44] for TiC-Mo_2C grains with Ni binder. It has been shown in experiments [43] that metallic interfaces are soft enough that failure mechanism is ductile. Considering [23, 24] we note that in the WC/Co CMs, the most of the fracture energy is dissipated via the ductile failure of the soft Co binder. In the failure, the two fracture mechanisms are engaged, namely, dimple rupture mechanism [64] and microcracks development in the Co binder/ WC grain interface, [65]. Co binder ductile failure seems to be the main mechanism of fracture resistance in the WC/Co CM, [24].

Up to now, the evaluation of equivalent mechanical constants of CMs has been carried out under static load, for example, [33], [44-49]. The loading models depict continuous modifications of the material. The changes of the internal structure concern rotations of grains, peaks of plastic strains in metallic binders, microcracks development [28, 50-51]. In the mentioned papers, the authors investigate ductile failure of the Co intergranular thin layer with I fracture initiation mechanism. It means, the coalescence of micro-voids that further cause cracking nucleation and then their propagation. Description of the II fracture initiation mechanism of the Co binder is given in [23, 24]. The ductile failure of the Co binder is caused by phenomena that initiate in the WC/Co interface. It has been found that the role of the interfaces is important and depending on the phenomena in interfaces may trigger different ways of cracks advancement. For example, the phenomenon has been observed in the CMC bi-material interfaces [52-54]. In this case, the nucleated cracks can be arrested for a moment or can advance along the stiff bound barrier.

WC/Co is widely purchased for fabrication of cutting tools. Therefore, the understanding of the microcracks development that is vital for the durability of the tools is of high importance. The mirocracking process is revealed in experiments for repeatable compressive load [55], and also for repeated impacts [74]. Layered composites under dynamic load are described in [66-68, 76]. Examples of impacts and blasts exerted on such structures are shown, for example, in [67, 69-73, 75]. Nevertheless, we still see a gap in the wide literature concerning WC/Co behaviour in the case of impacts.

To extend previously developed models for the quasi-static loading of the WC/Co [28, 47–51] we analyse microcracks development during compressive impact of the CM sample into rigid wall. It is justified by the fact that short impacts appear during hard materials machining. The numerical model under consideration takes into account dimensions and structure of the components. The constitutive laws are considered for the components, as well. In consequence, the factors that affect internal structure of the CM variation are as follows: (a) positions of CM components, (b) presence of grain/binder transition zones (interfaces) that are the microcracks triggers, (c) finite deformations of the Co intergranular layers (binders), (d) plastic strains in binders, (e) grains rotation. The other vital factor is the velocity of the impact. It was shown by means of damage parameter that has been pointed out peaks of microcracks, [56].

2. Physical and numerical model

The CM polycrystal consists of the elastic WC grains of very high strength and Co binders that are elastic-plastic. The constitutive material model is in agreement with experiments that are shown in [24, 43]. While the grains are discretized with hexahedral 3D elements only the binders are discretized with assemblage of 3D hexahedrals and 3D interface elements. The interface elements are present in the binders, in the binders joints and in their structure. They allowed introducing discontinuities into the model.

The finite element model stands for the Representative Volume Element (RVE) of the CM. The calculation scheme of the system that includes CM sample, rigid wall and initial and boundary

IOP Conf. Series: Materials Science and Engineering **416** (2018) 012088 doi:10.1088/1757-899X/416/1/012088

conditions is shown in figure 1. The sample impacts the rigid wall along *x*-axis of the model. The individual discretizations of the binders and the grains are presented in figure 2 and figure 3.

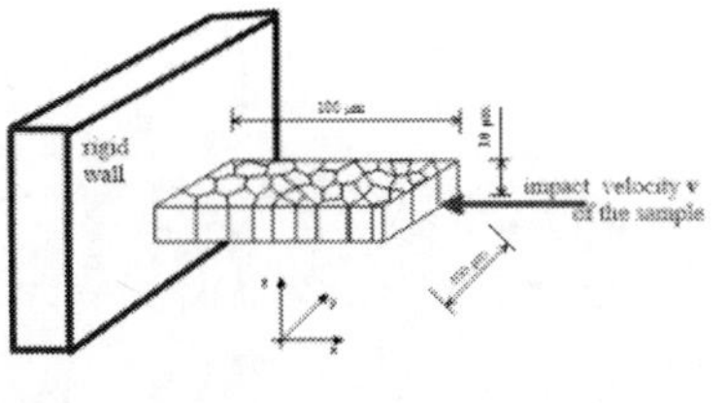

Figure 1. Scheme of the structure.

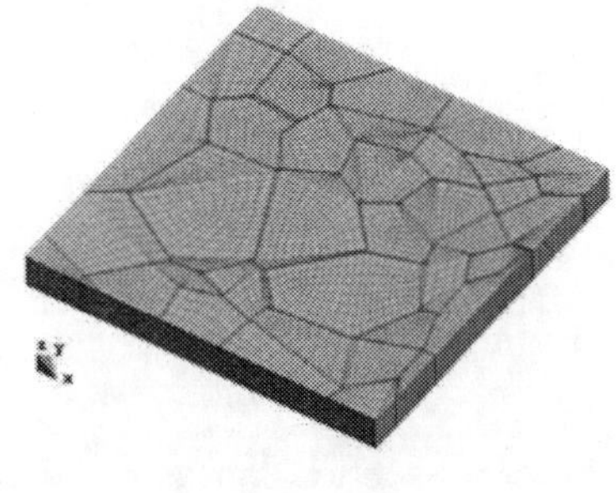

Figure 2. Discretization of the sample.

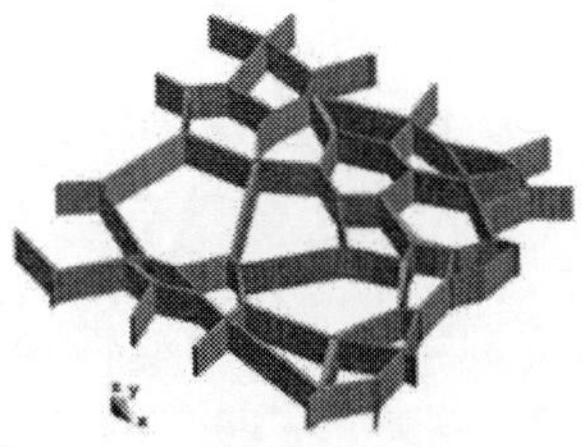

Figure 3. Discretization of the interfaces.

We consider two models, continuous (M1) and discontinuous (M2). The potential discontinuities are allowed by the interface elements that are inserted into the binders. The model M1 counts 34572 linear C3D8 hexahedrals and 41216 nodal points. The model M2 is additionally enriched with 47407 linear 8-nodes interface elements that cause the increase of the number of nodal points up to 152169. We apply MSC Patran [58] for geometrical modelling, Abaqus program for the numerical simulations [57] and program GiD for the postprocessing phase [59]. The material properties of the grains read: Young's modulus is $4.1\text{x}10^{11}$ Pa, and Poisson's ratio is 0.25. The intergranular binders are elastic-plastic with Young's modulus $2.1\text{x}10^{11}$ Pa, Poison's ratio 0.235, yield stress $2.97\text{x}10^{8}$ Pa. The material obey the isotropic hardening rule.

The created finite element model allows for the brick elements separation in the cobalt binders. It is achieved by the inserting of the cohesive interface elements that surrounds each of the hexahedral element in the binder. The properties of the binders are established considering the work [26], and adopting the traction-separation constitutive law in the interface elements, [60, 61].

The constitutive parameters of the constitutive law in the interfaces read: elastic modulus normal to the midsurface of the interface element (*n*) is $2.1\text{x}10^{11}$ Pa, both shear moduli in the directions that are tangent to the midsurface of the interface element (*s, t*) read $1.373\text{x}10^{11}$ Pa. The damage initiation condition is postulated quadratic:

$$\left\{\frac{\varepsilon_n}{\varepsilon_n^o}\right\}^2+\left\{\frac{\varepsilon_s}{\varepsilon_s^o}\right\}^2+\left\{\frac{\varepsilon_t}{\varepsilon_t^o}\right\}^2=1 \qquad (1)$$

In the equation above, ε_n is the nominal strain in the normal direction *(n)* to the midsurface of the interface element, and ε_s, ε_t are the nominal strains in the two orthogonal *(s)* and *(t)* tangent directions to the midsurface of the interface element. The variables ε_n^o, ε_s^o, ε_t^s are the maximum nominal strains at the normal mode (*n*) and the shear modes in the directions (*s*) and (*t*) with the postulate of their independence. The maximum nominal strains are $2.5\text{x}10^{-4}$, $6.65\text{x}10^{-4}$ and $6.65\text{x}10^{-4}$, respectively. Considering the isotropic damage, the maximum reduced stress is given as follows, [62]:

$$\sigma_{\max}^D=\sigma_{\max}(1-D) \qquad (2)$$

The reduced stress refers to the damage variable $D=A/A_o$ with the damaged cross-section area *A*, and the area of the pristine cross-section A_o. The damage variable *D* is in the range of 0 and 1. The D variable The evolution of the damage variable follows an exponential rule. The damage variable reflects the deterioration of the cross-section because of microcracks showing up in the course of loading path. The evolution equation of damage variable is as follows, [60]:

$$D = 1 - \left(\frac{\delta_m^o}{\delta_m^{\max}}\right)\left(1 - \frac{1 - \exp\left(-\alpha\left(\frac{\delta_m^{\max} - \delta_m^o}{\delta_m^f - \delta_m^o}\right)\right)}{1 - \exp(-\alpha)}\right) \tag{3}$$

The $\sigma_{\max}$ is the stress referred to the cross-section that is not damaged. The stress $\sigma_{\max}^D$ is the stress referred to the pristine cross-section. It is the nominal stress. Further on, δ_m^o is the displacement at which the stress touches the maximum strength $\sigma_{\max}$ at the start of damage, $\delta_m^{\max}$ is the maximum displacement attained in the loading path, δ_m^f is the failure displacement, and α is the damage rate.

3. Numerical results

We note qualitative dissimilarity in displacement fields in the models M1 and M2 that undergo the impact velocity of 50 m/s. While comparing figure 4(a) and figure 4(b) we observe significant influence of the discontinuities in the binders. We mind a wedge that builds along the grains bounds, figure 4(a). Another observation concerns the displacement fields, namely, we find that the displacement field in the case M1 is smoother than in the case of model M2, figure 4(b).

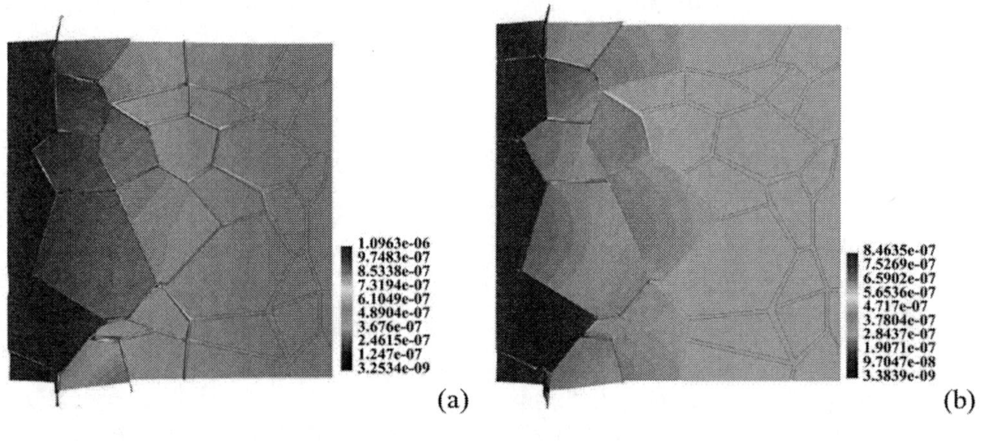

Figure 4. Displacement fields for two models, top view: (a) case M1; (b) case M2.

In contrast to model M1, the wedge is not formed in case M2. In both cases we can see thinning of the binders and squeezing their material off. However, we note that the squeezing off of the interfaces material in case M1 is more intense than in the case M2. In model M2, the discontinuities are sharper than in case of M2. In cases M1 and M2, the maximum displacements are 1.10E-06 m and 8.46E-07 m with the difference 23.1%. The squeezing off develops along the formed wedge, figure 5(a) and ranges deeper into the sample than in case of model M2, figure 5(b). The squeezed-out material is more distributed in the model with discontinuities.

IOP Conf. Series: Materials Science and Engineering **416** (2018) 012088 doi:10.1088/1757-899X/416/1/012088

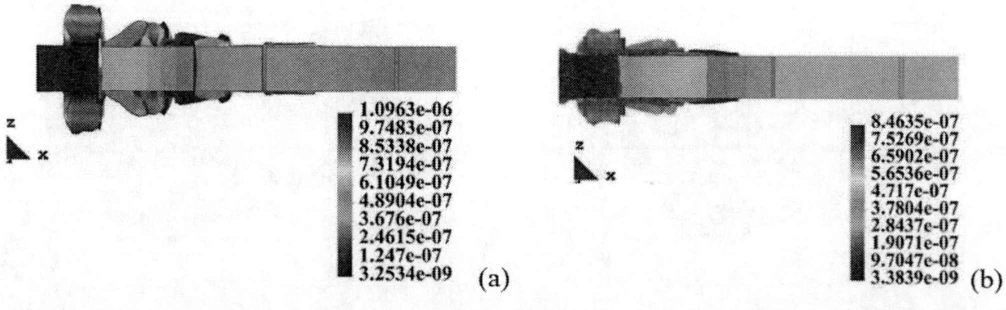

Figure 5. Displacement fields for two models, side view: (a) case M1; (b) case M2.

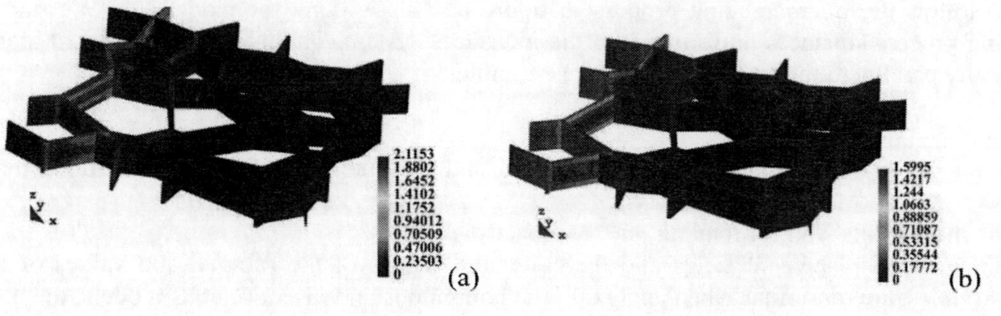

Figure 6. Equivalent plastic strain distribution in binders, time instant 1.0E-08 s: (a) case M1; (b) case M2.

Gross equivalent plastic strain is qualitatively similarly distributed in binders in both models, figure 6. However, the plastic deformation is higher in case M1 than in case M2 with the difference of 134%. We note that plastic strains in interfaces start to develop mostly at the edges of the interfaces. It concerns both M1 and M2 models. The effect can be seen particularly well in the binders that are placed close to the striking edge of the sample.

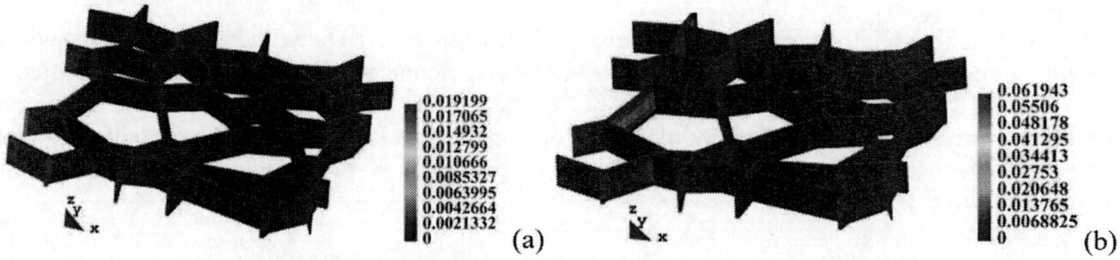

Figure 7. Damage parameter development: (a) time instant 2.0E-09 s; (b) time instant 1.0E-08 s.

The damage parameter distribution is given in figure 7. Damage appears in interfaces at the attacking edge, figure 7 (a). When the process advances, damage propagates into the interfaces system. We note that damage is higher in the middle of interfaces than at their edges, figure 7(b). We recall here figure 6 that illustrates spatial distribution of plastic strains. We note a qualitative difference between damage parameter distribution and equivalent plastic strain distribution. Plastic strains have appeared on the edges of interfaces while damage parameter is high inside interfaces.

Figure 8. Microcracks development: (a) time instant 2.0E-09 s; (b) time instant 1.0E-09 s.

We follow the microcracking process in figure 8. This is done for model M2. We find that the cracking process advances uniformly into the interfaces system. The front of the cracked interfaces is practically parallel to the attacking edge of the sample.

4. Final remarks

We point out main attributes of the new model and the summary of our observations that are as follows:

- the model suits well for loading due to impact,
- since the displacement fields and thereafter the shapes of the sample, and values of the other variables, are dissimilar when analyzing the continuous and discontinuous models, the model that includes the cracks effect should be considered,
- in the continuous model, the grains move and create a visible wedge,
- concerning the discontinuous model the grains do not build the wedge but they move along the binders that are about parallel to the striking boundary of the sample,
- plastic strains and damage parameter give their presence at the start of the impact process,
- plastic strains shows up in the bounds of binders and develop towards their interior,
- damage parameter starts to grow in the interior of the interfaces.

The model is intended to develop RVE [63, 77] of larger size and more precise.

Acknowledgments

This work was financially supported by Ministry of Science and Higher Education (Poland) within the statutory research (IPPT PAN) and National Science Centre (Poland) project No 2016/21/B/ST8/01027 (Lublin University of Technology). The calculations were done at the Interdisciplinary Centre for Mathematical and Computational Modelling, University of Warsaw, Poland. The licenses for the MSC Patran and Abaqus programs were provided by Academic Computer Centre in Gdańsk, Poland.

References

[1] Fett T and Munz D 1999 *Mechanical Properties, Failure Behaviour, Materials Selection, Ceramics* (Berlin Heidelberg New York: Springer)
[2] Rice R W 1998 *Porosity of ceramics* (New York: Marcel Dekker Inc.)
[3] Gömze L A and Gömze L N 2009 *Ěpitöanyag: JSBCM* **61** 38-42
[4] Sadowski T and Samborski S 2003 *J. Am. Cer. Soc.* **86** 2218-21
[5] Gömze L A and Gömze L N *IOP Conf. Ser.: Mater. Sci. Eng.* **47** 012033
[6] Sadowski T and Samborski S 2003 *Comp. Mat. Sci.* **28** 512-17
[7] Espinoza H D and Zavattieri P D 2003 *Mech. Mater.* **35** 333-64
[8] Espinoza H D and Zavattieri P D 2003 , *Mech. Mater.* **35** 365-94
[9] Sadowski T and Marsavina L 2011 *Comput. Mat. Sci.* **50** 1336-46
[10] Sadowski T 2012 *Comput. Mat. Sci.* **64** 209-11

[11] Sadowski T and Samborski S 2008 *Comp. Mater. Sci.* **43** 75-81
[12] Ghosh D, Banda M, Akurati S, Kang H and Fakharizadeh O 2017 *Scripta Mater.* **138** 139-44
[13] Sadowski T and Pankowski B 2016 *Comp. Struct.* **143** 388-94
[14] Sadowski T and Golewski G 2008 *Comp. Mat. Sci.* **43** 119-26
[15] Winter M 2002 *Nanocrystalline ceramics, synthesis and structure* (Berlin: Springer)
[16] Koch C C, Ovid'ko I A, Seal S and Veprek S 2007 *Structural nanocrystalline materials: fundamentals and applications* (Cambridge: Cambridge Univ. Press)
[17] Suresh S and Mortensen A 1998 *Fundamentals of functionally graded materials* (Cambridge: The University Press)
[18] Sadowski T, Marsavina L, Peride N, Craciun E M 2009 *Comput. Mat. Sci.* **46** 687-693
[19] Marsavina L, Linul E, Voiconi T, Sadowski T 2013 *Polymer Testing* **32** 673-680
[20] Sadowski T and Golewski P 2012 *Comput. Mat. Sci.* **52** 293-97
[21] Sadowski T and Golewski P 2012 *Comput. Mat. Sci.* **64** 285-88
[22] Siegl L S and Exner H E 1987 *Metall. Trans. A.* **18A** 1299-308
[23] Siegl L S and Fischmeister H F 1988 *Acta Metall.* **136** 887-97
[24] Ravichandran K S 1994, *Acta Metal. Mater.* **42** 143–50
[25] Hönle S and Schmauder S 1998 *Comput. Mater. Sci.* **13** 56-60
[26] Felten F, Schneider G and Sadowski T 2008 *Int. J. Ref. Mat. Hard. Mat.* **26** 55-60
[27] Li W, Wang H, Wang L, Hou C, Song X, Liu X and Han X 2017 *Mat. Res. Letters* **5** 55-60
[28] Postek E and Sadowski T 2011 *Comp. Interf.* **18** 57-76
[29] Nemat-Nasser S and Horii M 1999 *Micromechanics: overall properties of the heterogeneous materials* (Amsterdam – New York – Oxford – Tokyo: Elsevier)
[30] Kachanov M 1993 *Advances in Appl. Mech.* **30** 259-445
[31] Gross D andSeelig T 2006 *Fracture mechanics with an introduction to micromechanics*, (Berlin- Heidelberg: Springer)
[32] de Borst R and Sadowski T (Eds) 2008 *Lecture notes on composite materials* (Wien-New York: Springer)
[33] Schmauder S and Mishnayevsky L *Micromechanical and nanosimulation of metals and composites* (Berlin Heidelberg: Springer)
[34] Sadowski T and Trovalusci P (Eds) 2014 *Multiscale Modelling of of Complex Materials, CISM Courses and Lectures No. 556 International Centre for Mechanical Sciences* (Wien – NewYork: Springer)
[35] Altenbach H and Sadowski T (Eds) 2015 *Failure and Damage Analysis of Advanced Materials, CISM Courses and Lectures No. 560 International Centre for Mechanical Sciences* (Wien – NewYork: Springer)
[36] Flavata A, Trovalusci P and Masiani R 2016 *Comput. Mater. Sci.* **116** 22-31
[37] de Borst R 2008 *Comput. Mater. Sci.* **43** 1-15
[38] Kassner M L, Nemat-Nasser S, Suo Z, Bao G, Barbour J C, Brinson L C, Espinosa H, Gao H, Granick S, Gumbsh P, Kim K S, Knauss W, Kubin L, Langer J, Larson B C, Mahadevan L, Majumdar A, Torquato S and van Swol F 2005 *Mech. Mat.* **37** 231-59
[39] Hughes T J R 1995 *Comput. Meth. Appl. Mech. Eng.* **127** 387-401
[40] Lippmann N, Steinkopff T, Schmauder S and Gumbsh P 1997 *Comput. Mater. Sci.* **9** 28-35
[41] Soppa E, Schmauder S, Fisher G, Brollo J and Weber U 2003 *Comput. Mat. Sci.* **28** 574-86.
[42] Dalgleish B J, Lu M C and Evans A G 1998 *Acta Metall.* **36** 2029-35
[43] Dalgleish B J, Trumble K P and Evans A G 1989 *Acta Metall.* **37** 1923-31
[44] Zarikson J, Larsson A and Andren H O 2001 *Micron.* **32** 707-12
[45] Sadowski T and Nowicki T 2008 *Comput. Mat. Sci.* **43** 235-41
[46] Tkalich D, Cailletauld G, Yestrebov V Y and Kane A 2017 *Mech. Mater.* **105** 166-87
[47] Sadowski T, Hardy S and Postek E 2005 *Comput. Mat. Sci.* **34** 46-63
[48] Sadowski T, Hardy S and Postek E 2006 *Mat. Sci. Eng. A.* **424** 230-38
[49] Sadowski T, Postek E and Denis C 2007 *Comput. Mat. Sci.* **39** 230-236.

IOP Conf. Series: Materials Science and Engineering **416** (2018) 012088 doi:10.1088/1757-899X/416/1/012088

[50] Dębski H and Sadowski T 2014 *Comput. Mat. Sci.* **83** 403-11
[51] Dębski H and Sadowski T 2017 *Comp. Struct.* **159** 121-27
[52] Marsavina L and Sadowski T 2009 *Comput. Mat. Sci.* **45** 693-97
[53] Marsavina L and Sadowski T 2009 *Comput. Mat. Sci.* **44** 941-50
[54] Golewski G, Sadowski T 2014 *Const. Build. Mat.* **51** 207-14
[55] Krüger L, Mandel K, Krause R and Radajewski M 2015 *Int. J. Refrac. Met. Hard. Mat.* **51** 324–31
[56] Zhang H, Lu Q, Zhang L and Fang Z Z 2010 *Int. J. Refr. Met. Hard. Mat.* **28** 434-40
[57] Abaqus 6.13. User's Manual.
[58] http://www.mscsoftware.com/product/patran (accessed 20 December 2017)
[59] http://www.gidhome.com/ (accessed 20 December 2017)
[60] Camanhi P P and Davila C G 2002 *Mixed-Mode Decohesion Finite Elements for the Simulation of Delamination in Composite Materials* NASA/TM-2002-211737
[61] Zavattieri P B, Hector L G and Bower A F 2008 *Eng. Fract. Mech.* **75** 4309-32
[62] Kachanov L M 1986 *Introduction to Continuum Damage Mechanics* (Dordrecht: Martinun Nijhoff Publishers)
[63] Jimenez-Pique E, Turon-Vinas M, Chen H, Trofinov T, Fair J, Tarres E and Llanes L 2017 *Int. J. Refr. Met. Hard. Mat.* **67** 9-17
[64] Liu X, Wang H, Wang L, Hou Ch, Song X, Liu X and Han X 2017 In situ study of fracture behavior of ultrafine WC–Co cemented carbide, *Mater. Res. Lett.* **5** 55-60
[65] Sigl L S and Exner H E 1987 *Metall. Trans. A.* **18** (1987) 1299-308
[66] Kärger L, Baaran J, Gunnion A and Thomson R, 2009 *Compos. Part B.* **40** 71-6
[67] Vo T P, Guan Z W, Cantwell W J and Schleyer G K 2013 *Compos. Part B.* **44** 141-51
[68] El Sayed S and Sridharan S 2001 *Compos. Part B.* **32** 545-53
[69] Krishnan K, Sockalingam S, Bansal S and Rajan S D 2010 *Compos. Part B*, **41** 583-93
[70] Corbett G G, Reid S R and Johnson W 1996 *Int. J. Impact Eng.* **18** 141-230
[71] Grady D 1999 *Int. J. Impact Eng.* **23** 307-17
[72] Kawai N, Tsusru K, Shind D, Motoyashiki Y and Sato E 2011 *Int. J. Impact Eng.* **38** 542-45
[73] Compton B G and Zok F W 2013 *Int. J. Impact Eng.* **62** 75-87
[74] Huang Z Q and Gang L 2017 *Eng. Fail. Anal.* **80** 273-77
[75] Allix O 2001 *Compos. Sci. and Tech.* **61** 2193-205
[76] Guinard S, Allix O, Guédra-Degeorges D and Vinet A 2002 *Compos. Sci. and Tech.* **62** 585-89
[77] Karamnejad A and Sluys L J 2014 *Comp. Meth. Appl. Mech. Eng.* **278** 423-44

IOP Conf. Series: Materials Science and Engineering **416** (2018) 012089 doi:10.1088/1757-899X/416/1/012089

The Mechanical Properties of Waste Tire Cords Reinforced Geopolymer Concretes

M Łach[1,*], A Kiszka[1], K Korniejenko[1] and J Mikuła[1]

[1] Institute of Materials Engineering, Faculty of Mechanical Engineering, Cracow University of Technology, Jana Pawła II 37, 31-864 Cracow, Poland.

*Corresponding author: mlach@pk.edu.pl

Abstract. The article aim is to determinate the possibilities of recycling the waste textile cord from used tires in geopolymer composites. The matrix was based on fly ash from power station located in city named Skawina (Poland) and fine sand in the ratio of 1:1. The process of activation was made by 10M sodium hydroxide solution combined with the sodium silicate solution. In order to manufacture the composites the additions of 5, 10, 20 and 30% of waste textile cord by volume were applied. The research methods used: microstructure research, density, flexural strength and compressive strength tests as well as analysis of breakthroughs. The results show the influence of addition of waste textile cord form used tires on the composite properties. The addition of textile fibres causes a reduction in the weight and density of the geopolymer. For samples with 20% fibre noticed the lowest density and the highest bending strength. The highest compressive strength was obtained for the geopolymer with 5% textile fibres.

1. Introduction

Global tire manufacturing was estimated to be over 17 million tonnes in 2016 and it is growing about 4% per year through 2022 [1,2]. In European Union the tire sales was 4.89 million tonnes in 2016 [3, 4]. Disposal of waste tire rubber is one of the most important problems in waste management in all parts of the world [5]. Disposal of waste tire can be potentially harmless for environment. Not only because of fire thread on the landfills, but also due to their heavy metal and other pollutant content and risk for the leaching of toxins into the groundwater [6]. In some parts such as in European Union, disposal of waste is prohibited by regulations [7]. They should be recovered and recycled. One of the possibility is thermal depolymerisation, also known as pyrolysis [7, 8]. The main problem is that this kind of solution required very expensive infrastructure and during the process we recover only energy not the raw material. According to the newest EU regulation the pyrolysis with energy recovery is one of the less desirable methods of waste treatment.

Nowadays, the material recycling of tire products is still on very low level. It is estimate that in Poland it is only 15% by weight of used tires. It is caused by a lot of problems during processing used tires. They are multicomponent construction. It mainly consists of elastomer (rubber), metal and textile fibres called textile cord [8, 9]. The exemplary scheme for line for used tires processing is presented in the figure 1 [9, 10].

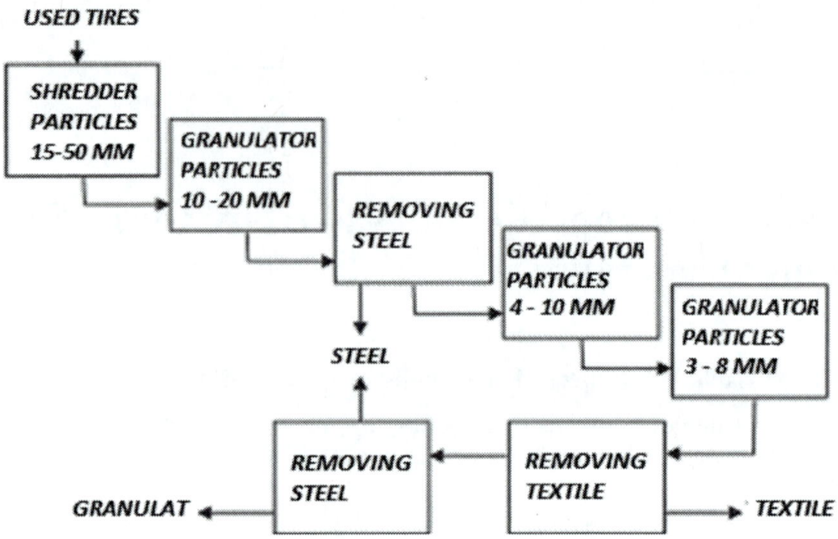

Figure 1. Diagram of a line for grinding tires [9, 10].

In the first step, tires are cut and crushed or dissolved. The rubber obtained in this way is subject to further fragmentation and thermal treatment. In the mechanical grinding process, rubber particles of various sizes, steel and textile waste are also obtained. This process is very energy-consuming and causes high noise emission. It is also possible to carry out the process in low temperatures, it happens that the rubber becomes brittle and much easier to crush.

A lot of components causes difficulties in the tire recycling process. The components such as rubber and steel can be relatively easy recovered and find the application for example in concrete industry [11, 12]. Obtained rubber part are used depending on their size. The large one as: grits and granules are used most often as surfaces of playgrounds, sports fields and as a material for sound and vibrations absorbing. The rubber dust is added to rubber mixtures, which are used for car mats, rubber mats, etc. The main problem is the recycling of the textile cord, especially because of its usually high degree of contamination with rubber material, which limits possibilities of use it after recovery [13, 14].

The problem of the re-use of textile fibres from the recycling of car tires is still not sufficiently explored. The first attempts of using a textile cord in modified mortars had a place in 2000. These tests gave positive results by improving the mechanical properties of the mortar, however, economic reasons prevented their application in practice [15, 16] Others were using textile waste as reinforcement materials for the production of polypropylene (PP) and made use of this composite in the production of car bumpers. In this case, the fibres also gave positive results, increasing bending strength, modulus of elasticity and acceptable impact strength compared to conventional PP [17]. Nowadays there are attempts to entirely utilize the waste tires granulates in geopolymers [18]. The research shows that it can be a promising way for utilization this kind of waste, but it still requires further research work.

2. Experimental procedure

Specimens based on geopolymer matrix with additions of 5, 10, 20 and 30% of waste textile cord by volume were investigated. The variables include different amount of reinforcement and reference samples based on matrix material without additions.

2.1. Materials

2.1.1. Geopolymer matrix. The geopolymer matrix were made from fly ash from the CHP plant in Skawina (Poland) and sand in the ratio 1:1. This fly ash is rich in oxides such as SiO_2 – 55.89% by mass and Al_2O_3 – 23.49% by mass. The morphology of the particles of fly ash was typical of such by-products of coal combustion and suitable for the process of alkali-activation. The process of activation has been made by 10M sodium hydroxide solution combined with the sodium silicate solution.

2.1.2. Waste tire cords. Textile fibres from used tires are a by-product obtained in the process of grinding a tire into a granulate. During subsequent shredding steps, the textile cord is separated from the rubber granulate. The weight of textile fibres makes up about 6% of the total weight of the tire, both in the case of passenger cars and trucks [19, 20]. The diameter of a single fibre is several tens of micrometres. The textile cord consists of numerous strands of twisted thin fibres made of cellulose, poly(butylene terephthalate) - PBT, polyethylene terephthalate - PET and polyamide-6,6 - PA66. The caravan's tire carcass consists of approximately 1500-1800 threads. They are arranged in a parallel manner and appropriately impregnated and joined with rubber [17, 20]. Fibres used for research were recovered by the Polish company Orzeł S.A.

2.2. Specimens
The samples were prepared using sodium promoter, fly ash and fibres (5, 10, 20 and 30% by volume of the composite). In the table 1 there is presented the addition according to volume and mass percentage.

Table 1. The content of fibres in the composite given in percent by volume and by mass.

Addition of waste textile cord by volume	Addition of waste textile cord by mass
5%	0.488
10%	0.864
20%	1.707
30%	2.530

The fly ash, sand, alkaline solution and fibres were mixed for about 20 minutes by using mixing machine Varimixer Bear (to receive the homogeneous paste). Next, it was poured into sets of plastic moulds. They have following dimensions: 20 x 5 x 10 cm. Then, the samples were subjected to vibratory removal of air bubbles. Tightly closed moulds were heated in the laboratory dryer for 24h at 75 °C. Then, the samples were unmoulded. Next, the samples was cured in laboratory condition during 28 days. Before the tests they were cut for samples proper for flexural and compressive strength tests (figure 2).

a) b) c)

Figure 2. Exemplary samples: a) geopolymer matrix without reinforcement, b) with 5% fibers addition, c) 30% fibers addition.

3. Analytical procedure

3.1. Microstructure research.
SEM type JEOL JSM 820 has been performed for microstructure investigations. The research has been made for the samples that previously had been broken while strength tests. The samples were covered with a thin layer of gold with JEOL JEE-4X vacuum sputter. The investigations were made for composite with 30% textile cord addition by volume at various magnifications.

3.2. Compressive strength test
Compressive strength tests were carried out according to the methodology described in the standard EN 12390-3. ('Testing hardened concrete. Compressive strength of test specimens'), because of absence of separate standards for geopolymer materials. The tests involved 2 cubic samples: 50 x 50 x 50 mm. Tests were performed on an concreate press - MATEST 3000kN with speed 0,5 [MPa/s].

3.3. Flexural strength test
Flexural strength tests were carried out according to the methodology described in the standard EN 12390-5. ('Testing hardened concrete. Flexural strength of test specimens'), because of absence of separate standards for geopolymer materials. The tests involved 2 cuboid samples: 50 x 50 x 150 mm. Tests were performed on an universal testing machine - MATEST 3000kN with speed 0,05 MPa/s.

3.4. Density
The density of the material was count on the basis of volumetric measurements and the weight of the analysed samples. The composites were treated as homogeneous material. The density was calculated by divided the total mass by total volume. The investigations were made on two specimens.

4. Experimental results and discussion

4.1. Microstructure research
The SEM observations were made for geopolymer with 30% textile cord addition. The exemplary results are presented in the figure 3.

Figure 3. SEM images of fly ash based geopolymer with 30% textile cord addition.

SEM investigations give a preliminary information about fibers distribution and the coherency between waste cord fibers and the geopolymer matrix. The observations in low magnification (20x and 50x) allow to estimate that the fibers distribution was regular. The investigations in high magnification (100x and more) allow to evaluate the fibers coherency to matrix as good (for the further research mechanical tests are needed).

4.2. Compressive strength test

Results for compressive strength test are presented in table 2.

Table 2. The average values for compressive strength tests.

The content of fibres by volume [%]	d_1 [mm]	d_2 [mm]	force [kN]	Compressive strength [MPa]
0%	50.31	50.34	67.551	26.671
5%	50.27	48.80	79.974	32.592
10%	50.73	49.30	59.068	23.640
20%	50.87	50.92	47.894	18.534
30%	49.62	54.01	51.468	19.207

The best result for compressive strength tests are for composites with small amount of fibres. The addition of 5% textile fibres give -32.59 MPa. The lowest compressive strength value is observed for material containing 20% of the cord, it is 18.53 MPa. Significant reduction of compressive strength values for samples with 20% and 30% fibre content could be influenced by the use of more sodium hydroxide during manufacturing process, which could lead to excessive reduction of the density of the tested material.

4.3. Flexural strength test

Results for flexural strength test are presented in table 3.

Table 3. The average values for flexural strength tests.

The content of fibres by volume [%]	d_1 [mm]	d_2 [mm]	force [kN]	Compressive strength [MPa]
0%	50.49	49.11	2.692	4.960
5%	51.28	49.45	2.443	4.389
10%	50.78	49.46	2.611	4.730
20%	52.33	49.92	2.951	5.085
30%	53.50	48.70	2.615	4.638

The addition of a textile cord slightly affects the value of bending strength of the tested composite. The most significant change is for 20% of fibers addition, it was increased by 0.13 MPa (for 5%, 10% and 30% a decrease in flexural strength were noted).

Additionally, after the flexural strength test the breakthroughs for the samples were examined. In the figure 4 the pores remaining after manufacturing process are visible, however, voids are much smaller in relation to voids visible on the geopolymer sample without additives. On the surface of the breakthrough, the textile cord fibers are visible.

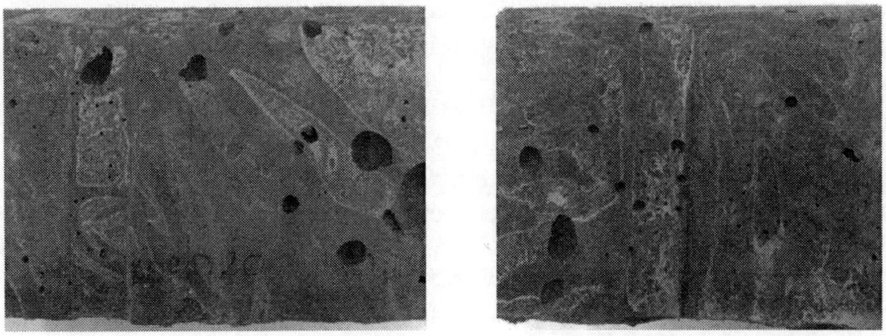

Figure 4. Breakthroughs of samples with 20% and 30% of textile fibers

IOP Conf. Series: Materials Science and Engineering **416** (2018) 012089 doi:10.1088/1757-899X/416/1/012089

Comparing with the material without the reinforcement the breakthroughs have more ductile character. The crack is slightly visible as a small line in the middle of samples.

4.4. Density

The results of density calculations are presented in table 4.

Table 4. The average values for density.

The content of fibres by volume [%]	a [mm]	b [mm]	h [mm]	volume [cm^3]	weight [g]	density [g/cm^3]
0%	52.072	50.246	50.972	133.3643	223	1.672112
5%	48.994	49.841	51.759	126.3920	208	1.645673
10%	49.114	49.221	51.079	123.4823	205	1.660157
20%	47.891	50.200	52.483	126.1759	206	1.632642
30%	49.515	49.589	53.758	131.9980	217	1.643965

On the basis of the presented data, we clearly notice a decrease in the density of the material with increasing the amount of the textile cord. The lowest density was noted for the material with the addition of 20% fibres and it amounted to 1.633 g/cm3. The highest result for 30% amount of fibres is probably the results of voids in materials (pores during manufacturing the sample). The decrease in density is related with the mass reduction. It is a beneficial phenomenon for many applications, because for example reduction of the cost of transportation.

5. Conclusions

The composites with addition of 5, 10, 20 and 30% of waste textile cord by volume were investigated. The conducted research shows that the use of these waste textile cord as an additive to a geopolymer-based composite material is possible. The addition of textile fibres reduces the weight and density of the geopolymer. For the geopolymer with the addition of 20% by volume, the lowest density was connected with the highest flexural strength. Additionally, introduction of reinforcement change the character of breakthrough from brittle to ductile. The highest compressive strength was obtained for a geopolymer with an addition of 5% by volume of fibres.

The results show that the addition textile fibers recovered from end-of-life tires to geopolymer can be a promising way of their utilization. On the basis of waste materials it is possible to receive the composite with good mechanical properties that could be used in construction industry. The industrial application required further research such as: water absorption, durability, cost-effectiveness and corrosion resistance.

6. References

[1] Sienkiewicz M, Kucinska-Lipka J, Janik H, Balas A 2012 *Waste Management* **32** 1742–1751
[2] Smithers Rapra 2017 *The Future of Tire Manufacturing to 2022*, https://www.smithersrapra.com/market-reports/tire-industry-market-reports/the-future-of-tire-manufacturing-to-2022 (access: 23.11.2017)
[3] Molanorouzi M, Mohaved SO 2016 *Polymer Degradation and Stability* **128** 115
[4] ETRMA, Statistics - Edition 2017 2017 *The European Tyre and Rubber Industry in figures*, http://www.etrma.org/library-2 (access: 23.11.2017)
[5] Thomas BS, Gupta RC 2016 *Renewable and Sustainable Energy Reviews* **54** 1323
[6] Derakhshan Z, Ghaneian MT, Mahvi AH, Conti GO, Faramarzian M, Dehghani M, Ferrante M 2017 *Environmental Research* **158** 462
[7] Jouhara H, Czajczyńska D, Krzyżyńska R, Spencer N 2017 *Energy* **134** 1121
[8] Hossain MS, Islam MR, Rahman MS, Kader MA, Haniu H 2017 *Energy Procedia* **110** 453
[9] Gronowicz J, Kubiak T 2007 *Problemy eksploatacji* **2** 5
[10] Parasiewicz W, Pyskło L, Magryta J 2005 *Recykling zużytych opon samochodowych*

IOP Conf. Series: Materials Science and Engineering **416** (2018) 012089 doi:10.1088/1757-899X/416/1/012089

(Instytut Przemysłu Gumowego 'STOMIL', Piastów)
[11] Thomas BS, Gupta RC, Panicker VJ 2016 *Journal of Cleaner Production* **112 (1)** 504
[12] Hesami S, Hikouei IS, Ali Emadi SA 2016 *Journal of Cleaner Production* **133 (1)** 228
[13] Kiszka A, Łach M, Korniejenko K, Mikuła J 2017 The Possibilities of Use Tire Cord as An Additive to Geopolymer Concretes, *Development of Eco-Friendly Composite Materials Based on Geopolymer Matrix and Reinforced with Waste Fibers*, eds. Mikuła J, Korniejenko K (Wyd. Instytutu Inzynirii Materiałowej Wydz. Mech. PK, Cracow) pp. 39-41
[14] Landi D, Vitali S, Germani M. 2016 *Procedia CIRP* **48** 508
[15] Bignozzi MC, Saccani A, Sandrolini F 2000 *Composites: Part A* **31** 97
[16] Bignozzi MC, Saccani A, Sandrolini F 2002 *Composites: Part A* **33** 205
[17] Czvikovszky T, Hargitai H 1997 *Nuclear Instruments and Methods in Physics Research B* **131** 300
[18] Mucsi G, Szenczi A, Nagy S 2018 *Journal of Cleaner Production* **178** 429
[19] Parres F, Crespo-Amorós JE, Nadal-Gisbert A 2009 *Construction & Building Materials* **23 (10)** 3182
[20] Landi D, Gigli S, Germani M, Marconi M 2018 Investigating the feasibility of a reuse scenario for textile fibres recovered from end-of-life tyres *article in press* Waste Management

Acknowledgments
This work was supported by the ERANet-LAC 2nd Joint Call (http://www.eranet-lac.eu) and funded by National Centre for Research and Development, Poland, under grant: "Development of eco-friendly composite materials based on geopolymer matrix and reinforced with waste fibers".

Organic Polymers Reinforced Inorganic Polymers - An Overview

K Korniejenko[1,*], B Tyliszczak[2], M Łach[1], J Mikuła[1], M Hebdowska-Krupa[1] and D Mierzwiński[1]

[1] Institute of Materials Engineering, Faculty of Mechanical Engineering, Cracow University of Technology, Jana Pawła II 37, 31-864 Cracow, Poland.
[2] Department of Chemistry and Technology of Polymers, Faculty of Chemical Engineering and Technology, Cracow University of Technology, Jana Pawła II 37, 31-864 Cracow, Poland.

*Corresponding author: kinga.korniejenko@mech.pk.edu.pl

Abstract. Nowadays hybrid materials started to be more and more popular. The main attention is focused on materials containing organic–inorganic components. Different combinations of organic and inorganic components allow the achievement of exceptional properties, such as electrical, catalytic, antibacterial, radiation-resistant, optical, excellent thermal, and mechanical properties. It gives a potential application for tis materials in completely new fields. The main objective of the article is to analyse challenges and future trends for new composites - inorganic-organic polymers as well as shows the results of preliminary research. The article based on: critical analysis of literature sources and laboratory research: visual assessment and compressive strength test.

1. Introduction

Inorganic polymers called also as geopolymers are chemically alkaline aluminosilicates [1,2]. They are obtained in the reaction of polycondensation of ortosilicans (comprising Si and Al atoms in their structure) with activated NaOH or KOH. Structurally, the obtained materials are all aluminosilicates made from Si and Al tetrahedra linked alternately by the oxygen atoms. In this network, there are also present positive ions (Na +, K +, Li +, etc.) that balance the electrical charge of aluminium, which is in tetrahedral coordination. Cross-linked in this way, silicates consist of chains which may include such entities as: sialan, silokso-sialan, or disilokso-sialan [3]. Geopolymers are attractive for industrial application taking into consideration several perspectives [1,4]: environmental (CO_2 emission and energy efficiency during the production process) [5,6], effectiveness of the manufacturing process [4,7], simplicity of application: low shrinkage and adherence to such materials as: concrete, steel, glass, ceramics [8,9], attractive mechanical properties [1,10], possibilities of application in different condition because their chemical resistant to atmospheric conditions and a variety of acids [11] and salts [12, 13] and fire resistant [1,14].

The one of the most important limitations of geopolymers is brittle behavior. They have relatively low tensile and flexural strength. To improve this mechanical properties it is possible to introduce to the matrix some additives [1,6]. The best results are achieved through the mixed organic and inorganic polymers. Because of that the effect of polymer admixtures on geopolymers is one of the up-to-date research topic on the optimization of the mechanical properties [5,15].

Recently, the research in this area are focused on the development of mechanical properties , especially brittleness reduction [5, 16]. A lot of research on reinforcement for geopolymers matrix such as fibres has been made [4,7]. Incorporation such kind of additives provide the control of cracking by bridging action during both micro and macrocracking of the matrix thus increasing the fracture toughness [16, 17].

Nowadays hybrid materials started to be more and more popular. The main attention is focused on materials containing organic–inorganic components. They have gained increased interest due to their exceptional multifunctional properties [18, 19]. Different combinations of organic and inorganic components allow the achievement of exceptional properties, such as electrical, catalytic, antibacterial, radiation-resistant, optical, excellent thermal, and mechanical properties. It gives a potential application for tis materials in completely new fields [5, 17].

Synthesis of organic–inorganic components together could change the chemical composition of the components, it is feasible to realize different kinds of materials, with different properties, whose applications depend on the ratio between organic and inorganic phases [18].

The literature review shows that attempt with some different materials have been made, especially with: polyvinyl acetate [5], polyacrylic acid (PAA) [20], sodium polyacrylate (PAA_{Na}) [20], polyethylene glycol (PEG) [20, 21], polyvinyl alcohol (PVA) [5, 20], polyacrylamide (PAAM) [20], Polydimethylsiloxane (PDMS) [22], and different kind of resins, including melamine resins [5, 23]. The investigations show different influence additives on geopolymers properties. Exemplary, the mechanical properties of geopolymers can be improve by admixture of polyacrylic acid (PAA) and sodium polyacrylate (PAA_{Na}) and they can be decreasing by admixture of polyacrylamide (PAm) and polyvinyl alcohol (PVA) [20]. However, additive of PAm and PVA have negative influence for results of the compressive strength tests in the same time it is beneficial for other material properties. By proper composition two or more organic polymers used together, it is possible to improve simultaneously the mechanical property, water-resistant property and the specific strength [20]. The promising results are given by epoxy resins, such as melamine resins makes the novel materials, thanks to isolating properties, are very attractive for the manufacturing thermo-resistant and thermo-insulating panels [5]. The results of the research show that incorporating content of 1 % by mass resin to geopolymer composites increasing significantly mechanical properties trough the limitation of the microcracks on the surface of geopolymers, The admixture prevent the cracking growth through filling effect, immobilizing water molecules and effectively postpone the water evaporation [23, 24].

This novel modification by organic polymers for inorganic polymers are required. The composition of several organic polymers is supposed to be suitable for wide range of geopolymer based on alkali activated materials, such as: kaolinite-based, fly ash-based, slag-based and waste-based geopolymers, because there is only minor difference between the chemical reactions of these systems [20].

2. Experimental procedure

Specimens based on geopolymer matrix with 5, 10 and 15% of different organic polymers were investigated. The variables include: different materials and different amount of additions as well as two ways of samples preparation. The composites in series I – the organic polymers were added as dry material and preliminary mixing with fly ash, then the alkaline solution was added. In series II – the organic polymers were prepared as water solution and then added during the mixing process together with the alkaline solution.

2.1. Materials and methods

2.1.1. Inorganic polymer. The base for inorganic polymers was fly ash form CHP plant in Skawina, Poland, which is rich in oxides such as SiO_2 – 55.89% by mass and Al_2O_3 – 23.49% by mass. The morphology of the particles of fly ash is typical for by-products of coal combustion. The process of alkali activation has been made by 8M sodium hydroxide solution combined with the sodium silicate solution.

IOP Conf. Series: Materials Science and Engineering **416** (2018) 012090 doi:10.1088/1757-899X/416/1/012090

2.1.2. Organic polymers. The seven different organic polymers have been chose to preliminary research: polyvinyl alcohol - PVA, starch (amylum), chitosan, gelatin (gelatine), polyvinylpyrrolidone - PVP, polyethylene glycol - PEG, and gum Arabic (also known as acacia gum). The polymers was chosen based on literature review, exemplary PEG and PVA and taking under consideration also lack of information about investigation with particular composition, exemplary chitosan.

PVA (polyvinyl alcohol) is a water-soluble synthetic polymer, made from polyvinyl acetate through hydrolysis. It has been used during the first half of the 20th century worldwide. PVA is easily degradable by biological organisms and it is a solubilized crystalline structure polymer in water. It is used in papermaking, textiles, food packing, and a variety of coatings, including surgical threads [25, 26].

Starch (amylum) is a polysaccharide carbohydrate consisting of a large number of glucose units joined together by glycosidic bonds. Starch is produced by all green plants as an energy store. It has some application in the food industry, especially after chemical modification [27].

Chitosan is a linear polysaccharide composed made by treating the chitin (usually shells of shrimp) with an alkaline substance, exemplary sodium hydroxide. It has commercial application in biomedical area, agriculture (including winemaking) and as a component of coatings [28].

Gelatin (gelatine) is a mixture of peptides and proteins produced by partial hydrolysis of collagen extracted from various animal body parts. It is applied in food and pharmaceutical industry as well as for cosmetic manufacturing.

PVP (polyvinylpyrrolidone) is a water-soluble polymer. It is PVP is used in wide range of medical and in many technical applications and also in cosmetic and food industry [29].

PEG (polyethylene glycol) produced by the interaction of ethylene oxide with water, ethylene glycol, or ethylene glycol oligomers. It is also known as polyethylene oxide (PEO) or polyoxyethylene (POE), depending on its molecular weight. It has numerous of applications including: medicine, chemistry, biological laboratories, in industry [30, 31].

Gum Arabic, also known as acacia gum [32] is an edible biopolymer obtained from trees of *Acacia senegal* and *Acacia seyal* which grow principally in the Africa. Chemically, it is a complex mixture of macromolecules of different size and composition, mainly carbohydrates and proteins. It is applied in many industrial sectors such as: textiles, ceramics, cosmetics and pharmaceuticals, food [32, 33].

2.2. Specimens

The samples were prepared using sodium promoter, fly ash and additives of various organic polymers (5%, 10% and 15% by mass). The process of alkali activation has been made by 8M sodium hydroxide solution combined with the sodium silicate solution (liquid glass at a ratio of 1:2.5). To produce geopolymers flakes of technical sodium hydroxide were used and an aqueous solution of sodium silicate (R-145) whose molar module was 2.5 and density was about 1.45 g/cm^3. The tap water was used instead of the distilled one. The alkaline solution was prepared by means of pouring the aqueous solution of sodium silicate over the solid sodium hydroxide. The fly ash, alkaline solution and organic polymer (series I – organic polymer as dry material, and series II - organic polymer as water solution) were mixed about 15 minutes. The homogeneous dispersion of the organic particles was obtained just by hand mixing. Next, it was poured into sets of plastic moulds for compressive strength test. Tightly closed molds were cured in ambient temperature (20^{0}C) trough 28 days. Then, the samples were unmoulded.

2.3. Analytical procedure

Compressive strength tests were carried out according to the methodology described in the standard EN 12390-3 'Testing hardened concrete. Compressive strength of test specimens', because of the lack of separate standards for geopolymer materials. The tests involved at least 3 samples. Samples used to the compressive strength test had cylindrical shape with dimensions: Ø = 36.5 mm and h = 70 mm. Test was made on the concreate press - MATEST 3000kN).

3. Experimental results and discussion

The results of the compressive strength test are shown in the table 1.

IOP Conf. Series: Materials Science and Engineering **416** (2018) 012090 doi:10.1088/1757-899X/416/1/012090

Table 1. The results of the compressive strength test - series I (samples preparing without solvent).

Sample	Compressive strength [MPa]	Standard deviation
Inorganic polymer	11.396	1.293
Inorganic polymer with chitosan (5%)	7.586	1.161
Inorganic polymer with chitosan (10%)	6.111	1.122
Inorganic polymer with chitosan (15%)	3.539	0.314
Inorganic polymer with gelatine (5%)	4.835	0.385
Inorganic polymer with gelatine (10%)	5.373	0.113
Inorganic polymer with gelatine (15%)	3.722	0.462
Inorganic polymer with gum Arabic (5%)	8.085	1.149
Inorganic polymer with gum Arabic (10%)	5.716	0.939
Inorganic polymer with gum Arabic (15%)	4.009	0.791
Inorganic polymer with PEG (5%)	6.411	0.380
Inorganic polymer with PEG (10%)	4.944	0.991
Inorganic polymer with PEG (15%)	4.580	0.502
Inorganic polymer with PVA (5%)	7.034	1.195
Inorganic polymer with PVA (10%)	3.056	0.743
Inorganic polymer with PVA (15%)	1.777	0.372
Inorganic polymer with PVP (5%)	6.499	1.876
Inorganic polymer with PVP (10%)	8.397	2.688
Inorganic polymer with PVP (15%)	7.759	0,748
Inorganic polymer with starch (5%)	6.964	1.896
Inorganic polymer with starch (10%)	7.171	1.262
Inorganic polymer with starch (15%)	5.959	1.291

The best results were achieved for the matrix without reinforcement. The additives of organic polymers decreased the compressive strength results. The relatively good results were achieved for 10% PVP, 5% Arabic gum and 5% chitosan.

Table 2. The results of the compressive strength test - series II (samples preparing with solvent).

Sample	Compressive strength [MPa]	Standard deviation
Inorganic polymer	11.396	1.293
Inorganic polymer with chitosan (5%)	8.147	2.738
Inorganic polymer with chitosan (10%)	10.074	1.813
Inorganic polymer with chitosan (15%)	6.352	2.167
Inorganic polymer with gelatine (5%)	8.519	1.481
Inorganic polymer with gelatine (10%)	5.458	1.471
Inorganic polymer with gelatine (15%)	3.515	0.680
Inorganic polymer with gum Arabic (5%)	8.366	2.410
Inorganic polymer with gum Arabic (10%)	8.053	1.938
Inorganic polymer with gum Arabic (15%)	6.310	0.116
Inorganic polymer with PEG (5%)	8.136	1.209
Inorganic polymer with PEG (10%)	7.245	1.397
Inorganic polymer with PEG (15%)	6.961	1.326
Inorganic polymer with PVA (5%)	8.318	0.340
Inorganic polymer with PVA (10%)	4.892	0.822
Inorganic polymer with PVA (15%)	3.872	0.152

Inorganic polymer with PVP (5%)	10.562	2.543
Inorganic polymer with PVP (10%)	9.640	1.036
Inorganic polymer with PVP(15%)	7.129	0.419
Inorganic polymer with starch (5%)	8.638	2.205
Inorganic polymer with starch (10%)	7.493	0.840
Inorganic polymer with starch (15%)	6.230	0.640

The samples preparing with solvent give better results that the samples in series I (samples preparing without solvent). The best results were archived for 10% of chitosan addition and 5% of PVP addition. The manufacturing process have a significant influence for final properties of the material.

4. Conclusions

The literature research as well as the conducted preliminary research show that inorganic-organic composites can be interesting material for different applications. However the compressive strength decreased, the other properties have been changed, especially samples surface. It can happened increasing of resistant of the material, but it required future research.

The planned further research should be connected with improving the matrix properties, especially flexural strength. It can be made by introducing to the matrix some additives investigated above, especially in amount of 5-10% and optimization the process. The most promising additives seem to be chitosan and PVP, because of the mechanical properties. This materials have not been investigated in world literature yet. The further research should also include the method of manufacturing. The application the same amount of organic polymers to inorganic matrix by using different processing methods gives significantly changes with the results of compressive strength test.

5. References

[1] Davidovits J 2015 *Geopolymer Chemistry and Applications*, 4th ed. (Saint-Quentin, France: Geopolymer Institute) p 644

[2] Olivia M. and Nikraz H. 2013 *Proc. 13th East Asia-Pacific Conf. on Structural Eng. & Construction*, (Sapporo, Japan: EASEC-13) D-6-6

[3] Jasiczek M, Kroisova D, Louda P, Sobczyk-Guzenda A and Szymanowski H 2010 *Inżynieria Materiałowa* **1/4** 993

[4] Korniejenko K, Mikuła J and Łach M 2015 *Prec. International Conference on Bio-Medical Engineering and Environmental Technology /BMEET-15* (London, UK: UREBE), p 208

[5] Roviello G, Ricciotti L, Ferone C, Colangelo F and Tarallo O 2015 *Cement and Concrete Composites* **59** 89

[6] Korniejenko K, Łach M and Mikuła J 2018 Mechanical Properties of Raffia Fibres Reinforced Geopolymer Composite *Advances in Natural Fibre Composites* eds. R. Fangueiro S Rana (Springer, Cham) pp. 135-144

[7] Korniejenko K, Mikula J and Łach M. 2015 *Proc. ICCST/10: 10th International Conference on Composite Science and Technology* (Lisboa, Portugal: Instituto Superior Tecnico) p 1 (ID177)

[8] Aguirre-Guerrero AM, Robayo-Salazar RA and de Gutiérrez RM 2017 *Applied Clay Science* **V, 135** 437

[9] Huseien GF, Mirza J, Ismail M, Ghoshal SK and Hussein AA 2017 *Renewable and Sustainable Energy Reviews* **V, 80** 54

[10] Zhang Z, Yao X, Zhu H, Hua S and Chen Y 2009 *Journal of Central South University of Technology*, **16** 49

[11] Bakharev T 2005 *Cement and Concrete Research* **V, 35 (4)** 658

[12] Zhang Z, Yao X and Zhu H 2010 *Applied Clay Science* **V, 49 (1-2)** 7

[13] Rashidian-Dezfouli H and Rao Rangaraju P 2017 *Construction and Building Materials* **V, 153** 996

[14] Cheng-Yong H, Yun-Ming L, Al Bakri Abdullah MM and Hussin K 2017 *Sci. Rep.* **7**, 45355
[15] Mikhailova O and Rovnanik P 2016 *World Academy of Science, Engineering and Technology, International Journal of Chemical and Molecular Engineering*, **10 (12)** 1472
[16] Ferone C, Roviello G, Colangelo F, Cioffi R, Tarallo O 2013 *Applied Clay Science* **73** 42
[17] Yan L, Kasal B and Huang L 2016 *Composites - Part B-Engineering* **92** 94
[18] Korniejenko K, Frączek E, Pytlak E and Adamski M 2016 *Procedia Engineering* **151** 388
[19] Jankauskaite V, Lazauskas A, Gri Konis E, Lisauskaite A and Ukiene K 2017 *Molecules* **22 (6)** 964
[20] Zhang S, Gong K and Lu J 2004 *Materials Letters* **58** 1292
[21] Mikhailova O and Rovnaník P 2016 *Procedia Engineering* **151** 222
[22] Roviello G, Menna C, Tarallo O, Ricciotti L, Ferone C, Colangelo F, Asprone D, Maggio R, Cappelletto E, Prota A and Cioffi R 2015 *Materials and Design* **87** 82
[23] Zhang YJ, Wang YC, Xu DL and Li S 2010 *Materials Science and Engineering, A*, **527** 6574
[24] Zhang YJ, Li S, Xu DL, Wang BQ, Xu GM, Yang DF, Wang N, Liu HC and Wang YC 2010 *Journal of Material Science* **45** 1189
[25] Gaaz TS, Sulong AB, Akhtar MN, Kadhum AAH, Mohamad AB and Al-Amiery AA 2015 *Molecules* **20** 22833
[26] Demerlis CC and Schoneker DR 2003 *Food Chem. Toxicol.* **41** 319
[27] Abbas KA, Khalil SK and Hussin ASM 2010 *Journal of Agricultural Science* **2 (2)** 90
[28] Younes I and Rinaudo M 2015 *Marine Drugs* **13 (3)** 1133
[29] Haaf F, Sanner A and Straub F 1985 *Polymer Journal* **17** 143
[30] Israelachvili J 1997 *PNAS* **94 (16)** 8378
[31] Kahovec J, Fox RB and Hatada K. 2002 *Pure and Applied Chemistry* **74 (10)** 1921
[32] Montenegro AM, Boiero ML, Valle L and Borsarelli CD 2012 Gum Arabic: More Than an Edible Emulsifier *Products and Applications of Biopolymers* ed. J Verbeek (InTech, Rijeka) pp. 3-26
[33] Wyasu G and Okereke NZJ 2012 *J. Nat. Prod. Plant Resour.* **2 (2)** 314

Acknowledgments
This work was supported by the ERANet-LAC 2nd Joint Call (http://www.eranet-lac.eu) and funded by National Centre for Research and Development, Poland, under grant: "Development of eco-friendly composite materials based on geopolymer matrix and reinforced with waste fibers".

IOP Conf. Series: Materials Science and Engineering **416** (2018) 012091 doi:10.1088/1757-899X/416/1/012091

Modelling Drilling Experiments on Al7075 Using the Response Surface Methodology

P Kyratsis[1,*], N Efkolidis[1] and K Kakoulis[1]

[1]Western Macedonia University of Applied Sciences, Department of Mechanical Engineering and Industrial Design, Kila Kozani GR50100, Greece

[*]Corresponding author: pkyratsis@teiwm.g

Abstract. The current research deals with the mathematical simulation of the drilling manufacturing process using the response surface methodology (RSM). The workpiece used was made of aluminium alloy 7075 and the tools used were manufactured from Kennametal. A number of different drilling tool diameters were used, with variable cutting speeds and feed rates. The thrust force (Fz) and the cutting torque (Mz) were measured using a Kistler dynamometer, while the mathematical models in both cases were proved to be sufficient, in order to describe the effect of the drilling parameters on both the thrust force and the cutting torque.

1. Introduction

Drilling is an extremely popular manufacturing process in a variety of industries i.e. structural, automotive, aerospace, manufacturing. Its simulation and the development of mathematical models are used for the prediction of its result i.e. surface quality, cutting force and torque, geometrical characteristics [1, 2]. A number of different methodologies have been applied (response surface methodology, genetic algorithms, fuzzy logic) based on experimental results for different drilling tools and workpiece materials, or validated simulations with finite elements and CAD based methods [3, 4]. They resulted in documenting the drilling manufacturing process outputs for use in industry and in academia. Those results are used for predicting the drilling behavior in different cases studies and can lead in drilling optimizations [5, 6].

The current paper aims to present mathematical models for the thrust force and the cutting torque based on the response surface methodology (RSM), when drilling Al7075 with solid carbide tools commercially available. For this reason, a full factorial set of 27 experiments were used and their results were handled following the RSM methodology and the analysis of variance (ANOVA). The derived mathematical models are very accurate and can be used directly for predicting both the thrust force and the cutting torque within the limits of the manufacturing parameters used.

2. Experiments set up

A HAAS VF1 machining center was used for performing a number of experiments. The material of the specimen used was Al7075, since it is one of the most popular materials in industry i.e. automotive, aerospace, constructions. As a result, the outcome of the present research is expected to be adopted from a variety of researchers and industry engineers.

A Kistler type 9123 dynamometer was used for acquiring the thrust force (Fz) and the cutting torque (Mz), while both signal ware processed by a multichannel signal conditioner (5223) and data acquisition unit (5697). Fig.1 depicts the research workflow followed for delivering the mathematical models.

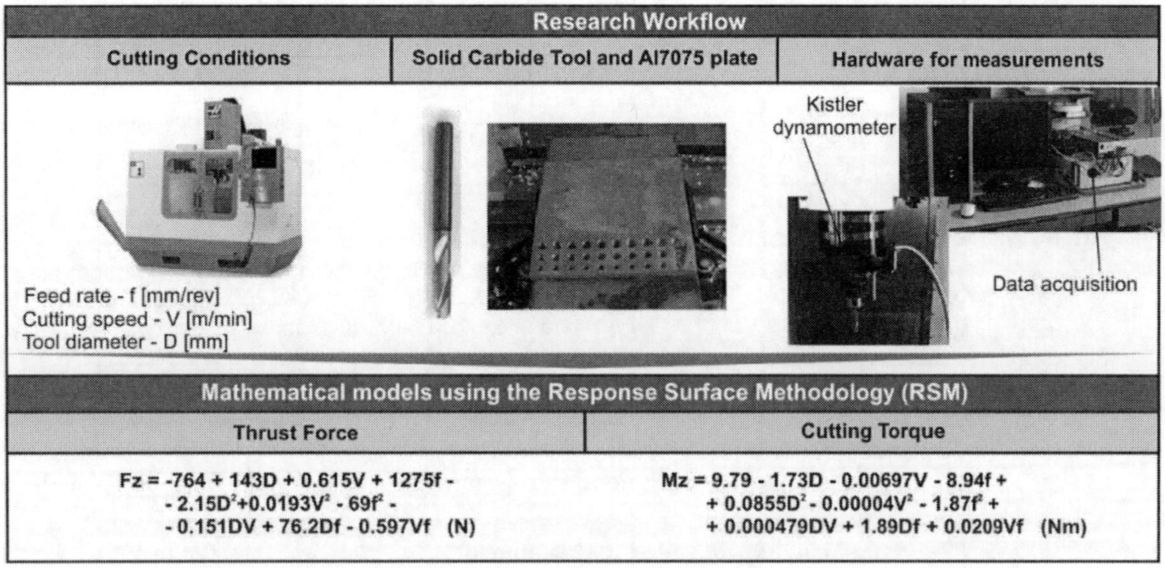

Figure 1. Research flowchart

Kennametal is the manufacturer of the drilling tools used (solid carbide -KC7325). Their diameters were 10mm, 12mm and 14mm, while the feed rates and cutting speeds selected were 0.20mm/rev, 0.40mm/rev and 0.60mm/rev and 10m/min, 40m/min and 70m/min, respectively. Table 1 depicts these drilling parameters together with their notations and values. The levels were selected to be relative wide in order to be able to increase the derived models usability.

27 experiments were performed, while all the combinations of the drilling parameters were achieved. As a result a full factorial set of data for both the Fz and Mz were measured and used for the calculation of the mathematical models based on the response surface methodology and the analysis of variance.

Table 1. Cutting parameters and tools

Manufacturing parameter	Notation	Unit	LEVELS		
Tool diameter	**D**	**mm**	**10**	**12**	**14**
Feed rate	**f**	**mm/rev**	**0.20**	**0.40**	**0.60**
Cutting speed	**V**	**m/min**	**10**	**40**	**70**

Figure 2 offers an overview of the total set of measurements acquired for the thrust force and the cutting torque. When the tool diameter increases, both the thrust force and the cutting torque increase to a great extent as expected. The same is the case every time the feed rate is increased. On the other hand, increasing the cutting speed does not affect substantially the required thrust force and cutting torque. This is expected to affect the statistical definition of the cutting speed significance during the application of the response surface methodology.

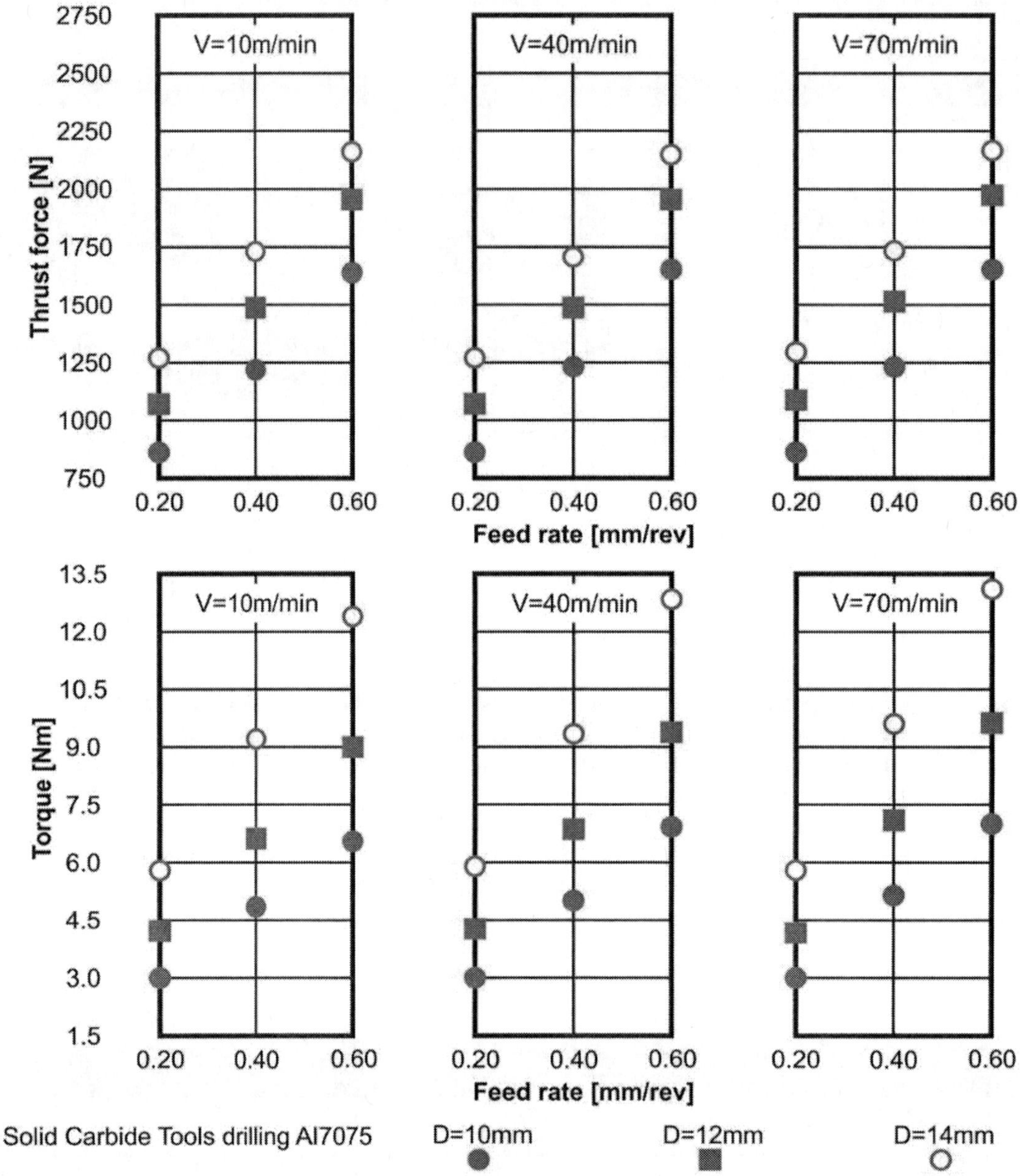

Figure 2. Drilling experiments in Al7075

3. Mathematical models proposed for both the Fz and Mz

A full factorial set of experiments was used in order to develop mathematical models for both the thrust force and the cutting torque during this research. All 27 experiments were performed in order to be able to describe the influence of the cutting parameters and the tool diameter to the thrust force and cutting torque measured. A second order polynomial model was the basis in both cases and the equations derived are presented:

$$F_z = -764 + 143D + 0.615V + 1275f - 2.15D^2 + 0.0193V^2 - 69f^2 - 0.151DV + 76.2Df - 0.597Vf\ (N) \quad (1)$$

and

$$M_z = 9.79 - 1.73D - 0.00697V - 8.94f + 0.0855D^2 - 0.00004V^2 - 1.87f^2 + 0.000479DV + 1.89Df + 0.0209Vf\ (Nm) \quad (2)$$

where:
Fz is the thrust force (N),
Mz is the cutting torque (Nm),
D is the tool diameter (mm),
f is the feed rate (mm/rev) and
V is the cutting speed (m/min).

Table 2. Analysis of variance for the thrust force mathematical model

Source of variation (Fz)	DF	SS	MS	F	P
Regression	9	4182029	464670	2729,14	0
Residual error	17	2894	170		
Total	26	4184924			
R-sq(adj)	99,90%				

Predictor	Coef.	SE Coef.	T	P
Constant	-763,60	198,100	-3,86	0,001
D	142,69	32,320	4,42	0,000
V	0,62	0,930	0,66	0,517
f	1275,00	158,100	8,07	0,000
D^2	-2,15	1,332	-1,62	0,124
V^2	0,02	0,006	3,26	0,005
f^2	-69,40	133,200	-0,52	0,609
D*V	-0,15	0,063	-2,41	0,027
D*f	76,25	9,417	8,10	0,000
V*f	-0,60	0,628	-0,95	0,355

Table 3. Analysis of variance for the torque mathematical model

Source of variation for M_z	DF	SS	MS	F	P
Regression	9	217,386	24,154	3533,56	0,000
Residual error	17	0,116	0,007		
Total	26	217,503			
R-sq(adj)	99,9%				

Predictor	Coef.	SE Coef.	T	P
Constant	9,79300	1,25500	7,80	0,000
D	-1,72650	0,20480	-8,43	0,000
V	-0,00697	0,00589	-1,18	0,253
f	-8,94100	1,00200	-8,93	0,000
D^2	0,08547	0,00844	10,13	0,000
V^2	-0,00040	0,00004	-1,06	0,304
f^2	-1,86530	0,84380	-2,21	0,041
D*V	0,00048	0,00040	1,20	0,245
D*f	1,88646	0,05967	31,62	0,000
V*f	0,02093	0,00398	5,26	0,000

The adequacy of all mathematical models is provided at a level of significance of 5%. The analysis of variance provided proofs of validity for the developed mathematical models. Based on the methodology principles, the calculated values of the F-ratio (Table 2 for the Fz and Table 3 for the Mz), are significantly high compared to the tabulated values provided by the F-table for 95% confidence level (2729.14 for the Fz and 3533.56 for the Mz). The P values are equal to zero, which proves that both the developed mathematical models are adequate at a 95% confidence level.

The validity of the fit of the models can also be proved, by the adjusted R-sq coefficient, which provides a measure of variability of the output. The coefficient in both cases is 99.9 %, which means that the mathematical models have adequate predictive capacity. In addition, the significant terms of the models can be recognized by the magnitude of their P-values. They should be less than 0.05 due to the confidence level used. With respect to the Fz model the following factors are statistically significant: D (P-value= 0.000), f (P-value= 0.000), V^2 (P-value= 0.005), D×V (P-value= 0.027), D×f (P-value= 0.000), while for the Mz model the significant factors are: D (P-value= 0.000), f (P-value= 0.000), D^2 (P-value= 0.000), f^2 (P-value= 0.041), D×f (P-value = 0.000) and V×f (P-value= 0.000).

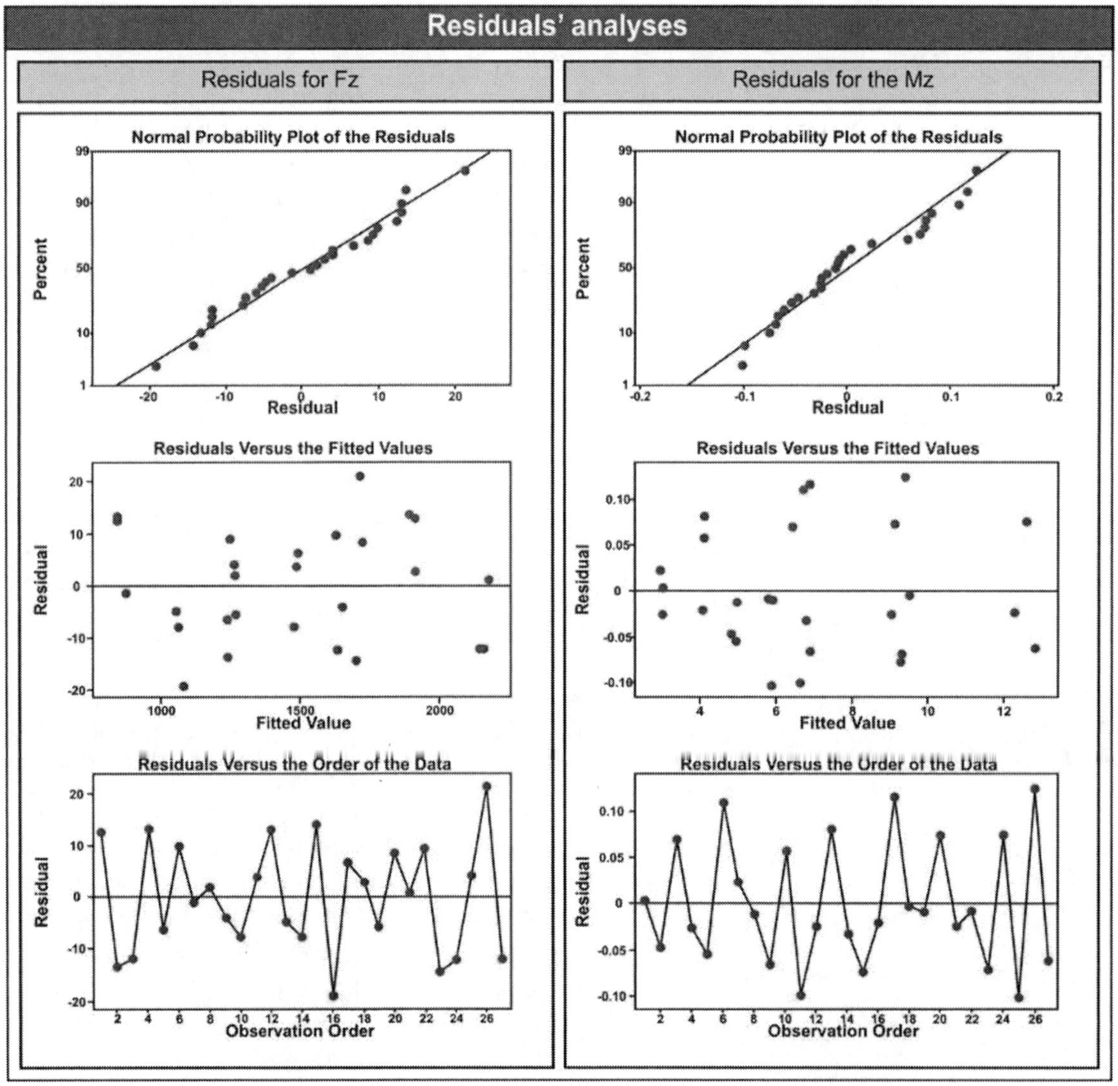

Figure 3. Residuals analysis graphs for the Fz and Mz

IOP Conf. Series: Materials Science and Engineering **416** (2018) 012091 doi:10.1088/1757-899X/416/1/012091

Figure 3 depicts the normal probability plots of the calculated residuals for the Fz and the Mz. The residuals closely follow straight lines (blue line) and their distribution follows the normal distribution pattern. Both the scatter diagrams of the residuals vs the fitted values provide evidence that the residuals are evenly distributed on both sides of the reference line and randomly scattered. The same is the case for the residuals vs the order of the experiments, which proves that the order of performing the experimental work does not affect the quality of the mathematical models. All the statistical results were acquired with the use of the statistical software Minitab. The regression analysis was accompanied with the residuals' analysis in order to secure that all the hypotheses of the theory are met.

The above mentioned methodology provided a solid basis for concluding that the prediction models sufficiently explain the relationship between the Fz and Mz with the drilling parameters used (diameter, feed rate and cutting speed) in the case of drilling Al7075 with the abovementioned drilling tools.

4. Conclusion

The current paper presents mathematical models for both the thrust force and cutting torque, when drilling Al7075 with solid carbide tools. Those tools were selected by the catalogue of one of the greater tool manufacturers and commercially available. The response surface methodology was followed and the proposed models were established to be accurate to a great extent. All the significant manufacturing parameters were identified in both cases and the high R-sq(adj) coefficient lead to excellent fitting results.

5. References

[1] Kyratsis P, Bilalis N and Antoniadis A 2011 Comput Aided Des **43** 1879
[2] Patra K, Anand RS, Steiner M and Biermann D 2015 Mater Manuf Processes **30** 248
[3] Karnik SR, Gaitonde and Davim JP 2008 Int J Adv Manuf Technol **38** 868
[4] Kumar BS and Baskar N 2013 Int J Adv Manuf Technol **65** 1501
[5] Badan I, Oancea G, Vasiloni M 2012 Bull Transilvania Univ Brasov-Ser I Eng Sci **5**(54):31
[6] Myers RH, Montgomery DC and Anderson-Cook CM 2009 Response surface methodology: process and product optimization using designed experiments, 3rd edn. Wiley, New York

Comparative Study of Plastic Strain Accumulations at Thermal Cycles for Solder Alloys

C Vălean[1,*] and E Linul[1]

[1]Department of Mechanics and Strength of Materials, Politehnica University of Timisoara, 1 Mihai Viteazu Avenue, 300 222 Timisoara, Romania

*Corresponding author: cristina_valean@yahoo.com

Abstract. This paper presents an investigation of plastic strain accumulation at thermal cycles for 4 different solder alloys used in electronics. The investigation was performed on an assembly which is part of an electronic packaging. The behavior of the solder was numerically analyzed by means of the finite element method, using Ansys program. This work was carried out for a thermal variation of 165°C, in the range of -40...+125°C. The first material (60Sn40Pb) was described using Elastic-Plastic, Chaboche and Anand models, which combine the plasticity with creep. Following the investigation of the first solder alloy, a correlation between the Anand model and the Elastic-Plastic one was proposed, this being obtained by using a numerical constant. It was observed that the proposed correlation factor is closely related to hardening constant (a parameter of the Anand model). Following the first investigation, another 3 types of solder alloys were also analyzed. Each new material was described with both the Anand model and the Elastic-Plastic model. The possibility of correlation between the two models, as well as the verification of the relationship between the used correlation factor and hardening constant, were investigated. Following these investigations, it was concluded that, for the studied thermal field, the material models can be correlated. Also, for 50% of investigated cases, the correlation factor is closely related with a parameter of the Anand model.

1. Introduction

In electronics are used many different soldering materials. The major problem occurs when the material is subjected to repeated thermal cycles, and these lead to plastic strain accumulations, initiation of cracks in tin and then fracture. Thus, damage to soldering material occurs.

In the past three decades a lot of analytical, numerical and experimental investigations, regarding the viscoplastic behavior of different kind of materials have been performed. In 1985, Anand [1], develops constitutive equations which describe the elastic-viscoplatic behavior of metal materials, considering small elastic stretches and large plastic deformations. This developed material model does not require a yield conditions and attendant loading criteria. Moreover, the model has defined only one scalar parameter which is isotropic resistance to plastic flow. In a word, the model proposed by Anand combine the plasticity with creep, and its constitutive equation is a combination of 9 material parameters. Afterwards, Brown and coworkers [2], develop constitutive equations which model large elastic-viscoplastic deformations of metals at high temperatures. Cheng and coworkers [3] apply the Anand model to study the plastic deformations for solder alloys, also performing experimental tests to determine the parameters of model. For studied cases (4 types of solder), constitutive equations accurately reproduce the basic tests. Busso and coworkers [4] analyze a solder alloy by applying thermal cycles to predict the shape of plastic strain and material microstructure evolution. Wang and coworkers

[5] use the Anand model to study the plastic deformation behavior of 4 soldering materials used in a series of electronic packaging. Liu and coworkers [6] investigate the inelastic behavior for 2 solder alloys, at both high temperature and different strain rates. Chen and coworkers [7] use a modified Anand model to describe plastic deformation for solder materials used in electronics, at temperatures between -40 and 125°C. Also, in this article is performed a comparison between analytical and experimental results.

Considering all these problems which can occur during the operation of a product (plastic strain accumulation, initiation and propagation of cracks in material and brake of it, respectively), it is necessary to perform in-depth investigations in terms of material viscoplasticity and in their comportment at thermal cycles. The main objective of this paper is the comparative investigation of the behavior of an electronic assembly, subjected to temperature cycles. Various material models, such as elastic-plastic with kinematic and isotropic hardening, the Chaboche model and the Anand model were used.

2. Material models

In order to describe the solder alloys, three material models (Elastic-Plastic model, Chaboche model and Anand model, respectively) were used.

In the case of Elastic-Plastic model, material properties for each solder alloy are shown in Table 1. For first round of analyses, 6 models of elastic-plastic model were described, with two types of hardening. Therefore, for both hardenings (isotropic and kinematic) 3 different tangent modules, Et, namely: (i) Et = 0 [MPa], (ii) Et = E/10 000 = 3 [MPa] and (iii) Et = E/100 = 300 [MPa], were used.

Table 1. Material data for Elastic-Plastic model

Solder Alloy	ρ [kg/m^3]	α [C^{-1}]	Rp_{02} [MPa]	ν [-]	E [MPa]	E_t [MPa]	Hardening
60Sn40Pb	8 600	$2.4 \cdot 10^{-5}$	32	0.4	$3 \cdot 10^4$	0 3 300	Iso. & Kin.
62Sn36Pb2Ag	8 600	$2.45 \cdot 10^{-5}$	38	0.35	$3.44 \cdot 10^4$	0	Isotropic
96.5Sn3.5Ag	7 360	$2.185 \cdot 10^{-5}$	34	0.4	$5.27 \cdot 10^4$	0	Isotropic
92.5Pb5Sn2.5Ag	11 110	$2.95 \cdot 10^{-5}$	17	0.33	$1.38 \cdot 10^4$	0	Isotropic

In Table 1, the parameters are: ρ – density; α – coefficient of thermal expansion; Rp_{02} – yield stress; E – Young modulus; Et – tangent modulus.

Material model proposed by Chaboche [8] defines a combined behavior for both kinematic and isotropic hardening. It is used for evaluating the effects of cycle loadings and for model both of cycling hardening and softening. To define the behavior of material described by Chaboche model, 3 material constants are necessary: (i) yield point, Rp_{02} = 32 [MPa], (ii) linear kinematic hardening, C = 13.6 [MPa] and (iii) dynamic recovery coefficient, γ = 457.9 [-]. This last term introduces the nonlinearity between the back stress and the actual plastic strain. The values for C and γ, in case of 60Sn40Pb, were taken from paper of Basaran and coworkers [9].

As specified above, the material model performed by Anand is based on a 9-parameter equation. For the study investigated in this paper, the values for all parameter were taken from Ref. [5].

In Table 2, the following parameters were used: s_0 – resistance to initial deformation; A – pre-exponential factor; ξ – multiplier of stress; m – strain rate sensitivity; h_0 – hardening/softening constant; $\hat{s}$ – saturation value of deformation resistance; n – strain rate sensitivity for the saturation value of deformation resistance; a – strain rate sensitivity of hardening/softening; Q/R – activation energy/gas constant.

Table 2. Anand parameters for all used materials.

Parameter	Solder alloy			
	60Sn40Pb	62Sn36Pb2Ag	96.5Sn3.5Ag	92.5Pb5Sn2.5Ag
s_0 [MPa]	56.330	42.320	39.090	33.070
A [Hz]	$1.49 \cdot 10^7$	$2.3 \cdot 10^7$	$3.3 \cdot 10^6$	$1.05 \cdot 10^5$
ξ [-]	11	11	6	7
m [-]	0.303	0.303	0.182	0.241
h_0 [MPa]	2 640.8	4 121.31	3 321.15	1 432
$\hat{s}$ [MPa]	80.420	80.790	73.810	41.630
n [-]	0.0231	0.0212	0.018	0.002
a [-]	1.34	1.38	1.82	1.3
$\frac{Q}{R}$ [K]	10 830	11 262	8 900	11 010

3. Finite element analysis

The study presented in this paper was performed on an assembly (part of an electronic packaging) presented in Figure 1. The assembly is composed by a plastic skeleton (1) which is over molding to a pin (2). The connection between pin and PCB (3) are made by a solder alloy (4).

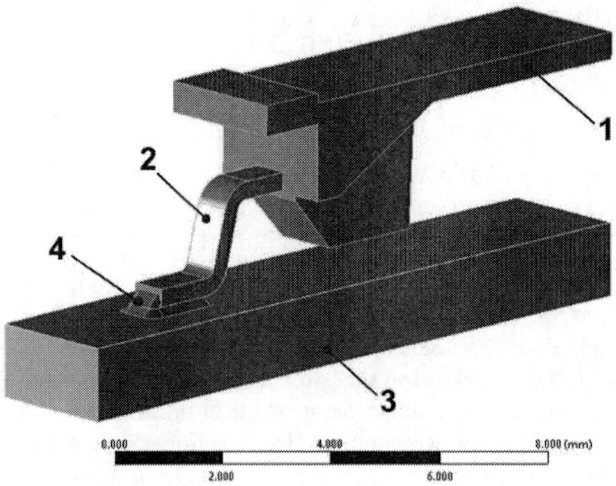

Figure 1. CAD model for analysis

Boundary conditions were chosen considering the missing parts of assembly. Therefore, both for lateral faces and for rear sides it was used a frictionless support, which allow the sliding in its plane. To block the rigid body motion, the displacement on Z axis of a horizontal edge was suppressed (Figure 2a). The contacts between all parts (1-2, 2-4, 3-4) were defined as bonded (Figure 2b).

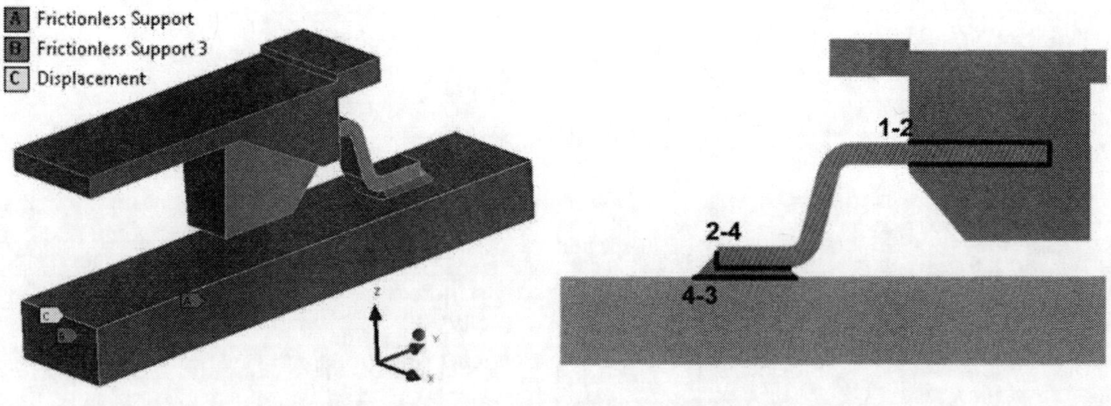

(a) (b)

Figure 2. Boundary conditions and contacts

Due to the fact that the main subject of this investigation is the solder alloy, which has small dimensions, it was used a fine meshing for it (elements with an approximate dimension of 0.1 mm). Considering time reduction of analysis, all other parts was meshed with a larger dimension than solder alloy's elements. All meshed assembly is presented in Figure 3.

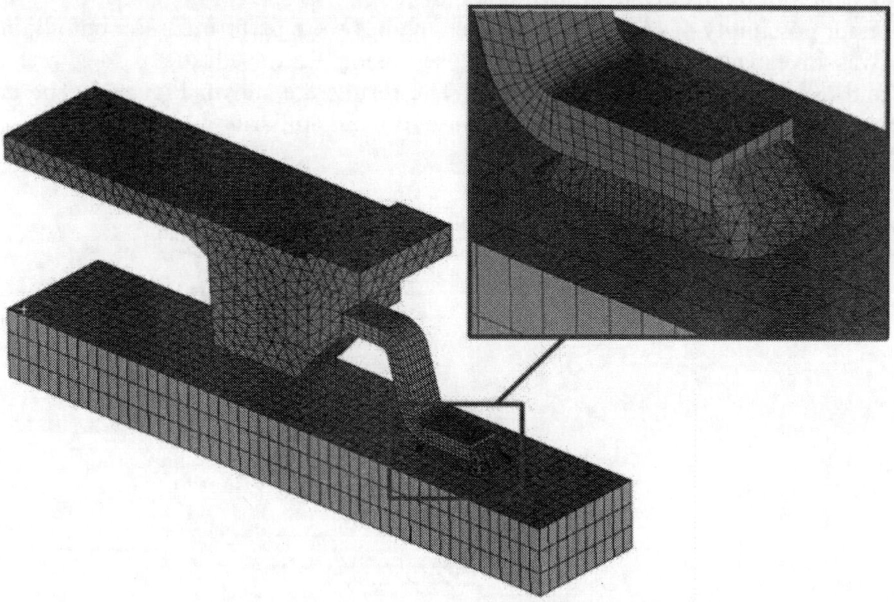

Figure 3. Meshed assembly

The whole assembly was subjected to a thermal condition with the charging scheme shown in Figure 4. The temperature varies between -40 and 125 ° C, having a charging time of 300 s (5 minutes), and a maintenance of 1 500 s (25 minutes).

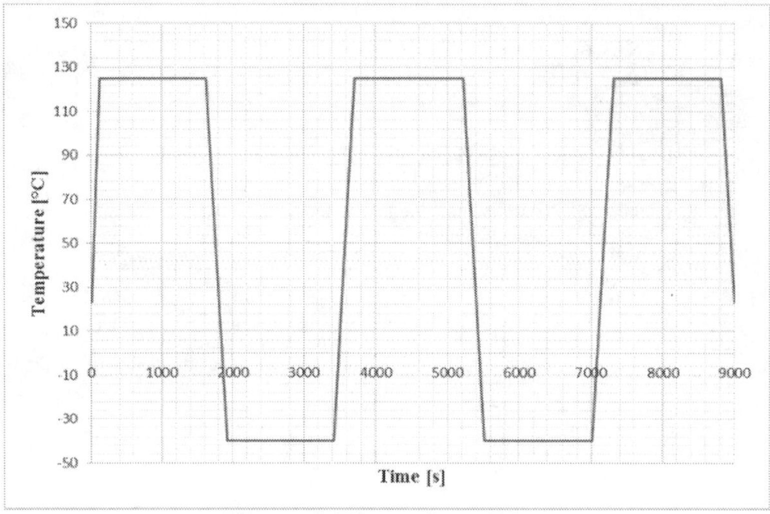

Figure 4. Thermal condition

4. Results

It must be mentioned that the running time for an analysis in Ansys [10], using a material described with Anand model, is approximately 8 times higher than for the other models. Due to this reason, the investigation for possibility of correlation between models were performed. First of all, the behavior of 60Sn40Pb was investigated by running 6 analyses using Elastic-Plastic model, one analysis with Chaboche and one with Anand material model. The results are shown Figure 5. The curves for two Elastic-Plastic models were not represented because of their similarity with the others.

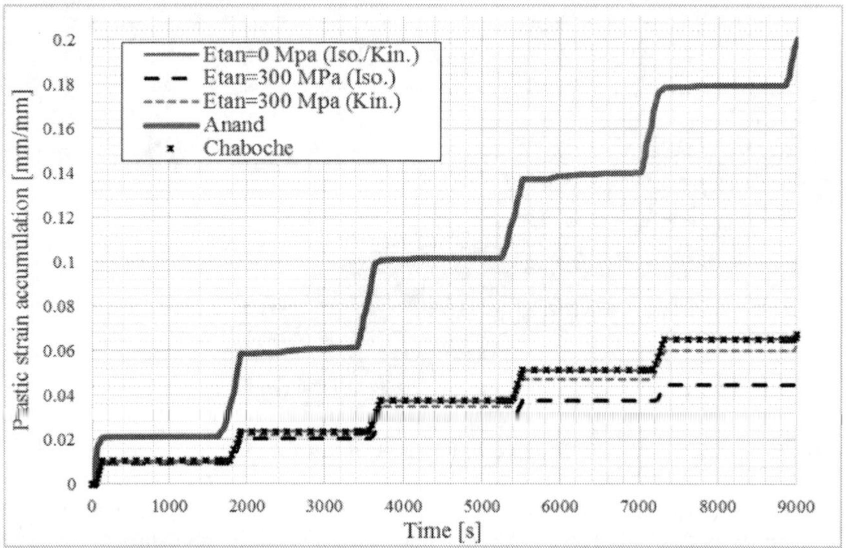

Figure 5. Plastic strain accumulation for studied models

After running the analyses with Anand (the reference model), it was obtained a plastic strain accumulation of 0.2 [mm/mm], for an interval of 9 000 s. Also, it was proposed a correlation between Chaboche and Anand, and Elastic-Plastic (Et=0 MPa) and Anand, respectively, for the studied time period.

After the investigation and the interpretation of results, it was observed that if the values of models are multiplied with a constant equal with 2.7, both of Elastic-Plastic (Figure 6a) and Chaboche (Figure 6b) are closely similar with reference model, Anand, on the investigated domain. On the other hand, this thing is not happening for the other material models; in case of these can be observed significant differences (up to 40%) between them and Anand (Figure 6c, 6d).

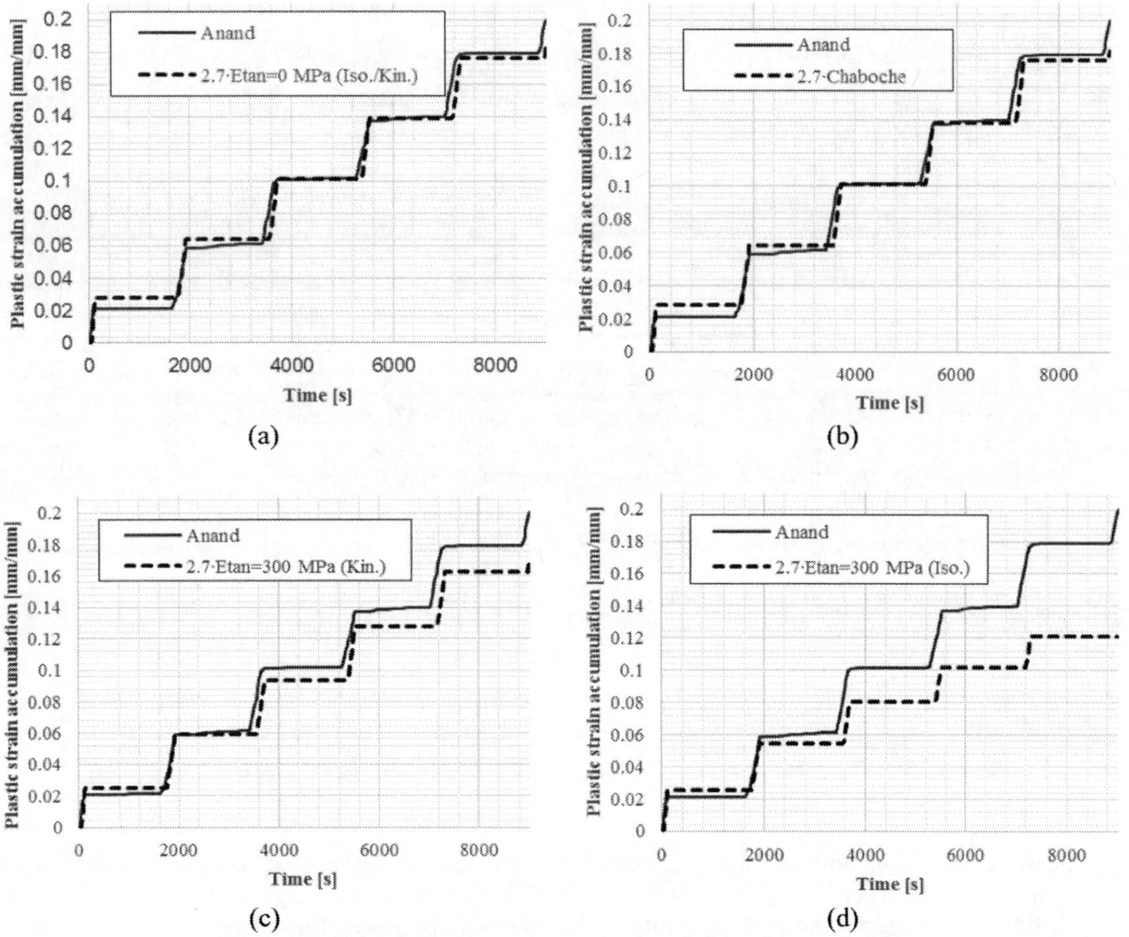

Figure 6. Correlation between Anand and the other models

There can be observed a close relationship between the correlation factor (2.7) and the value of hardening constant, h_0. It can be easy to see that the correlation factor is approximatively equal with hardening constant divided with 1 000. Due to this reason, the investigations continue for another 3 solder alloys. But, for these, were used just two material models: Anand (because it is the reference one) and Elastic-Plastic with Et=0 MPa, in expense of Chaboche, due to its simplicity of parameters determination.

The results for the other 3 materials (62Sn36Pb2Ag, 96.5Sn3.5Ag and 92.5Pb5Sn2.5Ag) are shown in Figure 7. For 62Sn36Pb2Ag was used 2.9 as correlation factor, the hardening constant being 4 121.31 MPa; for 96.5Sn3.5Ag was used 3.3, while h_0 =3 321.15 MPa; for 92.5Pb5Sn2.5Ag was used 2.3 and its hardening constant has a value of 1 432 MPa.

IOP Conf. Series: Materials Science and Engineering **416** (2018) 012092 doi:10.1088/1757-899X/416/1/012092

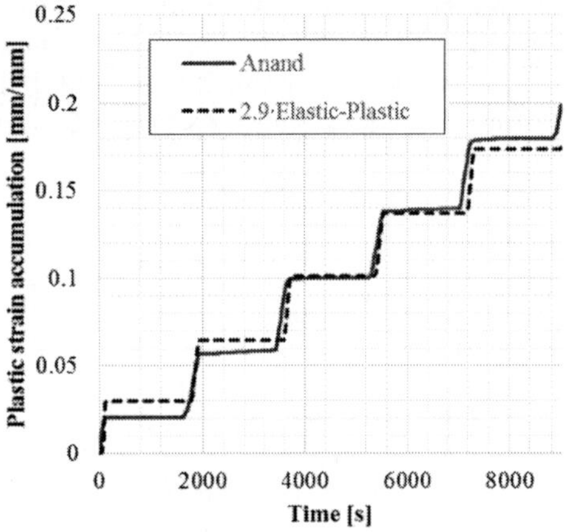

(a)

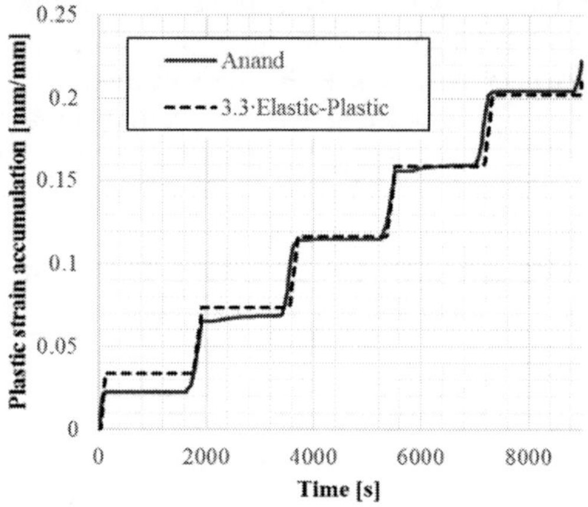

(b)

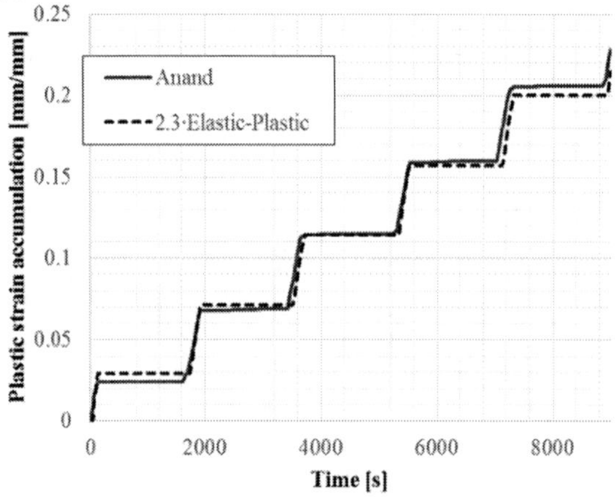

(c)

Figure 7. Correlation between Anand and Elastic-Plastic for the other 3 materials

It can be easy to see that just in one case of the last three, the correlation factor is closely related to Anand parameter, hardening constant. But it can be said that for all investigated cases, in 50% of them, the correlation between these two constants is possible, with proposed formula.

5. Conclusions

In this paper are presented results of investigations effectuated to characterize the behavior at thermal cycles of 4 solder alloys, namely 60Sn40Pb, 62Sn36Pb2Ag, 96.5Sn3.5Ag and 92.5Pb5Sn2.5Ag. During the study, the plastic stain accumulation at thermal cycles was investigated, by using 3 material models. The study was performed for a thermal variation of 165°C, in range of -40...+125°C (shown in Figure 5).

Time for running an analysis with finite elements, for each kind of material model, is different. In case of Anand model, time for running is 8 times higher than in case of the other 2 material models. Due to this problem, the possibility of correlation between models was studied.

It was proposed a constant to correlate the material models. Between correlation factor and value of hardening constant was observed a close relationship, according to formula $\frac{h_0}{1\,000}$.

In Table 3 are shown the results for all investigates materials.

Table 3. Hardening constant vs. correlation factor

Solder Alloy	Hardening constant h_0 [MPa]	Correlation factor
60Sn40Pb	2 640.8	2.7
62Sn36Pb2Ag	4 121.31	2.9
96.5Sn3.5Ag	3 321.15	3.3
92.5Pb5Sn2.5Ag	1 432	2.3

After all investigations, can be concluded that for studied thermal domain, the material models can be correlate. Also, for 50% of investigated cases, the correlation factor has a close relationship with a parameter of Anand model, hardening constant.

6. References

[1] Anand L 1985 *Int J Plasticity* **1** 213-231
[2] Brown SB, Kim KH and Anand L 1989 *Int J Plasticity* **5** 95-130
[3] Cheng ZN, Wang GZ, Chen L, Wilde J and Becker K 2000 *Solder Surf Mt Tech* **12(2)** 31-36
[4] Busso EP, Kitano M and Kumazawa T 1994 *J Electron Packaging* **116** 7-15
[5] Wang GZ, Cheng ZN, Becker K and Wilde J 2001 *J Electron Packaging* **123** 247-253
[6] Liu JC, Yu HJ, Zhang G, Wang ZH and Ma JS 2014 Constitutive behavior and Anand model of novel lead-free solder Sn-Zn-Bi-In-P. *Proc of the Int Conf on Electronics Packaging (ICEP)*, Toyama International Convention Center, Japan, 23-25 Apr 2014, 156-161
[7] Chen X, Chen G and Sakane M 2005 *IEEE T Compon Pack T* **28** 111-116
[8] Chaboche JL 1986 *Int J Plasticity* **2** 249
[9] Basaran C, Zhao Y, Tang H and Gomez J 2004 *J Electron Packaging* **127(3)** 208-214
[10] ANSYS® Academic Research, Release 17.1

Acknowledgment
This work was supported by the department of Powertrain – Sensors & Actuators from S.C. Continental Automotive Romania S.R.L.

IOP Conf. Series: Materials Science and Engineering **416** (2018) 012093 doi:10.1088/1757-899X/416/1/012093

Classical and Modern Methods of Manufacturing on CNC Milling Centers

A But[1,*] and L Gal[2]

[1] Politehnica University of Timişoara, Mechanical Faculty, Department of Materials and Manufacturing Engineering Arad 310130, B-dul Mihai Viteazu nr. 1, Timişoara 300222 , ROMANIA
[2]"AUREL VLAICU" University of Arad, B-dul Revolutiei nr 77. Arad 310130, ROMANIA

[*]Corresponding author: adrian.but@upt.ro

Abstract. The milling operation is widely used and the results are directly influenced by the milling strategies adopted. The paper presents two processing methods: one classical and one actual, both methods with wide industrial applications. Was made the comparison of these methods in manufacturing, are presented the results that were obtained on a real application. Knowing the milling processes, the new strategies developed by the CAM software, bring more substantial results and efficiency in manufacturing process.

1. Introduction

More milling strategies was developed by the Computer Aided Manufacturing (CAM) software's with the target to reduce the time of manufacturing and to up grate the surface quality.

Some of this new milling strategies are :

1.1. "Helical Finish"

The Helical Milling strategy is especially useful for milling all around steep cavities or cores, especially in manufacturing the electrodes. (Fig.1)

The benefits using this strategy are : higher surface quality and a shorter machining time when we have more complicate surfaces to manufacturing .(Fig. 2)

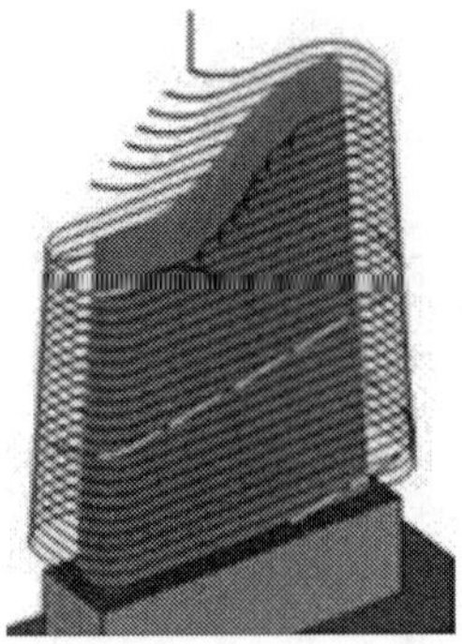

Figure 1 Helical Finish

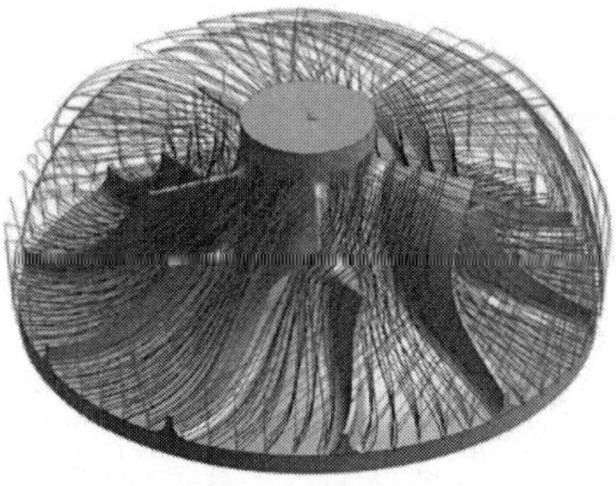

Figure 2 5X Multi-Blade

1.2. "5X Multi-Blade"
5X Multi-Blade milling strategy is a dedicate application for programming of multi-blade parts for aerospace area , and is recommend for 5X-contiuous milling applications for: roughing, finish floor and finish walls. (Fig.2)

The benefits of this milling strategy are:

-Fast and simple programming
-Safe, high quality tool path
-Finish surface quality

1.3. "Prevent waterfall"
New capabilities for better results when machining near a surface's edge:

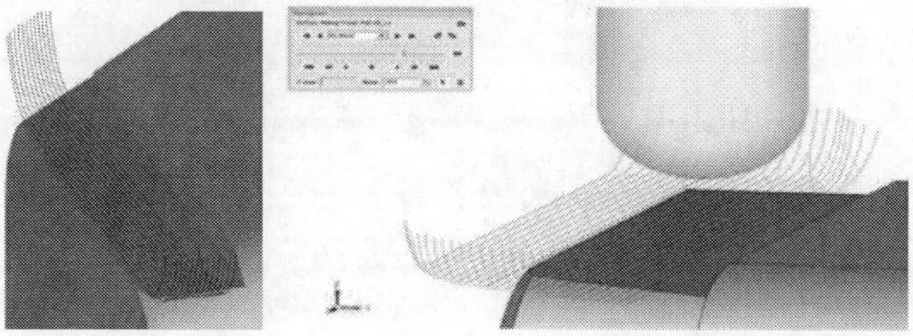

Figure 3. Prevent waterfall strategy

Air extension is smooth and have a natural extension of the tool path outside the part in all directions. (Fig.3)

The benefits are : extensions, along path, extra passes.

1.4. Rough "VoluMill"
Is a new strategy for ultra high volume removal and generate an high speed machining (HSM) all-rounded tool-path to maintain constant material removal rate.

This milling strategy consider updated stock, take in consideration the holder of the cutting tools and is execute between layers.

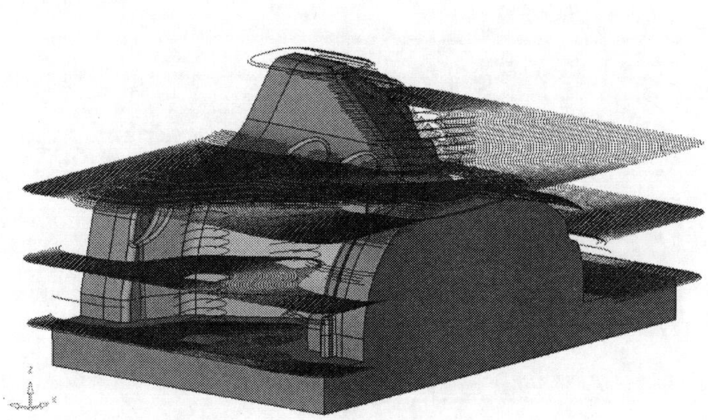

Figure 4. "VoluMill" strategy

IOP Conf. Series: Materials Science and Engineering **416** (2018) 012093 doi:10.1088/1757-899X/416/1/012093

2. State of art

Regarding the "Volumill" strategy (Fig.4) in the article was presented the comparison with the classical milling method - "Z constant" milling strategy.

Was manufacturing one part with one part with a geometry not so complicate but with some sharp shapes with narrow channels and with sudden changes of direction(Fig.5)

Figure 5. The part what was manufacturing

The NC programs was generated by the CAM software who use the cutting tools parameters what was recommended by the cutting tool catalogue.

The raw part is made from steel C45 with the dimensions 100 x 100 x 50 mm and the thickness what will be manufacturing is 5mm and was use one end mill cutting tools 10 mm diameter with four teeth.

2.1. Constant Z milling strategy

The thickness (15mm) was divided in 3constant pases, so the a_p will be 5mm and the radial angagment will be 60% from the cutting tool diameter: a_e=6mm. Starting with this technical informations we will have :

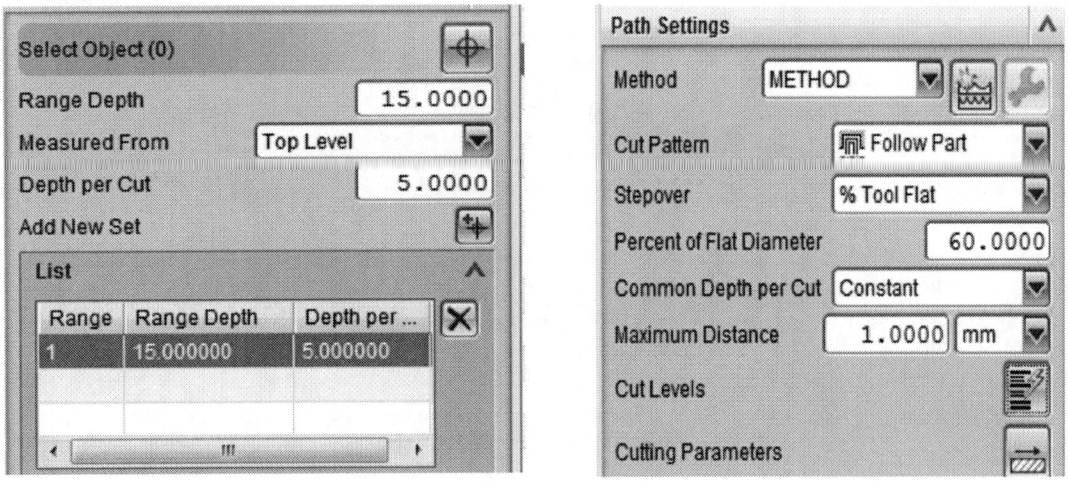

Figure 6. Axial and radial engagements

IOP Conf. Series: Materials Science and Engineering **416** (2018) 012093 doi:10.1088/1757-899X/416/1/012093

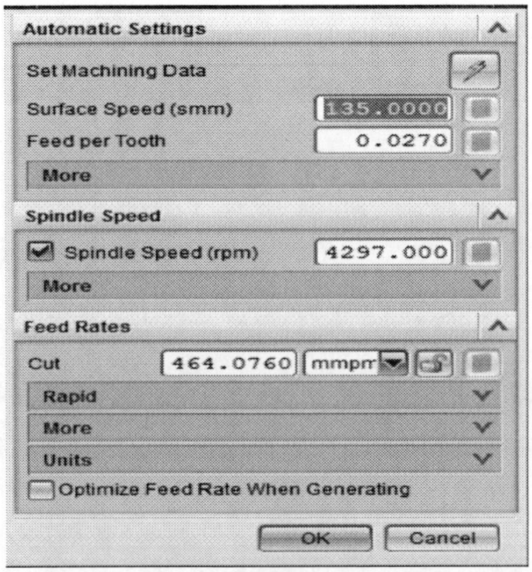

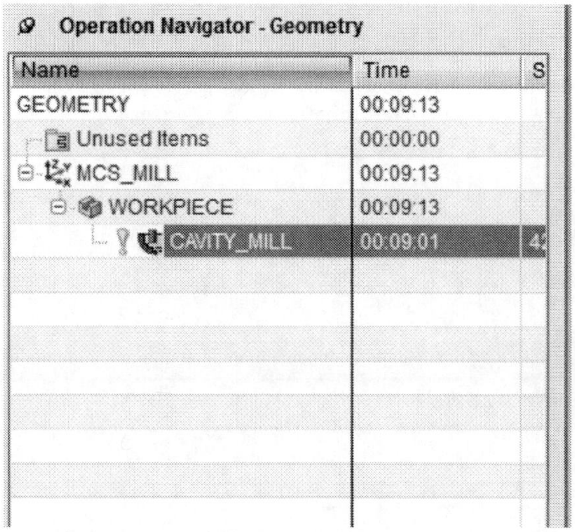

Figure 7. The cutting speed , feed by teeth and the time of manufacturing

The cutting speed of the tool will be V_c=135m/min and the feed per tooth is f_z=0,027mm/tooth, and we will use cooling liquid . (Fig.7)

The trajectory of the cutting tool is present in (Fig.8)

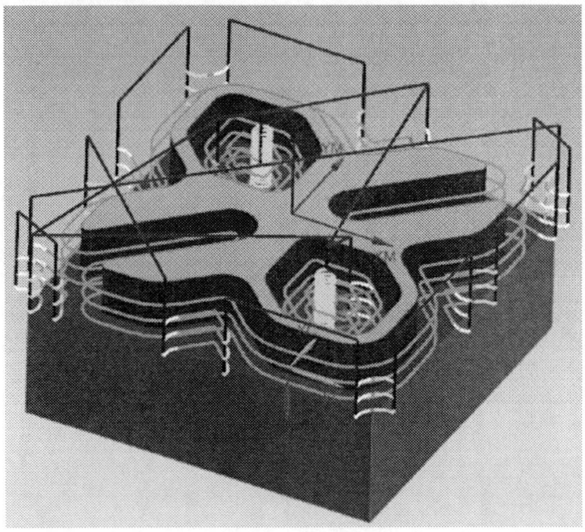

Figure 8. The trajectory of the cutting tool

Figure 9. The cutting tool and chips

In Fig.9 we can see the chip that is around the cutting tool what affect the life of the cutting tool and the surface quality of the part .

IOP Conf. Series: Materials Science and Engineering **416** (2018) 012093 doi:10.1088/1757-899X/416/1/012093

2.2. The Volumill strategy

The same raw part was manufacturing with "Volumill strategy". Now all the thickness 15mm will be manufacturing in one single pass: a_p=15mm and the radial engagement is 10%/D_c The value is a_e=1mm. (Fig.10)

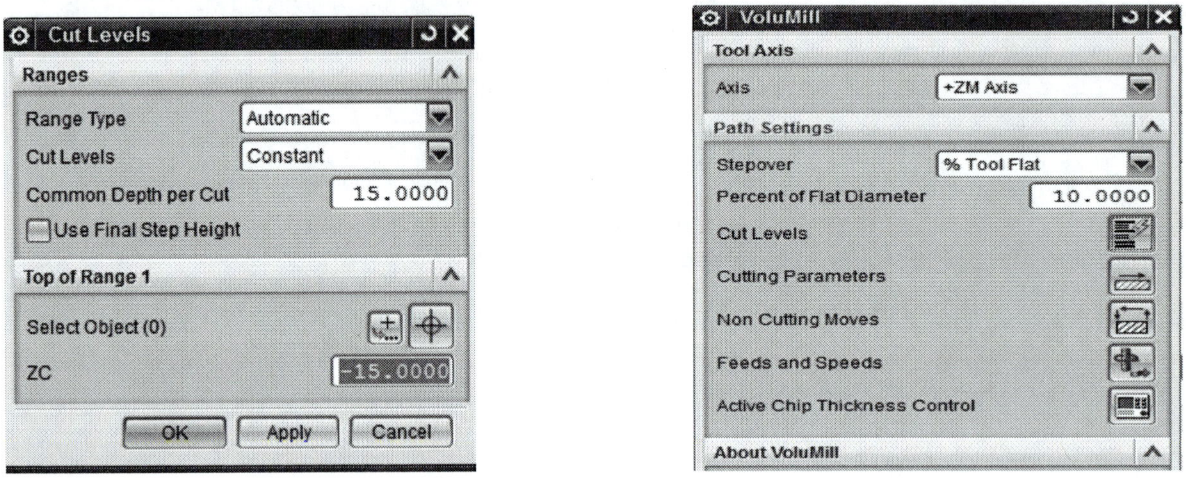

Figure 10. Axial and radial engagement

The radial engagement is small, the cutting tools trajectory are tangents, the chip volume is minimal so will be possible to use higher cutting parameters: spindle speed 9708 rpm. and the feed per tooth is 0.1 mm (Fig. 11).

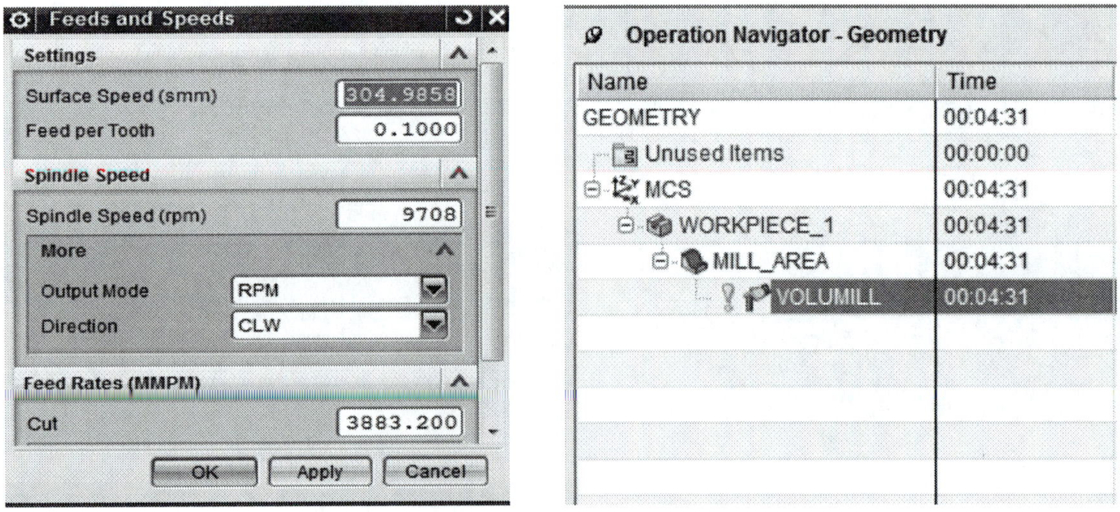

Figure 11. The cutting tool parameters and the processing time

The cutting tool is very short time in contact with the material, the heating transfer is very law, so isn't necessary to use cooling - the chips are not coloured. (Fig.13.)

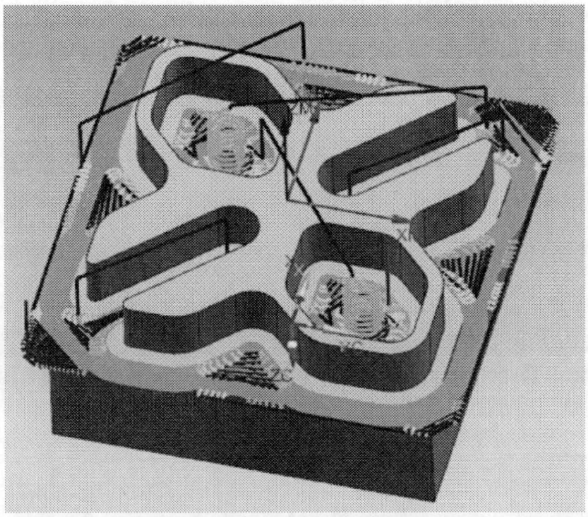

Figure 12. The cutting tool trajectories in CAM

Figure 13. The cutting tool and the chips

3. Conclusions

After the tests we conclude that "Volumill" is much more productive, the quality of the surface is much better and the cutting tool live was increase.

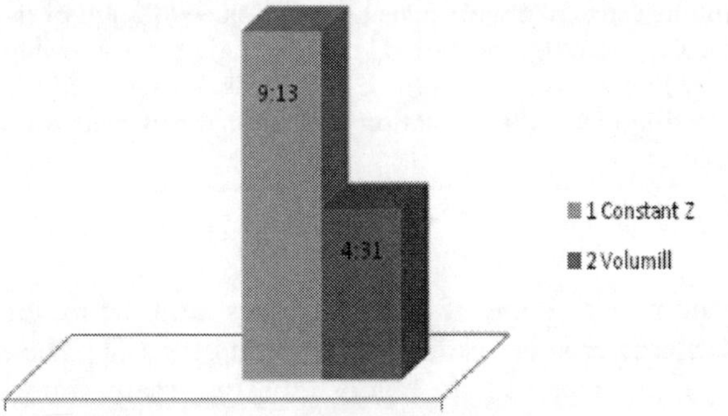

Figure 14.The manufacturing time

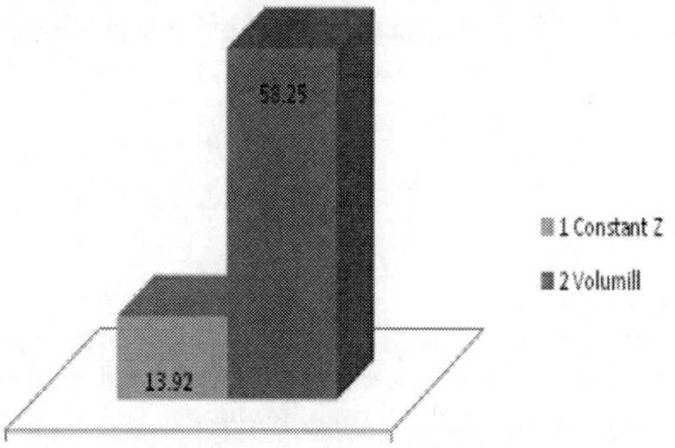

Figure 15. The chip volume diagram

The time of manufacturing using the " Z constant " strategy was 9 minutes and 13 seconds (Fig.14) in comparison with "Volumill" strategy when was 4 minutes and 31 seconds and the quantities of the chips evacuation is three times more. (Fig.15) This experimental test data's show us a high productivity and using the all highest cutting parameters of the tool demonstrate the productivity of this new " Volumill " milling strategy.

4. References

[1] But A-(2003) "*Advanced machine tools and manufacturing systems*" Editura Politehnica Timisoara 2009
[2] *Metal Cutting Technology* Training Handbook Sandvik Coromant
[3] *Metal Cutting Technology* Technical guide

IOP Conf. Series: Materials Science and Engineering **416** (2018) 012094 doi:10.1088/1757-899X/416/1/012094

Experimental Analysis of Tribological Behaviour of Organic Composites Materials

AL Craciun[1], C Pinca-Bretotean[1,*], ID Utu[2] and A Josan[1]

[1]University Politehnica Timisoara, Department of Engineering and Management, 5 Revolutiei Street, 331128 Hunedoara, Romania
[2]University Politehnica Timisoara, Materials and Manufacturing Department, 1Mihai Viteazu Street, 3300222 Timişoara, Romania

*Corresponding author: camelia.bretorean@fih.upt.ro

Abstract. The friction material plays an important role in ensuring the quality of the brake pads. Friction and wear characteristics of friction material are the decision makers in new formulations of composites materials developed for the brake systems. In this work, it was developed four recipes of frictional materials using coconut fibres reinforcement in aluminium matrix, using powder metallurgy technique. For these new materials developed are presented the physico-mechanical characteristics in accordance with various relevant international standards. These materials for brake pads are tested to determine their tribological behaviour in a standard pin on disc equipment. In this sense, the wear resistance and the evolution of friction coefficient of these composites materials were determined. The comparative study reveals that proposed composites have acceptable level of physic-mechanical and tribological characteristics. The study allows selecting the best formulation of frictional materials for use in manufacturing of brake pads for vehicles.

1. Introduction

Brake pads play an important role in braking system. These are of different types, and their choice is based on: vehicle type, engine type and general characteristics of the standards in effect [1].The main components of the brake pads are: friction material, intermediate layer, adhesive, main frame and damping layer. Of all these components, the friction material plays an important role in ensuring the quality of the brake pads in the braking system assembly [1]. This is a mix of materials in which each component has a certain concentration for each type of vehicle. The ideal brake friction material should have constant coefficient of friction under various operating conditions and it should also has various desirable properties such as: resistance to heat, water and oil, low wear rate, high thermal stability, low level of noise, and does not damage the brake disc [2].

In general, each formulation of friction material has its own frictional behaviours and wear-resistance characteristics. Friction material is a heterogeneous material and is composed of a few elements and each element has its own function. Changes in element types or weight percentage of the elements in the formulation may change the physical, mechanical and tribological properties of the brake friction materials [1], [2], [3]. Increasing braking performance can be achieved by improving the physical, mechanical and tribological characteristics of the friction material used in the manufacture of the brake pads [1], [3].

Earlier researchers have concluded that there is no simple correlation between friction and wear properties of a friction material with the physical and mechanical properties [4], [5], [6]. Therefore, each new formulation developed needs to be subjected to a series of tests to evaluate its friction and wear properties. Friction and wear characteristics of friction material play an important role in deciding which new formulations developed are suitable for the brake system. Composition and formulation of brake pads also play a big role on the friction behaviour, and since composition-property relationship are not known well enough [7]. Typical values for friction coefficient are 0.3 to 0.6 taking into account the operating conditions. Studies release that the friction coefficient fluctuate between 0.3 and 0.5 through a pressure range 1 to 2 MPa [1], [2], [8].

Friction materials have also been produced in the literature consisting of organic materials, for example: eggs shells, bananas peels, palm fibres, periwinkle shells, cocoa beans shells [6],[8],[9],[10]. Current trend in the research field of automotive industry is to utilization of different wastes as a source of raw materials for composite development due to the carcinogenic effects of asbestos [5], [6].The papers in the field has also shown composite materials made with coconut fibre, especially in tropical countries [8],[9]. Coconut fiber composite materials have a wide range of applicability in these countries, such as: building materials, marine ropes, fishing gear, furniture, household appliances [8].Regarding coconut fiber in the literature, it has been shown that it has properties that approve its application in the manufacture of friction materials [4], [7], [9].

In this work, were developed four recipes of frictional materials using coconut fibres reinforcement in aluminium matrix using power metallurgy technique. Each sample was subjected to determine the physical, mechanical, friction and wear characteristics, in accordance with various relevant international standards. Pin-on disc equipment TR-20 is used to evaluate the friction and wear characteristics of the organic composite materials used for brake pad applications. The comparative study reveals that proposed composites has acceptable level of physic-mechanical and tribological character. The study allows selecting the best formulation of composite for use in manufacturing of brake pads for vehicles. Considering that in our country coconut is an exotic fruit, its price is relatively high, this being an initial disadvantage for the friction materials produced in this paper.

2. Experimental Methodology

In this work, it has been made four recipes of composite materials using coconut fibres reinforcement in aluminium matrix using powder metallurgy technique. Their purpose is to make brake pads for small and medium vehicles. The choice of constituents was made following the study of the literature in the field and according to their role in meeting the requirements for the brake pads [4], [5], [7], [8], [9], [10]. The recipes of every composite contain eight ingredients; six ingredients were maintained constant and two ingredients were varied, aluminium and coconut fibre in complementary manner as shown in Table 1, and 2.

Table 1. Composite recipes for Sample 1 and Sample 2

No. of sample	Aluminium (%)	Graphite (%)	Zirconia oxide (%)	Silicon carbide (%)	Titanium oxide (%)	Phenolic resin (%)	Hexametyl tetramine (%)	Coconut fibre (%)
Sample 1	20	5	2	11	11	40	6	5
Sample 2	15	5	2	11	11	40	6	10

Table 2. Composite recipes for Sample 3 and Sample 4

No. of sample	Aluminium (%)	Graphite (%)	Zirconia oxide (%)	Silicon carbide (%)	Aluminium oxide (%)	Phenolic resin	Hexametil tetramine (%)	Coconut fibre (%)
Sample 3	20	5	2	11	11	40	6	5
Sample 4	15	5	2	11	11	40	6	10

In order to produce the composite materials in the laboratory, the following steps are taken: weighing, blending, cold pressing, oven storing, hot pressing, hot and cold cooling. After producing the samples according to the four recipes, these will be characterized and tested. All laboratory equipment used take placed at Engineering and Management Department, Faculty of Engineering Hunedoara.

The constituent elements of the recipes were mixed with a mechanical stirrer at 3000 rpm, respecting a certain order of introduction of the elements into the mixture. Thus, for the first time, aluminium and silicon carbide were mixed for 10 minutes. Separately was mixed, for 5 minutes, zirconium oxide and titanium oxide (for Sample1 and Sample2), respectively zirconium oxide and aluminium oxide (for Sample3 and Sample4). This mixture was added over the first one, after which stirring was continued for 4 min at a speed of 1500 rpm. Graphite and coconut fibre was added and mixed at the same speed for another 10 min. The phenolic resin and hexamethyltetramine were mixed separately. The solid mixture was introduced over the liquid mixture, the entire composition being mixed for 10 minutes at a speed of 1000 rpm to obtain the homogeneity of the composition. At the end of the mixing process, the composition, having the appearance of a viscous paste, was introduced into the mold for sintering. The mold allows making samples of 95 mm diameter. The mold base and its piston have been protected with aluminium foil. Before to the introduction of the composition into the mold, a thin graphite layer was sprinkled on, for easy extraction of the sample at the end of the sintering process. After insertion of the mixture into the mold, the screws are tightened and a cold pressing was performed, at a pressure of 20 MPa for 5 minutes. The sample mold assembly was introduced into the heated oven at 200^0 C, where the sample was held for one hour. After this time a hot pressing was performed with a pressure of 50 MPa, after which the sample was reintroduced into the furnace at 180^0C where it was held for 4 hours. It is mentioned that the pressing was intermittent, taking pauses at equal intervals in order to eliminate the gases into sample. Cooling was done in two steps, in the first step in the oven at 100 ° C for 8 hours, for convert the resin from thermoplastic to thermosetting and in the second step, in ambient air for 48 hours [4], [5], [6], [7], [8]. The stages and parameters of the sintering technology of composite materials have been established as a result of several experimental tests that allowed critical analyses to be performed and successive improvements.

Figure 1 shows the samples obtained from the four recipes proposed at the end of the sintering process.

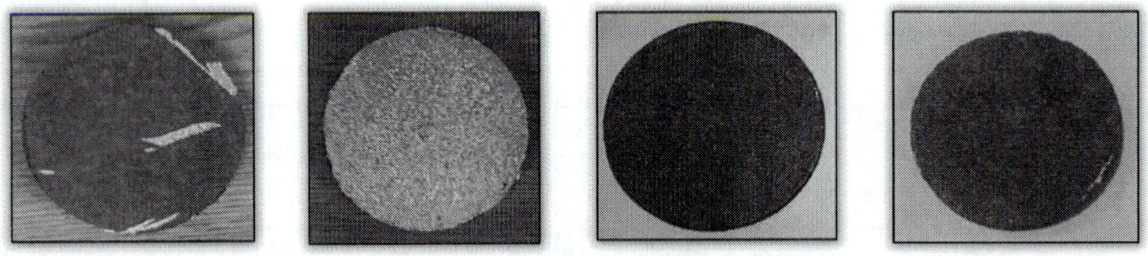

Sample 1 Sample 2 Sample 3 Sample 4

Figure 1. Samples obtained from the four recipes proposed at the end of the sintering process.

The samples for each test were carried out by cutting in concordance with actual standards, in order to determine the physico-mechanical and tribological characteristics.

3. Discussion and results

Table 3 presents the results of physical and mechanical characteristics of the produced composites. The density values for the obtained composite materials are relatively low compared to the metal densities. The highest densities are obtained by the specimens manufactured with 5% coconut fibre recipes (Sample 1 and Sample 3), due to the fact that these recipes contain the largest amount of metal. Density depends on the percentage of ingredients included in recipes, given that a metallic element has a density greater than an organic element [2], [4] this justifying the values obtained for density according to the percentage of metal introduced into the recipe.

In Table 1, it can be seen that all composite materials have higher porosity in water than porosity in oil. Similar results were obtained in works [6]. The smallest porosity has Sample 2, followed by Sample 4, both of them are made with 10% coconut fibre. Specialty literature states that the brake pads must have a certain porosity to minimize the effect of water and oil on the friction coefficient and to reduce the noise during braking [4], [5]. The obtained values were compared with those obtained by other researchers who studied the behaviour of composite materials with coconut fibre and similar destination and [4], [6].

Table 3. Physico-mechanical characteristics for the samples

Samples	Density (g cm^{-3})	Porosity in water (%)	Porosity in oil (%)	Hardness HBS	Breaking force (N)	Compressive strength (N mm^{-2})	Modulus of longitudinal elasticity (N mm^{-2})
Sample 1	1.54	0.75	0.29	128	11886	(-)27.6141	262999.142
Sample 2	1.37	0.55	0.14	176	14577	(-)35.7912	32537.454
Sample 3	1.46	0.88	0.32	109	8280	(-)25.6162	19406.212
Sample 4	1.29	0.62	0.17	150	12992	(-)33.0730	26248.412

The lowest hardness has Sample 3 (109 HBS), followed by Sample 1 (128 HBS) both having the same amount of aluminium and coconut fibre. Different values of hardness are explained by the fact that Sample 1 contains titanium oxide, and in Sample 2 aluminium oxide. The hardness values are higher than of other materials used for brake pad applications produced from other researchers [4], [6], [9].

In order to perform a qualitative analysis of the recipes produced on the basis of the established sintering technology and considering the destination of these materials, the compressive strength for each sample Sample1-4 was determined. Compression test was performed with universal testing machine. The compressive strength has increased with an increase in coconut fibre content, due to the correct dispersion of the filler particles with the binder, which has led to an appropriate interaction between these elements. Proper particle dispersion has improved the interaction between coconut fibres and binder. Table 1 shows the average values of mechanical characteristics of composite materials produced after four recipes. The highest compressive resistance has Sample 2, followed by Sample 4. The ultimate strength of the Sample 2 corresponds to 35.7912 N/mm^2. The breaking of the Sample 4 composite corresponds to a stress of 33.0730 N / mm^2, and breaking force of 12992N / mm^2. The highest value of the longitudinal elastic modulus was recorded for Sample 2.

The friction coefficient was determined at the University Politehnica Timişoara, Department of Materials and Manufacturing Engineering. Tribological researches of composite material developed have the following objectives: to study the evolution of wear and the friction coefficient. A pin-on-disc equipment, TR-20 was used to investigate the dry sliding wear characteristics of organic composite materials as per ASTM G99-95 standards. The equipment has four adjustable bays that allow the fitting of various shapes and sizes. The pin of equipment is a steel ball of 6 mm diameter. The pin on disc equipment is programmable to study friction against speed, load and wear. The specimens for the tribological tests are rectangular with the dimensions (25x7x7mm) and they were polished for friction and wear tests. The tests were conducted for five hours at load of 15N. Wearing diameter was 15 mm, speed is 150 rpm and test distance is 2200 m. Figure 2 shows a sample mounted in the pin on disc equipment prepared to determine wear and friction coefficient.

The initial weight of the specimen was measured with an electronic weighing machine, model Sortorius CP2202S-OCE with a least count of 0.0001mg. After running through a fixed time period, the specimen were removed, cleaned with acetone, dried and weighed to determine the weight loss due to wear, [4], [5], [10].

IOP Conf. Series: Materials Science and Engineering **416** (2018) 012094 doi:10.1088/1757-899X/416/1/012094

Figure 2. A sample prepared to determine wear and friction coefficient

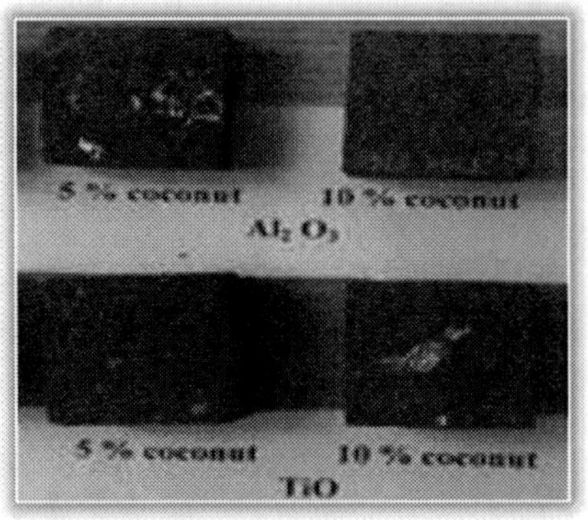

Figure 3. The specimens at the end of the tribological tests, with highlighting the friction trace

The difference in the weight measured before and after the test, gives the wear of the samples [5]. The wear rates were determined using the weight loss method and the results are presented in Table 4. Figure 3 shows the specimens at the end of the tribological tests, with highlighting the friction trace.

Table 4. Tribological characteristics of the samples

Composites	Initial weight (g)	Final weight (g)	Wear mass (g)	Wear mass relative (%)	Mass wear on the rest distance (g m^{-1})	Max. friction coefficient μ_{max}
Sample 1	7.7657	7.6057	0.16	2.06	0.00007272	0.358
Sample 2	7.5917	7.5517	0.04	0.52	0.00001818	0.415
Sample 3	8.2419	8.1119	0.13	1.57	0.00005909	0.331
Sample 4	8.7389	8.6789	0.06	0.68	0.00002727	0.452

Figure 4 presents the variation of mass wear for the organic composite materials and figure 5 presents the evolution of the coefficient of friction during the test. The wear resistance of the material is the inverse of the specific wear rate [5]. In this context, Sample2 has the highest wear resistance. The wear resistance of Sample 4 is close to the wear resistance of Sample 2.

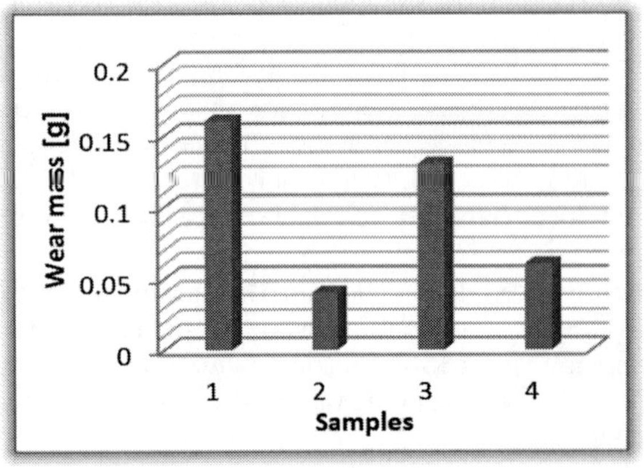

Figure 4. Variation of mass wear for the organic composite materials

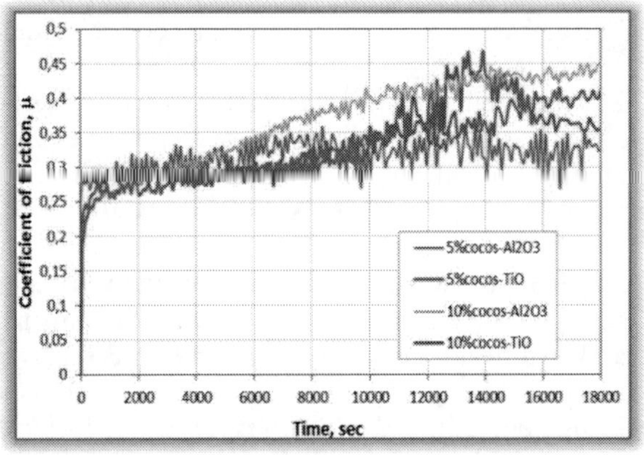

Figure 5. Evolution of the coefficient of friction during the test

The initial running of each sample is associated with a slow increase of the friction coefficient. The friction coefficient touches the value of 0.3 for all samples. After a certain trial, the friction coefficient became stable. With increased friction coefficient, the mass wear was lower.

4. Conclusions

From the results and discussion of this study it can be observed:

- the fabrication of organic composites materials depends on mixing sequences, speed and time of mixing;
- the density of composite materials produced by own recipes, depends on the percentage of metallic component introduced in the recipe, which confirms the data from the technical literature;
- the smallest porosity both in water and oil have Sample 2 and Sample 4 that are made with 10% coconut fibre;
- the compressive strength has increased with an increase in the coconut fibre content;
- the higher coconut fibre content, led to the better friction performance in the pin on disc procedure;
- for all samples after a certain trial time, the friction coefficient became stable;
- Sample 2 and Sample 4 is more efficient than other compositions, since it has a good coefficient of friction with less wear rate;
- Sample 2 and Sample 4 made with 10% coconut fibre content have the best physico-mechanical and tribological characteristics;
- Sample 2 and Sample 4 showed similar properties and could be used in fabrication of non-asbestos friction materials for brake pad application;
- the disadvantages that exist with newly created composites consist in: coconut fibre has a relatively high price, the fruit being an exotic one and the organic fibre has non uniform filler size.

5. References

[1] Blau P J 2001 Compositions, functions and testing of friction brake materials and their additives *Contract DE-AC05-00OR22725* Tennesse USA
[2] Cho MH, Kim SJD and Jang H 2005 *Wear* **258** 1682
[3] Anderson AE 1992 *Friction Lubrication and Wear Technology* **18** 569
[4] Maleque MA, Atiqah A, Talib RJ and Zahurin H 2012 *International Journal of Mechanical and Materials Engineering* **7(2)** 166
[5] Vijai R, Jees SMA and Thiyagarajan V 2013 *Advances in Tribology* **2013** 165859
[6] Olabisi AI, Adam AN and Okechukwu OM 2016 *International Journal of Materials Science and Applications* **5(2)** 66
[7] Lawal SS, Bala KC, and Alegbede AT 2017 *Leonardo Journal of Sciences* **30** pp. 47-56
[8] Sapuan SM and Harimi M 2003 *The Arabian Journal for Science and Engineering* **28(2B)** 171
[9] Sutikno M, Marwoto P and Rustad S 2010 *Carbon* **48** 3616
[10] Craciun AL and Pinca-Bretotean C 2016 *Mater. Sci. Eng.* **163** 012014

IOP Conf. Series: Materials Science and Engineering **416** (2018) 012095 doi:10.1088/1757-899X/416/1/012095

The Dispersion of the Heat Flow in the Engine Compartment. Case Study on Ford Fiesta 1.4i

C Birtok Băneasă[1,*], A Budiul Berghian[1], S A Rațiu[1] and D M Stoica[1]

[1]Politehnica University of Timisoara, Department of Engineering and Management, 5 Revolution Street, Hunedoara, 331128, Romania

[*]Corresponding author : corneliu.birtok@fih.upt.ro

Abstract. This paper presents a study concerning the influence of the dispersion of the heat upon the air filter and the intake manifold. The data was registered in real time with the help of thermo imaging camera at different operating regimes, route and traffic installed on a FORD Fiesta car, equipped with a gasoline engine with a cylindrical capacity of 1.4l. The propeller group is mounted transversely, a front constructive solution. The gas circulation is a inline cross flow type with the intake manifold placed in front of the engine and the exhaust manifold behind the engine. The air circulation through the air filter is a loop flow type having the in/out port orientated towards the front of the vehicle. This study wants to determine the influence of the materials which were used for the construction of the air filter and of the intake manifold upon the performance of the internal combustion engine. The results obtained after the tests allow us to determine the optimum parameters on the efficiency of the heat transfer.

1. Introduction

Because of the position in the engine compartment the intake is highly expose to thermal radiations from the cooling radiator, exhaust pipe and engine itself which is a disadvantage, the intake air it heated and the result is a lower density, thus containing less oxygen per volume unit than cold air. Air intake temperature is important to ensure higher amount of fuel involves in the combustion process. We can say that the fuel consumption is improved with the decrease of air intake temperature regardless of engine speed. Also, carbon monoxide and unburned hydrocarbons were decreased with the decrease of air intake temperature, again regardless of engine speed. So, based on the specialized studies, the lower air intake temperatures resulted in lowest fuel consumption and reduced exhaust emissions regardless of engine speed [1].

The study focused on Ford Fiesta1.4i, which engine is characterized by having a crossflow type gas circulation with the intake manifold placed in front and the exhaust manifold behind the engine. The air filter is placed in the upper part of the engine, the shell being made from plastics in a prismatic shape. The air circulation through the air filter is a loop flow type having the in/out port orientated towards the front of the vehicle.

2. Conditions and measurements

The measurements were made in the engine compartment of the FORD Fiesta car, equipped with a gasoline engine with a cylindrical capacity of 1.4i (figure 1.a), gasoline, multi point injection. The air filter is placed in the upper part of the engine, the shell being made from plastics in a prismatic shape.

IOP Conf. Series: Materials Science and Engineering **416** (2018) 012095 doi:10.1088/1757-899X/416/1/012095

The air circulation through the air filter (figure 1.b) is a loop flow type having the in/out port orientated towards the front of the vehicle.

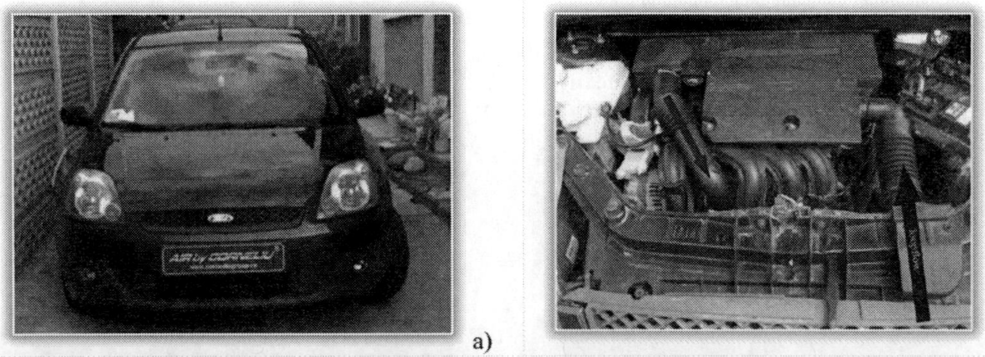

a) b)

Figure 1. a) FORD Fiesta car– hood surface, b) engine compartment

The comparative measurements were made after reaching the regime temperature of the engine, at the level of the engine compartment, from the air filter shell , of the intake manifold and the in/out connections hoses of the air filter [2].

The temperatures were monitorized depending on the functioning regime (normal traffic, crowded traffic) and route (altitude difference, ramp, downhill) as it follows:

» A – cold stopped engine– reference state
» B – traffic mode highway
» C – traffic mode normal urban
» D – traffic mode mountain (ascending)
» E – traffic mode mountain (discending)
» F – traffic mode crowded urban

The air filter (figure 2) is of panel type filter element (from microporous cardboard) placed in the upper part of the engine, the framework is made from plastics in a prismatic shape and it covers 100% the engine. The air circulation is loopflow type, having the direction from front to front [3].

Figure 2. Air filter

Figure 3. Intake manifold

The intake manifold (figure 3) is placed in the front part of the engine. On the intake manifold there are placed the throttle module and the multipoint injection ramp. The material used in the construction of the intake manifold is PA66 GF35 [4].

The heat sourced of the air filter and of the intake manifold are the following, in the case of the Ford car: exhaust manifold, engine, and cooling radiator.

3. The heat flow dispersion highlighted by the thermographic measurements

Comparative temperature measurements are presented next from the engine compartment (overview), from the air filter shell, of the intake manifold and of the in/out connections of the air filter for the above mentioned situations. Using the thermal imaging camera one can highlight the areas influenced by the heat transfer as a result of the dispersion through the engine compartment [5].

The comparative measurements on the surface of the hood are presented in Figure 4.

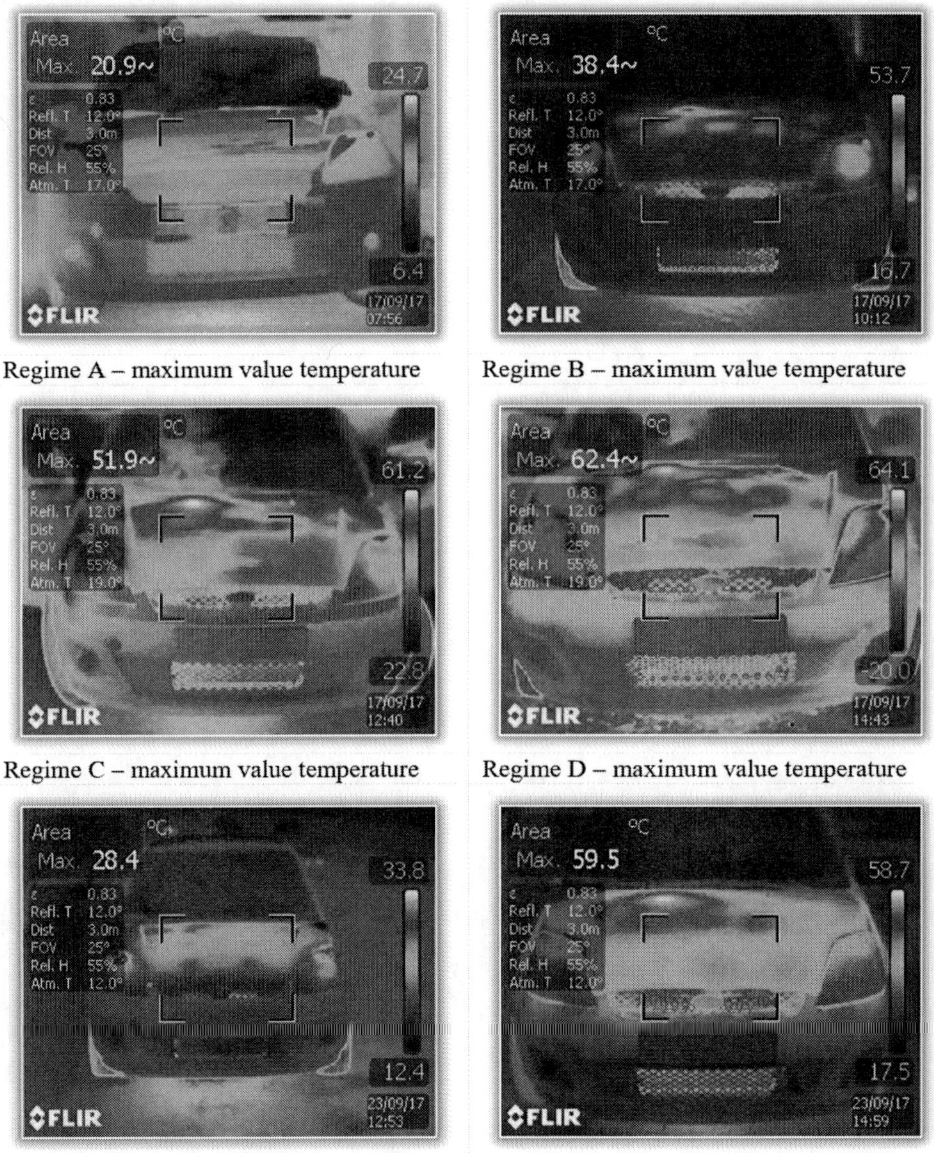

Regime A – maximum value temperature | Regime B – maximum value temperature

Regime C – maximum value temperature | Regime D – maximum value temperature

Regime E – maximum value temperature | Regime F – maximum value temperature

Figure 4. Measurements made on the surface of the hood

The comparative measuremetns in the engine compartment are presented in Figure 5. The comparative measurements of the shell of the air filter are presented in Figure 6. The comparative measurements on the surface of the intake manifold are presented in Figure 7.

IOP Conf. Series: Materials Science and Engineering **416** (2018) 012095 doi:10.1088/1757-899X/416/1/012095

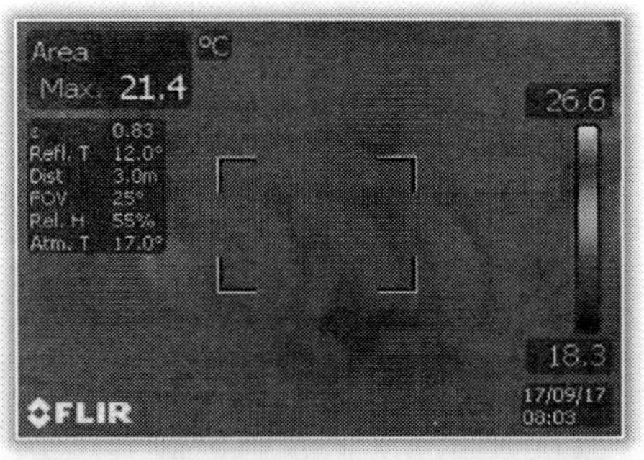

Regime A – maximum value temperature

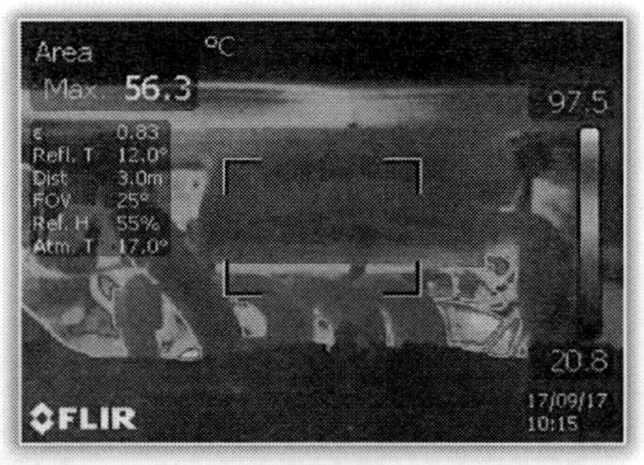

Regime B – maximum value temperature

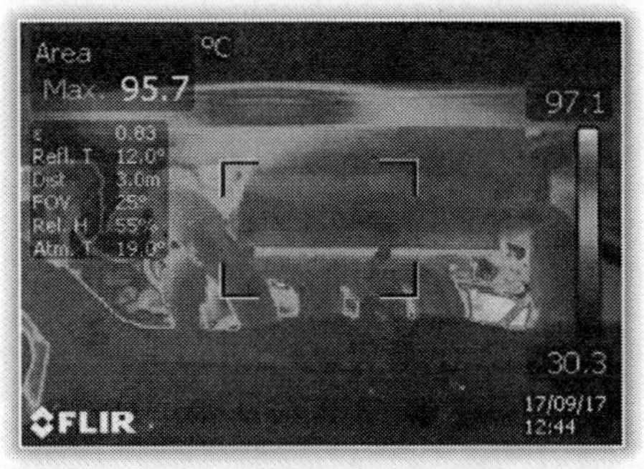

Regime C – maximum value temperature

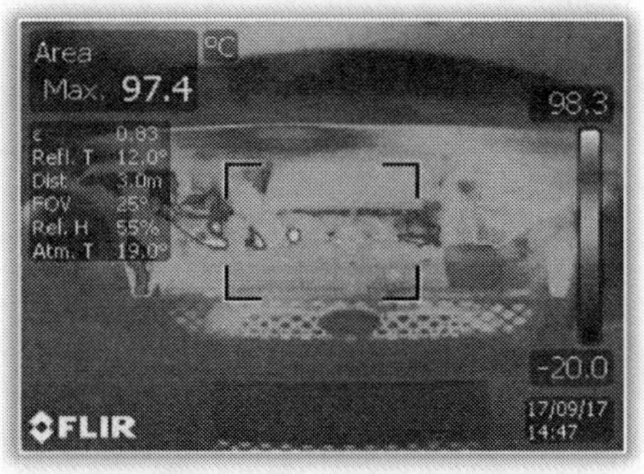

Regime D – maximum value temperature

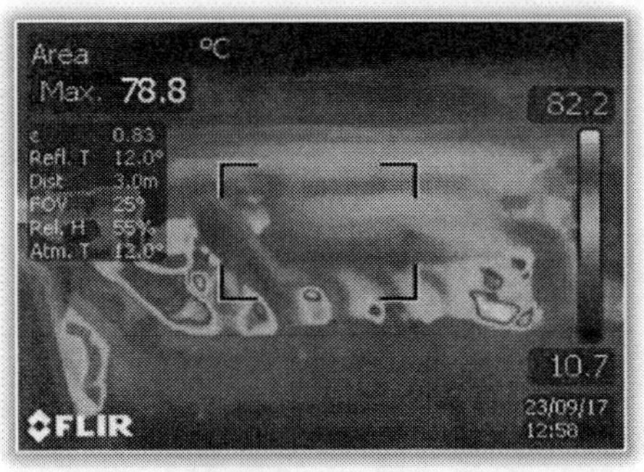

Regime E – maximum value temperature

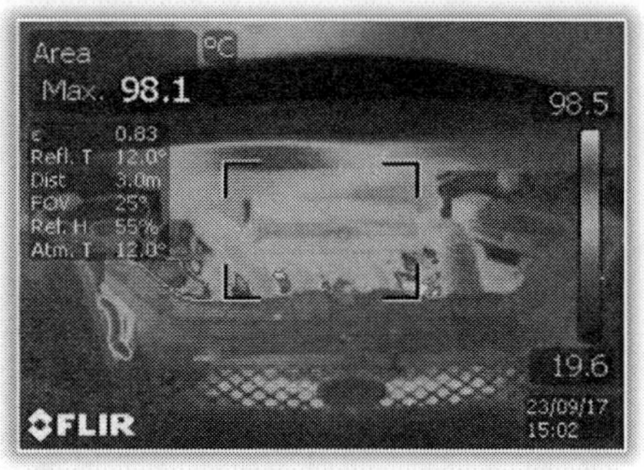

Regime F – maximum value temperature

Figure 5. Measurements made in the engine compartment

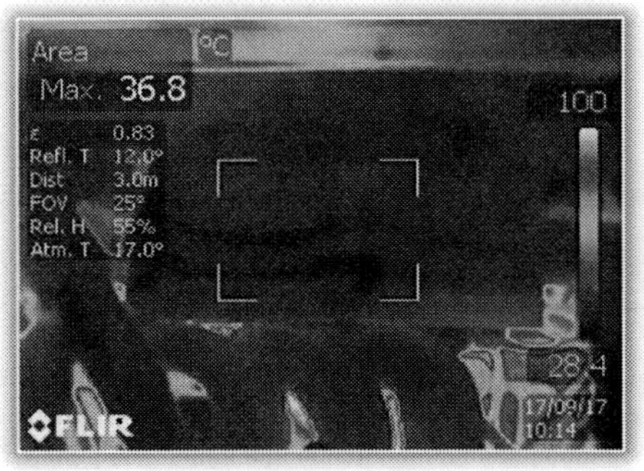

Regime B–maximum value temperature

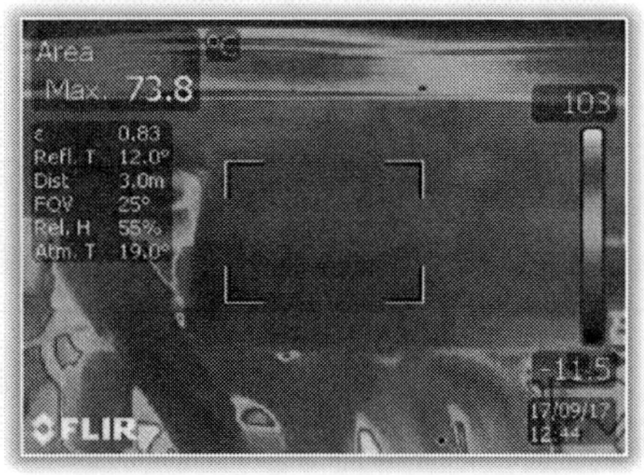

Regime C–maximum value temperature

Figure 6. Measurements made on the surface of the air filter

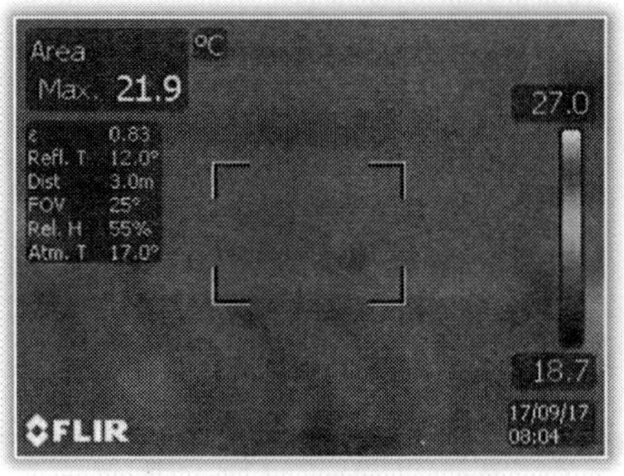

Regime A–maximum value temperature

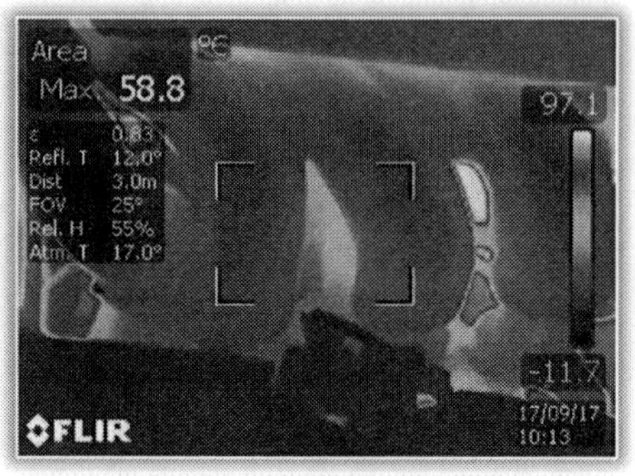

Regime B – maximum value temperature

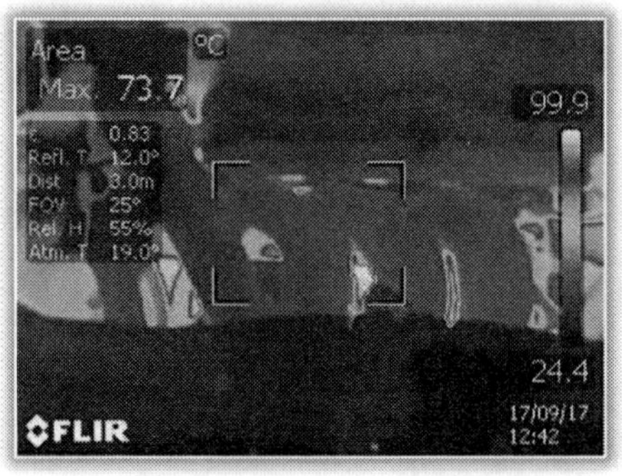

Regime C – maximum value temperature

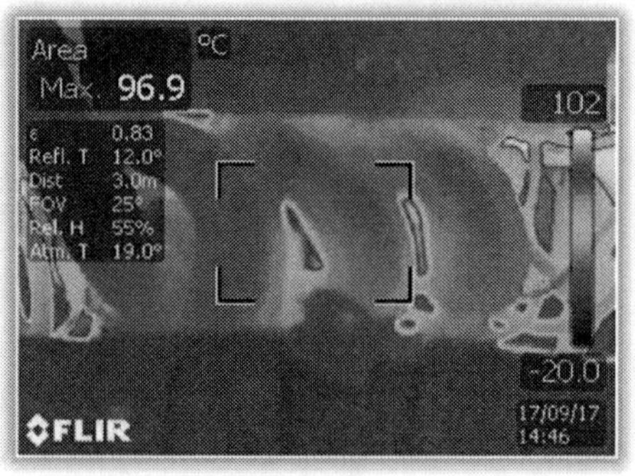

Regime D – maximum value temperature

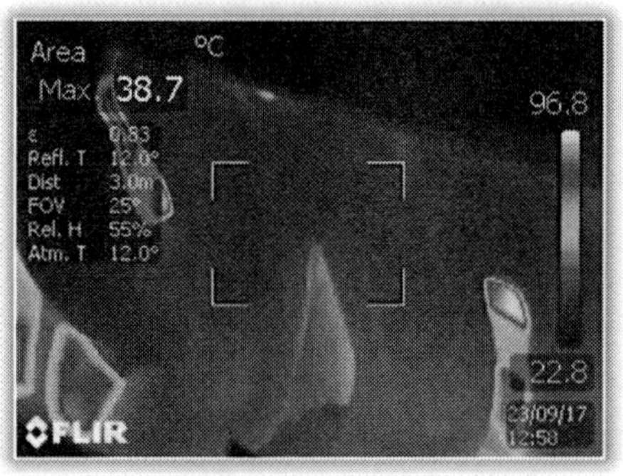

Regime E – maximum value temperature

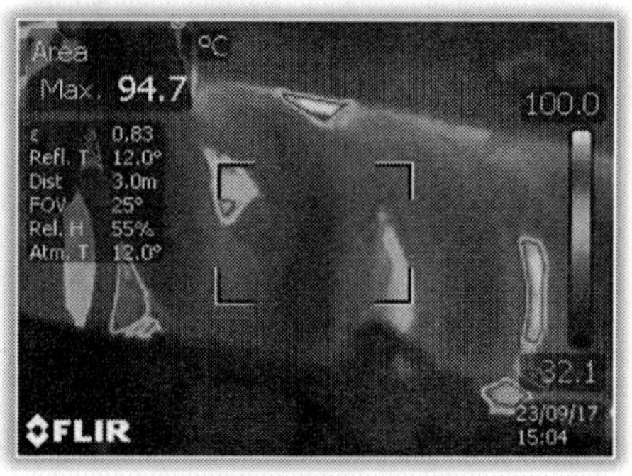

Regime F – maximum value temperature

Figure 7. Measurements made on the surface of the intake manifold

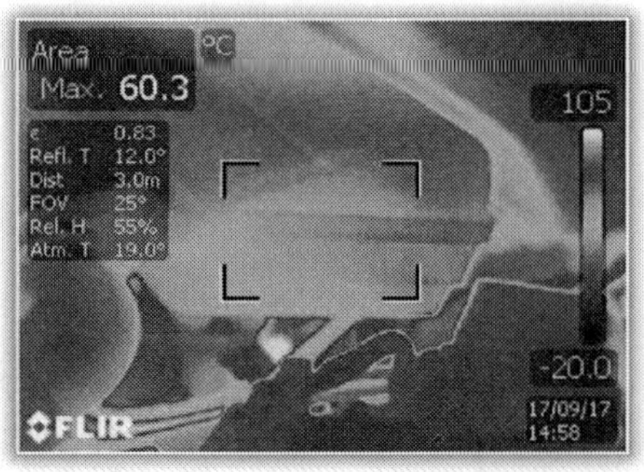

Regime D – minimum value temperature

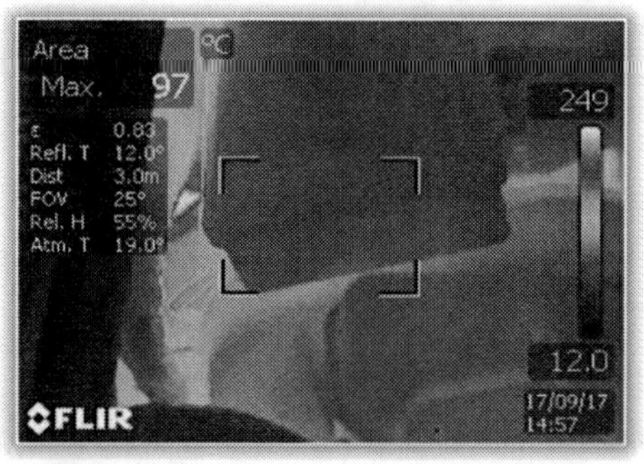

Regime D – maximum value temperature

Figure 8. Measurements made in critical areas

Area which presents critical temperatures at the entrance and exit from the air filter shell are presented in figure 8.b.

The comparative measurements on the entrance/exit connections hoses of the air filter are presented in figure 8.b.

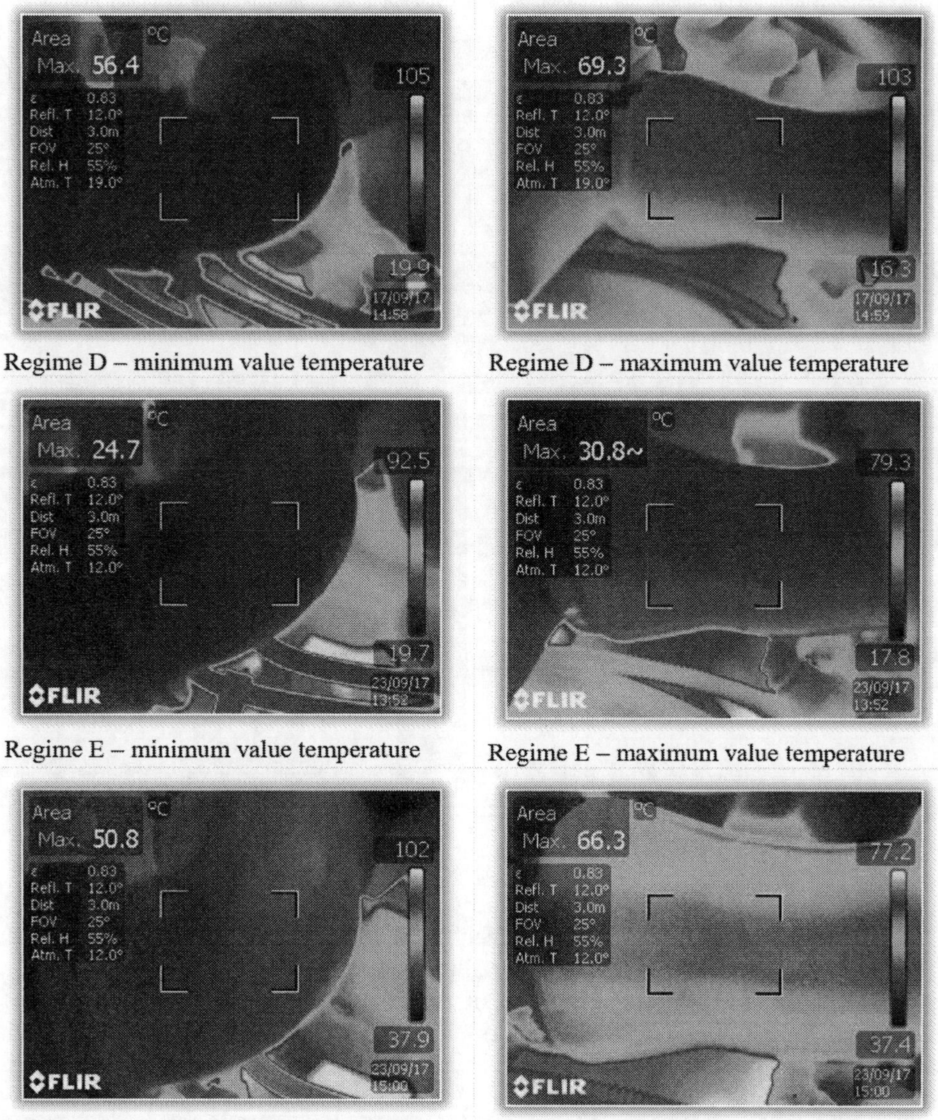

Regime D – minimum value temperature | Regime D – maximum value temperature

Regime E – minimum value temperature | Regime E – maximum value temperature

Regime F – minimum value temperature | Regime F – maximum value temperature

Figure 9. Measurements on the entrance/exit connections hoses

Temperature measurements were made on the surface of the intake manifold for all the considered functioning regimes of the engine (Graphics 1, 2) and in the case of traffic mode mountain (D) data was collected referring to the temperature of all the components (Graphics 3, 4), a modeling was realized using the method of the smallest squares. The programme used to generate the graphics was realized in the Matlab language.

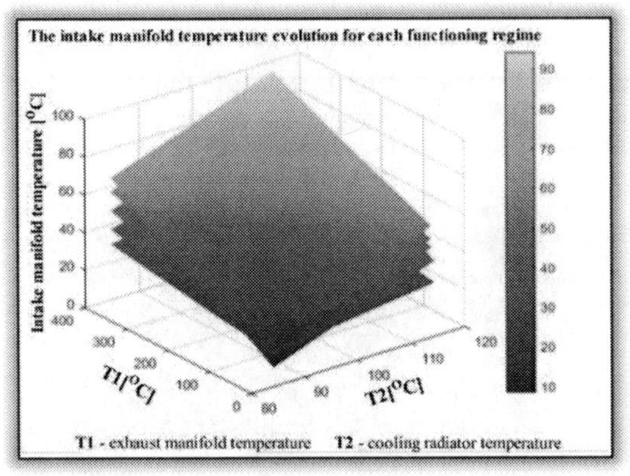

Graphic 1. The intake manifold temperature evolution for each functioning regime

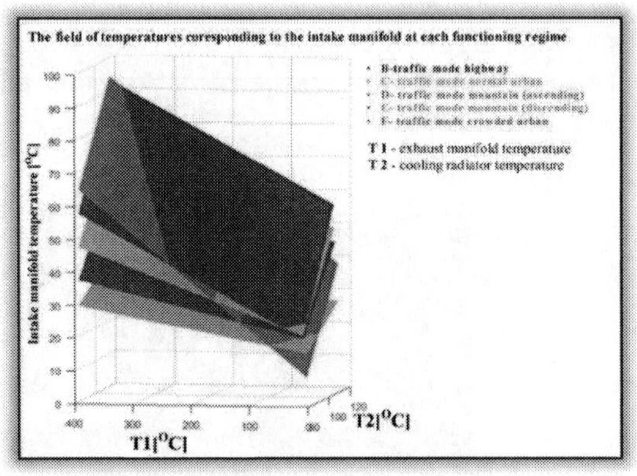

Graphic 2. The field of temperatures coresponding to the intake manifold at each functioning regime

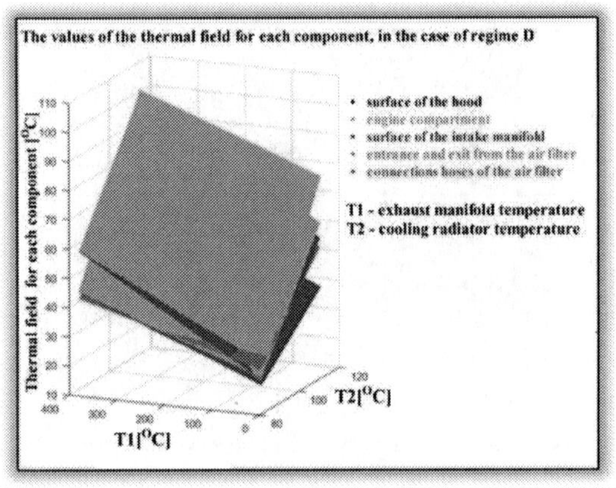

Graphic 3. The values of the thermal field for each component,in the case of regime D

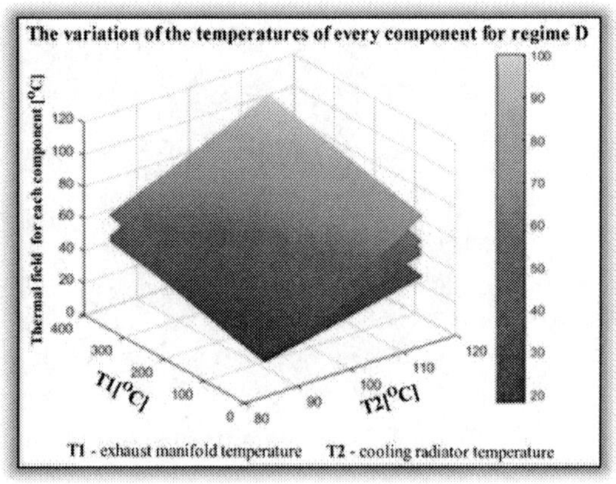

Graphic 4. The variation of the temperatures of every component for regime D

4. Final results and conclusions

At the end of the measurements it is emphasized the fact that the constructive solution chosen for the FORD Fiesta 1.4l car, namely placing the air filter above the engine, and its dimensions, lead to an excessive heat of the air for the functioning of the engine, favoring the appearance of overheating phenomena of the engine, of detonation, or normal wear, etc., with influences on reducing the filling coefficient of the cylinders,with consequences on reducing the power of the engine, growing of the fuel consumption and of the emissions.

Also from figure 8 can notice, on the air filter shell, the existence of some critical areas which present temperatures up to 30% higher than the maximum temperature recorded on the surface of the air filter, this cand be explained through the gathered influence of the 3 heat sources (exhaust manifold, engine and cooling radiator).

5. References

[1] Birtok–Băneasă C, Rațiu S A and Hepuț T, 2017 *Calculation of thermal conductivity for new materials used in intake systems of internal combustion engines,* International Conference on Numerical Analysis and Applied Mathematics (ICNAAM), Rhodes (Greece), SEP 19–25, 2016, AIP Conference Proceedings **1863** pp 1–6

[2] Birtok–Băneasă C 2017 The dispersion of the heat flow in the engine compartment. case study fiat panda 1.2i *Annals of Faculty Engineering Hunedoara* **XV**(4) pp 215–219
[3] Birtok–Băneasă C and Raţiu S 2011 *Air intake in internal combustion engines – superaspirant air filters– dinamycs transfer systems* Timişoara: Politehnica Publishing House
[4] Raţiu S, Birtok–Bãneasã C, Alic C and Mihon L 2009 *New concepts in modeling air filters for internal combustion engines* 20th International DAAAM SYMPOSIUM "Intelligent Manufacturing & Automation: Theory, Practice & Education" Vienna (Austria) **20** pp 1–6
[5] www.corneliugroup.ro

Recovery and Valorification of Iron From Industrial Waste

O Lupu[1,*], L Zgripcea[1], A Socalici[1] and E Ardelean[1]

[1]Politehnica University of Timisoara, Department Engineering and Management, Hunedoara, Romania

[*]Corresponding author: oana.lupu@yahoo.com

Abstract. In order to apply the concepts regarding sustainable development a balance needs to be established between the raw materials used and the volume of capitalized wastes, the aim being to reduce the level of pollution. An emphasis must be made on the technologies that allow the capitalization of waste that can revaluated after the manufacturing and storage processes. This paper presents the results obtained after laboratory testing the processing of ferrous waste with the aim of obtaining raw material that can be reused in the ferrous metallurgy. Processing and capitalizing such waste may result in recovering iron.

1. Introduction

Currently, the issue of handing waste through capitalization is an ecological and economical priority. Circular economy (fig.1) stimulates the growth of economic value at the same time that the consumption of natural resources decreases [1].

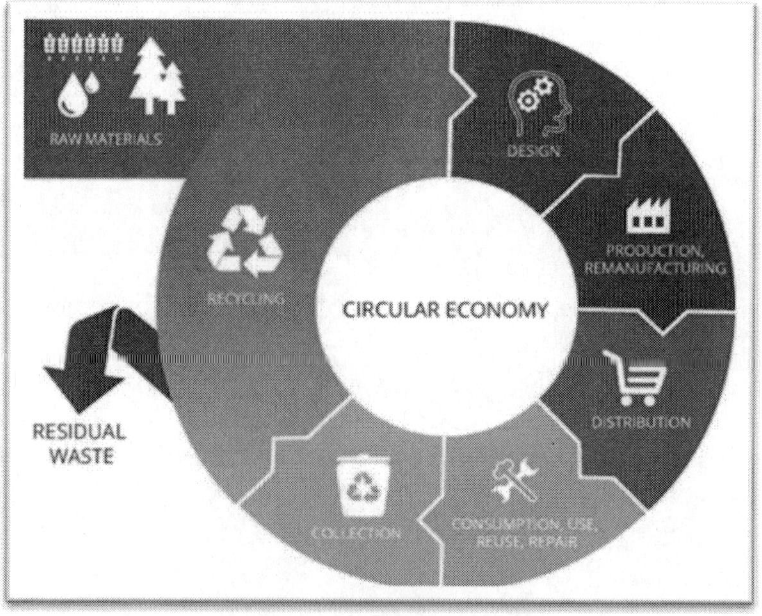

Figure 1. Circular Economy [1]

The capitalisation of waste is an essential part of recycling and it answers the connection between the vital and final point of the process in the economic circuit of waste. This dynamic offers circularity for economy. The capitalisation of industrial waste leads to the minimization of natural resource consume, the development of businesses, the optimization of costs and the creation of new jobs. The European Union must transform the linear economical model into a circular one which favours reuse and recycling of materials and existing products. By generating less waste the economy becomes more competitive, reducing the pressure of limited natural resources and the impact of the environment [2].

The European Commission integrated aspect related to circular economy in the Reference Documents about best techniques available (BREF). This thing will contribute to the reduction of waste generation, will stimulate recycling and the reduction of resource use thus leading to a greater durability and competivity in industrial sectors. Among the best available techniques for the metal industry we have: techniques of best row material use and techniques of reduction the generation of compost by alternating process residue [3].

2. Experimental researches

The paper presents the capitalization of pulverous ferrous waste [4]:

- Electric steel plant dust (10.02.08 – solid waste from gas clearance);
- Ferrous sludge (10.02.12 – waste from cooling water clearance).

The pulverous ferrous waste come from a plant of making and producing metallic powders. Choosing the procedure and technology for capitalization must have in view both the characteristics of waste and the destination of the waste, plus the processing installation existing in the west area.

It has been determined the chemical composition (table 1 and 2) and granulometric classes (figure 2 and 3) of the waste used in laboratory testing.

Table 1. Chemical composition of ferrous sludge

Chemical composition, [%] ASTM E 1479/2016												
Al	Cd	Cu	Cr	Mg	Mn	Ni	Pb	Fe	Sn	Sb	Zn	Other elements
0.003	0.003	0.70	0.54	0.003	0.079	0.046	0.008	84.40	0.016	0.003	0.003	14.196

Table 2. Chemical composition of steel plant dust

Chemical composition, [%] ASTM E 1479/2016															
Al	Cd	Cr	Cu	Fe	Mg	Mo	Mn	Ni	P	Pb	Si	Sn	Ti	Zn	Other elements
0.7	0.01	0.04	0.085	65.5	2.3	0.005	0.87	0.03	0.013	0.1	3.45	0.006	0.02	3.1	23.77

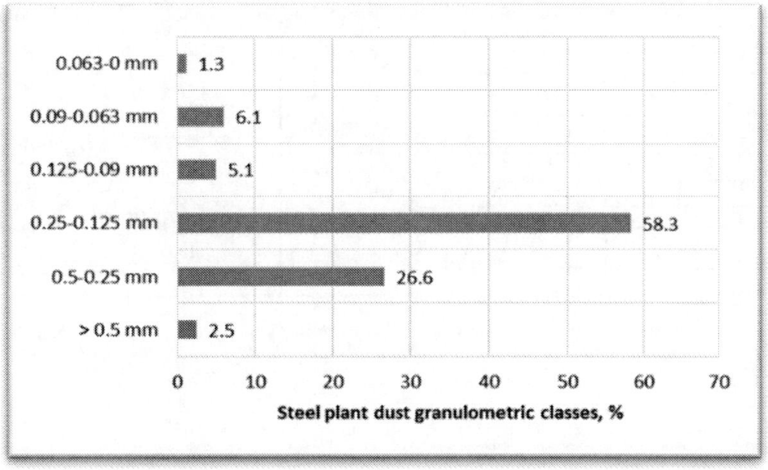

Figure 2. Steel plant dust granulometric classes

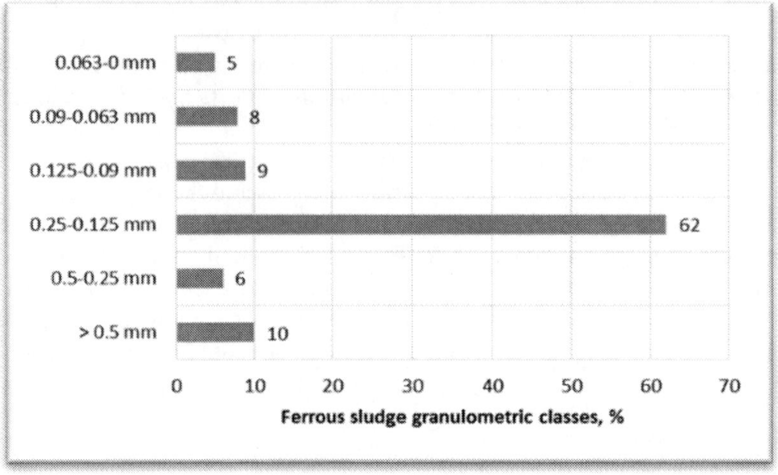

Figure 3. Ferrous sludge granulometric classes

In this case, we chose to process small wastes through briquetting processing with different binders adding so that we could their capitalization as raw material for steel making furnace [5,6]. At choosing the processing way we had in view the granulometric class determined in laboratory conditions, the balance of size between 0.25 mm being 87% from steel plant dust and 77% from ferrous sludge.

Also, at recipes composing we tried using exclusive waste with small add-ons of brinder (bentonite, cement and slag LF).We obtained briquettes after 8 recipes, the componence of experimental recipes are presented in table 3. Technological flux of briquettes processing is presented in figure 4.

Table 3. Component of recipes

Components, %	Recipes							
	R1	R2	R3	R4	R5	R6	R7	R8
Steel plant dust	0	0	0	0	35	40	40	40
Ferrous sludge	95	95	100	90	60	60	50	50
Bentonite	5	0	0	0	0	0	0	0
Cement	0	0	0	0	5	0	10	0
Slag LF	0	5	0	10	0	0	0	10

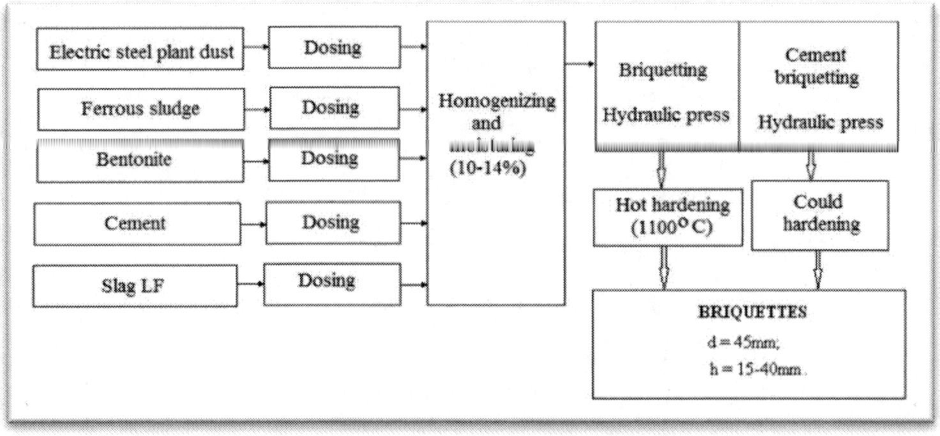

Figure 4. Technological flux

According to this, wastes are proportioned, smoothened and slightly wet: after words the mix is briquetted with the help of the installation from the laboratory of Energetic and raw material basis of the Faculty of Engineering from Hunedoara. The briquetting was made in a cylinder mould with forces varying between 200-350kN. The experimental briquettes that have bentonite component and slag as binder were hardened in the oven according to figure 5. The aspect during waste processing are presented in figure 6.

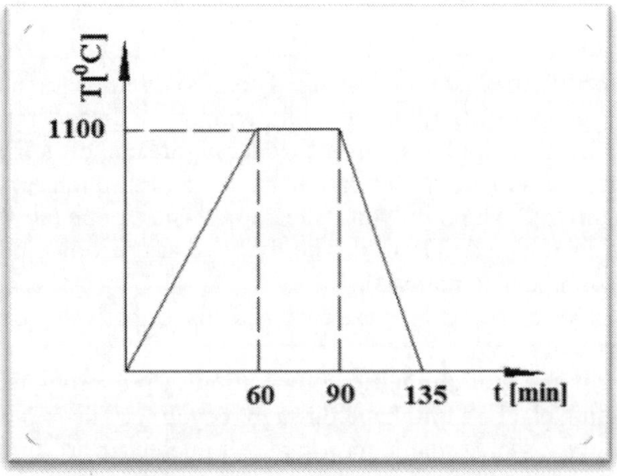

Figure 5. The briquettes hardening diagram
(T- hardening temperature, [^{0}C]; t- hardening time,[min])

Figure 6. Material and equipment used for experiments, and the resulting briquettes

The chemical composition of usual briquettes is presented in table 4.

Table 4. Chemical composition of briquettes

Chemical composition, [%]												
Fe_2O_3	SiO2	CaO	Al_2O_3	MgO	MnO	ZnO	Na_2O	SO_3	K_2O	MoO_3	Cr_2O_3	Other oxides
85.09	9.60	0.80	1.85	0.78	0.25	0.33	0.44	0.15	0.18	0.06	0.05	0.42

3. Conclusions

After preliminary examination of waste processing of some pulverous ferrous waste with the purpose of reintroducing them in the economical circuit: the fallowing were discovered:

- Analyzing the chemical and granulometric components of ferrous waste from a metallic powders plant results show that this can be capitalized through recycling;
- The procedures through which pulverous ferrous waste can be recycled in the iron industry are: briquetting, aggregation and pelleting. The first two are recommended for wastes with a varied granulometric componence material;
- This makes the experimental procedure easy and cost effective by companies which process and recycle these types of waste.

Analyzing the chemical and granulometric composition of pulverous ferrous waste resulted from a metallic powders production can be capitalized through recycling. The experimental briquettes obtained are destined to use as raw material at manufacturing steel in electric arc furnace.

An important interest is shared in non-conventional recycling of waste in order to obtain a product with high metal iron composite. The experimental briquetting obtained have a high amount of iron 60%Fe. It is necessary to intensify the capitalization process of these waste for both economical (they represent a source of iron), technological and ecological reasons.

4. References

[1] http://ec.europa.eu/environment/circular-economy
[2] Platon V 2006 Sustainable development and the recycling of materials *Expert* București
[3] Constantin N 2002 Unconventional methods to obtain ferrous materials *Printech* Bucuresti
[4] www.mmediu.ro
[5] Project no. 31-098/2007 *Prevention and fighting pollution in the steel making, energetic and mining industrial areas through the recycling of small-size and powdering wastes* Program PN2 Consortium CO Responsable Teodor Heput, Beneficiary CNMP Romania
[6] Crisan A, Vilceanu L, Ardelean M and Putan V 2013 Research regarding the compression behaviour of ferrous briquettes *Technical Gazette* **20**(4): 581-586

IOP Conf. Series: Materials Science and Engineering **416** (2018) 012097 doi:10.1088/1757-899X/416/1/012097

Use of Aqueous NaOH Solutions to Recover Silver from Radiographic Films

L Cânda[1,*], E Ardelean[1], T Heput[1] and M Ardelean[1]

[1]Politehnica University of Timisoara, Departament Engineering and Management, Revolutiei street no.5, Hunedoara, Romania

[*]Corresponding author: letitia.canda@fih.upt.ro

Abstract. Radiographic films fall into the category of medical waste, and their management is subjected to special regulations. Still widely used in Romania, radiographic films, are subsequently stored in large quantities; at minimum values of 2g Ag recovered per kilogram of radiographic film, a silver recovery of at least 300kg/ year is estimated. In the literature, numerous methods of silver recovery from exposed radiographic films have been approached, of which: leaching by sodium hydroxide solution, using reducing agents in a two steps process and silver nitrate cementation with more active metals.

The paper presents the results obtained under laboratory conditions on the recovery of silver using aqueous NaOH solution, running a complete flow of processing, from radiographic films to metallic silver production. There are also presented correlations between the monitored parameters, expressed both analytically and graphically.

1. Introduction

The varying but relatively high content of silver on radiographic films and effluent used in their processing (content ranging from 2–9g / kg, sometimes even higher: 15g / kg) requires its recovery in the context of sustainable development.

In the literature, numerous methods of silver recovery have been approached from exposed radiographic films, of which:

– leaching into sodium hydroxide solution [1,2];
– using of two–step reducing agents [3];
– silver nitrate cementation with more active metals [4,5].

The migration of the emulsion from the radiographic film is a relatively simple process, especially if particular attention is paid to the favorable factors, which are usually: the dissolution temperature, the concentration of the solutions and the speed of rotation if magnetic stirring is used. In recent years, although difficult to achieve, besides recovering silver itself, a particular emphasis has been put on the purity and form of recovered particles.

The starting point of the research was the laboratory production of the highest purity silver from the X–rays films based on the classical and modern methods of recovery or purification, respectively by combining old or new technologies and by making the effectively decisive parameters of these processes. For this purpose, the study of national and world–wide researches has been carried out over time to successfully apply new ways of synthesizing silver [6,7]. Therefore, it was considered research study conducted nationally and globally over time to the successful implementation of new ways of synthesizing silver.

2. Laboratory experiments

In our own research we used radiographic films waste from Municipal Hospital "Alexander Simionescu" Hunedoara, in compliance with legislation relating to privacy, which are kept in the hospital archives.

The basic element of the leaching process is the separation of the inorganic component from the polymeric layer of the radiographic film by leaching with NaOH. Our own preliminary researches [8] have been deepened and expanded through the study of several technological parameters resulting in graphical and analytical correlations.

For quantitative evaluation of leaching phenomenon based on NaOH concentration, aqueous solutions were used to extract silver from the polymer substrate. The film was sequentially treated with NaOH solutions of various concentrations: 1M, 1.25M, 1.5M and 2M respectively – the technological processes diagram A, B, C and D, respectively, as shown in Figure 1, 12 determinations being carried out, each 3 per series. 250g of radiographs were used for each 250g solution. Moreover, laboratory glassware with larger base areas were also used, so that the dimensions at which the film was cut could be larger, 5x5cm^2 respectively [8].

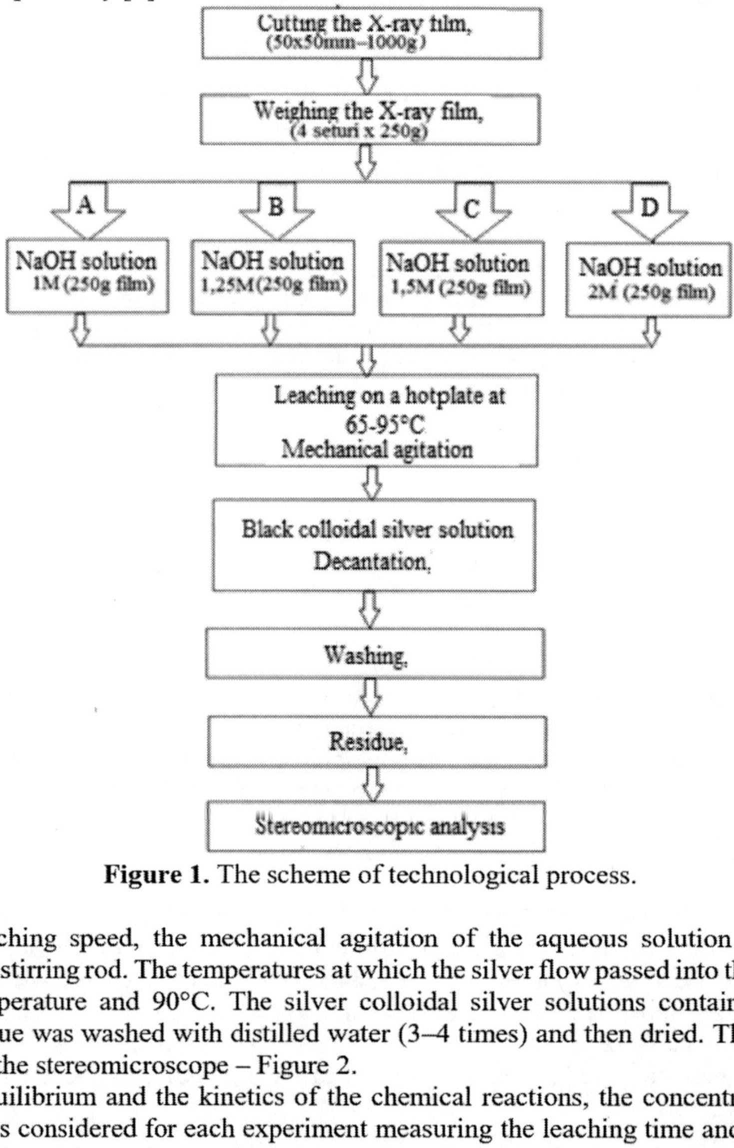

Figure 1. The scheme of technological process.

To increase leaching speed, the mechanical agitation of the aqueous solution was carried out continuously with a stirring rod. The temperatures at which the silver flow passed into the solution varied between room temperature and 90°C. The silver colloidal silver solutions containing NaOH were decanted. The residue was washed with distilled water (3–4 times) and then dried. The silver particles were visualized on the stereomicroscope – Figure 2.

To study the equilibrium and the kinetics of the chemical reactions, the concentration of aqueous NaOH solutions was considered for each experiment measuring the leaching time and temperature. As to the rate of agitation, this can be considered a constant factor in the experiments performed, since

IOP Conf. Series: Materials Science and Engineering **416** (2018) 012097 doi:10.1088/1757-899X/416/1/012097

mechanical stirring was performed to reduce the leaching time. Figure 3 shows the dependence between the leaching time and temperature at different concentrations of aqueous NaOH [9].

Figure 2. Aspects highlighted by stereomicroscope (20x).

For establishing correlations between the experimental data, they were processed in the Excel (Simple Correlations) and Matlab (Multiple Correlations) programs, the results being presented both in analytical and graphic form [10].

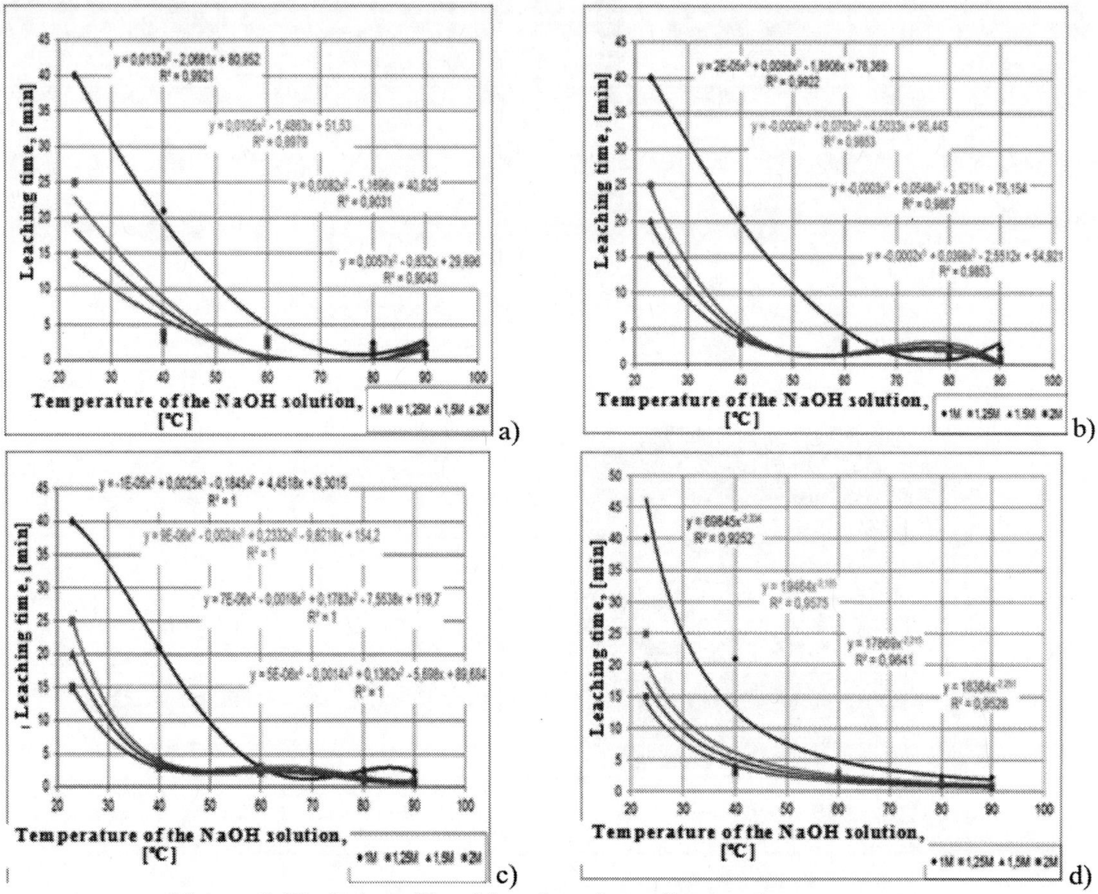

Figure 3. Variation of leaching time depending on the temperature of the solution aqueous NaOH and its concentration
a) polynomial correlations, degree II; b) polynomial correlations, degree III; c) polynomial correlations, degree IV; d) correlations, power function.

3. Interpretation of results

Regarding the simple correlations, these are expressed by II, III and IV degree polynomial functions as well as the power function (Figure 3). Considering the values for the correlation coefficient, it can be considered that, regardless of the type of correlation, they are representative. Increasing the temperature of the solution leads to a decrease in the leaching time, which is intense within the temperature limits of 20–60°C for the 1M concentration solutions, the time decreasing from 40 min to 5 min. For solution solutions of 1.25M; 1.5M and 2M respectively decrease in time is intense in the temperature range of 20–40°C, the reduction of time being from 20 to 10min. At temperatures above 60°C, it can be considered that the reduction of the leaching interval is insignificant. It can be also considered that an optimal leaching time is obtained at temperatures above 60°C and is of technological interest the average value obtained by applying the four correlations.

Furthermore, the graphical dependencies of Figure 3 show that an increase in the concentration of the solution from 1M to at least 1.25M results in a significant decrease in the silver transition time of 80–90°C, thus increasing the concentration to 2M or above is not justified (both for economic and environmental reasons). The minimum leaching time at 80–90°C takes place at a concentration of 1.5M.

Referring to double correlations performed in the Matlab calculation program, we considered the temperature of the aqueous solution of NaOH (T_{NaOH}) and its concentration (c_{NaOH}), as independent parameters, and as a dependent parameter: the leaching time (t_l), correlations shown in Figure 4 [10].

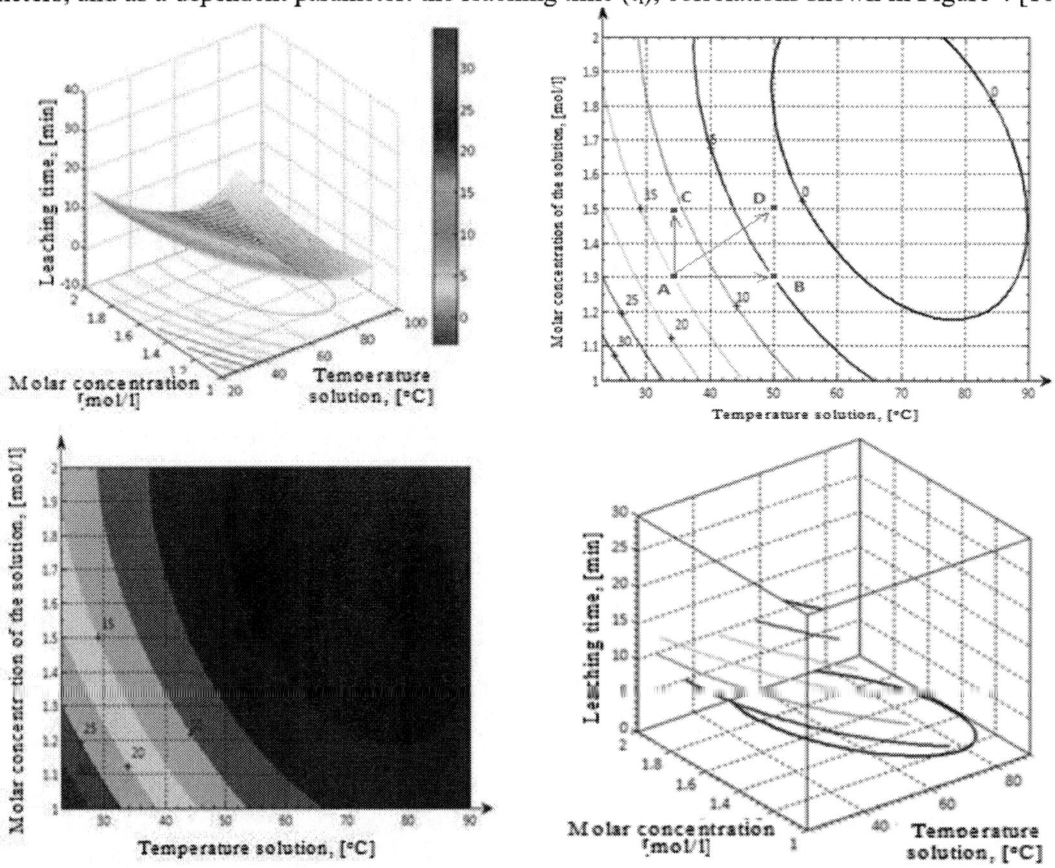

Figure 4. Dependence of leaching time on the temperature and molar concentration of the solution

The multiple regression equation is given by relation (1), the correlation coefficient being $R^2 = 0.954$.

$$t_l = 0.009 \cdot T^2_{NaOH} + 16.6 \cdot c^2_{NaOH} + 0.329 \cdot T_{NaOH} \cdot c_{NaOH} - 1.862 \cdot T_{NaOH} - 77.954 \cdot c_{NaOH} + 126.312 \quad (1)$$

Figure 4 shows the correlation surface, the plane projection of the level curves, the leaching time variation subdomains and the spatial projection curves respectively. Knowing the values for the level curves in the plane projection or the subdomains allows, for any temperature and concentration, the graphical determination of the leaching time, or the matching of the desired leaching time, respectively, the values of the independent parameters can be determined. For example, we consider point A (35, 1.5, 15) – Figure 4. Increasing the temperature to 50°C while maintaining the constant concentration leads to a decrease of the leaching time to 5min (i.e. by 0.66min/°C) – the point B (50, 1.3, 5) respectively. The reduction of the leaching time can also be achieved by increasing the concentration of the leaching solution; an increase of 0.2M at the same temperature determines leaching times of 11min – point C(35, 1.5, 11), thus 20min/mole solution. A simultaneous increase of the two independent parameters: temperature or molar concentration of the solution, also leads to a decrease of the leaching time. Thus, starting at point A (35, 1.5, 15) at point D (50, 1.5, 2.5) a decrease of the leaching time is obtained from 15min to 2.5min i.e. 0.83min/°C or 62.5 min/mole solution.

After complete drying in the oven, the residue was introduced into a ceramic crucible, covered with an amount of about 30 g of borax and introduced into a silicon bar furnace, in which the thermal regime of fig.5 has been respected. Cooling of the crucible was done with the oven, to avoid thermal stresses in the ceramic crucible. At extraction of the crucible from the furnace pieces of silver metal caught in a glass borax table could be seen – "borax pearl" – Figure 6. The recovery of the pieces of silver – Figure 7 was done after the crucible was reheated in the presence of water which favored the decomposition of the borax mass, the working temperature in this case being 100–150°C.

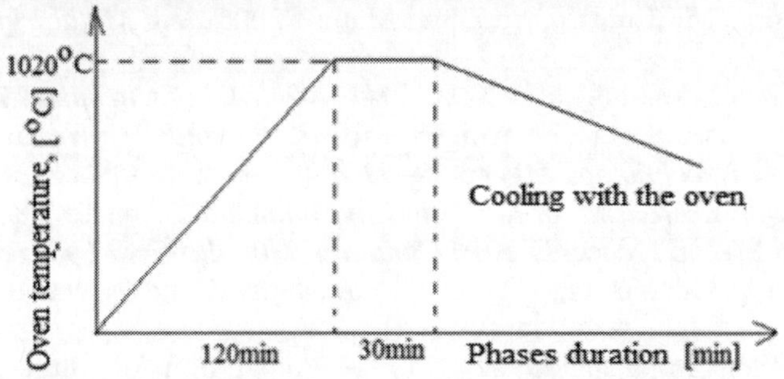

Figure 5. Diagram of the melting process in the oven

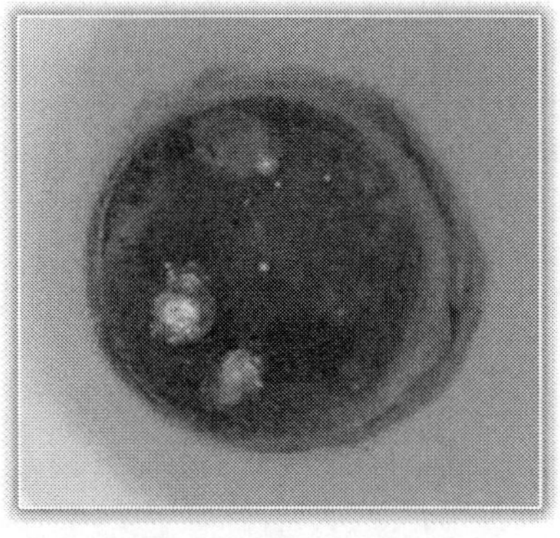

Figure 6. The ceramic crucible with borax pearl

Figure 7. Silver pieces result

4. Conclusions

As a result of the experimental approach to the leaching of radiographic films with aqueous NaOH, the following conclusions were drawn:

- The optimal working concentration of the aqueous NaOH solution is 1.5M – technological variant B, which was maintained constant by supplementing it with distilled water quantity to ensure the same volume of the leaching vessel;
- The leaching process was initiated at the ambient temperature of the laboratory and the working temperature ranged between 65–90°C;
- The length of the leaching process varied between 40–50min for each technological variant;
- The total residue resulting from the complete evaporation of the aqueous solution was 16.21g;
- The amount of silver resulted after melting the residue in the furnace in the presence of borax was 6.54g;
- The washing steps (especially the use of distilled water) and the drying of the films (natural or even more in the oven), which leads to a reduction of the processing time and economic efficiency of the process (reduction of water costs and energy), are no longer necessary;
- For economic reasons, the use of current water was used in the process of silver solution transfer from radiographic films, as well as the washing of the filtrate instead of the double distilled or triple distillate, with the same efficiency of the process;

Recycling of the polymer support on the plastic circuit was also considered, concurrently with the recycling of the residual solution in the recovery process, to avoid pollution of the ecosystem to final disposal.

5. References

[1] Marincovic J, Korac M, Kamperovic Z and Matic I 2006 *Acta Metalurgica Slovaca* **12** pp 262–268
[2] Nakiboğlu N, Toscali D and Nişli G 2003 *Turkish Journal of Chemistry* **27**(1) pp 127–133
[3] Shankar S, More SV, Seeta Laxman R 2010 *Kathmandu University Journal Of Science, Engineering And Technology* **6**(I) pp 60–69
[4] Jaskuła M 2009 *Jordan Journal of Earth an Environmental Sciences* **2**(1) pp 84–95
[5] Aktas S, Morcali MH and Yucel O 2010 *Canadian Mettalurgical Quarterly* **49**(2) pp 147–154
[6] Li G, He D, Qian Y, Guan B, Gao S, Cui Y, Yokoyama K and Li W 2012 F *Int J Mol Sci.* **13**(1) pp 466–476
[7] https://www.google.com/patents/US6290749 – Preparation of ultra–pure silver metal US 6290749 B1
[8] Cânda L, Heput T, Ardelean E 2017 *IOP Conference Series: Materials Science and Engineering* **106** pp 1–9
[9] Cânda L, Ardelean E, Hepuţ T 2018 *IOP Conference Series: Materials Science and Engineering* **294** pp 1–8
[10] Cânda L *Research on the recovery of silver from radiographic films and effluents – PhD Thesis* 2018 University Politehnica Timişoara

IOP Conf. Series: Materials Science and Engineering **416** (2018) 012098 doi:10.1088/1757-899X/416/1/012098

Consolidated Composites with Natural Textile Fabrics

MP Todor[1], C Bulei[1], T Heput[2] and I Kiss[2,*]

[1]University Politehnica Timisoara, Doctoral School, Timisoara / Hunedoara, Romania
[2]University Politehnica Timisoara, Faculty of Engineering Hunedoara, Department of Engineering and Management, Hunedoara, Romania

[*]Corresopnding author: imre.kiss@fih.upt.ro

Abstract. The problems raised by the usage and manufacturing of composite materials are numerous, starting from the prediction of the properties of a new material and going to the determination of some procedures, methods or materials that will eliminate some disadvantages. Because there is a wide variety of the materials already used, with a great diversity of characteristics and properties in a wide range of combinations involving various effects (mechanical, physical, chemical and so on), there have also been a number of issues that need to be addressed differently. The general properties of the composites are decisively influenced by the nature and proportions of constituents, by the nature of the matrix–reinforcement interface, the volume fraction of reinforcement in the composite and its orientation, and the reinforcing component architecture. This paper addresses a major research area of great interest and relevant in the present context, the emphasis is increasingly on environmental protection, saving material resources by developing and creating new manufacturing and execution technologies to be used in composite materials. The paper presents the ways of producing composite materials reinforced with textile materials made of natural fibres, as matrix being used polyester resins. Considering the quantitative availability of these textile fabrics, the experiments take into consideration variants of the bast fibers (flax, hemp and/or jute) for the reinforcement component.

1. Introductory notes

Within the generation of new materials that replace classical and deficient materials, considering their characteristics and prospects for the future, a special attention should be given to composites, recently referred to as reinforced materials.[1] Representing the most well–known category and marking the beginning of an industrial scale use of new materials, fiber composites are mainly imagined as long fiber–reinforced polymer matrix, which for a long period of time had been the only type of wide–spread composites. This explains the fact that, even today, this class of materials is often identified in specialty literature with the very notion of composites. [1,2] By definition, the concept of "composite" is assigned to a complex system composed of many different kind of materials. This is due to the fact that the possibilities of modifying the basic constituents, of the techniques of "assembly" and manufacturing, the level of the performance and the cost are practically infinite. [1–5]

The problems that have arisen in trying to define as much possible exactly the composite materials are a proof of the extremely wide field of this type of material, a field in a continuous and rapid expansion. As a general definition, composite materials are systems of two or more components whose properties complement each other, resulting in a material with superior properties to those specific to each component.[1–5] Thus, these components will cooperate, the deficiencies of some being supplemented by the qualities of others, conferring to the assembly, properties that no component can have alone. As a result, composites are a category of multi–phasic materials, characterized by the fact that the qualities of each component are combined. [1–5]

For fiber composite materials in a polymeric matrix, the following constituents are considered:[1–5]

— the reinforcement, which constitutes the reinforcement or skeleton of the composite, providing mechanical resistance (it is of a lamellar nature, made up of fibers or fiber cloth);
— the matrix, linking the reinforcement, distributing the efforts and ensuring both the protection of the reinforcement and the shape of the piece (it is by definition a polymer or an organic resin);
— the interface, assures the reinforcement–to–matrix compatibility, ensuring the transmission of the efforts from one to the other without a relative displacement (having a good thin layer adhesion, of the micron order).
—

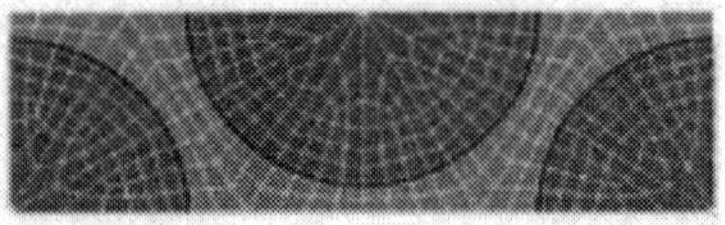

Figure 1. The composite's constituents

In a broad sense, a composite material is a combination of two or more chemically different materials with an interface between them (Figure 1).[1–5] The fiber can be inserted and mixed into matrices or it can be impregnated with a susceptible binder in order to solidify after the parts are configured. Technically, they are considered to be composite materials, continuous or discontinuous fiber arrangements disposed in a form of reinforcement which is immersed in a matrix whose mechanical strength is much lower. The possibility to use very diverse combinations of raw materials in a great variety of combinations results in a wide range of composites. [2,4]

The design of fiber composite materials is based on the concurrent use of two criteria, namely:[1–5]

— the geometrical features of the reinforcement, and
— the way to orient or arrange it in the matrix.

Composite materials "associate" chemically different materials in the same mass that exhibit a range of performance, either in terms of the ease of machining or mechanical, physical or chemical resistance. This produces "demand" materials with good stiffness and low density. The association between a matrix and an armature cannot be coincidental and it depends on: [1–5]

— the chemical compatibility of the components which influence the transfer of effort through the interface;
— the production technology that is chosen according to the geometry of the piece;
— the obtained technological characteristics and properties;
— the cost of manufacturing.

The matrix– reinforcement mixture does not acquire the properties of the composite materials except in the last manufacturing phase: hardening of the matrix (polymerization in the case of composite materials with the matrix of polyester resin). After hardening, the properties of composite materials can no longer be altered as with metallic alloys by thermal treatments. The connection between the armature and the matrix is created during the composite material manufacturing phase. It has a determining influence on the mechanical properties of the composite material. [1–5]

2. Technical textile materials for reinforcement

Textiles find their applications as reinforcement in composites. The study of some technologies has revealed that they could be applied on an industrial scale for the economic advantages, performance and simplicity of design.[5–9] Textile fibers are the basic raw material of the textile industry but, at the same time, they are also important material for many other areas of the industry. In the modern period, pre–consumer waste textile materials, like those resulting from technological processes, but also from post–consumer waste, i.e. those collected, have been used to obtain a variety of textile products such as yarns, fabrics, knitwear, nonwovens, beds and so on, products used in many sectors. [5] The fibers resulting from the recovery of waste, which serve to obtain nonwoven products, are used in the form of fibrous layers. These are textile supports, consist of fiber–oriented agglomerations or multidirectional. Practically, all types of fiber, including fibers recovered from reusable materials, can be used as starting materials for technical textiles. Reusable textile materials, also known as textile waste, come primarily from textile processing (spinning, weaving, knitting and so on), fabrication, machining in other textile processing sectors or as a result of physical or moral wear of textile products.

Textiles have the advantage of mass / strength ratio, reducing the time to achieve, eliminating waste resulting from the production and processing process, a better control of the final shape of the product and superior quality. [2,5–9] At the same time, textile materials allow direct control over the material characteristics from the design stage. In the case of textile materials, the architecture is favored by their extreme deformability being possible to obtain knitted fabrics of special complexity. This feature has made textile materials to be regarded as a possible alternative to the production of preforms for advanced composite materials recently. [5–9] Textile materials can also have three–dimensional architectures and the production of some of the most complex fittings eliminates the intermediate stages of preforms (with implications over production time and costs) and the possibility of controlling the final shape from the material design stage. [5–9]

3. The architecture of textiles

The form of textiles / fabrics is an extremely important factor in the design phase as it influences the raw material, structure and technology options. The sum of the qualities of a fabric, which includes flexibility and maneuverability to model, is manifested by the layout of the material to generate a pattern or to obtain a pattern. Textile materials can be defined according to the essential dimensions to the way they are constructed: [2,5,9]

— materials with one–dimensional architectures, for which the length is decisive, in this category being included the textile fibers and yarns (Figure 2a);
— materials with two–dimensional architectures, with a geometry characterized by two dimensions (length and width), the case of knits, fabrics, braids and non–woven materials (Figure 2b);
— materials with three–dimensional architectures, built along the three axes, with knits, fabrics and knitted fabrics with three–dimensional complex shapes (Figure 2c).
—

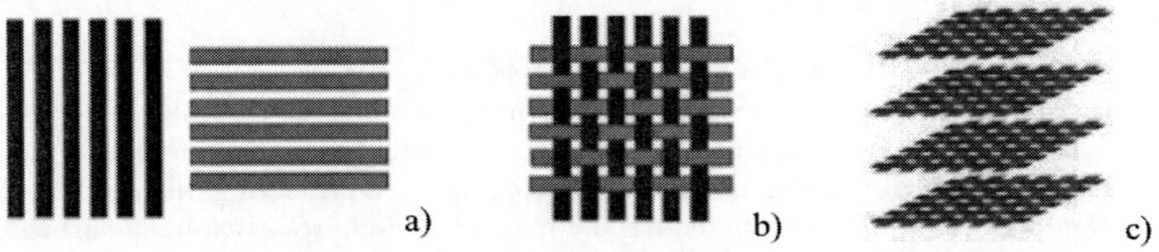

Figure 2. The architecture of textiles: a) vertical fibers (warp)/ horizontal fibers (weft); b) two–dimensional shapes (plain fabrics); c) three–dimensional shapes

The process of increasing the complexity of the shape of the products required the adaptation of the textile materials and led to the emergence of the concept of materials with three–dimensional architecture. [2,5] In the specialty literature, the concept of three–dimensional textile material is still being discussed, with several views on what types of materials can be considered as three–dimensional. In broad sense, three–dimensional architectures are defined as: [2–5]

— multiaxial (multilayer), having a structure in layers of wire systems deposited at different angles, independent of each other, assembled in a unitary knitted structure. The threads are arranged in layers at angles of 0^0, 90^0, +/– 30^0 and +/– 45^0
— sandwich, consisting of two layers of fabric, knitted or non–knitted.
— three–dimensional contours, the 3D layout of which is achieved by using contour lines, which define structures on incomplete rows, with complex shapes similar to the shape of the finished product.

Reinforcing with threads arranged at different angles improves the mechanical behavior of multiaxial knits on the reinforcement directions as well as the shear properties (Figure 3). In contrast, increased stiffness significantly reduces the formability of these architectures. The shape of composite materials reinforced with sandwich structures can be controlled by varying the structure of the outer layers or layout and structure parameters of the bonding layers.[2–5,9]

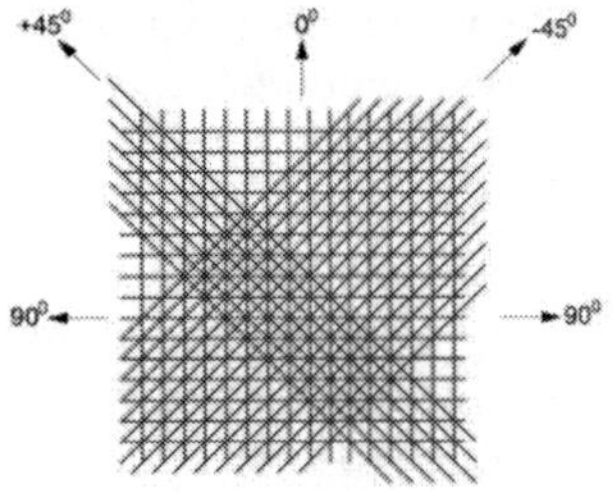

Figure 3. The shape of composite materials reinforced with sandwich structures

Three–dimensional preforms are less studied and used, especially due to the following problems related to their production and properties: [2–5]
— the development of these types of architecture is still in the laboratory stage;
— the specific properties and methods of their prediction are not yet well determined;
— the correlation of technological parameters – properties of composite material is not completely defined;
— pre–stretching the fabric before resin impregnation determines the unevenness of the composite behavior due to the random change in the arrangement of the yarns in the fabric.

Concluding, the main advantages of textile materials are related to:
— special formability, due mainly to the draping properties;
— the high degree of complexity of the obtained forms;
— the possibility of obtaining composites with superior behavior to impact demands.

These things recommend technical textiles for applications using complex architecture materials, including reinforcements for composite materials.

4. Technologies for obtaining consolidated composites with fabrics

The first problem with the manufacturing of fiber–reinforced composites is the choice of component materials depending on the conditions of use. It is very important that in a composite the mechanical stresses are supported by fibers, the matrix having the role of protecting fibers against oxidation, corrosion, hitting and friction. It is obvious that the manufacturing processes must ensure the exercise of this role from the moment the composite is formed.[2–5] The second problem is the incorporation of fibers into the matrix, which must respond to three fundamental requirements:
— the uniform distribution of fibers
— their alignment in a common direction and
— making a connection between fibers and matrix.

In the case of the fabrics usage, the first two requirements are solved by choosing the type of fabric, the fiber distribution and their alignment being solved by the manufacturing technology. Making a fiber–matrix link remains a difficult problem. The difficulty consists in ensuring that the fibers are "wetted" by the matrix without forming diffusion layers at the interface to alter the characteristics of the fabrics in the fabric.[2,5] This problem is solved by using low surface tension matrix relative to the textile material. Contact formation is the easiest method in the manufacturing of fiber reinforced polymer composite components (fabrics).[1–5] The matrix (resin) and the reinforcement (the textile material) are deposited in layers, and after each deposition it is ensured pressing and adhering the previous layer to the next with a contact roller. Thus, the resin is "buffered" over the fabric which is properly impregnated. In the case of several layers of fabric, for proper adherence, the next layer was laid before the previous one was fully hardened.

The reinforcement systems for composite textiles refer to a wide range of textile surfaces that can be obtained using all textile technologies (weaving, knitting, and sewing) or obtaining nonwoven materials. The characteristics of the resulting composite material depend on the ratio between the volumes and the reinforcement masses and the matrix.[1–5] To these there can be added the processes characterized by the production of reinforcement and composite material at the same stage, such as filament winding and polishing. When choosing the optimal variant of the composite preforms, the strengths and weaknesses of each composite preform must be taken into account. The criteria used in choosing the technological process for the reinforcement manufacture refer to the dimensional stability, to the imposed mechanical properties, and to the draping / forming properties of the reinforcement system.[2–5] For reinforcing we can meet uni–, two– or three–dimensional armatures as it follows:

— reinforcement of yarn (textile fibers) + matrix;
— fabric (textile surfaces) + matrix;
— fabric systems (3D textile materials) + matrix;

Preparation of the reinforcement is an operation that takes a very important place in the process of obtaining composites and involves knowing all the factors that compete for the building of an appropriate architecture of the reinforcement to a particular type of landmark.[2–5] There are usually used the following basic armatures:

— unidirectional (support from textile surfaces, in different layouts, in one layer);
— non–woven (non–assembled textile support, layered on several textile layers);
— woven (supported from reinforced textile surfaces, disposed on several textile layers);

Textile supports are textile being made of plain or composite fabrics (stratified supports). Textile fibrous layers and layered supports may be subjected to pre–consolidation processes by simple interlacing and then to mechanical consolidation processes and, more particularly, physic–chemical, with the aid of adhesives.[2,5]

The fabric is the textile product obtained by the right–angled crossing of two yarn systems: a longitudinal system and a transverse system. The obligatory condition for fabrication is the alternate passage of longitudinal threads over and under the transverse threads, and vice versa. Fabric elements are considered to be the crosses of the yarns. The structural characteristics and properties of the fabrics are determined by the links used to obtain the fabrics (basic, derived, combined, multiple or special), the characteristics of the component yarns (nature of the raw material, the process of production, the weaving method) and on the transverse system), the thickness, the degree of corrugation of the yarns in the fabric and its mass.

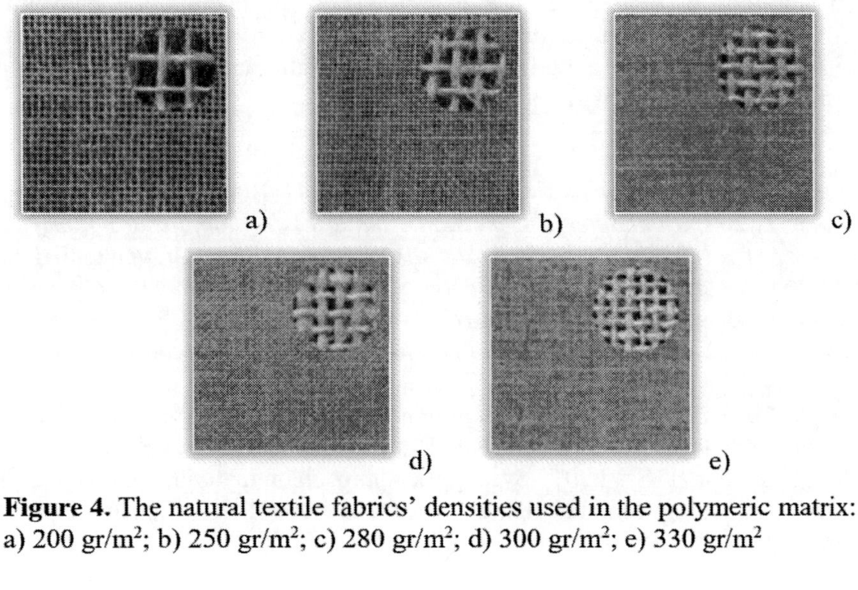

Figure 4. The natural textile fabrics' densities used in the polymeric matrix: a) 200 gr/m^2; b) 250 gr/m^2; c) 280 gr/m^2; d) 300 gr/m^2; e) 330 gr/m^2

Fabric of jute, hemp or sack cloth is an ecological and biodegradable natural raw material. For textile inserts in the polymeric matrix we use fabrics with densities between 200–330 gr/m^2, obtained from the selective collection of textile waste (from textile products that have completed their life cycle) or technological debris from textile factories (Figure 4).

The polyester resin used is an unsaturated, pre–accelerated one getting hardened at the ambient temperature. From the point of view of processability, the polyester resin is easier to use due to higher viscosity and low density. A hardener is added to the base component in a proportion that can vary between 1–3%, depending on the temperature and speed of the desired hardened. The hardener acts quickly and is highly concentrated, so a 2% addition is sufficient in most cases. The addition of hardener depends on the amount of base that is prepared at a single casting.

The working temperature must be between 18–25°C, the higher the temperature, the harder the hardening. Practically, at a temperature of 0°C, hardening does not take place. Working time may vary between 20–40 minutes depending on the amount of the hardener added.

5. Concluding remarks and future works

The main advantage of using composite materials is both qualitative and economical. The natural fibers used as a textile reinforcement favor the improvement of the properties of the made products, which makes them considered a successful alternative. The choice of technical textiles for a domain is made according to the functional characteristics, the reliability and the destination of the products being made. Achieving new performance materials to meet the widest possible range of requirements is a must–have nowadays. The idea of creating new techniques and technologies to improve the traditional and composite materials, the best use of secondary resources, the increase of recycling and the use of waste is becoming increasingly important.

Today, the concept of development means sustainable competitive growth, based on the innovation of technologies, products and processes, under conditions of increased environmental efficiency. In order to find solutions for waste recovery, it is taken into account that technological progress can no longer only follow economic criteria. The objectives include eliminating the concept of waste in product creation using materials, energy and resources that can be immediately recycled, reused and introduced into the circuit.

In the category of waste, an important place belongs to the textile waste, also to the knowledge and studying of the problem regarding their valorization in the industry (own, by recuperation of proper waste or in other industries, in composite products with textile friability). The development of alternative waste disposal solutions in high added value products by producing non–conventional products containing recovered fibers constitutes an ecological alternative to the capitalization of textile waste. Thus, fiber–reinforced composite materials are promising alternatives for reducing environmental problems regarding lightweight construction area.

6. References

[1] Ispas St 1987 *Materiale compozite / Composite materials* (Bucuresti: Editura Tehnica)
[2] Theodorescu H 2007 *Fundamente si mecanica materialelor composite polimerice / Basics and mechanics of polymeric composite materials* (Brasov: Editura Universitatii Transilvania)
[3] Carcea I 2008 *Materiale composite – Fenomene la interfata / Composite materials – Interface phenomena* (Iasi: Editura Politehnium)
[4] Bolunduţ I–L 2012 *Materiale şi tehnologii neconvenţionale / Materials and unconventional tehnologies* (Chişinău: Editura Tehnica–Info)
[5] Coterlici RF 2015 *Eco–compozite sustenabile cu proprietăţi performante / Sustainable green composite with performance properties* (Braşov: University Transilvania)
[6] Todor MP, Bulei C and Kiss I 2016 Systematic approach on materials selection in the automotive industry for making vehicles lighter, safer and more fuel–efficient, *Applied Engineering Letters*, **1**(4), pp 91–97

[7] Todor MP, Bulei C and Kiss I 2017 Inducing the biodegradability of polymeric composite materials using bioranforts, *Applied Engineering Letters,* **2**(2), pp 84–90

[8] Todor MP, Bulei C and Kiss I 2017 An overview on fiber–reinforced composites used in the automotive industry, *ANNALS of Faculty Engineering Hunedoara – International Journal of Engineering,* **XV**(2), pp. 181–188

[9] Todor MP, Bulei C, Heput T and Kiss I 2018 Researches on the development of new composite materials complete / partially biodegradable using natural textile fibers of new vegetable origin and those recovered from textile waste, *IOP Conf. Series: Materials Science and Engineering,* **294**, pp. 1–9

IOP Conf. Series: Materials Science and Engineering **416** (2018) 012099 doi:10.1088/1757-899X/416/1/012099

Recovering Aluminium for Recycling in Reusable Backyard Foundry that Melts Aluminium Cans

C Bulei [1], MP Todor [1], T Heput [2] and I Kiss [2,*]

[1]University Politehnica Timisoara, Doctoral School, Timisoara / Hunedoara, Romania
[2]University Politehnica Timisoara, Faculty of Engineering Hunedoara, Department of Engineering and Management, Hunedoara, Romania

[*]Corresponding author: imre.kiss@fih.upt.ro

Abstract. In a world which is increasingly demanding sustainability, nonferrous metal recycling has become a very important practice. Millions of tons of nonferrous scrap are recovered annually and used by smelters, refiners, ingot makers, foundries, and other manufacturers. Secondary materials are essential to the industry's survival because even new metals often require the combined use of recycled materials. The recovered materials are melted down in a furnace, poured into casters and shaped into ingots. These ingots are either used in the foundry industry or they can be transformed into flat sheets and other wrought products, which are then used to manufacture new products. Aluminium offers intelligent and practical solutions to recovering for recycling. The main activities described in this article are carried out during the experimental phase of the research and they are oriented towards the recovering aluminium for recycling. The objective of the research is to develop some recycling facilities for re–smelt and re–cast of aluminium by using charcoal briquettes into a mini metal foundry, designed and manufactured in laboratory. In this research we experimented a reusable backyard foundry that melts aluminium cans (soda and beer) easily and safely.

1. Introductory notes regarding the recovery of recycling metal resources

Recycling is a modern concept in the waste management, emerged as one of the possibilities to limit waste and to use resources more efficiently.[1–3] It has become increasingly clear that the industrialization and the sustained growth of the population have led to the consumption of increasingly large amounts of resources. An economic development without metal resources is inconceivable in the 21st Century and it is not possible to be in the global competition, with about 75–85% of the world economy depending on these resources. Recovery and reusing of recyclable resources are means of solving the contradiction between the requirements of the economic growing process and the restrictive nature of natural resources. [1–6] In this broad context, non–ferrous metals and their alloys are in the center of modern life and in many developments of high technology, especially in high–top industries.

The policy concerning waste aims for reducing resource consumption and favors the practical application of the "waste hierarchy", which classifies the various waste management options from the best to the least good for the environment. [1–3] According to this hierarchy (Figure 1), the priority is to prevent waste generation, followed by their preparation for reuse and recycling and, lastly, disposal. Although the waste hierarchy is not a rigid rule, the aim of moving to a recycling and recycling society means that materials should be preferred to disposal by storage. Thus, products considered being waste because they have lost their value of use can be reintroduced into the economic circuit through re–use or recycling. [1–9]

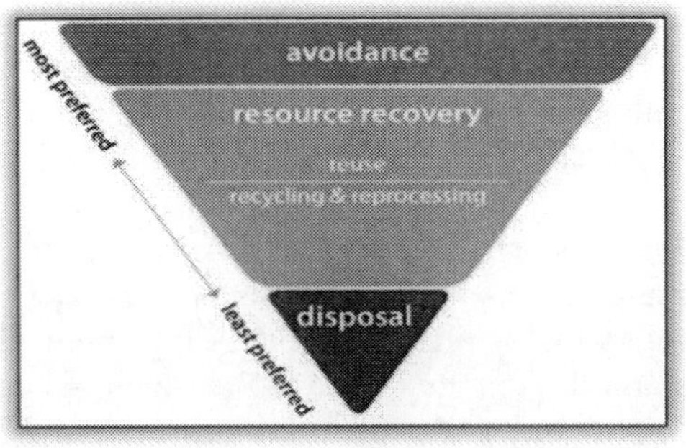

Figure 1. The "waste hierarchy" policy concerning waste

Researches on identifying solutions, designing and implementing innovative technologies for recycling and reuse of non–ferrous metal based raw materials are numerous. [1–3] The specific strategic objective of capitalizing recyclable metal resources – including the non–ferrous metal raw materials area – is the implementation of new concepts and new technologies for sustainable processes in the non–ferrous metals industry and environmental protection (standards and best practices in the non–ferrous metals industry) through new technologies and innovative solutions on: [1,2]

– substitution of critical metals;
– the efficient use of natural resources;
– improving metal recovery by recycling and reuse of raw materials;
– replacing raw materials with attractive commercial alternatives with low impact on the environment;
– the development of new materials through material capitalization;
– techniques for the manufacture of metallic materials;
– prevention and minimization of metal waste;
– rehabilitation of polluted areas through metallurgical and / or mining activities;

According to the principles of the "waste hierarchy", the management of non–ferrous metal waste includes activities such as: [1–3]

– providing the best options for waste collection in order to make the most efficient use of waste;
– reusing, recycling and treatment for the proper recovery of non–ferrous metals from waste;
– preventing uncontrolled disposal, thus preventing the loss of potentially useful materials;
– implementation of measures to prevent the generation of waste, ensuring its recovery and recycling;

Recovery of non–degradable waste is a pressing issue of mankind and in the last century the concern for this subject has increased steeply. [1–3] The depletion of natural resources is generated outside the natural transformation circuit, which can cause major imbalances occurring when human activities create artificial movement outside the ecosystem, producing waste. Thus, the concept of recycling, from the point of view of recycling and recycling of reusable resources, comprises two categories: [1,2]

– the open–circuit transformation, possible when the environment is able to accept waste for recycling without causing imbalance in the environment. This type is not proper for any metal waste;
– the closed circuit transformation when some waste from human activity does not naturally decompose into the ecosystem and cannot integrate into it, artificially created products must be artificially transformed. This type is often identified as the actual recycling of metal waste.

Aluminum is one of the few materials (along with steel and copper) for which recycling costs are exceeded by the sales price of the recycled product. Aluminum is the most expensive recycled product and recycling of aluminum is economic: it uses less energy and recycling is self–sustained due to the high value of the aluminum used. In other words, it is worth recycling it from all points of view: it is easy to carry, recyclable, infinitely reusable, it does not rust. For this reason, aluminum is rarely lost.

2. Recycling metal recipients

Aluminum is the most precious metal that is used in packaging, because this package has many advantages for both beverage and a consumer manufacturer, which neither PET nor glass has. [1–3] For example, the aluminum can for beverages is a lightweight, unbreakable package, cools quickly in the refrigerator, protects the content of factors such as light and air, has long life, maintains the effervescence and freshness of beverages, requires no labeling, and not last it is easy to recycle. [4,5,8,9] Aluminum cans are recycled in a closed circuit after the recycling process, and aluminum can be re–injected again or can be used for other products. Recyclable aluminum can come from a wide range of sources, including industries and private households. They also include metal containers used as packaging by large beverage manufacturers (juice and beer). [1–3,4,5,8,9]

Aluminum beverage cans are, like PETs, omnipresent, but similarities stop here. Unlike plastic, aluminum can be recycled to infinity without degrading at all. The process is simple: the boxes are sorted, then washed to remove the contaminants and then melted. The molten metal is converted into ingots which are then transferred to another processing unit and pulled into thin sheets, which can then take the form of beverage cans. On the whole, the process is very fast, a beer can deposited at a waste collector is recycled and reaches the market again. [4,5,8,9]

There are many initiatives involved in improving the return of aluminum waste and the industry is an active player in this area. Recycling empty beverage cans is only an example. Aluminum dosages are an important and extremely valuable resource, representing an important source of secondary aluminum. Once the beer or juice is consumed, if the empty can is properly collected and subsequently recycled, it can be revalued indefinitely without losing anything. Thus, aluminum cans can become raw material. Aluminum packaging has an intrinsic quality and cannot be said about it as a waste, but as a resource, being 100% recyclable. [1–9] Thus, out of all recyclable packaging, aluminum cans have become the most innovative and durable packaging. Aluminum can be recycled indefinitely without losing its properties unlike other recyclable materials such as plastic, paper or glass, which can be recycled for a limited number of times. The aluminum industry has all the interest to promote recycling as part of the industrial strategy. In Europe, about 50% of the total aluminum used to produce new beverage cans and other aluminum packaging products comes from recycled aluminum. [1–3]

However, a large proportion of packaging waste is generated in households, which requires a separate waste collection infrastructure from metallic cans which are mixed with other household waste or separately through locally developed networks. [1–3,7–9] The volume / weight ratio for aluminum metal dosages is quite high: one aluminum box has 13–15 grams, the boxes are made of 0.17 millimeter aluminum sheet. Thus, a pound of aluminum cans consists of approx. 60–70 pieces, which can be collected, flattened or even perforated. That is why the sustainable solution for fulfilling the recycling obligations and, implicitly, increasing the recycled quantities is the further extension of the population's access to the selective waste collection services. Creating and promoting the market for recycling and developing recycled materials from these packaging are fundamental. Thus, aluminum cans are an important resource that is also extremely valuable.

Figure 2. Recycling aluminum cans by melting

The aluminum can is the most recycled beverage packaging. [1–5,8,9] Obviously, this is only possible if they are collected and (re)introduced into the recycling circuit. If a can is left in nature, obviously it cannot fulfill its recycling potential. All the cans are 100% recyclable once they have been collected and reached a recycling point, even the smallest piece of aluminum is not lost. There are several forms to recycle beer or soft drinks cans. [1–3,6] As a rule, they can be melted and reused by an infinite number of times without losing quality (Figure 2).

In the European Union the recovery rate of these wastes varies between 40–70%, the latest statistical data indicating that the rate of recycling of aluminum cans reaches approx. 65%. [1,2] Worldwide, two out of three cans are recycled, but only one in three is recycled in Romania, or about 30–35%. [1,2] In 2007, the Recycling Standard for Romania required recycling at least half of aluminum waste by 2020. The recycling rate was then 3%, then rose to 5% and only in 2010 the growth was more significant, reaching 10%. [1–3] Nowadays a recycling rate of approx. 35% of the quantity placed on the market due to the sustained increase of the industry's contribution to a coherent and efficient system of selective collection of packaging waste, capable of meeting these obligations through the development of waste management services for recycling purposes. This is due to the efficiency of the partnership with over 200 management companies that carry out waste collection, sorting and transport services for recycling purposes. [1–3]

In Romania there are only small collection centers for aluminum cans, but there are also small units processing them. [1–3] Our country is lacking in recycling facilities, so the 33 or 50 centiliters cans are taking the road to Europe, to European recyclers that actually do the recycling of can bales, the aluminum being melted in factories with special technologies. This raw material is brought back in the form of an aluminum sheet and is transformed into another can. [1–4,8,9]

3. Construction and exploitation of the holding aggregate

Recycling is the reprocessing of materials in new products. [1–9] Thus, in order to carry out the recycling process of collected non–ferrous metal waste, aluminum waste melting facilities are required. This enables cost efficiency and faster recycling of waste from aluminum. [1–9] In this direction we propose a micro station for melting of aluminum wastes from the beverage cans. Research refers to a process for obtaining secondary aluminum from waste, by directly melting it into an experimental aggregate designed in the Faculty of Engineering.

To achieve the "melting unit" we need a cylindrical support with refractory lined walls. A gas cylinder was used to carry the aggregate, which was previously cut to the top, an oxyacetylene flame operation. (Figure 3) At the top two handles were welded. The cover of the oven was made of steel sheet fixed by the aggregate by means of a hinge, which allows its sliding. The following materials were used to make the inner linings of the aggregate: chamotte brick, refractory mortar, silicon sulphate. Due to the small size of the chamotte brick aggregate, it was cut into 57.5mm / 30mm strips to be able to reproduce the radius inside the aggregate. (Figure 4, Figure 5)

Figure 3. The "melting unit" – The achievement of the metallic "body"

Figure 4. The "melting unit" – The materials preparation for the refractory walls

The crucible was made of a 90 mm diameter steel pipe. At 10 cm from the bottom of the furnace a hole of 12 mm diameter was made where an air blowing installation will be installed. (Figure 6) To fix the crucible, a steel mounting plate was drilled in the middle so that the crucible's mouth would pass through it. On the fixing plate, holes were made to prevent air during the melting process. Through a hole, a pipe is inserted at the end of which we mount a vacuum cleaner to increase combustion. In this experimental melting aggregate, the crucible is placed in the center, over a layer of coal briquettes. A lot of charcoal briquettes are placed around the crucible until they are filled. The charcoal is ignited and when the crucible is hot, the aluminum cans are introduced. Melting takes place at a temperature of 680–750°C. (Figure 7) Aluminum melting slags are periodically evacuated, avoiding entrainment of molten aluminum.(Figure 8)

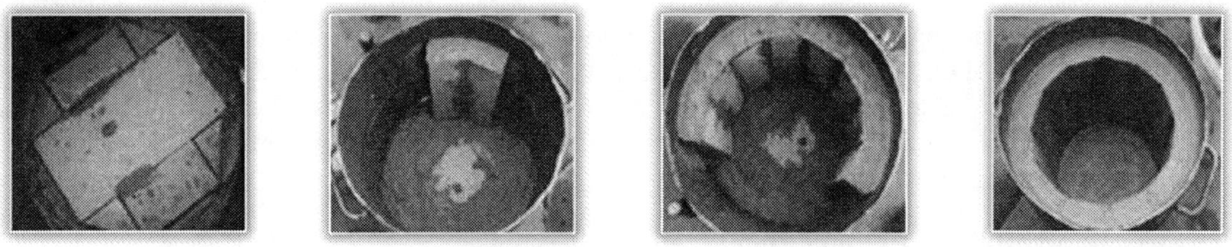

Figure 5. The "melting unit" – The achievement of the refractory walls

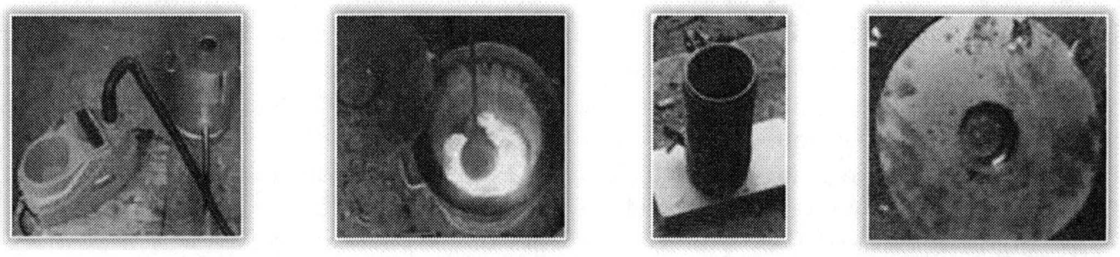

Figure 6. The experimental melting aggregate, the air blowing installation and the crucible

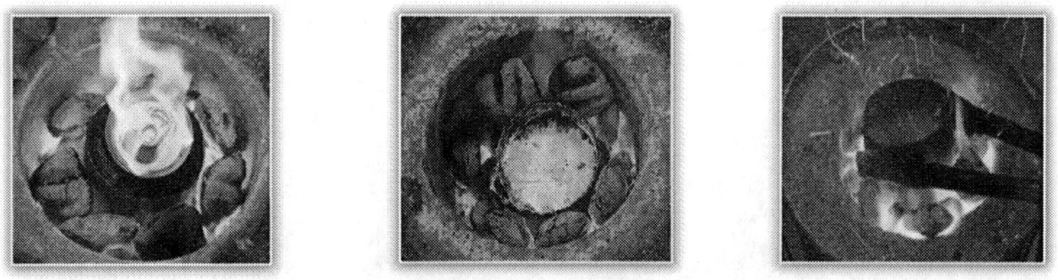

Figure 7. The melting of the aluminum cans

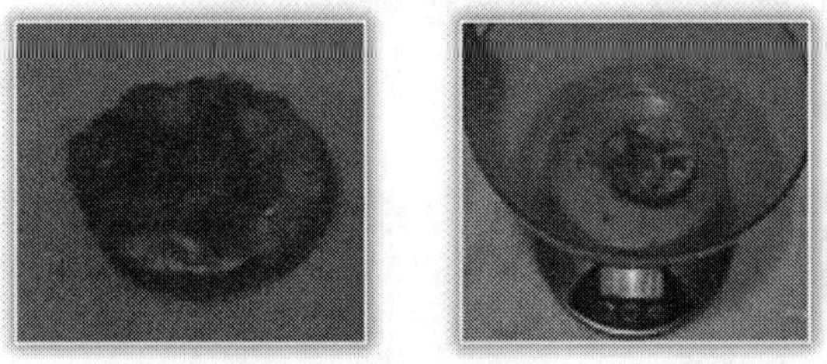

Figure 8. The aluminum melting slags

Figure 9. The melted aluminum

When the crucible is filled with molten aluminum, it is gently picked up using a pliers and poured the pure aluminum content carefully in the prepared forms. (Figure 9) Thus, small ingots or secondary aluminum chips are obtained. The aluminum remained will overturn on a concrete board and the crucible can be used for a new melting. The resulting melt may be transferred, prior to molding, to a holding furnace where corrections of the composition can be made. From 40–45 cans, an amount of approx. 500 grams of fused aluminum can be obtained. Varnishes and paints for marking the packaging can be removed by injecting a hot air stream at approx. 500°C, directly onto the waste before it is introduced into the melting crucible. Thus, the volume of slag involved in the melting process can be reduced.

4. Concluding remarks

Since many of the metals are widely used and natural reserves are for some, less and less important, the importance of recycling is increasing. Depending on the quality and the composition of the materials procured for recovery, the decision is made as to how the waste is to be processed. Recycling must include the following:

- the flow of raw materials must be managed in such a way as to facilitate optimal reuse and recycling, thus avoiding the waste and the depletion of natural resource deposits.
- the waste management must play a very important role, as it also represents a possible source of secondary raw materials.

The urban and industrial development of the cities, as well as the general increasing of the standard of living for the population, lead to the production of increasingly large amounts of waste – this is well known and undisputed. Waste of any kind, resulting from many human activities, is a nowadays theme, both due to the increasing in the quantities and their types, as well as to the significant quantities of reusable materials that can be recovered and introduced into the economic circuit. The big advantage of recycling is that old metal can be recycled many times without losing its attributes. Non–ferrous recycling, especially aluminum, offers smart and practical solutions to modern life, having many properties that do not alter by turning it into a product.

Containers and cans (beverages and beer), all made of aluminum, are so widely used worldwide that we cannot even imagine life without this light metal. However, not everyone is aware of the value of a single empty can. That's why, even the wild places in the country are suffocated by these cans that do not take the right path after they're used. Beyond the volume of recycled aluminum cans, it is more important to realize the benefits of recycling and transforming the collected aluminum cans into a routine. The extensive liability of the manufacturer is effective in the field of recycling, the interest in collecting and recycling packaging waste being constantly increasing.

5. References

[1] Centrul Naţional pentru Dezvoltare Durabilă / National Centre for Sustainable Development 2007 *Stategia Naţională pentru Dezvoltare Durabilă a României: Orizonturi 2013–2020–2030 / National Sustainable Development Strategy for Romania: Horizons 2013–2020–2030* Bucuresti http://www.mmediu.ro

[2] Capră G S and Stan A-C 2013 *Strategia Naţională de Gestionare a Deşeurilor 2014–2020 / National Waste Strategy 2014–2020*, Bucuresti http://www.mmediu.ro

[3] Ganea O, Simionescu A, Boţa I and Popa M 2011 *Reciclarea deşeurilor, o piaţă emergentă – Raport privid piaţa muncii în domeniul gestiunii deşeurilor în România* Bucureşti www.eco–social.ro

[4] Niero M and Olsen S I *Resources, Conservation & Recycling* **114** pp. 18–31

[5] Wernick I K and Themelis N J 1998 Recycling metals for the environment *Annual Review of Energy and the Environment* **23**(1) pp 465–497

[6] Verran G O and Kurzawa U 2008 An experimental study of aluminum can recycling using fusion in induction furnace, *Resources, Conservation and Recycling* **52**(5) pp 731–736

[7] Thomas M P and Wirtz A H 1994 *Resources, Conservation and Recycling* **10**(1–2) pp 193–204

[8] Das S K and Yin W 2007 The worldwide aluminum economy: The current state of the industry, *JOM*, **59**(11) pp 57–63

[9] Millbank P 2004 Aluminium recycling vital to global supply chain, *Aluminium International Today* **16**(5) pp 44–49

IOP Conf. Series: Materials Science and Engineering **416** (2018) 012100 doi:10.1088/1757-899X/416/1/012100

Researches Regarding Heterogeneous Welding between Carbon and Stainless Steel used in Fabrication of Cylinders' Rod for Magnesium Alloy Extruders

L Zgripcea[1,*], O Lupu[1], M Ardelean[1] and V Putan[1]

[1]Politehnica University of Timisoara, Engineering and Management Department, 5 Revolution Street, Hunedoara, 331128, Romania

[*]Corresponding author: laurentiu.zgripcea@gmail.com

Abstract. Magnesium alloys are used more and more in automotive industry due to specific strength (ratio between tensile strength and density) of 158kNm/kg versus 46kNm/kg in case of structural steel. Another advantage of magnesium alloys is machinability, aspect and automated caste/extrusion. With this last procedure is possible to achieve high rate of productivity for complicate pieces, needed in automotive sector. Technologically, a big issue of magnesium alloy is its reactivity in contact with carbon steel, accentuated by high temperature and pressure from extrusion chambers.

1. Introduction

Welding austenitic stainless steels to carbon and low alloy steels are used in extrusion of magnesium alloys for automotive industry.[1,2]

In the process of extrusion, magnesium alloy is pushed by a piston rod inside cylinder. The most stressed element is the piston rod, charged with high compression loads, under relatively high temperature (400 Celsius) and high pressure (300...700MPa) as is shown in figure 1.

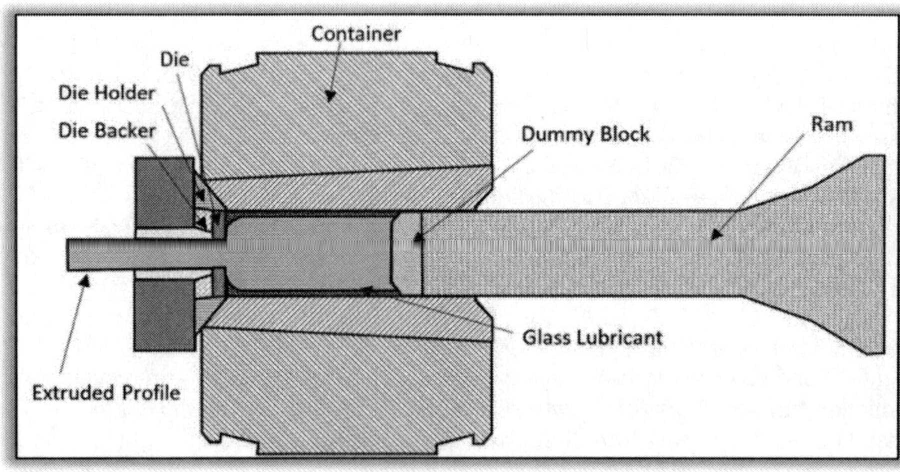

Figure 1. Typical process application. [1]

From this reason, piston rod (ram) is covered by a screwed bronze cap (dummy block). Very often rod's material does not resist and bend under stress (figure 2).

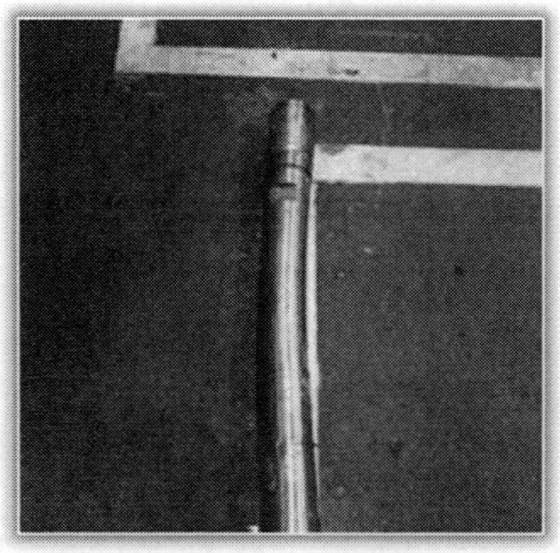

Figure 2. Bent rod. [2]

Study was pointed to realisation of dummy block – ram assemble using new materials and assemble to solve the next process issues [3–5]:

- Ram material was changed to high strenght medium alloyed steel.
- Dummy block amterial was changed to austenitc stainless steel
- Welding technology was developed to realize a strong and compact dummy block – ram assemble

Another problem during production process is the direct contact between high reactive magnesium alloy and dummy block. This lead in time to extensive wear and short life time.

For this reason, we implement a new design and technology, by welding a stainless steel dummy block. Welding of stainless steel dummy block can be done using specific welding procedures and filler materials carefully tested before.

Selection of filler material is key factor of the weld, with major impact over mechanical properties of assemble, stability at high temperature and corrosion resistance. From metallurgical point of view, welded joint is considered rather stainless than carbon steel, especially that over alloyed filers is used to obtain resistant microstructures when alloy with carbon steel.

2. Realization of prototype

2.1. *Choosing of parent materials*

From mechanical point of view, ram must have high mechanical properties at high temperature. Various steel grades were tried, and best results had 34CrNiMo6 (1.6582 EN 10027–2). From corrosion resistance and weldability, for dummy block the best results had austenitic stainless steel AISI304 (1.4301EN 10027–2).

2.2. *Welding conditions*

Although stainless steel does not require special condition, rod material has a high equivalent carbon (1.02%) and has high hardenability, susceptible to hydrogen cracking.

Welding joint to carbon steel must have slow cooling rate and reduce hardness in the heat–affected zone (HAZ), which creates a weld that is less brittle and more ductile. These characteristics are especially important for materials more susceptible to hardness at elevated temperatures, such as cast iron, medium– and high–carbon steel, or high– carbon–equivalency steel.

Another factor which requires preheating is dimension of welded parts (90mm diameter).

IOP Conf. Series: Materials Science and Engineering **416** (2018) 012100 doi:10.1088/1757-899X/416/1/012100

Taking into account these factors, result a preheating temperature of 360^{0}C.

According wit EN1011 standard, we have the next equivalent carbon for ram material [2]:

$$CE = (C + Mn)/6 + (Cr + Mo + V) + (Ni + Cu)/15 = 1.02\% \quad (1)$$

$$Tp = 679 \times CE - 160 \times \tanh(\frac{d}{35}) + 62 \times HD^{0.35} + (53 \times CE - 32) \times Q - 328 = 360\ °C \quad (2)$$

where:

- CE is equivalent carbon 1.02%;
- d is assemble diameter 90mm;
- HD is hydrogen content of the weld metal 5ml/100g;
- Q is the heat imput 2 kJ/mm.

Welding environment and filler material must be as dried as possible, at welding preheating temperature carbon steel has a much more hydrogen affinity and its related cracks versus austenitic steel.

2.3. Selection of filer material

Usually, stainless steel filer type ER308 can be used for this application. To improve weld quality, trials were made with ER309 filer which gives better results due to higher chromium and nickel.

From microstructure point of view, ER309 gives a higher ferrite concentration which is favourable.

From temperature point of view ER309 has higher nickel, so better results at working temperature of 400^{0}C.

Figures 3 and 4 show the ferrite calculation for ER308 and ER309.

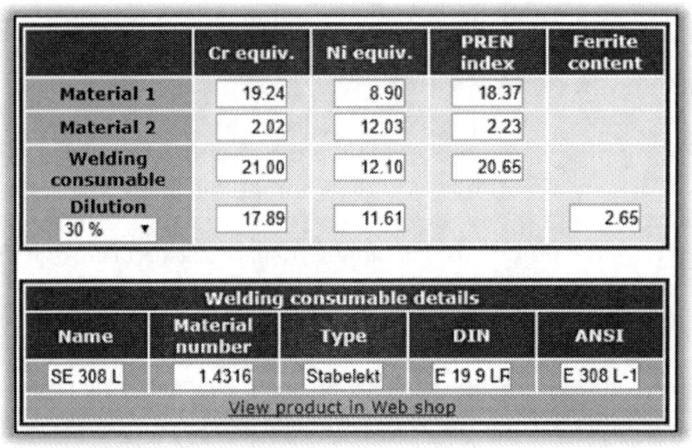

	Cr equiv.	Ni equiv.	PREN index	Ferrite content
Material 1	19.24	8.90	18.37	
Material 2	2.02	12.03	2.23	
Welding consumable	21.00	12.10	20.65	
Dilution 30 %	17.89	11.61		2.65

Welding consumable details				
Name	Material number	Type	DIN	ANSI
SE 308 L	1.4316	Stabelekt	E 19 9 LR	E 308 L-1
View product in Web shop				

Figure 3. Cr and Ni equivalent for ER308

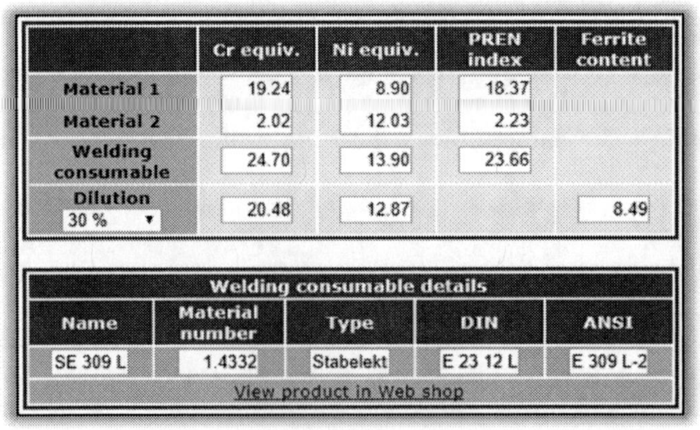

	Cr equiv.	Ni equiv.	PREN index	Ferrite content
Material 1	19.24	8.90	18.37	
Material 2	2.02	12.03	2.23	
Welding consumable	24.70	13.90	23.66	
Dilution 30 %	20.48	12.87		8.49

Welding consumable details				
Name	Material number	Type	DIN	ANSI
SE 309 L	1.4332	Stabelekt	E 23 12 L	E 309 L-2
View product in Web shop				

Figure 4. Cr and Ni equivalent for ER309

For selected filler grade ER309 another check was made with Schaeffler diagram (figure 5).

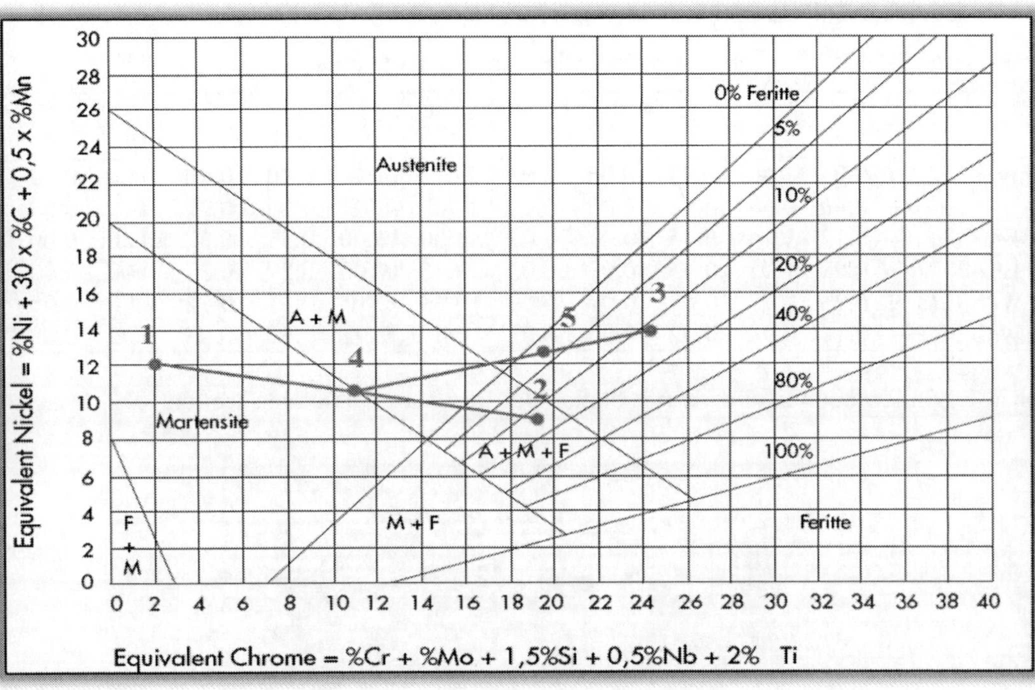

Figure 5. Schaeffler diagram: – point 1 represents rod material, carbon steel, grade; – point 2 represents cap material, stainless austenitic steel grade; – point 3 represents filer material grade ER309; – point 4 represents the mixture of rod and cap material; – point 5 represents the mixture of parents materials and filer material at a dilution (leverage) of 30% means 30% of parents material will participate in weld joint, rest being filer material.

3. Results of welding trial and conclusions

Trial was made using circular weld, using different filler materials, for filler materials half ER308 and for next half ER309. Welding procedure was MMA with 3.2mm electrodes at a current of 140 amps. Bevel angle was at 30 degree and height 10 mm.

For both filler materials weld joint was satisfactory and in transversal section no porosity or defect were observed (figure 6 and 7).

Figure 6. Joint preparation

Figure 7. Assemble welded

Chemical analyses of both welded joints shown difference in chemical composition, the most relevant is for carbon and chromium (table 1).

Table 1. Chemical compositions.

	Chemical composition, [%]											
	C	Mn	Si	S	P	Cr	Ni	Cu	Mo	Ti	V	Al
Dummy block	0.05	1.46	1.53	0.00	0.02	19.15	11.10	0.10	0.00	0.00	0.12	0.01
Ram	0.40	0.66	0.16	0.00	0.01	0.85	1.73	0.25	0.21	0.00	0.00	0.01
Filler ER309	0.12	1.75	0.45	0.03	0.03	24.00	13.00	0.37	0.37	0.00	0.00	0.00
Filler ER308	0.08	1.75	0.45	0.03	0.03	20.75	10.00	0.37	0.37	0.00	0.00	0.00
Weld with ER309	0.23	0.79	0.53	0.01	0.00	10.93	6.01	0.15	0.12	0.00	0.05	0.00
Weld with ER308	0.11	0.98	0.92	0.00	0.01	16.49	8.84	0.12	0.08	0.02	0.08	0.00

For ER309 filler, dilution of carbon is higher. Figure 8 shows assemble section.

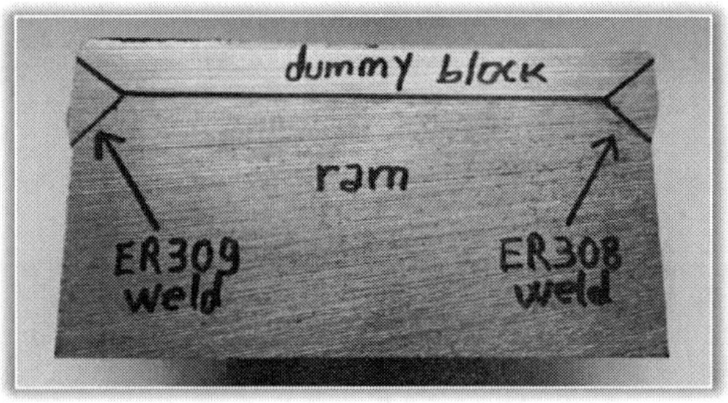

Figure 8. Assemble section

4. Conclusions and further researches

Both experimental welded joints are visually free of defects. Regarding chemical composition, for joint made with ER309 filler participation of base material is higher, indicated by carbon diluted from ram steel. That indicates a stronger weld compared with classical ER308 filler.

As a result of chemical composition difference between the two joints, we have a difference also in steel microstructure. This difference is in ferrite content: 2.65% for ER308 and 8.49% for ER 309.

The research is continued in the sense of process experimentation, key factors recorded are:

- Lifetime of ram;
- Aspect of dummy block and weld (corrosion).

5. References

[1] https://www.engineering.com/AdvancedManufacturing/ArticleID/13324/New–Extrusion–Method–Makes–Magnesium–More–Practical–for–the–Auto–Industry.aspx
[2] *** Procedures for fabrication of magnesium extruded parts 2014 TRW Factory Timisoara
[3] Dehelean D 1997 *Sudarea prin topire* Sudura Timisoara
[4] Miclosi V and Dehelean D 2014 International Welding Engineer
[5] Solomon Gh and Cicic DT 2010 T*eoria proceselor de sudare* BREN Bucuresti

IOP Conf. Series: Materials Science and Engineering **416** (2018) 012101 doi:10.1088/1757-899X/416/1/012101

The Influence of the Cutting Tools Parameters on Inconel CNC Manufacturing

A But[1,*] and L Gal[2]

[1]Politehnica University of Timişoara, Mechanical Faculty, Department of Materials and Manufacturing Engineering, B-dul Mihai Viteazu nr.1, Timişoara 300222, Romania
[2]"Aurel Vlaicu" University of Arad, B-dul Revolutiei nr 77. Arad 310130

[*]Corresponding author: adrian.but@upt.ro

Abstract. There are several types of super alloys: Titanium, Inconel. Titanium alloys are typical for aerospace products. They are attractive because of their low specific weight, and their resistance. Inconel alloys are corrosion and oxidation resistant materials, well suited for work in extreme environments who have a good resistance at high pressure and heat. Inconel alloys retain their properties in a wide temperature range and are attractive for high temperature applications where aluminum and steel parts will not resisted. But the processing of these super alloys is complex and requires dedicate cutting tools and special cutting tools parameters. In the paper will be present solutions at this requirements: new cutting tools and new solutions to manufacturing aerospace parts from Inconel.

1. Introduction

Inconel alloy 718 can be readily machined, but its high strength and work-hardening characteristics must be considered in the selection and use of proper tool materials and design, operating speeds, and coolants.

Until today Ti 6Al 4V (Ti64) is still the most employed of all Ti-alloys both for engine and air-frame applications. Ti64 is characterized by an optimum combination of properties: Uncritical in processing, high strength at low temperatures, excellent machinability and good, weldability. Ti64 is available as forging, sheet and since about 1975 as investment casting for stator components, for example compressor casings. Ti 6242 and Ti 6246 have higher strength and temperature capabilities. At present the most advances Ti-alloy for disk applications (up to 550^{0}C) is IMI 834. Above 450^0 C this alloy is superior to any other Ti-alloy. Its price, however, is about twice the price of Ti64 with respect to its complex metallurgy and the thermo mechanical processing parameters needed for optimum microstructure and mechanical properties. [2]

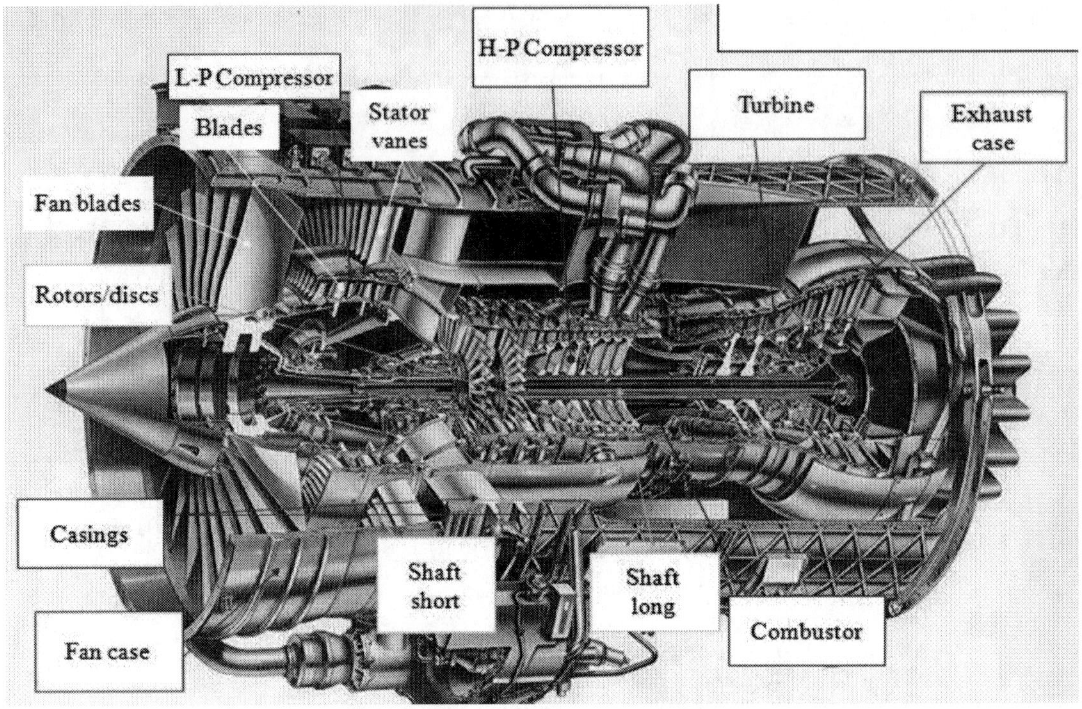

Figure 1. The aerospace components

When machined in the age-hardened condition the alloy will have a slightly better finish; chip action on chipbreaker tools will be better. The use of annealed material, however, provides easier machining and longer tool life.

Because of its strength, INCONEL alloy 718 is more resistant than most materials to deformation during hot forming. Its relative resistance is shown by pressures developed in the roll gap at 20% reduction .

It is readily hot-worked if sufficiently powerful equipment is used.

Hot forming is performed in the 900°-1400°C temperature range. In the last operation, the metal should be worked uniformly with a gradually decreasing temperature, finishing with some light reduction in the 900°-950°C range.[3]

This procedure is necessary to ensure notch ductility in stress-rupture applications when material has been annealed and aged.

In heating for hot working, the material should be brought up to temperature, allowed to soak a short time to ensure uniformity, and withdrawn.

2. State of art

Take in consideration the new materials of the aerospace components(Fig.1) who use super alloy and the complexity of the parts, the aerospace industrial needs are:1.Educated scientists and engineers 2.Globally sourced 3."Clearable" for defence-related projects 4.Mathematical and analytical capability 5.Probability and statistics 6.Communications across various disciplines: Not just interdisciplinary studies, but interaction.

Industrial aerospace research directions: 1.Increased global sourcing of subcomponents and materials 2.Global sourcing of assembly 3.Networked manufacturing 4."Materials by Design"5. Lean manufacturing 6."Green" materials and processes7.Networked aircraft, satellites.[8]

Regarding the manufacturing we can conclude:

IOP Conf. Series: Materials Science and Engineering **416** (2018) 012101 doi:10.1088/1757-899X/416/1/012101

The cutting conditions are totally different when we manufacturing Inconel in comparison with general steel manufacturing. The cutting speed is lower, the depth of cutting is much more less because the wear of the cutting tool is much higher.

The cutting tool parameters in the milling process have connections between : cutting tool speed V/"V_c" [m/min], the feed "s" /"f_z" [mm/rot], and the depth of cut "t" / „ a_p" [mm] with the setting parameters: rotational speed "n" [rot/min], feed by teeth "f_z" [mm], and the axial and radial depth of cut a_p, a_e [mm] (Fig.2).[1]

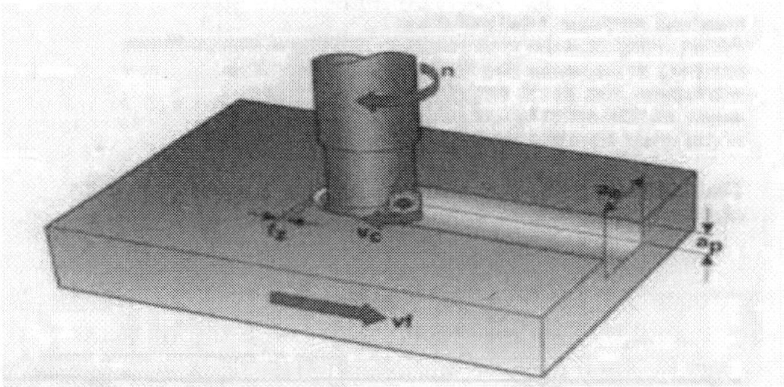

Figure 2. Milling process with the cutting tool parameters

Take in consideration this inconveniences, the cutting tools was adapted to this hard conditions of Inconel manufacturing. Was develop new geometries of the inserts (this we can see this, we can measure it) and new thermal coatings (this procedures are not presented by the companies - are their protected patents). All this new technical solutions must be tested and after that, to be confirmed.

In the paper was tested new cutting tools with new round rotate insert K0-90 in manufacturing Inconel super alloy.

2.1. The roud indexable inserts K0-90°

This have a new design with a chip breker who assure a good stability of the cutting process.

This round inserts will be fix on the plate suport (Fig.3.b) with 7 ° pozitiv inclination (Fig.3.a) or with 0 ° inclination (Fig.3.c). The cutting angle is recomanded for the martensitic materials X38CrMo17, austenitic steels (316Ti), where was obtain very good rezults.[3]

a) Positiv axial b) The Cutting plate support c) Negativ axial

Figure 3. Position of the cutting plate in support

The plate geometry is different for Inox (Fig 4a) , general milling, (Fig 4b) and super alloy (Fig 4c).

IOP Conf. Series: Materials Science and Engineering **416** (2018) 012101 doi:10.1088/1757-899X/416/1/012101

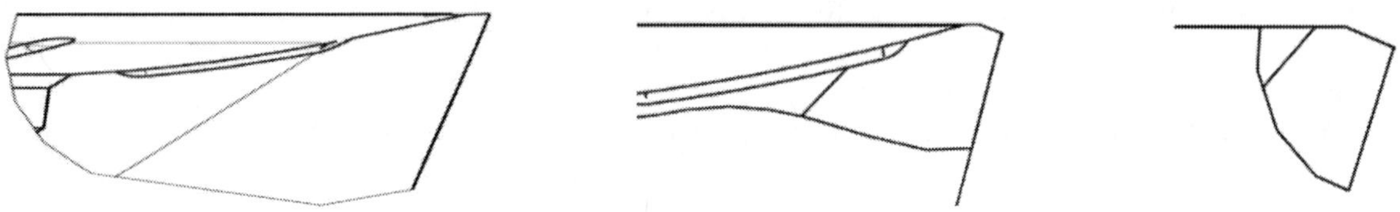

a) Inox b) General milling c) Hard roughing, superalloy

Figure 4. The plate geometry

2.2. New System

The new innovative tooling system change an old conception of the cutting process : "the insert has to be rigidly coupled to the cutter body-if it lose in its insert seat it would break soon or later ".

So this theory was unconfirmed by new research in the cutting tools manufacturing where the insert is rotating in the milling process.[4]

So this theory was unconfirmed by new research in the cutting tools manufacturing where the insert is rotating in the milling process.

With this new technical solution it is not necessary to stop the machine and to make a manual re-positioning with the change the wear area of the inserts. The results is that we will minimize the set - up time and will eliminate the un- productive time of manufacturing.

More that then, we will obtain the same wear along the circumferences of the round plate, we will use 100% of the total insert periphery and we have a multiple increase of tool life.

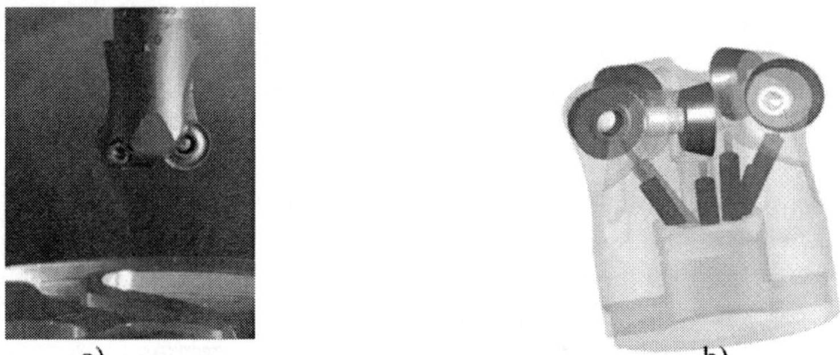

a) b)

Figure 5. a) The new system-with rotating plate b) the internal cooling systems

The geometry of this rotating plate have a dedicated design for the cutting edge. (Fig. 4). The clearance angle W1 reinforces the cutting edge while the clearance angle W2 provides sufficient space to eliminate the chips and to grow up the cooling liquid in the cutting tool area.

With this geometry was stabilized the cutting edge and form the limiting feature for the wear mark.

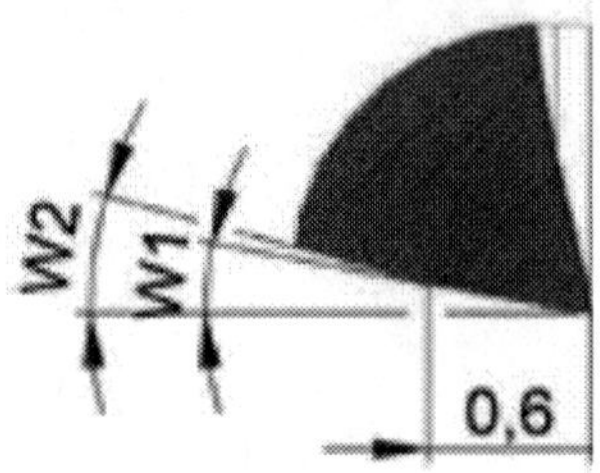

Figure 6 The geometry of the cutting edge

IOP Conf. Series: Materials Science and Engineering **416** (2018) 012101 doi:10.1088/1757-899X/416/1/012101

2.3. Milling Application areas

Titanium requires specific considerations due to its high strength to weight ratio which results in higher cutting temperatures.

It is also chemically reactive which can result in adhesion, welding and smearing when machining. The combination of mechanical, thermal and chemical loads will create a typical wear pattern known as edge chipping. (Fig.7)

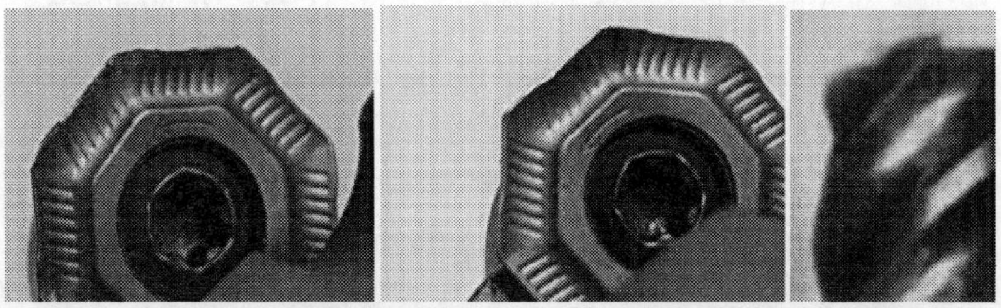

a) b) c)

Figure 7. Wear types a) Roughing, b) Semi-finishing c) Finishing

2.4. Recommendations in programming

1.Entering the component

When the cutter is programmed to enter straight into the workpiece, thick chips will be produced at the exit until the cutter is fully engaged. This will dramatically reduce tool life.

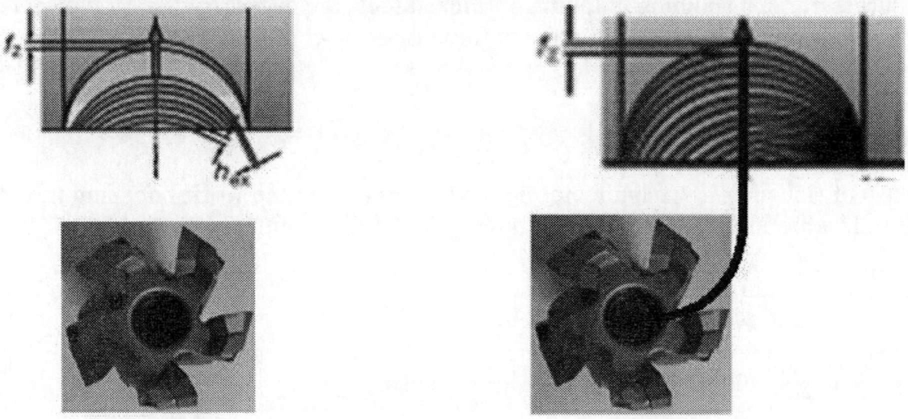

a) b)

Figure 8. a) enter straight to cut b) Roll in to cut

The main way to solve this problem is:

Roll in to cut (Fig. 8 b)

Program a roll into cut in a clockwise motion. By rolling into cut, the chip thickness on exit is always zero, allowing for higher feed and longer tool life.

Keep cutter constantly engaged

Sharp changes of direction in a cut will cause the same problem as occurs when entering straight into the workpiece(Fig.8a). Rolling around all corners should always be applied as a key step to provide a robust, optimized process. Program around interruptions and holes when possible.[3]

2.5. Programming Centre line or periphery feed

A machine is designed for either centre line feed, v_f (without radius compensation) or periphery feed, v_{fm} (with radius compensation).

IOP Conf. Series: Materials Science and Engineering **416** (2018) 012101 doi:10.1088/1757-899X/416/1/012101

If the machine requires a centre line feed, and periphery feed is programmed instead (on machines without radius compensation), the f_z value will become too high, with the subsequent risk of insert breakdown.[3]

Centre line feed programming

The NC codes generated will program the centre of the cutter rather than the periphery.

For straight line cutting (G1), the feed at the wall of the component, v_{fm} is the same as the programmed feed, v_f while the periphery feed around a radius (G2) will be higher than the tool centre feed.

Therefore the table feed, v_f needs to be reduced to maintain the feed per tooth, f_z and keep constant hex.

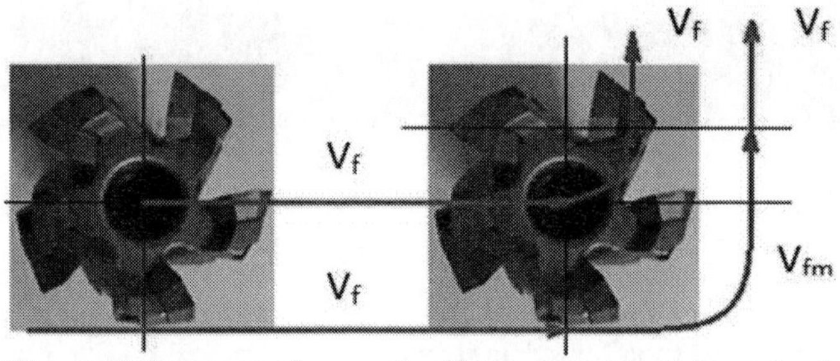

Figure 9 Programming centre line versus periphery feed

To be successful in machining titanium, optimized tools are necessary.

However, to achieve an optimized process these must be combined with application 'know how' to deliver a robust machining process.[5]

3. Conclusions

After the test of Inconel 718 manufacturing that was made in the workshop, sing the "New System" with roud indexable inserts (the rotating round plate) versus with the " Standard-System" we can conclude:

Cutting data recommendations:

Roughing v_c 50 m/min Depth of cut : > 3mm
Semi-finishing v_c 70 m/ Depth of cut : min 2 mm
Finishing v_c 100 m/ Depth of cut : min 0.5-1 mm

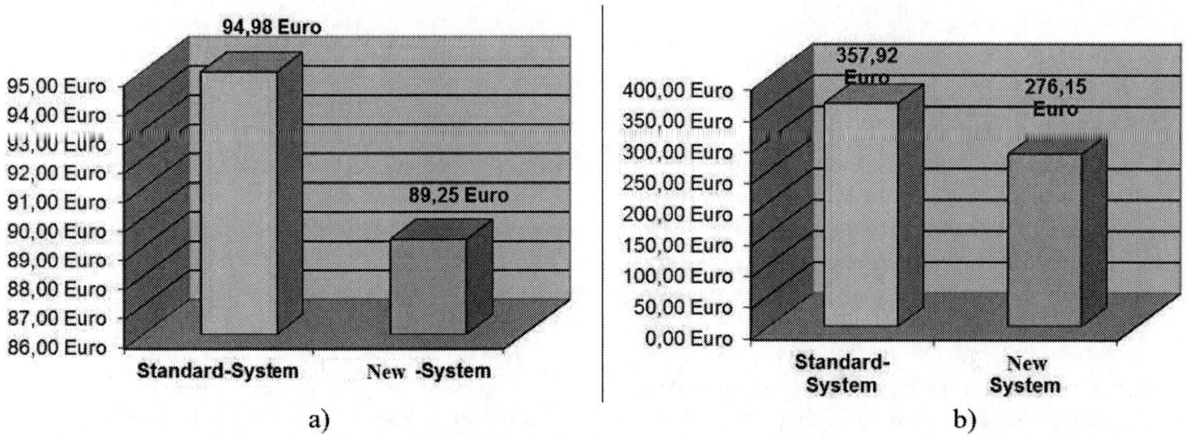

a) b)

Figure 10. a) the price for the inserts "Standard -System" /" New System" b) the price for manufacturing Standard -System" /" New System"

In the paper was presented the new trend in Inconel manufacturing, was tested and validate the "New System" - using the revolutionary solution in cutting tool construction - the round plate is rotating in the cutting process, and was manufacturing a part with establish the optimise cutting tools parameters for better quality and the time manufacturing.

Making the comparison between the " Standard -System" - the fixed round plate with the roud indexable inserts (the rotating round plate)- " New System", we made one significant reduce price for the cutting plate used: with 7% and with 23 % reduce the price for manufacturing the part with " New Systems".

4. References

[1] But A 2009 "*Advanced machine-tools and manufacturing systems*." Ed. Politehnica-Timisoara
[2] *Metal Cutting Technology* Technical guide
[3] *** SPINWORKS -Pokolm cutting tools catalog- 2016
[4] Kumar C and Deb S 2012 *Journal of Advance Manufacturing Systems* **11**(01) 67-80
[5] Sormaz DN, Arumugam J and Ganduri C 2007 Integration of rule-based process selection with virtual machining for distributed manufacturing planning, Process planning and scheduling for distributed manufacturing, *Springer*, 61-90
[6] Nikolaus K 2014 "Manufacturing : Self-Organizing Factories" *Siemens-Digital Factories*
[7] Bahemia H and SquireB 2010 *Int. J. Innovative Management* **14**(4) 603-627

IOP Conf. Series: Materials Science and Engineering **416** (2018) 012102 doi:10.1088/1757-899X/416/1/012102

Quasi-Static Mechanical Characterization of Lightweight Fly Ash-Based Geopolymer Foams

E Linul[1,*], K Korniejenko[2], D A Şerban[1], R Negru[1], L Marşavina[1], M Łach[2] and J Mikuła[2]

[1]Department of Mechanics and Strength of Materials, Politehnica University of Timisoara, 1 Mihai Viteazu Avenue, 300 222 Timisoara, Romania
[2]Institute of Materials Engineering, Faculty of Mechanical Engineering, Cracow University of Technology, Poland

[*]Corresponding author: emanoil.linul@upt.ro

Abstract. This paper investigates the compressive and flexural behavior of a closed-cell fly ash-based geopolymer foam with a density of 240 kg/m^3. The influence of loading speed (0.1, 10, 100, 250 and 500 mm/min) and anisotropy (in-plane and out-of-plane directions) on the compressive mechanical properties at room temperature were investigated. Under compressive loads, the stress-strain curves shows a cellular material typical behavior with three different regions: linear-elastic, plateau and densification region. On the other hand, the three-point bending tests were carried out on both un-notched and notched specimens (for fracture toughness determination). In compression, the foam progressively collapsed with a complete destruction of the successive rows of cells, while in the case of bending it exhibited a brittle behavior without plastic deformation, the fracture occurring by the propagation of a single crack.

1. Introduction

Porous materials such as polymeric [1-4] and metallic [5-8] foams have gained a lot of attention in recent years due to their excellent strength properties [9-12], energy absorption capabilities [13-15] and other significant features [16-18]. Many efforts have been made in last three decades in order to characterize experimentally [19-22], analytically [23, 24] ad numerically [25-27] the above mentioned foams. All of these studies have been conducted since cellular materials [28-30] begin to gain more applicability compared to fully dense solid materials [31-34].

In recent years, the interest in geopolymer-based cellular materials has grown significantly. The reduction of geopolymer density through foaming is increasingly being reported in the literature as it has been shown to be effective in improving their properties [35], cellular geopolymers exhibiting very good physical and mechanical properties, energy absorption and fireproofing [36, 37]. The foamed geopolymers have attracted a lot of attention and are promising alternative to other cellular materials [38], mainly in civil engineering, these may replace in the future conventional concrete based on Portland cement (which is responsible for a significant amount of global CO_2 emissions). Geopolymer foams use high-volume industrial waste with a significant reduction in CO_2 emissions during production [35, 36].

A lot of studies were focused on microstructure [39], different foaming methods [35, 40], thermal properties [36, 37], fire performances [41] of geopolymer foams, while only few papers have been focused on the determination of mechanical properties [39, 42, 43]. This paper investigate the effect of

IOP Conf. Series: Materials Science and Engineering **416** (2018) 012102 doi:10.1088/1757-899X/416/1/012102

loading speed and loading direction (effect of anisotropy) on fly ash-based geopolymer foam samples under quasi-static compression tests. Also the three point bending behavior on notched and un-notched samples was investigated.

2. Materials and methods

The presented research are based on geopolymer foams made from the fly ash provided by the local Polish CHP plant (located in city called Skawina) and microspheres. As an activator, an alkaline solution of 14M aqueous solution of NaOH + liquid glass at a ratio of 1:2.5 was used. The cellular structure was achieved thanks to the addition the hydrogen peroxide to manufacturing process.

Large fly-ash based geopolymer foam plates were foamed (Fig. 1a) and subsequently processed as cubic specimens (20 mm × 20 mm × 20 mm) for compression tests (Fig. 1b), while prismatic samples (10 mm × 20 mm × 80 mm) were used for 3PB tests (Fig. 1c). The samples used for the determination of fracture toughness have shown a notch of 10 mm in length and notch thickness of 0.6 mm. Figure 1d presents the macrostructure morphology of the investigated foam. The obtained average foam density was 240 kg/m^3.

a) b) c) d)

Figure 1. Foamed plate (a), compression sample (b), 3PB samples (c) and foam microstructure (d)

In order to investigate the anisotropy effect, all cubic samples were cut from the same foam plate under three different loading directions, named direction (1), direction (2) and direction (3). Quasi-static compression tests were carried out at room temperature (~23°C) on a 5 kN Zwick Roell 005 testing machine. A constant crosshead speed of 10 mm/min was used for all experimental tests, while 0.1, 10, 100, 250 and 500 mm/min was used for determining the effect of loading speed. Five samples were provided for each test condition.

3. Results and discussion

3.1. Compression tests

Figure 2 present a typical stress-strain curve obtained from compression tests carried out on a cubic fly-ash based geopolymer foam. The obtained curve is similar to those studied by other researchers on different cellular materials [44-47], highlighting three regions of deformation: a linear-elastic region followed by a large plateau region and ends with the onset strain of densification. The serrated character of the stress-strain curves is mostly due to the brittleness of the matrix material used for the foam production.

IOP Conf. Series: Materials Science and Engineering **416** (2018) 012102 doi:10.1088/1757-899X/416/1/012102

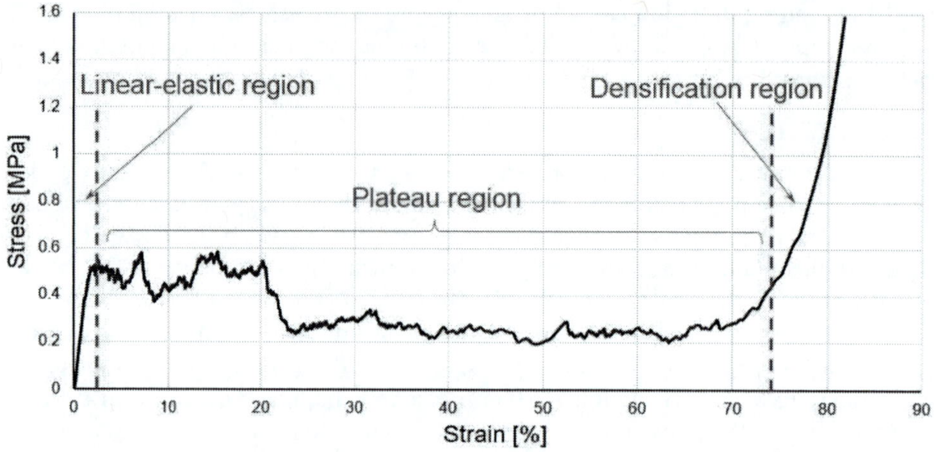

Figure 2. Detailed regions of fly-ash geopolymer foam stress-strain curve

The following compressive mechanical properties were investigated in this paper: compressive modulus (E), yield stress (σ_y), strain corresponding to yield stress (ε_y), stresses at 20% ($\sigma_{20\%}$) and 40% ($\sigma_{40\%}$), plateau stress (σ_{pl}), densification stress (σ_d) and densification stress (ε_d).

Figure 3 presents the stress-strain curves of investigated foams under three different loading directions. As it can be seen, all curves highlights the same compressive behavior with above mentioned regions. No large differences were observed between directions (1) and (2), while a decrease of compressive modulus and strength properties was obtained for direction (3) of loading. Therefore, the investigated foam does not have a strong anisotropic behavior, like other types of cellular materials [47].

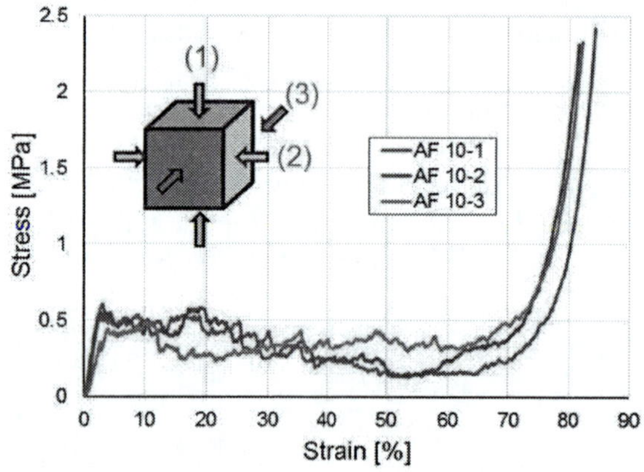

Figure 3. Stress-strain curves. Influence of loading direction

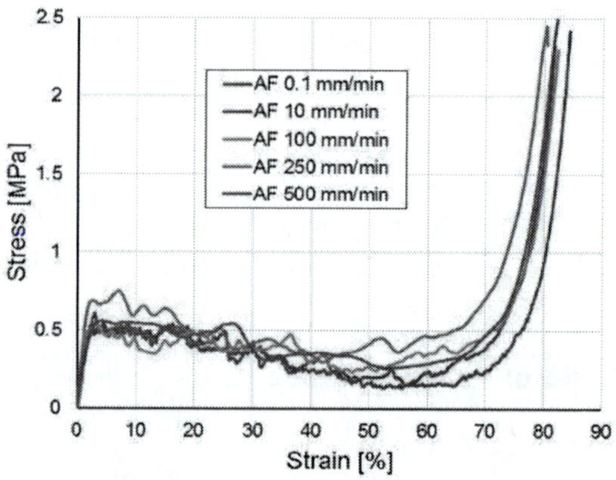

Figure 4. Stress-strain curves. Influence of loading speed

The main mechanical properties of investigated geopolymer foams samples depending on the loading directions are presented in Table 1, while Table 2 shows the same properties according to different loading speeds.

On the other hand, the loading speed (Fig. 4 and Table 2) present significantly higher elastic properties (about 50%) only at 500 mm/min, while for the other loading rates, approximately the same values are obtained (except for 22.63 MPa value at 100 mm/min).

Table 1. The mechanical properties of geopolymer foams. Influence of loading direction

Loading direction	Linear-elastic region			Plateau region			Densification region	
	E [MPa]	σ_y [MPa]	ε_y [%]	$\sigma_{20\%}$ [MPa]	$\sigma_{40\%}$ [MPa]	σ_{pl} [MPa]	σ_D [MPa]	ε_D [%]
1	33.92	0.60	3.15	0.41	0.24	0.325	0.15	64.99
2	34.73	0.54	2.96	0.57	0.24	0.405	0.18	67.24
3	19.19	0.35	2.75	0.26	0.37	0.315	0.38	66.64

Table 2. The mechanical properties of geopolymer foams. Influence of loading speed

Loading speed [mm/min]	Linear-elastic region			Plateau region			Densification region	
	E [MPa]	σ_y [MPa]	ε_y [%]	$\sigma_{20\%}$ [MPa]	$\sigma_{40\%}$ [MPa]	σ_{pl} [MPa]	σ_D [MPa]	ε_D [%]
0.1	30.21	0.50	1.92	0.40	0.31	0.355	0.24	62.96
10	33.92	0.60	3.15	0.41	0.24	0.325	0.15	64.99
100	22.63	0.55	3.53	0.49	0.30	0.395	0.36	64.92
250	42.53	0.69	2.74	0.48	0.35	0.415	0.48	63.35
500	68.91	0.55	3.14	0.47	0.35	0.410	0.34	66.21

The main strength properties (σ_y and σ_{pl}) are not influenced by the loading speed, only densification stress (σ_d) is slightly different. Also, the onset strain of densification show a deviation of ±1.5% between all investigated loading speeds.

3.2. Three point bending tests

Figure 5 presents the stress-strain curve of foam loaded under three point bending fixture on samples without notch. The flexural modulus was calculated from the initial slope of the curve, while the maximum stress at failure represents the flexural strength of foam. Table 3 lists the obtained bending mechanical properties.

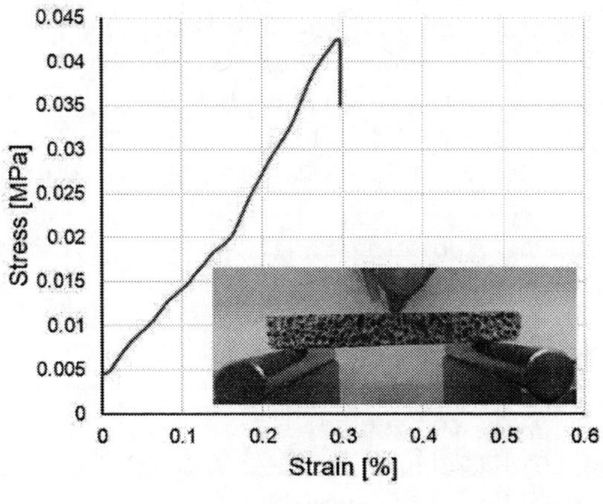

Figure 5. 3PB stress-strain curve of un-notched sample.

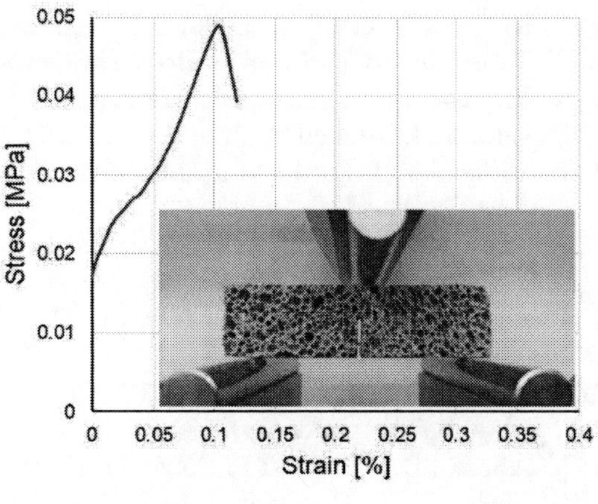

Figure 6. 3PB stress-strain curve of notched sample.

Table 3. The main 3PB mechanical properties of un-notched geopolymer foam

Geometrical parameters			Mechanical properties		
Sample high h [mm]	Sample width b [mm]	Span length l [mm]	Flexural modulus E [MPa]	Flexural strength σ_f [MPa]	Strain corresponding to σ_f ε_f [%]
10	20	64	14.96±1.76	0.042±0.004	0.3±0.02

On the other hand, Fig. 6 presents the stress-strain curve of investigated foam under 3PB test on notched samples. For all tested samples plan strain condition was fulfilled [48]. Brittle fracture was observed and the linear-elastic behavior was confirmed during the test when no cushioning occurs and no plastic deformation remain after the test [49-52]. It has been found a fracture toughness value of 0.17±0.02 MPa·m$^{0.5}$. The fracture occurs by the initiation and propagation of a single crack.

4. Conclusions

Quasi-static compressive and 3PB mechanical properties (compressive and flexural modulus, compressive and flexural strength, fracture toughness and densification strain) of closed-cell fly ash-based geopolymer foams are experimentally investigated.

The following conclusions can be drawn:

- The research results confirm that the investigated geopolymer foams have good mechanical properties for their specific weight;
- Investigated foam emphasize slightly anisotropic compressive behavior;
- Loading speed does not significantly affect the main mechanical properties;
- All 3PB notched and un-notched samples showed brittle failure without plastic deformation.

References

[1] Gibson, L.J. and Ashby, M.F., Cellular Solids-Structures and properties-Second edition. Published by the Press Syndicate of the University of Cambridge, UK, 1997.
[2] Linul E and Marsavina L 2015 *P. Romanian Acad. A* **16(4)** 522-530
[3] Şerban DA, Weissenborn O, Geller S, Marşavina L, Gude M 2016 *Polym. Test.* **49** 121-127
[4] Aliha MRM, Linul E, Bahmani A, Marsavina L 2018 *Polym. Test.* **67** 75-83
[5] Ashby MF, Evans A, Fleck NA, Gibson LJ, Hutchinson JW, Wadley HNG, Delale F 2000 *Metal Foams: A Design Guide.* Butterworth-Heinemann, USA
[6] Linul E, Marsavina L, Kováčik J 2017 *Mat. Sci. Eng.-A Struct.* **690** 214-224
[7] Kovácik J, Jerz J, Mináriková N et al. 2016 *Frattura ed Integrita Strutturale* **36** 55-62
[8] Taherishargh M, Linul E, Broxtermann S, Fiedler T 2018 *J. Alloy. Compd.* **737** 590-596
[9] Movahedi N, Linul E, Marsavina L 2018 *J. Mater. Eng. Perform.* **27(1)** 99-108
[10] Linul E, Movahedi N, Marsavina L 2018 *J. Alloy. Compd.* **740** 1172-1179
[11] Voiconi T, Linul E, Marsavina L, Sadowski T, Knec M 2014 *Solid State Phenomena* **216** 116-121
[12] Marsavina L, Linul E, Voiconi T, Constantinescu DM, Apostol DA 2013 *Frattura ed Integrita Strutturale* **34** 444-453
[13] Linul E, Şerban DA, Marsavina L, Kovacik J 2017 *Fatig. Fract. Eng. Mater. Struct.* **40(4)** 597-604
[14] Movahedi N and Linul E 2017 *Mater. Lett.* **206** 182-184
[15] Linul E, Movahedi N, Marsavina L 2017 *Compos. Struct.* **180** 709-722
[16] Apostol DA, Stuparu F, Constantinescu DM et al. 2016 *Materiale Plastice* **53(4)** 685-688
[17] Marsavina L, Constantinescu DM, Linul E et al. 2015 *Eng. Fail. Anal.* **58** 465-476
[18] Marsavina L, Kovacik J, Linul E 2016 *Theor. Appl. Fract. Mech.* **83** 11-18
[19] Şerban DA, Voiconi T, Linul E, Marsavina L, Modler N 2015 *Materiale Plastice* **52(4)** 537-541
[20] Negru R, Şerban DA, Pop C, Marşavina L 2018 *Theor. Appl. Fract. Mech.* DOI:10.1016/j.tafmec.2018.01.016

[21] Apostol DA, Stuparu F, Constantinescu DM et al. 2016 *Materiale Plastice* **53(2)** 280-282
[22] Marsavina L, Linul E, Voiconi T, Negru R 2016 *IOP Conf. Ser: Mater. Sci. Eng.* **123(1)** 012060
[23] Marsavina L, Berto F, Negru R et al. 2017 *Theor. Appl. Fract. Mech.* **91** 148-154
[24] Birsan M, Sadowski T, Marsavina L et al. 2013 *Int. J. Solids Struct.* **50** 519-530
[25] Şerban DA, Linul E, Voiconi T, Marsavina L, Modler N 2015 *Iran. Polym. J.* **24** 515-529
[26] Linul E, Serban DA, Marsavina L 2018 *Phys. Mesomech.* **21(2)** 178-186
[27] Linul E and Marsavina L 2011 *Iran. Polym. J.* **20(9)** 736-746
[28] Voiconi T, Negru R, Linul E et al. 2014 *Frattura ed Integrità Strutturale* **30** 101-108
[29] Linul E, Serban DA, Voiconi T, Marsavina L, Sadowski T 2014 *Key Engineering Materials* **601** 246-249
[30] Voiconi T, Linul E, Marsavina L et al. 2014 *Key Engineering Materials* **601** 254-257
[31] Şerban DA, Marşavina L, Silberschmidt VV 2012 *Comp. Mater. Sci.* **52** 139-146
[32] Şerban DA, Marşavina L, Modler N 2015 *Fatig. Fract. Eng. Mater. Struct.* **38(11)** 1383-1394
[33] Marsavina L, Cernescu A, Linul E, Scurtu D, Chirita C 2010 *Materiale Plastice* **47(1)** 85-89
[34] Şerban DA, Marşavina L, Silberschmidt VV 2012 *Comp. Mater. Sci.* **64** 116-121
[35] Masi G, Rickard WDA, Vickers L et al. 2014 *Ceram. Int.* **40** 13891-13902
[36] Łach M, Korniejenko K, Mikuła J 2016 *Procedia Engineering* **151** 410-416
[37] Cheng-Yong H, Yun-Ming L, Al Bakri Abdullah MM, Hussin K 2017 *Sci. Rep.* **7** 45355
[38] Movahedi N and Linul E 2018 *Mater. Lett.* **217** 194-197
[39] Provis JL, van Deventer JSJ 2009 *Geopolymers: structure, processing, properties and industrial applications*, first ed., Woodhead Publishing, Cambridge
[40] Zhang Z, Provis JL, Reid A, Wang H 2014 *Constr. Build. Mater.* **56** 113-127
[41] Zhao Y, Jow J, Cai X, Lai SY 2015 *World of Coal Ash (WOCA) Conference* in Nasvhille, TN-May 5-7
[42] Sanjayan JG, Nazari A, Chen L, Nguyen GH 2015 *Constr. Build. Mater.* **79** 236-244
[43] Skvara F, Sulc R, Tisler Z et al. 2014 *Ceram.-Silikaty* **58** 188-197
[44] Linul E, Marsavina L, Kovacik J, Sadowski T 2017 *Proceedings of Romanian Academy-Series A* **18(4)** 361-369
[45] Linul E, Şerban DA, Marsavina L, SadowskiT 2017 *Archive of Civil and Mechanical Engineering* **17(3)** 457-466
[46] Linul E, Movahedi N, Marsavina L 2018 *Materials* **11(4)** 554
[47] Linul E, Voiconi T, Marsavina L, Sadowski T 2013 *J. Phys. Conf. Ser.* **451** 012002
[48] Marsavina L, Linul E, Voiconi T, Sadowski T 2013 *Polym. Test.* **32** 673-680
[49] Marsavina L, Constantinescu DM, Linul E, Apostol DA, Voiconi T 2014 *Eng. Fract. Mech.* **129** 54-66
[50] Negru R, Marsavina L, Voiconi T et al. 2015 *Theor. Appl. Fract. Mech.* **80** 87-95
[51] Marsavina L, Constantinescu DM, Linul E, Stuparu FA, Apostol DA 2016 *Eng. Fract. Mech.* **167** 68-83
[52] Linul E, Marsavina L, Sadowski T, Kneć M 2012 *Solid State Phenomena* **188** 205-210

Acknowledgment
This work was supported by the ERANet-LAC 2nd Joint Call (http://www.eranet-lac.eu) and funded by National Centre for Research and Development, Poland, under grant: "Development of eco-friendly composite materials based on geopolymer matrix and reinforced with waste fibers" and funded by the Romanian National Authority for Scientific Research and Innovation, CCCDI - UEFISCDI, project number ERANET- LAC-FIBER 17/2017, within PNCDI II.

IOP Conf. Series: Materials Science and Engineering **416** (2018) 012103 doi:10.1088/1757-899X/416/1/012103

Evaluation of the Elastic Properties of Highly Porous Alumina Foams using Finite Element Analysis

D A Şerban[1,*], N Tessier-Doyen[2], J Absi[2], L Marşavina[1] and R Negru[1]

[1]Department of Mechanics and Strength of Materials, Politehnica University of Timisoara, 1 Mihai Viteazu Avenue, 300 222 Timisoara, Romania
[2]Université de Limoges, IRCER, UMR CNRS 7315, 12 Rue Atlantis, 87068 Limoges, France

[*]Corresponding author: dan.serban@upt.ro

Abstract. In this work, a two-dimensional approach is considered in the evaluation of the elastic properties of an alumina foam exhibiting a porosity of around 87%. Preliminary experimental procedures were performed for the determination of the Young's modulus. The stiffness was evaluated with the help of microstructural models consisting of regular structures composed of tessellated patterns of different cell shapes (hexagons, circles, octagons/squares). The experimental data was compared with the numerical results.

1. Introduction

Cellular materials are used in a wide array of applications (aerospace, automotive, terrestrial and naval transportation, biomedicine, civil engineering etc.) due to their particular characteristics: good mechanical properties at reduced specific mass, good thermal and sonic insulation, high energy absorption capabilities [1, 2].

The microstructure of porous ceramics influences their properties: open-cell ceramic foams are used as filters, bioreactors, cancellous bone tissue substitutes etc. while closed-cell ceramic foams have applications in sonic and thermal insulation or as structural components [3].

The prediction of the mechanical properties of cellular structures through finite element analysis represents a useful tool in understanding and designing structural foams [4]. Both two-dimensional [5, 6] and three-dimensional [7] structures were considered in literature for the estimation of various mechanical characteristics.

The first step in the assessment of the mechanical behavior of the designed microstructures is the evaluation of the strength and stiffness in comparison with the experimental data [6]. Further studies can be extended to the studying the structures' damping and dynamic response [8]. The development of accurate microstructures can also aid in the numerical prediction of the material fracture toughness [9].

This study investigates the mechanical characteristics of high-porosity alumina foam, which has applications in various fields such as thermal insulation, biomedical engineering or filtration [10].

The investigated ceramic foam exhibits a closed cell structure and has a porosity of 87%, a density of 465 kg/m^3, the cell size varying between 30 and 250 μm, with cell wall thicknesses of around 5 μm (Figure 1). Three structures (hexagonal lattice, octagonal/quadratic lattice and squared cell lattice with circular orifices) were considered in predicting the mechanical behavior of the aforementioned foam structure.

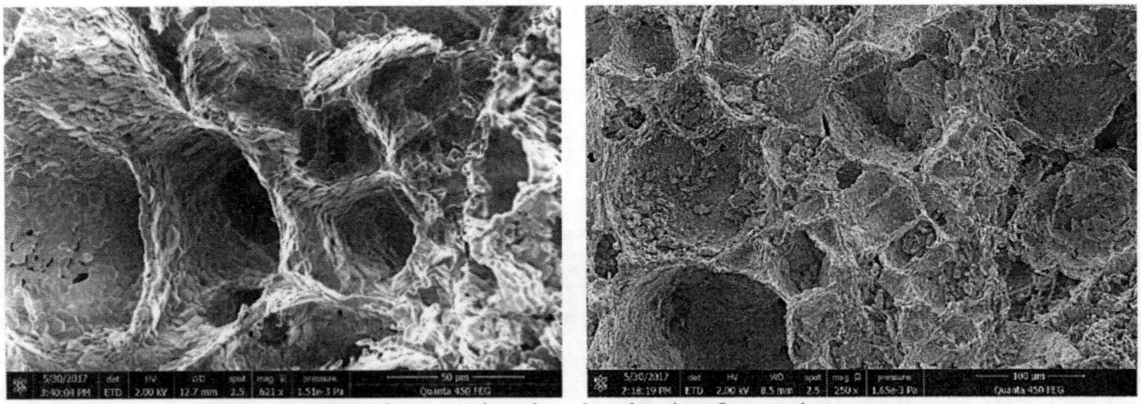

Figure 1. SEM images showing the alumina foam microstructure

2. Experimental results

The ceramic foam's Young's modulus and Poisson ratio were determined using ultrasonic echography tests with infinite medium transmission. 15mm x 13mm x 11mm specimens were cut from bricks provided by the manufacturer.

The principle of ultrasonic echography is based on acoustic wave emission in a material of a given thickness by means of two transducers: one sending longitudinal waves and the other sending transversal waves [11]. The contact between the transducers and the specimen is assured through an intermediary coupling agent. After the emission of the first waves, the values of the transversal (V_T) and longitudinal (V_L) propagation speeds through the thickness of the specimen is calculated through the delay propagation times, determined with the help of an oscilloscope. Knowing the density of the material ρ, the values of the Young's modulus E, shear modulus G and Poisson ratio ν are calculated with the relations:

$$E = \rho \cdot \left(\frac{3V_L^2 - 4V_T^2}{\frac{V_L^2}{V_T^2} - 1} \right) \qquad G = \rho \cdot V_L^2 \qquad \nu = \frac{E}{2G} - 1$$

The ultrasonic echography tests yielded a value of Young's modulus of 13.23 ± 2.62 GPa and a value for the Poisson ratio of 0.24 ± 0.01.

3. Development of structures

The regular structures investigated in this work consist of tessellations of certain cell geometries: hexagonal cells, square cells with octagonal and rectangle orifices (equivalent to an octagon/square lattice) and square cells with circular orifices.

For each structure, several parameters were considered: cell wall length l, cell wall thickness t, chamfer radius r (for the hexagonal lattice and the octagonal/quadratic lattice), cell length L and orifice radii d and δ (for the square cell with circular orifices). Considering the fact that, for regular structures, scaled versions of the same geometry behave identically, the structure variables were chosen as a ratio of the "primary" parameters (t, r, d) with respect to the "secondary" parameters, which determine the size of the cell (cell wall length l for the honeycomb and octagonal/quadratic lattice and cell length L

for the square cell with circular orifices). The geometries of each structure were generated using *SolidWorks*™.

The porosity Π of a material is defined by the ratio of the volume of voids V_{void} to the volume of the solid or the bulk material V_{solid}:

$$\Pi = \frac{V_{void}}{V_{solid}} \; [-]$$

For two-dimensional structures, the thickness can be equaled to the unit, thus, the porosity becomes a ratio of the surface of the voids A_{void} to the surface of the circumscribed figure A_{solid}. In order to determine the porosities of each structure, the CAD software was used in evaluating the surface area of the structure A_{str}, the resulting porosity being calculated with the relation:

$$\Pi = \frac{A_{solid} - A_{str}}{A_{solid}}$$

a) Hexagonal lattice (honeycomb)

The honeycomb geometry represents a tessellation of regular hexagons, presented in Figure 2 for two models (with and without chamfer). The model has two variable parameters: the cell wall thickness to cell wall length ratio t/l and the chamfer radius to cell wall length ratio r/l.

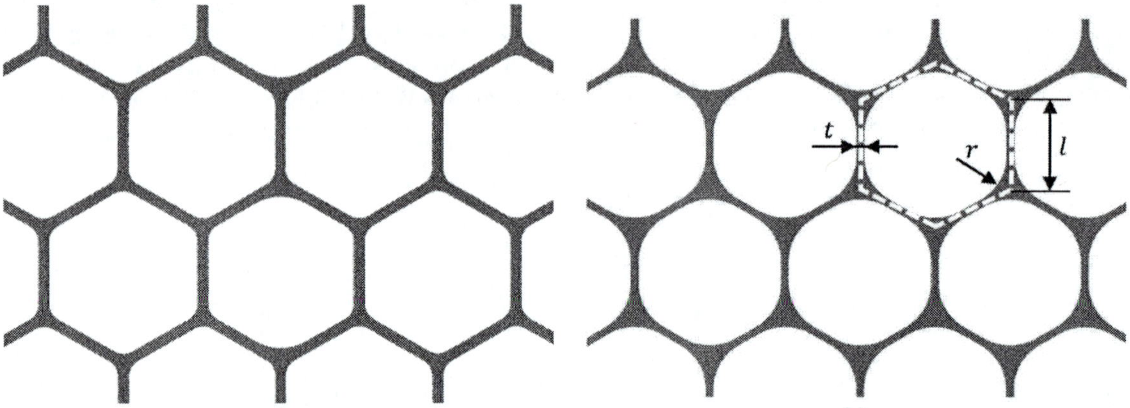

a) b)

Figure 2. Hexagonal lattice with no chamfer (a) and with a chamfer to cell wall length ratio of 0.6 (b)

In order to determine the variation of the porosity of the structure with the variables, the two parameters t/l and r/l were varied in five steps. The values of the variables along with the resulting porosities are presented in Table 1 while the resulting surface is presented in Figure 3.

Table 1. Variation in porosity with t/l and r/l for the hexagonal lattice

	t/l [-]	r/l [-]				
		0	0.1	0.2	0.4	0.6
Π [%]	0.1	0.782	0.781	0.777	0.762	0.738
	0.08	0.824	0.822	0.819	0.804	0.779
	0.06	0.866	0.865	0.861	0.846	0.821
	0.04	0.909	0.908	0.905	0.89	0.865
	0.02	0.954	0.953	0.949	0.942	0.909

A surface fitting equation was performed using the commercial software TableCurve 3D with the resulting date, obtaining in the relation:

$$\Pi\left(\frac{t}{l},\frac{r}{l}\right) = 1.33\left(\frac{t}{l}\right)^2 - 2.31\frac{t}{l} - 0.12\left(\frac{r}{l}\right)^2 + 1$$

Solving the equation for $\Pi = 0.87$ and setting three values for the r/l ratio resulted in the pairs of parameters presented in Table 2.

Table 2. Value of the parameters t/l and r/l for $\Pi = 0.87$ for the hexagonal lattice

Π [%]	r/l [-]	t/l [-]
	0	0.582
0.87	0.2	0.553
	0.6	0.39

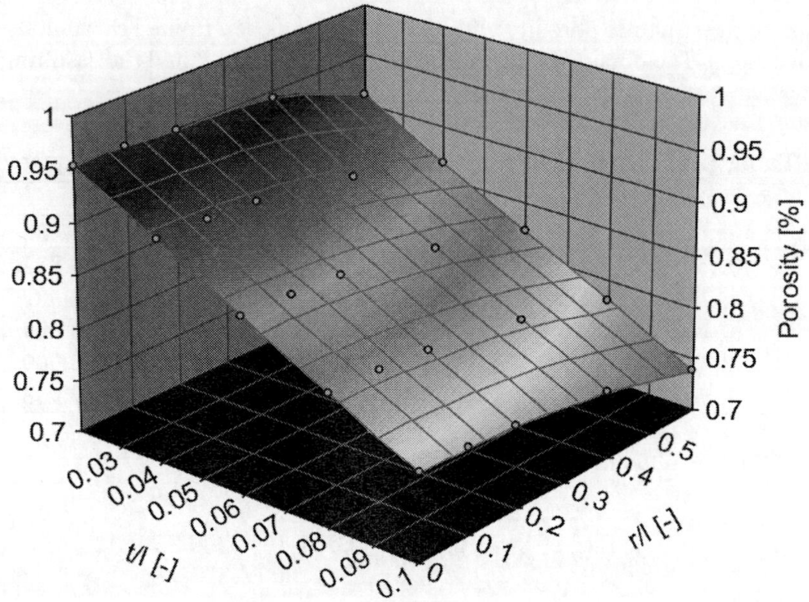

Figure 3. Graphical variation of porosity with t/l and r/l for the hexagonal lattice

b) Octagonal/quadratic lattice

The second investigated structure consists of a tessellated pattern of regular octagons and squares, presented in Figure 4 for models with and without chamfer. Similar to the honeycomb, the octagonal/quadratic lattice has two variable parameters: the cell wall thickness to cell wall length ratio t/l and the chamfer radius to cell wall length ratio r/l.

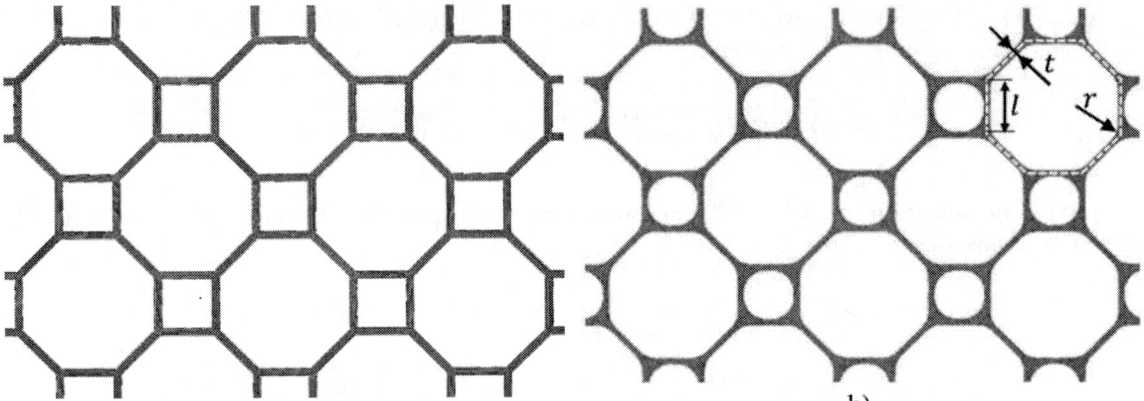

Figure 4. Octagonal/quadratic lattice with no chamfer (a) and with a chamfer to cell wall length ratio of 0.4 (b)

The variation of the structure porosity with t/l and r/l was determined through the variation of each parameter in five steps. The obtained values are presented in Table 3 and the resulting surface in Figure 5.

Table 3. Variation in porosity with t/l and r/l for the octagon/square lattice

	t/l [-]	r/l [-]				
		0	0.1	0.2	0.3	0.4
Π [%]	0.15	0.853	0.851	0.845	0.837	0.824
	0.125	0.876	0.874	0.869	0.86	0.848
	0.1	0.9	0.898	0.893	0.884	0.872
	0.075	0.924	0.923	0.917	0.909	0.896
	0.05	0.95	0.947	0.942	0.933	0.921

The surface fitting resulted in the equation:

$$\Pi\left(\frac{t}{l},\frac{r}{l}\right) = -0.97\frac{t}{l} - 0.001\left(\frac{r}{l}\right)^2 + 0.99$$

Solving the equation for $\Pi = 0.87$ and setting three values for the r/l ratio resulted in the pairs of parameters presented in Table 4.

Table 4. Value of the parameters t/l and r/l for $\Pi = 0.87$ for the octagon/square lattice

Π [%]	r/l [-]	t/l [-]
	0	0.131
0.87	0.2	0.124
	0.4	0.102

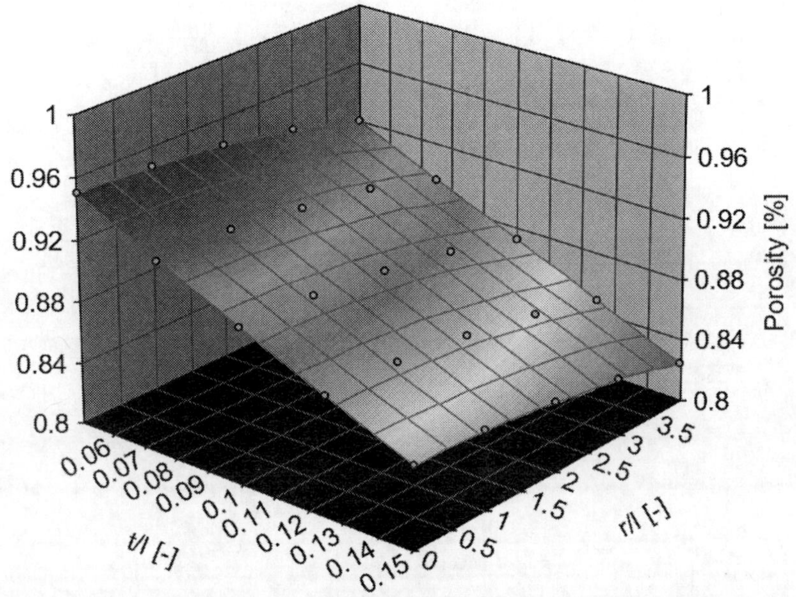

Figure 5. Graphical variation of porosity with t/l and r/l for the octagon/square lattice

c) Square cell with circular orifices

The third regular structure consists of a tessellation of a square cell with circular orifices, presented in Figure 6. In order to obtain high levels or porosity, two circle diameters were chosen. To simplify the generation of the models, a fixed ratio of 0.3 was set between the two diameters. Thus, the only variable for this geometry was the ratio between the cell length L and the diameter of the large circles d.

The variation in structure porosity with d/L was determined by varying the parameter in five steps. The obtained values are presented in Table 5 and the resulting curve in Figure 7.

Table 5. Variation in porosity with d/L

d/L [-]	4.9	4.8	4.7	4.6	4.5
Π [%]	0.8969	0.8607	0.8252	0.7905	0.7564

The equation that describes the variation is:

$$\Pi\left(\frac{d}{L}\right) = 3.57\left(\frac{d}{L}\right)^2 + 0.15\frac{d}{L} - 0.036$$

Solving the equation for $\Pi = 0.87$ resulted in a d to L ratio of 0.482.

IOP Conf. Series: Materials Science and Engineering **416** (2018) 012103 doi:10.1088/1757-899X/416/1/012103

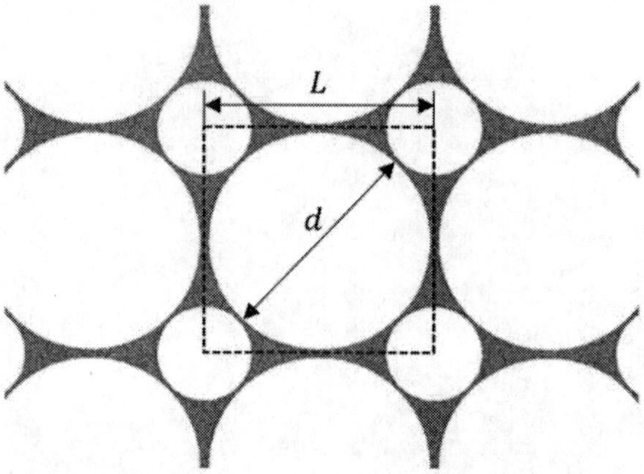

Figure 6. Square cell/circular orifice lattice

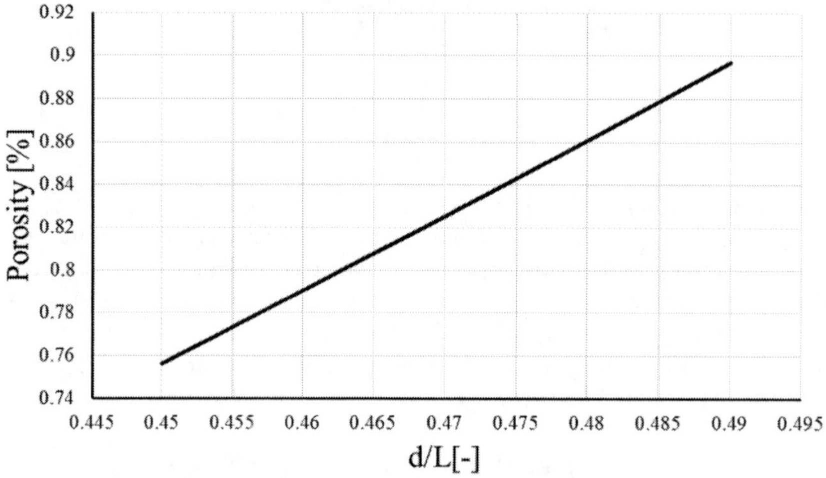

Figure 7. Graphical variation of porosity with d/L

4. Numerical results

The structures described in the previous paragraphs (with the corresponding parameters for $\Pi = 0.87$) were analysed in compression using the commercial software *Abaqus*™. The material model for the solid was chosen as isotropic linear elastic, with a Young's modulus of 350 GPa and a Poisson ratio of 0.24 (parameters for solid Alumina) [9]. The mesh was generated using second order quadratic elements in plane stress formulation (CPS8). The average size of the element was chosen so that a minimum of three elements per strut thickness would be obtained. As boundary conditions, an x-axis symmetry was imposed on the vertical edges of the model, a y-axis symmetry on the bottom edge and a displacement along the y-axis on the top edge.

The size effect was investigated using three models: one with 4x4 cells, one with 8x8 cells and one with 16x16 cells. The results, presented in Table 6, show very little variation between resulting stiffness.

Table 6. Relative error for stiffness results in size effect study

Number of cells [-]	Relative error [%]
16	0.045
64	0.009
256	0

The simulation results for a 4x4 lattice for each structure, along with the experimental results is presented in Table 7.

Table 7. Simulation results and comparison with the experimental data

	t/l [-]	r/l [-]	E Simulation [MPa]	E Experimental [MPa]	Error [%]
Hexagon	0.582	0	12566.64		-5.014
	0.553	2	12600		-4.762
	0.39	6	11625.41		-12.13
	t/l [-]	r/l [-]	E [MPa]		
Octagon with squares	0.131	0	11170.62	13230	-15.56
	0.124	2	11077.51		-16.27
	0.102	4	9897.95		-25.18
	d/L [-]		E [MPa]		
Square cell with circular orifices	0.482		9306.92		-29.65

5. Discussions and conclusion

In this study, the prediction of the elastic properties of high-porosity alumina foams was investigated for three different parametric geometries: hexagonal lattice, octagonal/quadratic lattice and squared cell lattice with circular orifices. For all the studied parameters, the stiffness resulted from the numerical analyses is lower than the value determined experimentally, with errors ranging from around 5% to 30%.

The structure that yielded the most accurate results was the one based on the hexagonal lattice. It was also observed that the lowest errors were determined for low values of the fillet radius to strut length ratio r/l, a value of 0.2 resulting in an error of -4.76%.

6. Bibliography

[1] Gibson L and Ashby M 1997 Cellular solids. Structure and properties, Cambridge University Press

[2] Şerban D, Weissenborn O, Geller S, Marşavina L and Gude M 2016 "Evaluation of the mechanical and morphological properties of long fibre reinforced polyurethane rigid foams," *Polymer Testing* **49** 121–127

[3] Sciamanna V, Nait-Ali B and Gonon M 2015 "Mechanical properties and thermal conductivity of porous alumina ceramics obtained from particle stabilized foams," *Ceramics International* **41** 2599-2606

[4] Daxner T 2010 "Finite Element Modeling of Cellular Materials," in *Cellular and Porous Materials in Structures and Processes*, H. Altenbach and A. Oechsner, Eds., Springer Link, 47-106

[5] Ren X and Silberschmidt V 2008 "Numerical modelling of low-density cellular materials," *Computational Materials Science* **43** 65–74

[6] Şerban D, Linul E, Voiconi T, Marsavina L and Modler N 2015 "Numerical evaluation of two-dimensional micromechanical structures of anisotropic cellular materials: case study for polyurethane rigid foams," *Iranian Polymer Journal* **24** 515-529

[7] Şerban D, Linul E, Sărăndan S and Marsavina L 2016 "Development of parametric Kelvin structures with closed cells," *Solid State Phenomena* **254** 49-54

[8] Şerban D, Voiconi T, Linul E, Marşavina L and Modler N 2015 "Viscoelastic Properties of PUR Foams: Impact excitation and dynamic mechanical analysis," *Materiale Plastice* **52** 537–541

[9] Linul E, Şerban D and Marşavina L 2018 "Influence of Cell Topology on Mode I Fracture Toughness of Cellular Structures," *Physical Mesomechanics* **21** 178-186

[10] Studart A, Gonzenbach U, Tervoort E and Gauckler L 2006 "Processing Routes to Macroporous Ceramics: A Review," *Journal of the American Ceramic Society* **89** 1771-1789

[11] Tessier-Doyen N, Glandus M, Huger M 2007 "Experimental and numerical study of elastic behavior of heterogeneous model materials with spherical inclusions", *Journal of Materials Science* **42**(14) 5826 – 5834

IOP Conf. Series: Materials Science and Engineering **416** (2018) 012104 doi:10.1088/1757-899X/416/1/012104

Cracking Risk Analysis of Mass Concrete Structures Produced at Constant and Changing Ambient Temperature

E Domonyi*

Óbuda University, Bánki Donát Faculty of Mechanical and Safety Engineering, Hungary

*Corresponding author: domonyi.erzsebet@bgk.uni-obuda.hu

Abstract. The low-heat cured cement-based mass concrete structures need to satisfy some special requirements to avoid crack forming. In this study, the influence of constant and changing ambient temperature on the crack formation is investigated and compared using a finite-element based numerical analysis.

1. Introduction

A real problem is that the quality and the cracking sensitivity of mass concrete structures depend on the changing ambient temperatures and the relative ambient humidity determined by the actual weather conditions. The risk of cracking in concrete structures can be important during summer concreting when thermal stresses are induced due to the hydration heat generation. In the case of winter concreting the low ambient temperature has a decisive influence on the temperature differences developed between the surface and the core of concrete parts. At the beginning of the solidification process in mass concretes, the early-age phase is characterized by an exothermic chemical reaction. During this reaction, the temperature and the volume modification can be observed. These phenomena determine the stress conditions forming between the surface and the core of concretes. Cracking initiates once stresses developed exceed the tensile stress capacity of the concrete.

As it is known, the difference between the surface temperature and the highest temperature of the core influences strongly the risk of crack formation. According to the Eurocode used in the engineering design of concrete parts, the maximum temperature difference is limited, and it cannot exceed 20 °C. Because the design of high performance concrete mixtures is a complex procedure, the technologist should be taken into consideration simultaneously the heat generation, the material properties, and the curing parameters, such as curing temperatures and durations.

During hydration, cement reacts with water and heat is generated. The curing temperature affects strongly the progress of hydration: when the curing temperature is increased, the cement hydrates more rapidly [1]. The amount of heat generated is a function of time and temperature.

Qingbin Li [2] et al. produced concrete specimens in summer and in winter and examined the change of mechanical parameters. It has been observed that the mechanical properties of concrete poured in different time are strongly affected by the external weather conditions (different temperatures). More exactly, fracture parameters are influenced by the coupling effect between temperature and age. According to the fracture test results, they concluded that the fracture properties of concrete poured in summer were better than that of concrete poured in winter at the same age.

Zhou Yunchuan [3] et al. used different curing methods for studying the cracking risk reduction. They applied finite element analysis to describe the solidification process. According to experiments performed at constant ambient temperatures they concluded that the cracking risk depends on the temperature difference forming between the surface and the core of the concrete. The authors recommended the use of a curing procedure to obtain an optimised body temperature field.

Hordijk and Reinhardt [4] performed experiments to investigate the effect of curing conditions on the fracture parameters of concretes. The results obtained indicate that the fracture characteristics are influenced by the applied curing conditions.

Kim et al. [5] studied the age effects, their results demonstrate that the fracture parameters of concretes can be improved with age.

In what follows, to simulate the temperature field in mass concrete structures, experiments by using a finite element software are presented. Based on results obtained the impact of the daily temperature change on the cracking risk in concrete structures is demonstrated.

2. Material and method

As it is known, the composition of the cement belongs to the major contributors to heat hydration. The examined material was a usual concrete with density 2400 kg/m^3, it has been made using CEM III B 32,5 N cement. Figure 1 illustrates the heat generation in the concrete as a function of time. The diagram has been determined by calorimetry.

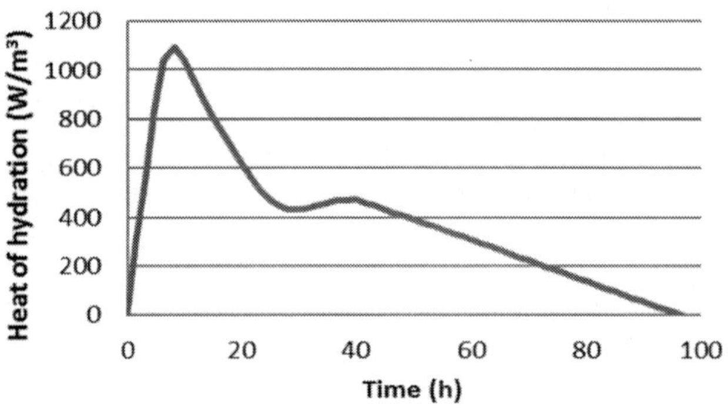

Figure 1. Heat generation curve of the concrete as a function of time

The thermal properties of the material investigated are summarized in Table 1. These parameters values are considered to be constant during the simulation.

Table 1. Material properties of the concrete

property	value
specific heat	1228 J/(kgK)
isotropic thermal conductivity	3.5 W/(mK)
temperature of fresh concrete	20 °C

As usual in the finite element calculations, the geometry of the specimen investigated was represented by a regular cube with size 1m x 1m x 1m [6]. The non-linear finite element model is based on the Fourier law including the internal heat generation due to the exothermic hydration reaction.

Concerning the boundary conditions, in the simulation experiments two cases have been investigated: In the first case it was supposed that the surface temperature is maintained at a uniform constant temperature (at a constant ambient temperature T_S= 20 °C), in the second case we assumed the surface temperature is changing periodically in the interval (15 °C - 25 °C). In this latter case, the changing ambient temperature T(t) was defined by equation (1).

$$T(t) = T_{amp} \sin\left(\frac{2\pi}{24} t\right) + T_{avg} \tag{1}$$

where

- T_{amp} is the amplitude of the temperature modification in °C,
- t is the time in hour,
- T_{avg} is the daily average temperature in °C.

The periodical temperature function (1) is depicted in Figure 2.

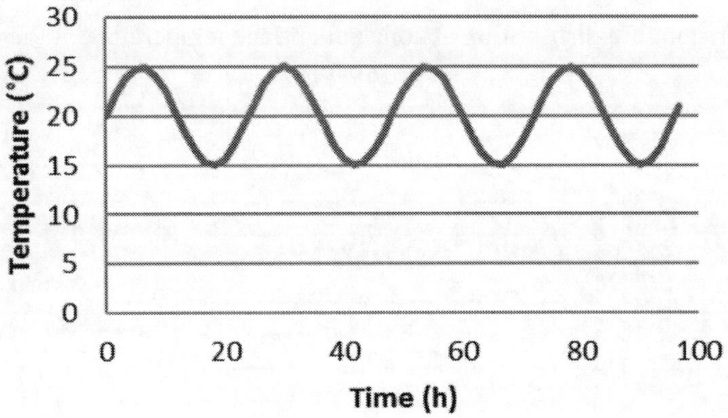

Figure 2. The changing surface temperature defined by Eq.(1)

In the model the heat transfer coefficient was considered to be constant: 5 W/(m²K). The simple specimen geometry and the symmetric heat transfer conditions (initial and boundary conditions) enables to take into account only the one-eighth of the unit cube. (See Figure 3.)

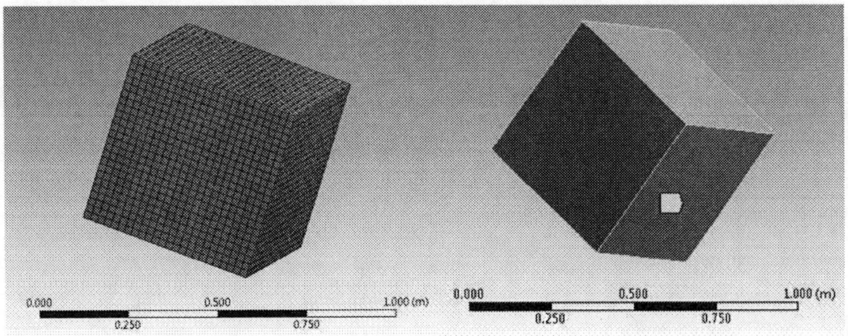

Figure 3. The selected meshing and the convection definition (with yellow)

3. Results

The results obtained by finite element analysis are demonstrated in Figures 4-6. As can be seen in Figure 4, the minimum values of temperature can be observed at the corners of the cube, while central point represents the maximum temperature.

IOP Conf. Series: Materials Science and Engineering **416** (2018) 012104 doi:10.1088/1757-899X/416/1/012104

According to diagrams depicted in Figures 5 and 6, due to the hydration heat the minimum and maximum temperatures increase simultaneously.

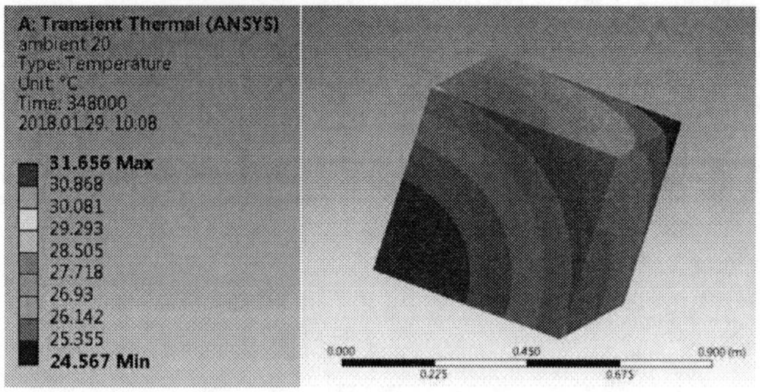

Figure 4. The temperature distribution at constant surface temperature at the end of the hydration process

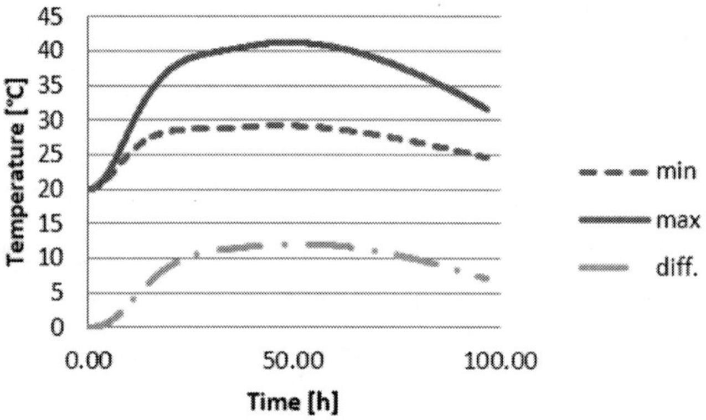

Figure 5. The variation of temperature in the cube as function of time in the case of constant surface temperature

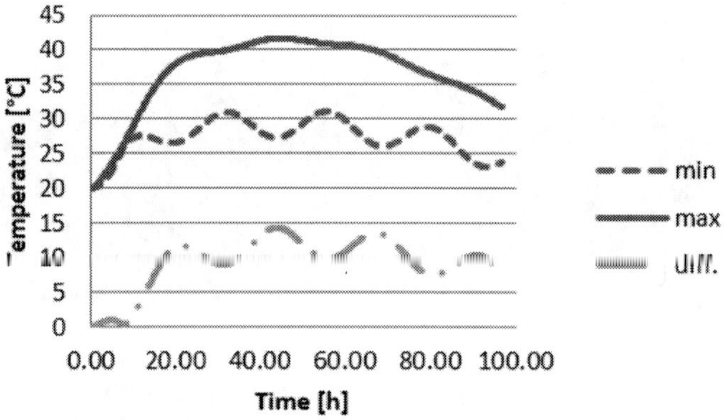

Figure 6. The variation of temperature in the cube as function of time in the case of periodically changing surface temperature

The computed results are given in Table 2.

Table 2. The minimum and the maximum temperatures, the maximum temperature-difference ΔT, and the time t_{max} when the temperature-difference obtains its maximum value

	Tmin [°C]	Tmax [°C]	ΔT [°C]	t_{max} [h]
T(t)= const	29.24	41.30	12.11	51.2
T(t)≠const	31.05	41.65	14.36	44.07

Based on the results of the finite element analysis, it can be concluded that the minimum and the maximum core temperatures are different in the two cases of the selected boundary conditions.

If the surface temperature is constant, lower minimum and maximum temperatures are observed in the body. In the case of periodically changing surface temperature, the difference between the minimum and maximum temperatures in the body is larger. In this latter case it can be observed that there is a shift in times t_{max} characterizing the maximum temperature differences (See values 51.2 h and 44.07 h). It should be noted that the maximum value of heat generation is detected at 8 hours, as it is indicated in Figure 1.

For comparison, the results obtained are demonstrated in Figure 7.

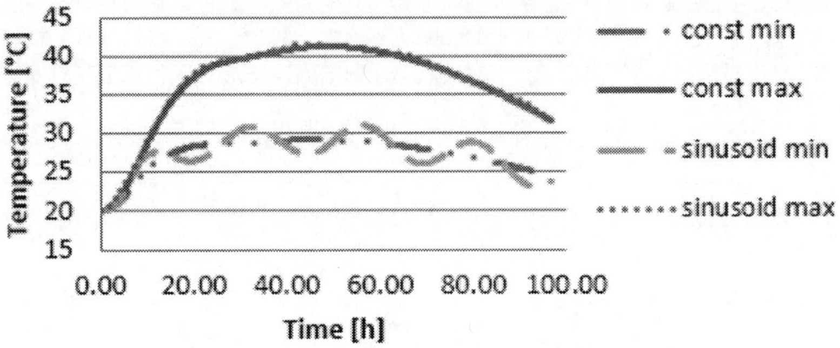

Figure7. Comparison of computed results

From diagram depicted in Figure 7, it can be observed that the maximum temperature difference ΔT appears at time 44 h in the case of constant surface temperature, while the maximum ΔT temperature is reached at time 51.2 h when the surface temperature is periodically changing.

4. Conclusions

The control and the reduction of hydration heat in concrete structures belong to the technological problems of primary importance. Based on the application of a simplified finite-element based model, from our preliminary results it can be concluded that the cracking sensibility can be decreased by an appropriate selecting of the starting time of concreting and taking into consideration the actual weather conditions.

5. References

[1] J. L. Meadows, 2007 Early-age cracking of mass concrete structures, Thesis, Auburn University, Auburn, Albama

[2] Qingbin Li, Junfeng Guan, Zhimin Wu, Wei Dong, Shaowu Zhou, 2016 Equivalent maturity for ambient temperature effect on fracture parameters of site-casting dam concrete, *Construction and Building Materials* **120** pp 293-308

[3] Zhou Yunchuan, Bai Liang, Yang Shengyuan, Chen Guting, 2012 Simulation Analysis of Mass Concrete Temperature Field *Procedia Earth and Planetary Science* **5** pp 5-12

[4] D. A. Hordijk, H. W. Reinhardt, 1990 Fracture of concrete in uniaxial tensile experiments as influenced by curing conditions, *Engineering Fracture Mechanics,* **35** pp 819-826
[5] J. K. Kim, Y. Lee, S. T. Yi, 2004 Fracture characteristics of concrete at early ages, 2004 *Cement and Concrete Research,* **34** pp 507-519
[6] Domonyi Erzsébet, Csizmadia Béla, Telekes Gábor 2016 Tömegbetonok repedési hajlamának elemzése a hőmérséklet és a hőátadási tényező függvényében, *GÉP LXVII* (5-6), pp 44-47

Acknowledgments
This article supported by the Únkp-17- 3 New National Excellence Program of the Ministry of Human Capacities

IOP Conf. Series: Materials Science and Engineering **416** (2018) 012105 doi:10.1088/1757-899X/416/1/012105

Effect of Nano Admixture of CSH on Selected Strength Parameters of Concrete Including Fly Ash

B Szostak[1,*] and G L Golewski[2]

[1] MSc Eng., Lublin University of Technology, Lublin, Poland
[2] Assoc. Prof., Ph.D., D.Sc. Eng., Lublin University of Technology,Lublin, Poland

[*]Corresponding author: b.szostak@pollub.pl

Abstract. Nowadays, various mineral additives and chemical admixtures are more and more frequently used in concrete mixtures. It is done in order to improve specific parameters of concrete. One of the most frequently used additives are siliceous fly ash (FA). These material are by-products of hard coal combustion processes. Utilization of FA has an ecological aspect (due to reducing the deposit areas of troublesome industrial wastes) as well as the economic aspect (a possibility to replace part of cement binder in the concrete mixture). Research, that carried out so far indicates a significant decrease in strength parameters of concrete with the addition of FA in early ages of curing. On the other hand, numerous experiments prove that using Calcium Silcate Hydrate (CSH) seeds causes an increase in mechanical features of composites with cement matrix. Therefore, it seems possible to strengthen the weakened structure of concrete containing FA through the use of a mixture of active modifiers included in the admixture. Based on the above, the authors of the study decided to carry out laboratory tests on concrete mixtures with the addition of FA and nano admixture of CSH. This study presents the results of tests on concrete with the addition of FA and CSH nanocrystals used as an admixture. Mixtures with 0 and 20% of FA and 4% admixture of CSH were prepared. All experiments were carried out after: 4, 8, 12, 24, 72h, and 7,28 days of curing.

1. Introduction

In recent years, the area of nanotechnology in the field of concrete technology has been developed considerably, e.g., [1-6]. In order to improve the physical and mechanical parameters of concretes, a modification occurs not only on the macro or micro scale but also on the nano scale.

The necessity of producing concretes with new and improved parameters is caused by a huge amount of investments in the area of construction and a fast rate of their growth. Currently, prefabricated elements are used more and more often during buildings' erection. This solution has many advantages, including:

- significant shortening of technological breaks,
- a limitation of the "wet processes" performed on the construction site,
- independence from weather conditions,
- the possibility of obtaining concrete composites with better parameters in the production process of precast products,

- the ability to reduce production costs as a result of a careful care of the concrete mixture, and the use of mineral additives and chemical admixtures.

The basic binder of the cementitious concrete, i.e., Portland cement, is replaced with various mineral additives, e.g., [7-10]. Such treatments are referred to as the green concrete technology in civil engineering, e.g., [11-13]. One of the most popular cement substitutes are siliceous fly ash (FA). These materials are industrial wastes produced as a result of the combustion of hard coal in power plants or combined heat and power plants, and they are safe from the radiological point of view [14]. Nevertheless, the storage of FA is costly and burdensome for the natural environment, so the use of this waste is important both from the economic [15] and ecological [16] point of view. In addition, FA used in an amount of up to 20% of cement mass significantly improve the mechanical parameters of concrete, including their fracture toughness after 28 days and in longer periods of curing [11, 17-21]. The disadvantage of this solution, however, is the reduction of an early concrete strength, e.g., [22], which plays a key role in prefabrication.

Due to the appropriate strength in the first hours of maturation and the faster, than traditional, hardening process of the concrete mixture, the efficiency of the prefabrication plant increases considerably. For this purpose, manufacturers decide on technological treatments such as autoclaving, low pressure steam curing, electro-heating, and/or using chemical admixtures.

The paper presents test results presenting an attempt to accelerate the binding processes in a concrete mixture with the FA addition, using a chemical admixture – the Calcium Silicate Hydrate (CSH). This admixture is an aqueous suspension of nucleuses in the form of nanoparticles. Thanks to nucleation, the growth of CSH crystals is definitely accelerated. The use of nucleuses as activators leads to the construction of a microstructure with improved physical and mechanical properties, e.g., [23,24]. This means increasing the initial strength and the possibility of earlier demoulding and transporting of prefabricates.

2. Experimental program

2.1. Scope of the studies

The tests were carried out for mixtures with the FA addition and the CSH nano-admixture. Concretes' compressive strength (fcm) was tested after: 4, 8, 12, 24 and 72 hours as well as 7 and 28 days of curing. All tests were carried out in accordance with the EN 12390 series standards.

2.2. Materials

The following materials were used to prepare the cubic test samples:

- CEM I 32.5R Portland cement from Ożarów - OPC
- Siliceous fly ash - from Puławy, combined heat and power plant
- CSH nanoparticles - Master X-SEED 100
- Water - from the municipal water supply
- Sandof 0 – 2 mmfraction - from the sand mine inNiemce
- Gravel of 0 – 8 mm fraction - from the Rokitno deposit
- Plasticizer – Stacheplast 125
- Superplasticizer – MasterGlenium ACE 430

2.3. Methodology

The tests were carried out for the following mixtures:

- 0% FA + 0% CSH (FA-0+CSH-0),
- 0% FA + 4% CSH (FA-0+CSH-4),
- 20% FA + 0% CSH (FA-20+CSH-0),

- 20% FA + 4% CSH (FA-20+CSH-4).

The composition of the mixtures tested and their designations are presented in Table 1. They were selected in such a way as to be able to compare the CSH modified concretes (FA-0+CSH-4, FA-20+CSH-4) with reference concrete (FA-0+CSH-0, FA-20+CSH-0).

Table 1. Details of concrete mixtures (kg/m^3)

Constituent	Mixtures			
	FA-0 CSH-0	FA-0 CSH-4	FA-20 CSH-0	FA-20 CSH-4
Opc	352	352	282	282
Siliceous Fly Ash	0	0	70	70
Csh Nanoparticles	0	14	0	14
Water	141	127	141	127
SAND OF 0 – 2 Mm FRACTION	676	676	676	676
GRAVEL OF 0 – 8 Mm FRACTION	1205	1205	1205	1205
Plasticizer	2	–	2	–
Superplasticizer	-	2	-	2

When composing concrete mixtures, it was assumed that in each case a constant value of the water-binder ratio will be maintained – w/b = 0.4. The use of this proportion of water to the binder reduces the pore content in the cement matrix structure, which reduces the subsequent impact of the cement matrix on the damage processes in the concrete.

Figure 1. Preparing cubic samples

42 cubic samples with 150 mm side (6 samples for each maturing period) were made for all types of the mixtures. After 15 minutes from mixing the ingredients, a consistency test was carried out for each mixture according to the Vebe method, in accordance with [25]. All mixtures were made with the V2 consistency class. The highest Vebe time in seconds was observed for the FA-0+CSH-0 mixture (17 s.), while the lowest for the FA-20+CSH-4 mixture (13 s.).

Compressive strength tests were carried out in the laboratory on the Controls Advantest 9 hydraulic press. The load increase was set at a level of 0.5 MPa/s.

IOP Conf. Series: Materials Science and Engineering **416** (2018) 012105 doi:10.1088/1757-899X/416/1/012105

Figure 2. Test stand – Control Advantest 9

Figure 3. Specimen during the test.

All samples, tested up to 24 hours, were demoulded immediately before the test. Samples tested after 3, 7 and 28 days were demoulded after 24 hours and then cured in water at 20°C. The cubes were removed from the water one hour before the test. For the study after 28 days, the samples matured in a full immersion for 14 days.

3. Research results and their analysis

Fig. 4 shows the average strength f_{cm} of concretes from each series determined on the basis of results from 6 samples tested.

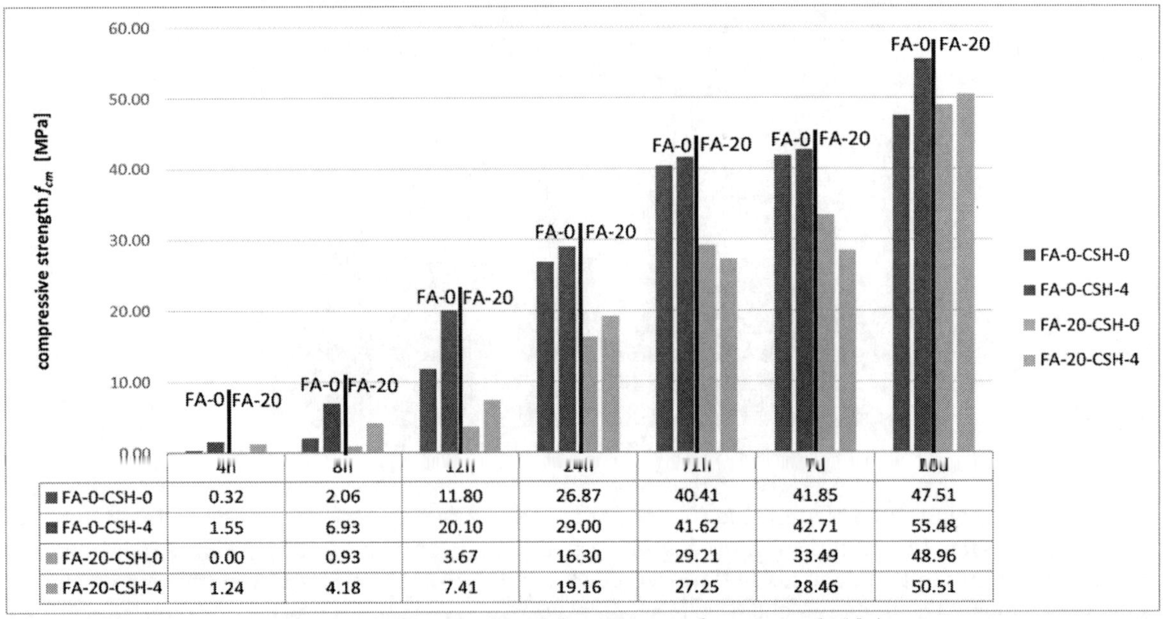

	4h	8h	12h	24h	72h	7d	28d
FA-0-CSH-0	0.32	2.06	11.80	26.87	40.41	41.85	47.51
FA-0-CSH-4	1.55	6.93	20.10	29.00	41.62	42.71	55.48
FA-20-CSH-0	0.00	0.93	3.67	16.30	29.21	33.49	48.96
FA-20-CSH-4	1.24	4.18	7.41	19.16	27.25	28.46	50.51

Figure 4. The results of the compressive strength (f_{cm})

On the basis of the tests conducted, a significant increase in the strength of concrete in the early curing periods (up to 24 hours) and after 28 days was noted. After 4 hours in the reference concretes, without FA additive, the admixture caused over 4 times the strength increase. In concretes with FA (FA-20-CSH-0), without the admixture, it was not possible to demould the mixture after 4 hours. After 8 hours, the increase in strength was over 3 times in both sets compared. With each subsequent study, the differences in the strength of the reference mixtures and mixtures with the CSH were significantly reduced. After 24 hours from the execution of the mixtures, the strength parameters of the mixtures compared were very similar to each other. However, after 3 and 7 days in the case of concretes with a 20% addition of FA (FA-20) the f_{cm}parameter was larger in the reference mixture, without the admixture used. After 28 days, an increase in the concrete compressive strength was observed for all concretes with the CSH nano-admixture. The concretes modified were characterized by higher compressive strength than reference concretes (not modified with admixture).

Based on the tests conducted, it can be concluded that the significant influence of the admixture is visible in the first 24 hours of the mixture's aging. Between 24 hours and 28 days, concretes with the admixture may obtain lower values of mechanical parameters, however their strength after 28 days is similar to the reference concretes and according to [26] the concretes would be marked with the same strength class.

4. Conclusions

After analyzing the results obtained from the research, the following conclusions can be drawn:

1) Modification of concretes in the nano-structure due to admixtures is a favorable solution during the production of prefabricated elements.
2) The use of the admixture significantly improves the early parameters of concretes both with and without the addition of FA. Thanks to this, there is a chance to shorten the process of prefabrication of elements and increase its efficiency.
3) Using the admixture, there is a chance to improve the parameters of concrete mixtures with the addition of FA both in early periods (up to 24 hours) and later periods (28 days). In the case of all mixtures compared after 4 hours, the strength increase was over 4 times, and after 8 hours more than 3 times. In research periods of over 8 hours, the impact of the CSH nano-admixture is insignificant, and after 3 days it disappears completely.
4) An increase in the strength of the CSH modified concretes after 28 days in relation to concretes without the nano-admixture was noted. Compressive strength results for the CSH modified mixtures without FA were more than 15% higher, while with the addition of FA increased by nearly 3%.
5) According to the [26] standard, all mixtures after 28 days receive the same class of the concrete strength.

5. References

[1] Zhang P, Wan J, Wang K and Li Q 2017 Influecne of nano-SiO2 on properties of fresh and hardened high performance concrete: A state-of-the-art view *Constr. Build. Mater.* **148** 648–658.

[2] Niewiadomski P., Hola J., Cwirzen A., Study on properties of self-compacting concrete modified with nanoparticles, 2018, *Arch. Civ. Mech. Eng.* **18** 877–86.

[3] Jayapalan A.R., Lee B.Y. Kurtis, K.E., 2013, Can nanotechnology be "green"? Comparing efficacy of nano and microparticles in cementitious materials, 2013, *Cem. Concr. Compos.* **36** 16–24.

[4] Kawashima S., Hou P., Corr J.D., Shah S.P., 2013, Modification of cement-based materials with nanoparticles, *Cem. Concr. Compos.,* **36**, 8–15.

[5] Horszczuk E., Sikora P., Lukowski P., 2016, Application of nanomaterials in production of self-sensing concretes: contemporary developments and prospects, *Arch. Civ. Eng.* **LXII**(3) 61–73.

[6] Szelag M., Mechano-Physical Properties and Microstructure of Carbon Nanotube Reinforced Cement Paste after Thermal Load, 2017, *Nanomaterials 7(9),* **267**.

[7] Giergiczny E., Giergiczny Z., Influence of siliceous fly ash variable quality on the propeties of cement-fly ash composties, 2010, *Cem. Wapno Beton,* **3**, 157–163.

[8] Konkol J., Prokopski G., Fracture toughness and fracture surfaces morphology of metakaolinite-modified concrete, 2016, *Constr. Build. Mater.* **123**, 638–648.

[9] Sosdean C., Marsavina L., De Schutter G., 2016, Experimental and numerical determination of the chloride penetration in cracked mortar specimens, Eur. J. Env. Civ. Eng. **20** (2) 231–249.

[10] Garbacz A., Sokołowska J.J., 2013, Concrete-like polymer composites with fly ashes – Comparative study, *Constr. Build. Mater.* **38** 689-699.

[11] Golewski G L 2018, Green concrete composite incorporating fly ash with high strength and fracture *J. Clean. Prod.* **172** 218–226.

[12] Blaszczynski T., Krol M., 2015, Usage of green concrete technology in civil engineering, *Proc. Eng.* **122** 296–301.

[13] Liew K.M, Sojobi A.O., Zhang L.W., 2017, Green concrete: Prospects and challenges, *Constr. Build. Mater.* **156**, 1063-1095.

[14] Golewski G.L., 2015, Studies of natural radioactivity of concreto with siliceous fly ash addition, *Cem. Wapno Beton*, 106-114.

[15] Alwaeli M., 2011, Economic calcus of the effectiveness of waste utilization processed as substitutes of primary materials, *Env. Prot. Eng.* **37** 51-58.

[16] Alwaeli M., 2013, Application of granulated lead-zine slang in concrete as an opportunity to save natural resources, *Rad. Phys. Chem.* **83** 54-60.

[17] Golewski G L 2017 Improvement of fracture toughness of green concrete as a result of addition of coal fly ash. Characterization of fly ash microstructure, *Mater. Charact.* **134** 335-346.

[18] Golewski G L 2018 Effect of curing time on the fracture toughness of fly ash concrete composites *Compos. Struct.* **185** 105-112.

[19] Golewski G L 2017, Generalized fracture toughness and compressive strength of sustainable concrete including low calcium fly ash, *Materials* **10**(12) 1393.

[20] Sadowski T and Golewski G L 2018 A failure analysis of concrete composites incorporating fly ash during torsional loading *Compos. Struct.* **183** 527-535.

[21] Golewski G L 2017 Effect of fly ash addition on the fracture toughness of plain concrete at third model of fracture J. Civ. Eng. Manag. **23** 613-620.

[22] Golewski G L and Sadowski T 2012 Experimental investigation and numerical modeling fracture processes under Model II in concret composites containing fly-ash additive at early age, Sol Stat. Phenom. **188** 158-163.

[23] BASF – MasterX-SEED 100 – Hardening accelerating admixture for concrete - (https://www.master-builders-solutions.basf.ng/en/products/master-x-seed/535- on 01.02.2018r.).

[24] Antonovic V., Pundiene I., Stonys R., Cesniene J., Keriene J., 2010, A review of the possible applications of nanotechnology in refactory concrete *J. Civ. Eng. Manag.* **16** 595–602.

[25] EN 12390-3:2009 - Testing hardened concrete. Compressive strength of test specimens.

[26] EN 206:2013 + A1:2016 - Concrete - Specification, performance, production and conformity.

IOP Conf. Series: Materials Science and Engineering **416** (2018) 012106 doi:10.1088/1757-899X/416/1/012106

Multi-Scale Numerical Viscoelastic Simulation of Fatigue Behavior of Asphalt Mixtures Modified with Polyphosphoric Acid

M Khodadadi[1,*], A Khodaii[1], P Hajikarimi[1], F Fakhari Tehrani[2,3] and J Absi[4]

[1]Hafez Avenue, Department of Civil & Environmental Engineering, Amirkabir University of Technology, Tehran, Iran

[2]Centre Universitaire de Génie Civil – Université de Limoges, boulevard Jacques Derche, Egletons, France

[3]Conservatoire national des Arts et métiers, Paris, France

[4]University of Limoges, SPCTS, UMR CNRS, 7315, 12 Rue Atlantis, 87068 Limoges, Cedex, France

[*]Corresponding author: m-khodadadi@aut.ac.ir

Abstract. Fatigue cracking is one of the main distresses affects the service life of asphalt mixture. Recently, polyphosphoric acid (PPA) widely used as a modifier to improve mechanical and rheological characteristics of asphalt binders and mixtures especially at high and low service temperatures. While there are several studies devoted to investigate high and low service temperature of PPA modified asphalt binder and mixture, its fatigue behaviour needs to be studied in more detail. The main objective of this paper is to develop a multiscale heterogeneous numerical finite element (FE) model as well as experimental program to investigate the effect of PPA modification on fatigue behaviour of different bituminous composites from neat asphalt binder to asphalt mixture. The experimental program includes frequency sweep test at test temperature of 25°C by implementing dynamic shear rheometer (DSR) for neat and PPA modified asphalt binder and also indirect tensile fatigue (ITF) at frequency of 1 Hz (0.1 s loading time and 0.9 s rest time) under constant stress of 300 kPa at test temperature of 25°C by using a UTM-25kN. Three different contents of PPA is used for modifying neat asphalt binder including 0.5, 1 and 1.5 wt.% of neat binder to consider the variation of modifier extent on fatigue behaviour of neat asphalt binder and mixture. The mixture specimens were prepared with granite aggregate and asphalt binder of 85/100 penetration grade. The asphalt mixture is modeled as a biphasic medium composed of granular inclusions with linear elastic properties and bituminous matrix exhibiting linear viscoelastic behavior at small strain values. The generalized Maxwell rheological model was used to simulate viscoelastic behavior of the asphalt binder and PPA modified asphalt binder. In order to create the aggregate skeleton of the bituminous composites, a custom software (called "MOA", French acronym for Random Object Modeler) developed in University of Limoges, has been used to generate random inclusions of various sizes and shapes. The experimental results of asphalt binder and mixture indicated that modification with PPA has a significant effect on improving asphalt mixture's fatigue life especially in dosage of 1%. On the other hand, results of numerical simulation have shown a good agreement with experimental results.

IOP Conf. Series: Materials Science and Engineering **416** (2018) 012106 doi:10.1088/1757-899X/416/1/012106

1. Introduction

Asphalt mixture as a heterogeneous structure is subjected to numerous loading and unloading conditions during its service life which causes crack distresses in this structure. This material consists of bitumen, aggregate and air void in which aggregate has a wide range of particle size.

Although several studies have been considered asphalt mixture as a humongous material, but the fact is that this assumption makes an error in the behaviour simulation of asphalt mixture, because of its microstructure affection on the mechanical performance. The reason of this difference come out from the aggregate content, orientation and contacts etc. since it is necessary to consider the asphalt mixture as a heterogeneous material in order to investigate its performance and find more reliable response from the simulation [1, 2]. For this purpose, several simulation programs were proposed. One of these programs which developed by University of Limoges is MOA (French acronym for Random Object Modeler) which can generate a random distribution of aggregate particles based on aggregate curve for importing into commercial finite element software like ABAQUS.

It is well known that fatigue cracking is one of the major distresses of asphalt mixture which can be investigated by considering the heterogonous combination. Using modifiers is a common approach for improving asphalt mixture's resistance against the fatigue crack. Polymers such as polyphosphoric acid (PPA), styrene Butadiene Styrene (SBS), PVE etc. are some instances of these modifiers that have been used recently in industry [3-7]. The PPA causes an improvement of high and low temperature performance of bitumen and asphalt mixture. While several studies have been also devoted during the recent decades [8, 9] to the use of this modifier for evaluating the asphalt mixtures distress, its fatigue behaviour needs to be studied in more detail. So, this research focuses on both experimental and numerical approach for evaluating the effect of PPA on fatigue behaviour of asphalt mixture. For this purpose, the fatigue behaviour of asphalt mixture as a viscoelastic heterogeneous material was investigated by performing experimental tests including the dynamic shear rheometer in frequency sweep mode on bitumen and asphalt mastic samples and the indirect tensile fatigue test on asphalt mixture and also applying a multi-scale numerical simulation. In this study, asphalt mixture is modelled as heterogeneous material containing irregular aggregate skeleton bonded with asphalt mastic.

2. Experimental program

In this paper, the fatigue behaviour of bitumen, asphalt mastic and asphalt mixture were studied. The bitumen with penetration grade of 85/100 (PG 58-22) and modified bitumen with PPA (by registry code 8017-16-1 Produced by Merck Company) were used. In this research, three different percentages of PPA including 0.5, 1 and 1.5% (based on weight of asphalt binder) was added to original bitumen. The PPA was blended with original asphalt binder implementing a low shear mixer at 150°C for 45 min. Table 1 depicts conventional properties of PPA used in this study based on producer's data sheet. Also, a well-graded Limestone filler was selected to construct asphalt mastic with volume relative of 18% (V_{filler} /V_{mastic}=18%) according to aggregate curve and particle size distribution. Finally, asphalt mixtures were designed according to ASTM D1559 standard in which the optimum percentage of bitumen and air void was 4.7% and 5%, respectively. Figure 1 depicts aggregate curve of asphalt mixture. Considering the fact that PPA does not affect the optimum bitumen percentage, 4.7% of bitumen was assumed constant for modified asphalt mixtures.

Table 1. Conventional properties of polyphosphoric acid (PPA)

Property Title	Value
Boiling point	530°C (1013 hPa)
Density	2.06 gr/cm^3 at 20°C
Melting Point	-20°C
Vapor pressure	2 hPa at 20°C
Assay (acidimetric, calc. as P_2O_5)	83 – 87%

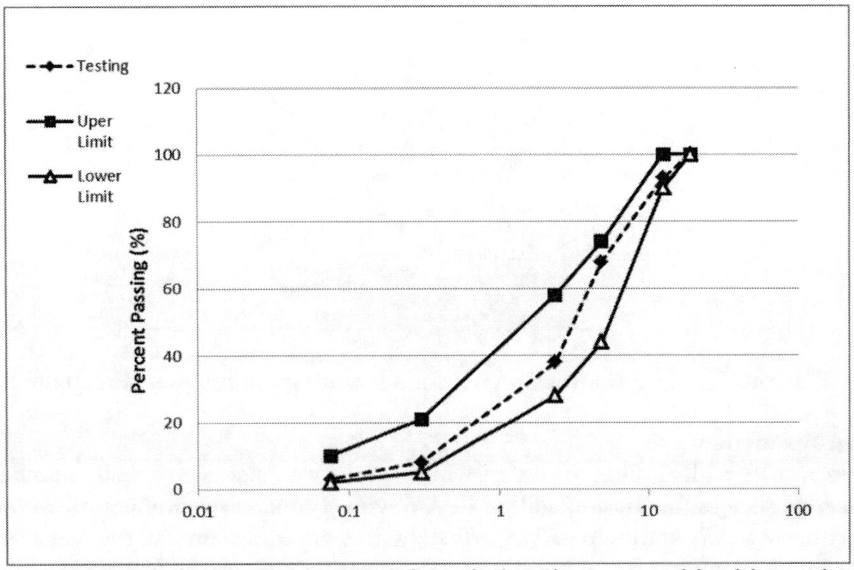

Figure 1. Aggregate Curve of Asphalt Mixture used in this study

The frequency sweep test at 25°C was conducted for bitumen and asphalt mastic to determine the rheological properties and also indirect tensile fatigue (ITF) at frequency of 1 Hz under constant stress of 300 kPa at 25°C was performed on asphalt mixture.

3. Methodology

The commercial software for numerical simulation is unable to develop randomly distributed particles. To overcome such a shortcoming, original software called MOA was written in C++ language based on the Delaunay triangulation method. This software can generate virtual models containing a matrix with a prearranged percentage of pores (volume filling rate) and given number of inclusions with different shapes including circle and polygon in 2D. It should be noted that MOA is capable of automatically generating samples derived from a matrix with a given percentage of air voids and containing a known number of inclusions featuring different shapes, extending from spherical to polyhedral and randomly distributed in the volume defined by the matrix in the form of a rectangular prism or cylinder.

In this research, the MOA is used to randomly generate aggregate inclusions and build up heterogeneous numerical model using the four following steps: 1) generating the main cadre of numerical model as a circle in two-dimensional domain to simulate ITF test, 2) generating inclusions arrangement with respect to gradation curve considering coarse and fine aggregate particles, 3) applying the Boolean operation to create the matrix and 4) assembling generated matrix and aggregate inclusions into one model by using function "Tied" in the ABAQUS software to tie the surfaces of aggregates and asphalt binder together. Figure 2 shows an example of two-dimensional models developed using MOA and ABAQUS.

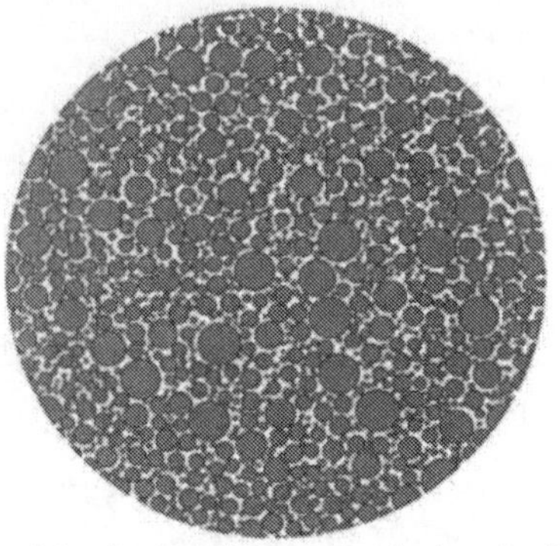

Figure 2. A 2D cylindrical model for simulating indirect tensile fatigue test

4. Results and Discussion

Figure 3 shows results of IDT fatigue test in which strain was sketched versus number of cycles to failure. As it can be seen in this figure, adding PPA obviously increases number of cycles to failure for all asphalt mixtures which shows positive effect of this modifier on fatigue behaviour of asphalt mixtures at moderate temperatures. According to these results, using 1% of PPA has more significant effect on improving the fatigue performance of asphalt mixture even in comparison with dosages of 1.5% of PPA. Therefore, 1% was obtained as the optimum dosage of PPA. Also using the dosage of 2% of PPA caused an extreme stiffness in bitumen during sample preparation process which makes some problems such as increasing in mixing temperature and compaction temperature of asphalt mixture. since, the usage amount of 2% was removed from selections.

In order to verify numerical model which is proposed for multi-scale approach, the loading and boundary conditions of experimental test was applied to numerical model which was developed for asphalt mixture. Constructing such a numerical model was done by using three following procedures:

1) Constructing asphalt mastic: considering asphalt binder as matrix and combining filler particles with binder as inclusions based on volumetric ratio of filler particles,

2) Constructing asphalt mortar: considering asphalt mastic as matrix and combining fine particles (< 1 mm) with asphalt mastic as inclusions based on volumetric ratio of fine aggregate particles,

3) Constructing asphalt mixture: considering asphalt mortar as matrix and combining coarse particles (> 1 mm) with asphalt mortar as inclusions based on volumetric ratio of coarse aggregate particles.

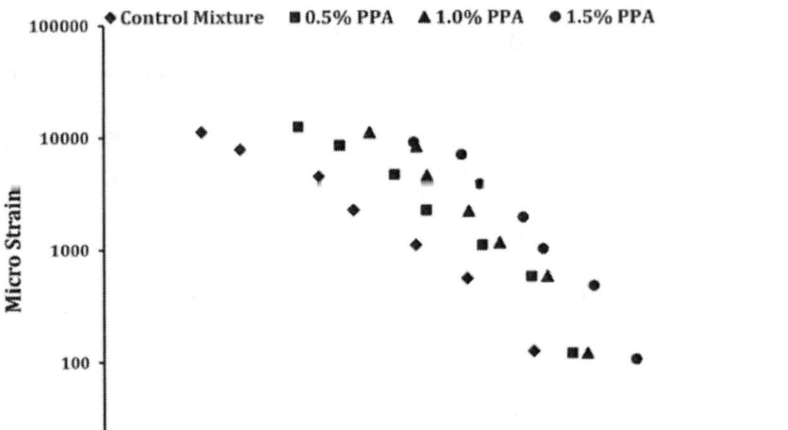

Figure 3. Strain versus number of cycles to failure for original and modified asphalt binders

To start numerical modelling with original and PPA modified asphalt binders, a generalized Maxwell model was used which is depicted in Figure 4 by implementing the Eq. (1) as follows:

$$G(t) = G_\infty + \sum_{i=1}^{m} G_i \exp(-t/\tau_i) \tag{1}$$

In which, G_∞ and G_i are the equilibrium modulus and the shear relaxation strength, respectively and τ_i is relaxation time. Also, m is number of the generalized Maxwell model element. In this study, the number of components of the generalized Maxwell model is selected equals to 22 to provide accurate material characterization.

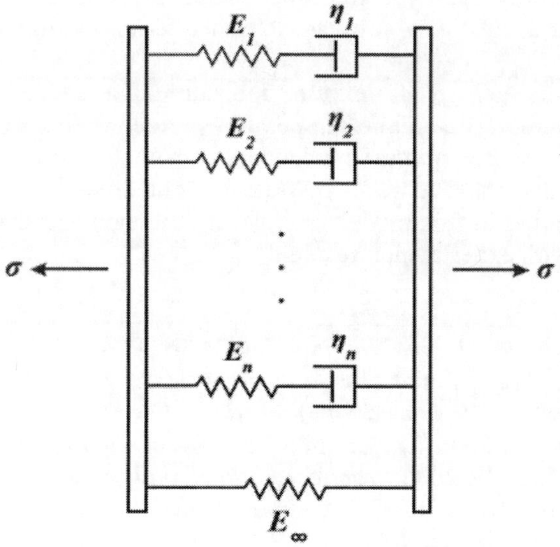

Figure 4. The generalized Maxwell model

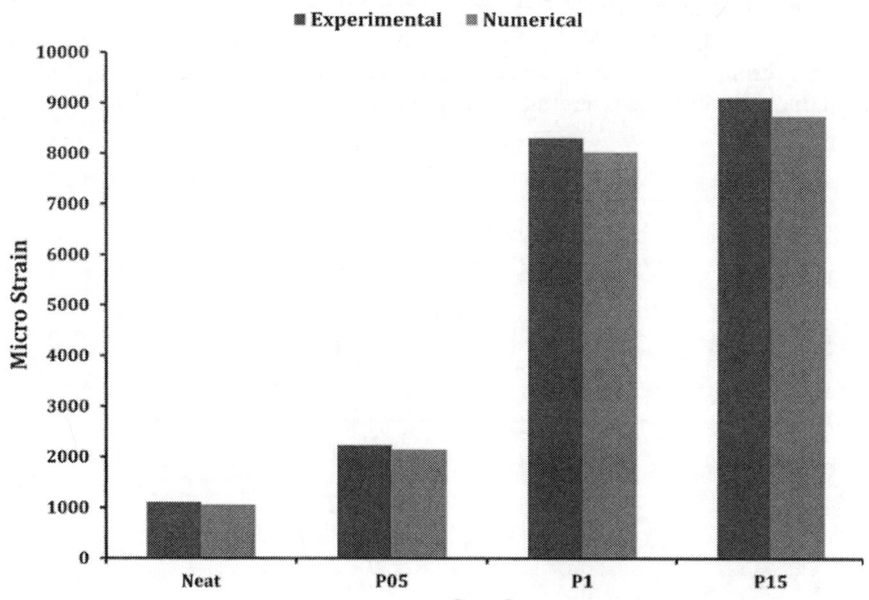

Figure 5. Micro strain of experimental and numerical approach for number of cycles of 10000

IOP Conf. Series: Materials Science and Engineering **416** (2018) 012106 doi:10.1088/1757-899X/416/1/012106

Figure 5 shows results of numerical model with experimental test results. As it can be seen in this figure, using linear viscoelastic parameters of asphalt binder and then extending these parameters to asphalt mastic and asphalt mortar phase, it is possible to predict fatigue behaviour of asphalt mixtures. The numerical models are run for 10000 cycles due to calculation cost and limitation of running time in this research. However, it can simply be done to increase number of cycles for numerical running and compare experimental results with numerical ones.

5. Conclusion

This paper, investigates the effect of PPA modifier on fatigue performance of asphalt binder and mixture including experimental tests and numerical simulation. The experimental program includes frequency sweep test using DSR at 25°C for all asphalt binder and also indirect tensile fatigue (ITF) for all specimens of asphalt mixture. On the other hand, the aggregate skeleton of the bituminous composites was performed using MOA software to generate random inclusions of various sizes and shapes.

The experimental results of asphalt mixture indicated that adding PPA improve the fatigue behaviour of asphalt mixtures at moderate temperatures. According to the experimental result, the dosage of 1% was obtained as the optimum percentage of asphalt binder modifier. On the other hand, the numerical model results approve the experimental outcomes by considering the generalized Maxwell model which applied in this model. So, it can be concluded that numerical simulation has shown a good agreement with experimental results.

6. References

[1] Chen J et al. 2015 *Applied Thermal Engineering* **84** 368–374
[2] Tehrani FF, Absi J, Allou F, Petit CH 2013 *Comput Mater Sci* **69** 186-196
[3] Panda M, Mazumdar M 1999 *Journal of Materials in Civil Engineering* **11**(2) 131-137
[4] Polacco G, Muscente A, Biondi D, Santini S 2006 *European Polymer Journal* **42**(5) 1113- 1121
[5] Chen JS, Liao MC, Tsai HH 2002 *Practical Failure Analysis* **2**(3) 75-83
[6] Polacco G, Berlincioni S, Biondi D, Stastna J, Zanzotto L 2005 *European Polymer Journal* **41**(12) 2831-2844
[7] Giavarini C, De Filippis P, Santarelli ML, Scarsella M 1996 *Fuel* 1996 **75**(6) 681-686
[8] Kodrat I, Sohn D, Hesp SAM 2007 *Journal of the Transportation Research Board* 47–55
[9] Orange G, Dupuis D, Marin JV, Farcas F, Such C, Marcant B 2004 "Chemical Modification of Bitumen through Polyphosphoric Acid: Properties-Microstructure Relationship", 3rd Euraphalt & Eurobitume Congress, Vienna, Paper 334, book 1, p. 733-745.

IOP Conf. Series: Materials Science and Engineering **416** (2018) 012107 doi:10.1088/1757-899X/416/1/012107

A Comparative Study of Linen (Flax) Fibers as Reinforcement of Fly Ash and Clay Brick Powder Based Geopolymers

G Silva[1], S Kim[1,*], A Castañeda[2], R Donayre[2], J Nakamatsu[2], R Aguilar[1], K Korniejenko[3], M Łach[3], J Mikuła[3]

[1]Engineering Department, Pontificia Universidad Católica del Perú PUCP, Av. Universitaria 1801, Lima 32, Lima, Peru

[2]Science Department, Pontificia Universidad Católica del Perú PUCP, Av. Universitaria 1801, Lima 32, Lima, Peru

[3]Cracow University of Technology, Faculty of Mechanical Engineering, Institute of Materials Engineering, Warszawska 24, 31-155 Cracow, Poland

[*]Corresponding author: skim@pucp.pe

Abstract. Geopolymers have great potential for the development of sustainable building materials due the utilization of industrial by-products and waste materials for their production. The introduction of natural fibers to geopolymer matrices has shown interesting results for producing alternative building materials with improved mechanical properties. This research explores the use of linen (flax) fibers as reinforcement of clay brick powder based-geopolymers and fly-ash based geopolymers. Linen fibers were obtained from the Institute of Natural Fibres and Medicinal Plants while clay brick powder were obtained from construction waste in Lima (Peru) and fly ash were extracted from Skawina (Poland). The influence of the addition of linen fibers on the mechanical properties of geopolymer matrix was studied by unconfined compression tests and three-point bending tests. The results show that addition of flax fibers significantly enhanced both compressive and flexural strengths of fly ash and clay brick powder based geopolymers.

1. Introduction

The search for eco-friendly building materials to replace traditional ones is a global concern due to all the environmental issues and unsustainability regarding construction industry. Recycling Construction & Demolition Wastes (C&D) for the production of alternative building materials will have significant benefits, from the reduction of energy and raw materials usage, to the decrease of use of available land for extracting resources or for landfills [1]. Clay brick powder, a C&D product, has shown great potential as raw material for geopolymer production. Baronio and Binda [2] claimed that clay brick powder has pozzolanic activity potential due to the destruction of the crystalline network when structural hydroxyl groups of clay minerals are lost by high temperatures during production. On the other hand, as global electricity production shows an important dependency on coal sources, enormous quantities of fly ash are produced around the world, e.g. only US produced 51 million tons in 2014 [3]. The percentage of fly ash used in other applications is low, so the major part of these ashes are stored in landfills and storage lagoons resulting in potential land, environmental and human health issues [4]. The chemical composition of fly ash varies depending of the coal source and the combustion process but is mainly

IOP Conf. Series: Materials Science and Engineering **416** (2018) 012107 doi:10.1088/1757-899X/416/1/012107

composed by SiO_2 and Al_2O_3 [4]. Therefore, fly ash is also suitable for geopolymer production since it contains high amounts of silica and alumina.

Unfortunately, unreinforced geopolymers exhibit a brittle behavior, similar to Portland cement pastes, which makes geopolymers inadequate for many applications. However, addition of synthetic and natural fibers to the geopolymer matrix has shown promising results as it enhances mechanical properties e.g. increases fracture toughness, improves flexural and impact strength, provides ductility and suppresses brittle behavior [5,6]. Natural fibers are eco-friendly and have lower density, higher specific strength and lower cost than synthetic fibers, and do not require a high energy consuming process [7]. Since a composite material based on recycled clay brick powder or an industrial by-product as fly ash reinforced with natural fibers would be a good option for the development of an eco-friendly building material, the objective of this article is to explore the use of untreated linen fibers as a reinforcement of fly ash and clay brick powder based geopolymers.

2. Experimental plan

2.1. Materials

Fired clay bricks from construction wastes in Lima (Peru) were subjected to a milling process consisting of four phases with different equipment. First, the entire brick was placed in an impact crusher to obtain pieces smaller than 25 mm. These pieces were then placed in a jaw crusher to reduce their size. After that, a rolling mill was used to produce brick particles smaller than 0.5 mm, which were finally placed in a ball mill for 3 hours. Resulting brick powder was passed through ASTM sieve No. 200 and particles smaller than 0.075 mm were used as a raw material in the production of clay brick powder-based geopolymer. Analysis of elemental composition of brick powder was performed by energy dispersive spectroscopy (EDS) with an Ametek EDAX TEAM system coupled to a FEI Quanta 650 scanning electron microscope. The brick powder showed high content of O (~35%), Fe (~29%), Si (~17%) and Al (~11%).

Geopolymer matrices were also made from fly ash from the CHP plant in Skawina (Poland) and sand in ratio 1:1. This fly ash form CHP plant in Skawina (Poland) is rich in oxides such as SiO_2 (~56%), Al_2O_3 (~23%) and Fe_2O_3 (~6%) [8].

The flax fibers were after ratting process. The fibers were cut to a length of 5 mm approximately and were added to the geopolymer matrix in the production process. The addition of fibers in geopolymer mixture was 1% by weight of both clay brick powder as well as fly ash based matrix.

2.2. Preparation of samples

Unreinforced and fiber reinforced geopolymer samples were prepared to study the effect of linen fiber reinforcement on the mechanical properties of clay brick powder-based geopolymers. In both cases, geopolymer matrices with an alkaline solution to clay brick powder ratio of 0.4 were considered. Alkaline solution consisted of a mixture of 10 M sodium hydroxide (NaOH) and commercial sodium silicate (Na_2SiO_3) in a NaOH:Na_2SiO_3 ratio of 1:2.5 by volume. First, clay brick powder and linen fibers were dry-mixed to ensure homogenous distribution (see Figure 1a). Then, alkaline solution was gradually added and manually mixed until a homogenous paste was obtained (see Figure 1b and c). The geopolymer paste was hand-formed into prismatic silicon moulds and subject to vibration in order to remove trapped air bubbles. Specimens were kept in an oven at approximately 70°C for 24 hours, after which specimens were left at ambient temperature until the mechanical tests. All specimens were tested after 1, 7, 14 and 28 days of age.

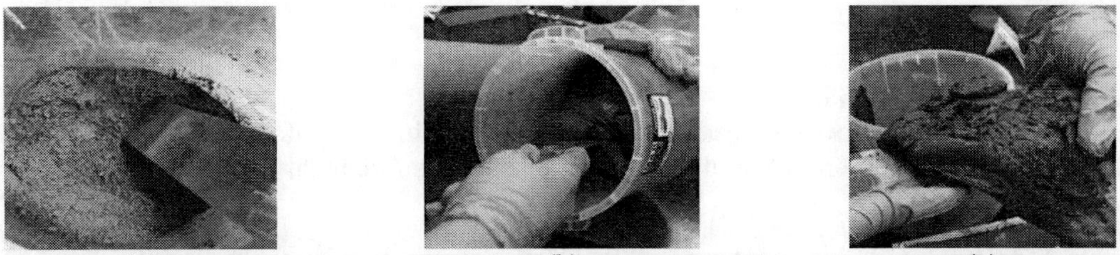

(a) (b) (c)

Figure 1. Clay brick powder-based geopolymer production: (a) Dry-mixing of clay brick powder and linen fibers; (b) Hand-mixing after solution addition; (c) Geopolymer paste.

In case of fly-ash based samples the manufacturing process was very similar. The process of activation has been made by 8 M sodium hydroxide solution combined with the sodium silicate solution (liquid glass at a ratio of 1:2.5). Technical sodium hydroxide was used and an aqueous solution of 2.5 M sodium silicate (R-145) with density about 1.45 g/cm^3. Tap water was used instead of the distilled one. The alkaline solution was prepared by means of pouring the aqueous solution of sodium silicate over the solid sodium hydroxide. The solution was mixed and left until its temperature became stable and the concentrations equalized about 2 hours. The fly ash and sand, alkaline solution and fibers were mixed about 20 minutes by using low speed mixing machine (to form a homogeneous paste). Next, samples were hand-formed into two sets of plastic moulds and then subjected to vibratory removal of air bubbles. Tightly closed moulds were heated in the laboratory dryer for 24h at 75 °C. Then, the samples were demoulded and investigated after 28 days.

2.3. Research procedures

Scanning electron microscopy (SEM) type JEOL JSM 820 with EDS has been performed for microstructure research. The research has been made for fibers and the samples made from fly ash-based geopolymer previously broken from compressive or flexural strength tests. Samples were covered with a thin layer of gold with JEOL JEE-4X vacuum sputter. The investigations were made at various magnifications (between 20 – 100x).

To evaluate the influence of linen fibers as a reinforcement of clay brick powder-based geopolymers, three-point bending tests and compression tests were performed according to ASTM C348 [9] and ASTM C349 [10], respectively. Three-point bending tests involved at least testing 3 prismatic samples with approximate dimensions of 40 mm x 40 mm (cross section) x 160 mm (length). Then, the two resultant broken parts from each bending test (cubic specimens with approximate dimension of 40 mm x 40 mm x 40 mm) were tested in compression. An electromechanical testing machine MTS Exceed 45.105 [11] was used for all mechanical testing. The displacement rate of the load frame was set at 0.3 mm/min for three-point bending tests and 0.5 mm/min for compression tests. Displacements of the load frame were recorded in all tests and were taken as global deformations. Furthermore, a digital image correlation technique was also used to measure localized deformations in the middle third of geopolymer specimens during mechanical tests by means of an ARAMIS 5M system with a strain precision of 0.005% [12].

The fly ash-based geopolymers were tested according to EU standards. Compressive strength tests were carried out according to the standard method EN 12390-3 ('Testing hardened concrete. Compressive strength of test specimens') [13], because of the lack of separate standards for geopolymer materials. The tests involved at least 6 samples and were performed on a universal testing machine - single-point load (Instron type 4465). Samples used for compressive strength test had cubic shape and dimensions (approx.): 50 mm × 50 mm × 50 mm. Flexural strength tests were carried out according to the standard EN 12390-5 ('Testing hardened concrete. Flexural strength of test specimens') [14]. Samples used to the flexural strength test had cuboid shape and dimensions (approx.): 200 mm × 50 mm × 50 mm. Tests were performed with a universal testing machine - single-point load (Instron type 4465).

IOP Conf. Series: Materials Science and Engineering **416** (2018) 012107 doi:10.1088/1757-899X/416/1/012107

3. Results

3.1. Microstructure investigation

SEM observation of microstructure for composites gives a preliminary information about the coherency of flax fibers with the geopolymer matrix. Figure 2 shows exemplary images that presented the composite morphology.

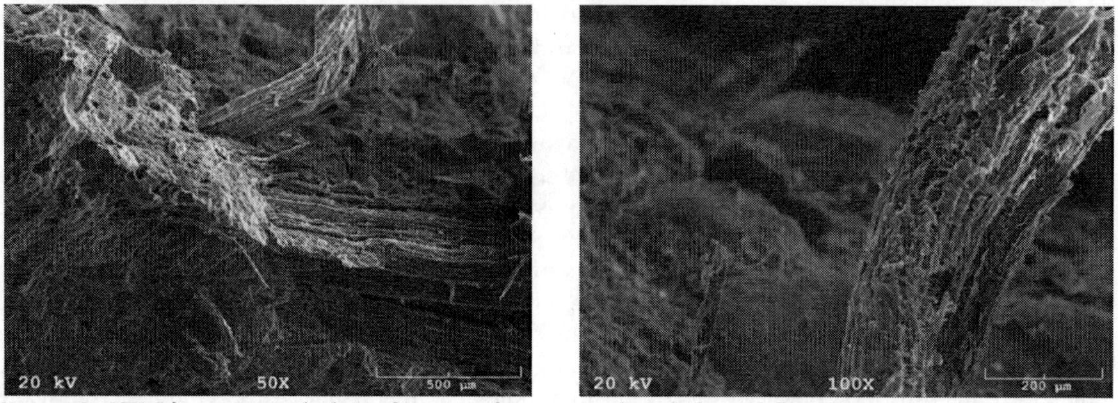

Figure 2. SEM image morphology for geopolymer matrix reinforced flax fibers

SEM images also showed that fiber distribution is regular. Fiber distribution in the matrix influences the properties of the specific composite, especially aggregation of fibers can decrease its mechanical properties.

3.2. Three-point bending tests

Three-point bending tests were used to study the flexural behavior of unreinforced geopolymer (URG) and fiber reinforced geopolymer (FRG) by means of applying a concentrated load in the middle of a simply supported prismatic sample. The history of applied force and mid-span vertical displacements of the load cell during three-point bending tests were used to generate the curves that relate bending stress and displacement. The bending stresses (σ_f) were calculated using equation 1.

$$\sigma_f = \frac{3FL}{2BH^2} \qquad (1)$$

where F is the applied force while L, H and B denote the space between supporting points (for conducted tests:120 mm), height and width of the prismatic samples, respectively.

Figures 3a, b, c and d show the curves and envelopes resulting from the three-point bending tests for URG and FRG samples at ages of 1, 7, 14 and 28 days, respectively. A notorious difference in flexural response between UGR and FGR samples due to the reinforcement of linen fibers is evidenced in Figure 3. FGR samples could withstand higher bending stresses as a result of the linen fiber's presence. Likewise, the abrupt reduction of resistance after reaching the maximum stress observed in URG samples is not seen in the FRG curves, which indicates that the linen fibers help to avoid a fragile failure.

IOP Conf. Series: Materials Science and Engineering **416** (2018) 012107 doi:10.1088/1757-899X/416/1/012107

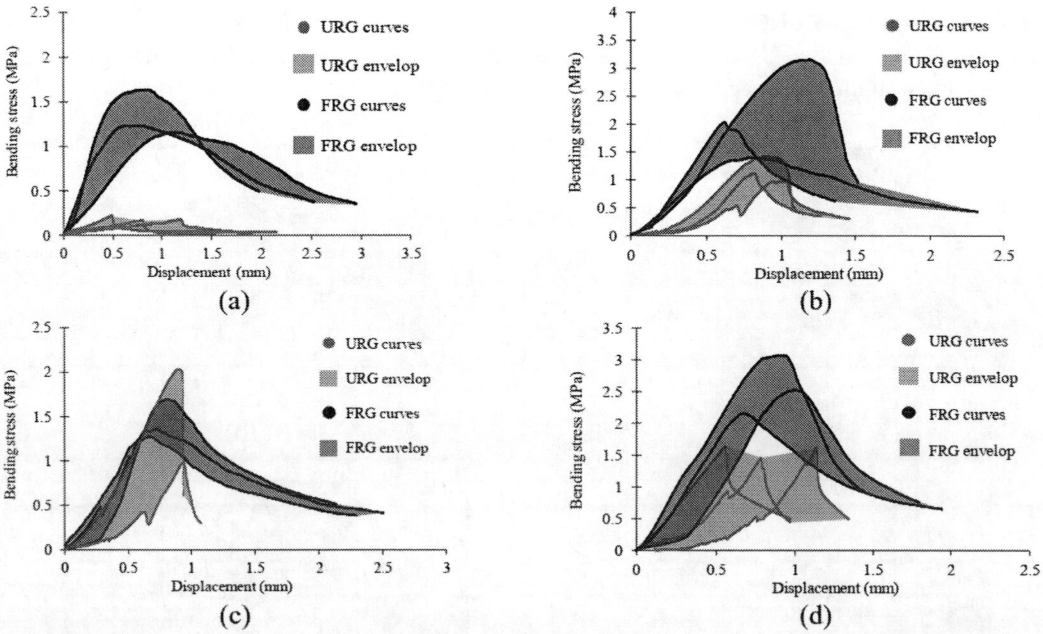

Figure 3. Bending stress vs. mid-span vertical displacements curves and envelopes for clay brick powder-based geopolymer specimens tested at different ages: (a) 1 day age test; (b) 7 days age test; (c) 14 days age test; (d) 28 days age test.

The research for samples based on fly ash was performed after the 28 days. For the calculation of flexural strength of fly ash-based geopolymer (f_{cf}) has been applied following equation 2:

$$f_{cf}=\frac{Fl}{d_1 d_2^2} \tag{2}$$

where F is maximal load while l, d_1 and d_2 denote the space between supporting points (for conducted tests:140 mm), height and width of the prismatic samples, respectively. The results of flexural strength test are presented in table 1.

Table 1. Flexural strength of fly ash-based geopolymer after 28 days.

Sample	MPa	StD
Geopolymer without reinforcement	5.22	0.251
Geopolymer with flax fibers (1%)	8.39	0.098

The results for samples with reinforcement are significantly better that for geopolymer matrix. Additionally, similar as in case of clay brick powder-based geopolymers, the fiber addition helps avoid the brittle failure.

3.3. Compression tests

Compression stress-strain curves were obtained by means of uniaxial compression tests on the two resultant portions from each three-point bending test. Compression stresses were calculated as the ratio of the applied load recorded during the test and the transversal area of the 40 mm x 40 mm metal plate. Local deformations measured in the middle third of the lateral face in URG and FRG samples were considered as the strains to calculate compression stress-strain curves. Figures 4a, b, c and d show the stress-strain curves and envelopes resulting from the uniaxial compression tests for URG and FRG samples at ages of 1, 7, 14 and 28 days, respectively. Again, FRG specimens exhibited higher compressive strength than URG samples at all studied aging times. Also, FRG samples evidenced higher capacity of deformation than URG presenting higher level of ultimate strains. This behavior is associated with an increment in ductility capacity of the FRG specimens.

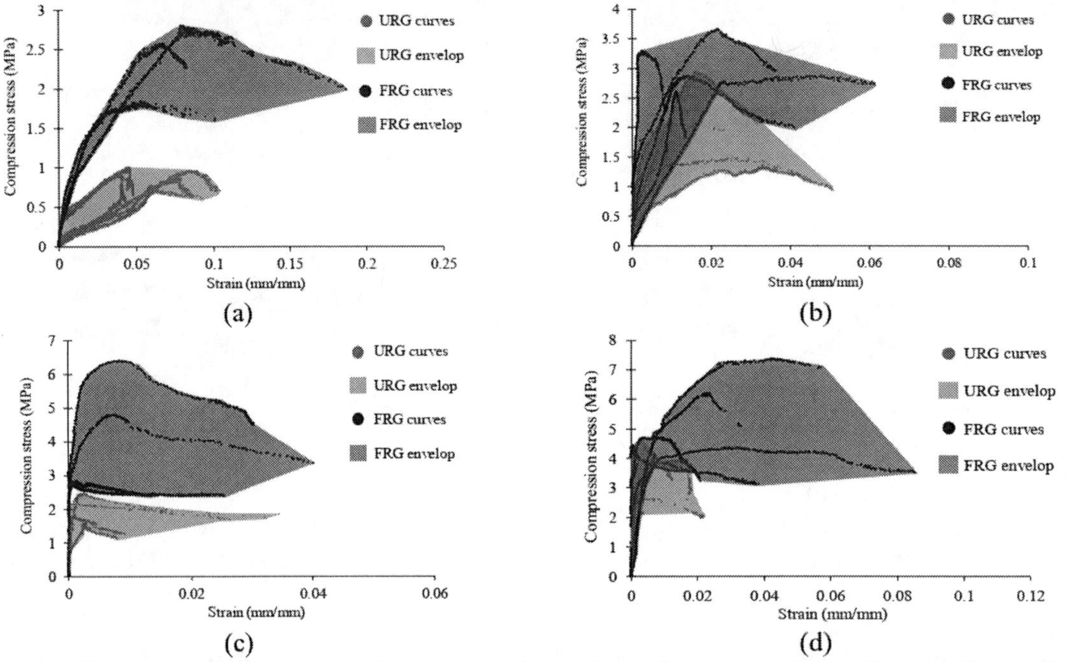

Figure 4. Compression stress vs. strain curves and envelopes for clay brick powder-based geopolymer specimens tested at different ages: (a) 1 day age test; (b) 7 days age test; (c) 14 days age test; (d) 28 days age test.

The research for samples based on fly ash was performed after the 28 days. The results of compressive strength tests are presented in table 2.

Table 2. Compressive strength of fly ash-based geopolymer after 28 days.

Sample	MPa	StD
Geopolymer without reinforcement	42.70	3.381
Geopolymer with flax fibers (1%)	44.64	4.047

The results for samples with reinforcement and without are very similar. Even though results for samples with reinforcement are slightly better, there are no statistically significant differences.

4. Discussion of results

In order to illustrate the influence of flax fiber reinforcement regarding flexural and compressive strengths, two columns graphs with error bars are given in Figure 5a and b. Values of flexural and compressive strengths presented in Figure 5 are the arithmetic mean of peak values for each bending stress vs. displacement and compression stress vs. strain curves shown in Figure 3 and 4, respectively. For data processing, the Interquartile Range (IQR) criterion was used to discard outliers and work with more representative values. Even though data were filter with IQR criterion, error bars display high values of standard deviation. High variability of data could be explained by differences during manual production i.e. hand-mixing and hand-operated vibration of geopolymer paste. Both column graphs also show the development of flexural and compressive strengths over time, reaching their maximum values at 28 day. However, FRG samples at 14 days exhibited a flexural strength much lower than 7-day-age

FRG, thus, it did not follow the marked trend (see Figure 5a) so 14-day-age FRG flexural strength were taken as an inconsistent value. Column graphs in Figure 5 reveal the large influence of the linen fiber reinforcement on the mechanical properties of geopolymer matrices. Addition of linen fibers to produce FRG composites caused an increase in flexural strength respect to URG of 605%, 89% and 64% at 1, 7 and 28 days, respectively. This improvement in flexural strength due to natural fiber reinforcement has been reported by other researches [5,7]. Despite the fact that the main objective of fiber reinforcement is to provide post-cracking ductility [15], compressive strength also increased significantly by the presence of linen fibers. FRG composites evidenced an increase in compressive strength of 170%, 39%, 86% and 53% in comparison to samples with no fibers at 1, 7, 14 and 28 days, respectively.

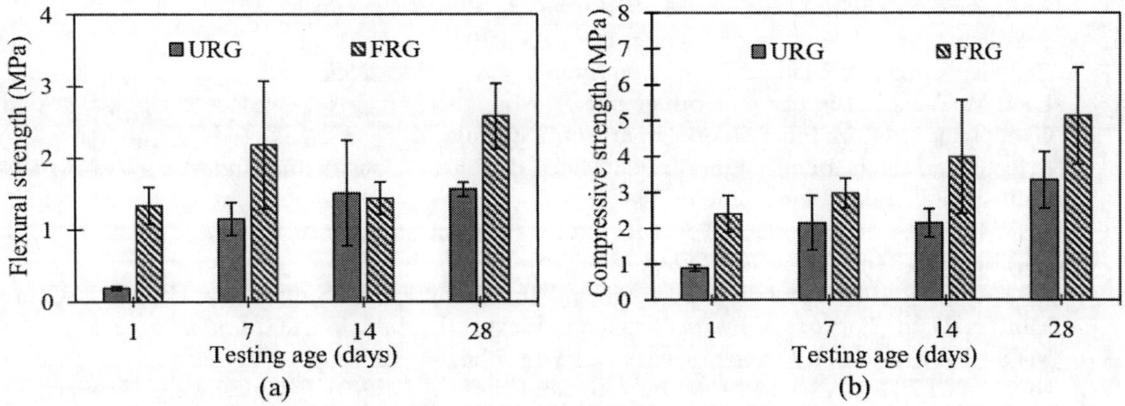

Figure 5. Summary of mechanical test results for clay brick powder-based geopolymer specimens tested at different ages: (a) Flexural Strength; (b) Compressive Strength.

In case of composites based on fly ash it is possible also observe the positive influence of flax fibers. The composites with fibers have better mechanical properties flexural strength (table 1) as well as compressive strength (table 2). Additionally the fragile is more ductile. The SEM observation confirms the good coherence linen fibers to matrix.

A comparative study of flax fibers as reinforcement of fly ash and clay brick powder based geopolymers shows that in both cases linen fibers have positive influence for composite properties. The differences of final properties of the composites, in this case, are depending on the matrix properties.

5. Conclusions

This work focused on the study of the reinforcement of clay brick powder-based geopolymers and fly ash geopolymers with the addition of natural linen fibers. Mechanical characterization consisted on three-point bending tests and compression tests showed that their addition to the mixture significantly enhanced flexural and compressive strengths when compared with the unreinforced matrices. Fiber reinforcement increased flexural strength by 605%, 89% and 64% at 1, 7 and 28 days, respectively. While compressive strength was upgraded by 170%, 39%, 86% and 53% at 1, 7, 14 and 28 days, respectively. The increasing the mechanical properties for fly ash-based geopolymers was: 161% in case of flexural strength and 104% in case of compressive strength.

Besides strength improvements, bending stress vs. displacement and compression stress vs. strain envelopes evidenced that fiber reinforced geopolymer samples did not exhibit an abrupt decrease of resistance after peak stress was reached, and samples had the capacity to withstand larger deformations and presented a more ductile failure mode than unreinforced matrices. Therefore, the use of natural linen fibers to improve the mechanical properties of geopolymer matrix based on waste materials as clay brick powder or fly ash allows the development of an eco-friendly building composite material with relatively good mechanical properties.

The research results confirms the possibility of modification of geopolymers and alkali activated materials by short natural fibers addition. It is supposed this kind of additive will be suitable also for geopolymers based on, others materials, exemplary slags or natural ashes, because there is only minor difference between the chemical reactions of these process.

6. References

[1] Allahverdi A and Najafi Kani E 2009 *Int. J. Civ. Eng.* **7**(3) 249–257
[2] Baronio G and Binda L 1997 *Constr. Build. Mater.* **11**(1) 41–46
[3] ACAA 2014 "American Coal Ash Association," no. April, 1–36
[4] Diaz EI, Allouche EN, Eklund S 2010 *Fuel* **89**(5) 992–996
[5] Alomayri T, Shaikh FUA, Low IM 2014 *Compos. Part B Eng.* **60** 36–42
[6] Sakulich AR 2011 *Sustain. Cities Soc.* **1**(4) 195–210
[7] Chen R, Ahmari S, Zhang L 2014 *J. Mater. Sci.* **49**(6) 2548–2558
[8] Łach M, Wajs J, Tempka C, Korniejenko K, Mikuła J 2015 Fibre-reinforced Composites with Geopolymer Matrix: Fiber-Geopolymers, in: J. Mikuła, K. Korniejenko (eds.), Innovative, Cost-Effective and Eco-friendly Fibre-based Materials for the Construction Industry, Wydawnictwo Politechniki Krakowskiej, Cracow 41–50
[9] ASTM C348 1998 "Standard Test Method for Flexural Strength of Hydraulic-Cement Mortars ASTM C348" *Annu. B. ASTM Stand.* **4** 2–7
[10] ASTM C349 1998 "Standard Test Method for Compressive Strength of Hydraulic-Cement Mortars (Using Portions of Prisms Broken in Flexure)1" *Annu. B. ASTM Stand.* 4–6
[11] MTS, "MTS Criterion ® Series 40 Manual Title," 2014.
[12] GOM mbH 2010 "ARAMIS User Information Hardware," 49(0). Brauncshweig, Germany
[13] British Standards Institution BSI 2009 "12390-3 Testing hardened concrete," **3**(1) 420–457
[14] British Standards Institution BSI 2009 "Testing hardened concrete (Flexural)" **3**(1) 420–457
[15] Bentur A, Mindness S, Bentur S 2007 *Civ. Eng.* 625

Acknowledgements

This work was supported by the ERANet-LAC 2nd Joint Call (http://www.eranet-lac.eu) and funded by National Centre for Research and Development, Poland, and the CONCYTEC (FONDECYT), grant no. J108-2016, under project: 'Development of eco-friendly composite materials based on geopolymer matrix and reinforced with waste fibers', under the ERANet- LAC: Latin America, Caribbean and European Union funded by the European Commission, within the 7th Framework Programme for Research and Technology Development (FP7), Topic #02: Waste management, recycling and urban mining (project no. ELAC2015/T02- 0721). The authors would also like to thank the fellowship funding for postgraduate studies for the first author by CONCYTEC, PERU (Contract N° 232-2015-FONDECYT).

IOP Conf. Series: Materials Science and Engineering **416** (2018) 012108 doi:10.1088/1757-899X/416/1/012108

A Parametric Study of the Mechanical Properties of Open-Cell Kelvin Structures

D A Şerban[1,*], S Sărăndan[1], R Negru[1], G Belgiu[2] and L Marşavina[1]

[1]Department of Mechanics and Strength of Materials, Politehnica University of Timisoara, 1 Mihai Viteazu Avenue, 300 222 Timisoara, Romania
[2] Department of Management of Production and Transportation, Politehnica University Timişoara, Romania

[*]Corresponding author: dan.serban@upt.ro

Abstract. This study investigates the elastic response of a tessellated structure based on cylindrical struts generated along the wireframe of a truncated octahedron (the Kelvin cell). The structure has two variable dimensions (strut thickness and strut length), the relative density being plotted as a function of the given parameters and ranged between $6.7 \cdot 10^{-4}$ and 0.217. The variation of the relative stiffness of the structure was determined as a function of relative density. The size effect was also investigated by determining the elastic response for different cell numbers.

1. Introduction
Cellular solids represent a class of materials which consist of interconnected matrices of solid layers having a well-defined shape of the edges and facets [1]. Due to their good mechanical properties (good impact resistance and high energy absorption) relative to their specific mass [2], cellular structures are used in a large variety of applications (from aerospace and automotive to personal protection or packaging [1]), either as stand-alone materials [3, 4] or as components in composite materials, especially in sandwich structures [5, 6].

Inspired by natural cellular structures such as honeycombs, xylem tissue or trabecular bone, a new class of materials was developed, called metamaterials (beyond materials) [7]. They represent human developed structures with increased mechanical, thermal or optical properties, that have no equivalent in nature [7]. This paper focuses on a class of mechanical metamaterials consisting of a Kelvin cell tessellation.

Considering the deformation of cellular structures in uniaxial loadings, Maxwell proposed two types of microtruss lattice structures: stretch dominated structures and bending dominated structures [8]. Bending dominated structures were considered as locked joints connecting bar frames that underwent bending when the structure is loaded. On the other hand, for the stretch dominated structures, the bars would be subjected primarily to tension or compression. In order to determine the behaviour of such lattices, Maxwell came up with a simple equation which takes into account the number of bars b and the number of joints j. For a three-dimensional space, the relation is presented in Equation (1).

$$M = b - 3j + 6 \qquad (1)$$

If $M < 0$, the structure is bending dominated. Otherwise, the structure is stretch dominated.

Considering the slenderness of thin bars that make up such structures, it was observed that stretch dominated structures present higher efficiency than bending dominated structures (as bars are generally stiffer when stretched than when bent) [9]. The variation of relative stiffness with relative density is presented in Equation (2a) for the ideal stretched dominated structure and in Equation (2b) for the ideal bending dominated structure.

$$\frac{\tilde{E}}{E_s} \approx \frac{\tilde{\rho}}{\rho_s} \qquad (2a)$$

$$\frac{\tilde{E}}{E_s} \approx \left(\frac{\tilde{\rho}}{\rho_s}\right)^2 \qquad (2a)$$

where $\tilde{E}$ and $\tilde{\rho}$ are the structure stiffness and density while E_s and ρ_s represent the solid stiffness and density.

For the Kelvin structure, $M = 36 - 3 \cdot 24 + 6 < 0$, determining a bending dominated behaviour. Though less efficient than stretch dominated structures, The Kelvin structure is one of the most efficient foams with equal volume cells (a structure with the least area of surface between cells), surpassed only by the Weaire–Phelan structure with a margin of 0.3% [10]. The advantage that the Kelvin structure holds over the Weaire–Phelan structure lays in its simplicity. The Kelvin structure consists of tessellated truncated octahedrons while the Weaire–Phelan structure consists of two types of cells: an irregular dodecahedron and a truncated hexagonal trapezohedron [10].

The advantages of parametric structures consist in the ease of designing the required mechanical characteristics: the relative density/stiffness needed for a specific application can be obtained through the input of parameter values (such as strut thickness, strut length or chamfer radius) [11]. In addition, the regular structure avoids defects that occur in natural cellular material; the behaviour of structures with reduced number of cells can be accurately extrapolated to structures with large and very large number of cells, significantly reducing computation times in predicting the mechanical response.

The aim of this study is to determine the stiffness variation of the kelvin structure with the cell wall thickness and cell wall length as well as with the relative density of the structure, for small deformations. Future work will investigate the yielding and post-yielding behaviour of the lattice. Considering that such structures will be manufactured with rapid prototyping technology, the mechanical behaviour of the polymers must be taken into account for complex loading scenarios such as dynamic loadings, high deformations fatigue or fracture. In such cases, the viscoelastic nature of the polymers [12-14] will play an important role in the behaviour of the structure [15].

2. Structure development

The Kelvin cell represents a regular geometrical entity obtained by truncating the tips of an octahedron. The design of the cell began with the generation of a 3D sketch, resulting in a wireframe structure (Figure 1 a). Considering that all the sides of the cell have equal length, the dimension was considered the first structure variable. The solid structure was obtained through the generation of cylinders of equal radius (the second structure variable) along the wireframe (Figure 1 b).

In order for the cell to be tessellated into a structure, the cylinders (cell struts) had to be cut, as seen in Figure 2 a. The Kelvin structure was obtained by multiplying and positioning the cells accordingly, merging the solids and cropping the structure in the shape of a prism with a square base. The resulting 3 x 3 x 2.5 cell structure is presented in Figure 2 b.

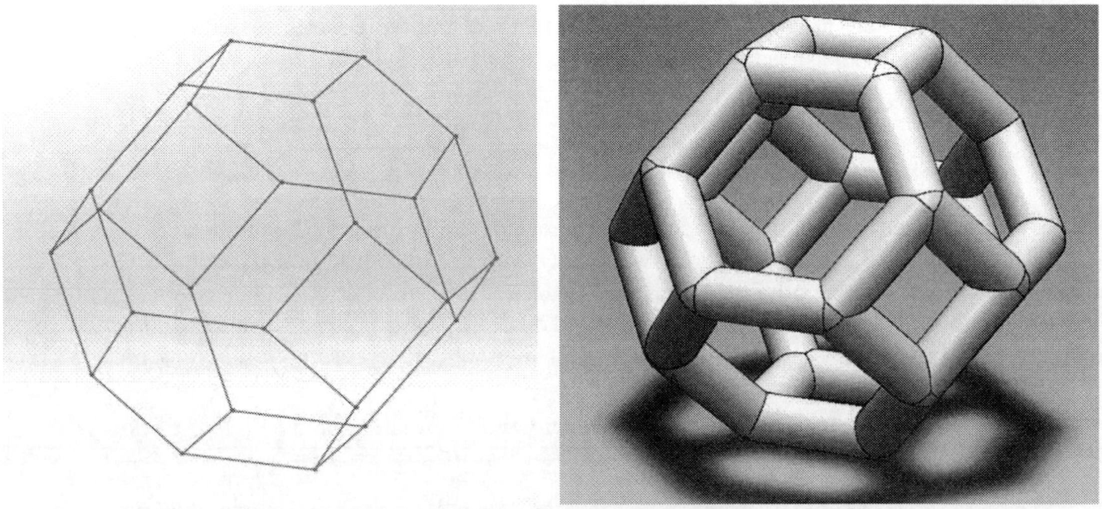

a) b)

Figure.1 Kelvin cell wireframe (a) and solid structure (b)

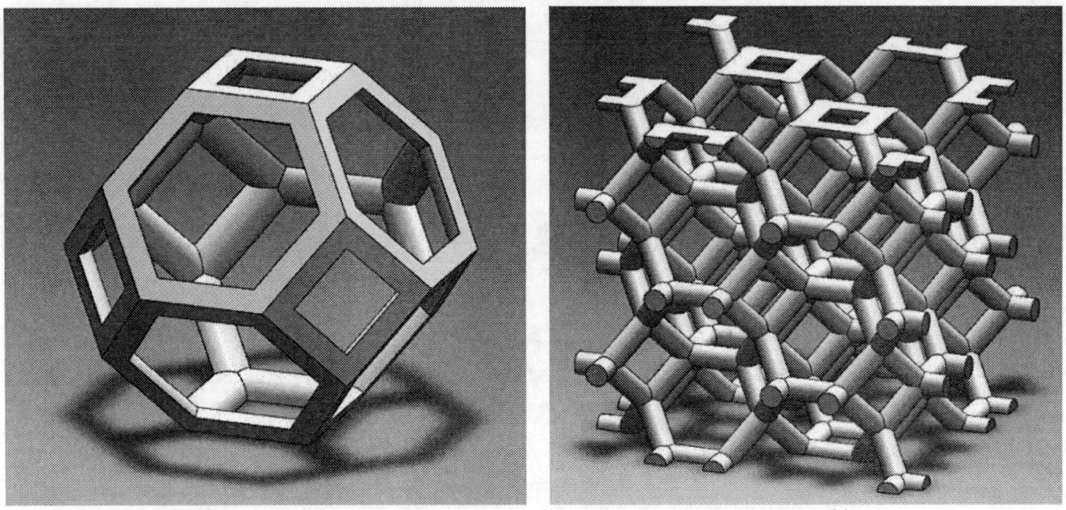

a) b)

Figure 2. Kelvin cell with cut struts (a) and tessellated structure (b)

The relative density ρ_{rel} of a cellular material is defined as:

$$\rho_{rel} = \frac{\rho_{cell}}{\rho_s} \ [-] \tag{3}$$

where ρ_{cell} represents the density of the cellular material and ρ_s represents the density of the solid material. Considering that the densities are calculated for the same unit volume, the relative density can be expressed as the cellular structure mass m_{cell} divided by the solid mass m_s.

$$\rho_{rel} = \frac{m_{cell}}{m_s} \ [-] \tag{4}$$

As both the cellular structure and the solid are made of the same material (with the same density ρ_s), the relative density of the structure can be expressed as:

$$\rho_{rel} = \frac{V_{cell}}{V_s} \ [-] \quad (5)$$

where V_{cell} represents the cell volume and V_s represents the volume of the circumscribed prism.

For the study of the variation in relative density with the two structural parameters, five values were chosen for the cell wall length l and 7 values for the cell wall thickness t. The cellular structure volume and the circumscribed prism value were evaluated by *SolidWorks*™. The variation of the relative density with the two parameters is presented in Table 1. The graphical representation is presented in Figure 3.

Table 1. Variation of relative density with cell length and cell wall thickness

		Cell length [mm]				
		15	20	25	30	35
Cell wall thickness [mm]	1	0.0036	0.002	0.0013	0.00091	0.00067
	2	0.014	0.0079	0.0051	0.0036	0.0026
	3	0.0306	0.0176	0.0114	0.0079	0.0059
	4	0.0529	0.0306	0.0199	0.014	0.0103
	6	0.1119	0.0659	0.0433	0.0306	0.0228
	8	0.1863	0.1119	0.0743	0.0529	0.0395
	10	0.2714	0.1665	0.1119	0.0802	0.0602

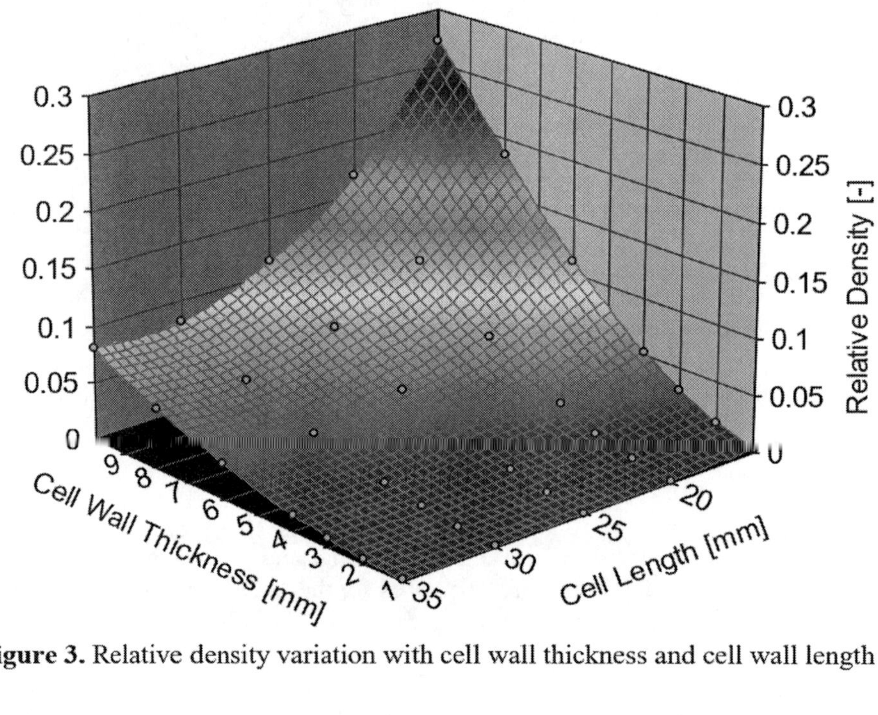

Figure 3. Relative density variation with cell wall thickness and cell wall length

IOP Conf. Series: Materials Science and Engineering **416** (2018) 012108 doi:10.1088/1757-899X/416/1/012108

A surface fitting was performed for the data presented in Table 1, resulting in the relation presented in Equation (6), having an error $r^2 = 0.9993$:

$$\rho_{rel}(l,t) = e^{9.32+1.8\ln(l)+\frac{15.01}{\sqrt{t}}} \tag{6}$$

3. Numerical analyses

The structures presented in chapter 3 were subjected to numerical analyses in the commercial software *Abaqus*™. The geometries were meshed using second order tetrahedral elements (C3D10), the average element size varying with the cell wall thickness, with an average of 20 elements per cross section. As boundary conditions, a symmetry was imposed on the bottom surface (to avoid stress concentration that would occur in a fixed condition) and a displacement on the top surface, along the normal direction. The material model chosen was isotropic linear elastic. A linear correspondence was observed between the stiffness of the solid material and the structure, so the actual value of the Young's modulus is irrelevant when considering the relative stiffness. The simulation results for the relative stiffness variation with cell wall length and thickness is presented in Table 2.

Table 2. Variation of relative stiffness with cell length and cell wall thickness

		Cell length [mm]				
		15	20	25	30	35
Cell wall thickness [mm]	1	$1.075\cdot10^{-5}$	$3.34\cdot10^{-6}$	$1.35\cdot10^{-6}$	$6.48\cdot10^{-7}$	$9.67\cdot10^{-7}$
	2	$1.82\cdot10^{-4}$	$5.61\cdot10^{-5}$	$2.26\cdot10^{-5}$	$1.07\cdot10^{-5}$	$5.74\cdot10^{-6}$
	3	$9.51\cdot10^{-4}$	$2.94\cdot10^{-4}$	$1.18\cdot10^{-4}$	$5.6\cdot10^{-5}$	$2.98\cdot10^{-5}$
	4	0.0031	$9.53\cdot10^{-4}$	$3.83\cdot10^{-4}$	$1.82\cdot10^{-4}$	$9.65\cdot10^{-5}$
	6	0.0151	0.0048	0.0019	$9.53\cdot10^{-4}$	$5.03\cdot10^{-4}$
	8	0.0431	0.0148	0.0062	0.003	0.00162
	10	0.0914	0.0342	0.0147	0.0073	0.00399

The graphical representation of the variation of the relative stiffness with cell wall thickness and length is presented in Figure 4.

Similar to the case of relative density, a surface fitting was performed, yielding the relation presented in Equation (7), with an error $r^2 = 0.9994$.

$$\rho_{rel}(l,t) = e^{13.56-3.35\ln(l)+\frac{-19.98}{\sqrt{t}}} \tag{7}$$

All the values for the determined relative stiffness were plotted as a function of the relative density, along with the ideal behaviour for the stretch dominated and the bending dominated structures, the resulting graph in double logarithmic scale being presented in Figure 5.

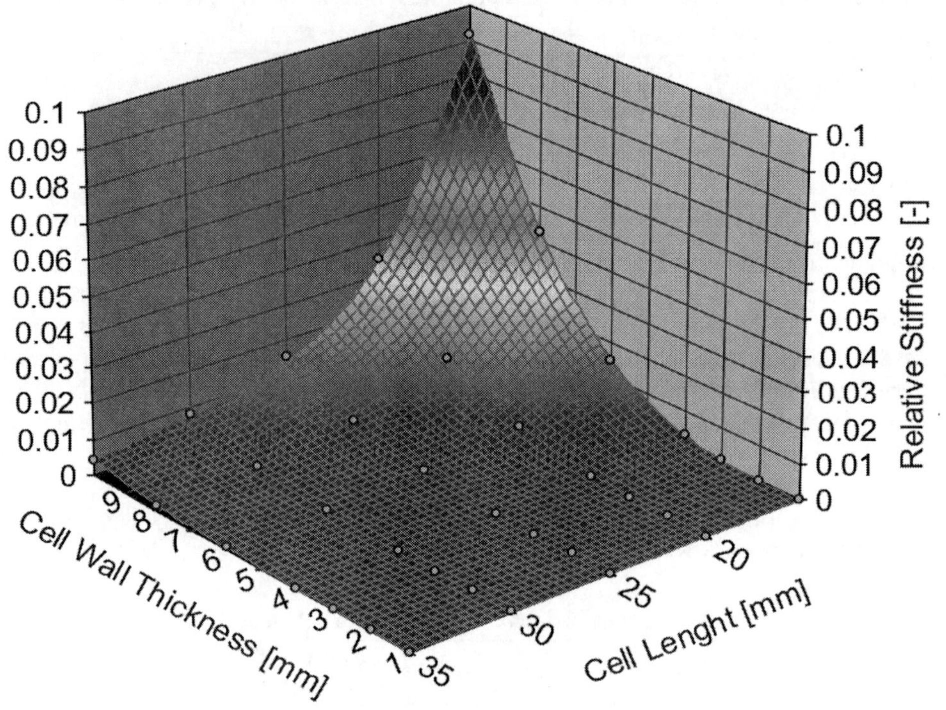

Figure 4. Relative stiffness variation with cell wall thickness and cell wall length

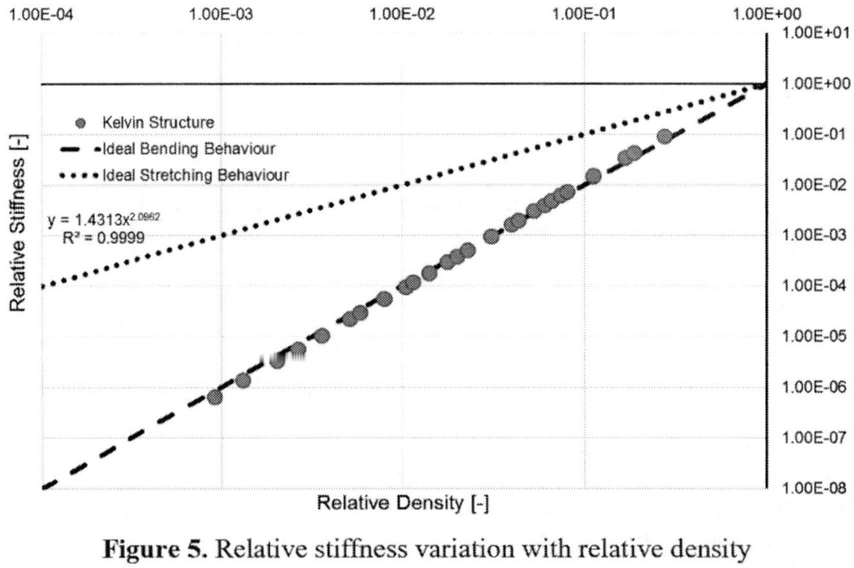

Figure 5. Relative stiffness variation with relative density

The curve fitting procedure resulted in the relation presented in Equation (8), with an error $r^2 = 0.9999$.

$$E_{rel} = 1.43 \cdot \left(\frac{\tilde{\rho}}{\rho_s}\right)^{2.096} \tag{8}$$

4. Discussions and conclusions

This work investigated the elastic response of a parametric Kelvin structure with two variables: cell wall length and cell wall thickness. It was observed that the variation in relative stiffness with relative density is very close to that of an ideal bending dominated structure, which recommends the use of the structure in engineering applications.

The development of the three equations (relative density and stiffness variation with cell wall thickness and length and relative stiffness variation with relative density) will aid in designing Kelvin structures for specific applications that require particular mechanical properties.

Future work on this topic will focus on the large deformation behaviour of the structure. Experimental investigations will be performed on prototyped structures and the results will be compared to the numerical ones in order to validate the designed models.

5. Bibliography

[1] Gibson L and Ashby M 1997 Cellular solids. Structure and properties, Cambridge University Press

[2] Zhang J and Ashby MF 1994 "Mechanical selection of foams and honeycombs used for packaging and energy absorption," *Journal of Materials Science* **29**(1) 157-163

[3] Avalle M, Belingardi G and Montanini R 2001 "Characterization of polymeric structural foams under compressive impact loading by means of energy-absorption diagram," *International Journal of Impact Engineering* **25** 455-472

[4] Mills N, Fitzgerald C, Gilchrist A and Verdejo R 2003 "Polymer foams for personal protection: cushions, shoes and helmets," *Composites Science and Technology* **63** 2389-2400

[5] Henaoa A, Carreraa M, Miravete A and Castejon L 2010 "Mechanical performance of through-thickness tufted sandwich structures," *Composite Structures* **92** 2052–2059

[6] Şerban D, Weissenborn O, Geller S, Marşavina L and Gude M 2016 "Evaluation of the mechanical and morphological properties of long fibre reinforced polyurethane rigid foams," *Polymer Testing* **49** 121–127

[7] Zheng X, Lee H, Weisgraber T, Shusteff M, DeOtte J, Duoss E, Kuntz J, Biener M, Ge Q, Jackson J, Kucheyev S and Fang N 2014 "Ultralight, Ultrastiff Mechanical Metamaterials," *Science* **344**(6190) 1373-1377

[8] Maxwell J 1864 "On the calculation of the equilibrium and stiffness of frames," The London, Edinburgh, and Dublin Philosophical Magazine and Journal of Science **27** 294-299

[9] Scheffler M and Colombo P 2005 Cellular Ceramics: Structure, Manufacturing, Properties and Applications, WILEY-VCH Verlag GmbH

[10] Weaire D and Phelan R 1994 "A counter-example to Kelvin's conjecture on minimal surfaces," *Philosophical Magazine Letters* **69** 107-110

[11] Şerban D, Linul E, Voiconi T, Marsavina L and Modler N 2015 "Numerical evaluation of two-dimensional micromechanical structures of anisotropic cellular materials: case study for polyurethane rigid foams," *Iranian Polymer Journal* **24** 515-529

[12] Şerban D, Weber G, Marşavina L, Silberschmidt V and Hufenbach W 2013 "Tensile properties of semi-crystalline thermoplastic polymers: Effects of temperature and strain rates," *Polymer Testing* **32** 413–425

[13] Şerban D, Marşavina L and Silberschmidt V 2012 "Response of semi-crystalline thermoplastic polymers to dynamic loading: A finite element study," *Computational Material Science* **64** 116–121

[14] Şerban D, Marşavina L and Modler N 2015 "Low-cycle fatigue behaviour of polyamides," Fatigue and Fracture of Engineering Materials and Structures **38** 1383–1394
[15] Şerban D, Voiconi T, Linul E, Marşavina L and Modler N 2015 "Viscoelastic Properties of PUR Foams: Impact excitation and dynamic mechanical analysis," *Materiale Plastice* **52** 537–541
[16] Şerban D, Linul E, Sărăndan S and Marsavina L 2016 "Development of parametric Kelvin structures with closed cells," *Solid State Phenomena* **254** 49-54

Institute of Physics
Dirac House, Temple Back
Bristol BS1 6BE UK

ISSN: 1757-8981
ISBN 978-1-5108-7380-3